29th International Laser Radar Conference (ILRC 29)

EPJ Web of Conferences Volume 237 (2020)

Hefei, China
24 - 28 June 2019

Editors:

Dong Liu
Yingjian Wang
Yonhua Wu
Barry Gross
Fred Moshary

ISBN: 978-1-7138-1442-9

Printed from e-media with permission by:

Curran Associates, Inc.
57 Morehouse Lane
Red Hook, NY 12571

Some format issues inherent in the e-media version may also appear in this print version.

For additional information, please contact EDP Sciences – Web of Conferences at the address below.

EDP Sciences – Web of Conferences
17, Avenue du Hoggar
Parc d'Activité de Courtabœuf
BP 112
F-91944 Les Ulis Cedex A
France

Phone: +33 (0) 1 69 18 75 75
Fax: +33 (0) 1 69 28 84 91

contact-edps@webofconferences.org

Additional copies of this publication are available from:

Curran Associates, Inc.
57 Morehouse Lane
Red Hook, NY 12571 USA
Phone: 845-758-0400
Fax: 845-758-2633
Email: curran@proceedings.com
Web: www.proceedings.com

TABLE OF CONTENTS

The 29th International Laser Radar Conference (ILRC29)

June 24-28, 2019, in Hefei, Anhui Province, China

ILRC29 Committees

ICLAS Chairman

- Alex Papayannis* (National Technical University of Athens, Greece)

ILRC29 Co-Chairmen

- Yingjian Wang (Hefei Institutes of Physical Science, Chinese Academy of Sciences）
- Dong Liu*（Anhui Institute of Optics and Fine Mechanics, Chinese Academy of Sciences）

Programme Committee

- ***SPACE LIDARS***
  Pierre Flamant (Inst Pierre Simon Laplace, CNRS, France)
  Upendra N. Singh** (NASA Langley Research Center, USA)
  Anne Grete Straume* (European Space Agency)
  George Tzeremes* (European Space Agency)
  Dave Donovan* (Royal Netherlands Meteorological Institute The Netherlands)
  Itsushi Uno (Kyushu University, Japan)
  David M. Winker (NASA Langley Research Center, USA)
  Weibiao Chen (Shanghai Institute of Optics and Fine Mechanics, Chinese Academy of Sciences)

- ***LIDAR NETWORKS***
  Eduardo Landulfo* (Instituto de Pesquisas Energeticas e Nucleares, Brazil)
  Gelsomina Pappalardo (Consiglio Nazionale delle Ricerche, Istituto di Metodologie per l'Analisi Ambientale (CNR-IMAA), Italy)
  Makato Abo* (Tokyo Metropolitan University, Japan)
  Anatoli Chaikovsky (Institute of Physics of the National Academy of Sciences, Belarus)
  Kevin B. Strawbridge* (Air Quality Processes Research Section, Canada)

Michael J. Newchurch (University of Alabama in Huntsville, USA)
Yingjian Wang（Hefei Institutes of Physical Science, Chinese Academy of Sciences）

- ***LIDAR DATA ANALYSIS and MODELS***

Laurent Menut (Inst Pierre Simon Laplace, CNRS, France)
Michael Schulz (University of Oslo,Dept. of Geosciences, Norway)
Stavros Solomos (Inst Astron Astrophys Space Applicat & RemoteSen, Greece)
Mark A. Vaughan (NASA Langley Research Center, USA)
Jianping Huang (Lanzhou University, China)

- ***STRATO-, MESOSPHERE and UPPER ATMOSPHERE***

Xinzhao Chu* (University of Colorado Boulder, USA)
Sophie Godin-Beekmann (UVSQ Université Paris-Saclay, UPMC Univ. CNRS, France)
Thomas J. McGee** (Goddard Space Flight Center, USA)
Makato Abo* (Tokyo Metropolitan University, Japan)
Takuji Nakamura (Kyoto University, Japan)
Faquan Li (Wuhan Institute of Physics and Mathematics, Chinese Academy of Sciences)

- ***AEROSOLS***

Lucas Alados-Arboledas (Universidad de Granada, Spain)
Dimitris Balis* (Aristotle University of Thessaloniki, Greece)
Edwin Eloranta (University of Wisconsin-Madison, USA)
Ulla Wandinger (Leibniz Institute for Tropospheric Research, Germany)
Silke Gross (Deutsches Zentrum f¨ur Luft- und Raumfahrt DLR,Germany)
Igor Veselovskii (Physics Instrumentation Center, Russia)
Zhengqiang Li (The Institute of Remote Sensing and Digital Earth, Chinese Academy of Sciences)

- ***BOUNDARY LAYER, POLLUTION, GREENGOUSE and TRACE GASES***

Sergey Bobrovnikov* (V.E. Zuev Institute of Atmospheric Optics, Russia)
Fabien Gibert* (Institut Pierre Simon Laplace, France)
Martin Wirth (Institut für Physik der Atmosphäre, DLR, Germany)
Alex Papayannis* (National Technical University of Athens, Greece)
Christoph Senff (NOAA Earth System Research Laboratory, USA)
Nobuo Sugimoto (National Institute for Environmental Studies, Japan)
Fred Moshary* (City College of New York, USA)
Dong Liu* (Anhui Institute of Optics and Fine Mechanics, Chinese Academy of Sciences)

- ***WIND and TURBULENCE***

Andreas Behrendt (Universität Hohenheim, Germany)
Bruce M. Gentry (NASA Goddard Space Flight Center, USA)
Michael Hardesty (National Oceanic and Atmospheric Administration, USA)
Oliver Reitebuch (Deutsches Zentrum für Luft- und Raumfahrt DLR, Germany)

Shoken Ishii* (National Institute of Information and Communications Technology, Japan)
Songhua Wu (Ocean University of China, China)

- ***NEW LIDAR TECHNOLOGIES***

Jorg Neumann (Laser Centrum Hannover, Germany)
Boyan Tatarov (University of Hertfordshire, UK)
Andreas Fix* (German Aerospace Centre, DLR, Germany)
Haris Riris (NASA Goddard Space Flight Center, USA)
Petros Mulugeta (NASA Langley Research Centre, USA)
Wei Gong (Wuhan University, China)

- ***LIDAR MEASUREMENTS COMBINATIONS, OCEAN-BIO APPLICATIONS***

Mikkel Brydegaard (Lund University, Sweden)
Reichardt Jens (Deutsches Zentrum für Luft- und Raumfahrt DLR, Germany)
Dengxin Hua (Xi'an University of Technology, China)
Liang Mei (Dalian University of Technology, China)

Travel Grant Committee

- Tom McGee (NASA Goddard Space Flight Centre, USA)
- Milton Huffaker (Coherent Investments. Inc., USA)
- Hiroaki Kuze (Chiba University, Japan)
- Eduardo Landulfo* (Instituto de Pesquisas Energeticas e Nucleares, Brazil)
- Adolfo Comeron (Technical University of Catalonia, Spain)
- Dong Liu* (Anhui Institute of Optics and Fine Mechanics, Chinese Academy of Sciences)

Local Organizing Committee

- Dong Liu* (Anhui Institute of Optics and Fine Mechanics, Chinese Academy of Sciences)
- Chenbo Xie (Anhui Institute of Optics and Fine Mechanics, Chinese Academy of Sciences)
- Huige Di (Xi'an University of Technology, China)
- Jiqiao Liu (Shanghai Institute of Optics and Fine Mechanics, Chinese Academy of Sciences)
- Yingying Ma (Wuhan University, China)
- Xiaoquan Song (Ocean University of China, China)
- Zhongwei Huang (Lanzhou University, China)

Paper Review Committee

George Tzeremes, Dave Donovan, Weibiao Chen, Anne Grete Straume, Eduardo Landulfo, Kevin Strawbridge, Michael Newchurch, Yingjian Wang, Stavros Solomos, Mark Vaughan, Gelsomina Pappalardo, Jianping Huang, Sophie-Godin Beekmann, Thomas McGee, Makato Abo, Faquan Li, Dimitris Balis, Silke Gross, Igor Veselovskii, Dong Liu, Fabien Gibert, Martin Wirth, Christoph Senff, Fred Moshary, Andreas Behrendt, Oliver Reitebuch, Shoken Ishii, Songhua Wu, Jorg Neumann, Boyan Tatarov, Andreas Fix, Wei Gong, Dengxin Hua, Liang Mei, Adolfo Comeron, Alex Papayannis

*ICLAS member

** ICLAS Executive Committee member

29th INTERNATIONAL LASER RADAR CONFERENCE

Date : 24-28 June, 2019

Location : Hefei, Anhui Province, China

Venue : The International Hall
at the Swan Lake Hotel

Website : http://ilrc29.csp.escience.cn
http://www.ilrc29.cn

ILRC29

Conference Topics

- New lidar technologies and techniques
- Atmospheric boundary layer and air pollution
- Atmospheric water vapor and temperature measurements
- Wind, turbulence and gravity waves
- Greenhouse and trace gases monitoring
- Aerosol, cloud and atmospheric radiation interactions
- Tropospheric and lower stratospheric dynamics and transport
- Stratosphere and upper atmosphere physics and chemistry
- Lidar networks and applications
- Space-based lidar, validation and global observations
- Lidar measurements combined with other sensors and techniques
- Lidar data analysis and data assimilation in models
- Ocean, land surface and biological applications

ICLAS Chairman

Prof. Alex Papayannis
National Technical University of Athens
Laser Remote Sensing Unit
Physics Department
15780 Zografou, GREECE
apdlidar@central.ntua.gr

Co-Chairmen

Prof. Yingjian Wang
Director General
Hefei Institutes of Physical Science
Chinese Academy of Sciences
350 Shushanhu Road, Hefei, Anhui 230031, CHINA
wyj@aiofm.ac.cn

Prof. Dong Liu
Anhui Institute of Optics and Fine Mechanics
Hefei Institutes of Physical Science
Chinese Academy of Sciences
350 Shushanhu Road, Hefei, Anhui 230031, CHINA
dliu@aiofm.ac.cn

Deadlines

Feb. 28, 2019	Deadline for paper submission
	Deadline for travel grant request
Mar. 31, 2019	Notification for paper acceptance
	Notification for travel grant award
Apr. 30, 2019	Deadline for early registration
May 31, 2019	Deadline for hotel reservation
Jun. 23, 2019	Registration and Welcome party

edp sciences

EPJ Web of Conferences

Statement of Peer review

In submitting conference proceedings to EPJ Web of Conferences, I certify to the Publisher that I adhere to the **Policy on Publishing Integrity** of the journal in order to safeguard good scientific practice in publishing.

1. All articles have been subjected to peer review administered by the proceedings editors.
2. Reviews have been conducted by expert referees, who have been requested to provide unbiased and constructive comments aimed, whenever possible, at improving the work.
3. Proceedings editors have taken all reasonable steps to ensure the quality of the materials they publish and their decision to accept or reject a paper for publication has been based only on the merits of the work and the relevance to the journal.

Title, date and place of the conference

the 29th International Laser Radar Conference, 24-28 June, 2019, in Hefei, Anhui Province, China

Proceedings editor(s):

Dong Liu, Yingjian Wang, Yonhua Wu, Barry Gross, and Fred Moshary

Date and editor's signature

June 21, 2020

Preface

On behalf of the International Co-ordination group for Laser Atmospheric Studies (ICLAS) I would like to welcome the readers of the volume containing the scientific Abstracts of the 29th International Laser Radar Conference held in Hefei (China) on 24-28 June 2019.

For 51 years now, the International Laser Radar Conference (ILRC) traditionally remains the world recognized venue where scientists and engineers from all over the world working in the field of the optical remote sensing applied to the atmospheres of the Earth and other planets and to the oceans, meet together to report on new results and developments and obtain a comprehensive and state-of-the-art knowledge at a vast range of topics (technology, devices, applications, education) associated with the laser radar (lidar) technique. The ILRC remains an enjoyable and stimulating meeting at which the international lidar community can get together to discuss and even argue over controversial topics and future directions. A primary focus of the Conference is to encourage young scientists to attend and present their work, ensuring the vitality of the field of laser remote sensing.

Finally, I would like to acknowledge the suppport of all sponsors of the 29th ILRC who facilitated the realisation of this Conference.

Prof. Dr. Alex Papayannis
ICLAS President

The 29th International Laser Radar Conference (ILRC29) was successfully held on June 24-28, 2019 in Hefei, China. The conference has more than 550 registrants from 35 countries/regions. Submitted papers were peer reviewed by the Paper Review Committee and 280 papers were accepted of which 86 papers including 16 invited talks were selected for oral presentations and 194 papers for poster presentations. The papers were organized in 8 sessions:

I. Space Lidars; II .Aerosol; III. Boundary Layer, Pollution, Greenhouse and Trace Gases; IV. Strato-, Mesosphere and Upper Atmosphere; V. Lidar Networks; VI. Wind, Water Vapor and Temperature Measurements, VII New Lidar Technologies and Applications; VIII. Lidar Data AnalysisaAnd Models

We appreciate the work of all the participants and the reviewers in making the conference a great success.

Review Committee: George Tzeremes, Dave Donovan, Weibiao Chen, Anne Grete Straume, Eduardo Landulfo, Kevin Strawbridge, Michael Newchurch, Yingjian Wang, Stavros Solomos, Mark Vaughan, Gelsomina Pappalardo, Jianping Huang, Sophie-Godin Beekmann, Thomas McGee, Makato Abo, Faquan Li, Dimitris Balis, Silke Gross, Igor Veselovskii, Dong Liu, Fabien Gibert, Martin Wirth, Christoph Senff, Fred Moshary, Andreas Behrendt, Oliver Reitebuch, Shoken Ishii, Songhua Wu, Jorg Neumann, Boyan Tatarov, Andreas Fix, Wei Gong, Dengxin Hua, Liang Mei, Adolfo Comeron, Alex Papayannis.

In this proceedings, 172 papers are presented which cover a broad range of topics including:

https://doi.org/10.1051/epjconf/202023700001

the initial assessment of the performance of the ESA's space-borne Doppler lidar AEOLUS, the proposed next-generation of the laser remote sensors for NASA's future earth and space science, the airborne lidar measurement for atmospheric composition and meteorological parameters, the latest research on lidar networks such as EARLINET and LALINET, and new lidar technologies and applications. The proceedings show the great progresses made by lidar scientists in every area.

We would like to thanks the 18 enterprise's sponsors (http://www.ilrc29.cn/) and all volunteers who supported this excited event

Let's meet again in the next ILRC in USA.

Co-Chairmen of the 29th ILRC
Prof. Yingjian Wang
Prof. Dong Liu

A COMPARISON OF SURFACE SLOPES EXTRACTED FROM ICESAT WAVEFORM DATA AND HIGH RESOLUTION DEM

Huan Xie [1, *], Hong Tang [1], Wenjia Du [1], Xiaohua Tong [1]

[1] *College of Surveying and Geo-informatics, Tongji University, 1239 Siping Road, Shanghai 200092, P. R. China*

**Email: huanxie@tongji.edu.cn*

ABSTRACT

Surface slope is an important topographic variable, accurate surface slope can support many research appliacations. Large footprint full waveform data has been used to estimate the surface slope and performes well. In this paper, surface slope within laser footprint is calculated using the Ice, Cloud, and land Elevation Satellite (ICESat) Geoscience Laser Altimeter System (GLAS) full waveform data and a high resolution Digital Elevation Model (REMA, the Reference Elevation Model of Antarctica). A comparison is done between two extracted surface slopes, the results show that the slopes extracted from full waveform data are close to slopes extracted from DEM, and the width of waveform can be used to extract surface slope in moderately sloping surface.

1. INTRODUCTION

The Ice, Cloud, and land Elevation Satellite (ICESat) was designed to measure ice sheet mass balance, change of polar climate, cloud and aerosol heights, land topography and vegetation characteristics[1]. From 2003 to 2009, the Geoscience Laser Altimeter System (GLAS) aboard ICESat provided 15 different kinds of products around the globe, including global elevation data, global atmosphere data, global land surface altimetry data and so on. These data have been widely used in polar ice sheet change detection, biomass information extraction, land elevation changes, oceanographic and sea-ice thickness studies, cloud height determination and vertical distribution of aerosol[2]. Accurate surface slope information is important to monitor the elevation and mass changes of ice sheet using laser altimeters[3]. For vegetation research, GLAS waveform data can be used to extract vegetation parameters such as vegetation height, fractional coverage, biomass, etc. However, the accuracy of these parameters are affected by sloping terrain, and surface slope corrections should be considered in vegetation parameters extraction[4]. The width of waveform from sloping surfaces are broader than those from flat surfaces[5], and the echo widths can be used to estimate surface slope. In this paper, a model based on triangle geometry is used to extract the surface slope from ICESat waveform data, and the surface slope is also extracted via a high resolution DEM (REMA). Finally, a comparison between the slope extracted from the waveform data and those estimated using the REMA DEM data is presented.

2. EXPERIMENTAL DATA

2.1 ICESat/GLAS data

The National Snow and Ice Data Center (NSIDC) provides 15 Level-1 and Level-2 data products from GLAS acquired between 2003 to 2009. In this paper, we used GLAH01(version 33), GLAH05 and GLAH14 data from version 34. GLAH01 contains the echo waveform signal and other waveform shape parameters, GLAH05 provides the size, shape and orientation parameters of footprints on the Earth's surface, and GLAH14 stores position information of laser footprints.

2.2 REMA

REMA (The Reference Elevation Model of Antarctica) is a high resolution, time-stamped Digital Elevation Model (DEM) of Antarctica with an 8-meter spatial resolution. REMA is extracted from satellite images including WorldView-1, WorldView-2, WorldView-3, and GeoEye-1, which were acquired between 2009 and 2017. Each individual DEM was vertically registered to satellite altimetry data from Cryosat-2 and ICESat, and the absolute uncertainties are less than 1 m over most of its area, and the relative uncertainty is on the order of decimeters[6].

3. METHODOLOGY

3.1 Surface slope extracted from waveform

GLAS mounted on ICESat transmits bell-shaped laser pulses and records the full return signals which we refer to as full waveform data. The transmitted bell pulse is modeled as a Gaussian function with a certain peak position, width and amplitude. After the pulse interacts with the earth surface, the peak position, width and amplitude of the echo waveform will be altered and the waveform of echo can be modeled as a single (or sum of several) Gaussian function(s)[7]. Generally, the echo waveform has a lower amplitude and a wider width when compared to the transmitted pulse due to the effects of attenuation, surface sloping and roughness[8]. This phenomenon allows a possibility for extracting surface slope from the broadened waveform.

Based on triangle geometry, the surface slope was described as follows[9]:

$$S = \theta + \frac{\beta_t}{2} + arctan(t_\Delta \times 10^4/\sigma_\Delta)$$

The surface slope S is calculated by the off-nadir laser pointing direction θ, the laser divergence angle β_t, the echo waveform width t_Δ, and mean pulse delay σ_Δ. From Ref. [9] we know that the median of θ is 0.009 degree, the β_t is 0.0086 degree, the σ_Δ is 4.02×10^7 ns. After decompression, smoothing and Gaussian decomposition, the width of waveform t_Δ is extracted from preprocessed waveform. Then the surface slope is calculated by the formula above.

3.2 Surface slope extracted from REMA

To validate the performance of surface slope extracted from GLAS waveform data, the surface slope is also calculated using REMA (The Reference Elevation Model of Antarctica) mosaic tiles with 8-meter spatial resolution. The location of footprint is obtained from GLAH14, and the size, shape and orientation of footprint are extracted from GLAH05. The mean slope within the footprint is derived from REMA data. Figure 1 shows the footprint on DEM, θ is the azimuth of the major axis, a is the semi-major axis of footprint, and b is semi-minor axis of footprint, which is calculated via semi-major axis and the footprint eccentricity.

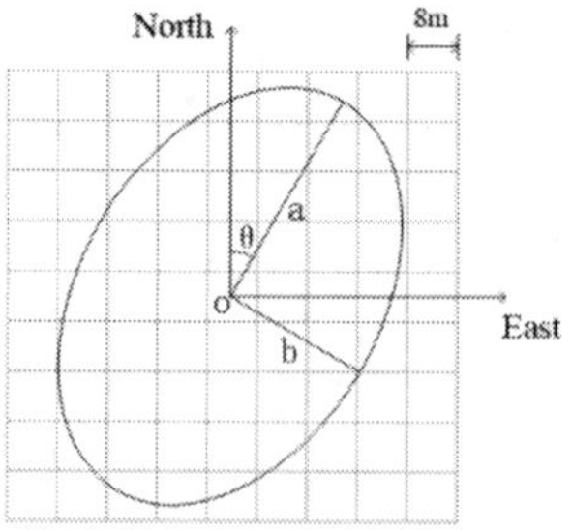

Figure 1. Footprint on DEM. The black ellipse is footprint, and the grid is DEM with 8-meter spatial resolution

4. RESULTS AND DISCUSSIONS

A comparison between slopes extracted from ICESat waveform data and REMA is shown in Figure 2. Quantitative statistics derived from this comparison are shown in Table 1. We find that the bias between two kinds of slopes is larger for areas where the surface slope is smaller than 1 degree. Hence, the slope extracted from waveform cannot be used in flat area. In flat area, the error may bigger than the true value.

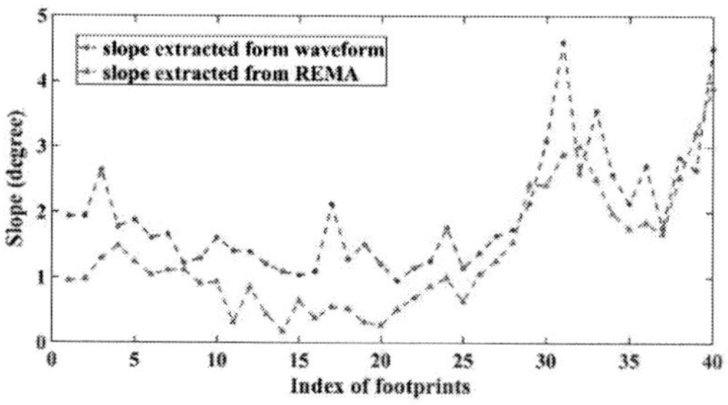

Figure 2. Comparison of waveform slopes and DEM slopes

Table 1. Statistical result of two slope types

Slope Type	Mean of Bias	Standard Deviation of Bias
Slope < 1°	0.76	0.32
Slope > 1°	0.55	0.42

In order to estimate a statistical relationship between the two slope data sets, a linear regression analysis is performed. We find that the regression result is good, the RMSE is 0.4497, and the R-square is 0.7429. There are two reasons for the bias. Firstly, the width of waveform for calculating slope is not accurate. Secondly, the waveform data and satellite images for extracting DEM were observed in different time.

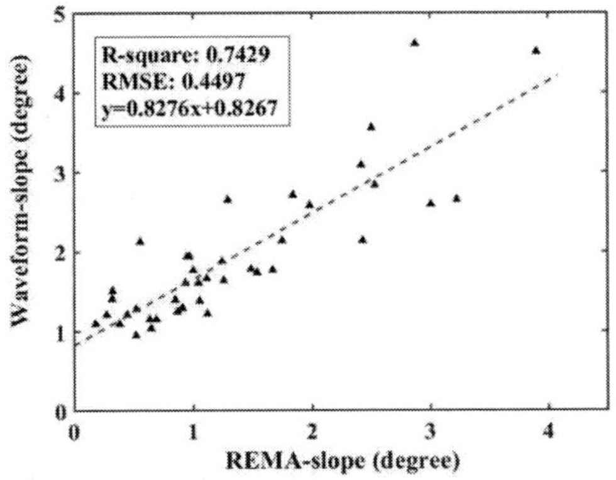

Figure 3. Scatter plot of the REMA slope versus the waveform slope and the corresponding regression model

To summarize, this paper extracts the surface slopes using ICESat/GLAS waveform data and REMA DEM data. A comparison is done between two sets of extracted slopes. The results show that there is a systematic bias between two extracted slope. The limitation of the method which extract slope from the waveform data is that it can only be used in moderately sloping surface. In the future, with the improvement of waveform data processing and simulating, as well as the improvement of the model for calculating surface slope, the results will be more accurate.

ACKNOWLEDGEMENTS

This work was supported by the National Natural Science Foundation of China under Grant No. 41822106 and 41571407, National High Resolution Ground Observation System of China (GF-7) under Grant No. 11-Y20A12-9001-17/18, Science and Technology Innovation Action Plan Program of Shanghai under Grant No. 18511102100, Dawn Scholar of Shanghai under Grant No. 18SG22, and the Fundamental Research Funds for the Central Universities of China.

REFERENCES

[1] H. J. Zwally *et al.*, "ICESat's laser measurements of polar ice, atmosphere, ocean, and land," *Journal of Geodynamics,* 34(3-4): 405-445 (2002)

[2] X. W. Wang et al., "Earth science applications of ICESat/GLAS," *International Journal of Remote Sensing* 32(23): 8837-8864 (2011)

[3] S. Nie *et al.*, "A Novel Model for Terrain Slope Estimation Using ICESat/GLAS Waveform Data," *Ieee Transactions on Geoscience and Remote Sensing,* 56(1): 217-227 (2018)

[4] C. Mahoney *et al.*, "Slope Estimation from ICESat/GLAS," *Remote Sensing,* 6(10): 10051-10069 (2014)

[5] D. H. Yi *et al.*, "ICESat measurement of Greenland ice sheet surface slope and roughness," *Annals of Glaciology,* 42: 83-89 (2005)

[6] I. M. Howat *et al.*, "The Reference Elevation Model of Antarctica," *Cryosphere,* 13(2): 665-674 (2019)

[7] V. H. Duong, "Processing and Application of ICESat Large Footprint Full Waveform Laser Range Data," (2010)

[8] A. C. Brenne *et al.*, "Geoscience laser altimeter system algorithm theoretical basis document: Derivation of range and range distributions from laser pulse waveform analysis. Algorithm Theoretical Basis Documents (ATBD)," (2003)

[9] X. L. Li *et al.*, "Terrain slope estimation within footprint from ICESat/GLAS waveform: model and method," *Journal of Applied Remote Sensing,* 6: 24 (2012)

AEOLUS L2 ACTIVITIES AT KNMI

D. P. Donovan*, Gert-Jan Marseille, Jos de Kloe and Ad Stoffelen.

Royal Netherlands Meteorological Institute (KNMI), PO Box 201,
3730 AE de Bilt, the Netherlands
**Email: Donovan@knmi.nl*

ABSTRACT

KNMI has had a long involvement with the recently launched Aeolus mission. Aeolus is an ESA Earth-Explorer mission and embarks the Atmospheric Laser Doppler Instrument (ALADIN). ALADIN is the first space-borne Doppler Wind Lidar (DWL) and was successfully launched in late August 2018. KNMI has had a long association with this mission and is sharing the software development of the L2B processor used to generate L2B wind products at ECMWF, develops related scene classification algorithms and performs calibration/validation activities. In addition, KNMI conducts more fundamental work for improving the retrieval of atmosphere optical properties, taking into account multiple scattering effects.

1. INTRODUCTION

ALADIN is a high spectral resolution Doppler lidar operating at 355 nm mounted on the Aeolus satellite. Aeolus orbits in a sun-synchronous dusk/dawn orbit at 320 km height. ALADIN currently emits 40-60 mJ pulses of circularly polarized light, slightly less than the opted 80 mJ, at a PRF of 50 Hz. The backscattered light is collected using a 1.5 m diameter telescope. The return signal is separated into two channels; the so-called "Mie" and "Rayleigh" channels.

The Mie channel uses a Fizeau interferometer coupled to an ACCD array to spectrally resolve and enable the separation of the spectrally narrow elastic backscatter (Mie) peak from clouds/aerosol and the spectrally broader return from atmospheric molecules. By measuring the spectral position of the Mie peak, the Doppler shift relative to the emitted signal frequency can be estimated and thus the line-of-sight velocity of the cloud/aerosol particles deduced. About 20% of the atmospheric samples have, in general, a cloud/aerosol return of sufficient signal-to-noise ratio to retrieve accurate winds, depending on the vertical sampling strategy.

The Rayleigh channel is built around a dual Fabry–Pérot interferometer. The etalon is split into two zones, which are imaged separately on to an ACCD detector. The two resulting sub-channels are offset spectrally either side of the center frequency such that the spectrally broad Doppler shift of the atmospheric backscatter can be deduced using the ratio of the two branches of the Rayleigh channel returns. For more information see [1]. The Rayleigh channel is unique in that it provides wind information in clear air and it provides generally winds for about 80% of the samples, again depending on vertical sampling settings.

ALADIN is tilted 35degrees off nadir in the direction perpendicular to the orbit track enabling the estimation of horizontally-projected line-of-sight (HLOS) atmospheric wind vector components with a typical resolution of 87 km along-track and a targeted accuracy of 2 to 3 m/s throughout the troposphere and lower stratosphere. The vertical resolution is fixed to 24 bins in each channel with bin sizes equal to multiples of 250 m , typically ranging from 250 m in the lowest km to 1 to 2 km up to an altitude of about 30 km for the Rayleigh channel.

The line-of-sight wind profiles (which are being assimilated into numerical weather prediction models and used to monitor atmospheric dynamics and circulation) are the primary Aeolus products. However, cloud and aerosol optical properties are also being produced with a horizontal resolution down to 3 km, the basic Aeolus horizontal integration length, as secondary product.

2. KNMI ACTIVITIES

As part of the ESA Aeolus DISC project, KNMI is leading the development of the so-called chain-of-processors, an end-to-end cal/val and Aeolus product R&D tool, operated in an internet cloud and shared within the Aeolus Cal/Val team. It is used for end-to-end sensitivity tests in the wind processing chain and accumulates new calibration settings, algorithm and tool development and product monitoring and evolution. This tool is relevant for all scientific activities towards Aeolus product improvement.

2.1 Cal/Val

To support Cal/Val activities of wind products, ECMWF and KNMI are involved in monitoring and diagnosing the optical properties and wind products from Aeolus, using prior weather information from the ECMWF model as reference. These fields are available at all Aeolus measurement locations and their availability facilitated a swift identification of anomalies in the Aeolus data processing, the resolution of some major anomalies and the beneficial assimilation of early and preliminary Aeolus data in the ECMWF model.

In addition to model data, ground-based data is used for validation. In fact, KNMI combines all data types in triple collocation for error attribution. In particular, over land Europe sees a large volume of tropospheric wind data emerging through Mode-S airplane winds [2]. These winds are collocated with Aeolus and ECMWF winds for error analysis at each vertical level from the ground to 11 km.

Finally, KNMI launches radiosondes from Paramaribo in Suriname in order to obtain high-resolution tropical wind, humidity and temperature profiles collocated with Aeolus overpasses. Elsewhere, this is augmented with other in-situ continuous profiling networks, such as Cabauw in the Netherlands.

2.2 Scene classification and optical properties

Scene classification is an important component of the Level-2B wind processor (L2Bp). Its main objective is to detect cloud and aerosol layers along the Aeolus track. The result is a binary map, also denoted feature mask, used in the L2Bp classification procedure as input for the accumulation of Mie and Rayleigh signals from measurement (2.88 km along track) to observation level (87 km along track).

Two options for scene classification have been implemented in the L2Bp, (i) based on the application of a threshold value to the scattering ratio estimate calculated by the Level-1B processor and (ii) from the Optical Properties Code (OPC), which is part of the L2Bp. Both options are operated on measurement level, which poses a challenge because of the high levels of measurement noise on top of the atmospheric return signal.

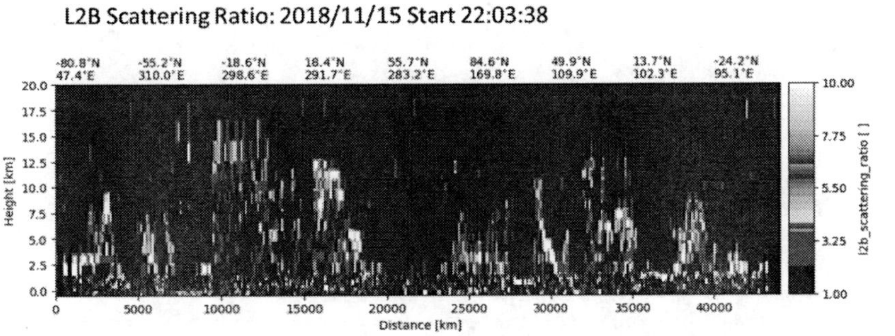

Figure 1: Sample preliminary L2B scattering ratios estimates (derived from Rayleigh channel)

Sample L2B scattering rato preliminary results are shown in Figure 1. Both L2B and L1B options are complementary because the L1B scattering ratio is inferred from the Mie channel signal only, while OPC uses the Rayleigh channel signal only. OPC also contributes to a mission risk reduction in case the Mie channel performs less than anticipated. Since scene classification is a crucial procedure in the L2Bp, the quality of produced winds strongly relies on the performance of the classification procedure in case of cloud and/or aerosol presence.

The L2 processor development in the ESA DISC project and in collaboration with ECMWF is being further optimized for to optimize the NWP impact.

2.3. Impact of Lidar Multiple Scattering on ALADIN retrievals.

For much of the ALADIN processing it is implicitly assumed that lidar multiple scattering effects are negligible. It is true that the receiver field-of-view (fov) of Aeolus is small (0.019) mrads), a value which is an order of magnitude smaller than those commonly used by terrestrial lidar systems. On this basis it may be supposed that lidar multiple-scattering effects present in Aeolus return signals will be small and can safely be ignored. This assumption, however, is not always valid.

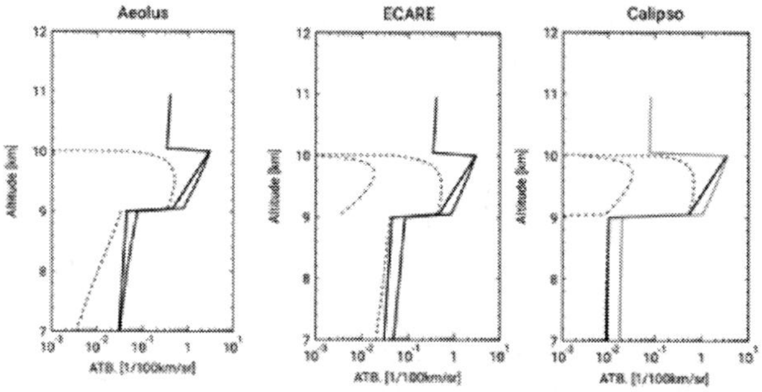

Figure 2: Simulated relative (total Ray+Mie) attenuated backscatter profiles corresponding to an idealized cirrus cloud with an optical thickness of 1 and an effective particle size of 20 microns. The Black lines correspond to the single-scatter only return, the Red lines show the 2nd and 3rd order scattering respectively (3rd order return for the Aeolus case is too small to be shown) and the Blue and Green lines represent the total single scattering (SS)+multiple scattering (MS) returns.

The relative strength of Multiple-Scattering (MS) in lidar returns depends on the laser divergence, the receiver field-of-view (fov), the distance from the target, the target extinction coefficient and the scattering phase function of the target. Of particular importance is the relative strength of the forward-scattering lobe of the phase function which increases with increasing particle sizes. The more forward scattering that occurs, the more chance multiple-scattering photons will stay within the receiver fov. This can be relevant even when the optical thickness of the target is small. This means that, e.g., cirrus clouds whose typical particle effective radii can be several -tens of microns are general targets where MS effects are particularly evident. Figure 1 shows an example of the results of an approximate analytical model [3] to an idealized cirrus cloud of optical thickness 1.0 with an effective particle size of 20 microns. Cases corresponding to CALIPSO, ATLID (the EarthCARE lidar) and Aeolus are shown. Here it can be seen that within the cloud the MS contribution can approach (and in the case of CALIPSO) exceed that of the SS return. Below the cloud, the return from the molecular Rayleigh atmosphere also contains a significant MS component arising from forward scattering events within the cloud above. In the case of CALIPSO a slight decay in the below-cloud MS return is present. For ATLID, the decay is somewhat more apparent and is obvious for Aeolus.

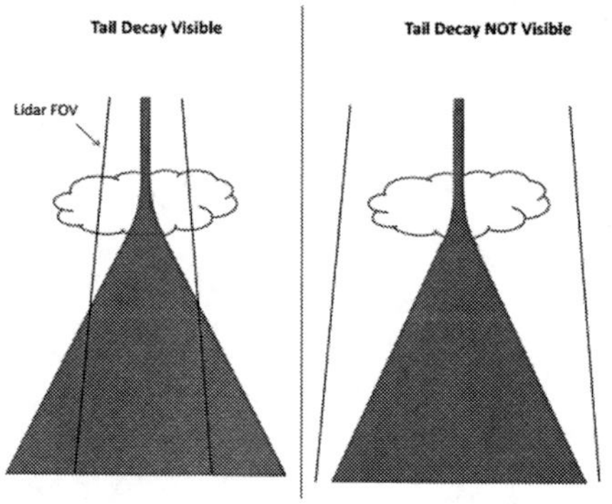

Figure 3: Schematic illustrating the basic mechanism behind the appearance of below cloud "tails" in space-based lidar returns.

The mechanism behind the appearance of these tails is illustrated schematically in Figure 2. Basically the scattering events occurring within the cloud serve to broaden the angular-spatial extend of the lidar beam. This beam then emerges from the cloud-base with a certain effective diameter and divergence. If the effective divergence is small compared to the lidar fov then no decay will be visible, but if this is not the case, then these type of "tails" will be apparent. Such tails indeed can be observed for CALIPSO but only for cases of relatively small cirrus particles (e.g. see Appendix A of [4]).

The approach used in [3] is limited in its accuracy and cannot accurately address the effects of MS on the spectral-polarization state of the lidar return. However, accurate lidar Monte-Carlo calculations can be used for this purpose. In this work the lidar radiative transfer model and instrument model of the EarthCARE simulator will be applied to the case of Aeolus. As well as exactly treating the lidar multiple-scattering also calculates the spectral-polarization state of the return signal [4,5].

Implications for Aeolus

Neglecting lidar MS effects in the retrieval algorithms may lead to a number of issues in the L2 products if not ameliorated in some manner.

The retrieval of cloud optical depth will be underestimated by up to several tens of percent if MS effects are not accounted for. For example, if the method of comparing the directly above and below cloud Rayleigh returns would be applied to the case shown in Figure 1 then the cloud transmission would be off by a factor of more than 2.

The polarization-spectral state of the MS return will not be identical to that of the SS only return. In particular, the greater angular distribution of the MS returns would be expected to alter the spectra of the in-cloud and even below cloud returns. The impact on the spectra shape are expected to be small though due to the small laser divergence and receiver field-of-view and not important for wind-retrievals.

In order to facilitate the investigation of anomalies and the possible effects of multiple-scattering on Aeolus products the EarthCARE simulator lidar modules [5] is being extended and tested in order to accurately handle Aeolus parameters and viewing geometry. This investigation contributes to the broader assessment of the feature detection/classification and optical properties retrievals with a particular effort directed to the quantification of multiple-scattering effects on the instrument performance (e.g., calibration, cross-talk) and on the feature detection/classification and optical properties and wind retrievals.

3. SUMMARY

KNMI is supporting the Aeolus mission in several areas. The activities at KNMI include leading the development of the so-called chain-of-processors,, participating in data monitoring and various Cal/Val activities and developing the L2B Scene classification algorithm (an important component of the Level-2B wind processor). In addition, more fundamental research relevant for Aeolus (e.g. the impact of MS on Aeolus signals) is also be conducted.

ACKNOWLEDGEMENTS

This work is supported in by ESA under the Aeolus DISC contract no. 40000126336/18/I-BG.

REFERENCES

[1] "ADM-Aeolus (Atmospheric Dynamics Mission)". https://directory.eoportal.org/web/eoportal/satellite-missions/a/aeolus ESA. Retrieved 18 Feb, 2019.

[2] Haan, S. de, and L. J. Bailey, and J. E. Konnen, 2013: Quality assessment of Automatic Dependent Surveillance Contract (ADS-C) wind and temperature observations from commercial aircraft, *Atmos. Meas. Tech.*, **6**, 199 – 206.

[3] E.W. Eloranta, 1998: Practical model for the calculation of multiply scattered lidar returns, *Appl. Opt.* **37**, 2464-2472 (1998).

[4] Reverdy, M., H. Chepfer, D. Donovan, V. Noel, G. Cesana, C. Hoareau, M. Chiriaco, and S. Bastin , 2015: An EarthCARE/ATLID simulator to evaluate cloud description in climate models, *J. Geophys. Res. Atmos.*, **120**, 11,090–11,113.

[5] Donovan, D. P., Klein Baltink, H., Henzing, J. S., de Roode, S. R., and Siebesma, A. P. 2015: A depolarisation lidar-based method for the determination of liquid-cloud microphysical properties, *Atmos. Meas. Tech.*, **8**, 237-266.

AN ADAPTIVE FILTER BASED ON LOCAL SLOPE BY ADJUSTING ELLIPTIC FILTER KERNEL FOR ICESAT-2 PHOTON COUNTING LASER ALTIMETRIC DATA

Huan Xie [1,*], Dan Ye [1], Gang Hai [1], Xiaohua Tong [1]

[1] *College of Surveying and Geo-informatics, Tongji University, 1239 Siping Road, Shanghai 200092, P. R. China*

**Email: huanxie@tongji.edu.cn*

ABSTRACT

The Ice, Cloud and Land Elevation Satellite-2 (ICESat-2) was launched on September 15th, 2018, which continues to perform measurement tasks as the successor to ICESat. Unlike full waveform technology of ICESat, ICESat-2 employs micropulse photon counting technology, which provides higher accuracy, but produces plenty of noise. This paper proposes an adaptive filter based on local slope by adjusting elliptic filter kernel. The general approach is 1) data preprocessing, 2) Gaussian density calculation, 3) OTSU adaptive threshold calculation. This method is seen to be robust in identifying signal points from high background noise points and suitable for low density data caused by slope.

Index Terms— photon counting, noise filtering, Gaussian density, ICESat-2

1. INTRODUCTION

Follow-up mission to ICESat, the main science objectives of ICESat-2 are measuring ice sheet changes, sea ice thickness and vegetation biomass[1]. ICESat-2 is equipped with the Advanced Topographic Laser Altimeter System (ATLAS), which collects elevation data using photon counting technology. The photon counting LiDAR has high level of sensitivity and high repetition frequency, but needs low pulse energy, which makes it possible to run at a higher altitude and provide larger coverage[2]. ICESat-2 is also the first time to equip a photon counting LiDAR on a satellite for Earth observation.

The ATLAS instrument adopts a laser pulse in the green (532 nm), which is spilt into 6 beams organized in 3 pairs. The distance is 3.3 km between each pair and 90 m inside each pair. Each pair consists of a strong and a weak beam, and the strong is 4 times as the weak. The sensor uses a 10 kHz repetition rate (~0.7 m on the ground) at the altitude of ~500 m and produce ~17 m overlapping footprint[3]. The data is discrete point clouds, which mainly come from target reflection, atmospheric scattering, solar background noise and the instrument dark count[4]. An efficient filter method for photon counting LiDAR data is extremely important to obtain ground information.

In 2014, a density-based clustering method is proposed[5]. This method changes the shape of search area from a circle to an ellipse and calculated local area density. Subsequently the signal points were extracted according to the threshold of density. It could effectively extract ground and canopy, but did not consider the impact of surface slope. For inclined surface, the direction of ellipse consistent with the slope can get better clustering results. This paper changes the direction of ellipse according to surface slope. Since near-ground noise points are difficult to distinguish from ground points, we give a greater weight in the direction of the track when calculating the density, and adopt the Gaussian density. Because the size of the laser footprint varies with the slope, the size of the elliptical nucleus also varies depending on the slope. Finally, the OTSU adaptive threshold method[6] is used to obtain a threshold of density for filter.

2. METHODOLOGY

In this paper, an improved adaptive filter method is proposed, as shown in Figure 1.

2.1 Preprocessing

Due to the edge effect, the density of points in the bounding area are smaller, the edge points are supplemented by mirror image method[7].

2.2 The elliptical Gaussian density calculation

Since the signal points are tightly distributed, like a line, and the density is greater than the noise, the filter kernel has changed to an ellipse[5]. The size of semi-major axis a and semi-minor axis b of the ellipse at p point were determined by the topographical features and the ground reflectivity.

$$dist(p,q) = \left[\frac{\Delta X_\theta{}^2}{a^2} + \frac{\Delta h_\theta{}^2}{b^2}\right]^{\frac{1}{2}}$$

$$\begin{cases} \Delta X_\theta = cos\theta(X_p - X_q) + sin\theta(h_p - h_q) \\ \Delta h_\theta = cos\theta(h_p - h_q) - sin\theta(X_p - X_q) \end{cases}$$

Where: θ is the slope of the center p, X_p and X_q are the distances along the track, and h_p and h_q are the elevations. If $dist(p,q)$ is less than 1, the point is inside the ellipse, else outside the ellipse.

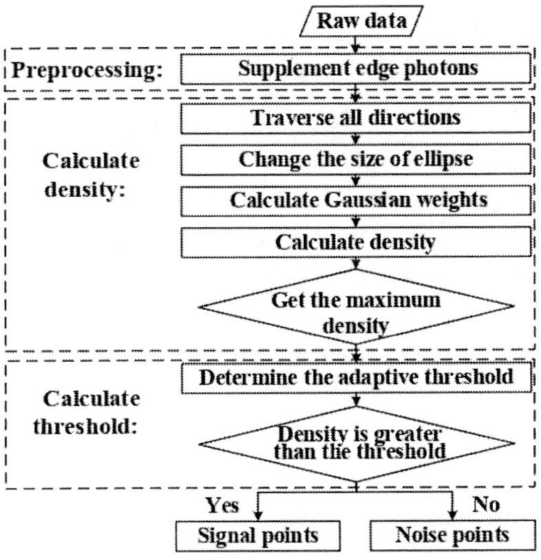

Fig 1. The flow chart of the elliptical Gaussian density adaptive filter method

Since the laser light irradiate on the slope and the distribution of laser energy varies, the size of the ellipse also varies accordingly.

Centering on each point, the points within the ellipse are given Gaussian weights[8]. The sum of the weights of all points in the ellipse is the density of the center point. There is the highest density along the tangential direction of the ground for ground points. Traversing all directions, the direction θ_i in which has the maximum density value is considered to be the slope direction, and the direction of filter kernel is set to θ_i.

2.3 OTSU adaptive threshold calculation

The OTSU method is an algorithm for determining the image binarization threshold. In principle, the method is called the maximum inter-class variance method. Because the image binarization is divided according to the threshold obtained by the OTSU method, the variance between the foreground and the background image is the largest. In this paper, we use the OTSU method to obtain the density adaptive threshold and extract the signal points.

3. EXPERIMENT AND RESULTS

Because the ICESat-2 data has not yet been released, this paper use its onboard simulation data MABEL (the Multiple Altimeter Beam Experimental Lidar) instead. In order to verify that the algorithm is suitable for terrain with slope, we chose the MABEL data with a steep terrain gathered on August 2, 2014.

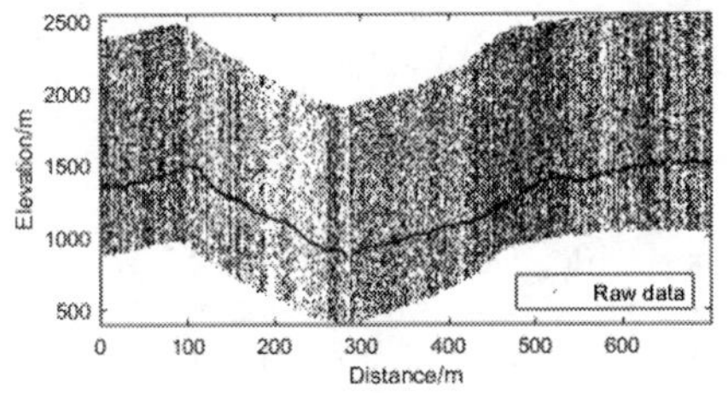

Fig 2. Profile of raw data, the x-coordinate is the distance along the track, and the y-coordinate is the elevation

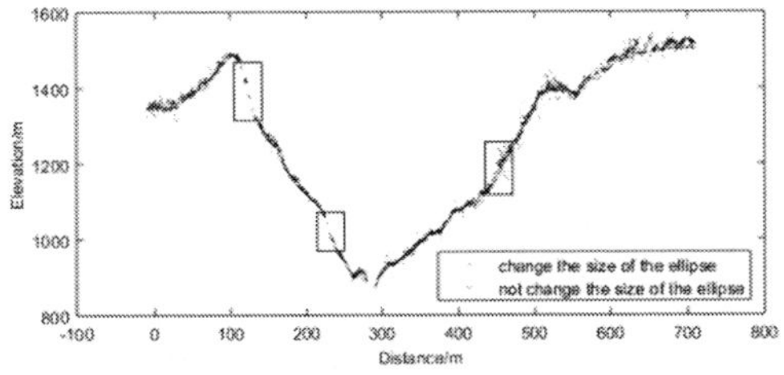

Fig 3. Results of signal points extraction (the green dots are the result of the ellipse size according to the slope, and the blue don't change)

The results of signal points extraction are shown in Figure 3. Parameters used in this paper are: $a = 10$, $b = 2$, Gaussian weight variance $\sigma = 2$. Here the blue dots represent the result without changing the size of the cluster ellipse, and the green dots represents the filter result of the ellipse size according to the gradient. The results within the red rectangle show that the method is more effective for extracting signal points on the slope surface.

4. CONCLUSION

In this paper, the existing elliptical cluster filter method is improved. In terms of density calculation, the near-ground noise points and the ground points are considered very close. This paper uses the Gaussian density to give a greater weight in the direction of the track. Secondly, since the footprint and energy of the laser vary in the slope, the density becomes smaller, and we change the ellipse size according to the slope. The OTSU method is used to determine an adaptive threshold. The results show that the filter method in this paper have some improvement in the slope.

ACKNOWLEDGEMENTS

This work was supported by the National Natural Science Foundation of China under Grant No. 41822106 and 41571407, National High Resolution Ground Observation System of China (GF-7) under Grant No. 11-Y20A12-9001-17/18, Science and Technology Innovation Action Plan Program of Shanghai under Grant No. 18511102100, Dawn Scholar of Shanghai under Grant No. 18SG22, and the Fundamental Research Funds for the Central Universities of China.

REFERENCES

[1] W. Abdalati et al., "The ICESat-2 Laser Altimetry Mission," Proceedings of the IEEE, 98(5): 735-751 (2010)

[2] J. J. Degnan, "Photon-counting multikilohertz microlaser altimeters for airborne and spaceborne topographic measurements," Journal of Geodynamics, 34(3): 503-549 (2002)

[3] L. A. Magruder et al., "Performance Analysis of Airborne Photon- Counting Lidar Data in Preparation for the ICESat-2 Mission," IEEE Transactions on Geoscience and Remote Sensing, 56(5): 2911-2918 (2018)

[4] F. Xie et al., "An adaptive directional filter for photon counting Lidar point cloud data" (in Chinese), Journal of Infrared and Millimeter Waves,1: 107-113 (2017)

[5] J. Zhang et al., "A clustering approach for detection of ground in micropulse photon-counting LiDAR altimeter data," in Geoscience and Remote Sensing Symposium, 177-180 (2014)

[6] N. Ohtsu, "A Threshold Selection Method from Gray-Level Histograms," IEEE Transactions on Systems Man & Cybernetics, 9(1): 62-66 (1979)

[7] S. Nie et al., "Estimating the vegetation canopy height using micro-pulse photon-counting LiDAR data," Opt Express, 26(10): A520-A540 (2018)

[8] U. C. Herzfeld et al., "Surface-Height Determination of Crevassed Glaciers-Mathematical Principles of an Autoadaptive Density-Dimension Algorithm and Validation Using ICESat-2 Simulator (SIMPL) Data", Ieee Transactions on Geoscience and Remote Sensing, 55(4): 1874-1896 (2017).

CO_2 PROFILING BY SPACE-BORNE RAMAN LIDAR

Paolo Di Girolamo[1*], Carmine Serio[1], Volker Wulfmeyer[2], Andreas Behrendt[2], Davide Dionisi[3]

[1]*Scuola di Ingegneria, Università della Basilicata, Potenza, Italy*
[2]*Institut fuer Physik und Meteorologie, Universitaet Hohenheim, Stuttgart, Germany*
[3]*ISMAR,CNR, Roma, Italy*
**Email: paolo.digirolamo@unibas.it*

ABSTRACT

As clearly reported in the IPCC fifth Assessment Report, CO_2 emissions are already producing destructive effects to the plant ecosystem through the alteration of soil-atmosphere interaction mechanisms.

Although the space and ground network for CO_2 monitoring has regularly expanded over the past 50 years, it does not guarantee the necessary spatial and temporal resolution needed for a quantitative analysis of sources and sinks. For the purpose of estimating forests' carbon capturing capabilities, accurate measurements of CO_2 gradients between the forest floor and the top of the canopy, which ultimately translates into the capability to measure CO_2 concentration profiles. Space sensors provide CO_2 measurements above forest canopies, which do not allow to properly estimate Gross Primary Production (GPP).

These observational gaps could be addressed with an active remote sensing system in space based on the vibrational Raman lidar technique. CO_2 profile measurements are possible, together with simultaneous measurements of the temperature and water vapour mixing ratio profile and a variety of additional variables (aerosol backscatter profile, aerosol extinction profile, PBL depth, cloud top and base heights, cloud optical depth). An assessment of the expected performance of the system has been performed based on the application of an analytical simulation model developed at University of Basilicata.

1. INTRODUCTION

Global warming is a consequence of the alteration of Earth's natural greenhouse effect. More than 75 % of this effect is dominated by water vapour, while the remaining 25 % primarily by CO_2. While human activities modify the Earth's hydrological cycle in a minor extent, they substantially alter the carbon cycle: CO_2 mixing ratio in the atmosphere has substantially increased from values around 300 ppm in the fifties to a current value exceeding 405-410 ppm, with an annual increasing rate of approximately 2 ppm.

Approximately half of CO_2 amount produced through fossil fuel combustion and other human activities is injected in the atmosphere and accumulates in it. The remaining part is absorbed by the oceans and the terrestrial biosphere (e.g. [1]). An appropriate quantitative assessment of the various components of the carbon cycle requires accurate measurements of the different sources and sinks and, ultimately, of chlorophyll vegetation photosynthesis, i.e. the Gross Primary Productivity or GPP.

Forests cover ~30% of the Earth's global land area [2]. Forests store large amounts of carbon captured from the atmosphere and retained in living and dead biomass and soil. The estimated amount of carbon dioxide (CO_2) in the atmosphere is equivalent to an amount of carbon of 8×10^{14} kg, while the amount stored in terrestrial biomass is 5×10^{14} kg, 60% of which (3×10^{14} kg) being stored in forest systems [3]. The missing balance between the carbon released in the atmosphere through the combustion of fossil fuels and deforestation on one side, and the uptake by sinks in oceanic and terrestrial systems on the other side has led to the 2 ppm per year increase in atmospheric CO_2 concentration mentioned at the beginning of this section.

Based on the above considerations, it appears clear that accurate measurements of CO_2 gradients between the forest floor and the top of the canopy, and their temporal variations, are urgently needed. These measurements are of paramount importance for an appropriate assessment of the potential impact of increased atmospheric CO_2 on ecosystems and for the characterization of coupling processes between the forest and the convective boundary layer.

Our current knowledge about atmospheric CO_2 concentrations and surface fluxes at regional scales over the globe comes primarily from ground-based in situ measurements from air

https://doi.org/10.1051/epjconf/202023701004

sampling networks and tall towers [4]. However, due to the sparseness of such measurements, there are still large uncertainties especially on the surface fluxes [5].

CO_2 measurements based on the application of the vibrational Raman lidar technique have been carried out since the early nineties [6-7]. Despite its high potential impact, the Raman lidar technique for CO_2 measurements has received limited attention in the last 25 years both at theoretical and experimental level, mainly because of its limited precision resulting from the small cross-section of vibrational Raman scattering and the low CO_2 concentration.

However, in the last decade, considerable technological advances have been achieved in the design and development of high-power solid-state laser sources, large-aperture telescopes and high gain/quantum efficiency detectors. These advances allow today reaching a new level of performance in CO_2 Raman lidar measurements from both ground and space platforms.

In the present paper, system performance in terms of measurement precision of space-borne CO_2 Raman lidar has been estimated based on the application of an analytical simulator developed at Univ. of Basilicata [8,9].

2. METHODOLOGY

2.1 CO_2 vertical variability

The seasonal and annual mean of CO_2 vertical profile reflect the combined influences of surface fluxes and atmospheric mixing. In the Northern Hemisphere, during the summer season, midday atmospheric CO_2 concentrations are generally lower near the surface than in the free troposphere, reflecting the higher impact of terrestrial photosynthesis over industrial emissions [5]. Conversely, during the winter, respiration and fossil-fuel sources lead to elevated low-altitude atmospheric CO_2 concentrations. The gradients are comparable in magnitude in both seasons, but the positive gradients persist for a larger portion of the year, leading to annual-mean gradients also showing higher atmospheric CO_2 concentrations near the surface than aloft [5]. The average midday differences in the Northern Hemisphere between 1 and 4 km, expressed in terms of CO_2 mixing ratio, is –2.2 ppm in summer, +2.6 ppm in winter, while instantaneous differences may be as large as 5-10 % [10]. The Southern Hemisphere locations show relatively constant CO_2 profiles in all seasons, with slightly higher values in the free troposphere [5]. Similar gradients are found in the NOAA's Carbon Tracker version 2010 [4].

CO_2 gradients between the forest floor and the top of the canopy range between 75 and 100 ppm at night, compared to 10-50 ppm during the day. Both ranges represented spring conditions when canopy leaf area development is not completed [11].

2.2 The lidar system

The Raman lidar technique has proved to be very effective in the profiling of atmospheric thermodynamic and compositional properties [12,13,14]. The Raman lidar BASIL has been deployed in a variety of international field experiments [15,16,17,18,19,20,21].

The lidar setup considered in the present research effort relies heavily on the ATLAS mission concept [9] which was submitted to the European Space Agency as part of the Earth Explorer-10 Call. Simulations consider a sun-synchronous low Earth orbit, with an orbiting height and speed of 450 km and 7 km/s, respectively. A dawn-dusk orbit with overpasses at 6/18 h local time has been selected for the simulations.

The lidar transmitter consists of a frequency-tripled, diode-laser pumped Nd:YAG laser, with an an average power in the UV (at 354.7 nm) of 250 W (single pulse energy = 2.5 J, repetition rate = 100 Hz). An electrical-to-optical efficiency > 5 % is assumed in order to achieve such high laser power based on the available on-board electrical power. The receiver consists of a large-aperture telescope, with a primary-mirror diameter of 4 m, and a high efficiency receiving unit for the collection of the Raman signals. Receiving field-of-view (FWHM) is 10 μrad. The present mission concept considers a Raman lidar system collecting five primary lidar signals: the CO_2 vibrational and water vapour roto-vibrational Raman signals, $P_{CO2}(z)$ and $P_{H2O}(z)$, respectively, the high- and low-quantum number rotational Raman signals, $P_{LoJ}(z)$ and $P_{HiJ}(z)$, and the elastic backscatter signal at λ_0, $P_{\lambda 0}(z)$.

The spectral selection is performed through the use of narrow-band interference filters (IFs). Specifically, the IFs are used to select the Q-branch of the two bands ν_1 and $2\nu_2$ both in the Stokes and anti-Stokes branches of CO_2

vibrational Raman band. IFs have central wavelengths (CWLs) of 338.1, 339.2, 371.71 and 373.1 nm, bandwidths of 0.1 nm, transmissions at CWLs of 80%, out of band rejections of OD6 @ 200-1200 nm, OD12 @ 354.7 nm and OD7 @ 325-340 and 370-390 and nm. Reported simulations consider photon detection by accumulation charge-coupled devices (ACCDs), with UV quantum efficiencies of 85%.

The central wavelength of the CO_2 band is almost coincident with the twenty-first anti-Stokes roto-vibrational O_2 line, which represents a potential source of contamination for CO_2 Raman lidar measurements [12]. Riebesell [6] and Ansmann et al. [7] came to the conclusion that CO_2 Raman lidar measurements were hardly feasible due to the difficulties in accurately estimating the magnitude of the interference by O_2 rotational lines. These authors also argued that fluorescence generated by either the receiver optics or atmospheric particles could potentially prevent from reaching a measurement accuracy at the 1 ppm level.

However, these early research efforts were conducted using excimer laser sources (Xe:Cl mixture), which have an emission spectrum typically spanning over a spectral interval of ~0.4 nm. Such broad spectrum makes separation of CO_2 and O_2 lines more difficult with respect to what is presently achievable based on the use of narrowband interference filters (0.1 nm) and injection-seeded Nd:YAG laser sources (typical spectral width of ~0.01 nm). Whiteman et al. [12] demonstrated that the bias affecting CO_2 measurements as a result of the contamination by the 21st O_2 rotational line is not exceeding 1 % if a 0.1-nm wide IF and a narrow-band Nd:YAG laser source (<0.02 nm) are used. Based on an appropriate rotational strength modeling as a function of temperature and relying on independent temperature profile measurements provided by the same lidar system, the magnitude of the interference can be properly estimated and subtracted from the CO_2 signal so that a residual systematic uncertainty of the order of 0.1-0.2 ppm is left in CO_2 mixing ratio measurements.

3. RESULTS

3.1 Input data for the Simulator

An appropriate input CO_2 mixing ratio profile was generated which accounts for different vertical gradients associated with the conflicting contributions of terrestrial photosynthesis and industrial emissions, and CO_2 capturing within forest canopy.

This input profile includes a 5 ppm increase at an altitude of 2 km (from a value of 405 to one of 410 ppm), introduced in order to simulate the daytime CO_2 depletion taking place within the mixed layer. In addition to this, a 50 ppm decrease is considered at an altitude of 50 m above the surface level, which is intended to represent CO_2 capturing within forest canopy.

Atmospheric quantities considered in the simulation include vertical profiles of pressure, temperature, and humidity from the U.S. Standard Atmosphere 1976 atmospheric reference model. Aerosol optical properties are simulated based on the use of the median aerosol extinction data from the ESA ARMA Model [22].

3.2 Simulation results

Considering vertical and horizontal resolutions of 200 m and 200 km, respectively, the statistical uncertainty affecting CO_2 mixing ratio profile measurements is not exceeding 2 ppm at night and 10 ppm in daytime from the surface up to an altitude of 5 km, with slightly higher values in the vertical region from the forest floor to the top of the canopy (Fig. 1). Ground-based demonstrations of the present instrument concept are under implementation both at Univ. of Basilicata and Univ. of Hohenheim.

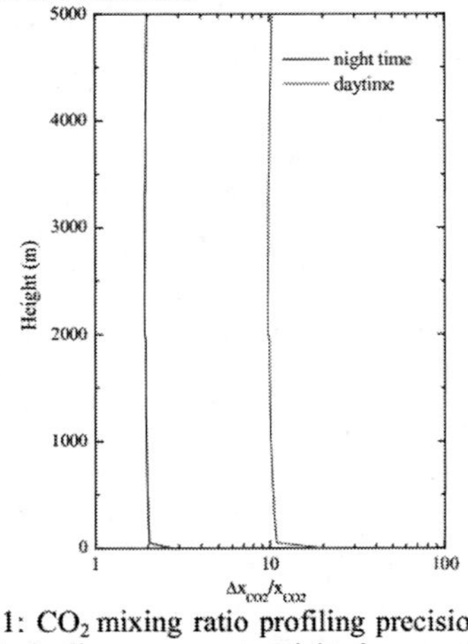

Figure 1: CO_2 mixing ratio profiling precision (in %). The black line represents night-time performance, while the red line represents daytime performance.

ACKNOWLEDGEMENTS

This work was supported by the Italian Ministry for Education, University and Research under the Grant OT4CLIMA.

REFERENCES

[1] C. Le Quéré *et al.*, The global carbon budget 1959–2011, Earth Syst. Sci. Data, 5, 165–185 (2013).

[2] G.J. Nabuurs et al., Forestry. In Climate Change 2007: Mitigation. Contribution of Working Group III to the Fourth Assessment Report of the Intergovernmental Panel on Climate Change, Cambridge University Press, Cambridge, UK and New York, NY (2007).

[3] D.C. McKinley *et al.*, A synthesis of current knowledge on forests and carbon storage in the United States, Ecol. App., doi:10.1890/10-0697.1 (2011).

[4] M. Reuter *et al.*, A simple empirical model estimating atmospheric CO_2 background concentrations, Atmos. Meas. Tech., 5, 1349–1357 (2012).

[5] B. B Stephens *et al.*, Weak Northern and Strong Tropical Land Carbon Uptake from Vertical Profiles of Atmospheric CO_2, Science, 316, 1732-1735 (2007).

[6] M. Riebesell, Raman lidar for the remote sensing of the water vapor and carbon dioxide profile in the troposphere (in German). Ph.D. thesis, GKSS document 901/F/13, University of Hamburg, 127 pp. (1990).

[7] A. Ansmann *et al.*, Combined Raman Elastic-backscatter lidar for vertical profiling of moisture, aerosol extinction, backscatter, and lidar ratio, Appl. Phys., 55B, 18-28 (1992).

[8] P. Di Girolamo *et al.*, Spaceborne profiling of atmospheric temperature and particle extinction with pure rotational Raman Lidar and of relative humidity in combination with differential absorption Lidar: performance simulations. Applied Optics, 45, 2474-2494, doi: 10.1364/AO.45.002474 (2006).

[9] P. Di Girolamo *et al.*, Space-borne profiling of atmospheric thermodynamic variables with Raman lidar: performance simulations, Optics Express, 26(7), 8125-8161, doi: 10.1364/OE.26.008125 (2018).

[10] Gatti et al., Vertical profiles of CO2 above eastern Amazonia suggest a net carbon flux to the atmosphere and balanced biosphere between 2000 and 2009, Tellus, 62B, 581–594 (2010).

[11] Buchmann et al., Global Change Biology, 2, 421-432, 1996.

[12] P. Di Girolamo *et al.*, Rotational Raman Lidar measurements of atmospheric temperature in the UV. Geophysical Research Letters, 31, doi: 10.1029/2003GL018342 (2004).

[13] D. N. Whiteman *et al.*, Demonstration Measurements of Water Vapor, Cirrus Clouds, and Carbon Dioxide using a High-Performance Raman Lidar, Journal of Atmospheric and Oceanic Technology, 24, 1377-1388 (2007).

[14] P. P. Di Girolamo *et al.*, Multiparameter Raman Lidar Measurements for the Characterization of a Dry Stratospheric Intrusion Event, Journal of Atmospheric and Oceanic Technology, 26, 1742-1762, doi: 10.1175/2009JTECHA1253.1 (2009).

[15] Bhawar *et al.*, Spectrally Resolved Observations of Atmospheric Emitted Radiance in the H_2O Rotation Band, Geophysical Research Letters, 35, L04812, ISSN: 0094-8276, doi: 10.1029/2007GL032207 (2008).

[16] Serio *et al.*, Retrieval of foreign-broadened water vapor continuum coefficients from emitted spectral radiance in the H2O rotational band from 240 to 590 cm-1, Optics Express, 16/20, 15816-15833, doi: 10.1364/OE.16.015816 (2008).

[17] Wulfmeyer *et al.*, Research campaign: The convective and orographically induced precipitation study - A research and development project of the World Weather Research Program for improving quantitative precipitation forecasting in low-mountain regions, Bulletin of the American Meteorological Society, 89, 1477-1486, ISSN: 0003-0007, doi: 10.1175/2008BAMS2367.1 (2008).

[18] Bennett *et al.*, Initiation of convection over the Black Forest mountains during COPS IOP15a, Quarterly Journal of the Royal Meteorological Society, 137, 176-189, doi: 10.1002/qj.760 (2011).

[19] Kiemle *et al.*, Latent heat flux measurements over complex terrain by airborne water vapour and wind Lidars. Quarterly Journal of the Royal Meteorological Society, 137, 190-203, ISSN: 0035-9009, doi: 10.1002/qj.757 (2011).

[20] Steinke et al., Assessment of Small-Scale Integrated Water Vapour Variability During HOPE, Atmospheric Chemistry and Physics, 15(5), 2675-26929, doi:10.5194/acp-15-2675-2015 (2015).

[21] Macke *et al.*, The HD(CP)2 Observational Prototype Experiment HOPE - An Overview, Atmospheric Chemistry and Physics, 17, 4887–4914, doi:10.5194/acp-17-4887-2017 (2017).

[22] ESA, "ARMA Reference Model of the Atmosphere," in Technical Report APP-FP/99-11239/AC/ac (European Space Agency, 1999).

DEVELOPMENT OF ATLID RETRIEVAL ALGORITHMS

D.P. Donovan[*,1], G-J van Zadelhoff[1], J. E. Williams[1], U. Wandinger[2], M. Haarig[2], Z. Qu[3]

[1]*Royal Netherlands Meteorological Institute (KNMI), Utrechtseweg 297, PO Box 201, 3730 AE de Bilt, the Netherlands*
[2]*Leibniz Institute for Tropospheric Research. Permoserstraße 15. 04318 Leipzig, Germany.*
[3]*Environment and Climate Change (MRD), Downsview, Ontario, Canada.*
**Email: Donovan@knmi.nl*

ABSTRACT

ATLID ("ATmospheric LIDar") is the lidar to be flown on the multi-instrument Earth Clouds and Radiation Explorer (EarthCARE or ECARE) joint ESA/JAXA mission now scheduled for launch in 2022. ATID is a 3 channel linearly polarized High-Spectral Resolution (HSRL) system operating at 355nm. Cloud and aerosol optical properties are key ECARE products. This paper will provide an overview of the ATLID L2a (i.e. single instrument) retrieval algorithms being developed and implemented in order to derive cloud and aerosol optical properties.

1. INTRODUCTION

EarthCARE[1] is a cloud-aerosol-precipitation-radiation process oriented mission. In addition to ATLID, the instrument suite includes, a 94 GHz, Doppler Cloud Radar supplied by Japan (CPR), a Multispectral Cloud/Aerosol imager for narrow-band TOA radiances (MSI) and a 3-view Broad-Band Long- and Short-Wave Radiometer for TOA radiance (BBR) measurements and flux estimation.

ATLID is a linearly polarized three channel lidar operating at 355nm. The vertical resolution of the return signal is 100 m throughout most of the atmosphere. The PRF is 51 Hz and nominally two shots will be averaged on-board giving a horizontal resolution on the order of 275 m. The lidar delivers profiles of the parallel, Mie (or particulate) attenuated backscatter, the parallel Rayleigh (molecular) attenuated backscatter and the corresponding total (Mie+Rayleigh) cross-polarized return. More details can be found in [2].

A comprehensive suite of retrieval algorithms is being developed for EarthCARE As outlined in [1], both (primarily) single-instrument (L2a) and synergistic multi-instrument (L2b) algorithms are being developed. In this paper, the focus is on the L2a ATLID algorithms.

2. THE L2a ATLID ALGORITHMS

In this Section, the ECARE L2a lidar algorithms are briefly described in turn.

2.1 ATLID Featuremask (A-FM).

A-FM uses a combination of image processing techniques in order to identify regions of clouds/aerosols, surface returns, clear air, or attenuated regions. The detected aerosol/cloud regions are separated into cloud phase and aerosol type later in subsequent processing steps.

2.2 Aerosol oriented extinction and backscatter retrieval routine (A-AER).

This procedure uses conventional HRSL retrieval methods for determining extinction and backscatter at the 50km+ horizontal scale (e.g. deriving the extinction based on the log derivative of the Rayleigh signal). In order to do this the lidar signals must be appropriately masked and averaged to achieve a target SNR. The averaging mask originates from the A-FM output which is used to avoid averaging "strong" and "weak" features together.

2.3 Cloud and aerosol Extinction, Backscatter and Depolarization procedure (A-EBD).

This routine retrieves the aerosol and cloud extinction and backscatter profiles at the 1-km horizontal scale. At this scale the SNR of the molecular scattering channel return is too low to enable the techniques employed by the A-AER approach. Instead, the method relies on finding the optimal profile of the extinction-to-backscatter ratio (S) that allows one to invert the total lidar signal to produce an extinction profile consistent with the observed Rayleigh channel signal. Multiple-scattering (MS) effects which are important for accurate cloud (and in some cases aerosol) retrievals are included in the procedure which is cast in an optimal-estimation framework..

As a-priori information, the S estimates produced by A-AER are used.

2.4 Target classification procedure (A-TC).

A-TC uses extinction, backscatter and depolarization ratio, as well as auxiliary inputs such as ECMWF forecast temperature, in order to classify targets into classes such as water or ice cloud or aerosol type. The aerosol typing scheme is based primarily on using the S and particle depolarization ratio to assign a class to the aerosol [3]. The cloud phase determination scheme uses backscatter and depolarization in a manner similar to that employed for the CALIOP retrievals.

2.5 ATLID Cloud Top Height (A-CTH)

The A-CTH procedure actually detects all recoverable significant boundaries. The cloud top height itself is defined as the upper geometrical boundary of the uppermost cloud layer in the atmosphere. The A-CTH product is derived from the Mie co-polar signal by searching for characteristic signal gradients using a wavelet covariance transform approach [4]. Depending on the cloud optical thickness and the signal-to-noise ratio, the boundaries are detected with a horizontal resolution of 1-km (thick clouds) or 11 km (thin clouds). The product contains a simplified classification of the uppermost cloud as well as quality indicators in terms of level of confidence of the detection and level of consistency with the A-TC product.

2.6 ATLID Aerosol Layer Descriptor (A-ALD)

The A-ALD product contains geometrical and optical information on aerosol layers. The same wavelet transform algorithm as used in the A-CTH retrieval is used to derive aerosol layer boundaries from the Mie co-polar signal, averaged over about 10 km horizontally. Appropriate threshold settings allow for the discrimination of clouds and aerosol. The A-ALD product provides the upper and lower geometrical boundaries of significant aerosol layers, the optical thickness of each layer (ALOT), the column and the stratospheric AOT at 355 nm. ALOT and AOT are calculated by integrating the extinction profile taken from the A-EBD product. In addition, layer-mean values of extinction and backscatter coefficient, lidar ratio and particle linear depolarization ratio are calculated from the A-EBD product. The A-ALD product is defined for cloud-free conditions only. It contains quality indicators in terms of level of confidence for the aerosol layer detection and level of consistency with the A-TC product.

3. ALGORITHM DEVELOPMENT PROCESS

A-Train observations are playing important roles within the ECARE algorithm development process. However, given the wavelength difference and the non-HSRL nature of the instrument, CALIPSO observations are not sufficient for ATLID algorithm development purposes. Accordingly, a number of simulated realistic scenes have been developed and simulated L1 data sets have been created using the multiple-scatter lidar models contained within the EarthCARE simulator framework [5].

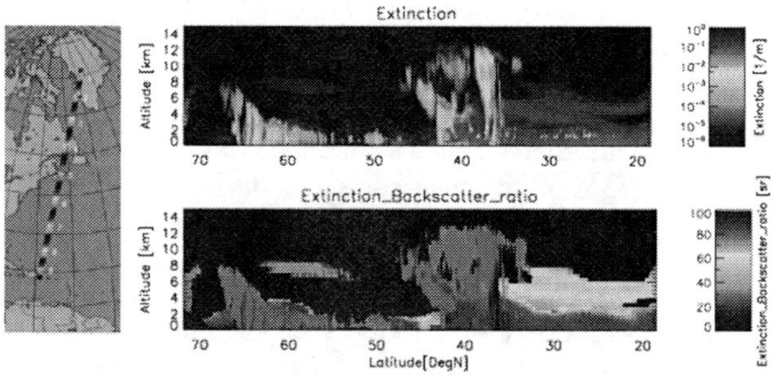

Figure 1: Illustration of the Halifax scene swath and the seven high resolution (0.25 km horizontal resolution) sections comprising the scene (Left). Corresponding derived vertical nadir slices of the 355nm extinction and backscatter-to-extinction ratio fields (Right).

One such scene (the so-called Halifax scene) is depicted in Figure 1. The scene is built around high-resolution segments of the environment Canada GEM forecast model [6]. The GEM model uses a dual moment micro-physical cloud scheme covering multiple hydrometers (cloud ice, snow, cloud water rain, graupel and hail). Using these fields as a foundation and imposing realistic particle habits and size distributions, the cloud/hydrometer fields necessary for the lidar simulations were created. Aerosol information did not come from the GEM model data. Rather, co-located aerosol CAMS aerosol fields [7] were used to construct the necessary aerosol fields.

Simulated fields of L1 attenuated backscatter for the three ATLID channels are shown in Figure 2.

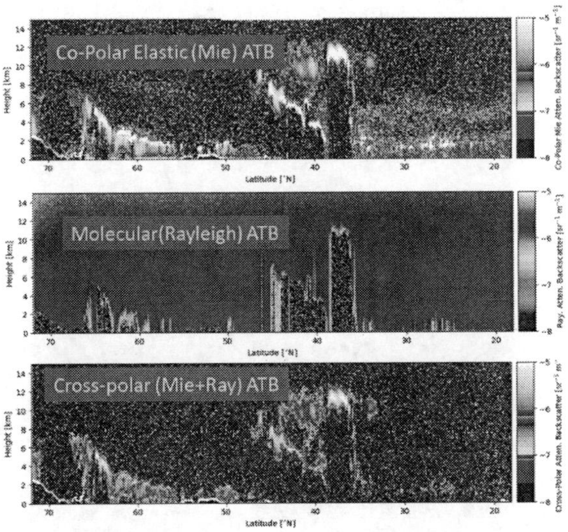

Figure 2: Simulated Attenuated backscatter for the Co-polar elastic (Mie), co-polar molecular (Rayleigh) and total cross-polar ATLID channels corresponding to the Halifax scene.

4. EXAMPLE RESULTS

In this section various illustrative sample results from the algorithms previously introduced will be presented.

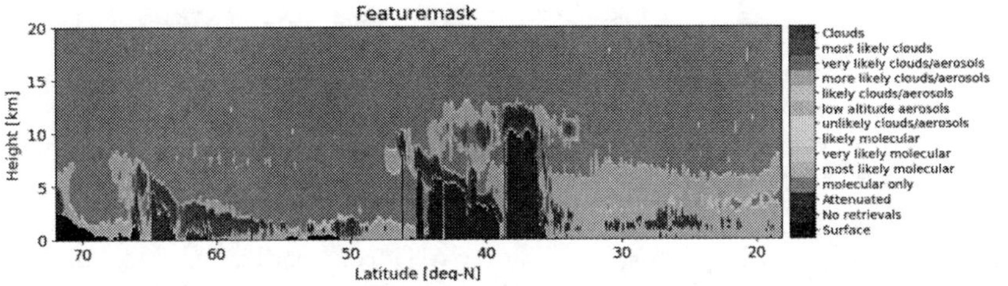

Figure 3: A-FM output for the Halifax Scene.

4.1 A-FM

The main output generated by A-FM for the Halifax scene is shown in Figure 3. Here the Red colors correspond to confidently detected strong features while the light-Blue areas correspond to confidently detected clear air regions. At this stage, the feature classes ((likely cloud, aerosol, likely clear etc.) assigned to the level of target detection confidence are preliminary and are intended to be interpreted in a qualitative sense.

A more robust and accurate classification is performed in the following processing steps.

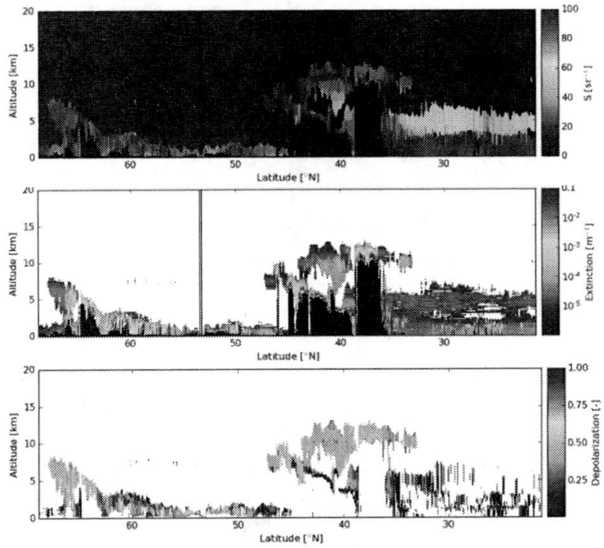

Figure 4: A-EBD results for the Halifax scene. Shown are the retrieved backscatter-to-extinction ratios (Top), the cloud and aerosol extinction fields (Middle) and the cloud/aerosol depolarization ratio (Bottom).

4.2 A-EBD

Sample output fields from the A-EBD algorithm are shown below. As discussed earlier the A-EBD algorithm uses low resolution input aerosol and "thin" ice clouds S fields from the A-AER algorithm (not shown) as well as a-priori values for the cloud S values in order to produce merged results for both clouds and aerosols. The data shown are actually a composite of two horizontal resolutions. For the "strong" features the horizontal resolution is 1-km while for the "weak" features it is 121 km.

The S and extinction fields are, generally, well retrieved (an exception being the high S values at around 14 km at about 35 °N). Quantitative statistical evaluations show that e.g. that typical aerosol extinction profiles with a horizontal resolution of 121 km and a vertical resolution of 0.1 km are retrieved with an RMS accuracy within about 20%.

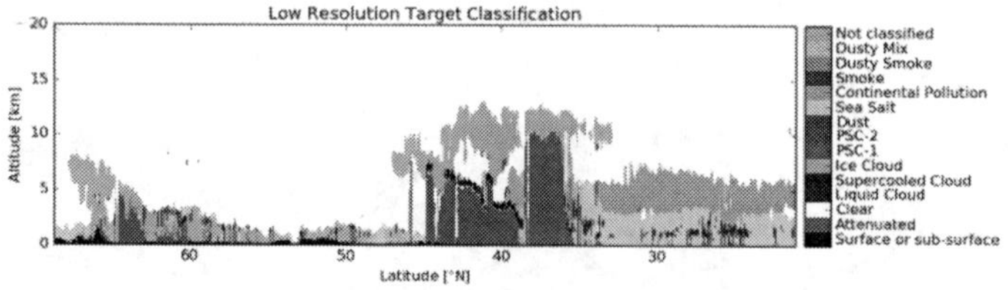

Figure 5: Target classification corresponding to the Halifax scene.

4.3 A-TC

Compared to A-FM, A-TC is able to draw on more information (i.e. attenuated backscatter only in the case of A-FM while attenuated backscatter, particle depolarization, particle extinction and extinction-to-backscatter are all used by A-TC) to produce a more accurate and detailed classification of the clouds and aerosols present.

The A-TC output corresponding to the Halifax scene is shown in Figure 5. As was the case for the data shown in Figure 4 the cloud classification has a horizontal resolution of 1 km while the aerosol classification field corresponds to a horizontal resolution of 120 km. Here it can be seen that the cloud/aerosol discrimination is robust and a good separation between the boundary-layer marine aerosol field and the lofted aerosol layer is achieved.

4.4 A-CTH

Sample A-CTH output is shown in Figure 6. Here a "zoomed in" cloud dominated section of the Halifax scene is presented in order to better illustrate the output. A-CTH uses a wavelet technique to detect significant layers and so compliments the A-FM based approach.

One of the main intended uses of A-CTH data is in synergy with the MSI also mounted on EarthCARE. To this end, A-CTH provides not only the detected cloud-top-height but also information on the number of detected significant layers. This is illustrated by the various colored lines present in Figure 6.

3. SUMMARY

The launch of EarthCARE at this point remains in the future. However, given the complexity of the processing chain and the novel nature of ATLID data the intervening time is being put to good use. There is no doubt that the real ATLID measurements will confront the algorithm development teams with surprises. However, the extensive realistic simulations will have certainly helped ensure that high-quality products from EarthCARE will be generated quickly after real observations have been acquired.

ACKNOWLEDGEMENTS

The work presented in this paper is performed under the APRIL project funded by ESA (contract 4000112018/14/NL/CT).

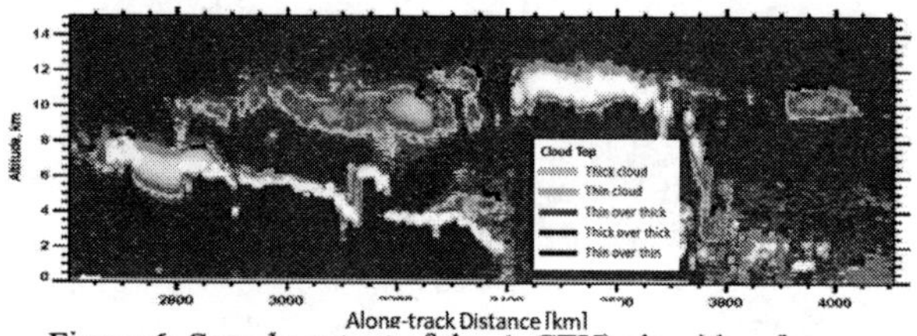

Figure 6: Sample output of the A-CTH) algorithm for a sub-section (approx. 33 to 47 °N) of the Halifax scene.

REFERENCES

[1] Illingworth, A., et al., 2014: THE EARTHCARE SATELLITE: The next step forward in global measurements of clouds, aerosols, precipitation and radiation. *Bull. Am. Met. Soc.* **96**, 1311–1332.

[2] do Carmo, J.P., et al., 2016: ATLID, ESA Atmospheric LIDAR Developement Status, Proc. ILRC27,DOI:https://doi.org/10.1051/epjconf/201611904003

[3] Wandinger U., et al., 2016: HETEAC: The Aerosol Classification Model for EarthCARE, Proc. ILRC27, DOI:10.1051/epjconf/201611901004

[4] Brooks, I., 2003: Finding Boundary Layer Top: Application of a wavelet covariance transform to lidar backscatter profiles, *J. Atmos. Ocean. Tech.* **20**, 1092–1105.

[5] Donovan, D. P., et al., 2015: A depolarisation lidar-based method for the determination of liquid-cloud microphysical properties, *Atmos. Meas. Tech.*, **8**, 237-266, https://doi.org/10.5194/amt-8-237-2015.

[6] Côté, J., etl al., , 1998: The operational CMC-MRB

Global Environmental Multiscale (GEM) model: Part I - Design considerations and formulation, *Mon.Wea. Rev.*, **126**, 1373-1395.

[7] Inness, A., et, al., 2019: The CAMS reanalysis of atmospheric composition, *Atmos. Chem. Phys.*, **19**, 3515-3556, https://doi.org/10.5194/acp-19-3515-2019.

ESA'S LIDAR MISSIONS AEOLUS AND EARTHCARE

Thomas Kanitz[1*], Alessandra Ciapponi[1], Alessia Mondello[4], Alessandro D'Ottavi[4], Ana Baselga Mateo[1], Anne-Grete Straume[1], Christoph Voland[1], Didier Bon[3], Elena Checa[1], Emilio Alvarez[1], Ida Bellucci[4], Joao Pereira Do Carmo[1], John Brewster[2], Jon Marshall[2], Marc Schillinger[3], Mark Hannington[2], Michael Rennie[7], Oliver Reitebuch[6], Olivier Lecrenier[3], Paolo Bravetti[5], Valentina Sacchieri[4], Valeria De Sanctis[4], Alain Lefebvre[1], Tommaso Parrinello[8], and Denny Wernham[1]

[1]ESA-ESTEC, Noordwijk, The Netherlands
[2]ADS, Airbus Defense and Space, Stevenage, United Kingdom
[3]ADS, Airbus Defense and Space, Toulouse, France
[4]Leonardo, Pomezia, Italy
[5]ADS, Airbus Defense and Space, Rome, Italy
[6]DLR, Institute of Atmospheric Physics, Oberpfaffenhofen, Germany
[7]ECMWF, Reading, United Kingdom
[8]ESA-ESRIN, Frascati, Italy
**Email: thomas.kanitz@esa.int*

ABSTRACT

ESAs Earth Explorer Aeolus was launched in August 2018. Aboard the first spaceborne wind lidar ALADIN (Atmospheric LAser Doppler INstrument) was switched on in early September 2018 and demonstrated the capability to provide atmospheric wind profiles globally from particle and molecular backscatter. In doing so, it will contribute to the improvement in numerical weather prediction (NWP) and the understanding of global dynamics. At the same, it is a major step for powerful and frequency stabilized ultraviolet (UV) lasers for space applications. In parallel, ESA and its partners continue the development of this technology by setting up further ground tests based on Aeolus, and preparing the next milestone with ATLID (ATmospheric LIDar) for the Earth Cloud, Aerosol and Radiation Explorer (EarthCARE) mission. ATLID is currently fully integrated and getting prepared for its on-ground testing.

1. INTRODUCTION

ALADIN is the single payload of ESA's wind lidar mission Aeolus within the Earth Explorer family of the ESA Living Planet Programme [1]. Aeolus shall demonstrate the potential of space-based Doppler Wind lidars to improve NWP and advance the understanding of atmospheric dynamics and climate processes. In order to do so, it shall provide vertical profiles of one component of the horizontal wind vector along the laser line-of-sight (LOS) from ground up to the lower stratosphere (to 30 km). Aeolus was launched from Kourou (French Guiana) on 22 August 2018 and is in operation since then with only minor interruptions.

ATLID is part of the payload of the EarthCARE mission, the sixth Earth Explorer of ESA [2]. The EarthCARE payload consists of a cloud Dopper radar, an imager, and a radiometer. In a synergetic manner, the four instruments will retrieve vertical profiles of clouds and aerosols, their radiative characteristics, and their micro-physical properties in order to determine top of atmosphere radiance and flux profiles. ATLID is gearing up for its ground performance and qualification tests in 2019.

2. AEOLUS – ALADIN

ALADIN is a direct-detection Doppler wind lidar operating at the UV wavelength of 354.8 nm [3]. It is composed of a 1.5 m-diameter telescope, a highly sensitive receiver setup with an imaging Fizeau spectrometer for narrowband Mie backscatter from aerosol and cloud particles, a double-edge Fabry-Perot spectrometer for broad Rayleigh-Brillouin molecular backscatter and photon counting accumulation charge coupled device (ACCD) detectors. The frequency-stabilized UV-laser was designed to emit 80 mJ pulses at a repetition frequency of 50.5 Hz. The return signal is acquired with a vertical resolution between 0.25 and 2 km and an accumulation of typically 20 pulses, resulting in a horizontal raw data resolution of 3 km per measurement. Further averaging (600 pulses) yields data at the so-called observation resolution which meets the wind precision requirement of 1 to 2.5 m/s [4].

2.1 From Launch to Operation

Aeolus was launched on 22 August 2018. After the successful launch and early orbit phase, the commissioning was started 48 hours later. Before ALADIN could be turned on, the in-situ cleaning system (ICS) was used to purge ALADIN twice to minimize the risk of laser induced contamination (LIC). After these flushes, the ICS was set to nominal mode to operate ALADIN in a residual oxygen pressure environment starting from 31 August 2018 [5].

For the first in-orbit operations, ALADIN's transmitter Flight Model A (FM-A) was selected, rather than the most powerful and stable Model B (FM-B). FM-A switched to its warm up mode allowing the assessment of the general status of the transmitter itself, but also the thermal and power performance of the instrument, e.g., telescope and transmitter thermal regulation. In parallel the detection chain was enabled and first scientific data became available to characterize the dark signal on the ACCDs, and allowing running the data chain ground processors and dissemination facility up to Level 1 [6]. On 2 September 2018, the operation of the transmitter FM-A started in defined intermediate energy level steps. The first light through ALADINs internal path could be measured at low emit energy, followed by the first frequency scans, laser beam monitoring and spectral coregistration of the Mie and Rayleigh spectrometers. Subsequently, the first profile wind measurements were able to begin on 4 September 2018. The following unbroken four month of operation were dedicated to instrument characterization with weekly routine calibrations for, e.g., spectral characterization over different frequency ranges and special operations for targeted investigations. Within this period the existing seven instrument characterization modes were extended by three procedures. The automated mission planning facility took over in week twelve of the mission and the data was assessed fit for distribution to the Aeolus Calibration and Validation teams by the data analysis core team in mid-December 2018 [7,8].

Two main instrument related issues dominated the first seven months of operation. The initial UV emit energy of FM-A was below recorded levels during last on-ground tests, respectively 65mJ compared to 80mJ (Fig. 1). Thermal conditions resulting in an unexpected gradient at the interface of laser and cooling systems are considered the likely root cause. During the following weeks and months, the emitted energy decreased continuously over the span of weeks with an ever increasingly shallower slope (Fig.1). Special operations including sensitivity tests of thermal adjustments of the laser cold plate temperature, and pre- and power amplifiers, as well as adjustments to the main currents were performed to keep the emit energy above 50mJ. This (preliminary) value was derived from NWP considerations. After a spurious switch to safe mode on 14 January, Aeolus returned to operation on 15 February 2019, still using FM-A, and a emit energy of 50mJ since 6 March 2019. No signs of LIC have been observed so far.

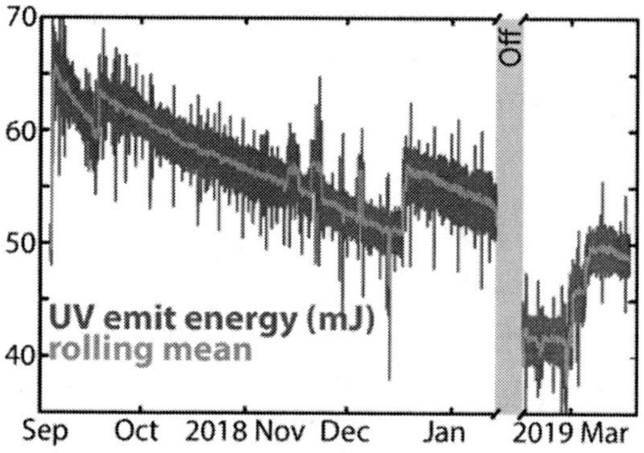

Figure 1 Aeolus inflight ultraviolet emit energy from housekeeping telemetry at power laser head level.

The second issue are appearing defects on a limited set of pixels of the ACCD (5 pixels on each channel), resulting in higher dark current rates and consequently biases in the final wind product [7,8]. After activation of a single pixel, the increased dark current decreases to a base offset. A dedicated calibration has been added to the routine operations and forward corrections are developed for the ground processing. Apart from these two issues, Aeolus has shown remarkable thermal and spectral stability.

Figure 2 The Laser Operation and Verification Facility at ESTEC.

2.3 Laser Operation and Verification Facility

ESA is currently setting up a Laser Operation and Verification Facility at ESTEC. The aim of this facility is to verify results in laser performance induced by operational commands before ultimately applying these to the operational lasers in-flight. To achieve this goal, the Flight Model spare laser (FM-C) with equal power and reference laser heads will be operated with electrical, thermal and telemetry harness in a flight-like configuration. This includes running the laser with a low pressure oxygen supply in a vacuum chamber in order to simulate the orbital environment.

To guard against contamination inside of the vacuum chamber and on the optical ground support equipment, the major parts of the facility are installed underneath an ISO7 clean tent. The setup is placed on a seismic block to decouple the vibration sensitive measurements from environmental disturbances. Temperature stability is assured by an air condition under the tent and a shield inside the vacuum chamber. The laser will be wrapped in multi-layer insulation to minimize temperature variation that could affect the laser performance. The main outcomes of this facility are:

- Problem solving of in-flight issues by replicating the observations on ground, e.g., the impact of vibration on output performance
- The mitigation of risks associated with the change of critical laser parameter changes by applying them first on ground
- Opportunity to investigate parameters not covered by pre-flight tests and in-flight data

The operation of this facility will help to improve the knowledge of high UV energy lasers with demanding frequency stability requirements and their operation in the space for future missions.

2.4 Next step for In-Flight Monitoring Sensors

In-Flight Monitoring Sensors (IFM) implemented in ALADIN and ATLID flight model lasers were limited to measuring temperatures and energies. Lessons learnt from on-ground testing and in-flight operation indicate the need of more extensive monitoring, e.g., the spatial beam characteristics (beam near field profiles, fluence, divergence, and pointing) and pulse duration.

Leonardo is currently developing a set of innovative and more sophisticated sensors for future spaceborne applications. Each sensor is mainly composed of a classical power and a new imaging section in order to fulfill the identified needs.

A complementary metal–oxide–semiconductor (CMOS) material was selected for the imaging sensor. The challenge here is to realize a detector working at both 532 and 355 nm and possessing at least 0.5 MPixels, a 40% active area filling-factor and a 5 µm pixel-size size while being constrained to 50x50x30 mm interface cube size.

For the energy section the photodiodes already customized for ALADIN/ATLID lasers were selected, where the light is scattered with a diffusing ceramic element on a photodiode.

In order to reduce the signal sensitivity, the beam impinging on the photodiodes can be scattered by an integrating sphere achieving total collection of light, spatial integration, and insensitivity to beam motions.

Figure 3 The IFM sensors (image and energy) bread boarding with PM-D at Leonardo.

The characterization of the IFM sensors will be done on the Proto Model (PM-D), a laser based on all rejected parts from ALADIN FMs. based on rejected parts from ALADIN FMs. Despite its origins, the laser performs at a high level. It emits more than 110 mJ UV emit energy, beam dimensions of 5.4x4.4 mm, divergence of 270x660 µrad, and pulse width of 21.5 ns.

3. EarthCARE - ATLID

ATLID's task is to provide vertical profiles of optical properties of thin cloud and aerosol layers, as well as the altitude of cloud boundaries. ATLID includes a laser similar to ALADIN [9], but pressurized and with only one pump unit. It emits laser pulses with 40mJ energy in the UV, at a repetition rate of 51 Hz. The backscatter signal is collected by a 620 mm telescope, and is then filtered and redirected through the optical focal plane assembly, allowing independent measure of atmospheric Mie and Rayleigh return signals.

Following the manufacturing, qualification and delivery of all ATLID units, the optical and electrical integration has been conducted in parallel in order to assemble the Optical Flight Model and the Electrical Flight Model. Following the initial performance and functional tests of these models [10], and with the integration of the flight laser cooling system, the instrument assembly approaches its final flight configuration. After the completion of additional instrument functional verification tests, the full instrument performance and qualification test campaigns are planned to start second quarter 2019, followed by the instrument delivery to the satellite prime.

4. CONCLUSIONS

ESAs lidar missions with ALADIN and ATLID have had a long road of development. ALADIN is technological pioneer for powerful frequency-stable UV lasers in space. The launch of ALADIN and its successful operation covering several months to date has demonstrated the potential of space-borne lidars to deliver wind profiles and other important geophysical parameters. ESA, together with its partners, are establishing a valuable ground test facility to further improve the technological knowledge, which adds to the experience of inflight data and will be useful for ATLID operation in the future, too.

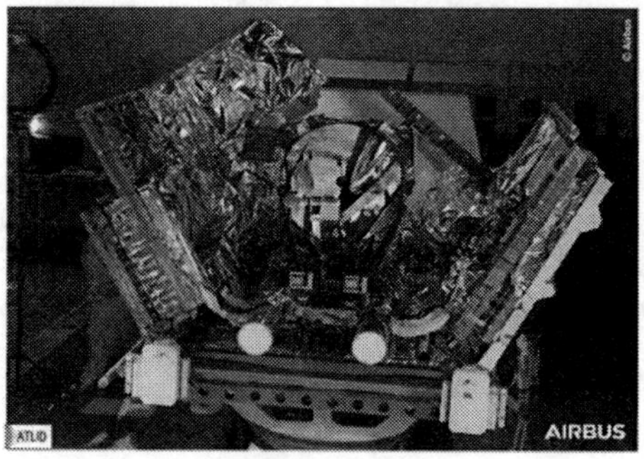

Figure 4 Optical Flight Model inside the housing structure of the Electrical Flight Model.

ACKNOWLEDGEMENTS

ESA wants to thank the science community and its partners for the continuous support during the development of these challenging Earth Explorers.

REFERENCES

[1] ESA, 2008: ADM-Aeolus Science Report, *ESA report SP-1311*, 121 pp.

[2] http://esamultimedia.esa.int/docs/SP_1279_1_EarthCARE.pdf

[3] Morancais, D., et al., 2004: ALADIN: The first European lidar in space. Proceedings of the 22nd ILRC

[4] European Space Agency ESA, 2016: ADM-Aeolus Mission Requirements Documents, AE-RP-ESA-SY-001, Issue 2, 16/11/2016, 57pp.

[5] Kanitz, T., et al., 2018: Aeolus First Light – First Glimpse. Proc. Int. Conf. Space Optics, ICSO.

[6] Reitebuch, O., Huber, D., Nikolaus, I., 2018: Algorithm Theoretical Basis Document ATBD: ADM-Aeolus Level 1B Products, AE-RP-DLR-L1B-001, V. 4.4, 20.04.2018, 117pp.

[7] Reitebuch, O. et al., 2019: Initial assessment of the performance of the first wind lidar in space on Aeolus, Proceedings of the 29th ILRC.

[8] Straume, A.G. et al., 2019: ESA'S space based Doppler mission Aeolus – First Results of the Wind and Aerosol Product Assessment, Proceedings of the 29th ILRC.

[9] Cosentino, A., et al., 2012: Spaceborne Lasers Development for ALADIN and ATLID Instruments, Geoscience and Remote Sensing Symposium, IGARSS.

[10] de Villele, G. et al., 2018: ATLID, ESA Atmospheric Lidar: integration of instrument and tests, Proc. Int. Conf. Space Optics, ICSO.

https://doi.org/10.1051/epjconf/202023701007

ESA'S SPACE-BASED DOPPLER WIND LIDAR MISSION AEOLUS – FIRST WIND AND AEROSOL PRODUCT ASSESSMENT RESULTS

A.G. Straume[1*], M. Rennie[3], L. Isaksen[3], J. de Kloe[4], G.-J. Marseille[4], A. Stoffelen[4], T. Flament[5], H. Stieglitz[5], A. Dabas[5], D. Huber[6], O. Reitebuch[7], C. Lemmerz[7], O. Lux[7], U. Marksteiner[7], F. Weiler[7], B. Witschas[7], M. Meringer[8], K. Schmidt[8], I. Nikolaus[9], A. Geiss[10], P. Flamant[11], T. Kanitz[1], D. Wernham[1], J. von Bismarck[2], S. Bley[2], T. Fehr[1], R. Floberghagen[2], T. Parinello[2]

[1]European Space Agency (ESA) ESTEC, Noordwijk, Netherlands
[2]European Space Agency (ESA) ESRIN, Frascati, Italy
[3]European Centre for Medium-Range Weather Forecasts (ECMWF), Reading, United Kingdom
[4]Royal Dutch Meteorological Institute (KNMI), de Bilt, Netherlands
[5]MétéoFrance, Toulouse, France
[6]DoRIT, Oberpfaffenhofen, Germany
[7]DLR, Institute of Atmospheric Physics, Oberpfaffenhofen, Germany
[8]DLR, Remote Sensing Technology Institute, Oberpfaffenhofen, Germany
[9]University of Applied Sciences, Munich, Germany
[10]Ludwig-Maximillians-University, Meteorological Institute, Munich, Germany
[11]LATMOS, Institut Pierre Simon Laplace (IPSL), Paris, France
**Email: Anne.Straume@esa.int*

ABSTRACT

The European Space Agency (ESA) wind mission, Aeolus, hosts the first space-based Doppler Wind Lidar (DWL) world-wide. The primary mission objective is to demonstrate the DWL technique for measuring wind profiles from space, intended for assimilation in Numerical Weather Prediction (NWP) models. The wind observations will also be used to advance atmospheric dynamics research and for evaluation of climate models. Mission spin-off products are profiles of cloud and aerosol optical properties. Aeolus was launched on 22 August 2018, and the Atmospheric LAser Doppler INstrument (Aladin) instrument switch-on was completed with first high energy output in wind mode on 4 September 2018 [1], [2]. The on-ground data processing facility worked excellent, allowing L2 product output in near-real-time from the start of the mission. First results from the wind profile product (L2B) assessment show that the winds are of very high quality, with random errors in the free Troposphere within (cloud/aerosol backscatter winds: 2.1 m/s) and larger (molecular backscatter winds: 4.3 m/s) than the requirements (2.5 m/s), but still allowing significant positive impact in first preliminary NWP impact experiments. The higher than expected random errors at the time of writing are amongst others due to a lower instrument out- and input photon budget than designed. The instrument calibration is working well, and some of the data processing steps are currently being refined to allow to fully correct instrument alignment related drifts and elevated detector dark currents causing biases in the first data product version. The optical properties spin-off product (L2A) is being compared e.g. to NWP model clouds, air quality model forecasts, and collocated ground-based observations. Features including optically thick and thin particle and hydrometeor layers are clearly identified and are being validated.

1. INTRODUCTION

The European Space Agency (ESA)'s Living Planet Programme includes two types of complementary user driven missions: the research oriented Earth Explorer missions and the operational service oriented Earth Watch missions. The Earth Explorer missions are proposed, assessed and recommended by the European Scientific community, and address the scientific challenges set out in the ESA Living Planet Programme and the ESA Earth Observation Science Strategy [7]. Following the recommendations at the ESA User Consultation Workshop in Granada, Spain in 1999, ESA implemented Aeolus as an Earth Explorer Core mission. The Aeolus mission industrial prime is Airbus Defense and Space (ADS) United Kingdom, and the Aladin instrument prime is ADS France. The Aladin instrument lasers are built by Leonardo, Pomezia, Italy.

2. THE AEOLUS MISSION

2.1 Mission motivation and objectives

The Aeolus mission selection was motivated by the need for more abundant direct wind profile measurements in the World Meteorological Organization (WMO) Global Observing System (GOS), which is used as input to weather forecast models world-wide [3]. Currently, the global distribution of conventional wind profile measurements (e.g. radiosondes, aircraft winds, ground-based wind lidars and radars) is not homogenous and mainly located over land in the Northern Hemisphere. Winds are also inferred from temperature soundings for the large-scale wind fields in the extra-tropics. Scatterometers provide surface winds over the oceans (single level), and Atmospheric Motion Vector (AMV) winds deduced from cloud, aerosol, humidity or ozone tracking are limited in vertical coverage and have less accurate height-assignments. The lack of 3D homogenous wind field sampling, in particular in the tropics, over the oceans and in polar areas, hence leads to difficulties in constraining atmospheric models and limits the studying of key processes in coupled climate systems. Aeolus winds will contribute to mitigate the current wind observation deficit.

The primary mission objective is to demonstrate the DWL technique for measuring wind profiles from space, intended for Near-Real-Time (NRT) assimilation in NWP models. Meteorological Centers world-wide are currently preparing to ingest Aeolus winds near-real-time in their weather models as soon as the data quality is of sufficiently good quality. Aeolus is expected to allow for significant forecast improvements. The wind observations will also be used to advance atmospheric dynamics research and for evaluation of climate models.

2.2 Instrument concept

Aeolus embarks a single payload, the direct detection Doppler Wind Lidar instrument Aladin. It is a pulsed UV lidar (355 nm, circularly polarized, 50 Hz), measuring continuously along the orbit (Figure 1, left panel). Its high spectral resolution concept allows for the detection of the parallel polarized molecular (Rayleigh) and particle (Mie) backscattered signals from the circularly polarized emitted light in two separate channels, each sampling the backscatter in 24 vertical height bins (Figure 1, right panel). This allows the instrument to measure winds both in clear and (partly) cloudy conditions down to optically thick clouds. The height of the wind measurement is detected by time gating. A quasi-global coverage is achieved daily (~ 16 orbits per day) and the orbit repeat cycle is 7 days (111 orbits). The orbit is sun-synchronous with a local equatorial crossing-time of 6 am/pm. A more detailed description of the instrument design and operation can be found in [3] and [6].

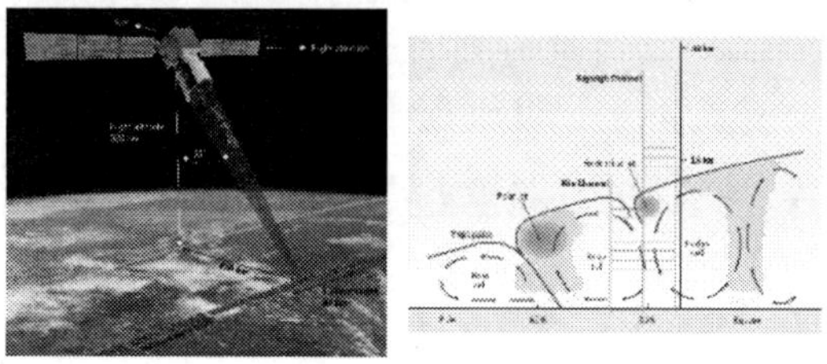

Figure 1: The Aeolus orbit, pointing and sampling characteristics (left) and an example of the vertical measurement binning by the two channels measuring winds from particle (Mie) and molecular (Rayleigh) returns (right).

2.3 Products and product requirements

The main product from Aeolus is horizontally projected line-of-sight (HLOS) wind profile observations (approximately zonally oriented) from the surface up to 25-30 km altitude (L2B product) [4]. The vertical sampling and range is configurable, and is being optimized in the operational phase of the mission. The atmospheric backscattered signal for the individual laser pulses are averaged on-board to yield ~3 km measurements along-track. These measurements are further averaged on-ground to observations, representing horizontal scales up to ~88 km. The vertical resolution of the winds varies from 0.25 to 2 km, adapted as a function of the underlying topography and/or climate zone. The HLOS wind observation random error (precision) requirement is 1 m/s in the Planetary Boundary Layer (PBL), 2.5 m/s in the free troposphere, and 3-5 m/s in the stratosphere, and the bias (systematic error) requirement is 0.7 m/s. The wind profiles are classified into cloudy and cloud-free observations.

The Aeolus Level 2A product contains profiles of particle and molecular parallel polarized

https://doi.org/10.1051/epjconf/202023701007

backscatter and extinction coefficients, scattering ratios, and backscatter-to-extinction ratios [5]. From these parameters it is possible to derive particle layer height, multi-layer cloud/aerosol stratification, cloud/aerosol optical depths, and some information on cloud/aerosol type.

Other products will be developed during the mission operational phase. ESA issues invitations to tender (ITTs) calling for further mission spin-off products and scientific applications (http://emits.sso.esa.int/).

3. RESULTS

First results of the Aeolus instrument characterization and calibration are presented in [1] and [2]. This paper shows first results of the Aeolus Level 2 product assessment.

3.1 First results of the Aeolus L2 wind product assessment

The initial assessment of the Aeolus primary product, the L2B wind profile observations, has been done by ECMWF and KNMI under ESA contract. ECMWF, KNMI and MétéoFrance have developed the Aeolus L2B processor and processing facility, including product quality monitoring using the ECMWF weather model. ECMWF hosts the Aeolus L2B processing facility, is currently performing NWP impact experiments and will actively assimilate the L2B product as soon as the quality is adequate. KNMI maintains and develops the L2B processor, with a special emphasize on the scene classification. Results from the first data quality assessment is shown in Figure 2, where Aeolus winds from one orbit are compared with ECMWF model winds projected onto the Aeolus line-of-sight.

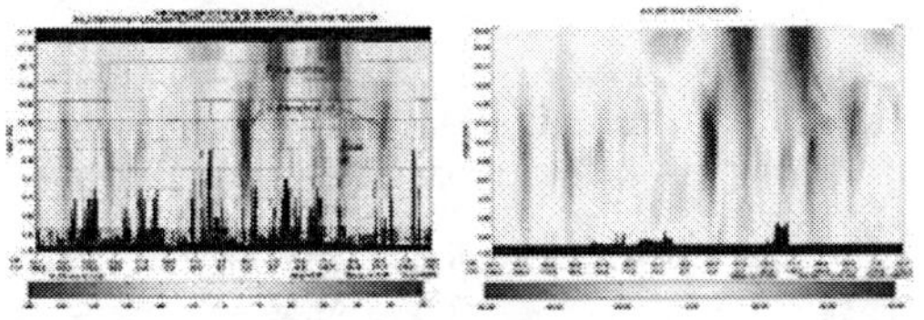

Figure 2: Aeolus L2B product winds combining observations from the Mie and Rayleigh channels over one orbit (September 15 2018), and collocated ECMWF model winds along the same orbit (M. Rennie).

The plots show a very good correlation between the large-scale wind features in the Aeolus observations and ECMWF model winds, such as sub-tropical jets and polar stratospheric jets. In Figure 3, Aeolus observation minus model background wind (o-b) statistics from 7 orbits are shown.

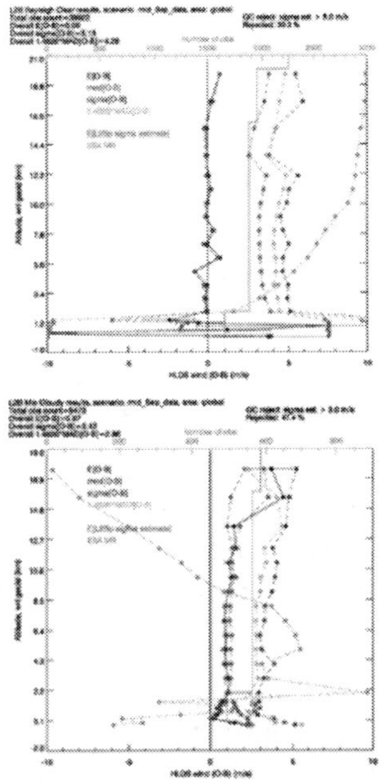

Figure 3: Aeolus Rayleigh (molecular return, top panel) and Mie (particle return, lower panel) wind observation monitoring statistics, 15 September 2018 (M. Rennie). Green: observation random error requirement, Blue: mean (observation (o) – ECMWF model background (b)), Red: σ(o – b), Pink: median value (o – b), Magenta: scaled medium absolute deviation (o - b), Purple: L2B processor estimated random error.

The Rayleigh wind (from molecular backscatter) deviation from the model background (Fig. 3 top panel, expressed as a robust estimate of the Aeolus observation standard deviation) is 4.3 m/s between 2 and 16 km for this dataset, which is larger than the mission random error requirement (2.5 m/s, [6]). The mean bias is 0.1 m/s which is lower than the mission bias requirement (0.7 m/s [6]). In the PBL (defined as 0-2 km altitude), the random errors are larger than the requirement, largely due to high-resolution vertical sampling leading to too low signal per height bin. PBL vertical sampling was intentionally optimized for ground detection investigations in this dataset. After the initial instrument calibration activities (first ~5 months), the vertical binning was changed to optimize the wind quality in the PBL. The Mie wind (from particle and hydrometeor backscatter) deviation from the model background (Fig. 3 lower panel, expressed as a robust estimate of the observation standard deviation) are 2.1 m/s in the free troposphere and hence meet the mission requirements. The mean bias is 1 m/s and higher than the mission requirements, because the bias correction using the instrument calibration measurements in the data processing was disabled. This was because the instrument characterization and calibration needed to be completed before the calibration processing could be switched on. With the upcoming processor deliveries and data reprocessing, it is expected that the product biases will be compliant with the requirements. First NWP impact experiments by M. Rennie at ECMWF show positive impact of the Aeolus observations in the ECMWF model tropical troposphere and southern hemisphere wind, humidity and temperature forecast at short-range relative to other observation types, which looks very promising. Further results from the Aeolus NWP impact assessment is expected to be available for the 29th ILRC conference.

3.2 First results from the Aerosol L2 optical properties products assessment

The initial assessment of the Aeolus L2A backscatter and extinction coefficient profile observations, has been done by MétéoFrance under ESA contract. MétéoFrance has developed the Aeolus Calibration Suite (for L2B Rayleigh-Brillouin scattering correction and L2A specific calibration) and the L2A atmospheric optical properties processors. An example of Aeolus L2A data from one orbit is shown in Figure 4, where cloud features up to 14 km in the tropics are visible. White areas indicate total signal attenuation. The assessments of the data quality are on-going at MétéoFrance, comparing the products to ground and space-based Atmospheric Optical Densities (AODs), ECMWF model clouds and CAMS air quality model aerosol analysis fields. Further results will be shown in the 29th ILRC presentation.

4. CONCLUSIONS

First results from the Aeolus L2 wind and atmospheric optical properties product assessment show that the Aeolus Doppler wind lidar measurement principle has been proven from space. The Aeolus wind product quality is close to the mission requirements, and the activation of the instrument calibration processing is expected ensure the winds to meet the bias requirements. The Aeolus product quality is assessed by scientific teams world-wide with collocated ground-based, airborne and space-based measurements, weather and air quality models. Results from the Aeolus CAL/VAL and Science workshop in March 2019 will be shown at the 29th ILRC.

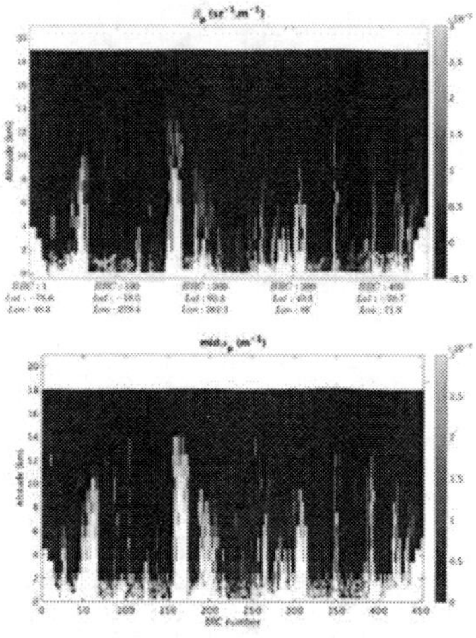

Figure 4: Aeolus L2A product particle backscatter (top panel) and particle extinction (lower panel) coefficient profiles for one orbit, 2 January 2019 (T. Flament, H. Stieglitz).

REFERENCES

[1] T. Kanitz *et al.*, ESA'S Lidar Missions Aeolus and EarthCARE, Proceedings of the 29th ILRC, this issue.
[2] O. Reitebuch *et al.*, Initial assessment of the performance of the first wind lidar in space on Aeolus, Proceedings of the 29th ILRC, this issue.
[3] ESA, 2008, ADM-Aeolus Science Report. ESA SP-1311, 121 pp., available on http://www.esa.int/aeolus
[4] D. Tan *et al.*, 2017, Aeolus Level-2b Algorithm Theoretical Basis Document, available on https://earth.esa.int/aos/AeolusCalVal
[5] P. Flamant *et al.*, 2017, ADM-Aeolus L2A Algorithm Theoretical Baseline Document, available on https://earth.esa.int/aos/AeolusCalVal
[6] ESA, 2016, ADM-Aeolus Mission Requirements Document, available on http://www.esa.int/aeolus
[7] http://www.esa.int/Our_Activities/Observing_the_Earth/The_Living_Planet_Programme

FIRST RESULTS FROM THE GERMAN CAL/VAL ACTIVITIES FOR AEOLUS

Holger Baars[1*], Alexander Geiß[2], Ulla Wandinger[1], Alina Herzog[1], Ronny Engelmann[1], Johannes Bühl[1], Martin Radenz[1], Patric Seifert[1], Albert Ansmann[1], Anne Martin[2], Ronny Leinweber[3], Volker Lehmann[3], Martin Weissmann[4], Alexander Cress[4], Maria Filioglou[5], Mika Komppula[5], and Oliver Reitebuch[6]

[1]Leibniz Institute for Tropospheric Research (TROPOS), Leipzig, Germany
[2]Ludwig-Maximilians-University Munich, Meteorological Institute, Munich, Germany
[3] Deutscher Wetterdienst, Richard-Aßmann-Observatorium, Tauche, Germany
[4]Deutscher Wetterdienst, Offenbach, Germany
[5]Finnish Meteorological Institute, Atmospheric Research Centre of Eastern Finland, Finland
[6]Deutsches Zentrum für Luft- und Raumfahrt e.V., DLR, Institute of Atmospheric Physics, Oberpfaffenhofen, Germany

**Email: baars@tropos.de*

ABSTRACT

On 22nd August 2018, the European Space Agency (ESA) launched the first direct detection Doppler wind lidar into space. Operating at 355 nm and acquiring signals with a dual channel receiver, it allows wind observations in clear air and particle-laden regions of the atmosphere. Furthermore, particle optical properties can be obtained using the High Spectral Resolution Technique Lidar (HSRL) technique. Measuring with 87 km horizontal and 0.25-2 km vertical resolution between ground and up to 30 km in the stratosphere, the global coverage of Aeolus observations shall fill gaps in the global observing system and thus help improving numerical weather prediction. Within this contribution, first results from the German initiative for experimental Aeolus validation are presented and discussed. Ground-based wind and aerosol measurements from tropospheric radar wind profilers, Doppler wind lidars, radiosondes, aerosol lidars and cloud radars are utilized for that purpose.

1. INTRODUCTION

ESA launched the Earth Explorer Mission Aeolus on 22 August 2018. This mission aims to demonstrate significant improvements in weather forecasting by measuring height-resolved wind profiles in the troposphere and lower stratosphere. ALADIN (Atmospheric Laser Doppler Instrument) is the first Lidar (Light Detection and Ranging) instrument on a European satellite. It is also the first space-borne instrument capable of measuring vertical profiles of wind on a global basis. Next to wind measurements, aerosol properties can be obtained as a spin-off product via the HSRL technique, which is a space-borne novelty as well.

Within the German initiative EVAA (Experimental Validation and Assimilation of Aeolus observations), Calibration/Validation (Cal/Val) activities of this space mission have been performed by Ludwig-Maximilians-University of Munich (LMU), Leibniz Institute of Tropospheric Research (TROPOS), German Weather Service (DWD) and DLR (Deutsches Zentrum f. Luft- und Raumfahrt). The aim of EVAA is to validate the wind and aerosol products of Aeolus and to quantify the benefits of these new measurements for weather forecasting by assimilation experiments.

For this purpose, validation measurements with aerosol lidar, Doppler cloud radar, wind lidar and radar wind profiler have been performed since Aeolus launch. Four radar wind profiler stations in Germany have been used (e.g. [1]) as well as aerosol and wind lidar observations at Leipzig, Germany and Punta Arenas, Chile. Furthermore, lidar data at the PollyNET [2] locations of Haifa, Israel and in the United Arab Emirates (UAE) are used.

2. WIND LIDAR MISSION AEOLUS

The instrument onboard Aeolus is called ALADIN, a High Spectral Resolution Lidar operating at a wavelength of ca. 355 nm. The laser pulses are emitted with a frequency of 50.5 Hz and have a circular polarization. More technical details can be found e.g. in [3] and [4].

Aeolus has a weekly repeating polar, sun-synchronous orbit at an inclination of 97° and a mean altitude of 320 km. Aeolus flies along the day/night border facing towards the night to minimize the background radiation. Thus, the measured Line-of-Sight (LOS) wind is mainly oriented either to the west or to the east.

The mission requirements for the Aeolus Mission are a vertical resolution of 500 m in the planetary boundary layer (PBL), 1 km in the troposphere and 2 km in the lower stratosphere. The precision of the horizontal component of LOS (HLOS) was required to be within 1 m/s for the PBL, 2.5 m/s for the troposphere and 3 m/s for the lower stratosphere.

As ALADIN is a high-spectral resolution lidar, having a Rayleigh and Mie channel, it can obtain the extinction and the backscatter coefficient independently as spin-off products. It's the first time that the lidar ratio can be directly measured from space. However, due to the optical architecture of ALADIN which has the same telescope for the outgoing and the receiving signal, only the co-polar component of the circular polarized outgoing light which was backscattered in the atmosphere can be detected. For strongly depolarizing particles, like ice crystals, desert dust or volcanic ash, a significant cross-polar component will thus not be detected. This results in an underestimation of the backscatter coefficient and therefore an overestimation of the lidar ratio [5].

3. FIRST CAL/VAL EFFORTS

After first promising results soon after launch, calibration and validation of Aeolus products becomes an essential part of the satellite mission. In order to determine and understand observation biases and random errors as well as possible instrument degradation, independent measurements of co-located instruments are necessary.

3.1 WIND PRODUCTS

As part of EVAA, ground-based wind measurements have been performed with tropospheric radar wind profilers (RWP), Doppler wind lidars, radiosondes and cloud radars. Especially radar wind profilers using frequencies between 400 and 500 MHz, which are able to measure wind up to 16 km with high accuracy in clear air as well as in regions with particles and clouds, are very well suited for Aeolus comparisons. The German weather service operates 482 MHz radar wind profilers at four sites in Germany on a 24/7 basis which is enough data for first performance statistics. Figure 1 shows a map with the locations of these four wind profilers and the corresponding Aeolus ground tracks from overpasses of one week.

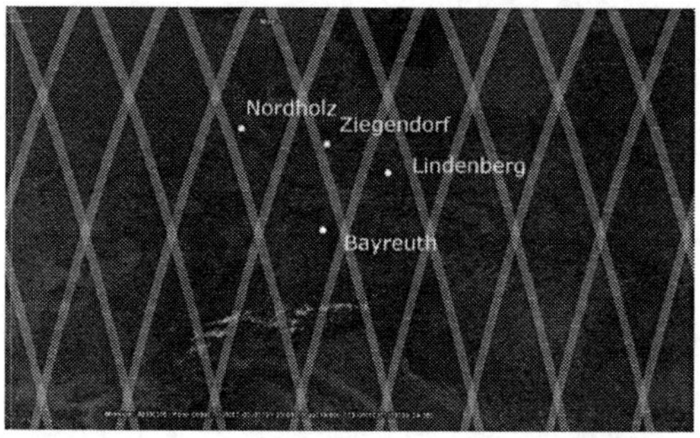

Fig. 1: Map of Central Europe showing the predicted Aeolus ground tracks of one week (magenta) and the four radar wind profiler stations of DWD (white) with a surrounding radius of 100 km (red).

However, in a comparison of instruments with different spatio-temporal resolution and measurement geometry, thus probing different volumes of the atmosphere, special attention must be paid to the representativeness of both measurements.

First results from comparisons of preliminary Aeolus Level 2B Rayleigh wind observations with radar wind profiler measurements are shown in Figure 2 for the first four months of operation. There is good agreement between both instruments showing a bias of 1.82 m/s and a scaled median absolute deviation of 4.75 m/s.

From these first preliminary results, it was concluded that the observed Rayleigh wind bias is continuously increasing with time. This is related to the use of ALADIN calibrations from September 2018 for the processing, and thus no instrument alignment drifts are taken into account. With the next product updates, the systematic error characteristics will be improved.

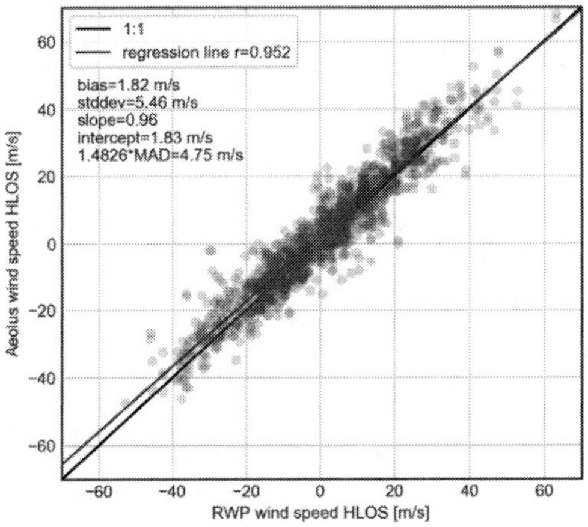

Fig. 2: Comparison of HLOS Rayleigh clear wind speed measured by Aeolus and HLOS projected wind speed from RWP measurements between 04 September 2018 and 15 January 2019. 1630 valid measurements are shown.

3.2 AEROSOL PRODUCTS

Aeolus Level 2A data were evaluated against ground-based multiwavelength Raman polarization lidars of type PollyXT [6]. These lidar systems are comprised in a voluntary, scientific network called PollyNET [2]. Fig. 3 shows the current status map of PollyNET. Green circles are active lidar sites while orange points are past campaigns. The yellow circles indicate the locations which have been used for the Cal/Val efforts.

For Aeolus Cal/Val, a first focus was laid on homogenous and temporally stable aerosol conditions, best with lofted and high altitude aerosol layers above 2 km. This was requested by ESA during the first data release to the Cal/Val teams. For that reason, PollyNET lidar measurements at Haifa, Israel run by TROPOS and near Dubai in UAE run by the Finish Meteorological Institute (FMI) were considered at first glance.

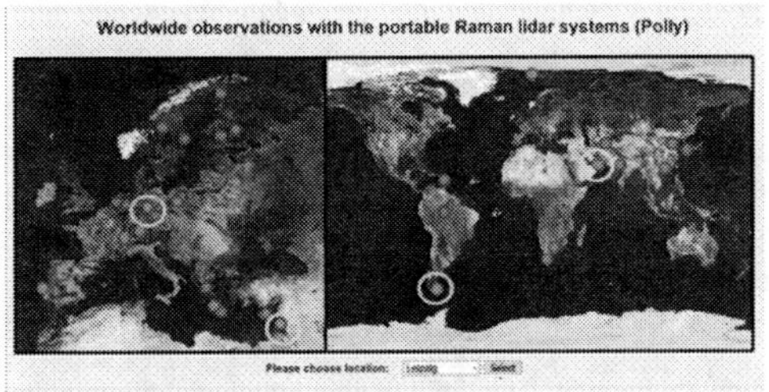

Fig. 3: Map of PollyNET showing active (green) and past (orange) locations. Stations for Aeolus Cal/Val are marked with a yellow circle.

Overpasses in a distance of 100 km to the ground-based lidar site were analyzed. From the four months of operation until mid of January, about five dedicated scenes with mostly dusty atmosphere at each of the locations could be found, so that in total about 10 suitable comparisons could be made. For the analyses, the prototype Level 2A data have been used, which is so far only accessible for Cal/Val teams.

Figure 4 shows one example overpass at Haifa, Israel on 29th September 2018, thus shortly after launch. The PollyXT observations at 355 nm (thin, black line) were transformed to the co-polar component (thick black line) with use of the measured linear depolarization ratio following the method outlined by [5].

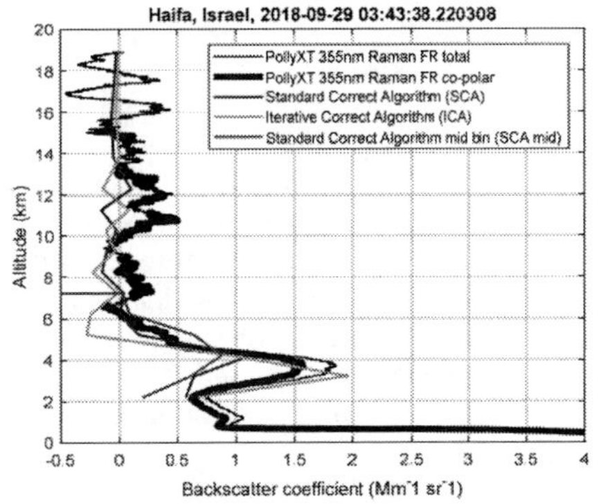

Fig. 4: Comparison of PollyXT at Haifa, Israel during Aeolus overpass on 29 September 2018.

For Aeolus, all three available products, namely the standard correction algorithm (SCA), the standard correction algorithm mid bin (using 2 bins to reduce noise) and the iterative correction algorithm (ICA) are shown while knowing that the SCA is the most developed algorithm [7]. Aeolus products below 2 km are not yet usable due to operational range-bin settings in the early orbit phase. Besides, a good agreement between Aeolus and the ground-based lidar with respect to the co-polar backscatter coefficient is found for this specific case, considering the long averaging length of Aeolus (87 km) and the distance to the lidar (around 100 km). However, also strong deviations have been found for other cases, when e.g. Cirrus are present or the scenes are horizontally heterogeneous. Then the representativeness error of Aeolus has to be taken into account.

3. CONCLUSIONS AND OUTLOOK

First results with respect to Aeolus Level 2A (aerosol) and 2B (wind) have been discussed. These comparisons already have yielded valuable insights into the performance of Aeolus. For the Level 2B Rayleigh wind product, a mean bias of 1.82 m/s and a scaled median absolute deviation of 4.75 m/s were observed. The preliminary investigation of the observed Rayleigh wind bias (wind obtained from backscattering of molecules only) in clear air showed an increase with time. By including updated calibrations in the next product versions, the systematic errors are aimed to be reduced.

Aeolus does not provide valid aerosol products in the lowermost 2 km yet. Thus, the focus for validation was laid on locations in Israel and the United Arab Emirates. About 10 cases have been found, which are suitable for a direct comparison. For all other overpasses, the large observational integration length (87 km) of Aeolus together with the distance to the ground-based lidar makes a direct comparison very difficult. Therefore, a change in the focus of interest and the application of new methods needs to be done in future.

ACKNOWLEDGEMENTS

This work was funded by the German Federal Ministry for Economic Affairs and Energy (BMWi) under grant no. (FKZ) 50EE1721C and 50EE1721A. Authors acknowledge also support through ACTRIS-2 under grant agreement no. 654109 from the European Union's Horizon 2020 research and innovation programme. The lidar data collected from United Arab Emirates are based on work supported by the National Center of Meteorology & Seismology, Abu Dhabi, UAE, under the UAE Research Program for Rain Enhancement Science. The Analysis is based on preliminary, non-public Aeolus data provided by ESA specifically to Cal/Val teams.

REFERENCES

[1] Illingworth, A.J., D. Cimini, C. Gaffard, M. Haeffelin, V. Lehmann, U. Löhnert, E.J. O'Connor, and D. Ruffieux: Exploiting Existing Ground-Based Remote Sensing Networks to Improve High-Resolution Weather Forecasts. Bull. Amer. Meteor. Soc., 96, 2107–2125, 2015.

[2] Baars, H., et al.: An overview of the first decade of Polly^{NET}: an emerging network of automated Raman-polarization lidars for continuous aerosol profiling, Atmos. Chem. Phys., 16, 5111-5137, 2016.

[3] European Space Agency ESA, 2008: ADM-Aeolus Science Report, ESA SP-1311

[4] P. Ingmann and A.G. Straume. ADM-AEOLUS mission requirements document. AE-RPESA- SY-001 EOP-SM/2047, 2016.

[5] U. Wandinger et al., Validation of ADM-Aeolus L2 aerosol and cloud products employing advanced ground-based lidar Measurements (VADAM), ADM-Aeolus Science and CAL/VAL Workshop, 2015.

[6] R. Engelmann et al., The automated multiwavelength Raman polarization and water-vapor lidar PollyXT: the next generation. Atmospheric Measurement Techniques, 9(4):1767–1784, 2016.

[7] Flamant, P. H. et al., ADM-Aeolus L2A Algorithm Theoretical Baseline Document Particle spin-off products, AE-TN-IPSL-GS-001 Version 5.5, 2017.

FIRST STEPS OF MATURATION TOWARDS SPACE OF NESTED CAVITY OPTICAL PARAMETRIC OSCILLATOR AND AMPLIFIERS FOR DIAL BASED ON PERIODICALLY POLED NONLINEAR MATERIALS

JEAN-BAPTISTE DHERBECOURT[1], JEAN-MICHEL MELKONIAN[1], VINCENT LEBAT[1], NICOLAS TANGUY[1], SOPHIE DUZELLIER[2], CEDRIC BLANCHARD[1], RIAN S. COETZEE[3], CARLOTA CANALIAS[3], VALDAS PASISKEVICIUS[3], ANTOINE GODARD[1], MYRIAM RAYBAUT[1]*

[1] *DPHY, ONERA, Université Paris Saclay, F-91123 Palaiseau, France*
[2] *ONERA/DPHY, Université de Toulouse, F-31055 Toulouse, France*
[3] *Department of Applied Physics, Royal Institute of Technology, Roslagstullsbacken 21, 10691 Stockholm, Sweden*
**Email: myriam.raybaut@onera.fr*

ABSTRACT

We report on the first steps of maturation towards space of a nested cavity optical parametric oscillator and amplifiers, based on periodically poled nonlinear materials, emitting in the 2μm range for multi species differential absorption lidar (DIAL) .

1. INTRODUCTION

Emitter and detectors technologies, for future spaceborne DIAL instruments for monitoring of the main green-house gases, have been the subject of several developments over the last decades. As far as the emitter is concerned, developments of bulk solid state laser sources, fiber laser sources, parametric sources have been actively pursued, including in the frame of the first DIAL spaceborne mission, MERLIN [1-5].

Here, we report on the results obtained in the frame of GENUIN (for Generic Frequency Converter Unit for Spaceborne Lidar Instruments) ESA Technical Readiness Program. In this study, we worked on the first steps of maturation towards space of a nested cavity optical parametric oscillator and amplifiers, based on periodically poled non-linear materials, emitting in the 2μm range for multi species DIAL.

Indeed, a few years ago, we began to work on the development of a generic instrument dedicated to the DIAL measurement of the three major greenhouse gases (CO_2, H_2O, CH_4) with a single instrument based on a single versatile emitting device [6]. In the prospect of future space missions, it shows high promises since 1/ such amplified optical parametric oscillator scheme can benefit from the latest developments of 1μm lasers for space, for which the maturation process was difficult [7], 2/ a mission targeting several species (for instance both CO_2 and H_2O ...) can be envisioned, 3/ a single spatialization process of the emitter for different missions dedicated to different gas measurements (CO_2, CH_4, H_2O) can be performed. Such capability is offered by the combination of i) the injection-free nested cavity optical parametric oscillator architecture, and ii) the periodically poled materials properties allowing frequency tuning by temperature change rather than angle tuning in classical birefringence phase matched non-linear materials.

We detail in this paper the first opto-mechanical design for the NesCOPO, which was realized taking into account the specifications for space operation, and on which preliminary vibration testing was performed, and the overall amplified NesCOPO performances, while pumped by a 100mJ laser. Radiation testing results on the specifically developed rubidium doped, periodically poled KTP crystals are also presented.

2. METHODOLOGY

2.1 The GENUIN emitter architecture

The emitter set-up is depicted in Fig.1.

The pump laser used here is a high energy (100 mJ), 30 Hz, commercial single frequency Nd:YAG laser, delivering 12 ns pulses. In this work, the goal was not to work on the pump laser itself, given the work perform in the frame of other projects for spaceborne 1μm lasers.

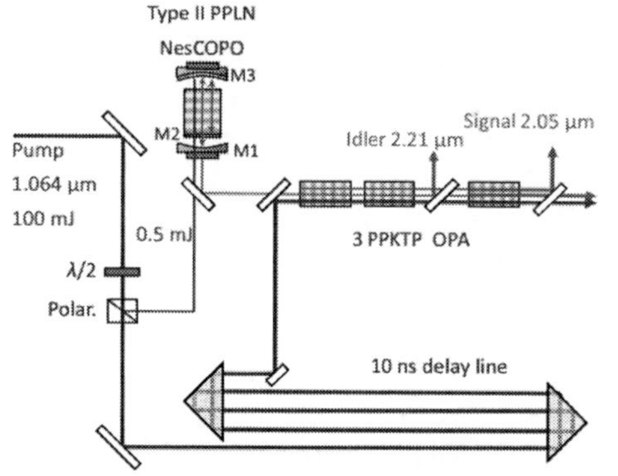

Fig. 1. GENUIN emitter optical layout.

Fig. 2. GENUIN NesCOPO picture.

GENUIN NesCOPO, depicted in Fig.2., is realized in zerodur glass material. The copper crystal oven shape was optimized through thermomechanical modelling, and heated via a Kapton encapsulated resistor. The NesCOPO is pumped by 0.5 mJ. The remaining pump energy is delayed to optimize the temporal OPO output with the pump pulse with amplifier stage, composed of three Rb:PPKTP crystals of 12mm length. A filtering element placed between crystals 2 and 3 aviods saturation of the amplifiers.

2.2 GENUIN amplifier non linear crystals

The non linear crystals are high aperture (5x7 mm^2) Rb:PPKTP crystals specifically developed for this project. They were exposed to gamma radiation, and tested in terms of transmission and poling period structuration, as detailed in [8].

3. GENUIN emitter main performances

The NesCOPO crystal temperature was set as to emit a single frequency signal wavelength in the 2.05μm range. With a single poling period, at a given temperature, it is possible to address both the CO_2 R30 line and the adjacent H_2O line. The 2.06μm H_2O line can be targeted by elevating the crystal temperature of around 15°C. The methane 2.29 μm line can be addressed by using an adjacent poling period of the NesCOPO crystal and the same mirrors as shown in [5]. Given the chosen opto-mechanical set-up, translation of the crystal in the cavity was not implemented here. However, given the NesCOPO low oscillation threshold, an additional NesCOPO targeting methane could be added.

We managed to generate up to 10,7 mJ of signal only, at 2.05μm, and measured an amplifier conversion efficiency of 37% towards signal plus idler (i.e. a 19,8% conversion efficiency towards signal alone). Addition of another crystal could increase again the conversion efficiency, or reducing the pump beam (with the potential damage risk).

Preliminary vibration testing was performed using a vibration pod available in our lab, with the applied acceleration described in Fig.3

	Test	Description
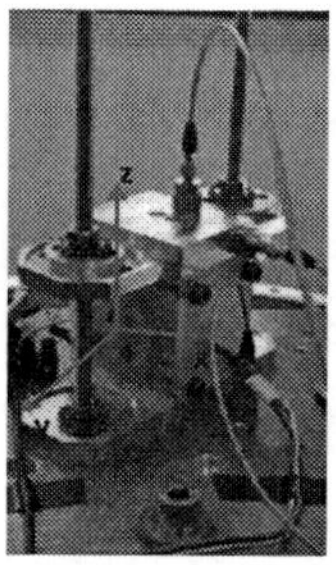	SIN2.5	2.5 g accelerations will be applied from 75 to 140 Hz, rate : 2 octaves / minute along vertical axis
	SIN10	10 g accelerations will be applied from 20 to 75 Hz rate : 2 octaves / minute along vertical axis
	RAND	Random vibration test with the level profile described in TN4, Along vertical axis, for 2 min.

Fig. 3. GENUIN NesCOPO on the vibration pod and applied levels

Additional details will be given concerning the frequency stability, tuning procedure, and perspectives for more energy output with pumping with a higher energy laser.

ACKNOWLEDGEMENTS

This development was supported by ESA in the frame of Contract n°4000107439/13/NL/SFe.

REFERENCES

[1] Tamer F. Refaat, Upendra N. Singh, Jirong Yu, Mulugeta Petros, Ruben Remus, and Syed Ismail, "Double-pulse 2-μm integrated path differential absorption lidar airborne validation for atmospheric carbon dioxide measurement," Appl. Opt. 55, 4232-4246 (2016)

[2] F. Gibert, D. Edouart, C. Cénac, F. Le Mounier, and A. Dumas, "2-μm Ho emitter-based coherent DIAL for CO2 profiling in the atmosphere," Opt. Lett. 40, 3093-3096 (2015)

[3] A. Amediek, G. Ehret, A. Fix, M. Wirth, C. Büdenbender, M. Quatrevalet, C. Kiemle, and C. Gerbig, Appl. Opt. 56, 5182-5197 (2017)

[4] K. Numata et al., Journ of Applied Remote sensing vol.6 (2012).

[5] H. Ishizuki et al., Opt. Express. 20, (2012).

[6] J. Barrientos Barria, D. Mammez, E. Cadiou, J.-B. Dherbecourt, M. Raybaut, T. Schmid, A. Bresson, J.-M. Melkonian, A. Godard, J. Pelon, and M. Lefebvre, Opt. Lett. **39**, 6719–22 (2014).

[7] D. Wernham, J. Alves, F. Pettazzi, and A. P. Tighe "Laser-induced contamination mitigation on the ALADIN laser for ADM-Aeolus", Proc. SPIE 7842, Laser-Induced Damage in Optical Materials: 2010, 78421E

[8] R..S. Coetzee, S. Duzellier, J.B. Dherbecourt, A. Zukauskas, M. Raybaut, V. Pasiskevicius, Opt. Mat. Express **7**, 4138-4146 (2017).

https://doi.org/10.1051/epjconf/202023701010

INITIAL ASSESSMENT OF THE PERFORMANCE OF THE FIRST WIND LIDAR IN SPACE ON AEOLUS

Oliver Reitebuch[1], Christian Lemmerz[1], Oliver Lux[1], Uwe Marksteiner[1], Stephan Rahm[1], Fabian Weiler[1], Benjamin Witschas[1], Markus Meringer[2], Karsten Schmidt[2], Dorit Huber[3], Ines Nikolaus[4], Alexander Geiss[5], Michael Vaughan[6], Alain Dabas[7], Thomas Flament[7], Hugo Stieglitz[7], Lars Isaksen[8], Michael Rennie[8], Jos de Kloe[9], Gert-Jan Marseille[9], Ad Stoffelen[9], Denny Wernham[10], Thomas Kanitz[10], Anne-Grete Straume[10], Thorsten Fehr[10], Jonas von Bismarck[11], Rune Floberghagen[11], Tommaso Parrinello[11]

[1]DLR, Institute of Atmospheric Physics, Oberpfaffenhofen, Germany
[2]DLR, Remote Sensing Technology Institute, Oberpfaffenhofen, Germany
[3]DoRIT, Fürstenfeldbruck, Germany
[4]University of Applied Sciences, Munich, Germany
[5]Ludwig-Maximillians-University, Meteorological Institute, Munich, Germany
[6]Optical & Lidar Associates OLA, Buckinghamshire, United Kingdom
[7]Météo-France, CNRM, Toulouse, France
[8]ECMWF, Reading, United Kingdom
[9]KNMI, de Bilt, The Netherlands
[10]ESA-ESTEC, Noordwijk, The Netherlands
[11]ESA-ESRIN, Frascati, Italy
**Email: oliver.reitebuch@dlr.de*

ABSTRACT

Soon after its successful launch in August 2018, the spaceborne wind lidar ALADIN (Atmospheric LAser Doppler INstrument) on-board ESA's Earth Explorer satellite Aeolus has demonstrated to provide atmospheric wind profiles on a global scale. Being the first ever Doppler Wind Lidar (DWL) instrument in space, ALADIN contributes to the improvement in numerical weather prediction (NWP) by measuring one component of the horizontal wind vector. The performance of the ALADIN instrument was assessed by a team from ESA, DLR, industry, and NWP centers during the first months of operation. The current knowledge about the main contributors to the random and systematic errors from the instrument will be discussed. First validation results from an airborne campaign with two wind lidars on-board the DLR Falcon aircraft will be shown.

1. INTRODUCTION

The Aeolus satellite was successfully launched from Kourou (French Guiana) on August 22, 2018. Its single payload ALADIN is the first European lidar and the first DWL in space [1]. It is considered as a technology demonstrator for future operational wind lidar missions. Aeolus provides profiles of one component of the horizontal wind vector along the laser line-of-sight (LOS, projected to the horizontal HLOS) from ground up to the lower stratosphere (20 km to 30 km) with 250 m to 2 km vertical resolution. A wind profile is obtained from horizontal averaging along track in order to achieve the precision requirement (random error) of 1 m/s to 2.5 m/s (HLOS) depending on altitude [2]. The horizontal averaging length of typically 90 km might vary depending on signal-to-noise ratio (SNR) and atmospheric scene. Systematic errors which cannot be corrected by a-priori knowledge (e.g., calibration or instrumental corrections) – also called "unknown biases" – are very detrimental for NWP. Thus, the requirement for the maximum bias of 0.7 m/s (HLOS) is more stringent than the random error requirement.

2. THE ALADIN LIDAR

The lidar ALADIN is based on a direct-detection DWL operating at an ultraviolet (UV) wavelength of 354.8 nm [3]. The optical receiver consists of two spectrometers to determine the Doppler shift from the spectrally broad Rayleigh-Brillouin molecular backscatter and the spectrally narrow Mie backscatter from aerosols and cloud particles. Accumulation charge coupled device (ACCD)

https://doi.org/10.1051/epjconf/202023701010

detectors are used for both the Mie and Rayleigh spectrometer. The minimum vertical resolution of the ALADIN instrument is limited by the ACCD detectors to 250 m and 24 different altitude levels (range bins). The vertical resolution can be varied between 250 m and 2000 m within one vertical profile. The number of laser pulse returns accumulated directly on the detector is a configurable parameter and determines the minimum along-track horizontal resolution for the raw data, which is typically 3 km for a number of 20 accumulated laser pulses (called "measurement"). Both the Mie and the Rayleigh spectrometer data are stored with this temporal resolution on-board. Further averaging of 600 laser pulse returns to so-called "observations" is performed on-ground for the Level 1b (L1b) products [4]. A grouping algorithm is applied for Level 2b (L2b) wind products which classifies the measurements in clear air Rayleigh winds and cloudy Mie winds [5]. The L2b Rayleigh wind retrieval uses temperature and pressure profiles from NWP models for taking into account the influence onto the molecular lineshape broadening arising from Rayleigh-Brillouin scattering [6].

3. RESULTS

Only nine days after launch on August 31st ALADIN was switched-on and first backscattered light from the atmosphere was sampled on September 5th showing typical lidar signals including molecular returns, cloud tops and extinction below the clouds (Fig. 1). The first atmospheric Doppler frequency shifts and atmospheric winds could be retrieved already three weeks after launch [7] and first results regarding the performance were reported during ICSO in October 2018 [8].

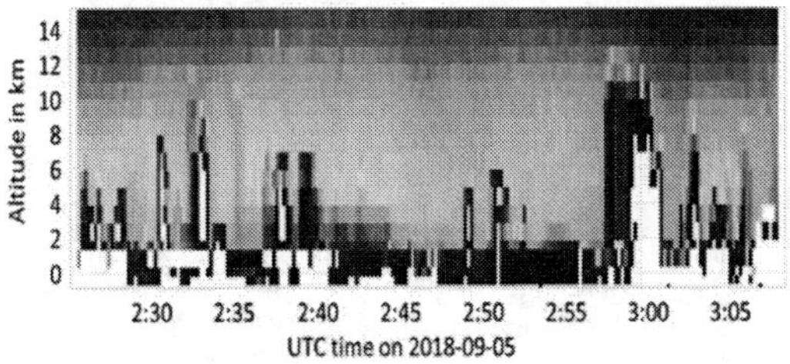

Figure 1 First atmospheric signal from the Rayleigh spectrometer received on 5 September 2018 along ½ orbit (visualized with VirES[9])

3.1 Random error performance

The random errors for ALADIN are mainly influenced by SNR and spectral widths of the backscattered signals. The dominating noise source for both Mie and Rayleigh spectrometer winds is arising from signal photon shot noise. The electronic noise in the detection chain is significantly lower due to the accumulation principle of the ACCD allowing photon counting. Apart from the range of the satellite to the atmosphere and the range resolution, the detected signal levels are mainly determined by both the emitted energy as well as the optical receive path transmission.

The evolution of the laser energy from the first switch-on until mid-January 2019, where the ALADIN instrument went into a standby mode, is shown in Fig. 2. A maximum energy of 65 mJ (UV) was measured initially which decreased to 50 mJ until mid-January, hence, being lower than the anticipated 80 mJ. By modifying the operating parameters (e.g. temperatures or currents) during the first months in order to increase the emitted laser energy, both the UV and IR energies were affected. A stable IR energy level was achieved by mid-December, which might indicate a thermo-mechanical settling in the laser. It was demonstrated that a UV laser could be operated in space over several months proving the successful deployment of an oxygen cleaning system to prevent laser induced contamination.

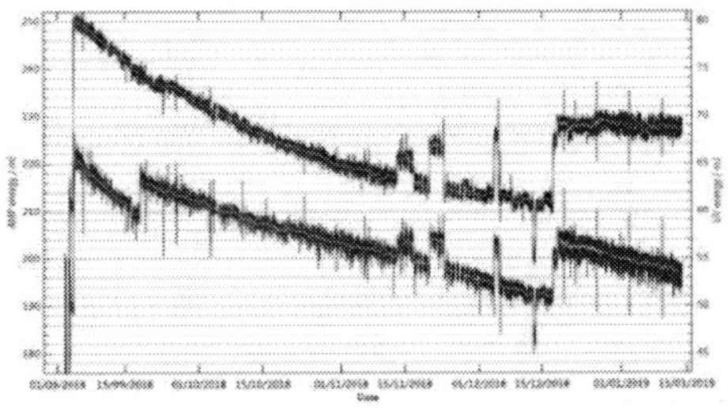

Figure 2 Evolution of laser energy from initial switch-on in September 2018 until mid-January 2019.

As the atmospheric molecular Rayleigh backscatter for high altitudes and above clouds depends only on atmospheric density and hence atmospheric temperature and pressure, the signal levels (i.e. the radiometric performance) were first assessed using the Rayleigh spectrometer signals.

Comparison of signal levels using an end-to-end simulator (E2S) with measured atmospheric signals showed lower signals by a factor of 2.5 to 3 consistently for different observing geometries (off-nadir, nadir pointing), over one orbit and over the period from September to December. A factor of 1.25 to 1.6 can be explained with lower laser energy, as the pre-launch simulations were performed with 80 mJ and only 50 mJ to 65 mJ were achieved in-orbit. A factor of 1.5 to 2 is presumably arising from further signal losses in the optical transmit-receive path. It is hereby assumed that signal clipping at the instrument field stop occurs, which limits the field-of-view to only 18 µrad. This leads to an increase of the Rayleigh wind errors from 2.5 m/s (expected pre-launch with 80 mJ) to about 4 m/s, which is confirmed by comparison with NWP models.

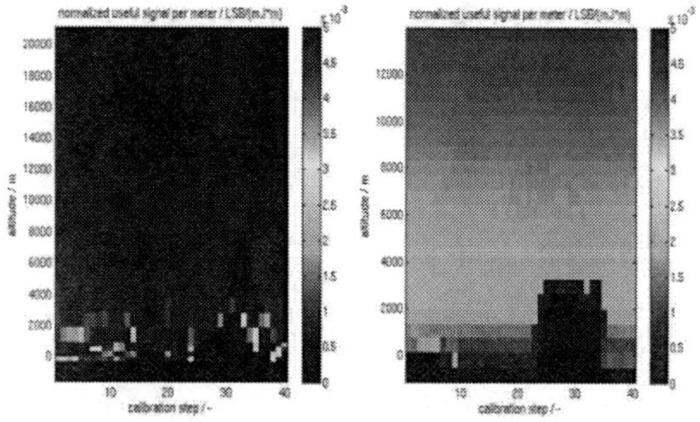

Figure 3 Comparison of normalized Rayleigh useful signal (least significant bit LSB/mJ m) from observations (left) and E2S (right) over Antarctica in nadir-pointing during a response calibration (step# corresponds to distance); high signals (red colors) show ground-return signals over ice.

3.2 Systematic errors

Already in mid-September 2018, a systematic error in Rayleigh winds for one specific range gate was spotted by comparison of several orbits of Aeolus data with the ECMWF model. This bias in Rayleigh winds could be traced back to one suspicious pixel on the Rayleigh ACCD which showed enhanced dark current levels compared to the other pixels by only a few digitizer counts. The number of these "hot" pixels increased continuously over time affecting five range gates for Mie and five range gates for Rayleigh winds (as of mid-March 2019). A specific instrument mode to characterize the dark current levels and occurrence of the "hot" pixels was introduced, and correction algorithms are currently tested.

Furthermore, slow drifts of the incidence angles on the spectrometers were observed which lead to slow drifts in the systematic errors. A strategy for correcting these drifts is currently being refined. It will be based on a combination of spectral characterization using internal measurements, response calibrations performed in nadir pointing, and the utilization of ground-return velocities to correct for remaining linear and harmonic drifts in wind speed bias.

3.2 First airborne validation results

Already several years before the launch, an airborne prototype of the Aeolus payload, the ALADIN Airborne Demonstrator (A2D), was developed at DLR. Owing to its representative design and operating principle, the A2D provides valuable information on the wind measurement strategies of the satellite instrument as well as on the optimization of the wind retrieval and related quality-control algorithms [10, 11]. In November 2018, the first airborne validation campaign after launch was carried out from the airbase in Oberpfaffenhofen, Germany. Four underflights of the Aeolus satellite could be realized and provided first comparisons between the two airborne wind lidars (A2D and a coherent 2-µm wind lidar) and the satellite instrument (Fig. 4).

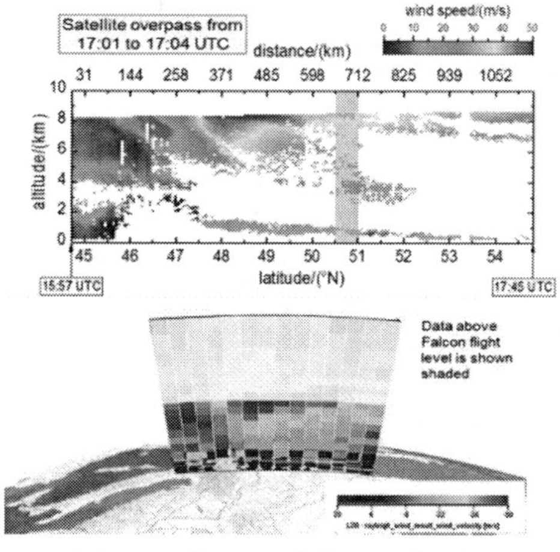

Figure 4 First airborne validation of L2b Rayleigh winds with 2-µm wind lidar (top) from 17 November 2018. Aeolus data (bottom) is visualized by VirES [9].

4. CONCLUSIONS

From a lidar-instrument perspective the ALADIN measurements of the first half year have demonstrated that a space-borne wind lidar can measure atmospheric winds by analyzing molecular Rayleigh and cloud Mie backscatter. It has been shown that a powerful UV laser can be operated in space with high frequency stability. This particularly proves the approach of using oxygen cleaning for mitigation the risk of LIC on optics in the UV for the first time.

The Rayleigh wind random errors are about 4 m/s and therefore higher than the expected value of 2.5 m/s from pre-launch simulations. This is mainly due to the significantly lower Rayleigh signal levels (factor 2.5-3) due to enhanced signal losses in the optical receiver (presumably at the field stop) and the lower laser energy between 50 mJ and 65 mJ compared to the target of 80 mJ. As it could be demonstrated, that no significant additional noise sources are present, the expected gain in signal would be a factor of 1.6 with 80 mJ, which is anticipated for the redundant laser.

The ongoing characterization of instrumental drifts and calibration will allow correction for the known biases with the upcoming algorithm, processor and data products updates. Correction for the enhanced dark signal levels of some ACCD pixels are currently being implemented and will almost completely remove the biases observed for single range gates for both the Mie and Rayleigh spectrometer output. A strategy for correction of instrumental drifts is being refined, based on a combination of spectral characterization using internal measurements, response calibrations performed in nadir pointing, and the use of ground-return velocities to correct for remaining linear and harmonic drifts in wind speed bias.

ACKNOWLEDGEMENTS

This work was supported by ESA under contract no. 18366/04/NL/MM and the Aeolus DISC contract no. 40000126336/18/I-BG. The airborne campaigns were funded by DLR and ESA under contract no 4000114053/15/NL/FF/gp.

REFERENCES

[1] European Space Agency ESA, 2008: ADM-Aeolus Science Report, ESA SP-1311, 121 pp.

[2] European Space Agency ESA, 2016: ADM-Aeolus Mission Requirements Documents, AE-RP-ESA-SY-001, Issue 2, 16/11/2016, 57pp.

[3] Reitebuch, O., 2012: The space-borne wind lidar mission ADM-Aeolus. in Atmospheric Physics – Background, Methods, Trends. U. Schumann (Ed.) *Springer Series on Research Topics in Aerospace*, 815-827.

[4] DLR: Reitebuch, O., Huber, D., Nikolaus, I., (2018). Algorithm Theoretical Basis Document ATBD: ADM-Aeolus Level 1B Products, AE-RP-DLR-L1B-001, V. 4.4, 20.04.2018, 117pp.

[5] ECMWF: Tan, D., M. Rennie, E. Andersson, P. Poli, A. Dabas, J. de Kloe, Gert-Jan Marseille, Ad Stoffelen 2018: ADM-Aeolus Level-2B Algorithm Theoretical Baseline Document. AE-TN-ECMWF-L2BP-0024, Version 3.1, 31 Jan 2018, 133 pages.

[6] Dabas, A., Denneulin, M.-L., Flamant, P., Loth, C., Garnier, A., Dolfi-Bouteyre, A., 2008: Correcting winds measured with a Rayleigh Doppler lidar from pressure and temperature effects. *Tellus* **60A**, 206-215.

[7] ESA 2018: Aeolus wows with first wind data, press release on 12 September 2018 https://www.esa.int/Our_Activities/Observing_the_Earth/Aeolus/

[8] Kanitz, T. J. Lochard, J Marshall, P. McGoldrick, O. Lecrenier, P. Bravetti, O. Reitebuch, M. Rennie, D. Wernham, and A. Elfving, 2018: Aeolus First Light – First Glimpse. Proc. Int. Conf. Space Optics ICSO, 9-12 October 2018, Chania, Greece.

[9] VirES for Aeolus 2018: https://aeolus.services/

[10] Lux, O., Lemmerz, C., Weiler, F., Marksteiner, U., Witschas, B., Rahm, S., Schäfler, A., Reitebuch, O. 2018: Airborne wind lidar observations over the North Atlantic in 2016 for the pre-launch validation of the satellite mission Aeolus. Atmos. Meas. Tech., 3297-3322.

[11] Marksteiner, U., Ch. Lemmerz, O. Lux , S. Rahm, A. Schäfler , B. Witschas, O. Reitebuch 2018: Calibrations and Wind Observations of an Airborne Direct-Detection Wind LiDAR Supporting ESA's Aeolus Mission. Remote Sensing, 10, 2056.

EPJ Web Conferences **237**, 01011 (2020)
ILRC 29
https://doi.org/10.1051/epjconf/202023701011

LASER REMOTE SENSORS FOR NASA'S FUTURE EARTH AND SPACE SCIENCE MISSIONS

Upendra N. Singh

NASA Engineering Safety Center, NASA Langley Research Center, Hampton, Virginia 23681, USA
**Email: Upendra.n.singh@nasa.gov*

ABSTRACT

Active optical (Laser/Lidar) measurement techniques are critical for the future National Aeronautics and Space Administration (NASA) Earth, Planetary Science, Exploration, and Aeronautics measurements. The latest science decadal surveys recommend a number of missions requiring active optical systems to meet the science measurement objectives and the aeronautics community continues to use Laser/Lidar technologies to meet the aeronautics measurement objectives. This presentation will provide an overview of NASA efforts in developing and maturing state-of-the-art advanced solid-state flight laser/lidar systems for airborne and space-borne remote sensing measurements. The presentation will also provide details of a strategic approach for active optical technologies and techniques to meet the NASA's future Earth and Space Science measurements/missions needs and requirements for space-based applications.

1. OVERVIEW

NASA's Strategic Plan calls for NASA "to improve prediction of climate, weather, and natural hazards," and to provide services to the Nation including "weather forecasting; climate prediction; natural hazard assessment, prediction, and response; and environmental management, including air quality forecasting and land use assessment." These assignments will require the use of Light Detection And Ranging (lidar) (or laser radar) remote sensing systems. Lidar sensors provide the excellent vertical resolution needed to do these jobs. NASA also is working to advance radar, laser, and light detection and ranging technologies to enable monitoring of such key Earth system parameters as land surface, oceans, ice sheet topography, and global tropospheric winds that could lead to advances in weather and severe storm prediction. After 35 years of passive and then microwave radar remote sensing, NASA began to utilize lidar sensors in the 1990's. However, many difficulties were encountered with the lidar missions. Part of the cause of these difficulties arose from the large gap between the technology maturities after Instrument Incubator Program (IIP) completion (~ \$3M), and the desired maturities to begin an Earth System Science Pathfinder (ESSP) flight projects (~ \$300M). Too often, the laser and lidar technologies were still being developed during the flight projects, leading to schedule delays and large labor bills from the project personnel. These concerns led to NASA commissioning an Earth Science Independent Laser Review Panel during the early part of 2000. The Panel report cited critical deficiencies in NASA's laser strategy and recommended that NASA should address the cited deficiencies and put together a Laser Risk Reduction Program [1]. Through the Panel report, it was clear that development teams needed to mature laser and lidar technologies before the approval of flight projects. For this reason, NASA approved the Laser Risk Reduction Program (LRRP), which began in 2002, had enjoyed many successes, and ended during 2010 [2-3].

2. RATIONALE FOR ACTIVE OPTICAL SENSING

Many physical phenomena critical to improving our understanding of Earth and planetary systems can only be measured practically using lidar techniques. Lidar remote sensing is analogous to radar remote sensing, except lidar employs electromagnetic radiation with wavelengths 4 to 5 orders of magnitude shorter than radar. This wavelength difference causes proportionally less beam divergence; profound differences in usable targets, atmospheric transmission, and attainable resolution. Lidar also requires different beam generation, handling, and detection technologies. Therefore, lidar sensors maintain the advantages of active radar sensors with the self-contained light source, choice and knowledge of exciting signal properties, pointing at desired target, and range-resolved return signal, while providing greatly improved horizontal and vertical spatial resolution, and access to new atmospheric and surface targets. In summary, lidar sensors offer:

- Excellent vertical and horizontal spatial resolution
- Relative independence from natural light sources, their angles, and time of day
- Choice of optical wavelength, pulse duration, pulse shape, and polarization to optimize science return (i.e., the ability to target specific chemical and physical phenomena by choosing the appropriate wavelength)
- Choice of pointing direction

• Choice of several lidar techniques: basic, altimetry, Doppler, DIfferential Absorption Lidar (DIAL), and polarization

An active optical remote sensing technology development and maturation strategic approach will mitigate lidar system technology development risks by coordinating activities with multiple NASA Centers and government agencies. This will move NASA into the next logical era of lidar remote sensing by enabling such critical space-based measurements as winds, ozone, CO2, moisture, clouds, aerosols, surface topography, ice thickness, vegetation, trace atmospheric species, biogenic traces gases and materials, and oceanic properties.

Lidar is also crucial to improving the safety, reliability, and affordability of solar system exploration by providing atmospheric information for spacecraft aerocapture, entry and descent; and guidance and hazard avoidance data to the Planetary Landers. Furthermore, lidar techniques are also regarded as enabling for a variety of homeland security, aviation, and environmental pollution applications. NASA's leadership in this area would further cement its public perception as a national asset and "crossover" entity encompassing both civil and DoD functions. This effort will bring all lidar remote sensing technologies together, including planetary, earth viewing, aerospace, and aeronautics.

3. KEY MEASUREMENTS ENABLED

An active optical strategic approach enables several key measurements that are likely to open new frontiers of research that practitioners cannot obtain in any other way. Especially critical measurements from the Earth Science measurement roadmaps include:

• Global tropospheric winds, which are acknowledged as the greatest unmet observational need for improving global weather forecasts on anything more than an incremental basis. Doppler lidar techniques are recognized as the only means for acquiring such data with the required precision (1 m/s, 25-100-km horizontal resolution).

• Similarly, the location, quantification, and extent of CO2 sources and sinks; and their variability are prima facie requirements for balancing the carbon budget in global climate models. High precision profiling of tropospheric CO2 (0.3% mixing ratio, 2-km vertical scale) is essential to understanding the global carbon cycle, greenhouse warming and sustainability of life on Earth can only be obtained through the use of differential absorption lidar (DIAL) lidar techniques.

• Measurement of the global distributions of tropospheric ozone and aerosols using a space-based DIAL system will help address the causes and impacts of long-term climate variability and will distinguish natural from human-induced drivers. This measurement is essential to the modeling and assessment of chemical, radiative, convective, dynamical, and transport processes in the troposphere. Again, precise, vertical measurements (1 to 2 km vertical) are only available through the use of lidar techniques.

• Lidar is the only method for measuring particle profiles in the oceans' mixing layer, necessary to understand how oceanic carbon storage and fluxes contribute to the global carbon cycle

• DIAL lidar is the only technique for global moisture profiles at high resolution (0.5 km vertical by 50 km horizontal) in the boundary layer. This is an essential need for understanding the convective processes and severe storm development, and to improve our understanding of the global hydrological and energy cycle.

• Backscatter lidar is the only method for high vertical resolution (30 m) measurements of optical properties of clouds and aerosols including planetary boundary height, cloud base, cloud top, cloud amount, cloud depolarization, and aerosol microphysical properties needed in climate modeling and research.

• Altimetry lidar is the only technique for directly profiling a surface level changes of less than 1 cm/year, essential for studying terrestrial land cover vegetation, land surface topography, volcano monitoring, global sea level, and polar ice sheet level changes caused by climatic changes.

Especially critical measurements from the Space Science measurement roadmaps include:

• Lidar techniques offer versatile planetary surface topography measurements and other planetary surface deformation information with sub-meter vertical accuracy laser altimetry to study geological, volcanic, climatic, and tidal effects in planetary bodies including Mars, the Moon, Venus, Titan, and Jupiter's Icy Moons.

• As with terrestrial applications, backscatter lidar is the only method for determining detailed vertical profiles of the Martian and Jovian atmospheres including dust storms and surface eddies that can provide information about their aerosol and clouds needed to study the structure, dynamical, and radiative properties of the atmospheres of these planetary bodies.

• DIAL lidar enables measurement of the variability of water vapor, ozone, CO2, and other trace gas species in planetary atmospheres (e.g., Mars) to study the chemical and radiative properties and obtain

https://doi.org/10.1051/epjconf/202023701011

information about water and water vapor that are of keen interest as an indispensable ingredient for life.

• Doppler wind lidar is the only technique to study the three-dimensional dynamical and transport properties, dust storms, and turbulence in the atmospheres of Mars, Jupiter, Venus, Titan, and other planetary systems that can shed new insight into knowledge about the state of these atmospheres

• Surface-based lidar is ideal for measuring tropospheric winds on Mars as a boundary condition in global climate models, but also to ensure design safety of next-generation entry, descent, and landing systems.

• Lidar with high power lasers can be used as remote sensing mass spectrometers by performing laser ablation of distant planetary bodies including Jupiter's Icy Moons, as well as direct compositional information about primitive objects including asteroids and comet nuclei.

• Lidar Raman Spectroscopy for detection of organic phases at local and remote scales (e.g., on Mars)

• Lidar spectroscopy for Biogenic Trace Gas sensing (i.e., detection of methane and other species) in the Martian atmosphere.

4. ACTIVE OPTICAL CAPABILITY ASSESSMENT AND STRATEGY

NASA Sensors and Instrumentation (S&I) Capability Leadership Team (CLT) was recently tasked by the NASA Agency Program Management Council (APMC) to determine if the Agency has the necessary expertise and capabilities to execute successfully the active optical-based systems necessary to make the required measurements for Science, Exploration, and Aeronautics. Given the charter from the APMC, the S&I CLT leadership assembled a core team to realize the assessment. The Active Optical assessment process utilized a multi-directorate teaming with representatives of the Science, Human Exploration and Operations, Aeronautics Research, and Space Technology Mission Directorates to provide guidance and list the priority measurements requirements to meet their critical future needs.

An Active Optical Tiger team consisting of mission directorate representatives, S&I capability and technical leads/co-leads, and an external assessment team worked together to seek national and international community inputs to formulate an integrated Agency level strategy. Given the cross-cutting synergies in critical Active Optical measurements for NASA Science, Exploration and Aeronautics, a strategic approach is needed to identify areas where NASA should "lead, leverage or collaborate" with existing national and international industries to meet its future needs. An essential part of the assessment strategy was to engage with other US Government entities, as well as industry and academia, to leverage common Active Optical interests and complementary skills/expertise resident in those sectors. To seek the input from broader national entities, such as NASA, industry, academia, Federally Funded Research and Development Centers (FFRDC) and Other Government Agencies (OGA), the S&I CLT organized a Technical Interchange Meeting (TIM) during July 31 through August 2, 2018 in Columbia, Maryland, USA. Participants in the TIM exchanged perspectives on the current state of the discipline's technologies and the direction NASA needs to take in the future to raise the TRL of the measurement technologies to meet these measurement needs in the applications domains. The TIM aimed at focusing NASA's directions to attain the necessary technology TRL levels to meet the Agency-level priority Active Optical measurements in Space and Aeronautics.

5. PRIORITY ACTIVE OPTICAL BASED MEASUREMENT AREAS

The NASA TIM organized focused technical sessions in the priority active-optical based measurements area for NASA critical needs in Science, Exploration, and Aeronautics. These sessions were:

5.1 Mission Area I: Earth Science Focused Measurement Topic Areas

Atmospheric Composition sounding measurements in the troposphere including aerosol and cloud properties and profiling, water vapor profiling, and Planetary Boundary Layer height

• Earth's Surface & Interior via geodetic imaging including wide-swath laser altimetry

• Climate Effect Monitoring including ice mass changes, global ice characterization, snow depth, and snow-water equivalent measurements

• Weather including global tropospheric Lidar wind measurements for planetary boundary layer energy, momentum, and mass exchanges; forecasting; convection; fluxes of heat, momentum, water vapor and other gases; and model validation and improvement

• Carbon, Ecosystems, & Biogeochemistry including measurements of CO_2 and CH_4 fluxes and trends, vegetation 3D structure, biomass, and disturbance, ocean mixing layer

5.2 Mission Area II: Space Science and Exploration Focused Measurement Topic Areas

• Lunar Exploration including landing navigation and hazard avoidance, and surface exploration and characterization

• Jovian Moons including surface features and characterization, and tidal signatures

• Jovian Environment including atmospheric cloud and wind structure, atmospheric composition

• Mars Exploration including atmospheric winds, density, and dust profiling, surface topography, surface exploration and characterization, search for organic signatures, moisture distribution, and cloud characterization

• Martian Mission Support including precision landing, navigation, and hazard avoidance, sample return rendezvous and capture, atmospheric characterization to support entry, descent, and landing

• Titan and Venus including balloon-borne active optical instrumentation

• Situational awareness of space weather phenomena

• Laser interferometry, active –optical sensing of telescope and satellite swarm elements to enable observations

5.2 Mission Area III: Aeronautics Focused Measurement Topic Areas

• Turbulence detection on flight vehicles

• Situational awareness detection of atmospheric hazards including volcanic ash, ice, and snow

• Sense and avoidance for autonomous UAVs and future autonomous urban mobility aircraft

• Airport ground control management for taxiing, takeoff, and landing including wake vortex detection

• UAV-based target imaging

• Sensing and avoiding hazards in the National Air Space including miniaturized, low-power active optical sensors and instrumentation for UAV and personal air vehicles

• Quantification of flow field and surface parameters in ground-based wind tunnel testing of aero vehicle

To include the international active optical capability as a part of the NASA strategy, the S&I CLT coordinated efforts with other space agencies and organized an International IEEE-GRSS Workshop on Space-based Lidar Remote Sensing Techniques and Emerging Technologies during June 4-8, 2018 at Milos Conference Center, Adamas, Milos Island, Cyclades, Greece. The NASA and ESA leadership jointly organized similar technical sessions as in the TIM. This Workshop brought together Active Optical experts and policy makers from NASA, ESA, JAXA, CNES, and DLR engaged in advancing space missions/measurements in the Active Optical area for Science and Exploration.

6. CONCLUSIONS

To address the critical Agency needs in the area of active optical measurements, the NASA Sensors and Instrumentation Capability Leadership Team worked with experts and leaders from NASA, industry, academia, FFRDCs, OGAs, and international Space Agencies to assess existing active optical capabilities and opportunities for leveraging and collaboration for addressing NASA's critical needs for Science, Exploration, and Aeronautics. The S&I CLT hosted TIM and international workshop to capture national and international technical inputs from experts. Inputs gathered were provided to an independent and external Active Optical Assessment and Recommendation Team (AOART). In coming months, AOART team will complete a final report and present its findings, observation, and recommendation to the APMC, including a suggested strategy to address the Agency's needs in a crosscutting, synergistic and cost effective manner. The conference presentation will summarize these results and recommendations.

REFERENCES

[1] S. Alejandro, et al., NASA Internal Report, November 2000.
[2] U. Singh, et al., Proc. of SPIE 4882-81 (2002).
[3] F. Peri, et al., Proc. of SPIE 4893-24 (2002).

EPJ Web Conferences **237**, 01012 (2020)
ILRC 29
https://doi.org/10.1051/epjconf/202023701012

LIDAR CONCEPT OF "GUANLAN" MISSION FOR SPACE OCEANOGRAPHY

Songhua Wu[1, 2 *], Weibiao Chen[3], Junwu Tang[2], Chaofang Zhao[1, 2], Ge Chen[1, 2],

[1] Ocean Remote Sensing Institute, College of Information Science and Engineering, Ocean University of China, Qingdao 266100, China
[2] Laboratory for Regional Oceanography and Numerical Modeling, Pilot National Laboratory for Marine Science and Technology (Qingdao), Qingdao 266237, China
[3]Shanghai Institute of Optics and Fine Mechanics, Shanghai 201800, China
**Email: wush@ouc.edu.cn*

ABSTRACT

Among the various challenges in spaceborne radar observation of the ocean, the following two issues are probably of a higher priority: inadequate dynamic resolution, and ineffective vertical penetration. Two highly anticipated breakthroughs in the coming decade are likely to be associated with radar interferometry and ocean lidar technology, which are expected to make a substantial contribution to a sub-mesoscale-resolving and depth-resolving observation of the ocean. The planned "Guanlan" science mission comprises a dual-frequency (Ku and Ka) interferometric altimetry (IA) and a near-nadir pointing ocean lidar (OL). The spaceborne active OL will ensure a deeper penetration depth and an all-time detection which leads to a layered characterization of the optical properties of the subsurface ocean. The simultaneous functioning of the OL and a dual-frequency (Ku and Ka) interferometric altimetry system will allow an enhanced understanding of contributions of the atmosphere and the air-sea interface which in turn considerably reduce the error budgets of the two sensors. The OL payload is expected to partially reveal the marine food chain and ecosystem with 10-m vertical interval in the euphotic layer, moving a significant step down to the oceanic mixed layer both dynamically and bio-optically.

1. INTRODUCTION

Since the advent of ocean radar satellite in 1970s [1], its development has followed three generic trends: (i) better sampling, (ii) higher accuracy, and (iii) more variables. Better sampling of the ocean normally refers to a broader coverage, a deeper penetration, a higher spatiotemporal resolution, and a longer time series. Upgraded sensors with refined algorithms (both theoretical and empirical) are major approaches to improve the accuracy of remotely sensed parameters. Meanwhile, tremendous efforts have been devoted to testing new instrument (e.g., ocean lidar) and deriving new variables from spaceborne measurements. Ideally, the ultimate goal of ocean remote sensing is towards a quantitative derivation of all required variables with a satisfactory accuracy and a perfect sampling in near-real time. In reality, however, this can never be fully realized, but can always be gradually approached. In this paper, the concept design of the ocean lidar (OL) for Chinese Guanlan science mission is described.

As a pioneer of spaceborne lidar, the US space agency NASA initiated the Lidar In-Space Technology Experiment (LITE) in 1988 [2]. Following a successful feasibility study with an aerosol and cloud lidar onboard the Space Shuttle of Discovery in 1994, NASA and the French space agency CNES launched their joint mission of CALIPSO (Cloud Aerosol Lidar and Infrared Pathfinder Satellite Observation) in 2006 [3], which transformed the satellite lidar technology from experimental to operational. In the years that followed, a dozen of spaceborne lidars have been proposed by several national and international space agencies such as NASA and ESA (European Space Agency).

The CALIPSO mission has provided valuable by-products for oceanographic studies during the past decade (see Hostetler et al. [4] for a recent review). Behrenfeld et al. [5] use CALIOP measurements to quantify global ocean phytoplankton biomass and total particulate organic carbon stocks. Lu et al. [6] find significant relationships between integrated subsurface backscatter and chlorophyll-a concentration, as well as particulate organic carbon, which indicate a potential use of CALIPSO lidar to estimate global chlorophyll-a and particulate organic carbon concentrations. Lu et al. (2016) introduce an approach for estimating the ocean subsurface layer-integrated backscatter and particulate backscattering coefficient from CALIOP 30° off-nadir lidar measurements. Behrenfeld et al. [7] report a decade of uninterrupted polar phytoplankton biomass cycles, and find that polar phytoplankton dynamics are categorized by "boom–bust" cycles resulting from slight imbalances in plankton predator–prey equilibria.

Compared to historical passive optical remote sensing of the ocean, spaceborne lidars have two unique advantages: First, vertical penetration into the mixed layer through which profiles of optical and even oceanographic properties can be quantitatively obtained; Second, all-time measurement independent of solar radiation can be carried out. But they also have a critical limitation that all measurements are vertically integrated without depth information. A dedicated spaceborne lidar such as the one proposed for the Guanlan science mission is therefore an urgent need for a better understanding of the truly three-dimensional (3-D, rather than 2.5-D) structures in the subsurface ocean.

2. THE OCEAN LIDAR SYSTEM

2.1 The lidar system

As an active optical detection instrument capable of providing time-resolving and spectral-resolving measurements, the Guanlan ocean lidar is built to perform all-time profiling of the attenuated coefficients, the backscatter coefficients and the particle linear depolarization ratios from the sea surface to the mixed layer. Additionally, the OL is also designed to make possible measurements of the ocean fluorescence and the sea level elevation. The specifications of the OL is listed in Tab 1.

Tab 1. The specifications of the OL

Parameters	Specification
Wavelength	1064.4±0.1nm 532.2±0.05nm 486.13±0.05nm
Single pulse energy	≧0.2J@1064.4nm ≧0.15J@532.2nm ≧0.08J@486.1nm
Beam divergence	≤0.2 mrad
Pulse width	~10 ns
Telescope diameter	1.2m
FOV	~ 0.3 mrad
Receiver optical bandwidth	≤0.2 nm
Height resolution	≤10 cm
Depth resolution	≤5 m
Penetration depth of ocean water	150m~300m (low orbit) 130m~240m (high orbit)

The Guanlan OL has two fully redundant lasers, each with a narrow-linewidth laser that transmits simultaneous, co-aligned multi-wavelength (including the blue and green bands) pulses (Fig. 2). The total output pulse energy is at joule level. The two lasers are expected to cover the full mission lifetime. The primary laser is scheduled to be fired since the launch of Guanlan, while the second one serves as a backup. For the receiver, a Cassegrain telescope with lightweight silicon carbide is designed to collect backscatter signals. The diameter of the telescope is approximately 1.2 m. For accurate measurements of the troposphere and subsurface ocean, a data acquisition subsystem with a high sensitivity and a large dynamic range is required. Therefore, an analog mode along with a photon counting mode will be operating simultaneously. To ensure data integrity and processing flexibility, high-speed waveform acquisition and variable digital discrimination techniques are adopted. The heat generated by lasers is dissipated via a radiant cooling panel. The OL is installed to the satellite platform in the overall mounting framework.

The technical route of the laser system is a solid-state main oscillation power amplifier combined with a nonlinear frequency conversion technology. Multistage laser amplifiers are used to amplify the seed laser pulses from a 1064-nm main oscillator. The frequency of the amplified high energy 1064-nm laser is doubled to obtain 532.2-nm green laser pulses. A 355 nm ultraviolet laser is generated by a frequency sum of the 532.2-nm laser and the remaining 1064-nm laser. The 486.1-nm blue laser pulses will be generated by the 355-nm laser pumped optical parametric oscillator. Considering the scale and light damage in the actual space environment, multiple lasers are designed to meet with the parameters described above and obtain the 486.1-nm blue laser and the 532.2-nm green laser with single pulse energy greater than 1 joule each. A 1°-3° incidence angle will be adopted to avoid ocean surface reflection at nadir.

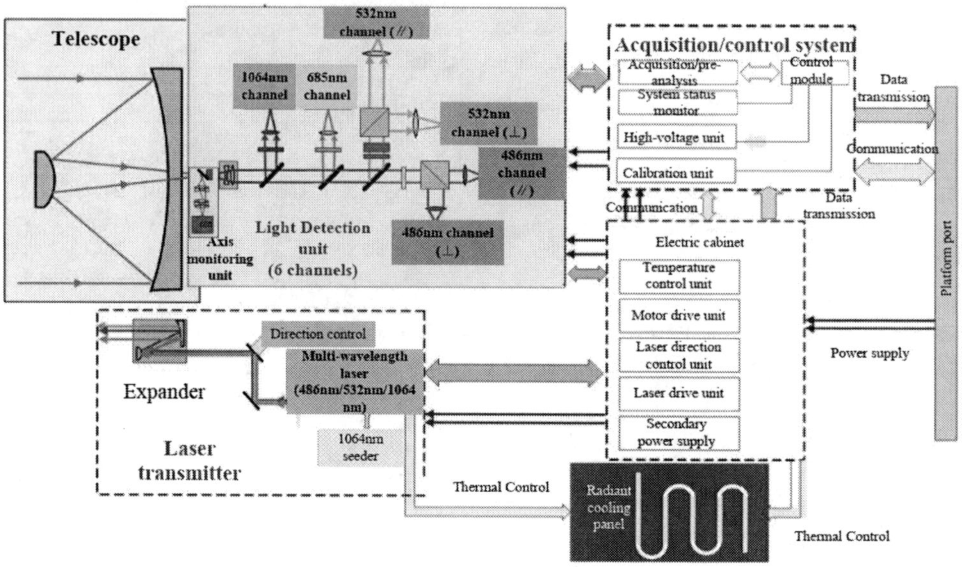

Fig 2. Technology scheme of the spaceborne ocean lidar

The receiving telescope adopts a Cassegrain structure, consisting of a primary mirror and a secondary mirror. The mirror-coated metal reflective film has a reflectivity of more than 95% in the blue-green band. The telescope has an effective receiving aperture of 1.2 m and a field of view of 0.2 mrad. Under the premise of ensuring the rigidity and stability of the mirror body, the weight of the telescope is to be reduced as much as possible. The primary and secondary mirrors are made of silicon carbide materials with high rigidity, excellent thermal conductivity and high stability. The weight of the optimized telescope will be limited to less than 200 kg.

A single photon counting technique is used to detect laser echoes in multiple channels. With this technology, the echo signal of a single photon can be detected with a count rate of 100 MHz. A good signal-to-noise ratio can be obtained by time integration during the detection. In the case of receiving a large number of photons at the same time, the analog signal output can be realized by the accumulation of multi-channel signals. The analog signal is received by the acquisition card with a sampling rate of 1 GSps, and the continuous signal of the underwater vertical profile can be collected. The hybrid method of analog sampling and photon counting will improve detection sensitivity meanwhile provide a high dynamic range.

2.2 Calibration and validation activities

Synchronous calibration/validation of the OL sensors onboard the Guanlan satellite is a challenging task. The spaceborne ocean lidar will be mainly calibrated by an airborne lidar. Based on satellite orbit and laser emission direction, the trajectory of laser footprint at the sea surface can be calculated. The airborne lidar verification experiment is then carried out synchronously with satellite measurement in the trajectory direction. A flight altitude of about 4 km results in a ground path of greater than 50 km, and the trajectory deviation between the footprints of airborne and spaceborne lidars at the sea surface is expected to be less than 2 km. The time deviation between measurements from the two platforms will be less than 0.5 hour. Aside from aircraft based measurements, the calibration method based on the comparison between ocean surface signal and theoretical results (e.g., Cox-Munk theory) using wind speed or surface cross section derived from the radar is also feasible.

The validation of the Guanlan OL requires simultaneous measurements of the subsurface ocean by the laser satellite and an integrated optical buoy. To do so, an optical multi-parameter profiling system is developed using an automatic winch, along with a ship-borne lidar. The optical buoy will be deployed at the footprint of the satellite orbit, and continuous profiling observations will be carried out when the satellite overflies it. The maximum depth of the profile will exceed 300 m (almost twice as much as the maximum penetration depth of an ocean lidar) and the vertical resolution will be less than 1 m. Multiple parameter measurements of temperature, salinity, chlorophyll concentration, backscattering coefficient and so on, will be collected and transmitted to the data center in near real time, where an integrated calibration and validation of the Guanlan OL will be implemented in order to reach the level of precision and accuracy.

3. UNIQUE CHARACTERISTICS AND POTENTIAL BENEFITS

As one of the first spaceborne ocean lidars and an oceanic counterpart of CALIPSO, the Guanlan OL will be used as an experimental sensor to test a number of new ideas such as a replacement of the nadir-pointing conventional altimeter, a complementary sensor for measuring the ocean component of the global carbon cycle, a new sensor for demonstrating the feasibility of optical-acoustical conversion in the ocean, and an extended sensor for tracing the paths of terrestrial dust into the ocean as well as its biogeochemical effects.

4. OUTLOOK AND RECOMMENDATIONS

The ocean has entered the big data stage at the surface while still remains in the small data stage at the bottom. The coexistence of big and small data in terms of ocean measurement is expected to last for a long time. As such, persistent availability of global ocean optical profile data from the Guanlan OL down to the depth of the maximum penetration appears to be extremely valuable, making unique contribution to the observation of oceanic mixed layer. It will also benefit from combination with concurrent in situ surface drifters and Argo floats to routinely provide a systematic monitoring of the upper ocean between 0-2000 m.

In summary, the key objectives of the Guanlan science mission is to extend the penetration depth of space-borne ocean lidar to the vicinity of thermocline, moving a critical step forward towards an ultimate goal of "transparent ocean" which relies on fine resolution, full spectrum, and multidisciplinary oceanographic sensing and observation. Meanwhile, an integrated physical-biological measurement of the ocean will be achieved, serving as China's novel contribution to the international ocean remote sensing and space oceanography communities.

ACKNOWLEDGEMENTS

This research is supported by the Pilot National Laboratory for Marine Science & Technology (Qingdao), China, under grants No. 2018ASKJ01 and 2015ASTP-OS15.

REFERENCES

[1] L. Fu, et al., Oceanography 23: 14-25 (2010)
[2] D. Winker, et al., Proceedings of the IEEE 84: 164-180 (1996)
[3] D. Winker, et al., Bulletin of the American Meteorological Society 91: 1211–1229 (2010)
[4] C. Hostetler, et al., Annual Review of Marine Science 10: 121-147 (2018)
[5] M, Behrenfeld, et al., Geophysical Research Letter 40: 4355-4360 (2013)
[6] L. Fu, et al., Journal of Geophysical Research 119: 4305-4317 (2014)
[7] M, Behrenfeld, et al., Nature Geoscience 10: 118-122 (2017)

EPJ Web Conferences **237**, 01013 (2020)
ILRC 29
https://doi.org/10.1051/epjconf/202023701013

MCT APD DETECTION SYSTEM FOR ATMOSPHERIC PROFILING APPLICATIONS USING TWO-MICRON LIDAR

Tamer F. Refaat*, Mulugeta Petros, Ruben Remus and Upendra N. Singh

NASA Langley Research Center, Hampton, Virginia 23681, USA
**Email: tamer.f.refaat@nasa.gov*

ABSTRACT

An advanced detection system, based on mercury cadmium telluride avalanche photodiodes array, was implemented within a 2-µm lidar. Detection system characterization was conducted for performance evaluation including settling time, noise-equivalent-power (NEP) and dynamic range. Results indicated an average NEP of 1.4 $fW/Hz^{1/2}$ per pixel. Lidar range resolved profiling demonstrated the dynamic range capability by measuring near-field aerosol scattering and far-field clouds reflection simultaneously.

1. INTRODUCTION

An airborne triple-pulse integrated path differential absorption (IPDA) lidar has been developed at NASA Langley Research Center [1-3]. This instrument measures both atmospheric carbon dioxide (CO_2) and water vapor (H_2O) simultaneously. The IPDA transmitter produces high-energy laser pulses at 2-µm wavelength with a repetition rate of 50 Hz. Each laser shot consists of a burst of three 50 ns pulses, separated by 200 µs, with a different wavelength setting for each pulse [4]. The operating wavelengths of the three pulses were selected, with respect to the CO_2 R30 line center, to target H_2O absorption, CO_2 absorption and minimum absorption (off-line), for the first, second and third pulses, respectively [1]. The IPDA receiver consists of a 0.4 m Newtonian telescope that focuses the return radiation onto a 300 µm spot size. The return radiation is collimated and filtered, before being split (90%-10%) into high and low signal channels. The high signal channel is focused onto a 300-µm diameter commercial extended-range InGaAs PIN photodetector. The low signal channel is used to extend the detection dynamic range for high returns without saturation. In addition, the low signal channel can be used to test other 2-µm detection technologies [3].

Recent developments of advanced single-charge-carrier Mercury Cadmium Telluride (HgCdTe or MCT) electron-initiated avalanche photodiodes (e-APD) are a breakthrough in lidar detection technology [5-6]. Compared to other detector materials, MCT is a mature and reliable technology amenable to space qualification. Several IPDA lidar validations were conducted using MCT e-APD detection systems. These validations were focused on the 1.6 µm wavelength targeting CO_2, methane and oxygen using the IPDA lidar technique [7-9]. Recently, an MCT e-APD detection system was integrated and characterized for the 2-µm wavelength [10]. In this paper, atmospheric testing of the detection system using the 2-µm triple-pulse IPDA lidar is presented. Focusing on the off-line return, the MCT e-APD direct detection system exhibited a dynamic range and sensitivity suitable for atmospheric profiling within the boundary layer, while observing far-field clouds.

2. MCT e-APD DETECTION SYSTEM

The MCT e-APD consists of 16 pixels arranged in 4×4 pixel-array with an 80×80 μm^2 pixel-size. This device operates at 77 K and exhibits IR sensitivity from 1-µm up to 4-µm [5]. The e-APD is connected to a readout integrated circuit (ROIC) that accesses each pixel through individual, variable gain, trans-impedance amplifiers (TIA). An output summing amplifier (SUM) produces the summation of a specific number of pixels. Pixel activation is selected by software, which controls device operation and allows changing the bias voltage and the TIA gains. A fixed gain buffer amplifier, for driving 50 Ω loads, follows the SUM. The TIA and SUM are coupled through either dc or ac coupling using hardware connectors. The detection system output is limited to 1.4 V [6, 10].

The 2-µm e-APD detection system includes a vacuum system, which allows cooling the device using a cryo-cooler to reduce dark current and noise. A cold narrow band-pass optical filter was integrated inside the chamber to transmit 2-µm

https://doi.org/10.1051/epjconf/202023701013

and reduce background radiation. Additional custom designed aft-optics couples the input radiation to an optical fiber, and focuses it onto the center 2×2 pixels. Limiting the number of active pixels reduces the noise of the detection system. The whole system is integrated in a rack mountable chassis at NASA Goddard Space Flight Center [6]. The detection system bandwidth is set to 6 MHz. An advanced, 12-bit, 1 GS/s digitizer is connected to the output for data recording.

2.1 Impulse Response and Coupling

Generally, backscatter lidar detection systems are required to acquire both high and low signals. High frequency, high amplitude signals, in the form of pulses, are generated from near-field scattering and hard targets such as ground and cloud returns. Low frequency, low amplitude signals are generated from aerosol scattering, which is characterized by a slowly varying spatial distribution, close to the background signal. Therefore, dc coupling is typically used for backscatter lidar applications. It is important for the detection system to recover quickly from high frequency pulses to avoid bias introduction to the background signal [11]. Thus, pulse short time recovery is an important aspect of the impulse response of the detection system. IPDA is a special lidar technique only focused on hard target returns. In this case, ac coupling can be implemented for IPDA only if the transmitted pulse-width is short compared to the resistance-capacitance (RC) time constant of the coupling.

An experimental setup was used to measure the impulse response of the detection system using different pulse-widths and couplings. A radiation pulse was simulated using a 2-μm continuous-wave laser and an acousto-optic modulator. The modulator is driven by a pulse generator in order to turn on and off the radiation toward the detector. The pulse-width was varied using a pulse generator. Figure 1 shows the impulse response results of the detection system output at different pulse widths using ac and dc couplings. For longer pulse-width (100 μs) ac coupling results in pulse deformation, compared to dc coupling, with long recovery time. The recovery time is measured as the difference between the pulse peak time and the time required to achieve a signal level equal to the pulse pre-trigger. For 1 μs pulses recovery times are 1.7 and 0.9 μs for dc and ac coupling, respectively. These reduce to 1.0 and 0.6 μs for the 10 ns pulse. Thus, ac coupling may be favorable for IPDA applications, which results in shorter recovery time for short pulses. In addition, ac coupling reduces low-frequency noise (e.g., 1/f noise) and preserve detection dynamic range by eliminating dc offsets. This results in an enhanced signal-to-noise ratio (SNR). Besides, ac coupling may underestimate pulse amplitude, as indicated in figure 1, causing IPDA biases.

2.2 Dynamic Range

For a fixed bias voltage and gain setting, the detection system dynamic range is defined by the minimum and maximum detectable optical powers in Watt. Minimum detectable power corresponds to the system noise-equivalent-power (NEP in W/Hz$^{1/2}$) multiplied by the square root of the bandwidth. From the NEP definition, this power results at unity SNR. Maximum detectable power is equal to the highest radiation intensity that can be applied to the detection system without causing saturation or clipping in either e-APD detector, electronics or digitizer. This corresponds to the digitizer maximum output voltage divided by gains and responsivity [10].

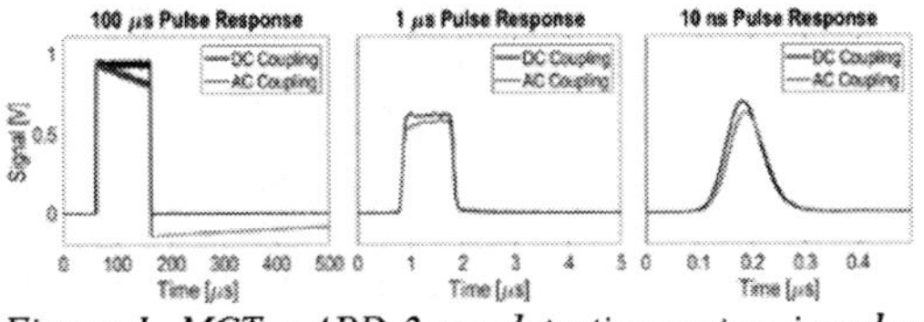

Figure 1. MCT e-APD 2-μm detection system impulse response using dc and ac coupling at different radiation pulse-width. Results obtained using gain setting of 0, 77.6K and 12 V bias.

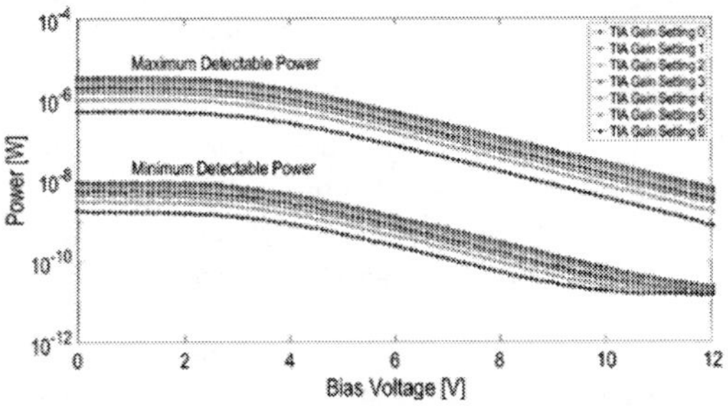

Figure 2. MCT e-APD 2-μm detection system dynamic range versus bias voltage using different gains. Results are obtained for the 2×2 center pixels at 77.6K.

 https://doi.org/10.1051/epjconf/202023701013

These procedures were repeated for different bias voltages and TIA gain settings. Figure 2 shows the dynamic range of the MCT e-APD detection system. These results indicate more than two orders of magnitude input dynamic range, which can be controlled by either the bias or the TIA gain. The absolute minimum detectable power, observed as lower saturation at high bias, is 14 pW, which is independent of TIA gain. This results in a total detection system NEP of 5.7 $fW/Hz^{1/2}$ for four pixels. This translates to about 1.4 $fW/Hz^{1/2}$ average NEP per pixel.

3. ATMOSPHERIC PROFILING

The 2-μm triple-pulse IPDA lidar is installed inside a mobile trailer for ground testing [12]. The instrument is nadir pointing similar to an airborne setup. Two, 0.6 m diameter, 45° mirrors are used to convert the IPDA pointing to zenith through a rooftop IR coated window. The objective of this test was to use clouds as hard target to assess the capability of the new MCT e-APD detection system. In these experiments, the detection system was dc coupled and the APD was cooled to 77 k.

3.1 Near-Field Cumulus Clouds

On December 12, 2017, the IPDA lidar was operated to target near-field cumulus clouds at around 2 km altitude. Figure 3 shows a sample profile of the off-line return, seeded at 2051.1915 nm, and the range-corrected signal. The MCT e-APD was biased using 7 V, with six active pixels and a TIA gain setting of four. Signal processing included pulse time alignment with respect to the initial window reflection. This is followed by 100 shot averaging (2 s), background subtraction, and range correction. Figure 4 shows a false color diagram for successive signals acquired during a 30 minute record of the cloud structure.

3.2 Far-Field Cirrus Clouds

On December 14, 2017, the IPDA was focused toward high-altitude (6 km) cirrus clouds. The objective was to test the instrument measurement capability using far-field targets. In this record, four pixels of the MCT e-APD were activated. The device was biased with 8 V and a TIA gain setting of five. Signal processing included pulse alignment, 50 shot averaging (1 sec) and background subtraction followed by range correction. Figure 5 shows a sample profile for an off-line return signal with the cloud return pulse occurring at 40 μs. This is equivalent to 6 km altitude as shown in the range corrected signal in the same figure. Focusing on the near-field range corrected signal reveals aerosol scattering from the boundary layer, as shown in the same figure.

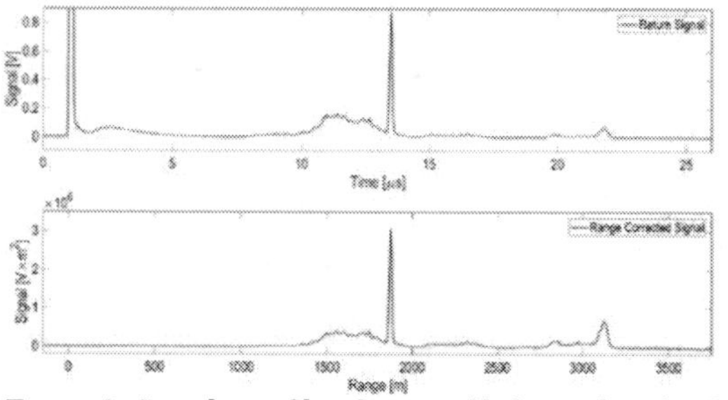

Figure 3. Sample profile of a near-filed cumulus cloud return at 2-μm using MCT e-APD detection system (top) and the range corrected signal (bottom).

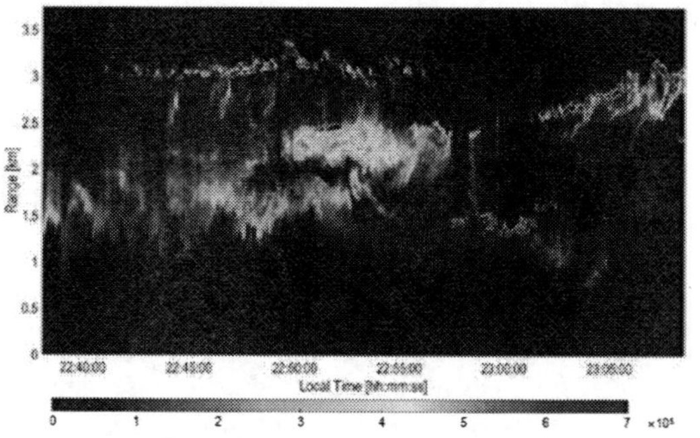

Figure 4. False color diagram for near–field record of cumulus clouds structure. Colors represent the 2-μm rang corrected signal strength.

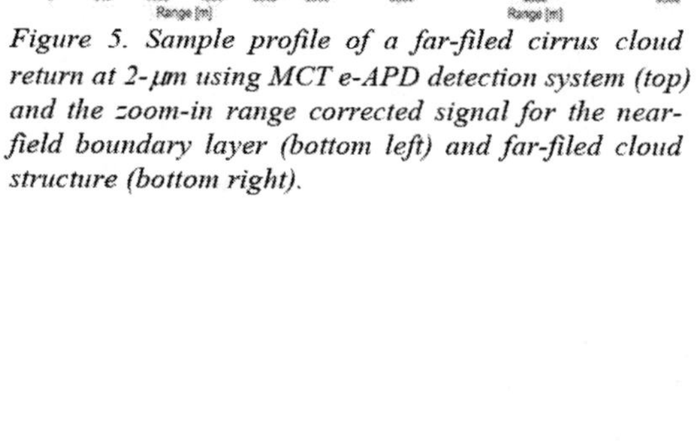

Figure 5. Sample profile of a far-filed cirrus cloud return at 2-μm using MCT e-APD detection system (top) and the zoom-in range corrected signal for the near-field boundary layer (bottom left) and far-filed cloud structure (bottom right).

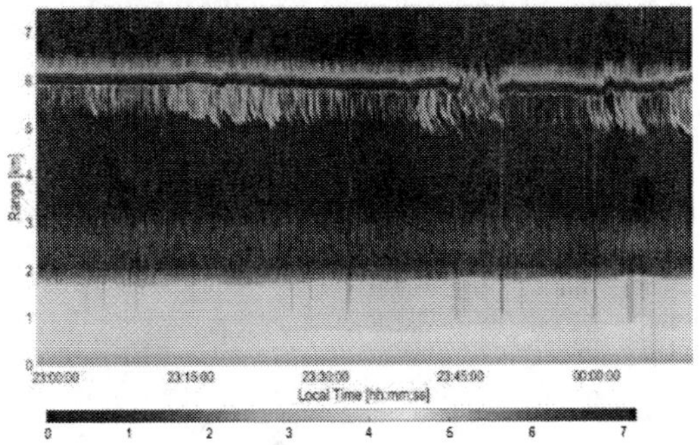

Figure 5. False color diagram for far–field record of cirrus clouds and boundary layer structures. Colors represent the logarithmic scale of the 2-μm rang corrected signal strength.

Figure 6 shows a false color diagram for successive signals acquired during a 30 minute recording of cirrus clouds features (including precipitating cirrus) and boundary layer structure. The color scale represents the strength of the range corrected backscatter signal, presented in logarithmic scale, to capture the large dynamic range. Figure 7 illustrates the boundary layer including the altitude distribution of different airmasses. In this figure, false color represents a linear scale of the range corrected signals. MCT e-APD sensitivity and detection system dynamic range enable observing near-field boundary layer structure and far-field clouds simultaneously at the 2-μm wavelength. These results, demonstrating both large dynamic range and high sensitivity, enable the potential for space-based CO_2 and H_2O atmospheric profiling using the range-resolved differential absorption lidar (DIAL) technique.

4. CONCLUSIONS

NASA Langley Research Center has been developing 2-μm lidar transmitter technologies for Earth science applications. In coordination with NASA's Earth Science Technology Office, NASA Goddard Space Flight Center integrated a 2-μm MCT e-APD lidar detection system. Similar detection systems demonstrated breakthroughs in lidar applications, with maturity amenable to space applications. Combining 2-μm transmitter and MCT e-APD detection technologies in a single lidar results in an instrument that could meet the necessary requirements for space-based CO_2 measurements.

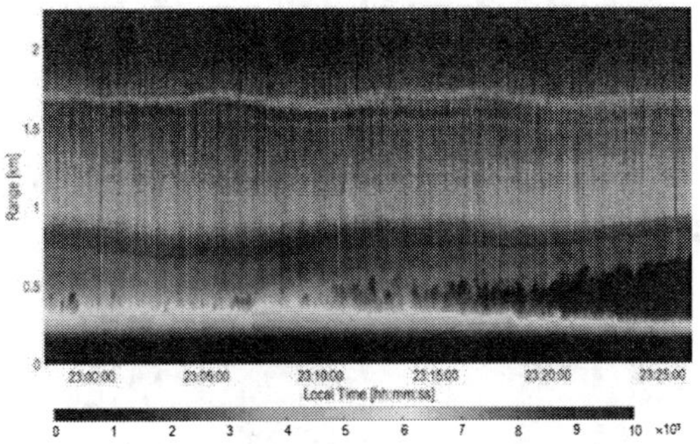

Figure 6. False color diagram for zoom-in near–field boundary layer structure at 2-μm. Colors represent range corrected signal strength.

Detection system characterization included the impulse response and dynamic range. AC coupling enhances SNR, while limited by the RC time constant. AC coupling can underestimate signals amplitudes, which may introduce biases for IPDA applications. Results indicated an average NEP of about 1.4 $fW/Hz^{1/2}$ per pixel. IPDA zenith operation enables assessment of the MCT e-APD sensitivity and detection system dynamic range. The system was capable of observing near-field boundary layer structure and far-field clouds simultaneously at the 2-μm wavelength. These results are encouraging for CO_2 DIAL profiling applications.

ACKNOWLEDGEMENTS

This work was funded and supported by NASA's Earth Science Technology Office. The authors acknowledge NASA Goddard Space Flight Center support. Thanks go to Dr. S. Ismail for his input.

REFERENCES

[1] T. Refaat, et al. Appl. Opt., 54, 1387-1398 (2015).
[2] U. Singh, et al. Proc. of SPIE, 10406-03 (2017).
[3] M. Petros, et al. ILRC 28, 176, 01009 (2018).
[4] U. Singh, et al., Optical Materials Express, 5(4), 824-837 (2015).
[5] J. Beck, et al., J. Elec. Mat., 43, 2970-2977 (2014).
[6] X. Sun, et al. Proc. SPIE, 9114-0K (2014)
[7] J. Abshire, et al., Atmos. Meas. Tech., 11, 2001-2025 (2018).
[8] H. Riris, et al., J. Appl. Rem. Sen., 11, 034001 (2017).
[9] H. Riris, et al., Opt. Exp., 25, 29307-29327 (2017).
[10] T. Refaat, et al., IEEE IGARSS, 1865-1868 (2018)
[11] A. Dumas, et al., Appl. Opt., 56, 7577-7587 (2017).
[12] T. Refaat, et al. Proc. of SPIE, 10779-02 (2018).

PARTICULATE OPTICAL PROPERTIES IN THE MEDITERRANEAN AND BLACK SEAS THROUGH CALIPSO SPACEBORNE LIDAR MEASUREMENTS

Davide Dionisi[1*], Vittorio Brando[1], Gianluca Volpe[1], Simone Colella[1] and Rosalia Santoleri[1]

[1]*Institute of Marine Sciences (ISMAR), Italian National Research Council (CNR), Rome - Tor Vergata, Italy*

**Email: davide.dionisi@cnr.it*

ABSTRACT

New applications on global-scale plankton retrievals using the CALIOP (Cloud-Aerosol Lidar with orthogonal Polarization) lidar measurements on the CALIPSO (Cloud-Aerosol Lidar and Infrared Pathfinder Satellite Observations) satellite recently suggested that space-based lidars could provide information about the depth distribution of optical scattering. Assessing the oceanic surface layer's optical properties through CALIOP is one of the reasons of the extension of the CALIOP mission for another 3 years (2018-2020). The objective of this work is the evaluation of the potential CALIOP ocean products in the Mediterranean and Black seas using the ocean color products provided by the Copernicus Marine Environment Monitoring Systems (CMEMS).

1. INTRODUCTION

The Lidar active remote sensing technique constitutes a highly reliable tool of investigating the marine environment. This technique allows detecting and determining the spatial and quantitative distributions of several optical and physical parameters of seawater with good resolution and accuracy at long distances. Lidar sensors operate in a range of conditions that are highly complementary to ocean color passive observations: under any lighting conditions; between broken cloud and through tenuous cloud; and through significant overlying aerosol loading with no need for complex correction algorithms.

Though originally designed for retrieving spatial and optical properties of clouds and aerosols, new applications on global-scale plankton retrievals using CALIOP (Cloud-Aerosol Lidar with Orthogonal Polarization) measurements suggested that space-based Lidars can provide information about the depth distribution of optical scattering [1-3]. In particular, the surface hemispheric backscattering coefficients (b_{bp}) at 532 nm (CALIOP emission wavelength) was estimated using the ratio between the cross-polarization and co-polarization channels. Despite these results, this new application for ocean retrievals has limitations and uncertainties that need to be further assessed and estimated.

The objective of this work is the evaluation of the potential CALIOP ocean products at basins scale. Within this frame, the ocean column-integrated depolarization ratio measurements at 532 nm were derived through CALIOP measurements in the Mediterranean (MED) and Black sea (BS). This parameter allowed deriving the backscattering coefficients that have been compared to the b_{bp} at 443 nm and to the chlorophyll-a (Chl-a) derived by the multi-sensor L3 product (MODIS-AQUA and NPP-VIIRS data) in MED and BS provided by the Copernicus Marine Environment Monitoring Systems (CMEMS). Analyses of annual and seasonal distributions of CALIOP and CMEMS b_{bp} are presented for both basins.

2. DATA

2.1 CALIOP dataset

To derive the ocean parameters, the CALIOP version 4.1 (V4.1) level 1 (L1) data products are used. Substantial improvements in the calibration accuracies and in the ancillary data used for the surface detection and calibration have been made for this data version. In particular, a new algorithm that estimates the likelihood of the saturation of the signal returning from the planetary surface was included in the analysis. This information, crucial for the retrieval of ocean product is reported as a flag value assuming the integer values of 0, 1 or 2 for a not saturated, possibly saturated or certainly saturated detection channel, respectively [4]. More information about the data products can be found at the Atmospheric Science Data Center (ASDC) web site (https: //eosweb.larc.nasa.gov/project/calipso/calipso_table, last access: 29 march 2019).

2.2 CMEMS Ocean product

The Copernicus Marine Environment Monitoring Service (CMEMS) is one of the six services of the Copernicus programme. It provides regular and systematic reference information on the physical state, variability, and dynamics of the ocean, ice, and marine ecosystems for the global ocean and the European seas. For this work, we used, as reference products, the multi-sensor L3 product (MODIS-AQUA and NPP-VIIRS data) of b_{bp} at 443 nm and Chl-a in the MED and BS. The near-real-time multi-sensor processing chain set up to derive these products takes care of the band shifting, the inter-sensor bias correction, and sensor merging. Further details can be found in [5].

3. METHODOLOGY

CALIOP retrieval algorithm for ocean applications has been developed accordingly to the main retrieval procedures introduced by [6, 1, 2]. The primary data set comprise the vertical profiles of co-polarized and cross-polarized component of the attenuated backscatter coefficients at 532nm (β_{co} and β_{cr}, respectively), with a vertical resolution of 30 meters (22.5 in water) and a single-shot along-track horizontal resolution of 330 m. Firstly, to have a reliable S/N for the 4th bin below the surface at around 78 m depth, the shots along-track are averaged into 10 km profiles. Secondly, the following criteria are adopted:

-1) only profiles with surface peaks located within ±120 m (±4 bins) of the surface Digital Elevation Model (DEM) parameter recorded in the CALIOP L1 data are considered.

-2) The surface backscatter signal for both β_{co} and β_{cr} should not be saturated (i.e. Surface_Saturation_Flag_532 = 0)

-3) The Integrated Attenuated Backscatter (IAB) should be lower than 0.017 sr^{-1}.

This preliminary filtering procedure reduces the original data sample of around 67%.

The subsurface β_{co} and β_{cr} profiles are contaminated by the strong signal return from the ocean surface that is spread by CALIOP 532 nm photomultiplier tubes (PMTs) transient response over several adjacent range bins, due to the PMT's noise tail effect and low-pass filter's broadening effect [7-8]. These effects are removed following the de-convolution method proposed by [9] to water clouds and applied by [2] to water surface return. After this correction, only the corrected attenuated backscatter coefficient profiles (β_{co_corr} and β_{cr_corr}, respectively) that have all the first four bins below the surface with values greater than zero are selected. The corrected profiles, which corresponds to 20% of the original data sample, are used to compute the integrated total (δ_T) and the ocean subsurface depolarization ratio (δ_w), defined as follows:

$$\delta_T = \frac{\sum_{i=p}^{i=p+5} \beta_{cr\_corr}(z_i)}{\sum_{i=p}^{i=p+5} \beta_{co\_corr}(z_i)} \tag{1}$$

$$\delta_w = \frac{\sum_{i=p+1}^{i=p+5} \beta_{\perp}^{corr}(z_i)}{\sum_{i=p+1}^{i=p+5} \beta_{\parallel}^{corr}(z_i)}. \tag{2}$$

Finally, the column-integrated ocean subsurface lidar backscatter (γ) is obtained combining (1), (2) and the ocean surface mean squares wave slopes (σ^2) estimated using the wind speed from scatterometer ASCAT, OSCAT and from ECMWF analysis:

$$\gamma = \frac{0.0209}{4\pi\sigma^2 cos^4\theta} exp\left[-\frac{tan^2\theta}{2\sigma^2}\right] \frac{\delta_T}{1-\delta_T/\delta_w}. \tag{3}$$

The associated relative uncertainties ($\Delta\gamma/\gamma$) is around 0.15 [2]. The γ final horizontal resolution is 0.25°.

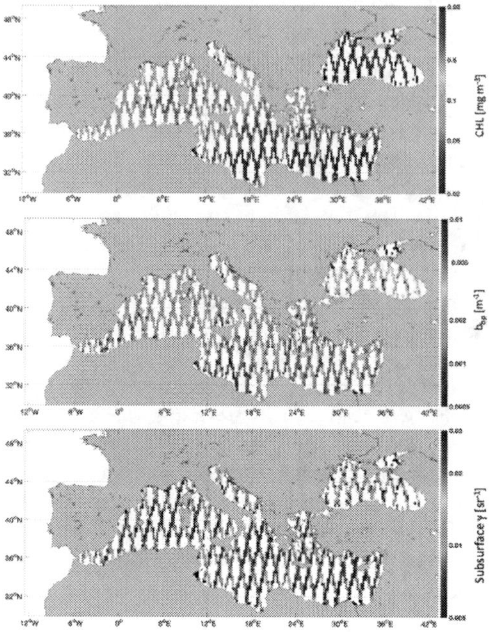

Fig.1 Distribution of Chl-a, b_{bp} and γ along the CALIOP groundtracks (high, middle and low panels, respectively) obtained from CMEMS computation and from eq. (3), respectively.

4. RESULTS

The CMEMS b_{bp} and Chl-a products are compared to the γ parameter obtained for Mediterranean (MED) and Black Sea (BS) (Fig.1). According to different 'trophic regimes' [10], MED is divided in four bioregions: West MED (1), East MED (2), Tyrrhenian Sea (3), Adriatic Sea (4). The correlation coefficients and the linear fit parameters (*a, b* of logY=*a*logX+*b*) between γ and b_{bp} and γ and CHL for the different sea regions are resumed in Table 1 and 2, respectively. For γ and b_{bp}, an high correlation ($R \geq 0.8$) is observed for the BS and the Adriatic Sea. A significant correlation is also calculated for West Med, whereas low signal and no correlation characterize no bloom regions (2 and 3). Conversely, no correlation is found between γ and CHL except for the Adriatic Sea.

Table 1. Results of correlation coefficients and linear fit parameters (*a, b* of logY=*a*logX+*b*) between γ and b_{bp}.

γ vs b_{bp}	1	2	3	4	MED	BS
N° of pts	2678	7920	955	520	10905	1105
R	0.35	0.35	0.50	0.57	0.32	0.09
a,b [logy = alogx+b]	0.83 0.55	0.73 0.20	0.73 0.30	0.65 0.33	0.77 0.34	-0.01 -0.47

Figure 2 depicts the mean annual cycles and the associated standard deviation of Chl-a, b_{bp} and γ (high, central and low panels, respectively) for MED and BS (blue and red colors). The summertime minima and wintertime maxima of the Chl-a signals (high panel of Fig.2) highlight that MED and BS have a similar phytoplankton biomass annual cycle. On the contrary, the b_{bp} patterns show semi-annual and annual cycles for BS and MED, respectively. In terms of γ, there is a good accordance between γ and b_{bp} for BS, whereas for MED the γ annual cycle seems to agree with Chl-a.

Table 2. Results of correlation coefficients and linear fit parameters (*a, b* of logY=*a*logX+*b*) between γ and Chl-a.

γ vs CHL	1	2	3	4	MED	BS
N° of pts	2678	7920	955	520	10905	1105
R	0.35	0.35	0.50	0.57	0.32	0.09
a,b [logy = alogx+b]	0.83 0.55	0.73 0.20	0.73 0.30	0.65 0.33	0.77 0.34	-0.01 -0.47

5. CONCLUSIONS AND PERSPECTIVES

Particulate optical properties in the Mediterranean and Black seas were estimated through CALIPSO spaceborne lidar measurements. The obtained results show that the derived CALIOP ocean product (γ) presents good correlation with the standard b_{bp} ocean color product in optically complex waters (coastal and/or dominated by terrigenous load). Furthermore, γ demonstrated to be as efficient as b_{bp} and Chl-a parameters for the phytoplankton dynamics at seasonal and basin scales. Further studies are planned to validate not only the column integrated but also the subsurface profile of depolarization ratio. Within this frame, the analysis of ADM-AEOLUS measurements, with its observing geometry of 30° off-zenith angle, could be used to evaluate the sub-surface signal contribution.

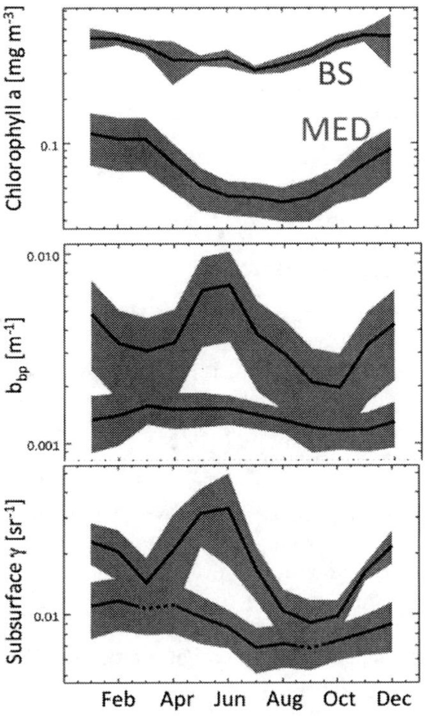

Figure 2. Mean annual cycles and the associated standard deviation of Chl-a, b_{bp} and γ (high, central and low panels, respectively) for MED and BS (blue and red colors).

REFERENCES

[1] Behrenfeld M.J., Hu Y., Hostetler C.A., Dall'Olmo G., Rodier S.D., et al.: Space-based lidar measurements of global ocean carbon stocks. Geophys. Res. Lett. 40:4355–60, 2013.

[2] Lu X., Hu Y., Trepte C., Zeng S., and Churnside, J.H: Ocean subsurface studies with the CALIPSO spaceborne lidar. *J. Geophys. Res.* 119, 4305-4317. doi: 10.1002/2014JC009970, 2014.

[3] Behrenfeld, M. J., Hu, Y., O'Malley, R. T., Boss, E. S., Hostetler, C. A., Siegel, D. A., Sarmiento, J. L., Schulien, J., Hair, J. W., Lu, X., Rodier, S., and Scarino, A. J.: Annual boom-bust cycles of polar phytoplankton biomass revealed by space-based lidar, Nat. Geosci., 10, 118–122, https://doi.org/10.1038/ngeo2861, 2017.

[4] Lu, X., Hu, Y., Yang, Y., Vaughan, M., Liu, Z., Rodier, S., Hunt, W., Powell, K., Lucker, P., and Trepte, C.: Laser pulse bidirectional reflectance from CALIPSO mission, Atmos. Meas. Tech., 11, 3281-3296, https://doi.org/10.5194/amt-11-3281-2018, 2018.

[5] Volpe, G., Colella, S., Brando, V. E., Forneris, V., La Padula, F., Di Cicco, A., Sammartino, M., Bracaglia, M., Artuso, F., and Santoleri, R.: Mediterranean ocean colour Level 3 operational multi-sensor processing, Ocean Sci., 15, 127-146, https://doi.org/10.5194/os-15-127-2019, 2019.

[6] Churnside, J., McCarty, B., and Lu, X.: Subsurface ocean signals from an orbiting polarization lidar, Remote Sens. 5(7), 3457–3475, 2013.

[7] Hu, Y., Powell, K., Vaughan, M., Tepte, C., Weimer, C., Beherenfeld, M., Young, S., Winker, D., Hostetler, C., Hunt, W., Kuehn, R., Flittner, d., Cisewski, M., Gibson, G., Lin, B., and MacDonnell, D.: Elevation information in tail (EIT) technique for lidar altimetry, Opt. Express, 15, 14504–14515, https://doi.org/10.1364/OE.15.014504, 2007.

[8] McGill, M. J., Vaughan, M., Trepte, C. R., Hart, W. D., Hlavka, D. L., Winker, D. M., Kuehn, R.: Airborne validation of spatial properties measured by the CALIPSO lidar, J. Geophys. Res., 112, D20201, doi:10.1029/2007JD008768, 2007.

[9] Li, J., Hu, Y., Huang, J., Stamnes, K., Yi, Y., and Stamnes, S.: A new method for retrieval of the extinction coefficient of water clouds by using the tail of the CALIOP signal, Atmos. Chem. Phys., 11, 2903-2916, https://doi.org/10.5194/acp-11-2903-2011, 2011.

[10] D'Ortenzio, F. and Ribera d'Alcalà, M.: On the trophic regimes of the Mediterranean Sea: a satellite analysis, Biogeosciences, 6, 139-148, https://doi.org/10.5194/bg-6-139-2009, 2009.

 https://doi.org/10.1051/epjconf/202023701015

RAYLEIGH WIND RETRIEVAL FOR THE ALADIN AIRBORNE DEMONSTRATOR OF THE AEOLUS MISSION USING SIMULATED RESPONSE CALIBRATION

Xiaochun Zhai[1,2], Uwe Marksteiner[1], Fabian Weiler[1], Christian Lemmerz[1], Oliver Lux[1], Benjamin Witschas[1], Oliver Reitebuch[1]*

[1]Deutsches Zentrum für Luft- und Raumfahrt e.V. (DLR), Institut für Physik der Atmosphäre, Oberpfaffenhofen 82234, Germany
[2]Ocean University of China, College of Information Science and Engineering, Ocean Remote Sensing Institute, Qingdao 266100, China
**Email: Oliver.Reitebuch@dlr.de*

ABSTRACT

Aeolus, launched on 22 August 2018, is the first ever satellite to directly observe wind information from space on a global scale. An airborne prototype called ALADIN Airborne Demonstrator (A2D) was developed at the German Aerospace Center (DLR) for validating the Aeolus measurement principle based on realistic atmospheric signals. However, atmospheric and instrumental variability currently limit the reliability and repeatability of the A2D instrument response calibration. In this study, a simulated Rayleigh response calibration (SRRC) is presented for resolving the limitations of A2D instrument response calibration.

1. INTRODUCTION

Continuous global wind measurement is of highest priority for improving the accuracy of numerical weather forecast as well as our knowledge of atmospheric dynamics [1]. Among the various techniques such as radiosonde, radar wind profiler and so forth, a spaceborne Doppler wind lidar is considered as the most promising one to meet the need of near-real time observations of global wind profiles. Aeolus, launched on 22 August 2018, is the first ever satellite to directly retrieve wind information from space on a global scale. The unique payload, the Atmospheric LAser Doppler INstrument (ALADIN), is a direct detection wind lidar operating at 355 nm which can probe the lowermost 30 km of the atmosphere from a 320 km orbit. It provides one component of the wind vector along the instrument's line-of-sight (LOS) with a vertical resolution of 0.25 km to 2 km and wind speed precision of 2 m/s to 4 m/s depending on altitude. The backscatter signals from particles (Mie channel) and molecules (Rayleigh channel) are independently received by two different interferometers, that is, a Fizeau interferometer and double-edge Fabry-Perot interferometers (FPIs), respectively. The novel combination of these two techniques, which were not implemented in a wind lidar before, enlarges the observational altitude range compared to the coherent Doppler wind lidar which only relies on particulate scattering. In the frame of the Aeolus program, an airborne prototype of ALADIN called ALADIN Airborne Demonstrator (A2D) was developed at the German Aerospace Center (DLR) [2]. Due to its representative design and operating principle, the A2D has provided valuable information on the wind measurement strategies of the satellite instrument as well as on the optimization of the wind retrieval and quality control algorithms [3,4]. The A2D was deployed in several ground and airborne campaigns over the last 12 years.

As opposed to coherent Doppler wind lidars where the Doppler frequency shift can be directly determined via the beat signal between the backscattered and emitted laser signal, the measured quantities from a direct detection wind lidar cannot be directly related to the Doppler frequency shift. The response calibration indicating the relationship between measured quantities and Doppler frequency shift is thus a prerequisite to retrieve wind information accurately. As for A2D, a small portion of transmitted laser radiation, referred to as internal reference signal, is collected and coupled into a multi-mode fiber firstly, and then entered the spectrometer optics [3], this optical path, called internal reference path, is used for internal reference laser frequency measurements. While

https://doi.org/10.1051/epjconf/202023701015

atmospheric backscattered signal is firstly collected by telescope and then guided to the front and spectrometer optics in turn, the atmospheric path is used for the measurement of received atmospheric frequency. Therefore, the response calibrations for both internal reference and atmospheric path are prerequisite for accurate wind retrieval. In order to calibrate the ALADIN or A2D Rayleigh channel, different instrument calibration approaches have been carried out using both measurements and simulations [4,5]. In this paper, a simulated Rayleigh response calibration (SRRC) is presented to resolve the reliability and repeatability limitations of the A2D instrument response calibration caused by atmospheric and instrumental variability.

2. METHODOLOGY

Fig.1 presents the measurement principle of the A2D Rayleigh channel. The emitted laser spectrum is depicted using a Gaussian function with a full width at half maximum (FWHM) of 50 MHz, and the molecular Rayleigh backscatter spectrum is modeled using an analytical fit of the Tenti S6 line shape model [6-8] at $T = 270$ K and $P = 700$ hPa. In this case, it is assumed that there is no Mie contamination on the Rayleigh channel. The transmitted signals through each filter (I_A, I_B), as indicated by light blue and magenta filled areas respectively, are proportional to the convolution of the respective filter transmission function and the line shape of the atmospheric backscatter signal. Therefore, the contrast between I_A and I_B represents a measure of the frequency shift between emitted laser pulse and molecular backscattered signal. Herein, the frequency-dependent Rayleigh response function is defined as:

$$R(f) = \frac{I_A(f) - I_B(f)}{I_A(f) + I_B(f)} \quad (1)$$

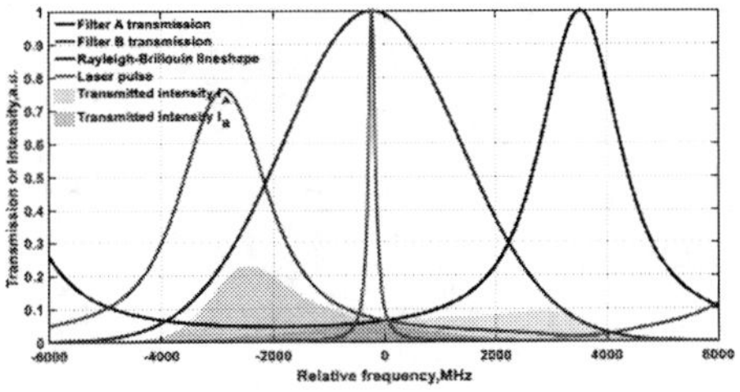

Fig. 1 Modeled spectral distribution of the transmitted laser pulse (pink line) and pure molecular backscatter (blue line) at $T = 270$ K, $P = 700$ hPa. The Rayleigh channel transmission spectra of two FPIs are shown in black $T_A(f)$ and red $T_B(f)$ lines, respectively. The transmitted integrated intensities through FPI A and B are marked with light blue and magenta filled areas.

A frequency scan of the laser transmitter over a wide frequency range of 1.7 GHz with step size of 25 MHz is carried out to obtain the response function during A2D instrument response calibration. During the instrument response calibration, the contribution of Doppler frequency shifts related to the motion of molecular or particle along the instruments' LOS direction needs to be eliminated. In practice, this is accomplished by flying a curve with a roll angle of 20°, resulting in approximately nadir pointing of the instrument and thus nearly zero wind contribution, provided that the vertical wind is negligible.

Following the procedure of the A2D instrument response calibration, SRRC can be carried out to obtain simulated response function based on Eq. (1). A linear least square fit is firstly applied to the simulated response function, and the sensitivity β_x and intercept α_x are obtained from:

$$\beta_x = \frac{\partial R_x(f)}{\partial f}, x = ATM, INT \quad (2)$$

$$\alpha_x = R_x(f = 0), \quad (3)$$

where x represents the simulated result derived from the internal reference or atmospheric path, respectively with f being the relative frequency (wrt. frequency at the filter crosspoint). The non-linearity $\gamma_x(f)$ is defined as the difference between $R(f)$ and the linear fit of simulated response function. Then a 5th order polynomial fit is used to model $\gamma_x(f)$. The difference between $R(f)$ and the fitted simulated response function is defined as the response residual. The result shows a periodic fluctuation of the response residual for both the internal reference and the atmospheric path with quite small response value less than 1.5×10^{-4}.

As mentioned above, the realization of SRRC needs not only knowledge of the transmission

https://doi.org/10.1051/epjconf/202023701015

characteristics of the FPIs, but also atmospheric temperature and pressure profiles for calculation of the molecular spectrum. These can be obtained from radiosondes, dropsondes or numerical weather prediction models. In the frame of the North Atlantic Waveguide and Downstream Experiment (NAWDEX) campaign in autumn of 2016, four aircraft equipped with diverse payloads were employed to investigate the influence of diabatic processes on the midlatitude weather [9]. The DLR Falcon 20 was deployed with the A2D and a well-established 2 μm coherent Doppler wind lidar, offering an ideal platform to demonstrate the feasibility of A2D in complex dynamic conditions. The other three aircraft were equipped with dropsonde dispensers to provide temperature, pressure, wind and humidity profiles, offering essential atmospheric temperature and pressure profiles for SRRC.

Since the transmission characteristics of the FPIs are different for the internal reference and atmospheric path due to slightly different illumination conditions of the different optical paths, different FPI transmission functions from different ground and airborne campaigns were used to calculate different simulated response functions. Each combination has been tested and analyzed. After comparison with the measured Rayleigh response function obtained during the NAWDEX campaign, it was found that the combination where the internal reference and atmospheric response functions were simulated using FPI parameters derived from the internal reference path during NAWDEX airborne campaign (INTA) and the atmospheric path during the BRAINS ground campaign (ATMG) [10] provides the best consistency with the measured one, which are used for initial SRRC determination. Fig. 2 shows the comparison of response functions between measurement and simulation, showing the best match for INTA and ATMG, respectively.

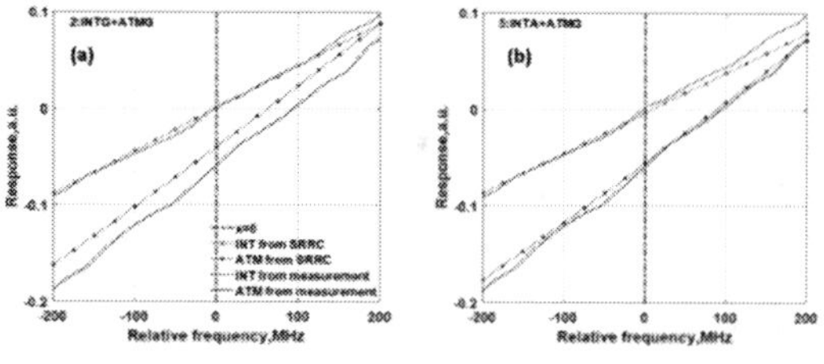

Fig. 2 The response functions of INT and 8th range-bin ATM from calibration measurement (red dot and blue dot lines, respectively) and SRRC using different combinations of FPIs. (a) INTG+ATMG and (b) INTA+ATMG for INT and ATM response simulation, respectively.

As slightly different illumination of the FPIs during Rayleigh response calibrations could yield large errors in retrieved LOS wind speed, an optimization of the FPI parameters is needed. Considering the optical path of the A2D Rayleigh channel, the FPI center frequency of the atmospheric path is sensitive to the incidence angle of the interferometer. Thus, the FPI transmission function for the atmospheric path could be adjusted for the center frequency of filter A or B. Assuming that the center frequencies of filters A and B have the same offset Δf_0 compared to the values obtained from ATMG and the parameters from INTA are regarded as ideal, Fig. 3 (a) (b) present the effect of Δf_0 on atmospheric response sensitivity and intercept, respectively. It can be seen the increase of center frequency of filter A and B results in a decrease of the response sensitivity and intercept. A cost function $F(\Delta f_0)$ is defined to determine optimized frequencies as follows:

$$F(\Delta f_0) = \sum_{i=1}^{N} | V_{LOS,SRRC}(i) - V_{LOS,dropsonde}(i) | \quad (4)$$

where $V_{LOS,SRRC}$ and $V_{LOS,dropsonde}$ are LOS wind speeds derived from SRRC and a dropsonde dataset used as a reference, while $i=1$ to N denote the atmospheric range gates of the A2D. Fig. 3 (c) shows that $F(\Delta f_0)$ has its minimum when the center frequencies of both filters A and B increase by 20 MHz, corresponding to the optimization case for LOS velocity retrieval using SRRC.

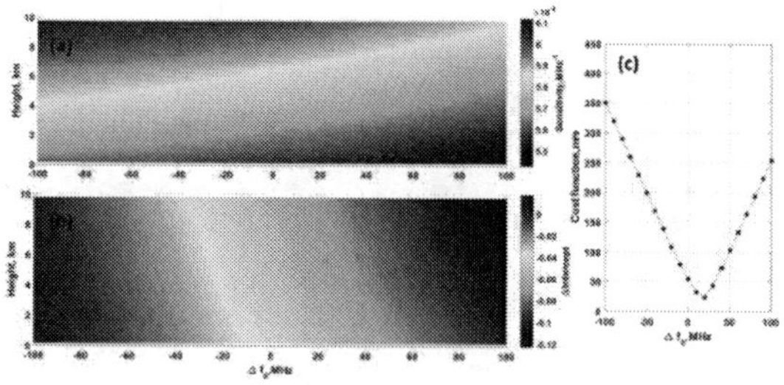

https://doi.org/10.1051/epjconf/202023701015

Fig. 3 The effect of the center frequency offset of filters A and B on the ATM response (a) sensitivity, (b) intercept, and (c) corresponding cost function.

3. RESULTS

In order to assess the accuracy of the LOS wind velocity retrieval from SRRC after FPIs optimization, Fig. 4 shows a case study comparison between measurement and simulation. Fig. 4 (a) (b) indicates that the sensitivity and intercept of the atmospheric path derived from SRRC have the same trend as the measured one, indicating the effect of atmospheric temperature and pressure on the response calibration. Because of the ground elevation limitation during A2D instrument response calibrations, the measured response below 2 km cannot be obtained, thus the response function at an altitude of 2 km is used for LOS wind retrieval below 2 km, causing larger discrepancies compared to the dropsonde and coherent Doppler wind lidar dataset, as shown in Fig. 4 (c). As the SRRC is not affected by this limitation it improves the accuracy of A2D wind measurements especially at lower altitudes, as shown in Fig. 4 (c).

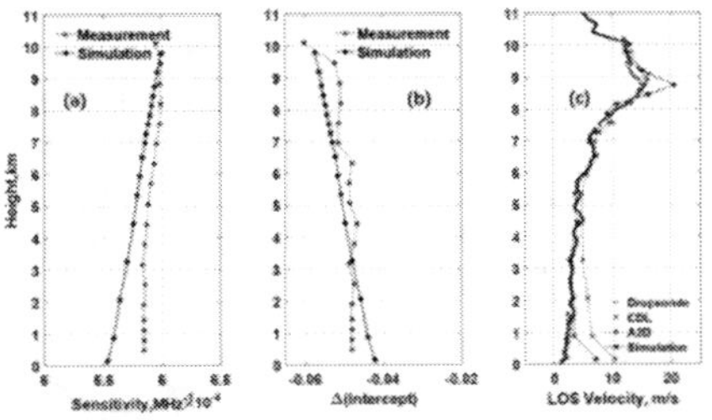

Fig. 4 Case study comparison of (a) sensitivity (b) intercept (c) LOS velocity between results from A2D Rayleigh channel measurement (red) and SRRC after FPIs optimization (blue). The LOS wind velocity from dropsonde (black) and coherent Doppler lidar (pink) are also presented.

4. CONCLUSIONS

The SRRC can resolve the reliability and repeatability limitations caused by atmospheric and instrumental variability and constraints during A2D instrument response calibrations. Unlike the A2D instrument response calibration, the SRRC is not affected by possible ground elevation limitations and can hence improve the accuracy of A2D wind measurements especially at lower altitudes. Further studies based on A2D SRRC will be performed with a focus on the atmospheric temperature and pressure effect, the Mie contamination correction and the optical properties of particles. From this, new sights into A2D and Aeolus measurements and wind results are expected.

ACKNOWLEDGEMENTS

Xiaochun Zhai appreciates the support from the Chinese Scholarship Council (CSC) to conduct the research at DLR. DLR and ESA provided funding for the airborne campaigns mentioned in the text.

REFERENCES

[1] A. Stoffelen, et al. Bulletin of the American Meteorological Society, 86.1 (2005): 73-88.
[2] O. Reitebuch, et al. Journal of Atmospheric and Oceanic Technology, 26.12 (2009): 2501-2515
[3] O. Lux, et al. Atmospheric Measurement Techniques, 11.6 (2018): 3297-3322.
[4] U. Marksteiner, et al. Remote Sensing, 10.12 (2018): 2056.
[5] A. Dabas, et al. Tellus A: Dynamic Meteorology and Oceanography, 60.2 (2008): 206-215
[6] G. Tenti, et al. Canadian Journal of Physics, 52.4 (1974): 285-290.
[7] B. Witschas, et al. Applied Optics, 50.3 (2011): 267-270
[8] B. Witschas, et al. Applied Optics, 50.29 (2011): 5758-5758.
[9] A. Schäfler, et al. Bulletin of the American Meteorological Society, 99.8 (2018): 1607-1637.
[10] B. Witschas, et al. Applied Optics, 51.25, (2012): 6207-6219.

 https://doi.org/10.1051/epjconf/202023701016

RESEARCH ON THE CONTROL POINT EXTRACTION METHOD OF CHINESE GAOFEN 7 SATELLITE USING ALTIMETER AND FOOTPRINT CAMERA

Binbin Li[1], Huan Xie[1,*], Xiaohua Tong[1], Yinqiao Cai[2], Zhijie Zhang[1,3]

[1] *College of Surveying and Geo-informatics, Tongji University, Shanghai 200082, China*
[2] *Shanghai Institute of Technical Physics, Chinese Academy of Sciences, Shanghai 200083, China*
[3] *School of Surveying and Land Information Engineering, Henan Polytechnic University, Jiaozuo 454000, China*
**Email: huanxie@tongji.edu.cn*

ABSTRACT

Satellite laser altimetry is one of the most advanced information acquisition technologies in Earth observation system. It can provide high-accuracy elevation information, however, due to the lack of detail intensity information, its planimetric accuracy is usually worse than the elevation accuracy. Gaofen 7 (GF-7) satellite scheduled for launch in 2019 will be equipped with laser altimeter, footprint camera, stereo mapping camera, etc. The laser altimeter together with the footprint camera was designed to provide high accuracy ground control point of satellite mapping. The laser altimeter can provide the high-accuracy elevation information and the joint processing of footprint camera and stereo mapping camera can provide high-accuracy planimetric information. Therefore, this paper mainly studies the technology of extracting high-accuracy control points based on GF-7 satellite's altimeter, footprint camera and stereo mapping camera using a simulated dataset extracted from Quickbird image and ICESat altimetric data.

1. INTRODUCTION

Satellite laser altimetry has been used in deep space exploration since 1960s because of its high-accuracy range detection capability. In the field of earth science, besides some experimental loads, the laser altimeter GLAS of ICESat satellite[1] launched by NASA in 2003 is the first satellite-borne laser altimeter system in the world and was retired in 2010. As its second generation satellite, ICESat-2 satellite[2] was successfully launched in 2018 and is currently in the testing and calibrating stage.

In addition to the experimental laser loads carried by Ziyuan 3 satellite (ZY-3) launched by China in May 2016, there is a plan to launch the Gaofen 7 (GF-7) satellite in 2019. GF-7 is a high resolution stereo mapping satellite and its main payloads include laser altimeter, footprint camera and stereo mapping camera. Among them, stereo mapping camera obtains the stereo image of the earth surface under flight conditions and the laser altimeter is used to provide the control point in the stereo image with the assistance of the footprint image imaged by footprint camera. The terrain data onto stereo mapping camera will be corrected by these control points, which can improve the elevation accuracy of the stereo image so as to satisfy the requirement of 1:10000 stereographic mapping under the condition of few control points. Therefore, this paper focuses on the research of the technology of extracting high accuracy control points based on GF-7 load characteristics using a simulated dataset by Quickbird image and ICESat altimetric data.

2. METHODOLOGY

The elevation information of control points can be obtained from laser altimetric data, and the planimetric information of control points can be obtained by processing the footprint image and the stereo image.

2.1 Elevation extraction using laser altimetric data

At first, the initial distance between the laser altimeter and the laser footprint can be obtained by multiplying the speed of light in a vacuum and half of the time difference between the centroids of the transmitted pulse and the received pulse[3].

And then, the initial laser footprint location can be calculated by using the satellite laser footprint location model[4][5] combined with the initial distance, the satellite position, attitude, the rotation matrix from the International Celestial Reference Frame (ICRF) coordinate system to the International Terrestrial Reference Frame (ITRF) coordinate system, the rotation matrix from satellite body coordinate system to ICRF coordinate system and so on.

In order to eliminate the spacecraft velocity influence on laser pointing, relevant resolution strategy[6] that the satellite attitude data at the time of laser launching and the satellite position data at the time of laser reaching the ground are used to participate in the calculation of the laser footprint positioning model has been adopted.

Finally, related geophysical corrections including atmospheric delay correction[7] and tidal correction[8] are made to correct the initial laser footprint location. Through the above process, the elevation accuracy is generally about 14 cm for ICESat[9][10].

2.2 Planimetric position extraction by combining footprint image and stereo image

A large footprint will be formed on the ground when satellite laser altimeter emits laser pulse to the ground with an altitude of approximately 600 km. For example, the diameter of laser footprint produced by ICEsat is approximately 70 m. Since laser altimeter cannot produce the continuous ground intensity information, its planimetric accuracy is usually worse than the elevation accuracy. Different from previous laser altimetric satellite, GF-7 will be equipped with footprint camera. It means that laser footprint planimetric information can obtained by matching footprint image with stereo image, which is a new model of extracting the planimetric information of laser footprint.

Footprint camera images the ground at the time of laser emission and ideally, the center of mass of the footprint in footprint image is the position of the laser footprint. Base on traditional gray weighted centroid algorithm, the centroid of the laser footprint recorded by the footprint camera can be extracted. And then, the stereo image will be re-sampled to the same resolution as the footprint image in order to extract and match feature points between two images. The centroid of the laser footprint in two images can be matched according to their matched relation, as shown in Figure 1. Finally, the planimetric information of control point can be extracted from the position of the centroid of the laser footprint in stereo image.

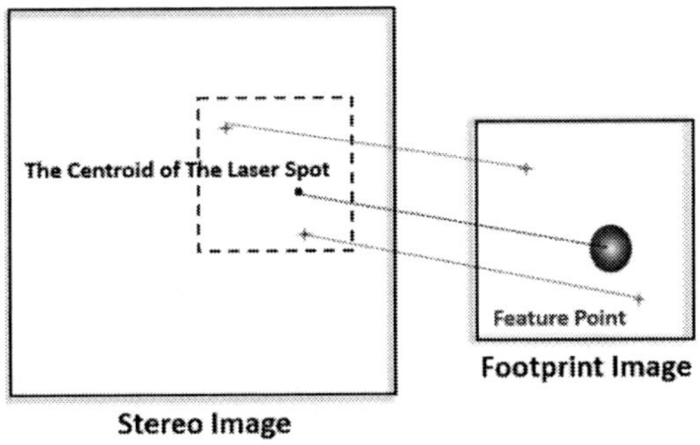

Fig. 1 The matching process of footprint image and stereo image

At present, the accuracy of image registration can reach sub-pixels level, which means that the planimetric information of laser footprint can be obtained at meter level in meter-scale stereo image. Because GF-7 has not yet been launched, this paper used Quickbird image as simulation data to generate stereo image and footprint image. By processing 30 groups of simulation images, image registration accuracy can reach sub-pixel level and the laser footprint planimetric accuracy can reach meter-level.

3. SUMMARY

Different from other traditional satellite laser altimeters which are used to observe the earth, GF-7 altimeter will further improve the accuracy of control points for satellite optical stereo image adjustment and stereo mapping because of the assistance of footprint camera. To achieve the goal, this paper researched the process of control point extraction technology for GF-7 with the assistant of footprint image, which will be of some reference value to GF-7 stereo mapping with high accuracy.

ACKNOWLEDGEMENTS

This work was supported by the National Natural Science Foundation of China under Grant No. 41822106 and 41571407, National High Resolution Ground Observation System of China (GF-7) under Grant No. 11-Y20A12-9001-17/18,

Science and Technology Innovation Action Plan Program of Shanghai under Grant No. 18511102100, Dawn Scholar of Shanghai under Grant No. 18SG22, and the Fundamental Research Funds for the Central Universities of China.

REFERENCES

[1] B. E. Schutz, et al. Overview of the ICESat Mission. Geophysical Research Letters, 32(21):L21S01. (2005)

[2] W. Abdalati, et al. The ICESat-2 Laser Altimetry Mission, Proceedings of IEEE, 98(5), 735-751. (2010)

[3] A. C. Brenner, et al. Derivation of Range and Range Distributions from Laser Pulse Waveform Analysis for Surface Elevations, Roughness, Slope, and Vegetation Height. GLAS Algorithm Theoretical Basis Document, Version 3.0. (2011)

[4] B. E. Schutz. Laser Footprint Location (Geolocation) and Surface Profiles. GLAS Algorithm Theoretical Basis Document, Version 3.0. (2002)

[5] X. M. Tang, et al. The rigorous geometric model of satellite laser altimeter and preliminarily accuracy validation. Acta Geodaetica Et Cartographica Sinica. 45(10):1182-1191.(2016)

[6] S. Bae, et al. Precision Attitude Determination (PAD). Geoscience Laser Altimeter System (GLAS) Algorithm Theoretical Basis Document, Version 2.2. [S. l.]: NASA. (2002)

[7] J. C. Owens, Optical refractive index of air: Dependence on pressure, temperature, and composition, Appl. Opt., 6, 51-59. (1967)

[8] G. Petit, et al. IERS conventions (2010). IERS Technical Note No.36. (2010)

[9] C. A. Shuman, et al. ICESat Antarctic elevation data: Preliminary precision and accuracy assessment. Geophysical Research Letters, 33(7):L07501. (2006)

[10] X. Wang, et al. Earth science applications of ICESat/GLAS: a review. International Journal of Remote Sensing, 32(23):28. (2011)

AEROSOL INVESTIGATION DURING THE ARCTIC HAZE SEASON OF 2018: OPTICAL AND HYGROSCOPIC PROPERTIES

Kim Janka Müller [1,2*], **Christoph Ritter**[1], **Konstantina Nakoudi**[1,3]

[1]*Alfred Wegener Institute, Helmholtz Centre for Polar and Marine Research, Potsdam, Germany*
[2]*Carl von Ossietzky University, Oldenburg, Germany*
[3]*Institute of Physics and Astronomy, University of Potsdam, Potsdam, Germany*
**Email: kim.janka.mueller@alum.uol.de*

ABSTRACT

In the beginning of March 2018, Lidar measurements were performed on Svalbard, Arctic Ocean, in order to analyse the optical and hygroscopic properties of Arctic aerosol. In this study, aerosol backscatter showed significant higher values in lower altitudes. The analysis of the Colour Ratio (CR) revealed smaller particles in lower altitudes, with larger particles appearing only above $5km$. Investigation of the hygroscopic character was done by applying the growth parameter introduced by Gassó et al. (2000). It was found that the method of Zieger et.al. (2010) can be successfully extended to backscatter and CR data from Lidar measurements. Ice nucleation was examined in ice supersaturation conditions, with no ice cloud formation observed. This indicated that the role of Arctic aerosol as ice nuclei is still a poorly understood issue.

1. INTRODUCTION

During springtime, aerosol containing air masses are transported into the Arctic troposphere, forming the Arctic Haze. The main sources are suggested to be anthropogenic, originating from i.e. biomass burning, industry, heating, and forest fires from Eastern Europe or Siberia. [1] *Arctic Haze* lasts from February to April with its peak in March [2]. Polluted particles are mainly found below $2km$ [3]. Arctic is a sensitive area experiencing significant changes in the radiation budget [4]. The phenomenon of *Arctic Amplification* describes the increase of near-surface temperature in Arctic regions compared to lower latitudes [5]. Increasing temperature leads to an earlier melting of sea ice and hence to a smaller albedo [6]. Since aerosol contribution to *Arctic Amplification* is not fully understood, it is important to analyse the optical, microphysical and hygroscopic properties of Arctic aerosol especially during the Haze period [7], [8] and [9].

2. INSTRUMENTS

KARL

The Lidar system used to collect the data for aerosol analysis is called KARL (**K**oldewey **A**erosol **R**aman **L**idar). This Lidar is operated at the AWIPEV station in Ny-Ålesund, Svalbard in the European Arctic ($78.9°N, 11.9°E$). KARL is a *"3β+2α+2δ+2wv"* Raman Lidar using a Nd:YAG laser, which emits radiation at the wavelength of $1064nm$. By frequency doubling and tripling the wavelengths of 532 and $355nm$ become available, which are subsequently being detected in parallel and perpendicular polarising states [10]. Aerosol backscattering was determined by the numerical stable method of Klett [11] with $30\,m/10\,min$ resolution. The backscatter coefficients can be used to define the Colour Ratio (CR, Eq. 01) [1].

$$CR(z,\lambda_1,\lambda_2) = \frac{\beta_{\lambda 1}^{Aer}(z)}{\beta_{\lambda 2}^{Aer}(z)} \qquad \text{(Eq. 01)}$$

λ_1 denotes the shorter wavelength and λ_2 the longer wavelength, z the distance travelled by laser light. Taking into account the spectral dependency of aerosol interaction with light, the CR provides an estimation of the aerosol size [12]. Throughout the collection of this study's data, KARL mainly measured during night time in order to provide a clear measurement set without the impact of day stray light.

Radiosondes

During the Lidar measurement period, radiosondes were launched four times a day in March 2018 at $05{:}00, 11{:}00, 17{:}00$, and $23{:}00$

UT in the framework of the *year of polar prediction*. The radiosondes were *Vaisala RS-41* type. Various meteorological parameters as air-pressure, temperature, relative humidity (RH) and wind speed were obtained. RH over water was taken from the radiosonde as well as the RH over ice was determined, using the water vapour partial pressure (P) in *hPa* and the saturation pressure of ice (SP_{ice}):

$$RH_{ice} = \frac{P}{SP_{ice}} \qquad \text{(Eq.02)}$$

This work considers Lidar and radiosonde data taken between 1st and 13th of March 2018.

3. METHODOLOGY

Some aerosol types exhibit a hygroscopic character, which means that they attract water. After a certain threshold of humidity, aerosol particles potentially start growing with their scattering efficiency changing [13]. Zieger et al. (2010) determined a growth factor ($g(RH)$) as in Eq. 03 using the aerosol's diameter in wet (D_{wet}) and dry (D_{dry}) surroundings [14].

$$g(RH) = \frac{D_{wet}(RH)}{D_{dry}} \qquad \text{(Eq.03)}$$

Dry surroundings were defined in a humidity range of $RH_{dry} = 44 - 45\%$ and wet surroundings within $RH_{wet} = 84 - 86\%$ [1]. In this study, the growth factor was extended by the growth function ($f(RH)$ as in Eq. 04, with γ denoting a hygroscopic growth parameter. High values indicate strong hygroscopic growth, whereas smaller show weak hygroscopicity [15].

$$f(RH) = (1 - RH)^{-\gamma} \qquad \text{(Eq. 04)}$$

Since the particle's size was not available by Lidar data, the application of the growth function was extended to backscatter and CR data. This modification was associated with the dependency of backscattering and CR on RH. Eq. 03 was transformed into Eq. 05 and Eq. 06.

$$g_{\beta}(RH) = \frac{\beta_{wet}(RH)}{\beta_{dry}} \qquad \text{(Eq.05)}$$

$$g_{CR}(RH) = \frac{CR_{wet}(RH)}{CR_{dry}} \qquad \text{(Eq. 06)}$$

4. RESULTS

4.1 Statistical Overview of Optical Properties

Histograms (Fig. 01) provide a statistical overview of aerosol backscatter during the measurement campaign. Four different height ranges were set to distinguish the behaviour of aerosol backscatter between different altitudes. The analysis of backscatter (Fig. 01) showed that lower altitudes feature higher backscattering, indicating that due to the haze higher number of aerosols were observed compared to other seasons e.g. summer.

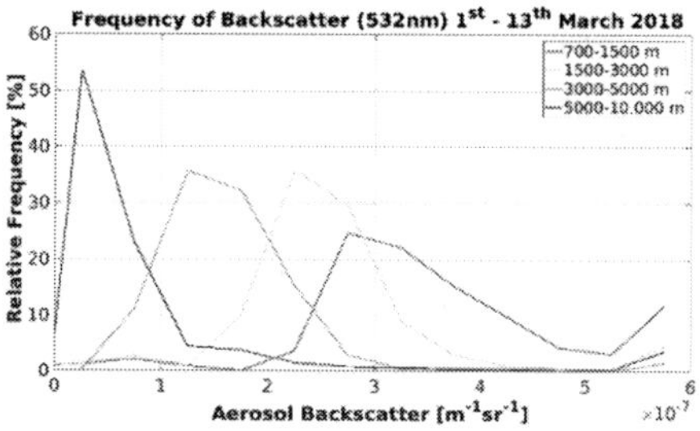

Fig.01: Distribution of aerosol backscatter in different altitude intervals.

Additionally, the analysis of CR showed high values for lower altitudes, indicating that mainly smaller particles were present. Only in an altitude higher than $5km$, smaller CR values were found, indicating bigger particles. Statistical analysis of the aerosol linear depolarisation ratio revealed that most of the observed aerosol were characterised by low depolarisation values, indicating spherical particles (only 2% of the data showed values of $\delta > 10\%$ indicating aspherical particles).

4.2 Hygroscopic Growth

The hygroscopic character of aerosols allows investigating the dependency of their properties on the ambient humidity. Taking the growth function and parameters as described in Eq. 04 and 05, the dependency of backscattering (Fig. 02) and CR (Fig. 03) on humidity were investigated. Additionally, the number of data points for each growth parameter were determined and are listed together with the distribution of particles (Tab. 01) for data points with $RH > 75\%$. Concerning aerosol backscattering, a weak hygroscopic behaviour was dominating (23% revealed a weak hygroscopic parameter γ1). Nevertheless, strong hygroscopicity was apparent for a number of cases (11% for γ_3; non-valid data 26%).

Considerable is the distribution of all measured data (backscattering and CR), showing that the majority (60%) was found in a range of RH lower than 45%. In the range of $RH = 45 - 60\%$ lay 13%, $RH = 60 - 90\%$ contained 25%, while in a very humid surrounding only 2% was found.

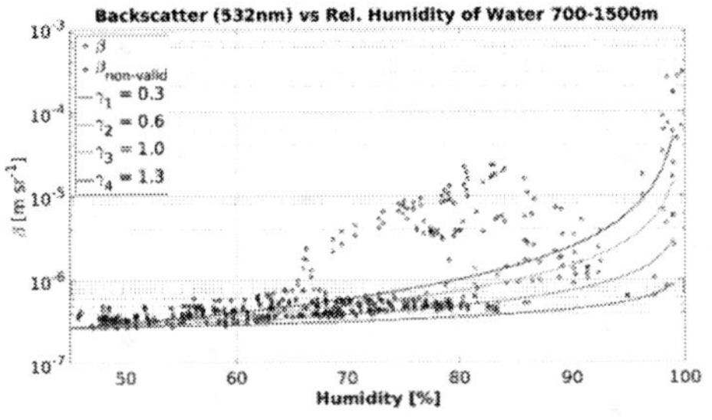

Fig.02: Backscattering versus humidity with fitted growth functions for different growth parameters.

Concerning CR (Fig. 03), its dependency on relative humidity was less pronounced compared to that of backscattering. The same formalism as in Eq.04 was applied but with a different sign for γ. It should be noted that the majority (21%) showed strong hygroscopicity ($\gamma_3 = 1.0$). This showed that Zieger's method can be applied to CR data as well (non-valid data 24%).

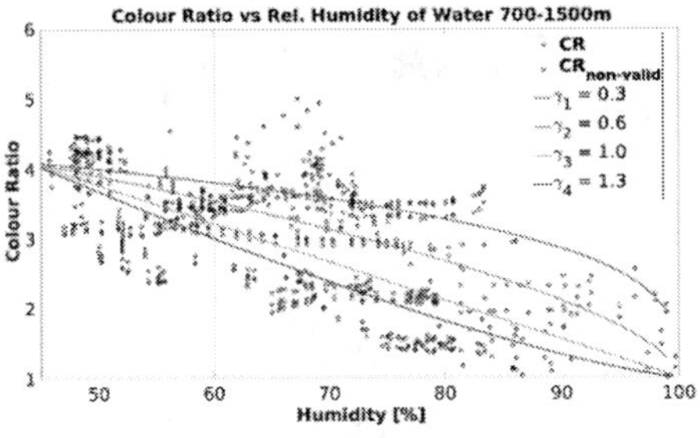

Fig.03: Same as Fig. 02, except for Colour Ratio.

Tab. 01: Distribution of backscatter and CR for different growth parameters (for $RH > 75\%$).

Hygroscopicity	**β [%]**	**CR [%]**
Weak ($\gamma_1 = 0.3$)	23	20
Moderate ($\gamma_2 = 0.6$)	30	19
Strong ($\gamma_3 = 1.0$)	11	21
Very strong ($\gamma_4 = 1.3$)	10	16

4.3 Case Study: Clouds in Ice Supersaturation?

A case study examining the aerosol ice nucleation efficiency was analysed.

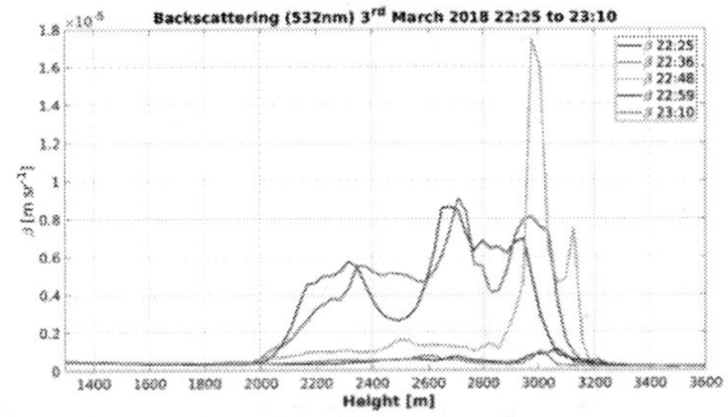

Fig.04: Backscattering in a time range of ±25min where the radiosonde (22:45UTC) was launched.

Depolarisation was investigated whenever high RH_{ice} was found, so as to determine whether an ice cloud was present or not. β^{Aer} (Fig. 04) maximised at $3km$ ($\beta^{Aer} = 1.7x \cdot 10^{-5} m^{-1} sr^{-1}$at 22:48 UTC, while in the time range of ca. ±10min significant lower values were recognised. Simultaneously, RH_{ice} (Fig. 05) indicated supersaturation at $3.5km$, whereas depolarisation (Fig. 06) was considerably low ($\delta_{max} = 3\%$) around $3km$.

Despite supersaturation over ice, no ice cloud was observed. Water cloud formation could have happened within the time range of $10min$. Taking into account the dependency between humidity and temperature, temperature fluctuations must have taken place to confirm this hypothesis.

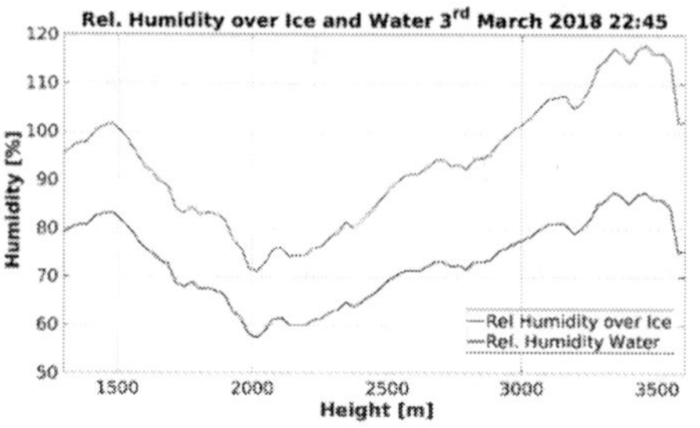

Fig.05: Vertical profile of RH_{ice}. For comparison RH_{water} is shown.

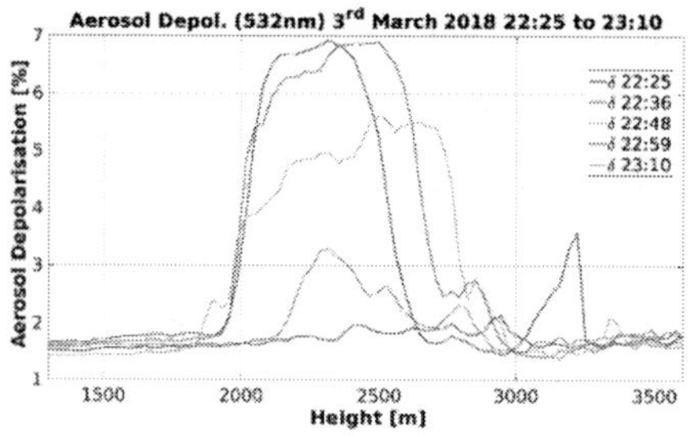

Fig.06: Same as Fig.04, except for depolarisation.

5. CONCLUSIONS

In this study the aerosol optical and hygroscopic properties have been investigated during the Arctic Haze season 2018. The presented work revealed higher aerosol concentration in the first $1500m$ from ground, resulting from the presence of Arctic Haze. CR analysis indicated that most of the particles in lower altitudes were relatively small and only above $5km$ bigger particles appeared. The application of the method of Zieger et al. (2010) [14] was extended to backscatter and CR data originating from Lidar measurements. The growth functions and their parameters were fitted into the present data set, with this adaption yielding satisfying results.

A case study concerning the aerosol ice nucleating efficiency indicated that not every aerosol type serves as ice nuclei. Thus, the indirect aerosol effect is still a poorly understood issue, since ambient conditions and aerosol type are potentially crucial factors.

APPENDIX

The aerosol depolarisation can be defined with Eq. 07 using the aerosol backscattering in parallel (∥) and perpendicular (⊥) polarisation as a function of the distance (z) and wavelength (λ).

$$\delta_{(z,\lambda)} = \frac{\beta_{\perp}^{Aer}(z,\lambda)}{\beta_{\parallel}^{Aer}(z,\lambda)} \qquad \text{(Eq. 07)}$$

Depolarisation is determined by the particle's shape, thus conclusion about e.g. ice particles can be drawn [16].

REFERENCES

[1] M. Stock et al, Atmos. Env. 52, 48-55 (2011)
[2] P. Tunved et al. Atmos. Chem. Phys, 13, 3643-3660 (2013)
[3] W.R Leaitch et al. J. Clim. Appl. Met. 23, 916-928 (1984)
[4] IPCC, Working Group 1, The Physical Science Basis (2007)
[5] S.R. Hudson, J. Geophys. Res. Atmos., 116, D16102 (2011)
[6] M.C. Serreze et al. Science, 315, 1533-1536 (2011)
[7] C. Ritter, et al. Atmos. Env. 141, 1-19 (2016)
[8] J. Lisok et al. Atmos. Env. 140, 150-166 (2016)
[9] K. Nakoudi et al. 29th International Laser Radar Conference (ILRC 2019), Hefei, China (2019)
[10] A. Hoffmann, Reports on Polar and Marine Research, 630, ISSN:1866-3192 (2011)
[11] J.D. Klett Appl. Opt. 31, No. 33, 211-220 (1981)
[12] H.C. van de Hulst, *Light Scattering by Small Particles* (1981)
[13] climatescience, *Aerosols and Climate*, retrieved from http://oxfordre.com/climatescience/view/10.1093/acrefore/9780190228620.001.0001/acrefore-9780190228620-e-13?rskey=FGAWKq&result=1 (2018)
[14] P. Zieger et al. Atmos. Chem. Phys., 10, 3875-3890 (2010)
[15] S. Gassó et al. Tellus B, 53, 546-567 (2000)
[16] J. Biele, Optics Express 7(12), 427-435 (2000)

AEROSOL INVESTIGATION DURING THE ARCTIC HAZE SEASON OF 2018: OPTICAL AND MICROPHYSICAL PROPERTIES

Konstantina Nakoudi[1, 2*], Christine Böckmann[3], Christoph Ritter[1], Vasileios Pefanis[4, 5], Marion Maturilli[1], Astrid Bracher[4, 5], Roland Neuber[1]

[1]*Alfred Wegener Institute, Helmholtz Center for Polar and Marine Research, Potsdam, Germany*
[2]*Institute of Physics and Astronomy, University of Potsdam, Potsdam, Germany*
[3]*Institute of Mathematics, University of Potsdam, Potsdam, Germany*
[4]*Alfred Wegener Institute, Helmholtz Center for Polar and Marine Research, Bremerhaven, Germany*
[5]*Institute for Environmental Physics, University of Bremen, Bremen, Germany*
**konstantina.nakoudi@awi.de*

ABSTRACT

In this work, optical and microphysical properties of Arctic aerosol as well as their radiative impact are investigated. Air-borne Lidar observations along with ground-based measurements are evaluated for the Arctic Haze season of 2018. Aerosol abundance as inferred from particle backscatter was typical for this period of the year, with nearly spherical and large particles. The inversion of microphysical properties yielded high Refractive Index (RI) together with low Single-Scattering Albedo (SSA), suggesting absorbing particles. A fitted lognormal volume distribution revealed a fine mode with effective radius (r_{eff}) of 0.23 μm and a coarse mode with r_{eff}=0.75 μm. The total radiative balance on ground was positive (12 Wm^{-2}).

1. INTRODUCTION

Aerosol affect significantly the radiation budget, especially in the Arctic, where climate change is substantially accelerated as compared to lower latitudes, a phenomenon described as Arctic Amplification [1]. During Arctic springtime, aerosol concentration in the accumulation mode exhibits an annual maximum, forming the Arctic Haze [2]. The aim of this work is to investigate the optical and microphysical properties of Arctic aerosol and assess their radiative impact during spring.

2. METHODOLOGY AND INSTRUMENTS

2.1 Lidar Systems

The multi-wavelength system Koldewey Aerosol Raman Lidar (KARL) is installed in Ny-Ålesund, Svalbard Archipelago (78.9 °N, 11.9 °E). KARL is a "3β+2α+2δ+2wv" system equipped with a Nd:YAG laser emitting at 1064, 532 and 355 nm after frequency doubling and tripling, respectively [3]. The receiver consists of a 70cm-diameter telescope with a 1.75 mrad field of view (FOV). Photomultiplier Tubes (PMTs) are used for the detection of backscattered photons in all channels, except for 1064 nm, where Avalanche Photodiodes (APDs) are used. Signal acquisition is performed in both Photon Counting (PC) and Analog (A) mode, allowing for large dynamical detection range.

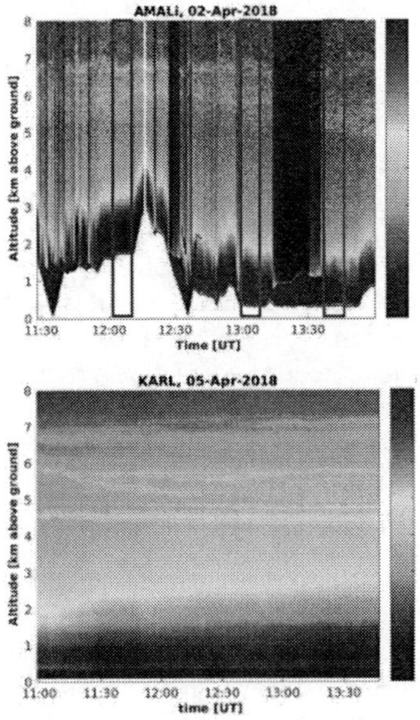

Figure 1: Lidar range-corrected signal (arbitrary units) at 532 nm over Fram Strait (top) and Ny-Ålesund (bottom). Rectangulars indicate temporal windows, for which optical properties were evaluated.

The elastic system Air-borne Mobile Aerosol Lidar (AMALi) is installed on board of the research aircraft Polar 5, operated by the Alfred Wegener Institute (AWI) [4]. A Nd:YAG laser emits pulses at 15 Hz and a 10 cm off-axis telescope mirror with a 3.1 mrad FOV collects backscattered light. PMTs are used for the detection of elastically backscattered photons at 532 and 355 nm. Additionally, the cross-polarized component is recorded at 532 nm. Licel Transient Recorders allow for high temporal resolution (1sec) in both PC and A mode. During Polar Air-borne Measurements and Arctic Regional Climate Model Simulation Project (PAMARCMiP) campaign in spring 2018 AMALi was in zenith measuring configuration.

Aerosol backscatter coefficient (β) profiles were derived by the Klett method [5], while vertically-resolved retrievals of the aerosol extinction coefficient (α) were performed with the Raman technique [6]. KARL profiles were obtained with 2 min-7.5 m resolution. AMALi data were evaluated on a 5 min-7.5 m resolution, with restrictions regarding the flight altitude change. More specifically, after testing several combinations of vertical displacement and the corresponding vertical gradients, we concluded that a shift of 30 m, has a negligible effect on the resulting profiles. Additionally, the profiles obtained by AMALi were vertically smoothed with a 37.5 m gliding average. In this study, we independently retrieved the aerosol optical properties from 3 AMALi profiles and, subsequently, estimated the average optical properties and their standard deviation within the altitude range of 5-6.9 km. Analogous approach was followed to retrieve the aerosol optical properties from KARL in altitudes of 4.4-6.9 km.

2.2 Measurement Site

Ny-Ålesund is the northernmost permanent research settlement, located on Spitsbergen Island, Svalbard Archipelago. Ny-Ålesund is affected by the warm West-Spitsbergen current, which is a sub-branch of the Gulf Stream, and thus experiences quite milder temperatures for the given latitude [7]. Furthermore, Ny-Ålesund is characterized by complex orography, since it is surrounded by Kongfjorden fjord in the North and mountain ranges (up to 1 km) in the South, East and North direction. Strong topography is, hence, a challenge in closure of aerosol properties at the surface and in the atmospheric column [8].

2.3 Microphysical Properties Inversion Scheme

The model relating the optical parameters Γ(λ) with the volume size distribution v(r) is described by a Fredholm integral operator of the first kind with a kernel function K(r,λ,m) (Equation 1),

$$\mathrm{K}(\mathrm{r},\lambda,\mathrm{m}) = \frac{3}{4r}\, Q(r,\lambda,m)$$

$$\Gamma(\lambda) = \int_{r_{min}}^{r_{max}} K(\mathrm{r},\lambda,\mathrm{m})v(r)dr \quad (1)$$

where λ denotes the wavelength, r the particle radius, Γ(λ) stands for either the aerosol extinction or backscatter coefficient, while $Q(r,\lambda,m)$ stands for either the extinction or backscatter Mie efficiency (dimensionless). The input for microphysical properties retrieval was obtained by KARL system. More specifically, 5 discrete values were provided i.e. aerosol β at 3 wavelengths (β_{355}, β_{532} and β_{1064}) and aerosol α at 2 wavelengths (α_{355}, α_{532}). Identifying Γ(λ) as our measurement data and v(r) as the unknown volume distribution, the problem reduces to the inversion of Eq. (1). Knowing the volume distribution, we can then extract the following microphysical parameters

Total surface-area concentration ($\mu m^2 cm^{-3}$)	$s_t = 3\int \frac{v(r)}{r}\,dr$
Total volume concentration ($\mu m^3 cm^{-3}$)	$v_t = \int v(r)\,dr$
Total number concentration (cm^{-3})	$n_t = \frac{3}{4\pi}\int \frac{v(r)}{r^3}\,dr$
Effective radius (μm)	$r_{eff} = 3\,\frac{v_t}{s_t}$

 https://doi.org/10.1051/epjconf/202023702002

Additionally, the complex RI and $SSA_{355, 532}$ can be retrieved. More details about the Lidar inversion code can be found in [9], [10] and [11].

2.4 Radiation Measurements

A BSRN (Baseline Surface Radiation Network) station is in operation since 1992 in Ny-Ålesund [12]. A CMP22 Pyranometer is measuring broadband (0.2-3.6 μm) diffuse, global and reflected radiation. A CHP1 Pyrheliometer is used for monitoring broadband (0.2-4 μm) direct fluxes, while a Pyrgeometer is measuring downward and upward fluxes (3.5-50 μm).

3. RESULTS

3.1 Aerosol Optical Properties

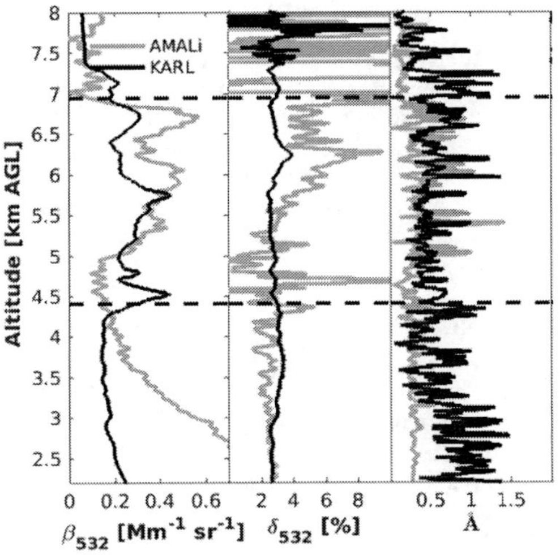

Figure 2: Vertically resolved optical properties as inverted from AMALi and KARL.

The investigated layers were observed over Fram Strait (2 April) and Ny-Ålesund (5 April) at approximately 4.5-7 km (Fig. 1). The inversion of aerosol optical properties (Fig. 2) yielded a bit higher β coefficient over Fram Strait (β_{532}^{AMALi} = 0.39 ± 0.17 Mm^{-1}sr^{-1}) compared to Ny-Ålesund (β_{532}^{KARL} = 0.28 ± 0.09 Mm^{-1}sr^{-1}). Regarding backscattering at 355 nm (not shown here), this also exhibited higher values over Fram Strait (β_{355}^{AMALi} = 1.25 ± 0.4 Mm^{-1}sr^{-1}; β_{355}^{KARL} = 0.62 ± 0.24 Mm^{-1}sr^{-1}). Nearly spherical particles were indicated by the linear depolarization ratio, with average values (± 1 σ) δ_{532}^{AMALi} = 4.5 (± 2.6) % and β_{532}^{KARL} = 2.8 (± 0.6) % over Fram Strait and Ny-Ålesund, respectively. Ångstrom (Å) exponent (β_{355}/β_{532}) indicated aerosol with a size parameter close to unity. These particles are expected to exhibit low spectral dependency in their interaction with light. The average Lidar ratio (LR) at 355 nm was 42 sr, with polluted continental aerosol expected to have similar optical impact [13].

3.2 Aerosol Microphysical Properties

The retrieved RI (Tab. 1) was quite higher compared to other transport events into the Arctic, indicating higher absorption [14]. Besides, SSA was somewhat lower in comparison to typical Haze periods, suggesting more absorbing aerosol [15]. The fine aerosol mode was characterized by high surface-area and volume concentration and, consequently, exhibited higher number concentration compared to the coarse mode (Fig. 3). However, coarse particles have high potential of dominating the total scattering efficiency of the layer. In validating the latter hypothesis, Mie code calculations are necessary.

Table 1: Aerosol preliminary microphysical properties for the fine and coarse aerosol mode.

Microphysical Properties 355 nm	**Mean (± 1 σ)**	
RI	1.63 + i · 0.03 (± 0.01 + i · 0.002)	
SSA	0.83 (± 0.0003), 355 nm 0.87 (± 0.0002), 532 nm	
	fine	**coarse**
r_{eff} (μm)	0.23	0.75
n_t (cm^{-3})	50.3	0.04
s_t (μm^2 · cm^{-3})	23.7	0.24

$v_t\,(\mu m^3 \cdot cm^{-3})$	1.8	0.06

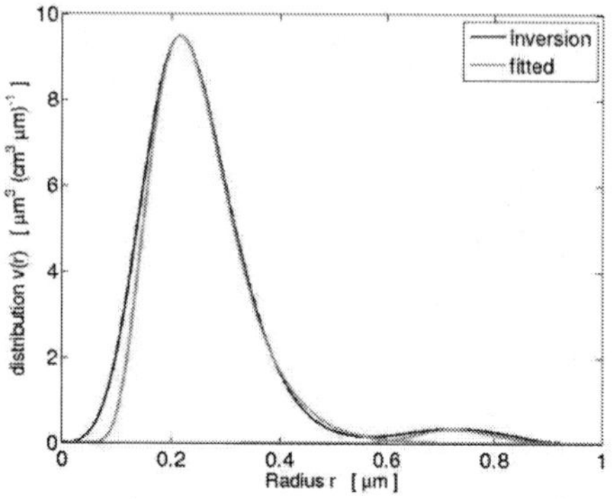

Fig 3: Retrieved volume distribution and log-normal fits for the fine and coarse aerosol mode.

3.3 Radiative Overview

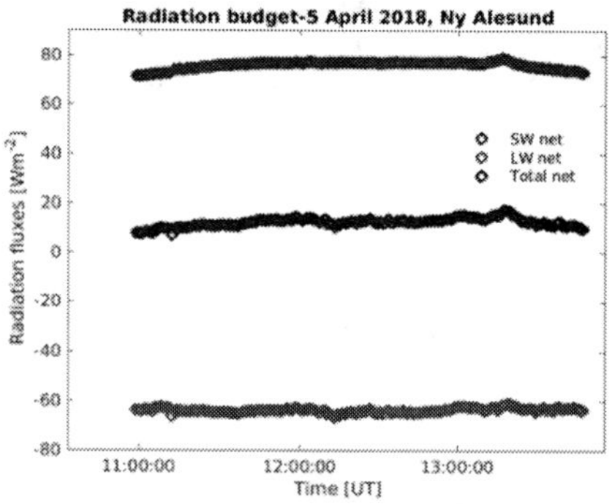

Fig 4: Net radiation budget on ground level at Ny-Ålesund, 5 April 2018.

SW fluxes resulted in an average positive budget of 76 Wm^{-2}, while a negative LW balance (-64 Wm^{-2}) was found. The combined impact of broadband LW and SW radiation was positive (12 Wm^{-2}), suggesting heating at ground level (Fig. 4).

4. CONCLUSIONS AND FUTURE WORK

In this, study, aerosol layers extending over a large spatial area in the Arctic were investigated. The retrieved optical properties, indicated quite large and nearly spherical particles. The layers under investigation were persistent, with the evolution of their optical properties not being pronounced. Additionally, their optical properties exhibited weak spectral dependency, especially over Ny-Ålesund, where the layers consisted of more aged particles compared to those observed over Fram Strait. A bimodal volume distribution was derived from the inversion of aerosol microphysical properties, with higher RI and lower SSA compared to other transport events into the Arctic. The inferred microphysical properties can be utilized in closure schemes with in-situ measurements. Hemispheric radiation measurements yielded moderate total heating at the surface of Ny-Ålesund. In the next steps of our research, we will assess the radiative impact of this event at ground and at different atmospheric levels from the perspective of Radiative Transfer Model for SCIAMACHY (SCIATRAN) simulations [16].

ACKNOWLEDGEMENTS

We gratefully acknowledge the funding by the Deutsche Forschungsgemeinschaft (DFG, German Research Foundation) – Project Number 268020496 – TRR 172, within the Transregional Collaborative Research Center "ArctiC Amplification: Climate Relevant Atmospheric and SurfaCe Processes, and Feedback Mechanisms **(AC)³**", subprojects BO6, CO3 and BO2.

REFERENCES

[1] M.C. Serreze et al. Glob. Planet. Change 77, 85-96 (2011)
[2] A. Stohl, JGR 111, 11306 (2006)
[3] A. Hoffmann, PhD Thesis Uni.Potsdam (2011)
[4] I. Stachlewska et al. Atmos. Chem. Phys. 10, 2947-2963 (2010)
[5] J. D. Klett, Appl. Opt. 20, 211-220 (1981)
[6] A. Ansmann et al. Appl. Opt. 31, 7113-7131 (1992)
[7] S. Dahlke et al. Adv. Meteorol. 2017, 4928620 (2017)
[8] C. Ritter, et al. Atmos. Env. 141, 1-19 (2016)
[9] C. Böckmann et al., Appl. Opt. 40, 1329-1342 (2001)
[10] S. Samaras et al. J. Comput. Phys. 299, 156-174 (2015)
[11] D. Müller et al. Atmos. Meas. Tech. 9, 5007-5035 (2016)
[12] M. Maturilli, et al. Theor. Appl. Climatol. 120, 333-339 (2015)
[13] D. Müller et al. JGR 112 D16202 (2007)
[14] C. Ritter et al. TellusB Chem. Phys. Meteorol. 70:1, 1-24 (2018)
[15] J. Lisok et al. Atmos. Env. 140, 150-166 (2016)
[16] V.V. Rosanov et al. J. Quant. Spectrosc. Radiat. Transf. 133, 13-71 (2014)

https://doi.org/10.1051/epjconf/202023702003

AEROSOL PARAMETERS DURING WINTER AND SUMMER SEASONS AND METEOROLOGICAL IMPLICATIONS

Pradeep Kumar[1*], Arti Choudhary[2], Abhay Kumar Singh[1], Rajendra Prasad[3], Anuradha Shukla[2]

[1]*Department of Physics, Institute of Science, Banaras Hindu University, Varanasi, India*
[2]*Tansport Planning and Environment Division, CSIR-Central Road Research Institute, New Delhi, India*
[3]*Department of Physics, Indian Institute of Technology (BHU), Varanasi, India*
**Email: pradeepph84@gmail.com*

ABSTRACT

The MICROTOPS II aerosol optical depth (AOD) and Moderate Resolution Imaging Spectroradiometer (MODIS) Aqua AOD and angstrom exponent (AE) were collected during December 2016 to May 2017. Higher AOD values were recorded during winter (December - February) and summer (March - May) Months. These values were observed by MICROTOPS II (0.35 - 1.279) and MODIS (0.222 - 1.904) during winter season. During summer AOD values were recorded by MICROTOPS II (0.272 – 1.744) and MODIS (0.227 – 1.33). Whereas MODIS AE (0.218 – 1.799) values were found high during winter season indicates about the dominance of fine particulates. During summer months MODIS AE (.001 – 1.648) values are indicating about the mixing of the aerosol particles. The relative humidity values during December and January months were found to be high, while its values were found decreasing during late February due to transition phase. Relative humidity values were recorded low during summer (36 – 86%) months in compare to the winter (50 - 100%) months.

1. INTRODUCTION

The continuously increasing air pollutant concentrations are not only an environmental issue within the cities, while it has also regional and global environmental effects [1, 2]. The rapidly increasing population, vehicles, industrial developments and the resulting anthropogenic activities are the major concerns for the air, water and land pollution in most of the developing Asian countries [1-6]. So, it becomes very essential to the management of near surface air quality due to its possible implications for the human health, visibility, precipitation and agricultural crop yield [7-9]. Indo Gangetic Plain (IGP) in India, experiences a very high aerosol loading with a strong intra-seasonal to inter-annual variability [10]. IGP region traversed by the Ganga river and its tributaries is one of the largest, densely populated, industrialized and developing region of the world where aerosol not only affect the Indian monsoon but also the global climate system [11]. The high mortality estimates from satellite observations in comparison to the ground-based databases were found by [8]. All the anthropogenic activity, including various festivals, affects the environment especially in India.

2. METHODOLOGY

2.1 Study area and sampling

Samples were collected in Varanasi (25.2677° N, and 82.9913° E) over the middle IGP during December 2016 – May 2017. The AOD (550 nm) samples were collected on daily basis using MICROTOPS II Sunphotometer. MODIS Aqua AOD (550 nm) and AE (412 – 470 nm) were collected from NASA Giovanni website (http://giovanni.gsfc.nasa.gov/giovanni/). The meteorological parameter (relative humidity) data was collected from the India Meteorological Department (IMD) established center at Banaras Hindu University (BHU), Varanasi.

3. RESULTS

3.1 AOD and AE for the atmospheric aerosols during winter season

During winter months AOD values were in the ranges (0.35 - 1.279) collected by MICROTOPS II. The maximum limit is not very large because the AOD data by MICROTOPS II was available from late January 2017 to February 2017. The AOD values were found very high during December month, can be clearly seen from the MODIS collected AOD (0.222 - 1.904) data. The temporal variation of MODIS aqua AOD and AE, MICROTOPS II AOD and relative humidity are shown in Figure 1. The decreasing trend of AOD

https://doi.org/10.1051/epjconf/202023702003

was observed from December 2016 to February 2017. AE values were found high during December and January months while its values were observed decreasing at the late February month. The AE>1 values are indicating about the dominance of fine particles during winter season. During winter season the aerosols were originated from biomass/biofuel burning and due to vehicular pollution over the site [12].

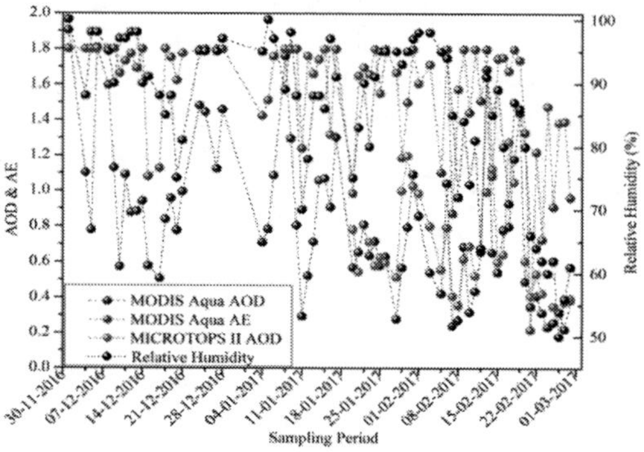

Figure 1: Temporal variation of MICROTOPS II AOD, MODIS Aqua AOD and AE and relative humidity during winter season.

The Five days back trajectories using HYSPLIT model over Varanasi for 6 January 2017 with huge aerosol loading is shown in Figure 2.

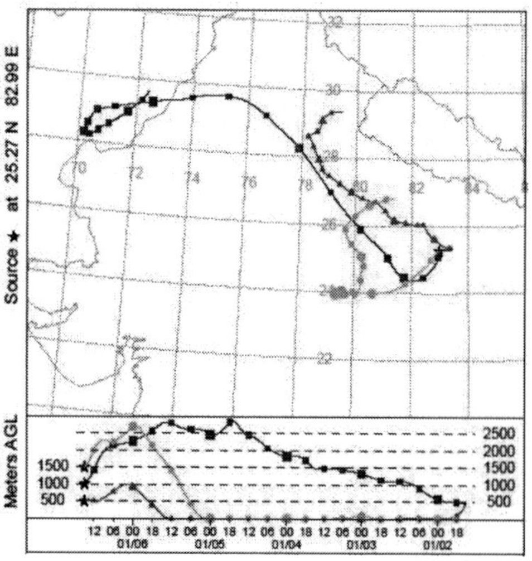

Figure 2: The Five days back trajectories using HYSPLIT model over Varanasi for 6 January 2017 with huge aerosol loading.

During 6 January 2017 it was observed that some of the aerosols are coming from Pakistan country. However mostly aerosols are arising from the states of Uttarakhand, Uttar Pradesh and Madhya Pradesh.

3.2 AOD and AE for the atmospheric aerosols during summer season

The AOD values varies in the different ranges during summer months by MODIS (0.227 – 1.33) and MICROTOP II (0.272 – 1.744). Good agreement between MODIS and MICROTOPS II AOD was observed. The agreement can clearly be seen by Figure 3.

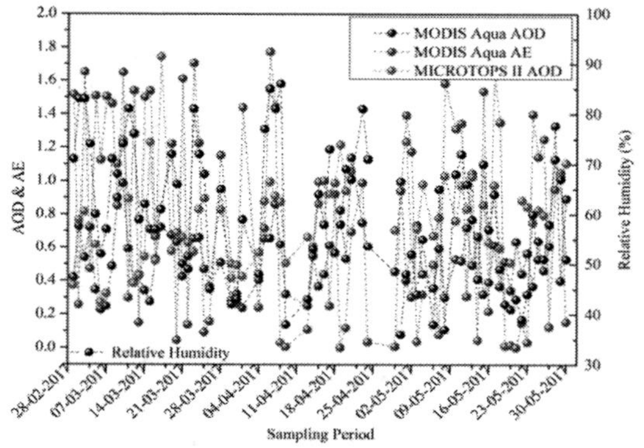

Figure 3: Temporal variation of MICROTOPS II AOD, MODIS Aqua AOD and AE and relative humidity during summer season.

While larger range of AE (.001 – 1.648) values are indicating about the mixing of the different size particles. Some of the AE values were observed very low while some were very high. For the most of the cases the AE<1 values during summer months is the indication of the dominance of the coarser particles. Long range transportation of the mineral dust during summer months are the major concern for the dominance of coarser particles. The coarse-mode aerosol concentrations were enhanced because of IGP is strongly affected by frequent and intense dust storms through long-range transport from Arabian Sea, southwest Asia, and the Thar Desert [13]. The Five days back trajectories using HYSPLIT model over Varanasi for 12 May 2017 with huge aerosol loading is shown in Figure 4.

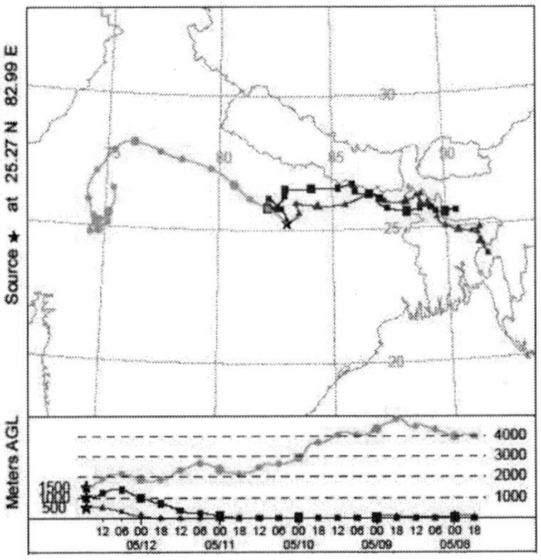

Figure 4: The Five days back trajectories using HYSPLIT model over Varanasi for 12 May 2017 with huge aerosol loading.

During 12 May 2017 it was observed that some of the aerosols are coming from Bangladesh country. However some of the aerosols are coming from Rajasthan and North Eastern states over the site of Varanasi.

3.3 Meteorological implication over the site

Meteorological parameters play a crucial role during observation of aerosol particles and its parameters. During winter months relative humidity was observed very high. While relative humidity values were found to be decreasing in the late February due to transition phase. It's ranging between (50 - 100%) during winter months. The exponential increase in AOD in highly humid condition (relative humidity>85%) is mostly attributed to the hygroscopic growth of fine particulates over the region [14]. During summer months relative humidity ranges between (36 – 86%) indicating about the dominance of the coarser particles. Lower values of relative humidity were observed during summer months in compare to the winter months.

ACKNOWLEDGEMENTS

Present study is financially supported by the SERB, New Delhi, India, under the scheme of SERB National Post-Doctoral Fellowship (PDF/2017/001898). Metrological data were taken from IMD centre BHU, Varanasi. Authors are also thankful to MODIS team and NASA (https://giovanni.gsfc.nasa.gov/giovanni/) for the satellite data acquisition. The back-trajectories are generated from website https://www.ready.noaa.gov/HYSPLIT.php.

REFERENCES

[1] A. Choudhary, et al. Variability in emission rate of Auto-Rickshaw based on real world driving profile: A case study in Guwahati city. 2019 URSI Asia-Pacific Radio Science Conference (AP-RASC). IEEE, (2019)

[2] A. Choudhary, et al. Urban mobility associated ambient air quality and policies for Environmental implication. Atmospheric Pollution and their complications: From regional to global scale. Cambridge Scholars Publishing 2019 (In Press)

[3] S. Menon, et al. Climate effects of black carbon aerosols in China and India. Science 297:2250–2253 (2002)

[4] H. Akimoto, Global air quality and pollution. Science 302:1716–1719 (2003)

[5] Prasad et al. Seasonal climatology of aerosol optical depth over Indian subcontinent: trend and departures in recent years. International Journal of Remote Sensing 27:2323–2329 (2006)

[6] S. Tiwari et al. Identification of aerosol types over Indo-Gangetic Basin: implications to optical properties and associated radiative forcing. Environmental Science and Pollution Research 22:12246–12260 (2015)

[7] S. Han et al. Effect of aerosols on visibility and radiation in spring 2009 in Tianjin, China. Aerosol and Air Quality Research 12, 211–217 (2012)

[8] J. Evans et al. Estimates of global mortality attributable to particulate air pollution using satellite imagery. Environmental Research 120:33–42 (2013)

[9] A. Mhawish, Remote sensing of aerosols from space: retrieval of properties and applications. In: Remote Sensing of Aerosols, Clouds, and Precipitation. Elsevier Inc, pp. 1–38. http://dx.doi.org/10.1016/B978-0-12-810437-8.00003-7 (2018)

[10] R. P. Singh et al. Variability of aerosol parameters over Kanpur, Northern India. Journal of Geophysical Research109: D23206,doi:10.1029/2004JD004966 (2004)

[11] S. K. Guttikunda et al. The contribution of mega cities to regional sulphur pollution in Asia. Atmospheric Environment, 37(1): 11-22 (2003)

[12] P. Kumar, et al. Temporal variation of atmospheric aerosols and associated optical and metrological parameters. 2019 URSI Asia-Pacific Radio Science Conference (AP-RASC). IEEE, 2019

[13] P. Kumar, et al. Atmospheric aerosols over Indo Gangetic Plain: characteristics, sources and climate

implications. Atmospheric Pollution and their complications: From regional to global scale. Cambridge Scholars Publishing (2019) (In Press).

[14] O. Altaratz et al. Relative humidity and its effect on aerosol optical depth in the vicinity of convective clouds. Environmental Research Letters 8 (3) 034025 (2013)

AEROSOLS AND CLOUDS INTERACTIONS IN AN URBAN ATMOSPHERE

Indira Gunaseelan[1], Vijay Bhaskar[2]
[1,2]Department of Bio-Energy
School of Energy, Environment and Natural Resources
Madurai Kamaraj University, Madurai, India-21
Email: gunaindra17@gmail.com

ABSTRACT
Aerosols create great uncertainties in studying climate change under global warming and atmospheric dynamics. To understand the impacts of aerosols on cloud properties in Madurai, we have analyzed an extensive collection of aerosol and cloud properties, obtained from the Moderate resolution Imaging Spectroradiometer (MODIS) data, over the study site during 2012-2013. Monthly, seasonal and annual variations of aerosols and clouds studied along their interactions and impacts on climate. Considering annual averages for all these parameters, most often the year 2012 was dominated with a higher presence of AOD, COD, CER, CTT, CTP whereas rainfall and CF were found to be dominated in 2013. The presence of higher CF in 2013 may be a cause for the higher rainfall and the lower level of CF in 2012 may be a cause for less rainfall. High aerosol loading in this area is due to biomass burning and urban air pollution which may significantly suppress precipitation. Increased aerosols and the local aerosol emissions may reduce the precipitation efficiency, which is responsible for the precipitation reduction and vice-versa.

1. INTRODUCTION

Indian summer monsoon is the biggest source of freshwater resource and more than 70 % of the annual precipitation over India occurs during the monsoon season. Indian monsoon rainfall is critical to food and water security. Aerosol is known to impact the formation and the life cycle of clouds and sometimes it is known to influence the rainfall. Higher aerosol concentrations may decrease the cloud droplet size, suppressing rainfall and enhancing the retention of liquid water by the cloud. Hence increasing the lifetime of the cloud and also the total cloud cover (Albrecht, 1989). Smaller cloud droplets are less efficient at producing rainfall than larger ones; an enhanced aerosol population will lead to a longer cloud life (Rosenfeld, 2000). Longer cloud life results in a lower rate of surface evaporation, a more stable and drier atmosphere, and consequently a reduction in cloud formation (Hansen *et al.*, 1997). The interaction between aerosols, cloud and precipitation can be in two ways; (i) presence of more number of aerosols leads a smaller droplet formation with less coalescence and contributes cooling effect and (ii) abundant smaller cloud droplets have a tendency to reduce precipitation they alter cloud cover with increase cloud lifetime and albedo (Devara and Manoj, 2013). Changes in aerosol, cloud properties cause an uneven distribution of rainfall associated with flash flood or prolonged drought. The changes resulted in major loss of human lives and damages to crops and properties with devastating societal impacts. The studies of aerosols are very important because due to lack of understanding of aerosol properties and their spatial and temporal variations, their effects will be indefinite and vast aerosols withstand in the atmosphere (Gunaseelan *et al.*, 2014).

2. METHODOLOGY

2.1 MODerate resolution Imaging Spectroradiometer (MODIS)

MODerate resolution Imaging Spectroradiometer (MODIS) is a key instrument aboard the Terra - Earth Orbiting System (EOS AM) and Aqua (EOS PM) satellites. The MODIS aerosol algorithm has two components, one adapted for retrieval over land (Kaufman *et al.*, 1997), and the other for retrieval over ocean (Tanré *et al.*, 1997). The daily mean of aerosol optical depths (AOD) derived from the MODIS with level 3 and the collection version of 5.1 with grid size of 1° x 1° at 550 nm aerosol data product from the Terra platform is used in the present work. AODs which are above 0 and less than 1.0 are only considered. This limit is imposed on the assumption that AOD value greater than 1.0 would have resulted most likely due to cloud contamination (Chung *et al.*, 2005).

 https://doi.org/10.1051/epjconf/202023702004

2.2 Study area/ site description

Madurai (9.98° N, 78.21° E) is a semi-arid region on the banks of river Vaigai. It has an average elevation of 100 m. Madurai has the typical climate of the Deccan plateau and remains hot and humid during most of the year with very bright sun shines during summer season manifested by the temperature ranging from 27 °C to 40 °C. Therefore, the climate is quite hot during the summer, winter starts from December and lasts till February with the temperatures ranging between 20 °C and 30 °C. Madurai experiences similar monsoon pattern with Northeast monsoon and Southwest monsoon, with the most rain during October to December. Rainfall is very frequent in the city and it receives the major share of rainfall between the months of July and October. The city is mainly utilized for agricultural activities and the major crop is paddy. Agricultural and biofuel burning practices, heavy transportation and high dense population are the main source of the environmental pollution in this Madurai city.

3. RESULTS

3.1 AEROSOL AND CLOUD INTERACTIONS

The aerosol, cloud and rainfall interaction play an important role in atmospheric dynamics, climate change, radiation budget. Troposphere clouds and scattering aerosols tend for cooling the earth's surface, whereas clouds in the high altitude and absorbing aerosols tend to arming effect (IPCC, 2013). AOD during 2012 and 2013 at 550 nm were taken to study the cloud and rainfall interactions in the Madurai city. AOD oscillated between 0.27 to 0.79 for 2012, whereas for 2013 the values fluctuate from 0.27 to 0.60 and the annual averages are 0.46 ± 0.16 (2012) and 0.39 ± 0.10 (2013.) Maximum AOD peak throughout the two years was noticed in May and minimum AOD was in January (2012) and in February (2013). Maximum concentration in the study site is due to the high temperature and also due to a local religion festival which takes place in a particular month. The city covered with mushroom population, generator usage for the festival, traffic related issues with lots of fireworks. Usage of crackers and also people from outside of the festival use open burning for their usages. Differences in temperature are also attributed to aerosol loadings in the different seasons because the circulation shifts are associated with the monsoon system. Minimum AOD may be due to cold temperature and winter manifested reversal of the spatial gradient of aerosol loading. During this period spatial gradient of aerosol shows comparatively much low levels over entire Indian peninsula (Indira *et al.*, 2013).

Monthly, seasonal and annual variations of cloud parameters and rainfall along with Pearson correlation were studied. Cloud Parameters such as Cloud Optical Depth (COD), Cloud Fraction (CF), Cloud Effective Radius (CER), Cloud Top Temperature (CTT) and Cloud Top Pressure (CTP) data were collected from the satellite data.

COD during 2012, varies from 2.18 to 11.08, maximum is noticed in October whereas minimum is in May. In 2013, the COD values set as 3.50 to 7.50, upper limit values are noticed on December and lower limit values are noticed on May. Considering the seasonal variations of the two years, during 2012 ceiling of COD was in post monsoon and low limit was in the monsoon. In 2013, the ceiling limit was noticed in winter and lower limit was in summer. By Studying the annual variation between the two years 2012 dominated with huge COD than 2013. Correlation analysis gives that in both the years the AOD and COD show a strong negative relationship. Moisture density is more important for the COD, the heavy rain in 2013 might have produced more moisture content in the city with significant COD.

The variations of CF was studied, in 2012 the values swings from 0.36 to 0.92, the highest peak was noticed in August and low peak was in January meanwhile in 2013 the values rise and fall as 0.49 to 0.98 and the maximum durations were noted in June, July along the minimum variations in January. Considering the seasonal variations, presence of higher and lower CF was observed on the monsoon and winter for both the years. Comparing the both years for annual variations more CF was noted in 2013 than in 2012. CF and AOD correlation demonstrates that the relationship between them throughout the study period is a strong positive correlation. Presence of number of particles in the aerosols act as CCN and they are capable to produce more new clouds i.e. directly increases the cloud fraction.

Monthly CER for 2012 oscillates from 12.71 μm to 23.04 μm, it dominates in the November and least was in September. In 2013, the monthly CER values rolled among 14.31 μm to 19.60 μm, bare minimum was in February and utmost CER peak noticed on November. Huge CER whereas minimum CER was noticed on monsoon in 2012 and in 2013 it was in summer. Annual variations resulted that CER dominated in 2012 when compared to that of 2013. The correlation between CER and AOD during 2012 was positively correlated whereas in 2013 it was correlated negatively. Reduction of CER during monsoon and summer may be due to the microphysical effects of aerosols over clouds and also due to the changes in the atmospheric circulation patterns.

CTT values were found to be ceiling in January and found to be small in October and their level diverges from 244.99 °C to 278.02 °C for 2012. CTT during 2013 ranged between 244.09 °C to 275.61 °C and higher and lower variations are noticed in February and July. The seasonal variations outcome shows that elevated CTT was perceived during winter for both the years (2012 and 2013). Least amount of CTT was in the post monsoon during 2012 whereas in 2013, minimum CTT was in the monsoon. CTT for annual variations showed that minimum crest noted in 2013 and the maximum crest was in 2012. CTT and AOD relationship during 2012 was quite strongly negative relation whereas in 2013 these parameters were moderately correlated with a negative relationship. Two physically possible mechanisms can relate AOD with CTT and rainfall. During 2013, the invigoration effect might have taken place in Madurai. This effect is most significant for the deeper clouds and has the ability to enhance the convection and may produce heavy rainfall. Microphysical effect may take place in the study region during 2012 which cause for the size reduction of the cloud particle, rainfall rate and further leads for precipitation suppression.

Monthly CTP values are varied from 244.99 hPa to 278.02 hPa, the elevated level was in November and bare minimum was in September. CTP monthly values for 2013 vary from 378.34 hPa to 722.37 hPa the monthly maxima and minima are noted in February and July. Seasonal deviation results show that throughout the study period, CTP was found maximum in winter and minimum in monsoon. Yearly consideration of CTP outcome represents that CTP is highly dominated during 2012 when compared with 2013. Low CTP is necessary for most prominent clouds for producing rainfall than high CTP. Low CTP in 2012 and 2013 monsoon in the study site helps for mature and well-formed clouds for precipitation formation and it could be the likely reason for elevated rainfall in 2013. A strong negative relationship was noticed between AOD and CTP throughout the research period, increasing in aerosol enhancement may decrease the CTP.

The statistical approach Pearson's correlation was carried was out between AOD, cloud parameters and rainfall during the study periods. During 2012, AOD is positively correlated with CF and CER it suggests that this relationship is due to an inverse aerosol indirect effect. AOD values are negatively correlated with COD, CTT, CTP and with rainfall and the results are correlated significantly. Since the year receives less rainfall and also the wet removal will be negligible and hence aerosols are dominated. For the period 2013, a significant positive correlation was noticed between AOD and CF. Negative correlation was noticed between AOD and CER, COD, CTT, CTP, rainfall and the indirect radiative effects of aerosols are the most probable cause. During this year, the study site might dominate by giant aerosols, it can lead to form larger droplets rapidly than CCN does and it accelerates the onset of precipitation. Considering annual averages for all these parameters, most often the year 2012 was dominated with a higher presence of AOD, COD, CER, CTT, CTP whereas rainfall and CF were found to be dominated in 2013. The presence of higher CF in 2013 might be a cause for the higher rainfall and the lower level of CF in 2012 may be a cause for less rainfall. Throughout both the years, AOD was correlated in a same manner, but the changes in CER between the two years might be a probable reason for the drought and monsoon conditions. In Madurai during 2012 and 2013, variations in CF and CER play an important role in the formation and obstruction of rainfall. The possible reason for low rainfall is due to the abundance of aerosols in 2012. Apart from aerosols, clouds and meteorological parameters, location of the study site, the dominance of the aerosol types and other parameters also play an important role. Since various natural and anthropogenic activities may happen at any cost of time and it is impossible to predict the situation in any location. Man-made emissions induce the brighter clouds with less rainfall and affect the particle size distribution. Increase of aerosol in the atmosphere changes the radiative balance, reduction in visibility (Zhou *et al.*, 2014) and health impacts. Smoke aerosols can be a possible reason to suppress precipitation and decrease cloud droplets. The open burning of crop wastes and forests contribute to the aerosol concentration over this region. It was concluded that the role of aerosols with cloud and rainfall are extremely complex phenomena and also unpredictable.

ACKNOWLEDGEMENTS

The author is gratefully acknowledging the GESDISC's interactive online visualization and analysis infrastructure (Giovanni) as a part of NASA's Goddard Earth Sciences (GES) Data and Information Services Centre (DISC). We acknowledge MODIS, NCEP/NCAR and TRMM for providing the datasets.

REFERENCES

1. B. Albrecht. Science 245:1227-1230 (1989).
2. C. Chung, et al. Journal of Geophysical Research Letters 110, D24207 (2005).
3. P.C.S. Devara and M.G. Manoj. Particuology 11 (1), 25-33 (2013).
4. J. Hansen, et al. Journal of Geophysics Research, Atmosphere 102(D6), 6831-6864 (1997).
5. G. Indira, et al. Environmental Science and Pollution Research 21, 13, 8188-8197 (2014).
6. G. Indira, et al. Aerosol and Air Quality Research 13: 1608–1618 (2013).
7. IPCC, Summary for Policy makers. (2006).
8. Y.J. Kaufman, et al. Journal of Geophysical Research Letter 102: 17051–17067 (1997).
9. D. Rosenfeld. Science 287 (5459), 1793-1796 (2000).
10. D. Tanre, et al. Journal of Geophysical Research 102, 16 971–16 988 (1997).
11. G. Q. Zhou, et al. Aerosol Air Quality Research 14:3, 783-792 (2014)

AIRBORNE POLLEN OBSERVED BY POLLY^{XT} RAMAN LIDAR AT FINOKALIA, CRETE

Xiaoxia Shang[1,*], Stephanie Bohlmann[1], Maria Filioglou[1], Elina Giannakaki[1,2], Mikko R.A. Pitkänen[1], Annika Saarto[3], Vassilis Amiridis[4], Maria Kanakidou[5], Mika Komppula[1]

[1]Finnish Meteorological Institute, P.O. Box 1627, 70211 Kuopio, Finland
[2]Department of Environmental Physics and Meteorology, University of Athens, 15784 Athens, Greece
[3]Biodiversity Unit, University of Turku, 20014 Turku, Finland
[4]IAASARS, National Observatory of Athens, 15236 Athens, Greece
[5]ECPL, Department of Chemistry, University of Crete, 70013 Heraklion, Greece
**Email: xiaoxia.shang@fmi.fi*

ABSTRACT

In order to document and study airborne pollen in the Mediterranean region, a pollen measurement campaign was performed during February-May 2018, at the Finokalia station in Crete. A ground-based multi-wavelength Raman polarization lidar Polly^{XT} performed continuous measurements, together with a Hirst-type Burkard pollen sampler. The optical properties of pollen layers with presence of airborne pollen are retrieved and presented. Dust-free condition is applied for pollen study, using the dust models.

1. INTRODUCTION

Airborne pollen is recognized as one of the major agents of allergy-related diseases such as asthma, rhinitis, and atopic eczema. Pollen is also a form of biogenic air pollutant which affects both the solar radiation reaching the Earth and cloud optical properties [1]. Various networks are built to monitor pollen concentrations at ground level using in situ instruments [2]. One of the most allergenic tree pollen is produced by birch (*Betula*) in north, central, and Eastern Europe. In the Mediterranean region, olive (*Olea europaea*) pollen is considered as one of the most important causes of respiratory allergic disease, whereas cypress (*Cupressus*) releases an enormous amount of anemophilous pollen and has been recognized to be responsible for a large part of total annual amount of airborne pollen [3].

Recently an increasing interest in pollen has arisen in the aerosol lidar community. It has been observed that the non-spherical pollen grains generate strong depolarization [4][5]. Our previous study on pollen shows that lidar measurements can detect the presence of pollen grains in the atmosphere, and lidar derived parameters (e.g. depolarization ratio and Ångström) provide the possibility to identify different pollen types (e.g. birch and pine pollen) [6]. In order to get more information on the distribution of different pollen types in the atmosphere, Finokalia station, located at a remote coastal site in the Eastern Mediterranean Sea, was selected for the pollen campaign. Marine and dust aerosols in this area make the pollen detection and separation more challenging.

2. METHODOLOGY

2.1 Measurement site

A pollen measurement campaign was performed during February-May 2018, at the Finokalia station (35°20'N, 25°40'E) on the north coast of Crete, Greece. The Finokalia station is located at the top of a hilly elevation (238 m above sea level) facing the sea within a sector 270° to 90°. The site is ideal for studying natural aerosols such as mineral dust, marine aerosols, and airborne pollen, since it is not affected by anthropogenic emissions [7]. The nearest large urban center, to the station, is Heraklion with 150 000 inhabitants located 70 km west of Finokalia.

2.2 Instruments

The pollen concentration, for two periods between 23 February and 26 March, and 10 April to 12 May 2018, was retrieved by the Burkard 7-Day Recording Volumetric Spore Sampler [8], placed 4 meters above ground level (agl). This Hirst-type pollen sampler enables microscopical identification of pollen types and their concentration with a 2-hour time resolution.

A ground-based multi-wavelength Raman polarization lidar Polly^{XT} [9] performed continuous measurements. The lidar is equipped with three

https://doi.org/10.1051/epjconf/202023702005

elastic channels (355 nm, 532 nm and 1064 nm), two rotational vibrational Raman channels (387 nm and 607 nm), two linear depolarization channels (355 nm and 532 nm), and one water vapor channel at 407 nm. The combined use of near and far field telescopes provides reliable vertical profiles of aerosol optical properties from 0.25 km to 10 km agl [10].

2.3 Dust models

The island of Crete is very often affected by windblown dust originating from the Sahara, due to its proximity to the African coastline. In order to avoid the dust effects on pollen properties retrieval, only dust-free periods are considered in this study. The periods are defined using two models.

The NMMB/BSC-Dust (Non-hydrostatic Multiscale Model / Barcelona Supercomputing Center) [11] is an online multi-scale atmospheric dust model designed to accurately describe the dust cycle in the atmosphere, and intended to provide short to medium-range dust forecasts for both regional and global domains. It provides vertical profiles (from 0 to 12 km agl) of dust concentration every 6 hours, with a horizontal resolution of 0.3°×0.3°.

The second aerosol product was the Copernicus Atmosphere Monitoring Service (CAMS) near-real-time dust mass mixing ratio forecast [12][13]. The model is based on the global ECMWF-IFS NWP model, and is extended with CAMS atmospheric composition features including data assimilation from satellites and other observational data. The 3-hourly short range dust forecast data used in this study are provided at 25 pressure levels in a 0.4° regular grid.

2.4. Lidar data processing

During cloud-free day-time, the aerosol optical depth derived from the AERONET sun-photometer station of Finokalia (http://aeronet.gsfc.nasa.gov/) is used as a constraint to retrieve the altitude-independent lidar ratio (LR). The Klett method [14] is then applied to retrieve the vertical profiles of aerosol extinction coefficient and backscatter coefficient. During night-time, the standard Raman inversion algorithm is used, which provides additional LR profile [15][16]. LR is considered an important criterion to analyse atmospheric aerosols, as it depends on their single scattering albedo and backscatter phase function, and is thus a function of size distribution and chemical composition.

The volumetric depolarization ratio (VDR) and linear particle depolarization ratio (PDR) at 355 nm and 532 nm are retrieved [17]. The PDR provides information on the shape of the scattering particles. Multi-wavelength measurements allow the retrieval of Ångström exponents (Å), which is related to the particle nature, mostly the size.

A temporal averaging of 2 hours is used to increase the signal-to-noise ratio and to match the Hirst-type pollen sampler data.

3. RESULTS

3.1 Airborne pollen

During the campaign, 30 types of pollen were measured by Burkard Sampler. The five most abundant airborne pollen types were: *Olea* (Olive), *Cupressus* (Cypress), *Platanus* (plane trees), *Quercus* (Oak), and *Ulmus* (Elm), their microphotographs are shown in Figure 1. Their shapes are almost spherical, with size ranging from ~15 to ~30 μm. Sensitisation rates to olive and cypress pollen in Greece are 31.8% and 12.7%, compared to 8.2% to plane pollen and 7.6% to oak pollen. Elm pollen appears in the air in fairly small quantities [18]. The pollination periods for several pollen types were often overlapped, thus a mixture of pollen types were always observed in the atmosphere.

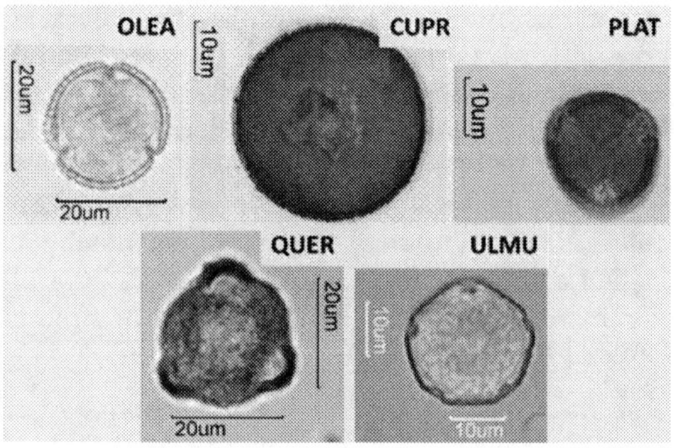

Figure 1. Microphotographs of pollen grains: ***Olea*** *(Olive),* ***Cupressus*** *(Cypress),* ***Platanus*** *(plane tree),* ***Quercus*** *(Oak),* ***Ulmus*** *(Elm). Source: http://www.microlabgallery.com/ and https://pollen.tstebler.ch/.*

3.2 Aerosol optical properties

A "pollen layer" is defined as the lowest observed layer; we assume that pollen grains are distributed in this layer, and that anthropogenic emissions were not present. "Clean atmosphere" and "non-typed aerosols" were removed from our analysis following the criteria of aerosol typing given in [19]. The top heights of pollen layers during the campaign ranged from 0.4 to 3 km agl. Mean values of the optical properties in the pollen layer within 2 hours were used.

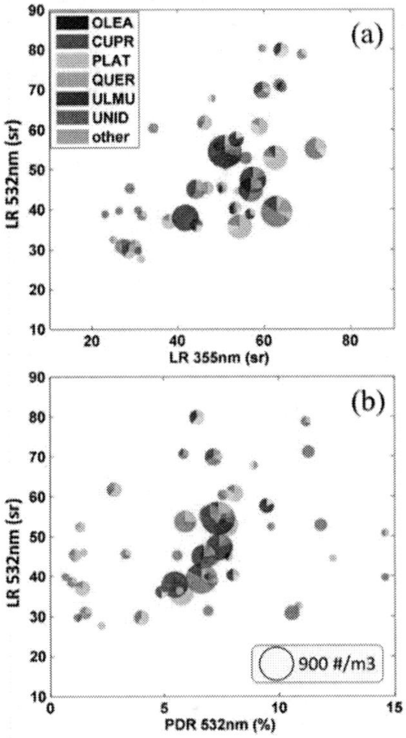

Figure 2. (a) Lidar ratio (LR) at 355 nm, and (b) linear particle depolarization ratio (PDR) against LR at 532 nm for night-time measurements. Mean values within 2 hours in the defined "pollen layer" are used. The size of the symbols is relative to the total pollen concentration measured by the Burkard sampler, the main pollen types are shown by colors, UNID means unidentified pollen.

3.2.1 Night-time observations

Night observations were obtained and analyzed during the period. Mean values in the pollen layers of (a) LR at 355 nm, and (b) PDR at 532 nm against LR at 532 nm are shown in Figure 2. LR in both wavelengths ranged between 20 and 80 sr, with a mean value of 47.1 ±13.0 sr at 355 nm and 48.1 ±13.9 sr at 532 nm. Unlike our previous study of pine pollen in Finland with high depolarization, the maximum PDR was found to be ~15%, with a mean value of 6.5 ±3.5%. This indicates that pollen observed during our campaign in this Mediterranean area are almost spherical. The backscatter-related Ångström exponent (Å) at 355/532 nm is 1.0 ±0.4, which indicates relatively small pollen particles.

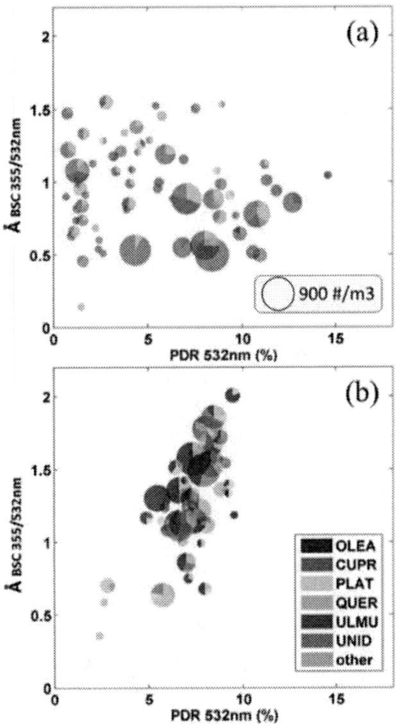

Figure 3. Lidar-derived linear particle depolarization ratio (PDR) at 532 nm against backscatter-related Ångström exponents (355 nm/532 nm), for the periods of (a) 23 Feb to 26 Mar., and (b) 10 Apr. to 12 May, 2018. Mean values within 2 hours in the defined "pollen layer" are used. Legend is same as in Figure 2.

3.2.2 All-day observations.

The cypress tree shed profuse amounts of pollen 6-7 months of the year. *Cupressus* is observed for the whole campaign, with the highest pollination period at the beginning of March. Olive flowers from mid spring to early summer. *Olea* pollen was observed from April to May in this campaign, with highest pollination period at the end of April. Two periods are considered separately in this study: (i) 23 Feb. to 26 Mar. 2018, *Cupressus* is the dominant pollen; (ii) 10 Apr. to 12 May. 2018, *Olea* is dominant. Mean values in the pollen layers of the backscatter-related Ångström exponent (Å) at 355 / 532nm and PDR at 532 m are shown in Figure 3. For the period (i), PDR values for the pollen layers range from very small to about 15%, with a mean value of 5.6 ±3.9%. The Å is found to be 1.0 ±0.3. For the period (ii), PDR values are more concentrated at 7.3 ±1.6%, and Å is bigger compared to the period (i), with the mean value of 1.3 ±0.4. The smaller Å of period (i) is related to the bigger size of the dominate pollen grain (*Cupressus*). With these retrieved values of optical properties, the conventional classification algorithms will always identify these particle as marine mixtures, pollution mixtures, or dust mixtures [20].

CONCLUSION

We characterized the optical properties of airborne pollen in a remotely located site in the Mediterranean region, using a multi-wavelength Raman polarization lidar and a Hirst-type Burkard pollen sampler. We found that the linear particle depolarization ratio of pollen layer was relatively small, with a maximum value of ~15%, since the shape of the majority of pollen types in this region is quasi-spherical. Optical properties of airborne pollen retrieved in this study are classified as "mixture" using conventional classification algorithms in literature. Future work will be focused on the aerosol separation of dust, pollen and marine aerosols.

ACKNOWLEDGEMENTS

This work was supported by the Academy of Finland (project no. 310312). The authors would like to thank the AERONET network for sun-photometer products. Dust data used in this study is from the NMMB/BSC-Dust model, operated by the Barcelona Supercomputing Center (http://www.bsc.es/ess/bsc-dust-daily-forecast/), or generated using Copernicus Atmosphere Monitoring Service Information (2019).

REFERENCES

[1] A. L. Steiner, et al., Geophys. Res. Lett., 42, 3596–3602 (2015)
[2] T. Giesecke, et al., Veg. Hist. Archaeobot., 19, 247–258 (2010)
[3] G. D'Amato, et al., Allergy, 62 (9), 976-990 (2007)
[4] M.Y. Noh, et al., Atmos. Environ., 69, 139-147 (2013)
[5] M. Sicard et al., Atmos. Chem. Phys., 16, 6805-6821 (2016)
[6] X. Shang, et al., in Proceedings of European LiDAR Conference (ELC), (2018)
[7] S. Apostolopoulou, et al., in Proceedings of the 27th International Laser and Radar conference (ILRC27), (2017)
[8] J. M. Hirst, Ann. Appl. Biol., 39, 257-265 (1952)
[9] R. Engelmann, et al., Atmos. Meas. Tech., 9, 1767–1784 (2016)
[10] S. Solomos, et al., Atmosphere, 9, 240 (2018)
[11] C. Pérez, et al., Atmos. Chem. Phys., 11, 13001-13027 (2011)
[12] Up-to-date online documentation of the global CAMS-IFS forecasts https://atmosphere.copernicus.eu/global-production-system (valid on 2019-03-22)
[13] Quarterly quality assurance documents of the CAMS-IFS atmospheric composition forecasts https://atmosphere.copernicus.eu/node/325 (valid on 2019-03-22)
[14] J. D. Klett, Appl. Optics, 20, 211–220 (1981)
[15] A. Ansmann, et al. Opt. Lett., 15, 746 (1990)
[16] X. Shang, et al., Atmos. Meas. Tech., 11, 6525-6538 (2018)
[17] P. Chazette, et al., Atmos. Chem. Phys., 12, 7059–7072 (2012)
[18] D. Gioulekas, et al., Allergy, 59: 174–184. (2004)
[19] H. Baars, et al., Atmos. Meas. Tech., 10, 3175–3201 (2017)
[20] B. Heese, et al., Atmos. Chem. Phys., 17, 6679-6691 (2017)

ANALYSIS OF A POLLUTION TRANSMISSION PROCESS IN HEFEI CITY BASED ON MOBILE LIDAR

Shuai Zhang[1,*] , Zhaoming Zhou[1], Conglei Ye[1], Jibing Shi[1], Peng Wang[1], Dong Liu[2]

[1]Hefei GBo-Qua.Technology Co.,LTD, Hefei, Anhui 230088, China

[2]Key Laboratory of Atmospheric optics, Anhui Institute of Optics and Fine Mechanics, Chinese Academy of Sciences, Hefei, Anhui 230031, China

**Email: zhangshuai@gbqtech.com*

ABSTRACT

The air pollution has been regional in China with the development of economy. To monitoring the air pollution transmission, a new technique, mobile lidar system (GBQ-S01), was introduced. In this paper, a pollution transmission process happened on October 26th, 2017, was analyzed with the use of mobile lidar, air quality monitoring stations data, and Hysplit backward trajectories. The results showed that the polluted air mass was transferred from northeast under the force of air pressure. Under the influences of air pollution transmission and bad meteorological diffusion conditions, The PM_{10} quality concentrations in Hefei increased a lot within 5 hours; among all the 10 national air quality monitoring stations, the Luyang District (the northernmost one) and Changjiang Middle Road (the easternmost one) received the most serious impact with PM_{10} concentration reached up to 252 μg/m³ and 219 μg/m³ at 22:00 (Beijing Time).

1. INTRODUCTION

The air pollution in China has been extremely serious, with rapid development of economy but less concern of environment[1]. Atmospheric environment problem, as one of the hottest topics in scientific researches, has attracted attentions of the whole society recently. The main characteristic of air pollution in our country is the deterioration of air quality in urban areas, especially in large cities as Beijing, Shanghai and Guangzhou[2]. Lidar, combining traditional radar technology with modern laser technology, is one of the most advanced active optical remote sensing instruments. It uses laser as light source, and remotely detects the targets by receiving the back-scattering signals produced by the interaction between laser and targets. Because of its high resolution, good direction, and small volume, it shows outstanding prospects in air quality monitoring and meteorological elements measurement[3].

With further demand of environment monitoring, mobile lidar was manufactured by integrating the lidar into a motor vehicle. It inherits the advantages of fixed-point lidar in observation, and the mobility of vehicle at the same time, making the mobile monitoring of atmospheric environment come true.

A mobile lidar experiment was conducted in Hefei, combining with the data measured in the air quality monitoring stations and the backward trajectories retrieved by Hysplit model, to understand the trends of pollution transmission in October, 2017.

2. METHODOLOGY

2.1 The mobile lidar system

The mobile lidar system (GBQ-S01, Fig. 1), designed and manufactured by Hefei GBo-Qua Technology Co., LTD., having the utility of three-dimensional observations of the atmospheric environment, was used in this project. This system integrates the aerosol lidar on a vehicle platform with a special data analysis software to present the real-time distribution of particle pollution, from near ground to over ten kilometers, along the navigation path.

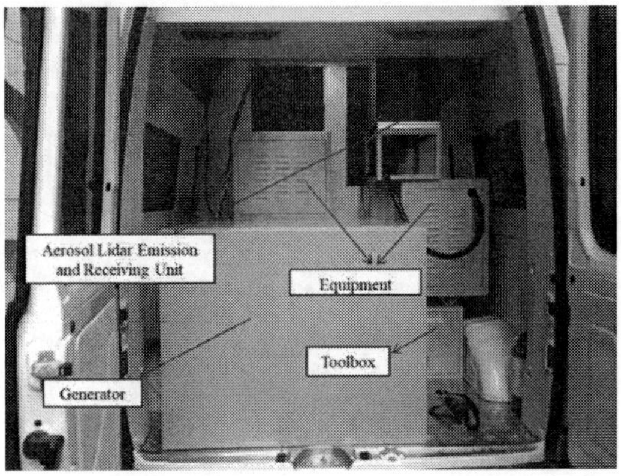

Fig. 1 Structure of mobile lidar system

2.2 Aerosol lidar

https://doi.org/10.1051/epjconf/202023702006

Aerosol lidar, the observation-device of mobile lidar system, contains two principal components: laser and telescope. The laser is used to emit pulsed laser light and the light will be expanded through a beam expander before it enters into atmosphere and reacts with molecules and aerosol particles. Some of the laser photon will be absorbed and also some of them will be scattered. The Cassegrain telescope is used to receive the backscattered signals and the signals will be filtered through a convergent lens before it enters into the signal acquisition channel. The polarization prism in the signal acquisition channel will polarize the received signals into vertical and parallel components, which will then be converted into electrical signals respectively by photodetectors in two designed channels. Extinction coefficient and depolarization ratio profiles, which indicate the vertical distribution of aerosol particles, can then be calculated from these electrical signals by using the Fernald method[4-5]. The schematic diagram of aerosol lidar is shown in Fig. 2, and the main parameters of mobile lidar system are listed in Table 1.

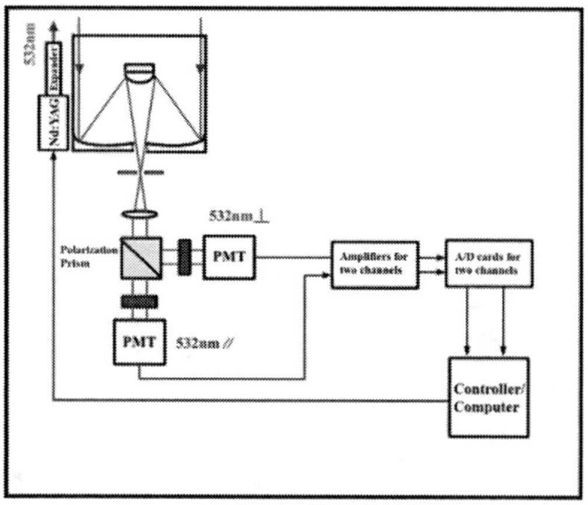

Fig. 2 Schematic diagram of aerosol lidar

Table 1 Main parameters of mobile lidar system

Parameters	Value
Wavelength	532 nm
Pulse energy	> 1 mJ
Repetition Frequency	⩾ 3 kHz
Observation altitude	⩾ 15 km
Speed limits	0 – 120 km/h
Temporal resolution	3 s – 30 min (adjustable)
Spatial resolution	7.5 m

2.3 Data Acquisition

2.3.1 Mobile lidar experiment set up

To track the pollution transmission channel of Hefei in a pollution process transmitted from northern areas, a mobile lidar observation experiment was designed and implemented on October 26th, 2007. The navigation routs are shown in Fig. 3.

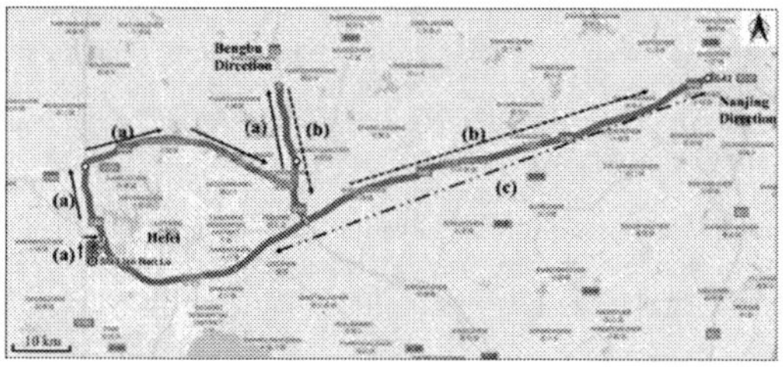

Fig. 3 Schematic chart of navigation path layered on google map

2.3.2 Air quality data

Air quality data are monitored, audited and released by the National Atmospheric Control Station. The detail introductions about the measuring instruments and methods are available in [6].

2.3.3 Backward trajectories

Three backward trajectories at different altitude (500m, 1000m and 1500m) starting from Hefei were calculated by Hybird Single Particle Lagrangian Integrated Trajectory Model (HYSPLIT) [7-8].

3. RESULTS

3.1 Mobile lidar observation results

The mobile lidar observation results showed that aerosols in Hefei and its surrounding areas during navigation were mainly concentrated within 1.5 km altitude. At the northeast corner of Hefei, a pollution zone was detected at the altitude around 1.3 km as shown in Fig. 4(a) circled by a solid circle; and when we went northern (in the Bengbu direction), we also found a pollution zone at the same height as shown in Fig. 4(a) as dashed circle, but it disappeared on the way back (as shown in Fig. 4(b) as dashed circle). Meanwhile, the air was clear on the way to Nanjing and from Nanjing direction back to Hefei. In conclusion, the air pollution could transfer from the north-eastern direction under the force of air pressure.

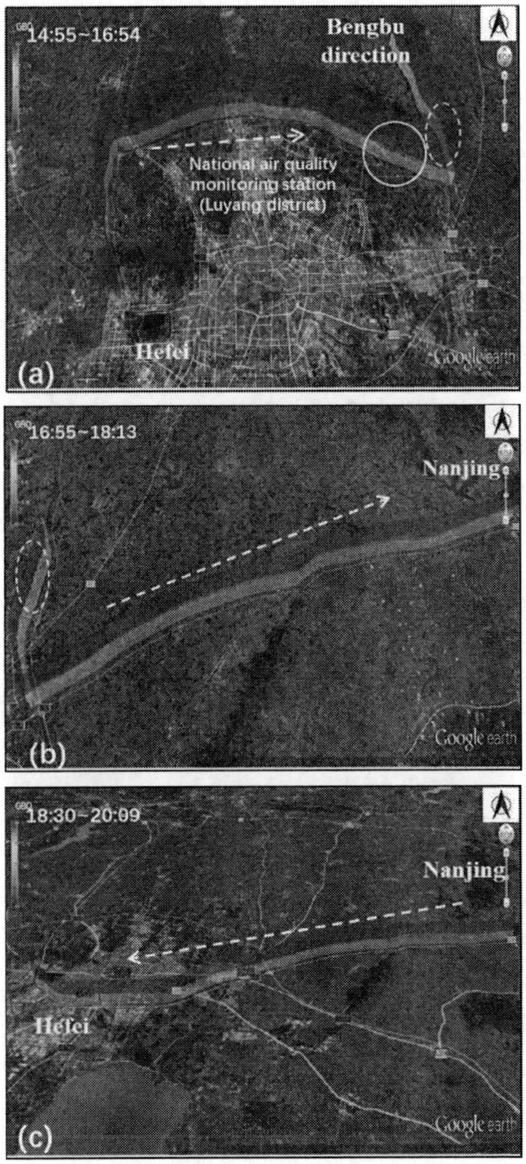

Fig. 4 Mobile lidar observation results

3.2 Air quality data

Measured PM_{10} quality concentrations, at 10 national air monitoring stations in Hefei, showed that Hefei was not affected in this reginal air pollution process until 17:00 (Beijing Time, BT). From 17:00 to 22:00 (BT), except for High-tech zone (the westernmost one) and Dongpu Reservoir (the northwesternmost one), the PM_{10} quality concentrations at other eight stations increased up to 100 μg/m³ or more. Among them, the Luyang District (the northernmost one) and Changjiang Middle Road (the easternmost one) received the most serious impact with PM_{10} concentration reached up to 252 μg/m³ and 219 μg/m³ at 22:00 (BT). Thus, the same conclusion as mobile lidar observation can be gotten that the source of this regional air pollution is from the north-eastern direction.

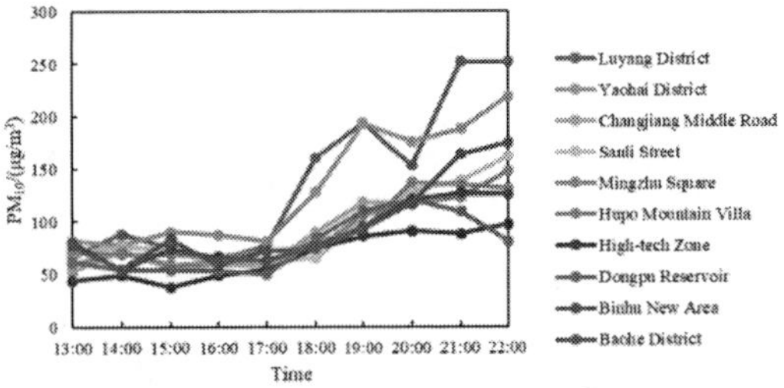

Fig. 5 Temporal variations of PM_{10} concentration measured at each air quality monitoring stations in Hefei on October 26th, 2017

Fig. 6 Schematic map of the positions of air quality monitoring stations in Hefei layered on google map

3.3 Backward trajectories

The backward trajectories ending at 22:00 (BT) at lower altitude (100 m and 500 m) also showed that the air pollution in Hefei was from northeastern direction and the wind direction changed a lot in the last 24 hours. At higher altitude (1000 m), the air particles were transfer from northwest, that means the predominant wind direction at higher altitude was different from the ones at lower altitudes. Vertical wind shear was not good for the spread of polluted air mass that cased the serious pollution in Hefei.

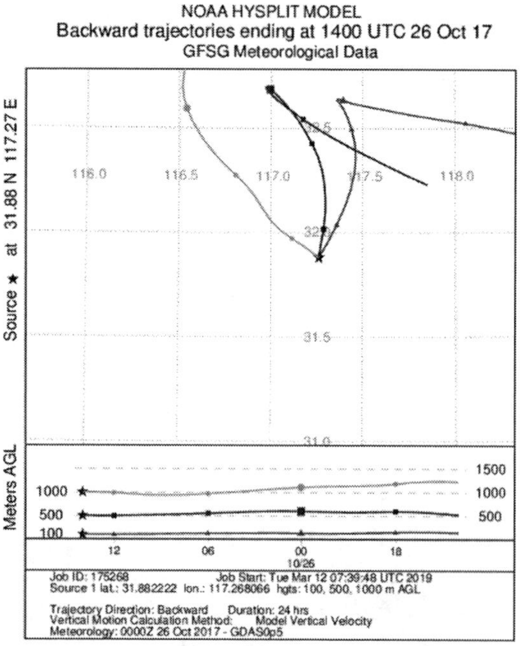

Fig. 7 Hysplit model simulated 24-hours backward trajectories of Hefei ending at 22 pm (Beijing Time) or 14 pm (Coordinated Universal Time, UTC) on October 26th, 2017

ACKNOWLEDGEMENTS

The authors gratefully acknowledge the NOAA Air Resources Laboratory (ARL) for the provision of the HYSPLIT transport and dispersion model and/or READY website (http://www.ready.noaa.gov) used in this publication.

REFERENCES

[1] J. Hao, et al. Air pollution and its control in China. Frontiers of Environmental Science & Engineering in China 1(2): 129-142 (2007).

[2] C. Chan, et al. Air pollution in mega cities in China. Atmospheric Environment 42(1): 1-42 (2008).

[3] Q. Yin, et al. Application of lidar in monitoring meteorological and atmospheric environment. Journal of Meteorology and Environment 25(5): 48-56 (2009).

[4] J. Shi, et al. Tropospheric aerosol measurements by lidar in Northern suburbs of Nanjing. Laser and Progress Optoelectronics, (10):73-81 (2012).

[5] F. G. Fernald. Analysis of atmospheric lidar observation: some comments. Appl. Opt., 23:652-653 (1984).

[6] X. Zhao, et al. Analysis on the spatial-temporal distribution characteristics of O_3 and its influencing factors in Hefei City. Acta Scientiae Circumstantiae 38(2): 649-660 (2018).

[7] A. F. Stein, et al. NOAA's HYSPLIT atmospheric transport and dispersion modeling system. Bull. Amer. Meteor. Soc. 96, 2059-2077 (2015).

[8] G. Rolph, et al. Real-time Environmental Applications and Display system: READY. Environmental Modelling & Software 95, 210-228 (2017).

ANALYTICAL RELATIONSHIPS BETWEEN BULK MICROPHYSICAL PARAMETERS OF NUCLEATION, AITKEN, ACCUMULATION AND COARSE MODE PARTICLES AND EXTINCTION AND BACKSCATTER COEFFICIENTS MEASURED WITH LIDAR

Alexei Kolgotin[1*], Detlef Müller[2], Vadim Griaznov[1], Igor Veselovskii[1], Mikhail Korenskiy[1], Albert Ansmann[3]
[(1)] A.M. Prokhorov General Physics Institute, Moscow, Russia, Email: kolgotin@pic.troitsk.ru*
[(2)] University of Hertfordshire, Hatfield, Hertfordshire, AL10 9AB, UK
[(3)] Leibniz Institute for Tropospheric Research, Leipzig, Germany

ABSTRACT

We derive analytical relationships between bulk microphysical parameters of nucleation, Aitken, accumulation and coarse mode particles and extinction and backscatter coefficients measured at wavelengths 355, 532 and 1064 nm. The bulk parameters are represented by number concentration, mean (effective) radius, variance and complex refractive index. Analytical relationships hold true for arbitrarily shaped particles and complex refractive indices $m=m_R-im_I$, where $m_R>>m_I$. The accuracy is above 60%.

1. INTRODUCTION

Retrieval of particle microphysical parameters from lidar measurements of extinction coefficients at 355 and 532 nm and backscatter coefficients at 355, 532 and 1064 nm is a complex task which can be solved in terms of parametric and direct, inversion (as, for example, least-squares approaches) and inversion with regularization methods. Direct methods of solving this task became popular recently. The direct approach uses a sample function, as for example a logarithmic-normally distributed function that approximates atmospheric particle size distributions.

The most common case permits the use of up to 5 aerosol components [1]. In this contribution we further simplify this concept. We describe an arbitrary external mixture as superposition of 4 components, i.e. particles in the nucleation, Aitken, accumulation and coarse mode, respectively. We present standard algebraic equations that describe the interdependencies between backscatter and extinction coefficients and particle bulk parameters, i.e. number concentration, mean (effective) radius, variance and complex refractive index for each mode separately.

In section 2 we present the methodology. In section 3 we carry out a numerical test with synthetic optical data. Section 4 summarizes our results.

2. METHODOLOGY

Extinction $\alpha(\lambda)$ and backscatter $\beta(\lambda)$ coefficients measured with lidar at wavelengths λ =355, 532 and 1064 nm are related to the particle size distribution $f(r)$ via integral equations of the form

$$\int_0^\infty K_g(\lambda,m,r)f(r)dr = g(\lambda), \qquad g=\alpha,\beta \quad (1)$$

where the kernels of the integral equation $K_g(\lambda,m,r)$ are extinction ($g=\alpha$) or backscatter ($g=\beta$) cross-sections of particles of radius r and complex refractive index $m=m_R-im_I$. The real and imaginary parts of the complex refractive index fulfill the conditions

$$m_R\in[1.3, 1.8] \quad m_I\in[0, 0.1] \quad (2)$$

For example, if the function $f(r)$ describes number particle size distribution we can compute mean (effective) radius r_m (r_{eff}), standard deviation σ, number (N), surface-area (S) and volume (V) concentrations. We can split the radius domain $r\in(0;\infty)$ into 4 intervals [in µm]

$$r\in(0, 0.005) \quad [a] \qquad r\in(0.005, 0.1) \quad [b]$$
$$r\in(0.1, 0.5) \quad [c] \qquad r\in(0.5, \infty) \quad [d] \quad (3)$$

and formally consider Eq. (1) on each interval separately. This approach leads us to the representation of optical data as a superposition of optical data of the nucleation [Nu, interval (3a)], Aitken [Ai, (3b)], accumulation [Ac, (3c)] and coarse [Co, (3d)] mode particles, i.e.

$$g_{Nu}(\lambda)+g_{Ai}(\lambda)+g_{Ac}(\lambda)+g_{Co}(\lambda)=g(\lambda) \qquad g=\alpha,\beta \quad (4)$$

We aim to express the optical data $g(\lambda)$ of each mode $f_{mo}(r)$ on the basis of its own particle microphysical parameters $\mu_{mo}(r_{eff,mode})$, σ_{mo}, N_{mo} and m_{mo} where each number mode is approximated by a lognormal distribution, i.e.

$$f_{mo}(r) \approx \frac{N_{mo}}{(2\pi)^{1/2} r \ln\sigma_{mo}} \exp\left[-\frac{(\ln r - \ln\mu_{mo})^2}{2(\ln\sigma_{mo})^2}\right]$$

$$\text{mo=Nu, Ai, Ac, Co} \quad (5)$$

2.1. Case of large particles (coarse mode)

We start our study with the largest particles. In that case $r\to\infty$. In this asymptotic case the extinction kernel $K_\alpha(\lambda,m,r)$ converges to the particle cross-section πr^2 with factor ~2.1. As a result, the left-hand side of Eq. (1) coincides with the definition of surface-area concentration [2], i.e.

$$\int_{0.5}^{\infty} K_\alpha(\lambda,m,r)f(r)dr \approx \frac{2.1}{4}S \qquad g=\alpha,\beta \quad (6)$$

A statistical analysis of our reference look-up table shows that extinction coefficient and surface-area concentration of large particles are linearly correlated with a high correlation coefficient close to 1 [2]. For particles with effective radius $r_{eff}\geq 0.5$ µm and measured with lidar at λ=532 nm the equation

$$\alpha_{Co}(532)=0.635S\pm 25\% \qquad (7)$$

is fulfilled. In view of approximation (5) this equation can be written in terms of number concentration, mean (effective) radius and variance $\alpha_{Co}(532)$ =

$$= 0.635\times 4\pi N_{Co} r_{eff,Co}^2 \exp(-3\ln^2\sigma_{Co}) \pm 25\%$$
$$= 0.635\times 4\pi N_{Co} \mu_{Co}^2 \exp(2\ln^2\sigma_{Co}) \pm 25\% \qquad (8)$$

Eq. (8) can be directly used for the analytical estimation of the coarse-mode extinction coefficient from particle microphysical parameters and conversely. The estimation accuracy ε_{Co} =± 25% in Eq. (8) is defined as

$$\varepsilon_{mo}(\lambda)= g_{true}(\lambda) / g_{mo}(\lambda) - 1 \qquad g=\alpha,\beta;$$
$$mo=Nu, Ai, Ac, Co, \qquad (9)$$

wherer term $g_{true}(\lambda)$ is the true value. Eq. (8) neither depends on particle shape nor complex refractive index.

2.2. Nucleation mode and Aitken mode

We now consider the smallest particles r_{lim}<0.1 µm. In this case the dimensionless size parameter is $x_{lim}=2\pi r_{lim}/\lambda \approx 1$ at λ=532 nm. We need to use the 1st and 2nd member of the series expansion that asymptotically approximates the kernels $K_g(\lambda,m,r)$ in Mie theory. The use of the 2nd member of the series expansion provides us with a sufficiently accurate approximation, if the magnitude $x_{lim}\approx 1$ [3]. This 2nd order approximation allows us to find the kernels $K_g(\lambda,m,r)$ and one can show that

$$K_\alpha(\lambda,m,r)=A_1(\lambda,m)r^3+ A_2(\lambda,m)r^5 + A_3(\lambda,m)r^6 \quad (10)$$
$$K_\beta(\lambda,m,r)=B_1(\lambda,m)r^6+ B_2(\lambda,m)r^8 + B_3(\lambda,m)r^{10} \quad (11)$$

where

$$A_1(\lambda,m) = -\frac{8\pi^2}{\lambda}\mathrm{Im}(\frac{m^2-1}{m^2+2}); \qquad (12a)$$

$$A_2(\lambda,m) = -\frac{32\pi^4}{15\lambda^3}\mathrm{Im}\left[(\frac{m^2-1}{m^2+2})^2\frac{m^4+27m^2+38}{2m^2+3}\right]; \qquad (12b)$$

$$A_3(\lambda,m) = -\frac{128\pi^5}{3\lambda^4}\mathrm{Re}\left[(\frac{m^2-1}{m^2+2})^2\right]; \qquad 12c)$$

$$B_1(\lambda,m) = B_1(\lambda,m_R) = (\frac{2\pi}{\lambda})^4(\frac{m_R^2-1}{m_R^2+2})^2; \qquad (13a)$$

$$B_2(\lambda,m) = B_2(\lambda,m_R) = 2(\frac{2\pi}{\lambda})^6(\frac{m_R^2-1}{m_R^2+2})^2\kappa(m_R); \qquad (13b)$$

$$B_3(\lambda,m) = B_3(\lambda,m_R) = (\frac{2\pi}{\lambda})^8(\frac{m_R^2-1}{m_R^2+2})^2\kappa^2(m_R); \qquad (13c)$$

$$\kappa(m) = \kappa(m_R) = \frac{3}{5}\frac{m_R^2-2}{m_R^2+2} - \frac{1}{30}(m_R^2+2) - \frac{5}{30}\frac{m_R^2+2}{2m_R^2+3} \qquad (13d)$$

The equalities $B_j(\lambda,m)=B_j(\lambda,m_R)$ in Eq. (13) hold true because $m_R>>m_I$, see the conditions (2).

We now come back to Eq. (1) in which the kernels $K_g(\lambda,m,r)$ are determined with Eqs. (10) and (11). Eq. (1) can be analytically integrated if we use approximation (5) for the function $f(r)$ [4]

$$\alpha_{mo}(\lambda)=N_{mo}[A_1(\lambda,m)\mu_{mo}^3\exp(4.5\ln^2\sigma_{mo})+$$
$$A_2(\lambda,m)\mu_{mo}^5\exp(12.5\ln^2\sigma_{mo})+$$
$$A_3(\lambda,m)\mu_{mo}^6\exp(18\ln^2\sigma_{mo})] \qquad (14)$$
$$\beta_{mo}(\lambda)=N_{mo}[B_1(\lambda,m_R)\mu_{mo}^6\exp(18\ln^2\sigma_{mo})+$$
$$B_2(\lambda,m_R)\mu_{mo}^8\exp(32\ln^2\sigma_{mo})+$$
$$B_3(\lambda,m_R)\mu_{mo}^{10}\exp(50\ln^2\sigma_{mo})] \qquad (15)$$
$$mo=Nu, Ai$$

The estimation accuracies that follow for the extinction and backscatter coefficients with these analytical expressions are close to ε_{mode}=0 for nucleation and Aitken modes if $f_{mo}(\mu_{mo}) >> f_{mo}(r_{lim}) = f_{mo}(0.1)$.

Eqs. (14) and (15) still can be used with an accuracy up to $\varepsilon_{mo}\approx 15\%$ if $\mu_g \leq r_{lim}$=0.1 µm where

$$\mu_g = \frac{\int_0^\infty rK_g(\lambda,m,r)f_{mo}(r)dr}{g_{mo}(\lambda)} = \frac{M_{g,mo}(\lambda,m)}{g_{mo}(\lambda)}$$
$$g=\alpha,\beta; \qquad mo=Nu, Ai \qquad (16)$$

and

$$M_{\alpha,mo}(\lambda,m)=N_{mo}[A_1(\lambda,m)\mu_{mo}^4\exp(8\ln^2\sigma_{mo})+$$
$$A_2(\lambda,m)\mu_{mo}^6\exp(18\ln^2\sigma_{mo})+$$
$$A_3(\lambda,m)\mu_{mo}^7\exp(24.5\ln^2\sigma_{mo})] \quad (17)$$
$$M_{\beta,mo}(\lambda,m)=N_{mo}[B_1(\lambda,m_R)\mu_{mo}^7\exp(24.5\ln^2\sigma_{mo})+$$
$$B_2(\lambda,m_R)\mu_{mo}^9\exp(40.5\ln^2\sigma_{mo})+$$
$$B_3(\lambda,m_R)\mu_{mo}^{11}\exp(60.5\ln^2\sigma_{mo})] \quad (18)$$

Standard multiwavelength lidar delivers 3 backscatter and 2 extinction coefficients. If we consider the five expressions (14) and (15) in the spectral range λ=355-1064 nm we obtain a system of 5 nonlinear algebraic equations and 5 unknown parameters N_{mo}, μ_{mo}, σ_{mo} and m_{mo}. This system can be solved numerically, see Ref. [5], for the case of single modes and small particles in the Aitken mode.

We stress that Eq. (14) and (15) can be modified with a correction factor r_{lim}/μ_g so that

$$\alpha_{mo}^{corr}(\lambda)=(r_{lim}/\mu_\alpha)\alpha_{mo}(\lambda) \qquad mo=Nu, Ai \ (19)$$
$$\beta_{mo}^{corr}(\lambda)=(r_{lim}/\mu_\beta)\beta_{mo}(\lambda) \qquad mo=Nu, Ai \ (20)$$

This correction allows us to use Eq. (14) and (15) for large μ_{mo} and σ_{mo} but condition μ_g=1…5r_{lim}

in Eq. (16) for these μ_{mo} and σ_{mo} still holds true. In this case the estimation accuracy of the analytical expressions is better than $\varepsilon_{mo} \approx 60\%$.

In contrast to large particles, extinction and backscatter coefficients of small particles [see analytical expressions (14), (15) and (19), (20)] significantly depend on their complex refractive index. Still, these analytical expressions can be used for arbitrarily shaped particles because their size is much less than the measurement wavelengths used in lidar applications [3].

2.3. Intermediate size of particles

Our analysis of Eq. (8) and (14) shows that the extinction coefficient of large particles converges to their surface-area concentration; note that we can describe this parameter in terms of volume concentration normalized by effective radius, see Ref. [2]. The extinction coefficient of particles in the nucleation mode (3a) is proportional to volume concentration normalized by wavelength. The statistical analysis of our reference look-up table shows that the extinction coefficient at 532 nm and surface-area concentration are still linearly correlated but with a lower correlation coefficient

$$\alpha_{Ac}(532) = 0.57 S_{Ac} \pm 60\% = \\ = 0.57 \times 4\pi N_{Ac} \mu_{Ac}^2 \exp(2\ln^2 \sigma_{Ac}) \pm 60\% \quad (21)$$

That means we can use the linear regression equation, but the estimation uncertainty grows to $\varepsilon_{Ac} = \pm 60\%$ in the worst case. Again, we note that Eq. (21) does not depend on the complex refractive index and particle shape; see Eq. (8).

The estimation accuracies of the analytical expressions that can be used for the extinction coefficient, see (8), (14), (19) and (21) do not exceed 60%. The highest uncertainties are related to particles with a radius of approximately 0.1 μm which is approximately at the border between Aitken and accumulation mode particles. We can improve the accuracy of estimation if we combine the analytical expressions with best accuracies ε_{mo} [see, for example, Eqs. (8)] and (14)]. In fact, any lognormal distribution (5), e.g. denoted as f_1, can be approximately represented as superposition of two other lognormal distributions, i.e. f_2 and f_3 [6]

$$f_1 \to f_2 + f_3$$

In that case the concentrations of the particle size distributions fulfill the conditions

$$N_1 = N_2 + N_3 \qquad S_1 = S_2 + S_3 \qquad V_1 = V_2 + V_3 \quad (22)$$

These conditions ensure the same effective radius r_{eff} as well as similar optical data $g(\lambda)$ that are produced by f_1 and by a superposition of f_2 and f_3. We consider the conditions (22) and determine preliminary (1st order) estimates of the parameters N_1, μ_1, σ_1 and N_2, μ_2, σ_2. Details on how this step is done exactly will be provided in a publication. We find the parameters for the 3rd distribution which can be described by the following equations

$$N_3 = N_1 - N_2 \quad (23a)$$

$$\mu_3 = \frac{[N_1 N_3^{-1} \mu_1^2 \exp(2\ln^2 \sigma_1) - N_2 N_3^{-1} \mu_2^2 \exp(2\ln^2 \sigma_2)]^{0.5}}{\exp(\ln^2 \sigma_3)} \quad (23b)$$

$$\sigma_3 = \exp\left\{\left[\frac{2}{3}\ln \ldots\right]^{0.5}\right\} \quad (23c)$$

where $\cdots$ denotes

$$\frac{N_1 \mu_1^3 \exp(4.5\ln^2 \sigma_1) - N_2 \mu_2^3 \exp(4.5\ln^2 \sigma_2)}{\left[N_1 N_3^{-1/3} \mu_1^2 \exp(2\ln^2 \sigma_1) - N_2 N_3^{-1/3} \mu_2^2 \exp(2\ln^2 \sigma_2)\right]^{3/2}}$$

Thus we can always (1) fit the parameters (N_2, μ_2, σ_2) and (N_3, μ_3, σ_3) [which simultaneously fulfill Eq. (22) and (23)], (2) find respective optical data $g_2(\lambda)$ and $g_3(\lambda)$ with, for example, Eq. (8) and/or (14) and finally (3) estimate the optical data produced by the size distribution f_1:

$$g_1(\lambda) = g_{mo}(\lambda) \approx g_2(\lambda) + g_3(\lambda) \quad mo = Ai, Ac \quad (24)$$

The estimation accuracies of $g_2(\lambda)$ and $g_3(\lambda)$ determine the estimation accuracy of $g_1(\lambda)$.

3. NUMERICAL TEST

We compare the extinction and backscatter coefficients $g(\lambda)$ estimated with the analytical expressions (8), (14), (15), (19)-(21), (24) and computed with a Mie-scattering code [7]. Besides that comparison, we also compare the kernels $K_g(\lambda,m,r)$ that result from the use of Eq. (10) and (11) and which are described by the strict theory of Mie scattering [7].

We use for the comparisons lognormal size distributions with parameters $\mu \in [0.001;2.5]$ μm, $\sigma \in [1.35;2.55]$ and complex refractive indices in the intervals (2) to estimate the synthetic optical data. We split the parameter domain (μ,σ) into 5 regions (see Fig. 1b,c). We use Eq. (8) for region I, Eq. (14) and (15) for region II, Eq. (19), (20) for region III, Eq. (21) for region IV and Eq. (24) for region V. The results of the estimated values of the kernels and effective radius are shown in Fig. 1a and 1b, respectively. The estimation accuracy (9) for the extinction coefficient, i.e. $\varepsilon_{\alpha(532)}$ at λ=532 nm and m=1.7-i0.05 is shown in Fig. 1c.

Fig. 1a shows that the kernels computed with the analytical expressions (dotted lines) and with the Mie-scattering code (solid lines) coincide for all particle sizes r<0.1 μm. This result shows why we obtain such a high accuracy of the estimated extinction (Fig. 1c) and backscatter (not shown)

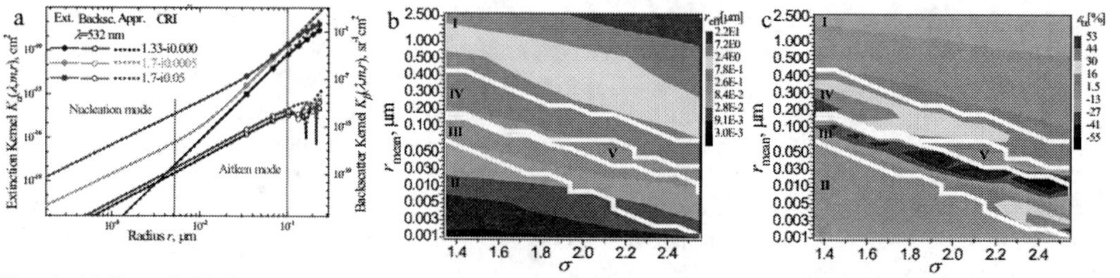

Fig. 1. (a) Kernels $K_g(\lambda,m,r)$ computed with the analytical expressions (dotted lines) and with a Mie-scattering code (solid lines). (b) Effective radius r_{eff} and (c) estimation accuracy of the extinction coefficient $\varepsilon_{\alpha(532)}$ at m=1.7-i0.05 vs mean radius and variance ($r_{mean}=\mu$, σ). Regions I and IV are used for the correlation relationships; see text. Regions II and III are needed for the 2nd order approximation of the kernel functions. Region V means that we use the combination of correlation (I, IV) and approximation (II-III).

coefficients, i.e. $\varepsilon_g \approx 0$ in region II where Eq. (14), (15) can be used. We also find a high estimation accuracy of $|\varepsilon_{\alpha(532)}| \approx 20\%$ based on the correlation relationship used in region I. the correlation relationship (21) works, but the accuracy degrades up to 60%. Region III shows an estimation accuracy of $\varepsilon_{\alpha(532)} \approx$ -55% (overestimation). This result indicates the limit where the use of the 2nd order approximation of kernel functions still can be used. The region V where we use the combination of the correlation relationships and the 2nd order approximation provides the accuracy to approximately $|\varepsilon_{\alpha(532)}| \approx 50\%$.

4. CONCLUSION

We derive analytical expressions that allow us to estimate extinction and backscatter coefficients at wavelengths 355, 532 and 1064 nm. We derive these expressions and thus estimates of these coefficients based on the following approaches:

a. 2nd order approximation of kernel functions can be described by Mie-scattering theory for the case of small particles with r<0.1 μm (nucleation and Aitken modes),

b. correlation relationships between extinction coefficient and surface-area concentration are used if particles are large and intermediate sizes, i.e. r>0.1 μm (accumulation and coarse modes),

c. combination of approaches a and b in the case of small particles where $r \approx 0.1$ μm (Aitken and/or accumulation modes).

The expressions directly link the backscatter and/or extinction coefficients through the underlying particle bulk parameters such as number concentration, mean (effective) radius, variance and complex refractive index to standard algebraic equations. The expressions hold true for arbitrarily shaped particles and complex refractive indices $m=m_R-im_I$, where $m_R \gg m_I$. Minimum uncertainties of the estimates of the parameters are 0-20% in the case of nucleation and coarse mode particles, respectively. Maximum uncertainties are 60% at worst in the case of Aitken and accumulation mode particles.

We will show how these analytical expressions can be applied for estimating microphysical parameters and the contributions of nucleation, Aitken, accumulation and coarse mode particles to extinction and backscatter coefficients measured with lidar. A paper that describes the results is in preparation.

REFERENCES

1. I. Veselovskii, et al, "Characterization of smoke/dust episode over West Africa: comparison of MERRA-2 modeling with multiwavelength Mie-Raman lidar observations", *Atmos. Meas. Tech.*, **11**, pp. 949-969 (2018).
2. A. Kolgotin, et al, "Improved identification of the solution space of aerosol microphysical properties derived from the inversion of profiles of lidar optical data, part 1: theory", *Appl. Opt.*, **55**, pp. 9839 – 9849 (2016).
3. H.C. van de Hulst, *Light scattering by small particles*, John Wiley&Sons, New York, 1957, 537 p.
4. I.S. Gradshtein, I.M. Ryzhik, *Tables Of Integrals, Series And Products* (https://archive.org/details/GradshteinI.S.RyzhikI.M.TablesOfIntegralsSeriesAndProducts)).
5. A. Kolgotin, et al, "Particle microphysical parameters and complex refractive index from 3β+2α HSRL/Raman lidar measurements: retrieval uncertainties versus measurement errors", manuscript in *preparation*, 2019.
6. N.C. Beaulieu, et al, "An optimal lognormal approximation to lognormal sum distributions", *Transactions on Vehicular Technology*, **53**, No. 2, pp. 479 – 489 (2004).
7. F. B. Bohren and D. R. Huffman, *Absorption and scattering of light by small particles,* John Wiley, New York, 1983, 530 p.

APPLICATION OF REGULARIZATION ALGORITHM TO HSRL-2 OBSERVATIONS DURING ORACLES CAMPAIGN: COMPARISON OF RETRIEVED AND *IN SITU* PARTICLE SIZE DISTRIBUTIONS AND SINGLE SCATTERING ALBEDO

Alexei Kolgotin[1*], Igor Veselovskii[1], Mikhail Korenskiy[1], Detlef Müller[2]
[(1)] *A.M. Prokhorov General Physics Institute, Moscow, Russia, Email*: kolgotin@pic.troitsk.ru*
[(2)] *University of Hertfordshire, Hatfield, Hertfordshire, AL10 9AB, UK*

ABSTRACT

Data obtained from HSRL-2 observations carried out on 20 September 2016 during the ORACLES campaign are publicly accessible. In our presentation we invert 3β+2α data into (1) particle size distributions with a regularization algorithm, and subsequently compute (2) single scattering albedo. We carry out a first comparison to the same particle characteristics measured with airborne *in-situ* instruments. We find good agreement of the data products. However, a more detailed study is needed as correction factors and sources of retrieval and measurement uncertainties need to be tested.

1. INTRODUCTION

Numerous approaches have been suggested for the retrieval of particle size distributions from multiwavelength lidar observations during the past 2 - 3 decades. Multiwavelength lidar based on a tripled Nd:YAG laser is capable to provide us with aerosol extinction coefficients (α) at wavelengths λ=355 and 532 nm and backscattering coefficients at 355, 532 and 1064 nm; we denote this data set as 3β+2α data. The retrieval of particle size distributions from 3β+2α data is a complex problem because the number of input optical data is limited and characterized by high measurement errors. Moreover, the range of possible variations of radii (r) and complex refractive indexes ($m=m_R - im_I$) is very wide. This makes the task of retrieving particle microphysical properties from 3β+2α measurements very challenging. Still, the results presented in [1] demonstrate that inversion with regularization is capable to provide estimates of microphysical parameters such as effective/mean radius (r_{eff}/r_{menan}), variance (σ), number (n), surface-area (s) and volume (v) concentration.

Validation of this approach can be carried out by comparing the retrieved particle size distributions with *in situ* measurements [2]. The data acquired in the framework of the campaign *Observations of Aerosols Above Clouds and Their Interactions* (ORACLES) is publicly accessible (https:// espoarchive.nasa.gov/ archive/ browse/ oracles). We took the opportunity of carrying out a comparison study based on the accessible data. The ORACLES data contain measured *in situ* particle size distributions together with time series of 3β+2α data measured with NASA Langley Research Center's high-spectral-resolution lidar (HSRL-2).

In this study we retrieve profiles of particle microphysical parameters and single scattering albedo from HSRL2 data with our regularization algorithm [1] and compare the inversion results to airborne *in situ* measurements.

2. METHODOLOGY

2.1 HSRL-2 Optical Data

During the ORACLES campaign HSRL-2 provided 3β+2α data with 300 m vertical resolution and 1 min (i.e. about 6 km at nominal aircraft speed) horizontal resolution (https:// espoarchive.nasa.gov/archive/browse/oracles/id8 /ER2) [3]. In this study 3β+2α data from the observations on 20 Sep 2016 between minutes 820 and 839 on that day, i.e. between 13:40 UTC and 13:59 UTC were averaged. During that time the aircraft travelled from −17.13° S to −19.30° S and from 8.24° W to 8.73° W (see table 1).

2.2 Regularization Algorithm

The HSRL-2 3β+2α measurements were inverted to the particle size distributions with a regularization algorithm [1]. Since 2004 the algorithm was updated with new options that allow for the processing of large volumes of data in an automated, unsupervised mode for a wide range of particle radii and complex refractive indices [4]:

$$r\in[0.03;10]\ \mu m;\ m_R\in[1.3;1.8];\ \ m_I\in[0;0.1] \quad (1)$$

2.3 *In situ* data

The particle microphysical and optical properties were measured via *in situ* sampling with an isokinetic, low-turbulence inlet mounted on the port side of the P-3B aircraft [3]. The samplings with 1 sec temporal resolution are available at https://espoarchive.nasa.gov/ archive/ browse/ oracles/P3/mrg1. In this study we analyze measurements taken between 5.2-5.5 km height on 20 Sep 2016 between 13:40 and 13:41 UTC. The P-3B aircraft travelled from −18.96° S to −19.02° S and from 9.78° W to 9.83° W (see table 1) during that time.

Table 1. HSRL-2 and in situ data measured in the framework of the ORACLES campaign

	$3\beta+2\alpha$	*in situ*
Source	.../oracles/id8/ ER2	.../oracles/P3/ mrg1
File	HSRL2-Microphysics_ ER2_20160920_ R1.h5	mrg1_P3_ 20160920_R2 5.nc
Date	20 Sep 2016	20 Sep 2016
Latitude,°deg S	-17.13...-19.30	-18.96...-19.02
Longitude,°deg W	8.24...8.73	9.78...9.83
Height, m	959.1... 5680.9	5173...5514
Time, min	820...839	820...821

3. CASE STUDY

The $3\beta+2\alpha$ profiles (20 Sep 2016) are shown in Fig. 1a-b. We obtained them by averaging the data (taken between 820 and 839 minutes since start of 20 Sep) into 20-minute intervals. The time intervals allow us to decrease the measurement uncertainties of lidar ratios and Ångström exponents (backscatter- and extinction-related) in the height range from 1.9 - 5.7 km (above sea level). Below 1.9 km height the data oscillate and cannot be used for the retrieval. For that reason, we only invert the $3\beta+2\alpha$ profiles in the 1.9 - 5.7 km height range.

We also analyzed profiles of particle linear depolarization ratios measured by HSRL-2 at 355, 532 and 1064 nm. The depolarization ratios slightly decrease with height from 0.1 (at 1.3 km) to 0.02 (at 5.7 km). Thus, we used the spherical particle approximation in data inversion. 0.02 is well within the range of values (up to 0.05) in which the spherical particle model (Mie's light-scattering theory) can be used. Although 0.1 is already a bit too far outside that safe range we know from a few studies, e.g. [5, 6] that some particle parameters may be retrieved with acceptable accuracy. Thus, we also used these data in our study.

To retrieve the particle size distributions $dv(r)/dr$ the $3\beta+2\alpha$ profiles from Fig. 1a-b were used as input in the regularization algorithm [1] The retrieval routine is carried out in an automated unsupervised fashion on the domains (1) without the use of any constraints on the solution space in consecutive order from the lowest height bin (l=1) at 1.9 km to the topmost one (l=13) at 5.7 km height above sea level. The analysis of the solution space at each height bin shows that the individual solutions are retrieved with a regularization parameter γ=0, i.e., regularization of the individual solutions is not needed. This result means that the quality of the HSRL-2 data used in this study is high. The final solution space of the particle size distribution is an average over those 100 individual solutions that are nearest (with regards to their own individual discrepancy values) to the minimum discrepancy, as shown in Fig. 2.

The particle size distributions retrieved below 4.4 km height have mean (effective) radii of approximately 0.1 μm for the fine and approximately 4 μm for the coarse mode. The only variation in the particle size distributions below 4.4 km is the particle concentration in the fine mode, which varies from 70 to 120 $\mu m^2 cm^{-3}$ (red and blue curves in Fig. 2). However, above 4.4 km the coarse mode decreases and cannot be identified anymore by the regularization algorithm at 5.7 km (solid black, gray and green curves).

We use the retrieved particle size distributions for our estimation of the particle microphysical parameters, i.e. effective/mean radius (r_{eff}/r_{mean}), variance (σ), number (n), surface-area (s) and volume (v) concentrations. The extensive microphysical parameters are proportional to the extinction coefficient $\alpha(355)$. Surface-area concentration varies from 80 to 500 $\mu m^2 cm^{-3}$, number concentration varies from 500 to 4600 cm^{-3}, and volume concentration varies from 5 and 55 $\mu m^3 cm^{-3}$; see Fig. 1e-f. The total effective radius, which correlates with the extinction-related Ångström exponent, varies from 0.17 to 0.4 μm.

The analysis of the retrieval results shows that s, v/r_{eff}, n ($r_{mean}^2+\sigma^2$) and $\alpha(355)$ are linearly correlated. Correlation coefficients are R>0.93. The correlation coefficient for effective radius versus the extinction-related Ångström exponent is R>0.8. The regression coefficients that describe the linear correlation relationships (not shown) agree with those that were used in [7]. This fact again may support our assumption regarding the high quality of the HSRL-2 $3\beta+2\alpha$ data in the 1.9-5.7 km height range.

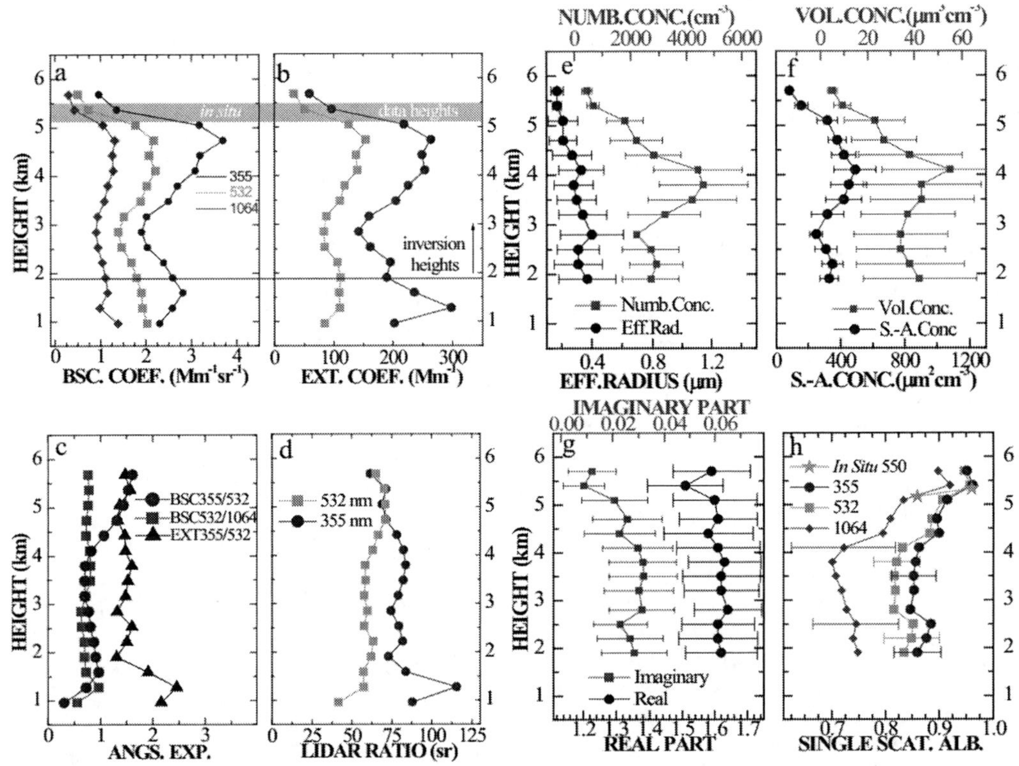

Fig.1. Profiles of the backscattering (a) and extinction (b) coefficients measured with HSRL-2 at 355, 532, and 1064 nm together with the backscatter- and extinction-related Ångström exponents (c) and the lidar ratios (d). The backscattering and extinction profiles describe 20-minute averages of data taken between 820 (equivalent to 13:40 UTC*) and 839 minutes (equivalent to 13:59* UTC*) since start of 20 Sep 2016. Profiles of particle effective radius (e, circles), number (e, squares), surface-area (f, circles) and volume (f, squares) concentrations, real (g, circles) and imaginary (g, squares) parts of the complex refractive index, and single scattering albedo(h) were retrieved with the regularization algorithm from the profiles of the 3β+2α data, see Fig.* 1*a-b. The green stars correspond to in situ measurements at 550 nm.*

The retrieval algorithm also allows us to estimate the complex refractive index from the optical data, see Fig. 1g. The real part does not depend on height and is approximately 1.6; the exception is the height bin at 5.4 km. The imaginary part is approximately 0.03 and does not change significantly from this value below 4.4 km height above sea level. It decreases to 0.01 above 4.4 km height above sea level. This decrease of the imaginary part may (at least in part) explains the increase of the backscatter-related Ångström exponent (at 355/532 nm) with height; see blue curve in Fig. 1c. This behavior also explains the increase of particle single scattering albedo at 355 and 532 nm from 0.90 to 0.95 above 4.4 km height (see Fig. 1h).

Finally, we compare the retrieval results with *in situ* data. We averaged into size bins the particle concentrations $n(r)$ measured *in situ* and retrieved from the HSRL-2 observations. The averaging time is approximately 1 minute for the P-3B aircraft data. We note that the aircraft travelled above 5.5 km height after 13:41 UTC during the time we use for this comparison. We note that the lidar data degraded above 5 km height above sea level during that time. Data are not available above 5.7 km height above sea level after minute 822 on that day. Thus, we need to take the comparison with some caution. We converted the *in-situ* value of $n(r)$ ["# particles per cm^3 per log10"] into units of a volume particle size distribution $dv(r)/dr$ [$\mu m^2 cm^{-3}$]. For that purpose we used the equation

$$dv(r_i)/dr = \frac{4}{3ln(10)lg\frac{r_{i+1}}{r_i}} \pi r_i^3 n(r_i) \quad i=1,2,\ldots \qquad (2)$$

The parameter r_i denotes the radius bins that were used for the *in-situ* measurements of the particle size distributions (see black dotted curve in Fig. 2). Unfortunately, the *in-situ* measurements of the particle size distributions covered only the particle radius range $r<0.5$ µm. Thus, we cannot compare the regularization results and the *in-situ* measurements for particles above 0.5 µm. It also remains unclear if particle loss effects caused by the inlet system were significant and how they were corrected for.

Finally, we note that relative humidity measured *in situ* at 13:40 UTC had decreased from 60% to 10% (not shown). This low value makes it unnecessary to carry out a humidity correction of the radius of the observed particles.

Fig. 2 shows that the particle size distributions retrieved with the regularization algorithm at 5.1 km (black solid line) and at 5.4 km (gray line) height above sea surface. The *in-situ* data show the average value between 5.2 and 5.5 km height (black dot). The data cover the radius range from 0.03 to 0.5 µm. We find reasonable agreement, i.e. position (radius bin) of the maximum value of volume particle size distributions in these heights. The maximum value measured *in-situ* is between respective values retrieved with regularization algorithm at 5.1 and 5.4 km. Single scattering albedo retrieved at 532 nm and measured *in situ* at 550 nm coincide if we take account of the involved uncertainties. We find a value of ~0.96 at 5.4 km height in both approaches.

4. CONCLUSION

We inverted 3β+2α data taken with HSRL-2 on 20 Sep 2016 during the ORACLES campaign. We identified two aerosol layers and analyzed two height bins for our study. The lower aerosol layer, below 4.4 km height above sea level, is characterized by a bimodal particle size distribution. In the upper aerosol layer above 4.4 km height the coarse mode disappears. Simultaneously, the imaginary part of the complex refractive index decreases to 0.01 in the upper aerosol layer. A first comparison of the results obtained with the regularization algorithm (between 5.1-5.4 km height) and measured with *in-situ* instruments (between 5.2-5.5 km height) agree reasonably well.

A future, more detailed study of the ORACLES data needs to include a sensitivity study on the uncertainties of the optical input data and uncertainties of the data products we inferred from the regularization algorithm and measured with the *in-situ* instruments.

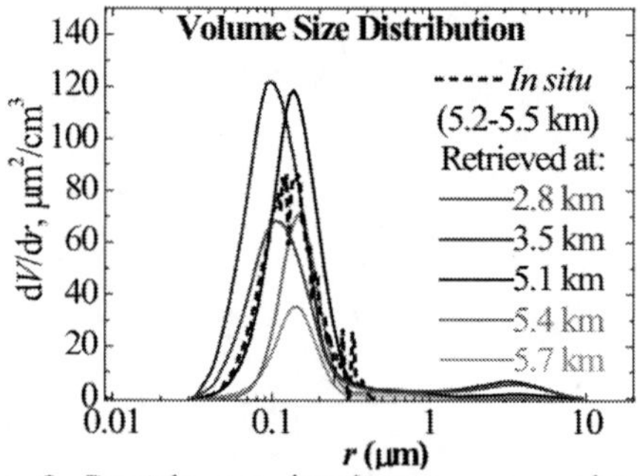

Fig. 2. Particle size distributions measured in situ (see dotted black curve and details in Table 1) and retrieved with the regularization algorithm (solid curves) from the 3β+2α data (see Fig. 1) at different height levels (see legend).

REFERENCES

1. I. Veselovskii, et al., "Inversion of multiwavelength Raman lidar data for retrieval of bimodal aerosol size distribution", *Appl. Opt.* **43**, 1180-1195 (2004).

2. D. Müller, et al., "Vertical profiles of microphysical particle properties derived from inversion with two-dimensional regularization of multiwavelength Raman lidar data: experiment", *Appl. Opt.*, Vol. **50**, No 14, pp. 2069-2079 (2011).

3. P. Sawamura, et al., "HSRL-2 aerosol optical measurements and microphysical retrievals vs. airborne *in situ* measurements during DISCOVER-AQ 2013: an intercomparison study", *ACP*, **17**, pp. 7229–7243 (2017).

4. D. Müller, et al., "Automated, unsupervised inversion of multiwavelength lidar data with TiARA: assessment of retrieval performance of microphysical parameters using simulated data", *Appl. Opt.*, **58**, No 6, in press (2019).

5. I. Veselovskii, et al, "Characterization of smoke/dust episode over West Africa: comparison of MERRA-2 modeling with multiwavelength Mie-Raman lidar observations", *AMT*, **11**, pp. 949-969 (2018).

6. M. Tesche et al., Vertically Resolved Separation of Dust and Smoke Over Cape Verde Using Multiwavelength Raman And Polarization Lidars During Saharan Mineral Dust Experiment 2008, *JGR, 114*, D13202, 2009.

7. A. Kolgotin, et al., "Improved identification of the solution space of aerosol microphysical properties derived from the inversion of profiles of lidar optical data, part 3: case studies", *Appl. Opt.*, **57**, No 10, pp. 2499-2513 (2018).

CHARACTERIZATION OF AEROSOL SIZE AND MICROPHYSICAL PROPERTIES FROM MULTI-WAVELENGTH RAMAN LIDAR MEASUREMENTS: INTER-COMPARISON WITH IN SITU SENSORS ONBOARD THE ATR 42 IN THE FRAMEWORK OF HYMEX-SOP1

Benedetto De Rosa[1] , Paolo Di Girolamo[1] , Donato Summa[1] Dario Stellitano[1]

[1] *Scuola di Ingegneria, Università degli Studi della Basilicata, Potenza, 85100, Italy*
**Email: bundit@hotmail.it*

ABSTRACT

This extended abstract reports measurements that were carried out by the Raman lidar system BASIL in the frame of the Hydrological Cycle in the Mediterranean Experiment – Special Observation Period 1 (HyMeX-SOP1). A specific case study was selected revealing the presence of variable aerosol properties at different altitudes. Specifically, Raman lidar measurements on 02 October 2012 reveal the presence of two distinct aerosol layers, a lower one extending up to ~3 km and an upper one extending from 3.5 km to 4.7 km. Aerosol and size microphysical properties are determined from multi-wavelength measurements of particle backscattering and extinction profiles based on the application of a retrieval scheme which employs Tikhonov's inversion with regularization. Inversion results suggest a size distribution with the presence, in both the lower and upper aerosol layer, of two particle modes (a fine mode, with a radius of ~0.2 μm, and a coarse mode, with radii in the range 2-4 μm), volume concentration values of 2-4 mm^3cm^{-3} and effective radii in the range 0.2-0.6 μm.

This effort benefited from the dedicated flights of the French research aircraft ATR42, equipped with a variety of in situ sensors for measuring aerosol/cloud size and microphysical properties. Aerosol size and microphysical properties retrieved from multi-wavelength Raman lidar measurements were compared with simultaneous and co-located in-situ measurements.

1. INTRODUCTION

Tropospheric aerosols are a fundamental atmospheric component of the Earth's radiation budget. Increased aerosol concentrations intensify atmospheric scattering and absorption processes for both solar and planetary radiation. Additionally, increased tropospheric aerosol concentrations may lead to enhanced nucleation and intensified cloud formation processes The above mentioned direct and indirect aerosol effect have a strong impact on the Earth's radiation budget. A quantitative assessment of these effects is difficult to obtain and estimates are typically affected by large uncertainties.

Despite the recognized importance of having vertically resolved measurements of aerosol size and microphysical properties, a limited number of techniques can provide this information..

Airborne in situ sensors may be effective. However, their exploitation is complex and costly, and consequently their use is limited. Remote sensing techniques have strong potential, but their applicability and the reliability of results from need to be verified against independent measurements. Among others, multi-wavelength Raman lidar measurements appear to be very promising. The dataset reported and discussed in the present research effort represents a unique opportunity to verify the quality of the results in terms of size and microphysical properties obtained through the application of obtained the retrieval scheme based on Tikhonov's inversion with regularization.

Reported measurements were carried out in the frame of HyMeX-SOP1, which took place in the period September–November 2012 over the North-Western Mediterranean Sea and its surrounding coastal regions in France, Italy and Spain. In the frame of HyMeX-SOP1, the University of BASILicata ground-based Raman Lidar system (BASIL) was deployed in the Cévennes-Vivarais site (Candillargues, Southern France, Lat: 43°37' N, Long: 4° 4' E, Elev: 1 m).

As part of HyMeX-SOP1, HyMeX-SOP1 the French research aircraft ATR42, equipped with a variety of in situ sensors for turbulence and aerosol/cloud microphysical measurements, performed more than 60 flight hours, 8 of which as part of the EUFAR project "WaLiTemp. A specific flight pattern was defined for this project, with the aircraft making spirals up and down around a central location, originally aimed to be the atmospheric supersite in Candillargues, but, because of air-traffic restrictions, 20 km eastward of the supersite.

2. METHODOLOGY

2.1 The lidar system

The lidar transmitter is developed around a Nd:YAG laser source, including both second and third harmonic generation crystals.. Laser pulses at 1064, 532 and 355 nm are simultaneously transmitted in the atmosphere along the zenith. BASIL is capable to perform high-resolution and accurate measurements of atmospheric water vapour and temperature, both in daytime and night-time, based on the application of the vibrational and rotational Raman lidar techniques, respectively, in the UV [1, 2, 3, 4, 5]. Besides water vapour and temperature, BASIL, in its HyMeX configuration, also performed measurements of the particle backscattering coefficient at 355, 532 nm and 1064 of the particle extinction coefficient at 355 nm and 532, and of particle depolarization at 355 nm. In addition to HyMeX-SOP1, BASIL has been deployed in several international field campaigns [6-9].

A retrieval algorithm, exploiting Tikhonov's inversion with regularization, is applied to multi-wavelength particle backscattering and extinction measurements in order to retrieve particle size and microphysical parameters. The retrieval scheme makes use of kernel functions for spherical particles. Veselovskii et al. [10-11] demonstrated that an input data set consisting of particle backscattering coefficient profiles at 355, 532 and 1064 nm and particle extinction coefficient at 355 and 532 is sufficient to allow retrieving specific aerosol size and microphysical parameters, such as number, surface, volume density, effective radius and complex refractive index. In order form this to be possible the statistical uncertainty affecting particle backscatter measurements must not exceed 5 %, while the uncertainty affecting particle extinction measurements must not exceed 10 % [10-11].

3. RESULTS

3.1 Case of study 2 October 2012

We have focused our attention on the case study on 2 October 2012. The color map in figure 1 illustrates the time evolution of particle backscattering coefficient at 355 nm covering a period of approx. 4 h from 19:20 to 23:40 UTC. On this day the ATR 42 flew from 19:20 to 22:20 UTC.

For the purpose of retrieving particle size distribution parameters as a function of altitude, we focused on a time interval when aerosol concentration was higher. In order to get high enough signals-to-noise ratios, and consequently low statistical uncertainty, for the retrieval scheme to be applicable, we considered a 2-h averaging for both particle backscattering and extinction coefficient, i.e. the time interval from 20:30 to 22:30 UTC.

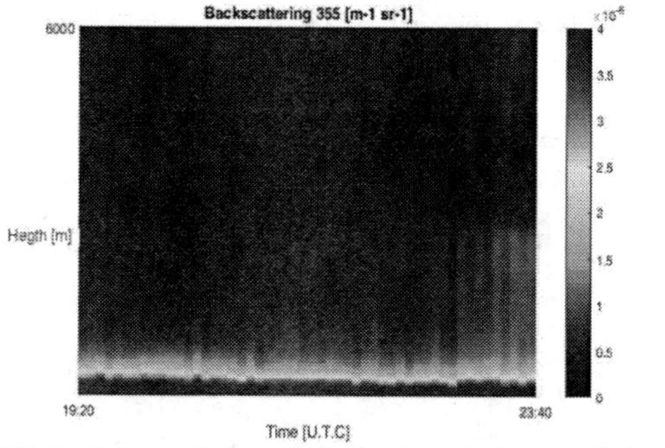

Fig.1 Time evolution of the backscattering coefficient at 355 nm in the time interval 19:20-23:40 UTC on 2 October 2012.

Figure 2 shows the 5-days back-trajectory analysis as determined with NOAA Lagrangian back-trajectory model HYSPLIT. The analysis reveals that air masses reaching the measurement site at an altitude of 600 m were originated over the Northern Atlantic Ocean, South of Iceland, and passed at low altitudes (500-600 m) over continental areas with high anthropic impact (Ireland, England and Northern France). The chemical sensors onboard the ATR 42 show high concentration values for NO_3 SO_4 NH_4, which are typical of a polluted aerosol (figure 3).

The back-trajectory analysis also reveals that air masses at 4000 m were originated over the Northern Atlantic Ocean, off-shore the Canadian coast, and overpassed the Northern coast of Spain before reaching the measurement site. This is a marine aerosol, with a high concentration of sulfates.

Both particle backscattering and extinction coefficient profiles, shown in figure 4, reveal the presence of the two different aerosol layers. The lower polluted layer extends up to ~3 km, while the upper marine layer extends from 3.5 to 4.7 km. To reduce the statistical noise the backscattering coefficient was determined with a vertical resolution of 300 m, while the extinction coefficient with a vertical resolution of 600 m. The backscatter coefficient at 355 and 532 was determined based on the application of the Raman technique, while at 1064 the Klett-modified method was applied [12-13]. The extinction at both 355 and 532 nm was calculated using the approach proposed by Ansmann [14].

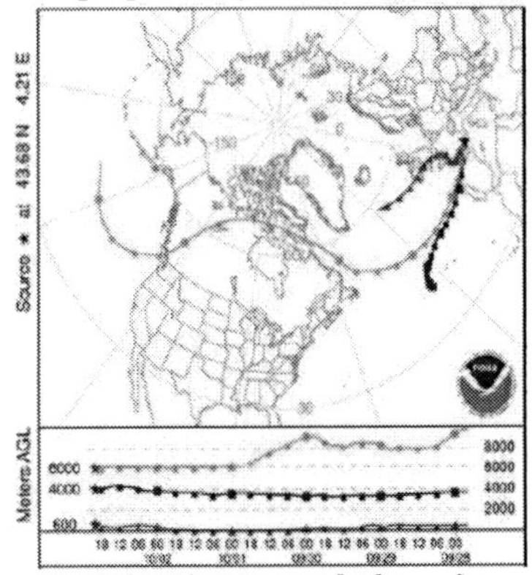

Figure 2 Back-trajectory analysis as determined with NOAA HYSPLIT model. The three illustrated trajectories are those ending over the lidar site at 600 m, 4000 m and 6000 m (in red, blue and green, respectively) at 20:00 UTC on 2 October 2012.

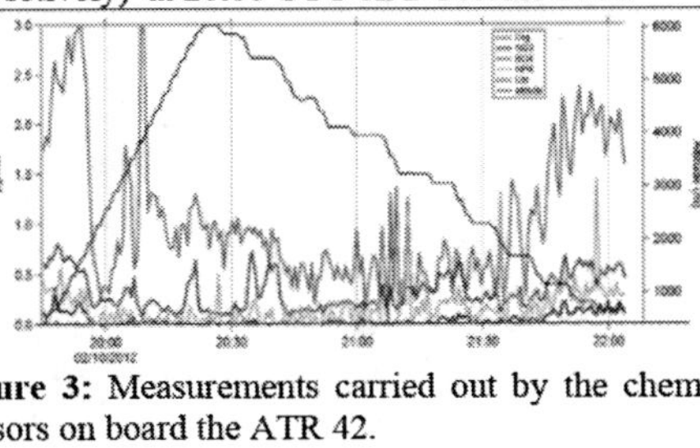

Figure 3: Measurements carried out by the chemical sensors on board the ATR 42.

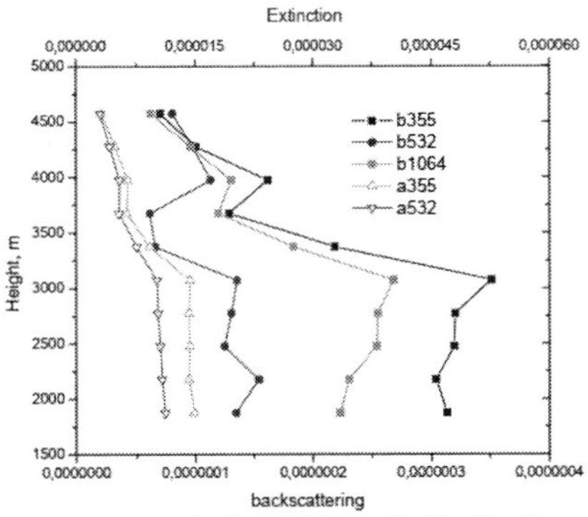

Figure 4: Vertical profile of the backscattering coefficient at 355, 532 and 1064 nm and the extinction coefficient at 355 and 532nm.

Aerosol size and microphysical properties obtained from the multi-wavelength Raman lidar data based on the application of the above mentioned inversion algorithm are compared with those simultaneously measured from in-situ sensors on-board the ATR42. Figure 5 illustrates the comparison expressed in terms of volume concentration measurements, with the two instruments showing similar values both in the lower (2-4 mm^3cm^{-3} for the Raman lidar and 2-5 mm^3cm^{-3} for the in-situ sensors) and upper layer (2-3 mm^3cm^{-3} for the Raman lidar and 1-3.5 mm^3cm^{-3} for the in-situ sensors).

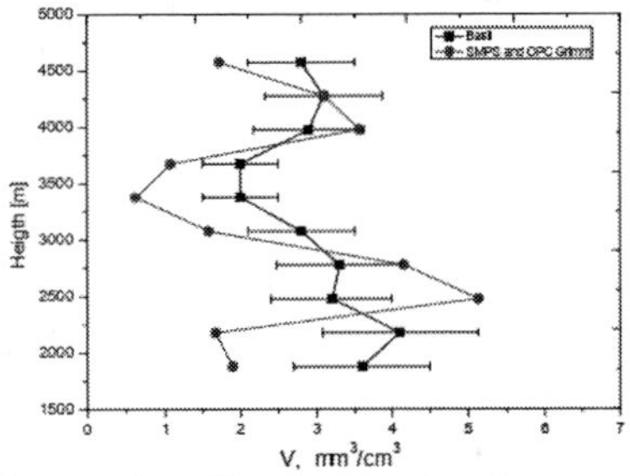

Figure 5: Comparison in terms of volume concentration measurements between BASIL and the on-board in situ sensors.

Figure 6 shows the variability with altitude of the effective radius, with values in the range 0.2-0.6 μm. The agreement between BASIL and the in-situ sensors is good at all altitudes, with values measured by the in-situ sensors always within the error bar of the corresponding measurements by the Raman lidar, with the only exception of the values at ~4 km. As expected, values in the upper layer are slightly larger than those in the lower layer, as in fact marine aerosols are typically larger than urban continental aerosols.

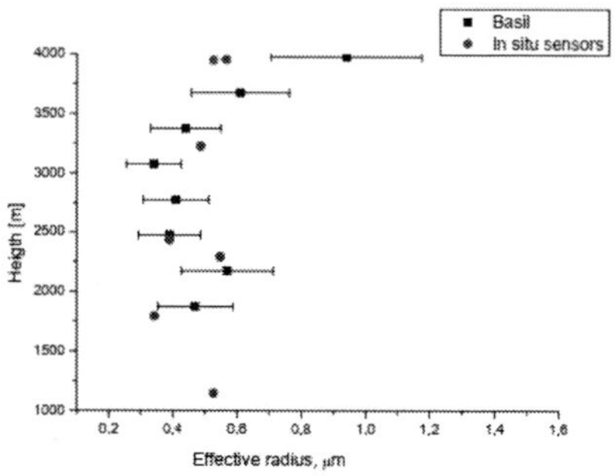

Figure 6: Comparison in terms of effective radius measurements between BASIL and the on-board in situ sensors.

Figures 7, 8 and 9 show the aerosol size distribution at 2, 2.7 and 4 km. The agreement between Raman lidar and in-situ sensors is good at all altitudes. It is to be specified that the in-situ sensors are not sensitivity to particle with diameters in excess of 2 μm.

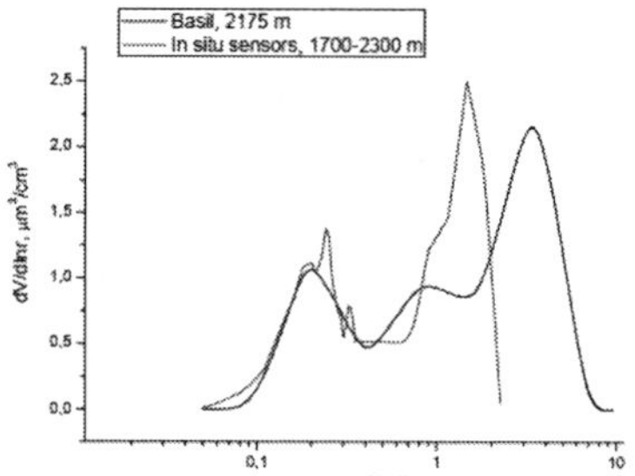

Figure 7: Comparison in terms of size distribution at 2175 m measurements between BASIL and the on-board in situ sensors.

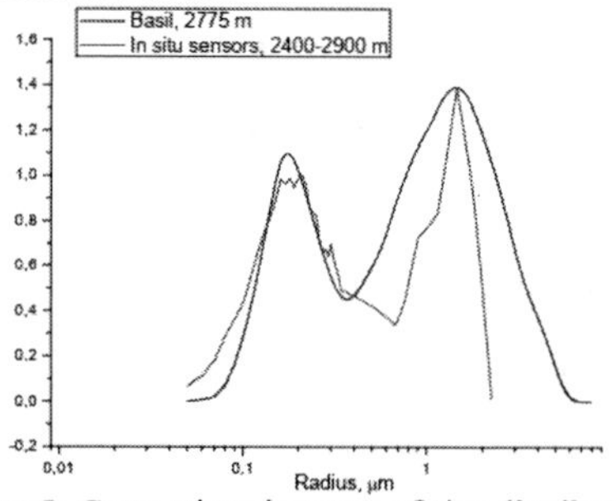

Figure 8: Comparison in terms of size distribution at 2775 m measurements between BASIL and the on-board in situ sensors.

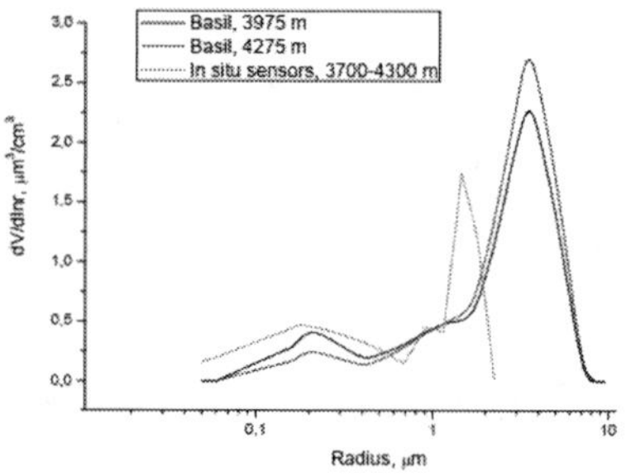

Figure 9: Comparison in terms of size distribution at 3975 m measurements between BASIL and the on-board in situ sensors.

At 2 and 2.7 km the size distribution is dominated by the continental aerosols, while at 4 km most aerosols have a marine origin and have crossed the Atlantic before reaching the site.

REFERENCES

[1] A. Behrendt, J. Reichardt, Appl. Optics 39, 1372–1378, doi: 10.1364/AO.39.001372 (2000).

[2] P. Di Girolamo, et al. Geophys. Res. Lett., 31, doi: 10.1029/2003GL018342 (2004).

[3] P. Di Girolamo, et al. Appl. Opt. 45, 2474-2494, doi:10.1364/AO.45.002474 (2006).

[4] R. Bhawar, et al. Quarterly Journal of the Royal Meteorological Society 137, 325–348, doi: 10.1002/qj.697 (2011).

[5] A. Behrendt, Weitkamp, C., Springer, New York, (2005).

[6] R. Bhawar, et al. Geophysical Research Letters 35, L04812, doi: 10.1029/2007GL032207 (2008).

[7] C. Serio, et al. Optics Express 16/20, 15816-15833, doi: 10.1364/OE.16.015816 (2008).

[8] N. Kalthoff, et al. Atmospheric Research 93, 680-694, doi: 10.1016/j.atmosres.2009.02.010 (2009).

[9] A. Macke, et al. Atmos. Chem. Phys. 17, 4887-4914, doi:10.5194/acp-17-4887-2017 (2017).

[10] I. Veselovskii, et al. Inversion with regularization for the retrieval of tropospheric aerosol parameters from multiwavelength lidar sounding, Appl. Opt., AO, 41, 3685–3699, doi:10.1364/AO.41.003685 (2002).

[11] I. Veselovskii, et al. J. Geophys. Res., 115, D21203, doi: 10.1029/2010JD014139, (2010).

[12] P. Di Girolamo, et al. Journal of Aerosol Science 26, 989-1001, doi: 10.1016/0021-8502(95)00025-8 (1995).

[13] P. Di Girolamo, et al. Appl. Opt 38(21), 4585–4595, doi: 10.1364/AO.38.004585 (1999).

[14] A. Ansmann, et al. Measurement of atmospheric aerosol extinction profiles with a Raman lidar, Opt. Lett., OL, 15(13), 746–748, doi:10.1364/OL.15.000746 (1990).

CONTINUOUS LIDAR OBSERVATION OF NEAR SURFACE AEROSOL USING OPTICAL AND SAMPLING DATA FROM GROUND-BASED INSTRUMENTS

Jamrud Aminuddin[1], Ilham Alimuddin[3], Laras Tursilowati[4], Naohiro Manago[5], and Hiroaki Kuze[2]

[1]Department of Physics, Faculty of Mathematics and Natural Science, Universitas Jenderal Soedirman, Jl. dr. Suparno 61 Karangwangkal-Purwokerto, Jawa Tengah, 53123, Indonesia.
[2]Center for Environmental Remote Sensing, Chiba University, 1-33 Yayoi-Cho, Inage-Ku, Chiba, 2630043, Japan
[3]Department of Geology, Faculty of Engineering, Universitas Hasanuddin, Jl. Perintis Kemerdekaan km 10, Makassar, Sulawesi Selatan, 90245, Indonesia.
[4]Space Science Center, Indonesian Institute of Aeronautics and Space (LAPAN), Jl. Dr. Djunjunan No.133, Bandung, Jawa Barat, 40173, Indonesia.
[5]Institute of Space and Astronautic Science, Japan Aerospace Exploration Agency, 3-1-1 Yoshinodai, Chuo-Ku, Sagamihara, Kanagawa, 2525210, Japan.
**Email: jamrud.aminuddin@unsoed.ac.id*

ABSTRACT

Monitoring of near-surface aerosol is important for both public health issues and radiation budget studies. In this study, we report a continuous observation method of aerosol particles by means of a vertical Mie-scattering lidar in combination with other optical and sampling instruments operated at the ground level. In the Fernald method used for processing the lidar signal, the most appropriate value of lidar ratio at 532 nm is estimated from the Mie-scattering calculation. The input parameters, namely, the mode radius, variance, and both real and imaginary parts of refractive index, are so determined as to reproduce the data from ground-based sampling instruments. Instead of the far-end boundary condition, the extinction coefficient at the surface level is used for constraining the retrieved aerosol extinction profile. The correction of the truncation and relative humidity (RH) effects on the scattering data from the sampling is made with the help of the optical data from a visibility-meter. We discuss the observed features in both low and high RH cases. Such a capability will be useful for uninterrupted lidar observations of near-surface aerosols irrespective of the presence of clouds that often hinders signal observations at higher altitudes where the aerosol-free atmosphere is assumed for the conventional Fernald analysis.

1. INTRODUCTION

The effects of atmospheric aerosols have been discussed in the context of both public health and the Earth's radiation budget in relation to global climate change [1]. Since most of the sources and sinks are located on the ground, the precise monitoring of aerosol behavior near the surface level is indispensable for better understanding of aerosol characteristics. Mie-scattering lidars have provided a valuable tool for studying aerosol profiles as well as the aerosol-cloud interaction. The problem in the conventional analysis based on the Fernald method [2] is that first, the value of lidar ratio (S_1) has to be assumed, and second, lidar signals at the far-end boundary is needed so that the analysis can be started at the nearly aerosol-free altitude, 5~6 km above the surface. Occasionally in the lower troposphere, the high variability of aerosol property leads to the substantial change in the lidar ratio. Besides, the presence of lower clouds often makes it impossible to observe lidar signals at higher altitudes. These problems can potentially be solved by combining the lidar data with sampling and optical data from instruments continuously operated in the observatory where the lidar setup is installed [3], [4].

Various remote sensing instruments and aerosol sampling instruments have routinely operated at the Center for Environmental Remote Sensing (CEReS), Chiba University to study the detailed properties of urban aerosol [5]–[7]. In the present paper, we report on the analysis of the data from a vertical lidar with ancillary information on aerosol observed with an integrating nephelometer (scattering coefficient), an aethalometer (absorption coefficient), and an optical particle counter (size distribution). The extinction coefficient obtained from these sampling data is compared with the in-situ optical data from a visibility-meter in order to carry out the truncation and relative humidity (RH) corrections of the sampling data [8], [9]. Then the Mie-scattering calculation is implemented to reproduce the observed aerosol properties at the surface level, and this provides us with the value of S_1 as well.

2. METHODOLOGY

2.1 Sampling instruments and Mie-scattering calculation

Near-surface values of aerosol extinction coefficient (AEC) are obtained from the ground-based sampling instruments. An integrating nephelometer (TSI 3563) provides the scattering coefficient measured at the three wavelengths of 450, 550, and 700 nm. An aethalometer (Magee, AE31) measures the values of black carbon (BC) concentration at the seven wavelengths (370, 470, 520, 590, 660, 880, and 950 nm). Then, the temporal change of AEC at the lidar wavelength (532 nm) is calculated by carrying out the interpolation based on the Angstrom exponent [6]. The value of lidar ratio can be estimated by means of the Mie-scattering calculation using aerosol parameters of the mode radius, variance, and the real and imaginary parts of refractive index. The mode radius and variance can be directly obtained from the record of a particle counter (Rion, KC22B). The values of refractive index are estimated from the condition that the observed wavelength dependence of both AEC and absorption coefficient is reproduced [6], [7].

2.2 Optical measurement

The in-situ, optical data on the ambient aerosol is obtained using a visibility-meter (Vaisala, PWD 52). This instrument yields the value of meteo-rological visibility, V, every 10 min. The AEC value at 550 nm, α_{ext}, is obtained using the Köschmeider equation of $\alpha_{ext}=\ln(20)/V-\alpha_{Ray}$ [6], where α_{Ray} is the extinction due to the Rayleigh scattering of air molecules [10].

2.2 Truncation and RH corrections

Normally the scattering coefficient derived from the nephelometer is smaller than the ambient value because of the truncation error [8]. The causes of this error are first, the loss of larger particles in the sampling process, and second, the limited angular coverage (7° to 170°) of the scattering measurement. We employ the AEC value from the visibility-meter to carry out the correction of this error. When the ambient RH is high (>~50%), an additional correction of $f(RH)$ is applied to consider the RH difference between the ambient (wet) and instrumental (dry) conditions. Here $f(RH)$ stands for the ratio of aerosol scattering cross-sections between the ambient and dry conditions [9].

2.2 Lidar system and Fernald analysis

In the present work, we use the 532 nm lidar data of the National Institute for Environmental Studies (NIES) lidar [11] operated on the campus of Chiba University. The vertical profile of AEC can be obtained every 15 min by analyzing the data using the Fernald analysis. In this analysis, the lidar ratio for air molecules (S_2) is fixed at 8.52 sr, whereas the parameter for aerosol (S_1) is determined through the Mie scattering calculation as mentioned above. Instead of the far-end boundary, near surface values of AEC is obtained from the ground-based instruments to constraint the derived profile of AEC.

3. RESULTS AND DISCUSSION

Figures 1(a) and (b) show the temporal change of AEC at 550 nm recorded during March 19-20 and May 19-21, 2017, respectively, exemplifying the lower and higher RH cases. It is seen that in both cases, the original, sampling-based AEC value is notably small as compared with the optically (visibility-meter) measured value. In the low RH case (with minimum RH of ~20%) of Fig. 1(a), the AEC value after correcting the truncation error is in good agreement with the ambient value. In the high RH case (with minimum RH of ~60%) in Fig. 1(b), on the other hand,

https://doi.org/10.1051/epjconf/202023702010

deviations are seen between the sampling and optical data when only the truncation correction is applied. A better agreement can be seen after further applying the $f(RH)$ correction.

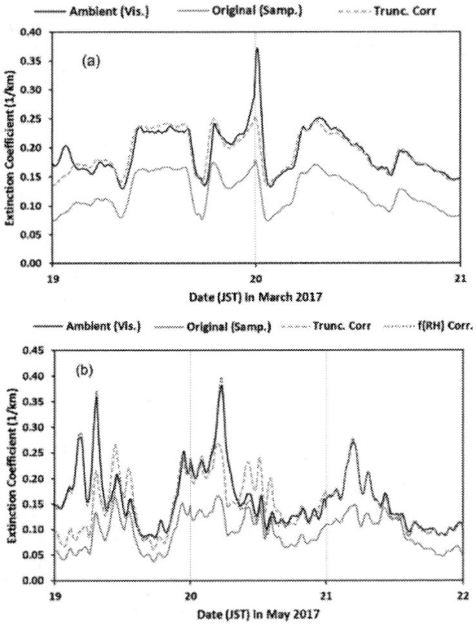

Fig. 1 Temporal change of AEC observed under a relatively (a) low RH and (b) high RH conditions. In (b), both the truncation and $f(RH)$ corrections are applied to the value from the sampling measurement.

Figure 2 shows an example of Mie calculation applied to the data from sampling instruments. The size distribution measured with the optical particle counter is fitted as shown in Fig. 2(a). The wavelength dependence of the normalized scattering coefficient and absorption coefficient is plotted in Fig. 2(b) and (c), respectively. The single scattering albedo is plotted in Fig. 2(d). The resulting Mie-scattering parameters for this case (10:00 JST on March 20, 2017) are as follows: mode radius = 0.0268 μm, variance = 0.3323, and refractive index =1.593-0.023i. Since the scattering phase function $f(\theta)$ is readily obtained from the Mie calculation, the value of S_1 can be derived as the value of $f(\pi)^{-1}$.

Figures 3(a) and (b) show the temporal change of lidar ratio at 550 nm for the case of relatively low and high RH case, respectively. Then, the lidar ratio is employed in the Fernald analysis for retrieving the AEC profiles from the NIES lidar data.

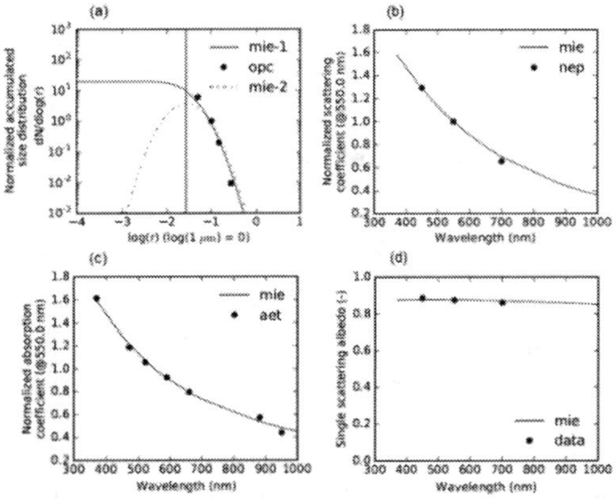

Fig. 2 Fitting of observed aerosol parameters with Mie scattering calculation for the case of data observed at 10:00 JST on March 20, 2017.

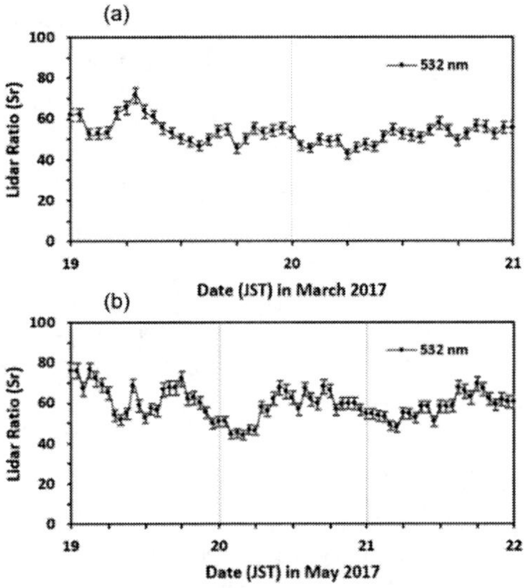

Fig. 3 Temporal change of lidar ratio in the case of (a) low RH observed during March 19-20, 2017 and (b) high RH observed during May 19-21, 2017.

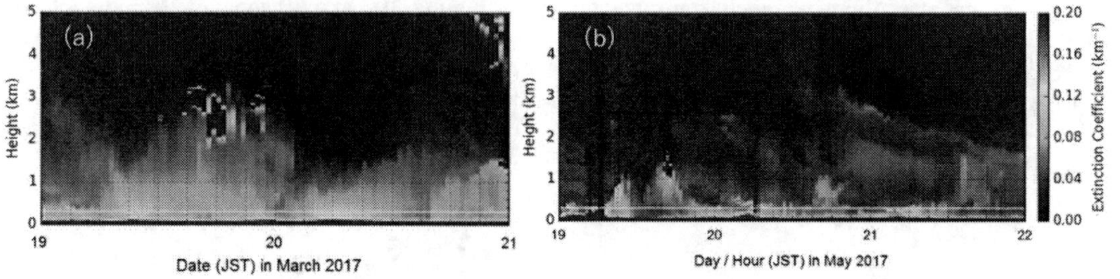

Fig. 4 Time-height indicator plot of the lidar data observed at 532 nm: (a) low RH case during March 19-20, 2017, and (b) high RH case during May 19-21, 2017.

Figures 4(a) and (b) show the time-height indicator plot of the AEC profile from the 532-nm observation data of the NIES lidar. In the May case with high RH, significant increase of AEC is seen in the morning of each day. This is caused by the increase in RH that leads to the onset of the *f*(*RH*) effect in the atmosphere of Chiba City. Similar variation can also be seen in the surface data illustrated in Fig. 1(b). Another important feature seen in Fig. 4 is that continuous profiles are observed even when relatively thick clouds appear, and hence, far-end boundary values are not available from the lidar data. Obviously, this is the advantage of our approach in that the constraint of the AEC retrieval is imposed at the surface level by utilizing the data obtained from the ground-based instruments. This method will be useful for the detailed study of aerosol-cloud interaction occurring in the lower troposphere.

ACKNOWLEDGEMENTS

The first author (J.A.) would like to thank to Universitas Jenderal Soedirman, Chiba University, ICLAS, IRC, and ILRC29 Committee for supporting in ILCR 2019.

REFERENCES

[1] IPPC, "Climate Change 2007: The Physical Science Basis," Geneva-Switzerland, 2007.

[2] F. G. Fernald, "Analysis of atmospheric lidar observations: some comments.," *Appl. Opt.*, vol. 23, no. 5, p. 652, 1984.

[3] H. Kuze, "Characterization of tropospheric aerosols by ground-based optical measurements," *SPIE Newsroom*, pp. 2–4, 2012.

[4] Y. Mabuchi, N. Manago, G. Bagtasa, H. Saitoh, N. Takeuchi, M. Yabuki, T. Shiina, and H. Kuze, "Multi-wavelength lidar system for the characterization of tropospheric aerosols and clouds," in *International Geoscience and Remote Sensing Symposium (IGARSS)*, 2012, pp. 2505–2508.

[5] H. Kinjo, H. Kuze, T. Takamura, M. Yabuki, and N. Takeuchi, "Determination of aerosol extinction-to-backscattering ratio from multiwavelength lidar observation," *Jpn. J. Appl. Phys.*, 40(1), pp. 434–440, 2001.

[6] J. Aminuddin, S. Okude, N. Lagrosas, N. Manago, and H. Kuze, "Real time derivation of atmospheric aerosol optical properties by concurrent measurements of optical and sampling instruments," *Open J. Air Pollut.*, vol. 07, no. 02, pp. 140–155, 2018.

[7] J. Aminuddin, B. Purbantoro, N. Lagrosas, N. Manago, and H. Kuze, "Landsat-8 satellite and plan position indicator lidar observations for retrieving aerosol optical properties in the lower troposphere," *Adv. Remote Sens.*, vol. 07, no. 03, pp. 183–202, 2018.

[8] T. L. Anderson and J. A. Ogren, "Determining aerosol radiative properties using the TSI 3563 integrating nephelometer," *Aerosol Sci. Technol.*, vol. 29, no. 1, pp. 57–69, 1998.

[9] P. Zieger, R. Fierz-Schmidhauser, E. Weingartner, and U. Baltensperger, "Effects of relative humidity on aerosol light scattering: Results from different European sites," *Atmos. Chem. Phys.*, vol. 13, no. 21, pp. 10609–10631, 2013.

[10] B. A. Bodhaine, N. B. Wood, E. G. Dutton, and J. R. Slusser, "On Rayleigh optical depth calculations," *J. Atmos. Ocean. Technol.*, vol. 16, no. 11 PART 2, pp. 1854–1861, 1999.

[11] A. Shimizu, N. Sugimoto, and I. Matsui, "Detailed description of data processing system for lidar network in East Asia," in *25th International Laser Radar Conference*, 2010, pp. 911–913.

DIURNAL BEHAVIOR OF AEROSOL OPTICAL PROPERTIES STUDIED WITH LIDAR AND GROUND-BASED INSTRUMENTS

Prane Mariel Ong[1, 2*], Nofel Lagrosas[2], Tatsuo Shiina[1], Hiroaki Kuze[1,2]

[1]*Graduate School of Science and Engineering, Chiba University, 1-33 Yayoi-cho, Inage-ku, Chiba, 263-8522, Japan*
[2]*Center for Environmental Remote Sensing (CEReS), Chiba University, 1-33 Yayoi-cho, Inage-ku, Chiba 263-8522, Japan*
**Email: prane.ong@chiba-u.jp*

ABSTRACT

The combined use of remote sensing and in-situ monitoring instruments could help improve the assessment of near-surface aerosol properties. In this paper, we analyze the diurnal behavior of aerosol extinction coefficients, $\alpha_{Ext}(\lambda)$, at λ=349 and 550 nm using a lidar and a present weather detector, respectively. We utilize the aerosol optical thickness (AOT), single scattering albedo (SSA), and Ångström exponent (AE) from SKYNET sky radiometer, and AE from aethalometer, and the number distribution from optical particle counter to evaluate the effect of relative humidity (RH) on aerosol extinction coefficients. It is found that although $\alpha_{Ext}(\lambda)$ often exhibits a positive correlation with the ambient RH, this relation is obscured when both the number distribution and particle size change simultaneously. Moreover, α_{Ext} at 349 nm is more sensitive to this change than at 550 nm.

1. INTRODUCTION

Aerosol plays an important role in the Earth climate change. One parameter that controls visual air quality is the aerosol extinction coefficient. Numerous studies have been performed on relatively higher altitudes, but studies in the near-range are still lacking. In this report, we examine the diurnal behavior of aerosol extinction coefficient by focusing on the lower troposphere where most of the sources and sinks of aerosol are located. The values of aerosol extinction coefficient, α_{Ext} (λ), are derived from a lidar system with a low elevation angle and a visibility meter. Then, they are related to other aerosol optical properties, particle size and number distribution. Synergetic use of the data from these remote sensing and sampling instruments could help improve the assessment of aerosol radiative forcing to climate [1-2].

2. METHODOLOGY

A campaign was conducted in the whole month of November 2017 using a near-horizontal lidar system, present weather detector, sky radiometer (SKYNET), aethalometer, optical particle counter, and weather monitor. With these simultaneous measurements, it enables us to investigate the behavior of near-ground aerosol optical properties and particle size. In this paper, we will limit the analysis to the diurnal behavior of aerosols observed under clear and cloudy sky conditions.

Most of the instruments are located on the rooftop of the Faculty of Engineering Research Building in Chiba University in Japan (35.62°N, 140.10°E), about 65 m above seal level. The campus is in the urban area of Chiba City, along the east side of the Tokyo Bay.

2.1 Lidar system

The lidar system is pointed in the direction of 10° east of north with an elevation angle of 4°. Although the system is originally a plan position indicator (PPI) lidar that can perform 360° azimuthal rotation [3], it was operated in a static mode during this November 2017 campaign. Figure 1 shows the schematic diagram of the lidar system. It is based on a diode-laser pumped Nd:YLF laser emitting at 349 nm wavelength with 60 μJ pulse energy and 1 kHz pulse repetition rate. A photomultiplier tube (PMT, Hamamatsu H10304-00-NN) is used as the receiver sensor. The PMT is connected to a transient recorder (Licel, TR20-160) for data recording of the backscattered signal every 5 min. The obtained data are used to calculate the aerosol extinction coefficient (α_{Ext}) using the adaptive slope method [4-5]. This method is used on the premise that a well-mixed layer of aerosol exists below the maximum covered altitude of 350 m (at 5 km range).

2.2. Other Instruments

Meteorological optical range (MOR) data are routinely obtained every minute using a visibility-meter (Vaisala, PWD52). The measurement is based on the detection of forward scattering at 45° scattering angle. The visibility, V, which is automatically converted by the system to the 550 nm wavelength equivalent, is used to calculate the aerosol extinction coefficient at 550 nm using the relation of $\alpha_{Ext} = \ln(20)/V$, assuming a contrast ratio of 5% for the meteorological visibility.

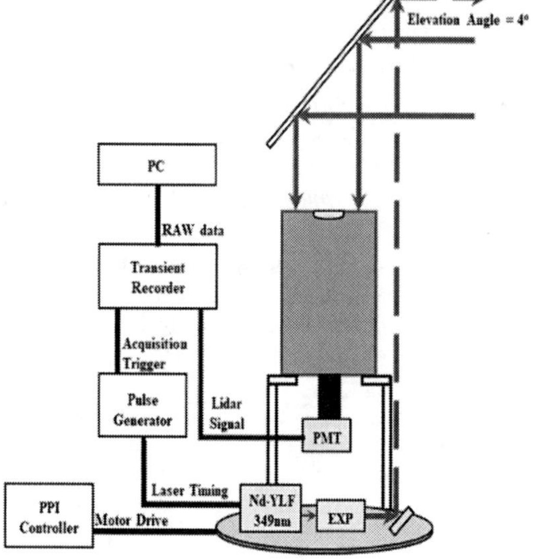

Fig. 1. Schematic diagram of the lidar system.

Table 1. Summary of instruments with their list of products, wavelengths and time resolution.

List of Instruments	List of Products	Wavelengths	Time Resolution
Aethalometer (Magee AE31)	Equivalent Black Carbon (EBC)	**370**, 470, **520**, 590, 660, 880, 950 nm	5 min
Aethalometer (Magee AE31)	Ångström Exponent (AE)	470-880 nm	5 min
Lidar	Extinction Coefficient (α_{Ext})	349 nm	5 min
Present Weather Detector (PWD52)	Visibility Range, Extinction Coefficient (α_{Ext})	550 nm	5 min
Optical Particle Counter (OPC, Rion KC-01E)	Count/L and dN/dlnR (0.3, 0.5, 1.0, **2.0, 5.0** µm particle size)	-	10 min
Sky Radiometer (POM-01, SKYNET)	Aerosol Optical Thickness (AOT)*	**340**, 380, 400, **500**, 675, 870, 1020 nm	10 min
Sky Radiometer (POM-01, SKYNET)	Ångström Exponent (AE)*	400-870 nm	10 min
Sky Radiometer (POM-01, SKYNET)	Single Scattering Albedo (SSA)*	**340**, 380, 400, **500**, 675, 870, 1020 nm	10 min
Weather Monitor (Davis Vantage Pro2)	Relative Humidity (RH), Wind Speed and Velocity, Rain Rate	-	5 min

*in clear sky condition

A ground-based, seven-wavelength sky radiometer (POM-02, Prede) of SKYNET (http://atmos3.cr.chiba-u.jp/skynet/chiba/chiba.html) is routinely operated on the same rooftop along with the lidar and the PWD52 systems. It measures the intensities of the direct solar irradiance and the diffuse solar radiance distribution. Among its products are the aerosol optical thickness (AOT), the single scattering albedo (SSA) at 340, 380, 400, 500, 675, 870, 1020 nm) and the Ångström exponent (AE, 400-870 nm). For AOT and SSA, we only consider the wavelengths 340 nm and 500 nm, which are close to the lidar (349 nm) and PWD52 (550 nm) wavelengths.

The maximum value of SSA is unity, which indicates pure scattering, while the value less than one represents the contribution of absorption. For 400-870 nm, AE values below ~1.0 indicate mainly coarse particles (radius, $r > 1$ µm), while AE values above ~1.5 indicate mainly fine particles ($r < 1$ µm). Fukagawa et al. [5] analyzed the seasonal variation of AOT and AE in Chiba during 1999-2005. They reported that the AOT in winter is relatively small (< 0.2), though occasionally the formation of inversion layer leads to air pollution inside the atmospheric boundary layer [6].

Sampling instruments such as an aethalometer (Magee, AE31) and an optical particle counter (OPC; Rion, KC-01E) are also utilized to observe the absorptive property of particles and particle size distributions, respectively. The former is a seven wavelength (370, 470, 520, 590, 660, 880, 950 nm) aethalometer that measures the elemental black carbon (EBC) concentration, which then is used to estimate the absorption coefficient (α_{Abs}) as

$$\alpha_{\text{Abs}} = \text{EBC}\left(\frac{14625}{C_{ref}\lambda}\right). \tag{1}$$

Here, C_{ref} (= 2.14) is the correlation factor due to multiple scattering effects in the fiber filter inside the instrument. For wavelength range between 470 and 880 nm, the AE can be computed using the following linear regression between $\ln(\alpha_{\text{Abs}})$ and $\ln(\lambda)$ [7]:

$$\text{AE} = -\text{dln}[\alpha_{\text{Abs}}(\lambda)]/\text{dln}(\lambda)\ . \tag{2}$$

The site is also equipped with a weather monitor (Davis, Vantage Pro 2) that constantly measures the ambient weather parameters including temperature, relative humidity (RH), rain rate, wind speed and direction every 5 min. Table 1 summarizes all these instruments with their products and temporal resolutions.

3. RESULTS AND DISCUSSION

Figure 2 shows examples of the diurnal variation of aerosol optical properties with respect to a (a) clear sky, and (b) cloudy sky condition observed on November 29 and 20, respectively. On both days, low relative humidity (RH) was observed in mid-afternoon and high RH in the nighttime to early morning. The AOT values at 340 and 500 nm were below 0.35 and 0.21, respectively. In both cases, the wind speed was limited (< 2 m·s^{-1}).

 https://doi.org/10.1051/epjconf/202023702011

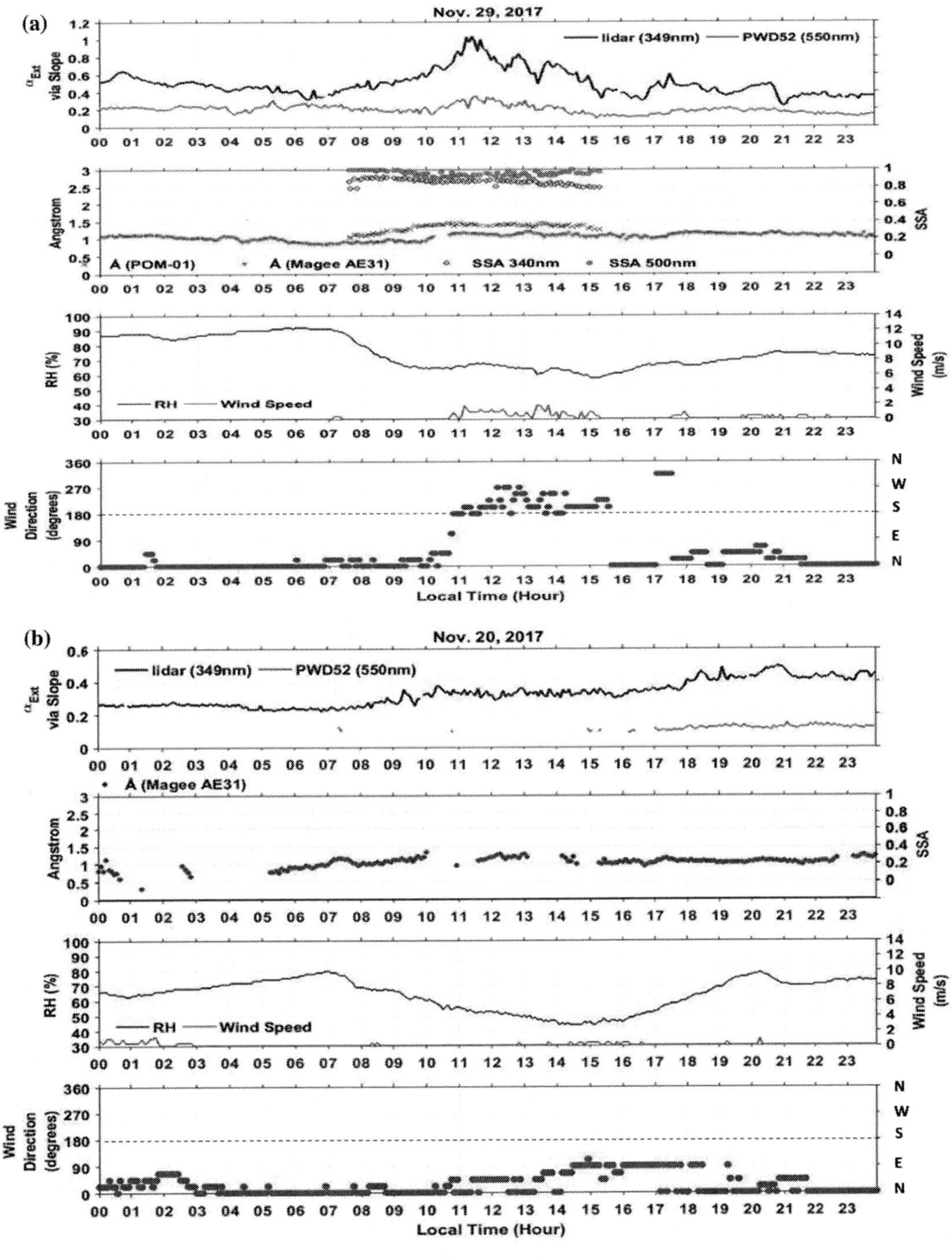

Fig. 2. Example of the diurnal cycle of aerosol optical properties in (a) clear with low visibility sky condition on November 29, and (b) cloudy sky condition on November 20.

3.1 Clear Sky (November 29, 2017)

In the early morning with a constant high RH (~90%), a decreasing trend is seen for $\alpha_{Ext,Lidar}$ (349 nm), while almost no change in $\alpha_{Ext,PWD}$ (550 nm). From the OPC measurement, on the other hand, there is an apparent decrease in the coarse particles while the fine particles do not vary that much. At around 07:00 JST, as the RH starts to decrease, $\alpha_{Ext,PWD}$ decreases as well, but $\alpha_{Ext,Lidar}$ increases. This opposite reaction between the visibility-meter and lidar data may be attributed to the increase in the EBC concentration and hence, the increase in AE. Moreover, traffic usually increases in the surrounding area of the measurement site during this time.

Another interesting observation is during the shift in the wind direction from north to south (11:00 JST), resulting in decreases in SSA and AE over time. A decrease of fine particle concentration and constant high concentration of coarse particles are also observed. Although both $\alpha_{Ext,Lidar}$ and $\alpha_{Ext,PWD}$ have negative slopes, $\alpha_{Ext,Lidar}$ is more negatively sloped.

3.2 Cloudy Sky (November 20, 2017)

There are no SKYNET products during this day due to the cloud cover. The diurnal cycle of $\alpha_{Ext,Lidar}$ is almost in-phase with the increase/decrease in RH. The only exception of this trend is around 07:00 JST when a sudden increase of fine particles is observed in conjunction with the onset of the decrease in RH.

Figure 3 shows the change of the α_{Ext} observed with the lidar (349 nm) and present-weather detector (550 nm) plotted against the number concentrations at a geometric mean radius of (a) 0.387 μm for fine particles and (b) 3.162 μm for coarse particles, both for clean and cloudy conditions. This result also shows that more sensitive change of α_{Ext} is seen at the UV wavelength, for both fine and coarse particles under both sky conditions.

4. SUMMARY AND CONCLUSION

We have derived the aerosol optical properties near the surface region from different instruments and analyzed their relationship. We observed that the change in α_{Ext} is brought about either by the change in the ambient RH, or that in the particle size and number distribution. Moreover, for the cloudy sky condition without the SKYNET products, we can still infer similar observation from aethalometer and OPC. In addition, α_{Ext} at 349 nm is more sensitive to this change than at 550 nm. Thus, the present approach based on the near-horizontal lidar observation will be useful for characterizing near-surface aerosols under cloudy or nighttime conditions, in which the conventional methodology based on the sunphotometer or sky radiometer cannot be undertaken.

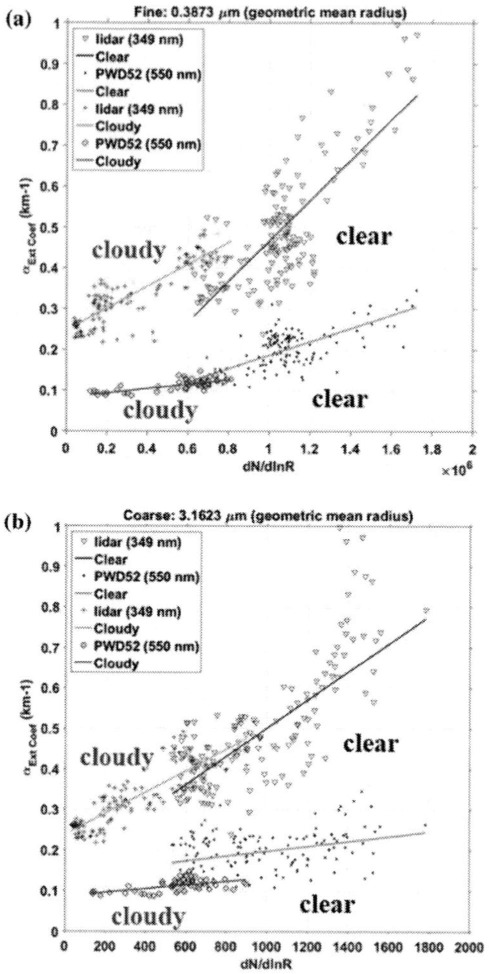

Fig. 3. Extinction coefficient at 349 nm (lidar) and 550 nm (PWD52) versus the number concentration (OPC) for (a) fine and (b) coarse particles.

REFERENCES

[1] IPCC: In: Climate change 2013: The Physical Science Basis, Cambridge, United Kingdom: Cambridge University Press; 2013.
[2] Gebhari, K.A., et al.: Atmos. Environ. 35 pp5177-5191, 2001.
[3] Kamiya, Y., Mabuchi, Y., Haji, S., Saito, H., Bagtasa, G., Manago, N., Irie, H. and Kuze, H., ISRS2013, P1-06 1462, Shin-Narashino, May 2013.
[4] Kunz, G.J., and Leeuw G.D., Applied Optics, 32(18) pp3249-3256, 1993.
[5] Yu-Bing, P., et al., Atmos. Sci. and Oceanic Lett., 7(3) pp203-209, 2014.
[6] Fukagawa, S. el al., Atmos. Environ. 40 pp2160-2168, 2006.
[7] Ångström, A., Tellus, Ser. A, 16(1), 64–75, 1964, doi:10.1111/j.2153-3490.1964.tb00144.x.

DUAL-WAVELENGTH HIGH-SPECTRAL-RESOLUTION LIDAR FOR PROFILING OPTICAL PROPERTIES OF AEROSOL AND CLOUD

Xue Shen[1], Nanchao Wang[1], *Dong Liu[1], Da Xiao[1], Yuhang Rong[1], Tianfen Zhong[1], Chong Liu[1], Yupeng Zhang[1], Yudi Zhou[1], Sijie Chen[1]

[1] *State Key Laboratory of Modern Optical Instrumentation, College of Optical Science and Engineering, Zhejiang University, Hangzhou 310027, China*
*Email: *liudongopt@zju.edu.cn*

ABSTRACT

A dual-wavelength high-spectral-resolution lidar (HSRL) based on an iodine absorption filter and a field-widened Michelson interferometer (FWMI) has been developed to profile backscatter and extinction coefficients of aerosols and clouds accurately. This instrument was tested and calibrated on multiple observations in Hangzhou and Zhoushan, respectively, from August 2018 to April 2019. This paper discusses the design and the internal calibration method of the lidar system in detail, with several typical cases of observations and the analysis of these data products. The optical properties of urban aerosols in Hangzhou and the evolvement of clouds in Zhoushan are presented, respectively.

1. INTRODUCTION

There is a strong need to make more observations on the vertical distribution and optical properties of aerosols and clouds. Both aerosols and clouds play significant roles in influencing the Earth's climate, but our limited understanding of them leads to very large uncertainties in our ability to estimate the climate changes. The vertical distribution and optical properties of clouds contribute significantly to the climate models. As for aerosols, more accurate estimates of their impact on climate require global monitoring of aerosol scattering and absorption during their production and transportation. Furthermore, the vertical distribution of the aerosols can induce aerosol-cloud interaction. Unfortunately, these parameters are difficult to estimate from passive satellite data. On the other hand, Lidar is widely recognized as a necessary component in any strategy to provide the information on aerosol and cloud distribution and optical properties required to address these outstanding issues [1, 2].

Using a conventional backscatter lidar, the optical properties cannot be measured directly, only with assumption of the extinction-to-backscatter ratio, the so-called lidar ratio. However, the lidar ratio is a highly variable quantity in atmosphere, so that large errors in the retrieval are expected. Using a high spectral resolution lidar (HSRL) is a good way to solve this problem. HSRL takes advantage of the different spectral broadening of light, backscattered by molecules and aerosols. By means of a narrow bandwidth optical filter the aerosol contribution is separated from the molecular backscatter. Thus, backscatter and extinction coefficients can be measured directly and no assumption about the lidar ratio is required [3]. Furthermore, our preliminary study shows that the parameters such as spectral depolarization ratio and color ratio can be obtained by adding a lidar of another wavelength on the basis of a single-wavelength HSRL, which can reduce the discrimination error of component and intensity of aerosols [4]. Therefore, a dual-wavelength HSRL is of great significance for the determination of the vertical structure and evolution of clouds and aerosols.

This paper introduces the first dual-wavelength HSRL independently developed by Zhejiang University. The system design and calibration method are introduced at first. Some typical results of the initial observation are subsequently presented and analyzed, which demonstrate the system's performance.

2. METHODOLOGY

2.1 The lidar system

The developed dual-wavelength HSRL system mainly includes transmitting part and receiving part. The transmitter used for the HSRL measurements is a injection seeded Nd: YAG laser with a narrow-linewidth CW laser. It has pulse energies of 100 mJ at 532 nm, 150 mJ at 1064 nm, and a laser repetition rate of 10 Hz. The atmospheric backscatter is collected by means of a 280 mm Cassegrain telescope. A field stop within the focal plane of the telescope limits the field of view to 0.36 mrad. Dichroic beam splitters are used to spectrally separate the backscatter signals at 1064 nm and 532 nm. The 532 nm subsystem is filtered with an interference filter of 0.3 nm bandwidth to suppress sunlight, and then use an iodine filter to suppress particle-dominated components for the HSRL measurement. At 1064 nm, a pressure-tuned field-widened Michelson interferometer (FWMI) is used. Figure 1(a) shows the picture of HSRL's receiving system. The entire lidar system is deployed in an experimental container, as Fig. 1(b) shown, which can provide good working conditions for HSRL.

Fig.1. (a) The dual-wavelength high spectral resolution lidar, (b) the experimental container deployed in Zhoushan to accomplish the first field observation.

2.2 Calibration approach

In the polarized HSRL, several calibrations should be made before the instrument operation. The first one is the gain ratio of polarization channels. We use a common approach called the ' ±45° calibration method', i.e., rotating the polarization direction of laser beam with respect to the polarized beam splitter, and recording the signals of the detector in the parallel channel and perpendicular channel [5].The second calibration is on the gain ratio of molecular channel with the background transmittance of narrowband discriminators. We prepare a laser beam has the same properties as the emitted laser but with multiple attenuations of energy, straight into the telescope. The frequency of the laser beam should be scanned on, and the largest ratio of molecular signal to combined signal is the target value. At the same time, the spectrum transmittance of the discriminator can be measured. The third calibration is signal transmittance of Rayleigh and Mie can be calculated with the spectrum transmittance measured. This method was inspired by NASA's calibration method using backscattered signal of hard target [6]. The fifth calibration is the overlap function of the HSRL. Since the HSRL instrument we built cannot observe the atmosphere horizontally, the commonly used 'horizontal calibration' for overlap factor would be unavailable. We add an extra backscatter lidar receiver besides HSRL to measure system's overlap function [7]. This is an ingenious design because there is no need to calibrate the difference between the two systems according to HSRL's algorithm. Overall, lidar calibration is an interesting and challenging topic, which needs more researches to be done in order to improve the retrieval accuracy.

3. RESULTS

3.1 Aerosol optical properties

Figure 2 shows the retrieval results of depolarization ratio and lidar ratio of aerosol measured in Hangzhou. These images have a temporal resolution of 100s and a spatial resolution of 7.5m. The aerosol layer extends to 3km, the depolarization ratio value is around 0.1, and the lidar ratio is around 40 sr, which fits the range of urban aerosol optical properties in the literature [4].

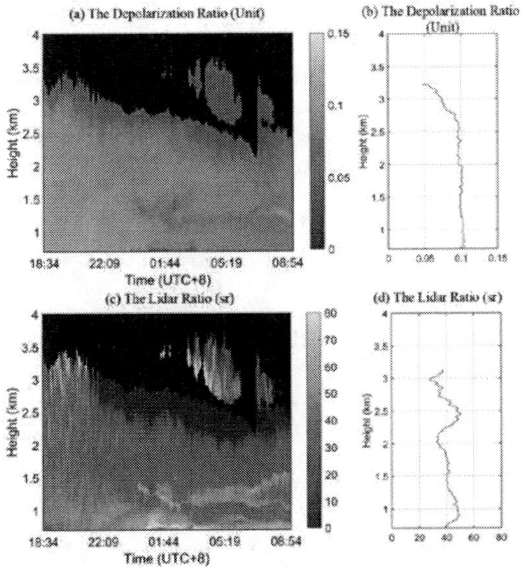

Fig.2. Measured aerosol's depolarization ratio (a & b) and lidar ratio (c & d) in Hangzhou from 18:34 pm (UTC + 8), 12 March 2019 to 8:54 am (UTC + 8), 13 March 2019.

3.2 Cloud optical properties

Figure 3 shows the retrieval results of the cloud evolution process before a rainfall at Dongshawan beach in Zhoushan. This image has a temporal resolution of 100s and a spatial resolution of 7.5m. As the figure shows, there is a relatively obvious cloud signal below 14km. As the rain approaches, the cloud gradually falls, and the air humidity can be detected to gradually increase. When the rain falls at 8:23 am on the second day, the cloud drops to the vicinity of 3km, then the lidar system is shut down.

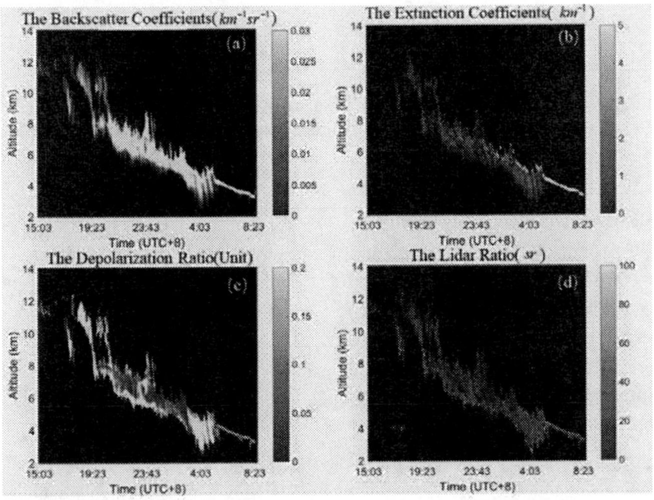

Fig.3. The retrieval results of cloud evolution in Zhoushan from 15:03pm, 15 April 2019 to 8:23am, 16 April 2019. These four pictures are: (a) the backscatter coefficients; (b) the extinction coefficients; (c) the depolarization ratio; (d) the extinction-to-backscatter ratio.

 https://doi.org/10.1051/epjconf/202023702012

Figure 3 presents data products for the field experiment, which demonstrate the power of the HSRL for observing the spatial distribution of cloud, quantifying cloud optical parameters, and inferring cloud type. The extensive parameters measured with the high spectral resolution technique are backscatter and extinction coefficients at 532nm, and are shown in Figs. 3(a) and 3(b). The backscatter data is averaged over 100s and 7.5m in altitude, and the extinction is analyzed over 100s and 7.5m in altitude. The intensive parameters measured by the HSRL instrument are shown in Figs. 3(c) and 3(d) and include the depolarization ratio at 532nm (100s, 7.5m), the extinction-to-backscatter ratio (Sa) (100s, 7.5m). As depicted by Fig.3, in the evening of April 15, 2019, rain-producing clouds form below the cirrus clouds, along with the depolarization ratio of cloud layer tends to decrease, and the backscatter coefficient increases gradually [8]. By the morning of April 16, 2019, just before the rain, clouds were gathering around 3km above the ground. The backscattering of rain-producing clouds is extremely large, and the lidar ratio is between 15 sr and 20 sr. The results shown in Fig. 3 are reasonable. Figure 4 shows the profile of cirrus clouds. At this time, there is a strong backscattering signal at an altitude of 10-12km. The backscattering coefficient reaches up to $0.02km^{-1}sr^{-1}$, and the maximum scattering ratio is up to 60, while the depolarization ratio is as much as 0.3, the lidar ratio varies from 5 sr to 10 sr, which conforms to the reported range of cirrus clouds' optical properties.

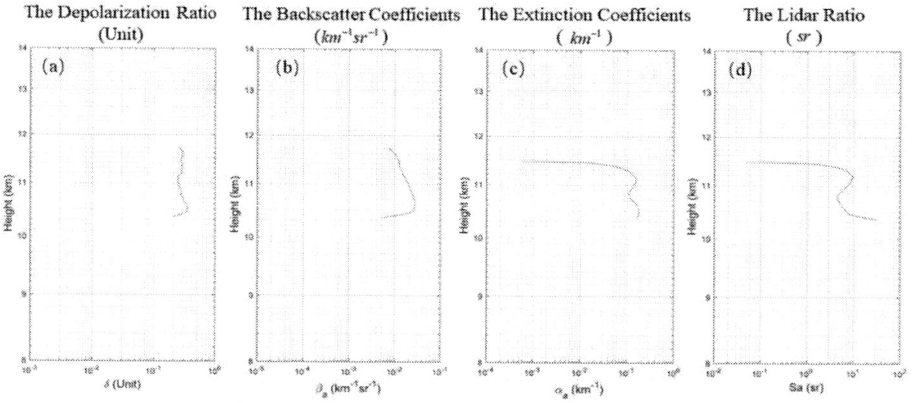

Fig.4. Retrieval profiles of cirrus clouds at an altitude of 10-12km around 18:35pm, 15 April 2019. These four pictures are: (a) the depolarization ratio; (b) the backscatter coefficients; (c) the extinction coefficients; (d) the extinction-to-backscatter ratio.

ACKNOWLEDGEMENTS

This work was supported by the National Key Research and Development Program of China (2016YFC1400900, 2016YFC0200700); National Natural Science Foundation of China (NSFC) (41775023); Excellent Young Scientist Program of Zhejiang Provincial Natural Science of China (LR19D050001); Fundamental Research Funds for the Central Universities; State Key Laboratory of Modern Optical Instrumentation Innovation Program.

REFERENCES

[1] Hair J W, Hostetler C A, Cook A L, et al. Applied Optics, 47(36):6734-52 (2008).

[2] Francois-Marie Bréon. Science, 313(5787):623-624 (2006).

[3] Esselborn M, Wirth M, Fix A, et al. Applied Optics, 47(3):346-358 (2008).

[4] Dong Liu, et al. Journal of Applied Remote Sensing, 9(1):096006 (2015).

[5] B. Liu, and Z. E. Wang, Optics express **21**, 14583-14590 (2013).

[6] S. P. Burton, C. A. Hostetler, et al. Appl. Opt. 57, 6061-6075 (2018)

[7] Juan Luis Guerrero-Rascado, et al. Opt. Express 18, 20350-20369 (2010)

[8] S. Groß, et al. Atmospheric Chemistry and Physics 13.5(2013).

FEASIBILITY STUDIES OF THE THREE-WAVELENGTH MIE-SCATTERING POLARIZATION SCHEIMPFLUG LIDAR TECHNIQUE

Zheng Kong[1], Teng Ma[1], Zhenfeng Gong[1], Kun Liu[1], Liang Mei[1,*]

[1]*School of Optoelectronic Engineering and Instrumentation Science, Dalian University of Technology, Dalian 116024, China*
**Email: meiliang@dlut.edu.cn*

ABSTRACT

A three-wavelength Mie-scattering polarization Scheimpflug lidar system, utilizing 808-nm, 520-nm and 405-nm multimode laser diodes as light sources and two CMOS sensors as detectors, is developed for the studies of the aerosol extinction coefficient, depolarization ratio and the Ångström exponent. Atmospheric monitoring has been carried out on a near horizontal path from 23:00 January 14th to 06:00 January 15th, 2019 at Dalian, which is a coast city in Northern China. By studying the depolarization, aerosol extinction coefficient and Ångström exponent, it has been found out that a strong north wind blew away local spherical haze particles and brought external non-spherical large-size particles. The measurement results indicated a promising future of employing the present three-wavelength polarization Scheimpflug lidar system in the applications of atmospheric remote sensing.

Keywords: Three-wavelength, Scheimpflug lidar, laser diode, aerosol, polarization, remote sensing.

1. INTRODUCTION

Mie-scattering atmospheric lidar is a very useful remote sensing tool and has been extensively used to investigate clouds, aerosol and boundary layer for decades. The conventional lidar technique is mainly based on the principle of photon time of flight, measuring the atmospheric backscattering signal of the transmitted laser pulse according to the time-of-arrival. The most common laser source used in the pulsed lidar systems is the Nd:YAG laser, which has a fundamental emission at 1064 nm and can be frequency doubled and tripled to 532 nm and 355 nm, respectively [1-3]. The recently developed Scheimpflug lidar (SLidar) technique has also shown a great potential in the applications of atmospheric remote sensing [4-6]. The Scheimpflug principle reveals that the laser beam transmitted into atmosphere can be in-focused on a tilted image sensor while employing a large-aperture telescope. In the present work, we developed a three-wavelength Mie-scattering polarization Scheimpflug lidar system based on 405-nm, 520-nm and 808-nm high-power continuous-wave laser diodes and two CMOS cameras. Atmospheric measurements have been performed to investigate the performance and feasibility of the lidar system in the applications of atmospheric aerosol monitoring in Dalian, which is a coast city in Northern China.

2. EXPERIMENTAL SYSTEM

The schematic of the three-wavelength Mie-scattering polarization SLidar system is shown in Fig. 1. Two 808-nm laser diodes, a 405-nm laser diode and a 520-nm laser diode were used as light sources. Each laser diode was mounted on a copper sheet with a thermoelectric cooler to control the temperature.

The depolarization ratio measurement was achieved based on a time-multiplexing detection scheme, by employing two 808-nm laser diodes (LD I, LD II) with the same polarization states as light sources and a single CMOS camera as the detector. A half-wave plate was used to rotate the plane of polarization of LD II by 90°. The two orthogonal polarized 808-nm laser beams were combined by a polarization beam-splitter cube. A photodetector was utilized to monitor the intensities of residual beams, in order to monitor the variation of the two 808-nm diode lasers.

As laser beams emitted by the 405-nm and 520-nm laser diodes (LD III and LD IV) have much larger divergences, a F6 lens will truncate the laser beam significantly, leading to a very low geometrical transmission efficiency of about 25%. Thus, a cylindrical lens pair was designed to reduce the divergence along the fast axis and placed in front of the 405-nm and the 520-nm laser diode to improve the transmission efficiency [7,8]. The 405-nm and 520-nm laser beams were first combined by a dichroic mirror and then coupled together with the 808-nm laser beams.

https://doi.org/10.1051/epjconf/202023702013

The joint laser beams were reflected by an elliptical mirror and then transmitted into the atmosphere through a F6 refractor. Each laser diode was mounted on an x-y-z linear stage for focal and position adjustments.

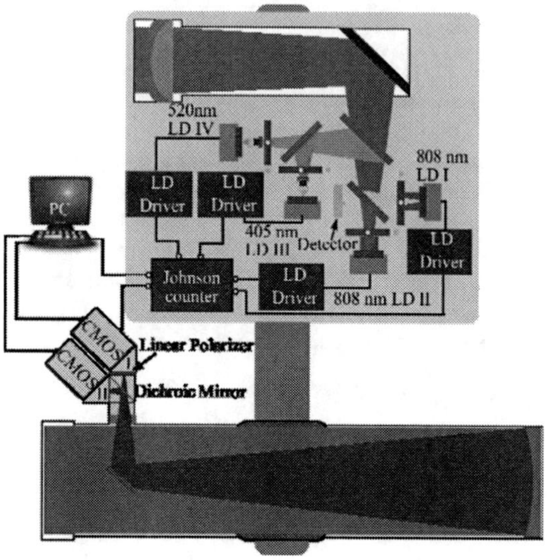

Fig. 1. Schematic of the three-wavelength Mie-scattering polarization Scheimpflug lidar system. LD I: 808 nm, 4 W; LD II: 808 nm, 4 W; LD III: 405 nm, 0.6W; LD IV:520 nm, 1W; Laser diode collimator: f=600 mm, Ø104 mm; Newtonian receiver: F4, *f*=800 mm, Ø200 mm; Separation between the receiver and the collimator is about 806 mm, the tilt angle of the CMOS sensor is 45°.

The backscattering signals were collected by a Newtonian telescope and detected by two 45° tilted CMOS cameras. The short-wavelength backscattering signals (405 nm and 520 nm) were separated from the near infrared signals by a dichroic mirror. The CMOS camera I was employed to detect the 808-nm parallel and perpendicular polarized backscattering signals based on a time-multiplexing detection scheme [9]. A linear polarizer, with the polarization axis parallel to the polarization state of LD I, is mounted in front of the CMOS camera I. As the two 808-nm laser diodes are turned on and off sequentially, the CMOS camera I is able to detect the parallel and perpendicular polarized backscattering signals successively. A dark frame was also acquired, when the two laser diodes were both turned off, to eliminate background signals. The CMOS II camera detected the 405-nm and 520-nm backscattering signals based on the time-multiplexing detection scheme.

The raw lidar signals were obtained by performing vertical pixel binning, background subtraction and median averaging. The total measurement time for all lidar curves is 1 minute, when performing 100-time signal averaging with 200-ms exposure time.

3. MEASUREMENT AND RESULTS

The three-wavelength polarization SLidar system was placed on the 8th floor of the Graduate School of Education Building, Dalian University of Technology (DLUT) and the laser beam was about 27-m height above the outside ground. The lidar system was pointed to the south direction on a near horizontal path. Since no interference filters have been employed to suppress the sunlight background, the lidar system is only able to operate during nighttime. Thus, the measurement campaign was carried out from 23:00 January 14th to 06:00 January 15th, 2019 in Dalian, China (38.886°N, 121.532°E).

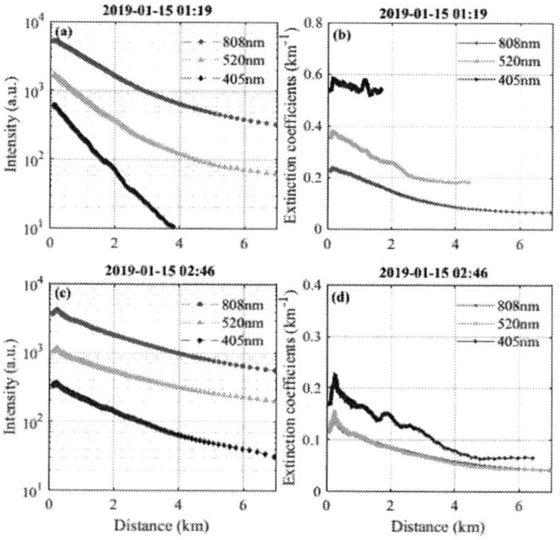

Fig. 2. (a) Atmospheric backscattering signals and (b) the aerosol extinction coefficients measured at 01:19 on January 15th, 2019; (c) Atmospheric backscattering signals and (d) the aerosol extinction coefficients measured at 02:46 on January 15th 2019.

The backscattering signals measured under two typical atmospheric conditions as well as the corresponding aerosol extinction coefficient retrieved based on the Fernald method are shown in Fig. 1 [10]. The lidar ratios for 808 nm, 520 nm and 405 nm are set to 40 sr, 50 sr and 50 sr, respectively [11-14]. Fig. 1(a) and (b) were obtained during the haze while Fig. 1 (c) and (d) were acquired when the haze has been dissipated. Among all the backscattering signals, the 808-nm channel had the strongest intensity while the 405-nm channel was the weakest one, mainly attributed to the power differences of laser diodes. In the meanwhile, the atmospheric extinction coefficient was generally larger in short wavelengths. Figure 3 shows the space-time evolution of the extinction coefficient at 808 nm, which clearly illustrated the whole process of the transportation and the dissipation of the haze.

The depolarization ratio $\delta(z)$ is given by

$$\delta(z)=k\frac{P_s(z)}{P_p(z)} \qquad (1)$$

Here k is a polarization-dependent system constant of the lidar system and its value was 1.02, calibrated by measuring backscattering signals without the linear polarizer. $P_s(z)$ and $P_p(z)$ are the s- and p-polarized lidar signals detected by the polarization Scheimpflug lidar system.

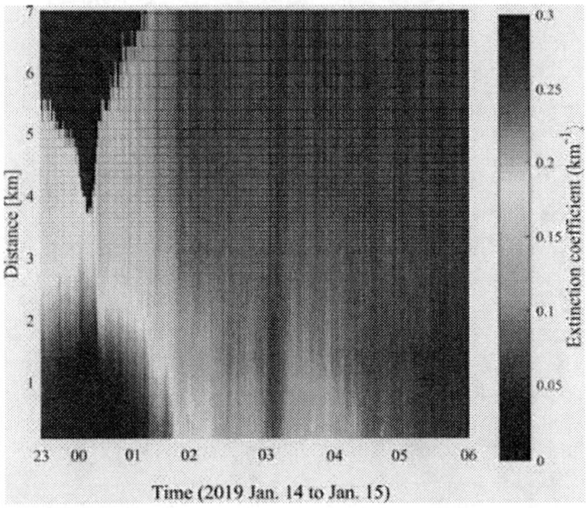

Fig. 3. Time-range map of aerosol extinction coefficients retrieved by the Fernald method at 808 nm.

The time-space map of the depolarization ratio is shown in Fig. 4. The spatial variation of the depolarization ratio was not significant since it was measured on a near horizontal path, but the temporal change was obvious. The Ångström exponent, often used as a qualitative indicator of aerosol particle size [15-17], is also evaluated based on the lidar signals measured at 808 nm and 520 nm. The spatial-temporal variation of the Ångström exponent is shown in Fig. 5.

According to the space-time variation of the 808-nm extinction coefficient, a severe haze can be clearly observed during 23:00 January 14th to 01:20 January 15th. The PM2.5 and PM10 concentrations were up to 90 μg/m^3 and 148 μg/m^3, respectively, reported by the Qixianling national pollution monitoring station. The RH was larger than 70%. The depolarization ratio was 0.09±0.02 and the Ångström exponent was about 1.0.

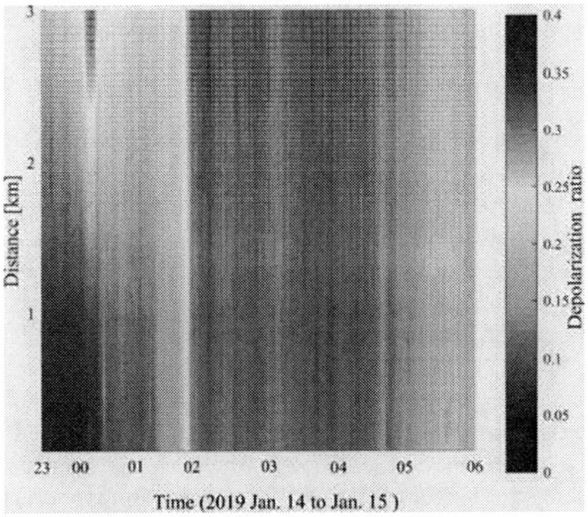

Fig. 4. Time-range map of depolarization ratio.

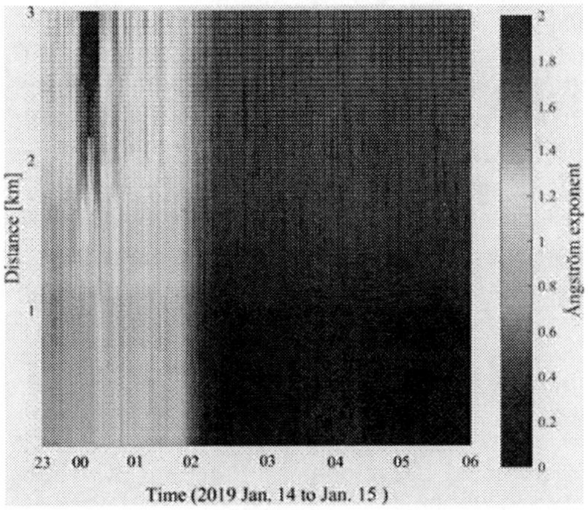

Fig. 5. Time-range map of the Ångström exponent at 808 nm and 520 nm.

From 01:20 January 15th to 02:00 January 15th, the local haze started to dissipate while the external pollution appeared due to the strong north wind. Correspondingly, the value of the extinction coefficient was decreasing, the depolarization ratio increased from 0.1 to 0.3 and the Ångström exponent decreased from 1.0 to 0, which implied that the relative concentration of spherical particles gradually decreased and the atmosphere was dominated by non-spherical large-size dust particles [18-22]. During 02:00 to 06:00 on January 15th, the concentration of PM10 was beyond 100 $\mu g/m^3$ and the concentration of PM2.5 drops to below 20 $\mu g/m^3$, indicating that large-size particles dominated the atmosphere, which was in good agreement with the Ångström exponent.

4. CONCLUSION

This work has developed a three-wavelength polarization Scheimpflug lidar system for atmospheric remote sensing by employing four high-power continuous-wave multimode laser diodes as the laser sources and two CMOS sensors as the detectors. Atmospheric measurements have been performed from 23:00 January 14th to 06:00 January 15th, 2019. The dissipation of local haze and the occurrence of dust particles due to external transportation have been evaluated based on the extinction coefficient, depolarization ratio, Ångström exponent as well as other atmospheric parameters. During this period, the depolarization ratio increased from 0.1 to 0.3 and the Ångström exponent decreased from 1 to nearly zero. The PM10 concentration was always beyond 100 $\mu g/m^3$, while the PM2.5 concentration decreased from 90 $\mu g/m^3$ to 20 $\mu g/m^3$ due to the strong north wind, which blew away local spherical haze particles and brought external non-spherical large-size particles. Experimental measurements are planned in the near future for in-depth studies of the system performance as well as application explorations.

ACKNOWLEDGEMENTS

This work was supported by National Key Research and Development Program of China (2016YFC0200600); National Natural Science Foundation of China (61705030); Natural Science Foundation of Liaoning Province, China (201602163); Fundamental Research Funds for the Central Universities (DUT18JC22).

REFERENCES

1. Y. Mabuchi, N. Manago, G. Bagtasa, H. Saitoh, N. Takeuchi, M. Yabuki, T. Shiina, and H. Kuze, 2012 Ieee International Geoscience and Remote Sensing Symposium (Igarss), 2505-2508 (2012).
2. J. F. Shao and J. D. Mao, Aopc 2015: Optical and Optoelectronic Sensing and Imaging Technology **9674**, (2015).
3. Y. Hara, T. Nishizawa, N. Sugimoto, K. Osada, K. Yumimoto, I. Uno, R. Kudo, and H. Ishimoto, Remote Sens-Basel **10**, 937 (2018).
4. L. Mei, P. Guan, and Z. Kong, Opt. Express **25**, A953-A962 (2017).
5. L. Mei and M. Brydegaard, Opt. Express **23**, 1613-1628 (2015).
6. G. D. Sun, L. A. Qin, Z. H. Hou, X. Jing, F. He, F. F. Tan, and S. L. Zhang, Opt. Express **26**, 7423-7436 (2018).
7. L. Mei, P. Guan, Y. Yang, and Z. Kong, Opt. Express **25**, A628-A638 (2017).
8. L. Mei, Z. Kong, and P. Guan, Opt. Express **26**, A260-A274 (2018).
9. L. Mei and P. Guan, Opt. Lett. **42**, 3562-3565 (2017).
10. F. G. Fernald, Appl. Opt. **23**, 652-653 (1984).
11. J. Su, Z. Y. Liu, Y. H. Wu, M. P. McCormick, and L. Q. Lei, Atmos. Environ. **79**, 36-40 (2013).
12. C. Cordoba-Jabonero, J. A. Adame, J. R. Campbell, E. Cuevas, J. P. Diaz, F. Exposito, and M. Gil-Ojeda, EPJ Web Conf. **119**, (2016).
13. H. Z. Ji, Y. C. Zhang, S. Y. Chen, H. Chen, and P. Guo, Opt. Commun. **416**, 54-60 (2018).
14. H. Zhao, J. D. Mao, C. Y. Zhou, and X. Gong, J. Quant. Spectrosc. & Rad. Transf. **217**, 224-228 (2018).
15. G. L. Schuster, O. Dubovik, and B. N. Holben, J. Geophys. Res. - Atmos. **111**, (2006).
16. N. G. Loeb and G. L. Schuster, J. Geophys. Res. - Atmos. **113**, (2008).
17. H. Moosmuller, R. K. Chakrabarty, K. M. Ehlers, and W. P. Arnott, Atmos. Chem. Phys. **11**, 1217-1225 (2011).
18. S. Shin, D. Muller, Y. J. Kim, B. Tatarov, D. Shin, P. Seifert, and Y. M. Noh, Asia-Pac J Atmos Sci **49**, 19-25 (2013).
19. Y. Song, B. Zhang, G. Shi, S. Li, H. Di, Q. Yan, and D. Hua, Particuology **40**, 62-69 (2018).
20. Y. Noh, D. Muller, K. Lee, K. Kim, K. Lee, A. Shimizu, S. W. Kim, I. Sano, and C. B. Park, Atmos. Chem. Phys. **17**, 6271-6290 (2017).
21. G. Y. Dai, S. H. Wu, and X. Q. Song, Remote Sens-Basel **10**, 378 (2018).
22. M. Haarig, A. Ansmann, H. Baars, C. Jimenez, I. Veselovskii, R. Engelmann, and D. Althausen, Atmos. Chem. Phys. **18**, 11847-11861 (2018).

EPJ Web Conferences **237**, 02014 (2020) https://doi.org/10.1051/epjconf/202023702014
ILRC 29

IDENTIFICATION OF AEROSOL SOURCES IN SIBERIA AND STUDY OF AEROSOL TRANSPORT AT REGIONAL SCALE BY AIRBORNE AND SPACE-BORNE LIDAR MEASUREMENT

Antonin Zabukovec[1], Gérard Ancellet[1], Jacques Pelon[1], J.D. Paris[2], Iogannes E. Penner[3], Grigorii Kokhanenko[3], Yuri S. Balin[3]

[1] *Laboratoire Atmosphere Milieux, Observations Spatiales (LATMOS), CNRS, Sorbonne Université, Université Versailles St Quentin, Paris, France*
[2]*Laboratoire des Sciences du Climat et de l'Environnement, CEA-CNRS-UVSQ, Gif sur Yvette, France*
[3] *Zuev Institute of Atmospheric Optics, Russian Academy of Sciences, Tomsk, Russia*

ABSTRACT

Airborne lidar measurements were carried out over Siberia in July 2013 and June 2017. Aerosol optical properties are derived using the Lagrangian FLEXible PARTicle dispersion model (FLEXPART) simulations and Moderate Resolution Imaging Spectrometer (MODIS) AOD. Comparison with Cloud-Aerosol Lidar with Orthogonal Polarization (CALIOP) aerosol products is used to validate the CALIOP aerosol type identification above Siberia. Two case studies are discussed : a mixture of dust and pollution from Northern Kazakhstan and smoke plumes from forest fires. Comparisons with the CALIOP backscatter ratio show that CALIOP algorithm may overestimate the LR for a dusty mixture if not constrained by an independent AOD measurement.

1. INTRODUCTION

Asian pollution and oil/gas flaring in Siberia have been identified as aerosol key sources, however the impact by these pollutants is poorly known. CALIOP or MODIS satellite observations provide valuable information about aerosol spatial distributions and optical properties may be compared with ground-based or airborne observations. Regional aerosol studies with CALIOP have been conducted for high latitudes [1], European Arctic [2], or the Arctic ice sheet [3], but no similar studies exist for Central Siberia.

2. METHODOLOGY

2.1 Objectives and methodology

Airborne backscatter lidar observations in Siberia were carried out and compared to CALIOP data using Level 2 Version 4 CALIOP aerosol data products [4] and Level 1 CALIOP attenuated backscatter processed at LATMOS [2] in order to assess CALIOP aerosol optical properties accuracy.

Aerosol types were characterized using the Lagrangian FLEXPART model, aircraft in-situ measurements, MODIS AOD and CALIOP aerosol vertical extent and depolarization in the source region. Biomass-burning emissions were identified using the daily fire radiative power (FRP) maps based on NASA Fire Information for Resource Management System (FIRMS) using MODIS [5] and the Visible Infrared Imaging Radiometer Suite (VIIRS) [6]. Only FRP > 0,05 GW were taken into account.

2.2 Aircraft campaigns

The aircraft campaigns took place in July 2013 and June 2017. The four flight tracks for each campaign are shown in Fig 1.

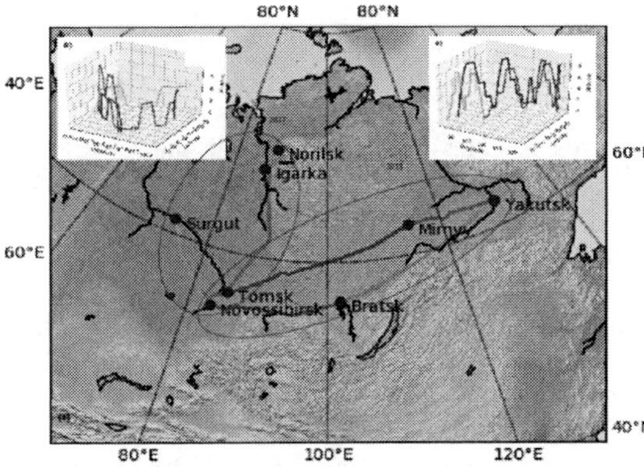

Fig. 1: Map of the 8 aircraft flight tracks carried out in 2013 (eastern loop) and in 2017 (northern

 https://doi.org/10.1051/epjconf/202023702014

loop). The aircraft altitude ranges are also shown for the 2017 (b) and 2013 (c) flights.

The flights were selected to fly over (i) the major Siberian cities (Novosibirsk, Tomsk, Krasnoyarsk, Yakutsk) (ii) the gas flaring fields of the Ob valley and the industrial city of Norilsk (iii) the Siberian taiga to sample long-range transport of forest fires and mid-latitude Eastern Asia emissions.

2.3 Airborne lidar system

The transmitter module is based on a solid state Nd-YAG laser emitting 8 ns laser pulse at 532 nm. The optical receiver is a 150 mm diameter reception lens coupled with a 1nm filter and two reception channels (co- and cross-polarization). The vertical sampling is less than 6 m. The full geometrical overlap is obtained between 80m and 150m. In practice the first 200 m are not used to account for lidar misalignment during the flights. The cross-polarization calibration is not good enough to characterize the aerosol type and is mainly used to discriminate cloud and aerosol layers.

2.4 Aerosol optical depth retrieval

The lidar calibration is performed several times during each flight using a normalization of the attenuated backscatter signal (PR2) to molecular backscatter in the range 200m-700m below the aircraft when the fit with the molecular density slope is better than 4%, only at altitudes above 3.5 km and when no aerosol/cloud layers are seen in the calibration range. The vertical profile of molecular backscatter was estimated from the 0.75° ERA-Interim ECMWF meteorological analysis.
Aerosol optical depth retrieval is based on the Fernald forward inversion of the calibrated PR2 [7], assuming a range independent value of aerosol lidar ratio (LR). The LR value is constrained using the distribution of 10 km MODIS collection 6 aerosol optical depth (AOD) in an area of ± 70 km and a time lag of ± 5 h around the aircraft observation.
The distribution of possible LR values was recursively tuned to obtain an airborne lidar AOD within the 25th and 75th percentile of the MODIS AOD distribution. To estimate the uncertainty of the retrieved backscatter ratio, 500 inversions were randomly performed using the LR distribution and using a backscatter coefficient distribution at the reference altitude according to the estimated calibration error.

3. RESULTS

3.1 Urban pollution case

During the 2017 campaign, between Novosibirsk and Surgut above the Ob Valley, the aerosol vertical distribution shows 100-200 km horizontal layers in the 0 and 2.5 km altitude range (Fig. 2).

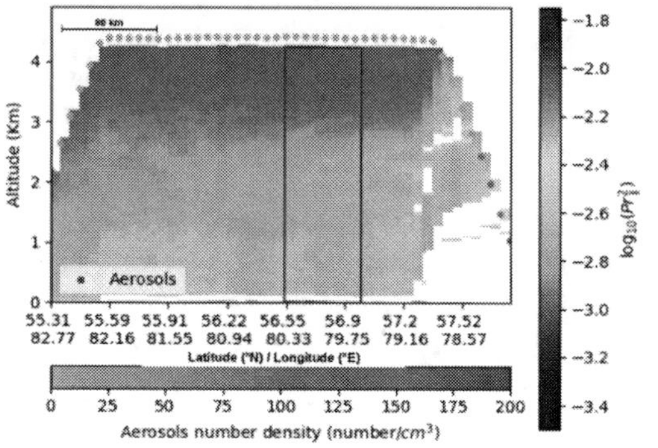

Fig. 2: Vertical cross-section of airborne lidar $\log_{10}(PR2)$ on June, 16 2017 . Calibration constant is 13458 ± 2%. Grimm aerosol concentrations in particle.cm^{-3} are shown at the aircraft altitude.

The FLEXPART simulation (Fig. 3) potential emission sensitivity (PES) is used to identify the possible aerosol sources (PES > 2000 s). Both aerosol emissions from the Novosibirsk area and from Kazakhstan can explain the layers seen by the airborne lidar.

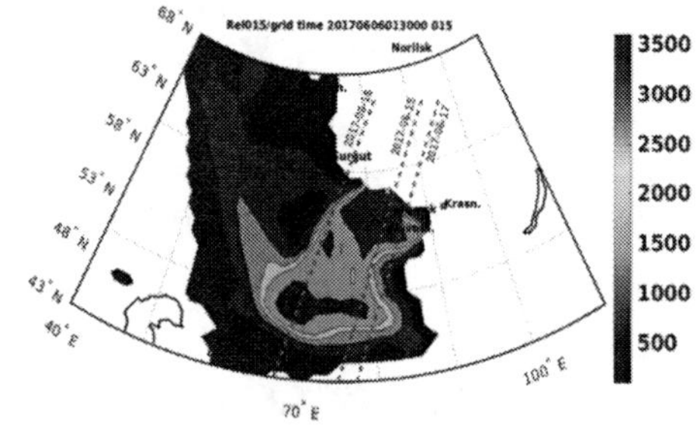

Fig. 3: Map of the PES distribution for FLEXPART backward simulation for the aerosol layers between 0-2 km at 56.6°N Fig. 2 (black

cross). The pink dotted lines are the selected CALIOP overpasses in the source area.

MODIS AOD and tropospheric CO column measured by IASI are however higher in Northern Kazakhstan (0.15-025 AOD, 2.0-2.5 $x10^{18}$ molecule.cm^{-2} CO column) than in the Novosibirsk area (0.05-0.12 AOD, 1.5-1.8 $x10^{18}$ molecule.cm^{-2} CO column).

No forest fires are located in the high PES areas according to the FIRMS data. A CALIPSO vertical cross-section above Northern Kazakhstan shows depolarizing aerosol layers (> 10%) up to 3km altitudes between 45°N and 50°N (Fig. 4). A mixture of dust and pollution source is then expected in the layers seen by the airborne lidar at 56°N, 80°E.

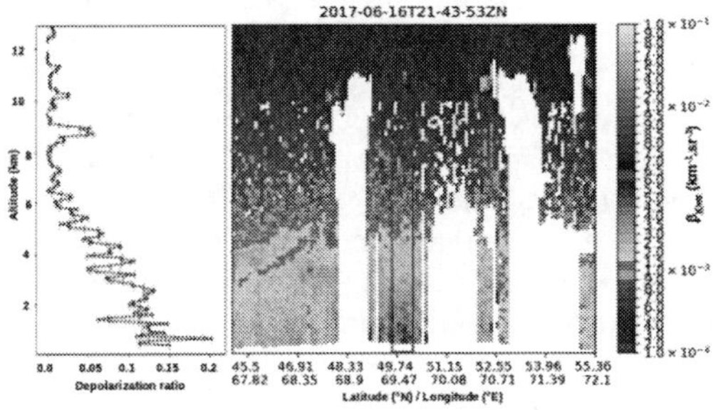

Fig. 4: Vertical cross-section of CALIOP 532 backscatter coefficient (right side) on June 16 in the Kazakhstan source region. CALIOP 532 depolarization ratio vertical profile at 49.74°N (left side).

The mean airborne lidar backscatter ratio at 56.7°N is 19 ± 0.1 below 2.5 km altitude with an AOD_{532} of 0.0895 ± 0.0225 and LR between 26 and 40, when constrained by MODIS observations (Fig. 5). The corresponding aerosol type would be "dusty mix" according to [8]. It is also in good agreement with the FLEXPART analysis.

Two CALIOP backscatter ratio profiles in the same area at 55.5°N, 82.5°E on June 15th, 21 UT and at 57°N, 87°E on June 17th, 20 UT are in good agreement with the airborne lidar observations. The corresponding AOD_{532} are 0.1 ± 0.04 (0.1 ± 0.04) and 0.168 ± 0.062 (0.139 ± 0.047) for the CALIOP L2 product (LATMOS processed L1 data) on June 15th and 17th respectively, with LR between 55 and 70 sr. A better match with the airborne lidar and MODIS AOD could be obtained by using LR values smaller than 40 as expected for a dusty mix type.

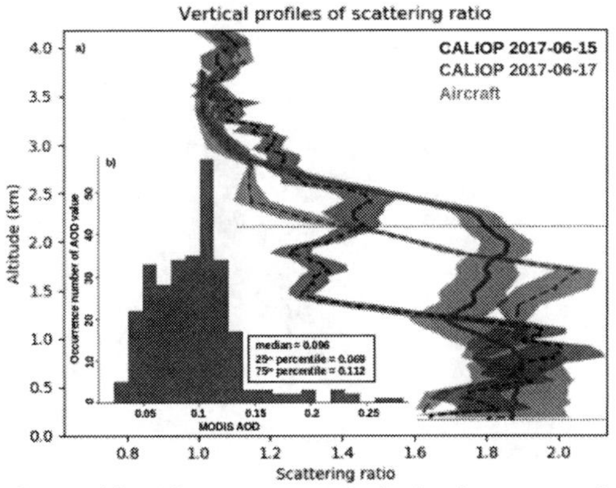

Fig. 5: (a) 50 km average of the backscatter ratio vertical profile for the airborne lidar on June 16th 05 UT (green), CALIOP on June 15th (blue) and on June 17th (red). (b) MODIS AOD_{550} distribution in the ± 70 km area around the aircraft.

3.1 Boreal forest fire case

During the 2013 campaign, a 500 m thick aerosol layer is observed by the airborne lidar at 2.0 km between Tomsk and Mirnyy (Fig.6).

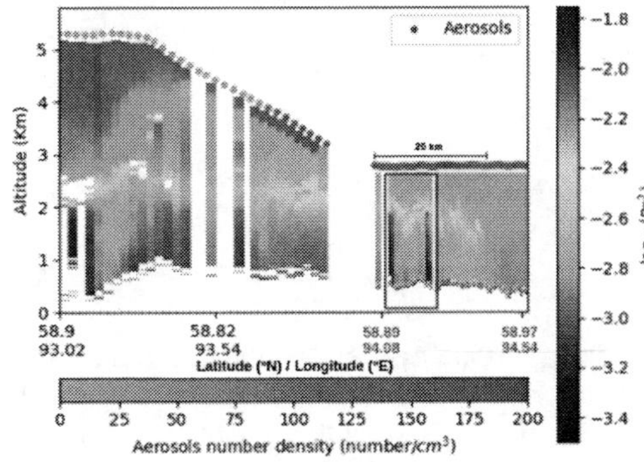

Fig. 6: Vertical cross-section of aircraft log_{10}(PR2) on July, 19 2013. Calibration constant is 491724 ± 10%. Grimm aerosol concentrations in particle.cm^{-3} are shown at the aircraft altitude.

According to the FLEXPART simulation, no aerosol transport is expected from large cities or pollution sources, while several forest fires were identified by the FIRMS data close to the aircraft position (Fig. 7). The mean airborne lidar backscatter ratio at 58°N (Fig. 6) is larger than 3.0

between 1.5 and 2.2 km altitude with an AOD_{532} of 0.122 ± 0.022 and LR between 53 and 56, when constrained by MODIS observations. Such values are consistent with a "fresh smoke" aerosol type according to [8].

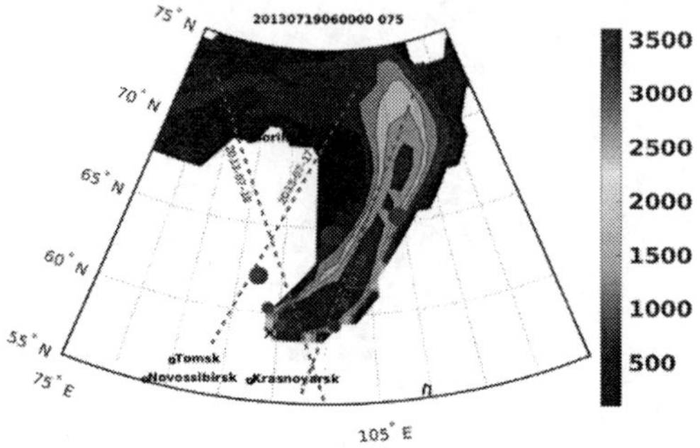

Fig. 7: Map of the PES distribution for FLEXPART backward simulation for the aerosol layer at 1.7 km Fig. 6 (black cross). The red dots are FIRMS forest fires and the pink dotted lines are the selected CALIOP overpasses in the source area.

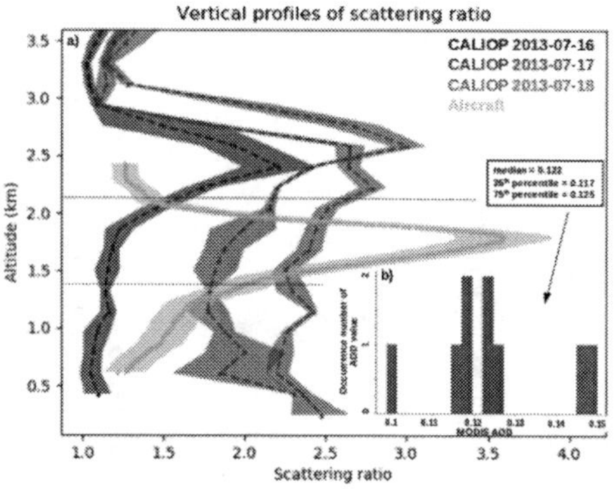

Fig. 8: (a) 20 km average of the backscatter ratio vertical profile from the airborne lidar on July 19th 6 UT (yellow), CALIOP on July 16th 19 UT (blue), on July 17th 19 UT (red) and on June 18th 5 UT (green). (b) MODIS AOD_{550} distribution in the ± 70 km area around the aircraft position.

This backscatter ratio profile can be compared with several CALIOP tracks observed near forest fires in the same area (Fig. 7). The CALIOP vertical profiles show smoke layers between 2km and 3 km altitude, but with backscatter ratio between 2 and 3 (Fig. 8). CALIOP AODs in the latitude band 60°N-62°N range from 0.1 to 0.3 with lidar ratio of 70 sr. AODs are also in good agreement with the range of MODIS AOD around the CALIOP tracks. While the airborne lidar is able to sample fresh smoke layer, CALIOP profiles did not overpass fires during the combustion phase and are thus mainly representative of fire plumes after transport and mixing. It explains both the vertical structure of the CALIOP aerosol layers (larger thicknesses and smaller peak values) and larger lidar ratio for aged biomass plume [8].

Conclusion and Perspectives

During the 2017 and 2013 aircraft campaigns, mixed layers of dust and pollution from Northern Kazakhstan and smoke layers in forest area were detected by the airborne lidar. Comparison with CALIOP backscatter ratio shows that CALIOP algorithm may overestimate the LR for a dusty mixture if not constrained by an independent AOD measurement. The use of low liquid water cloud reflectance from the Synergized Optical Depth of Aerosols product (SODA) or land surface reflectance from the Wide Field Camera (WFC) aboard CALIOP could be a step forward to derive the aerosol CALIOP extinction above Siberia.

ACKNOWLEDGEMENTS

This work was supported by Sorbonne Université and the French Centre National d'Etudes Spatiales.

REFERENCES

[1] M.Di. Pierro, and al. Atmospheric Chemistry and Physics 11.5, 2225-2243 (2011)
[2] G. Ancellet, and al. Atmospheric Chemistry and Physics 14.16, 8235-8254 (2014)
[3] C. Di Biagio, and al. Journal of Geophysical Research: Atmospheres 123.2, 1363-1383 (2018)
[4] M.H. Kim, and al. Atmospheric Measurement Techniques 11.11 (2018)
[5] L. Giglio, and al. Remote Sensing of Environment 87.2-3 , 273-282 (2003)
[6] W. Schroeder, and al. Remote Sensing of Environment 143, 85-96 (2014)
[7] F. Fernald, et al. Applied Optics 23.5, 652-653 (1984)
[8] S.P. Burton, et al. Atmospheric Measurement Techniques 6.5, 1397-1412 (2013)

IMPACT OF THE PLANETARY BOUNDARY LAYER HEIGHT ON THE SURFACE AEROSOL OPTICAL AND MICROPHYSICAL PROPERTIES

Salvatore Romano, Maria Rita Perrone

Mathematics and Physics Department, University of Salento, via per Arnesano, 73100 Lecce, Italy
**Email: salvatore.romano@unisalento.it*

ABSTRACT

Lidar, nephelometer, and aethalometer measurements at the surface, co-located in time and space with Particulate Matter (PM) measurements, have been performed to investigate the impact of the daily evolution of the Planetary Boundary Layer (PBL) height on the aerosol optical and microphysical properties. Measurements were performed at a coastal site of southeastern Italy characterized by a shallow (<1000 m) PBL height. The Standard Deviation technique applied to the vertical profiles of both the lidar range corrected signal (RCS) and the linear volume depolarization ratio (δ_r) has been used to determine the daily evolution of the PBL height and highlight benefits and limits of using RCS and δ_r vertical profiles. It is shown that the PBL height, which drives the particle dispersion at the surface, significantly affects the optical and microphysical properties of the surface particles since the particle dispersion varies with their size and, consequently, the mean optical and microphysical properties of the surface particles are affected. The impact of meteorological conditions on the daily trend of the PBL height and the surface particle properties has also been highlighted.

1. INTRODUCTION

The characterization of the Planetary Boundary Layer (PBL) is of primary importance for climate, meteorological forecasts, pollutant dispersion, and air quality studies. Processes occurring within the PBL control energy, water vapour, and pollutant exchanges between the surface and the free atmosphere. The PBL experiences a marked diurnal cycle that depends on both the synoptic and the local weather conditions [1]. The PBL tends to be lower in depth at night, while during the day it tends to have a higher depth because of the wind speed and the air thickness, which vary as a function of temperature. Strong wind speeds allow for more convective mixing, which will cause the PBL to expand. At night, the PBL contracts because of the reduction of rising thermals from the surface and since cold air is denser than warm air (e.g., [2]). Consequently, the PBL height can be estimated from measurements of the mechanical turbulence, the temperature enabling convection, or the concentration of the atmospheric constituents. Several techniques have been applied to trace the daily trend of the PBL height [3]. Active remote sensing instruments such as lidars present advantages over the more traditional use of radiosondes for PBL studies (e.g., [4], [5]). In this study, the Standard Deviation (SD) technique applied to the vertical profiles of the lidar range corrected signal (RCS) and the linear volume depolarization ratio (δ_r) has been used to determine the daily evolution of the PBL height and investigate its relationship with local meteorological parameters and aerosol properties at the surface. Main results will be presented and discussed by means of case studies.

2. METHODOLOGY

2.1 Site description and UNILE lidar system

Measurements have been performed in Lecce on the roof of the Mathematics and Physics Department of the University of Salento (40.33°N, 18.11°E, 30 m a.s.l.). The monitoring site is located on a narrow and flat peninsular area ~20 km away from both the Ionian and the Adriatic Sea and may be considered as representative of coastal sites of the Central Mediterranean away from large sources of local pollution [6]. The used ground-based lidar system, identified as UNILE (UNIversity of LEcce) lidar, operates within EARLINET (European Aerosol Research LIdar NETwork) since May 2000. It is currently composed by a Nd:YAG laser operating at 1064, 532, and 355 nm at a repetition rate of 30 Hz and a f/4 Newton telescope [7].

2.2 Nephelometer, aethalometer, and PM mass concentration measurements

A light-emitting diode-based integrating nephelometer (model Aurora 3000, ECOTECH, Knoxfield, Australia) was used to measure the aerosol scattering and backscattering coefficients (σ_s and β_s, respectively) at 450, 525, and 635 nm with a temporal resolution of 5 min. Then, an aethalometer model AE-31 (Magee Scientific, Berkeley, California, USA) was used to determine the aerosol absorption coefficient (σ_a) at seven wavelengths (370, 470, 520, 590, 660, 880, and 950 nm) with a temporal resolution of 5 min. Both instruments were equipped with a PM2.5 sampling head [8]. A PM sampler (model FH 62 I-R, Thermo ESM Andersen Instruments GmbH, Erlangen, Germany) fitted with a PM10 sampling head was used to investigate the daily evolution of the PM10 mass concentration.

2.3 PBL height identification

The turbulent nature of the mixing layer leads to a fluctuating boundary between the PBL and the free troposphere. Measurements of such fluctuations based on the standard deviation (SD) of lidar signals as a function of the altitude have been used to determine the PBL height (e.g., [9]). More specifically, the SD technique searches for the first maximum in the profile of the sum of the squares of the lidar signal deviations from the mean value over a selected time interval in order to detect the PBL height (e.g., [3]).

3. RESULTS

3.1 Daily trend of the PBL height and the meteorological parameters on 25 February 2016

Figure 1 shows by a color plot (a) the lidar RCS at 1064 nm and (b) δ_r at 355 nm retrieved from measurements performed from 09:05 to 14:00 and from 16:54 to 17:55 UTC of 25 February 2016, from the ground level up to 2.4 km. One observes that the aerosol particles were located at heights smaller than ~2 km above ground level (agl) and within different aerosol layers, whose respective vertical profiles varied with the day hours. Fig. 1b highlights that the δ_r values on average increased with the height, reaching values smaller than $\sim 7 \cdot 10^{-3}$ at heights smaller than ~950 m agl. Figure 2 shows by a color plot the SD of (a) the RCS and (b) δ_r signals. Black dots identify the PBL heights retrieved from the SD technique. The PBL height from the SD of the RCS increased with the day hours up to ~950 m agl around 13:15 UTC and then decreased up to ~450 m agl after sunset. The PBL heights with corresponding standard errors of the mean (SEM) retrieved from the RCS-SD (red dots) and δ_r-SD (blue triangles) are compared in Fig. 3 to better highlight the differences between the PBL heights from the two different signals. Figure 3 shows that the PBL heights retrieved from the RCS-SD were on average smaller than the corresponding ones obtained from the δ_r-SD. The mean difference was equal to ~120 m. The observed differences could likely be due to the very low values of δ_r, as well as to the increase of δ_r with the height, for altitudes larger than few hundred meters. Note also that we have observed a better agreement between the PBL height values retrieved from the SD of the two lidar signals at higher δ_r values. Therefore, the SD technique applied to the δ_r signals can lead to less reliable PBL heights if δ_r-values are rather small.

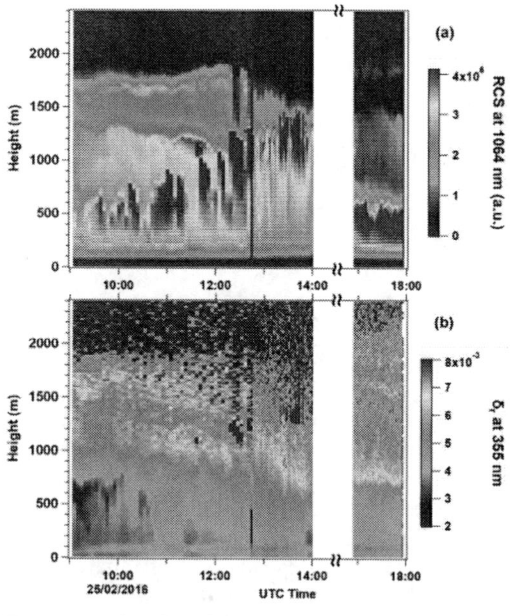

Figure 1 Color plot of the temporal evolution of (a) the range corrected lidar signal (RCS) at 1064 nm and (b) the linear volume depolarization ratio (δ_r) at 355 nm on 25 February 2016.

 https://doi.org/10.1051/epjconf/202023702015

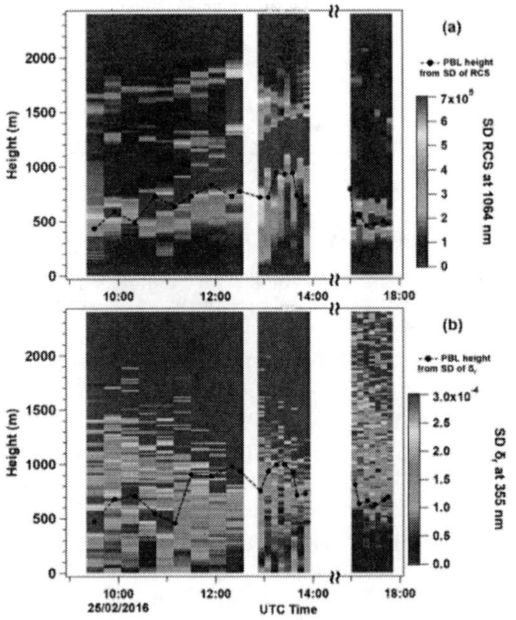

Figure 2 Color plot of the temporal evolution of the standard deviation (SD) of (a) the range corrected lidar signal (RCS) at 1064 nm and (b) the linear volume depolarization ratio (δ_r) at 355 nm on 25 February 2016. Black dots identify the determined PBL heights.

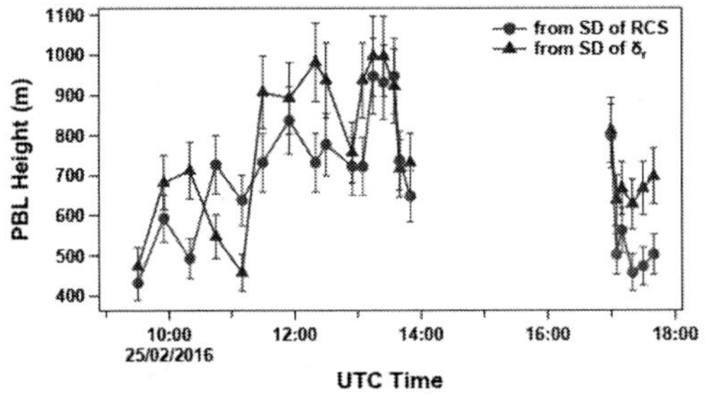

Figure 3 Time evolution of the PBL height (with the corresponding standard errors of the mean) retrieved from the RCS (red dots) and δ_r lidar signals (blue triangles).

Figure 4 shows the daily evolution of temperature (T), relative humidity (RH), and wind speed (WS) values from a local meteorological station in addition to the daily trend of the PBL height (black dots). One observes from Fig. 4 that the PBL height increase/decrease is mainly driven by the WS increase/decrease, which in turn depends on the T time evolution.

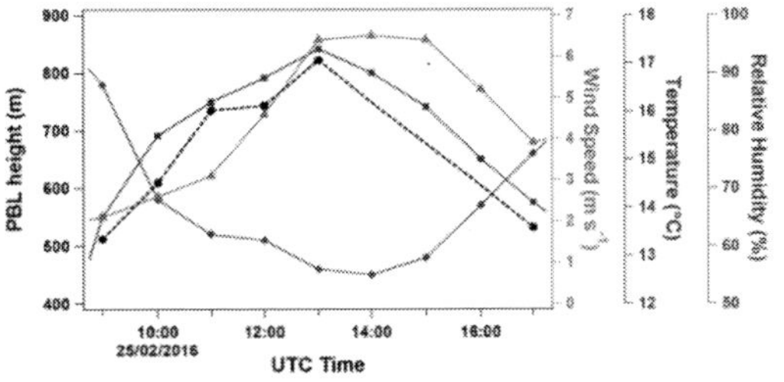

Figure 4 Temporal evolution of the hourly means of PBL height from RCS-SD (black dots), wind speed (green triangles), air temperature (blue boxes), and relative humidity (red diamonds).

3.2 PBL heights and surface aerosol optical and microphysical properties

The impact of the PBL height time evolution on the aerosol optical properties at the surface is clearly shown by Fig. 5, which shows the daily evolution of the aerosol scattering (σ_s) and absorption (σ_a) coefficients at 450 nm retrieved on 25 February 2016 from nephelometer and aethalometer measurements, respectively. The backscatter ratio ($\beta_R = \beta_s/\sigma_s$) calculated from the ratio of the aerosol backscatter coefficient (β_s) to σ_s, both at 450 nm, is also plotted in Fig. 5, in addition to the PBL heights (black dots). Figure 5 shows that σ_s and σ_a decrease with the increase of the PBL height, reaching the smallest values when the PBL height reaches its highest values. The backscatter ratio, which assumes lower values at higher PBL heights, indicates that the σ_s and σ_a decrease was likely due to the decrease of the fine-mode particle contribution at the surface. Indeed, the fine particles can be more efficiently dispersed than the coarse particles with the PBL height increase occurring during the central hours of the day [10]. The daily trend of the single-scattering-albedo SSA = $\sigma_s/(\sigma_s+\sigma_a)$, shown in Fig. 6, further supports this last comment. The increase of fine-mode particles, which on average are made by black carbon and/or carbonaceous species, is generally responsible for the decrease of the SSA. Therefore, Fig. 6 shows that the SSA increases with the PBL height up to ~ midday and, then, decreases with the PBL height decrease, likely because of the increase of fine-absorbing particle contribution at the surface. The hourly means of the PM10 mass concentration retrieved from the PM sampler measurements are also plotted in Fig.

https://doi.org/10.1051/epjconf/202023702015

6 (red diamonds). One observes that first the PM10 mass concentration assumed lower values (~30 µg m^{-3}) when the PBL height reached its maximum hourly mean (~820 m) at 13.00 UTC. Then, a decreasing trend of the PBL height was associated with an increasing trend of the PM10 mass concentration (up to 43 µg m^{-3}) until 17.00 UTC.

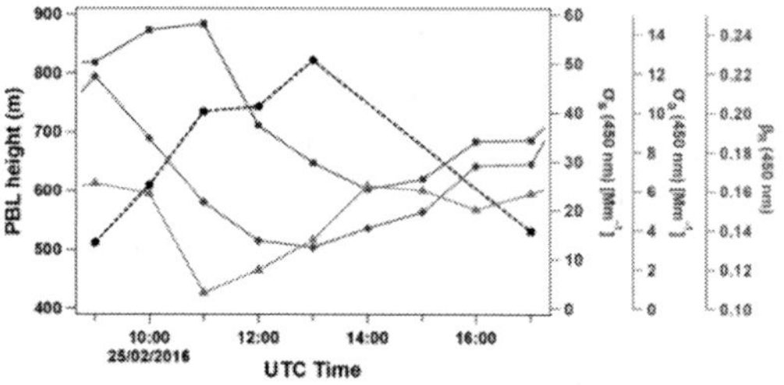

Figure 5 Temporal evolution of the hourly means of PBL height from the RCS-SD (black dots), aerosol scattering σ_s and absorption σ_a coefficient at 450 nm (blue boxes and red diamonds, respectively), and backscatter ratio β_R at 450 nm (green triangles) on 25 February 2016.

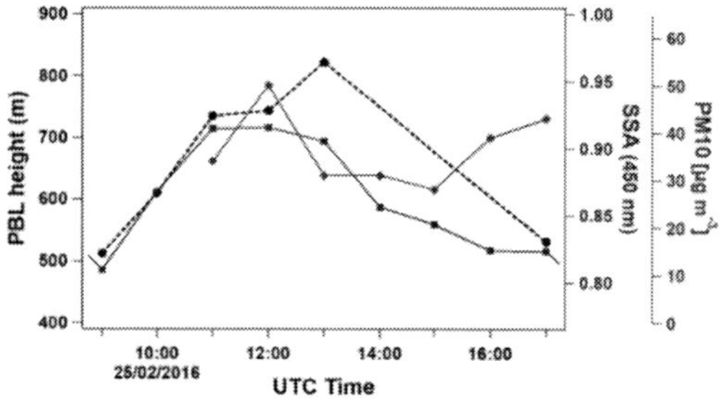

Figure 6 Temporal evolution of the hourly means of PBL height from the RCS-SD (black dots), single scattering albedo (SSA, blue boxes) and PM10 mass concentration (red diamonds) on 25 February 2016.

Figure 7 shows, as an example, the diurnal evolution of the hourly means of PBL height, σ_s, σ_a, and β_R calculated on 25 January 2016 to further highlight the relationships between the PBL height and the particle properties at the surface. Figure 7 also proves the inverse relation between the PBL height and σ_s, σ_a, and β_R. In particular, one can observe that σ_s, σ_a, and β_R reached minimum values around midday, when the PBL height reached its maximum value of ~550 m, likely because of the dispersion mainly of the surface fine-mode particles.

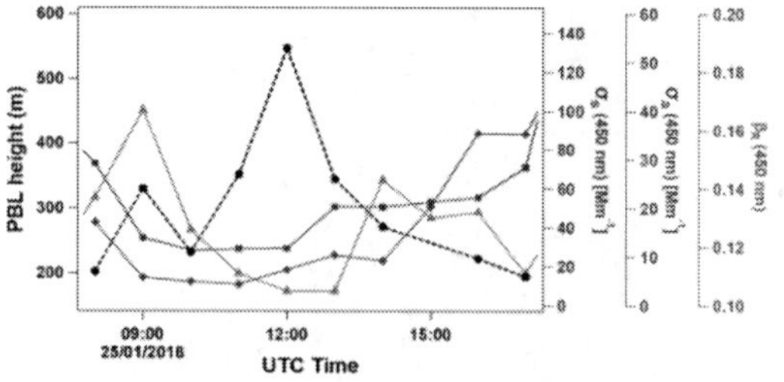

Figure 7 Temporal evolution of the hourly means of PBL height from the RCS-SD (black dots), aerosol scattering σ_s and absorption σ_a coefficient at 450 nm (blue boxes and red diamonds, respectively), and backscatter ratio β_R at 450 nm (green triangles) on 25 January 2016.

In conclusion, the results reported in this work have shown that the diurnal variations of the surface aerosol extensive (σ_s, σ_a, and PM10 concentration) and intensive (β_R and SSA) properties are strictly related to the corresponding evolution of the PBL height. The benefits of using the RCS instead of δ_r to calculate the PBL heights have also been highlighted.

ACKNOWLEDGEMENTS

S. Romano has carried out this work with the financial support of EARLINET as part of the ACTRIS Research Infrastructure Project by the European Union's Horizon 2020 research and innovation programme under grant agreements no. 654109 and 739530 (previously under grant agreement no. 262254) in the 7th Framework Programme (FP7/2007-2013).

REFERENCES

[1] R. B. Stull, Introduction to boundary layer meteorology. Kluwer, Dordrecht (1988).

[2] M. R. Perrone and S. Romano Atmos. Res. 213, 57-69 (2018).

[3] M. Collaud Coen, et al. Atmos. Chem. Phys. 14, 13205-13221 (2014).

[4] F. De Tomasi and M.R. Perrone Atmos. Res. 80, 86-103 (2006).

[5] J.A. Bravo-Aranda, et al. Atmos. Chem. Phys. 17, 6839-6851 (2017).

[6] S. Basart, et al. Atmos. Chem. Phys. 9, 8265-8282 (2009).

[7] M. R. Perrone, et al. Atmos. Chem. Phys. 14, 1185-1204 (2014).

[8] S. Romano, et al. Atmos. Environ. 203, 35-47 (2019).

[9] H. Baars, et al. Atmos. Chem. Phys. 8, 7281-7296 (2008).

[10] M. R. Perrone, et al. Environ. Sci. Pollut. R. 22, 16570-16589 (2015).

INVESTIGATIONS TO HYGROSCOPIC AEROSOL GROWTH WITHIN THE CONVECTIVE BOUNDARY LAYER

Dietrich Althausen[1*], Andrew Reigert[2], Ulla Wandinger[1], Jens Reichardt[3]

[1]Leipniz Institute for Tropospheric Research, 04318 Leipzig, Germany
[2]Bertrandt Ing-.Büro GmbH, 85080 Gaimersheim, Germany
[3]German Meteorological Service, Aßmann Observatory, Lindenberg, Germany
**Email: dietrich@tropos.de*

ABSTRACT

We present a new method to determine the hygroscopic growth of atmospheric particles. This method combines lidar measurements with high temporal resolution of the particle backscatter coefficient and water vapor mixing ratio with temperature measurements from radiosondes and a microwave radiometer. The hygroscopic growth is described by an equation that represents the two observed branches of the growth curve with different dependencies on the relative humidity. An example is presented to illustrate a first result from a continental air mass case.

1. INTRODUCTION

Hygroscopic particle growth capabilities are immanent properties of atmospheric particles. As particles grow and change their sizes, the lidar aerosol optical properties are influenced. This influence is deeply dependant on the different measured aerosols properties, especially their chemical composition.

These particle properties are only accessible by simultaneous measurements of their respective particle optical properties and the relative humidity with a high temporal resolution and within the same atmospheric volume. Especially, such investigations have to ensure that the measurements are performed for the same atmospheric particle type at different values of relative humidity.

The dependence of the particle growth on relative humidity has been investigated for a long time [1-5]. But these investigations were done by laboratory studies or by in-situ measurements. [6-7] investigated hygroscopic growth along a horizontal optical path in 30 m height above the ground. These comprehensive investigations were performed at unperturbed atmospheric humidity conditions. They showed the applicability of the parametrization scheme, but were limited to the ground and the current humidity conditions there.

This study investigates the hygroscopic growth of particles in the well mixed daytime planetary boundary layer. It is realized by combining vertical measurements of lidar aerosol optical properties and relative humidity.

2. METHODOLOGY

2.1 Enhancement factor of the particle backscatter coefficient

The hygroscopic growth is characterized by the increase of the particle backscatter coefficient β_{par} in dependence on the relative humidity ϕ and is described by the enhancement factor

$$f(\phi)=\frac{\beta_{\mathrm{par}}(\phi)}{\beta_{\mathrm{par}}(\phi=40\%)}.$$

This enhancement factor is determined by a no linear fit to the measurement data by using the following equation and the fit parameters ϕ_{thr}, a_0, a_1, and E_2:

$$f(\phi)=\begin{cases} a_0\left(1-\dfrac{\phi}{100}\right)^{-E_1} & \text{for } \phi\le\phi_{\mathrm{thr}} \\ a_0 a_1\left(1-\dfrac{\phi}{100}\right)^{-E_2} & \text{for } \phi>\phi_{\mathrm{thr}} \end{cases}$$

where $E_1=E_2-\dfrac{\ln(a_1)}{\ln\left(1-\dfrac{\phi_{\mathrm{thr}}}{100}\right)}$.

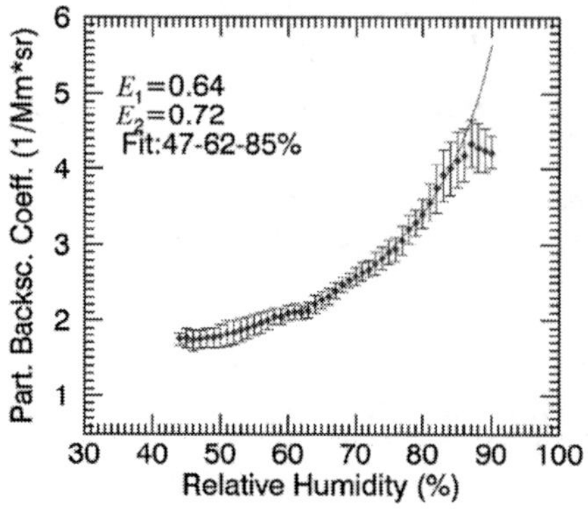

Fig. 1. Particle backscatter coefficient in dependence on relative humidity, 13 August 2015.

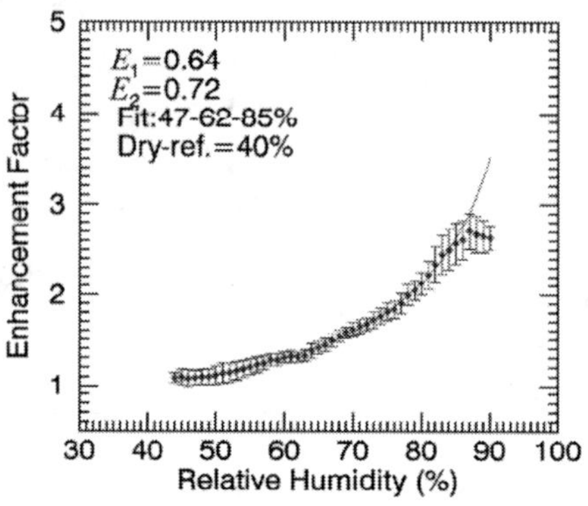

Fig. 2. Enhancement factor in dependence on relative humidity, 13 August 2015.

This approach takes into account that atmospheric aerosol is usually comprised of particles across several size ranges (size distribution) and different types (chemical composition). Hence, there is often no single deliquescence or fluorescence relative humidity identifiable.

2.2 The lidar system

We used the RAMSES (Raman lidar for atmospheric moisture sensing) system of the German Meteorological Service [8] for the determination of the particle backscatter coefficient and the water vapor mixing ratio. This ensures the observation of the same atmospheric column with varying humidity at a high temporal resolution. The RS-92 (Vaisala) radiosondes, launched every 6 hours, provided the data for the calibration of the lidar's water vapor mixing ratios and temperature profiles. The microwave radiometer MP-3039A from the Radiometrics Corporation (radiometrics.com) supplied temperature measurements between the radiosonde launches. By combining both temperature measurements using weighing functions and applying the result to the lidar's water vapor mixing ratio the relative humidity was determined. All measurements were taken at the Lindenberg Meteorological Observatory "Richard Aßmann" (52°12'29.29"N, 14° 7'11.01"E).

3. FIRST RESULTS

Figures 1 and 2 show the particle backscatter coefficient and the enhancement factor from 13 August 2015. The threshold relative humidity is 62% on this day. The height of the convective planetary boundary layer was approximately 1.8 km. Backward trajectories show an east-north-east air mass advection, which suggests a continental aerosol mixture with a strong regional influence.

REFERENCES

[1] H. Köhler, Trans. Faraday Soc., 32, 1152-1161 (1936)

[2] C. Orr et al., Journal of Atmospheric Sciences 15, 240-241 (1958)

[3] F. Kasten, Tellus, 21, 631-635 (1969)

[4] G. Hänel, Advances in Geophysics, 19, 74 (1976)

[5] P. Zieger et al., Tellus B, 66 (2014)

[6] A. Skupin et al. Atmospheric Measurement Techniques 7, 701-712 (2014)

[7] A. Skupin et al., Atmospheric Chemistry & Physics, 16, 1863-1876 (2016)

[8] J. Reichardt et al., Appl. Optics, 51 (34), 8111-8131 (2012)

https://doi.org/10.1051/epjconf/202023702017

IS NEAR-SPHERICAL SHAPE "THE NEW BLACK" FOR SMOKE?

Anna Gialitaki [1,2*], Alexandra Tsekeri [1], Vassilis Amiridis [1], Romain Ceolato [3], Lucas Paulien [3] Emmanouil Proestakis [1], Eleni Marinou [1,4], Moritz Haarig [5], Holger Baars [5] and Dimitris Balis [2]

[1] *National Observatory of Athens, IAASARS, Athens, Greece*
[2] *Laboratory of Atmospheric Physics, Physics Department, Aristotle University of Thessaloniki, Greece*
[3] *ONERA, The French Aerospace Lab, Toulouse, France*
[4] *Institute of Atmospheric Physics, German Aerospace Center (DLR), Oberpfaffenhofen, Germany*
[5] *Leibniz Institute for Tropospheric Research (TROPOS), Leipzig, Germany*
**Email: togialitaki@noa.gr*

ABSTRACT

We present smoke lidar measurements from the Canadian fires of 2017. The advected smoke layers over Europe are detected at both tropospheric and stratospheric heights, with the latter presenting non-typical values of the Particle Linear Depolarization Ratio (PLDR) with strong wavelength dependence from the UV to the Near-IR. Specifically, the PLDR values are of the order of 22, 18 and 4% at 355, 532 and 1064 nm respectively. In an attempt to interpret these results, we apply the hypothesis that smoke particles have near-spherical shapes. Scattering calculations with the T-matrix code support other findings in the literature ([1]- [2]), showing that the near-spherical shape (or closely similar shapes as in [2]), is the only shape that has been shown to reproduce the observed PLDR and Lidar Ratio (LR) values of the stratospheric smoke particles at the three measurement wavelengths.

1. INTRODUCTION

Recent studies show that the PLDR of smoke presents large variability ([3] - [7]). Since the PLDR is indicative of particle shape, this variability may be attributed to (a) smoke aerosol mixing with other aerosol types, (b) particles' aging through various atmospheric processes, (c) particle water uptake at different relative humidity conditions. These processes alter the smoke particle shape and composition. Indicatively, the typical PLDR values of aged and fresh tropospheric smoke varies from 2 to 10% at 532 and 355 nm ([3] - [5]), while this value can be larger when smoke is mixed with high depolarizing aerosol types (i.e. dust) ([6] - [7]). Lately, observational evidence of PLDR values that exceed the typical range has been reported. Burton et al. (2015) [8], showed air-borne HSRL measurements of smoke at 8 km altitude, originating from Canadian fires, revealing PLDR values of the order of 20, 9 and 1.8 % at 355, 532 and 1064 nm, respectively. These are also the first reported measurements of smoke PLDR spectral dependence. Other studies validate these high PLDR values for stratospheric smoke particles at 532 nm ([9] - [11]).

In the spotlight of the large-scale Canadian fires of 2017, this unique feature of high values of PLDR with unexpected high spectral dependence for smoke particles raises once again an interesting discussion. Here, we aim to seek for an explanation on the variation of PLDR and LR from UV to Near-IR under the hypothesis that the smoke particles have a near-spherical shape.

Our hypothesis and presented results are a continuation effort on previous work reported in the literature. Specifically, in Mishchenko et al. (2016) [1], the lidar measurements presented in [8] were reproduced considering near-spherical shapes for the smoke particles with an axial ratio 0.9 to 1.2. In Bi et al. 2018 [12], it was showed that the PLDR of near-spherical particles with refractive index of 1.3-1.7 + i0.001- i0.01, can take values up to 100%, depending also on the particle size parameter. However, the results of these studies are limited in terms of reproducing only the PLDR wavelength dependence. Here we make a step further, reproducing also the wavelength dependence of the LR considering near-spherical particles. Our methodology is summarized in Section 2, our results in Section 3 and we conclude in Section 4.

2. METHODOLOGY

https://doi.org/10.1051/epjconf/202023702017

2.1 The BERTHA lidar system

The lidar measurements presented in this study were conducted at the Leibniz Institute of Tropospheric Research (TROPOS), with BERTHA multi-wavelength lidar system [13]. The measurements have a 7.5 m vertical resolution, while for the specific case study presented here temporal resolution was 10 sec. The capability of BERTHA to provide independent measurements of the particle extinction coefficient and PLDR at three wavelengths constitutes this lidar system unique within the EARLINET network [14]. The BERTHA measurements used herein are not simultaneous for the particle extinction coefficient and PLDR at 1064 nm, since the 1064 cross-polarized channel had to be exchanged with the 1058 nm rotational Raman channel in order to measure the particle extinction at 1064 nm [15]. The overall duration of this procedure was 20-30 minutes [16]. This shortcoming is not expected to affect the study, since the scene analyzed is very stable in terms of aerosol layering, especially in the stratosphere (Figure 1).

2.2 Scattering calculations

In order to reproduce the unique values and spectral dependence of the PLDR and LR of the stratospheric smoke from the Canadian fires, we used light scattering calculations generated with the T-matrix code [17, 18]. The T-matrix code is based on the exact numerical solution of Maxwell's equations. For the calculations we considered a wide range of refractive indices and effective radii typical for aged smoke, employing near-spherical particles but also soot fractal aggregates.

3. RESULTS

Figure 1 presents the observations from Leipzig, on 22 August 2017, with the 3+3+3 polarization/Raman lidar BERTHA. As it can be seen from the Range Corrected Signal (RCS) at 1064 nm, the stratospheric smoke layer lies between 15 and 16 km, while there is also a smoke layer in the troposphere at 5-6 km. For deriving the smoke optical properties, lidar signals were averaged over a time window of 2.5 hours (20:45 – 23:15 UTC) to achieve satisfying signal to noise ratio (SNR), even at high altitudes. For deriving the PLDR at 1064 nm a time window of 40 min was used (23:50 – 00:30 UTC).

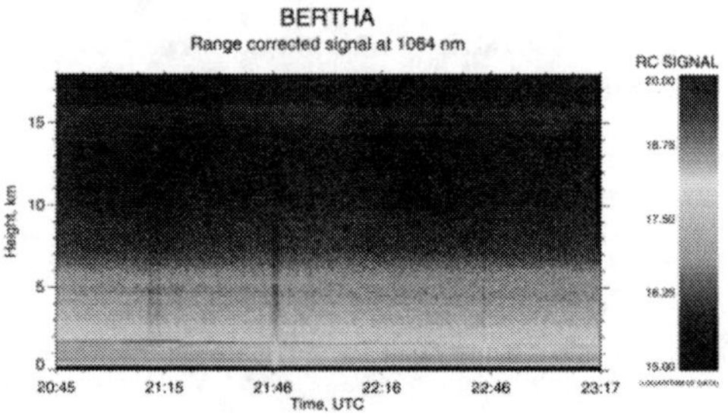

Fig.1. BERTHA RCS at 1064 nm, showing the smoke layers advected from Canada over Leipzig, in the troposphere (5 – 6 km) and in the stratosphere (15 – 16 km).

Figure 2 shows the mean PLDR and LR values of the stratospheric smoke layers and their errors. The LR values are found to be typical for aged smoke particles (40 ± 16) sr at 355 nm, (66 ± 12) sr at 532 nm, which are consistent with what is reported in the literature (i.e. [3] - [7], [19] - [20]), and (92 ± 27) sr at 1064 nm. No significant difference has been detected for the LR values between the tropospheric and stratospheric layers (see [16]), while this does not hold true for the PLDR values. In the case of stratospheric smoke the PLDR values are 22.5, 18.5 and 4% at 355, 532 and 1064 nm, respectively, while in the troposphere they are no higher than 3% at all wavelenghts [16].

Building on the previous studies of [1] and [12] we used near-spherical smoke particles to reproduce the measured PLDR and LR at 355, 532 and 1064 nm in the stratosphere. T-matrix simulations were conducted for a range of refractive index values: m = 1.4 – 1.75 (with step of 0.05) + i(0.005 – 0.045) (with step of 0.005), the particles are considered to follow mono-modal log-normal size distributions with mean effective radii of: r_{eff} = 0.1 to 0.7 μm (with step of 0.05 and a fixed effective variance of v_{eff} = 0.03) and mono-modal normal shape distributions with mean axial ratio: a/b = 0.6 to 1.55 (with step of 0.05 and fixed standard deviation σ_s = 0.05). The fitting of the measurements is performed on the basis of achieving the minimum sum of squares of relative differences between measurements and simulations as shown in Eq. (1), where M_λ^i represents the measured PLDR/LR values at wavelength λ_i, (i = 355, 532, 1064 nm) and S_λ^i the corresponding simulated values. Out of the results, the solution is accepted based on the requirement that each simulated value has to be within the measurement error (Eq. 2).

$$\sum_{i=1}^{n}\left(\frac{M_\lambda^i - S_\lambda^i}{M_\lambda^i}\right)^2 = min \quad (1)$$

$$S_\lambda^i \leq e(M_\lambda^i) \quad (2)$$

Figure 4 presents the best fit of the measured PLDR and LR at 355, 532 and 1064 nm, calculated for near-spherical smoke particles with mean a/b = 1.4, m = 1.55+ i0.025 and r_{eff} = 0.25 μm (Figure 3). Table 1 summarizes our results for near-spherical particles, in comparison with the Leipzig observations. The T-matrix-calculated PLDR values were found to be 21.8, 18.3 and 4.5 % at 355, 532 and 1064 nm respectively, while the LR values were 41.7, 64.2 and 98.5 sr at 355, 532 and 1064 nm.

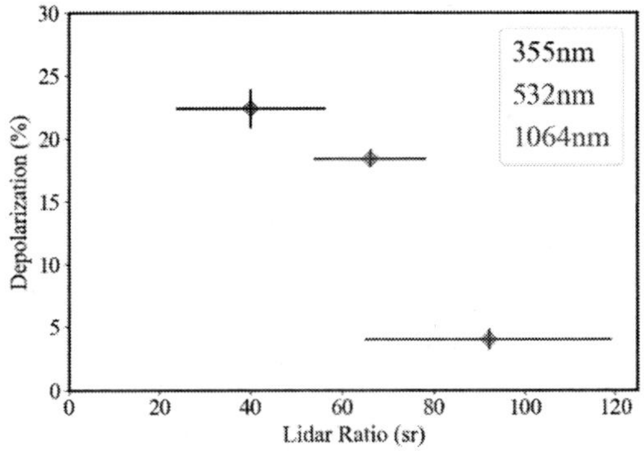

Fig.2 Intensive optical properties of the stratospheric smoke particles at 355, 532 and 1064 nm. All values correspond to the time window 22/8 20:45 – 23:15 UTC, except of the PLDR at 1064 nm which corresponds to the time window 23:50 – 00:30 UTC.

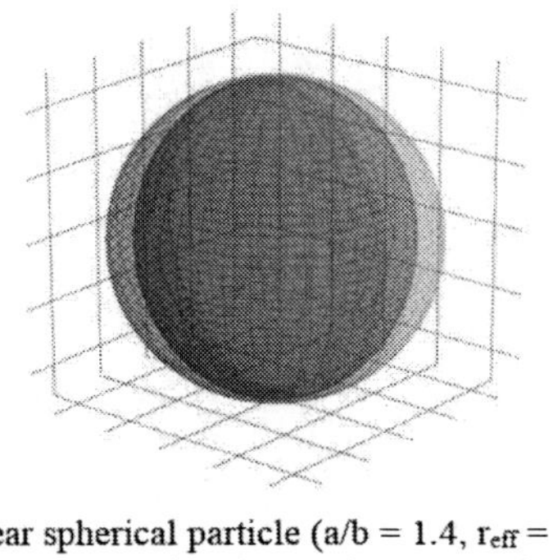

Fig. 3 Near spherical particle (a/b = 1.4, r_{eff} = 0.25 μm) found to reproduce the observed PLDR and LR values at all measurement wavelengths.

A slight deviation from the measured mean values is found at all wavelengths, however all calculated parameters are found to be within the measurement errors.

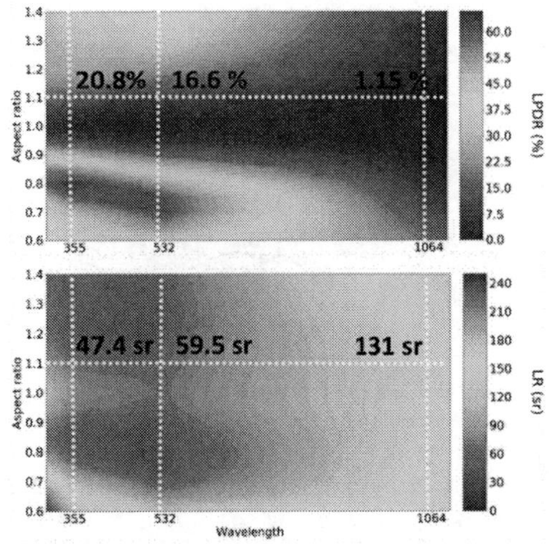

Fig. 4 a) PLDR and **b)** LR, calculated with the T-matrix code for near-spherical smoke particles with r_{eff} = 0.25 μm, v_{eff} = 0.03 and m = 1.55 + i0.025. The measured PLDR and LR at 355, 532 and 1064 nm for near-spherical smoke particles with mean axial ratio a/b = 1.4 (white dash lines).

Table 1. Leipzig measurements vs T-matrix simulations for near spherical particles with mean axial ratio of a/b = 1.4, m = 1.55 + i0.025 and r_{eff} = 0.25μm.

	355	*532*	*1064*
		PLDR	
Obs.	22.4 ± 1.5	18.4 ± 0.6	4.3 ± 0.7
Sim.	21.8	18.3	4.5
		LR	
Obs.	40 ± 16	66 ± 12	92 ± 27
Sim.	41.7	64.2	98.5

In an attempt to examine whether more complicated morphologies can reproduce the observations, we also performed scattering calculations with the Superposition T-Matrix Method (STMM) [21] for highly irregular smoke particles with a simple chain-like fractal morphology (Figure 5). Specifically, we considered a soot fractal aggregate with a fractal geometry provided by a diffusion-limited cluster-aggregation (DLCA) model [22]. Here, the fractal morphology of the particle is characterized by the fractal pre-factor $k_f = 1.3$, the fractal dimension $D_f = 1.8$, the number of monomers $N = 90$, and the volume-equivalent radius $R_V = 0.55\ \mu m$. The calculated LPDR of the fractal soot aggregate at 355, 532 and 1064 nm are presented in Table 2 for m = 1.42 + i0.02.

Fig. 5 Model morphology of soot fractal aggregate used in STMM simulations.

Although the use of this simple fractal geometry provides significantly large PLDR, it does not reproduce the measured values of both PLDR and LR, as it was also shown in greater detail in [2] (see Fig. 9 and 10). In order to reproduce the measurements for aged smoke particles the fractals need to be coated by other material, producing a shape that closely resembles the near-spherical shape, as shapes for Type-B, size 11, Vr = 20, found in [2], Fig. 4.

Table 2. Leipzig measurements vs superposition T – matrix simulations for a fractal aggregate with m = 1.42 + i0.02, $D_f = 1.8$, $k_f = 1.3$, $R_V = 0.55\ \mu m$.

	355	*532*	*1064*
		PLDR (%)	
Obs.	22.4 ± 1.5	18.4 ± 0.6	4.3 ± 0.7
Sim.	13	10	3

4. CONCLUSIONS

The use of near-spherical shapes explains at a large extent the exceptional PLDR values observed at stratospheric smoke layers over Leipzig, originated by Canadian wildfires. This hypothesis also reproduce the PLDR and LR spectral dependence found, when the aspect ratio is very close to unity for typical refractive index, and size of smoke particles.

ACKNOWLEDGEMENTS

We acknowledge support of this work by the project "PANhellenic infrastructure for Atmospheric Composition and climatE change" PANACEA (MIS 5021516), the Stavros Niarchos Foundation and the European Research Council under the European Community's Horizon 2020 research and innovation framework program / ERC Grant Agreement 725698 (D-TECT).

REFERENCES

[1] M. Mishchenko et al., Applied optics, 55, 9968–9973 (2016)
[2] H. Ishimoto et al., Atmos. Meas. Tech., 12, 107–118 (2019)
[3] D. Müller et al. J. Geophys. Res., 110, D17201, 5 (2005)
[4] D. Nicolae et al. J. Geophys. Res. Atmos., 118, 2956–2965 (2013)
[5] E. Giannakaki et al. Atmos. Chem. Phys., 15, 5429–5442 (2015)
[6] Tesche et al. J. Geophys. Res. 114, D13202 (2009)
[7] I. Veselovskii et al. Atmos. Meas. Tech., 11, 949-969 (2018)
[8] S. P. Burton et al. Atmos. Chem. Phys., 15, 13453 – 3473 (2015)
[9] M. Fromm et al. J. Geophys. Res. Atmos, 113 (2008)
[10] N. Sugimoto et al. SOLA, 6, 93–96 (2010)
[11] F. Dahlkötter et al. Atmos. Chem. Phys., 14, 6111–6137 (2014)
[12] L. Bi et al. Opt. Express, 26 (2018)
[13] M. Haarig et al. Atmos. Chem. Phys., 17, 10767–10794 (2017)
[14] G. Pappalardo et al. Atmos. Meas. Tech., 7, 2389–2409 (2014)
[15] M. Haarig et al. Atmos. Meas. Tech., 9, 4269–4278 (2016)
[16] M. Haarig et al. Atmos. Chem. Phys., 18, 11847-11861 (2018)
[17] P. C. Waterman, Phys. Rev. D, 3, 825 – 839 (1971)
[18] M. Mishchenko et al. J. Quant. Spectrosc. Radiat. Transfer, 60, pp. 309—324 (1998)
[19] T. Murayama et al. Geophys. Res. Lett., 31, L23103 (2004)
[20] M. Fiebig et al. J. Geophys. Res. Atmos., 107 (2002)
[21] D. W. Mackowski, J. Quant. Spectrosc. Radiat. Transfer 133, 264–270 (2014)
[22] S. Jungblut et al. J. Phys. Chem. C, 123 (1), 950–954 (2019)

LIDAR BASED SEPARATION OF POLLUTED DUST OBSERVED OVER WARSAW (CASE STUDY ON 09 AUGUST 2013)

Dominika Szczepanik[1], Eleni Tetoni[2], Dongxiang Wang[1], Iwona S. Stachlewska[1*]

[1]*University of Warsaw, Faculty of Physics, Pasteura 5, 02-093 Warsaw,*
[2]*German Aerospace Center, Institute of Atmospheric Physics, Muenchener Strasse 20, 82234 Wessling*
**Email: iwona.stachlewska@fuw.edu.pl*

ABSTRACT

This paper presents preliminary results of using an extended POLIPHON method for separation of dust and non-dust aerosol backscatter coefficient, applied on a case study of 9th August 2013. That day, long-range transport of mineral dust over EARLINET-ACTRIS lidar site in Warsaw was observed with the 8-channel PollyXT-UW lidar. The dust particles were also observed by the Cloud-Aerosol Lidar with Orthogonal Polarization (CALIOP) on board the CALIPSO satellite. The backward trajectories calculated using the HYSPLIT model confirmed the air-mass transport from Northern Africa. Results yield possible dust separation for the mixture of dust with other aerosol types, such as pollution, marine type, etc.

1. INTRODUCTION

Mineral dust is one of the major components of the atmospheric aerosol system that influences climatic and environmental conditions in several ways. It can affect climate directly by scattering and absorbing the incoming solar and outgoing terrestrial radiation fluxes. Indirectly, it can cause changes in the cloud cover, acting as nuclei for water vapor to create cloud droplets, which changes the optical properties and lifetime of clouds and may influence precipitation formation. The main sources of mineral dust observed over eastern and southern Europe are located in the North African and the Middle East deserts. Large amounts of mineral dust can be advected over long distances in the free troposphere, and therefore observed far from the source in different places over the globe [1]. In the case of climate modeling, different properties of the fine and coarse mode dust particles must be taken into consideration. The coarse mode covers the particle size spectrum with a diameter higher or equal 1 μm, whereas the fine mode is composed of particles with sizes below 1 μm. The particle linear depolarization ratios at 532 nm of 39±4% were reported for airborne mineral dust dominated by the coarse mode particles and of 16±5% for dust dominated by the fine mode particles [2].

In this study, the results of the separation of the contribution of the coarse dust, fine dust and non-dust within the particle backscattering coefficient profile for Raman lidar observations in Warsaw is presented.

2. METHODOLOGY

Data collected at the Remote Sensing Laboratory (RS-Lab) in Warsaw using the 8-channel PollyXT-UW lidar (details in [3]) were used. The quick look plots available via the PollyNET Website (*http://polly.tropos.de*) were examined to find the good-candidate case, which was defined as a day with strong aerosol load and depolarizing layer in the free troposphere that likely contained long-range transported mineral dust. The data provided by the CALIOP lidar onboard the CALIPSO satellite were searched for the corresponding overpass, defined as the nearest in time and distance from the RS-Lab site, with signatures of possible dust over Warsaw latitude. In this way, for analyses data on 9th August 2018 were selected. The HYSPLIT model [4] was used to calculate the backward trajectories of air-masses movement to estimate the possible source of aerosols with the use of the GDAS meteorological data for air-masses at 1.0, 2.5 and 3.8 km, lasting 168 h and ending at 01:00 UTC, 9th August 2018 in Warsaw.

The particle backscatter and extinction profiles at 355 and 532 nm were obtained independently, using the classical Raman approach [5] with an assumption of the Ångström exponent (AE=1) and the calibration of backscattering profiles at a range above 6 km. The particle depolarization ratio profiles were obtained at 355 and 532 nm, as described in [3], using the ±45° calibration method. All profiles were calculated with a coarse temporal resolution of 2 h for 7.5 m vertical resolution and smoothed with a running mean of 41 range-bins.

https://doi.org/10.1051/epjconf/202023702018

The separation algorithm was implemented in the way described in [1], using the threshold values on the particle depolarization ratios as provided in [1,2]. The separation algorithm was applied in two ways called the one- and two-step separation method. As a result of the one-step separation algorithm, the profiles of the particle backscattering coefficient due to dust and non-dust particles are separated within the particle backscattering coefficient profile. The dust particles contribution to the backscattering coefficient is obtained for the particle depolarization ratio δ_p between $0.05 \leq \delta_p \leq 0.31$. The non-dust contribution to the particle backscatter coefficient is obtained by subtraction of the dust backscattering contribution from the particle backscattering profile.

The two-step separation gives the contribution of dust coarse, dust fine and non-dust particles to the initial profile of the particle backscattering coefficient. Firstly, the coarse dust particles contribution to the backscattering coefficient is obtained for $0.24 \leq \delta_p \leq 0.39$ and subtracted from the particle backscattering profile. Secondly, from the remaining backscattering profile, the fine dust contribution is obtained for $0.05 \leq \delta_p \leq 0.24$, and again it is being subtracted. The extant particle backscattering profile is assumed to be the non-dust contribution to particle backscattering profile.

3. RESULTS

On 9th August 2013, a strong aerosol loading was observed up to 6 km in several layers, as depicted in the ground-based PollyXT-UW lidar quick-look (Fig. 1) and the satellite CALIOP lidar data product related to aerosol-typing (Fig. 2).

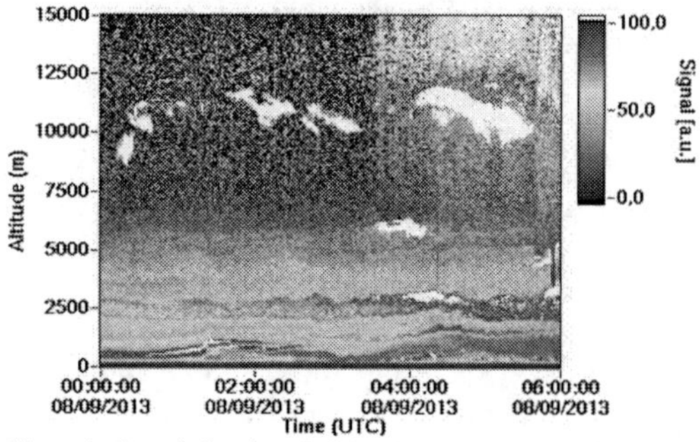

Fig. 1:Quick-look of the range corrected signal of PollyXT-UW lidar at EARLINET-ACTRIS site in Warsaw, Poland.
Source:http://polly.rsd.tropos.de

On the PollyXT-UW lidar quick-look in Warsaw aerosol layers can be seen, within the boundary layer below 1.0 km, at 2.5 km and 3.7 km. The CALIOP lidar also detected aerosols up to 6 km during the corresponding CALIPSO overpass. For the layers observed, the backward trajectories were calculated using the HYSPLIT model (Fig. 3).

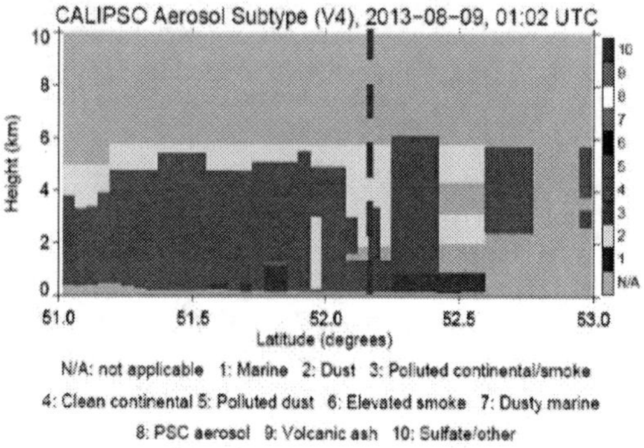

Fig. 2: CALIOP lidar aerosol subtype plot, with the latitude of Warsaw (52.22977) marked as red dashed line; dust (2) and polluted dust (5) particles were observed during this CALIPSO overpass in the nearest distance from the EARLINET-ACTRIS lidar site in Warsaw.

According to the backward trajectories, at both of those altitudes in the free troposphere (Fig. 3, in blue and green), one may expect aerosol load likely containing Saharan dust particles uplifted from over North Africa. As those warm and dusty air-masses were traveling for more than 6 days long before reaching Central Poland, it is highly probable that the mineral dust observed over Warsaw lidar site on that day, was composed of a mixture of different aerosol types; the air-masses mixing is indicated in Fig. 3 bottom panel. The aerosol carried by the air-mass marked in blue, from the area over the Mediterranean Sea that may contain marine aerosol mixed with the air-mass marked in green more likely containing pure Saharan dust. The dust observed over Warsaw is expected to be rather elderly, and therefore sorted due to dry deposition caused by gravitational sedimentation of coarse particles along its pathway. Hence, it shall have different optical properties than fresh/pure dust.

In Fig. 4, PollyXT-UW lidar-derived optical properties of the atmosphere are shown. The

https://doi.org/10.1051/epjconf/202023702018

profiles of the particle backscattering coefficients show peaks of an increased aerosol load in the layers between 2.0-3.0 km and 3.2-4.2 km. The peaks are less pronounced in the particle extinction coefficient, but still visible, especially for 355 nm. The noise in the extinction profiles increases with altitude, however, the obtained with Raman approach extinction profiles are rather high-in-range for the low smoothing applied.

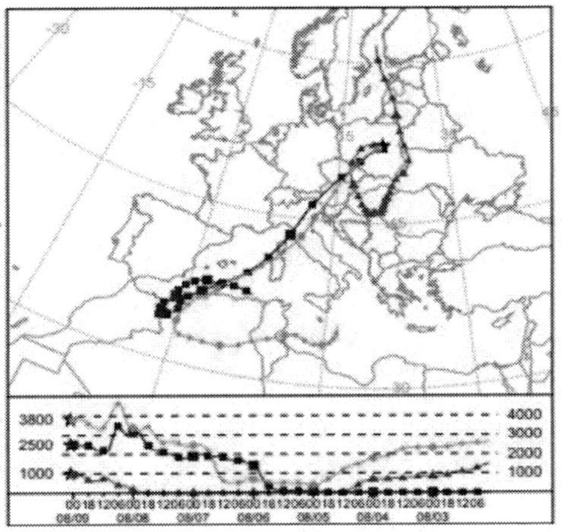

Fig. 3: HYSPLIT 168h-backward trajectories calculated for Warsaw, at 1.0, 2.5 and 3.8 km, ending at 01:00 UTC on 9th of August 2013.

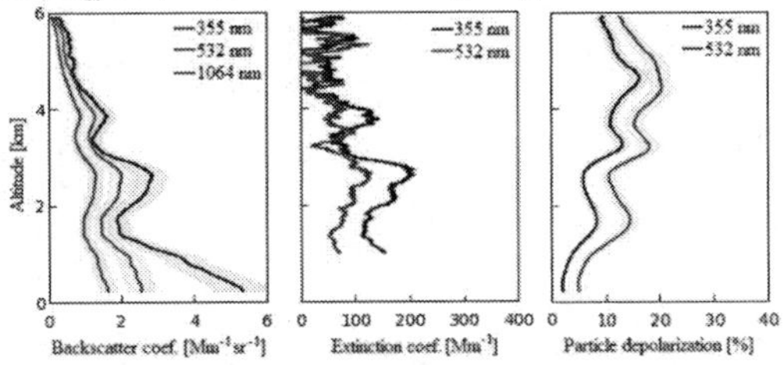

Fig. 4: PollyXT-UW profiles of the wavelength-dependent particle backscattering coefficient (left), particle extinction coefficient (center) and particle depolarization ratios (right), derived as average of 0:00-2:00 UTC on 9th August 2013 at the EARLINET-ACTRIS site in Warsaw.

The particle depolarization ratio increases with altitude, from 3-5 ±0.5% close to the ground to even 15-20 ±2% at 5 km and then starts decreasing. This indicates the existence of a high number of non-spherical particles. The particle lidar ratios (LR_{355} and LR_{532}) and Ångström exponent derived from the particle extinction AE_{α} (355/532) and backscattering coefficient profiles AE_{β}(355/532), AE_{β}(532/1064), AE_{β}(355/1064) were calculated. All of the AE_{β} were of 0.5±0.15. The AE_{α} was of 0.46±0.1 in the layer at 3.0-3.3 km and of 0.32±0.2 at 3.8-4.2 km. These values confirm the existence of large particles. The LR_{532} of 55±5 sr and LR_{355} was of 60±7 sr, hence within the range of the typical lidar ratios reported for Saharan dust between 50-60 sr [6]. For Leipzig, Saharan dust-related LR_{532} of 59 ± 11 sr and LR_{355} 65 ± 12 sr were reported [7]. Clearly, the values found for Warsaw correspond well with the latter.

The selected case was a great candidate to test the feasibility of the separation algorithm for Warsaw lidar data. The one- and two-step separation results are shown in Figs. 5 and 6, respectively. One can see that there is an interesting reverse relation between the particle backscattering coefficient and also non-dust contribution to the backscattering coefficient and particle depolarization ratio. When the peak in particle depolarization ratio is being observed, there is a local minimum in the backscattering coefficient profile. This kind of relation is not observed between dust contribution to the backscattering coefficient and particle depolarization ratio. It might be caused by the hydration of small aerosol particles, which causes the change of the shape of particles from non-spherical to more spherical ones. It also reveals that peaks of the particle depolarization ratio are likely to be caused by dust/coarse dust occurrence.

The one-step separation method results, depicted in Fig. 5, indicate that dust particles could have been present from 0.5 km up to 6 km. There was less dust than other aerosols only in the lower troposphere (<3 km). The same local maxima as in Fig. 4 are captured. The results lead to the conclusion that on 9th August 2013 a strong desert dust influx over Warsaw took place.

The two-step separation method results, depicted in Fig. 6, indicate that, if any, the coarse mode dust particles have been present only at 2.7-5.5 km (orange profile, left subplot) and at about 1.8 km. There was a quite large amount of fine mode dust particles (purple profile, right subplot), with several local maxima exceeding the non-dust contribution. The much lower load of the coarse dust than the fine dust particles can be explained as due to the gravitational fallout of coarse dust particles during the long-range transport from Sahara to Warsaw.

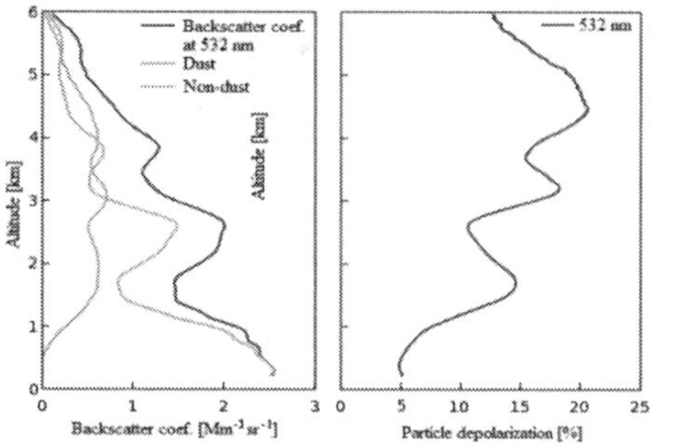

Fig. 5: The one-step separation algorithm applied to 532 nm particle backscattering coefficient profile for assessment of the dust and non-dust contribution and the particle depolarization ratio profile at 0:00-2:00 UTC on 9th August 2013 at Warsaw lidar site.

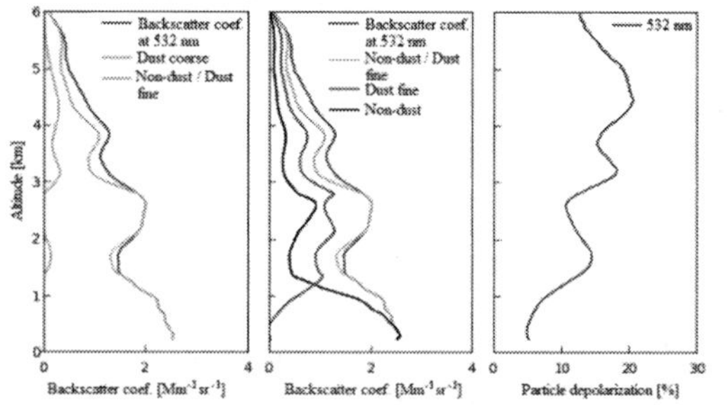

Fig. 6: The two-step separation algorithm applied to 532 nm particle backscattering coefficient for assessment of the coarse mode dust, fine mode dust and non-dust contribution at 0:00-2:00 UTC on 9th August 2013 at Warsaw lidar site.

4. CONCLUSIONS

The first results of a successful application of the separation algorithm proposed by [1] applied to the lidar data at the Warsaw EARLINET-ACTRIS site are presented. In the near future, adjustments of the evaluation procedure will be undertaken. For example, the profile obtained for testing the approach was derived for rather long temporal averaging, so as to work with a significantly high aerosol loading in order to ease the particle contribution separation. The threshold values currently used for testing of the separation algorithms ([1, 2]) are characteristic for pure mineral dust aerosol. In general, this shall not be the case for Warsaw affected by complex aerosol injections not only into the free troposphere [8] but also into the boundary layer [9, 10]. It is expected that the values of the particle depolarization thresholds must be different for mixtures of non-pure/non-fresh dust with other types of aerosol observed over Warsaw (e.g. local and advected anthropogenic pollution [10], long- and short-range transported biomass-burning [11, 12, 13],or Arctic marine air [9]). The higher number of dust observation episodes over Warsaw will be analyzed to refine the threshold values, which will be used in the separation algorithm.

ACKNOWLEDGEMENTS

The research leading to these results was founded by the ESA-ESTEC within the Technical assistance for Polish Radar and Lidar Mobile Observation System (POLIMOS 4000119961/16/NL/FF/mg).

The lidar development was financed by the FNITP, Poland (519/FNITP/115/2010).

We acknowledge the NOAA Air Resources Laboratory (ARL) for the provision of the HYSPLIT transport model (http://www.arl.noaa.gov/ready.php) used for the interpretation for the results obtained in this publication.

CALIPSO data were obtained from the NASA Langley Research Center Atmospheric Science Data Center.

REFERENCES

[1] R. E. Mamouri and A. Ansmann, Atmos. Meas. Tech., 7, 3717-3735 (2014)
[2] T. Sakai, et al. Appl. Opt., 49, 4441-4449 (2010)
[3] R. Engelmann, et al. Atmos. Meas. Tech., 9. 1767-1784 (2016).
[4] R.R. Draxler and G.D. Rolph, Silver Spring, (2012)
[5] A. Ansmann, et al. Opt. Lett., 15(13), 746-748 (1990)
[6] A. Nisantzi, et al. Atmos. Chem. Phys., 15, 7071-7084 (2015)
[7] A. Papayannis, et al. J. Geophys. Res., 113, D10204 (2008).
[8] L. Janicka et al. Atmos. Environ., vol. 169, pp. 162-174 (2017)
[9] D. Wang, et al. Atmos. Chem. and Phys. 19, 13097–13128 (2019)
[10] I. S. Stachlewska, Rem. Sens., vol. 9 (11), pp. Art. 1199 (2017)
[11] L. Janicka and I. S. Stachlewska, Atmos. Chem. and Phys. Disc. (2019)
[12] I. S. Stachlewska, et al. Rem. Sens., vol. 10(3), pp. Art. 412 (2018)
[13] I. S. Stachlewska, et al. Atmos. Res., vol. 194, pp. 258-267 (2017)

LIDAR DERIVED FINE SCALE RESOLUTION PROPERTIES OF TROPOSPHERIC AEROSOL MIXTURES

Lucja Janicka[1], Christine Böckmann[2], Iwona S. Stachlewska[1*], Dongxiang Wang[1]

[1]*University of Warsaw, Faculty of Physics, 02-093 Warsaw, Poland*
[2]*University of Potsdam, Institute of Mathematics, D-14476 Potsdam, Germany*
**Email: iwona.stachlewska@fuw.edu.pl*

ABSTRACT

Fine scale resolution analysis was applied to the complicated aerosol structures observed with the PollyXT-UW lidar over Warsaw during the night of 9/10 August 2015. The full sets of the particle optical properties profiles, so called 3β+2α+2δ+wv (3 backscattering, 2 extinction coefficients, 2 depolarization ratios and water vapour mixing ratio), were obtained to discriminate multiple aerosol height sectors, which were then used for the microphysical properties inversion. The statistical characterization of the main aerosol/mixture types was obtained.

1. INTRODUCTION

For studies on influence of atmospheric aerosol on the earth radiative budget it is crucial to know the optical and microphysical properties of aerosol particles related to particular aerosol type, the altitude of aerosol layers suspension, and the type of ground surface over which the layers are present. Additional challenge arises, when suspended aerosol does not represent pure aerosol type, but composes a mixture of different ones. Exploration of the aerosol mixtures properties is a difficult task and requires detailed analysis in the fine scale.

The presented study concerns multiple aerosol layers detected over Warsaw during a heat wave event over Poland, known as favoring the long-range transport of biomass burning aerosol [1]. Biomass burning particles advection (2-3 days old) from over France, Germany, and Czech Republic, was observed in majority of the layers over Warsaw, being sometimes mixed in with an anthropogenic pollution from sources in Poland [2].

2. METHODOLOGY

In the current study, the data from the 8-channel PollyXT-UW lidar [3], installed at the Remote Sensing Laboratory (RS-Lab) of the Institute of Geophysics, Faculty of Physics, University of Warsaw (52.21°N, 20.98°E, 112 m a.s.l) were analyzed. The RS-Lab operates within the European Aerosol Lidar Network EARLINET [4]. The quasi-continuous vertical lidar measurements with the 355, 532, 1064 nm laser light, provide the elastic backscattering signals at all three emitted wavelengths, the elastic depolarized signals at 355 and 532 nm, the non-elastic Raman signals on N_2 molecules at 387 and 607 nm and on H_2O at 407 nm. The raw signal resolution is 30 s and 7.5 m.

The sets of the particle optical properties profiles (3β+2α+2δ+wv) are derived using the classical Raman approach, according to the evaluation scheme given in [5], and then smoothed using a customized algorithm detailed in [2]. The profiles of linear particle depolarization ratio were derived with the ±45° calibration method [3]. The relative humidity profiles were derived as in [6].

The microphysical properties (complex refractive index (m), particle size distribution, effective radius (R_{eff}), surface (S) and volume (V) concentrations) were obtained using the inversion algorithm developed at University of Potsdam; details are given in [7],[8],[9]. For all calculations the spherical model was used, as the mineral dust contribution was not expected. The regularization method based on the truncated singular value decomposition (TSVD) was applied.

The optical properties were calculated for each consecutive 30 minutes from 19-02 UTC on the night of 9/10 August 2015. In total 9 sets of 3α+2β+2δ+wv profiles were obtained. In each set, multiple height sectors were discerned; based mainly on the extinction and relative humidity profiles. The mean values of optical properties were obtained in each height sector, and based on those the intensive aerosol properties calculation and microphysical inversion followed. Finally, the height sectors were grouped in the main aerosol/mixture layers and the optical properties within them were obtained in a statistical way. Extensive analysis of the results is given in [2].

3. RESULTS

3.1 Optical properties

Fine scale analysis of optical properties derived in all height sectors (for brevity only some are indicated in Fig.1) allows for discrimination of the same aerosol type layers in statistical way. In lowermost Layer 1 (Fig.1, in black), anthropogenic industry-driven pollution uplifted from boundary layer was likely mixed with biomass burning. Here the lidar ratios are being relatively low: LR_{355} of 55±6 sr and LR_{532} of 43±4 sr with Ångström exponent AE_{α} of 1.48±0.10. Above it, Layer 2 contained moderately-fresh biomass burning aerosol. Here the lidar ratios were significantly higher with the mean values of 76±7 sr for LR_{355} and 62±4 sr for LR_{532}. The AE_{α} of 1.35±0.18 was only slightly lower. For both layers, the particle depolarization ratios were similar at each wavelength: δ_{355} of 1.5-1.6 % and δ_{532} of 3.5-3.6 %.

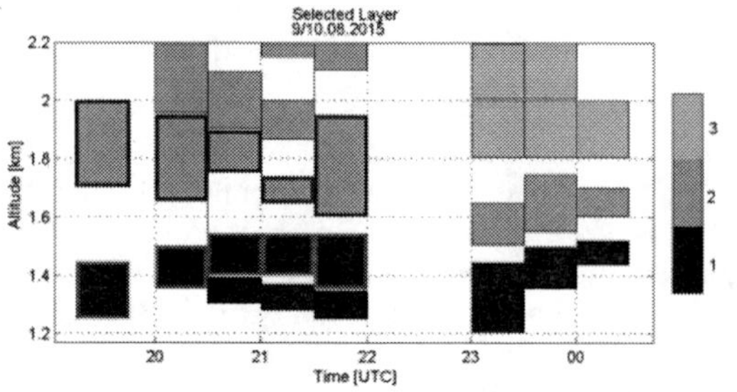

Figure 1: Schematic plot of aerosol layers derived on the night of 9/10 August 2016 from PollyXT-UW at the EARLINET-ACTRIS lidar site in Warsaw. Each color depicts different aerosol type. Blue and red rectangles mark layers for which microphysical inversion results are shown in Fig 2.

3.2 Microphysical properties

The microphysical properties derived in height sectors marked by red and blue rectangles in Fig.1, reveal very good agreement of the obtained results. The particle size distributions (Fig.2) have a fine mode dominating (~0.12 μm) and much smaller modes at larger radii (~0.75 and ~1.5μm). This consistency is clear in reference to individual height sectors and the mean values of each layer type, whereby the mean distributions for the biomass burning aerosol (Layer 2) and the anthropogenic pollution (Layer 1) differ only in terms of the 0.75 μm mode existing in the latter.

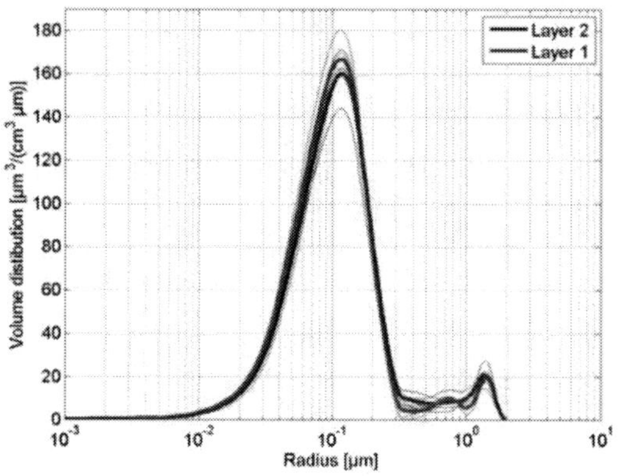

Figure 2: Particle size distributions obtained by the microphysical inversion of the PollyXT-UW lidar data at the night of 9/10 August 2016 at in Warsaw. Thin lines denote individual distributions for each height sector and the thick lines the mean distributions for each aerosol-type layer .

Table 1: Mean microphysical properties for aerosol layers derived in height sectors, depicted in rectangles in Fig. 1, with the standard deviation.

R_{eff} [μm]	*Re* m	*Im* m	S [μm²/cm³]	V [μm²/cm³]	SSA_{355}	SSA_{532}
Layer 1 mixed BB & AP						
0.15 ± 0.01	1.54 ± 0.01	0.013 ± 0.002	783 ± 34	40 ± 2	0.92 ± 0.01	0.90 ± 0.01
Layer 2 long-range transported BB						
0.17 ± 0.02	1.55 ± 0.02	0.028 ± 0.003	757 ± 52	43 ± 5	0.85 ± 0.02	0.83 ± 0.02

In Table 1, the mean microphysical properties for each aerosol layer are listed. The effective radius R_{eff}, the volume V and surface S concentrations are the same in the standard deviation limit. The real part of refractive index is also similar for both aerosol types. However, the imaginary part of 0.013±0.01 (Layer 1) and 0.028±0.02 (Layer 2) differ distinctly, indicating higher absorption for the biomass burning layer. The single scattering albedo (SSA) for this layer is very low ~0.84, as expected. In the anthropogenic pollution layer the SSA is of ~0.91, which is still indicating possible absorption, thus pollution mixed with biomass burning is confirmed.

4. CONCLUSIONS

In presented study, optical and microphysical properties of aerosol over Warsaw during a heat wave event were derived and analyzed in fine temporal and spatial scales. The analysis allowed for discrimination of aerosol layers dominated by the long-range transported biomass burning (lidar ratios higher than ~20 sr, twice higher imaginary part of the refractive index) and those being mixed with the anthropogenic pollution (additional mode at radius of ~0.75, higher SSA). This case study has a great potential for further use in aerosol type separation algorithms.

ACKNOWLEDGEMENTS

The PollyXT-UW lidar was developed within R&D collaboration of the University of Warsaw and the Leibniz Institute of Tropospheric Research (TROPOS), in the PollyXT development group of Dietrich Althausen. This development was financed by the Polish Foundation of Science and Technology (519/FNITP/115/2010).

The RS-Lab site provides the data products to the PolandAOD-NET (*http://www.polandaod.pl*), the EARLINET (*https://www.earlinet.org*), and the Polly.NET (*http://polly.tropos.de*) networks.

The algorithms used for lidar data evaluation were developed in the frame of ESA-ESTEC Contract 4000112373/14/NL/CT (MULTIPLY).

The EARLINET is supported by the ACTRIS-2 project, funded by the European Union Research Infrastructure Action under the H2020 specific program for Integrating and opening existing national and regional research infrastructures of European interest under Grant Agreement No. 654109 and No. 739530. The University of Warsaw and the University of Potsdam participate in ACTRIS-2 project as an associate partner without funding.

REFERENCES

[1] I. S. Stachlewska, et al.: Modification of local urban aerosol properties by long-range transport of biomass burning aerosol, Rem. Sens., 10, 412, doi:10.3390/rs10030412 (2018)

[2] L. Janicka, and I. S. Stachlewska: Properties of biomass burning aerosol mixtures derived at fine temporal and spatial scales from Raman lidar measurements: Part I optical properties, Atmos. Chem. Phys. Discuss. (2019)

[3] R. Engelmann, et al.: The automated multiwavelength Raman polarization and water-vapor lidar PollyXT: The next generation, Atmos. Meas. Tech., 9, 1767-1784, doi:10.5194/amt-9-1767-2016 (2016)

[4] G. Pappalardo, et al.: EARLINET: Towards an advanced sustainable European aerosol lidar network, Atmos. Meas. Tech., 7, 2389-2409, doi: 10.5194/amt-7-2389-2014 (2014)

[5] H. Baars, et al.: An overview of the first decade of Polly NET: An emerging network of automated Raman-polarization lidars for continuous aerosol profiling, Atmos. Chem. Phys., 16, 5111-5137, doi:10.5194/acp16-5111-2016 (2016)

[6] I. S. Stachlewska, et al.: Raman lidar water vapour profiling over Warsaw, Poland, Atmos. Res., 194, 258-267, doi:10.1016/j.atmosres.2017.05.004 (2017)

[7] C. Böckmann, et al.: Microphysical aerosol parameters from multiwavelength lidar, J. Opt. Soc. Am. A, 22(3), 518–528, doi:10.1364/JOSAA.22.000518 (2005)

[8] S. Samaras, et al.: Using Raman-lidar-based regularized microphysical retrievals and Aerosol Mass Spectrometer measurements for the characterization of biomass burning aerosols, J. Comput. Phys., 299, 156-174, doi:10.1016/j.jcp.2015.06.045 (2015)

[9] D. Müller, et al.: Microphysical particle properties derived from inversion algorithms developed in the framework of EARLINET, Atmos. Meas. Tech., 9, 5007-5035, doi:10.5194/amt-9-5007-2016 (2016)

LIDAR MEASUREMNT ON DUST TRANSPORT FROM THE SAHARAN DESERT TO THE IRAN PLATEAU

Hossein Panahifar [1] *, Ruhollah Moradhaseli[3], Hadi Bourzoie[4], Mahdi Gholami[1], Hamid Reza Khalesifard[1,2]

[1]*Department of Physics, and* [2]*Research Center for Climate Change and Global Warming, Institute for Advanced Studies in Basic Sciences, 4513766731 Zanjan, Iran*
[3] *Physics Department, Faculty of Science, Zanjan Branch, Islamic Azad University, Zanjan, Iran*
[4] *Faculty of Engineering, Sabzevar University of New Technology, 9615131113 Sabzevar, Iran*
**Email: h.panahifar@iasbs.ac.ir*

ABSTRACT

Optical properties of long-range Saharan dust particles transported to the Iran Plateau have been investigated. The results were derived from the measurements of a dual-wavelength Depolarized backscatter/Raman lidar and a Cimel CE318-2 sunphotometer. Observations were performed in Zanjan, Northwest Iran. The backward trajectory analysis show that the lofted dust plumes come from the Saharan desert and travel along Mediterranean Sea and Turkey toward Iran. The lidar ratio within the lofted dust layer has been found with mean values of 50 sr at 532 nm. For the depolarization ratio, mean values of 25% have been found.

1. INTRODUCTION

Atmospheric mineral dust can be transported more than tens of thousands of kilometers away from its sources in the free troposphere [1-3]. Dust is an important component of the atmospheric aerosol system and sensitively influences environmental and climatic conditions on the regional and global scales especially in northern hemisphere [4, 5]. The Saharan desert as a main part of the dust belt, is the world largest mineral dust source [6, 7]. The Iran Plateau (~26°-36° N, ~45°-61° E) which lies in the middle of the global dust belt is frequently affected by local, regional and trans-regional dust storm outbreaks.

Up to now, few works have been done to characterize the atmospheric aerosol properties as well as their transport scheme over the Iran Plateau. Khalesifard et al. installed a depolarized lidar to investigate the atmosphere above Tehran, the capital of Iran. They have shown that the urban pollution and polluted dust are the dominant contributors to the atmospheric aerosol burden most of the time, especially during the cold season. Dust particles are also the dominant atmospheric aerosol fraction during the warm season [8, 9]. Panahifar et al. have reported a long-range dust transport from Arabian Peninsula to Tehran, Iran [10]. Bayat et al. and Masoumi et al. used a sun-photometer and measured column integrated aerosol properties for the atmosphere over Zanjan. They characterized the Mesopotamian region as the most influencing dust source over the measurement site [11, 12]. Vishkaee et al. used a backscatter lidar to investigate some cases for transport of dust to Northwest of Iran during summer and winter times. They reported the Tigris and Euphrates basin and a dry region between the Aral and Caspian seas as sources for the mentioned cases [13]. There are also a few reports that are mentioning some parts of the Iran Plateau as sources for dust events. [3, 6].

This article provides a scenario of long-range dust transport from Northern Africa to Iran. In Sect. 2 the measurement site, the used lidar system, and the used auxiliary data are described. Sect. 3 gives an overview of the observations and two case studies are presented in detail. The conclusions are discussed in Sect. 4.

2. INSTRUMENTATION AND MEASUREMENTS

The **I**ASBS **R**emote sens**I**ng **S**tation (IRIS) in Zanjan is equipped with a Depolarized backscatter /Raman lidar and a Cimel CE318-2 sun-photometer (SPM) [11]. Both instruments are deployed at the IASBS Physics Department building (36.70° N, 48.51° E, 1800 m above mean sea level (amsl)). The lidar, has a frequency doubled pulsed Nd-YAG laser (pulse duration: 10 ns, energy per pulse: 150 mJ @ 1064 nm and 130 mJ @ 532 nm) with a linear polarized output, as its transmitter. The detection unit after a 11″ Schmidt–Cassegrain telescope can separate the parallel and cross-polarized signal components at 532 nm. Further measurement channels collect lidar return signals at 607 nm (nitrogen Raman channel) and 1064 nm (elastic backscatter). The basic temporal and spatial signal resolution, with which the raw signals are stored, is 60 s and 7.5 m, respectively. Figure 1 shows schematic diagram of the IRIS lidar setup.

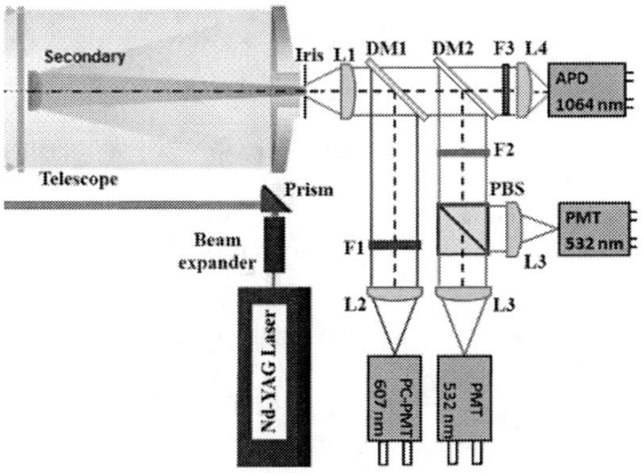

Figure 1 IRIS Polarization/Raman lidar setup

The particle backscatter coefficient at 532 nm, the particle extinction coefficient at 532 nm and the respective particle extinction-to-backscatter ratio (lidar ratio) at 532 nm are determined from the elastic and inelastic return signals in the way described by Ansmann et al. [14]. Aerosol Optical Depth (AOD) at 550 nm taken from the Moderate Resolution Imaging Spectroradiometer (MODIS) Deep Blue collection (MOD08, MYD08) available from the Giovanni Web portal (http://disc.sci.gsfc.nasa.gov/giovanni). The Back-trajectories are calculated with the NOAA HYSPLIT model [15].

3. RESULTS

In the following two cases which are representing trans-continental dust transport from the Saharan desert to Zanjan area have been presented.

3. 1 Case I, 20-22 January 2019

Figure 2 shows the time-height series of volume linear depolarization ratio (VLDR) at 532 nm recorded in Zanjan from 06:40 UTC, 20 January to 06:00 UTC, 22 January 2019. Fig. 2 demonstrates that the depolarization for the boundary layer varies between 10%-15% that can be due either to urban pollution or to temporally happening fog layers over the ground surface.

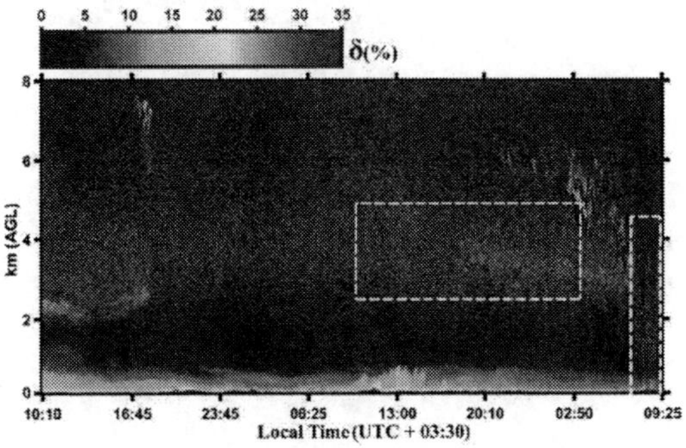

Figure 2 Time evolution of volume depolarization ratio between 06:40 UTC on 20 January 2019 and 06:00 UTC on 22 January 2019.

The Fig. 2 illustrate that at the last 2 hours of the time series (Dashed rectangle 1) the laser beam was obstructed by a dense layer of fog near the ground level and depolarization ratio increased probably due to multiple scattering (VLDR ~ 35%) [16]. A layer of lofted dust (20% ≤ VLDR ≤ 30%) can be seen around 4 km agl (Dashed rectangle 2). The layer spans from 09:30 UTC on 21 January 2019 until the early hour of next day.

The HYSPLIT 96 hour backward trajectories in Figure 3 demonstrate that the observed lofted dust layer (Fig. 2) is transported to Zanjan from North Africa. The backward trajectories that are starting from IRIS on 21 January 2019 at altitudes of 5 and 4.5 km agl, returning back to the south of Algeria and north of Niger respectively (Fig. 3). The air parcels travel over Mediterranean Sea, pass along Turkey-Syria border and eventually reach to the IRIS.

3. 2 Case II, 07-08 April 2013

Figure 4 represents the temporal development of the attenuated backscatter ratio at 1064 nm channel recorded from 09:30 UTC, 07 April to 06:30 UTC, 08 April 2013. Black area denotes periods where no measurements were performed.

The mean (Maximum) AOD variations at 440 nm, recorded on the SPM, during 07-08 April 2013 reaches to 0.25 (0.35).

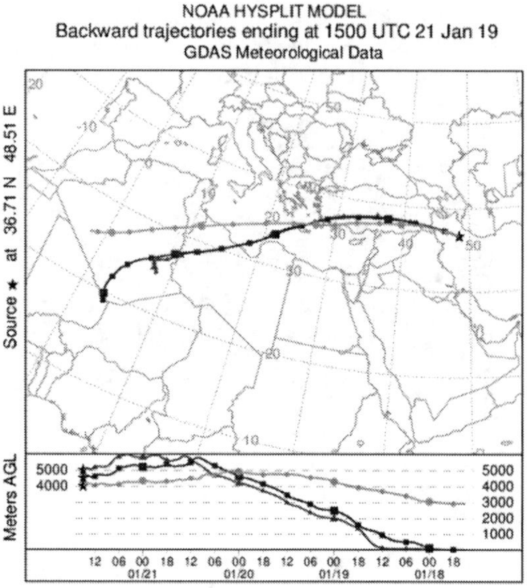

Figure 3 96 h HYSPLIT backward trajectories arriving to Zanjan on 15:00 UTC, 21 January 2019 at 5, 4.5 and 4 km heights agl.

The retrieved Ångström Exponent from the SPM data is obtained as 0.35 ± 0.1 for the mentioned date which corresponds to existence of particles in coarse mode. Fig. 4 depicts a lofted layer of Saharan dust that arrives over Zanjan at ~20:00 UTC on 07 April 2013 at altitudes 3-4 km agl. At the following times, beneath this layer, a thinner dust layer is recorded at altitudes around 2 km agl (Fig. 5c). Vertical profiles of the optical properties of these dust layers are shown in Figure 5. The average (maximum) value of the particle linear depolarization ratio at 532 nm for the recorded layer at ~ 4 km is obtained as 25% (30%).

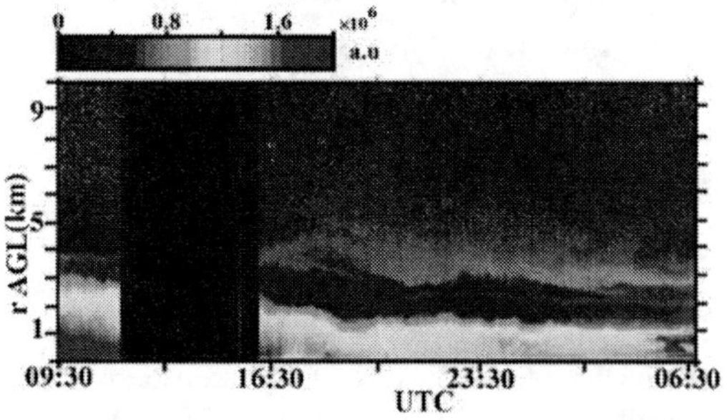

Figure 4 Time-height series of lidar attenuated backscatter profile recorded in Zanjan at 1064 nm between 09:30 UTC on 07 April 2013 and 06:30 UTC on 08 April 2013.

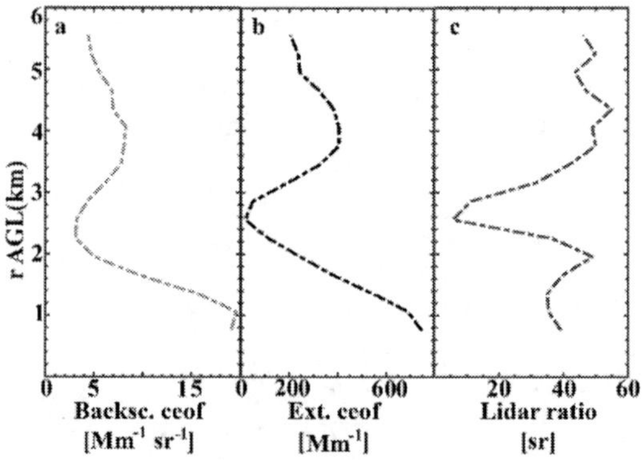

Figure 5 The averaged lidar profiles of a) Particle backscatter coefficient b) Particle extinction coefficient c) lidar ratio at 532 nm recorded during 19:40 UTC, 07 April 2013 to 01:30 UTC, 08 April 2013.

Fig. 5c depicts the mean lidar ratio values for this layer as 50.2 sr at 532 nm and its maximum reaches to 55 sr. These values are comparable with previously reported lidar ratios for transported Saharan dust to European regions [17].

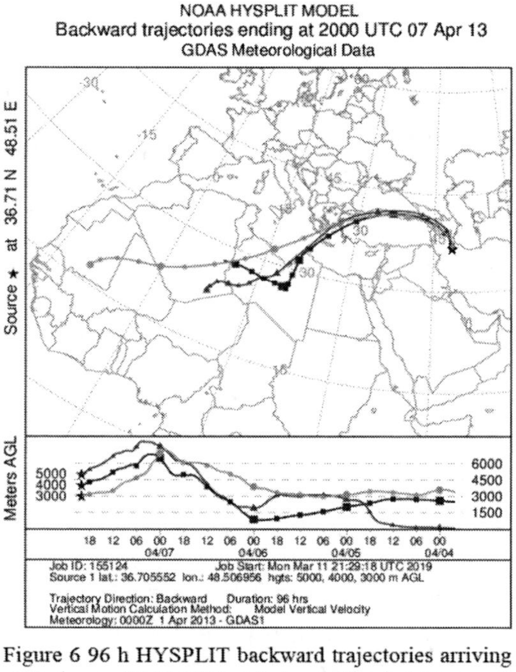

Figure 6 96 h HYSPLIT backward trajectories arriving at Zanjan on 20:00 UTC at 3, 4 and 5 km height.

Figure 6 shows the 96 hours HYSPLIT backward trajectories calculated for arrival heights of 5, 4 and 3 km agl starting from the IRIS. These heights are chosen based on the layer parameters that are represented in Figs. 5a-5c. Fig. 7 shows that the dust plume originates from Algeria and Niger, travels along the Mediterranean Sea, passes over Turkey and finally reaches to Zanjan. This behavior is very similar to Case I.

Figure 7 demonstrates MODIS derived daily Terra AOD Deep blue on 04 April 2013. Fig. 7 is corroborated by Fig. 6 and shows that AOD increased (AOD~1) over Niger.

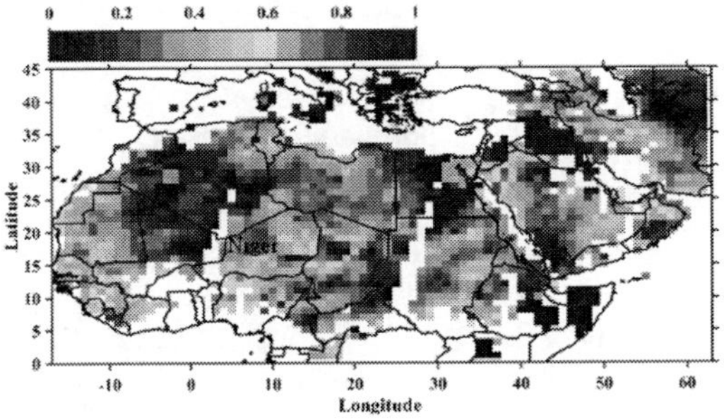

Figure 7 MODIS Terra daily AOD at 550 nm for 04 April 2013, Deep blue AOD for land surfaces.

4. CONCLUSION

We presented two cases where the Saharan dust transported across the Mediterranean Sea and reached to Northwest Iran. The extinction coefficient, the lidar ratio and the particle linear depolarization ratio have been calculated. The lofted Saharan dust layers are recorded at height between 3-5 km agl for both cases. The measured depolarization ratios for both cases range between 20% to 30% at 532 nm [18, 19]. Mean values of the lidar ratio of the long-range transported Saharan dust (07 April 2013) are 50 sr at 532 nm. These lidar ratio value is comparable with typical lidar ratios of the Saharan dust (50–60 sr) and higher than lidar ratios of dust from Middle East (35–45 sr) [17-19]. According to HYSPLIT backward trajectories, for both cases, the lofted dust layer originates from the Saharan desert and travel over Mediterranean Sea and could not mix with dust from Mesopotamia and the Arabian Peninsula.

ACKNOWLEDGEMENTS

The authors would like to appreciate the NOAA Air Resources Laboratory (ARL) for the provision of the HYSPLIT transport model. The MODIS mission scientists and associated NASA personnel is acknowledged for the production of the data used in this research effort.

REFERENCES

[1] Haarig et al., Atmospheric Chemistry and Physics 17, p.10767 (2017)
[2] Mamouri, Rodanthi-Elisavet, et al. Atmospheric Chemistry and Physics 16.21 (2016): 13711-13724
[3] Hofer, Julian, et al. Atmospheric Chemistry and Physics 17.23 (2017): 14559
[4] Myhre, G. and Stordal, J. Geophys. Res., 106, 18193–18204 (2001)
[5] Balkanski, Y., et al. Atmospheric Chemistry and Physics 7.1 (2007): 81-95
[6] Prospero, Joseph M., et al. Reviews of geophysics 40.1 (2002): 2-1
[7] Liu, Xiaohong, et al. Atmospheric Chemistry and Physics 11.15 (2011): 7781
[8] Khalesifard, Hamid Reza, and Hossein Panahifar, Optics and Photonics for Energy and the Environment. Optical Society of America, 2015
[9] Panahifar, Hossein, and Hamid Khalesifard. EPJ Web of Conferences. Vol. 176. EDP Sciences, 2018
[10] Panahifar et al. to be appear in Central Asian Dust Conference (CADUC 2019), Dushanbe, Tajikistan
[11] Bayat, Ali, et al. Atmospheric Measurement Techniques 6.10 (2013): 2659-2669
[12] Masoumi, A., et al. Atmospheric research 120 (2013): 343-355
[13] Abdi Vishkaee, Farhad, et al. Journal of Geophysical Research: Atmospheres 117.D3 (2012)
[14] Ansmann, Albert, et al. Appl. Opt., 31, 7113–7131, 1992
[15] Stein, A. F., et al. Bulletin of the American Meteorological Society 96.12 (2015): 2059-2077
[16] Weitkamp, Claus, ed. Springer Science & Business, 2006
[17] Nisantzi, Argyro, et al. Atmospheric Chemistry and Physics 15.12 (2015): 7071-7084
[18] Groß, S., et al. Atmospheric Chemistry and Physics 19 (2015): 11067-11080
[19] Tesche, Matthias, et al. Chemical and Physical Meteorology 61.1 (2009): 144-164

LIDAR OBSERVATIONS OF BIRCH AND SPRUCE POLLEN IN FINLAND

S. Bohlmann[1], X. Shang[1], E. Giannakaki [1,2], M. Filioglou[1], A. Saarto[3] and M. Komppula[1]

[1] *Finnish Meteorological Institute, P.O. Box 1627, 70211 Kuopio, Finland*
[2] *Department of Environmental Physics and Meteorology, University of Athens, 15784 Athens, Greece*
[3] *Biodiversity Unit of the University of Turku, 20014 University of Turku, Finland*
**Email: Stephanie.bohlmann@fmi.fi*

ABSTRACT

Pollen has various effects on human health and the environment. To understand phenomena behind atmospheric pollen transport and hence improve pollen forecasts, vertically resolved optical properties and geometrical characteristics of the pollen distribution need to be studied. Lidar measurements and especially the particle depolarization ratio have been found to be an excellent tool to track pollen grains. In this study we present first results of atmospheric pollen characterization based on a 11 days period of birch and spruce pollination events.

1. INTRODUCTION

Pollen is one type of primary biogenic atmospheric aerosol. In addition to the commonly known allergic impact, pollen also has various climatic and environmental impacts [1]. Concentration, types and sources are diverse, making the monitoring of pollen challenging. However, to improve pollen forecasts, the investigation of pollen properties is necessary. Using lidar it is possible to investigate the vertical distribution of aerosols. This information can be used for forecasting pollen concentration, their vertical distribution and transport. Recently lidar measurements have shown a strong depolarization signature caused by non-spherical pollen [2,3,4,5]. Thus, in the absence of other depolarizing non-spherical particles, the particle depolarization ratio can be used to track airborne pollen grains. To our knowledge previous studies are only based on single cases or short time periods. Our goal is to improve the existing knowledge of pollen detection using lidar measurements. We conducted a campaign from May to August 2016 at the rural forest station in Vehmasmäki (Finland) to measure different pollen types throughout the pollination season. Finland provides ideal conditions for the observation of pollen as 86% of Finland's total area is forestry land. Pine (*Pinus sylvestris*), spruce (*Picea abies*), and birch (*Betula pubescens, Betula pendula)* trees make up 97% of the growing stock on forest land [6].

During the 4-months campaign, the multi-wavelength Raman lidar Polly^{XT} was simultaneously operated with several in-situ measurements e.g. a Hirst-type volumetric air sampler. 21 different pollen types were detected. In this study we focus on the observation of birch pollen as those are considered to be one of the most allergenic pollen types.

2. METHODOLOGY

2.1 Measurement campaign

The measurement campaign took place from May-August 2016 at Vehmasmäki, Finland (62°44'N, 27°33'E, 190 m above sea level); a rural forest site 18 km from the city center of Kuopio, Eastern Finland. The combination of the multi-wavelength Raman polarization lidar Polly^{XT} and a Hirst-type pollen sampler enables simultaneous measurements of the vertical aerosol distribution and the pollen concentration on ground. Due to the location of the measurement site, far from major aerosol sources like dust and anthropogenic aerosol, the atmosphere is relatively clean and well suitable for pollen studies.

2.2 Instrumentation

Lidar measurements were conducted with the multi-wavelength Raman polarization lidar Polly^{XT} [7,8]. Polly^{XT} – FMI operates at three

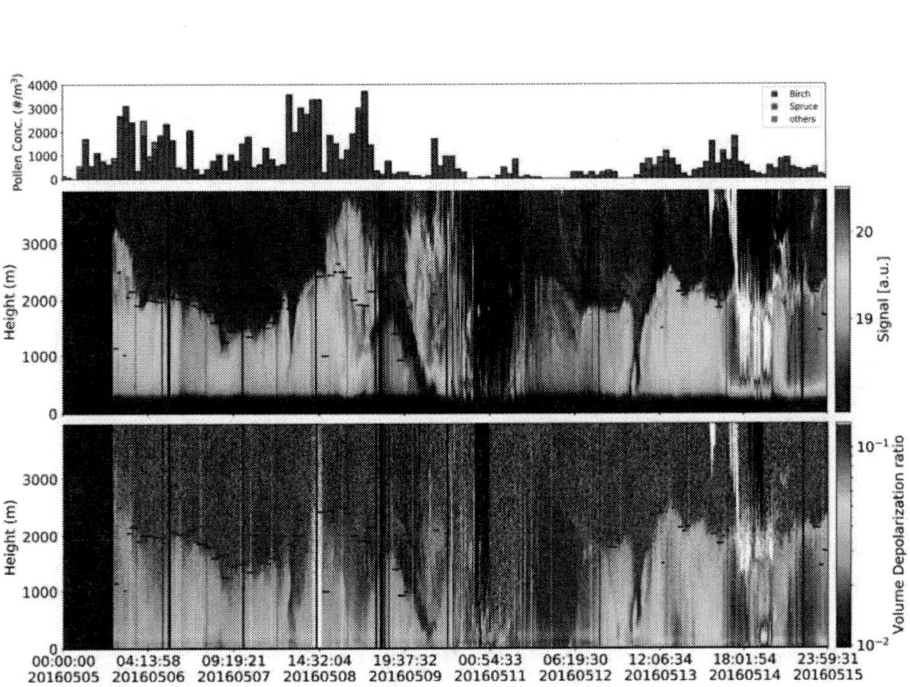

Figure 1 *Overview of the 11 days intense birch pollination period. Upper panel: pollen concentration measured with Hirst-type air sampler. Center panel: Range-corrected lidar signal at 1064 nm. Lower panel: Volume depolarization ratio at 532 nm. Black lines indicate top height of the layers used for pollen characterization.*

emission wavelengths (355, 532, 1064 nm) and seven detection channels. In addition to the three emitted wavelengths, the backscattered signals at the in-elastic Raman shifted wavelengths (387, 407 and 607 nm) and the cross-polarized signal at 532 nm are detected.

A Hirst-type volumetric air sampler located next to the lidar, 4 m above ground, monitored the pollen concentration. The sampling principle is based on the design described by Hirst [9]. This type of spore sampler enables 7-day collection of pollen grains with 2 hour time resolution.

3. RESULTS

Figure 1 shows the temporal evolution of pollen concentration (top panel), range-corrected signal at 1064 nm (middle panel) and volume depolarization ratio at 532 nm (lower panel) of the 11 days period from 5 – 15 May. This period has been selected to present the capability of the measurement setup to provide information on different pollen types. According to the pollen concentration measured by the Hirst-type sampler two periods can be defined. During the first period, from 5 – 10 May the measured birch pollen concentration is high and reaches concentrations of up to 3500 pollen grains per cubic meter. During the second period, from 12 – 15 May, a significant amount of spruce pollen mixed with birch was observed. The volume depolarization ratio during this second period is higher, which indicates the increasing presence of non-spherical particles. All lidar measurements with sufficient signal quality and suitable atmospheric conditions were analyzed in 2 hour time intervals to match the pollen collector resolution. From those measurements, layers with high aerosol load were selected using the gradient method [10], and further layers with a distance less than 500 m from each other were combined to one. For our pollen studies only the lowest layers were considered as those, most likely, contain the highest pollen load. The layer top of the presumed pollen layer is marked with black lines in Fig.1 (middle and lower panel). Mean values of lidar ratio (LR) at 532 nm (top), backscatter-related Ångström exponent (Å) at 532/1064 nm (middle)

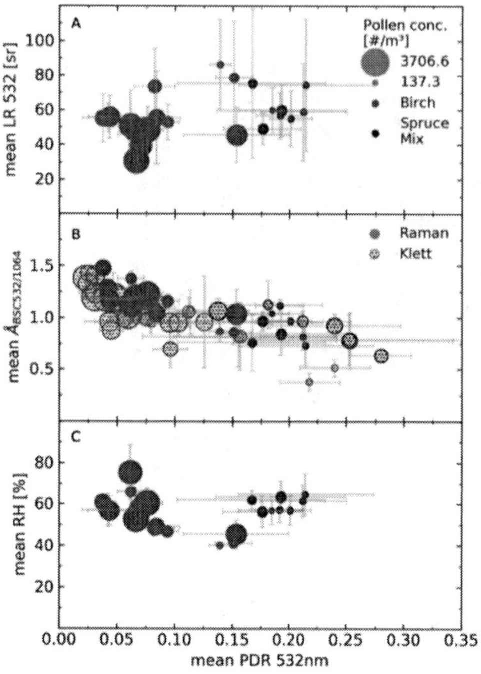

Figure 2 *Layer mean values of lidar ratio at 532 nm (top), backscatter-related Ångström exponent at 532/1064 nm (middle) and relative humidity (bottom) against mean PDR at 532 nm. Error bars denote the standard deviation, size the pollen concentration and color the measurement period.*

and relative humidity (RH, bottom) of the defined layers are shown against the particle depolarization ratio (PDR) at 532 nm in Fig. 2. Error bars denote the standard deviation. Cases marked in green have been measured during the first birch pollen dominated period. During the second period (black dots), birch and spruce pollen have been detected. The lidar ratio ranges in both periods between 20 and 80 sr, with a mean lidar ratio of around 55 sr for the birch period and 62 sr for the spruce mixed period. Figure 2b shows additionally all day-time measurements using the Klett method. A clear separation of the particle depolarization ratio is visible. Particle depolarization ratio in the second period ranges around 24% while it is around 10% during the pure birch period. This higher depolarization is caused by the non-spherical shape of spruce pollen. The Ångström exponent shows a tendency to smaller values with increasing depolarization. The lower Ångström exponent of about 0.8 during the second period is related to the bigger particle size of spruce pollen (~80 µm) compared to birch pollen grains (~20 µm). An external wetting of the pollen grains at high relative humidity (>89% [11]) could change the detected depolarization ratio of the pollen as the shape becomes more spherical. Figure 2c shows that the relative humidity has no effect on depolarization ratio in the range of our measurements as the measured relative humidity during our campaign is rather low.

CONCLUSION

Using lidar measurements it is possible to detect pollen grains in the atmosphere. Lidar-derived parameters provide the possibility of identifying different pollen types. In this study, birch and spruce pollen have been detected. Because of their non-spherical shape spruce pollen show a higher depolarization ratio than birch. In upcoming campaigns our dataset will be extended with more case studies and different pollen types, which will allow a more comprehensive characterization of optical pollen properties.

ACKNOWLEDGEMENTS

This work was supported by the Academy of Finland (project no. 310312).

REFERENCES

[1] A. L. Steiner et al., Geophys. Res. Lett., 42, 3596-3602 (2015)
[2] Y. M. Noh et al., Atmos. Chem. Phys., 13, 7619-7629 (2013)
[3] Y. M. Noh et al., Atmos. Environ., 69, 139-147 (2013)
[4] K. Sassen, Geophys. Res. Lett., 35, L18810 (2008)
[5] M. Sicard et al., Atmos. Chem. Phys., 16, 6805-6821 (2016)
[6] E. Vaahtera et al., Luonnonvarakeskus (2018)
[7] D. Althausen et al., J. Atmos. Oceanic Technol., 26, 2366–2378 (2009)
[8] R. Engelmann et al., Atmos. Meas. Tech., 9, 1767-1784 (2016)
[9] J. M. Hirst, Ann. Appl. Biol., 39, 257-265 (1952)
[10] C. Flamant et al., Bound.-Lay. Meteorol, 83, 247-284 (1997)
[11] P. T. Griffiths et al., Atmos. Sci. Lett., 13, 289-295 (2012)

LIDAR RATIOS OF DUST OVER WEST AFRICA MEASURED DURING "SHADOW" CAMPAIGN

Igor Veselovskii[1], Philippe Goloub[2], Qiaoyun Hu[2], Thierry Podvin[2], Michail Korenskiy[1]

[1]Prokhorov General Physics Institute, Moscow, Russia
[2]Univ. Lille, Laboratoire d'Optique Atmosphérique, Lille, France

ABSTRACT

The lidar ratios of Saharan dust at 355 and 532 nm (LR_{355} and LR_{532}) measured over West Africa during SHADOW field campaign are analyzed. Results demonstrate that even for pure dust, the lidar ratio may present strong height dependence. The possible reasons of height dependence of lidar ratios during strong dust events are considered.

1 INTRODUCTION

Atmospheric dust provides significant impact on the Earth system and this impact still remains highly uncertain. The Sahara desert is the largest source region of mineral dust, so quantification of the Saharan dust optical and microphysical properties is an important contribution to the study of the Earth radiative balance. In modeling the direct aerosol effect, the vertical profiles of the extinction coefficient belong to the basic inputs. When the extinction profiles are derived from the measurements of elastic backscatter lidar, the lidar ratio (extinction to backscattering ratio) is a key parameter in data inversion. The great advantage of Raman and HSRL lidars is in their ability to provide independent measurements of aerosol backscattering and extinction coefficients. The numerous lidar stations regular profile the properties of the Saharan dust layers during their transport to Europe and North America.

The properties of dust particles are modified during transport in atmosphere, so it is important to study the dust properties in the region of origin. Such observations were performed during SAMUM-1,2 experiments in 2006, 2008 (in Morocco and Capo Verde respectively), as well as during SHADOW campaign in Senegal in 2015-2016 in Senegal [1-5]. However, the dust is a mixture of various minerals whose relative abundance depends on dust origin, so even near the source of origin the dust properties may present significant variations. Atmospheric dust in Africa can also contain the products of biomass burning and the local pollutions. As a result, the comparison of lidar ratios observed during different campaigns my look controversy. The averaged values of LR observed during SAMUM-2 campaign, are LR_{355}=52.7±10.2 and LR_{532}=54.2±9.6 [3]. Corresponding averaged value of depolarization ratio at 532 nm is 31%, indicating that heavy dust episodes were predominant. In contrast to SAMUM, where lidar ratios at 355 and 532 nm were close, SHADOW measurements demonstrate that in many dust episodes the lidar ratio at 355 nm (LR_{355}) significantly exceeded LR_{532}, which was related by Veselovskii et al. [2] to increase of the imaginary part of the refractive index in UV. In our presentation we analyze vertical variation of the lidar ratios and consider relationship between LR_{355} and LR_{532} for different dust episodes.

2 EXPERIMENTAL SETUP AND RESULTS OF THE MEASUREMENTS

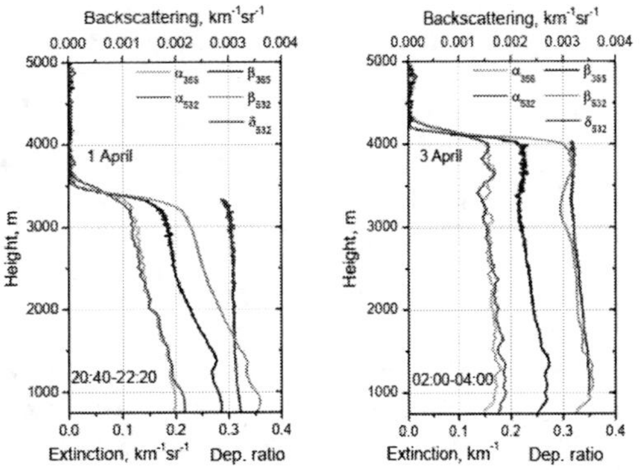

Figure 1. Vertical profiles of aerosol backscattering and extinction coefficients at 355 nm and 532 nm together with particle depolarization ratio δ_{532} for 1 April (20:40-22:20 UTC) and 3 April (02:00-04:00 UTC).

 https://doi.org/10.1051/epjconf/202023702022

The instrumentation site is located at Mbour, Senegal. The observations were performed during March-April 2015 and December 2015-January 2016. Detailed information regarding the SHADOW campaign and the instruments is presented in [2]. The LILAS lidar is based on a tripled Nd:YAG laser with a 20 Hz repetition rate, and pulse energy of 90/100/100 mJ at 355/532/1064 nm. The aperture of the receiving telescope is 400 mm. The lidar allows evaluation of three particle backscattering (355 nm, 532 nm, 1064 nm) and two extinction coefficients (355 nm, 532 nm). To improve the system performance at 532 nm, the rotational Raman channel was used instead of the vibrational one.

2.1. Dust episode on 1-4 April 2015

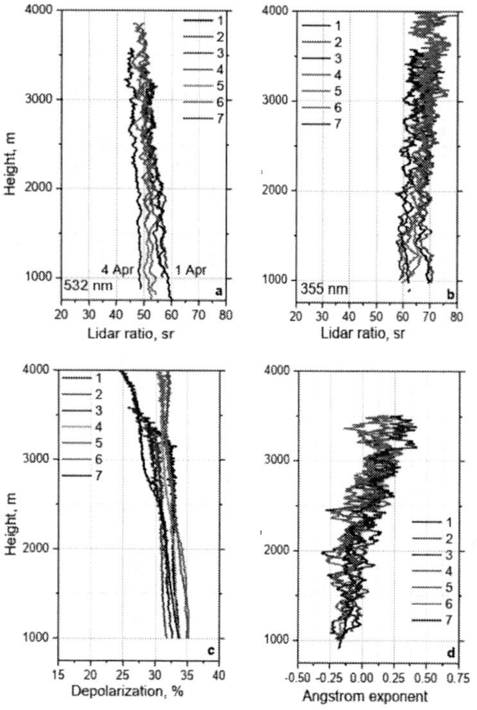

Figure 2. Vertical profiles of (a,b) the lidar ratios at 355 nm and 352 nm, (c) the particle depolarization ratio δ_{532} and (d) extinction Angstrom exponent at 355-532 nm for seven temporal intervals during 1-4 April period.

For period 1-4 April 2015 a strong dust episode occurred. Fig.1. shows vertical profiles of aerosol backscattering β and extinction α coefficients at 355 nm and 532 nm together with particle depolarization ratio δ_{532}, obtained during two measurements sessions on the nights 1-2 and 2-3 April. For both sessions δ_{532} exceeds 30%, and depolarization presents small decrease with height. The values α_{355} and α_{532} are close, while β_{355} exceeds β_{532}, which can be related to increase of the imaginary part m_I of the refractive index of dust at 355 nm, as discussed in [2].

Fig.2 shows profiles of lidar ratios at 355 nm and 532 nm for seven temporal intervals during 1-4 April sessions. At 1000 m height the LR_{532} decreases from 60 sr on 1 April to 50 sr by 4 April. Moreover, for all days lidar ratios at 532 nm decrease with height, reaching the value of 45 sr at 3500 m on 4 April. LR_{355} varies from 60 to 70 sr at 1000 m and, in contrast with LR_{532}, it rises with height, reaching 70 sr at 4000 m. The depolarization ratios for all sessions are above 30% and only on 4 April δ_{532} is below 30% near the top of dust layer. We should recall, that accuracy of calibration of depolarization measurements is about 15%, so day-to-day variations of depolarization can be partly due to uncertainty of calibration procedure. The Angstrom exponent is slightly negative (-0.15±0.1) at low heights, but it starts rising above 2000 m, indicating that particles become smaller.

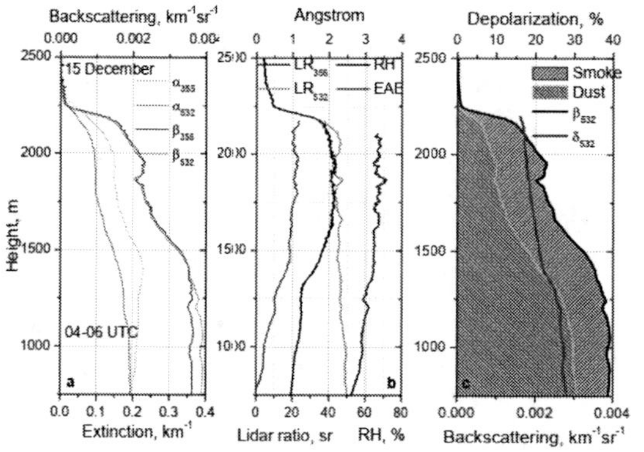

Figure 3. Vertical profiles of (a) extinction and backscattering coefficients at 355 and 532 nm, (b) lidar ratios, Angstrom exponent and relative humidity, (c) particle depolarization and decomposition of β_{532} for dust and smoke contributions.

One of possible explanations of observed height dependence of the particle intensive parameters is that at high altitudes the dust particles are mixed with another type of aerosol. The MERRA-2 aerosol transport model predicts, that for period considered the particles of organic carbon (main component of biomass burning aerosol) may occur above 2000 m [4]. Organic carbon (OC) is characterized by increased absorption in UV, which could explain the increase of LR_{355}. Presence of small OC particles could also explain the increase of the Angstrom exponent. However, for explanation of LR_{532} decrease with height we should assume that LR_{532} of OC is low. This may look unexpected, because lidar measurements usually provide high values of LR_{532} for the smoke (above 60 sr). We should keep in mind however, that our measurements were performed at condition of low relative humidity (below 30%).

2.2. Dust-smoke episodes

To analyze the values of lidar ratio of OC and their dependence on the relative humidity (RH), we have considered several smoke episodes during the campaign. The maximal value of RH in these episodes varied from 30% to 90%. The results obtained on 15 December 2015 are shown in Fig.3. The figure presents the lidar ratios at 355 and 532 nm together with profile of RH, calculated from lidar derived water vapor mixing ratio and radio sonde measured temperature. Maximal value of RH is below 40%. The LR_{532} and LR_{355} are close at 1000 m, but LR_{532} decreases with height from 50 sr to 40 sr, while LR_{355}, in opposite, increases up to 70 sr.

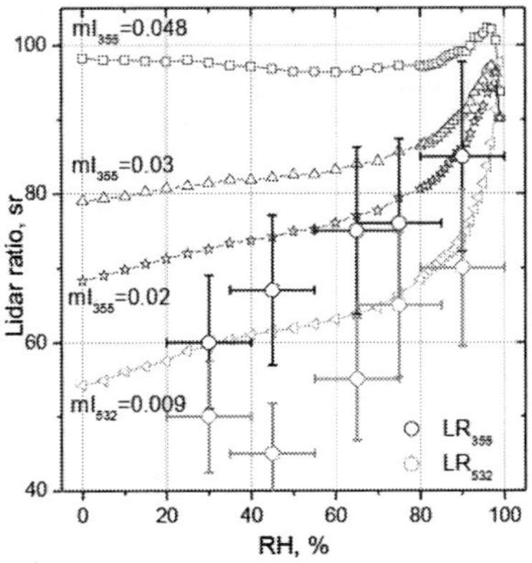

Figure 4. Lidar ratios LR_{355} and LR_{532} for five dust – smoke episodes (open symbols) together with simulated values (line + symbols). In simulation, three values of imaginary part of dry particles at 355 nm are considered: m_I =0.048, 0.03, and 0.02.

Assuming, that depolarization ratio of dust and smoke are known (35% and 7% correspondingly) the backscattering β_{532} can be decomposed for dust and smoke contributions [5], as shown in Fig.3c. The contribution of smoke to β_{532} becomes predominant above 2000 m, and corresponding value of LR_{532} is about 40 sr. For other dust - smoke episodes characterized by higher RH, the values of LR_{532} and LR_{355} are higher. Fig.4 shows dependence of lidar ratios on RH for five smoke episodes. Results are given for altitudes corresponding maxima of RH, where smoke contribution is predominant. The same figure shows the results of simulation, using parameters of OC from MERRA-2 model [4]. In simulation we considered three values of the imaginary part of dry particles at 355 nm: m_I =0.048, 0.03, 0.02. For 532 nm only one value m_I =0.009 was used. We should keep in mind, however, that smoke particles in different episodes may have different composition, so their properties depend not only on RH. For RH>70% the measured lidar ratios are in reasonable agreement with modeling (for mI_{355}=0.02), however for low RH the observed lidar ratios are smaller than modeled ones for both wavelengths.

2.3. Dust episode on 10-24 April 2015

Strong dust episode occurred also for period 10-24 April. Fig.5 shows extinction coefficients and lidar ratios at 355 and 532 nm for two sessions on 15-16 and 23-24 April. The depolarization ratio for both days is about 30%, so particles can be considered as a pure dust. Still, in contrast with 1-4 April episode, LR_{355} and LR_{532} are close, which can be an indication that absorption in UV is lower. To analyze the difference in dust origin, Fig.6 presents the back trajectories for 2 April (00:00 UTC) and 24 April (01:00 UTC). In first case, air mass arrives from North – East of Sahara. For second case, the origin of air mass above 1000 m is shifted more to the South. Close values of lidar ratios at 355 an 532 nm were observed for all sessions during this dust episode. It is possible, that difference between LR_{355} and LR_{532} for considered dust episodes is related to different mineralogy in the regions of dust origin.

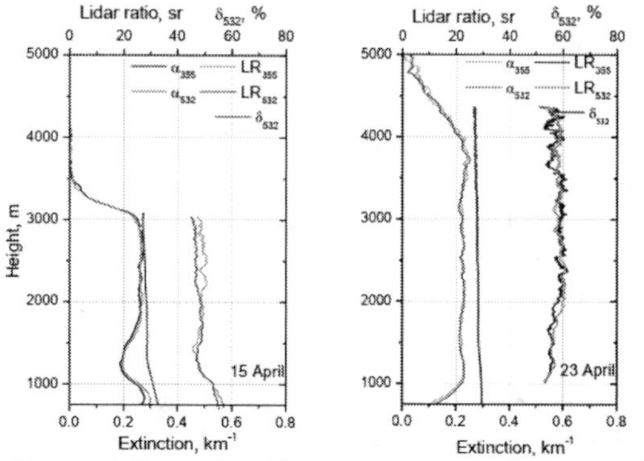

Figure 5. Vertical profiles of extinction coefficients and lidar ratios at 355 and 532 nm together with δ_{532} for the nights 15-16 and 23-24 April 2015.

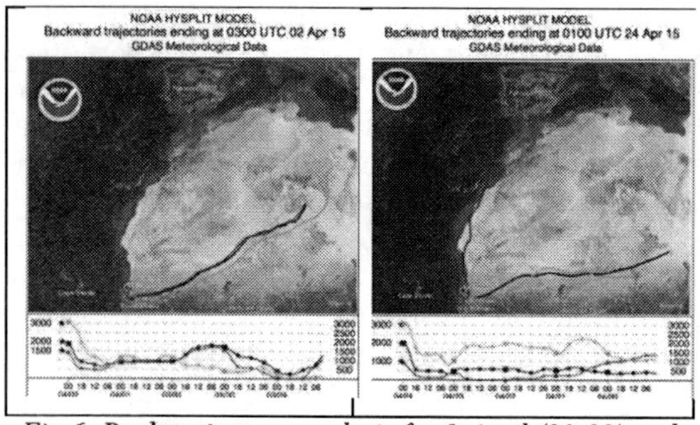

Fig.6. Back trajectory analysis for 2 April (00:00) and 24 April (01:00).

3 CONCLUSIONS

Presented results demonstrate that lidar ratio at 532 nm during dust episodes may decrease with height, which can be due to two reasons: presence of smoke near the top of the dust layer and height variation of the dust properties (in particular height variation of the imaginary part). Analysis of smoke episodes show that the lidar ratio of smoke depends on RH and varies from 40 to 70 sr at 532 nm. This confirms that smoke can contribute to the decrease of observed LR_{532} at condition of low humidity.

Observations demonstrate also that for some dust events the lidar ratios at 355 nm and 532 nm coincide, while for other episodes LR_{355} significantly exceeds LR_{532}. Such difference, in principle, can be the result if different mineralogy at the region of dust origin.

ACKNOWLEDGEMENTS

The authors are very grateful to the labex CaPPA for supporting this study. Development of the software for lidar data analysis was supported by the Russian Science Foundation (project no.16-17-10241).

References

[1] Ansmann, A., Petzold, A., Kandler, K., Tegen, I., Wendisch, M., Müller, D., Weinzierl, B., Müller, T., Heintzenberg, J.: Saharan Mineral Dust Experiments SAMUM–1 and SAMUM–2: what have we learned?, Tellus, 63B, 403–429, 2011.

[2] Veselovskii, I., Goloub, P., Podvin, T., et al.: Study of African dust with multi-wavelength Raman lidar during the "SHADOW" campaign in Senegal, Atm. Chem. Phys. 16, 7013–7028, 2016.

[3] Tesche, M., Groß, S., Ansmann, A., Müller, D., Althausen, D., Freudenthaler, V., and Esselborn, M.: Profiling of Saharan dust and biomass-burning smoke with multiwavelength polarization Raman lidar at Cape Verde, Tellus B, 63, 649–676, 2011.

[4] Veselovskii, I., P. Goloub, T. Podvin, et al.: Characterization of smoke/dust episode over West Africa: comparison of MERRA-2 modeling with multiwavelength Mie-Raman lidar observations, AMT 11, 949–969, 2018.

[5] Tesche, M., Ansmann, A., Müller, D., Althausen, D., Engelmann, R., Freudenthaler, V., and Groß, S.: Vertically resolved separation of dust and smoke over Cape Verde using multiwavelength Raman and polarization lidars during Saharan Mineral Dust Experiment 2008, J. Geophys. Res., 114, D13202, doi:10.1029/2009JD011862, 2009.

LONG-TERM ANALYSES OF AEROSOL OPTICAL THICKNESS USING CALIOP

Masahiro Fujikawa[1], Rei Kudo[2], Tomoaki Nishizawa[3], Eiji Oikawa[4], Akiko Higurashi[3], Hajime Okamoto[4]

[1] *Kyushu University Interdisciplinary Graduate School of Engineering Sciences, 6-1 Kasuga Park, Kasuga, Fukuoka 816-8580, JAPAN *Email: fujikawa@riam.kyushu-u.ac.jp*
[2] *Meteorological Research Institute, 1-1 Nagamine, Tsukuba, Ibaraki 305-0052, JAPAN*
[3] *National Institute for Environmental Studies, 16-2 Onogawa, Tsukuba, Ibaraki 305-0053, JAPAN*
[4] *Kyushu University Research Institute for Applied Mechanis, 6-1 Kasuga Park, Kasuga, Fukuoka 816-8580, JAPAN*

ABSTRACT

We developed an algorithm to derive extinction coefficients for four aerosol components (water-soluble, dust, sea salt, black carbon) from Cloud-Aerosol Lidar with Orthogonal Polarization (CALIOP) data. The algorithm was applied to the nine-year data for 2007–2015 and the results were compared to CALIOP standard product (CALIOP-ST) and MODerate resolution Imaging Spectroradiometer (MODIS) standard product (MODIS-ST). Comparisons of the total aerosol optical thickness (AOT) showed that MODIS-ST was the largest, followed by CALIOP-ST (Ver.4), and our product. CALIOP-ST (Ver.3) showed a similar magnitude to ours.

1. INTRODUCTION

Uncertainties of a climatic impact of aerosols are large. It is therefore highly demanded to have reliable observing systems and analyses of the global spatiotemporal distribution of optical and microphysical properties of aerosols. The observation by the two-wavelength polarization Mie scattering lidar CALIOP onboard Cloud-Aerosol Lidar and Infrared Pathfinder Satellite Observations (CALIPSO) satellite has exceeded 12 years and it provides long-term fluctuation of the global three-dimensional distribution of the aerosols.

The optical and microphysical properties of aerosols differ greatly depending on their chemical composition. Thus, identification of aerosol components has to be conducted to assess the climate effects of aerosols. A method to determine major aerosol species in the atmosphere using two-wavelength polarized Mie scattering lidar data has been developed using data from JAMSTEC research vessel MIRAI and ground-based lidar network AD-Net [1. 2, 3]. The algorithm was further extended to be applied to CALIOP data [4]. This method identifies four aerosol components, water-soluble particles (WS), mineral dust (DS), sea salt particles (SS), and internally mixed black carbon particles (LAC) and estimates vertical profiles of extinction coefficients for four aerosol components.

2. Method

We rely on the results from Aerosol detection and cloud screening in NASA-CALIOP ST products. However, our retrieval algorithm for aerosol microphysics is different from the NASA CALIOP standard algorithm. Our method uses all the measurement channels of the CALIOP: total attenuated backscattering coefficient (ß) at 532 and 1064nm and the depolarization ratio (δ) at 532 nm (CALIOP Level 1B Version 3 product). Note that the NASA standard algorithm uses ß and δ at 532nm. Maximum a posteriori (MAP) method is used to reduce estimation errors due to signal noise and to suppress the generation of meaningless solutions [5]. Following assumptions and modeling have been made for their shapes, size distributions, and refractive indexes to estimate extinction coefficients of four aerosol components from three-measurement channels. The shape of DS is assumed to be a spheroid; and the others are spherical. The particle size distribution is expressed as a log-normal distribution; the mode radii and variances are prescribed according to studies in [6] and [7]. As for LAC, Core-Gray-Shell (CGS) model [8] is applied where black carbon is assigned as the core and

https://doi.org/10.1051/epjconf/202023702023

WS as the shell. The dry mode radius of LAC is the same as WS. The mode radius of SS is estimated from the wind speed at sea level [9]. The complex refractive indices of WS and SS are adopted from the values of a database [6], and those of DS are taken from the values of Asian dust observation [10]. In addition, optical and microphysical properties of WS, SS, and LAC are modeled depending on humidity [6].

This algorithm is applied to the CALIOP data after removing the data contaminated by cloud using Cloud and aerosol masks (CALIOP Level 2 VFM Version 3 product).

3. Results

We applied the algorithm to the CALIOP data and evaluated annual mean global distributions of the clear-sky aerosol optical thickness (AOT) at 532 nm of four aerosol components (WS, SS, DS, and LAC) and total AOT (Fig.1). The total AOT (Fig.1e) is larger (around 0.3–0.5) in low to mid-latitudes, such as in China, India, the Middle East, Africa, and South America. It is found that the contributions of WS (Fig.1a) and DS (Fig.1c) are large.

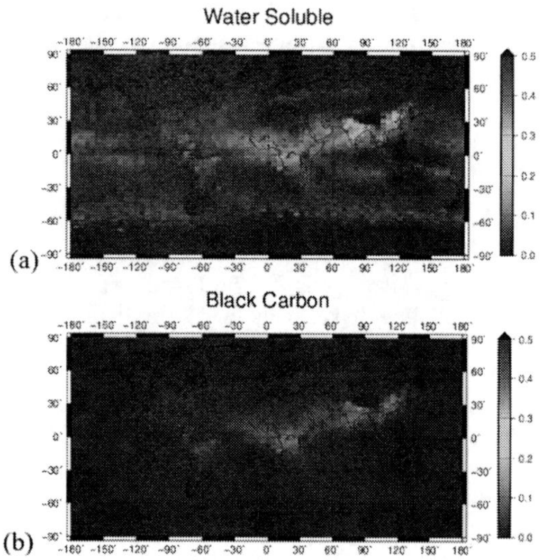

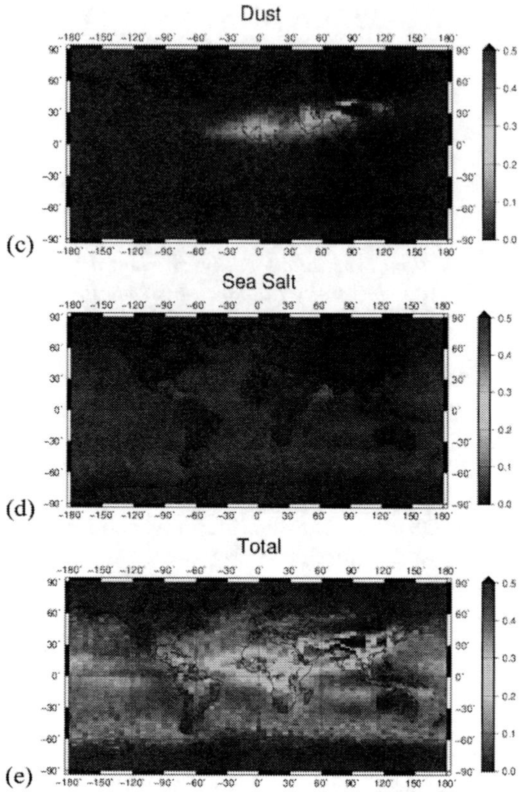

Figure 1. The annual mean clear-sky aerosol optical thickness (AOT) in 2010 during day-time from this study where the resolution of latitude 2° - longitude 5°. (a) Water Soluble, (b)LAC (Black Carbon), (c)Dust, (d)Sea Salt, (e)Total.

We compared these results with the other satellite products. Figure 2 shows total AOTs of CALIOP NASA standard products (CALIOP-ST) Version 3 (Ver.3) and Version 4 (Ver.4) and MODIS standard product (MODIS-ST). The geographical pattern of total AOT is similar to our product, CALIPSO-ST, and MODIS-ST. Total AOT of our product is comparable to those of CALIPSO-ST Ver.4 and MODIS-ST and larger than CALIPSO ST Ver.3 over the ocean. In Sahara and Arabian deserts, total AOT is smaller than 0.5 in our product, CALIPSO-ST Ver.3, and MODIS-ST, while that in CALIOP-ST Ver.4 is larger than 0.5.

Finally, 9 years of analyses of total AOT are conducted from 2007 to 2015 (Fig.3). CALIPSO

AOT for our product is smaller than MODIS AOT throughout the whole period. MODIS AOT tends to be overestimated by about 10 to 20%, because of cloud contamination of thin cirrus [11]. CALIOP AOT is smaller than MODIS AOT by 20–40% even after accounting for the cloud contamination effect of the passive sensor. In addition, it is found that the global mean AOTs of MODIS-ST and CALIOP-ST are relatively constant. Contrary to these two products, our CALIOP Ver.3 product shows large fluctuations from 2012 to 2013. When we use Ver.4 product as input, such fluctuation is found in the product. The possible reason for the time variation of our retrieved AOT is the minor update of the CALIOP Level 1B product from Version 3.02 to 3.30.

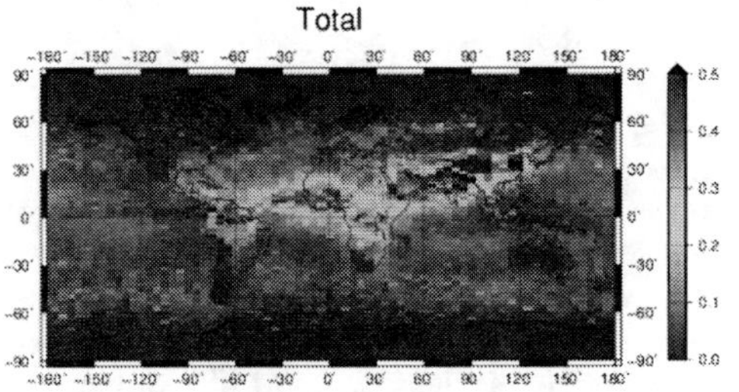

(a)

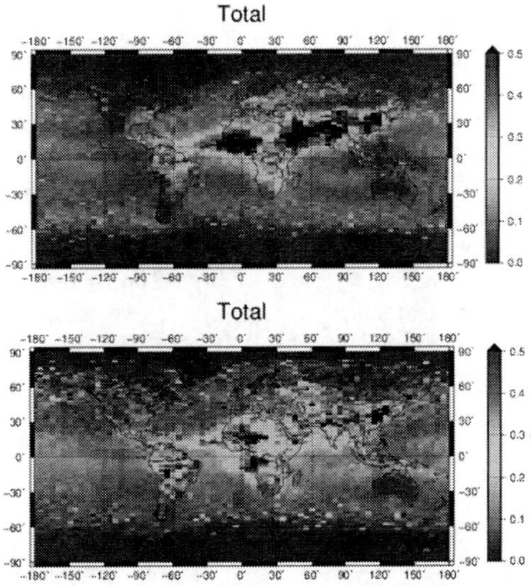

(b) (c)

Figure 2. Same as Fig.1e but (a) CALIOP NASA standard product Version 3 (CALIOP-ST Ver.3), (b) CALIOP NASA standard product Version 4 (CALIOP-ST Ver.4), and (c)MODIS standard (MODIS-ST) product (Level2, MAC04S0.002, "AOD_550_Dark_Target_Deep_Blue_Combined").

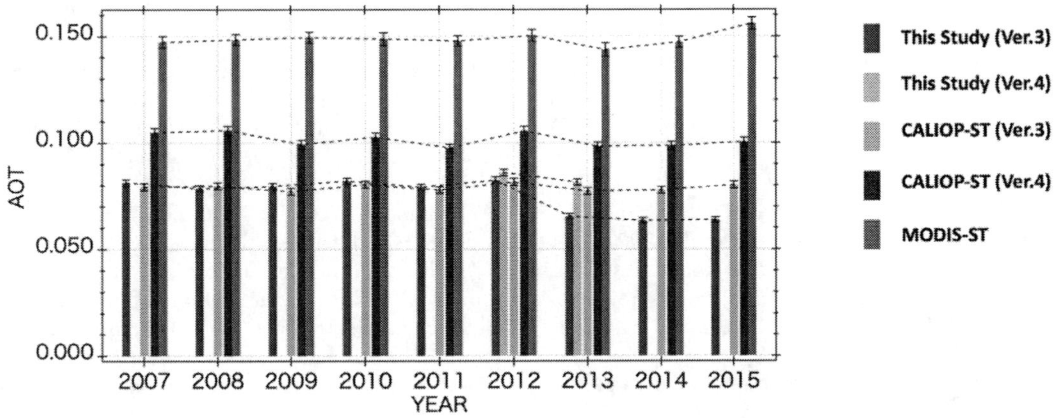

Figure 3. Changes in the annual global AOT average value of each product for 9 years (2007 – 2015, only day-time). Error bars represent standard error.

4. SUMMARY

The algorithm is developed to (1) distinguish four components of the aerosol by using CALIOP data (2 wavelengths, 1 polarization) and (2) retrieve their extinction coefficient. These results show that WS and DS account a large fraction of total AOT.

Comparisons of our products with other products are conducted for 9 years. The geographical pattern of total AOT is similar between our results, CALIPSO-ST, and MODIS-ST. The global mean values of our results are smaller than CALIPSO-ST(Ver.4) and MODIS-ST, while the interannual variation of AOT is less than 10% from 2007 to 2011 for all four products.

We plan to perform the comparison of total AOT between our results and the AERONET ground-based measurement to understand the reason for the underestimation of total AOT in our results.

ACKNOWLEDGMENTS

CALIPSO data were obtained from the Atmospheric Sciences Data Center at NASA's Langley Research Center. This study was supported by JSPS KAKENHI Grant Numbers JP17H06139, JP15H02808, JP15H01728, JP2524707, and by the Japan Aerospace Exploration Agency for EarthCARE Research Announcement.

REFERENCES

[1] Nishizawa, T., H. Okamoto, N. Sugimoto, I. Matsui, A. Shimizu, and K. Aoki, 2007, An algorithm that retrieves aerosol properties from dual-wavelength polarization lidar measurements, J. Geophys. Res., 112, D06212, doi:10.1029/2006JD007435.

[2] Nishizawa, T., H. Okamoto, T. Takemura, N. Sugimoto, I. Matsui, and A. Shimizu. 2008, Aerosol retrieval from two-wavelength backscatter and one-wavelength polarization lidar measurement taken during the MR01K02 cruise of the R/V Mirai and evaluation of a global aerosol transport model, J. Geophys. Res.. 113, D21201, doi:10.1029/2007JD009640.

[3] Nishizawa, T., N. Sugimoto, I. Matsui, A. Shimizu, and H. Okamoto, 2011, Algorithm to retrieve aerosol optical properties from two-wavelength backscatter and one-wavelength polarization lidar considering nonsphericity of dust, Journal of Quantitative Spectroscopy and Radiative Transfer Papers 112, 254-267.

[4] Kudo et al., 7th International EarthCARE science workshop (2018/6).

[5] Kudo, R., Nishizawa, T., and Aoyagi, T.: Vertical profiles of aerosol optical properties and the solar heating rate estimated by combining sky radiometer and lidar measurements, Atmos. Meas. Tech., 9, 3223-3243, https://doi.org/10.5194/amt-9-3223-2016, 2016.

[6] Hess, M., P. Koepke, and I. Schult, 1998: Optical properties of aerosols and clouds: The software package OPAC. Bull. Amer. Meteor. Soc., 79, 831–844.

[7] Smirnov, A., Holben, B. N., Kaufman, Y. J., Dubovik, O., Eck, T. F., Slutsker, I., Pietras, C., and Halthore, R. N.: Optical properties of atmospheric aerosol in maritime environments, J. Atmos. Sci., 59, 501–523, 2002.

[8] Michael Kahnert, Timo Nousiainen, and Hannakaisa Lindqvist, "Models for integrated and differential scattering optical properties of encapsulated light absorbing carbon aggregates," Opt. Express 21, 7974-7993 (2013).

[9] Erickson, D. J., and Duce, R. A.: On the global flux of atmospheric sea salt, J. Geophys. Res. 93, 14079-14088 (1988).

[10] Aoki, T., Y. Tanaka, A. Uchiyama, M. Chiba, M. Mikami, and J. R. Key, 2005: Sensitivity experiments of direct radiative forcing by mineral dust using spectrally detailed radiative transfer model, J. Meteorol. Soc. Japan, 83A, 315-331.

[11] Kaufman, Y. J., L. A. Remer, D. Tanr´e, R.-R. Li, R. Kleidman, S. Mattoo, R. Levy, T. Eck, B. N. Holben, C. Ichoku, J. Martins, and I. Koren (2005), A critical examination of the residual cloud contamination and diurnal sampling effects on MODIS estimates of aerosol over ocean, IEEE Trans. on Geoscience & Remote Sensing, 43(12), 2886–2897.

https://doi.org/10.1051/epjconf/202023702024

MOBILE OBSERVATIONS BY LIDAR, SUN PHOTOMETER AND IN SITU IN NORTH CHINA PLAIN

Ioana Elisabeta Popovici[1,2*], Philippe Goloub[1], Luc Blarel[1], Xiangao Xia[3], Zhaoze Deng[3], Hongbin Chen[3], Hongyan Chen[3], Yitian Hao[3], Nan Yin[3], Disong Fu[3], Christine Deroo[1], Augustin Mortier[4], Fabrice Ducos[1], Benjamin Torres[1], Oleg Dubovik[1], Stéphane Victori[2]

[1] *Univ. Lille, CNRS, UMR 8518 - LOA - Laboratoire d'Optique Atmosphérique, F-59000 Lille, France*
[2] *CIMEL Electronique, R&D Department, 75011 Paris, France*
[3] *CAS, LAGEO, Institute of Atmospheric Physics, Beijing, China*
[4] *Division for Climate Modelling and Air Pollution, Norwegian Meteorological Institute, 0313 Oslo, Norway*
**Email: ioanaelisabeta.popovici@univ-lille.fr*

ABSTRACT

A mobile laboratory integrating lidar, sun photometer and in situ instruments has been deployed to observe the aerosol spatial variability in North China Plain in May 2017. Results from the campaign are presented.

1. INTRODUCTION

MOABAI campaign (***M**obile **O**bservations of **A**tmosphere **B**y vehicle-borne **A**erosol measurement **I**nstruments*) has been carried out in the 5-23 May 2017 period in North China Plain (Fig. 1), one of the most populated and polluted regions of China, where long-standing heavy aerosol pollution episodes frequently occur. The campaign was purposed to elucidate and quantify the 4-D distribution of aerosols in a polluted region where few aerosol measurements are available, in order to get a comprehensive characterization of aerosol properties and their vertical distribution in variable atmospheric situations.

2. INSTRUMENTS AND METHODOLOGY

The mobile observations have been performed using a state-of-the-art instrumented van, which included both remote sensing and in situ instrumentation. The lidar and the mobile sun photometer from MAMS [1] (Mobile Aerosol Monitoring System) were transported and integrated on site in the existing van already equipped with in situ instruments.

2.1 Lidar

A commercial Cimel micropulse lidar (model CE370) has been integrated in the van. It is a one-wavelength elastic backscattering lidar operating at 532 nm with 20 µJ pulse energy and 4.7 kHz laser pulse repetition rate. The detection of backscattered light is done in photon counting mode using an APD. The lidar is a mono-axis system, consisting of a shared transmitter-receiver telescope connected to the control and acquisition unit through a 10 m optical fiber. The telescope has a diameter of 200 mm and a narrow field of view (55 µrad) in order to limit the background solar light. The lidar allows monitoring aerosols and clouds in the troposphere, typically up to 15 km with a vertical resolution of 15 m.

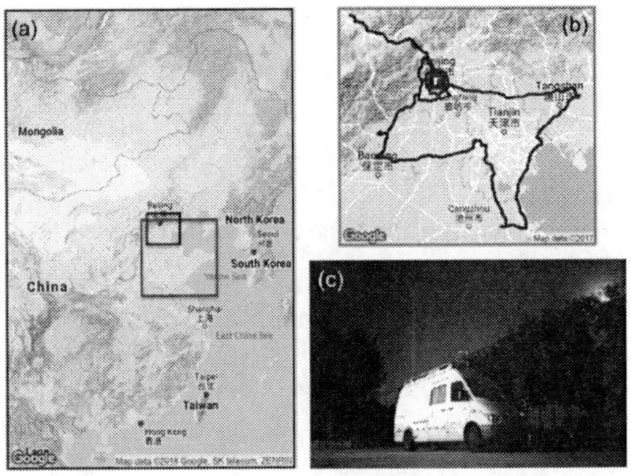

Figure 1. (a) North China Plain (red rectangle) and area investigated through mobile measurements (blue rectangle), (b) GPS track of all mobile transects during MOABAI and (c) IAP/LAGEO instrumented van used for mobile observations.

Vertical profiling of extinction coefficient is achieved using a Klett-based iterative inversion algorithm [2], constrained by coincident AOD from mobile sun photometer measurements. The mass concentration profiles are derived using an extinction-to-mass relationship, where additional information from in situ measurements at ground level and column-integrated volume size distribution are used in the calculations.

2.2 Sun photometer

The sun-tracking photometer PLASMA (*Photomètre Léger Aéroporté pour la Surveillance des Masses d'Air*) [3] is an instrument developed by LOA and SNO PHOTONS in Lille, France. Its unique features are the capability of performing direct solar radiation measurements during the vehicle motion and its light weight (~ 5 kg) and compactness (Ø 16 x H 23 cm). PLASMA has 9 spectral channels at same wavelengths as a standard Cimel CE318 photometer in AERONET. It is calibrated by SNO PHOTONS, French branch of AERONET, and undergoes the calibration protocol for the AERONET reference master instrument. Compared to CE318 sun photometer, the current PLASMA model performs only direct sun measurements.

PLASMA provides AOD measurements along the vehicle's track and column-integrated volume size distribution are derived from spectral AOD using GRASP-AOD algorithm [4].

2.3 In situ

The in situ instruments consisted of a polar nephelometer (Ecotech, model Aurora 4000), an aethalometer (Magee Scientific, model AE33), a Grimm Sky-OPC and trace gas analyzers for NO_2, SO_2 and O_3.

The measurements at ground level included particle number concentration, size distribution, particle scattering (450, 535 and 635 nm) and absorption (370, 470, 520, 590, 660, 880 and 950 nm) coefficients.

3. RESULTS

Ten days of mobile observations have been recorded during MOABAI campaign, six days in Beijing on the 4th, 5th and 6th ring-roads by daytime and nighttime and 4 days outside of Beijing, reaching Baoding, Tianjin and Tangshan. Four types of atmospheric situations were observed: two days of pollution and "background" (moderate pollution) situation in Beijing with a contribution of dust transported from Gobi desert (9 and 11 May), three heavy pollution days, when the air flow moved from south of China (18, 19 and 21 May), one day with moderate pollution (15 May) and one "clean" day, marked by lower aerosol loading (13 May), but with a contribution of long-range transported aerosols in altitude. In the case of pollution days, the height of the PBL extended up to 1 - 1.5 km altitude and the AOD at 440 nm was high (0.8 - 1.8), highest values being recorded when the air masses were moving from South direction. Angstrom Exponent (AE) values typical for fine mode particles predominance (1.2 – 1.6) were recorded for these days. For the "background" and "clean" situations, a contribution of both fine and coarse mode aerosols was evidenced, indicated by lower AE and by transported aerosol layers above 2 km. Three mobile observations were carried out outside Beijing (Fig. 2) : Beijing – Baoding - Tianjin (16 May), Tianjin - Tangshan (17 May) and Tangshan - Beijing (18 May). The AOD at 440 nm was in the range of 0.2 - 0.7, 0.3 - 0.79 and 0.43 - 1.35 for the transects on 16, 17 and 18 May, respectively. The values of AE were in the range of 0.38 - 1.5, 1.02 - 1.9 and 1.22 - 1.74 for 16, 17 and 18 May, respectively.

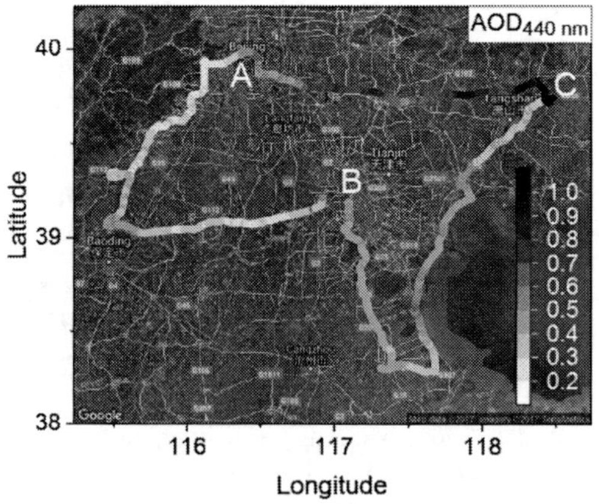

Figure 2. Spatial variability of AOD at 440 nm from PLASMA sun photometer measurements. The A, B and C denote the departure and arrival points for three transects: Beijing-Tianjin (AB) on 16 May 2017, Tianjin-Tangshan (BC) on 17 May 2017 and Tangshan-Beijing (CA) on 18 May 2017.

The lidar profiles showed higher particle concentrations when reaching polluted regions of Baoding, Tianjin, Binhai New Area and Tangshan and transport of desert dust from Inner Mongolia at about 2-3 km altitude.

Mapping of AOD, Angstrom Exponent and retrieved vertical profiles of aerosol extinction coefficients and column-integrated volume size distribution have been achieved. Results from some case studies in Beijing and along Tianjin coast will be presented.

3.1 Case study: 17 May 2017

The mobile observations on 17 May 2017 started from the city of Tianjin, along the Binhai New Area, on the coast of the Bohai Sea, and heading to Tangshan (Fig. 3). The weather was fair, with clear sky along the whole transect, with ambient temperatures in the 24° - 30° Celsius and RH in the 30 - 65 % range, with a noticeable increase along the coast of Bohai Sea, indicating the occurrence of a sea breeze event. Backward trajectories showed a flux from S-SE at ground level and up to 1 km and a flux from NW at 3 km, explaining the presence of fine dust layers transported from Inner Mongolia.

The Binhai New Area is a region marked by the presence of several industries, from machinery factories, petro-chemical manufacturing plants, automotive fitting factories and electronics facilities to sea salt production, shipbuilding and port activity. The area accounted 271 industrial enterprises in 2012, resulting in heavy pollution in the region [5, 6]. It is a region where the microphysical and optical properties of aerosols are not well characterized, much less at a fine scale. The situation is even more complex in spring season, as mineral dust transports frequently occur, adding to the local anthropogenic aerosols, which makes this region an interesting study area.

One previous study using mobile lidar measurements has been conducted in Tianjin, in different seasons in 2016 [7]. Nevertheless, this is the first time mobile observations with such rich instrumentation and methodology, combining remote sensing and in situ measurements, are conducted in Tianjin and in the Binhai New Area.

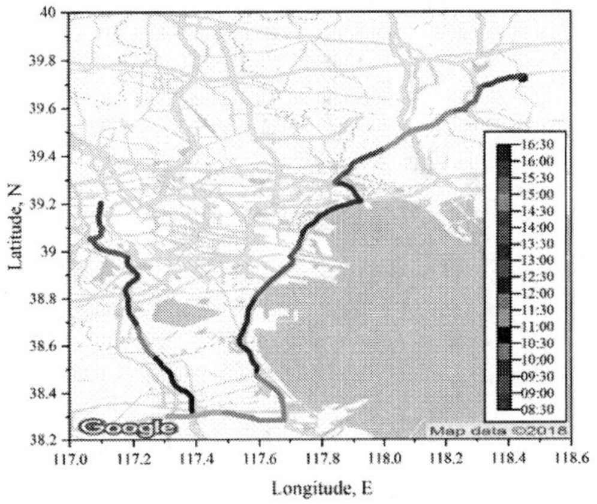

Figure 3. GPS track of the mobile transect from Tianjin to Tangshan on 17 May 2017, colour coded by 30 minutes time intervals (local time)

The spatial variability of the extinction coefficients profiles at 532 nm derived from the synergy of lidar and sun photometer measurements is presented in Figure 4. The mean extinction coefficient in the PBL, from the surface to about 2 km, was 0.14 ± 0.11 km^{-1} along the whole transect from Tianjin to Tangshan and a maximum of 0.57 km^{-1} was recorded at 650 m height at Tianjin. Above the PBL, a lofted aerosol layer in the 2.2 – 3.5 km altitude range was observed all along the mobile transect. The backward trajectories indicated it is most likely desert dust, transported from Inner Mongolia. The mean extinction coefficient of the dust layer was 0.05 ± 0.03 km^{-1}, and the mean AOD at 532 nm was 0.06 ± 0.01, representing around 18-20% of the total measured AOD.

The highest values of extinction coefficient were found at Tianjin and along the Binhai New Area. This shows both the urban print of a metropolis (the 13th world largest city) and the print of a coastal region with enhanced industrial activities and the presence of the port of Tianjin, the largest port in mainland China and one of the largest in the world.

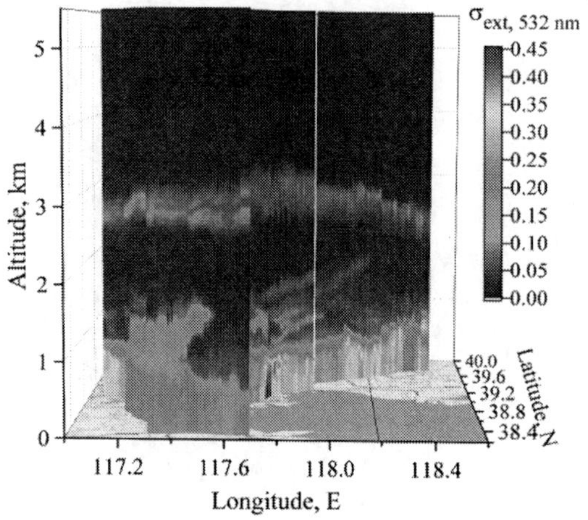

Figure 4. 3-D representation of the variability of lidar-derived extinction coefficients at 532 nm along the mobile transect from Tianjin to Tangshan

The profiles of aerosol mass concentration were estimated from the extinction coefficients profiles and using microphysical and optical properties of two defined aerosol models: an urban/industrial aerosol model in the PBL and a desert dust model for the elevated layer in the 2.2 – 3.5 km altitude range. The extinction profiles below 200 m are constrained by in situ extinction measurements at ground level.

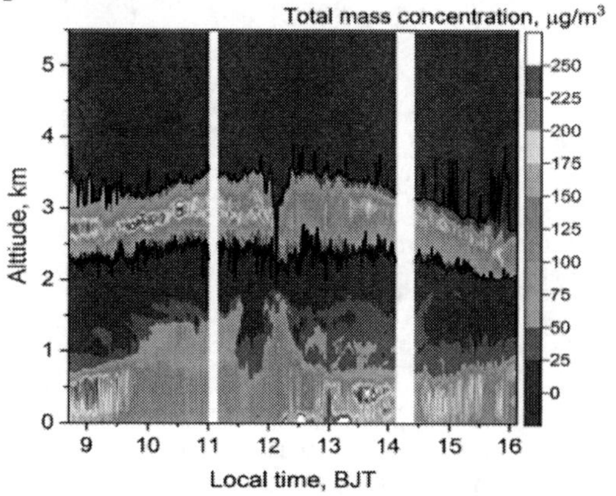

Figure 5. Particle mass concentration profiles derived from mobile lidar and sun photometer measurements on 17 May 2017 along the transect from Tianjin to Tangshan

The mean mass concentration in the dust layer was 95 ± 52 μg m^{-3} with a maximum of 284 μg m^{-3}. The mean mass concentration in the PBL was around 80 ± 62 μg m^{-3} for the whole transect. The standard deviations correspond to the variability along the transect. The highest particle mass concentrations were found near Tianjin, until 09:30, and when crossing the industrial coastal region, from 12:00 to 14:00 local time.

4. CONCLUSIONS

We have observed highly variable aerosol situations in North China Plain, proving the use of mobile observations for investigating the spatial variability of aerosol properties. The advantage of an instrumented van is that it can be easily deployed for the study of sudden aerosol events. The complementarity of in situ and remote sensing data is useful in the analysis of complex aerosol situations.

ACKNOWLEDGEMENTS

We wish to thank the CaPPA and CPER CLIMIBIO projects for the financial support and Service National d'Observation PHOTONS/AERONET from INSU/CNRS for the sun photometer data processing and calibration. The ACTRIS/AERONET Europe and ACTRIS-FR research infrastructure are also acknowledged.

REFERENCES

[1] I. E. Popovici et al. Atmos. Meas. Tech., vol. 11, no. 8, pp. 4671–4691 (2018)
[2] A. Mortier et al. Atmos. Chem. Phys., vol. 13, no. 7, pp. 3705–3720 (2013)
[3] Y. Karol et al. Atmos. Meas. Tech., vol. 6, no. 9, pp. 2383–2389 (2013)
[4] B. Torres et al. Atmos. Meas. Tech., vol. 10, no. 10, pp. 3743–3781 (2017)
[5] X. Su et al. Sci. Total Environ., vol. 584–585, no. April 2018, pp. 381–392 (2017)
[6] S. Kong et al.Sci. Total Environ., vol. 408, no. 20, pp. 4681–4694 (2010)
[7] L. Lyu et al. J. Meteorol. Res., vol. 32, no. February, 2018, pp. 60-68 (2018)

https://doi.org/10.1051/epjconf/202023702025

MONITORING ATMOSPHERIC AEROSOLS OVER THE URMIA LAKE BY CALIPSO AND A GROUND BASED DEPOLARIZED LIDAR

Hamid R. Khalesifard[1, 2]*, Hossein Panahifar[1], Fatemeh Ghomashi[1], Salar Alizadeh[1] and Ruhollah Moradhaseli[3]

[1]Department of Physics, and [2]Research Center for Climate Change and Global Warming, Institute for Advanced Studies in Basic Sciences, Zanjan 4513766731, Iran
[3]Physics Department, Faculty of Science, Zanjan Brach, Islamic Azad University, Zanjan, Iran
**khalesi@iasbs.ac.ir*

ABSTRACT

The Urmia Lake, a hypersaline lake in Northwest Iran is facing a severe drying scenario. We have installed an azimuthal scanning depolarized backscatter lidar in the coast of the lake to monitor the atmospheric aerosols that may originate from the dried lake bed. We also used the CALIPSO recordings to monitor the aerosol optical depth and particulate depolarization ratio just over the lake. Recordings of the lidar and CALIPSO both show that dry salt particles can be found in the atmospheric boundary layer over the lake especially in summer times. Also CALIPSO data in synergy with HYSPLIT model show that the lake is not an intense aerosol source comparing to neighboring sources like the Mesopotamia region but it is under their influence.

1. INTRODUCTION

The Urmia Lake (UL) in Northwest Iran (Figure 1) once was the second largest permanent hypersaline lake in the world. Now the lake is experiencing a serious drying challenge that has been started about twenty years ago. As a result of this environmental disaster, the water level of the lake has dropped significantly (Figure 2) in a way that its surface area has shrunk to just 12% of its value in 1995 [1]. Many reasons have been mentioned for the drought of the lake, including climate change [2, 3] and human factors like building dams, digging too many wells and excessive use of water resources in the catchment area of the lake [1, 4-6]. There is a debate that drying of the UL has turned it to an active source of salt or mineral particles [7]. In this work we are trying to find the types of the particles that may originate from the dried lake bed and investigate how active is the UL as a source for such particles. On doing this we have

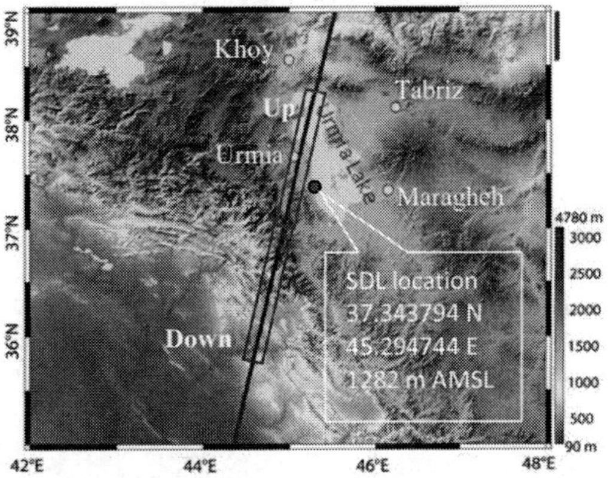

Figure 1 Orographic map of the Urmia Lake water catchment area in Northwest Iran and a night-time CALIPSO ground track over it. Blue spot at the west coast of the lake is the location of the installed lidar.

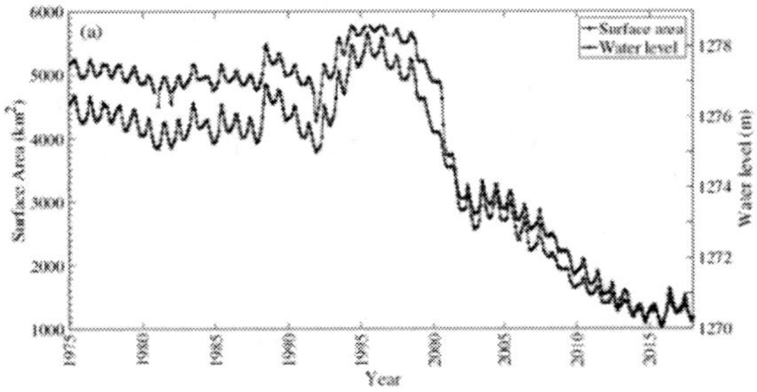

Figure 2 Variations of surface area and water level height of the Urmia Lake from 1975 to 2017.

used the recordings of the CALIPSO satellite over the UL since its launching time and a scanning depolarized elastic backscatter lidar (SDL) that we have installed at the western coast of the UL to monitor the atmospheric aerosols just in the vicinity of the lake. We are expecting that comparison of aerosol optical depth (AOD) and particulate depolarization ratio (PDR) from CALIPSO with SDL measurements on similar

https://doi.org/10.1051/epjconf/202023702025

parameters provides trustable tools to specify the atmospheric aerosols in the water catchment area of the UL and gives us an estimation about the activities of the UL as a source of atmospheric aerosols

2. METHODOLOGY

2.1 CALIPSO recordings

Figure 1 illustrates a nighttime CALIPSO ground track that is passing almost over the northern half of the UL. To compare the atmospheric particles inside the water catchment area of the lake with those that maybe found outside of it, the track is divided in two segments, Up and Down. The Up segment is almost inside the water catchment area but Down is outside of it and under the influence of dust events that are originating from the Mesopotamian region [8, 9]. To investigate the atmospheric aerosols over the mentioned segments, CALIPSO level 2 aerosol products have been used to extract AOD and PDR for atmospheric aerosols in these regions. The data set covers June 2006 to December 2017. Since the crest line of the surrounding mountains around the water catchment varies between 2000 m above mean sea level (amsl) to 3500 m amsl, the investigation has been carried out at two altitude regimes: i.e. altitudes from surface up to 2000 m amsl and altitudes higher than 3000 m amsl. From now on we label these two altitude regimes by letters *l* and *h*.

2.2 The scanning depolarized backscatter lidar, SDL

The SDL has a frequency doubled 10 ns pulsed Nd-YAG laser and a 5x beam expander as its transmitter. The laser has an energy of 100 (150) mJ/pulse at 532 (1064) nm. At the receiver, an 8" Cassegrain telescope collects the backscattered lights and then signals are recorded at 532 nm in parallel and perpendicular polarizations respect to that of the laser beam (Figure 3a). The scanning scheme of the SDL is shown in Figure 3b. The azimuthal angle, φ, varies between 0° to 180° in steps of 1.0° ± 0.1° degree by means of a stepper motor through a 1/90 reducing speed gear box. The elevation angle, θ, varies between 45° to 90° (±0.1°) continuously by means of a hand hold screw. The SDL also can aim the zenith direction (θ = 0 °). The spatial resolution of the lidar

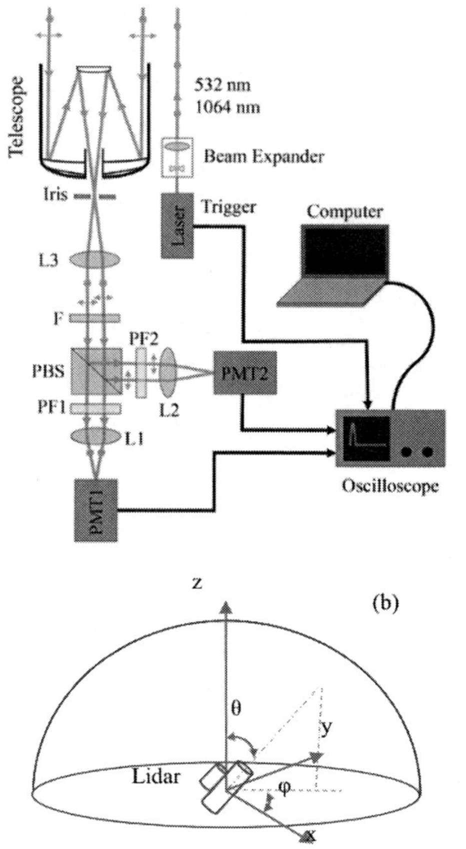

Figure 3 a) Optical setup of SDL; L-lens, PBS-polarizing beam splitter, PF-polarizing filter, b) the scanning scheme of SDL

signals is 12 m and its range is 12 km in zenith direction.

3. RESULTS AND DISCUSSIONS

3.1 CALIPSO recordings

Figure 4 shows variations of the aerosol layer height, h_{layer}, versus the aerosol layer optical depth, τ_{layer}, over Up and Down segments, obtained from the CALIPSO recordings. As the figure shows, the aerosol layers over the Up segment appear at higher altitudes. This can be seen more clearly, when an average would be performed over all recorded layer heights $\tau_{layer} \geq 0.2$. Dashed black (red) lines in Figure 4 depict this value for Up (Down) segment as 3.1 (2.5) km. In other words, aerosol layers over the UL mostly

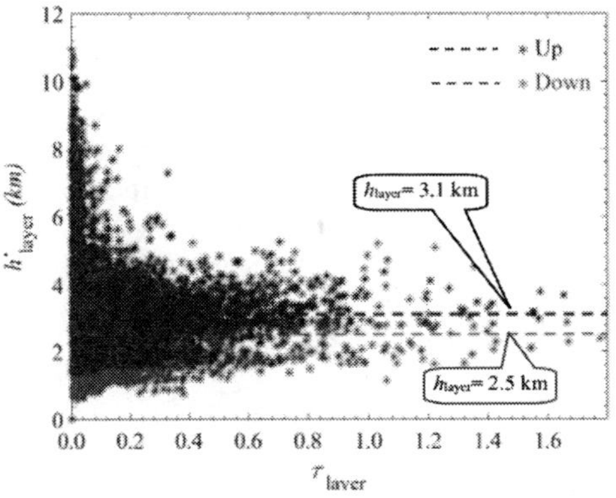

Figure 4 Variations of h_{layer}, versus τlayer, over Up and Down segments. Black (red) dashed lines corresponds to average height of the aerosol layers with $\tau_{layer} \geq 0.2$ for Up (Down) segment.

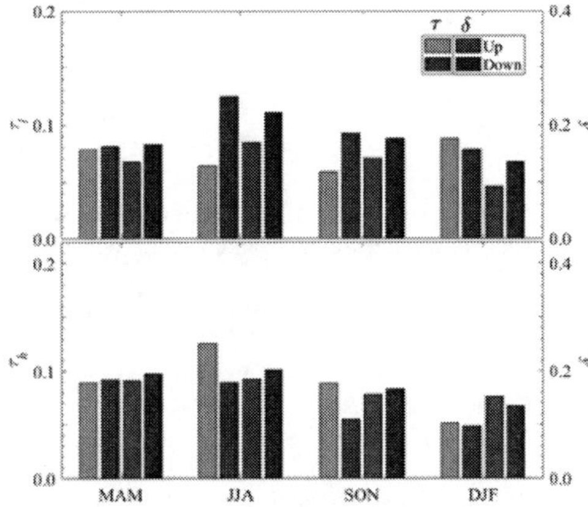

Figure 5 Seasonal variations of AOD and PDR over Up and Down segments, CALIPSO recordings, Jun. 2007 to Dec. 2017, a) and b) l and h altitude regimes.

are appearing above the crest lines of the mountains around the UL water catchment area. It is also interesting to look at seasonal variations of AOD and PDR over Up and Down segments for l and h altitude-regimes. These variations are shown in Figure 5a and 5b. As Figure 5a shows higher values of the AOD for altitudes lower than 2 km amsl, τ_l, in the Up (Down) segment are happening in winter (summer) times when the PDR, δ_l, has its lowest (highest) values. But for altitudes higher than 3.0 km, Figure 5b shows more or less similar behaviors for Up and Down segments. In other words, one can conclude that the water catchment area of the UL, like other regions in Northwest Iran, is very much under the influence of dust sources in its west and east sides, i.e. Mesopotamia and the dry region between the Caspian Sea and Aral Lake [8, 10]. This means that referring to such recordings, it is not easy to find the UL as an influencing source of mineral particles for the regions outside of its water catchment area.

3.2 SDL recordings

Since October 2018, the SDL has been installed at the western coast of the UL (37.343794 N, 45.294744 E, 1282 m AMSL) in Rashakan preserved area (Figure 1). Just before start of the first precipitations in October, when the lake bed in the vicinity of the installation site was quite dry, we performed few azimuthal scans (φ-scan in Figure 3b) at different θ values (Figure 3b). The scanning calibration of the lidar is in the way that, when φ=0, the lidar is just aiming the east. Figure 6 shows a recorded SDL scan at 11:30 UTC on 01 Oct. 2018, where φ is scanned from -70° to 70° and θ is set at 40°.

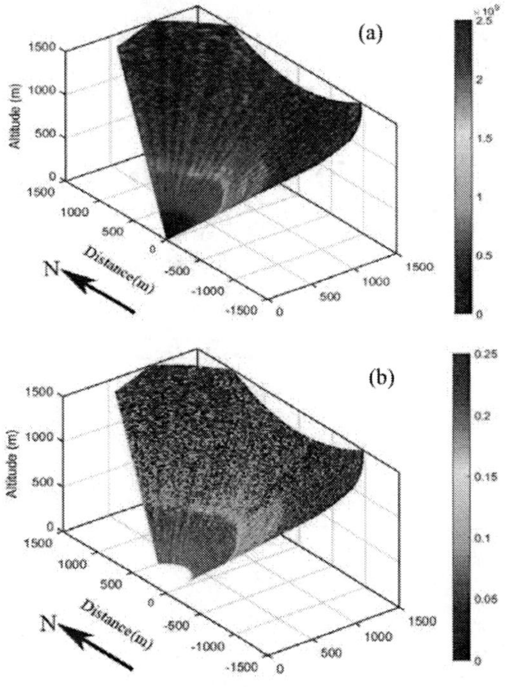

Figure 6 SDL signals for an azimuthal scan, φ = -70° to φ = 70° and θ = 40°, 11:30 UTC, 01 Oct. 2018, a) range corrected backscatter signals on parallel channel, b) depolarization ratio.

Figure 6a shows the range corrected lidar signals on parallel channel. An aerosol layer that is extending from 400 m to 800 m agl can be seen in this figure. Figure 6b depicts depolarization for this scan. Figure 6b shows that the depolarization ratio for this layer ranges from 0.1 to 0.15. This is in agreement with previous report on depolarization associated with dry salt particles [11]. It also confirms our previous work on monitoring the depolarization of dust layers in vicinity of the UL by CALIPSO recordings where we found it as: 0.16 ± 0.05 [12]. HYSPLIT backward trajectories in Figure 7 show that the aerosol layer in Figure 6 is originated from a local source, i.e. the Urmia Lake. Recoded time series of SDL in zenith direction during December 2018 (not shown here) also show similar or lower depolarization values for surface atmospheric layers. Considerable reduction of the depolarization at these time can be due to existence of anthropogenic particles.

4. CONCLUSION

We have used the recordings by the space lidar onboard the CALIPSO satellite and a ground based scanning backscatter depolarized lidar to investigate the type of atmospheric aerosols that may originate from the dried bed of the Urmia Lake. CALIPSO recordings on AOD and PDR don't show that the UL is an intense source of atmospheric aerosols for its water catchment area and outside of it. However, these recording indicate that the UL and its surrounding are under the influence of neighboring dust sources like Mesopotamia, especially during dry months of the year. These also show that during winter times the water catchment area of the UL mostly is contaminated with urban industrial particles. On the other side, SDL measurements over the UL are indicating that atmospheric aerosols that may be raised from the dried lake bed, mostly are dry salt particles

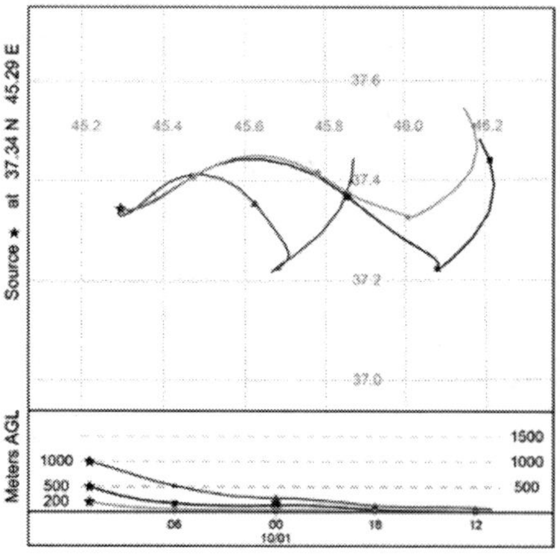

Figure 7 NOAA HYSPLIT backward trajectories ending in Rashakan, at 11:00 UTC, Oct. 01, 2018

ACKNOWLEDGEMENTS

This work was supported by Urmia Lake Restoration Program (ULRP) under Grant No. 96100201. The authors appreciate the NASA Langley Research Center for providing the CALIPSO data. We are also grateful to the NOAA Air Resources Laboratory (ARL) for the provision of the HYSPLIT transport model.

REFERENCES

[1] A. AghaKouchak, et al. Journal of Great Lakes Research **41** (1), 307-311 (2015).

[2] M. Abbaspour, et al. International Journal of Environmental Science and Technology **9** (2), 257-266 (2012).

[3] A. Delju, et al. Theoretical and applied climatology **111** (1-2), 285-296 (2013).

[4] F. Fathian, et al. Theoretical and Applied Climatology **119** (3-4), 443-464 (2015).

[5] E. Hassanzadeh, et al. Water Resources Management **26** (1), 129-145 (2012).

[6] A. Taravat, et al. Water **8** (11), 478 (2016).

[7] P. Ginoux, et al. Reviews of Geophysics **50** (3) (2012).

[8] F. Abdi Vishkaee, et al. Journal of Geophysical Research: Atmospheres **116** (D5) (2011).

[9] A. Masoumi, et al. Atmospheric research **120**, 343-355 (2013).

[10] F. Abdi Vishkaee, et al. Journal of Geophysical Research: Atmospheres **117** (D3) (2012).

[11] M. Haarig, et al. Atmospheric Chemistry and Physics **17** (23), 14199 (2017).

[12] F. Ghomashi and H. Khalesifard, to be appear in *Centeral Asian dust conferece* (Dushanbeh, Tajikistan, 2019).

PREPROCESSING OF RAMAN LIDAR SIGNAL OVER A HIGH ALTITUDE STATION IN INDIA: PRACTICAL CONSIDERATIONS

Jaswant[1,2], Shishir Kumar Singh[1,2], Radhakrishnan S.R.[1,2*], Devesh Shukla[1], Chhemendra Sharma[1,2]

[1]*Environmental Sciences and Biomedical Metrology Division, CSIR-National Physical Laboratory, Dr K.S. Krishnan Marg, New Delhi 110012, India*

[2]*Academy of Scientific and Innovative Research (AcSIR), CSIR Campus, CSIR Road, Taramani, Chennai 600113, India*

**Email: rkrishnan@nplindia.org*

ABSTRACT

Lidar can provide the range-resolved information about the vertical profiles of optical properties of aerosol content and clouds present in the atmosphere. However, before obtaining the information from received lidar signal we ought to use a few pre-processing techniques such as range-correction, temporal and spatial averaging, dead time correction of the photon counts (PC) signal, correction due to overlap effect, Simulation of molecular Rayleigh signal and gluing analog & PC signals. In this work, we have discussed some of the initial pre-processing techniques ought to be performed before lidar inversion.

1. INTRODUCTION

Lidar system has proved to provide range-resolved measurements of the vertical profiles of optical properties of aerosol content and clouds present in the atmosphere [1]. Before obtaining this information about aerosols and clouds through signal inversion, it is required to perform a number of preprocessing techniques so as to achieve better and improved signal such as signal averaging, dead time correction, bin shift correction, background correction, overlap correction, gluing both analog and photon counts (PC) signals etc. Here in this paper we have tried to provide a practical approach to some of these preprocessing techniques.

2. METHODOLOGY

A Raman Lidar system has been deployed at the CSIR-Institute of Himalayan Bioresource Technology (IHBT), Palampur, Himachal Pradesh, India. The station is situated at an altitude of 1347 m (amsl). This lidar system is currently being used to obtain the vertical distribution of clouds and aerosol over Palampur. Before deriving the optical properties from this data, a number of preprocessing techniques mentioned above can be performed. The data pre-processing includes the techniques that are applied on the raw lidar data before performing any inversion algorithm. The corrections or the techniques used in this work are mentioned below in sequential order.

2.1 Signal Averaging

The preprocessing of the raw lidar data starts with the temporal and spatial averaging. Both temporal and spatial averaging is done to increase the signal to noise ratio of the lidar measurements. The selection of time for averaging is chosen in such a way so that the variability in the lidar signal can be minimized. For our lidar data analysis we have taken the 30 minutes and 30m time and space averaged data respectively to increase the SNR.

2.2 Dead time Correction

The next correction that is applied on the averaged PC data is known as the dead time correction of PC signals. The dead time of any detector system is referred as the minimum amount of time that is required by the detector to distinguish two consecutive events or photon counts. If there is not enough time separation between two events, then second event may not be counted. Hence, the photons which have time separation more than dead time, in that case both the photons would be counted. In case of not enough separation, they may be counted as single photon. To correct this, we must apply the dad time correction on the received PC signals. For a non-paralyzable system the dead time corrected signal is determined as follows [2], [3]:

$$N_{true} = \frac{N_{observed}}{1 - \tau^* N_{observed}} \quad (1)$$

Here, N_{true} is the dead time corrected signal, $N_{observed}$ is the observed signal and τ is the dead time i.e. 4ns prescribed by the manufacturer [4].

2.3 Background Subtraction

After performing the dead time correction on the PC signals, the background noise or the solar/atmospheric radiation is to be removed from both analog and PC signals. The background noise is inherent in the received signal and it can be considered to be almost constant throughout the whole range in the signal. To remove the noise from the signals, first we have to calculate the noise offset value. For that we select a region in the higher heights where aerosol content can be considered to be negligible and the backscattered signal from such heights will mainly be due to the background radiation. The average of the lidar signal in this altitude range can be considered as noise and this is subtracted from the lidar signal throughout the whole range. For our study we have selected the area above 20km and calculated the noise value for both analog and PC channel and subtracted it from the respective signals.

2.4 Bin Shift Correction

After the noise removal from both signals there is a need to remove the time delay between both the signals. The analog signal is received at the detector side with some time delay as compared to the PC signal. This delay is caused mainly due to two reasons i). Due to the presence of anti-aliasing filter in analog channel and ii). ADC pipelining [5]. So before gluing we have to correct the analog signal for bin shift or we have to shift the analog signal by some memory bins. The amount of shift can be calculated by selecting a cloud profile and the peak of the signal should be at the same time in both the signals.

2.5 Overlap Correction

The lidar observations are significantly affected by the incomplete overlap found between the beam of the laser and the field of view of the telescope particularly at lower altitudes. To correctly measure the vertical profiles of the aerosol optical properties at lower altitude, it becomes extremely prominent to take overlap correction into account. To find out the overlap function and further correct it, we have utilised the approach mentioned by Wandinger and Ansmann [6] which demonstrates that the comparison of the inversion algorithms i.e. klett inversion (elastic inversion) and Raman inversion (inelastic inversion) can yield us the information about the incomplete overlap. The lidar signal is corrected for the overlap effect as mentioned in the equation 2 below,

$$P_{corrected}(R,\lambda) = \frac{1}{O(R,\lambda)} P_{measured}(R,\lambda) \quad (2)$$

here $O(R, \lambda)$ represents overlap function, $P_{measured}$ denotes the received power and $P_{corrected}$ is lidar signal after overlap correction.

2.6 Gluing of Analog and PC signals

The lidar data collected in analog mode exhibits high linearity at lower heights whereas the PC signal provides good sensitivity at higher altitudes. The SNR is high for analog at lower heights and for PC at higher heights. Due to this reason, there is a need to find a signal which shows higher SNR at both lower and higher heights. As a solution, we combine both the signals into a single signal through a process known as gluing. Although there are numerous strategies to achieve this but we have followed a method proposed by [7].

3. RESULTS

The lidar data of 1 month has been taken in this study after performing temporal (30 minutes) and spatial (30 meters) averaging on the raw lidar data to improve the SNR. The preprocessing of the signal has been shown from figure 1 to 4. The averaging was followed by the dead time correction of the raw PC signal with 4ns dead time. We can see from figure 1(b) that pulse pile-up effect was reduced in the dead time corrected PC signal (pink). Later, both analog and PC signals were corrected for the background noise caused due to atmospheric radiation. The results of the background corrected data for both analog and PC signal can be seen in figure 1(a) and 1 (b) respectively.

Figure 2 shows the cloud profile in both analog and PC mode, we can observe that there exists a shift of 9 bins between both analog and PC data as

 https://doi.org/10.1051/epjconf/202023702026

analog data is delayed by 9 bins compared to PC data because of the reasons mentioned in section 2.4. Results reveal that before gluing, the shift correction of the analog data becomes mandatory.

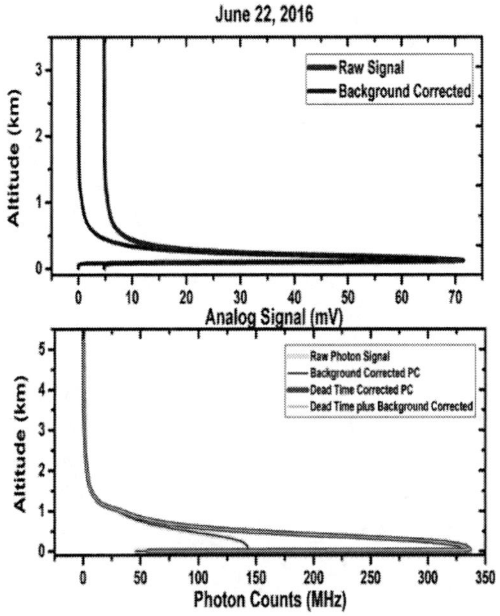

Figure 1(a) shows the raw analog signal and background corrected signal for the June 22, 2016. Figure 1(b) shows the PC signal, background removed PC signal, dead time corrected PC signal and dead time plus background corrected PC signal for June 22, 2016.

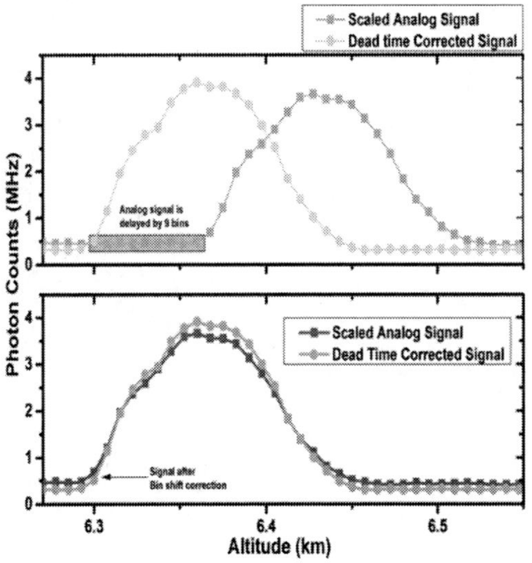

Figure 2(a) shows the comparison of scaled analog signal and dead time plus background corrected PC signal before bin shift correction, Figure 2(b) same as 2(a) after bin shift correction

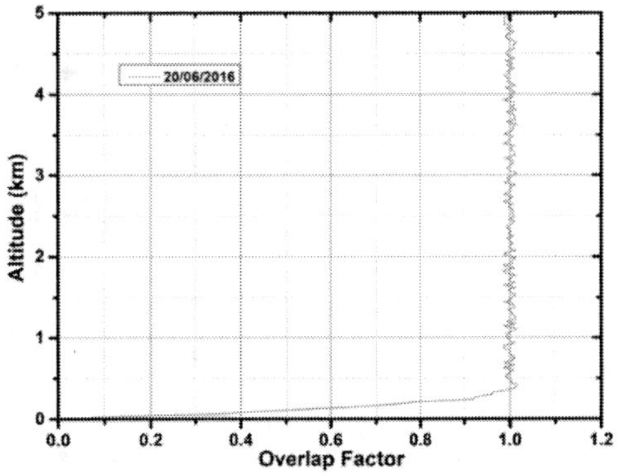

Figure 3 shows the overlap function calculated for the 20th June 2016 using iterative method.

Next, we determined the overlap function for the lidar system using analog data as analog signal shows good linearity at lower heights. We can see from figure 3(a) that there was incomplete overlap below 300 meters and the complete overlap was achieved beyond 300 meters. Below 300 meters, the value of overlap function was less than 1 and after 300 meters, the value was close to 1. Figure 3(b) shows the relative deviation of overlap factor calculated for each day with respect to mean overlap factor. The deviation above 300 meters was less than 5%, however for lower heights i.e. below 300 meters the deviation was high.

Figure 4 depicts the comparison of the scaled analog signal, dead time corrected PC signal and glued signal. It can be seen that the transition region for gluing between dead time corrected PC signal and scaled analog (red) was found from 1.7km to 3.2km. So, the scaled analog signal was completely replaced upto 1.7km by the glued signal (cyan) whereas dead time corrected PC signal was replaced beyond altitude of 3.2km by glued signal. So, after applying all these corrections this glued signal was used to further derive the optical properties of aerosols and clouds.

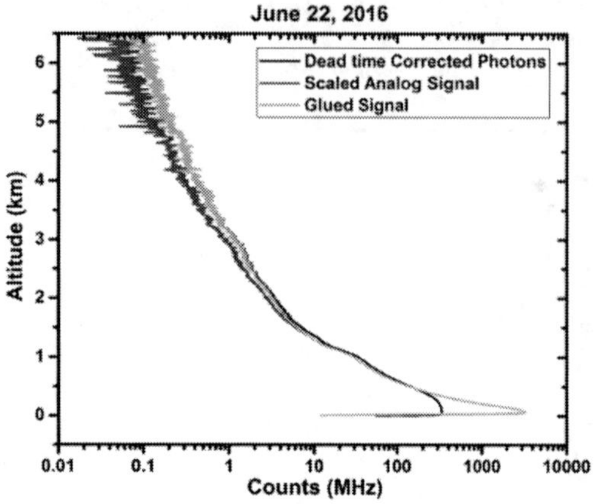

Figure 4 shows the comparison of the scaled analog signal, dead time corrected and glued signal for June 22, 2016.

4. CONCLUSIONS

This paper reports the techniques which we follow to preprocess the lidar signal before deriving optical properties from it. Results reveal that the lidar signal is improved to an optimal level after applying these preprocessing steps.

ACKNOWLEDGEMENTS

Authors are grateful to Director, NPL for necessary support. Authors are thankful to CSIR network project PSC-0112 for necessary financial support. Mr. Jaswant is thankful to Council of Scientific and Industrial Research (CSIR) for providing research fellowship and also to Academy of Scientific and Innovative Research (AcSIR) for facilitating as its PhD student.

REFERENCES

[1] A. Ansmann, P. Seifert, M. Tesche, and U. Wandinger, Atmospheric Chem. Phys. 12, 9399–9415 (2012)

[2] D. N. Whiteman, S. H. Melfi, and R. A. Ferrare, Appl. Opt. 31, p. 3068 (1992)

[3] Zhishen Liu, Zhigang Li, Bingyi Liu, and Rongzhong Li, Chin. Opt. Lett. 7, 1051–1054 (2009)

[4] "LR111-ESS-D200 Raman Depolarization LIDAR." User Manual by Raymetrics, (2016)

[5] "Licel Ethernet Controller – Installation and Reference Manual." Licel GmbH, 16-Jan-2014.

[6] U. Wandinger and A. Ansmann, Appl. Opt. 41, p. 511 (2002)

[7] C.-Z. Feng, B.-Y. Liu, J.-T. Liu, and S.-H. Wu, EPJ Web Conf. 119, p. 25021 (2016)

Profiling aerosol optical properties at the Central Asian site of Dushanbe, Tajikistan: pure dust cases

Julian Hofer[1,*], Dietrich Althausen[1], Sabur F. Abdullaev[2], Abduvosit N. Makhmudov[2], Bakhron I. Nazarov[2], Holger Baars[1], Ronny Engelmann[1] and Albert Ansmann[1]

[1]*Leibniz Institute for Tropospheric Research (TROPOS), Leipzig, Germany, *hofer@tropos.de*
[2]*Physical Technical Institute of the Academy of Sciences of Tajikistan, Dushanbe, Tajikistan*

ABSTRACT

Tajikistan is often affected by atmospheric mineral dust originating from various surrounding deserts. The direct and indirect radiative effects of that dust play a sensitive role in the Central Asian climate system and therefore need to be quantified. The Central Asian Dust Experiment (CADEX) provides for the first time an aerosol climatology for Central Asia based long-term aerosol profiling by ground-based lidar (PollyXT type) in Dushanbe, Tajikistan. For pure dust cases, mean depolarization(lidar) ratios of 0.23±0.03(44±3 sr) at 355 nm and 0.32±0.02(38±3 sr) at 532 nm wavelength have been measured. The mean extinction-related Ångström exponent was 0.18±0.15.

1 INTRODUCTION

The Central Asian country Tajikistan and its neighboring countries contain vast arid regions and are located within the global dust belt reaching from the Sahara to the Gobi desert. Thus, the occurrence of dust events with a broad range of intensities is frequent in Tajikistan and whole Central Asia [1]. In this region, which is highly affected by climate change [2, 3], this dust has impacts ranging from radiative effects to health hazards [4, 5]. Aerosol observations in Tajikistan are therefore highly important to understand regional and global transport of mineral dust and its effects on radiation budget, cloud formation etc. Therefore, long-term data on vertical profiles of aerosol optical properties are provided by the Central Asian Dust Experiment (CADEX).

2 INSTRUMENTS AND METHODS

The CADEX campaign was conducted at a field site of the Physical Technical Institute of the Academy of Sciences of Tajikistan in Dushanbe (38°33′34″ N, 68°51′22″ E, 864 m ASL) [6]. During the 18-month campaign (March 2015 to August 2016), a multiwavelength polarization Raman lidar PollyXT [7] was operated continuously for the lidar measurements. During the measurement period, strong dust events occurred several times and lofted dust layers at various altitudes were frequent, as well as dusty mixtures and pollution aerosol, especially in winter and spring [6, 8, 9].

This contribution, however, provides a closer look at the aerosol optical properties during periods when purest dust was observed. Such periods were identified by the layer-mean particle linear depolarization ratio at 532 nm wavelength plus one standard deviation (of the mean within the averaging height range) being larger than 0.31. The basis of this analysis are profiles of the optical properties retrieved at each possible night during the observation period. The profiles were manually evaluated with the Raman method [10, 11, 12]. 22 profiles fulfill the aforementioned criterion, but one case of a thin lofted layer was excluded, due to a small averaging height range and noisy extinction profiles. The mean averaging height range for the remaining 21 profiles was 1.6±0.5 to 2.7±0.7 km AGL. For comparison of the layer-mean extinction-related Ångström exponent, AERONET (Aerosol Robotic Network [13]) data of the collocated AERONET CIMEL sun photometer [14] was used. The FLEXible

PARTicle dispersion model FLEXPART [15] was used for the backward trajectory calculations.

3 RESULTS

The layer-mean particle linear depolarization ratio of the 21 pure dust cases ranged from 0.18±0.01 to 0.29±0.01 at 355 nm and from 0.30±0.02 (determined by the applied selection criterion) to 0.36±0.01 at 532 nm wavelength. The layer-mean lidar ratio ranged from 39±2 sr to 49±9 sr at 355 nm and from 32±2 sr to 43±3 sr at 532 nm wavelength. The layer-mean extinction-related Ångström exponent ranged from -0.12±0.05 to 0.46±0.14. The means of all the values are listed in Tab. 1.

Table 1: Mean and standard deviation of layer-mean intensive parameters (particle linear depolarization ratio, lidar ratio and extinction-related Ångström exponent) of 21 dust events during the CADEX campaign.

wavelength	355 nm	532 nm
depol. ratio	0.23±0.03	0.32±0.02
lidar ratio	44±3 sr	38±3 sr
Ångstr. exp.	0.18±0.15	

Figure 1 shows a comparison of layer-mean Ångström exponents from lidar with column Ångström exponents from AERONET as scatter plots. The 21 cases have a smaller range in the AERONET Ångström exponent of the 440-870 nm range (Fig. 1a) than the ones of the 380-500 nm range (Fig. 1b). The correlation is similar in both cases, but the AERONET Ångström exponents of the 380-500 nm range (more comparable to the lidar wavelengths) are obviously distributed closer along the identity line (Fig. 1b).

Figure 2 shows examples of 7-day FLEXPART backward trajectories, which were calculated for all averaged layers. Exemplary lofted layers (Figs. 2a,b) had high accumulated residence times over the Arabian Peninsula, Iran and partly Afghanistan and Uzbekistan. Exemplary lower level layers had high accumulated residence times mainly in Uzbekistan and Kazakhstan towards the Caspian and Aral Sea (Figs. 2b,c).

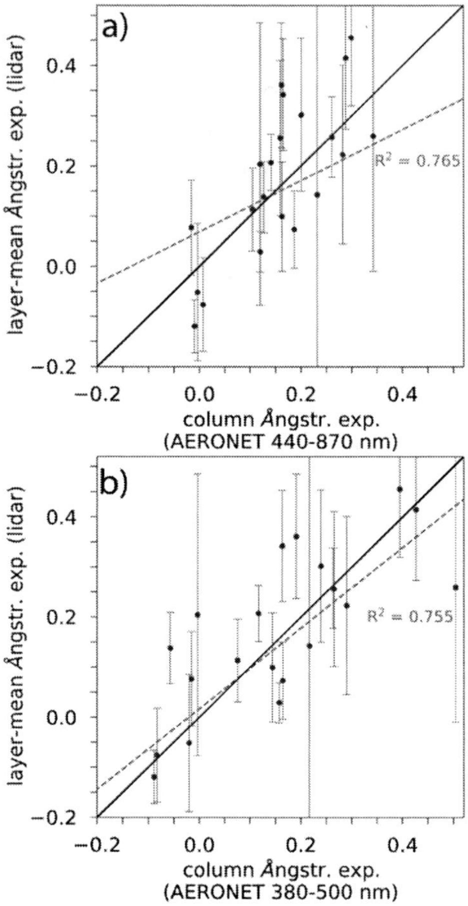

Figure 1: Comparison of layer-mean dust Ångström exponents from lidar with column Ångström exponents from AERONET from the 440-870 nm (a) and 380-500 nm (b) wavelength range. Errorbars indicate the standard deviation within the averaging height range.

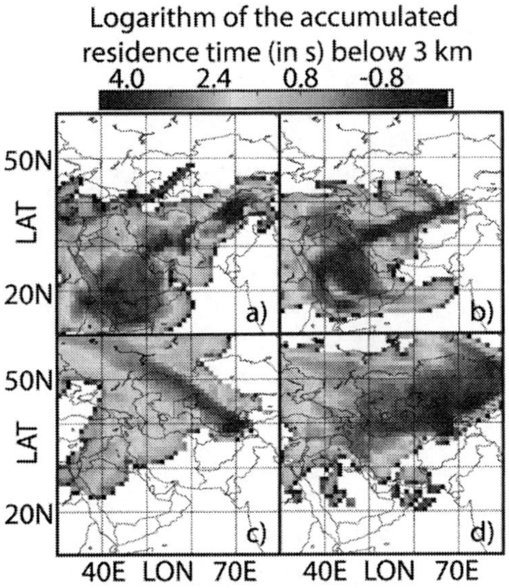

Figure 2: Accumulated residence times of 7-day FLEXPART backward trajectories arriving at Dushanbe at the following dates (UTC) and heights (AGL): a) 13 April 2015, 15:10-16:10, 2.8-3.8 km. b) 23 April 2015, 20:47-21:47, 3-5 km. c) 10 June 2015, 19:00-20:00, 1.4-2.5 km. d) 14 July 2016, 18:30-19:30, 1.4-1-8 km.

4 CONCLUSIONS AND OUTLOOK

The 18-month lidar observations in Dushanbe, Tajikistan, provided unprecedented data of vertical profiles of aerosol optical properties in Central Asia. The optical properties of pure dust cases observed during the CADEX campaign are presented in this study. Further analyses include the depolarization ratio, lidar ratio, Ångström exponent of all the different observed aerosol conditions as well as aerosol layer heights. The polarization-lidar photometer networking (POLIPHON) [16, 17, 18] method will be used to estimate dust and non-dust mass concentrations, as well as concentrations of cloud condensation nuclei and ice nucleating particles on a climatological basis. For further aerosol characterization, backward trajectory analyses and AERONET data will be used in addition to the lidar data. Furthermore, forward trajectory analysis can give insights on the transport of lofted dust layers further east.

ACKNOWLEDGEMENTS

The CADEX project was funded by the German Federal Ministry of Education and Research in the context of "Partnerships for sustainable problem solving in emerging and developing countries" under the grant number 01DK14014. This project has also received funding from the European Union's Horizon 2020 research and innovation programme under grant agreement No 654109.

References

[1] Indoitu, R., Orlovsky, L., and Orlovsky, N.: Dust storms in Central Asia: Spatial and temporal variations, *J. Arid Environ.* **85**, 62–70, (2012)

[2] Xu, H., Wang, X., and Zhang, X.: Decreased vegetation growth in response to summer drought in Central Asia from 2000 to 2012, *Int. J. Appl. Earth Obs. Geoinf.* **52**, 390–402 (2016)

[3] Lioubimtseva, E. and Henebry, G. M.: Climate and environmental change in arid Central Asia: Impacts, vulnerability, and adaptations, *J. Arid Environ.* **73**, 963–977 (2009)

[4] Shi, Z., Xie, X., Li, X., Yang, L., Xie, X., Lei, J., Sha, Y., Liu, X.: Snow-darkening versus direct radiative effects of mineral dust aerosol on the Indian summer monsoon onset: role of temperature change over dust sources, *Atmos. Chem. Phys.* **19**, 1605–1622 (2019)

[5] Wiggs, G. F. S., O'hara, S. L., Wegerdt, J., Van Der Meer, J., Small, I., Hubbard, R.: The dynamics and characteristics of aeolian dust in dryland Central Asia: possible impacts on human exposure and respiratory health in the Aral Sea basin, *Geogr. J.* **169**, 142–157 (2003)

[6] Hofer, J., Althausen, D., Abdullaev, S. F., Makhmudov, A. N., Nazarov, B. I., Schettler,

G., Engelmann, R., Baars, H., Fomba, K. W., Müller, K., Heinold, B., Kandler, K., Ansmann, A.: Long-term profiling of mineral dust and pollution aerosol with multiwavelength polarization Raman lidar at the Central Asian site of Dushanbe, Tajikistan: case studies, *Atmos. Chem. Phys.* **17**, 14 559–14 577 (2017)

[7] Engelmann, R., Kanitz, T., Baars, H., Heese, B., Althausen, D., Skupin, A., Wandinger, U., Komppula, M., Stachlewska, I. S., Amiridis, V., Marinou, E., Mattis, I., Linné, H., Ansmann, A.: The automated multiwavelength Raman polarization and water-vapor lidar Polly^{XT}: the neXT generation, *Atmos. Meas. Tech.* **9**, 1767–1784 (2016)

[8] Hofer, J., Althausen, D., Abduallaev, S. F., Nazarov, B. I., Makhmudov, A. N., Baars, H., Engelmann, R., Ansmann, A.: Aerosol layer heights above Tajikistan during the CADEX campaign, *Proceedings of the Central Asian Dust Conference (CADUC)*, Dushanbe, Tajikistan (2019)

[9] Hofer, J., Althausen, D., Abdullaev, S. F., Makhmudov, A. N., Nazarov, B. I., Schettler, G., Engelmann, R., Baars, H., Heinold, B., Müller, K., Fomba, K. W.: Central Asian Dust Experiment (CADEX): First Year Lidar Observations, in: *Light, Energy and the Environment*, EW2A.3, Optical Society of America (2017)

[10] Ansmann, A., Riebesell, M. A., and Weitkamp, C.: Measurement of atmospheric aerosol extinction profiles with a Raman lidar, *Opt. Lett.* **15**, 746–748 (1990)

[11] Ansmann, A., Riebesell, M. A., Wandinger, U., Weitkamp, C., Voss, E., Lahmann, W., Michaelis, W.: Combined raman elastic-backscatter LIDAR for vertical profiling of moisture, aerosol extinction, backscatter, and LIDAR ratio, *App. Phys.* **55**, 18–28 (1992)

[12] Ansmann, A., Wandinger, U., Riebesell, M. A., Weitkamp, C., Michaelis, W.: Independent measurement of extinction and backscatter profiles in cirrus clouds by using a combined Raman elastic-backscatter lidar, *Appl. Opt.* **31**, 7113–7131 (1992)

[13] Holben, B. N., Eck, T. F., Slutsker, I., Tanré, D., Buis, J. P., Setzer, A., E., V., Reagan, J. A., Kaufman, Y. J., Nakajima, T., Lavenu, F., Jankowiak, I., Smirnov, A.: AERONET A Federated Instrument Network and Data Archive for Aerosol Characterization, *Remote Sens. Environ.* **66**, 1–16 (1998)

[14] Abdullaev, S. F., Nazarov, B. I., Salikhov, T. K., Maslov, V. A.: Correlations of surface air temperature and optical thickness of arid aerosol according to AERONET data, *Atmos. Ocean. Opt.* **25**, 428–433 (2012)

[15] Stohl, A., Forster, C., Frank, A., Seibert, P., Wotawa, G.: Technical note: The Lagrangian particle dispersion model FLEXPART version 6.2., *Atmos. Chem. Phys.* **5**, 2461–2474 (2005)

[16] Ansmann, A. and Mamouri, R. E.: Estimated desert-dust ice nuclei profiles from polarization lidar: methodology and case studies, *Atmos. Chem. Phys.* **15**, 3463–3477 (2015)

[17] Ansmann, A. and Mamouri, R.-E.: Potential of polarization lidar to provide profiles of CCN- and INP-relevant aerosol parameters, *Atmos. Chem. Phys.* **16**, 5905–5931 (2016)

[18] Mamouri, R.-E. and Ansmann, A.: Potential of polarization/Raman lidar to separate fine dust, coarse dust, maritime, and anthropogenic aerosol profiles, *Atmos. Meas. Tech.* **10**, 3403–3427 (2017)

https://doi.org/10.1051/epjconf/202023702028

POLARIZATION LIDAR FOR DETECTING DUST ORIENTATION

Alexandra Tsekeri[1*], Volker Freudenthaler[2], George Doxastakis[3], Josef Gasteiger[4], Alexandros Louridas[3], George Georgoussis[3], Ioannis Binietoglou[5], Thanasis Georgiou[1], Zbigniew Ulanowski[6], and Vassilis Amiridis[1]

[1]*IAASARS, National Observatory of Athens, Athens, Greece*
[2]*Fakultät für Physik, Meteorologisches Institut, Ludwig-Maximilians-Universität, Munich, Germany*
[3]*Raymetrics S.A., Greece*
[4]*Faculty of Physics, University of Vienna, Vienna, Austria*
[5]*National Institute of R&D for Optoelectronics, Romania*
[6]*School of Physics Astronomy and Mathematics, University of Hertfordshire, Hatfield, UK*
**Email: atsekeri@noa.gr*

ABSTRACT

We present the preliminary design of a polarization lidar system for detecting oriented dust particles. The system emits linearly- and circularly-polarized light from two lasers and detects the backscattered light, using two telescopes. The emitted light by each laser is detected from both telescopes alternatively, providing in this way two times the signals recorded by a conventional system. The system has also the capability of performing measurements at different zenith angles. Its detection strategy is based on scattering calculations for mixtures of oriented and randomly-oriented dust particles in different atmospheric scenarios.

1. INTRODUCTION

Dust orientation is an ongoing investigation in recent years [1]. Its potential proof will be a paradigm shift for dust remote sensing, invalidating the currently used simplifications of randomly-oriented particles. Currently, the only signature of particle orientation comes from astronomical polarimetry measurements of dichroic extinction [1], however these measurements provide only an indication of the alignment and refer to column-integrated values, not being capable for vertically-resolved retrievals. Vertically-resolved measurements of dust orientation can be acquired with a polarization sensitive lidar designed to target the off-diagonal elements of the backscattering matrix which are non-zero only when the particles are oriented [2].

Previous studies have demonstrated the feasibility of using circular or linear depolarization lidar measurements to detect the orientation of ice crystals in clouds [3,4,5] and it has been theoretically shown that these techniques can be extended for oriented dust particles [2]. Geier and Arienti [2] showed that although the linear depolarization ratio is sufficient for discerning ice crystal orientation, this is not the case for smaller particles as dust. What they suggested is to use the linear or circular polarized light and detect the backscatter in different polarization planes. With this approach they were able to retrieve the off-diagonal elements of the backscatter phase matrix, which provide information on the asphericity and the orientation of the particles.

We will construct a polarization lidar that will emit linear- and circular-polarized light and detect the linearly- and circularly-polarized components of the backscattered light. The operating wavelength will be at 532 or 1064 nm for better probing of the dust coarse mode, and the design will follow the high quality EARLINET standards [6]. It will also employ the capability to measure at more than one zenith angle so as to provide more information on the dust orientation and microphysical properties, depending on the angle of the particle orientation.

To achieve highly accurate polarization measurements with high signal to noise ratio

(SNR), the proposed lidar system will use high power lasers, large aperture telescopes and small receiver fields of view. In order to achieve good SNR in short measurement times the system will be equipped with two laser sources emitting interleaved linear and circular polarized light and two telescopes for their detection. The linearly- and circularly-polarized components of the backscattered light will be measured utilizing a Polarizing Beam Splitter cube (PBS) in one telescope and a circular analyzer (i.e. a Quarter-Wave Plate (QWP) at 45° with a PBS) in the other.

The capability of the system in deriving the percentage of oriented dust particles in the atmosphere is investigated herein with a simulated dataset, considering different atmospheric scenarios. The corresponding backscatter matrices are calculated with the Amsterdam Discrete Dipole Approximation (ADDA) [7] for irregular-shaped oriented dust particles [8].

2. METHODOLOGY

2.1 System design

The system design is shown in Fig. 1. The two lasers (L1 and L2) shoot alternatively and their backscattered signals are collected by both Telescopes (T1 and T2). This "2-Lasers/2-Telescopes" setup allows the recording of eight signals from the four detectors of the system (D1, D2, D3 and D4 in Fig. 1). This cost-effective approach provides higher SNR (up to 4 times) and smaller measurement errors due to no moving optics in receiver and transmitter.

L1 emits 45° linearly-polarized light with respect to the plane of the PBS1. L2 emits linearly-polarized light which is transformed with QWP_{L2} with fast-axis angle of 45° to circularly-polarized light. The backscattered light collected by T1 is analyzed to its parallel and cross components of its linearly polarized component, recorded by D1 and D2, respectively. The circularly-polarized component of the backscattered light collected by T2 is transformed to linearly-polarized component, passing through the QWP2 with fast-axis angle of 45°. Further on, this linearly-polarized component is analyzed to the parallel and cross components through the PBS2 and measured at D3 and D4, respectively.

In our analysis we use the ratios of the recorded signals. Specifically, we use the natural logarithms of these ratios, so as to avoid large values in case of denominator values close to zero. The backscattered signals from L1, recorded at D1, D2, D3 and D4 are named as CH1, CH2, CH3 and CH4, respectively, and the backscattered signals from L2, recorded at D1, D2, D3 and D4 are named as CH5, CH6, CH7 and CH8, respectively. The ratios used are ln(CH1/CH2), ln(CH3/CH4), ln(CH5/CH6) and ln(CH7/CH8).

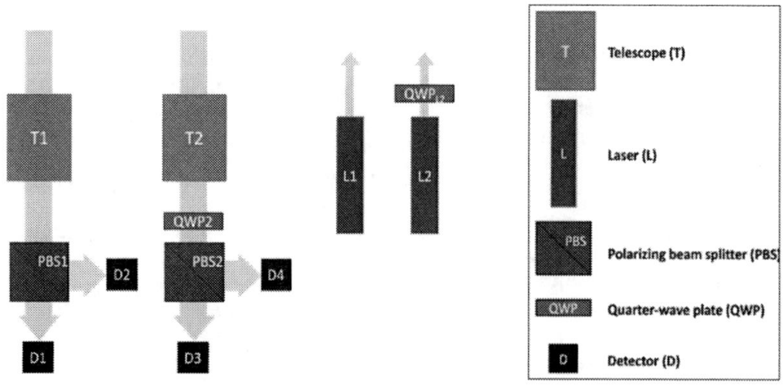

Figure 1: The polarizing lidar design.

2.2 Simulated dataset for system evaluation

The system capabilities and limitations in detecting the percentage of oriented dust particles are investigated using simulated atmospheric scenarios, generated from a system signal simulator and considering the scattering properties of dust particles with realistic irregular shapes and their mixtures as oriented and randomly-oriented particles. The scattering database is generated considering the irregular-shaped particles of [8] (Fig. 2), using the ADDA code and the extensive computational resources of MareNostrum HPC and ARIS HPC.

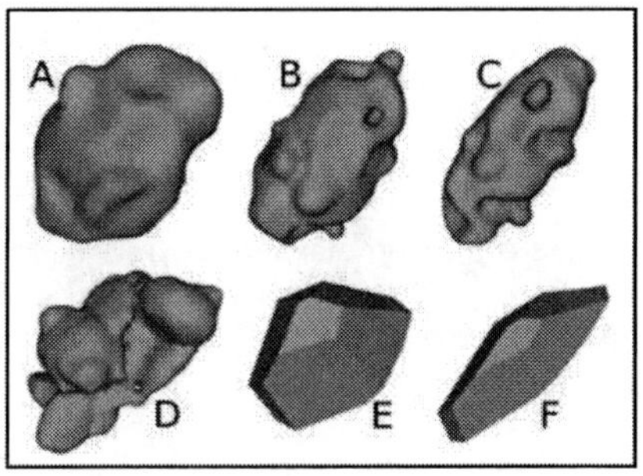

Figure 2: Particle shapes considered for dust particles (Source: [8]).

The maximum dust volume-equivalent size parameter considered is ~60 and the particle refractive indices are 1.6+i0 for "non absorbing" dust and 1.48+i0.002 for "absorbing" dust. We make the assumption that the particles are not oriented at fixed angles, but "wobble" over a range of angles along the vertical direction. Moreover, we consider various heights above ground and depths for the dust layers, and we omit the presence of other aerosol layers in the atmospheric scene. The AODs of the dust layers range from 0.05 to 1.5 at 532 and 1064 nm. The above assumptions may be over-simplifications of the realistic atmospheric scenes, but they are useful as a first step towards a more extensive analysis.

3. RESULTS

3.1 Information content for the percentage of oriented particles

Figure 3 presents a preliminary analysis of the information content for the percentage of oriented dust particles in the simulated signals considered. The analysis is done for a system emitting at 532 nm (Fig. 3a) and at 1064 nm (Fig. 3b) and performing measurements at 60° from zenith. It seems that the measurements at 1064 nm provide more information on the dust particle orientation. These results will be further analyzed considering the real SNR of 1064 and 532 nm detectors, taking also into account viewing angles at e.g. 30° and 45° from zenith. Finally, they will be used for constructing a retrieval scheme for the properties of oriented dust particles.

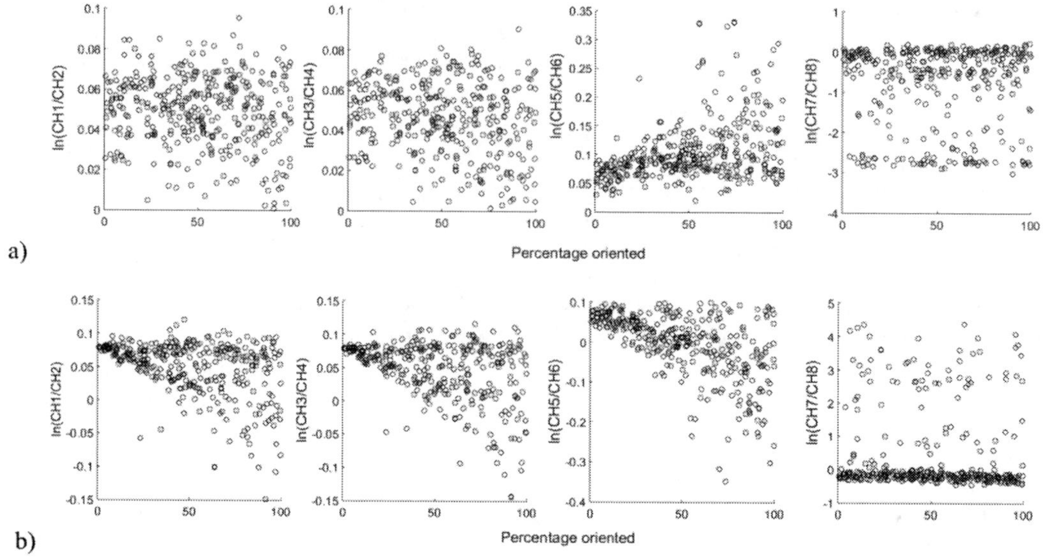

Figure 3: a) Percentage of oriented dust particles in the dust layers of the simulated atmospheric scenes, plotted versus the ratios of the logarithms of the signals (from left to right) ln(CH1/CH2), ln(CH3/CH4), ln(CH5/CH6) and ln(CH7/CH8), for backscattered light at 532 nm, viewed by the lidar with 60° from zenith. b) The same with (a), for backscattered light at 1064 nm.

ACKNOWLEDGEMENTS

This work was supported by the European Research Council under the European Community's Horizon 2020 research and innovation framework program/ERC grant agreement 725698 (D-TECT). The authors acknowledge PRACE for awarding access to MareNostrum at Barcelona Supercomputing Center (BSC), Spain, and GRNET for awarding computational time in the National HPC facility - ARIS - under project ID pa170906-ADDAPAS and pr005038-REMOD.

REFERENCES

[1] Z. Ulanowski, et al., Atmos. Chem. Phys. 7: 6161–6173 (2007)

[2] M. Geier and M. Arienti, J Quant Spectrosc Radiat Transf 149: 16–32 (2014)

[3] Y.Balin et al., Opt. Express, 21(11): 13408-13418 (2013)

[4] V. Bruno et al., Appl. Opt., OSA, 43: 6620-6628 (2004)

[5] S. N. Volkov et al., Appl. Opt., OSA, 54: 3095-3105 (2015)

[6] V. Freudenthaler and T. Berkoff, Proc. of 2nd GALION workshop, WMO, Geneva (2010)

[7] M. A. Yurkin and A. G. Hoekstra, J. Quant. Spectrosc. Radiat., 112: 2234–2247 (2011)

[8] J. Gasteiger et al., Tellus B, 63: 725–741 (2011)

POLARIZATION RAMAN LIDAR FOR ATMOSPHERIC MONITORING IN THE VIPAVA VALLEY

Longlong Wang[1*], Samo Stanič[1], William Eichinger[2], Xiaoquan Song[3,4], Marko Zavrtanik[5]

[1]*Center for Atmospheric Research, University of Nova Gorica, Nova Gorica 5270, Slovenia*
[2]*Department of Civil and Environmental Engineering, University of Iowa, Iowa City, IA 52242, USA*
[3]*Ocean Remote Sensing Institute, Ocean University of China, Qingdao 266100, China*
[4]*Laboratory for Regional Oceanography and Numerical Modeling, Qingdao National Laboratory for Marine Science and Technology, Qingdao 266237, China*
[5]*Jožef Stefan Institute, Ljubljana 1000, Slovenia*
**Email: longlong.wang@ung.si*

ABSTRACT

We report on the design, construction and performance of a polarization Raman lidar, built for atmospheric monitoring in the Vipava valley in SW Slovenia, a regional air pollution hot-spot where aerosols are expected to originate from a number of different sources. Its key features are automatized remote operation capability and indoor deployment, which provide high duty cycle in all weather conditions. System optimization and performance studies include the calibration of the depolarization ratio, merging of near-range (analog) and far-range (photon-counting) data, determination of overlap functions and validation of the retrieved observables with radiosonde data.

1. Introduction

Closed valleys in mountainous areas are often regional air pollution hot-spots due to local emissions, the main causes for air pollution being restricted vertical mixing and limited horizontal ventilation. In order to monitor the dilution of local pollutants (carbonaceous aerosols) and deposition effects of long range transport aerosols (mineral dust and sea salt) [1], experimental data on aerosol properties with good temporal and spatial resolution is needed. The town of Ajdovščina, which is located in the center of the Vipava valley is an ideal investigation site (Fig. 1). In addition to existing in-situ meteorological and environmental measurements (i.e. of black carbon concentration), remote sensing of atmospheric optical properties provides the opportunity to investigate atmospheric structures and estimate predominant aerosol types in these structures. The capability of continuous lidar profiling in all weather conditions is the key to revealing the dynamics of various poorly understood atmospheric phenomena, such as for example the effects of the strong, gusty downslope wind (Bora). Our goal was to develop a dedicated, high accuracy lidar system for long-term monitoring of a broad set of atmospheric optical properties (extinction coefficient, backscatter coefficients, depolarization ratio, backscatter Ångström exponent, lidar ratio and water vapor mixing ratio) above the Vipava valley.

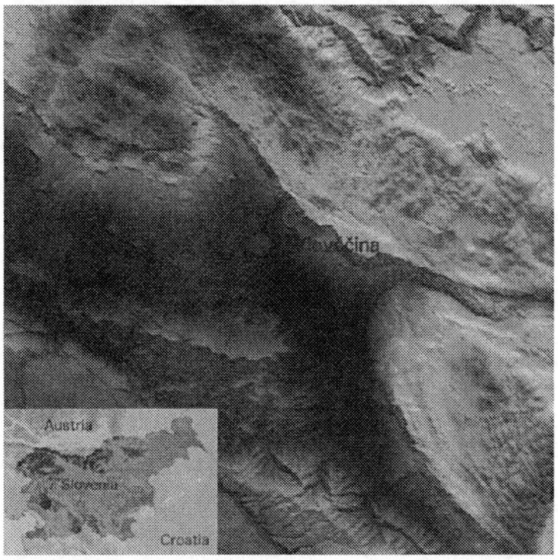

Fig. 1. Terrain configuration of the Vipava valley in the SW Slovenia, about 30 km from the Adriatic sea. To N, the valley is closed by a 1200 m high Trnovski gozd plateau and towards the Adriatic by the 300 m high Carst plateau.

2. System configuration

The biaxial polarization Raman lidar system consists of two independent transmitters, a receiver, spectroscopic filter with detectors for five separate channels and the data acquisition system, including signal amplifiers and digitizers. System setup is shown in Fig. 1.

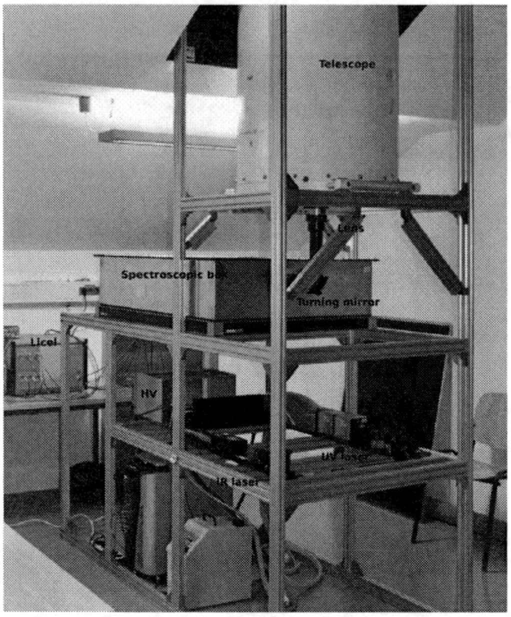

Fig. 1. Polarization Raman lidar system in its working environment. The atmosphere is accessed through a borosilicate rooftop window with > 85% light transmission. The telescope is mounted as close to the window as possible, leaving room for the spectroscopic filter and the transmitter below.

In order to facilitate operation in all-weather conditions, including very strong winds, the lidar is deployed indoors and accesses the atmosphere through a rooftop window. To achieve adequate maximum detectable range throughout whole troposphere of the system in the Raman channels, a large diameter telescope (600 mm) and a high efficiency spectroscopic system (18% to 42%, depending on the channel) was used. Using the four UV and one IR return channels, UV extinction coefficient, backscatter coefficients, depolarization ratio (DR), backscatter Ångström exponent (BAE), lidar ratio (LR) and water vapor mixing ratio were retrieved. The analysis of their uncertainties is presented in [2].

2.1 Transmitters

The transmitter part of the system consists of two high power Nd:YAG pulsed lasers (Q-smart 450 at 355 nm with 10 Hz and Big Sky Ultra at 1064 nm with 20Hz), beam expanders and turning mirrors, and an additional polarization beam splitter (PBS) for the 355 nm laser. For both lasers, the fixed full width divergence of the beams was reduced to less than 0.2 mrad using beam expanders. Laser pulses at 355 nm were more than 99.9% linearly polarized after passing the PBS.

2.2 Receiver

The receiver is an f/8 Cassegrain telescope with a 600 mm primary and 80 mm secondary apertures. A coupling lens was installed at the back plane of the telescope and adjusted to magnify focus at the far end of the filter box approximately 2 m from the end of the telescope. The diameter of the focus at the end of the box is approximately 2 cm. As we desire to make measurements in various weather conditions, including during strong winds, the lidar is installed indoors. It emits and receives signals through a rooftop window with special UV and IR-transparent borosilicate glass with the thickness of 5 mm and the dimensions of 850 mm × 980 mm. The total light received by the telescope is separated into five channels with specific wavelengths and polarizations (vibrational Raman nitrogen and water vapor channels at 386.7 nm and 408.4 nm, two Mie-

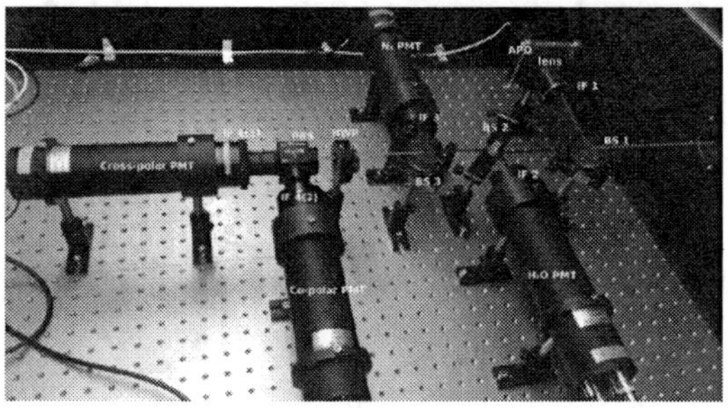

Fig. 2. The five channel spectroscopic system. Total optical efficiencies: Mie 1064 nm - 18.3%, N_2 - 41.7%, H_2O - 42.9%, P polarization Mie 355 nm - 33.5%, S polarization Mie 355 nm - 28.7%. IF: Interference Filter, BS: Beam Splitter, PBS: Polarization Beam Splitter, PMT: Photomultiplier Tube, APD: Avalanche Photodiode, HWP: Half Wave Plate.

https://doi.org/10.1051/epjconf/202023702029

Rayleigh channels at 355 nm with orthogonal polarization planes and a Mie-Rayleigh channel at 1064 nm). Layout of spectroscopic optical components and light detectors is shown in Fig 2.

2.3 DAQ system

Detector outputs were amplified and digitized by transient recorders (Licel TR-160) in both analog and photon counting detection chains. Digitized signals were read out by Linux based data acquisition (DAQ) computer via Ethernet link. The system uses our own C++ based DAQ software with graphical interface for detectors, digitizer and laser control, providing real-time lidar return information data quality monitoring [3]. Both system parameters and the digitized lidar return traces are stored into tree structured native binary files (CERN ROOT software toolkit).

2.4 Polarization correction

The retrieved depolarization ratio was calibrated according to [4] and is comparable to the results of EARLINET lidars. The calibration includes corrections for different gains of the detectors in the two polarization channels, diattenuation and retardance in optical components as well as for geometrical miss-alignment between polarization plane of the transmitter and polarization planes in the receiver. The misalignment of the polarization angle was found to be -18.2° with respect to the polarization plane of the emitter by minimizing

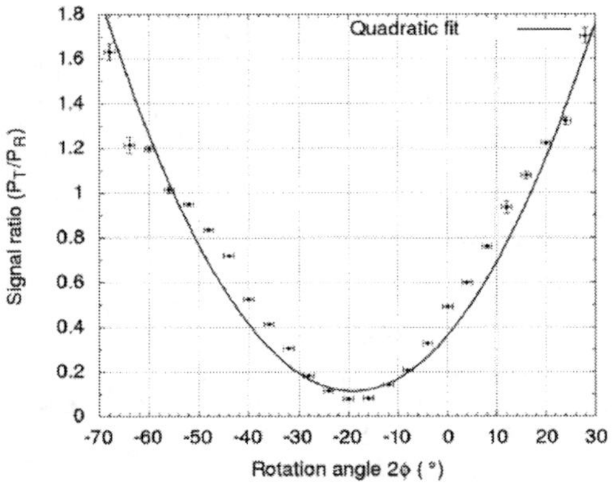

Fig. 3. The ratio between cross- and parallel polarization channels as a function of the orientation of the half wave plate. The offset angle of polarization plane was found to be at φ = -18.2°.

the ratio between the signal in the perpendicular and parallel polarization channel by varying the orientation of the half wave plate (Fig. 3).

2.5 Overlap correction

The correction assessment due to incomplete overlap of the transmitter and the receiver in the UV channels was performed during a very clear and stable night, which assured as constant lidar ratio throughout the incomplete overlap region as possible. The overlap function for the IR channel was estimated during the day. The planetary boundary layer was well mixed and no apparent aerosol sources were observed, so we assumed vertical homogeneity of aerosol concentration. The ranges for the complete overlap were found to be about 300 m for the UV and 200 m for the IR channels (Fig. 4).

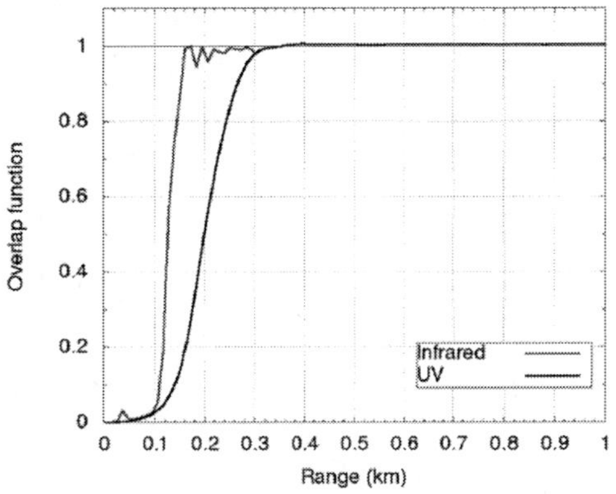

Fig. 4. Overlap functions of the receiver and the transmitter at 355 nm (black) and at 1064 nm (red).

2.6 Data homogenization

Systematic errors were estimated and range corrected UV backscattering scattering signals were normalized to the attenuated molecular backscatter coefficient calculated from radiosonde data at Udine (67 km away) in cloudless nighttime conditions using the Rayleigh fit [5]. Normalized signals from both (analog and photon counting) detection chains were merged in order to extend the detectable range. Systematic uncertainty for the return signal in the parallel polarization channel was found to be less than 5% for ranges below 10 km and about 20% for ranges below 30 km (Fig. 5).

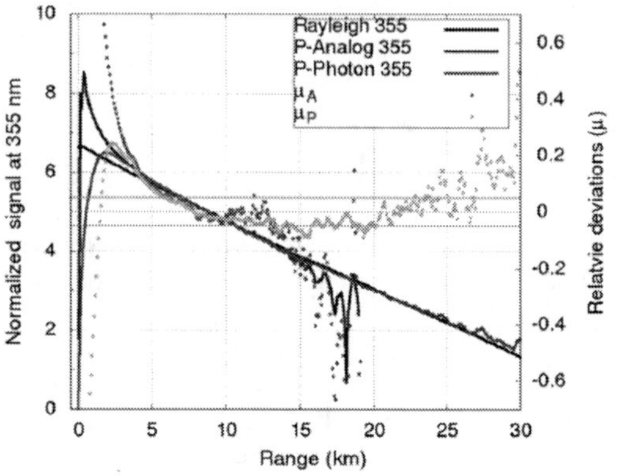

Fig. 5. Rayleigh fit (black) to the normalized range square corrected signals from the analog (blue) and photon-counting (red) detection chains for the UV signal. Pink and light blue dots indicate deviations from the attenuated molecular backscatter coefficient calculated from the radiosonde data at Udine, 67 km away. Horizontal green lines indicate 5% deviations.

aerosol optical property profiles (LR, DR, BAE) correspond well to this aerosol composition, as they indicate the presence of biomass burning aerosols in the PBL and marine aerosols in the EAL (Fig. 6).

4. Conclusions

We have demonstrated that the automatized polarization Raman lidar can provide reliable information on atmospheric optical properties in all weather conditions. Optical properties related to aerosol characteristics (LR, DR and BAE) indicated the same aerosol composition in stable conditions as the in-situ measurements, which suggests we could use them as a stand-alone tool for aerosol typing in non-stable conditions such as the onsets of Bora as well.

ACKNOWLEDGEMENTS

This work was supported by Slovenian Research Agency (grant P1-0385) and Fulbright Scholar

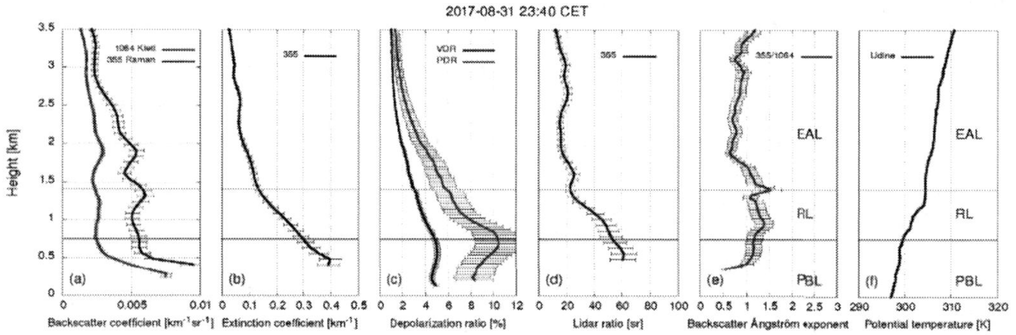

Figure 6. a-e) *An example of vertical profiles of aerosol optical properties in stable stratified atmosphere on 31 August 2017 retrieved from lidar data between 23:40 and 00:10+1 CET. Error bars represent total uncertainty. Horizontal lines (black and green) indicate the boundaries between different elevated aerosol layers.* ***f)*** *Potential temperature from the radiosonde measurement at 00:00 CET.*

3. Performance

Commissioning of the new lidar took place between August 2017 and August 2018 to assess its performance. In a total of 34 days of measurements in stable conditions, we observed good agreement between predominant aerosol types as determined from in-situ measurements and suggested by the retrieved lidar optical observables. As an example, in stable and stratified atmosphere on 1 September 2017 the aerosols in the PBL (below 0.7 km) were predominantly carbonaceous and those in the elevated aerosol layer (EAL, above 1.4 km) arrived from above the Adriatic. Lidar-retrieved

program. Sounding data was provided by Aeronautica Militare and OSMER-ARPA FVG.

REFERENCES

[1] L. Wang, *et al.*, Remote Sensing 11(2):106 (2019).
[2] L. Wang, *et al.*, Atmosphere 10, 128 (2019).
[3] S. Benzvi, *et al.*, Nucl. Instrum. Methods Phys. Res. A 574(1), 171-184 (2007).
[4] J. Bravo-Aranda, *et al.*, Atmos. Meas. Tech. 9(10), 4935 (2016).
[5] V. Freudenthaler, *et al.*, Atmos. Meas. Tech. Discuss. 1-35 (2018).

RECENT UPGRADES OF THE EOLE AND AIAS LIDAR SYSTEMS OF THE NATIONAL TECHNICAL UNIVERSITY OF ATHENS OPERATING SINCE 2000 IN ATHENS, GREECE

A. Papayannis*[1], P. Kokkalis[2], M. Mylonaki[1], R. Soupiona[1], C. A. Papanikolaou[1], R. Foskinis[1], and A. Giakoumaki[1]

[1]National Technical University of Athens, Laser Remote Sensing Laboratory, Physics Department, 15780 Zografou, Greece
[2]Kuwait University, Faculty of Science, Kuwait
**Email: apdlidar@central.ntua.gr*

ABSTRACT

The technical specifications and advances/ upgrades of the stationary (EOLE) and mobile (AIAS) lidars located at the National Technical University of Athens (NTUA) are presented in this paper. EOLE is a multi-wavelength combined backscatter/Raman lidar system, which is part of the EARLINET lidar network since May 2000. AIAS is a mobile 532 nm elastic depolarization lidar system. Both instruments have been upgraded during 2019, in the frame of PANACEA to be part of the Greek National Research Infrastructure for aerosol research, under the umbrella of the European Strategy Forum on Research Infrastructures (ESFRI).

1. INTRODUCTION

Since 2000, the Laser Remote Sensing Unit (LRSU) of NTUA [37.96°N, 23.78°E, 212 m asl.] is operating the EOLE lidar system. Through these years, the system has been gradually upgraded regarding its operational wavelengths and configuration. Nowadays, EOLE is an advanced 6-wavelength elastic-Raman lidar system (Fig. 1), equipped with two depolarization channels (355 and 532 nm), able to perform independent and simultaneous measurements of the vertical profiles of the aerosol backscatter coefficient at 354.93, 532 and 1064.2 nm, aerosol extinction coefficient and linear particle depolarization ratio at 354.93 and 532 nm, water vapor mixing ratio (using the H_2O Raman channel at 407 nm) and ozone in the troposphere [1-3]. Here we will limit our presentation on the aerosol measurement profiling.

2. EXPERIMENTAL SET-UP

The advanced elastic-Raman lidar system (EOLE) of LRSU is based on a pulsed Nd:YAG laser system (Spectra Physics Lab-170-10) (Fig. 1) which emits, simultaneously, pulses at 354.93-532-1064.2 nm, with energies of 240-310-260 mJ, respectively, at 10 Hz repetition frequency. The laser beams are horizontally polarized (>90%) at 532-1064.2 nm and vertically polarized (>90%) at 354.93 nm. The laser beam containing all three wavelengths is expanded by an achromatic Galilean telescope (x3), before being emitted in the atmosphere, thus its divergence remains <0.17 mrad (FWHM). A 300 mm diameter Cassegrainian telescope (focal length f=600 mm, FOV=1.5 mrad) collects all elastically backscattered lidar signals (354.93-532-1064.2 nm), as well as those generated by the spontaneous Raman effect (by atmospheric N_2 at 386.6-607.4 nm and by H_2O at 407.5 nm).

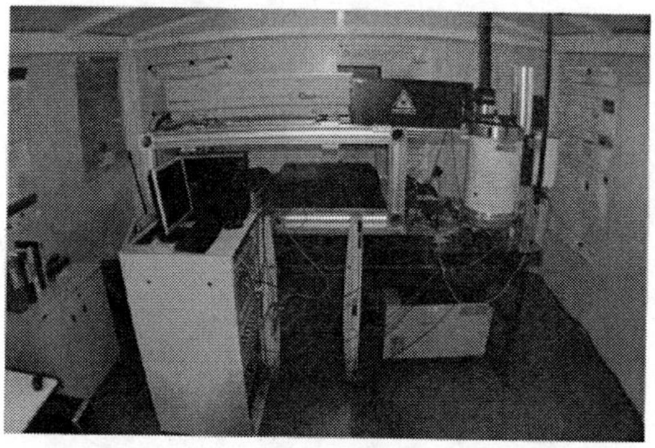

Figure 1. The LRSU elastic-Raman lidar (EOLE).

A second laser is used to emit vertically polarized beam at 355 nm (>99%), while an additional telescope (200 mm diameter, Dall-Kirkham Cassegrainian) with focal length f=1000 mm, collects the elastically backscattered lidar signals at 355 and 532 nm (at 2 polarization planes:

parallel and vertical), which are optically separated by polarizing beam splitter cubes.

At the entrance of the 300 mm diameter telescope a high grade fused silica optical fiber (N.A. =0.22±0.02 and 1.5 mm core diameter) is used to transfer the lidar signals to the 6-wavelength spectrometer, which is equipped with achromatic collimating lenses, dichroic beam splitters, as well as doublets, eye pieces and interference filters (IFF) placed in front of the detectors (Photomultiplier tubes-PMTs at 354.93-386.6-407.5-532-607.4 nm and Avalanche Photo Diode at 1064.2 nm). In Table 1 the technical characteristics of the receiving system of EOLE are summarized.

	354.9 nm	**386.6 nm**	**407.5 nm**	**532 nm**	**607.4 nm**	**1064.2 nm**
IFF bandwidth (nm, FWHM)	1.8	0.84	0.5	0.5	1.06	0.97
Transmission (%)	47.5	77.4	51	45	78.7	45.7
Out of band blocking (>OD)	6	8	10	4	6	4
Detector model-Hamamatsu	R7400P-03	R7400U-P06	R7400U-02	R7400-P02	R7400U-20	Si-APD-1.5
Eye-piece	YES	YES	YES	YES	YES	YES

Table 1. Technical characteristics of the optical components of the wavelength separation system of EOLE.

All optical components of the 6-wavelength spectrometer are mounted on micrometric positioners and rotators, fixed on an anti-vibrating optical table; therefore, they can be placed at an optimal position. The choice of the optical components, as well as their optimal position has been identified through advanced ray-tracing code (Zemax), by simulating the projection of the laser beam down to the PMTs' photocathodes. Narrowband IFFs are used to suppress the atmospheric background noise at the detected wavelengths (354.93-354.93-386.6-407.5-532-607.4-1064.2 nm) (cf. Table 1).

The PMTs (operating both in the analog and photon-counting mode) and APD (operating in the analog mode) output signals, are fed into fast transient recorders (Licel GmbH) working in both the analog (using 20 MHz Analog to Digital Converters-ADC at 12-bits resolution) and photon counting (maximum 250 MHz count rates) mode, at 8196 range bins. EOLE's overlap distance is of ~600 m, while the vertical raw resolution of the acquired lidar signals is 7.5 m.

3.1.2 Mobile two-wavelength depolarization elastic lidar system

Additionally, a Raymetrics S.A. (Greece) van-mounted 1064 nm (elastic) and 532 nm (elastic) depolarization lidar (AIAS) is available (Fig. 2), to provide the vertical aerosol linear depolarization ratio in the troposphere, thus evaluating the sphericity of the probed aerosols. AIAS is based on a pulsed Nd:YAG laser system (Litron Nano SG 150-10 Series) which emits pulses at 532 and 1064.2 nm, with energies of 95 and 55 mJ, respectively, at 10 Hz repetition frequency.

Figure 2. The 532 nm LRSU elastic depolarization lidar (AIAS lidar).

The laser beam at 532 nm is vertically polarized (>99%) using a λ-2 waveplate and is expanded by a Galilean telescope (x4), before being emitted in the atmosphere, thus its divergence remains <0.4 mrad.

A 200 mm diameter Dall-Kirkham Cassegrainian telescope (focal length f=1000 mm) collects the elastically backscattered lidar signals at 532 nm (at 2 polarization planes: parallel and vertical) and at 1064 nm. A secondary mirror is used to guide the backscattered light to wavelength separation unit (WSU), which is equipped with collimating lenses, dichroic beam splitters, polarizing cubes, as well as doublets and very narrow (0.6 nm

https://doi.org/10.1051/epjconf/202023702030

FWHM) interference filters (IFF) placed in front of the detectors (Photo-multiplier tubes-PMTs at 532 nm). A dichroic mirror is used to separate the 532 nm from the 1064.2 nm signals, while a polarizing cube beam splitter mounted in a rotating base, is used to separate the co-polarized and de-polarized light at 532 nm, and to calibrate the signals.

	532 nm (p and s-polarization)	**1064 nm**
IFF bandwidth (nm, FWHM)	0.6	0.97
Transmission (%)	44.24	45.70
Out of band blocking (>OD)	6	4
Detector model-Hamamatsu	R7400U-P02	Si-APD-3.0
Eye-piece	YES	YES

Table 2. Technical characteristics of the optical components of the wavelength separation system of AIAS.

Both EOLE [2] and AIAS [3] lidars follow the quality assurance protocols of EARLINET and have been validated through direct intercomparisons, both at hardware and software level, in the frame of the EARLINET, ASOS and ACTRIS projects [4-7].

4. The PANACEA PROJECT

The PANhellenic infrastructure for Atmospheric Composition and climatE change (PANACEA) is a Research Infrastructure (RI) where specialized atmospheric observations (aiming to adequately monitor the atmospheric composition, radiation disturbances and essential climate variables) will be performed. Thus, PANACEA is envisioned to be a distributed European RI (adopting the ACTRIS/ESFRI structure according to the EU Regulation 651/26.6.2014) to provide specific services to stakeholders, as well as a network of excellence at national level.

The specific objectives of PANACEA are to:

- create a novel RI of excellence in atmospheric sciences that will catalyze collaboration at national and international level, minimizing costs and enhancing networking and mobility through intra- and trans-national access;

- promote development of new technologies for atmospheric observation of aerosols, clouds, radiation, greenhouse and trace gases, in close partnership with national and European Union SMEs;

- provide access to tools and data, allowing various beneficiaries (e.g. scientific community, institutions, Regions, local/national authorities, large companies/industries and SMEs, NGOs) to exploit and develop synergistic atmospheric characterization techniques;

- improve the efficiency of current facilities, through the application of tailored harmonization processes; promote QA/QC via creation of specialized calibration centers;

- provide access to high-quality information and services to end-users (research community, education, local/regional/national authorities, large companies/industries/SMEs, environmental and health protection agencies, energy, tourism and transport sectors, international organizations and networks, NGOs, etc.);

- provide state-of-the-science expertise, assisting policy making on issues such as climate change, natural hazards, air quality and long-range transport of pollutants;

- create and use educational and training platforms for students and young scientists in the fields of air quality and climate change.

ACKNOWLEDGMENTS

The LRSU-NTUA elastic-Raman system is member of the EARLINET network (www.earlinet.org). We acknowledge support of this work by the project "PANhellenic infrastructure for Atmospheric Composition and climatE change" (MIS 5021516) which is implemented under the Action "Reinforcement of the Research and Innovation Infrastructure", funded by the Operational Programme "Competitiveness, Entrepreneurship and Innovation" (NSRF 2014-2020) and co-financed by Greece and the European Union (European Regional Development Fund) (https://panacea-ri.gr). The Biomedical Research Foundation of the Academy of Athens (BRFAA) is acknowledged for the provision of its mobile platform to host the NTUA AIAS lidar system.

REFERENCES

[1] Mamouri, R.E., A. Papayannis, G. Tsaknakis, et al., First water vapor measurements over Athens, Greece, obtained by a combined Raman-elastic backscatter lidar system, *Optica Pura Y Applicada* **41**, 109-116 (2008).

[2] Kokkalis, P., A. Papayannis, R. E. Mamouri, et al., The EOLE lidar system of the National Technical University of Athens, 629-632, *26th International Laser Radar Conference (26th ILRC)*, 25-29 June 2012, Porto Heli, Greece.

[3] Papayannis, A., D. Balis, P. Kokkalis, R. Mamouri, G. Tsaknakis, E. Giannakaki, N. Siomos, V. Amiridis, ARIADNE: The Greek Lidar Network, 349-352, *26th International Laser Radar Conference (26th ILRC)*, 25-29 June 2012, Porto Heli, Greece.

[4] Freudenthaler, V., The telecover test: A quality assurance tool for the optical part of a lidar system, *Proc. 24th International Laser Radar Conference*, 145-146 (2008).

[5] Pappalardo, G., A. Amodeo, U. Wandinger, et al., Aerosol lidar intercomparison in the frame of EARLINET: Part III: Aerosol extinction Raman lidar algorithm intercomparison, *Applied Optics* **43**, 5370-5385 (2004).

[6] Amodeo, A., et al., Proc. 28th International Laser Radar Conference, *EPJ Web of Conferences* 176, 09008 (2018).

[7] Müller, D., U. Wandinger, and A. Ansman, Microphysical particle parameters from extinction and backscatter lidar data by inversion with regularization: theory, *Applied Optics* **38**, 2346-2357 (1999).

SET OF ALGORITHMS AND TECHNIQUES FOR ACCURATE 3D, SINGLE BEAM – SINGLE POINTING, LIDAR MEASUREMENTS FOR SLANT RANGE VISIBILITY, PLANETARY BOUNDARY LAYER HEIGHT AND WIND SPEED RETRIEVAL, ATMOSPHERIC LAYERS SPATIAL DISTRIBUTION AND CATEGORIZATION IN REAL TIME

Alexandros Pantazis[1,*], Alexandros Papayannis[1]

[1]National Technical University of Athens, Laser Remote Sensing Laboratory, Physics Department, 15780 Zografou, Greece,
**Email: alexpant@mail.ntua.gr*

ABSTRACT

In this work, a full set of recently developed algorithms and techniques is presented, for a single beam-single pointing lidar to be able to perform operational and independent accurate 3 Dimensional (3D) measurements, for slant range visibility, wind speed retrieval, atmospheric layers spatial distribution and categorization, as well as Planetary Boundary Layer Height (PBLH) retrieval, in real or Near Real Time (NRT).The idea behind this development was for any single lidar to be able to perform a set of accurately measured products, either mobile or stationary, with or without network connectivity with other sensors for data-information exchange. The products were determined by the needs of lidar remote scientific and commercial community, in order to be even more attractive and valuable to atmospheric scientists, meteorologists, aviation and shipping safety operators, as well as to the Space lidar community.

1. INTRODUCTION

Following the work of [1, 2, 3] and Eqs 1-6 in these references, which also apply for this work, backscattering lidar signal received power (P(λ,R)), Range-squared Corrected lidar Signal (RCS), P′(λ,R) which is the received power after atmospheric and electronic noise BackGround (BG) correction and processing of lidar atmospheric coefficients like, aerosol extinction coefficient $\alpha_{aer}(\lambda,R)$ and aerosol backscattering coefficient $\beta_{aer}(\lambda,R)$, were used in order to produce a full set of algorithms and techniques for 3D lidar measurements. The original development of the algorithms used the above mentioned variables and Eqs 1-6, produced with Klett technique [4, 5], but they are also applicable using other retrieval techniques, like the Raman one [6].

The vertical, near vertical or even slant measurements that can achieve molecular values of $\beta_{aer}(\lambda,R)$ (and/or $\alpha_{aer}(\lambda,R)$), at a molecular height above the PBLH and at any wavelength, taken from [1, 2, 3, 7], ([8] p. 57, 58), which are recorded using the backscattering signal, are able to provide a clear view of the aerosol load located along the Line Of Sight (LOS) of the lidar beam. This aerosol load recorded as elastic backscattering signal (Eqs 1-6 already mentioned and Eqs 7-9 below), was used by these algorithms in order to have successive outcomes of processing.

Moreover, the aerosol optical depth (τ) is given below:

$$\tau_{C_n}(0,R) = \int_0^R a_{C_n}(\lambda,R')dR' \quad (7)$$

and $\tau_{total}(0,R_n) = \tau_{C_1} + \tau_{C_2} + \ldots + \tau_{C_n}$ (8)

where, R′ is the distance at which $\alpha_{aer}(\lambda,R)$ has been measured with a range resolution of 1 bin (i.e. 7.5 m), τ_{total} is the sum of τ_{Cn}, which is the individual optical depth (n=1,2,3,...) of a number (n) of atmospheric layers, decided to be used by the user in his measurement.

The relationship between the two coefficients $\beta_{aer}(\lambda,R)$ and $\alpha_{aer}(\lambda,R)$ also used is some cases, is the so-called lidar ratio, C(λ,R):

$$C(\lambda,R) = a_{aer}(\lambda,R)/\beta_{aer}(\lambda,R) \quad (9)$$

Flight and ground safety at airports, but also shipping safety has to do, in many ways, with visibility and meteorological conditions in the area. A major flight safety issue is to obtain visibility measurements especially from the pilot's point of view and to be able to have accurate

https://doi.org/10.1051/epjconf/202023702031

measurements of some meteorological parameters from a respected distance of few km, not only above the airport, but also in Slant Range (SR).

Also, accurate PBLH retrieval together with an estimation of the atmospheric layering and distribution, through the detection and tracking of the actual structure of $\beta_{aer}(\lambda,R)$ and/or $\alpha_{aer}(\lambda,R)$, RCS or $P'(\lambda,R)$ in 3D measurements, is also a main issue for today's scientific community, as no concrete technique could offer so far, solid outcomes in Real Time (RT), producing in this way a maximum contribution in present weather conditions estimation from a distance and 'help' in weather forecasting. The main difference in these algorithms is that they create a new kind of 'mathematical tools' (notional conditions) for processing, based on the backscattering lidar signal, the concordance of which is based upon the approximation of PBLH retrieval, actual layering and distribution according to the Bayes theorem ([9] pp. 1-3).

For wind measurements, the most known method is based on the Doppler effect which has been used for the retrieval of the PBLH through wind measurements [10]. A different technique, but the most correlated to the one proposed here, based on single beam–single pointing measurements, is the one mentioned in [7]; it uses the wind retrieval correlation method for backscatter lidar scanning schemes consisting of a few profiles along the LOS, which is mathematically formulated in a matrix solution form under different atmospheric conditions. This method assumes the frozen atmosphere model [7] and works with temporal cross-correlation functions that do not need to be maximized. It also applies for the case of slant scans based on the multiple-angle azimuth technique (horizontal wind retrievals) [7]. The main differences with our proposed wind algorithms and technique, is that even if we simultaneously use single beam measurements at each time, we can produce wind speed in 2 dimensions (2D), by staying in one single slant pointing angle and calculate 3D wind speed profiles by using a method similar to the 2-angle azimuth scan [7].

2. METHODOLOGY

The methodology used during the development of this set of algorithms was first to create some new denoising techniques like NOSUB-S/R [1] in order to reduce any noise from our backscattering signal and proceed to more accurate estimations of our variables, to be inputted for processing. In NOSUB-S/R some new filters-variables created in order to estimate the conditions where noise is taken over the actual signal returns and produces notional backscattering signal values. A subtraction occurs in this case of the actual noise, without 'harming' the valuable lidar backscattering signal, extending in this way the distance where valuable signal can contribute to our measurements and perform to the rest of the remaining valid signal, a subtraction also, of any added noise, in order to have handed for processing, actual and valid backscattering values of our signal.

Then, the way to Slant Range (SR) measurements became easier by creating LADDER or 3D Stepping Technique [1] (Fig. 1) where, we can retrieve $\beta_{aer}(\lambda,R)$ and/or $\alpha_{aer}(\lambda,R)$, RCS or $P'(\lambda,R)$ by performing measurements like 'stair steps' from vertical to horizontal direction, every 1° vertical angle, like "walking down a ladder".

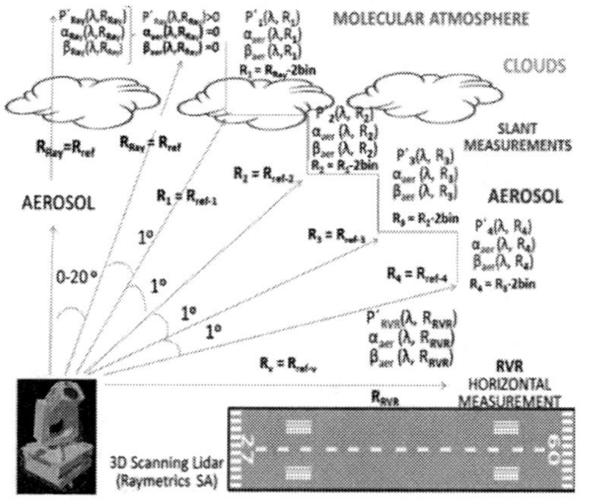

Fig. 1. The LADDER or 3D Stepping working principle [1].

In this way, we take as R_F at Eqs 1-6, the distance at which we abstract 1-2 slant - height bins (1 bin is equal to the spatial resolution ΔR) for every new measurement and we retain the last value of $\beta_{aer}(\lambda,R)$ and/or $\alpha_{aer}(\lambda,R)$, RCS or $P'(\lambda,R)$, at which R_F was previously taken, as the new calibrating values for $\beta_{aer}(\lambda,R)$ and/or $\alpha_{aer}(\lambda,R)$, RCS or $P'(\lambda,R)$. The results of such a technique when used properly were most effective (Table 1) in comparison with other techniques.

Method	α_{aer} error ($\Delta\alpha_{aer}$) %	Vis error (ΔVis) %	Comments
Klett	20-30	20-30	Homogeneous atmosphere
Raman	5-15	5-15	Homogeneous atmosphere
LADDER	0 (as the value of C(λ,R) is known)	0 (as the value of C(λ,R) is known)	Homogeneous/ Non homogeneous atmosphere
Scatterometer	N/A	±10 m or ±10 %, highest values applies	Homogeneous atmosphere

Table 1. Various methods and techniques used to derive/estimate the atmospheric extinction and/or visibility and their associated uncertainty, compared to the LADDER technique in homogeneous/non homogeneous atmospheres [1].

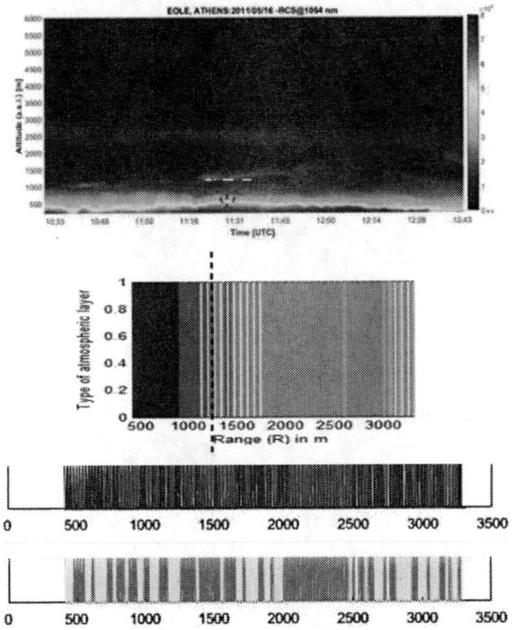

Fig. 2. (Upper) RCS signal acquired at 1064nm by the LRSU-NTUA lidar with METCON PBLH presentation in 3D, using the UNIFORM technique. The white dashed line denotes PBLH outcome with METCON-PBLH at 1372.5m while radiosonde data obtained by the HNMS around that day, at the same time, waved between 1340 to 1531m. (Middle) METCON algorithm [1] applied in 3D lidar data pointing vertically, up to 3000m. Where Blue is "Light Fog", Purple is "Thin Fog", Light Blue is "Light Haze" and Green is "Haze". The black dashed line denotes the PBLH retrieval with METCON PBLH, between layers change as supposed to happen. (Lower) Application of SIBESMEA with a case study of wind direction in 2D. Red and blue fill-ins are for the upwards and downwards wind direction and green and yellow fill-ins are for the east-west wind directions. The rest of the air mass missing between continuous measurements applies for 3rd dimension wind movement, by presuming high repetition frequency of pulsed signals. The speed of the 3rd dimension movement of the air mass could be calculated through the rate of air mass loss in that (+ or -) direction.

By solving SR measurements it was achievable to acquire SR visibility also from pilot's point of view, with NAVIS-T/P algorithm [1] and the use of Eqs 7-9. Towards this idea of SR visibility and PBLH retrieval, initially, METCON or WEATHER PHENOMENA algorithms [1] was created in order to produce equivalent layering and distribution in a 3D view from lidar's location, by means of atmospheric layering categorization mentioned in ([8] pp. 57, 58, 144) and ([11] pp. 178, 179) (Fig. 2, Middle).

As a next step, after being able to actually produce atmospheric layering and tracking in RT, the achievement was to create an algorithm being able to retrieve the PBLH in an area surrounding lidar's location. By using the principle of searching for a persistent appearance ([9] pp. 1-3) of a certain atmospheric layer inside the common PBLH area according to weather forecasting and time of the year at lidar's location, like fog or other similar atmospheric layer, in terms of close values calculation of backscattering variables measured with the ones produced by fog [1], ([8] pp. 57, 58, 144), ([11] pp. 178, 179) the algorithm acquires usually the upper limit of that atmospheric layer, as retrieved PBLH. In this effort, certain thresholds of first and second derivatives of a new kind of 'mathematical tools' (notional conditions) for processing used for VASPAT or MUSTI algorithms, VERDE or DECIS [2] and METCON PBLH [3] (Fig. 2, Middle) and succeeded a great efficiency in more than 94% of correct PBLH retrieval in more than 30 cases studied and compared with nearby radiosonde measurements, available by the Hellenic National Meteorological Service

https://doi.org/10.1051/epjconf/202023702031

(HNMS) at the same date and time [3] (Fig. 3). To this end, a new technique was born called UNIFORM technique [3] in order for 3D PBLH retrieval becomes possible (Fig. 2, Upper).

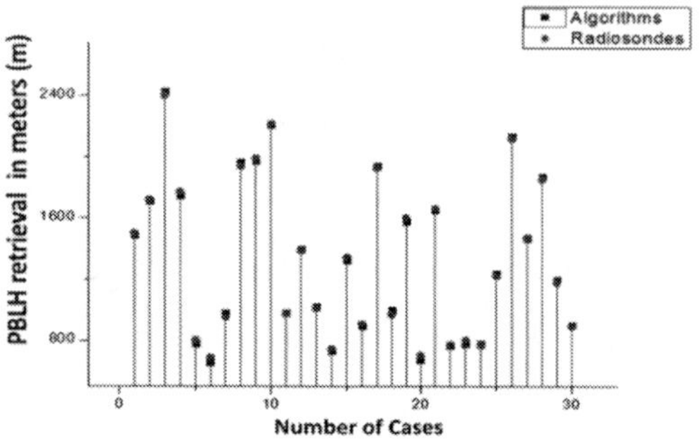

Fig. 3. PBLH retrieval with the single pointing EOLE NTUA lidar after the application of [2, 3] algorithms processing with appropriate settings (black squares), compared with HNMS radiosonde data (red cycles) at the same date and time over Athens, between the years of 2014 to 2016, showing an agreement inside 35m in all these cases.

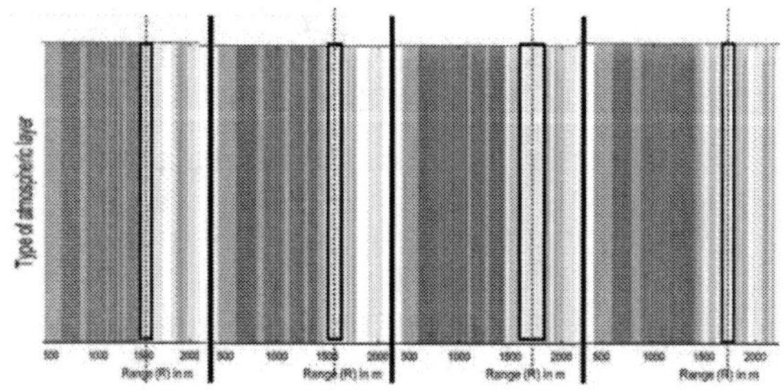

Fig. 4. WIND METCON algorithm [3] presentation of four sequential lidar data recording files every 10". The tracking of a certain atmospheric layer in Yellow color ("Light Haze" depicted with dashed line in its center) begins at ~1500m height and travelling up to 1750m in 30", presenting a layer tracking (wind) speed of 8.3m/s at that time and then it was lost as a specific layer. Air masses movement through atmospheric layer detection and the replacement of the "free" space in between by other atmospheric layer, is depicting the air masses movement and changes around PBLH and other atmospheric mixtures with time. The blue vertical lines represent distinguishes consecutive lidar shoots.

Finally, the goal was that once 3D atmospheric layering was possible and accurate, and through that, PBLH retrieval in a wider area in RT, tracking of these layers should become possible also for a 3D valid estimation of most of weather conditions presence, with a single lidar without the actual need of the expensive and heavy Doppler equipment or experienced user. So, SIBESMEA (Fig. 2, Lower) and WIND METCON (Fig. 4) algorithms developed [3] in order to overcome this obstacle with totally successful results, making it also possible for tracking atmospheric layers separately as well as tracking and visualizing the way they are transformed to other kind of layers through their movement in the atmosphere, even in PBLH level, were valuable case studies could be performed in the future (Fig. 2, Lower).

3. RESULTS

All these algorithms presented a unique performance after being developed and tested under 'real' lidar measurements, using EOLE lidar data from NTUA together with radiosonde data from the Hellenic National Meteorological service (HNMS).

ACKNOWLEDGEMENTS

We thank the active participation of Mrs E. Andrikopoulou in the Matlab organization, programming and implementation of the mathematical codes presented in this work. We also thank the reviewers for their valuable comments and corrections. Radiosonde data have been obtained from HNMS.

REFERENCES

[1] A. Pantazis, A. Papayannis, G. Georgousis, Appl. Opt., **56**, 6440-6448 (2017).

[2] A. Pantazis, A. Papayannis, G. Georgousis, Appl. Opt., **56**, 8199-8211, (2018).

[3] A. Pantazis, A. Papayannis, Appl. Opt., in press (2019).

[4] J. Klett, Appl. Opt. **20**, 211-220 (1981).

[5] J. Klett, Appl. Opt. **24**, 1638-1643 (1985).

[6] A. Ansmann et al. Appl. Phys. **B55**, 18 (1992).

[7] S. Tomás, F. Rocadenbosch, J. Geophys. Res. Atmos., **120**, 7758-7776 (2015).

[8] R.M. Measures, "Laser Remote Sensing. Fundamentals and Applications", Krieger, Sys. No 9247, MEA 621.3678 (1992).

[9] J. V. Stone, "Bayes' Rule: A tutorial analysis to Bayesian Anlysis", Sebtel Press (2013).

[10] Meng Huang, et al., Boundary-Layer Meteorol **162**, 503–522 (2017).

[11] W. L. Wright et al., NASA, CR-132724 C89 (1966).

SHIPBORNE AND GROUND-BASED LIDAR MONITORING OF THE ATMOSPHERE OVER THE LAKE BAIKAL IN 2018

Yurii S. Balin[1], Marina G. Klemasheva[1], Grigorii P. Kokhanenko [1], Sergey V. Nasonov[1*], Ioganes E. Penner[1]

**Email: nsvtsk@gmail.com*

[1]V.E. Zuev Institute of Atmospheric Optics, Rus. Acad. Sci., Tomsk 634021, Russia

ABSTRACT

The paper presents the results of studies aimed at the analysis and assessment of atmospheric pollution over the Lake Baikal in the summer. This information is necessary to create physical models of the formation and transfer of atmospheric aerosol fields, taking into account the physical and geographical features of the Baikal region. Measurements were carried out by a lidar «LOSA-A2» installed on the scientific-research vessel «Academician V.A. Koptyug». The vessel's route passed along the South, Middle and Northern Baikal, from July 15, 2018 to July 26, 2018. At the same time, observations were conducted using lidar «LOSA-M2». It was located in the background area at Boyarsky stationary site (51.84° N, 106.06° E), in the south-eastern part of the lake. The results of changes in the spatial structure of atmospheric aerosol fields in background conditions and during forest fires are shown.

1. INTRODUCTION AND METHODOLOGY

Interest in the Lake Baikal is associated with high requirements for the protection of nature of this unique object and the features of natural conditions [1-4]. Here, the formation and distribution of atmospheric admixtures in the summer period is influenced by mountain terrain and significant temperature contrasts between the cold surface of the lake and the atmosphere.

One of the most significant sources of pollution of the atmosphere of the Baikal region is smoke aerosol from forest fires, the number of which is increasing due to climate warming in the region.

The sounding of aerosol fields was carried out using the mobile aerosol-Raman lidar «LOSA-A2». It was for the first time tested in 2018 in this expedition. Lidar was located on the lower deck aft of the vessel. Thanks to this, it was possible to direct it vertically upwards or to tilt it over the horizon. The appearance of the «LOSA-A2» lidar is shown in Fig. 1a.

(a) (b)

Fig.1 Appearance of mobile aerosol-Raman lidars «LOSA-A2» (a) and «LOSA-M2» (b)

The lidar case is made of carbon fiber, which increases the stability of the disposition of optical elements under mechanical loads during transportation and operation of the lidar. In addition, the case has a low dependence from temperature.

The lidar uses as the transmitter LS-2131M laser (LOTIS TII, Belarus) operating at two wavelengths 1064 nm and 532 nm with a frequency of 20 Hz. The radiation passes a collimator, reducing the beam divergence angle to 0.5 mrad. A phase polarization plate is inserted between the laser and the collimator, which

https://doi.org/10.1051/epjconf/202023702032

changes the plane of linear polarization or creates circular polarization.

The lidar receiving part is made in the form of two identical lens objectives with a diameter of 110 mm and located symmetrically relative to the optical axis of the transmitter. One of these objectives is designed to register lidar signals in the visible range wavelengths (signals of elastic and Raman scattering - 532 nm and 607 nm, orthogonal polarization components at a wavelength of 532 nm). This receiving channel uses photomultipliers H11526-20-NF («HamamatsuPhotonics», Japan), operating in analog mode. The signal of Raman scattering on atmospheric nitrogen molecules is registered by the photodetector module H11706P-40-MOD («HamamatsuPhotonics», Japan) operating in the photon counting mode. The second receiving objective is designed to receive backscattered radiation at a wavelength of 1064 nm, using the developed original photodetector module based on the C30956E-EU avalanche photodiode («PerkinElemer», USA), operating in analog mode.

The probing of aerosol fields at the Boyarsky stationary site is carried out by the multi-frequency mobile aerosol-Raman lidar «LOSA-M2» (Fig. 1b) at two laser wavelengths of 1064 nm and 532 nm [5]. The detection of elastic backscattered radiation at these wavelengths was carried out in the analog mode, and for 532 nm also in the photon counting mode. In addition, a signal is detected at an offset wavelength of 607 nm of spontaneous Raman scattering in the photon counting mode.

The general preliminary interpretation of the data of sounding was performed on the basis of the analysis of the vertical cross-sections of the aerosol field of the troposphere represented in the units of the scattering ratio $R(\lambda_1,z)=(\beta^a(\lambda_1,z)+\beta^m(\lambda_1,z))/\beta^m(\lambda_1,z)$, where $\beta^a(\lambda_1,z)$ and $\beta^m(\lambda_1,z)$ are the aerosol and molecular backscattering coefficients at the wavelength of λ_1=1064 nm.

The main attention of the researches was focused on the study of the temporary changes in the vertical structure of aerosol fields and the identification of the physical mechanisms of their formation.

2. ANALYSIS OF LIDAR DATA

During the passage of the vessel along the Middle Baikal on July 19, 2018, the atmosphere was cleared by rains the previous night, the lidar recorded background concentrations of aerosol. An example of such results is shown in Fig. 2a, where you can see small values of the atmospheric scattering ratio parameter.

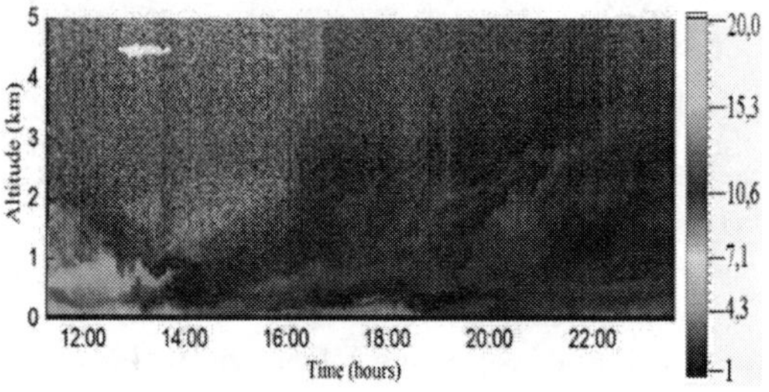

(a)

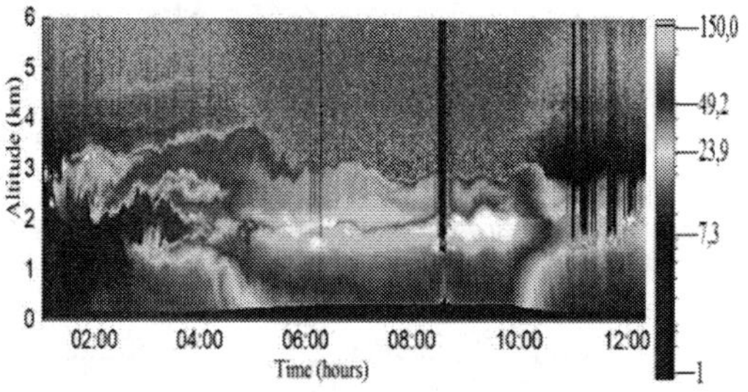

(b)

(c)

Fig.2 Spatiotemporal structure of aerosol field, obtained: (a) – by the lidar at the research vessel «Academician V.A. Koptyug» on July 19, 2018; (b) – by the lidar at the research vessel «Academician V.A. Koptyug» on July 22, 2018; (c) – by the lidar at the Boyarsky stationary site on July 22, 2018. Color scale is given in scattering ratio units.

According to laser-sensing data, the first traces of a smoke train from forest fires from the Krasnoyarsk Territory and Yakutia (Fig. 2b) were observed while the vessel was near Severobaykalsk. They appeared after midnight and were observed as independent aerosol formations with a thickness of 200–300 m (Fig. 2b). Further, there is an expansion of the smoke train in the altitude range of 2-3 km, up to a height of 3.5-4 km in the morning (04:00 – 05:00 a.m.). The aerosol content in the atmosphere exceeds the background values dozens of times. In the same time period, with a delay of about 1.5 hours (02:30 a.m.), the presence of a lower second smoke train at an altitude of 1.5 km was detected. This is probably due to the orographic features of the surrounding mountain terrain. These two smoke trains expanded during the observation time and merged into a single one at an altitude of about 2 km (04:20 a.m.). After that, within an hour there is a sharp filling of the lower layers of the atmosphere with a smoke aerosol, right up to the surface layer. Both smoke trains, joined to the total mass, have a pronounced border between them in the region of 2 km. This indicates a different concentration of aerosol at the boundaries of these smoke trains.

At 11:55 a.m. local time the vessel moved away from Severobaikalsk towards the eastern shore of the . The presence of smoke aerosol in the atmosphere, but of a lower concentration, was recorded according to laser sensing data along the route.

The increased aerosol content in the same period (July 22) was also recorded during observations at the Boyarsky stationary site (South Baikal). Here, measurements were carried out using a «LOSA-M2» lidar [5], in parallel with ship observations. On this day in the stationary site for weather conditions, the experiment began at 8:15 a.m. At this time, at an altitude of 250 m and 1500 m, layers with a high content of aerosol were observed (Fig. 2c). In the afternoon, the upper layer began to descend, while the lower layer weakened in aerosol concentration, and by 3 pm both layers lowered, and clouds appeared in the atmosphere for 5 km. According to the data of the meteorological complex on the Boyarsky stationary site, the experiment on July 22, 2018 took place under conditions of a decrease in atmospheric pressure (from 740 mm Hg to 713 mm Hg) and relative humidity (from 60% to 30%), with mostly northwesterly wind direction.

Similar data of lidar observations once again prove the complex processes of formation of the structure of aerosol fields above the lake.

Similar lidar studies of the formation of the structure of aerosol fields in mountain conditions over the Lake Baikal were made earlier [6]. In this work, ship studies were carried out of the specific features of the formation of longitudinal and transversal aerosol lidar transect depending on the lay of land (areas of South and Middle Baikal). Correlation analysis of the altitude distribution of the optical thickness and the altitude distribution of the mountain profile along the route of the vessel showed that the greatest correlation is observed in the lower layer of the atmosphere 0.5-1.5 km [6].

3. SUMMARY

The analysis of the spatial-temporal distribution of the aerosol in atmosphere over the Lake Baikal according to the results of ship measurements in July 2018 was conduct. Synoptic conditions and orographic features of the coastal zone of the Lake Baikal had a significant impact on the distribution of aerosol over the lake.

In this paper, we present the results of time dynamics of the formation of complex structures altitude airborne admixtures from forest fires in the coastal zone, which is due to features of airflow over mountain terrain.

ACKNOWLEDGEMENT

The authors thank the head of the Laboratory of Hydrochemistry and Atmosphere Chemistry Limnological Institute SB RAS Khodzher T.V. for organization the expedition «Baikal-18».

This work was supported by Russian Science Foundation grant № 18-77-10035.

REFERENCES

[1] Pancheko M. V., Domysheva V. M., Pestunov D. A., Shmargunov V. P. Spatial distribution of the carbon-containing gas fluxes direction in the "water - atmosphere" system in the littoral zone of Lake Baikal in summer // International Journal of

Hydrology Vol. 2, Issue 2 - 2018. P 195-196. DOI: 10.15406/ijh.2018.02.00068.

[2] Obolkin V., Khodzher T., Sorokovikova L., Tomberg I. Effect of long-range transport of sulphur and nitrogen oxides from large coal power plants on acidification of river waters in the Baikal region, East Siberia // International Journal of Environmental Studies. 2016. 73(3): P. 452-461.

[3] Arshinov M. Yu., Belan B. D., Ivlev G. A., Rasskazchikova T. M. Spatiotemporal characteristics of air circulation in the hollow of Lake Baikal // Atmospheric and oceanic optics, V. 14, No. 04, P. 263-266, (2001).

[4] Balin Yu. S., Klemasheva M. G., Kokhanenko G. P., Nasonov S. V., Novoselov M. M., Penner I. E. Vertical structure of the aerosol fields of the atmosphere in the period of forest fires over Lake Baikal in 2015 // Proc. SPIE 10035, 22nd International Symposium on Atmospheric and Ocean Optics: Atmospheric Physics, 100354D (29 November 2016) doi: 10.1117/12.2249294.

[5] Balin Yu. S., Bairashin G. S., Kokhanenko G. P., Klemasheva M. G., Penner I. E., and Samoilova S. V. "LOSA-M2" aerosol Raman lidar // Quantum Electron. 41 (10), 945–949 (2011). doi: 10.1070/QE2011v041n10ABEH014574.

[6] Balin, Y., Ershov, A., Penner, I. Shipborne LIDAR Investigations of Aerosol Fields in the Atmosphere Over Lake Baikal // 22nd Internation Laser Radar Conference (ILRC 2004), Proceedings of the Conference held 12-16 July, 2004 in Matera, Italy. Edited by Gelsomina Pappalardo and Aldo Amodeo. ESA SP-561. Paris: European Space Agency, 2004, P.665-668.

STRATOSPHERIC SMOKE PROPERTIES BASED ON LIDAR OBSERVATIONS IN AUTUMN 2017 OVER WARSAW

Dongxiang Wang[1] and Iwona S. Stachlewska[1*]

[1] *University of Warsaw, Faculty of Physics, Warsaw, 02-093, Poland*
**Email: iwona.stachlewska@fuw.edu.pl*

ABSTRACT

Smoke layers in the stratosphere were observed during autumn 2017 using PollyXT-UW Raman lidar at the European Aerosol Research Lidar Network in the frame of the Aerosol Cloud and Trace Gases Research Infrastructure, i.e. the EARLINET-ACTRIS site in Warsaw, Poland. The analysis was focused on discriminating very weak signatures of smoke layers in the stratosphere and investigating their optical properties. Preliminary results are presented and discussed. A decrease of the lidar-derived stratospheric aerosol optical depth contribution to the total optical depth was detected after the stratospheric smoke particles circled Northern Hemisphere.

1. INTRODUCTION

The role of the stratospheric aerosol burden in climate variability and global radiative budget causes increasing attention in recent years [1], [2]. Due to its variability, it is essential to perform long-term observations of stratospheric aerosol over the world [3]. While measurements from space are performed with a large diversity of techniques, also long-term ground-based observations of stratosphere are highly valuable, as they ensure continuity and coherence of the stratospheric aerosol record [4, 5].

In summer of 2017, the unusually hot and dry environment led to a wide spread of wildfires in western Canada and United States of America. The smoke aerosol traveled over the North Atlantic, Europe, northern Asia, and circled the globe within less than 20 days [6]. In August 2017, these Canadian wildfire smoke events were observed not only in the troposphere but also in the lower stratosphere (12-26 km) by several European lidar sites [7-10].

However, once injected into the stratosphere, the traces of the stratospheric smoke are likely to be observed for several months after the major summer event. Therefore, this paper focus was set on nighttime observations of any stratospheric aerosol signatures detected in September and October 2017 over Warsaw. Profiles obtained at the elastic total and polarized channels both at 355 and 532 nm were evaluated in search for any aerosol load at an altitude range of 12-30 km a.g.l. The corresponding optical properties of stratospheric smoke layers were derived and analyzed. A comparative study of this stratospheric smoke at several sites was recently published [11].

2. LIDAR INSTRUMENT

Since July 2013, quasi-continuous atmospheric observations are performed using PollyXT-UW lidar at the Remote Sensing Laboratory (RS-Lab, *http://www.igf.fuw.edu.pl/en/instruments*), at the Faculty of Physics of the University of Warsaw. Lidar observations in the frame of the ACTRIS-Poland activities are conducted at the European EARLINET (*www.earlinet.org*, [12]), the worldwide Polly.NET (*http://polly.tropos.de*, [13]) and Polish Aerosol Research Network PolandAOD-NET (*www.polandaod.pl*, [14]).

The next generation PollyXT-UW lidar [15,16], emits vertically to the atmosphere high-energy laser pulses (180 mJ at 1064 nm, 110 mJ at 532 nm, and 60 mJ at 355 nm) with a repetition rate of 20 Hz. The laser beam is characterized by a low divergence of 0.2 mrad. The signals are received by the narrow field of view (1 mrad) telescope of 300 mm diameter. The detection is performed at 8-channels (so-called $2\alpha + 3\beta + 2\delta$ + WV) enabling determination of the total and particle extinction and backscatter coefficient, the volume and particle linear depolarization ratio, and the water vapor mixing ratio profiles.

The lidar signals are recorded up to 48 km. The initial signals at all channels are acquired with a 7.5 m vertical resolution in temporal steps of 30 s.

3. METHODOLOGY OF EVALUATION

Study was focused on search for the smoke aerosol loading in the stratosphere. The data in the daytime were not used due to too high background light level affecting signals in stratosphere. All available lidar data for the overnight time period from 18:00 to 04:00 UTC were analyzed. In September and October 2017, for 20 days there were no observations due to malfunction of the lidar laser cooling system. 10 days were not analyzed because they provided less than 8 hours of observations in the given period. 24 days were excluded from analyses due to the evidence of thick impenetrable low and mid-level clouds in the signals. Finally, 7 days for the case study analysis were chosen. As it is challenging to identify the weak stratospheric smoke layers in the time-height matrix plots of the background and range corrected lidar signals, for each of the 7 cases a single average profile from 18:00 to 04:00 UTC was obtained. Stratospheric aerosol signatures can be seen in these averaged, background corrected, smoothed and range-corrected vertical profiles.

The particle backscatter coefficient profiles at 532 and 355 nm were computed by solving the elastic lidar equation (second order Bernouli equation solution for mathematically ill-posed problem) [17] with an assumption of the height-independent particle lidar ratio (LR). The choice of lidar ratio values (31 sr at 355 nm and 55 sr at 532 nm) was based on the sensitivity study, i.e. defined as the lowermost lidar ratio that, if applied, guarantee no negative values in the derived vertical profile of the particle backscatter coefficient. The selection of the calibration range (CR) was set at a range free of any aerosol signatures. Proposed in [17] forward and backward inversions were used respectively above and below the aerosol-free calibration range. The particle extinction coefficient profiles were estimated by multiplying the particle backscatter coefficient profiles with the constant lidar ratio (values as specified above). Similar methodology was used in [11] for Warsaw data.

The volume linear depolarization ratio (VDR) at 355 and 532 nm, defined as the ratio of the polarized-to-total elastic backscatter signal), were calculated with the mean calibration constants of V_{355} of $2.10*10^{-1}\pm3.79*10^{-2}$ and V_{532} of $5.37*10^{-2}\pm5.95*10^{-3}$, as in [14].

The aerosol optical depth (AOD) at 355 and 532 nm was derived by integrating the particle extinction coefficient profile for different height ranges: within the aerosol boundary layer, in the free troposphere, and in the lower stratosphere, as in [18]. The aerosol boundary layer (ABL) height and the tropopause (TRP) height were obtained using the wavelet covariance transform method (WCT), as in [19]. The advantage of the WCT method is that it is less affected by the signal noise than the other classical methods, such as the gradient method [20]. The WCT with the dilation of 30 m was applied on the three average overnight elastic signals, and the average height is marked as ABL/TRP height.

4. RESULTS

Figure 1 depicts the wavelength-dependent VDR of stratospheric smoke layers detected over several days in September and October 2017 in Warsaw. The smoke signatures mainly distribute within the range 14-22 km. Lack of data on many days is not that there was no layer; it was not possible to evaluate lidar data e.g. due to low clouds. The mean background (bluish) VDR values outside the layers were of 0.015±0.03 at 355 nm and 0.018±0.01 532 nm. The mean VDR values of stratospheric smoke layers was 0.031±0.04 (maximum 0.049) at 355 nm and 0.033±0.01 (maximum 0.059) at 532 nm. The uncertainty of the VDR values at 355 and 532 nm is < 10%.

This indicates that stratospheric smoke particles over were almost spherical in shape during observed cases in Warsaw. Ansmann [7] reported for Leipzig (the closest EARLINET site, around 700 km, SW off Warsaw), that on 21-22 August 2017 the VDR for the extreme smoke layer in the lower stratosphere was mostly between 0.1-0.2 at height from 12 to 16 km at 532 nm, indicating aspherical particles.

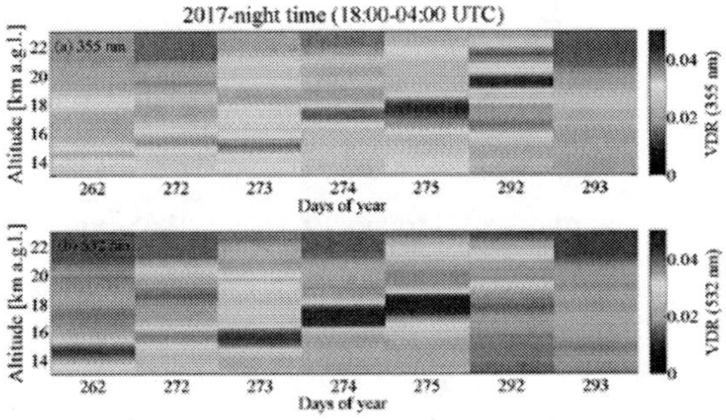

Figure 1. PollyXT-UW lidar-derived volume depolarization ratio (VDR) at 355 (a) and 532 nm (b) for stratospheric smoke layers on several nights in autumn 2017 over Warsaw.

Direct comparison of VDR values in the extreme smoke layer in the lower stratosphere is not possible due to laser failure in Warsaw in August 2017. Thus, the reason for the strong VDR contrast in August 2017 values in Leipzig and September-October values in Warsaw is due to both, the difference in each site location and the difference in observation period. The smoke aerosol can circle the globe within less than 20 days [7, 11], therefore during its transport smoke layers are expected to dilute and either dry and decrease their size or take up water – both will lower the VDR. Figure 2 depicts an example of the wavelength dependent mean vertical profiles of the particle extinction coefficient (α_p) averaged over the period from 18:00 UTC on 18 September to 04:00 UTC on 19 September 2017. The ABL height (separating the aerosol boundary layer and free troposphere) marks the red line and the TRP height (separating the troposphere and the stratosphere) the black line. The uncertainty in the derived particle extinction coefficient and the AOD values was less 20%. The variation of α_p profiles showed weak aerosol layers at 14-20 km, being indicative of an existence of the stratospheric smoke. In lower troposphere, at 6-10 km the aerosol load was significantly enhanced and the free troposphere AOD_{FT} was slightly higher than within aerosol boundary layer AOD_{ABL}, which indicates aerosol advection in the free troposphere (not of interest to this paper). The total aerosol optical depth (AOD_{TO}) was of 0.30 (355 nm) and 0.27 (532 nm). In lower stratosphere AOD_{STR} was of 0.06 and 0.02, contributing with 20% and 7% to AOD_{TO} at 355 and 532 nm, respectively. For the extreme smoke observed in August 2017 in Leipzig, the 532 nm smoke related AOD_{TR} was of 0.08 and as high as 0.2-0.25 for AOD_{STR} [7]. This is indicating a significant dilution of the stratospheric smoke aerosol load over Warsaw.

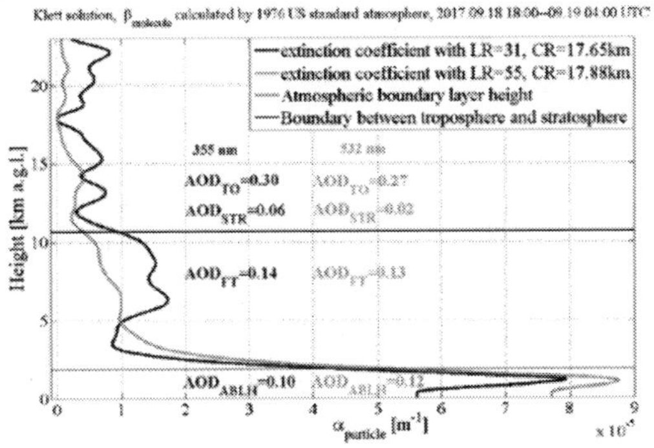

Figure 2. PollyXT-UW lidar-derived vertical profiles of particle extinction coefficient (α_p) at 355 nm and 532 nm, calculated as a mean of 18:00-4:00 UTC on 18/19 September 2017 in Warsaw. Aerosol optical depth within aerosol boundary layer (AOD_{ABL}), free troposphere (AOD_{FT}), stratosphere (AOD_{STR}) and total (AOD_{TO}) are given.

5. CONCLUSIONS

Well documented wildfires originating over North America in August of 2017 with smoke aerosol layers detected in the troposphere and stratosphere after long range transport over the globe, were explored in this study in view of the assessment of the change of their optical properties 2-3 months after the event.

The stratospheric smoke aerosol layers observed over Warsaw in September and October of 2017 were discussed in terms of the volume depolarization ratio (VDR) and aerosol optical depth (AOD). The VDR values of stratospheric smoke layers were in the range from 0.020 to 0.049 at 355 nm and from 0.020 to 0.059 at 532 nm. The AOD of the smoke layers in stratosphere contributed with 0.06 (20%) and 0.02 (7%) to the total AOD at 355 and 532 nm, respectively. The comparison of these results implies that the circulation of the smoke particles related to the biomass burning injection of North American wildfires smoke aerosol into the atmosphere in August 2017 [7] caused significant (10 times) dilution of the aerosol optical depth and indicate significant decrease of the volume depolarization ratio, which can be explained either by strong ageing (decreasing size and drying of particles) or by water uptake of the stratospheric aerosol.

ACKNOWLEDGEMENTS

The PollyXT lidar in Warsaw was developed in scientific collaboration of the Faculty of Physics, University of Warsaw (FUW) with the Institute of Tropospheric Research (TROPOS) financed by Polish Foundation of Science & Technology (519/FNITP/115/2010). We acknowledge work of colleagues of PollyXT Lidar Group lead by Dietrich Althausen.
The algorithms developed for this research were financed within the ESA-ESTEC Contract for Development of a European HSRL airborne facility (MULTIPLY) Contract No. 4000112373/14/NL/CT.

REFERENCES

[1] T. Deshler, et al. A review of global stratospheric aerosol: Measurements, importance, life cycle, and local stratospheric aerosol. Atmos. Res. 90, 223–232 (2008)

[2] S. Kremser, et al. Stratospheric aerosol – observations, processes, and impact on climate. Rev. Geophys. 54, 278–335 (2016)

[3] S. Solomon, et al. The Persistently Variable "Background" Stratospheric Aerosol Layer and Global Climate Change. Sci, 333, 866–870 (2011)

[4] Zuev, V. V, et al. 30-year lidar observations of the stratospheric aerosol layer state over Tomsk (Western Siberia, Russia). Atmos. Chem. Phys. 17, 3067-3081 (2017)

[5] S. Khaykin, et al. Variability and evolution of the midlatitude stratospheric aerosol budget from 22 years of ground-based lidar and satellite observations. Atmos. Chem. Phys. 17, 1829-1845 (2017)

[6] S. Khaykin, et al. Stratospheric Smoke With Unprecedentedly High Backscatter Observed by Lidars Above Southern France. Geophys. Res. Lett. 45, 1639–1646 (2018)

[7] A. Ansmann, et al. Extreme levels of Canadian wildfire smoke in the stratosphere over central Europe on 21–22 August 2017. Atmos. Chem. Phys. 18, 11831–11845 (2018)

[8] M. Haarig, et al. Depolarization and lidar ratio at 355, 532, and 1064 nm and microphysical properties of aged tropospheric and stratospheric Canadian wildfire smoke. Atmos. Chem. Phys. 18, 11847-11861 (2018)

[9] Q. Hu, et al. Long-range-transported Canadian smoke plumes in the lower stratosphere over northern France. Atmos. Chem. Phys. 19, 1173-1193 (2019)

[10] C Jimenez, et al. Polarization lidar: an extended three-signal calibration approach. Atmos. Meas. Tech. 12, 1077-1093 (2019)

[11] H. Baars, et al. The unprecedented 2017–2018 stratospheric smoke event: decay phase and aerosol properties observed with the EARLINET. Atmos. Chem. Phys. 19, 15183–15198 (2019)

[12] G. Pappalardo, et al. EARLINET: Towards an advanced sustainable European aerosol lidar network. Atmos. Meas. Tech. 7, 2389-2409 (2014)

[13] H. Baars, et al. An overview of the first decade of Polly NET: An emerging network of automated Raman-polarization lidars for continuous aerosol profiling. Atmos. Chem. Phys. 16, 5111-5137 (2016)

[14] K. Markowicz, et al. Study of aerosol optical properties during long-range transport of biomass burning from Canada to Central Europe in July 2013. J. Aeros. Sci. 101, 156-173 (2016)

[15] R. Engelmann, et al. The automated multiwavelength Raman polarization and water-vapor lidar PollyXT: The next generation. Atmos. Meas. Tech. 9, 1767-1784 (2016)

[16] I. S. Stachlewska, et al. Raman lidar water vapour profiling over Warsaw, Poland. Atmos. Res. 194, 258-267 (2017)

[17] J. D. Klett. Lidar inversion with variable backscatter/extinction ratios. Appl. Opt. 24, 1638–1643 (1985)

[18] I. S. Stachlewska, et al. Modification of local urban aerosol properties by long-range transport of biomass burning aerosol. Remote Sens. 10, 412 (2018)

[19] D. Wang, et al. Interrelations between surface, boundary layer, and columnar aerosol properties derived in summer and early autumn over a continental urban site in Warsaw, Poland, Atmos. Chem. Phys., 19, 13097–13128 (2019)

[20] I.M. Brooks. Finding boundary layer top: Application of a wavelet covariance transform to lidar backscatter profiles. J. Atmos. Ocean. Tech. 20, 1092-1105 (2003)

STUDIES ON AEROSOL OPTICAL PROPERTIES AT HIGH ALTITUDE STATION IN WESTERN HIMALAYAS USING RAMAN LIDAR

Shishir Kumar Singh[1,2], Jaswant[1,2], Radhakrishnan S.R.[1,2*], Davender Sethi[1,2] and Chhemendra Sharma[1,2]

[1]*Environmental Sciences and Biomedical Metrology Division*
[2] *CSIR-National Physical Laboratory, Dr. K. S. Krishnan Marg, New Delhi-110012, India*
**Email:* rkrishnan@nplindia.org

ABSTRACT

The aerosol optical properties have been investigated using the Raman lidar system for the month of November 2018 at the western Himalayan station of Palampur. Before deriving the optical properties, the lidar data has been applied with initial pre-processing such as Dead time correction, atmospheric noise correction, temporal and spatial averaging, range correction, gluing etc. The optical properties such as backscatter coefficient, extinction coefficient and linear depolarization ratio have been derived by using the inversion algorithm proposed by Fernald. The results show that the backscatter coefficient was found in the range from 9.00E-9 $m^{-1}sr^{-1}$ to 4.97E-6 $m^{-1}sr^{-1}$ and the extinction coefficient was found in the range from 3.16E-7m^{-1} to 1.74E-4m^{-1}. The Linear depolarization ratio was in the range from 0.0179 to 0.621 with lower values at near heights suggesting the dominance of spherical particles at the lower heights. We have also observed a cloud layer at a height of 9.5 km to 12.1 km with high depolarization ratio during the observation period on 22/11/2018.

1. INTRODUCTION

The effects of atmospheric aerosols on climate have large uncertainties including their direct and indirect effects related to aerosol-cloud interactions. The climatic effects of aerosol depend on its physical, chemical and optical properties along with its vertical distribution. This enables the need of advanced lidar systems which can provide the optical and microphysical properties of aerosols with its vertical distribution [1].

In India, only few measurements of aerosol vertical profiles by lidar have been carried out. Long-term measurements and continuous measurements of aerosol vertical distribution are even less. In this paper, we are presenting the measurement of aerosol vertical profiles in Palampur, Himachal Pradesh, India. The measurements are based on a Raman lidar which provides information on the vertical distribution of optical presence of aerosols and clouds. [2]

2. METHODOLOGY

2.1 Sampling Site

A Raymetrics Raman Lidar system, LR111-D200 is installed at the National Physical Laboratory's remote monitoring station at the campus of CSIR-Institute of Himalayan Bioresource Technology (IHBT), Palampur, Himachal Pradesh, India at an altitude of 1347 m above mean sea level to study the vertical profiles of physical and optical properties of aerosols and clouds. Palampur is located in the lower Himalayan region of Western Himalayas and lies within the latitude 32.1109° N and longitude 76.5363° E. Figure 1 shows the topographic map of Palampur. It is the first ground based lidar facility in the Western Himalayan region of India.

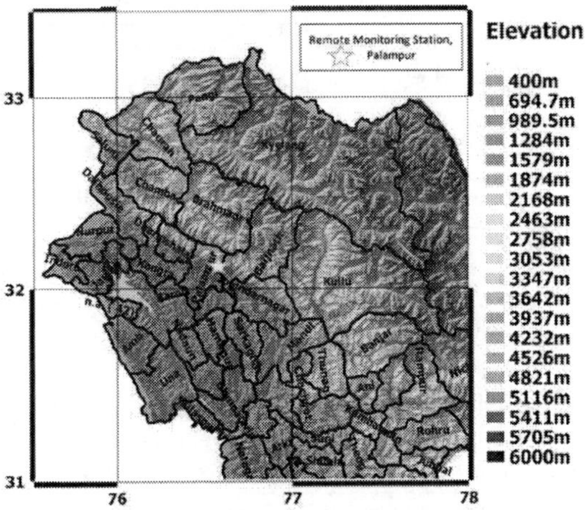

Figure 1: Topographic map of Palampur

https://doi.org/10.1051/epjconf/202023702034

2.2 The Lidar System

This lidar system is operated at 355 nm with Raman channel at 387 nm having a spatial resolution of 7.5 m. The return signal is analyzed by using the software provided by the manufacturer for further processing. The co-polarized and cross polarized Mie channels of the lidar system provides the linear depolarization ratio which is an indicator of the shape of the aerosol/cloud particle. The nitrogen Raman channel at 387 nm of the lidar system can be used to derive the lidar ratio, which is an important parameter for the identification of different types of aerosols in the troposphere. The technical specifications of the Raman lidar system are given in Table 1.

Table 1: Specifications of Raman Lidar

Pulsed Laser Source	Nd:YAG (Quantel ULTRA100 Series)
Wavelength	355 nm
Energy / Pulse	32.8 mJ
Repetition Rate	20 Hz
Pulse Width	5.60 nsec
Laser Beam Divergence	<0.8 mrad
Telescope Type	Cassegrain
Primary Diameter	200 mm
Secondary Diameter	48 mm
Transmitter–Receiver Distance	212 mm
Field of View	2.3 mrad (adjustable from 0.5 to 3mrad)
Detectors	3 PMTs
Data Acquisition Mode	Both Analog and Photon Counting

2.3 Lidar Data Analysis

The Raman lidar data for the 22nd November 2018 has been used for the study. The depolarisation ratio of the lidar system allows us to separate spherical particles from non-spherical. The lidar ratio gives us information about the properties and origin of particles by giving the ratio of the extinction and backscatter coefficient [2]. The Palampur has a very clear environment with very less anthropogenic activities, so a lidar ratio of 35sr has been taken at 355nm [6]. Before deriving optical properties from the raw lidar data, it has been temporally and spatially averaged. Other than this, the raw lidar data has been preprocessed with other corrections such as dead time correction of photon counting (PC) signal, background correction, bin shift correction, range correction, gluing etc. by the software provided by manufacturer. Figures 2 & 3 shows the aerosol backscatter coefficient, extinction coefficient and linear depolarization ratio at 355 nm wavelength for 22nd Nov. 2018.

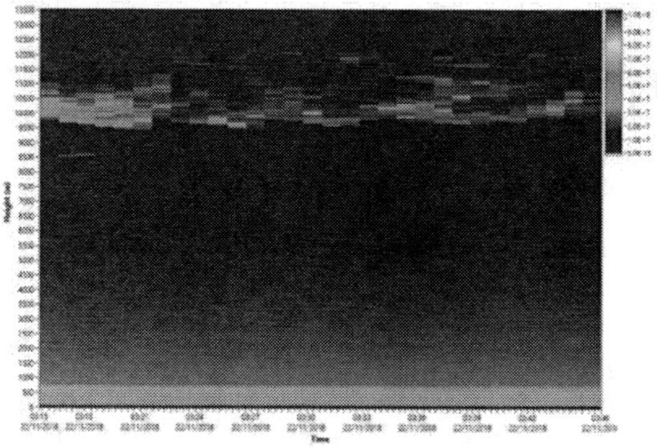

Figure 2 shows the temporal variation of Range Corrected Signal for 22nd November 2018

The figure 2 shows the temporal variation of range corrected signal (RCS) of 22nd November 2018. In this figure, we can see that there is high loading of aerosol over the Lidar station upto 1km and at height of 10-12km we can observe the clouds.

The backscatter coefficient was found in the range from 3.62E-8 $m^{-1}sr^{-1}$ to 1.76E-6 $m^{-1}sr^{-1}$ in low altitude and 9.00E-9 $m^{-1}sr^{-1}$ to 4.97E-6 $m^{-1}sr^{-1}$ in high altitude (9500m-12100m). The extinction coefficient was found in the range from 1.27E-6 m^{-1} to 6.15E-5 m^{-1} in low altitude (upto 800m) and 3.16E-7 m^{-1} to 1.74E-4 m^{-1} in high altitude (9500m-12100m). The value of the backscatter coefficient and extinction coefficient were high near the ground (0-1km) and then it decreased with the altitude. Figure 3 shows that beyond 1000 meters there were very few aerosols and the atmosphere can be considered clear.

The Linear depolarization ratio was in the range from 0.0179 to 0.621. The depolarization values were close to zero at the lower heights suggesting the dominance of spherical particles at the lower heights. We have also observed a cloud layer at a height of 9500m-12100m with high depolarization ratio.

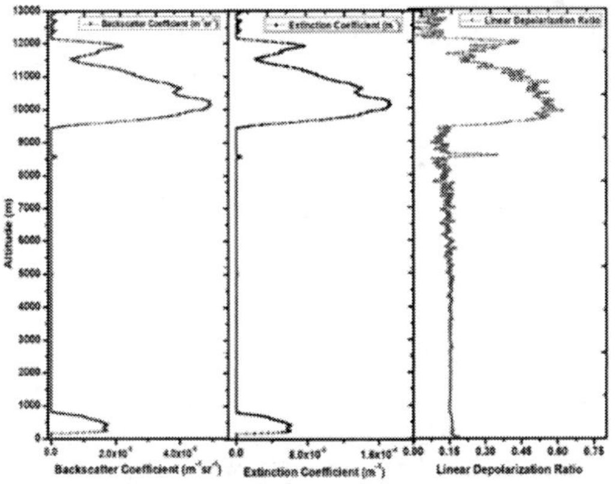

Figure 3 shows the backscatter coefficient, extinction coefficient and linear depolarization ratio of 22nd November 2018.

To trace the sources of air parcel, we have also plotted the HYSPLIT (Hybrid Single-Particle Lagrangian Integrated Trajectory) 3 days Back Trajectory for at three different levels, 500m agl (above ground level), 6000 m agl and 13000 m agl in figure 4.

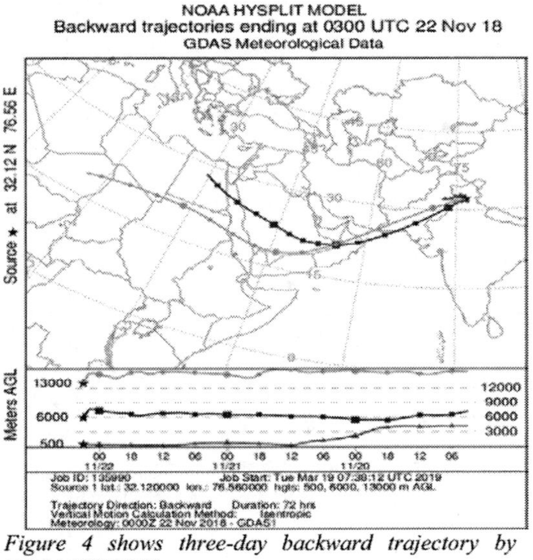

Figure 4 shows three-day backward trajectory by NOAA HYSPLIT

Results show that most of the aerosol loading was due to regional emissions at lower heights of the station. The air parcel arrived at the higher heights of station was from Africa and the western region of the India. This also shows that the cloud loading at higher altitude was originated from the north of Africa having high depolarization ratio.

4. CONCLUSION

The aerosol optical properties over a western Himalayan region Palampur have been investigated by a Raman lidar system for the month November 2018. The raw lidar signal has been pre-processed before deriving optical properties from it. The results show that the aerosol loading was high at the lower heights with the dominance of the spherical particles. A cloud layer was also observed at the height of 9.5 km to 12 km above the ground level.

ACKNOWLEDGEMENTS

Authors are grateful to Director, NPL for necessary support. Authors are thankful to CSIR network project PSC-0112 for necessary financial support. Mr. Shishir Kumar Singh is thankful to University Grant Commission (UGC)) for providing research fellowship and also to Academy of Scientific and Innovative Research (AcSIR) for facilitating as its PhD student.

REFERENCES

[1] M. Komppula et al. Atmos. Chem. Phys., 12, 4513-4524 (2012).
[2] M. N. Young et al. Atmospheric Research 86, 76–87 (2007)
[3] F. G. Fernald Appl. Opt. 23, 652-653 (1984)
[4] A. Ansmann et al. Optics Letter Vol. 15, Issue 13, 746-748 (1990)
[5] A. Ansmann et al. Applied Physics B 55(1): 18-28 1 (1992)
[6] K. Sasson et al. Geophysical Research Letters, Vol. 34, L08803 (2007)

SYNERGETIC OBSERVATIONS BY GROUND-BASED AND SPACE LIDAR SYSTEMS AND AERONET SUN-RADIOMETERS: A STEP TO ADVANCED REGIONAL MONITORING OF LARGE SCALE AEROSOL CHANGES

Anatoli Chaikovsky[1*], Andrey Bril[1], Oleg Dubovik[2], Anton Fedarenka[1], Philippe Goloub[2], Qiaoyun Hu[2], Anton Lopatin[3], Tatyana Lapyonok[2], Natallia Miatselskaya[1], Benjamin Torres[2], David Fuertes[3], Vladislau Peshcharankou[1], Thierry Podvin[2,36], Ioana Popovici[2], Dong Liu[4], Zhengqiang Li[5], Ourania Soupiona[6], Maria Mylonaki[6], Lucia Mona[7], Aldo Giunta[7], Nikolaos Papagiannopoulos[7], Maria Rita Perrone[8], Salvatore Romano[8], Dimitris Balis[9], Nikolaos Siomos[9], Kalliopi-Artemis Voudouri[9], Livio Belegante[10], Doina Nicolae[10], Dragos Ene[10], Nicolae Ajtai[11], Horatiu Stefanie[11], Vassilis Amiridis[12], Alexandra Tsekeri[12], Daniele Bortoli[13], Maria Joao Costa[13], Ina Mattis[14], Francesc Rocadenbosch[15,16], Alejandro Rodríguez-Gomez[15], Michael Sicard[15,16], Alfonso J. Fernandez[17], Francisco Molero[17]; Dietrich Althausen[18], Holger Baars[18], Juan Luis Guerrero Rascado[19], Pablo Ortiz-Amezcua.[19], José Antonio Benavent Oltra[19], Andrés Esteban Bedoya-Velásquez[19], Roberto Román[19], Lucas Alados-Arboledas[19], Yurii Balin[20], Grigorii Kokhanenko[20], Ivan Penner[20], Boris Chen[21], Leonid Sverdlik[21], Genadi Milinevsky[22,23], Nobuo Sugimoto[24], Atsushi Shimizu[24], Tomoaki Nishizawa[24], Rei Kudo[25], Itaru Sano[26], Teppei J. Yasunari[27], Hitoshi Irie[28], Toshihiko Takemura[29], Sang-Woo Kim[30], Nguyen Xuan Anh[30], Pham Xuan Thanh[31], Aleksander Pietruczuk[32], Iwona S. Stachlewska[33], Alessia Sannino[34], Xuan Wang[35], Antonela Boselli[7]

*[1]Institute of Physics, NAS of Belarus, Minsk, Belarus, *Email: chaikov@dragon.bas-net.by; [2]Laboratoire d'Optique Atmosphérique, CNRS – Université Lille, France; [3]GRASP-SAS, Villeneuve d'Ascq, France; [4]Anhui institute of Optics and Fine Mechanics, Chinese Academy of Sciences, Hefei, China; [5]Institute of Remote Sensing and Digital Earth, Chinese Academy of Sciences, Beijing, China; [6]National Technical University of Athens, Athens, Greece; [7]CNR - Istituto di Metodologie per l'Analisi Ambientale, Potenza, Italy; [8]University of Salento, Lecce, Italy; [9]Aristotle University of Thessaloniki, Greece; [10]National Institute of Research and Development for Optoelectronics, Magurele, Romania; [11]Babeș-Bolyai University, Cluj-Napoca, Romania; [12]IAASARS, National Observatory of Athens, Athens, Greece; [13]Instituto de Ciências da Terra, Escola de Ciências e Tecnologia, Universidade de Évora, Évora, Portugal; [14]Meteorological observatory Hohenpeissenberg, DWD, Germany; [15]COMMSENSLAB-Unidad María de Maeztu, Universitat Politècnica de Catalunya, Barcelona, Spain; [16]Institut d'Estudis Espacials de Catalunya (CTE-CRAE / IEEC), Universitat Politècnica de Catalunya, Barcelona, Spain; [17]Research Centre for Energy, Environment and Technology (CIEMAT), Madrid, Spain; [18]Leibniz Institute for Tropospheric Research, Leipzig, Germany; [19]Andalusian Institute for Earth System Research (IISTA-CEAMA), University of Granada. 18006, Granada, Spain; [20]V.E. Zuev Institute of Atmospheric Optics of the Russian Academy of Science, Tomsk, Russia; [21]Kyrgyz-Russian Slavic University, Bishkek, Kyrgyz Republic; [22]Kyiv National Taras Shevchenko University, Kyiv, Ukraine; [23]College of Physics, International Center of Future Science, Jilin University, Changchun, China; [24]National Institute for Environmental Studies, Tsukuba, Japan; [25]Meteorological Research Institute, Japan Meteorological Agency, Tsukuba, Ibaraki, Japan; [26]Faculty of science and engineering, Kindai University, Higashi-Osaka, Japan; [27]Arctic Research Center, Hokkaido University, Sapporo, Hokkaido, Japan; [28]Center for Environmental Remote Sensing, Chiba University, Chiba, Japan; [29]Research Institute for Applied Mechanics, Kyushu University, Fukuoka, Japan; [30]School of Earth and Environmental Sciences, Seoul National University, Seoul, Korea; [31]Institute of Geophysics, Vietnam Academy of Science and Technology, Hanoi, Vietnam; [32]Institute of Geophysics Polish Academy of Sciences, Warsaw, Poland; [33]University of Warsaw, Faculty of Physics, Warsaw, Poland; [34]Dipartimento di Fisica dell'Università Federico II di Napoli, Napoli, Italy; [35]CNR- Istituto Superconduttori, materiali Innovativi e dispostivi, Napoli, Italy; [36]CIMEL Electronique, Paris, France*

ABSTRACT

The paper presents the preliminary results of the lidar&radiometer measurement campaign (LRMC-2017), estimation of statistical relations between aerosol mode concentrations retrieved from CALIOP and ground-based lidar stations and case study of fire smoke events in the Eurasian regions using combined ground-based and space lidar and radiometer observations.

1. INTRODUCTION

Coordinated lidar and radiometer sounding (LRS technique [1]) and related algorithms for processing combined lidar and radiometer data (LidAlm, POLIPHON, LIRIC, GARRLiC, SKYLIDAR, [2-6]) provides estimates of the key aerosol optical parameters, which determines the aerosol layer contribution in the equation of radiation transfer, as well as aerosol mode concentration profiles.

However, the low number of coincident lidar and sun-radiometer stations restricts the scope of LRS for monitoring atmospheric aerosol.

New version of Lidar&Radiometer Sounding (LRS) technique, named LRS-S, takes data of ground-based lidar stations, as well as satellite lidar (CALIOP) in the vicinity of AERONET radiometer stations to increase about ten times the number of sites suitable for control of aerosol parameters [7].

Combined Lidar&Radiometer measurement campaign (LRMC-2017) was started in 2017 with the aim to verify an ability to obtain information on aerosol concentration profiles over a large region from coordinated radiometer and ground-based and space lidar measurements. The first measurement period lasted from 15 to 31 of May, 2017. Additional measurements were carried out in 2017 and 2018.

This paper presents the preliminary results of collocated measurements of ground-based and satellite lidars in the vicinity of AERONET stations and observations of some aerosol events over Eurasian continent to assess the performance of LRS-S technique and expose significant features.

2. MEASUREMENT PROCEDURE

To date, 26 lidar stations of EARLINET, AD-Net and CIS-LiNet lidar networks associated with AERONET sun-radiometers sites submitted measurement data to the general database.

Three types of coordinated lidar and radiometer measurements were carried out in the frame of LRMC-2017 (Figure 1):

- R&L, coordinated sun-radiometer and multi-wavelength lidar measurements at the ground-based complex radiometer and lidar stations; R&L measurements provide local information about aerosol mode concentration profiles averaged over a time interval, $\Delta T \approx 30-60$ min;
- R&C, AERONET radiometer measurements and CALIOP sounding in the vicinity of AERONET station, $\Delta S \approx 100km$; R&C measurements provide aerosol mode concentration profiles averaged over the trajectory segment $\Delta L \approx 100-200km$.
- R&L&C, combined radiometer, ground-based lidar and CALIOP measurements in the vicinity of AERONET station.

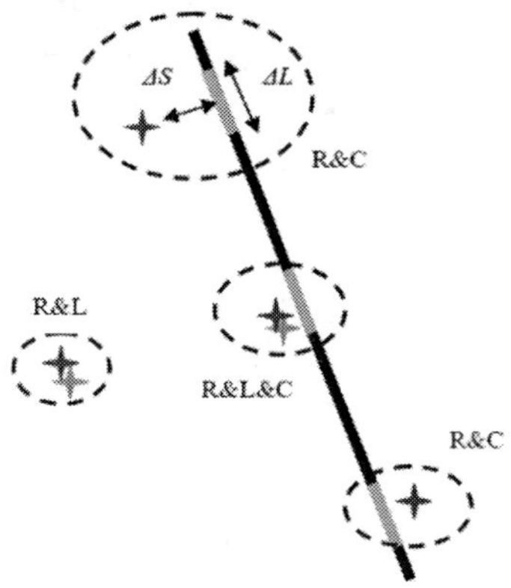

Figure 1. The scheme of measurement in the frame of LRMC-2017: black line is CALIOP ground track; blue segment is a part of trajectory of length, *ΔL*, used for retrieving aerosol parameters; *ΔS* is the distance between radiometer station and trajectory segment; red star is AERONET station; green star is multi-wavelength lidar system.

STUDYING OF STATISTICAL RELATIONS BETWEEN AEROSOL PARAMETERS

In general, the aerosol parameters retrieved from R&L and R&C measurements are different because they are calculated from different statistical ensembles of lidar data with the employment of time and/or space averaging procedures for R&L and R&C measurements, correspondingly. Coincidence of the results should be expected if the aerosol random field satisfies some conditions: horizontal spatial homogeneity and ergodicity.

Results of R&L&C measurements allow comparing the aerosol concentration profiles, $c_k^g(h)$ and $c_k^s(h)$, retrieved from ground-based and satellite lidar measurements to be compared. Index k is related to the aerosol mode.

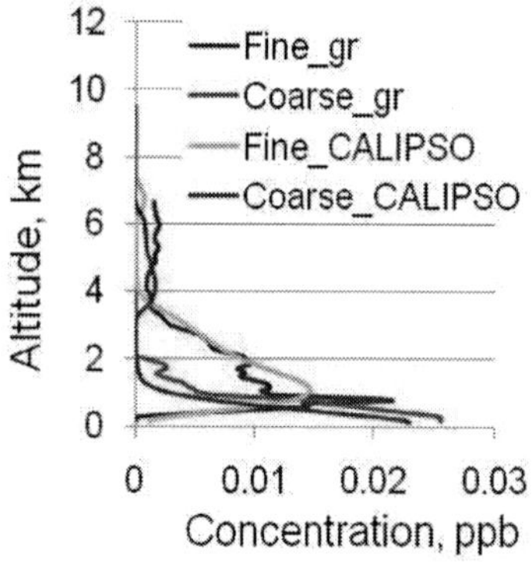

Figure 2. Implementation of R&L&C measurements at AD-Net stations: a) –scheme of measurements; b) – profiles of aerosol mode concentration in Fukuoka region from ground-based lidar measurements (gr) and satellite ones (CALIPSO); 17/05/2017; G(fine)=0.05, G(coarse)=0.1

Differences between profiles $c_k^g(h)$ and $c_k^s(h)$ were numerically estimated by parameter G:

$$G\left(c_k^g, c_k^s\right) = \frac{\sum_i \left(c_k^g(h_i) - c_k^s(h_i)\right)^2}{\sum_i \left(c_k^g(h_i)\right)^2 + \sum_i \left(c_k^s(h_i)\right)^2}; \quad (1)$$

where $G\left(c_k^g, c_k^s\right) \in [0,1]$; if $G\left(c_k^g, c_k^s\right)$ equals 0 or 1, it means the complete coincidence of the profiles $c_k^g(h)$ and $c_k^s(h)$ or the localization of aerosol in non-coincident atmospheric layers, correspondently.

Figure 2 shows the implementation scheme of R&L&C measurements at AD-Net stations and results of retrieving aerosol concentration profiles in Fukuoka region.

Probability distribution of the parameter $G\left(c_t^g, c_t^s\right)$ for total (fine and coarse) aerosol particles concentration are shown in figure 3. Average value of parameter $G\left(c_t^g, c_t^s\right)$ for all measured ensemble of data is 0.11.

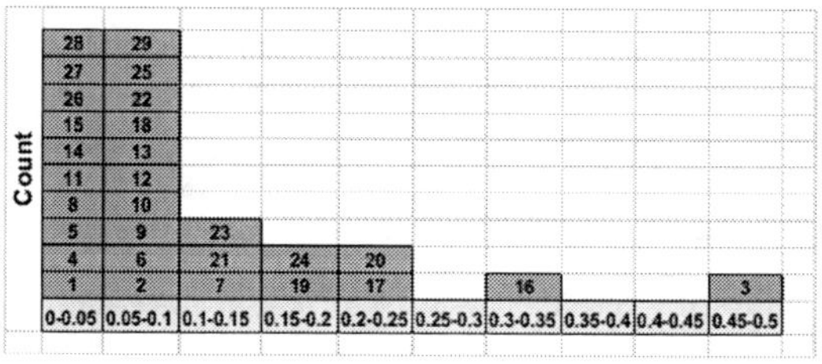

1- Athens *(EARLINET);* 2,3 - Barcelona *(EARLINET);* 4 - Granada *(EARLINET);* 5 - Hohenpeissenberg *(EARLINET);* 6 - Madrid *(EARLINET);* 7- Finakalia *(EARLINET);* 8 - Lille *(EARLINET);* 9 - Potenza *(EARLINET),* 10 - Lecce *(EARLINET);* 11 - Bucharest *(EARLINET);* 12-15 - Evora *(EARLINET);* 16,17 - Haifa *(TROPOS);* 18,19 - Minsk *(EARLINET,CIS-LiNet);* 20 - Tomsk *(CIS-LiNet);* 21,22 - Teplokluchenka *(CIS-LiNet);* 23- Tsukuba *(AD-Net);* 24- Chiba *(AD-Net;* 25 - Osaka *(AD-Net);* 26 - Fukuoka *(AD-Net);* 27,28 - Sapporo *(AD-Net);* 29-Seoul *(AD-Net)*

Figure 3. Chart of probability distribution of the parameter $G\left(c_k^g, c_k^s\right)$ for the height profiles of the total concentrations of aerosol particles, retrieved from ground-based and satellite lidar measurements

CASE STUDY OF DUST AND FIRE SMOKE EVENTS IN THE EURASIAN REGIONS

Eurasian continent includes regions with advance infrastructure for monitoring of environment as well as uninhabited and inaccessible areas where only satellite observations are available. LRS-S technique is a potentially optimal tool for studying large scale aerosol changes and monitoring the aerosol transport over Eurasian area.

In the frame of LRMC-2017 campaign we tested the algorithm and new software package LIRIC-2 for retrieving aerosol mode concentration profiles for some aerosol events in 2017 – 2018:

- transport of fire smoke to European regions from North America and East Russian regions and Ukraine in August, 2018;
- fire smoke in Vietnam, April 2017;

Figures 6 demonstrate the examples of retrieved aerosol concentration profiles. Intercontinental transport of fire smoke in the free troposphere creates a wide layer of small aerosol particles with a maximum at an altitude of 3-6 km in the East European region (Fig. 6, a-c). Local fire spots pollute boundary layer in Bac-Lieu, Vietnam (Fig. 6, d).

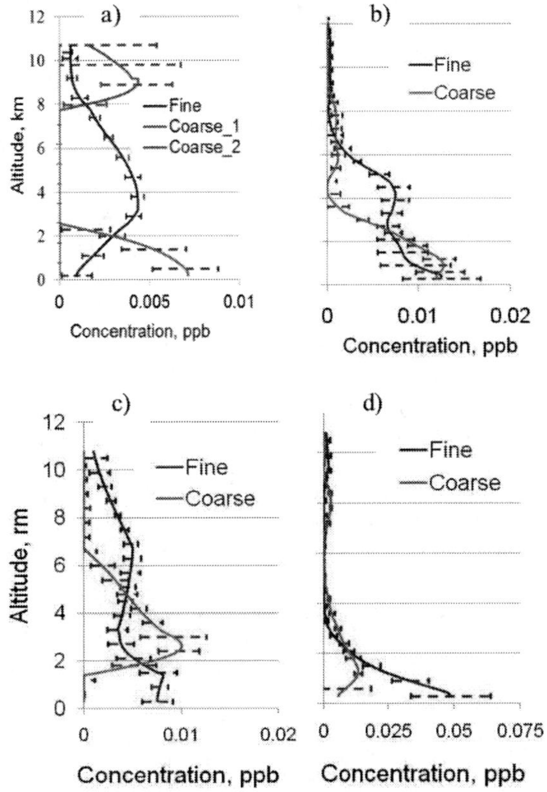

Figure 6. Profiles of aerosol mode concentration at: a)Belsk, Poland, 29/08/2018; b) Kyiv, Ukraine, 31/08/2018; c)Minsk, Belarus, 02/09/2018; d) Bac Lieu, Vietnam, 18/04/2017

SUMMARY

Implementation of LRS-S technique increases the number and geographical area of sites for monitoring of aerosol parameter profiles. Results of LRMC-2017 measurement campaign demonstrate that coordinated ground-based and satellite lidar and radiometer measurements with assimilation data of LRS-S observations for modeling aerosol transport are perspective research lines for development of regional large-scale aerosol monitoring.

ACKNOWLEDGMENTS

The financial support by the European Union's Horizon 2020 research and innovation program (ACTRIS-2, grant agreement no. 654109) is gratefully acknowledged. The investigation was supported by the Belarusian Republican Foundation for Fundamental, Research Agreement No. F18EA-011, by the Russian Foundation for Basic Research (grant No. 18-55-81001), and also by Vietnam Academy of Science and Technology (project code QTRU05.01/18-20). The Portuguese team acknowledges the support from the Portuguese Science Foundation, in the frame of the European Regional Development Fund - COMPETE 2020, under the projects UID/GEO/04683/2013 (POCI-01-0145-FEDER-007690)

We acknowledge the AERONET community, Atmospheric Science Data Center for the provision of the radiometer and CALIPSO data.

REFERENCES

1. A. Chaikovsky, et al. Proceeding of SPIE, 4678, 257-268 (2002).
2. J. Cuesta, et al. Appl. Opt., 47, 4598–4611 (2008).
3. A. Ansmann, et all. Atmos. Chem. Phys., 12, 9399–9415 (2012).
4. A. Chaikovsky, et al. Atmos. Meas. Tech., No. 9, 1181–1205 (2016).
5. A. Lopatin, et al. Atmos. Meas. Tech., 6, 2065–2088 (2013)
6. R. Kudo, et al. Atmos. Meas. Tech., No. 9, 3223-3243 (2016)
7. Chaikovsky, et al. EDJ Web of Conferences, 176, 08007 (2018)

WILDFIRE SMOKE IN THE STRATOSPHERE OVER EUROPE– FIRST MEASUREMENTS OF DEPOLARIZATION AND LIDAR RATIOS AT 355, 532, AND 1064 NM

Moritz Haarig[1*], Holger Baars[1], Albert Ansmann[1], Ronny Engelmann[1], Kevin Ohneiser[1], Cristofer Jimenez[1], Dietrich Althausen[1], Johannes Bühl[1], Patric Seifert[1], Rodanthi Mamouri[2], Argyro Nisantzi[2]

[1]*Leibniz Institute for Tropospheric Research (TROPOS), Leipzig, Germany*
[2]*Cyprus University of Technology, Dep. of Civil Engineering and Geomatics, Limassol, Cyprus*
**Email: haarig@tropos.de*

ABSTRACT

Canadian wildfire smoke was detected in the troposphere and lower stratosphere over Europe in August and September 2017. Lidar measurements from various stations of the European Aerosol Research Lidar Network (EARLINET) observed the stratospheric smoke layer. Triple-wavelength (355, 532, and 1064 nm) lidar measurements of the depolarization and the lidar ratio are reported from Leipzig, Germany. The particle linear depolarization ratio of the wildfire smoke in the stratosphere had an exceptional strong wavelength dependence reaching from 0.22 at 355 nm, to 0.18 at 532 nm, and 0.04 at 1064 nm. The lidar ratio increased with wavelength from 40±16 sr at 355 nm, to 66±12 sr at 532 nm, and 92±27 sr at 1064 nm. The development of the stratospheric smoke plume over several months was studied by long-term lidar measurements in Cyprus. The stratospheric smoke layers increased in altitude up to 24 km height.

1. INTRODUCTION

In summer 2017, a record-breaking amount of wildfires in western Canada emitted a huge amount of biomass burning smoke into the atmosphere. Pyrocumulonimbus events occurred due to the strong heat release and lifted the smoke up into the stratosphere [1]. Smoke transport from North American forest fires to Europe is known since a long time [2, 3]. However, for the first time a significant amount of smoke was detected in the stratosphere over Europe, where the light-extinction coefficients were 20 times stronger than after the Pinatubo eruption in 1991 [4]. The smoke in the stratospheric as well as in the troposphere reached Europe about 10 days after injection into the atmosphere over western Canada [5]. On 22 August 2017, the smoke layers were detected over Leipzig and measured with several lidar systems [6]. The stratospheric smoke layers remained for several months over Europe. In Cyprus the stratospheric smoke layers were detected till January 2018.

This abstract describes the first measurement of the lidar ratio and the depolarization ratio at all three typical lidar wavelengths (355, 532, and 1064 nm) to give a full characterization of the optical properties of stratospheric smoke. Furthermore, the stratospheric smoke as a long-term phenomenon is discussed.

2. LIDAR SYSTEMS

BERTHA (Backscatter Extinction lidar-Ratio Temperature Humidity profiling Apparatus) [7] is the first lidar which measures the extinction coefficient at 1064 nm [8, 9] and the depolarization ratio at 355, 532, and 1064 nm [10]. It takes 20-30 minutes to change from the extinction to depolarization measurement at 1064 nm. With the short delay, 3+3+3 measurements (3 backscatter coefficients, 3 extinction coefficients, 3 depolarization ratios) can be realized for the first time.

Long-term observations in Cyprus and other European sites were performed with an autonomous 3+2+2 Polly^{XT} lidar [11] in the framework of PollyNET [12] and CyCARE (Cyprus Clouds, Aerosol and pReciptation Experiment, October 2016 to March 2018, Limassol, Cyprus).

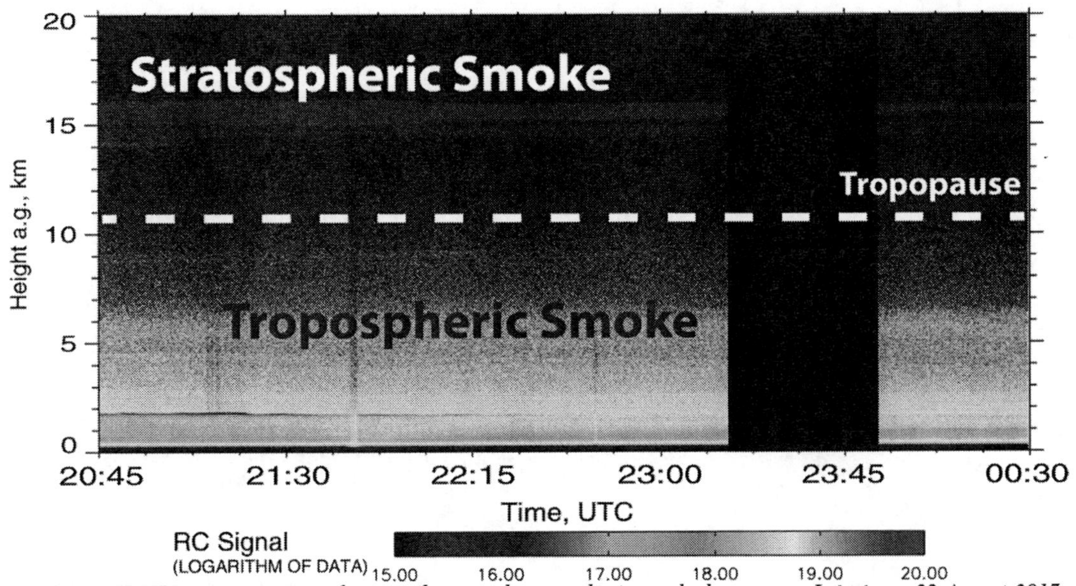

Figure 1 *BERTHA measurements of stratospheric and tropospheric smoke layers over Leipzig on 22 August 2017. The change from the 1064 nm extinction measurement to the 1064 nm depolarization-ratio measurement including calibration caused the gap in the measurements (23:15-23:50 UTC). The range corrected signal at 532 nm (cross polarized) is shown.*

3. RESULTS

3.1 3+3+3 measurements of stratospheric and tropospheric smoke

On 22 Aug 2017 smoke layers reached continental Europe and were measured in Leipzig with 3 lidar systems. Tropospheric smoke layers occurred mainly between 5 and 6.5 km height and a second much stronger smoke layer was found in the stratosphere between 15 and 16 km height (see Fig. 1) well above the local tropopause at 10.8 km height.

The extinction-to-backscatter ratio (lidar ratio) and the particle linear depolarization ratio (PLDR) were measured at 355, 532, and 1064 nm. The profiles of the optical properties are not shown here, but they can be found in Haarig et al. (2018) [6]. At this point, we focus on the main findings and the comparison of the smoke characteristics after the tropospheric and stratospheric transport way.

The lidar ratio of the long-range transported smoke increased with wavelength without a significant difference between the smoke in the troposphere and stratosphere as shown in Fig. 2a. A higher lidar ratio in the visible than in the UV is a clear indication for smoke in contrast to pollution [3]. The lidar ratio at 1064 nm was measured for the first time and could replace former estimates [e.g., 13]. The spectral dependence of the lidar ratio was a result of the strong wavelength dependence of the particle backscatter coefficient and a weak spectral dependence of the extinction coefficient [6].

The particle linear depolarization ratio shown in Fig 2b had values below 5% in the troposphere, which is in line previous observations of aged smoke [e.g., 14]. The particles were obviously compact and of almost spherical shape. In contrast, high values of the PLDR were observed in the stratosphere, 22% at 355 nm, 18% at 532 nm, and 4% at 1064 nm, indicating non-spherical particles. Observations of the same stratospheric smoke layer over Lille, France, confirmed this strong wavelength dependence [15]. A similar case was observed by [16] in the upper troposphere (around 8 km height).

The strong wavelength dependence of the PLDR of the stratospheric smoke compared to the tropospheric smoke points to significant different conditions during transport or already during the emission process. The fast lifting of the smoke particles during the pyrocumulonimbus event or the extremely dry conditions in the stratosphere

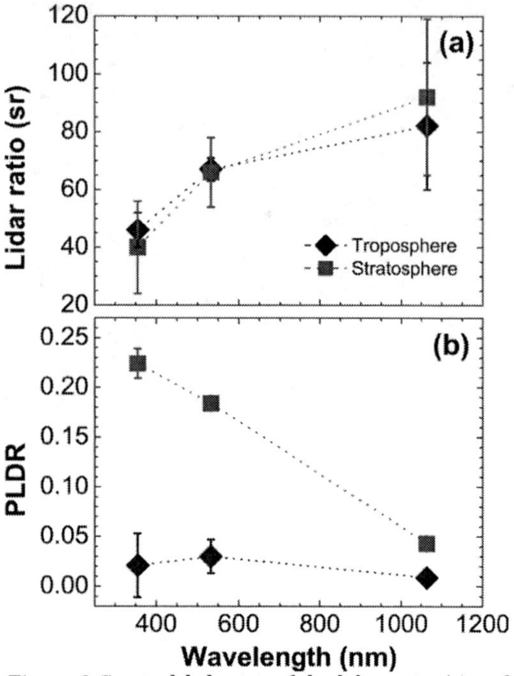

***Figure 2** Spectral behavior of the lidar ratio (a) and particle linear depolarization ratio (b) of smoke in the troposphere (black diamonds, 5 – 6.5 km height) and stratosphere (red squares, 15 – 16 km) over Leipzig on 22 August 2017.*

might be possible explanations for the less spherical smoke particles in the stratosphere.

The spectral behavior of the optical properties benefiting from the measurements at 1064 nm indicated the absence of coarse mode particles (radius > 500 nm). The particles must be small compared to the wavelength of 1064 nm. Lidar inversion calculations led to an effective radius of 0.17±0.06 µm (troposphere) and 0.32±0.10 µm (stratosphere) [6].

3.2 Evolution of the smoke layer with time

The injection of significant amounts of wildfire smoke into the stratosphere was a phenomenon of intercontinental scale. In August 2017 the smoke was released in Western Canada and traveled via Greenland to Europe. This process was studied from space with CALIPSO (Cloud-Aerosol Lidar and Infrared Pathfinder Satellite Observation) [5]. Most of the EARLINET stations between Norway and Crete observed the stratospheric smoke layer in late August and early September 2017 (Baars et al., 2019 in prep).

The stratospheric smoke layers were a long-lasting phenomenon. The measurements with a Polly^{XT} at Cyprus revealed that the smoke layers were present in the stratosphere for more than 5 months till January 2018. Figure 3 shows the vertical extend of the smoke layers over Limassol, Cyprus. Color-coded are the backscatter coefficient and the depolarization ratio.

In September 2017, the stratospheric smoke passed Limassol in several layers between 17 and 23 km height. These layers were characterized by a high depolarization ratio (> 16% at 532 nm) and a strong backscatter coefficient (Fig. 3). The smoke layers got diluted with time leading to a lower backscatter coefficient at a larger vertically

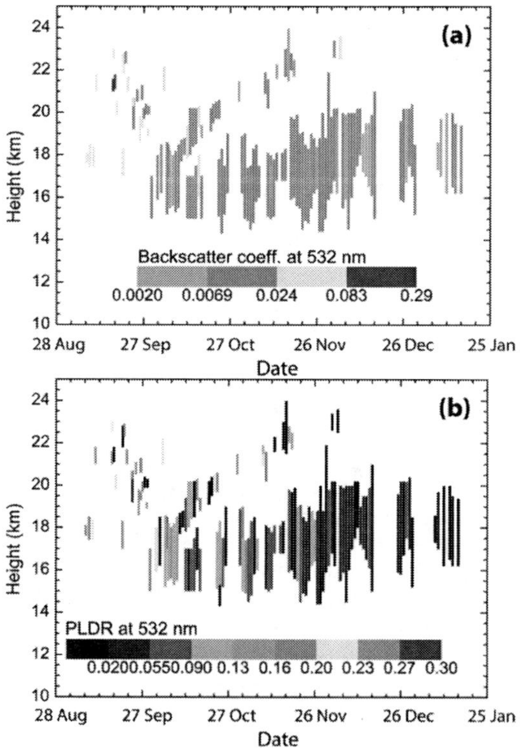

***Figure 3** The stratospheric smoke layer over Cyprus between August 2017 and January 2018. The vertical extend is marked with the bars, color coded with the particle backscatter coefficient (a) on a logarithmic scale and the particle linear depolarization ratio at 532 nm (b) on a linear scale.*

extend. The decrease of the particle depolarization ratio with time (Fig. 3b) is not easy to explain. Either the smoke particle got more spherical or their size decreased with time. Both, spherical particles and small particles cause low depolarization ratios. Or the later arriving smoke particles have a different history of emission and transport (slower transport speed).

Between September and November 2017, part of the stratospheric smoke increased in height up to 24 km. Compared to the tropospheric smoke which was washed out after a few weeks, the stratospheric smoke stayed for more than 5 months in the atmosphere and contributed significantly to the stratospheric aerosol load.

4. CONCLUSION AND OUTLOOK

Unique 3+3+3 lidar observations could be performed during the record-breaking stratospheric smoke event occurring over Europe in August 2017. The extension of the spectral range towards the near infrared (1064 nm) in terms of extinction and depolarization ratio measurements adds valuable information about the particle size and shape characteristics.

The observations at Cyprus show the advantage of continuous lidar measurements over longer time periods. The temporal evolution of the stratospheric smoke layers over the Eastern Mediterranean revealed that the smoke stays for several months in the stratosphere [17].

Wildfire smoke can have a major contribution to the stratospheric aerosol load in the same order of magnitude or even stronger than volcanic aerosol. The radiative fluxes, chemical processes, and even heterogeneous ice formation in the upper troposphere were influenced over a period of weeks to several months.

REFERENCES

[1] M. Fromm et al., B. Am. Meteorol. Soc., 91, 1193–1209 (2010)
[2] I. Mattis et al., J. Geophys. Res., 113, D20202 (2010)
[3] D. Müller et al., J. Geophys. Res., 110, D17201 (2005)
[4] A. Ansmann et al., Atmos. Chem. Phys. 18, 11831–11845 (2018)
[5] S.M. Khaykin et al., Geophys. Res. Lett., 45, 1639–1646 (2018)
[6] M. Haarig et al., Atmos. Chem. Phys. 18, 11847–11861 (2018)
[7] M. Haarig et al., Atmos. Chem. Phys. 17, 10767–10794 (2017)
[8] M. Haarig et al., Atmos. Meas. Tech. 9, 4269–4278 (2016)
[9] R. Engelmann et al., EPJ Web of Conferences 176, 01004 (2018)
[10] M. Haarig et al., EPJ Web of Conferences 119, 18009 (2016)
[11] R. Engelmann et al., Atmos. Meas. Tech. 9, 1767–1784 (2016)
[12] H. Baars et al., Atmos. Chem. Phys. 16, 5111–5137 (2016)
[13] M.-H. Kim et al., Atmos. Meas. Tech., 11, 6107–6135 (2018)
[14] L. Janicka et al., Atmos. Env., 169, 162–174 (2017)
[15] Q. Hu et al., Atmos. Chem. Phys., 19, 1173–1193 (2019)
[16] S.P. Burton et al., Atmos. Chem. Phys., 15, 13453–13473 (2015)
[17] H. Baars et al., Atmos. Chem. Phys., 19, 15183–15198 (2019)

 https://doi.org/10.1051/epjconf/202023703001

A RAMAN LIDAR WITH A DEEP ULTRAVIOLET LASER FOR CONTINUOUS WATER VAPOR PROFILING IN THE ATMOSPHERIC BOUNDARY LAYER

Masanori Yabuki[1*], Yuya Kawano[2], Yusaku Tottori[2], Makoto Tsukamoto[2], Eiji Takeuchi[2], and Toshitaka Tsuda[1]

[1]*Research Institute for Sustainable Humanosphere, Kyoto University, Uji, Kyoto 611-0011, Japan*
[2] *EKO Instruments Co., Ltd., Shibuya-ku, Tokyo 151-0072, Japan*
**Email: yabuki@rish.kyoto-u.ac.jp*

ABSTRACT

A Raman lidar with a deep ultraviolet laser was constructed to continuously monitor water vapor distributions in the atmospheric boundary layer for twenty-four hours. We employ a laser at a wavelength of 266 nm and detects the light separated into an elastic backscatter signal and vibrational Raman signals of oxygen, nitrogen, and water vapor. The lidar was encased in a temperature-controlled and vibration-isolated compact container, resistant to a variety of environmental conditions. Water vapor profile observations were made for twelve months from November 24, 2017, to November 29, 2018. These observations were compared with collocated radiosonde measurements for daytime and nighttime conditions.

1. INTRODUCTION

Water vapor, one of the most variable atmospheric constituents, plays an important role in atmospheric processes such as the atmospheric energy budget and the global water cycle. The distribution of water vapor is associated with that of clouds and rainfall through the vertical stability of the atmosphere caused by a large amount of latent heat related to the phase changes of water. Therefore, water vapor is a key parameter in understanding the localized extreme weather events associated with severe weather disasters such as torrential rain and floods. Information on the spatiotemporal distribution of water vapor is highly beneficial for improving the accuracy of weather forecasts made by mesoscale numerical weather prediction models.

The Raman lidar technique is a well-established tool for measuring the water vapor mixing ratio in the atmosphere[1]. While most of the Raman lidar previously used for the monitoring and field observation of water-vapor distribution employ a laser wavelength of 355 nm, Renaut et al. (1980, 1988) and Lazzarotto et al. (2001) demonstrated the Raman lidar with a 266 nm laser and a grating polychromator to conduct daytime observations during periods of high sky radiance[2,3]. The use of the deep ultraviolet wavelengths is convenient because of the low background noise during the daytime, since most of the solar radiation in the wavelength range below 300 nm is absorbed by the ozone layer in the stratosphere. Recently, the improved performance of UV optical components has led to the development of enhanced Raman lidar systems with a deep ultraviolet lidar. In this study, we constructed a Raman lidar by employing a laser at a wavelength of 266 nm and an interference-filter-based polychromator.

2. METHODOLOGY

2.1 Water vapor profiles from Raman lidar measurements

The water-vapor mixing ratio m at range R is obtained from the Raman backscatter signals of water vapor $P_{\lambda_{H_2O}}$, nitrogen $P_{\lambda_{N_2}}$, and oxygen $P_{\lambda_{O_2}}$ as follow:

$$m(R) = K_m \frac{P_{\lambda_{H_2O}}(R)}{P_{\lambda_{N_2}}(R)} \times \left[\frac{P_{\lambda_{O_2}}(R)}{P_{\lambda_{N_2}}(R)}\right]^{\frac{\sigma_{O_3}(\lambda_{H_2O})-\sigma_{O_3}(\lambda_{N_2})}{\sigma_{O_3}(\lambda_{N_2})-\sigma_{O_3}(\lambda_{O_2})}} \Delta T, \quad (1)$$

where K_m is the calibration constant and commonly evaluated by comparing the results of an independent measurement (e.g., radiosonde) of the water vapor mixing ratio; $\sigma_{O_3}(\lambda_x)$ is the ozone absorption cross section at Raman-shifted wavelength λ_x of species x; and ΔT indicates the difference in transmission of each Raman-shifted wavelength due to Rayleigh and Mie scattering processes.

2.2 System setup

The experimental setup of the Raman lidar is shown in Fig. 1. We employed the fourth harmonic output (266 nm) of an Nd:YAG laser (Continuum Surelite III-10, USA) and operated with the pulse energy of < 50 mJ. The beam divergence of the outgoing laser beam after passing the beam expander is 0.17 mrad. A custom-made Cassegrain telescope (Raymetrics, Greece) with a primary mirror diameter of 200 mm and a focal length of 1000 mm was used as the optical receiver. The backscattered radiation was separated by the dichroic beam splitters and the interference filters to elastic and inelastic wavelengths of 266 nm for the elastic scattering, and 277.5 nm, 283.6 nm, and 294.6 nm for the vibrational Raman scattering of oxygen, nitrogen and water vapor molecules, respectively. Three photomultiplier tubes (PMT) were used for detecting each backscatter light (R9880U, Hamamatsu, Japan). The data acquisition was performed using a four-channel transient recorder (TR 20-16bit, Licel, Germany) to record the analog and photon counts in the signal pulses. The lidar was encased in a temperature-controlled and vibration-isolated compact container with laser and quartz windows for transmitting the laser beam and the scattered light from the atmosphere, respectively.

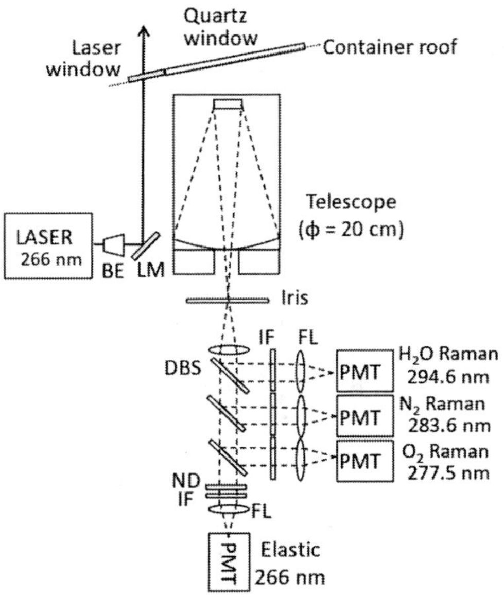

Figure 1. Schematic setup of water vapor Raman lidar. BE: beam expander, LM: laser mirror. FL: focusing lens; ND: neutral density filter; DBS: dichroic beam splitter; IF: interference filter; PMT: photomultiplier tube.

3. RESULTS

The Raman lidar container was installed at the Shigaraki MU radar observatory (34°51'N, 136°06'E, 385 m a.s.l.) of the Research Institute for Sustainable Humanosphere (RISH), Kyoto University, Japan, from November 24, 2017, to November 27, 2018. The MU radar observatory is located in Shigaraki in the central part of the Kinki Region of Japan and is surrounded by a mixed hardwood-coniferous forest.

The temporal variation of the calibration constant was evaluated by comparison with collocated measurements from a radiosonde that was launched 21 and 14 times during the daytime and nighttime, respectively, during the year-long lidar observation period. The calibration constants of both analog and photon counting signals indicated a monotonic variation within 3.5 % per year.

Water vapor mixing ratio profiles from the radiosonde and the Raman lidars for selected cases during nighttime and daytime are shown in Fig. 2. The trends shown by each observation agreed well up to about 1500 m, which is the top of the atmospheric boundary layer despite the low water vapor mixing ratio in winter. The height resolutions were determined in the range of 30–150 m every 15 minutes in the atmospheric boundary layer, considering statistical errors calculated from the signals.

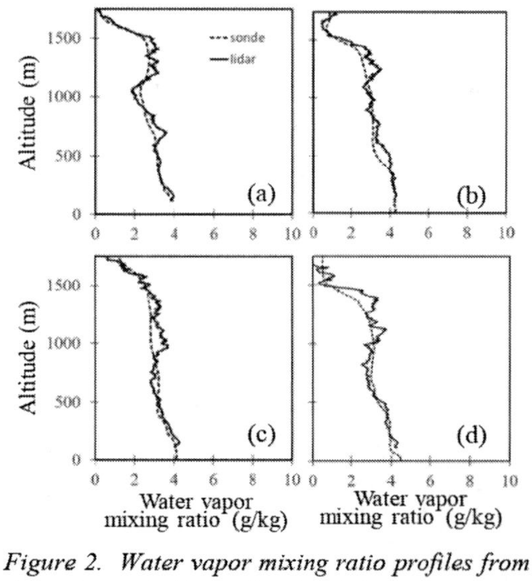

Figure 2. Water vapor mixing ratio profiles from the radiosonde (dashed line) and the Raman lidars (solid line) at Shigaraki, Japan at (a) 20:59–21:14 JST on December 22, 2017; (b) 01:15–01:30 JST; (c) 07:21–07:36 JST and (d) 10:41–10:56 JST on December 23, 2017. The range resolution of the Raman lidar changes in height from 30 m to 150 m, considering the Raman signal statistics.

Figure 3 shows time–height cross sections of the water vapor mixing ratio profiles observed by the Raman lidar. The lidar worked well throughout the twelve months under air temperature variations between −10 °C and 35 °C, except for the middle of the February and the beginning of the October during which time laser emission was stopped because of trouble with the air conditioner in the lidar container. In most cases where the observation range did not reach a height of 1500 m, there were optically thick clouds and fogs at the lower height level. Red arrows in Fig. 3 correspond to the time of replacing the flash lamp of the laser. The variations of the calibration constant as well as the derived water vapor profiles were small before and after replacing the flashlamp every 2–4 months.

4. CONCLUSIONS

We have constructed a Raman lidar for observing long-term monitoring of water vapor mixing ratio profiles in the atmospheric boundary layer for twenty-four hours. The Raman lidar employs a laser at a wavelength of 266 nm, characterized by the low background noise during daytime, and an interference-filter-based polychromator. The lidar was encased in a temperature-controlled and vibration-isolated compact container. We have demonstrated the potential of the developed lidar by collecting observations over a whole year at Shigaraki MU radar observatory in Japan. The lidar worked well throughout the twelve months. Water vapor mixing ratio profiles from the radiosonde and the Raman lidars agreed well up to about 1500 m which is the top of the atmospheric boundary layer. The variations of the calibration constant, as well as the derived water vapor profiles, were small before and after replacing the flashlamp every 2-4 months.

https://doi.org/10.1051/epjconf/202023703001

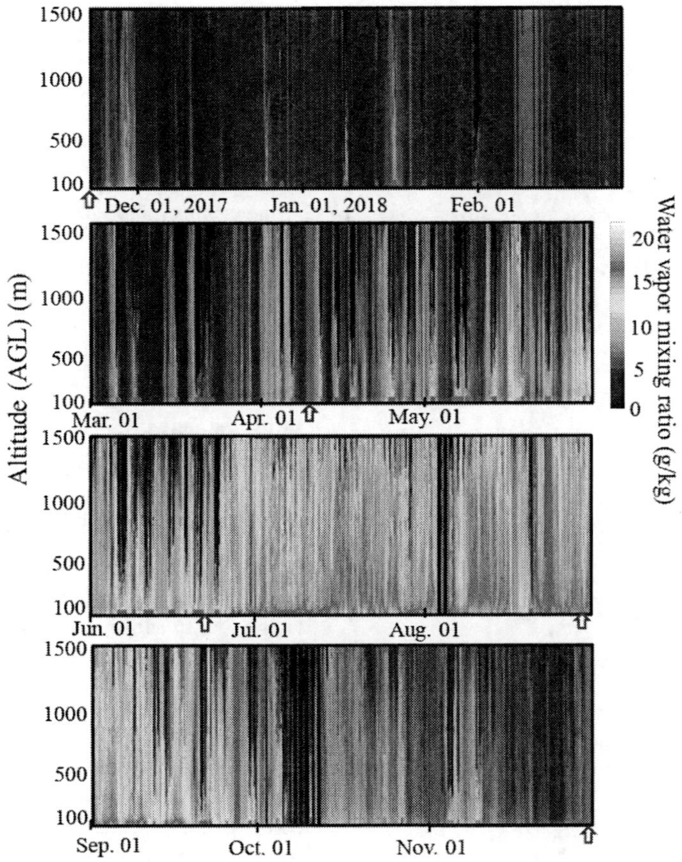

Figure 3. Time-height cross sections of the water vapor mixing ratio profiles observed by the Raman lidar at Shigaraki, Japan, from November 24, 2017, to November 29, 2018. Red arrows in x-axis correspond to the time of replacing the flash lamp of the laser.

ACKNOWLEDGEMENTS

A part of this study is supported by JSPS KAKENHI Grants 15H03724, 16H04999, and 26709042.

REFERENCES

[1] Whiteman, D., Melfi, S., and Ferrare, R., Raman lidar system for the measurement of water vapor and aerosols in the Earth's atmosphere, Appl. Opt., 31, 3068–3082 (1992).

[2] Renaut, D., Pourny, C., and Capitini, R., Daytime Raman lidar measurements of water vapor, Opt. Lett., 5, 232-235 (1980).

[3] Renault, D. and R. Capitini, Boundary-layer water vapor probing with a solar-blind Raman lidar: validations, meteorological observations and prospects, *J. Atmos. Oceanic Technol.*, 5(5), 585– 601 (1988).

[4] Lazzarotto, B., M. Frioud, G. Larchevêque, V. Mitev, P. Quaglia, V. Simeonov, A. Thompson, H. van den Bergh, and B. Calpini, Ozone and water-vapor measurements by Raman lidar in the planetary boundary layer: error sources and field measurements, Appl. Optics, 40 (18), 2985– 2997 (2001).

AEROSOL FIELD INFLUENCE ON THE RETRIEVAL OF THE OZONE VERTICAL COLUMN DENSITIES FROM PANDORA 2S MEASUREMENTS

Alexandru Dandocsi[1,2], Anca Nemuc[1*], Doina Nicolae[1], Livio Belegante[1], Alexander Cede[3], Martin Tiefengraber[3,4]

[1]*National Institute of Research and Development for Optolectronics INOE 2000, Remote Sensing Dept., Magurele-Bucharest, Romania*
[2]*University Politehnica of Bucharest, Faculty of Applied Sciences, Bucharest, Romania*
[3]*LuftBlick, Kreith 39A, 6162 Mutters, Austria*
[4]*Institute of Atmospheric and Cryospheric Sciences, University of Innsbruck, Innsbruck, Austria Romania*
**Email: anca@inoe.ro*

ABSTRACT

Total ozone and other trace gases are measured and reported regularly due to the increased interest started with the ozone hole discovery but the new satellites dedicated to worldwide observations of these species need both short-and long-term well calibrated ground based observation for validation procedures.

The ESA/NASA *Pandora* network established a sophisticated, automatic calibration procedure that utilizes a variety of narrow-line and broadband emission lamps with temperature control for their UV-Vis-NIR spectroradiometers. In this study, we describe additional calibration efforts for ozone retrievals.

In this paper we explore the local aerosol field influence on the retrieval of the ozone spectra from PANDORA 2S measurements using collocated lidar and sunphotometer measurements and proposed a methodology to be implemented in the calibration procedure of the instruments.

INTRODUCTION

Observations of ozone and trace gases during the last decades proved that anthropogenic activities do affect the environment on large scales and that the consequences are potentially very harmful to humans. But atmospheric processes are still far from being entirely understood and therefore there is a need for better, continuous and worldwide spread of both satellite and ground based observations.

Pandora instruments are field grade spectroscopic UV–Vis systems (Herman et al., 2009). They are part of the European Space Agency-sponsored Pandonia Global Network but also NASA (USA) sponsored network. The main goal of Pandonia is to provide consistent ground-based total O_3, NO_2, and HCHO columns for satellite validation, overlapping significantly with the spectral ranges of the S5p instrument as well as the MSI and ATLID on-board EarthCARE. The major advantages of networking are uniform instrument design and calibration, and centralized data monitoring, processing, and distribution. Direct-sun-observation geometry eliminates the need for atmospheric radiative transfer modeling and simplifies data interpretation (Tzourtiu, 2012). Pandonia currently operates 75 instruments and is expected to have about 300 instruments in the next 5 years. The quality assurance procedures are extremely important for satellite validation.

One issue not tackled by the comprehensive approach (Cede et al.) in the calibration fitting procedures is the influence of the aerosol local fields. Radiative effects of UV-absorbing aerosols produce spurious ozone column reductions. The magnitude of the total ozone underestimation depends on the aerosol type, optical depth, and aerosol layer height. The error is produced by complex interactions between ozone absorption (both stratospheric and tropospheric), aerosol scattering, and aerosol absorption. Tropospheric UV-absorbing aerosols such as dust and carbonaceous aerosols from biomass burning have been found to affect the retrieved ozone from radiance spectra.

At RADO (Romanian 3D Atmospheric Observatory) in Magurele, near Bucharest, Romania (26.029°E, 44.048°N, 93 m above sea level) several instruments for aerosol observations

 https://doi.org/10.1051/epjconf/202023703002

along with PANDORA 2S have been implemented during the last years and are performing regular measurements (Nicolae et al,2010). Therefore we took this opportunity and use lidar and sunphotometer measurements characterizing the vertical distribution and optical properties of aerosols.

In this paper we investigate how the retrieved ozone spectra from Pandora 2S measurements are influenced by the local aerosol fields and we propose a methodology for routinely using this by the stations from PANDONIA with collocated lidar and /or sunphotometer.

1 INSTRUMENTS AND METHODOLOGY

1.1 The lidar and sunphotometer

Advanced Lidar system with elastic and inelastic channels – such as our Multiwavelength Raman Lidar instrument -RALI– can be used into the inversion procedure to obtain all optical parameters of the aerosols. It emits and receives light at 1064, 532, and 355 nm. (Nicolae et al. 2010, Nemuc et al 2013)

Extinction profiles retrieved from the 355 nm and 532 nm channels (Figure 1) of the Lidar system are used to compute the aerosol optical depths by integrating the profile.

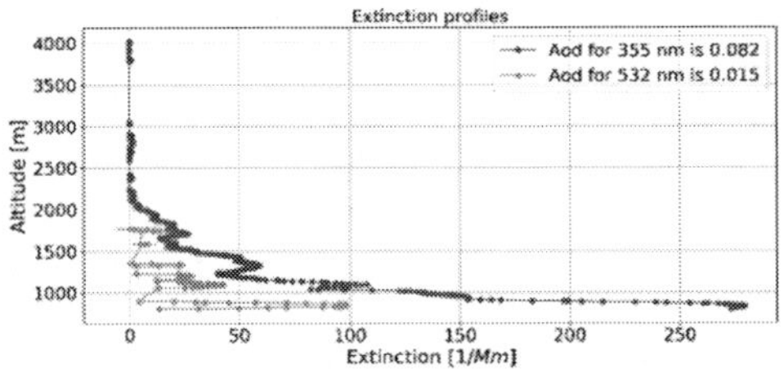

Figure 1 Extinction profiles from Lidar system

Column Aerosol properties are retrieved from automatic sun-photometer measurements (AERONET) at our site since 2007 (Carstea et al 2019). Standard measurements performed by the CIMEL CE-318 sun photometer are composed of solar spectral irradiance used to derive accurate spectral extinction AOD (±0.01) within the range 340 to 1640 nm as well as downward sky spectral radiances (440 to 1020 nm) in the solar principal plane and in the almucantar geometries. The instrument measures direct Sun with a 1.2° FOV at discrete channels from UV to NIR (i.e., 340, 380, 440, 500, 675, 870, 940, 1,020, and 1,640 nm). The direct-Sun measurements take about 10 s to scan all filter wheels in front of the detector for each wavelength. Data products used in this study were collected in cloud-free conditions, provided by the AERONET web site as "aerosol optical depth (V3) - solar" using the Smirnov et al. (2000) cloud-screening algorithm.

1.2 Pandora

A detailed description of the instrument has been provided by Herman et al. (2009). Pandora is a sun-viewing spectrometer that was initially developed for validation of the Ozone Monitoring Instrument (OMI) aboard the Aura satellite (Herman et al., 2009).

The instrument consists of a small Avantes low-stray-light spectrometer (280–525 nm with 0.6 nm spectral resolution with 5 times oversampling) and the second one (400-900 nm, 0.24nm step, 1.1nm resolution) connected to an optical head by 400 µm core diameter single-strand fiber optic cables. The optical head is attached to a small two-axis positioner, capable of accurate pointing to track the sun's center (±0.2°). A diffuser is included in the optical path to minimize the effect of small pointing errors. Direct-sun spectra are taken at variable integration times (2.5 ms to 4 s) with a total measurement duration of 40 s. Pandora measures unscattered solar photons in a narrow cone (2.1° field of view (FOV) full width at half maximum with a diffuser and 1.6° FOV without the diffuser) at a specific solar azimuth and zenith direction that changes from east in the morning to west in the evening.

Pandora spectra are automatically collected and submitted to LuftBlick servers for centralized uniform processing by the Blick Software Suite (Cede, 2018). All standard operational Pandora (instrument #111) data products are available live at http://live.pandonia.net/ and archived at http://pandonia.net/data (INOE site). (last access: 13 March 2019).

To achieve wavelength and radiometric calibration stability, the spectrometer is temperature stabilized inside an insulated enclosure using an actively coupled thermo-electric cooler and heater.

Total vertical column of O_3 is retrieved using direct sun measurements. As the reference spectrum, in this case, it is used an extraterrestrial spectrum from another source (i.e. not measured by the Pandora system), convoluted with the Pandora filter function. The reference spectrum is obtained in a similar way as described in Bernhard et al., 2004.

Algorithm and calibration related to Pandora Total Ozone Spectral Fitting Algorithm have been described in details by Tzortziou et.al. 2012.

Spectral fitting-data presented in this paper are from a 310 to 330 nm fitting window.

Briefly, spectral fitting is performed using laboratory measured absorption cross sections and implement shift squeeze functions to fit the observed spectra with the solar reference spectrum's Fraunhofer line structure (Figure 2) and spectrum from Ring effect is also considered. The fourth-order polynomial is applied for removal of aerosol and Rayleigh scattering effects.

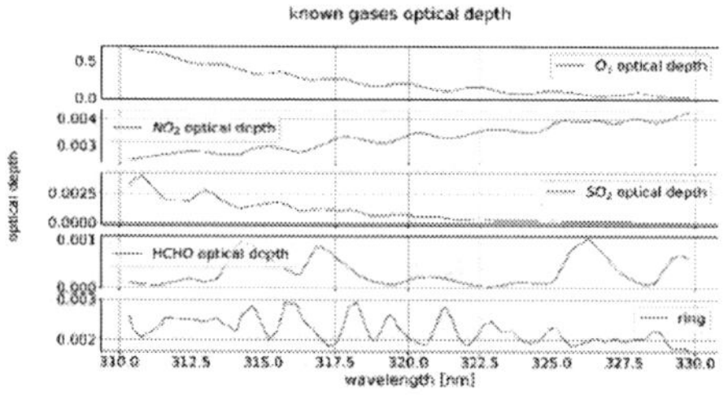

Figure 2 Known optical depths of species fitted for the retrieval of total vertical column of O_3

1.3 Methodology

During the first step of our proposed procedure the daily mean values of aerosol optical depths from Lidar and sunphotometer systems are evaluated. The Angstrom exponent is calculated from the 532 and 355 or 500 and 340 channels from the Lidar and sunphotometer respectively. Using the Lambert-Beer law, the aerosol optical depths for spectral range of 310 – 340 nm is extrapolated. Example of correction of the reference spectrum is shown in (Figure 3)

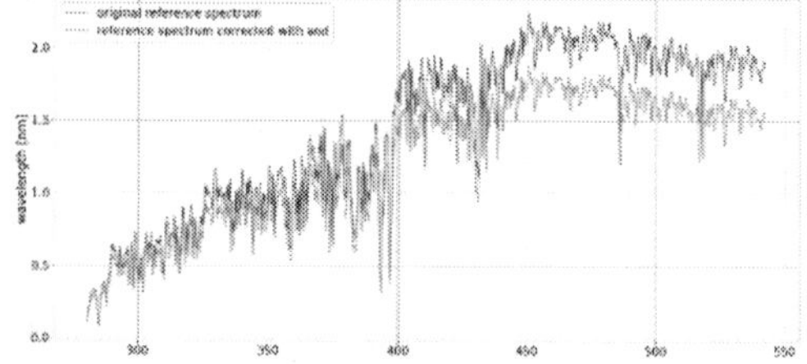

Figure 3 Reference spectrum modified based on aod from Lidar system

The vertical total column densities are evaluated using both original and corrected with the AOD spectra for several cases.

2. RESULTS

All analyzed cases were choose to be under cloud free conditions. Differences between total column densities retrieved using the original and modified reference spectrum ranges from 0.5% for lower values of AOD up to 5% for the cases where the calculated Angstrom exponent was higher.

An example of time series using both reference spectra is shown in Figure 4.

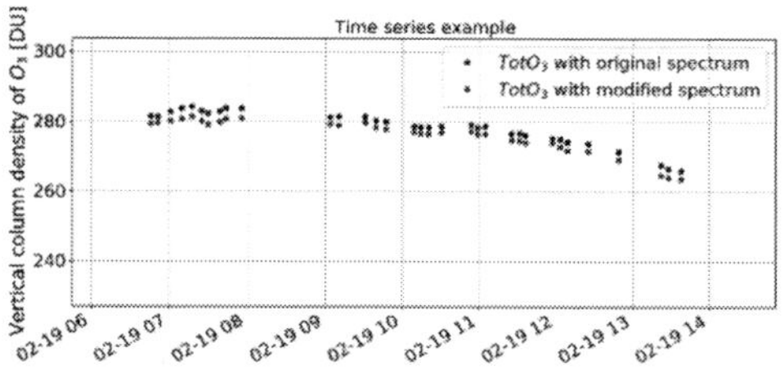

Figure 4 Example of vertical column density of O3 retrieved with original and modified reference spectrum

In this case the differences vary between 0.8% and 1.4%. Data points in Figure 4 are considered measurements with high confidence based on spectral fitting and measured spectra.

3. CONCLUSION

Aerosol information can be implemented in the retrieval of vertical column densities of trace gases based on differential optical absorption spectroscopy method.

In this paper the authors have implemented in the processing chain of the measured spectra from direct sun measurements the possibility of correcting the reference spectrum used in the retrieval of column density of O_3,. Collocated in space and time measurements of AOD have been considered.

For the analysed cases, differences using the Angstrom exponent to correct the reference spectrum ranges from 0.5% to 5% in the final product, the vertical column of O_3 expressed in Dobson Units.

Next steps will be dedicated to long term comparisons and also implementation and evaluation at other PANDONIA sites.

Through the implementation of this type of methodologies the well-calibrated ground-based PANDORA 2S will comprehensively complement spaceborne measurements in spectral and temporal domains.

ACKNOWLEDGEMENTS

The work performed for this study was funded by the Ministry of Research and Innovation through Program I - Development of the national research-development system, Subprogram 1.2 - Institutional Performance - Projects of Excellence Financing in RDI, Contract No.19PFE/ 17.10.2018 and by Romanian National Core Program Contract No.18N/2019. Also part of this study was supported through ASSESS (Atmospheric studies in support of ESA's sentinel 4 and 5 products) ROSA-STAR project, contract 150/2017

REFERENCES

[1] Cede, A.: Manual for Blick Software Suite 1.6, available at: http://pandonia.net/media/documents/BlickSoftwareSuite_Manual_v11.pdf (last access: 12 March 2019), (2018)

[2] Herman, J., A. Cede, E. Spinei, G. Mount, M. Tzortziou, and N. Abuhassan, NO2 column amounts from groundbased Pandora and MFDOAS spectrometers using the direct-sun DOAS technique: intercomparisons and application to OMI validation, J. Geophys. Res., 114, D13307, doi:10.1029/2009JD011848, 2009

[3] Herman, J., Evans, R., Cede, A., Abuhassan, N., Petropavlovskikh, I., and McConville, G.: Comparison of ozone retrievals from the Pandora spectrometer system and Dobson spectrophotometer in Boulder, Colorado, Atmos. Meas. Tech., 8, 3407-3418, https://doi.org/10.5194/amt-8-3407-2015, 2015.

[4] Tzortziou, M., J. R. Herman, A. Cede, and N. Abuhassan, High precision, absolute total column ozone measurements from the Pandora spectrometer system: Comparisons with data from a Brewer double monochromator and Aura OMI, J. Geophys. Res., 117, D16303, doi:10.1029/2012JD017814, 2012

[5] A. Smirnov, B. N. Holben, T. F. Eck, O. Dubovik,and I. Slutsker, Cloud-Screening and Quality Control Algorithms for the AERONET Database, REMOTE SENS. ENVIRON. 73:337–349, 2000

[6] D. Nicolae, J. Vasilescu, E.Carstea, K. Stebel, F. Prata, Romanian Atmospheric research 3D Observatory: Synergy of instruments, Romanian Reports in Physics 62/4, 838-853, 2010

[7] Nemuc, A., Vasilescu, J., Talianu, C., Belegante, L., and Nicolae, D.: Assessment of aerosol's mass concentrations from measured linear particle depolarization ratio (vertically resolved) and simulations, Atmos. Meas. Tech., 6, 3243-3255, https://doi.org/10.5194/amt-6-3243-2013, 2013

[8] Carstea, E., Fragkos, K., Siomos, N. et al., Columnar aerosol measurements in a continental southeastern Europe site: climatology and trends, Theor Appl Climatol. DOI: 10.1007/s00704-019-02805-z, 2019

[9] G. Bernhard, C. R. Booth, and J. C. Ehramjian. Version 2 data of the National Science Foundation's ultraviolet radiation monitoring network: South Pole. Journal of Geophysical Research (Atmospheres), 109(D21), 2004.

ATMOSPHERIC MERCURY IN CHINA STUDIED WITH DIFFERENTIAL ABSORPTION LIDAR

Guangyu Zhao[1], Zheng Duan[1], Ming Lian[1], and Sune Svanberg[1,2*]

[1]Center for Optical and Electromagnetic Research, South China Academy of Advanced Optoelectronics, South China Normal University, University City Campus, Guangzhou 510006, China

[2]Department of Physics, Lund University, SE 221 00 Lund, Sweden
**Email: sune.svanberg@fysik.lth.se*

ABSTRACT

Mercury is a very serious environmental pollutant which is intensely studied by many researchers. China is a main producer and consumer of mercury. We have pursued extensive monitoring of atomic mercury using the differential absorption lidar (DIAL) technique, based on a newly constructed mobile monitoring system. Mercury levels in major cities (Guangzhou, Zhengzhou, Xi´an) were measured, as well as in mercury mining areas in Wanshan, Guizhou province. Further, mercury emissions from the Emperor Qin Mausoleum in Xi´an were investigated together with archaeologists.

1. INTRODUCTION

Mercury is a very serious heavy metal pollutant, which is presently subject to strict international regulations [1]. All aspects of mercury in the environment are studied by a quite large community of researchers. China is a major producer and consumer of mercury [2]. Emissions of mercury are due to coal burning, metal processing, mining activities etc. An important aspect of mercury management is to characterize sources and quantify emissions. Mercury is also an interesting geophysical tracer gas, a fact which is related to its very high vapor pressure even at normal ambient temperatures [3]. Atmospheric mercury is predominantly in elemental, atomic form, making the differential absorption lidar (DIAL) method very sensitive and suitable for mercury mapping. The sensitivity is approaching the level of the Atlantic background, which is of the order on 2 ng/m^3. Mercury DIAL monitoring was pioneered by the Lund University lidar group in Sweden, and is now also available in China after the development of a mobile measurement system by the South China Normal University (SCNU) lidar group [4].

2. METHODOLOGY

The SCNU mobile laser spectroscopy laboratory, incorporating DIAL capability, is shown in Fig. 1, on site in the Wanshan mining area in the Guizhou province, China. The system incorporates a 20 Hz high-energy pulsed Nd:YAG laser, the 532 nm and 355 nm harmonics of which are used for pumping a narrow-band Sirah dye laser, which for the case of the atomic mercury resonance line close to 254 nm is operating with the dye Coumarine 307 at 508 nm, followed by frequency doubling. A vertically-looking 40 cm diameter Newtonian telescope is used in combination with a large roof-top folding mirror for the collection of back-scattered radiation. Recordings for two close-lying on- and off-resonance wavelengths with switching between every laser shot, is used for evaluation of the range-resolved mercury concentration profile, as illustrated in Fig. 2. Concentration profiles are evaluated from the DIAL curve, formed as the ratio between the on- and off-resonance recordings, as described in [5] and illustrated in the figure. We note, that a high slope on the DIAL

Fig. 1. The SCNU mobile differential absorption lidar system on site in the Wanshan mining area of China for atomic mercury monitoring.

https://doi.org/10.1051/epjconf/202023703003

curve corresponds to a high concentration value. A detailed description of the mobile laboratory and its capabilities is given in [4].

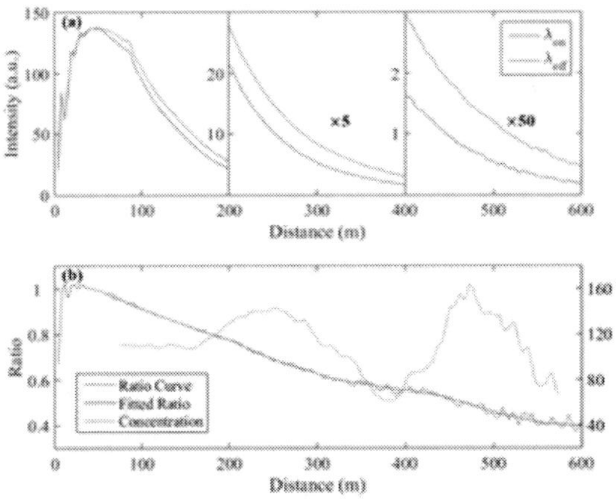

Fig. 2. a) Lidar recordings of atomic mercury for transmitted on- and off-resonance wavelengths. b) Divided DIAL curve, from which the range-resolved concentration profile is derived [6].

We have used the mercury monitoring capability in a number of projects on the Chinese scene. These include monitoring in the major cities of Guangzhou [7], Zhengzhou [8] and Xi´an [9].

Particularly polluted areas are found in connection to the mining areas in the Guizhou province, in particular in the mercury mining areas in Wanshan [10,11]. Although official mining has been closed down in 2002, artisanal mercury extraction from the cinnabar ore in still ingoing, and there are large deposits of roasted ore, still containing high concentrations of metal mercury. We have pursued a mercury monitoring campaign together with the Chinese Academy of Sciences Institute for Geochemistry, Guiyang [6].

As mentioned, mercury is an interesting geophysical tracer gas [3]. Emission are related to geothermal energy, volcanic activity and ore deposits. In ancient China, it seems mercury, at the time considered as an elexir of life, was deposited in tombs of high officials as a gift of the emperor [12]. We have performed an exploratory atomic mercury investigation at the maosoleum of the "Terracotta Army" emperor Qin in Xi´an [13].

3. RESULTS

We found, that the mercury concentration in the air of the major cities in China, where we performed DIAL measurements was in the range 5-10 ng/m^3, with occasionally higher values, especially during the very hot summer days in Xi´an.

In the mining areas in Guizhou we observed very high concentrations of atomic mercury, in spite of the temperature during the campaign being modest, around 7-14 degrees. The recording in Fig. 2 from the Wanshan area reaches concentrations of 150 ng/m^3. An overview of the measurement area in Wanshan is given in Fig. 3. Examples of vertical scans in the three measurement directions indicated in Fig. 2 a) are shown in Fig. 4. The highest concentrations, observed in the Fig. 4a) are associated with air flow from the Longtangping mercury plant.

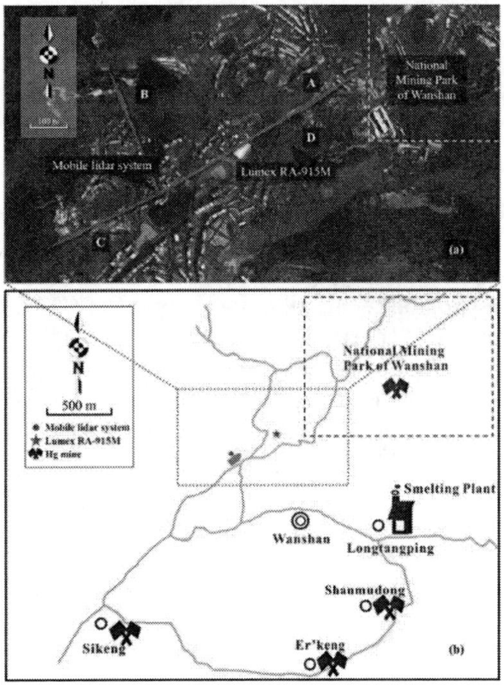

Fig. 3. Overview of the Wanshan area, with mining sites and the location of a cinnabar roasting plant indicated [6].

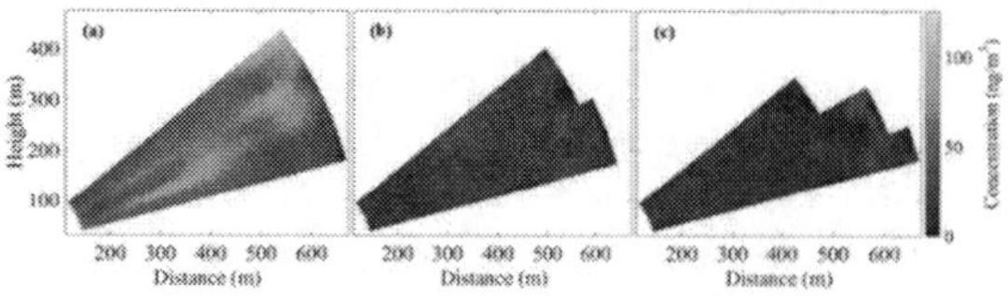

Fig. 4. a) Vertical scans recorded in the three directions indicated in Fig. 3a). Concentrations reaching 100 ng/m^3 are observed in a direction intersecting the air flow from a mercury processing plant [6].

We performed an intercalibration between our lidar system and a Lumex Zeeman modulation mercury instrument [6]. The lidar data were found to be 9 percent higher than the Lumex data, possibly indicating that the DIAL cross section, for which no full consensus exists, should be correspondingly adjusted.

A photo from our measurements at the Emperor Qin mausoleum in Lintong, 30 km from the city of Xi´an is shown in Fig. 5. Three different locations were chosen for the lidar system, which performed scans over the slopes of the mound, with the laser beam positioned about 5 meters above the top of the vegetation, covering the mound. The results of our campaign are reported in Ref. [13].

ACKNOWLEDGEMENTS

The authors gratefully acknowledge the support of Prof. Sailing He. A fruitful cooperation with further colleagues as SCNU, and at Henan Agricultural University, Zhengzhou, and the Institute of Geochemistry, Chinese Academy of Sciences, Guiyang, is much appreciated. This work was financially supported by a Guangdong Province Innovation Research Team Program under Grant No.201001D0104799318.

REFERENCES

[1] United Nations, Minamata convention on mercury. (2013)

[2] D. G. Streets, et al. Atmospheric Environment 39(40): 7789-7806 (2005)

[3] S. Svanberg, Optics and Lasers in Engineering 37, 245-266 (2002)

[4] G. Y. Zhao, et al. Applied Optics 56(5): 1506-1516 (2017)

[5] S. Svanberg, in Air Monitoring by Spectroscopic Techniques, (Eds: M. Sigrist), Wiley, N.Y. 1994, pp 85-161

[6] M. Lian, et al. Environmental Pollution 240: 353-358 (2018)

[7] G. Y. Zhao, et al. PIERS Guangzhou 2014, 2711-2714 (2014)

[8] M. Lian, G. Y. Zhao, Z. Duan, Y. Y. Li, S. Svanberg and J. D. Hu, Improvement of a Mobile Differential Absorption Lidar System and Measurements of Atmospheric Mercury in the City of Zhengzhou, accepted by Journal of Optoelectronics Lasers (2019)

[9] Z. Duan, G. Y. Zhao, S. M. Zhu, M. Lian, Y. Y. Li, W. X. Zhang, and S. Svanberg, Atmospheric mercury pollution in the Xi´an area, China (to appear)

[10] Z. H. Dai, et al. Atmospheric Chemistry and Physics 12: 6207–6218 (2012)

[11] G. L. Qiu, et al. Journal of Agricultural and Food Chemistry 56(7): 2465-2468 (2008)

[12] H. Z. Liu, et al. Geophysical and Geochemical Exploration 37(4): 670-674 (2013)

[13] G. Y. Zhao, W. X. Zhang, Z. Duan, M. Lian, N. B. Hou, Y. Y. Li, S. M. Zhu, and S. Svanberg, Mercury Emission from the Emperor Qin Tomb in Xi´an Studied by Differential Absorption Lidar, submitted (2019)

Fig. 5. Photograph from one of the lidar system measurement sites at the Emperor Qin Mausoleum area close to Xi´an.

AUTOMATIC ALERT SYSTEM FOR TROPOSPHERIC PARTICULATE POLLUTION MONITORING

Mariana Adam[1*], Konstantinos Fragkos[1], Ioannis Binietoglou[1]

[1]National Institute of Research and Development for Optoelectronics – INOE 2000, Magurele, Ilfov 077125, Romania

**Email: mariana.adam@inoe.ro*

ABSTRACT

An automatic alert system is implemented in order to detect atmospheric pollution layers and establish the degree of pollution load. When such layers are heavily loaded, automatic alerts are sent towards designed scientists in order to further analyze the event and inform the relevant stakeholders. The alert system is based on continuous measurements taken by a ceilometer supplemented by a photometer. Backtrajectory model is used to establish the pollution source.

1. INTRODUCTION

Atmospheric particulate pollution can impact human health [1] and safety, and to a larger extent, the biosphere. Heavily loaded pollution layers in troposphere can affect aviation [2], and when reaching the ground, they can diminish air quality [1], while they also affect agriculture areas or important infrastructure [3].

Within this context, we pursued to implement an automated system for pollution layers detection in near real time (NRT) where the degree of pollution is evaluated. When the pollution is high, we issue automated alerts towards designated scientists. The scientist can then perform dedicated measurements to gather additional information to get a thorough image of the pollution event. If the final assessment reveals a strong pollution event, alerts will be sent towards the national bodies in charge with emergency situations. Here we present the main methodology and the implementation of the system for Bucharest, Romania.

2. METHODOLOGY

2.1 Instruments

This methodology is based on measurements by continuously operating instruments, namely a CHM15k ceilometer from Lufft [4] and a CE318 photometer from Cimel [5]. In addition, Hysplit [6] back trajectory analysis is performed when strong pollution layer is detected, in order to gather information about the source of the event.

2.1 Procedure

The main steps of the procedure are the following:

- Ceilometer data are processed every 15 minutes. Time-height plots (quicklooks) based on range corrected signal (RCS), displaying the last 24h, are updated. Example images are shown on our website (http://liverali.inoe.ro/Ceilometer/). Furthermore, the lofted aerosol layers are detected. Currently, the algorithm developed by Lufft is used to identify up to three layers of aerosol layer height (ALH). Note that the algorithm does not perform above ~ 4000 m. An algorithm to evaluate ALH over the whole troposphere is under development in our department.
- When an aerosol pollution layer is detected by ceilometer over the last 15min, above 2500m, the system downloads the NRT Cimel sun photometer data to further characterize the aerosol situation. Specifically, aerosol optical depth (AOD) data and Spectral Deconvolution Algorithm (SDA) files are used. The current values of AOD, Ångström exponent (AE) at 440 nm / 870 nm, fine mode fraction (FMF) / course mode fraction (CMF) at 550 nm and the AOD for FMF/CMF at 550 nm are compared with the climatological values (monthly means) at our station [7]. The current photometer data refers to Version 3 level 1.5, which are cloud-screened and quality controlled and are provided by Aeronet [8].
- If one of the parameters exceed 90th percentile of the climatological values, the Hysplit backtrajectory is performed over 10 days. In addition, AE bellow 10th percentile is considered as well.

Specifically, the criteria to define the values outside the climatological range are:

- AOD > 90th percentile
- AE < 10th percentile or > 90th percentile
- AOD FMF, AOD CMF > 90th percentiles
- FMF > 90th percentile

A document is created containing overview information of the detected event, in order to help scientists quickly evaluate all available information. The document contains the RCS from ceilometer with the location of the ALH, the photometer values along with the monthly means from climatology and the Hysplit backtrajectory. An email is sent to designated scientists who will analyze the results. Also, the events are stores in a log file (contains the same information as that issued by email).

3. RESULTS

3.1 Current alert system

The implementation of the algorithm started in February 2018. The Hysplit feature was added at a later stage.

The following alerts types are issued:

a) When there is no photometer data, the aerosol load can not be evaluated through the optical properties given by photometer. An alert is sent by email where it is mentioned the presence of the ALH, while there is no photometer data available.

b) When the optical properties are evaluated w.r.t. climatological values but the current values are within expected limits.

c) When the optical properties are evaluated w.r.t. climatological values and the current values are outside the expected limits. A document is created showing the results.

An example of the information included in the document is shown in Figs 1-3. In Fig. 1 (slide 1) we observe that the current values (shown by black squares) are below the 90th percentiles (shown by filled triangles). However, AE is above the 90th percentile. Similar, the FMF and CMF AODs are below 90th percentile (Fig. 2). FMF is slightly larger than the 90th percentile. Figure 3 shows the RCS with overlapped ALH (also shown RCS with overlapped cloud base height – CBH). We can observe that in the last 15 min there are ALH (IInd or IIIrd layers) above 2500 m. The right plot shows the Hysplit backtrajectory. We observe that the backtrajectory reaches the ground level (shown by the black curve on the lower panel) after ~30 h back in time, close to Pristina in Kosovo. Apart from these images, the document includes numerical values of the parameters which are outside climatological values (in this case, AE). Large values of AE correspond to a small mean size of the aerosol and consequently to a larger FMF and AOD FMF. Small particles are associated with pollution due to smoke fires for example.

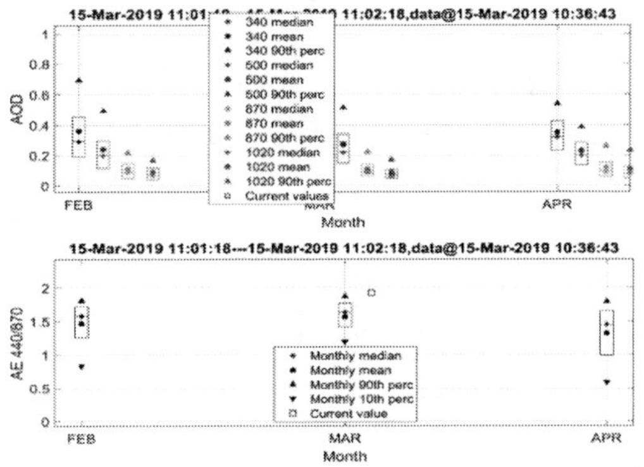

Fig. 1. AOD and AE (current values versus climatological monthly values).

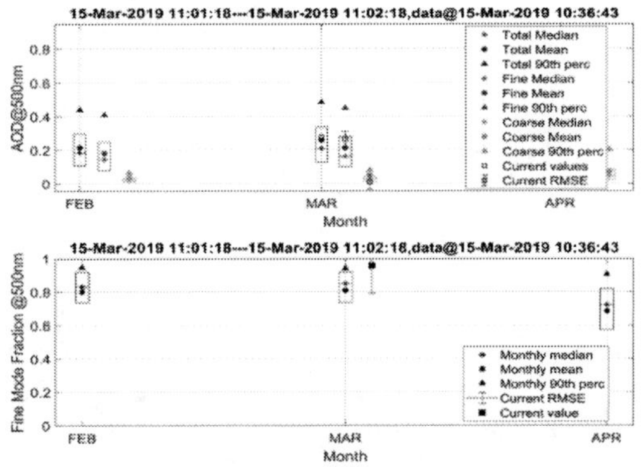

Fig. 2. AOD FMF and AOD CMF, FMF at 500nm.

A quick search in FIRMS [9] shows the fires which occurred 48h before. Figure 4 shows a screen shot of the area.

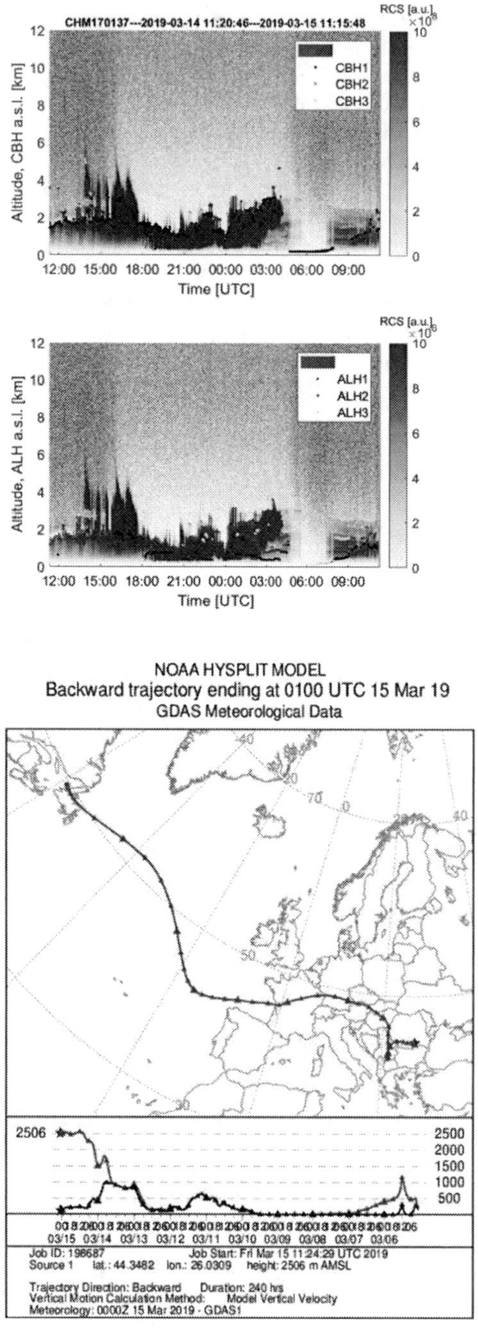

Fig. 3. RCS with overlapped CBH and ALH (upper) and Hysplit backtrajectory (lower).

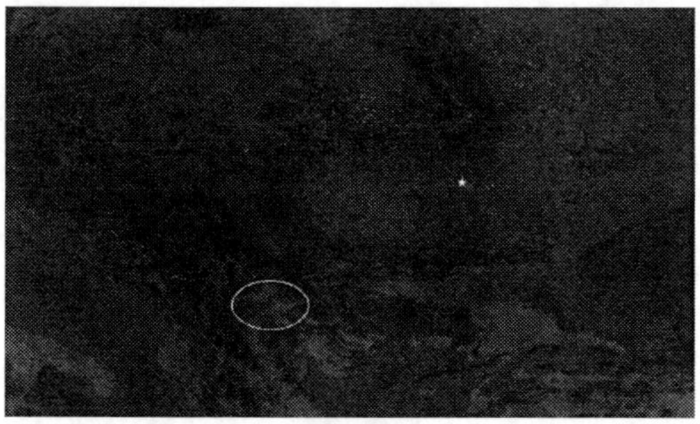

Fig. 4. Screenshot of the fires seen by MODIS 48h before the event run by Hysplit. The star shows INOE location and the eclipse shows the region with fires along backtrajectory.

CONCLUSIONS

The current alert system is implemented at INOE 2000 in Magurele (Romania). It is based on continuous measurements taken by a CHM15k ceilometer and a CE318 photometer. Climatological values (based on ten years measurements) of the aerosol optical properties revealed by photometer are used to determine the degree of the current pollution. Thus, alerts are sent when optical properties are above the 90th percentile of the climatological values plus value of AE below 10th percentile. Hysplit backtrajectory is performed to determine the pollution source.

Limitations:

- The algorithm works when photometer data is available (cloud free). Currently, only the day time measurements are used.
- ALH is determined by the manufacturer's algorithm. It focuses on ALH below ~ 4000m.

Further implementations:

- Use day and night photometer data. Use of Photons retrievals are envisaged [10].
- In-house algorithm to determine ALH in whole troposphere.

ACKNOWLEDGEMENTS

This work was supported by the Romanian National contracts 18N/08.02.2019 and 19PFE/17.10.2018.

REFERENCES

[1] V. Amiridis, et al., Atmos. Environ., 46, 536-544, (2012)
[2] J. R. Clarkson, et al., J. Aerospace Engineering, 230, 2274-2291 (2016)
[3] H. Craig, et al. J. Appl. Volcanol. 5:7, 2-31 (2016). (http://appliedvolc.springeropen.com/articles/10.1186/s13617-016-0046-1).
[4] https://www.lufft.com/products/cloud-height-snow-depth-sensors-288/lufft-ceilometer-chm8k-2405/
[5] https://www.cimel.fr/?instrument=sun-sky-lunar-multiband-photometer&lang=en
[6] A. F. Stein, et al., Bull. Amer. Meteor. Soc., **96**, 2059-2077, http://dx.doi.org/10.1175/BAMS-D-14-00110.1 (2015)
[7] E. Carstea, et al, Theoretical and Applied Climatology, https://doi.org/10.1007/s00704-019-02805-z (2019)
[8] O'Neill, N. T., Eck, T. F., Smirnov, A., Holben, B. N., and Thulasiraman, S., *J. Geophys. Res.*, 108, 4559, doi:10.1029/2002JD002975, (2003)
[9] https://firms.modaps.eosdis.nasa.gov/map/#z:7;c:26.7,43.9;d:2019-03-13..2019-03-15
[10] http://loaphotons.univ-lille1.fr/photons/data_monitor/AE/scp-week3-triple.php

https://doi.org/10.1051/epjconf/202023703005

CH_4 AND CO_2 IPDA LIDAR MEASUREMENTS DURING THE COMET 2018 AIRBORNE FIELD CAMPAIGN

Andreas Fix[1*], Axel Amediek[1], Christian Büdenbender[1], Gerhard Ehret[1], Christoph Kiemle[1], Mathieu Quatrevalet[1], Martin Wirth[1], Sebastian Wolff [1], Heinrich Bovensmann[2], André Butz[3], Michał Gałkowski[4], Christoph Gerbig[4], Patrick Jöckel[1], Julia Marshall[4], Jarosław Nęcki[5], Klaus Pfeilsticker[3], Anke Roiger[1], Justyna Swolkień[5], Martin Zöger[6], and the CoMet team

[1] *German Aerospace Center (DLR), Institute of Atmospheric Physics, Oberpfaffenhofen, Germany*
2 University of Bremen, Institute of Environmental Physics, Bremen, Germany
[3] *University of Heidelberg, Institute of Environmental Physics, Heidelberg, Germany*
[4] *Max Planck Institute for Biogeochemistry, Jena, Germany*
[5] *AGH University of Science and Technology, Kraków, Poland*
[6] *German Aerospace Center (DLR), Flight Experiments, Oberpfaffenhofen, Germany*
**Email: andreas.fix@dlr.de*

ABSTRACT

Installed onboard the German research aircraft HALO, the integrated-path differential-absorption (IPDA) lidar CHARM-F measures weighted vertical columns of both greenhouse gases (GHG) below the aircraft and along its flight track, aiming at high accuracy and precision. Results will be shown from the deployment during the CoMet field campaign that was carried out in spring 2018, with its main focus on one of the major European hot spots in methane emissions: the Upper Silesian Coal Basin (USCB) in Poland. First analyses reveal a measurement precision of below 0.5% for 20-km averages and also low bias, which was assessed by comparison with in-situ instruments. The measurements flights were designed to capture individual CH_4 and CO_2 plumes from e.g. coal mine venting and coal-fired power plants, respectively, but also to measure large and regional scale GHG gradients and to provide comparisons with the Total Carbon Column Observing Network (TCCON). Many other different instruments, both airborne and ground-based, complemented the lidar measurements to provide a comprehensive dataset for model analyses. CHARM-F also acts as the airborne demonstrator for MERLIN, the "Methane Remote Lidar Mission", conducted by the German and French space agencies, DLR and CNES, with launch foreseen in ~ 2024. In this context, the airborne lidar data are likewise important for mission support such as for e.g. algorithm development and improvement and, moreover, the CoMet mission was also an important step for MERLIN validation preparation.

1. INTRODUCTION

There are still significant deficits in the knowledge about the budgets of two major anthropogenic greenhouse gases, CO_2 and CH_4, which contribute to about 81% of the anthropogenic radiative forcing, such that the ability to accurately predict our future climate remains substantially compromised. Different feedback mechanisms which are insufficiently understood have significant impact on the quality of climate projections. In order to accurately predict future climate of our planet and support observing emission targets in the framework of international agreements (e.g. COP21 Paris Agreement), adequate knowledge of the sources and sinks of these greenhouse gases and their feedbacks is mandatory.

In spite of the recognized importance of this issue, our current understanding about sources and sinks of the gases CO_2 and CH_4 is still inadequate. In order to overcome such shortcomings, the CoMet (CO_2 and Methane) mission has been executed using the most advanced airborne, ground-based and satellite instruments for greenhouse gas monitoring to fill existing gaps and investigate new observation capabilities to support political entities towards raising their ambition to reduce GHG emissions. One of the key instruments of CoMet was the airborne Integrated Path Differential Absorption (IPDA) lidar [1] whose results are the focus of this contribution.

2. METHODOLOGY

2.1 The CoMet 1.0 mission

CoMet is a multi-aircraft mission performed in May-June 2018 deploying a suite of the most sophisticated airborne instruments to measure atmospheric CH_4 and CO_2 and supported by a variety of ground-based in-situ and remote sensing experiments. The German research aircraft HALO acted as the airborne flagship of that campaign. Figure 1 shows all 9 flight tracks of the scientific flights of HALO during CoMet out of Oberpfaffenhofen airport with a total distance of ~44.000 km. Among other instruments, HALO was equipped with the CHARM-F lidar [1] and in-situ instruments for the measurement of greenhouse gases [2]. Parts of the flights were concentrated on a European hot-spot region of methane emissions: the Upper Silesian Coal Basin (USCB).

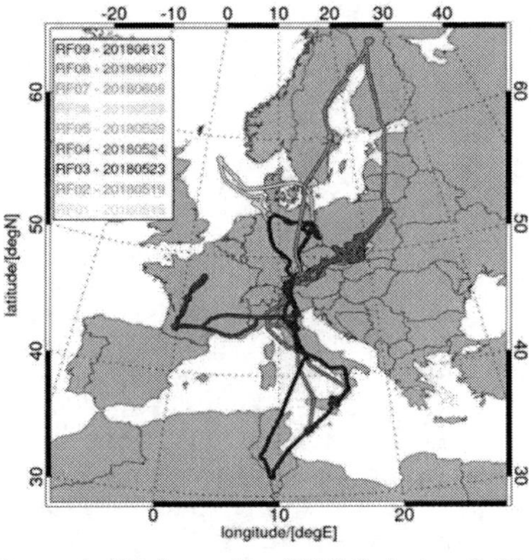

Figure 1: *Flight tracks of HALO during CoMet. On the 28 May, 29 May, 6 June and 7 June the HALO flights concentrated on the Upper Silesian Coal Basin and were coordinated with three propeller aircraft. The flight on 24 May was coordinated with the French Falcon operated by SAFIRE.*

Therefore, two other aircraft were performing extensive flight operations in that region with their base at Katowice Pyrzowice airport: a Cessna 208 Grand Caravan operated by DLR with a comprehensive in-situ payload [3] and a Cessna 207 operated by FU Berlin with the MAMAP passive spectrometer [4] as the main instrument. Additionally, two flights dedicated to survey exhaust plumes of coal mine ventilation shafts were performed with a Do228 aircraft equipped with a hyperspectral imager. In order to support the validation preparation activities of MERLIN [5], a coordinated flight of the French Falcon 20 operated by SAFIRE and HALO was performed in France. That flight was designed to provide comparison with airborne in-situ instruments, ground based remote sensing measurements (e.g. TCCON instruments), and balloons equipped with aircore samplers. On the ground, a variety of instruments supported the airborne measurements. In the USCB, an array of four ground-based and a mobile FTIR instruments were deployed. Three wind lidars accurately measured the local wind conditions to help inferring fluxes by means of e.g. mass balance approaches. A mobile backscatter lidar provide information about the planetary boundary layer evolution. In addition, in-situ measurements from mobile vans and small drones were made available through an in-kind contribution of the $MEMO^2$ network to provide near-surface information of GHGs and to quantify CH_4 and CO_2 emissions from individual coal mining shafts. All measurements were planned to provide best possible synergy with satellite measurements from GOSAT, Sentinel-5P and OCO-2. Further measurements activities such as in Sodankylä, Finland, supported CoMet. Finally, a model infrastructure (regional inverse modelling, chemistry-climate modelling with regional refinement) is employed in order to use the data streams of the individual instruments for modelling the greenhouse gas fluxes. Those models were also used in forecast mode to optimize flight strategies.

2.2 The CHARM-F lidar

In the past years, an airborne IPDA system (CHARM-F) has been developed and successfully tested at DLR [1]. To our knowledge, CHARM-F currently is the only airborne IPDA capable to measure both greenhouse gases, CH_4 (at 1645 nm) and CO_2 (at 1572 nm) at the same time [1].

The lidar transmitter (Fig. 2) is based on two optical parametric oscillators which are pumped by means of diode-pumped, injection seeded, and

Q-switched Nd:YAG lasers in a master-oscillator power-amplifier configuration. At both wavelengths, 1571 nm and 1645 nm, an energy of ~ 10 mJ is generated within a pulse length of ~ 20 ns.

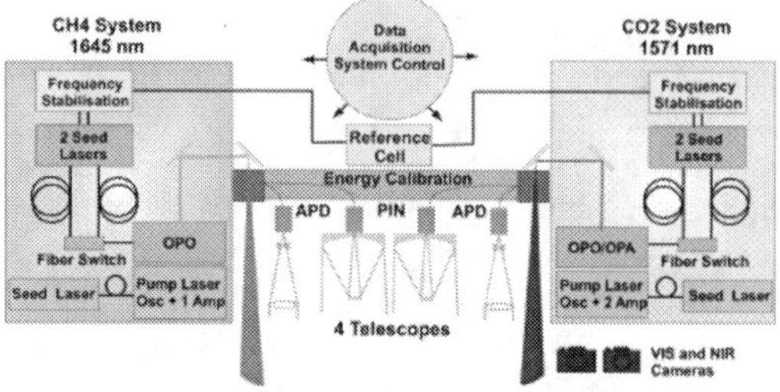

Figure 2: *Schematic set-up of the airborne CO_2 and CH_4 integrated path differential absorption lidar.*

Figure 3: *Photograph of the CO_2 and CH_4 IPDA lidar as installed into the cabin of HALO.*

The IPDA technique requires the generation of two wavelengths that act as the on-line and off-line. This is achieved by injection seeding cw radiation from two stabilized DFB lasers. In order to fulfil the stringent requirements on frequency stability for the on-line and off-line wavelengths, a sophisticated locking scheme has been developed that is based on DFB lasers referenced to a multipass absorption cell and offset locking techniques. In order to monitor the frequency stability of the seed system (at 1572 nm) with highest possible accuracy under in-flight conditions, a compact optical frequency comb (Menlo Systems) has been employed, to the best of our knowledge, for the first time onboard an aircraft. Previously, such measurements were only possible in the laboratory.

CHARM-F's receiving system consists of two telescopes for each wavelength range, one adapted for optimum operation onboard the aircraft and attached to an InGaAs pin diode, and another smaller telescope for mimicking the viewing geometry of a satellite system and equipped with an InGaAs avalanche photodiode (APD). This redundant measurement capacity proved to be very valuable for an independent quality assessment of the data. The received signals are sampled using fast digitizers and processed by means of a home-built data acquisition system. Housekeeping data monitoring as well as laser control is maintained by separate computer systems. Figures 2 and 3 show the schematic set-up and a photograph of the system installed on-board, respectively.

The prime quantities to be measured by CHARM-F are the column-integrated dry-air mixing ratio of corresponding GHG (denoted as XGHG) [6].

3. RESULTS

During previous test flights the precision of CHARM-F measurements of XCO_2 and XCH_4 have already been quantified to be below 0.5% for 20-km averages [1]. During CoMet, the accuracy of the lidar measurements has been carefully assessed by comparing the lidar data to measured profile data from a cavity ring-down spectrometer (CRDS). For this purpose, the in-situ profiles were integrated according to CHARM-F's weighting function and compared to the CHARM-F XGHG values from fly-bys in the vicinity of the respective descent. Figure 4 shows the comparison for the case of XCO_2 independently measured with the two different receivers. Given the fact that the CHARM-F data result from a simple cross section calculation by just using the HITRAN16 data base (www.hitran.org), the agreement is considered very good. The mean absolute bias is of the order of 0.6%. For XCH_4 the results are comparable.

These results lend confidence in quality of the CHARM-F data acquired during the research flights on all relevant spatial scales. Small sub-continental and regional greenhouse gas gradients

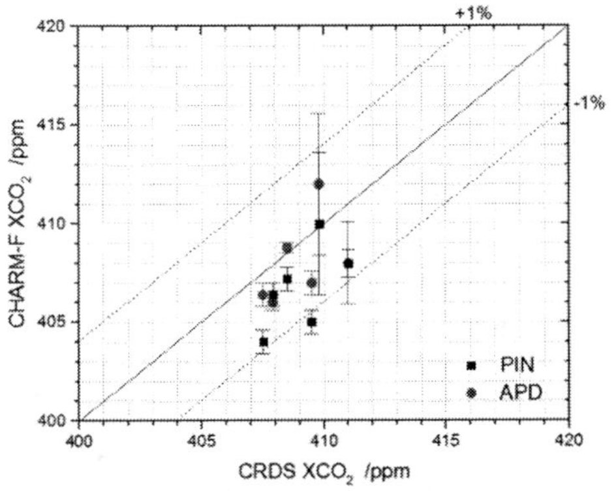

Figure 4: *CHARM-F column-integrated dry air mixing ratio of CO_2 compared to the same quantity derived from profiles of a cavity ring-down spectrometer (CRDS).*

were detected as well as plumes from individual sources such as coal-fired power plants or coal mine ventilation shafts. For the first time, emissions from such GHG hot spots were measured simultaneously by active and passive remote sensing techniques on aircraft and also in-situ instruments. The data will be exploited in conjunction with the models that are now going to be used to hindcast the campaign period to evaluate GHG fluxes and transport.

4. SUMMARY AND OUTLOOK

The CHARM-F lidar operated by DLR was successfully employed onboard HALO during the CoMet mission in May-June 2018. CoMet is a multi-aircraft mission comprising of a fleet of five aircraft that provided a comprehensive data set of active and passive remote sensing data as well as in-situ measurements. This innovative airborne payload together with the data from supporting ground-based instruments and modeling activities shows much prospect to determine greenhouse gas fluxes more precisely than previously possible and support activities to reduce GHG emissions.

While this first CoMet mission in 2018 focused on anthropogenic sources of the major greenhouse gases, successor missions are already in planning that will extend the CoMet study from Central Europe to the highly relevant natural fluxes of Arctic and Tropical wetlands. Beyond the scientific goals, those activities will continue the preparation of the validation activities of the German-French climate mission MERLIN [5] that is scheduled for launch in ~2024.

ACKNOWLEDGEMENTS

We would like to thank the following individuals as part of the CoMet team: T. Andersen, C. Bès, J. Borchardt, J. P. Burrows, H. Chen, C. Crévoisier, A. Dandocsi, A. Fiehn, F. Hase, P. Hochstaffl, K. Gerilowski, R. Kivi, R. Kleinschek, C. Köhler, P. Korben, J. Kostinek, S. Krautwurst, M. Kud, A. Kuze, J. Landgraf, A. Luther, H. Maazallahi, C. Mallaun, M. Menoud, A. Nickl, P. Poczta, M. Rapp, T. Röckmann, T. Ruhtz, J. Schmidt, M. Schmidt, I. Stachlewska, M. Stanisavljevic, C. van der Veen, D. Wang, and N. Wildmann. We would like to also acknowledge the support of B. Sprenger, B. Pröbster, and R. Holzwarth of Menlo Systems GmbH. CoMet was partly funded by German Research Foundation (DFG) within its Priority Program HALO-SPP 1294. The scientific data analysis is partly supported by Federal Ministry for Research and Education (BMBF) within its AIRSPACE project.

REFERENCES

[1] A. Amediek, et al., "CHARM-F—a new airborne integrated-path differential-absorption lidar for carbon dioxide and methane observations: measurement performance and quantification of strong point source emissions," Appl. Opt. 56, 5182 (2017).

[2] A. Filges, et al., "The IAGOS-core greenhouse gas package: a measurement system for continuous airborne observations of CO2, CH4, H2O and CO", Tellus B: Chemical and Physical Meteorology, 67:1, 27989 (2015).

[3] A. Fix, et al., "CoMet: an airborne mission to simultaneously measure CO2 and CH4 using lidar, passive remote sensing and in-situ techniques, EPJ Web Conf. Vol.176, The 28th International Laser Radar Conference (ILRC 28) (2018).

[4] Gerilowski, K. et al., "MAMAP – a new spectrometer system for column-averaged methane and carbon dioxide observations from aircraft: instrument description and performance analysis," Atmos. Meas. Tech., 4, 215 (2011).

[5] G. Ehret, et al., "MERLIN: A French-German Space Lidar Mission Dedicated to Atmospheric Methane," Remote Sens. 9, 1052 (2017).

[6] G. Ehret, et al., "Space-borne remote sensing of CO2, CH4, and N2O by integrated path differential absorption lidar: a sensitivity analysis," Appl. Phys. B90, 593 (2008).

CHARACTERISATION OF BIOMASS BURNING AEROSOLS IN THE SOUTHERN HEMISPHERIC MIDLATITUDES BY MULTIWAVELENGTH RAMAN LIDAR

Athina Avgousta Floutsi *,1, Holger Baars[1], Patric Seifert[1], Martin Radenz[1], Yin Zhenping[1,2,3], Ulla Wandinger[1], Boris Barja[4], Felix Zamorano[4], and Albert Ansmann[1]

[1]Leibniz Institute for Tropospheric Research, Permoserstraße 15, 04318 Leipzig, Germany
[2]School of Electronic Information, Wuhan University, Wuhan, China
[3]Key Laboratory of Geospace Environment and Geodesy, Ministry of Education, Wuhan, China
[4]Atmospheric Research Laboratory, University of Magallanes, Punta Arenas, Chile
**Email: floutsi@tropos.de*

ABSTRACT

Vertically resolved multiwavelength aerosol Raman lidar observations were conducted in the pristine environment of the Southern-hemisphere midlatitudes at Punta Arenas, Chile (53.1346°S, 70.8834°W). In contrast to the usually prevailing clean and pristine conditions at this site, two pronounced lofted aerosol layers were observed up to 4.2 and 4.4 km height on 4 and 5 February 2019, respectively. The layers mainly consisted of biomass burning aerosols originating from the region of Central Chile, where wildfires were also observed. Based on spectrally resolved backscatter and extinction coefficients, lidar ratios and depolarization ratio a detailed characterization of the aerosol optical properties is presented.

1. INTRODUCTION

Vertically resolved measurements of aerosol optical and microphysical properties are of high importance for climate research and the Light Detection And Ranging technique (lidar) is commonly used within the atmospheric research community for the retrieval of vertical aerosol profiles. On global scale, monitoring of the aerosol characteristics has been achieved through spaceborne missions, e.g. with the Cloud-Aerosol Lidar and Infrared Pathfinder Satellite Observations mission (CALIPSO, [1]) and the upcoming EarthCARE satellite [2], and by ground-based networks e.g. the European Aerosol Research Lidar Network (EARLINET [3]), the ceilometer network of the German Meteorological Service (DWD, [4]) and the portable Raman lidar systems network Polly^{NET} ([5], [6], [7]). More specifically, in the framework of worldwide field campaigns, Polly^{XT} systems have acquired unique lidar data sets in the Amazonian basin, India, China, South Africa, Finland, Korea, Chile and over the Atlantic and Arctic on board the research vessels Polarstern and Meteor. Since November 2018, the Leipzig Aerosol and Cloud Remote Observations System (LACROS) operates 24/7 in Punta Arenas, Chile (53.1346°S and 70.8834°W, 9 m a.s.l.) aiming to provide further information on aerosol, clouds and their interaction processes in the Southern Ocean environment.

2. METHODOLOGY

The lidar observations were conducted in the framework of the Dynamics, Aerosol, Cloud and Precipitation Observations in the Pristine Environment of the Southern Ocean (DACAPO-PESO). The Raman lidar Polly^{XT} [7] deployed on site allows the determination of vertically resolved profiles of volume extinction and backscatter coefficient (at 355, 532 and 355, 532, 1064 nm respectively), lidar ratio (at 355 and 532 nm), and depolarization ratio (at 355 and 532 nm). The extinction and backscatter coefficients were determined as described by Ansmann and Müller (2005) [8], while for the depolarization ratio the methodology as described in Freudenthaler et al. (2009) [9] was applied. The influence of noise to the lidar signals was reduced by means of a moving average filter with a window length of 742.5 m.

3. RESULTS

3.1 Aerosol optical properties

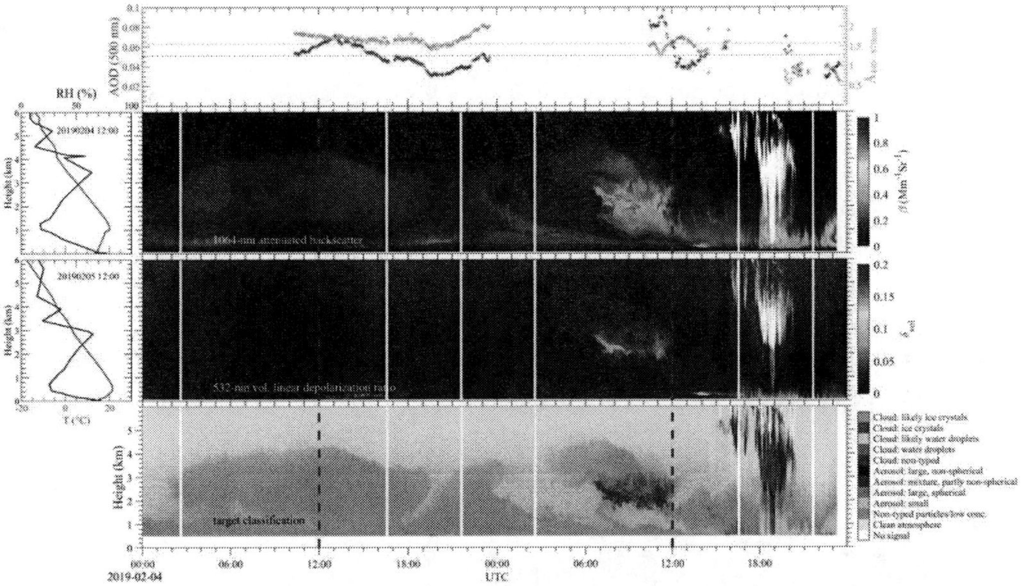

Figure 1. From the upper to the lower panel: AOD (500 nm) and Ångström exponent (440/870 nm), attenuated backscatter coefficient at 1064 nm (sr^{-1}m^{-1}), volume depolarization ratio (%), and target classification reflecting the atmospheric conditions over Punta Arenas from 04.02.2019 to 05.02.2019. Within the study period, two radiosonde launches took place (depicted with red dashed lines). The corresponding temperature and relative humidity profiles can be seen on the left panels.

During the DACAPO-PESO campaign, from 4 February to 5 February 2019, several lofted aerosol layers were observed between 1 and 5 km height. Figure 1 shows the two-day overview of the AOD (500 nm), attenuated backscatter coefficient (1064 nm), volume depolarization ratio (%), temperature (°C) and relative humidity (%) over Punta Arenas. As Figure 1 indicates, on 4 February 2019 between 02:00 and 19:30 UTC an extensive aerosol plume was present, characterized by low backscatter and extremely low depolarization values. Data from radiosondes (Fig. 1, left top) revealed a temperature inversion at 1 km height, while the relative humidity decreased rapidly up to 1 km and then increased with height up to 65% and up to the top of the layer at 4 km height. On 5 February 2019 between 06:00 and 12:00 UTC, the aerosol plume was characterized by higher backscatter values and, at some parts, higher depolarization ratios. A temperature inversion was present at 0.5 km height and up to the same altitude, the relative humidity decreased rapidly to values of 30%. Above that level the relative humidity increased up to 70% at 3 km above which it decreased strongly towards 20% throughout the clean middle troposphere (Fig. 1, left bottom).

According to the Aerosol Robotic Network (AERONET, [10]), the aerosol optical depth (AOD, which is a measure of the integrated extinction of solar radiation by atmospheric particles) reached values up to 0.07 and 0.09 at 500 nm on 4 and 5 February respectively. The AOD is relatively high (despite the values smaller than 0.1 observed) for the typical average AOD values observed over Punta Arenas of approx. 0.04 at 500 nm. The daily variability of AOD at 500 nm and Ångström exponent at the wavelength pair of 440/870 nm are shown in Fig. 1 (top panel).

The Hybrid Single-Particle Lagrangian Integrated Trajectory (HYSPLIT, [11]) model was used in order to identify the origins of the air masses observed. As shown in the 5-day backward trajectories (Fig. 2), the air masses arriving on 5 February 2019 at 08:00 UTC at 3, 2 and 1 km originate from the Southeastern Pacific Ocean, and on their descending route to Punta Arenas they pass over Central Chile (on 4 February 2019). Data from FIRMS (Fire Information for Resource Management System, [12]) show that above the aforementioned regions, where the air masses passed by on 3 and 4 February 2019, active fires and wildfires occurred (shown in Fig. 2 with red and yellow colors). Therefore, our observations are most likely biomass burning aerosols (4 February 2019) and biomass burning aerosols possibly mixed with some volcanic ash from the eruptions of Planchón-Peteroa and Nevados de Chillán volcanoes (5 February 2019).

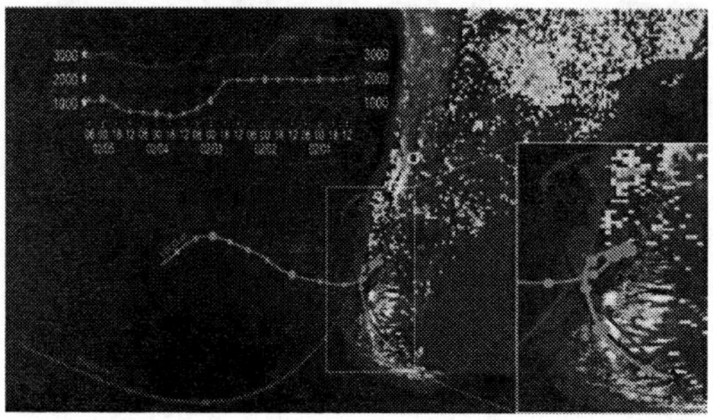

Figure 2. 5-day HYSPLIT backward trajectories arriving over Punta Arenas at 2 km (red line) and 3 km (blue line) on 05 February 2019, at 08:00 UTC. The underlying fire map derived from FIRMS shows all fires (yellow and red spots) detected during January 2019 and the locations of the two active volcanoes (white circles).

Figure 3 shows the aerosol optical properties observed over Punta Arenas on 4 (top panel) and 5 February (lower panel) 2019, respectively. As the backscatter coefficient profiles (Fig. 3a) indicate, the main aerosol load is located between 1.5 and 4 km height. For the same altitudes, the lidar ratio (Fig. 3c) at 355 nm is almost constant with height and it is roughly 50-60 sr, a value that indicates the presence of a biomass-burning dominated aerosol mixture. The backscatter-related Ångström exponents above 1 for both wavelength pairs (Fig. 3d), indicate the predominance of relatively small particles which are also spherical (low particle depolarization ratio, Fig. 3e). For the same period, the automated target classification (Fig. 1, [6]) was not able to assign a specific aerosol type due to the low aerosol scattering. The profiles of backscatter coefficient (Fig. 3f) along with the very low particle linear depolarization ratio (Fig. 3j) indicate that on 5 February 2019 the aerosol load, located between 1 and 4 km height, consisted mainly of smoke particles and a possible mix with

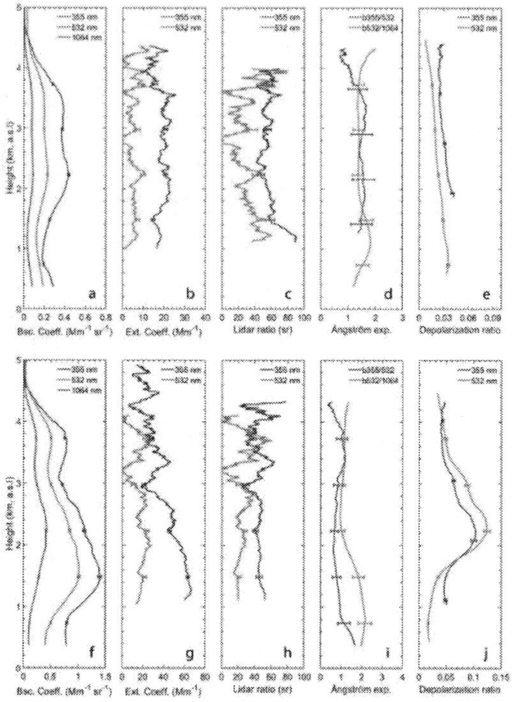

Figure 3. (a) and (f) mean particle backscatter coefficients (at three wavelengths), (b) and (g) extinction coefficients (at two wavelengths), (c) and (h) the corresponding extinction-to-backscatter ratios, (d) and (i) backscatter-related Ångström exponents, and (e) and (j) particle depolarization ratios measured at Punta Arenas (top) on 4 February 2019, 06:00-09:00 UTC and (bottom) on 5 February 2019, 08:00-09:00 UTC. Error bars (standard deviation) include systematic uncertainties and signal noise.

volcanic ash transported from the aforementioned erupting volcanoes. Below 2 km height, the lidar ratio is between 40 – 60 sr (at 355 nm, Fig. 3h) and the Ångström exponent values are higher than 1, indicating the predominance of small particles (Fig. 3i). At the same altitude, the particle linear depolarization ratio is smaller than 5% (for both wavelengths) indicating the presence of mainly spherical particles. Above 2 km height, the Ångström exponent values are decreasing, and the depolarization ratio increases up to 12 % (around 2.2 km) and then decreases with height again. From the aforementioned observations we can conclude the presence of relatively large, predominantly non-spherical particles above 2 km height. Our results are in agreement with the target classification (Fig. 1), which confirms the presence of small in size particles below 2 km height, whereas aerosols are classified as a mixture of partially spherical and non-spherical particles above that level.

4. SUMMARY AND OUTLOOK

Two lofted aerosol layers were observed by the Raman lidar Polly^{XT} in the pristine environment of the Southern Hemisphere Midlatitudes (Punta Arenas). Both the extensive and the intensive optical parameters of the layers were studied and it was found that the observed aerosol load was consisting of mainly biomass burning aerosols. In addition, this case study is intended to be used for validation purposes, as an input to a new aerosol typing algorithm that is currently under development at Leibniz Institute for Tropospheric Research (TROPOS). The algorithm will be applicable to both ground-based and spaceborne lidar and will exploit the multispectral information from sophisticated ground-based lidars (i.e., 3 backscatter, 2 extinction and 2 depolarization measurements at 355, 532 and 1064 nm), but will also be able to retrieve an aerosol typing from less information (e.g. EarthCARE). In brief, the aerosol mixing state (based on different aerosol types) will be estimated by means of optimal estimation by combining the lidar-measured quantities (lidar ratio, depolarization ratio and Ångström exponent) with the most representative microphysical description, in terms of size and shape distribution and spectral complex refractive index. Preliminary results from the algorithm show that separation between smoke, maritime, pollution and dust dominated aerosol mixtures and estimation of their mixing state are possible.

ACKNOWLEDGEMENTS

Parts of the research leading to these results have received funding from the European Union's Horizon 2020 research and innovation programme under grant agreement no. 654109 (ACTRIS).

REFERENCES

[1] D. M. Winker, et al. Journal of Atmospheric and Oceanic Technology 26.11: 2310-2323 (2009)
[2] A. J. Illingworth, et al. Bulletin of the American Meteorological Society 96: 1311-1332 (2015)
[3] G. Pappalardo, et al. Atmospheric Measurement Techniques 8: 2389-2409 (2014)
[4] H. Flentje, et al. Atmospheric Measurement Techniques Discussions 3: 3643-3673 (2010)
[5] D. Althausen et al. Journal of Atmospheric and Oceanic Technology 26.11: 2366-2378 (2009)
[6] H. Baars et al. Atmospheric Chemistry and Physics 16.8: 5111-5137 (2016)
[7] R. Engelmann et al. Atmospheric Measurement Techniques 9: 1767-1784 (2016)
[8] A. Ansmann, and D. Müller, Lidar and atmospheric aerosol particles, in LIDAR: Range-Resolved Optical Remote Sensing of the Atmosphere, edited by C. Weitkamp, pp. 105–141, Springer, New York (2005)
[9] V. Freudenthaler, et al. Tellus, Ser. B 61: 165–179 (2009)
[10] B. N. Holben, et al. Remote sensing of environment 66.1: 1-16 (1998)
[11] http://ready.arl.noaa.gov/HYSPLIT.php
[12] https://firms.modaps.eosdis.nasa.gov/firemap/

CHARACTERIZATION OF COMPLEX WATER VAPOUR FIELD STRUCTURES AND THEIR GENESIS BASED ON THE COMBINED USE OF RAMAN LIDAR MEASUREMENTS AND MESO-NH MODEL SIMULATIONS

Paolo Di Girolamo[1*], Marie-Noelle Bouin[2]

[1]*Scuola di Ingegneria, Università della Basilicata, Potenza, Italy*
[2]*Météo-France /CNRM, 29280 Plouzané, France*
**Email: paolo.digirolamo@unibas.it*

ABSTRACT

As part of the Cevennes-Vivarais site, the University of Basilicata Raman lidar system (BASIL) was deployed in Candillargues throughout the duration of HyMeX-SOP 1 (September-November 2012), providing high-resolution and accurate measurements, both in daytime and night-time, of atmospheric temperature, water vapour mixing ratio and particle backscattering and extinction coefficient at three wavelengths. Measurements carried out by BASIL on 28 September 2012 reveal a quite complex vertical structure of the water vapor field. Reported Raman lidar measurements were run in the time interval between two consecutive heavy precipitation events, from 15:30 UTC on 28 September to 03:30 UTC on 29 September 2012. Throughout most of this observation period, lidar measurements reveal the presence of four distinct humidity layers.

The present research effort aims at assessing the origin of the different humidity filaments observed by BASIL on this day. The analysis relies on the comparisons between Raman lidar MESO-NH model simulations. Back-trajectory analyses from the model reveal that air masses ending in Candillargues at different altitudes levels are coming from different geographical regions. Specifically, the analysis reveals that air masses within the surface humidity layer were originated over the Atlantic Ocean, while air masses within the elevated filamentary humidity layer, also coming from the Atlantic Ocean, overpassed the sea stretch North of Spain and Southern France at an altitude of ~1 km. In addition, air masses within the lower of the two upper layers are found to overpass Southern Spain and Marocco, descending from an elevation of 2-3.5 km, while air masses within the uppermost layer are found to overpass Algeria, descending from an elevation of 3.5-5 km.

1. HyMeX-SOP1

HyMeX was conceived with the goal of improving the comprehension of the water cycle in the Mediterranean [1]. Within this program a major field campaign, named Special Observation Period 1 (SOP1), took place in the period September–November 2012 over the northwestern Mediterranean Sea and its surrounding coastal regions in France, Italy and Spain, with a specific focus on the study of heavy precipitation and flashflood events [2,3].

As part of the Cévennes-Vivarais observational site, a set of research instruments was installed in an atmospheric 'supersite' located in Candillargues (43°37'N, 4°04'E, elevation: 1m), near the coast and south of the Massif Central. Among these instruments, a wind profiler, a radiometer, a radiosonde launching facility and a variety of radars and lidars (BASIL being one of these) were present.

2. BASIL

BASIL operated from 5 September to 5 November 2012, collecting more than 600 h of measurements, distributed over 51 measurement days and 19 Intensive Observation Periods (IOPs). BASIL uses a tripled Nd:YAG laser (single pulse energy power at 355 nm: 500 mJ, pulse repetition frequency: 20Hz, average power: 10 W) and a Newtonian telescope (primary mirror diameter: 45 cm, f/2.1). The major feature of BASIL is represented by its capability to perform high-resolution and accurate measurements of atmospheric temperature and water vapour, both in daytime and night-time, based on the application of the rotational and vibrational Raman lidar techniques in the ultraviolet [3-6]. Besides temperature and water vapour, BASIL is

https://doi.org/10.1051/epjconf/202023703007

also capable of providing measurements of particle backscatter at 355, 532 and 1064 nm, particle extinction at 355 and 532 nm, and particle depolarization at 355 and 532 nm [7-10]).

3. RESULTS

3.1 Lidar measurements

Raman lidar measurements considered in this research effort were run in the time interval between two consecutive heavy precipitation events, from 15:30 UTC on 28 September to 03:30 UTC on 29 September 2012. Figure 1 illustrates the time evolution of the range-corrected signal at 1064 nm as measured by BASIL over this time interval. The figure reveals the presence throughout most of the observational period of stratiform clouds around 4 km, with a very limited vertical extent (~0.5 km). Stratiform rain events are found to take place between 18:30 and 20:00 UTC on 28 September and between 00:30 and 03:30 UTC on 29 September 2019. One of these events, between 15:30 and 16:30 UTC on 28 September, appears in the form of Virga, with precipitating particles sublimating before reaching the ground. In this time interval, a lidar dark band is also observed at ~3.4 km, this phenomenon being associated with the variable scattering properties of melting hydrometeors in the snow flake to rain conversion process [10,11,12].

Figure 2 illustrates the time evolution of water vapour mixing ratio, m_{H2O}, as measured by BASIL over the same time period considered in Figure 1. This figure reveals the presence throughout most of the observation period of four distinct humidity layers: a surface layer extending up to 0.4-0.8 km a.s.l.; a filamentary structure, first observed around 17:30 UTC, with a progressively descending trend from ~1 km down to ~0.5 km and a variable vertical extent within the range 0.1-0.4 km; two upper humid layers, one extending from 1.3-1.5 km to 2.4-2.8 km, with values of m_{H2O} up to 7.5 g kg^{-1}, and one above extending up to approx. 4 km, with values of m_{H2O} not exceeding 6 g kg^{-1}. Relatively high values of m_{H2O} are observed throughout most part of the troposphere, with a residual value of 3 g kg^{-1} around 6 km.

3.1 Model simulations and back-trajectory analysis

For the purpose of this analysis the non-hydrostatic numerical research model MESO-NH was run over a 1446 x 1778 km^2 domain (35°-48° N, 8° W-16° E), with a horizontal resolution of 2.5 km. Atmospheric forcing are taken from AROME-WMED. The considered MESO-NH simulation started at 00:00 UTC on 27 September and ended at 00:00 UTC on 29 September 2012.

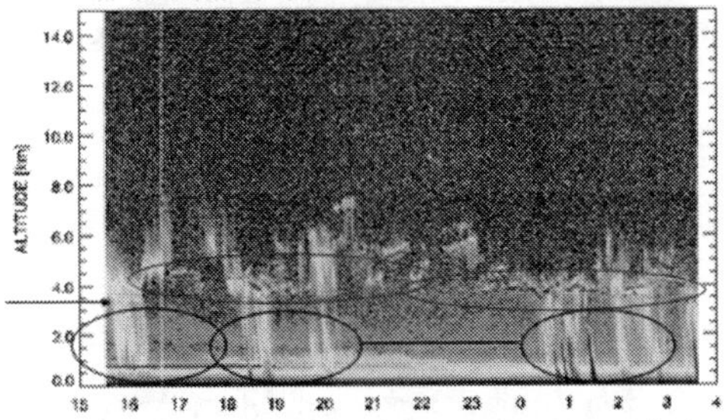

Figure 1: Time evolution of the range-corrected signal at 1064 nm (expressed in arbitrary units) as measured by BASIL over this time interval from 15:30 UTC on 28 September to 03:30 UTC on 29 September 2012. The red ellipses indicate the stratiform clouds, the red ellipses indicate the stratiform rain events, while the black arrow on the left portion of the figure reveal the presence of a dark band event.

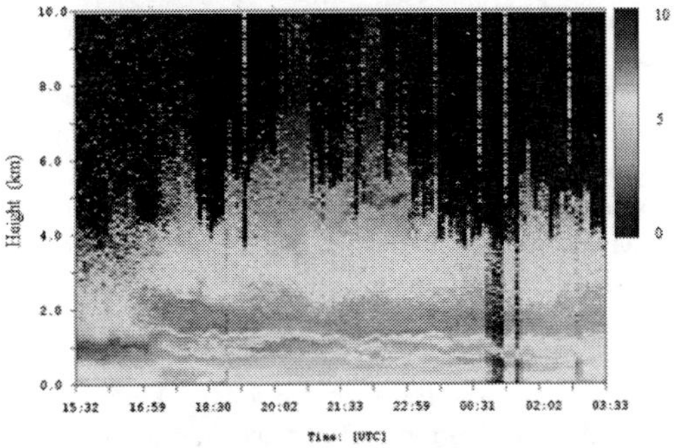

Figure 2: Time evolution of m_{H2O} as measured by BASIL over the same time period considered in Fig. 1.

Thus, back-trajectories can be extended back in time by 48 hours. Figure 3 illustrates the comparison between BASIL measurements and MESO-NH simulations in the time interval from 15:30-23:30 UTC on 28 September 2012. The four different humidity layers present in the Raman lidar measurements are well reproduced by Meso-NH, both in terms of timing of occurrence and mixing ratio values, with the only exception of the mixing ratio values within the elevated filamentary structure, with measured values not exceeding 7 g kg^{-1} and simulated values being as large as 8-9 g kg^{-1}.

 https://doi.org/10.1051/epjconf/202023703007

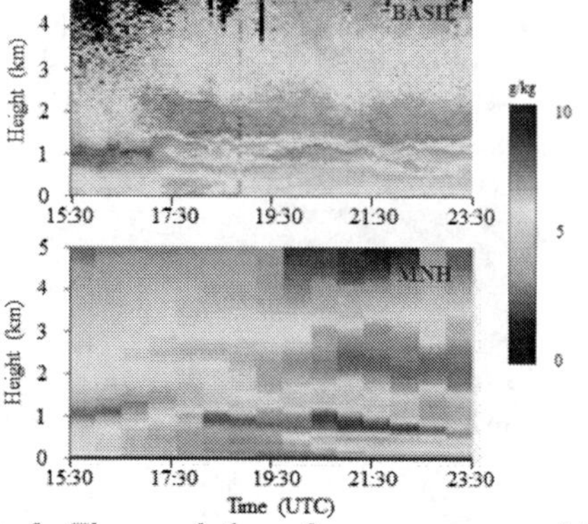

Figure 3: Time evolution of m_{H2O} as measured by BASIL (upper panel) and simulated by MESO-NH (lower panel).

An assessment of the origin of the different humidity layers observed by the Raman lidar on this day is also carried out based on the use of MESO-NH simulated data. Back-trajectory analyses from this model allow revealing that air masses reaching Candillargues at different altitudes levels are coming from different geographical regions. Figure 4 illustrates the 48h-backtrajectory analysis for the air masses ending in Candillargues at 15:30 UTC on 28 September 2012. The trajectory colors identify the altitude experienced by the air masses during their transport to Candillargues, while the numbers at the beginning of each trajectory identify the air masses ending altitude in Candillargues. The red ellipse in the figure highlights the back-trajectories of the air masses ending in Candillargues within the surface humidity layer. The analysis reveals that air masses within this layer were originated over the Atlantic Ocean. Figure 5 illustrates the 48h-backtrajectory analysis of the air masses ending in Candillargues at 18:00 UTC on 28 September 2012. The red ellipses identify the back-trajectories ending at the altitude of elevated filamentary humidity layer and within the two upper layers. The figure reveals that air masses within the elevated filamentary humidity layer are also coming from the Atlantic Ocean, overpassing the sea stretch North of Spain and Southern France at an altitude of ~1 km. The figure also shows that air masses within the lower of the two upper layers overpassed Southern Spain and Marocco, descending from an elevation of 2-3.5 km, while air masses within the uppermost layer overpassed Algeria, descending from an elevation of 3.5-5 km.

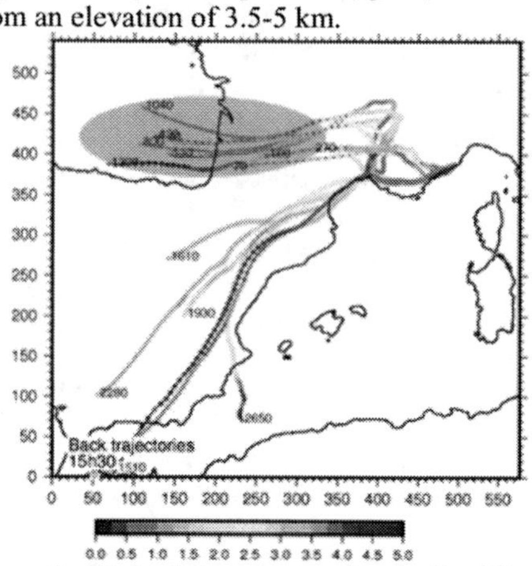

Figure 4: 48h back-trajectory ending in Candillargues at 15:30 UTC on 28 September 2012 as simulated by MESO-NH. Colors identify the altitude of the air masses during their transport to Candillargues, with the color legend being expressed in km. Numbers at the beginning of each trajectory identify the ending altitude in Candillargues. Red ellipse: back-trajectories ending in Candillargues within the surface humidity layer.

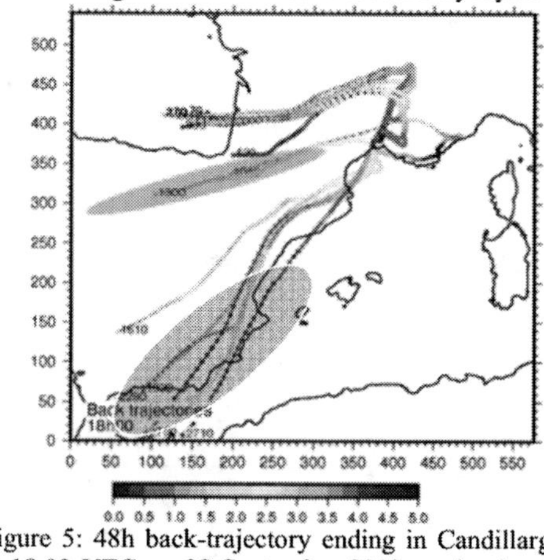

Figure 5: 48h back-trajectory ending in Candillargues at 18:00 UTC on 28 September 2012 as simulated by MESO-NH. Red ellipses: back-trajectories ending in Candillargues at the altitude of elevated filamentary humidity layer and within the two upper layers.

Back-trajectory analysis ending in Candillargues at 18:00 UTC on 28 September 2012 and starting 240 hours (10 days) earlier at 18:00 UTC on 18 September 2012 have also been determined through the application of the NOAA HYSPLIT

Lagrangian back-trajectory model [13]. The back-trajectory analysis extending back in time by 10 days reveals that air masses in the uppermost humid layer are originated over the Saharan desert in Central Africa (Mali).

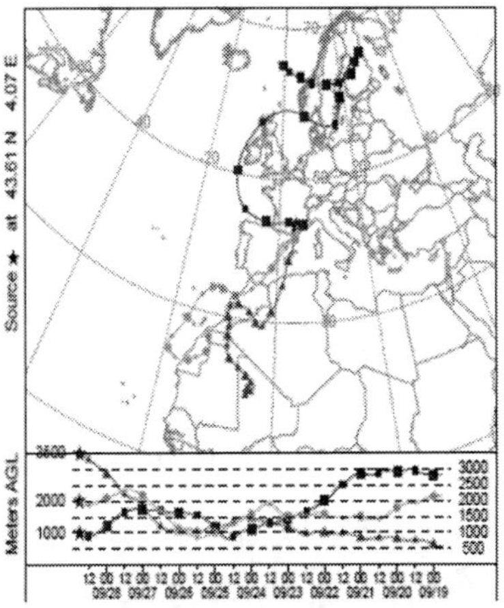

Figure 6: 240 h back-trajectory ending in Candillargues at 18:00 UTC on 28 September 2012 as simulated by NOAA-HYSPLIT back-trajectory model.

As a continuation of this research effort, we intend to investigate more precisely the origin of the sounded air parcels and their possible modifications along their route, also referring to the time series of potential temperature and other relevant parameters, as the aerosol properties.

ACKNOWLEDGEMENTS

This work was supported by the Italian Ministry for Education, University and Research under the Grant OT4CLIMA. The authors gratefully acknowledge NOAA Air Resources Laboratory for the use of the HYSPLIT transport and dispersion model.

REFERENCES

[1] P. Drobinski *et al.*, HyMeX, a 10-year multidisciplinary program on the Mediterranean water cycle, Bull. Am. Meteorol. Soc. 95: 1063–1082, doi: 10.1175/BAMS-D-12-00242.1 (2014).

[2] V. Ducrocq *et al.*, HyMeX-SOP1, the field campaign dedicated to heavy precipitation and flash-flooding in northwestern Mediterranean, Bull. Am. Meteorol. Soc. 95: 1083–1100, doi: 10.1175/BAMS-D-12-00244.1 (2014).

[3] F. Duffourg *et al.*, Offshore deep convection initiation and maintenance during HyMeX IOP 16a heavy precipitation event, Quarterly Journal of the Royal Meteorological Society 142 (Suppl 1): 259–274, doi: 10.1002/qj.2725 (2016).

[4] Di Girolamo et al., Rotational Raman lidar measurements of atmospheric temperature in theUV, Geophys. Res. Lett. 31: L01106, doi: 10.1029/2003GL018342 (2004).

[5] Di Girolamo P, Behrendt A, Wulfmeyer V., Spaceborne profiling of atmospheric temperature and particle extinction with pure rotational Raman lidar and of relative humidity in combination with differential absorption lidar: performance simulations, Appl. Opt. 45: 2474–2494, doi: 10.1364/AO.45.002474 (2006).

[6] P. Di Girolamo et al., Multiparameter Raman lidar measurements for the characterization of a dry stratospheric intrusion event, J. Atmos. Oceanic Technol. 26: 1742–1762, doi: 10.1175/2009JTECHA1253.1 (2009).

[7] Bhawar et al., The water vapour intercomparison effort in the framework of the Convective and Orographically-induced Precipitation Study: Airborne-to-ground-based and airborne-to-airborne lidar systems, Q. J. R. Meteorol. Soc. 137: 325–348, doi: 10.1002/qj.697 (2011).

[8] V. Griaznov et al., Spatial Distribution of Doubly Scattered Polarized Laser Radiation in the Focal Plane of a Lidar Receiver, Applied Optics 46, 6821-6830, doi: 10.1364/AO.46.006821 (2007).

[9] P. Di Girolamo et al., UV Raman Lidar measurements of relative humidity for the characterization of cirrus cloud microphysical properties, Atmospheric ChemistrY and Physics 9, 8799-8811, doi: 10.5194/acp-9-8799-2009 (2009).

[10] P. Di Girolamo et al., Raman Lidar observations of a Saharan dust outbreak event: Characterization of the dust optical properties and determination of particle size and microphysical parameters, Atmospheric Environment 50, 66-78 (2012).

[11] P. Di Girolamo et al., Lidar and radar measurements of the melting layer: observations of dark and bright band phenomena, Atmospheric Chemistry and Physics 12, 4143-4157, doi: 10.5194/acp-12-4143-2012 (2012).

[12] P. Di Girolamo et al., Model simulations of melting hydrometeors: a new Lidar bright band from melting frozen drops, Geophysical Research Letters 30, doi: 10.1029/2002GL016825 (2003).

[13] V. Griaznov et al., Numerical Simulation of Light Backscattering by Spheres With Off-Center Inclusion. Application for Lidar Case, Applied Optics 43, 5512-5522, doi: 10.1364/AO.43.005512.

[14] Draxler, R. R and Hess, G. D.: An overview of the HYSPLIT_4 modeling system for trajectories, dispersion and deposition, Australian Meteorological Magazine 47(4), 295–308 (1998).

 https://doi.org/10.1051/epjconf/202023703008

CHARACTERIZATION OF TRACE GASES AND GREEN HOUSE GAS IN MEGACITY NEW DELHI

Arti Choudhary[1*], Pradeep Kumar[2], Anuradha Shukla[1], Siddhartha Singh[3], Abhay Kumar Singh[2]

[1]*Tansport Planning and Environment Division, CSIR-Central Road Research Institute, New Delhi, India*
[2]*Department of Physics, Institute of Science, Banaras Hindu University, Varanasi, India*
[3]*India Meteorological Department, New Delhi, India*
**Email: choudharyarti12@gmail.com*

ABSTRACT

Air pollution and climate change is serious environmental concern due to its visible negative impact on human health. Around 14 Indian cities are placed among top 20 most polluted cities of the world. Trace gas like O_3, NO_x, CO and CO_2 are important pollutants which is associated with human health, climate change and adverse effect on growth and yield of crops. Stratospheric O_3 absorbs ultraviolet light and prevents it from reaching to the ground. Greenhouse effect of O_3 and CO_2 is prominent, O_3 in upper troposphere and ranked 3rd for its radiative potential after the carbon dioxide and methane. The amount of O_3 generated by photochemical reaction of air pollutants is much larger than the inflow from the stratosphere. This is indicating that trace gases and GHG are generated by anthropogenic activities. It is significantly high in urban area like megacity Delhi as compared to rural area due to excessive anthropogenic activity.

The ground level measurements of surface trace gas like O_3, NO_x, CO and CO_2 were conducted in Delhi-Mathura road near traffic intersection for year 2017, January to December. The daily mean concentration of O_3, NO_x, CO and CO_2 were 23.11±17.26ppb (range 58.38 to 6.42ppb), 26.41±4.24ppb (ranges 48.14 to 24.09ppb), 1.56±4.24ppm (ranges 6.6 to 0.69ppm) and 342.54±33.49 (ranges 508.23 to 323.33ppm), respectively. The mixing ratios of O_3 were highest of 32ppbv and lowest 17ppbv during the pre-monsoon and monsoon seasons, respectively. While the mixing ratios of both CO and NO_x showed highest and lowest values during the winter and monsoon seasons, respectively.

The analysis concluded seasonality of O_3, CO and NO_x were also governed by the long-range transport, mainly with the summer and winter monsoon circulations over the Indian subcontinent. The mixing ratios of CO and NO_x show strong correlations during winter and pre-monsoon seasons, while poor correlation in the monsoon season. The mixing ratios of CO and NO_x decreased with the increase in wind speed, while O_3 tended to increase with the wind speed.

Keywords: Trace gases, GHG, seasonal variability, mixing rate, vehicular pollution

1. METHODOLOGY

Major vehicles for transportation are buses, cars, two- and three-wheelers (motorbikes, scooter and auto-rickshaws). Due to rapidly increasing number of vehicles (about 10–15 % per year), the transport related activities are the major contributors of various pollutants [1-3]. The real time automatic data of ambient pollutant concentration and meteorological data were collected from Automatic Weather Station (AWS) and Automatic Rain Gauge (ARG) for the Year 2017, installed by India Meteorological Department (IMD, India) at Delhi-Mathura Road/National Highway 2 (NH2) passing through Delhi to Agra city. The site is geographically located at 28°37'39.99N 77°14'29.04E at 216 meters above mean sea level (MSL). The ground measured concentration of trace gases O_3 CO, NO_x and CO_2 will be considered for the analysis. The site has very high inflow and outflow of vehicles, about 170,000 vehicles per day but there are other major sources of pollution located in the vicinity of the site, such as sewage treatment plant, gasoline stations and automobile workshops. The major sources of gaseous pollutants in the surrounding areas of the site are vehicular exhaust and biofuel burning. The vehicle composition consists of mainly two-wheelers, three-wheelers, cars, buses and light commercial vehicles (LCVs) during day time and heavy-duty trucks during the night hours [4-6].

2. RESULTS

2.1 Diurnal variation of trace gases

The basic dataset in the present study were recorded for 30 min interval and averaging has

https://doi.org/10.1051/epjconf/202023703008

been done to derive hourly data. The amplitude of diurnal distribution varied from month to month. The diurnal distributions of O_3 in the pre-monsoon, post-monsoon and winters seasons show strong variability (Figure 1). However, the distribution of O_3 in the monsoon season shows weaker diurnal dependency. The maximum O_3 and minimum NO_x concentration were found during 14-17 hour and min CO, CO_2 concentration during 13-16 hr and maximum concentration were found during morning and evening traffic peak hour (7-11am and 17-20pm, Figure 1).

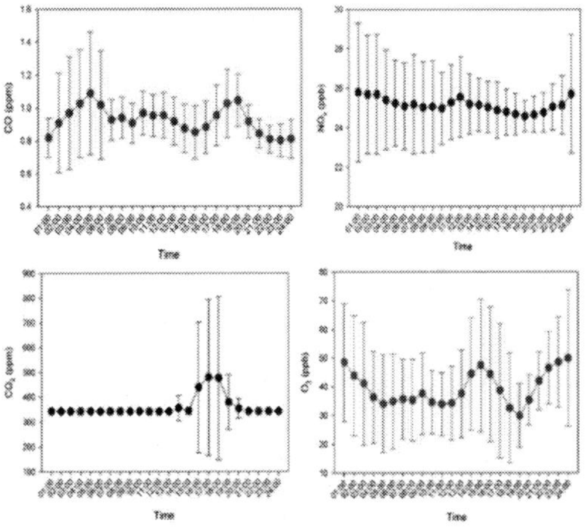

Figure 1: Diurnal variation of air pollutants.

The degree of diurnal variability in both CO and NO_x varied from month to month with strong variability in the pre monsoon, post-monsoon and winters seasons, while weaker diurnal dependency in the monsoon season. The mixing ratios of both CO and NO_x show a sharp peak in the morning hours between 07 hr and 10 hr. Due to elevated planetary boundary layer mixing ratios were observed in the afternoon hours (13–16 hr). The observations from night till early morning hours show high values mainly due to shallow nocturnal boundary layer depth, resisting the mixing of local emissions with the free tropospheric air. In other hand, during after noontime the higher PBL depth provides a larger mixing region and hence the pollutants get diluted. The morning and evening peaks are almost absent during monsoon months and lower concentrations of both NO_x and CO were observed with no significant differences between day and night.

2.2 Seasonal variation of trace gases and GHG gas

The higher O_3concentrationwas found in the pre-monsoon and winter seasons and it associated with the elevated levels of NO_x. The lower concentration of O_3 in the monsoon and post-monsoon seasons were associated with the lower concentration of NO_x. It is also evident that the highest O_3 mixing ratio was observed in the pre-monsoon season, while the maxima in precursor species were observed in the winter seasons (Figure 2).

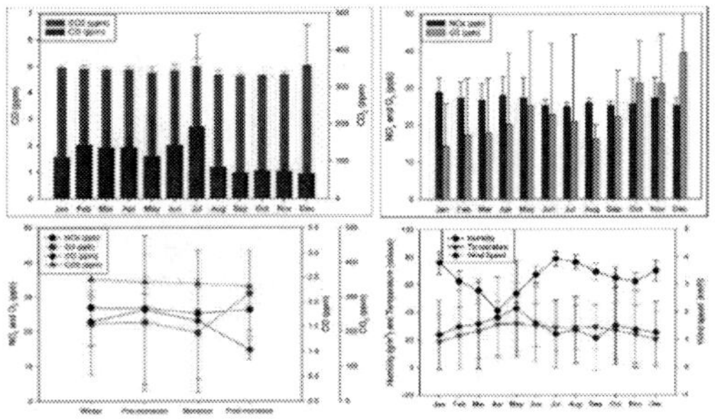

Figure 2: Monthly variation (top), seasonal variation (bottom) of pollutants.

However, CO and CO_2 concentration were found more or less similar trend throughout the year. It is indicating similar source and dominant through-out the year like roadway transportation. The seasonal changes in the long-range transport, emission, boundary layer height and photochemistry (via OH oxidation) play important roles in the observed seasonal variation of O_3 and NO_x concentration.

However, to stimulate the contributions of these processes require detailed model simulation dealing with these processes on local scale. During winter season, Indo-Gangetic Plain influenced the site leading to highest levels of NO_x.

ACKNOWLEDGEMENTS

Present study is financially supported by the SERB, New Delhi, India, under the scheme of SERB National Post-Doctoral Fellowship (PDF/2017/001284). Air Quality and Metrological

https://doi.org/10.1051/epjconf/202023703008

data were taken from India Meteorological Department (IMD), New Delhi.

REFERENCES

[1] Choudhary, A., & Gokhale, S. Urban real-world driving traffic emissions during interruption and congestion. Transportation Research Part D: Transport and Environment, 43: 59-70 (2016).

[2] Choudhary, A., & Gokhale, S. On-road measurements and modelling of vehicular emissions during traffic interruption and congestion events in an urban traffic corridor. Atmospheric Pollution Research, 10: 480-492 (2019).

[3] Choudhary, A., & Gokhale, S. Evaluation of emission reduction benefits of traffic flow management and technology upgrade in a congested urban traffic corridor. Clean Technologies and Environmental Policy, 21: 257-273 (2019).

[4] Choudhary, A., Kumar, P., Gaur, M., Prabhu, V., Shukla, A., & Gokhale S. Real world driving dynamics characterization and identification of emission rate magnifying factors for auto-rickshaw. Nature Environment and Pollution Technology (In Press) 2019.

[5] Choudhary, A., Gokhale, S., Shukla, A., Kumar, P., Singh, A. K. Variability in emission rate of Auto-Rickshaw based on real world driving profile: A case study in Guwahati city. 2019 URSI Asia-Pacific Radio Science Conference (AP-RASC). IEEE, (2019).

[6] Choudhary, A., Gaur, M., Shukla, A. Megacities of developing countries. Air Pollution: Sources, Impacts and Controls, Oxford publisher (CABI Press) pp. 150-162 (2019).

https://doi.org/10.1051/epjconf/202023703009

COMPARISON AMONG THE ATMOSPHERIC BOUNDARY LAYER HEIGHT ESTIMATED FROM THREE DIFFERENT TRACERS

Gregori de Arruda Moreira[1,2,3,4*], Fábio Juliano da Silva Lopes[3], Juan Luis Guerrero-Rascado[2], Pablo Ortiz-Amezcua[2], Alberto Cazorla[2], Amauri Pereira de Oliveira[1], Eduardo Landulfo[3], Lucas Alados-Arboledas[2]

[1]*Institute of Astronomy, Geophysics and Atmospheric Science, São Paulo, Brazil*
[2]*Andalusian Institute for Earth System Research (IISTA-CEAMA), Granada, Spain*
[3]*Institute of Energetic and Nuclear Research (IPEN), São Paulo, Brazil*
[4]*Federal Institute of São Paulo (IFSP), São Paulo, Brazil*

**Email: gregori.moreira@usp.br*

ABSTRACT

The Atmospheric Boundary Layer (ABL) is the lowermost part of the troposphere. In this work, we analysed the combination of ABL height estimated continuously by three different remote sensing systems: a ceilometer, a Doppler lidar and a passive Microwave Radiometer, during a summer campaign, which was held in Granada from June to August 2016. This study demonstrates as the combined utilization of remote sensing systems, based on different tracers, can provide detailed information about the height of ABL and their sublayers.

1. INTRODUCTION

The lowermost region in the troposphere is denominated Atmospheric Boundary Layer (ABL). This layer "*is directly influenced by the presence of the Earth's surface, and responds to surface forcings with a time scale of about an hour or less*" (Stull, 1988), so that this process originates an unstable layer denominated Convective Boundary Layer (CBL). Close to sunset, the reduction of incidence of solar radiation causes gradual suppression of the convective processes, resulting in a weak and sporadic turbulence. Due to this phenomenon, the CBL becomes two different layers: the Stable Boundary Layer (SBL), and the Residual Layer (RL).

The height of ABL ($ABLH$) is an important parameter for several research fields, like air quality, weather forecasting, etc. The $ABLH$ can be estimated from some atmospheric variables, e.g. vertical wind speed (w), potential temperature (θ), humidity (q) and vertical aerosol distribution, which does not necessarily provide the same results due to different $ABLH$ definitions adopted by each methodology. In this paper, we propose to estimate the $ABLH$ from three different remote sensing systems: microwave radiometer (MWR), ceilometer and Doppler lidar, which can provide an $ABLH$ estimation from the respective tracers: θ, vertical aerosol distribution and w, so that the combination of these different results can provide a detailed observation of ABL and their sublayers. The data acquire campaign was held in IISTA-CEAMA (Andalusian Institute for Earth System Research - Granada [37.16°N, 3.61°W, 680 m a.s.l.] - Spain) from June to August 2016.

2. INSTRUMENTATION AND METHODOLOGY

2.1 The MWR Method

The ground-based passive microwave radiometer (RPG-HATPRO G2, Radiometer Physics GmbH) is situated at IISTA-CEAMA and it is part of the MWRnet - (Rose et al., 2005). It measures the sky brightness temperature with a radiometric resolution between 0.3 and 0.4 K root mean square error at 1 s integration time and uses direct detection receivers within two bands: 22-31 GHz (water vapor - K band) and 51-58 GHz (oxygen - V band), which are used for deriving relative humidity and temperature profiles, respectively. The generated profiles have a range resolution varying between 10 and 200 m in the first 2 km and varying between 200 and 1000 km up to 10km (Navas-Guzmán et al., 2014). The algorithm applied to the MWR combines two methodologies

https://doi.org/10.1051/epjconf/202023703009

that are the parcel method - $[PM]$ (Holzworth, 1964) and the temperature gradient method $[TGM]$ - (Stull, 1988), in order to estimate the $ABLH_{MWR}$ in convective (CBL) and stable (SBL) situations, respectively. First of all the potential temperature profile, $\theta(z)$, is derived from the temperature $(T(z))$ profile provided by the MWR, using the following equation:

$$\theta(z) = T(z) + 0.0098 * z \quad (1) \text{ (Stull, 2011)}$$

where z is the height above the sea level, and 0.0098K/m is the dry adiabatic temperature gradient. The surface potential temperature $[\theta(z_0)]$ is obtained from the meteorological station co-located with the MWR. This $\theta(z)$ profile is analyzed in order to label the situation as stable or unstable. Such analysis is made by comparing the surface potential temperature ($\theta(z_0)$) with all points in the $\theta(z)$ profile below 5 km. If all points presents $\theta(z)$ values larger than $\theta(z_0)$ the situation is considered stable and TGM is applied, otherwise the situation is assumed as unstable and the PM is used instead (Figure 1).

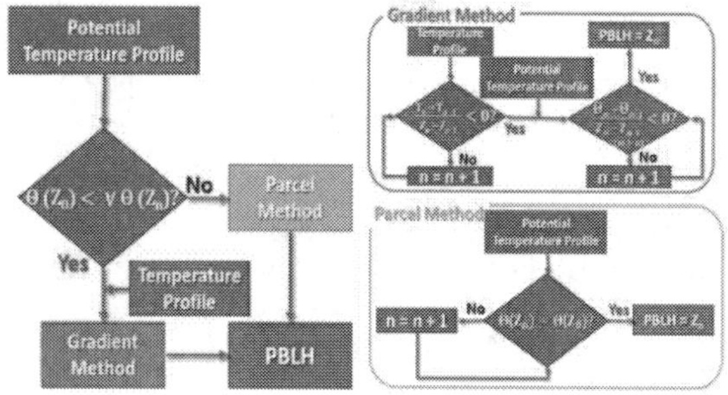

Figure 1 – Combination of two methods to estimate the *ABLH* based on Temperature Profile.

2.2 Doppler lidar: Variance threshold method

The coherent Doppler Lidar (Halo Photonics) model Stream Line is allocated at IISTA-CEAMA. This system uses heterodyne detection to measure the Doppler shift of backscattered light. It operates an eye-safe laser transmitter vertically pointing to zenith emitting at 1.5 µm with pulse energy and repetition rate of 100 µJ and 15 KHz, respectively. The system operates with 300 gates, where the range gate length is 30 m and its first gate is located at 60 m. The data acquisition is performed in Stare mode (only the vertical wind speed is measured) with a time resolution of 2 s.

The variance of vertical wind speed $({\sigma_w}^2)$ is used to estimate the vertical size of plumes growing due to homogeneous turbulent movement. Therefore, this variable is applied as an indicator of mixing layer height $(ABLH_{Doppler})$.The $ABLH_{Doppler}$ is adopted as the first height where ${\sigma_w}^2$ has a value lower than a predetermined threshold (th_{var}), which in this paper was adopted as 0.16 m²/s² (Moreira, 2018). The σ_w^2 is calculated using time intervals of 30 minutes.

2.3 Ceilometer: Gradient method

The ceilometer Jenopitk model CHM15k is operated at the IISTA-CEAMA station. It operates with a pulsed Nd:YAG laser emitting at 1064 nm and a telescope with a field of view of 0.45 mrad. The energy per pulse is 8.4 µJ with a repetition frequency in the range of 5–7 kHz. The laser beam divergence is less than 0.3 mrad. The spatial and temporal resolution used are 15 m and 15 s, respectively.

From ceilometer data the $ABLH$ is estimated by the aerosol backscatter using the gradient method $[GM]$ (Caicedo et al., 2017). This method consists in detecting the minimum of gradient in the $RCS(z)$. Due to a typical reduction of aerosol concentration in the free troposphere (FT) when compared to ABL, this transition region (corresponding to $ABLH_{ceilometer}$) is characterized by an abrupt reduction in Range Corrected Signal ($RCS(z)$ signal).

$$ABLH_{ceilometer} = \min\left(\frac{dRCS(z)}{dz}\right) \quad (2)$$

3. RESULTS

The figure 2 presents a box whisker plot of $ABLH_{ceilometer}$. From 01:00 to 08:00 UTC, the data present low variability and a median quite constant (1000 m). At 09:00 UTC the $ABLH_{ceilometer}$ begins to present greater variability and higher average values. The highest values are observed at 15:00 UTC, so that in this period, 75% of values are higher than 1800 m. After this period, the variability of data is reduced and the median value begins to decrease reaching around 1000 m at 00:00 UTC.

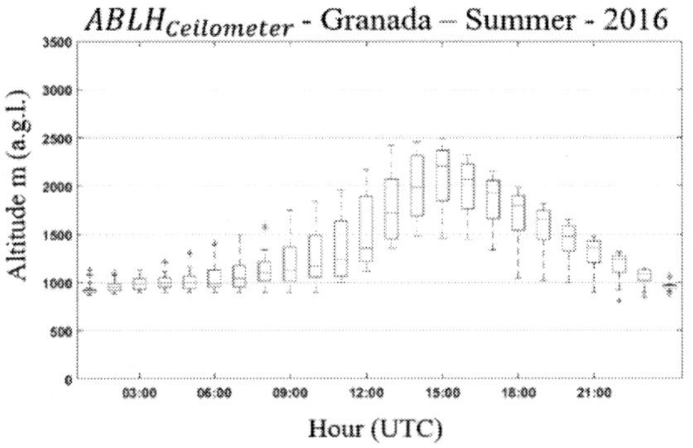

Figure 2 – Daily $ABLH_{Ceilometer}$ cycle for Summer 2016. Whiskers and boxes indicate 10, 25, 75 and 90% percentiles. The red lines represent the median.

In figure 3 a box whisker plot of $ABLH_{MWR}$ is presented. From 01:00 to 06:00 UTC the distribution presents low variability and a quite constant value of median (200 m). At 07:00 UTC the variability and the median begin to increase, reaching the maximum at 15:00 UTC (approximately 1600 m). Although the median begins to decrease at 17:00 UTC the high variability is maintained until 19:00 UTC, so that after this period the $ABLH_{MWR}$ decreases maintaining the same pattern observed during the dawn.

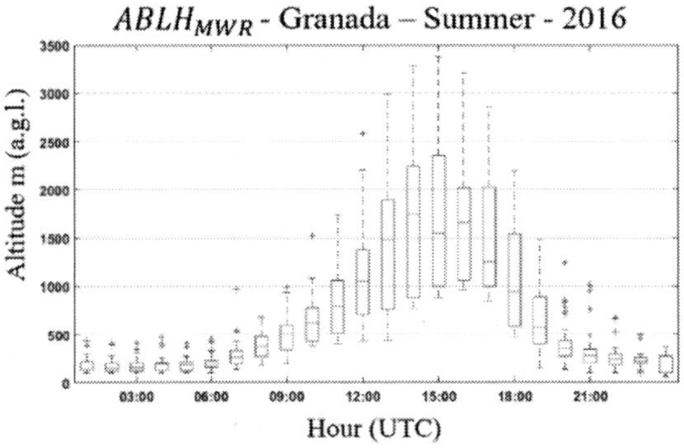

Figure 3 – Daily $ABLH_{MWR}$ cycle for Summer 2016. Whiskers and boxes indicate 10, 25, 75 and 90% percentiles. The red lines represent the median.

In figure 4 is presented a box whisker plot of the $ABLH_{Doppler}$. From 01:00 to 05:00 UTC the distribution presents low variability and a quite constant value of median (200 m). At 06:00 UTC the median value begins to increase, reaching its maximum at 15:00 UTC (1300 m) and maintaining it until 16:00 UTC. However, it is not observed an increase of the variability of distribution, as demonstrated in figures 3 and 4. At 17:00 UTC the median begins to decrease reaching at 00:00 UTC the same value observed during the dawn.

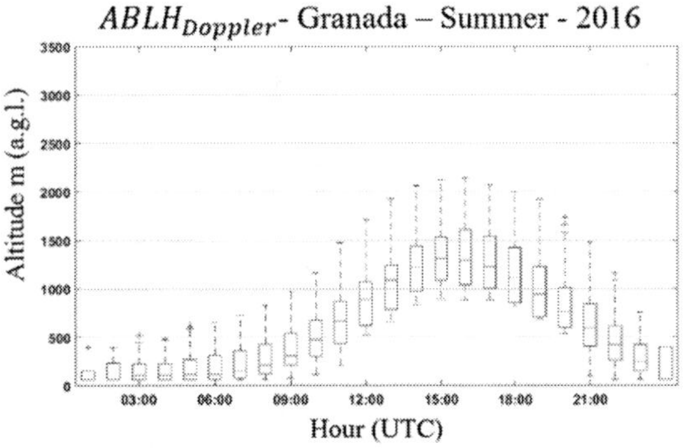

Figure 4 – Daily $ABLH_{Doppler}$ cycle for Summer 2016. Whiskers and boxes indicate 10, 25, 75 and 90% percentiles. The red lines represent the median.

The distribution of $ABLH_{Ceilometer}$ presents distinct behavior in comparison with $ABLH_{MWR}$ and $ABLH_{Doppler}$, due to the different ABL definition adopted by GM in comparison with the another ones, resulting in detection of distinct ABL sublayers.

The $ABLH_{Ceilometer}$ is based on vertical aerosol concentration, therefore it represents the full-developed CBL (often in the central region of the day) and the RL in the rest of the time. It justifies the higher values and low variability observed during the stable period.

On the other hand, $ABLH_{MWR}$ and $ABLH_{Doppler}$, although based on utilization of different tracers, have similar behavior estimating the height of CBL and SBL in convective and stable cases, respectively. It justifies the low values of median observed during the stable periods, as well as, the fast variations observed at 07:00 and 16:00 UTC. During the convective situations, $ABLH_{MWR}$ and $ABLH_{Doppler}$ have a similar behavior, although the $ABLH_{Doppler}$ has low variability in comparison with $ABLH_{MWR}$. It probably is caused by a reduction in the vertical resolution of MWR above 2000 m.

Considering the observation of different sublayers that compound the ABL, the joint observation of three $ABLH$ presented in previous figures can provide a detailed analysis of $ABLH$ behavior, as demonstrated in figure 5, where the green, red and blue line represents the average value of $ABLH_{Ceilometer}$, $ABLH_{MWR}$ and $ABLH_{Doppler}$, respectively.

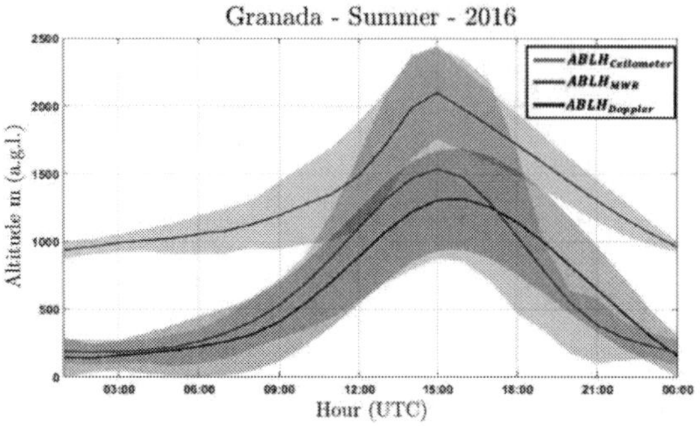

Figure 5 – Average values of $ABLH$ obtained during Summer 2016. Green, red and blue line represent $ABLH_{Ceilometer}$, $ABLH_{MWR}$ and $ABLH_{Doppler}$, respectively. The shadows represent the standard deviation of line with same color.

From this picture, we can observe the low variability of RL ($ABLH_{Ceilometer}$) and SBL ($ABLH_{MWR}$ and $ABLH_{Doppler}$) between 01:00 and 06:00 UTC. After this period the convective process increases, so that, the CBL ($ABLH_{MWR}$ and $ABLH_{Doppler}$) arises fastly, while the RL ($ABLH_{Ceilometer}$) ascends slowly. Approximately at 11:00 UTC the CBL breaks-up the RL and becomes full-developed. Consequently, all methods estimate the CBL height providing profiles with similar behaviors. After 16:00 UTC the difference between $ABLH_{Ceilometer}$ and other ones increase again because it returns to detect the RL other ones the SBL. The shadows (standard deviation) reinforce the similarity among the methods when the same sublayer is detected.

4. CONCLUSION

In this paper, we performed an estimation of $ABLH$ from three different tracers: w, Θ, and vertical aerosol distribution, which were detected from Doppler lidar, MWR and Ceilometer data, respectively, during a campaign executed during Summer 2016 at IISTA-CEAMA. The $ABLH_{MWR}$ and $ABLH_{Doppler}$ present similar behavior during all periods, although the first one has a higher variability during the convective period. On the other hand the $ABLH_{Doppler}$ presents a high similarity with the other ABL heights only in the central region of the day. It occurs, because each method is based on a distinct $ABLH$ definition, consequently during the stable period, $ABLH_{MWR}$ and $ABLH_{Doppler}$ estimates the SBL height, while $ABLH_{Doppler}$. However, the joint observation of these three variables can provide a detailed comprehension about the sublayers that compound the ABL and how they interact along the time.

ACKNOWLEDGEMENTS

This work was supported by the Spanish Agencia Estatal de Investigación, AEI, through projects CGL2016-81092-R, CGL2017-83538-C3-1-R and CGL2017-90884-REDT. We acknowledge the financial support by the EU Horizon 2020 research and innovation program through project ACTRIS-2 (grant agreement No 654109) and, CNPQ by the Projects 152156/2018-6 and 432515/2018 and, Post-Doctor Scholarship (154320/2018-8).

REFERENCES

[1] Stull, R. B. An Introduction to Boundary Layer Meteorology. vol. 13, Kluwer Academic Publishers, the Netherlands, Dordrecht/Boston/London, 1988.

[2] Rose, T., Creewll, S., Löhnert, U., Simmer, C. A network suitable microwave radiometer for operational monitoring of cloudy atmosphere. Atmospheric Research, 75, 183 – 200 (2005).

[3] Navas-Guzmán, F., Fernández-Gálvez, J., Granados-Muñoz, M. J, Guerrero-Rascado, J. L., Bravo-Aranda, J. A., and Alados-Arboledas, L. Tropospheric water vapor and relative humidity profiles from lidar and microwave radiometry. Atmospheric Measurement Techniques, 7, 1201-1211 (2104).

[4] Holzworth, C. G. Estimates of mean maximum mixing depths in the contiguous United States. Monthly Weather Review, 92, 235–242 (1964).

[5] Stull, R. B. Meteorology for Scientists and Engineers. 3rd Edition, Uni. Of British Columbia, 2011.

[6] Moreira, G. A, Guerrero-Rascado, J. L., Bravo-Aranda, J. A., Benvant-Oltra, J. A., Ortiz-Amezcua, P., Róman, R., Bedoya-Velásquez, A. E., Landulfo, E., Alados-Arboledas, L., Sutdy of the planetary boundary layer by microwave radiometer, elastic lidar an Doppler lidar estimations in Southern Iberian Peninsula. Atmos Resear., 213, 185-195 (2018).

[7] Caicedo, V., Rappenglück, B., Lefer, B., Morris, G., Toledo, D., Delgado, R. Comparison of aerosol lidar retrieval methods for boundary layer height detection using ceilometer aerosol backscatter data. Atmospheric Measurement Techniques, 10, 1609–1622, (2017).

CORRELATION STUDY OF PLANETARY-BOUNDARY-LAYER-HEIGHT RETRIEVALS FROM CL51 AND CHM15K CEILOMETERS WITH APPLICATION TO PM2.5 DYNAMICS IN NEW YORK CITY

Dingdong Li [1], Barry Gross [1,2], Yonghua Wu [1,2], Fred Moshary [1,2]*

[1] *Optical Remote Sensing Lab, City College of New York, USA*
[2]*NOAA – Cooperative Science Center for Earth System Sciences and Remote Sensing Technologies*
Email: moshary@ccny.cuny.edu

ABSTRACT

Planetary-boundary-layer-height (PBLH) plays a critical role in the study of urban air quality, weather and climate. Continuous observation is critical in understanding air pollution processes and evaluation of air quality/ weather models in the complex urban environment. In this study, we observe the PBLH variation using multiple ceilometers and lidar in New York City (NYC) during both the summer and winter time and explore the potential correlation with ground PM2.5. An automated quality control and quality assurance (QC/QA) method is developed to optimize the PBLH determination from the ceilometers (Vaisala CL51 and Lufft CHM15k) product. The PBLHs from the two ceilometers and lidar show good consistency (R^2=0.68~0.88) during the convective PBL period at 15:00-21:00 UTC (10:00-16:00 EST). We also investigate the seasonal variation and diurnal evolution of PBLH and demonstrate an inverse relation between the PBLH and PM2.5 during the morning transient period of PBLH growth. Further, the correlation between the ceilometer-attenuated backscatter and ground PM2.5 and its dependences on the vertical altitude are analyzed, showing that the aerosols in the PBL are more deeply mixed while also being influenced by the relatively high humidity variability during the summer.

1. INTRODUCTION

In conventional air quality monitoring of fine particulate matter (PM2.5) performed by the EPA, observations are generally limited to the surface level [1]. Although such surface monitoring is critical due to specific EPA NAAQS health standards for PM2.5 exposure [2], surface measurements alone do not provide insight into how ambient aerosols mix within the planetary boundary layer. In particular, diurnal cycle and season variation can have a dramatic effect on both the mixing height and the homogeneity of the aerosols. These mechanisms drive the venting of aerosol emission sources and allow pollution to vent through the PBL, resulting in a basic anti-correlation between PBLH and the surface pollution under well mixed conditions.

In prior researches at CCNY, Nd:YAG high power lidar was employed to extract PBLH information using WRF models [3]. Despite of high SNR, such lidar system can only measure a limited set of days due to manual operation and eye safety concerns and therefore is not suitable for long-term continuous studies. In this case, automatic lidar and ceilometer (ALC) instruments are advantageous for urban atmosphere observation given their continuous 24-hour measuring capacity of atmospheric profiling including cloud base height, aerosol attenuated backscatter coefficient and PBLH. In previous WRF-CMAQ PBLH and PM2.5 analyses [3], a Vaisala CL-31 ceilometer was used to retrieve the PBLH. However, the relatively weaker laser power of CL-31 leads to a lower signal-noise-ratio (SNR). And the lower SNR makes it more difficult to resolve PBLH > 1 km and find unique aerosol layer structure within the PBL.

To improve observational capacity, more powerful and robust ceilometers with higher SNR are desired for PBL and aerosol profile dynamics study. Compared to Vaisala CL-31, Vaisala CL51 and Lufft CHM15k have better SNR, higher detection range, but at the same time, suffer from longer optical non-overlap length. Under clear sky condition and in strong convective PBL period, the aerosols are well mixed, in which circumstance the PBLH often refers to mixing-layer-height (MLH), the PBLH results from different ALCs are highly consistent. But complex atmospheric conditions such as multiple aerosol layers, clouds or rainy weather will significantly affect ceilometer PBLH detection accuracy. Therefore, quality assurance/

quality control (QA/QC) post-process is necessary to exclude extraneous layer heights, outliers and data of rainy weather and thus improves the PBLH product quality and its applications in the air quality assessment.

2. METHODOLOGY

2.1 Observation instruments

In comparison study, the attenuated backscatters and PBLHs are retrieved by a Vaisala CL51 ceilometer, a Lufft CHM15k ceilometer and a 3-wavelength Elastic-Raman (Nd:YAG) lidar coupled with LICEL transient recorders (TR40-160) deployed on the roof of the Engineering School of City College of New York (CCNY). The distance between these three instruments is 5-10 m to avoid signal interference but close enough to collect data from the same atmospheric volume. Nearby is a standard surface air quality monitoring station by the New York State Department of Environment Conservation (NYSDE), collecting air quality data (PM2.5, ozone and carbon monoxide concentration) with 1-hour collection time. The Vaisala CL51 ceilometer operates at a wavelength of 910 ± 10 nm at 25 °C, and the measurement range is 0-10,000 m. The time/height resolution of the CL51 ceilometer is set to be 16 s /10 m. The center wavelength of Lufft CHM15k is 1064 nm and the measurement range is 5-15,000 m, with time/height resolution of 15 s/15 m, respectively. The lidar transmits three wavelengths (1064-, 532- and 355-nm) with a flash lamp-pumped Nd: YAG laser at a repetition rate of 30 Hz (Spectra-physics Quanta-Ray PRO-320). The two ceilometers make automatic 24-hour observation in all-weather conditions, while the lidar requires manual operation and thus only collects data in the daytime under appropriate weather condition. All data recording is based on UTC and the data analysis is done in MATLAB R2018b.

2.2 Data processing

To improve the SNR and align the timestamps of the data from two ceilometers, the attenuated backscatters of CL51 and CHM15k are averaged every 10 minutes. Along the range dimension, a 150 m sliding window is applied to the backscatters and the maximum range is set to be 4500m which captures all PBL aerosols but excludes high clouds and smoke plumes. Additionally, in PM2.5 and backscatter correlation evaluation (section 3.3), the attenuated backscatter coefficient of lower clouds is excluded based on the histogram of attenuated backscatter from that day.

Both ceilometers retrieve PBLH using negative gradient of the backscatter profile and provide up to three levels of PBLH attributions of each time. Here, the lowest detectable PBLH attribution is set to be the primary PBLH. To reduce the calculation consumption, the PBLHs are averaged every 10 minutes. Afterwards, a QA/QC post-process is implemented as follows:

- Initiate PBLH false positive flags (pbltfp) as 0 for each PBLH;
- Compare the PBLHs before sunrise ($PBLH_{bfsr}$), morning transition ($PBLH_{mt}$), afternoon ($PBLH_{aftn}$) with thresholds T_1, T_2, T_3, respectively, if PBLH greater than the corresponsing threshold, then pbltfp=1;
- Compare the PBLHs after sunset ($PBLH_{afst}$) with PBLH in the noon time, if $PBLH_{afst} > PBLH_{noon}$, then pbltfp=1;
- Find the shape peak/valley of PBLHs of an entire day using MATLAB built-in function 'findpeaks' (Copyright 2007-2016 The MathWorks, Inc.) and set their pbltfp =1;
- If PBLH > mean (PBLHs) $\pm$ 3std, pbltfp=1;

Apart from our QA/QC process of PBLH, the sky condition is evaluated by duration time of near surface backscatters which are greater than the threshold, indicating rain, fog or dust on the lens. If such duration time is more than 3 hours, the data file of this day will not be considered.

3. RESULTS and DISCUSSION

3.1 MLH correlation of lidar & ceilometers

Figure 1 shows the correlation of the MLHs from lidar with MLHs from CHM15k (Figure 1 (a)) and from CL51 (Figure 1 (b)). The MLHs are collected in 10:00-17:00 (EST) of 7 clear sky days (no rain). Generally, MLHs from both ceilometers are strongly correlated with MLHs from lidar measurement. But the CHM15k leads to a more accurate retrieval (R^2=0.81) with less root mean square error (RMSE=0.154). MLHs from CL51 are less consistent with lidar measurement (R^2=0.68, RMSE=0.204), which is mainly caused by lower SNR of CL51.

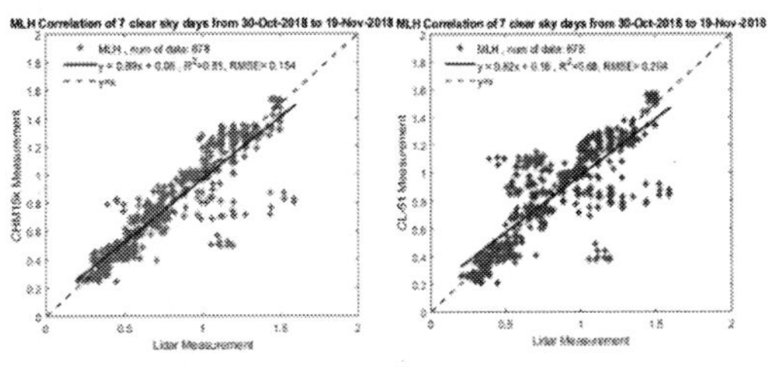

Figure 1. MLHs comparisons between lidar and ceilometers (a) CHM15k (b) CL51

3.2 PBLH correlation between the ceilometers

We calculated the correlation coefficient of PBLHs retrieved by both ceilometers for 24 hours over 17 clear sky days (no rain) in November and December 2018. Before QA/QC process, the correlation of PBLHs from both ceilometers is very poor with R^2 = 0.1~0.2 (not shown). After QA/QC process, the correlation between PBLHs from two ceilometers improves to R^2=0.76 with RMSE=0.172. Figure 2 shows the correlation coefficient R^2 of PBLHs from two ceilometers in different time periods. Strong correlation appears in 10:00-16:00 (EST) which is the convective PBL period. However, the correlation degrades in the nocturnal period.

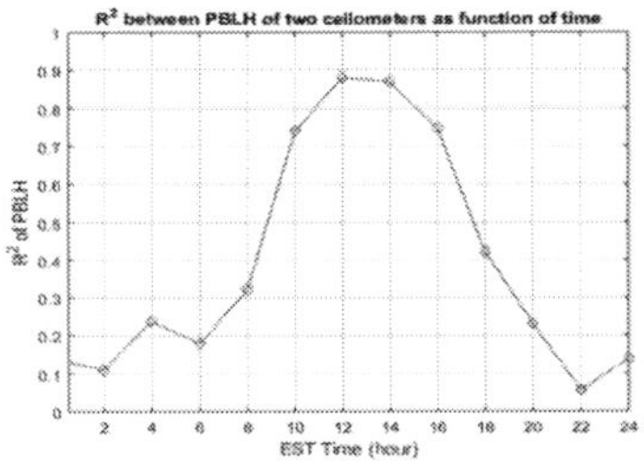

Figure 2. PBLH comparisons between CHM15k and CL51

Such discrepancy can be possibly explained by the following effects: 1. The influence from the residual layers. 2. The nocturnal PBLH during the fall/winter can be below the non-overlap range of CHM15k, so CHM15k detects higher aerosol layers. 3. The aerosol concentration is too low for ceilometers to capture the PBLH.

3.3 PBLH potential impact on PM2.5

With the PBLHs obtained from both ceilometers (no data from CHM15k during the summer), the averaged diurnal PBLHs in July, August and November, December 2018 under the clear sky condition (no rain) are calculated, in comparison with the averaged diurnal PM2.5 concentration of the same days. The hourly PM2.5 data are from air quality station in CCNY operated by NYSDEC.

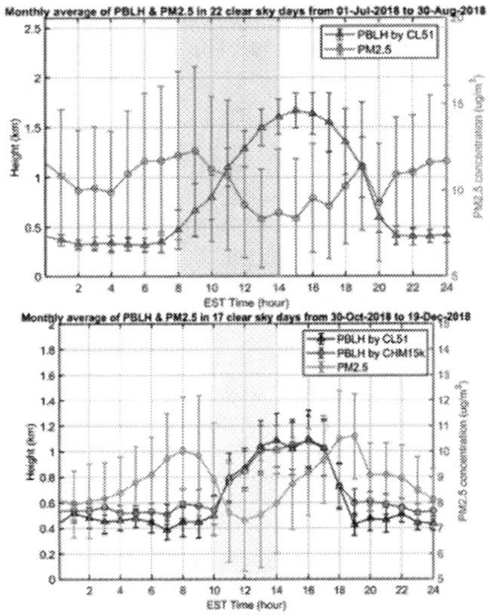

Figure 3. Inverse relationship exhibited between surface PM2.5 and PBL height during morning transition. (a) The diurnal averaged PBLHs measured by CL51 (no CHM15k data during summer) and the diurnal averaged PM2.5 in summer. (b) The diurnal averaged PBLHs measured by CL51 and CHM15k compared with the diurnal averaged PM2.5 in fall/winter.

In Figure 3 (a), the averaged diurnal PBLH demonstrates that PBLH starts to increase around 8:00 and reaches the maximum at 1670 m around 15:00 with growth rate ~ 174 m/hour. Meanwhile, PM2.5 reaches the peak around 9:00 and starts to decrease as the PBLH increases, then fluctuates but slightly increase in the rest time. Figure 3 (b) shows the averaged diurnal PBLHs calculated from CL51 measurements (blue line) and from CHM15k measurements (red line). Here we can see that the averaged PBLH from CHM15k is slightly higher than PBLH from CL51 during the nocturnal period, inferring that CHM15k tends to detect higher aerosol layer heights during the night time as mentioned before. In fall/winter, the morning transition of PBLH starts from 10:00, and PBLHs reach the peak at 1100 m around 2:00 (growth rate 147 m/hour). The PM2.5 diurnal cycle in fall/winter is similar to the summer, showing two peaks around 8:00 and 18:00 which is probably caused by car emission. Strong inverse relation between PBLH and PM2.5 also appears in the morning transition period, but slightly shifted forward compared to the result in summer. This inverse relation between PBLH and PM2.5 implies that the aerosol particles are diluted when the PBLH growth leads to larger volume of PBL.

3.4 Correlation of backscatter & PM2.5

Next, we explore the correlation coefficient of the attenuated backscatter at different heights with the surface level PM2.5 under clear sky condition. In particular, the result illustrates if the attenuated backscatter can be used as a proxy for fine particle matter and provides a better understanding of the mixing effect of PBL on the pollutants at surface level and upper atmosphere. Figure 4 left panel displays the correlation dependence on vertical altitude for both ceilometers in winter. The CHM15k correlation degradation below 100 m is due to its larger non-overlap range. The correlation between PM2.5 and PBLHs by CL51 in summer and fall/winter are shown in figure 4 right panel. Note that near field correlation is slightly better in winter which is probably due to the smaller changes of relative humidity in winter affecting the mass to optical coefficient relationship. Besides, the correlation decreasing rate with respect to the height is smaller for summer than that of the winter, implying the deeper mixing effect of aerosols during the summer.

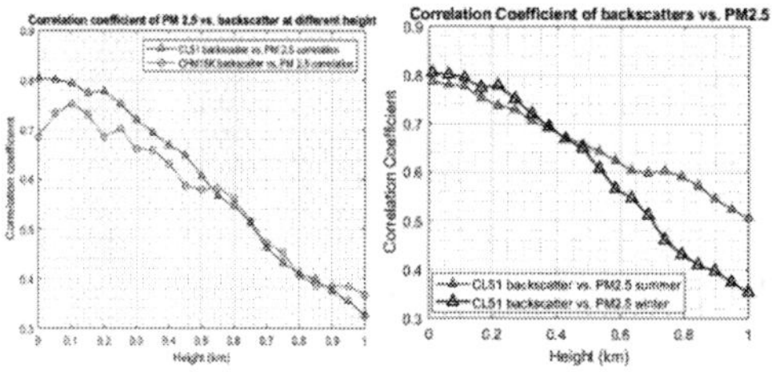

Figure 4. Left: Correlation as function of height between backscatter from two ceilometers and surface PM2.5 concentrations. Right: Correlation as function of height between backscatter from CL51 and surface PM2.5 concentrations in winter and summer.

4. CONCLUSIONS

This study demonstrates the potential of using ceilometers for PBL dynamics studies in complex urban environments. We compared and analyzed the PBLH results retrieved from lidar, CL51 and CHM15k ceilometers and optimized by an automated QA/QC method. Diurnal averaged PBLH were compared with PM2.5 during summer and fall/winter, showing strong inverse relation between them during the morning transition period. Finally, backscatter and PM2.5 exhibits strong correlation showing that the ALC measured attenuated backscatter has potential to serve as a proxy for particle mass, but the conversion coefficient of attenuated backscatters to PM2.5 need to be further addressed. In the future study, more climatic factors and larger date set should be assessed to fully understand the mechanism behind the PBLH and air quality and to assess PBL retrievals in Meteorological Models.

ACKNOWLEDGEMENTS.

This work is supported by the New York State Energy Research and Development Authority under Grant No.100415 and No.137482, and the he National Oceanic and Atmospheric Administration – Cooperative Science Center for Earth System Sciences and Remote Sensing Technologies under the Cooperative Agreement Grant #: NA16SEC4810008. We appreciate the data from NYS Department of Environmental Conservation. The statements contained within the manuscript are not the views of the funding agency or the U.S. government, but reflect the authors' views.

REFERENCES

[1]https://www3.epa.gov/ttn/amtic/contmont.html
[2]https://www.epa.gov/criteria-air-pollutants/naaqs-table
[3] C.-M. Gan, Y. Wu, B.L. Madhavan, B.Gross, F. Moshary, Application of active optical sensors to probe the vertical structure of the urban boundary layer and assess anomalies in air quality model PM2.5 forecasts, Atm. Environment 45, 6613-6621 (2012)
[4] Wu, Y., S. Chaw, B. Gross, et al., 2009: Low and optically thin cloud measurements using a Raman-Mie lidar, Appl. Opt., 48, 1218-1227.

DIURNAL VARIATIONS OF CO_2 MIXING RATIO IN THE LOWER ATMOSPHERE BY THREE WAVELENGTH DIAL

Yasukuni Shibata , Chikao Nagasawa, Makoto Abo

Tokyo Metropolitan University, Hino, Tokyo 1910065, Japan

**Email: sibata@tmu.ac.jp*

ABSTRACT

We have conducted the measurement of high accurate CO_2 mixing ratio profiles by measuring the temperature profiles simultaneously using the three wavelength CO_2 DIAL. The measurements of CO_2 diurnal variation in the lower atmosphere were carried out on sunny and cloudy days respectively. We find out that increasing of the CO_2 mixing ratio occurs over the altitude of about 2 km from the surface during nighttime. On the other hand, the CO_2 mixing ratio decreases over the lower atmosphere during daytime. In particular, the CO_2 mixing ratio decreases earlier on sunny days than on cloudy days after sunrise. This result suggests that CO_2 absorption by photosynthesis greatly contributes to the strength of the solar radiation.

1. INTRODUCTION

Validating and improving the global atmospheric transport model requires precise measurement of the carbon dioxide (CO_2) profile. In particular, knowledge of CO_2 diurnal variation in the lower atmosphere is important for the elucidation of sinks and sources. We have developed a ground based direct detection three wavelength 1.6 μm DIAL to achieve measurements of vertical CO_2 concentration and temperature profiles in the atmosphere[1-3]. In the DIAL observation, the temperature dependence of the CO_2 mixing ratio due to the temperature dependence of the absorption cross section is 1.6 ppm/K near the ground and the air density in the atmosphere. This means that the difference between the model and the actual temperature profile greatly influences the accuracy of the CO_2 mixing ratio. We carried out measurements of CO_2 diurnal variations on the sunny and cloudy day. In this paper, we report on the details of CO_2 diurnal variations and the differences in CO_2 profiles due to weather.

2. CO_2 DIURNAL VARIATIONS

The CO_2 DIAL consists of an optical parametric generator (OPG) transmitter that excited by the LD pumped Nd:YAG laser. The on-line DFB seed laser onto the line center (1572.992 nm) by the wavelength control unit including CO_2 reference cell. The off-line DFB seed laser (1573.137 nm) is operated by the free-run mode. The third DFB seed laser (1573.040 nm) for the temperature measurement is operated by the offset lock system. These seed lasers were connected to an optical fiber switch, and the switching speed is 2.5 ms. The atmospheric backscatters were collected by a 250 mm telescope and was connected into the near-infrared PMT module operated with an analog mode. We realized the CO_2 concentration measurement for daytime by using the narrowband interference filter with 1.0 nm FWHM.

Atmospheric CO_2 measurements were conducted from 0.39 to 2.5 km by the three wavelength DIAL at the Hino campus of Tokyo Metropolitan University (35.7N, 139.4E). Figure 1 shows the time-height cross section of the CO_2 mixing ratio with an integration time of 30 minutes. Japan was covered with high pressure during the observation period. It was fine and the wind was weak. A CO_2 gas sensor (LI-COR, LI-7500) was installed at the our lidar laboratory. CO_2 mixing ratio between the ground and 0.39 km altitude were interpolated. The sunrise was 5:00LT and the sunset was 18:23LT.

CO_2 mixing ratios increased from 22:00 LT on 21 April to 07:00 LT on 22 April at the altitude of 1.5 km or less. On the other hand, CO_2 mixing ratios decreased during daytime (especially in the afternoon). Furthermore, the increase or decrease in CO_2 mixing ratios occurs earlier in the lower layer than in the upper layer. Diurnal variations are attributable primarily to the natural photosynthetic and respiration cycle of vegetation and soils and to anthropogenic sources in urban areas.

https://doi.org/10.1051/epjconf/202023703011

Figure 2 shows the time-height cross section of the CO_2 mixing ratio on 30 - 31 October, 2018. It was cloudy during DIAL measurement and the clouds were about 5 km altitude and the wind was weak. The sunrise was 6:01LT. After 10:00LT, background level increased and we could not continue the DIAL measurement. As in the case of Fig. 1, the CO_2 mixing ratio at night time increased to an altitude of 1.5 km, and the value was 410 ppm or more. However, even after 2 hours from sunrise, the decrease in CO_2 mixing ratio as shown in Fig. 1 was not measured. We consider that the weaker photosynthetic activity on cloudy days compared to sunny days is due to the difference in the decrease of the CO_2 mixing ratio after sunrise.

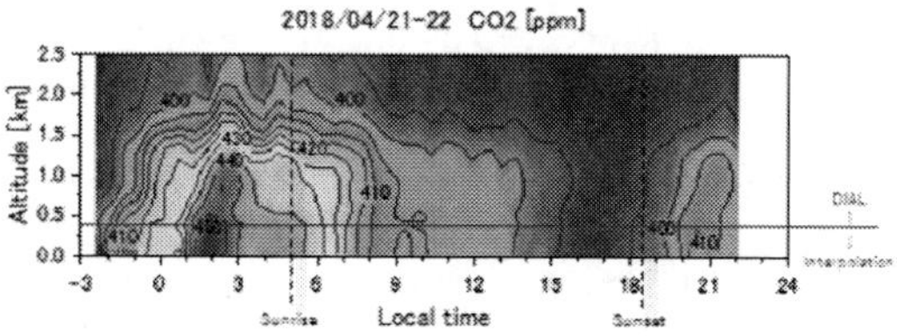

Figure 1 Time-height cross section of the CO_2 mixing ratio on 21-22 April 2018 with an integration time of 30 minutes. Between the ground and the altitude of 390 m, it is interpolated by the measurement of the ground CO_2 sensor (LI-7500) and the DIAL measurement.

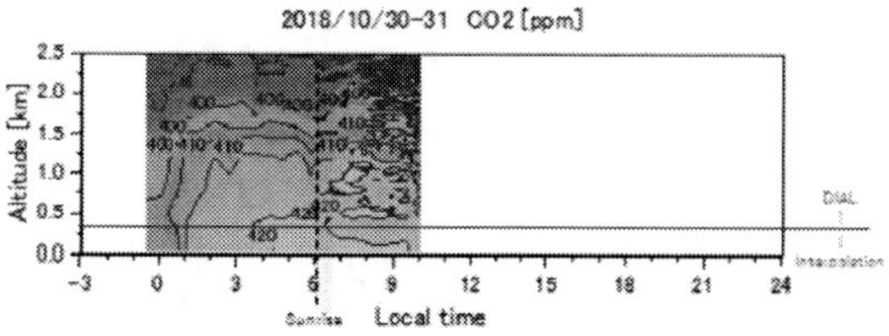

Figure 2 Time-height cross section of the CO_2 mixing ratio on 30 - 31 April 2018.

3. CONCLUSION

We have developed the three wavelengths 1.6 μm DIAL system for measurements of the CO_2 mixing ratio and the atmospheric temperature profiles in the troposphere. The CO_2 DIAL conducted measurements of a diurnal variation of CO_2 mixing ratio and temperature profile on 21-22 April 2018 and 30-31 October 2018 from 0.39 to 2.5 km altitude. We can see that the increasing of CO_2 mixing ratios occurs over the altitude of about 2 km from surface during nighttime. On a sunny day (April), the CO_2 mixing ratio decreased significantly for two hours after sunrise. On the other hand, on a cloudy day (October), the CO_2 mixing ratio did not decrease significantly even after two hours from sunrise. However, we could not continue the measurement except for four hours after sunrise due to the deterioration of the signal-to-noise ratio of increase of background radiation arising from clouds over the lidar site, so we could not know the behavior of the CO_2 mixing ratio over the entire daytime. Diurnal variations are attributable to the natural photosynthetic and respiration cycle of vegetation and soils and to anthropogenic sources in urban areas. However, we found out that CO_2 mixing ratio in the lower atmosphere attributes primarily to the photosynthesis of plants under the environment around our lidar site.

ACKNOWLEDGEMENTS

This work was financially supported by the System Development Program for Advanced Measurement and Analysis of the Japan Science and Technology Agency.

REFERENCES

[1] D. Sakaizawa, et al., Appl. Opt. 48, 748-757 (2009).
[2] Y. Shibata, et al., Appl. Opt.,56, 1194-1201 (2017).
[3] Y. Shibata, et. al., Sensors 18, 4064 (2018).

ENTRAINMENT AND MIXING OF TRANSPORTED OZONE LAYERS: IMPLICATIONS FOR SURFACE AIR QUALITY IN THE WESTERN U.S.

Christoph Senff[1]*, Andrew Langford[2], Raul Alvarez II[2], Tim Bonin[1], Alan Brewer[2], Aditya Choukulkar[1], Guillaume Kirgis[1], Richard Marchbanks[1], Scott Sandberg[2], Ann Weickmann[1], Michael Zucker[1]

[1]*University of Colorado CIRES at NOAA ESRL Chemical Sciences Division, Boulder, CO, USA*
[2]*NOAA ESRL Chemical Sciences Division, Boulder, CO, USA*
**Email: christoph.senff@noaa.gov*

ABSTRACT

Recently, two air quality campaigns were conducted in the southwestern United States to study the impact of transported ozone, stratospheric intrusions, and fire emissions on ground-level ozone concentrations. The California Baseline Ozone Transport Study (CABOTS) took place in May – August 2016 covering the central California coast and San Joaquin Valley, and the Fires, Asian, and Stratospheric Transport Las Vegas Ozone Study (FAST-LVOS) was conducted in the greater Las Vegas, Nevada area in May – June 2017. During these studies, nearly 1000 hours of ozone and aerosol profile data were collected with the NOAA TOPAZ lidar. A Doppler wind lidar and a radar wind profiler provided continuous observations of atmospheric turbulence, horizontal winds, and mixed layer height. These measurements allowed us to directly observe the degree to which ozone transport layers aloft were entrained into the boundary layer and to quantify the resulting impact on surface ozone levels. Mixed layer heights in the San Joaquin Valley during CABOTS were generally below 1 km above ground level (AGL), while boundary layer heights in Las Vegas during FAST-LVOS routinely exceeded 3 km AGL and occasionally reached up to 4.5 km AGL. Consequently, boundary layer entrainment was more often observed during FAST-LVOS, while most elevated ozone layers passed untapped over the San Joaquin Valley during CABOTS.

1. INTRODUCTION

Ground-level ozone (O_3) concentrations in the U.S. are often driven by local production, especially in and near urban areas. However, stratosphere-to-troposphere transport (STT) and long-range transport from Asia can also have significant impacts on surface O_3 levels, particularly in the western U.S. Ozone-rich air from the lower stratosphere can be carried into the lower free troposphere by stratospheric intrusions that are spawned by upper-level cyclones. This phenomenon peaks in late spring when cyclonic activity remains high and the O_3 content of the lower stratosphere is at a maximum. Also, many of these intrusions entrain pollution plumes transported across the Pacific Ocean from Asia as they descend through the troposphere and bring this ozone-enriched air toward the surface with them. Global model simulations [1,2] indicate that STT and Asian pollution contributions to U.S. surface O_3 in late spring are largest in the Southwest and the Intermountain West. The higher terrain there and deeper boundary layers compared to the rest of the country increase the likelihood that descending ozone-rich air reaches the surface [3]. O_3 produced in wildfire plumes is another potential source of increased surface O_3. High O_3 associated with STT, Asian pollution, and wildfire plumes is often concentrated in elevated layers in the lower free troposphere that need to be transported to the surface to affect ground-level O_3 concentrations. This can be accomplished by entrainment into the well-mixed boundary layer, isentropic transport directly to the surface in very deep intrusions, or terrain-modulated flow patterns.

To study the relative impact of these three sources on surface air quality compared to local O_3 production and to characterize the processes that transport the elevated O_3 layers to the surface, the CABOTS and FAST-LVOS air quality campaigns were conducted in May-August 2016 along the central California coast and the San Joaquin Valley

https://doi.org/10.1051/epjconf/202023703012

Figure 1: TOPAZ lidar truck at Visalia Municipal Airport during CABOTS.

(SJV) and in May-June 2017 in the Las Vegas Valley (LVV), respectively. Both, the SJV and the LVV often exceed the 8-hour O_3 National Ambient Air Quality Standard (NAAQS). However, the SJV suffers from more severe O_3 exceedances and has been designated as one of only two "extreme" non-attainment areas in the U.S.

2. METHODOLOGY

The NOAA ESRL Chemical Sciences Division (CSD) truck-based Tunable Optical Profiler for Aerosols and oZone (TOPAZ) lidar [4,5] (Figure 1) was a key component of both studies and was deployed at the Visalia Municipal Airport (VMA)

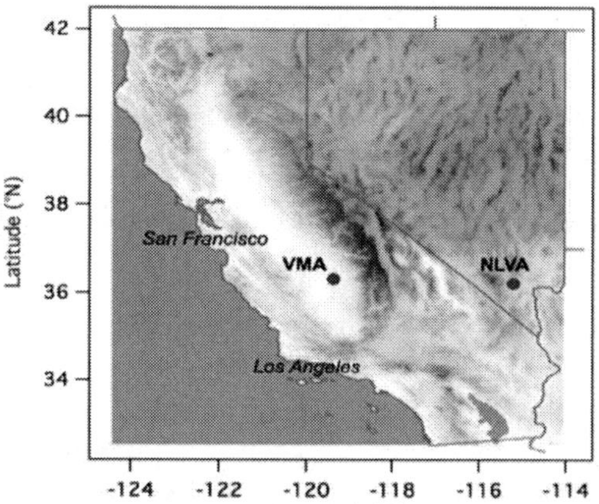

Figure 2: Locations of the NOAA lidars at Visalia, CA (VMA) during CABOTS and at Las Vegas, NV (NLVA) during FAST-LVOS.

during CABOTS and the North Las Vegas Airport (NLVA) during FAST-LVOS (Figure 2). TOPAZ uses the Differential Absorption Lidar (DIAL) technique to measure ozone as a function of distance from the lidar. A large scanning mirror on top of the TOPAZ truck permits scanning of the laser beam at a series of different elevation angles to measure ozone from 25 m to nearly 6 km above the surface during the day, and 8 km at night. TOPAZ is part of the NASA-sponsored Tropospheric Ozone Lidar Network (TOLNet). The autonomous, zenith-looking NOAA/ESRL/CSD MicroDoppler wind lidar was deployed next to TOPAZ during part of CABOTS (7/28 – 8/7/2016) and for the entire FAST-LVOS study to provide continuous observations of vertical wind speed, turbulence, and mixed layer height. The continuously operating San Joaquin Valley Air Pollution Control District (SJVAPCD) radar Radio Acoustic Sounding System (RASS) system, located at VMA, measures horizontal wind and virtual temperature profiles. The latter were used to compute hourly mixed layer height estimates during CABOTS.

3. RESULTS

The TOPAZ observations revealed a complex vertical ozone structure during the CABOTS and FAST-LVOS studies, including distinct high-ozone layers in the free troposphere associated with stratospheric intrusions, long-range transport from Asia, wildfire plumes, and regional pollution transport. The deployment locations of the NOAA lidars during the two studies (VMA during CABOTS and NLVA during FAST-LVOS, see Figure 2), were roughly at the same latitude and only ~400 km apart, however, the mixed layer height (MLH) characteristics were dramatically different. Afternoon MLHs were very shallow at Visalia, CA during CABOTS (200 – 1200 m AGL, Figure 3), which is due to strong subsidence associated with synoptic-scale high pressure and the valley-mountain circulation in California's Central Valley. In contrast, MLHs were much deeper and more variable at Las Vegas, NV during FAST-LVOS (1300 – 4500 m AGL, Figure 3). The different MLH regimes have important ramifications for boundary layer entrainment of transported O_3 layers and for potential impacts on surface O_3 concentrations.

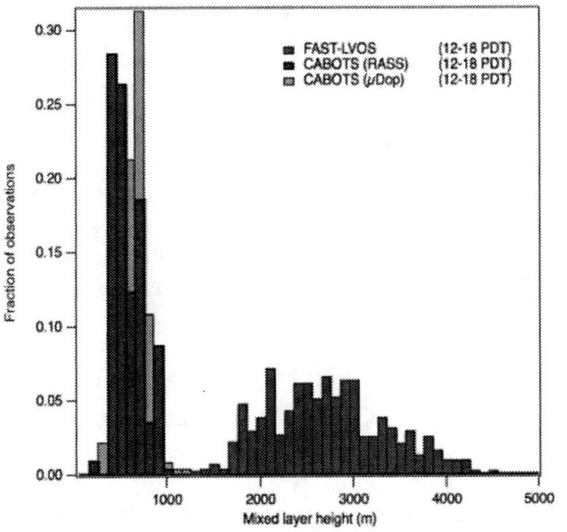

Figure 1: Distribution of observed afternoon mixed layer heights at VMA during CABOTS (from SJVAPCD radar RASS, dark blue; from MicroDoppler lidar, light blue, 7/28-8/7/2016) and during FAST-LVOS (from MicroDoppler lidar, red).

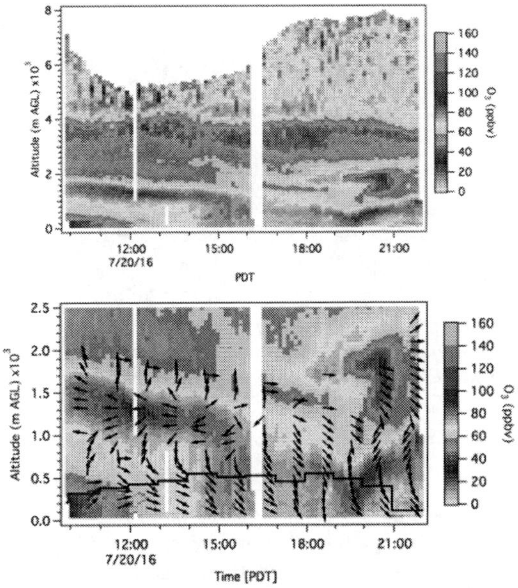

Figure 2: TOPAZ ozone profiles observed on 20 July 2016 during CABOTS. Radar RASS mixed layer heights (black line) and wind arrows are overlaid (bottom panel).

3.1 CABOTS

Figure 4 shows the O_3 distribution observed with TOPAZ on 20 July 2016 during CABOTS (full O_3 profiles up to 8 km AGL, top panel; O_3 profiles in the lowest 2.5 km AGL with horizontal wind and MLH observations overlaid, bottom panel). On this day, a low-altitude O_3 layer was entrained into the boundary layer in the early afternoon, which increased surface O_3 by several ppbv. This layer likely was formed by locally produced O_3 that had been transported into the Sierra Nevada Foothills by upslope flow the day before and then was re-circulated over the SJV in an easterly return flow, as indicated by the radar RASS measured winds in Figure 4 (bottom panel). The high-O_3 layer between 4 and 6 km AGL was of stratospheric and/or Asian origin and passed over the SJV, well out of reach of the shallow mixed layer. Similar observations on several other days during CABOTS confirmed that the mixed layer in the SJV was not nearly deep enough to entrain free-tropospheric O_3 layers associated with STT and transport from Asia, except for one exceptionally deep stratospheric intrusion event.

3.2 FAST-LVOS

During FAST-LVOS, we often observed partial or complete entrainment of elevated O_3 layers of stratospheric or Asian origin into the boundary layer. Figure 5 shows the O_3 profiles observed with TOPAZ on 28 June 2017 during the FAST-LVOS campaign. On that day, a narrow O_3 layer that likely was of Asian origin and contained O_3 concentrations in excess of 100 ppbv descended to 2.5 km AGL by midday, when it was completely engulfed by the growing mixed layer in the LVV. We estimate that the entrainment of this transported O_3 layer resulted in a ~ 5 ppbv increase in surface ozone concentrations. This rather small, but still significant O_3 enhancement may have contributed to O_3 NAAQS exceedances that were observed by the regulatory monitoring network in the LVV on that day. We observed several similar cases that resulted in estimated surface O_3 enhancements of up to 12 ppbv.

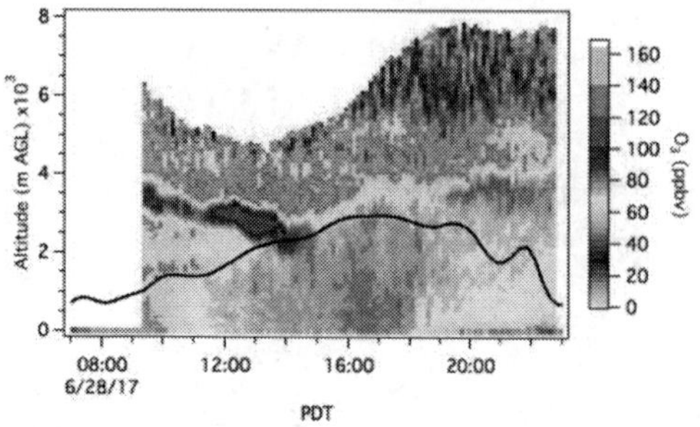

Figure 3: TOPAZ ozone profiles observed on 28 June 2017 during FAST-LVOS. MicroDoppler lidar mixed layer heights (black line) are overlaid.

4. CONCLUSIONS

Shallow mixed layers in California's San Joaquin Valley strongly limited entrainment of transported O_3 layers associated with stratospheric intrusions, Asian transport, and wildfire plumes. Most of these elevated O_3 layers passed over the San Joaquin Valley untapped. In contrast, deep mixed layers in the Las Vegas Valley frequently entrained transported O_3 layers from altitudes as high as 5 km above sea level. As a result, surface O_3 levels often increased, which likely contributed to exceedances of the O_3 NAAQS in the Las Vegas area.

ACKNOWLEDGEMENTS

This work was funded in part by the California Air Resources Board (CARB) under contract no. 15RD012, by the Clark County Department of Air Quality under contract no. CBE 604318-16, and by the NASA-sponsored Tropospheric Ozone Lidar Network (TOLNet).

REFERENCES

[1] Lin, M. Y., et al. (2012a), Springtime high surface ozone events over the western United States: Quantifying the role of stratospheric intrusions, *J. Geophys. Res.*, *117*, D00v22, doi:10.1029/ 2012jd018151.

[2] Lin, M. Y., et al. (2012b), Transport of Asian ozone pollution into surface air over the western United States in spring, *J. Geophys. Res.*, *117*, D00v07, doi:10.1029/2011JD016961.

[3] Langford, A. O., et al. (2017), Entrainment of stratospheric air and Asian pollution by the convective boundary layer in the southwestern U.S, *J. Geophys. Res.*, *122*(2), 1312-1337, doi:10.1002/2016JD025987.

[4] Alvarez, R. J., et al, 2011: Development and application of a compact, tunable, solid-state airborne ozone lidar system for boundary layer profiling, *J. Atmos. Ocean. Technol.*, doi: 10.1175/JTECH-D-10-05044.1

[5] Alvarez, R. J., Senff, C. J., A. M. Weickmann, S. P. Sandberg, A. O. Langford, R. D. Marchbanks, W. A. Brewer, and R. M. Hardesty, 2012: Reconfiguration of the NOAA TOPAZ Lidar for Ground-based Measurement of Ozone and Aerosol Backscatter, Proceedings of the 26th International Laser Radar Conference, Porto Heli, Greece, 25-29 June 2012, pp. 249-252.

https://doi.org/10.1051/epjconf/202023703013

FLIGHT DEMONSTRATION OF A 2-MICRON, DOUBLE PLUSED CO_2 IPDA LIDAR INSTRUMENT

Jirong Yu[1], Mulugeta Petros[1], Upendra Singh[1], Tamer Refaat[1], Karl Reithmaier[2], Ruben Remus[1], William Johnson[3]

[1] *NASA Langley Research Center, MS 468, Hampton, VA 23681 USA, j.yu@nasa.gov*
[2] *Science System &Applications, Inc, One Enterprise Parkway, Hampton, Virginia 23666 USA*
[3] *Michigan Aerospace Corporation, 1777 Highland Dr. # B, Ann Arbor, MI 48108 USA*

ABSTRACT

NASA Langley Research Center (LaRC) developed a double pulsed, high energy 2-micron Integrated Path Differential Absorption (IPDA) lidar instrument to measure atmospheric CO_2 column density. The 2-μm double pulsed IPDA lidar was flown ten times in March and April of 2014. It was determined that the IPDA lidar measurement is in good agreement with an in-situ CO_2 measurement by a collocated NOAA flight. The average column CO_2 density difference between the IPDA lidar measurements and the NOAA air samples is 1.48ppm in the flight altitudes of 3 to 6.1 km.

1. INTRODUCTION

NASA Langley Research Center (LaRC) has developed a double-pulsed, high energy, 2-micron direct detection IPDA lidar instrument (Yu et.al). Lidars operating in the 2 μm band offer high near-surface CO_2 measurement sensitivity due to the intrinsically stronger absorption lines (Menzies and Tratt 2003, Caron and Durand 2009). The objective of the airborne demonstration of the newly developed 2-micron pulsed IPDA lidar is to demonstrate the functionality and capability of the lidar instrument. The airborne IPDA lidar made measurements at different flight altitudes up to 8.3 km limited by aircraft capability and different ground target conditions such as vegetation, soil, ocean surface, snow and sand, and different cloud conditions. Strong lidar return signals were obtained for both on and off-line wavelengths at all flight altitudes. The CO_2 column dry mixing ratio is derived from the IPDA lidar measurement and available meteorological data profiles. This paper describes the measurement results of the 2-micron pulsed IPDA lidar instrument during this airborne campaign demonstration.

2. AIRBORNE DEMONSTRATION

2.1 Data Signal to Noise Ratio

Strong lidar return signals were obtained for both on and off-line wavelengths at different altitudes. Figure 1 shows lidar signal examples at two different lidar operating conditions. There are two pair of peaks in the return signal. The first pair of signal peaks are the on and off-line pulse signals reflected from the airplane window. The second pair of peaks are on and off-line pulse signals from the hard target echo. The off-line signals are intentionally offset from the on-line return signals in order to see the on and off-line return signals clearly. In fact, they are overlapped with each other in terms of range. Fig. 1a corresponds to a lidar signal with a flight at 1372 meters with pre-amplifier setting at 10^3. Since the airplane is at a low altitude, there is little absorption for the on-line pulse, thus, the amplitudes of the on-line and off-line returning signals are close to the amplitudes of the on and off-line laser pulses recorded by an energy monitor. Fig. 1b shows the lidar signal with an airplane flying at 6096 meters with pre-amplifier gain setting 10^5. Due to the longer absorption path length relative to the conditions in Fig 1a, the return signal amplitude from the on-line pulse is reduced. Therefore, the lidar signals from on and off-line pulses are comparable as shown in Fig. 1b. The inserts in Fig. 1 are the enlarged ground return signals. It clearly shows that the ground return signal is strong with high signal to noise ratio. The SNR is defined as the ratio of the area integrated under the return lidar signal waveform divided by baseline RMS noise, times the same time interval used to integrate the lidar signal.

 https://doi.org/10.1051/epjconf/202023703013

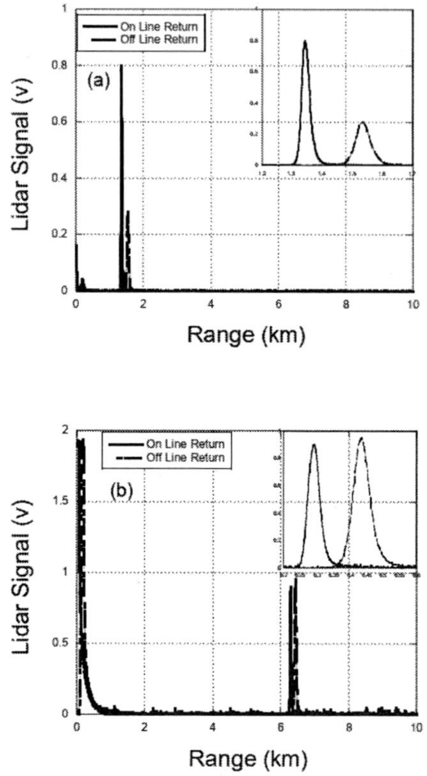

Fig. 1. IPDA lidar signal examples. The preamplifier gain settings are 10^3 and 10^5 for (a) and (b), respectively

Table 1. On and Off-line return signal SNR at various ground conditions

	Lidar S/N		Flight	Preamp
	On line	Off line	altitude (m)	gain
Soil/Veg.	287	478	6818	10^5
Sea Surface	90	138	6811	10^5
Cloud Tops	234	281	5631	10^5

2.2 DAOD Measurement Statistics

The lidar measures the backscattered signals from hard targets normalized to their emitted energy samples recorded by an energy monitor. The key measurement parameter is the differential absorption optical depth (DAOD), which is defined as the optical depth difference between the on and off-line frequency. DAOD can be calculated according to equation 1.

$$DAOD = \ln[(P_{off}\, t_{off} / E_{off}) / (P_{on}\, t_{on} / E_{on})] \qquad (1)$$

where the P_i is the lidar return signal power, E_i is the transmitted laser energy, and t_i is the effective pulse width of the return signal at the on or off-line frequency.

The accuracy of the DAOD measurements depends on the lidar signal and noise characteristics, and lidar system bias errors. Since the objective of the flights is to demonstrate functionality of the newly developed instrument, many lidar instrument settings were adjusted during the flights. Adjustments include the pre-amplifier gain, the on-line frequency shift from the R30 absorption peak, the receiver bandwidth, and the laser output energy. The instrument measured DAOD is compared with a model simulated DAOD value. The model used here is the US standard atmospheric model with an assumed atmospheric CO_2 concentration of 395 ppm.

Fig. 2 depicts an example of the DAOD measurement results. The data shown was taken in the morning on Mar. 27, 2014 over land at a flight altitude 6706 meters. The land condition varies between rural and residential areas.

Fig. 2a is a DAOD calculation based on a single shot return signal. The on line frequency is locked at 4 GHz from the R30 absorption peak. The preamplifier gain is at 10^5. The DAOD mean value for the single shot measurement is 1.0587 with a standard deviation of 0.0457. The model predicts the mean value of 1.0553. Random error can be reduced by shot averaging. Fig. 2b shows the result with 100 points moving average, which corresponds to 10 seconds average. The standard deviation is improved to 0.0123 for the lidar data as shown in Fig. 2b.

 https://doi.org/10.1051/epjconf/202023703013

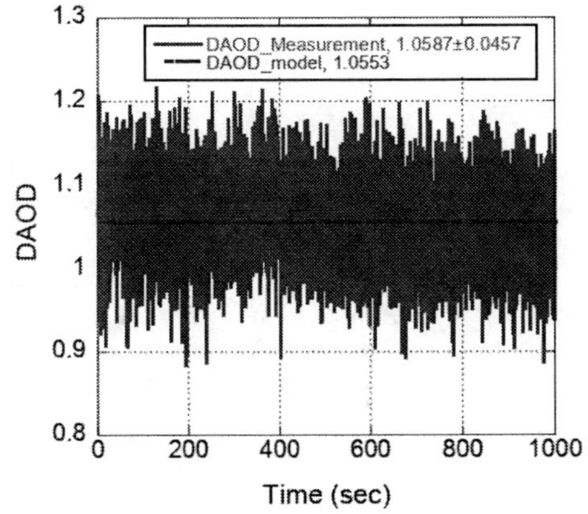

Fig. 2a. DAOD measurement by the IPDA lidar. The measurement is derived from a single shot return signal. It is compared to model predicted DAOD value.

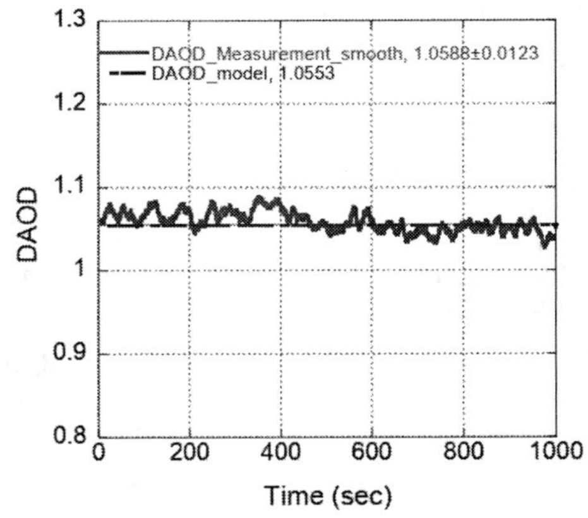

Fig. 2b. DAOD measurement by the IPDA lidar. The measurement is the result of 10s (100 shots) average.

2.3 CO_2 Mixing Ratio Measurement Validation

On 5 April 2014, NOAA conducted an air-sampling flight over the Atlantic Ocean off the coast of Cape May, New Jersey (CM; 38.83°N-74.32°W). The programmable multi-flask air sampling system provided very high precision for the CO_2 mixing ratio measurement per sample. It also measures temperature, pressure, relative humidity and other trace gases. This data can be used to make direct comparison and validation for the 2-micron IPDA lidar CO_2 mixing ratio measurements.

Flying the IPDA lidar over the ocean provides a target with near consistent surface reflectivity, which tends to reduce measurement uncertainty compared to elevated continental grounds that varies in both reflectivity and scattering surface elevation. The NOAA flight collected data at seven different altitudes, starting from 6126 meters and gradually descending to 912 meters. (6126, 5243, 3977, 3052, 2127, 1505, 912 m). It provided coarse vertical CO_2 and meteorological data profiles. Due to airspace restriction, our flight flew over the same location half an hour after the NOAA flight. The IPDA lidar flew at the same altitudes as the NOAA flight. The on-line frequency was set at 4 GHz from the R30 line absorption peak for the flight altitude above 3052 meters. The on-line frequency was changed to 3 GHz from the R30 line absorption peak below a flight altitude of 3052 meters because of less absorption due to shorter range. At an altitude of 3052 meters, the data with on-line frequency shift at both 3 and 4 GHz was taken.

The profiles of CO_2 mixing ratio xcd, temperature, pressure, and water vapor from the ground to 8 km can be obtained by linear extrapolation of the NOAA data. To make the direct comparison to the IPDA lidar column density measurement, the CO_2 weighted-average column dry-air volume-mixing ratio, Xcd, c, can be calculated. At a certain altitude, it is a weighted integration of xcd from that altitude to the surface.

IPDA lidar measures the DAOD according to equation 2. Using the NOAA measured meteorological data profile, the CO_2 weighted-average column dry-air volume-mixing ratio, Xcd, m, can be obtained with the lidar measured DAOD value. Then, the lidar measured Xcd, m can be directly compared to the NOAA measurement Xcd, c. The subscripts m and c represent the IPDA lidar measurement and the calculated result from them respectively. As shown in Fig. 2b there appears to be a small drift in the measured DAOD value, or a gradient in the CO_2 column value, due to land condition changes over the flight track.

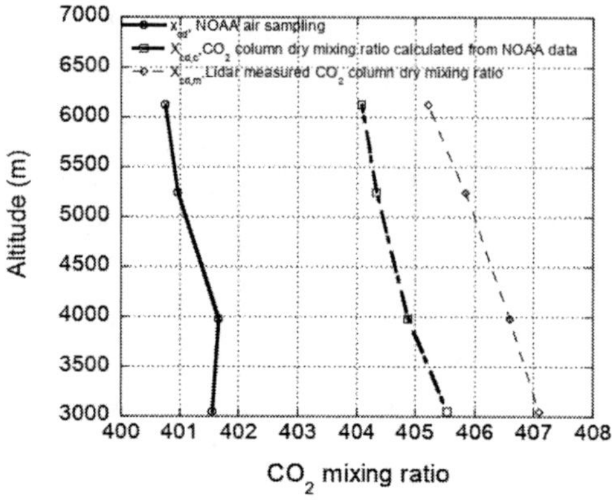

Fig. 3. CO_2 mixing ratio versus altitude calculated from NOAA flask sample data and from the IPDA lidar measurement.

Figure 3 shows the Xcd comparison between the IPDA lidar instrument measurement Xcd, m, and the model from the NOAA in-situ instrument, Xcd, c. The Xcd, m is the result of the 100 pulse average to reduce the error introduced from random noise. The direct comparison between Xcd, c, and Xcd, m, revealed that the column integrated CO_2 mixing ratio measured by the IPDA lidar instrument is higher than that derived from NOAA flask air sampling. The average difference is 1.4775 ppm, which corresponds to a 0.36% difference between the two instruments. This direct comparison between the two independent measurements validates the high precision measurement capability of the 2-μm double-pulsed IPDA lidar instrument.

3. CONCLUSIONS

NASA LaRC developed a double-pulse, 2-μm integrated path differential absorption (IPDA) lidar instrument for atmospheric CO_2 measurement. Advantages of the 2-micron high energy pulsed IPDA remote sensing technique include a high signal-to-noise ratio measurement with accurate ranging; favorable weighting function towards the ground surface to measure the source and sinks of the CO_2; and the capability to directly eliminate contaminations from aerosols and clouds to yield high accuracy CO_2 column measurements. IPDA CO_2 differential optical depth measurement results agree well with model prediction. With 10 s average, the standard deviation of the DAOD measurement is 0.0145. Compared to the CO_2 mixing ratio measured by NOAA flask sampling data, the 2-micron IPDA lidar provided an accurate measurement with 0.36% difference.

ACKNOWLEDGEMENT

The authors would like to thank the NASA Earth Science Technology Office for funding this project. The authors acknowledge support from the Engineering and Research Services Directorates at NASA Langley Research Center. Thanks also go to the dedicated efforts of the Research Systems Integration Branch personnel who made the airborne flight testing possible. Acknowledgements also go to the LaRC CAPABLE team and to NOAA for providing public information that is significant for the science validation process. The authors also would like to thank Dr. Robert Menzies at JPL, Dr. Syed Ismail and Anthony Notari at LaRC for very useful discussion, advice and help.

REFERENCES

Caron J., and Y. Durand, 2009: Operating wavelengths optimization for a spaceborne lidar measuring atmospheric CO_2. *Appl. Opt.*, 48, 5413-5422

Menzies, R. T. and D. M. Tratt, 2003: Differential laser absorption spectrometry for global profiling of tropospheric carbon dioxide: selection of optimum sounding frequencies for high-precision measurements. *Appl. Opt.*, 42, 6569-6577

T. Refaat, U. Singh, J. Yu, M. Petros, S. Ismail, M. Kavaya, K. Davis, 2015: Evaluation of an airborne triple-pulsed 2 μm IPDA lidar for simultaneous and independent atmospheric water vapor and carbon dioxide measurements, *Appl. Opt.* **54**, 1387-1398.

Yu J., M. Petros, U. N. Singh, T. Refaat, K. Reithmaier, R. Remus, W. Johnson: An Airborne 2-micron Double-Pulsed Direct Detection Lidar Instrument for Atmospheric CO_2 Column Measurements, *JTECH,* in press

INVESTIGATE WILDFIRE IMPACTS ON OZONE PRODUCTION BY VERTICAL OBSERVATIONS AND PHOTOCHEMICAL MODELING

Bo Wang[1], Michael Newchurch[1], Shi Kuang[1], Arastoo Biazar[1]

[1]The University of Alabama in Huntsville, AL 35806, USA

ABSTRACT

In troposphere, ozone is a toxic secondary pollutant produced when its precursors react in sunlight. An important source of ozone precursors is biomass burning. Here we investigate the impacts of 2016 Southeast U.S. Wildfires on ozone production by integrating vertical resolved ozone profiles and photochemical modeling. The results show that wildfires contributed to ozone lamina at the top of boundary layer and enhanced surface ozone up to about 10ppbv in Southeast U.S.. Ozone lidar observed a lower ozone change with respect to a fast growth of aerosol plume, of which the reason is also investigated. Current results indicate an effective integration of vertical observations and modeling for us to understand the ozone production from fires in troposphere.

1. INTRODUCTION

Tropospheric ozone (O_3) plays an important role in climate and affects air quality. It is transported from stratosphere or produced by photochemical reactions of its precursor gases that are emitted by multiple sources (e.g. vegetation, lightning, fossil fuel combustion, biomass burning (BB)) [1]. It is fairly well established that BB can release substantial O_3 precursors (nitrogen oxides, carbon monoxide, and hydrocarbons) that react in the presence of sunlight [2], and then affect regional air quality or far downwind by long-range transport [3, 4]. However, understanding the O_3 formation are complicated by fire emissions, photochemistry, meteorological patterns, and the mixing with extra airmasses [5]. Here we report a case study of wildfires. The main goal is to investigate the wildfire impacts on ozone production. Specific objectives include: (1) Identify and quantify fire impacts on vertical ozone profiles; (2) Estimate the potential impacts on surface ozone when smoke interacts with boundary layer; (3) Explore the dynamical and chemical driver for the observed ozone variations in smoke plumes.

2. METHODOLOGY

Here we use both ECC ozonesonde [6] and ground-based ozone lidar [7] at Huntsville station (34.725°N, 86.645°W) to obtain vertical resolved ozone profiles. The ozonesonde has vertical resolution of 100-150m, with precision better than ±5% and an accuracy better than ±10% for O_3 measurements. O_3 DIfferential Absorption Lidar (DIAL) determines O_3 profile by analyzing how rapidly the backscattered signals at the two UV wavelengths diminish with altitude [8]. The Rocket-city O_3 Quality Evaluation in the Troposphere (RO$_3$QET) lidar at Huntsville can retrieve O_3 between 0.1-12km with 2min temporal resolution, with 150m vertical resolution and ±10% precision in the lower troposphere [9]. As the retrieved O_3 number density is affected by large amounts of aerosol, we use the aerosol-correction O_3 in this study.

Vertical aerosol from ceilometer and DIAL as well as MODIS reflectance imagery are combined to identify the smoke plumes. CT25K ceilometer, in the University of Alabama in Huntsville (UAH), is a low-power pulsed lidar at high spatiotemporal resolution [10]. The backscatter signal consists of both molecular and aerosol components. We use the total backscatter intensity curtain as an indicator of aerosol structure. The aerosol extinction profiles have been retrieved from RO$_3$QET at 299nm with an iterative DIAL algorithm [11]. We have applied the newly retrieved DIAL 299nm aerosol extinction here.

We use a fully coupled meteorology-chemistry model, the Weather Research and Forecasting with Chemistry (WRF-Chem) model [12]—version 3.9.1, in this study. Gas and aerosols are calculated with the MOZCART chemical mechanism [13]. Other selected parameterizations include Morrison's microphysics scheme, the Rapid Radiative Transfer Model longwave and Goddard shortwave radiation schemes, the Re Monin-Obukhov surface layer, the Noah Land

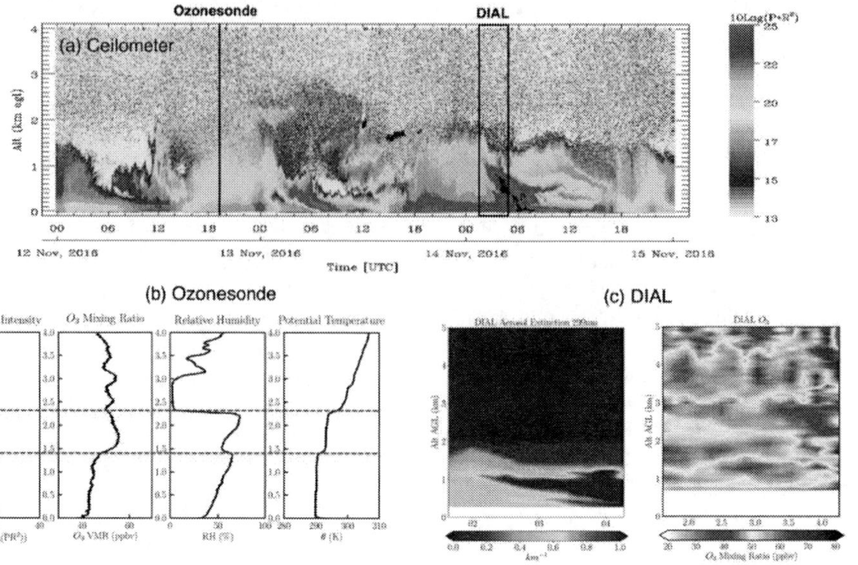

Figure 1. (a) UAH MIPS CT25K ceilometer backscatter intensity time-height curtain in Nov 12-14 (By courtesy of Kevin Knupp). (b) Ceilometer backscatter intensity (β), ozonesonde (Flight number: HU995) ozone volume mixing ratio (O_3 VMR), relative humidity (RH), and potential temperature (θ) on Nov 12 19UTC are displayed, respectively. Two red dashed lines indicate two inversion layers. (c) DIAL aerosol extinction coefficient at 299nm and aerosol-correction O_3 on Nov 14 01:37-04:16UTC.

Surface Model, the YSU PBL, the G3 Cumulus clouds, and the Madronich F-TUV photolysis. The initial and lateral conditions are provided by NCEP NAM 12km Analysis data for meteorology and MOZART-4 global model [14] outputs for chemistry. Fire emissions are calculated from the Fire Inventory from NCAR (FINN) [15] with an online plume rise model. Anthropogenic emissions from NEI-2011 and biogenic emissions calculated online by MEGAN module are also included. We performed the simulations at 16km and 4km horizontal resolution for two domains. We use the second domain output for analysis. The vertical configurations include 60 unequally spaced layers below 50hPa, with 12 layers below 2km altitude. The model performance has been evaluated by surface data, vertical observations, and aerosol optical depth (not shown here).

3. RESULTS

The 2016 Southeastern United States Wildfires are a series of wildfires along the Southern Appalachians during October-December in 2016. It approximately burned 127,555 acres (about 516 km^2) in Georgia, Tennessee, North Carolina, and South Carolina (Incident Information System). During Nov 12-14, MODIS (**Figure 3a-c**) suggests that wildfires continued to burn and emit a significant amount of smoke over southeast.

3.1 Vertical observations analysis

Figure 1a shows the continuous aerosol structure below 4km above ground level (AGL) indicated by UAH ceilometer backscatter in Nov 12-14. Several aerosol layers travelling over Huntsville, mostly in residual layer during nighttime. While an elevated aerosol plume around 2km persisted for nearly 24 hours, from Nov 12 12UTC to Nov 13 12UTC. Wildfires are supposed to affect this aerosol layer, as MODIS reflectance product shows overhead smoke plumes over Huntsville.

The ozonesonde was also launched at UAH campus at 19UTC on Nov 12 (local time 1pm). **Figure 1b** shows the profiles of O_3 VMR, RH, θ, as well as backscatter intensity from ceilometer at the sonde launch time. Upon the heating surface layer, θ is uniform across the convective boundary layer (CBL), indicating a high degree of vertical mixing. This corresponds to the uniform aerosol and O_3 distribution across CBL. Above that, two inversion layers at 1.4-1.5km and 2.3-2.4km AGL, respectively, can be identified with clear knees in the θ profile. Above 1.5km is free troposphere (FT). The upper inversion layer is likely caused by wind shear. Between the two inversion layers, enhanced (+12ppbv) O_3 lamina coexisted with that elevated aerosol plume. O_3 VMR increased from 43.8ppbv (on average in CBL) up to a peak of 55.6ppbv at 1.8km AGL. This coexistence of fire-impacted aerosol plume and enhanced O_3 indicates a fire-impacted ozone lamina in FT. To justify the hypothesis, we use sensitivity test by WRF-Chem to quantify the fire impacts later. In addition, this thick ozone lamina (nearly 1km thickness) was below a stratospheric intrusion layer, suggested by the extremely dry air (RH<5%) at 2.3-3.0km. As the fire-impacted O_3 lamina is wet (RH>50%), we could not claim any contribution from the stratospheric O_3 to this O_3 lamina through one sounding profile.

The most severe particulate air pollution occurred in Nov 13-14, indicated by the highest backscatter signal in ceilometer (**Figure 1a**). We carried out 3h DIAL measurement to observe the vertical O_3 and aerosol distribution under such a high aerosol loading condition, which covers the heavy aerosol subsidence process at nighttime. **Figure 1c** shows the time-height curtain of aerosol extinction coefficient retrieved at 299nm and O_3 after aerosol correction. The aerosol extinction below 2km is above 0.25km^{-1}, indicating heavy aerosol loading below the capping inversion layer. After 2UTC, a plume with over 1.0km^{-1} extinction presents at 1km. It gradually subsides during transport in residual layer. At 4UTC, extinction across the residual layer column has exceeded 1.0km^{-1}, with a thicker layer between two finer layers. This specified fine structure is highly consistent with that depicted by ceilometer. There are two obvious O_3 laminae illustrated by DIAL O_3 curtain, with O_3 VMR around 50-70ppbv. The 4km O_3 lamina on Nov 13 nighttime was elevated above the heavy aerosol column, indicating it is dominated by other O_3 source (e.g. STT) rather than fire emissions. Below that, another O_3 lamina was aloft around 2km--same altitude as on Nov 12--corresponding to about 0.2-0.6km^{-1} aerosol extinction. However, we notice a lower O_3 change with respect to the highest aerosol extinction over 1.0km^{-1}. It is unclear for the source of the ~2km O_3 lamina and the O_3 suppression within the heavy aerosol plume at night. To interpret those features in Nov 12-14, we use photochemical simulations to quantify the contribution of fire or extra sources to the observed ozone formation. Besides the local impacts, the impacts on a regional scale and surface ozone are also explored.

3.2 Model results

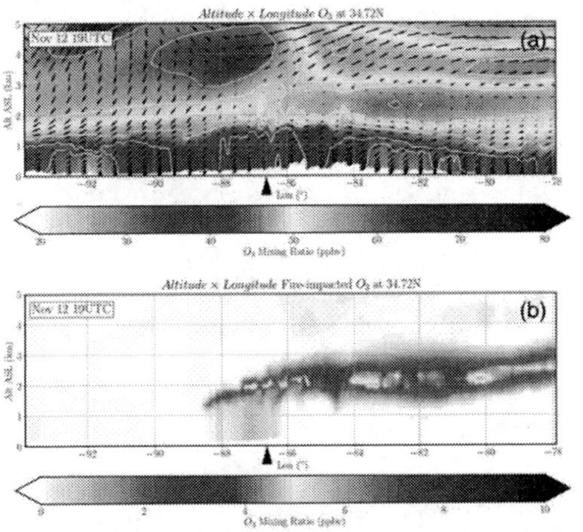

Figure 2. (a) Modeled longitude-altitude curtain of O_3 mixing ratio (ppbv) across 34.72°N on 1900UTC Nov 12-14, 2016 at 0-5km ASL altitude. Arrows indicate modeled horizontal wind direction. A black triangle marks the location of Huntsville. (b) Same as Figure 2a, but for fire-impacted O_3 mixing ratio. Note the colorbar range is different from Figure 2a.

Figure 2 shows the modeled vertical sections of O_3 and increased O_3 from fire along the latitude of Huntsville station from west to east over Southeast U.S. in Nov 12-14 at 19UTC. The curtain on Nov 12 shows that the enhanced O_3 lamina at 2km spread widely across 10° longitude (from 78°W to 88°W). This widely spreading layer is carried in light or calm wind field. As discussed before, this thick layer is aloft between two inversion layers: the bottom one is PBL entrainment zone and the upper one is likely formed by wind shear. The wind field simulation in **Figure 2a** and weather map (not shown here) confirms the formation of that wind shear, when a large-scale anticyclone circulation of winds around a high pressure blowing westward below 2km, while the jet stream above 3km (concurrent with stratospheric intrusion O_3) blowing eastward.

This structure is a key for capping the plume with enhanced O_3 above PBL height. On Nov 13, the stratification feature only persisted over west of 82°W, consistent with DIAL observation at Huntsville station (86.645°W), when the wind shear structure vanished at east coast. On Nov 14, as the wind field below 5km is dominated by westerly wind, there is no obvious laminae structure presented over southeast. **Figure 2b** shows the contribution from fire to the ozone lamina from a vertical viewpoint. The result shows that up to 5-6ppbv O_3 enhancement across 10° longitude is caused by fire emissions. The fire impacts on Nov 12 daytime across Huntsville latitude present a stratification feature, with obvious smoke-induced O_3 lamina above PBL but little impacts in PBL. It also suggests that the wildfires take responsibility for half of the ozone enhancement over Huntsville on Nov 12. It is worth discussing the extra contribution from stratospheric intrusion or anthropogenic emissions sources in the ongoing studies.

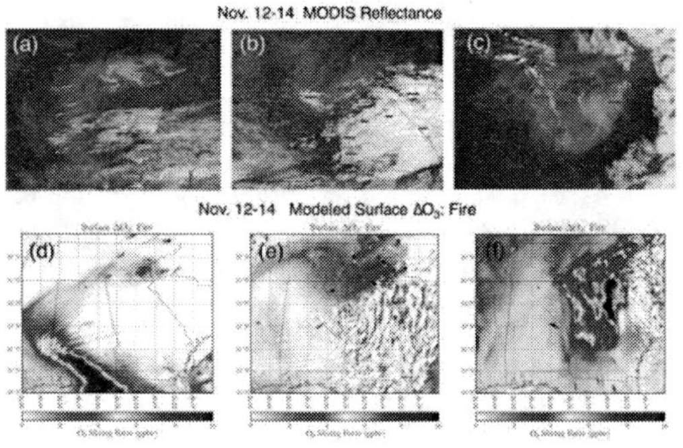

Figure 3. (a-c). MODIS true color reflectance in 2016 Nov 12-14 at 19UTC, respectively. (d-f) Spatial distribution of fire-impacted surface ozone.

Figure 3 shows the fire smoke in MODIS and the modeled fire impacts on surface O_3 in Nov 12-14 19UTC, respectively. On Nov 12, different from the elevated 2km O_3 enhancement near Huntsville (**Figure 2b**), the surface O_3 enhancement over 10ppbv is shown at southern AL and northern FL (**Figure 3d**), downwind of the fire locations (**Figure 3a**). The lower O_3 production near fires also explains the low CBL ozone observed over Huntsville. In Nov 13-14, when high pressure circulation accumulated smoke around fire sources (**Figure 3b, c**), model results show more concentrated fire impacts on ozone production around the Southern Appalachians (**Figure 3e, f**). As O_3 production from fires is lower on Nov 13 (**Figure 3e**), and the photochemical reaction is inactive at nighttime during DIAL measurement, a lack of ozone source likely caused the observed low ozone change during heavy aerosol loading at night. In addition, another sensitivity test with anthropogenic emissions on and off (not shown here) indicates NOx titration effect around urban and sub-urban area (Huntsville). Mixing of fire and anthropogenic pollution during the high pressure circulation needs to be studied.

4. CONCLUSIONS

Due to the multitude of tropospheric ozone sources and a lack of coincident measurements, coordinating observations and modeling are crucial for understanding O_3 production from fires. The vertical profiles and surface stations are essential to inform model, regional photochemical model will be good complementary to stationary and surface observations by its spatial coverage. It is critical to identify and exclude other O_3 sources for fire-impacts estimation in future studies.

ACKNOWLEDGEMENTS

This work was supported by Tropospheric Ozone Lidar Network (TOLNet). We also acknowledge the use of NCAR/ACOM MOZART-4 global model output and emission pre-processing tools.

REFERENCES

[1] Cooper, Owen R., et al. *Science*, *348*(6239), 1096-1097.
[2] Andreae, Meinrat O. and Pedro Merlet. *Global biogeochemical cycles* 15, no. 4 (2001): 955-966.
[3] Pfister, G. G., et al. *Geophysical Research Letters* 35, no. 19 (2008).
[4] Colarco, P. R., et al. *Journal of Geophysical Research: Atmospheres* 109, no. D6 (2004).
[5] Jaffe, Daniel A., and Nicole L. Wigder. *Atmospheric Environment* 51 (2012): 1-10.
[6] Newchurch, M. J., et al. *Journal of Geophysical Research: Atmospheres* 108, no. D1 (2003).
[7] Kuang, Shi., et al. *IEEE Transactions on Geoscience and Remote Sensing* 49, no. 1 (2011): 557-571.
[8] Browell, Edward V., et al. *Applied Optics* 24, no. 17 (1985): 2827-2836.
[9] Kuang, Shi., et al. *Journal of Geophysical Research: Atmospheres* 122, no. 2 (2017): 1293-1311.
[10] Knupp, Kevin. Journal of the atmospheric sciences 63, no. 8 (2006): 2016-2035.
[11] Kuang, Shi., et al. *the 29th ILRC (submitted)*.
[12] Grell, Georg A., et al. *Atmospheric Environment* 39, no. 37 (2005): 6957-6975.
[13] Pfister, G. G., et al. Atmos. Chem. Phys., 11, 7515-7532, 2011.
[14] Emmons, L.K., et al. Geosci. Model Dev., 3, 43–67, 2010.
[15] Wiedinmyer, C., et al. Geoscientific Model Development 4, no. 3 (2011): 625-41.

MODELING AND LIDAR STUDY ON OZONE OVER THE CHESAPEAKE BAY DURING OWLETS-2

Zhifeng Yang[1, 2*], Ruben Delgado[1, 2], Belay Demoz [1,2], John Sullivan[3], Guillaume Gronoff[4,5], Timothy Berkoff[4]

[1] *University of Maryland, Baltimore County; Baltimore, Maryland, 21250, USA*
[2] *Joint Center for Earth Systems Technology, Baltimore, Maryland, 21250, USA*
[3] *NASA Goddard Space Flight Center, Greenbelt, Maryland, 20771, USA*
[4] *NASA Langley Research Center, Hampton, Virginia 23681, USA*
[5] *Science Systems & Applications Inc., Hampton, Virginia 23666, USA*
**Email: zhifeng.yang@umbc.edu*

ABSTRACT

This study focuses on the distribution of ozone (O_3) concentration near the Chesapeake Bay, USA (hereafter CB) by integrating observations and model simulations. The motivation of this work is to understand reasons causing the horizontal and vertical distribution of pollutants (mainly O_3) near the CB. The O_3 exceedance over the CB happens very frequently during summer and the Maryland Department of Environment intends to find out the reasons in order to make policy-related decision. The observation data used in this study are from the Ozone Water-Land Environmental Transition Study-2 (OWLETS-2) field campaign, including observations from O_3 lidar, Doppler wind lidar, ozonesonde. The mesoscale model employed is Weather Research and Forecasting model coupled with chemistry (WRF-Chem) version 3.9.1. The anthropogenic emission dataset is from National Emission Inventory 2011 (NEI-2011), including various emission species, e.g., CO, NO_X, SO_2, NH_3, $PM_{2.5}$, PM_{10}, etc. The meteorological initial and boundary conditions are from the Northern American Regional Reanalysis (NARR) dataset, which is a high-resolution combined model and assimilated dataset from the National Centers for Environmental Prediction (NCEP). There are several findings of this study based on the model simulations and ground-based observations. Actually, at the beginning of study, we considered two different versions of anthropogenic emissions from NEI-2005 and NEI-2011 developed by the Environment Protection Agency (EPA). EPA added the anthropogenic emissions over CB from boats and ships while updating from NEI-2005 to NEI-2011. For model performance evaluation, we employed AirNow surface hourly O_3 mixing ratio diurnal variation and compared it with model simulations. For instance, at Essex site near Baltimore City, observed O_3 has a strong diurnal variation, with minimum (25 ppbv) just after sunrise (05:00 EST), and with maximum (75 ppbv) around afternoon (15:00 EST). Even the model simulation has a good agreement with the observation, it underestimates the mean O_3 mixing ratio by about 15-20 ppbv. Both the surface and 700 mb level horizontal spatial distribution of O_3 indicate the higher O_3 concentration over the north-middle CB, with surface O_3 mixing ratio of 40-50 ppbv and 700 mb level O_3 mixing ratio of 60 ppbv, which means the surface O_3 was lifted up after production. The vertical profiles of wind of both model and Doppler wind lidar match very well, indicating that the model captured the vertical variation of wind. However, the vertical profiles of O_3 from model simulation, ozonesonde, and O_3 lidar suggests that model simulation underestimated the O_3 from surface to 4.5 km. In addition, the model simulation captured the vertical mixing of O_3 from surface to 2 km, while misses the O_3 variation above 2 km. In order to study the influence of bay breeze on the O_3 small scale transport, three vertical cross sections through the CB from west to east at the northern, middle, and southern CB. The results show that higher O_3 concentration above the CB. The bay breeze over the southern CB is stronger than the northern CB. The planetary boundary layer height over the CB is dramatically lower than the surrounding land in the day, which contributes to the surface higher O_3 concentration over the CB.

1. INTRODUCTION

Tropospheric ozone (O_3), especially near the surface, has paramount significant environmental and human health impacts. Long-term exposure to high concentration of O_3 can induce serious health issues, for example, irritating lungs, aggravating bronchitis, emphysema and asthma [1-3]. In order to mitigate the surface pollution in the United States, the Clean Air Act requires the Environmental Protection Agency (EPA) to set National Ambient Air Quality Standards (NAAQS). The current NAAQS, in force since 2015, sets an ambient 8-hour O_3 mixing ratio criterion of 70 parts per billion by volume (ppbv) (National Ambient Air Quality Standards for Ozone, 2015, hereafter referred to as the Standard; [4]). However, the surface O_3 mixing ratio over some specific regions can exceed the Standard, for instance, as in the Chesapeake Bay (hereafter referred to as CB, [5, 6]). All these studies near water bodies surrounded by land and active research are ongoing on understanding the causes for exceedance of O_3. O_3, as a secondary pollutant, its concentration depends on its precursors, chemical reactions, and meteorological conditions [7, 8]. Understanding these processes helps to understand O_3 pollution sources, sinks and is useful for improving the accuracy of O_3 pollution forecasting.

As a study to better understand the behavior of O_3 and related trace gases across the water land transition O_3 in the CB, Ozone Water-Land Environmental Transition Study (OWLETS) field campaigns (OWLETS-1, 2017 [9]; OWLETS-2, 2018 [10]) integrates ground-based and airborne instruments from multiple institutes. Two key "super-sites" locate at University of Maryland Baltimore County (UMBC) and Hart-Miller Island (HMI) to observe pollutants simultaneously over land and water. All the measurements obtained from OWLETS serve for O_3 numerical forecast.

This study further investigates the influence of the CB on the local O_3 mixing ratio distribution, by employing a mesoscale model, the Weather Research and Forecasting model coupled with Chemistry (WRF-Chem; [11]). The model performance was evaluated by comparing model output with surface O_3 mixing ratio observations. Horizontal and vertical distributions were analyzed by studying the horizontal levels and vertical cross sections through the CB. In terms of novelty, this work utilizes the first-hand ground-based observations from various platforms during OWLETS-2 field campaign, i.e., NASA Goddard Space Flight Center TROPospheric Ozone (TROPOZ) DIAL lidar, NASA Langley Research Center Langley Mobile Ozone Lidar (LMOL), Doppler wind lidar, electrochemical concentration cell (ECC) ozonesonde, and coupled with mesoscale model simulations to study the impact of the CB on the local O_3 pollution. Future work includes employing the Data Assimilation Research Testbed (DART, [12]) developed by National Center of Atmospheric Research (NCAR) to assimilate the O_3 concentration 3-D distribution by employing both WRF-Chem model and ground-based O_3 observations.

2. METHODOLOGY

2.1 Model Configuration

In this study, WRF-Chem version 3.7 is employed to study the influences of CB on the local O_3 mixing ratio distribution in 3-D, along with the interaction between O_3 mixing ratio and meteorological fields. WRF-Chem is a fully coupled "online" model, which has the air quality component consistent with the meteorological components [11]. The gas-phase chemistry and aerosol module are based on the Carbon-Bond Mechanism Version Z (CBM-Z, [13]) and Model for Simulating Aerosol Interactions & Chemistry (MOSAIC, [14]). Radiation treatment utilizes the RRTMG short-wave and long-wave radiation schemes [15], without including the aerosol feedback. The radiation scheme setups are recommended for the model simulations over the continental U.S. by the WRF team [16].

The model study area (Fig. 3.1) is set up to start with the eastern U.S. The outer domain is further resolved by one two-way nested grid region that progressively focuses and centers on the study region, the CB. Thus, two domains are established, an outer domain (spatial resolution of 9×9 km, d01) and a nested domain (3×3 km, d02), selected for concentrating on detailed O_3 concentration and evolution investigation. Along the vertical, 35 levels are used with the model top located at 100 hPa and about 10-15 of the vertical levels set below 800 hPa, in order to resolve the planetary boundary layer (PBL). While there are several PBL schemes, the YSU scheme [17, 18] was selected for the runs shown here based on the performance evaluation through previous studies [19].

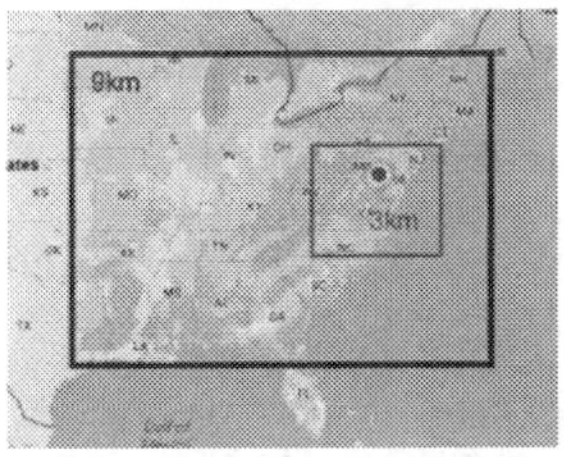

Fig. 3.1 Model domain setup with 9 km (red rectangle) outer domain and 3 km nested domain (pink rectangle).

In addition, EPA updated anthropogenic emissions over the CB when comparing NEI-2005 and NEI-2011 (Fig. 3.2). Several emission species were added over the CB, e.g., CO, NO_X, SO_2, NH_3, $PM_{2.5}$, PM_{10}.

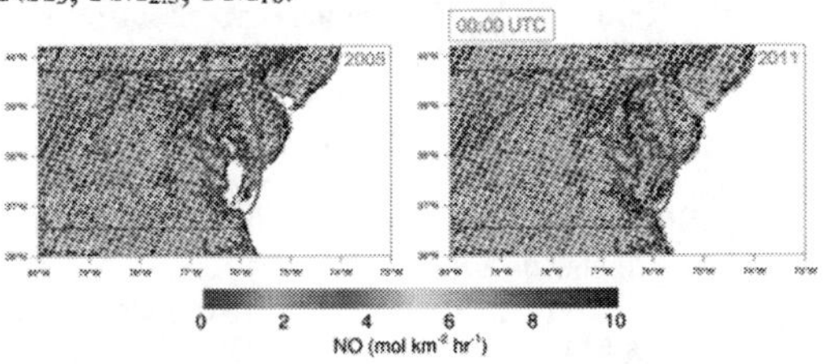

Fig. 3.2 NO emissions from NEI-2005 (left) and NEI-2011 (right). The pink ellipse shows NO emissions over the CB.

2.2 TROPOZ and LMOL ozone lidar

The NASA Goddard Space Flight Center TROPOZ DIAL lidar has been developed in a transportable trailer and used to take measurements of tropospheric ozone at UMBC (38.99N, 76.84W) [21]. The NASA Langley Research Center developed LMOL lidar and used to take measurements at HMI (39.24N, 76.36W) [22]. They are a part of the ground-based Tropospheric Ozone Lidar Network (TOLNet, https://wwwair.larc.nasa.gov/missions/TOLNet/index.html) which is designated to improve the current number of tropospheric ozone profiles [20]. TOLNet is a network of ozone lidar stations across the U.S. and Canada with the objective of improving the tropospheric O_3 vertical profile observation. Currently, TOLNet has seven stations and is designed to provide regular, high fidelity measurements of O_3 within the troposphere particularly the changing O_3 dynamics and laminae inside the PBL. We used the data observed by the GSFC TROPOZ DIAL lidar and LaRC LMOL lidar to evaluate the model performance on the O_3 vertical profile in this study.

3. RESULTS

3.1 O_3 Diurnal Variation

Surface O_3 diurnal variations from both AirNow surface observations and WRF-Chem simulations on June 29, 2018 are shown in Fig. 3.3. The observed O_3 has a strong diurnal variation, with minimum (25 ppbv) just after sunrise, and with maximum (75 ppbv) around afternoon. This is due to the large emissions during daytime and O_3 precursor photochemical reactions. Nevertheless, simulated O_3 also has similar diurnal variation as observation, albeit with significant underestimates about 15-20 ppbv. The explanation to this is that we did not include biogenic emissions in the model simulation.

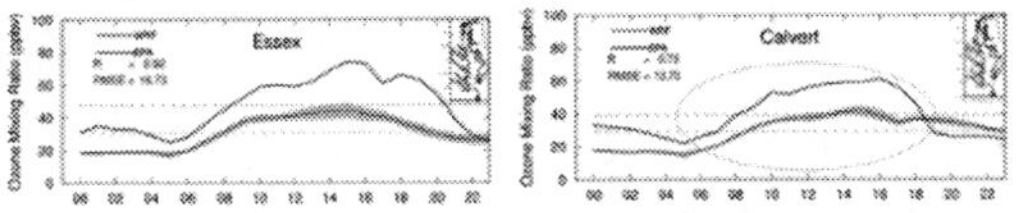

Fig. 3.3 Diurnal variation of O_3 mixing ratio at the surface from EPA AirNow and WRF-Chem simulation from two AirNow sites in Maryland (Essex and Calvert). The red shaded areas are standard deviations. O_3 surface mixing ratio daily averages are denoted as dashed lines, with blue for AirNow and red for WRF-Chem.

3.2 O_3 Vertical Profile

3.2.1 Ozonesonde

The O_3 vertical profiles from both ozonesonde and model simulation are shown in Fig 3.4. At UMBC, observed O_3 vertical profile from ascending (blue) has finer vertical resolution, with O_3 well-mixed below 2.5 km. Simulated O_3 captures this well mixture near the surface, but misses the trend above 2.5 km. At HMI, the observed O_3 increases rapidly from 60 to 85 ppbv from surface to 200 m. Note that the descending ozonesonde profile could be very far away from the original launch location, and the observation time could be hours later after launch.

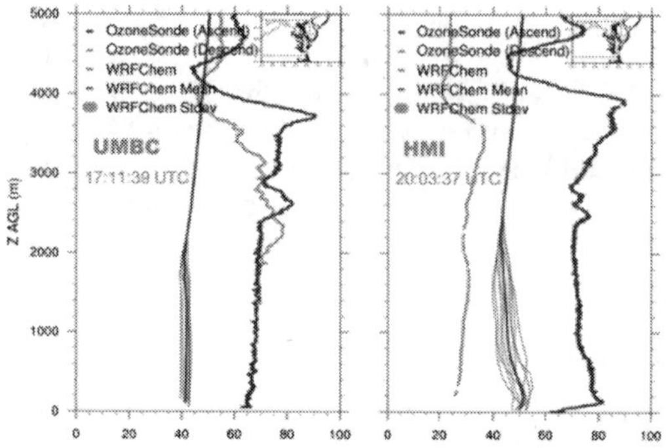

Fig. 3.4 O_3 vertical profiles from ozonesonde at UMBC (launch time at 17:11:39 UTC) and HMI (launch time at 20:03:37 UTC) sites on June 29, 2018. Blue is ascending, green is descending, and red is model simulation.

3.2.1 TROPOZ and LMOL O_3 lidar

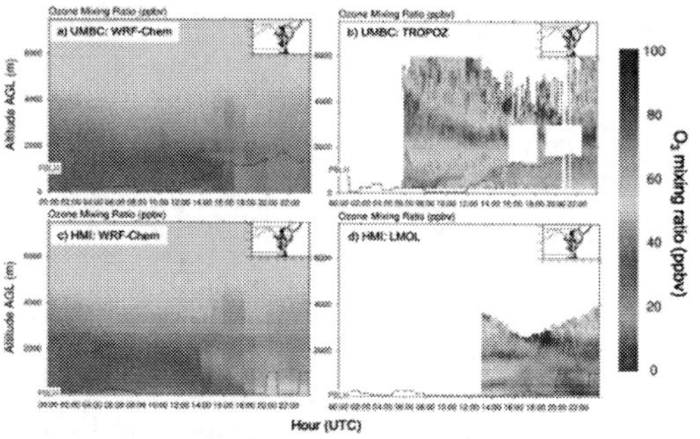

Fig. 3.5 O_3 vertical profile time series on June 29, 2018 observed by TROPOZ O_3 lidar at UMBC (b) and LMOL O_3 lidar at HMI (d). The corresponding model simulated O_3 vertical profile time series are shown in (a) and (c), respectively. PBLH from model simulation is denoted by pink curve.

From the O_3 vertical profile diurnal variation (Fig. 3.5), near-surface O_3 mixture is significantly influenced by the variation of the planetary boundary layer height (PBLH). As PBLH is shallow at night, there is lower O_3 to the surface. While PBLH increases in the day, O_3 is mixed down to the surface and well-mixed within 2 km. At UMBC, there are two layers with higher O_3 concentration from 2-4 km (Fig. 3.5 b). While at HMI, higher O_3 concentration was found near the CB surface (Fig. 3.5 d). The contribution of high O_3 concentration might be transported from long-range transport. In addition, PBLH fluctuates at HMI around 22:00 UTC. More evidence is needed to confirm it.

ACKNOWLEDGEMENTS

This work was supported by the Maryland Department of the Environment, USA. We also acknowledge NASA Tropospheric Composition Program for continued support to John Sullivan and Timothy Berkoff.

REFERENCES

[1] M. Lippmann, Environmental Science & Technology. 25, 1954-1962 (1991)
[2] I. S. Mudway & F. J. Kelly, Molecular aspects of medicine, 21(1-2), 1-48 (2000)
[3] M. H. Forouzanfar. et al. The Lancet, 388(10053), 1659-1724 (2016)
[4] O. R. Cooper, Science,348(6239), 1096-1097 (2015)
[5] C. P. Loughner, Journal of Applied Meteorology and Climatology, 53(7), 1697-1713 (2014)
[6] J. T. Sullivan, Bulletin of the American Meteorological Society, 100(2), 291-306 (2019)
[7] U. PROG, United Kingdom Photochemical Oxidants Review Group Report, Department of the Environment, London (1997)
[8] M. E. Jenkin, & K. C. Clemitshaw, Atmospheric Environment, 34(16), 2499-2527 (2000)
[9] J. T. Sullivan, et al., Bulletin of the American Meteorological Society, 100(2), 291-306 (2019)
[10] J. T. Sullivan, T. Berkoff, J. Dreessen, R. Delgado, G. Gronoff, L. Nino, B. Carroll et al. In AGU Fall Meeting Abstracts. 2018.
[11] G. A. Grell, et al., Atmospheric Environment, 39: 6957–6975 (2005)
[12] J. Anderson et al., Bulletin of the American Meteorological Society, 90(9), 1283-1296 (2009)
[13] R. A. Zaveri and L. K. Peters, J. Geophys. Res., 104, 30387–30415 (1999)
[14] R. A. Zaveri, et al., J. Geophys. Res., 113, D13204, doi:10.1029/2007JD 008782 (2008)
[15] M. J. Iacono, et al., J. Geophys. Res., 113, D13103, doi:10.1029/2008JD-009944 (2008)
[16] S. E. Peckham, WRF/Chem version 3.9 user's guide (2017)
[17] S. Y. Hong and H. L. Pan, Mon. Weather Rev. 124: 2322–2339 (1996)
[18] S. Y. Hong, et al., Mon. Weather Rev. 134: 2318–2341 (2006)
[19] X. Hu, et al., Journal of Applied Meteorology and Climatology, 49(9), 1831-1844 (2010)
[20] T. Leblanc, et al., Atmospheric measurement techniques, 11, 6137-6162 (2018)
[21] J. T. Sullivan, T. J. McGee, G. K. Sumnicht, L. W. Twigg, and R. M. Hoff., Atmospheric Measurement Techniques, 7(10):3529–3548, (2014)
[22] R. De Young, Carrion, W., Ganoe, R., Pliutau, D., Gronoff, G., Berkoff, T., & Kuang, S., Applied optics, 56(3), 721-730. (2017)

https://doi.org/10.1051/epjconf/202023703016

NAOMI GAZL: A MULTISPECIES DIAL TESTED ON THE TADI GAS LEAK SIMULATION FACILITY

JEAN-BAPTISTE DHERBECOURT[1*], JEAN-MICHEL MELKONIAN[1], ANTOINE GODARD[1], VINCENT LEBAT[1], NICOLAS TANGUY[1], CEDRIC BLANCHARD[1], STÉPHANIE DOZ[3], PIERRE-YVES FOUCHER[3], THIERRY HUET[3], XAVIER WATREMEZ[2], DOMINIQUE DUBUCQ[2], MYRIAM RAYBAUT[1]*

[1] *DPHY, ONERA, Université Paris Saclay, F-91123 Palaiseau, France*
[2] *TOTAL, France*
[3] *ONERA/DOTA, Université de Toulouse, F-31055 Toulouse France*
**Email: jean-baptiste.dherbecourt@onera.fr*

ABSTRACT

We report on a direct detection differential absorption lidar (DIAL), designed for remote detection of CH_4 and CO_2. The system is based on a single-frequency optical parametric oscillator/amplifier system, tunable in the 1.57-1.65 μm range. The DIAL system, called NAOMI GAZL, was tested on a controlled gas release facility in October 2018.

1. INTRODUCTION

Remote detection, localization, and quantification of gaseous species is a topic of high interest in the frame of various applications, from safety to environmental or industrial monitoring. In such context, differential absorption lidars (DIAL) are promising since they can be designed to address different species, especially using emitters in the infrared where most chemicals of interest display characteristic absorption bands.

In the frame of NAOMI project (for New Advanced Observation Method Integration), Onera and Total jointly investigated different remote sensing methods, to be able to probe gases such as CO_2 and CH_4 remotely. Several systems where developed, and tested on a specific gas leak simulation facility called TADI (for Transverse Anomaly Detection Infrastructure).

Here, we present a compact, range-resolved, multi-species DIAL for both CO_2 and CH_4, based a widely tunable, single frequency emitter, with a high conversion efficiency. Indeed, only few DIAL systems have been reported lately in the literature, for this kind of applications [1,2], and here we exploit a specific emitter architecture. The emitter is based on a pulsed, nanosecond, single frequency, nested cavity optical parametric oscillator (NesCOPO) [3], amplified to 16 mJ, and implemented in a direct detection lidar system. This lidar is called NAOMI GAZL.

2. METHODOLOGY

2.1 The NAOMI GAZL DIAL

Basically, as it is well known, in a DIAL, at least two laser pulses are emitted, at two wavelengths called ON (in coincidence with one of the targeted species absorption line) and OFF (on an absorption minimum). The laser pulses are backscattered by the atmosphere, collected and detected by the DIAL receiver, and a gas plume can be localized and quantified along the line of sight by comparing the backscattered ON and OFF signals.

NAOMI GAZL emitter is based on an amplified NesCOPO. The core of the NesCOPO is a multi-track periodically-poled $LiNbO_3$ (PPLN) crystal. Given the chosen poling periods, several lines in the 1.57-1.65 μm range can be addressed for CO_2 and CH_4 measurement (Fig.1). Water vapour lines are weak but could also be addressed.

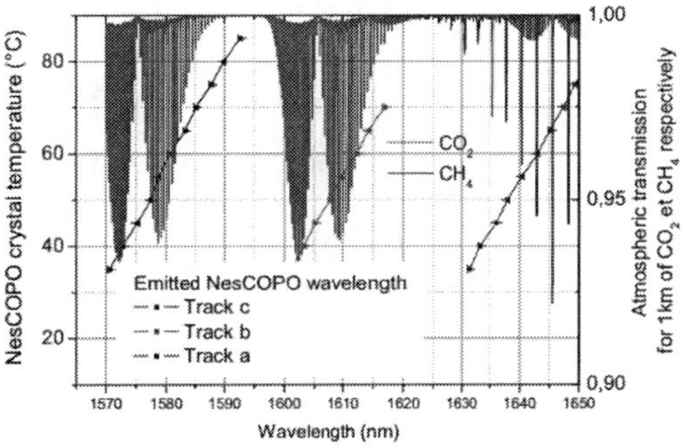

Fig. 1. NAOMI GAZL NesCOPO emitted wavelength as a function of the PPLN temperature, for three different poling periods a). The atmospheric transmission for CO_2 (for 400 ppm) and CH_4 (for 2 ppm) over 1 km are given.

The NesCOPO is pumped by less than 1 mJ of a 10 ns, 100 Hz, injection-seeded Q-switched Nd:YAG laser (Innolas). An optical parametric amplification stage, based on a PPLN crystal followed by KTA amplifiers allows to emit up to 16 mJ of single frequency radiation in the 1.57-1.65 µm range, while pumped by around 100 mJ.

NAOMI GAZL receiver unit is a custom Newtonian telescope, of 25 cm diameter, coupled to a commercial, 300 µm diameter, amplified InGaAs PIN photodiode, coupled to an 8-bits digitizer. Given the chosen amplifier, the maximum detection bandwidth is 14.5 MHz, corresponding approximately to 10 m spatial range resolution cells along the line of sight.

2.2 Preliminary testing on atmospheric CH_4

Prior to the system deployment on the TADI facility, the system was tested in ONERA facility in Palaiseau (France) in September 2018, on atmospheric CH_4. For this measurement, the photodiode alignment was optimized for long range measurement, resulting in a 200 m blind zone. A typical measurement result is displayed in Fig.2. The obtained statistical errors (due to the detection noise only) are around 3 ppm, at a 1 km range, for integration times of 250 s and a range resolution of 150 m, and less than 0.5 ppm at a 300 m range. These performances are sufficient given the simulated gas leak tests foreseen on TADI platform. For more accurate environmental monitoring, improvement of these performances via further averaging or better detection using a commercially available avalanche photodiode can be expected.

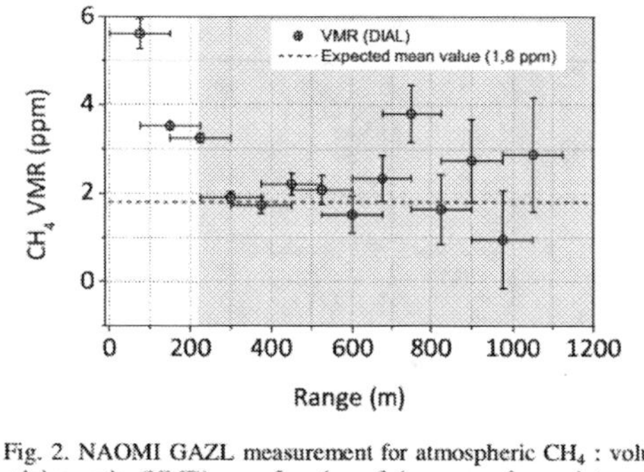

Fig. 2. NAOMI GAZL measurement for atmospheric CH_4 : volume mixing ratio (VMR) as a function of the range, for an integration time of 250 s and a measurement resolution of 150 m. The red dotted line corresponds to the expected 1.8 ppm atmospheric concentration, and the white area corresponds to the lidar blind zone.

3. OCTOBER 2018 TADI TEST CAMPAIGN

3.1 The TADI facility

Located in Lacq (France), TADI facility is an open-air site measuring 2,000 m² for qualifying systems designed to detect, localize and quantify remotely CO_2 and CH_4 releases. Equipped with surface facilities (pipelines, columns, recovered wellheads, etc.), this test area can reproduce around 30 scenarios of controlled emissions from 0.3 to 300 g/s of gas in an industrial environment.

NAOMI GAZL was deployed on the TADI facility during a test campaign in October 2018 (see Fig. 3).

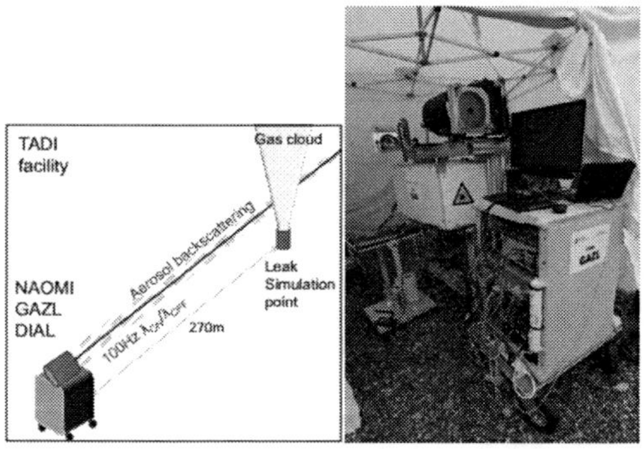

Fig. 3. NAOMI GAZL deployment scheme on TADI and picture of the system.

3.2 Example of measurement results

The NAOMI GAZL DIAL was located 250 m away from the gas release system. It was operated in a high resolution mode, i.e. with a spatial range resolution of 10 m (corresponding to the maximum detector amplifier bandwidth). Given the wind condition during these measurements, it was also operated on a high speed temporal mode with 10 s temporal averaging. Fig.4 a) and b) show an example of remote detection, quantification and localization of a CH_4 plume, generated by TADI, using GAZL. The system was tested for different fluxes of CH_4 and CO_2. Comparisons with integrated concentration in ppm.m retrieved from LWIR imaging hyperspectral data acquired during this campaign can also be used then to analyse and validate GAZL results (Fig.4 c).

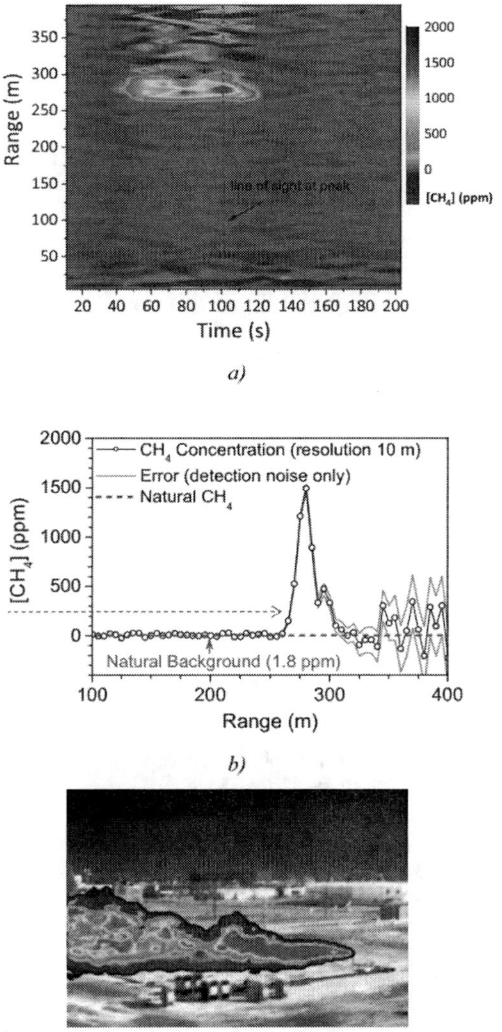

Fig. 3. a) Example of measurement performed using GAZL DIAL, on TADI facility: a gas plume is generated 250 m away from the DIAL. The estimated CH_4 plume concentration with a 10 m range resolution and a 10 s temporal resolution is shown on the left curve. In b), we can see a zoom of the CH_4 concentration along the line of sight, with error bars corresponding to the detection noise. c) Methane plume quantitative image (range: 0-200.000 ppm.m) from hyperspectral LWIR data at a distance of 100 m from the leak with a 5 s temporal resolution and a 15 cm spatial resolution.

ACKNOWLEDGEMENTS

This development was carried out in the frame of TOTAL and Onera NAOMI project.

REFERENCES

[1] G.A. Wagner, D.L. Plusquellic, Optics Express Vol.26, n°15,19420-19434 (2018)

[2] A. Amediek, G. Ehret, A. Fix, M. Wirth, C. Büdenbender, M. Quatrevalet, C. Kiemle, and C. Gerbig, Appl. Opt. **56**, 5182-5197 (2017)

[3] A. Godard, M. Raybaut, M. Lefebvre, Encyclopedia of Analytical Chemistry (2017)

New All-Solid-State KTA-Based DIAL for Tropospheric Methane Monitoring

Taieb Gasmi Cherifi
Division of Science and Engineering, Saint Louis University-Madrid campus,
Avenida del Valle, 34. Madrid 2803, Spain

ABSTRACT

We present an Optical Parametric Oscillator (OPO)-based differential absorption lidar (DIAL) that operates in the mid-infrared (3.0-4.5 um). Tunability of the OPO is achieved by the use of Potassium Titanyle Arsenate ($KTiOAsO_4$), or KTA crystal. The Dial measurements of methane are in good agreement with reported values and also exhibits diurnal fluctuations in concentration levels during atmospheric monitoring.

1. INTRODUCTION

We present a KTA crystal based-optical parametric oscillator pumped by a Q-switched Nd:YAG as a transmitter of a differential absorption lidar system. We will also conduct measurements of methane in an urban environment. The transmitter is a mid-IR laser source in a configuration based on a modified version of V. Vaicikauskas[1,2]. The singly resonant OPO, formed by two mirrors, contains a KTA crystal by Lasertec. In order to maximize the output energy, the cavity is carefully designed and its length is optimized to be 34 mm. The output laser pulse has the maximum energy of 6mJ, pulse duration of 6ns at a pulse repetition rate of 5-20 Hz. The KTA crystal of size 5×5×25 mm^3 cut along x-axis ($\theta = 90°$, $\varphi = 0°$) satisfies the type II non-critical phase matching NCPM condition for optical parametric conversion. The crystal is not cooled and is wrapped with the indium foil and mounted on a copper block. The crystal is coated on both faces to have high transmission at 1064 nm, 1.5 – 1.6 μm, and 3.0– 4.5 μm. The area monitored by the Lidar is a centric urban area where the main source of pollution can only be attributed to vehicle exhaust and to a very low extent to wind transport of methane from the further south industrial region.

2. METHODOLOGY, RESULTS & DISCUSSION

Figure 1 illustrates the backscattered signal intensity as a function of the Lidar wavelengths which span the fundamental methane absorptions under investigation.

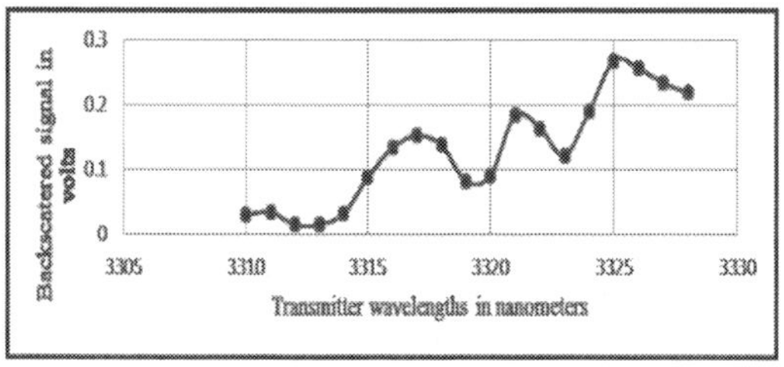

Fig.1: Backscattered signal intensity as a function of the Lidar wavelength for methane monitoring

The mid-infrared spectrum of methane is characterized by a strong vibration–rotational band [3,4], which is the Q branch ν_3 (3016 cm^{-1}), which is located within the tunable range of OPO laser (3–4 μm). The measurement of the urban concentration of the methane with high accuracy and sensitivity requires the prior knowledge of the transmission spectra of the atmosphere for a number of wavelengths from 3.0 to 3.45 μm. We have therefore made use of HITRAN2012 database[5] to identify the possible absorbing molecules in this band. This HITRAN edition has indeed witnessed major improvements within some hot bands for methane compared to HITRAN2008. It is worth mentioning that the mid-IR spectrum window contains absorption peaks for an extensive variety of trace gases[6], including H_2O, CO, CO_2, NO, N_2O, NO_2 and CH_4 which has a strong fundamental absorption at 3.3 μm. Moreover, the mid-infrared region 3.0-3.7 μm is predominantly significant because it contains fundamentally all the characteristic

https://doi.org/10.1051/epjconf/202023703017

stretching wavelengths of the molecules containing C–H bonds. Water vapor does not have significant absorption in the region 3.15 – 3.20 µm. Figure 2 shows the line intensity versus wavelength, generated based on the HITRAN, reveals clearly that the methane had strong absorption between 3.310 µm and 3.317 µm. Based on these studies, we have selected 3.316 µm as online wavelength (λ_{on}) and 3.326 µm as an offline wavelength (λ_{off}) to monitor atmospheric methane using our system with absorption1[7,8]cross section $\sigma_{on} \sim 3.497x10^{-19}$ cm^2 and $\sigma_{off} \sim 2.985x10^{-21}$ cm^2.

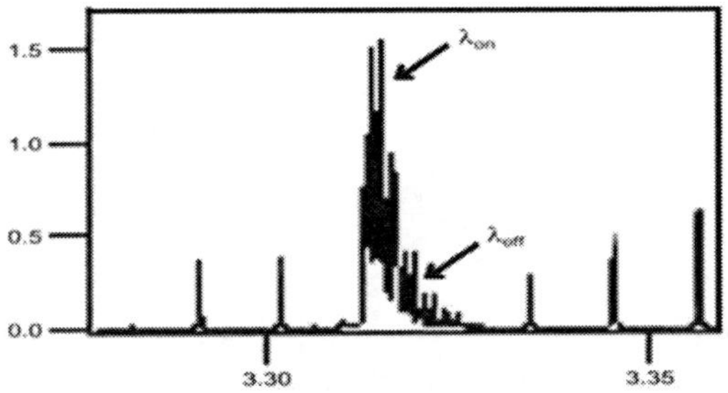

Fig. 2: Methane absorption spectra. Two selected lines – λ_{on} and λ_{off} –in the figure (source: HITRAN2012).

Figure 3 shows the experimental results of the variation of atmospheric methane as a function of time of the day. As it is well established that inversion layer is responsible for producing smog, trapping the pollutants produced by vehicles, fires and industrial activities. In our case, this effect is caused by the city topography that creates a temperature inversion since it can sometimes cause cold air to flow from the mountain peaks down into valleys where the city is located. The plateau concentration approached an average value of 1.8 ppm of methane which is slightly higher than that of background (1.75 ppm). This decrease in methane concentration is provoked by the decrease in vehicle traffic coinciding with the increase in the atmospheric mixing depth in the afternoon. However, the tendency of methane concentration reverses during the late hours of the day at around 17:00 which could be attributed to second traffic rush hour of the day. It is therefore important to note the existence of diurnal fluctuations in concentration levels during atmospheric monitoring of CH_4. This behavior is hence an additional evidence that urban areas are increasingly recognized as a globally important source of methane to the atmosphere; although, the location of urban methane sources and relative contributions of source sectors are not well known.

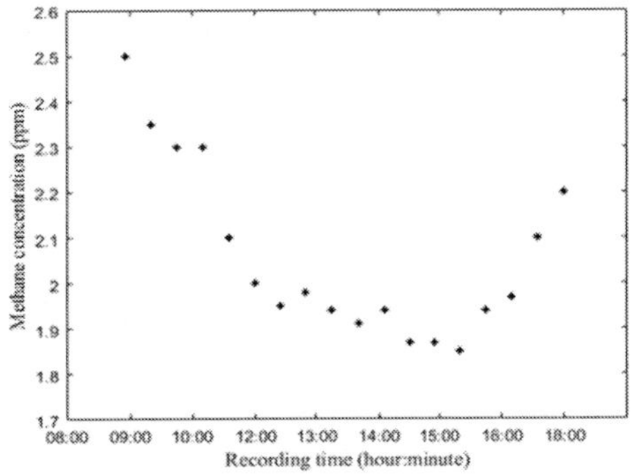

Fig. 3: Experimental results of the diurnal variation of atmospheric methane.

REFERENCES

[1] V. Vaicikauskasa, Z. Kuprionisb, M. Kaucikasa, V.Svedasa, V. Kabelkaa, "Mid-Infrared All-solid State DIAL for Remote Sensing of Hazardous Chemical Agents", *Laser Radar Technology and Applications XI, Proc. of SPIE, **6214**,* (2006).

[2] Q.B.Sun, H.J. Liu, N. Huang, C. Ruan, S.L. Zhu and W. Zhao " High energy and High efficiency 3.4 um extracavity KTA optical parametric oscillator", Laser Phys. Lett. 8, N° 1, 16-20 (2011).

[3] S.W. Lee, J. McNeil, T. Zenker, T.H. Chyba, "Methane Concentration Measurements with a Mid-infrared Optical Parametric Oscillator-based Differential Absorption Lidar System", *Proc. Lasers and Electro-Optics, CLEO/Pacific Rim '99*. Seoul, S-Korea, (1999).

[4] S.W. Lee, J. McNeil, T. Zenker, T.H. Chyba, "A Mid-IR OPO Based Lidar System for DIAL Measurements of Tropospheric Methane Concentration", *Proc. Conference on Lasers and Electro-Optics*, CLEO '99, Baltimore, MD, USA, (1999).

[5] "The HITRAN2012 Molecular Spectroscopic Database", *Journal of Quantitative Spectroscopy & Radiative Transfer*, **130**, 130, 4–50, (2013).
[6] "Silicon on silicon dioxide slot waveguide evanescent field gas absorption sensor", M.A. Butt, S:N: Khonina and N.L. Kazanskiy, Journal of Modern Optics, Vol. 65, N° 2, 174-178, (2018).
[7] S. Veerabuthiran, A. K. Razdan, M. K. Jindal, R. K. Sharma, V. Sagar. "Development of 3.0–3.45 um OPO Laser Based Range Resolved and Hard-target Differential Absorption Lidar for Sensing of Atmospheric Methane". *Optics & Laser Technology*, **73**, 1–5, (2015).
[8] R. J. Hargreaves, P. F. Bernath, J. Bailey, and M. Dulick, "Emperical Line Lists and Absorption Cross Sections for Methane at High Temperatures". *The Astrophysical Journal*, **813**, 12, (2015).

https://doi.org/10.1051/epjconf/202023703018

Observation of fast tropospheric dynamics and impacts on humidity and air pollution during a heatwave event in New York City, as observed by lidars, other profilers and surface measurements

Fred Moshary*, Adrian Diaz Fortich, Yonghua Wu, Mark Arend, Prathap Ramamurhty, and Barry Gross

NOAA Center for Earth System Science and Remote Sensing Technology, The City College of New York, New York, NY 10031, USA.
**Email: moshary@ccny.cuny.edu*

ABSTRACT

Exposure to heatwaves in densely populated urban areas are of major health concern. Interaction of heatwaves with urban areas are complex; not only driven by mesoscale meteorology, but also by the topology and thermal properties of urban surfaces, as well as anthropogenic thermal and pollution loadings. The impact of heatwaves on population in coastal urban areas such as New York City depend on interplay of these factors as well as the effects of sea/land breeze. These mechanisms can lead to rapid changes in pollution levels and heat index during the day. High time resolution synergistic observations with surface and profiling measurements are needed to capture and understand these dynamics. We report on fast tropospheric dynamics during a heatwave in New York City in summer 2018, resulting in a transient event during which dramatic reduction of pollutants and humidity in the tropospheric column is observed by lidars and other synergistic measurements.

1. INTRODUCTION

The United Stated National Weather Service defines heatwaves as three consecutive days with temperature highs above 90 F (32 C). On the average, in US, heatwaves are the deadliest severe weather events [1] and the US Centers for Disease Control estimates that annually, over 600 people die of extreme heat related complications. While globally, there are different definitions for heatwaves, the health impacts of heatwaves are of major concern across different communities. As a consequence of climate change, the frequency and intensity of heatwaves are expected to increase, representing a major threat to populations. This is especially exacerbated in urban areas, where surface roughness (urban topology), and thermal properties of surfaces (e.g. surface thermal potential and impervious surfaces) can lead to further elevated temperatures due to the Urban Heat Island effect [2]. Increasingly, air conditioning is used to alleviate the impact of heatwaves in private and public spaces, which can create further adverse effects in densely populated urban areas by increasing the heat load on the urban atmosphere [3]. Increased use of air conditioning also leads to increased electric demand. In certain cities like New York, this peak electricity demand is met by peaking generation plants within the city, leading to increased urban emissions. The combination of increased emissions together with ample heat and sunshine leads to efficient photochemical production of pollutants such as ozone and secondary aerosols (PM2.5), resulting in pollution exposure at levels exceeding the National Ambient Air Quality Standards (NAAQS) and posing additional health concerns for urban and downwind communities .

Convection driven turbulence and other related tropospheric dynamics during heatwaves can drive densities and fluxes of water vapor and other trace gases and particulates, and as such affect their impacts. We report on observation made during a heatwave episode in New York City during the multi-agency Long Island Sound Tropospheric Ozone Study (LISTOS) campaign in the Summer 2018. The specific episode lasted from June 29th to July 3rd. During the heatwave, ozone levels exceeded the NAAQS (8hr 70ppb) for June 30 to July 2nd where peak temperatures observed reached about 35 degrees each day. PM2.5 levels exceeding 35 $\mu g/m^3$ were observed but the 24 hr fine particulate matter NAAQS was not exceeded.

In this period of time, an extensive suite of surface

analyzers including measurement of meteorological parameters and trace gases, as well as column profiling instruments such as visible and microwave radiometers, and aerosol and wind lidars were deployed to characterize the surface and atmospheric conditions. During the day-time high temperature periods, fast and anomalous dynamics in atmospheric column concentrations of particulates and trace gases were observed and are presented in this report. Preliminary analysis suggest that these are due to free-troposphere-boundary layer exchanges and clear air subsidence.

2. METHODOLOGY

At CCNY-campus in northern Manhattan, NYC, a suite of ground-based remote sensing instruments are deployed on a building roof, which include: a 3-wavelength elastic-Raman lidar, a ceilometer (Vaisala CL51), a scanning mini-micropulse lidar (SigmaSpace mini-MPL) a scanning Doppler wind lidar (Leosphere 200S), thermodynamic microwave profiling radiometer (Radiometrics MP300A), and a CIMEL sun/sky radiometer (part of NASA AERONET). In addition, the New York State Department of Environment Conservation (NYSDEC) operates a sampling facility with O_3, CO, and $PM_{2.5}$ analyzers on the CCNY campus. Finally, a Campbell Scientific weather station measures surface level meteorological conditions.

3. OBSERVATIONS

Figure 1 shows the ceilometer observations of the boundary layer aerosol profiles during June 30.

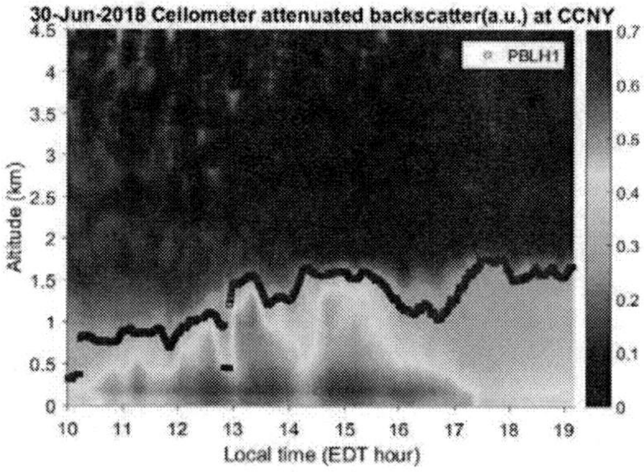

Fig. 1 Aerosol profiles observed by a Vaisala CL31 ceilometer on June 30.

During the daytime hours as the temperature increases, large pronounced oscillation in aerosol concentration within the boundary layer is observed. These features were also seen by the CCNY multiwaveleght lidar system and the observations were consistent with the similarly rapid changes in surface PM2.5 concentration and column aerosol optical depth as observed by NASA AERONET CIMEL instrument.

Later on during the heatwave on July 2nd, a more dramatic feature is observed during midday as shown in figure 2 This figure shows the ceilometer profile of aerosols within the boundary layer aat high time resolution to capture the rapid changes.

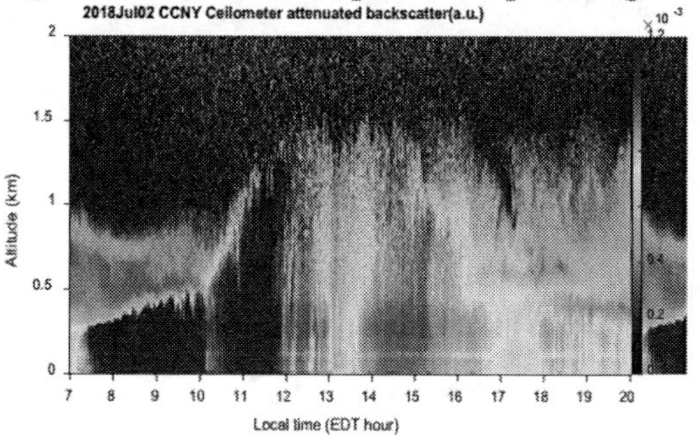

Fig. 2 Ceilometer aerosol profiles on July 2.

As can be seen from this figure, at approximately 1 pm local time, the atmospheric column clears all the way to the surface for about 1 hr. We particularly note the sharp features observed here. Again the same feature is observed by the multi-wavelength lidar.

Figure 3 shows the observed column optical depth consistent with the profiling observations.

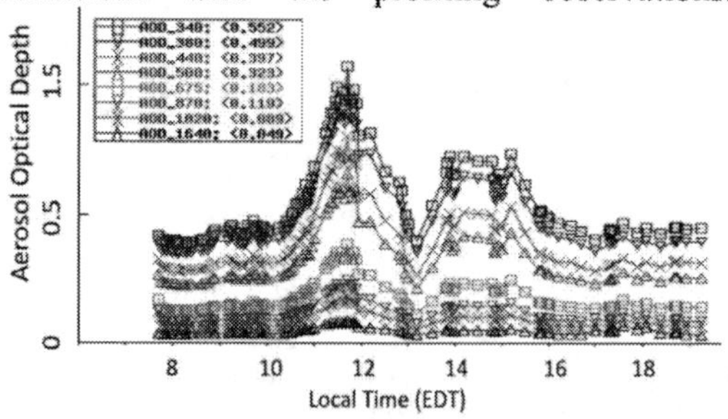

Fig. 3. AERONET AOD observations for July 12, 2018.

The near air temperature and humidity are shown in Figure 4 a and b. As the temperature climbs to its maximum of 35 C by 1 pm, there is a noticeable notch in the relative humidity at this time coincident with the clearing observed by the lidars.

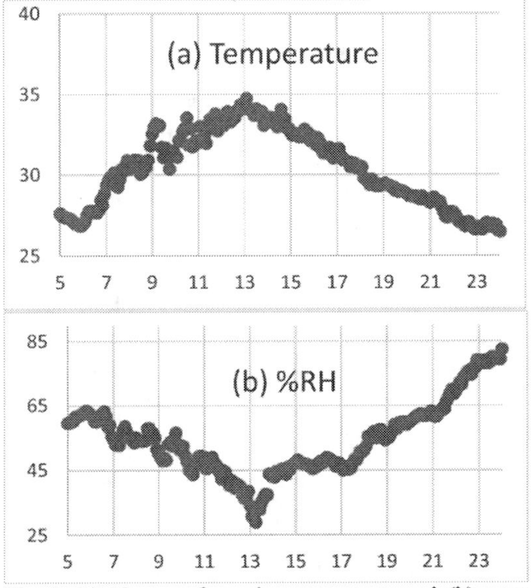

Fig 4. (a) Near surface air temperature and (b) near surface humidity measured for July 2, 2018.

Similarly, the microwave radiometer measurements of the column humidity profiler shows that the column dries out during this same event with dramatic decrease in relative humidity all the way to the surface as can be seen in figure 5.

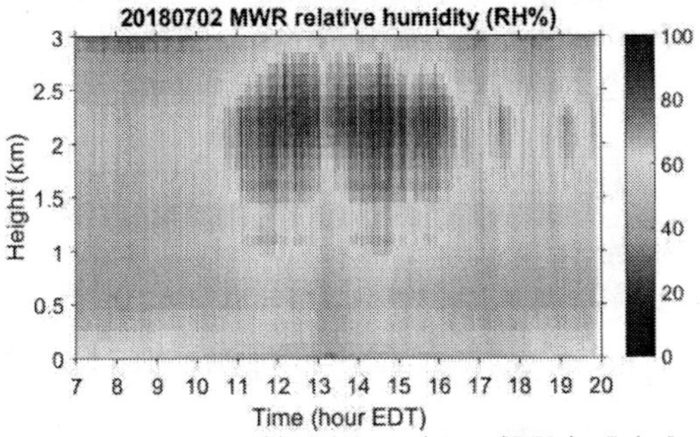

Fig 5. Microwave profiler observations of RH for July 2.

Figure 6a shows the surface PM2.5 measurement and 5b the surface ozone measurements for this day. At 11:30 am the PM2.5 concentration exceeds 40 µg/m^3 but by 13:30 the level have dropped to 3 µg/m^3. The surface ozone also exhibits a decrease from 117 ppb to 78 ppb.

The surface wind conditions in the morning leading up to the observed event are light and variable. The horizontal wind profiles measured with a wind lidar are consistent as seen in figures 7.

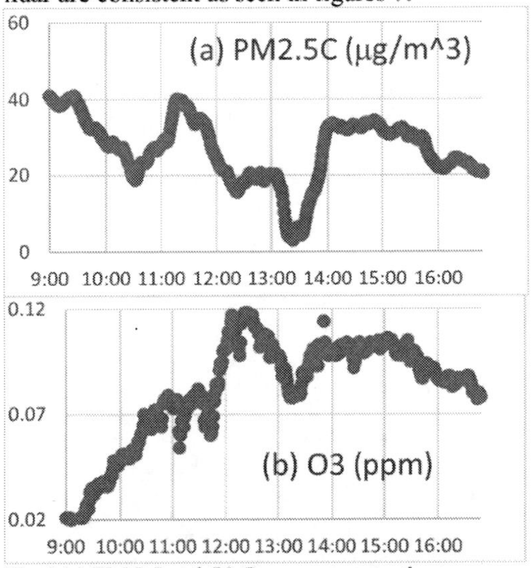

Fig 6 (a) PM 2.5 and (b) Ozone concentrations

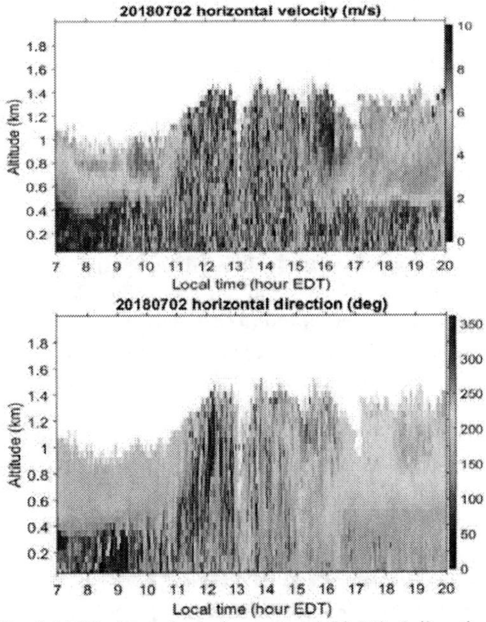

Fig. 7 (a) Horizontal wind speed and (b) Wind direction measured by LEOSPHERE 200S wind lidar.

During morning hours, we observe a low level jet from the south at 400-900 meter before 11 am.

After this time, the winds are variable for the next 5 hours across the entire column. At about 2pm, southerly near-surface winds evolve up to a 500 meter above which the winds are southwesterly to 1.2 km height. These are consistent with surface wind observations. The afternoon winds appear to gradually clear the atmospheric after about 3pm consistent with ceilometer and surface observation.

However, there is clear evidence of a dramatic updraft preceding the clearing of the atmospheric column at 1 pm.as can be seen in figure 8.

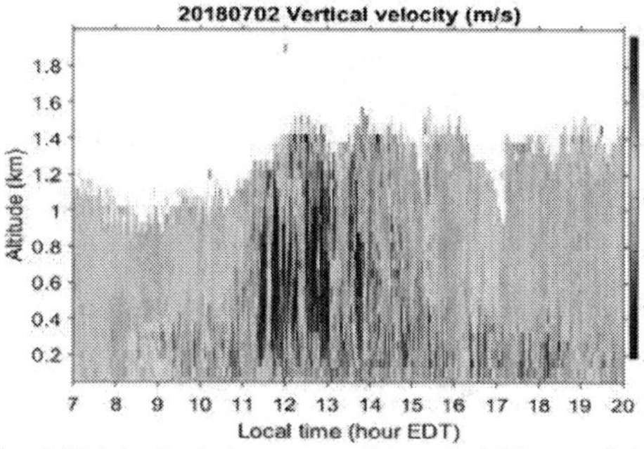

Fig. 8 Vertical winds measured by wind ldiar on July 2.

The strong updraft event lasts nearly 1.5 hrs staring at 11:30, during which time the mixing layer height grows to 1.5 km. However, the updraft event abruptly ends at 1 pm. After 1 pm, there is a turbulent period with both updrafts and down drafts, but these are reltatively short lived.

The observed event is limited in spatial extent. Observations from the horizontally scanning micropulse lidar pointed eastward at 100 meter elevation shows that the clearing extends at least 3.5 km east. Other lidar observations in the region show that the event is observed up to 7 km northeast of the CCNY, however it is not as dramatic at that distance. However, the event is not observed 14 km to the southeast or 20 km to the south, still within the boroughs of New York City.

4. DISCUSSIONS AND CONCLUSION

During a five day heatwave event in NYC, which lead to exceedance of ozone NAAQS, we report observation of rapid changes of aerosols and trace gase concentrations within the boundary layer column. The case of June 30th shows several oscillation while the case of July 2nd shows a single dramatic event that lasts for about 1 hr. These events seem to come about with sharp transitions on the order of minute. The July 2nd case is extensively studied with data presented from a number of ground-based analyzers and atmospheric column and profiling remote sensing instruments. Horizontal wind observations at the surface and above are not suggestive of transport being the reason for the observed clearing. However, as can be seen from figure 5, the event extends above the mixing layer all the way to about 3 km, ushering in dry air into the atmospheric column. This observation is suggestive of exchange with the free troposphere with clear air subsidence into the planetary boundary layer. Vertical wind profiles show some evidence of downdraft following 1 pm, however, not as dramatic as the updraft observed prior to the clearing of the atmospheric column. The column stays clear for about 1 hr, after which, the concentrations of water vapor and other trace constituents rapidly recover. The onset of horizontal winds after 3pm gradually clears the boundary layer, while the air temperatures cool into the evening. Based on remote sensing measurements spatial extent of the observed events is on the order of several kilometers. The dynamics observed, on the order of 1 hr or less on both June 30 and July 2 show that high time resolution is needed to capture these transients, while ground-based remote sensing probes the vertical and horizontal scale of the vent.

ACKNOWLEDGEMENTS. This study is supported by the New York State Energy Resources Development Authority (grant # 100415), NESCAUM (project #2411) and the NOAA CESSRST under the NOAA Agreement NA16SEC4810008.

REFERENCES

[1] https://www.nws.noaa.gov/om/hazstats.shtml

[2] Prathap Ramamurthy, Jorge Gonzalez, Luis Uriarte, Mark Arend, Fred Moshary, Impact of heatwave on a megacity: an observational analysis of New York City during July 2016, Environmental Research Letters 12, 54011 (May 2017).

[3] Xiaoyu Xu, Fei Chen, Shuanghe Shen, Shiguang Miao, Michael Barlage, Wenli Guo and Alex Mahalov, Using WRF-Urban to Assess Summertime Air Conditioning Electric Loads and Their Impacts on Urban Weather in Beijing, the J. Geophysical Res.: Atmospheres, 123, 5, (2475-2490), (2018).

OBSERVATION OF WILDFIRE SMOKE TRANSPORT AND PBL VARIATION DURING SUMMER 2018 LISTOS CAMPAIGN IN NEW YORK CITY

Yonghua Wu [1, 2*], Kaihui Zhao [3], Jianping Huang [4], Dingdong Li [1], Mark Arend [1, 2], Barry Gross [1, 2], Fred Moshary [1, 2]

[1] *the City College of New York, NY, NY 10031, USA. *Email: yhwu@ccny.cuny.edu*
[2] *NOAA – Cooperative Science Center for Earth System Sciences and Remote Sensing Technologies*
[3] *School of Environment and Energy, South China University of Technology, Guangzhou 510006, China*
[4] *NOAA-NCEP Environmental Modeling Center and IM System Group Inc., College Park, MD 20740, USA*

ABSTRACT

Air pollution associated with wildfire smoke transport and heat wave in summer pose serious public health concerns in the populated New York City (NYC) area. In this study, we present a synergistic lidar, ceilometer and in-situ observation for wildfire smoke transport and planetary-boundary-layer (PBL) variation in the NYC urban and coastal area during the summer 2018 Long Island Sound Tropospheric Ozone (O_3) Study (LISTOS). A dense smoke plume and mixing into PBL on August 15-17, 2018 was analyzed while the coincident enhancement of $PM_{2.5}$, CO and O_3 exceedance of NAAQS was demonstrated from both the observation and model. In addition, we show the temporal-spatial variation and difference of the PBL-height (PBLH) in the NYC urban and its coastal vicinity. We further evaluate the NAM-CMAQ model forecast of O_3, $PM_{2.5}$ and PBLH with the ground observations.

1. INTRODUCTION

Air pollution episodes frequently occur in summer in NYC area, particularly O_3 exceedance of national ambient air quality standard (NAAQS) in the urban and downwind coastal area of Long Island Sound (LIS) [1]. Extensive anthropogenic emissions, photochemical formation and urban heat island (UHI) magnify the air pollution impacts. In addition, long-range transport of wildfire smoke from northwest US and Canada can interact with PBL and affect the local air quality in NYC [2]. LIS area is subject to high-levels of O_3 in the summer associated with the urban pollution transport and metrological conditions (i.e. shallow, stable marine PBL and sea breeze) [1, 3]. To understand O_3 episodes, wildfire smoke transport effects, and assess model forecast, it is necessary to make comprehensive and simultaneous observation of time-height distribution of aerosols, O_3, and PBL in the NYC urban, coastal and marine region.

This abstract presents an integrated observation of elevated aerosol plumes, PBLH, ground $PM_{2.5}$, O_3 and its precursors in summer 2018 LISTOS campaign. The observations include a multi-wavelength elastic-Raman lidar, two ceilometers, a wind lidar and a microwave profiler to characterize diurnal vertical distribution and variation of aerosols and PBL. In particular, three-day ship-borne measurements for the PBLH, VOCs, O_3 and meteorological parameters are made in the LIS; and a coastal site facilitated with a ceilometer and O_3 monitor was operated from July 27 to Sep.18, 2018. With the measurements, an initial comparison to the NOAA NAM-CMAQ model forecast is evaluated [4].

2. METHODOLOGY

At CCNY-campus, a suite of ground-based remote sensing instruments is deployed on a building roof, which include a 3-wavelength elastic-Raman lidar, a ceilometer, a microwave radiometer and a CIMEL sun/sky radiometer (part of NASA AERONET). Three elastic-scattering and two Raman-scattering returns by nitrogen and water vapor molecules excited by 355-nm laser output are collected by a receiver telescope [5]. Two ceilometers (Vaisala CL-51 and CL-31) make automatic 24-hr/7-day observation at CCNY and a coastal site, respectively. We develop a quality-assurance (QA) method to optimize PBLH determination from raw aerosol-layer product. With the co-located sunphotometer-measured aerosol optical depth (AOD), aerosol extinction profiles and lidar-ratios are constrained in the lidar Fernald inversion [6]. In addition, the New York State Department of Environment Conservation

https://doi.org/10.1051/epjconf/202023703019

(NYSDEC) samples O_3, NO_x, $PM_{2.5}$ and its speciation in the NYC area [7]. Figure 1 shows ground-based sites used in this study.

Fig.1 Ground-site location (yellow 'x' set by CCNY)

3. RESULTS and DISCUSSION

3.1 Smoke plumes on Aug.16, 2018

Figure 2 (a) shows the multi-layer aloft plumes at 1.5 ~ 5 km altitude observed by CCNY-lidar on Aug. 16, 2018. Along with the PBLH growing and plume dispersion, the elevated plumes gradually mix into the PBL after 13:00 local time. The attenuated color-ratio in Fig.2 (b) can help discriminated the smoke from the scattered low clouds (large color-ratio at 15:30-18:00). The aerosol extinction coefficients in the aloft layer were larger than the ones in the PBL before 15:00 in Fig.2(c). According to the lidar-derived extinction profiles, the aloft-layers contribute 70% to the column AOD. In addition, Fig.1(d) shows the wildfire sources (red +) and smoke area (yellow) from the NOAA-HMS product, and the air backward trajectories (48hr long) with the end point at CCNY from the NOAA-HYSPLIT model. The results indicate that the main wildfire sources were located in the western Canada and northwest US, and the wildfire smoke was moving east across a large section of the northern US and southern Canada. Furthermore, we plot the vertical integrated smoke by the NOAA-HRRR smoke model on in Fig.2 (d); the dense smoke in the northwest US/Canada and transport to the northeast US can be clearly seen.

Figure 3 gives the temporal variation of ground $PM_{2.5}$, O_3 and carbon monoxide (CO) during Aug.12-31, 2018. Two significant enhancement processes of $PM_{2.5}$ are shown on Aug.15-17 (prevailing northwest wind) and 27-29 (southwest wind), respectively. The hourly O_3 are in exceedance of NAAQS (70-ppb). As a smoke indicator, the CO indicate concurrent increase. Such air pollution event associated with the wildfire smoke transport might be identified as the "exceptional event" according to EPA guidance.

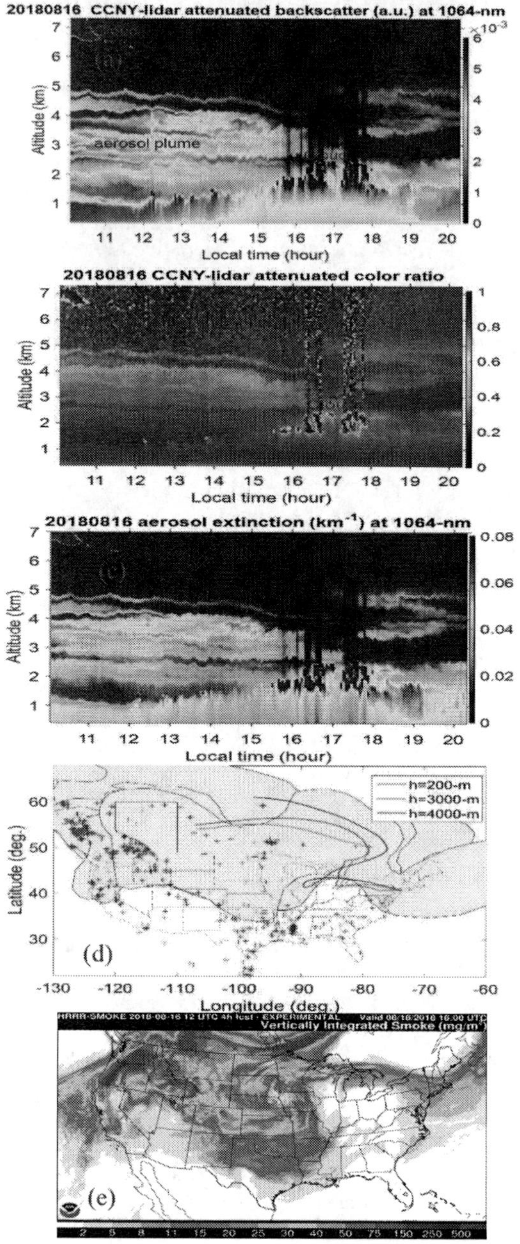

Fig.2 (a)-(c) CCNY-lidar attenuated backscatter, color-ratio, aerosol extinction (S_{1064}=55-sr), (d)-(e) NOAA-HYSPLIT backward trajectory, HMS fire points (red+)-smoke (yellow), HRRR model smoke on Aug.16, 2018.

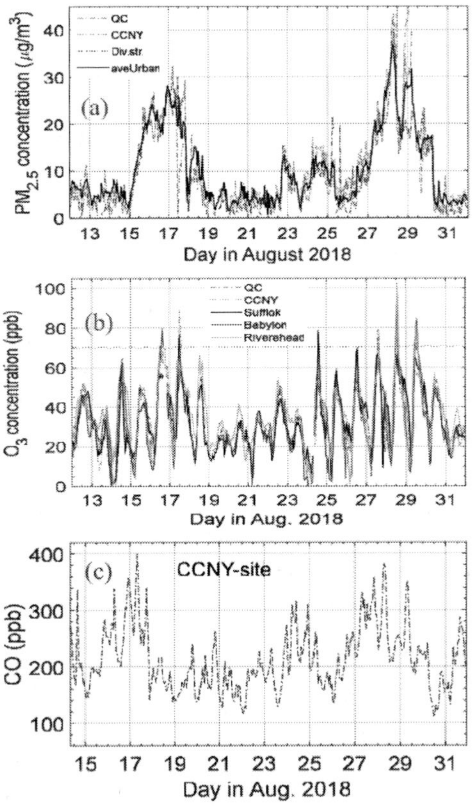

Fig.3 (a)-(c) Ground-level $PM_{2.5}$, O_3 and CO variation in NYC during Aug.12-31, 2018.

3.2 Mixing-layer-height (MLH) variation in the urban, coastal and marine area

Figure 4 gives the MLH comparison measured by the ceilometer and lidar in the NYC urban, coastal and marine area on Aug. 29, 2018. In Fig.4 (a), the ship-track with the local time in the LIS is given. In the marine and costal area, the mixing-layer shows a low and stable top at 0.3 km before 11:30, then a consistent increase up to 1.5 km at 14:00, following with a high top. In contrast, in the urban area as shown in Fig.4(c), the MLHs start to increase at 10:00 am, then reach up to maximum at ~12:30, and remains consistently high, in agreement with MLH in the coastal and marine area (Fig.4d). Overall, there is around a 2-hr delay of the MLH morning transition in the coast and marine area in comparison to the urban area, which might be due to different surface temperatures and roughness over the land and water that affect the PBL development. The low MLH in the morning corresponds to high ground $PM_{2.5}$ in Fig.3 (a). In addition, the PBLHs from the NAM model (4-km grid) indicate much lower values in the LIS than the ones in the NYC urban area after 14:00 EDT, which is not consistent with the observations.

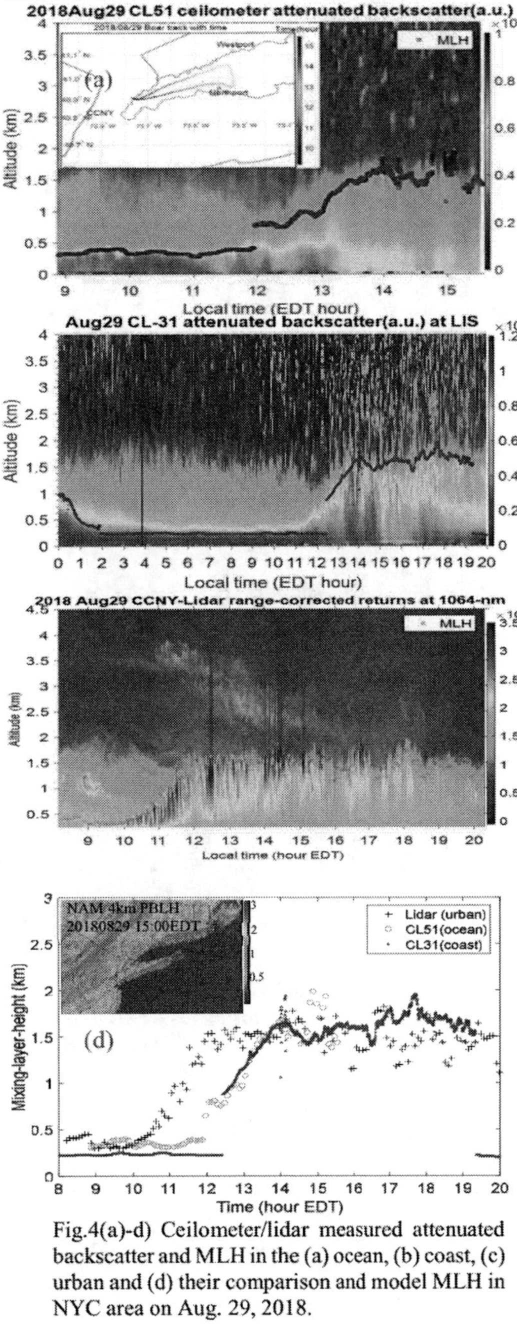

Fig.4(a)-d) Ceilometer/lidar measured attenuated backscatter and MLH in the (a) ocean, (b) coast, (c) urban and (d) their comparison and model MLH in NYC area on Aug. 29, 2018.

3.3 Evaluation of model forecast

Figure 5 gives the comparison of PBLH, ground $PM_{2.5}$ and O_3 between the NAM-CMAQ model and the observations at CCNY-site, respectively. First, Fig.5 (a) indicates consistent diurnal evolution of PBLH, particularly for the convective PBL. But, the model shows an underestimate in the morning and night. On the other hand, the model $PM_{2.5}$ generally agree with the observations on Aug. 27-28, but show large biases on Aug. 29. The model O_3 agree with the observation in term of temporal variation and peak values at noon, but are underestimated in the morning and night.

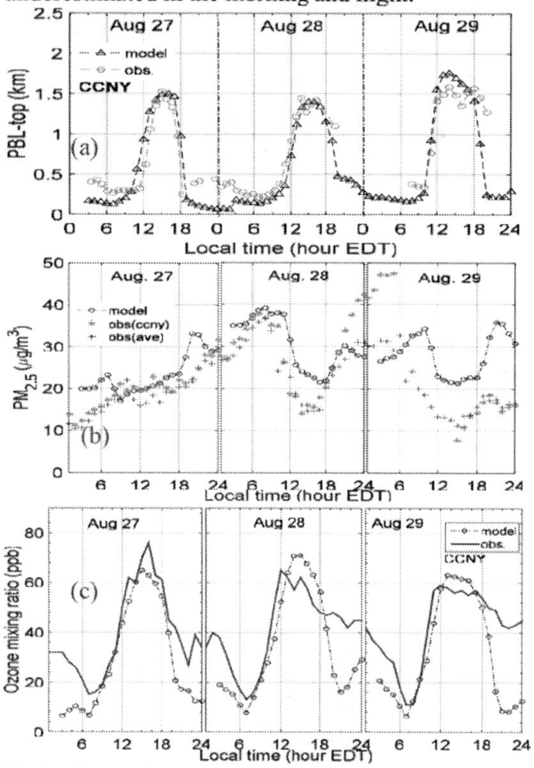

Fig.5 (a)-(c) Comparison of PBLH, $PM_{2.5}$ and O_3 between the NAM-CMAQ model and observation.

4. CONCLUSION

This abstract considers a case study and investigation of the wildfire smoke transport and PBLH spatial variation in NYC area in summer 2018. The optical properties of aloft aerosols, mixing into the PBL and potential impacts on air quality are indicated from the lidar and in-situ measurements. The elevated plumes contribute as much as ~70% to the column AOD whereas the ground $PM_{2.5}$ and CO show dramatic increase (e.g. $PM_{2.5}$ from 5 to 30 $\mu g/m^3$) in the NYC area. The sources and transport paths from the northwest US/Canada to the northeast US are verified by the NOAA-HMS, HYSPLIT backward trajectory and HRRR-smoke model. In addition, we present a case study of temporal and spatial variation of MLH in the urban, coastal and marine area. The simultaneous observations by a CCNY-lidar and two ceilometers indicate around 2-hr delay for the morning transit of MLH in the coast and marine area in comparison to the urban area, but consistently high MLH or PBLH from noon to afternoon. The initial comparison of NOAA NAM-CMAQ forecast with the observations indicate the consistent diurnal variation of PBLH and O_3, but biases in the early morning and night, which may be associated with the vertical mixing process, PBLH and chemical formation.

ACKNOWLEDGEMENTS

The New York State Energy Research and Development Authority (grant # 100415 and 137482), NESCAUM (grant # 2411) and the NOAA-Cooperative Science Center for Earth System Sciences and Remote Sensing Technologies under the Cooperative Agreement Grant #NA16SEC4810008 support this study. We appreciate the data from NYSDEC, NASA and NOAA-HMS, HYSPLIT and HRRR model.

REFERENCES

[1] Miller, Paul J., 2017: Retrospective and Future Analysis of Air Quality In and Downwind of New York City, DRAFT White Paper, available at ww.nescaum.org/documents/listos.

[2] Wu, Y., et al., 2018: Intra-continental wildfire smoke transport and impact on local air quality observed by ground-based and satellite remote sensing in New York City, *Atmos. Environ.*, 187, 266-281.

[3] Angevine, W.M., et al., 2004: Coastal Boundary Layer Influence on Pollutant Transport in New England. *J. Appl. Meteor.*, 43, 1425–1437.

[4] Huang, J., et al., 2017: Improving NOAA NAQFC $PM_{2.5}$ predictions with a bias correction approach, *Wea. and Forecasting*, 32, 407-421.

[5] Wu, Y., et al., 2009: Low and optically thin cloud measurements using a Raman-Mie lidar, *Appl. Opt.*, 48, 1218-1227.

[6] Holben, B.N., et al., 1998: AERONET A federated instrument network and data archive for aerosol characterization, *Rem. Sens. Environ.*, 66, 1-16.

[7] Rattigan, O. V., et al., 2010: Multi-year hourly $PM_{2.5}$ carbon measurements in New York: diurnal, day of week and seasonal patterns, *Atmos. Environ.*, 44 (16), 2043-2053.

https://doi.org/10.1051/epjconf/202023703020

OBSERVATIONS OF AEROSOL SPATIAL DISTRIBUTION AND EMISSIONS IN NEW YORK CITY USING A SCANNING MICRO PULSE LIDAR

Adrian Diaz Fortich*[2], Victor Dominguez[2], Yonghua Wu[1], Barry Gross[1], Fred Moshary[1]

[1]NOAA Center for Earth System Science and Remote Sensing Technology
Electrical Engineering Department, City College of New York, New York, NY 10031, USA
[2]NOAA EPP Earth System Science and Technology Fellow
**Email: adiaz001@citymail.cuny.edu*

ABSTRACT

In order to better understand the behavior of particulate pollution and atmospheric dynamics in New York City, it is of great importance to analyze the spatial distribution of aerosols. A scanning lidar system allows for horizontal range-resolved observations of aerosol backscatter with high space and time resolution. A challenge to analyzing the lidar returns is to disentangle extinction over the range of the observations to retrieve the backscatter coefficient with distance. This work presents horizontal measurements taken with a scanning eye-safe Micro Pulse Lidar in New York City. The measurements are analyzed using the Slope Method to get an estimate of the range-resolved aerosol backscatter coefficient. The results are presented as backscatter coefficient maps that display the aerosol spatial distribution within the field of view of the scanning pattern deployed. These observations clearly resolve aerosol dynamics and emission sources within the urban areas.

1. INTRODUCTION

Particulate pollution is a great concern in densely populated urban areas, such as New York City, as studies have shown that high exposure to fine particulate matter (particulates with a diameter of less than 2.5 μm, also known as $PM_{2.5}$) can be associated with adverse health effects on human respiratory and cardiovascular systems [1-2]. Lidar imaging of the spatial distribution of aerosols can shed light on particulate pollution and atmospheric dynamics. Vertical distribution of aerosols above the CCNY campus have been observed using a Raman-Mie lidar and ceilometer and have produced important studies on particulate pollution and plume transport [3-5]. However, vertically pointing instruments are unable to obtain multi-dimensional views of the atmosphere, and aerosol dynamics across the cityscape, hence scanning lidars are needed.

Previous implementations of scanning lidars at other locations have combined transmitter sources in different regions of the spectrum with different lidar techniques, some of them include elastic-backscatter lidar in the UV [6], differential absorption lidar in the IR [7] and Raman-shifted and polarization lidar in the IR [8-9]. Scanning lidars are required to perform rapid scans with moderate range, which can be achieved with high pulse-energy sources that allow measurements with reduced averaging time. However, these are usually not eye-safe and must be operated as established by the strict guidelines set forth by the US Federal Aviation Administration, and American National Standards Institute (ANSI) guidelines for Safe Use of Lasers Outdoors [10]. On the other hand, micro-pulse lidars (MPL) are simple, low cost and more importantly, meet the necessary eye-safety requirements for scanning. MPL consists of a transmitter source with kHz repetition rate and pulse energy on the order of micro Joules. The energy transmitted by this source meets the eye-safety requirements at all ranges when appropriately expanded in the beam size. Moreover, the receiver employs an effective photon counting detection using a compact avalanche photo diode detector [11]. Currently, MPLs are used in NASA's Micro-Pulse Lidar Network (MPLNET), which continuously measures boundary layer heights as well as aerosol and cloud vertical structure from multiple sites around the globe [12].

In this study, a commercial scanning eye-safe mini-MPL is used to perform horizontal measurements from which backscatter coefficient maps are derived. These maps display the aerosol spatial distribution and emission sources present within the field of view of the scanning pattern used.

2. METHODOLOGY

The instrument used, the Mini-MPL by Sigma Space Corp., is a scanning eye-safe elastic-backscatter lidar, located on top of the Grove School of Engineering at City College of the City University of New York.

Transmitter	
Wavelength	532 nm
Repetition rate	2500 Hz
Energy	4 μJ
Receiver	
Diameter	8 cm
Field of view	220 μrad
Detector	APD photon counting
Range resolution	5/15/30/75 m

Table 1. Mini-MPL specifications.

As defined by Campbell et al. [13], MPL measured raw photon counts take the form:

$$n(r) = \frac{\frac{CEO(r)\beta(r)T(r)^2}{r^2}+n_b+n_{ap}(r)}{D[n(r)]} \qquad (1)$$

Where $n(r)$ equals the measured signal return in photoelectron counts per microsecond at range r. C represents a dimensional system calibration constant. E is the transmitted laser pulse energy. $O(r)$ is the overlap correction as a function of range caused by the field of view of the transceiver system. $\beta(r)$ is the backscatter coefficient due to all types of atmospheric scattering. $T(r)$ is the atmospheric transmittance. n_b is the background contribution from ambient light. $n_{ap}(r)$ is the contribution from afterpulse, and $D[n(r)]$ is the detector deadtime correction factor as a function of raw counts. Rearranging the terms:

$$N(r) = C\beta(r)T(r)^2 = \frac{\{n(r)D[n(r)]-n_b-n_{ap}(r)\}r^2}{EO(r)} \qquad (2)$$

In this new form, $N(r)$ represents the Normalized Relative Backscatter (NRB), which is the primary product of MPL as it includes corrections for the different artifacts present in the raw measurements. The NRB is a quantity proportional to the attenuated backscatter as it is directly affected by the atmospheric transmittance term $T(r)$. This implies that the NRB values on the far end will tend to be small due to path attenuation. In order to correct for this and have a better representation of the aerosol distribution, it is necessary to further operate on the NRB values looking to separate the atmospheric transmittance term from the backscatter coefficient.

$$T(r) = e^{-\int_0^r \sigma(r')dr'} \qquad (3)$$

The atmospheric transmittance is defined in terms of the extinction coefficient, $\sigma(r)$ (Equation 3). Therefore, it is necessary to know the extinction coefficient at all ranges in order to calculate the atmospheric transmittance. The slope method of inversion defines the extinction coefficient for a homogeneous atmosphere, σ_{hom}, in terms of the signal slope, as follows [14]:

$$\sigma_{hom} = -\frac{1}{2}\frac{d\, ln[N(r)]}{dr} \qquad (4)$$

The accuracy of this method increases with increasing aerosol concentration, favoring its application to the lidar measurements conducted in the boundary layer where the highest aerosol concentrations are usually observed [15]. Typically, the slope of the least squares straight line fit to $ln[N(r)]$ over any interval where it appears to be nearly a straight line is used as the best estimate of $\frac{d\, ln[N(r)]}{dr}$. In other words, the slope method can be applied to determine the extinction coefficient only in certain parts of the signal. Since it is necessary to calculate the atmospheric transmittance $T(r)$ at all ranges, the extinction coefficient obtained using the slope method, σ_{hom}, is used as an estimate for the extinction coefficient $\sigma(r)$. Estimating the extinction coefficient allows us to perform a first-order correction by removing the atmospheric transmittance from the NRB. As a result, we are able to obtain an approximation to the backscatter coefficient that can be linked directly to the aerosol number density given that the molecular contribution is insignificant as it generally is within the boundary layer where aerosols prevail.

Please note that using σ_{hom} as an estimate for the extinction coefficient $\sigma(r)$ at all ranges introduces error particularly in regions where inhomogeneities occur. Nevertheless, we found the simplicity and stability of this approach to yield the best results. The solution proposed by Fernald [17] was considered as well, however, difficulties were encountered since the results produced were affected by inaccuracies on the far-end condition and error propagation while performing backward integration.

3. RESULTS

3.1. Emissions

The local utility companies in New York City, operate several peaking power generation stations throughout the city in which electricity is generated using fossil fuels (mostly natural gas). These plants are used to meet peak electric demands, especially during heat-wave days where air-conditioning electric loads exceed the power transported to New York City by long distance transmission lines. When the plants are in operation, they contribute to local emissions and resulting pollution within the City.

Our location at the City College of the City University of New York, is within 7 km of a number of such plants located in South Bronx and the Astoria Neighborhood in the Queens borough of New York City. In order to observe and capture emissions resulting from the generation of electricity taking place at these stations, continuous horizontal scans were carried out between the azimuth angles of 134° to 141° with respect to north, during daytime hours on July 1st, 2018. From these measurements, backscatter coefficient values were obtained as it was explained in section 2. Figure 1 shows a map in which the backscatter coefficient values obtained from one of the scans are displayed. 4 different sources of emissions can be clearly identified within this narrow field of view. One of them is located in South Bronx at about 3.5 km away from our position. Two more are found 5 km away in the neighborhood of Astoria, Queens. And the last one, at 6.2 km, is also located in Astoria, Queens. Monitored emissions are used to estimate plant activity during heatwave and non-heatwave days.

3.2. Spatial Distribution

Low income neighborhoods tend to have higher concentrations of vulnerable population to air pollution and often are in the vicinity of emission point-source and area sources [16]. A considerable amount of these neighborhoods are located in the Bronx borough of New York City which makes it an area of great interest for air pollution studies. On May 10th, 2017, measurements over the Bronx area were taken for up to 4 km from our location between 40° and 80° azimuth angles with respect to north. A map with backscatter coefficient values from one of the scans is shown on Figure 2. In this particular case, a high load of aerosols is observed to concentrate over an area that corresponds to central Bronx.

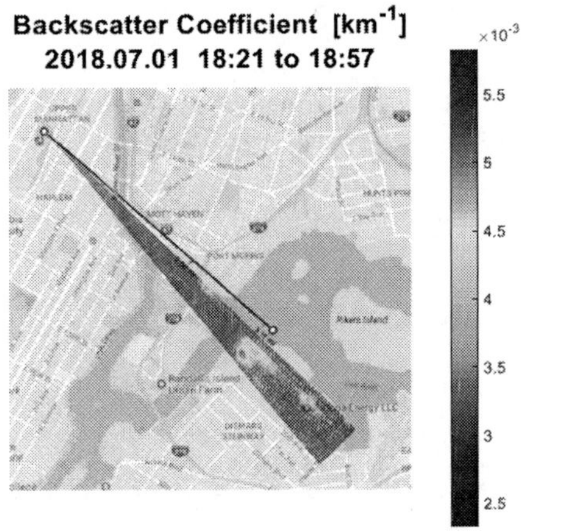

Figure 1. Backscatter coefficient map for July 1st, 2018 between 18:21 and 18:57 EDT.

Figure 2. Backscatter coefficient map for May 10th, 2017 between 16:54 and 17:26 EDT.

4. CONCLUSIONS

Aerosol spatial distribution maps have been derived by applying the slope method to horizontal measurements taken with an eye-safe scanning MPL. Using the backscatter coefficient to represent the aerosol distribution provides an important improvement in terms of accuracy since far-end measurements can be better analyzed once the attenuation term is removed. Two case studies are shown looking at different regions within New York City. In both cases, measurements on the far end are of great importance as they show emissions and high concentrations of aerosols.

ACKNOWLEDGEMENTS

This study is supported by The National Oceanic and Atmospheric Administration – Cooperative Science Center for Earth System Sciences and Remote Sensing Technologies (NOAA-CESSRST) under the Cooperative Agreement Grant #: NA16SEC4810008, and by the New York State Energy Research and Development Authority (NYSERDA) Agreement No. 100415. Its contents are solely the responsibility of the authors and do not necessarily represent the official views of the funding agencies.

REFERENCES

[1] C. A. Pope III, et al. Journal of the air & waste management association 56(6), 709-742 (2006).
[2] S. A. Weber, et al. Environmental Research 151, 399-409 (2016).
[3] C. Gan, et al. Atmospheric Environment 45 (37), 6613-6621 (2011).
[4] Y. Wu, et al. Atmospheric Environment 63, 32–42 (2012).
[5] Y. Wu, et al. Atmospheric Environment 106, 358-368 (2015).
[6] A. Behrendt, et al. Atmospheric environment, 45(16), 2795-2802 (2011).
[7] F. Späth, et al. Atmospheric Measurement Techniques, 9(4), 1701-1720 (2016).
[8] S. F. De Wekker, et al. Journal of Applied Meteorology and Climatology, 48(7), 1483-1499 (2009).
[9] S.D. Mayor, et al. Optical Engineering, 46(9), 096201-096201 (2007).
[10] American National Standard for Safe Use of Lasers Outdoors, American National Standards Institute, New York (2015).
[11] J. D. Spinhirne, et al. IEEE Transactions on Geoscience and Remote Sensing, 31(1), 48-55. (1993).
[12] E. J. Welton, et al. Second International Asia-Pacific Symposium on Remote Sensing of the Atmosphere, Environment, and Space, 151-158 (2001).
[13] J. R. Campbell, et al. Journal of Atmospheric and Oceanic Technology 19(4), 431-442 (2002).
[14] J. D. Klett, et al. Applied optics, 20(2), 211-220 (1981).
[15] T. Dreischuh, et al. Aerosols-Science and Case Studies (2016).
[16] I. Kheirbek, et al. The New York City Community Air Survey: neighborhood air quality 2008–2014 (2017).
[14] F. G. Fernald, Applied optics, 23(5), 652-653 (1984).

OBSERVATIONS OF THE LOWER-TROPOSPHERIC TEMPERATURE PROFILES USING THREE WAVELENGTH CO_2-DIAL

Yasukuni Shibata , Nagasawa Chikao, Makoto Abo

Tokyo Metropolitan University, Hino, Tokyo 1910065, Japan

**Email: sibata@tmu.ac.jp*

ABSTRACT

The eye-safe lower-tropospheric temperature profiler with three wavelength differential absorption lidar (DIAL) technique which can perform the continuous temperature profile observation through daytime and nighttime is conducted. The DIAL consists of a Nd:YAG laser pumped an OPG tuned around 1573 nm of an CO_2 absorption line with 2 mJ/pulse at 400 Hz repetition rate and a receiving telescope of 25cm diameter. In this paper, we show the result of continuous temperature profile observations over 25 hours from 0.39 to 2.5 km altitude in the lower-troposphere. We can see temporally the generation and disappearance of the temperature inversion layers in the planetary boundary layer.

1. INTRODUCTION

Continuous observations of the temperature profile of lower-troposphere are important for studies in the atmospheric science. The lidar is the most effective instruments for temperature profiling with high temporal and spatial resolutions. Different lidar techniques for measuring the atmospheric temperature profile have been developed, namely the rotational Raman lidar (RRL), the high-spectral resolution lidar (HSRL) technique and the DIAL[1]. In lower-troposphere observation, achievement of the high rejection rate for the Mie scattering signal is difficult for the RRL and the HSRL. In daytime observation, as the solar background noise is higher in UV and visible wavelength, high power laser is required.

The 1.6 μm CO2 absorption band has the reasonable absorption intensity for DIAL observations in the troposphere[2-4]. We added a third wavelength for temperature measurement to the conventional two-wavelength DIAL. In addition to λ_{on} (1572.992 nm) and λ_{off} (1573.137 nm), the third wavelength λ_T (1573.040 nm) for temperature measurement is adopted to the off-center absorption line. Temperature profiles by the DIAL technique can be obtained from the ratio of receiving signals λ_T and λ_{on}.

$$\frac{S_T(z)}{S_{on}(z)} = \frac{E_T}{E_{off}} exp\left\{-2\int_0^z n_{co2}(Z)\Delta\sigma dZ\right\} \quad (1)$$

$$\Delta\sigma = \sigma_{on}(T,z) - \sigma_T(T,z) \quad (2)$$

The atmospheric temperature $T(z)$ is calculated from the CO_2 density $n_{CO2}(z)$ which obtained from conventional DIAL technique and the differential absorption cross section $\Delta\sigma$.

2. TEMPERATURE OBSERVATIONS

This DIAL system is composed of an optical parametric generator (OPG) transmitter excited by a laser-diode pumped Nd:YAG laser. The partial power of λ_{on} and λ_T injection seeders were as split off and directed through wavelength-controlled units, respectively. Three injection seeders, which are DFB lasers, are connected to an optical fiber switch. The output seeder wavelength is switched for every shot in turn of λ_{on}, λ_{off} and λ_T. The atmospheric backscatters are collected by 25 cm diameter telescopes and bandwidth of the interference filter is 1 nm FWHM. The optical receiver includes a near-infrared PMT operating in the analog mode.

Atmospheric temperature observations were conducted from 0.39 to 2.5 km by the three wavelength DIAL in 21 - 22 April 2018 at the Hino campus of Tokyo Metropolitan University (35.7N, 139.4E). It was covered with high pressure during the observation period, it was fine and the wind was weak. Figure 1 shows temperature profiles with an integration time of 30 minutes. Temperatures between the ground and 0.39 km of the DIAL minimum altitude were interpolated. There is the stable boundary layer at 0.75 km altitude and the inflection point of the temperature gradient at an altitude of 1.5 km until 6:00 LT. The temperature on the ground is lower than 0.39 km altitude between 2:00LT and 6:00LT. The surface inversion layer is appearing in this time. After sunrise (5:00 LT), the surface temperature increased, and the inversion layers disappeared. After sunset (18:23 LT), an inversion layer was formed again at an altitude of 0.6 km. Typical diurnal variations of temperature profiles in fine weather were observed in detail.

3. DISCUSSION AND CONCLUTION

We have developed three wavelengths 1.6 μm DIAL system for measurements of the atmospheric temperature profiles with the CO_2 mixing ratio in the lower-troposphere. This DIAL conducted measurements of a diurnal variation of temperature profile on 21 - 22 April, 2018 from 0.39 to 2.5 km altitude. At an altitude of 1.5 km or less, the inflection point of the temperature gradient and the existence of a clear inversion layer were observed. Usually nocturnal surface inversion layer height is lower than 100 m, but observed inversion layer height is higher than this. This result will be explained following mechanism. Our observation site is located at the west edge of the Kanto plain. The north and west boundaries of the plain consist of high mountains while the other sides face the Pacific Ocean. Numerical results show cold air accumulation in the western parts of the plain[5]. The accumulated cold air generated a deep surface inversion layer.

ACKNOWLEDGEMENTS

This work was financially supported by the System Development Program for Advanced Measurement and Analysis of the Japan Science and Technology Agency.

REFERENCES

[1] A. Behrendt, Range Resolved Optical Remote Sensing of the Atmosphere, Springer, New York, (2005).
[2] D. Sakaizawa, et al., Appl. Opt. 48, 748-757 (2009).
[3] Y. Shibata, et al., Appl. Opt.,56, 1194-1201 (2017).
[4] Y. Shibata, et. al., Sensors 18, 4064 (2018).
[5] H. Kondo, J. of Appl. Meteor., 34, 1439 (1994)

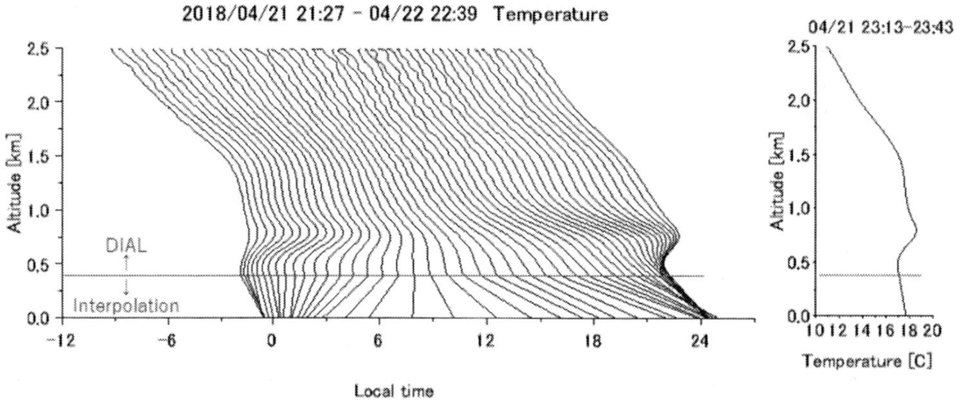

Figure 1 Atmospheric temperature profiles with an integration time of 30 minutes obtained from three wavelength CO_2 DIAL on 21-22 April 2018.

OZONE LIDAR OBSERVATIONS IN THE CITY OF PARIS: SEASONAL VARIABILITY AND ROLE OF THE NOCTURNAL LOW LEVEL JET

Gérard Ancellet[1*], François Ravetta[1], Jacques Pelon[1], Andrea Pazmino[1], Amélie Klein[1], Elsa Dieudonné[2], Patrick Augustin[2], Hervé Delbarre[2]

[1] *Laboratoire Atmosphere Milieux, Observations Spatiales (LATMOS), CNRS-INSU, Sorbonne Université, Université Versailles St Quentin, Paris, France*
[2]*LPCA Université du Littoral-Côte d'Opale, Dunkerque, France*
**Email: gerard.ancellet@latmos.ipsl.fr*

ABSTRACT

Ozone lidar measurements have been carried out in Paris during 4 years to characterize the seasonal variability of the vertical gradient within the urban planetary boundary layer (PBL). The interaction between NO_x emission and thermal stability of the PBL is the main driver of the winter strong positive O_3 gradient, while summer neutral gradient is related to weaker thermal stability and photochemical ozone production at the regional scale. Simultaneous lidar measurements of ozone and wind vertical profiles during 36 hours in September 2014 also show that the nocturnal low level jet (NLLJ) plays a significant role in the early morning ozone increase.

1. INTRODUCTION

Understanding the behavior of ozone in the urban boundary layer is necessary to develop effective air pollution control strategies in cities throughout the world. A key uncertainty in understanding ground level ozone is knowledge of the role of vertical and regional transport processes in controlling ozone and ozone precursors mixing and dispersion Chemical and physical processes at night and during transitions (night-to-day and day-to-night) as well as lack of vertically resolved measurements limit our ability to forecast "next-day" ozone concentrations [1].

2. METHODOLOGY AND EXPERIMENTAL SETUP

2.1 Objectives and methodology

Assessment of the UV DIAL lidar observations to improve our knowledge of ozone transport and photochemistry in the urban boundary layer was made using (1) regular measurements of the O_3 vertical profile between 0.3km and 3km (4-8 times per month) in Paris city center to characterize the seasonal evolution of ozone vertical gradient within and above the urban planetary boundary layer (PBL) (2) continuous observations of O3 and wind vertical profiles and of PBL height during 48 hours. The seasonal cycle was studied using lidar measurements averaged over 1hour during daytime and with partial clear sky conditions up to 3km altitude. Such measurements are available during 4 years (2011-2014) with 4-8 profiles per month [2]. Regarding the analysis of the O3 vertical profile diurnal cycle, a specific campaign was held in Paris in September 2014 [3].

2.2 The ozone lidar observations

The ozone Differential Absorption Lidar (DIAL) instrument is an airborne lidar system [4,5] installed in a shelter at Sorbonne Université (SU) in the frame of the QUALAIR facility. The transmitter module is based on a solid state Nd-YAG laser and Stimulated Raman Scattering in deuterium to generate 3 wavelengths (266, 289, and 316nm). The wavelength pair 266-289nm is used for the ranges between 0.4km and 1.7km, while measurements with the pair 289-316nm provide data above 1.5km only. The Lidar vertical resolution varies from 100m at the 0.4km altitude to 1000m at the 3km altitude. The temporal resolution can be as low as 2 minutes, but only hourly profiles are considered for the analysis of the seasonal variability. In the altitude range 200m to 3000m the Lidar accuracy including the different error sources (signal to noise ratio, Lidar alignment, background signal correction) is of the order of $15 \mu g.m^{-3}$. At range below 500m, the DIAL measurement is mainly limited by the geometrical telescope overlap function uncertainty. To further reduce the error on the ozone measurements below 500m a new optical receiver has been added

 https://doi.org/10.1051/epjconf/202023703022

in October 2013 to the Lidar. It includes a 10cm telescope with a 3.3mrd field of view to reach a full overlap near 200m. Measurements are alternately made using the large 40cm telescope in the far field from 400m to 3km and using the small near field telescope from 200m to 900m. For the near field measurements, only two wavelengths (266 and 289nm) are necessary because the 266nm Lidar signal is still strong enough below 1km. The correlation between the surface measurements at the Eiffel Tower (310m ASL) and the Lidar data averaged between 250m and 500m high have been used to check the Lidar accuracy at 310m. The comparison with the far field ozone Lidar data (Fig.1a) shows a bias of 10µg.m-3 and a 14µg.m-3 standard deviation. When using the near field mode (Fig.1b) the bias is close to zero and the standard deviation is 7µg.m-3.

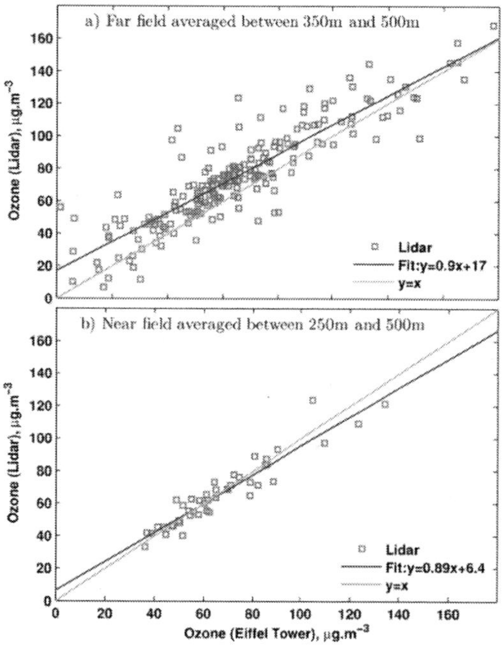

Fig. 1: Scatter plot of ozone concentrations measured by the Lidar and the Eiffel Tower measurements between 2011 and 2014. Top panel shows the Lidar far field (large telescope) averaged between 350m and 500m and the bottom panel shows the Lidar near field (small telescope) averaged between 250m and 500m.

2.3 The wind and backscatter lidar data

Wind fields during the September 2014 campaign were measured using an eye safe scanning wind Lidar (Leosphere Windcube WLS 100) operating at 1.54 µm with a repetition rate of 10kHz. The Doppler shift is measured with a heterodyne technique. The wind Lidar was located on the roof of a 60m tower providing wind fields in the range 0.19 to 1.5km with a 50m resolution [6]. The wind lidar accuracy is of the order of 0.5m.s^{-1}. The QUALAIR operational eye safe micropulse Lidar emitting at 532nm is also used to monitor the aerosol backscatter profile with a 15m vertical resolution and a 1s time resolution. The PBL height is calculated using the wind variance maximum during the night and the day-to-night and night-to-day transitions, while during the day the altitude is taken when the variance of the backscatter lidar is maximum [7].

2.4 Surface Data

The Paris air quality network provides O_3, NO_2 surface concentrations at about 70 stations. Two stations have been selected for our study: first at the top of the Eiffel Tower (275m AGL) and second at ground level away from direct traffic emission (Paris 13). Temperature lapse rate between the Montsouris station and the top of the Eiffel tower is also used to characterize the atmospheric thermal stability. The tropospheric column of NO_2 is measured by SAOZ (Système d'Analyse par Observation Zénithale) instrument. Although SAOZ was developed to monitor NO_2 total columns, it can also infer tropospheric columns in the polluted urban PBL [8,9].

3. RESULTS

3.1 The seasonal cycle of the ozone vertical gradient

The seasonal average and median of the vertical profiles of ozone measured by the LIDAR between 250m and 3km are shown in Fig.2 for the entire measurement period (2011-2014). The ozone gradients within the PBL are the largest in winter (60µg.m^{-3}.km^{-1}) and fall (40µg.m^{-3}.km^{-1}), while it is minimum in spring (27µg.m^{-3}.km^{-1}) and summer (20µg.m^{-3}.km^{-1}). The positive gradients are still visible using the quartiles, meaning that the result is robust. The seasonal variability of the

O_3 gradients in the urban layer near the surface shows a similar pattern when using the surface and Eiffel tower observations, but with lower values 25 and 14µg.m^{-3}.km^{-1} respectively in winter fall and less than 1µg.m^{-3}.km^{-1} in spring and summer. In winter, the amount of ozone destruction needed to explain an ozone positive gradient of 60µg.m^{-3}.km^{-1} between 60m and 590m is of the order of 8mg.m^{-2}, in good agreement with the winter NO_2 column measured by the SAOZ.

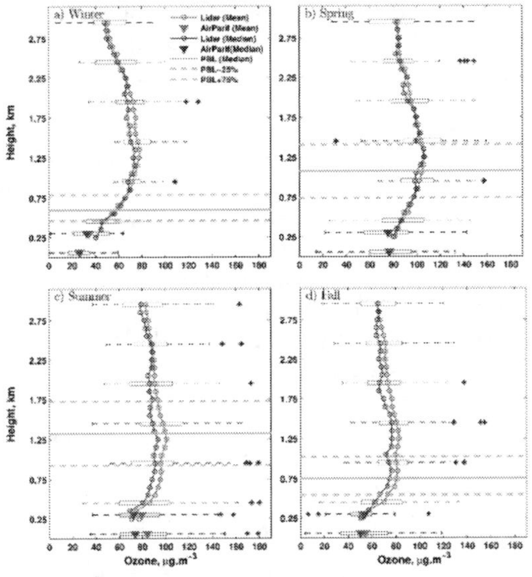

Fig. 2 Seasonal variation (a to d) of vertical profiles of ozone for the period 2011-2014. Blue (red) curves are the mean (median) of vertical ozone profiles for each season. Blue (red) triangles are the corresponding ozone seasonal mean (median) for in-situ measurements at the surface and at the Eiffel Tower. Whisker boxes limits show the interquartile range whereas outliers are showed by red markers. Planetary boundary layers (PBL) heights are represented by a green continuous line while the dashed green lines are for the percentiles 25% and 75%.

For spring and summer, ozone is well-mixed in the first 300m, but there is still a weak positive gradient above 500m. The spring and summer ozone gradients in the upper part of the PBL are however less visible when using the values of the 75th percentile. Considering the seasonal variability of the thermal gradient, we expect a neutral vertical ozone gradient in the urban layer up to 300m in summer. Considering that the level of NO_2 column 6mg.m^{-2}) is still significant in summer to induce a positive gradient across the PBL, it means that photochemical production at the scale of the Paris area and vertical mixing are strong enough to compensate for this NO_2 titration in summer.

3.1 The role of the nocturnal boundary layer jet on the surface ozone diurnal cycle

A campaign was held in Paris to study the role of the nocturnal low level jet (NLLJ) on the ozone diurnal cycle. Vertical ozone profiles were measured during a 36 hour intensive observation period (IOP) on 9-10 September 2014.

Ozone measurements from the Lidar are combined with in-situ measurements at the surface (Paris13) and at the Eiffel Tower (275m AGL) in one ozone contour plot (Fig.3a). Ozone Lidar measurements can be as low as 200m AGL, when using the near field telescope. The PBL height in Fig. 3 is calculated as described in Section 2.3. During the IOP, low nocturnal boundary layer (NBL) heights are observed, with values ~ 250m AGL. Ozone values are vertically homogeneous within the PBL during daytime due to strong vertical mixing during daytime as shown by the Doppler lidar vertical wind speeds (Fig. 3d). Ozone values are also larger in the afternoon in good agreement with the peak in the photochemical ozone production from NOx and VOCs emissions. During the night, O_3 concentrations are above 100µg.m^{-3} on 9 Sept., whereas they are as low as 40µg.m^{-3} late in the night/early morning on 10 Sept. This feature just above the NBL is co-located with a NLLJ, (Fig. 3b).The horizontal wind direction and air mass transport modelling (not shown) shows the NLLJ on 10 Sept. 2014 originates from a region with low ozone production northeast of Paris. Potential temperature and O_3 gradients have positive values during the night, corresponding to a stable NBL (Fig. 3b). Due to the stable NBL, ozone is trapped close to the surface and O_3 deposition combined with titration by NO result in a positive ozone gradient [10]. On 10 Sept.2014, the NBL lapse rate (+5K.km^{-1}) is however two times lower than the temperature gradient on 9 Sept., which favors vertical transport and mixing of the low ozone layer downwards. The

https://doi.org/10.1051/epjconf/202023703022

temporal evolution of NO_2 at the surface and the Eiffel tower (Fig. 3e) also shows that at early morning NO_2 gradient (06:00-08:00 LT) is higher on 9 Sept. than on 10 Sept. while emissions are expected to be the same. Increased vertical transport due to the NLLJ between the residual layer (RL) and the NBL is the most likely explanation for the ozone decrease in the RL. Enhanced nocturnal vertical mixing associated with NLLJs has already been observed in similar cases [1, 11]. The low ozone concentration in the RL has also consequences on the next day rate of ozone increase near the surface (Fig. 3a). Indeed the rate of increase between 10:30-12:30 LT is only 12µg.m^{-3}.h^{-1} on 10 Sept. while it is 26µg.m^{-3}.h^{-1} on 9 Sept. [3].

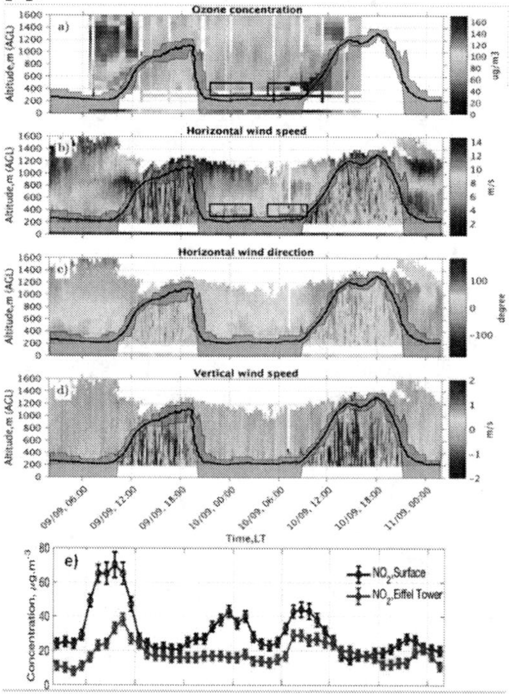

Fig. 3: Time evolution on 9 and 10 September of lidar O3 concentrations (a), of surface O3 (a) and NO_x (e) measured at the Eiffel Tower (275m AGL) and the Paris13 station (5m AGL), of the horizontal lidar wind speed (b) and direction (c), of vertical wind speed (d). Surface wind speed and direction measured at Montsouris are also shown in (b) and (c). The PBL height is shown in black with its uncertainty in grey.

Conclusion and Perspectives.

Our studies shows that regular and continuous observations of ozone and wind profiles in the urban boundary layer are mandatory to account for ozone surface variability. This is why the QUALAIR station (qualair.aero.jussieu.fr) was developed by SU in the center of Paris. The assimilation of ozone vertical profiles, PBL height and wind fields in chemical transport models will improve forecasts of "next-day" ozone pollution. While Doppler lidar, or urban canopy radar [12], are available to meet this goal, a new small size and autonomous UV DIAL lidar is still needed. It is currently under construction at LATMOS, using a diode pumped high repetition rate Nd-YAG laser and hollow core fiber for stimulated scattering.

ACKNOWLEDGEMENTS

This work was supported by SU and the OSU Ecce-TERRA. A. Klein and Elsa Dieudonné thesis were funded by Ministère de l'enseignement supérieur et de la recherche and supported by Ecole normale supérieure de Cachan (ENS Paris-Saclay).

REFERENCES

[1] J.T. Sullivan, et al. Atmospheric Environment 158, 160-171 (2017)
[2] A. Klein, et al. Atmospheric Environment 167, 603-615.(2017)
[3] A. Klein, et al. Atmospheric Environment 197, 92-102 (2019)
[4] G. Ancellet and F. Ravetta Appl. Opt. 37(24), 5509-5521 (1998)
[5] G. Ancellet and F. Ravetta J. Environ. Monit. 5, 47-56 (2003)
[6] I. Veselovskii, et al. Atmos. Chem. Phys 16, 7013-7028 (2016)
[7] L. Menut, et al. Appl. Opt. 38, 945–954 (1999)
[8] F. Tack, et al., Atmospheric Measurement Techniques 8~(6), 2417—2435 (2015)
[9] E. Dieudonné, et al. Geophysical Research Letters 40(6), 1247-1251 (2013)
[10]J. Stutz,et al. Journal of Geophysical Research: Atmospheres 109(D12), D12306 (2004)
[11] S. Kuang et al. Atmospheric Environment 45(33), 6078-6084 (2011)
[12] A. Oude Nijhuis et al. BAMS 99 (11), 2275-2293 (2018).

https://doi.org/10.1051/epjconf/202023703023

Pollution Monitoring Based on Vehicle-borne Particulate Quantum Lidar

XU WANG[1,] ,JING GAO[1],YUEFENG ZHAO[1]*

[1] *College of Physics and Electronic Sciences, Shandong Normal University, 250358, Jinan, China;*

**Email: yuefengzhao@126.com*

ABSTRACT

From February 7 to 10, 2018, due to unfavorable meteorological conditions, a process of air pollution occurred in Hefei and its surrounding areas, and moderate to severe pollution occurred in the municipal districts. Aiming at the pollution process in Hefei City, under the leadership of Hefei Environmental Protection Bureau and with the support of Hefei Environmental Protection Sub-bureau, four fixed observation points were selected to carry out all-weather environmental network monitoring pilot projects in Hefei municipal jurisdiction area. At the same time, a "mobile monitoring vehicle of atmospheric environmental pollution" was arranged to conduct all-weather walking observation to real-time monitor the spatial distribution and three-dimensional space of pollutants in Hefei urban area、inter-transport and space subsidence and diffusion. The RaySound Series Portable high-energy high-frequency lidar was used in the environmental networking monitoring in Hefei. The air quality in the observation area was evaluated comprehensively by using the mode of fixed vertical monitoring, plane scanning monitoring and mobile walking monitoring.

1. INTRODUCTION

As the main reason of air pollution, such as haze, aerosols or suspended particulates, especially fine particles, have an important impact on air pollution. For the measurement of particulate matter, some physical and chemical properties such as mass and concentration can be obtained by ground observation, but the spatial distribution information of particulate matter can not be obtained by ground measurement instruments. The vertical distribution of particulate matter can be mastered by means of sounding balloon and aircraft observation, but the operation cost is high and it is difficult to achieve data continuity [1]. Because the existing monitoring equipment in the station can not accurately judge the diffusion trend of atmospheric pollutants, in order to strengthen the capacity of building atmospheric pollution control monitoring and provide technical support for the action of air pollution control, it is urgent to add atmospheric particulate matter lidar and vehicle-borne particulate matter monitor. Atmospheric particulate matter lidar is mainly used for monitoring atmospheric transmission process, large-scale measurement in large areas, assisting vehicle-borne particulate matter monitor to realize the rapid moving monitoring of emission sources[2]. It monitors the main emission sources affecting atmospheric pollution in real time, including construction site dust, road dust, transport vehicle spraying, etc. and provides effective data support for pollution source analysis and related decision-making. In order to verify the feasibility of Lidar monitoring of pollutant sources, a Lidar Monitoring Experiment of pollutant sources was conducted in Hefei City by using vehicle-borne particulate matter quantum lidar.

2. METHODOLOGY

2.1 The lidar system

The basic principle of lidar detecting atmosphere is the mechanism of interaction between laser and atmosphere . Laser beams generated by lasers are emitted into the atmosphere. When they meet with air molecules, aerosols and other components in the atmosphere, they will scatter and absorb. A small part of the scattered energy (backscattered light) falls on the receiving telescope field of view and is received . Its basic principles can be expressed by formulas:

$$P(z) = \frac{\frac{CEO(z)\beta(z)T^2}{Z^2} + BG + AP}{DTC[P(Z)]}$$

In the formula, P (z) is the received signal, Z is the distance from the signal to the lidar, C is the constant of the lidar system, E is the energy of the laser pulse, the aperture is the backscattering coefficient, and R is the bidirectional transmittance. BG, AP, DTC and O (Z) represent background noise, residual pulse, detector delay correction and overlap factor correction respectively .

2.2 Inversion method of aerosol extinction coefficient

The inversion of aerosol extinction coefficient needs to solve the lidar equation.

$$P(z) = P_0 C z^{-2} [\beta_a(z) + \beta_m(z)] \exp[-2\int_0^z \alpha_a(z') + \alpha_m(z')dz']$$

Among them, P(z) is the receiving power at height z, C is the function of lidar system, B(z) is the backscattering coefficient, and A(z) is the extinction coefficient. Subscripts a and m are usually used to represent aerosols and atmospheric molecules. According to Rayleigh scattering theory, the backscattering coefficient of atmospheric molecules can be calculated from the standard molecular model. Assuming that the aerosol lidar ratio ($S_a = \frac{\alpha_a(z)}{\beta_a(z)} = 50sr$)and the atmospheric molecule ($S_m = \frac{\alpha_m(z)}{\beta_m(z)} = 8\pi/3$) , the aerosol extinction coefficient can be written as follows:

$$\alpha_a(z) = \frac{P(z)\exp[2\left(\frac{S_a}{S_m} - 1\right)\int_z^{z_c} \alpha_m(z)dz]}{\frac{p(z_c)}{\alpha_a(z_c) + \alpha_m(z_c)\frac{S_a}{S_m}} + 2\int_z^{z_c} P(z)\exp[2(\frac{S_a}{S_m} - 1)\int_z^{z_c} \alpha_m(z')dz'} - \frac{S_a}{S_m}\alpha_m(z)$$

The distribution of aerosol backscattering coefficients can be obtained by inversion of lidar m-scattering echo signal with the above formula [3] .

2.3 Monitoring Mode

Fixed vertical monitoring

Long-term vertical measurements at fixed stations can be used to analyze the transport of pollutants in space. Combining with the wind field lidar, the pollutant transport ratio can be calculated and the local/external ratio of pollutants can be analyzed. Vertical measurements can be used to obtain the spatial and temporal evolution data of vertical particulate matter and to understand the distribution and variation of local PBL layer, which is an important model data for local pollutant models and early prediction.

Mobile Navigation Monitoring

Vehicle-borne lidar mobile navigation monitoring technology can quickly obtain the atmospheric particulate matter profile in the monitoring area. Combining meteorological and topographic macro-environmental conditions, it can effectively analyze the generation and disappearance process and transboundary transport of particulate matter pollution, which has important application value in the detection of regional particulate matter pollution. The causes of urban pollution can be analyzed by combining navigation survey with meteorological conditions, and the pollution situation in various areas of the city can also be visually reflected. Under the combined mode of fixed navigation monitoring and navigation monitoring, Integrated data can be obtained comprehensively.

Planar Scanning Monitoring

Lidar scanning measurement can detect and record the changes of pollution sources (local and external sources) within the radius of 4KM on average, and find the sources of horizontal particulate matter emissions. Scanning data can be displayed directly on the map, that is to say, it can monitor the regional particulate matter in real time.

2.4 Monitoring factors and monitoring methods

According to the cruise monitoring of Hefei's surrounding environment and main pollution source components, the temporal and spatial evolution of particulate matter, PM10 and PM2.5 concentration was selected as monitoring factors. The cruise scanning monitoring was used to analyze the test data in coordination with conventional meteorological parameters such as wind direction, wind speed, low cloud, air pressure and temperature.

2.5 Data Selection Requirements

It is helpful to analyze the causes of Urban Haze by analyzing the pollution sources around the monitoring time in the heavily polluted weather of the city through radar data analysis. The monitoring point is chosen as a high place with wide field of vision and as far away from the water vapor weight as possible. Continuous and uninterrupted scanning was used to analyze the time of different non-polluting point sources, temporary pollution sources and fixed pollution sources. Because of the effect of urban wind, pollution will spread with the wind. Vehicle-borne lidar can be used for flexible monitoring and emergency monitoring of illegal discharge, leakage discharge and unknown sources of pollution in key areas or industrial parks.

3. RESULTS

3.1 Fixed vertical monitoring

Fig.1 Vertical Monitoring and Analysis Chart of Backscatter Lidar of Hefei City from 11:00 on February 8, 2018 to 10:00 on February 9, 2018

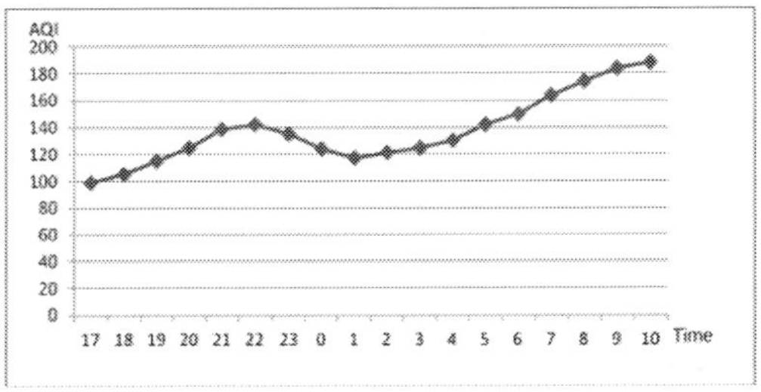

Fig.2 AQI Trend Map of State Control Points in Hefei from 17:00 on February 8, 2018 to 10:00 on February 9, 2018

According to Fig. 1, the inversion analysis map of fixed monitoring in Hefei from 19:03 on Feb. 8, 2018 to 10:13 on Feb. 9, 2018 shows that from 19:03 on Feb. 8, 2018 to early morning on Feb. 9, there exists diffusion and transmission of exotic pollutants about 600 meters above the air in Hefei, which layers clearly with the near surface layer, PBL layer gradually decreases, and the upper air pollution layer gradually subsides, until the early morning of Feb. 9. By the early morning of February 9, the upper pollution layer and the near-surface layer compounded, and the PBL layer gradually decreased from 19:00 on February 8 to about 600 m at high altitude until about 280 m in the early morning of February 9, and then remained basically unchanged. With the deposition of high-altitude pollution transmission and the decrease of mixing layer, while external pollution and local pollution are combined, the decrease of PBL layer is not conducive to the diffusion of pollution, resulting in the agglomeration of near-surface particulate pollution concentration.

Comparing with Figure 2, the trend of particulate matter basically represents the trend of AQI. The main reason for the increase of the monitoring index of the national control point on February 8 is the decrease of the mixed layer, which is not conducive to the diffusion of particulate matter. From the early morning of February 9, the monitoring data of the national control point began to rise, which is higher than that of February 8, Mainly in the case of lower mixed layer, the combination of external transmission pollution settlement and local pollution

3.2 Mobile Navigation Monitoring

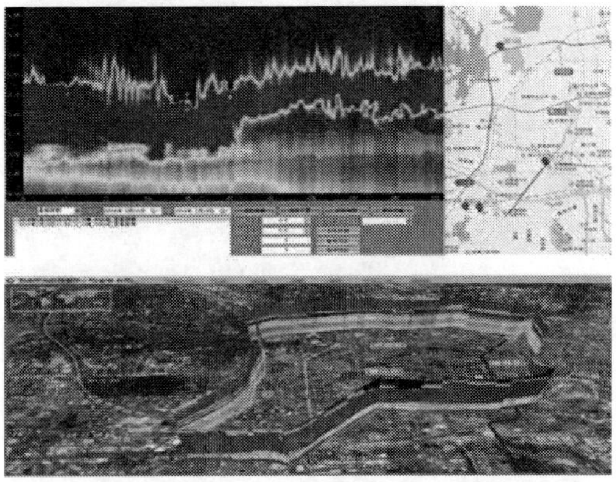

Fig.3 Inversion Chart of Navigation Monitoring at 10:05 on February 10, 2018 in Hefei City

Navigation monitoring was carried out at 10:05 on February 10, 2018. The wind speed was class 3, the direction was northeast ,and the air quality was moderately polluted. The navigation route was near the junction of Jade Road and Flourishing Avenue - Flourishing Avenue - Jinzhai South Road - Tongling Road - North Second Ring Road - West Second Ring Road - Jixian Road - Jixian Road and Luhua Road. From Fig. 3, it can be seen that there are thick clouds at about 2000m altitude during the voyage period. The atmospheric boundary layer changes with the change of monitoring road section. The influence of cloud height is greater. The thicker the cloud layer is, the lower the atmospheric boundary layer is, the worse the atmospheric diffusion condition is, and the concentration of particles near the ground increases. During the navigation monitoring period, there is a particulate matter pollution layer under the high-altitude clouds, the overall height is about 1000m. From the starting point of navigation to the intersection of Tongling Road and North Second Road, the regional clouds are relatively thick. , the layer height is lower and atmospheric boundary layer is about

500 m. Then from Tongling Road to North Second Road, the cloud layer thickness decreases, the atmospheric boundary layer rises to about 800m, the atmospheric diffusion condition improves, and the concentration of particles near the ground decreases.

3.3 Planar Scanning Monitoring

Fig.4 Feb. 10 21:43 Planar Scanning Analysis Chart

On February 10, 2018, 21:08-21:43, a plane scanning monitoring was carried out. The scanning angle was 5 degrees, the wind direction was northwest and wind speed was class 2, and the air quality index was in a good range. According to the results of plane scanning inversion of lidar ,the overall air quality is optimistic. It can be seen that in the spatial range of about 300 m horizontal distance and 100 vertical height at the scanning center, there is a higher concentration of particulate matter than other scanning areas, mainly near the ground. Under the influence of Northwest wind, the concentration of particles in the southeast direction of the scanning center is higher than that in the Northwest direction. It is suspected that there is a silent source near the scanning center with pollution source. Driven by the northwest wind direction, the particles diffuse to the southeast, resulting in higher concentration in the southeast region than that in the Northwest region[4].

3.4 Conclusion

In view of the long-term and mild pollution in this city, we can easily find that there are pollution mass transit over the city through radar data, and the pollution mass transit diffuses and settles. In addition, due to the influence of climate conditions, the PBL layer of the city is reduced, and the ability of pollutant diffusion is weak, which leads to the continuity of urban pollution[5]. The characteristics of vertical spatial boundary layer, pollutant capacity and pollutant input are obtained through integrated analysis of fixed monitoring, plane scanning and navigation monitoring. The spatial particle distribution in urban area is analyzed. At the same time, the monitoring results are basically consistent with the data released by the national control points, so as to get the precise sources and trends of pollution, and verify the feasibility and accuracy of lidar monitoring pollution sources.

ACKNOWLEDGEMENTS

The research work was supported by National Natural Science foundation of China under Grant No.81400285.And the authors would also like to gratefully acknowledge the support from the key research and development foundation of Shandong Province under Grant No.2016GGX101016.

REFERENCES

[1] He T Y, Stani S, Gao F, et al. Tracking of urban aerosols using combined lidar-based remote sensing and ground-based measurements [J]. Atmospheric Measurement Techniques, 2011,5(5):891-900

[2] WU Dexia, GONG Zhengyu, PAN Benfeng, et al. The applications particulate lidar in the stereo-monitoring for the complex atmospheric pollution ［J］. Environmental Monitoring in China, 2015, 31(5) : 156-162.

[3]Fernald F G. Analysis of atmospheric lidar observations: some comments[J]. Applied Optics, 1984, 23(5):652.

[4] Xie C, Zhao M, Wang B, et al. Study of the scanning lidar on the atmospheric detection [J]. Journal of Quantitative Spectroscopy&Radiative Transfer, 2015,150(150):114-120.

[5] ZHOU Yanqiu, NI Changjian, Liu Peichuan, et al. Analysis of mixed layer height by lidar detection over a heavy haze episode ［J］. Environmental Monitoring in China, 2016, 32(4) : 22-28.

POTENTIAL OF RAMAN LIDAR FOR PROFILING OF METHANE MIXING RATIO IN THE LOWER TROPOSPHERE.

Igor Veselovskii[1], Philippe Goloub[2], Qiaoyun Hu[2], Thierry Podvin[2], Mikhail Korenskiy[1]

[1]Prokhorov General Physics Institute, Moscow, Russia
[2]Univ. Lille, Laboratoire d'Optique Atmosphérique, Lille, France

ABSTRACT

The results of methane profiling in the lower troposphere by Raman lidar from Lille University observatory platform (France), are presented. The use of powerful DPSS tripled Nd:YAG laser allowed profiling of methane background mixing ratio of 2 ppm in the night time up to 4000 m with 100 m height and 1 hour temporal resolution. Enhancement of CH4 mixing ratio inside the boundary layer comparing to the free troposphere values was observed.

1 INTRODUCTION

Methane is currently the second (after carbon dioxide) most important greenhouse gas of anthropogenic origin. Methane is emitted from a variety of natural and anthropogenic sources (e.g. [1,2] and references therein) and on a per-molecule basis, methane is about 30 times more effective a greenhouse gas than carbon dioxide. Existing methane DIAL systems operate in the shortwave infrared (SWIR) spectral range between 1.65 and 2.3 μm, and due to low Raleigh scattering, methane profiling is possible only inside regions containing significant amounts of aerosol. Raman lidars, by contrast, can operate in aerosol free atmosphere.

Monitoring of the background CH_4 concentrations in the troposphere with airborne Raman lidar, based on powerful XeF laser, was reported by Heaps and Burris in 1996 [3]. However, the wideband radiation of excimer lasers requires the use of wideband interference filters in Raman channel, which, in turn, increases the sky background noise and possible contribution of aerosol fluorescence. Wideband detection also creates an additional complication related to interference from the oxygen Raman overtone (second Stokes shift). Significant progress in the development of the interference filters, detectors and laser sources during the last two decades provides, now the opportunity to develop the CH_4 Raman lidar based on a relatively compact tripled Nd:YAG laser. For narrowband 354.7 nm laser radiation the vibrational Raman line of methane is at 395.7 nm, while the oxygen Raman overtone (3089 cm^{-1}) is at 398.4 nm, which can be rejected by the interference filter.

In our presentation, we provide the first results of methane profiling in the lower troposphere using LILAS Raman lidar from Lille University observatory platform.

2 EXPERIMENTAL SETUP

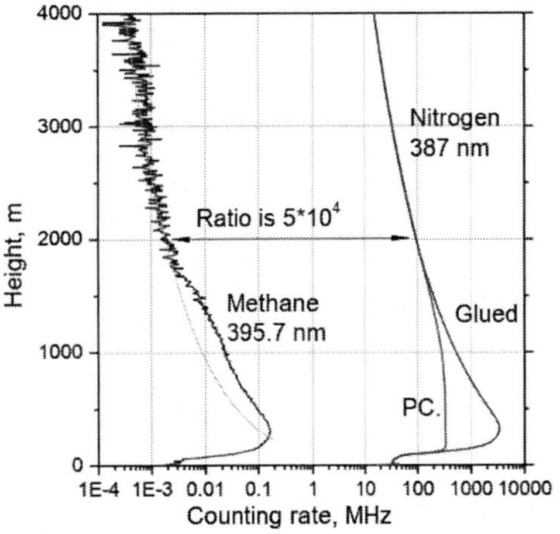

Figure 1. Raman signals corresponding to nitrogen (386.7 nm) and methane (395.7 nm) Raman scattering measured in June 2018. Signals accumulation time is 4 hours. Cian line corresponds to molecule scattering. In nitrogen channel the analog and photon count (PC) signals are glued.

The experiments were performed using multiwavelength Mie-Raman lidar from the Lille University (LILAS). The backscattered light is collected by a 40-cm aperture Newtonian telescope and the outputs of the detectors are

https://doi.org/10.1051/epjconf/202023703024

digitized using Licel transient recorders that incorporate both analog and photon-counting electronics. To perform CH_4 measurements shown in this paper, we modified the water vapor channel: 408 nm interference filter was replaced by the methane filter centered at 395.7 nm. The dichroic mirror in the receiver did not provide efficient selection of the methane (395.7 nm) Raman component, so it was replaced by the mirror with high reflectance at 395.7 nm. Only a small portion of the 387 nm component was transmitted to the nitrogen channel, while more than 95% of the 395.7 nm signal was reflected to the methane channel. Comparing nitrogen Raman backscatter intensity before and after the modification we find that the sensitivity of the 387 nm channel was degraded by a factor of 185.

The Alluxa interference filter in the methane Raman channel has a bandwidth of 0.3 nm with peak transmission greater than 80%. Suppression of 355 and 532 nm radiation is specified to be greater than 12 orders of magnitude. The Raman scattering cross section of methane is about 8 times higher than that of nitrogen, so for a 2 ppm methane concentration the ratio of intensities of Raman nitrogen to methane scattering is about $5*10^4$. Thus not only should the elastic backscatter components of the signal be suppressed sufficiently, but also the nitrogen and oxygen Stokes and anti-Stokes Raman lines should be as well. To eliminate the possible contribution of rotational anti-Stokes lines, the interference filter was combined with a notch filter, providing an additional OD4 blocking in the 348 – 360 nm range.

Two measurement sessions were organized. During the first session in May-June 2018, a flash lamp tripled Nd:YAG laser with a 20 Hz repetition rate, and pulse energy of 70 mJ at 355 nm was used. During second session in December 2018, it was replaced by the Merion MW 9-100 laser with a 100 Hz repetition rate, and pulse energy of 200 mJ at 355 nm. The results of first session were recently published in [4].

Fig.1 shows CH_4 and N_2 Raman lidar signals on the night of 14-15 June 2018. Aerosols are mainly located below 1700 m, in the same height range the methane backscatter is enhanced in respect to molecule profile. The ratio of nitrogen to methane lidar signals above the PBL top, estimated from our measurements of $5*10^4$, agrees well with prediction, which confirms that we observed methane Raman scattering.

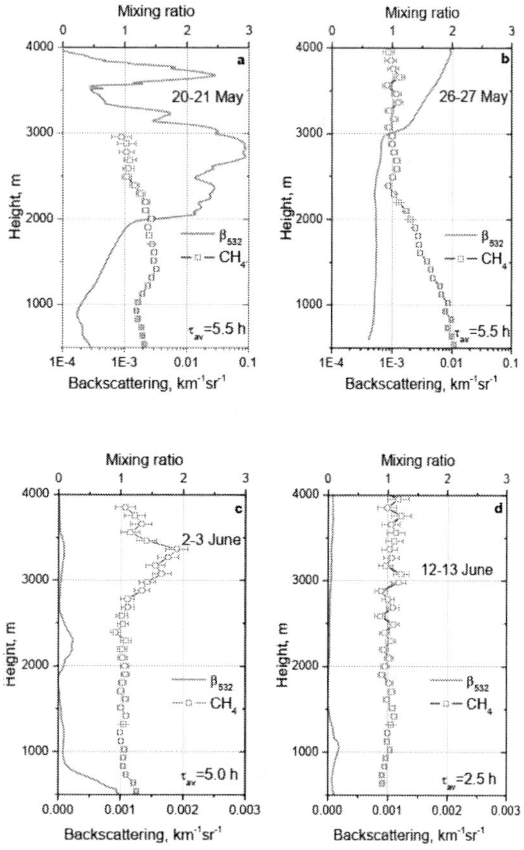

Figure 2. Vertical profiles of the methane mixing ratio and aerosol backscattering at 532 nm for four night measurement sessions. Mixing ratios are not calibrated and the value 1.0 corresponds to approximately 2 ppm. Signals averaging time τ_{av} is given in hours.

3. RESULTS OF THE MEASUREMENTS

The lidar derived mixing ratio of methane can be calculated from the ratio of CH_4 and N_2 lidar signals, corrected for the aerosol and molecular differential extinction.

The derived methane profiles exhibited strong night to night variation. Fig.2 shows the results of four measurement sessions, representing nights with different aerosol loading. The profiles of the methane mixing ratio are averaged over 100 m height bins. The same figure provides the profile of the backscattering coefficient at 532 nm. The mixing ratio is given in arbitrary units, assuming that the value of 1.0 corresponds to n_{CH_4}=2 ppm.

On the night of 20-21 May a scattering layer with peak value β_{532}=0.09 km^{-1}sr^{-1} occurs in the 2500 – 3100 m height range. Low lidar ratio (below 20 sr) and low depolarization ratio (below 5%) indicate that this layer is likely a water cloud. Strong cloud scattering demonstrates no influence on the mixing ratio, which is about 1.0 in the center of the cloud, proving that the interference filters provide sufficient rejection of elastic scattering. It should be mentioned also that the Raman band of the liquid water extends from 395 nm to 409 nm, so potentially it can be an interfering factor in the methane measurements. However, Fig.2a does not reveal a noticeable effect of liquid water Raman scattering on the methane profile due to the narrowband filter in the CH_4 channel. Cloud layers occurred also on 26-27 May (Fig.2b) at a height of approximately 4000 m with maximum values of backscattering coefficients of β_{532}=0.01 km^{-1}sr^{-1}. As in Fig.2a, the presence of clouds does not influence the methane measurements. By contrast, the night of 12-13 June (Fig.2d) was characterized by a low aerosol backscattering coefficient in the 500 – 4000 m range (β_{532} is below 2*10^{-4} km^{-1}sr^{-1}) and the mixing ratio shows no significant deviation from the 1.0 value in the whole height range. On 2-3 June, the increase of CH4 mixing ratio at 3400 m correlates with a weak aerosol layer (β_{532}<10^{-4} km^{-1}sr^{-1}) at the same height. It is interesting that a stronger aerosol layer at 2300 m is not accompanied by an increase in n_{CH_4}.

The use of powerful DPSS laser during session in December 2018 allowed to reduce the uncertainties of the measurements, as follows from Fig.3, showing nitrogen and methane Raman lidar signals on 25-26 December 2018. The DPSS laser allows to profile the methane mixing ratio with 1 hour resolution. Fig.4 presents twelve CH_4 profiles collected during the night 26-27 December for period 18:00 – 06:00 UTC. Statistical uncertainties are given for the first profile only (for the rest of profiles the uncertainties are similar). This figure demonstrates that profiling up to 4000 m with 1 hour resolution is possible. The time series of the methane ratio for three height intervals are given by Fig.5. The temporal variations of CH_4 exceed estimated statistical uncertainties.

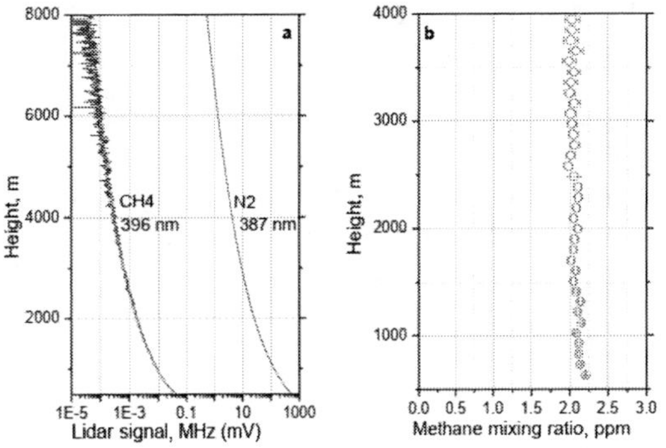

Figure 3. Raman lidar signals on 25-26 December 2018 corresponding to N_2 and CH_4 together with methane mixing ratio obtained with DPSS laser. Averaging time is 7 hours.

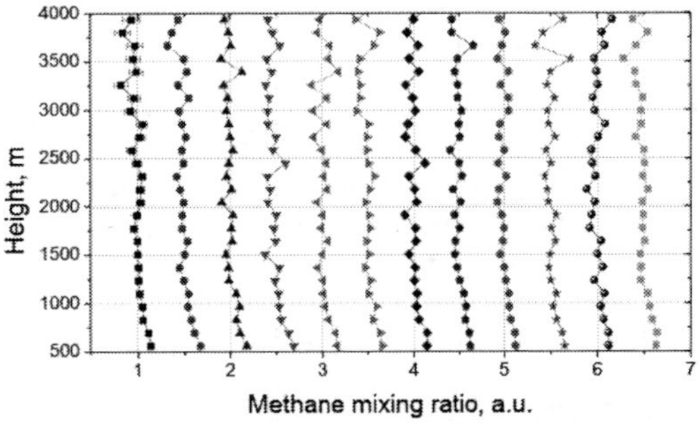

Figure 4. Vertical profiles of methane mixing ratio measured on the night 26-27 December 2018 for period 18:00 – 06:00 UTC. Profiles are averaged for 1 hour and shifted relatively to each other by 0.5. Mixing ratios are not calibrated and the value 1.0 corresponds to approximately 2 ppm.

4. CONCLUSIONS

The results presented here demonstrate the feasibility of profiling the background mixing ratios of methane in the lower troposphere using Raman lidar. In our measurements we observed enhanced concentrations of the methane inside the PBL, compared to aerosol free regions. The CH_4 profiles are not always correlated with aerosol backscattering, which corroborates the hypothesis that, in the PBL, not aerosol fluorescence but methane is measured. However, we cannot completely exclude fluorescence contribution. To measure and correct for it, if necessary, we plan to introduce an additional control channel close to 393 nm where no strong Raman lines exist and only fluorescence can be detected. Raman lidars for CH_4 monitoring cannot compete with IR airborne DIAL systems in sensitivity and accuracy. However, Raman lidars are capable to evaluate the vertical profile of methane concentrations through the boundary layer and in the aerosol free atmosphere.

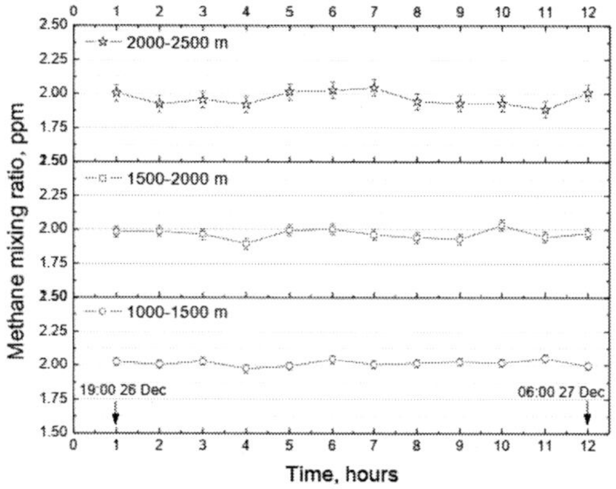

Figure 5. Time series of the methane mixing ratio in the height layers of 1000 – 1500 m, 1500 – 2000 m, 2000 – 2500 m measured on on the night 26-27 December 2018. Averaging time is 1 hour.

ACKNOWLEDGEMENTS

We acknowledge Emilie Colin and Melody Lardier from Lumibird, the new name of the QUANTEL-KEOPSYS group, for providing a Merion MW 9-100 laser and technical support. Development of the software for lidar data analysis was supported by the Russian Science Foundation (project no.16-17-10241).

References

[1] Baray, S., et al.: Quantification of methane sources in the Athabasca Oil Sands Region of Alberta by aircraft mass balance, Atmos. Chem. Phys., 18, 7361–7378, 2018.

[2] Kavitha, M., Nair, P. R.: Non-homogeneous vertical distribution of methane over Indian region using surface, aircraft and satellite based data, Atmospheric Environment 141, 174 – 185, 2016.

[3] Heaps, Wm. S. and Burris, J.: Airborne Raman lidar, Appl. Opt. 35, 7128-7135, 1996.

[4] Veselovskii, I., Goloub, P., Hu, Q., Podvin, T., Whiteman, D. N., Korenskiy, M., Landulfo, E.: Profiling of CH4 background mixing ratio in the lower troposphere with Raman lidar: a feasibility experiment., Atmos. Meas. Tech., 2019.

REAL-TIME POLLUTION ANALYSIS AND LOCATION BASED ON VEHICLE PARTICLE RADAR

Jing Gao[1], Xu Wang[1], Yuefeng Zhao[1*], Junyong Zhao[1]

[1]*School of Physics and Electronic Science, Shandong Normal University, 250014, Jinan*
**Email:yuefengzhao@126.com*

ABSTRACT

This paper first studies the advantages and main observation methods of atmospheric particulate matter lidar in Regional large-scale measurement, and then discusses the measurement principle of lidar and two commonly used methods for inversion of aerosol extinction coefficient.Finally, the mobile observation system consisting of lidar-assisted vehicle particle monitor is used. Anping County is used as the research area.The lidar data of Anping County on September 15 to September 20, 2017 are obtained by means of the combination of navigation and fixed vertical monitoring.The results of fixed detection showes that foreign pollutants begin to be imported around 10 pm on the 15th, and the superimposed pollutants peakes around 2 pm on the 16th.The concentration of particulate matter gradually decreased after 2 pm on the 16th, which is in line with the trend of air quality rising first and then falling in accordance with the trend of the Ministry of Environmental Protection.According to the particle trajectory tracking map of the navigation area, the contaminant transport was analyzed Combined with the meteorological conditions at that time.

1. INTRODUCTION

Suspended particulate matter or aerosols in the atmosphere, especially fine particles,lead to reduced visibility, frequent haze and other atmospheric pollution and cause damage to human health, making the monitoring, remediation and prevention of particulate matter a hot spot[1].For the measurement method of the particulate matter, some physical and chemical properties such as mass and concentration can be obtained from the ground observation, but the ground measurement instrument cannot obtain the spatial distribution information of the particulate matter[2]. Although the vertical distribution of particulate matter can be grasped by means of air balloon detection, airborne detection orsatellite remote sensing ,it is difficult to realize the continuity of the data because of the high operation cost.Atmospheric particulate matter lidar is mainly used for atmospheric transmission process monitoring and Regional large-scale measurement[3].The auxiliary vehicle particulate matter monitor realizes rapid movement monitoring of emission sources, and monitors the main emission sources that affect atmospheric conditions in real time, providing effective data support for pollution source analysis and related decision-making.

Common monitoring methods include: plane scanning, mobile navigation and fixed vertical monitoring.

2. METHODOLOGY

Lidar equation[4]:

$$P_L(z,\lambda_L)=K_L\frac{1}{z^2}\beta(z,\lambda_L)\exp\left\{-2\int_{z_0}^{z}\alpha(z',\lambda_L)dz'\right\} \quad (1)$$

Here, $P_L(z,\lambda_L)$ is the atmospheric backscattered echo power (W) at the height z received by the lidar; λ_L is the laser wavelength (nm); K_L is the lidar system constant $(W\cdot km^3\cdot sr^{-1})$; $\beta(z,\lambda_L)$ is the backscattering coefficient of the atmosphere at the height z, $\beta(z,\lambda_L)=\beta_m(z,\lambda_L)+\beta_a(z,\lambda_L)$,

$\beta_m(z,\lambda_L)$ and $\beta_a(z,\lambda_L)$ are the backscattering coefficients of air molecules and atmospheric aerosols respectively; $\alpha(z,\lambda_L)$ is the extinction coefficient (km^{-1}) of the atmosphere at the height z, $\alpha(z,\lambda_L)=\alpha_m(z,\lambda_L)+\alpha_a(z,\lambda_L)$,

$\alpha_m(z,\lambda_L)$ and $\alpha_a(z,\lambda_L)$ are the extinction coefficient of air molecules and atmospheric aerosols respectively; z_0 is the height where the lidar is located.

There are two main methods for solving the Mie- scattering lidar equation: the Klett method and the Fernald method[5-6].

Klett method: klett assumes that the atmospheric extinction coefficient and backscattering coefficient satisfy:

https://doi.org/10.1051/epjconf/202023703025

$$\beta(z,\lambda_L) = \frac{\alpha(z,\lambda_L)^k}{S_1},(2)$$

Here,k is a constant, the range of variation is 0.7 to 1.3, and the typical value is 1; S_1 is a constant.

Substituting equation (2) into equation (1), the atmospheric backscatter coefficient is eliminated, and a Bornoulli equation is obtained. Solve this equation and get the atmospheric extinction coefficient:

$$\alpha(z,\lambda_L)= \frac{\exp\{[S(z,\lambda_L)-S(z_c,\lambda_L)]/k\}}{[1/\alpha(z_c,\lambda_L)]-2/k\int_{z_0}^{z}\exp\{[S(z,\lambda_L)-S(z_c,\lambda_L)]/k\}dz},(3)$$

Here, $S(z,\lambda_L) = \ln\left[P_L(z,\lambda_L)z^2\right]$; z_c is the height at which the boundary value is selected. The Klett method is suitable for areas with high concentrations of atmospheric aerosols, such as atmospheric boundary layers, clouds, and aerosol layers with large optical thicknesses.

Fernald method: Fernald assumes the following relationship:

$$\alpha_a(z,\lambda_L) = S_1\beta_a(z,\lambda_L),(4)$$
$$\alpha_m(z,\lambda_L) = S_1\beta_m(z,\lambda_L),(5)$$

S_1 is the extinction backscattering ratio of atmospheric aerosols, also known as the lidar ratio. S_2 is the extinction backscattering ratio of air molecules, which can be derived from Rayleigh scattering theory.

Substituting equations (4) and (5) into equation (1) eliminates the extinction coefficient of atmospheric aerosols, and simplifies to obtain a Bernoulli equation. Solving this equation yields the backscattering coefficient of atmospheric aerosols:

$$\beta_a(z,\lambda_L)=-\beta_m(z,\lambda_L)+ \frac{X(z,\lambda_L)\exp\left[-2(S_1-S_2)\int_{z_c}^{z}\beta_m(z',\lambda_L)dz'\right]}{\frac{X(z_c,\lambda_L)}{\beta_a(z_c,\lambda_L)+\beta_m(z_c,\lambda_L)}-2S_1\int_{z_c}^{z}X(z',\lambda_L)\exp\left[-2(S_1-S_2)\int_{z_c}^{z'}\beta_m(z'',\lambda_L)dz''\right]dz'},(6)$$

Using the atmospheric aerosol backscattering coefficient obtained from the inversion and the assumed lidar ratio, the extinction coefficient of the atmospheric aerosol can be calculated according to formula (6).

3.Field monitoring

The purpose is to use the lidar navigation and fixed monitoring to analyze the trend of pollutants over the city and the causes of urban Pollution by

combining the meteorological conditions .

3.1 Urban air quality analysis

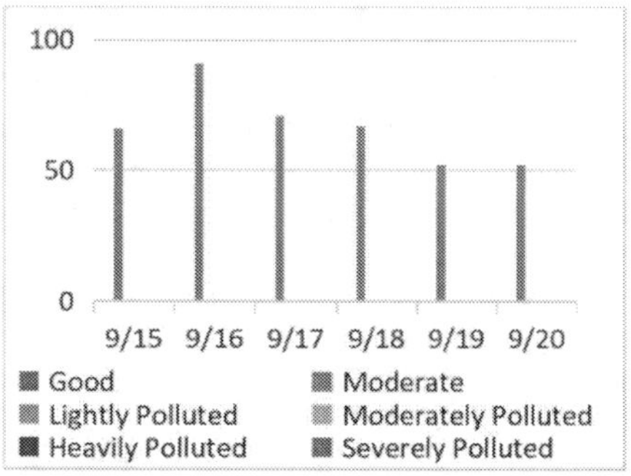

Figure 1 Anping County Air Quality Daily

Fig.1 shows that the air quality in Anping County is good in recent days, but the pollution index on September 16 is significantly higher than the pollution index in other days, and it shows a trend of rising first and then falling.

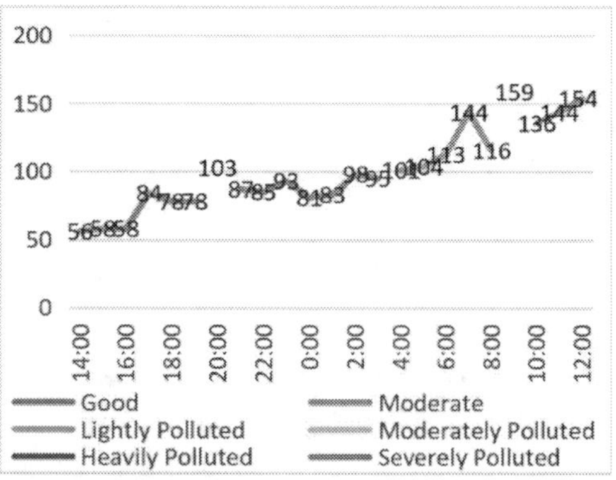

Figure 2 Air quality trend of Anping County in the past 24 hours

It can be seen from Fig.2that the air quality index began to rise at 2 am on September 16 and reached its maximum at around 9 am, and then floats up and down.This paper aims to combine the causes of this phenomenon with meteorological and environmental analysis based on radar detection data.

3.2 system parameters

The laser uses a pumped solid-state laser with a working wavelength of 532 nm, a single-pulse output energy of ≥1 mJ, and a pulse frequency of 2 kHz. The telescope has a diameter of

180mm. The particle monitor has a measurement range of 0-100 mg/m3 and an accuracy of less than ±10%. The wind speed and direction meter has two measurement range modes of 0 to 70 m/s and 0 to 360 degrees with an accuracy of ±3°.

4. Analysis of the overall urban pollution

4.1Fixed monitoring

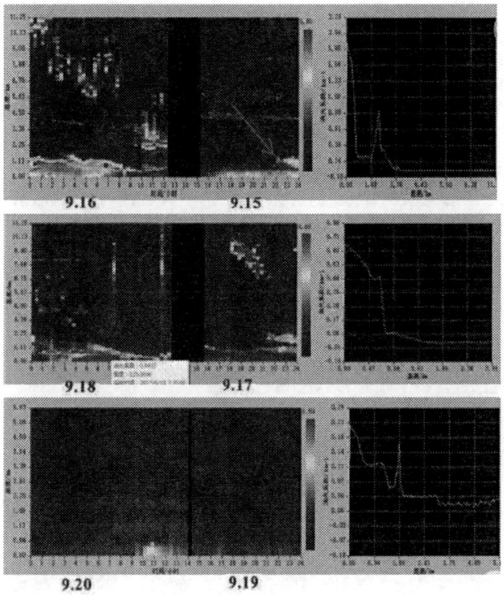

Figure 3Temporal and spatial evolution map of fixed particle concentration on September 15 to September 20, 2017

articulate matter concentration monitored from September 15 to September 20, 2017

As shown in Fig.3, the concentration of particulate matter was the highest on the 16th, and mainly the foreign input pollution. On the 19th, the concentration of particulate matter was the lowest and the air quality was the best. At around 10:00 on the 15th, foreign pollutants began to be transported.By about 2 pm on the 16th, the external pollutants and local pollutants were superimposed to reach a peak, and then the concentration of particulate matter gradually decreased.At around 21:00 on the 17th, there was another pollution and it was superimposed with local pollution to reach the peak at around 6:00 on the 18th.The particulate matter concentration then decreased and dropped to a minimum on the 19th.On the 19th and 20th, the concentration of particulate matter was lower and the air quality was better. On the 20th, around 10:00, there was a temporary increase in the concentration of particulate matter, which was caused by dust.

4.2Mobile navigation monitoring

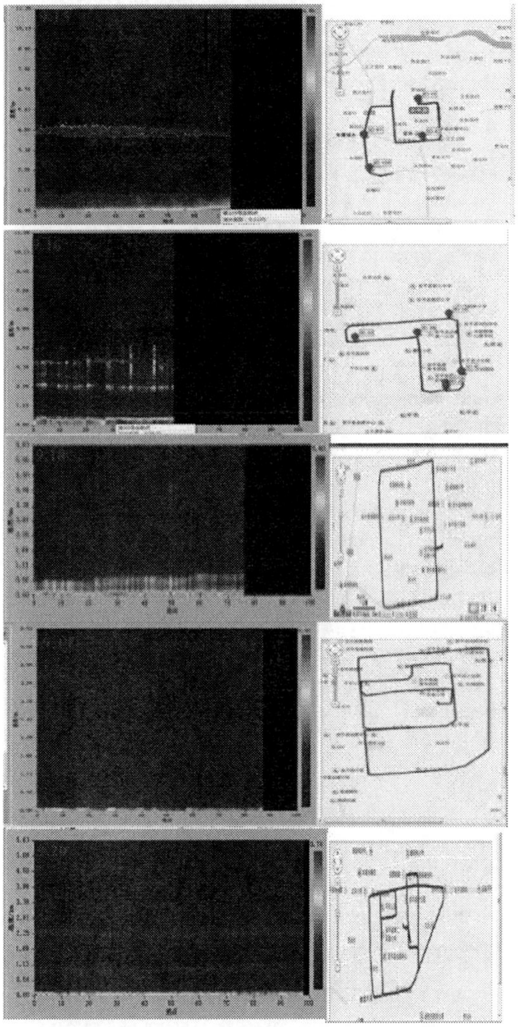

Figure4 Spatio-temporal evolution of particulate matter concentration in the navigation on September 15 to September 20, 2017

As shown in Fig.4, the concentration of particulate matter in Anping County was the lowest on the 19th and the highest on the 16th. The total inversion results show that from the 15th, the main pollution is some pollution near the surface, and the clouds in the air are thicker, which can be seen as cloudy days. According to the detection results, atmospheric particulate matter pollution began to aggravate on the 16th. At the beginning of the 18th, atmospheric particulate matter pollution was significantly weakened, and the inversion results from 19th to 20th remained basically unchanged.Compared with the urban air quality analysis report of Anping County from September 15 to September 20, the inversion results are consistent with the air quality forecast provided by the Ministry of Environmental Protection.

4.3analysis

By voyage, it can be found that the concentration of particulate matter in Hehuang Avenue is higher than that in other sections, and the concentration of particulate matter is significantly reduced from Hehuang Avenue to Hanwang Park. It can be inferred that the pollution source on Hehuang Avenue is mainly the discharge of motor vehicle exhaust and the dust generated by the motor vehicle when driving. The concentration of particulate matter in Yuhua Road and Hongqi Street is also high. Due to the laying of gas pipelines on Yuhua Avenue and Hongqi Street, the laying of gas pipelines causes dust, resulting in high concentration of particulate matter in this section. Finally, the EPA's particle concentration is also too large, there are construction sites around it, causing the spread of dust, and because of the surrounding buildings around the EPA, the diffused pollutants cannot be dispersed in time.

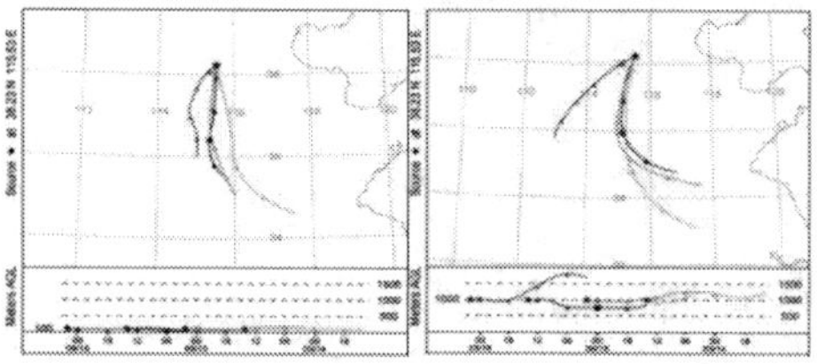

Figure5 particle backward trajectory tracking

It can be seen from Fig.5 that the wind direction of the navigation area is the southeast wind. The low-altitude diffusion is mainly at 100 meters, and the low-level diffusion is mainly the exhaust of motor vehicles. At 1000 meters, the main source of pollutants is high-altitude diffusion, while the high-level diffusion pollutants are mainly from the transmission of external pollution.At the same time, there are fewer vehicles on Hanwang Road, Hongqi Street and Beixin Avenue, and the concentration of particulate matter is relatively low. It can be speculated that the pollution source near the provincial road 301 is mainly the exhaust of the motor vehicle exhaust.

5. RESULTS

Based on the on-board particulate matter lidar mobile observation system, this paper monitors the weather conditions of Anping County from September 15 to September 20, 2017, and provides a comparison with the Ministry of Environmental Protection,confirming the accuracy and practicability of the system.

Vertical monitoring can directly reflect changes in urban pollution, while navigation monitoring can intuitively reflect regional pollution.The navigation detection adds the maneuverability of the detection based on the fixed detection, which is beneficial to better realize the rapid movement monitoring and real-time monitoring of the emission source, and provide effective data support for pollution source analysis and related decision-making.

ACKNOWLEDGEMENTS

The research work was supported by National N atural Science foundation of China under Grant No.81400285.And the authors would also like t o gratefully acknowledge the support from the k ey research and development foundation of Sha ndong Province under Grant No.2016GGX1010 16.

REFERENCES

[1]Inaba H.,*Opto-electronics*,*4*(2),101-123(1972).

[2]Reinhardt, A., In *2016 46th European Microwave Conference (EuMC)* (pp. 1509-1512). IEEE.

[3]Tao, J.,Science China Earth Sciences, 56(8), 1422-1433(2013).

[4]F. G. Fernald, Appl. Opt. 23, 652-653 (1984).

[5]Mao F.,et al.Optics express,21(7), 8286-8297(2013)

[6]Ji H, et al. Journal of Quantitative Spectroscopy and Radiative Transfer 188 (2017): 71-78.

A COMPREHENSIVE STUDY ON TROPICAL (10°N-15°N) MESOSPHERIC INVERSION LAYERS USING LIDAR AND SATELLITE (TIMED-SABER) OBSERVATIONS

K. Ramesh [1,*], S. Sridharan [2], K. Raghunath [2]

[1] *National Center for Atmospheric Research (NCAR), UCAR, Boulder, CO, USA.*
[2] *National Atmospheric Research Laboratory (NARL), Gadanki, A.P., India.*

Email: [*] *karanamram@gmail.com, susridharan@narl.gov.in, kraghunath@narl.gov.in*

ABSTRACT

One of the interesting and poorly understood features of mesosphere and lower thermosphere (MLT) region is the phenomenon of Mesospheric Inversion Layers (MILs). The poor understanding of MILs is due to limited access of their occurrence height region, however the lidars are more efficient tools which provide stratosphere and mesosphere nocturnal temperatures with high temporal and vertical resolutions. The state-of-the-art lidar system comprising Mie, Rayleigh lidars installed at National Atmospheric Research Laboratory (NARL), Gadanki (13.5°N, 79.2°E), India has provided an excellent opportunity to undertake this study. The Nd:YAG laser source with lower power (11W) has been replaced by the one with higher power (30W) in January 2007. As the laser power has been increased, the molecular back scatter signal is also increased and consequently the top height level of the temperature retrieval has been increased to ~90-95 km. In the present study, the role of dominant causative mechanisms for the occurrence of MILs has been discussed using mainly the lidar and satellite (TIMED-SABER) observations over Gadanki region.

1. INTRODUCTION

Mesospheric Inversion Layer (MIL) is the region of temperature gradient inversion from negative to positive in mesosphere, over several kilometers in thickness [1,2,3]. The study of MILs is essential for the comprehensive understanding of middle atmosphere/MLT energy budget. Alhough the MILs have been reported from several sites using lidar and space borne observations, their morphological characteristics are still unknown. However their causative mechanisms are quite complex, the MILs occur mainly due to Gravity Wave (GW) breaking, Planetary Wave (PW) critical level interaction, chemical heating and GW-Tidal interaction [2,3]. In the absence of damping, all the atmospheric waves *viz.*, GWs, PWs and Tides propagate vertically with exponentially ($e^{z/2H}$, z=altitude and H=scale height) increasing amplitudes in response to decreasing density up to MLT region. The waves deposit energy and momentum in the background when they break and consequently influence the thermal, dynamical structures and also the chemical composition. In addition to the above dynamical processes, the heating due to several exothermic reactions mainly among H, O, O_2, O_3, OH, HO_2 enhances the background temperature which is maintained as a MIL.

2. METHODOLOGY

2.1. The Gadanki Lidar System

The lidar system containing Mie, Rayleigh lidars installed at National Atmospheric Research laboratory (NARL), Department of Space, Gadanki (13.5°N, 79.2°E), India is operational since 1998. The system operates at 532 nm green laser with pulse repetition frequency (PRF) of 20 Hz and the relatively lower energy of ~550 mJ per pulse (PL8020, Continuum, USA). Later in 2007, the lidar system was upgraded with a high power laser of energy ~600 mJ/pulse and PRF of 50 Hz (PL9050, Continuum, USA) which is operated in monostatic biaxial mode. The 10X beam expander is used to expand the beam from its initial diameter of 9 mm to 90 mm so that the beam divergence decreases from ~0.45 mRad

https://doi.org/10.1051/epjconf/202023704001

to < 0.1 mRad and thus the beam remains within the field of view (FOV) of the receiving telescope at all height ranges. The Schmidt-Cassegrain (ø=35 cm) and Newtonian (ø=75 cm) telescopes serve as receivers to collect the atmospheric back scattered photons due to Mie (aerosols and clouds) and Rayleigh scattering (molecules) respectively. The received photons then allowed to fall on photo multiplier tubes (PMTs) of various gains (Hamamatsu, R3234-01). The PMT output signal will be collected and processed through the MCS-plus (Multi Channel Scalar) PC based photon count system with each two data acquisition channels of the receivers (Mie receiver:P, S channels & Rayleigh receiver : R, U channels). The photons are integrated for the time resolution of ~4 min corresponding to 12500 laser shots with dwell time of 2μsec for range resolution of 300 m. The laser source and the 90 mm expanded laser beam are shown in Figure 1(a-c).

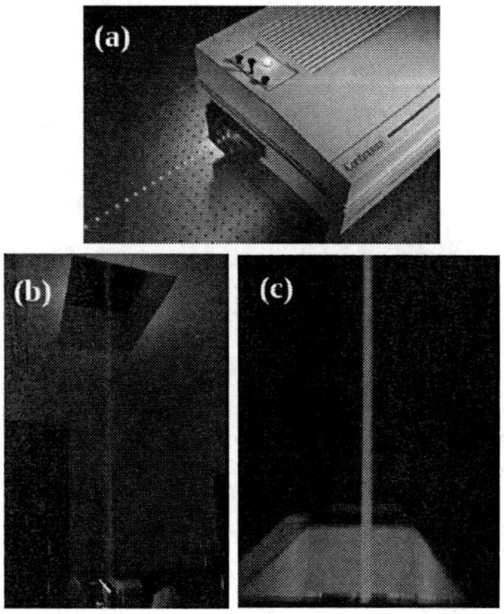

Figure 1. (a) Laser source from Continuum, USA (www.continuumlasers.com), (b) 90 mm beam steered upwards by a mirror, (c) beam entering into atmosphere.

The lidar equation for the backscattered photon counts can be given as follows.

$$N(z_i) = \frac{N_0 A K T^2(z_0, z_i)}{4\pi (z_i - z_0)^2} \left| n_r(z_i)\beta_r + n_m(z_i)\beta_m \right| \Delta z \quad \text{.........} (1)$$

where
z_i=Altitude of ith layer
N_i =Backscattered signal from i th altitude layer,
N_0=Number of emitted photons,
$n_r(z_i)$ & $n_m(z_i)$=Air molecules and aerosols concentrations,
β_r & β_m=Rayleigh and Mie backscattering cross sections,
z_0=Altitude of the lidar site,
A=Area of the receiver telescope,
$T^2(z_0,z_i)$=Atmospheric transmission,
K=Optical efficiency of the lidar system,
Δz=Thickness of the layer

The simulated lidar signal from equation (1) is as shown in Figure 2. The photon counts decreases with height in response to the decreasing atmospheric density.

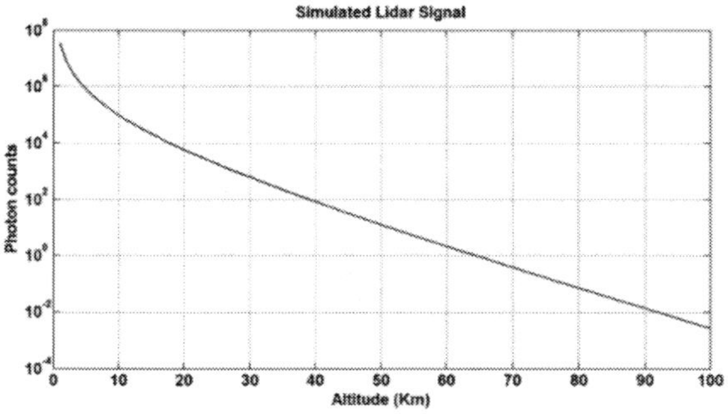

Figure 2. The simulated lidar signal for the lidar system at NARL, Gadanki.

2.2. Lidar Temperature Retrieval

The lidar temperature profiles are determined from the method given by *Hauchecorne and Chanin*,[1980] [4]. Above 30 km where the aerosol (Mie scattering) contribution is negligible and assuming the atmosphere is in hydrostatic equilibrium and obeys ideal gas law, the temperature at i[th] layer can be given by the following equation.

$$T(z_i) = \frac{M\, g(z_i)\Delta z}{R\, Log(1+X)} \quad \text{where } X = \frac{\rho(z_i) g(z_i)\Delta z}{P(z_i + \Delta z/2)} \quad \text{...............}(2)$$

and the statistical standard error in the temperature is given by the following equation.

 https://doi.org/10.1051/epjconf/202023704001

$$\frac{\delta T(z_i)}{T(z_i)} = \frac{\delta Log|1+X|}{Log|1+X|} = \frac{\delta X}{(1+X)\,Log(1+X)} \quad ....(3)$$

The scattering ratio is used to identify the aerosol and molecular layers and it can be defined as

$$R(z_i) = \frac{n_r(z_i)\beta_r + n_m(z_i)\beta_m}{n_r(z_i)\beta_r} \quad ....(4)$$

In general, the scattering ratio is assumed to be unity above 30 km so that the contribution from aerosol scattering is negligible.

2.3. TIMED-SABER Observations

The Sounding of the Atmosphere using Broadband Emission Radiometry (SABER) is one of the four space-borne instruments onboard NASA's Thermosphere-Ionosphere-Mesosphere Energetics and Dynamics (TIMED) satellite. It provides vertical profiles of temperature, ozone volume mixing ratio (O_3 vmr), chemical heating rates due to various exothermic chemical reactions among H, O, O_2 , O_3 , OH, OH_2 , and volume emission rates of different species. More details on SABER instrument and different data sets can be obtained from *Remsberg et al.* [2003] [5], *Mlynczak et al.* [2007] [6], and *Gan et al.* [2012] [7].

3. OBSERVATIONS AND RESULTS

The nightly (19:07-23:14 LT) mean Rayleigh lidar temperature (T) over Gadanki along with TIMED-SABER temperature profile for 12.5°N, 79.0°E at 23:58 LT are shown in Figure 3a. The lidar and SABER temperature profiles show large inversion layers above ~75 km with amplitude and thickness of ~93 K, ~85 K and ~8 km, ~5 km respectively. The MSIS-90 (Mass Spectrometer Incoherent Scatter-90) model temperature profile also included in the figure for reference. The profiles of lapse rate, Γ=-dT/dz and the Brunt-vaisala frequency square (N^2) calculated from equation (5) are shown in Figure 3b and Figure 3c respectively. The lapse rate and N^2 values represents the presence of turbulent layers and conditions for convective instability.

$$N^2(z) = \frac{g(z)}{T(z)}\left[\frac{dT(z)}{dz} + \frac{g(z)}{c_p}\right] \quad ....(5)$$

Here g(z), T(z) and c_p are acceleration due to gravity, mean temperature and specific heat at constant pressure (1004 J/K/Kg) respectively.

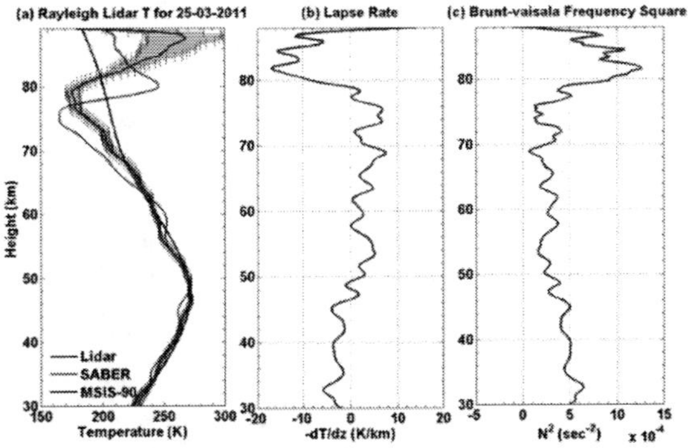

Figure 3. (a) lidar, TIMED-SABER and MSIS-90 temperatures, (b) lapse rate (-dT/dz), (c) Brunt-vaisala Frequency square (N^2) on the night of 25-03-2011 over Gadanki region.

It is evident from Figure 3b that the turbulent layers are not present as the lapse rate (Γ) is not equal to the adiabatic lapse rate (Γ_d ~10 K/km). Further it can be observed from Figure 3c that the condition for convective instability (N^2 <0) is not satisfied above ~75-80 km which indicates the negligible role of gravity wave breaking for the occurrence of inversion layer. Further the SABER chemical heating rates among H, O, O_2, O_3, OH and HO_2 for 25-03-2011 are shown in Figure 4. The reactions are listed in Table 1.

Table 1. Potential Exothermic Reactions for Mesospheric Heating Rates.

S.No.	Reaction
R_1	$O+OH \rightarrow H+O_2$
R_2	$H+O_2+M \rightarrow HO_2+M$
R_3	$H+O_3 \rightarrow OH+O_2$
R_4	$O+HO_2 \rightarrow OH+O_2$
R_5	$O+O+M \rightarrow O_2+M$
R_6	$O+O_2+M \rightarrow O_3+M$
R_7	$O+O_3 \rightarrow O_2+O_2$

 https://doi.org/10.1051/epjconf/202023704001

The heating rates increase above ~77 km and the reaction between H and O_3 dominates all the reactions with ~10 K/day. Also the total heating rates are about 20 K/day in the inversion region at ~80 km.

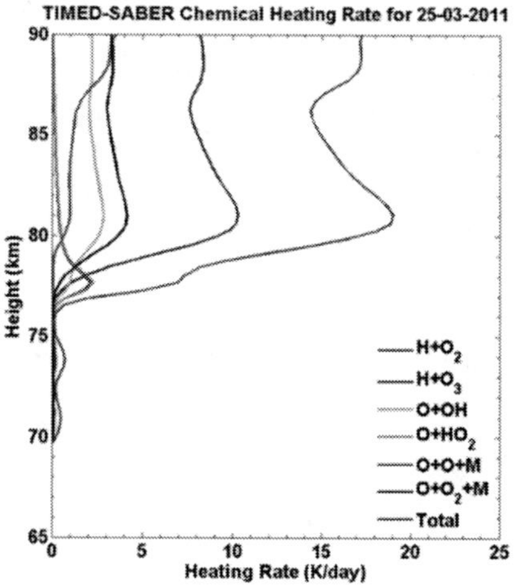

Figure 4. SABER chemical heating rates among H, O, O_2, O_3, OH, HO_2 for 25-03-2011.

4. SUMMARY AND DISCUSSION

The increased laser power of the NARL lidar system enabled to achieve the temperature profiles even up to ~90 km with the considerably lower uncertainty values. *Sridharan et al.* [2010] [8] observed a considerable reduction in the standard error with the new laser (PL9050) at heights ~5 km higher than the same obtained with the low power laser (PL8020). As an example, they obtained the standard error as 20 K at 85 km with low power laser and the same standard error is obtained above 90 km with the high power laser. They concluded that it is best suitable to study upper mesospheric temperature inversions, which takes place at altitudes above 85 km. In this study, it is clearly evident that the dynamics in connection with gravity wave breaking is negligible that can be observed primarily from the vertical profiles of lapse rate and Brunt-vaisala frequency square values. However the total chemical heating due to the exothermic reactionsamong H, O, O_2, O_3, OH, HO_2 is dominant for the occurrence of the MIL with the total heating rate of ~20 K/day above ~75-80.

5. ACKNOWLEDGEMENTS

The authors thank the technical staff for operating and maintaining the Mie-Rayleigh lidar system at NARL, Gadanki, India. They also gratefully acknowledge the NASA's TIMED-SABER team for providing the data used in this study through the website: http://saber.gats-inc.com/. One of the authors, K. Ramesh acknowledges NCAR/UCAR for NASA's Jack-Eddy Postdoctoral Fellowship.

6. REFERENCES

[1] T. Leblanc and A. Hauchecorne, Recent observations of mesospheric temperature inversions, J. Geophys. Res., 102, 19471-19482 (1997).

[2] J. W. Meriwether, and A.J. Gerrard, Mesosphere inversion layers and stratosphere temperature enhancements, Rev. of Geophys., 42, RG3003 (2004).

[3] W. J. Meriwether, and C.S. Gardner, A review of the mesospheric inversion layer phenomenon, J. Geophys. Res., 105, 12405– 12416 (2000).

[4] A. Hauchecorne, and M.L. Chanin, Density and temperature profiles obtained by lidar between 35 and 70 km, Geophys. Res. Lett., 7, 565–568 (1980).

[5] E. Remsberg et al., On the verification of the quality of SABER temperature, geopotential height, and wind fields by comparison with Met Office assimilated analyses, J. Geophys.Res., 108(D19), 4628, doi:10.1029/2003JD003720 (2003).

[6] M.G. Mlynczak et al., Sounding of the Atmosphere using Broadband Emission Radiometry observations of daytime mesospheric $O_2(^1\Delta)$ 1.27 µm emission and derivation of ozone atomic oxygen, and solar and chemical energy deposition rates, J. Geophys. Res., 112, D15306, doi:10.1029/2006JD008355 (2007).

[7] Q. Gan et al, TIMED/SABER observations of Lower mesospheric inversion layers at low and middle latitudes, J. Geophys. Res., 117, D07109, doi:10.1029/2012JD017455 (2012).

[8] S. Sridharan et al., Near simultaneous lidar observations of upper mesospheric inversion layers and sporadic sodium layers over Gadanki (13.5°N, 79.2°E, Asian J. Phys., 19, 4, 13-18 (2010).

https://doi.org/10.1051/epjconf/202023704002

DYNAMIC DRIVERS OF TIFE DIURNAL CYCLE IN ANTARCTICA

Zhibin Yu[1,2*], Xinzhao Chu[2], Xian Lu[2], and Cao Chen[2]

[1]*Harbin Institute of Technology (Shenzhen), Guangdong 518055, China. *Email: yuzb@hit.edu.cn*

[2]*University of Colorado Boulder, Colorado 80309, USA*

ABSTRACT

The discovery of the thermosphere-ionosphere Fe (TIFe) layers has opened a door to exploring the least understood thermosphere and ionosphere region between 100 and 200 km with ground-based lidar instruments. The characteristics of the polar TIFe layers, and the impacts of the atmosphere neutral dynamics, electrodynamics, and metallic chemistry on the formation of TIFe layers deserve further investigation, especially the diurnal cycles of TIFe layers observed by lidar. This paper aims at investigating the major driving forces with 1-D Thermosphere-Ionosphere Fe/Fe$^+$ (TIFe) model. A main question to answer is whether neutral dynamics like tidal winds or electrodynamics like the convection electric fields and currents in the magnetosphere and ionosphere are responsible for the diurnal cycle of TIFe layers.

1. INTRODUCTION

As the lower boundary of space weather regime, the region between 100 and 200 km is one of the key regions for the coupling among the neutral-ion chemistry, neutral dynamics, and ionospheric electrodynamics. Atmospheric waves of upward propagation transport energy and momentum and cause strong disturbances in the 100–200 km region [1, 2]. In polar region, energetic particle precipitations have substantial impacts on the properties of the neutrals and plasma in the upper atmosphere and ionosphere via direct ionization. However, due to lack of observations, 100–200 km region is still one of the least understood regions in the upper atmosphere.

The first TIFe layers up to 155 km discovered by Prof. Xinzhao Chu's group at Antarctica exhibit clear gravity wave structures with ~1.5 h period [3]. These thermospheric metal layers provide excellent traces for studying the coupling among neutral dynamics, chemistry, and electrodynamics in the E and lower F regions of ionosphere. Numerous studies based on lidar observations and numerical simulations have contributed to our understandings of the chemical and dynamical processes of the main metal layers distributing between around 70 km and 110 km [4]. The fundamental differences in the characteristics of the main metal layers and thermospheric metal layers give rise to non-available theory that could be directly applied to explain the formation of the thermospheric metal layers.

Chu et al. [3] originally hypothesize that the thermospheric Fe layers are formed through the direct recombination of the electrons with Fe$^+$ ions. These ions are uplifted by the polar electric field from the main deposition region to E-F region and vertically converged by neutral wind and electric field. This qualitative explanation is supported by the numerical simulations using a thermosphere-ionosphere Fe/Fe$^+$ (TIFe) model [Chu and Yu, 2017], which provide a detailed quantitative explanation of the gravity-wave-shaped Fe layers. The vertical wind shear generated by gravity waves appears to be the dominant factor to converge Fe$^+$ ions redistributed from the main deposition region to the upper regions by the convective electric field. The energetic particles originated from the magnetosphere substantially increase the electron density, enhancing the speed of direct recombination between Fe$^+$ and electrons.

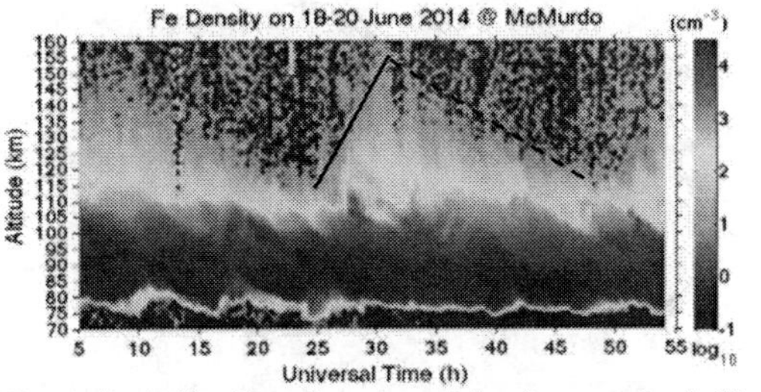

Fig. 1. Periodic diurnal cycles of the thermospheric Fe layers during a continuous 50 h of lidar observations at McMurdo, reproduced from Chu et al. [2016].

Besides the gravity-wave-shaped thermospheric metal layers, tidal-shaped thermospheric metal layers have been observed globally by the recent lidar studies [6, 7, 8, 9]. Unlike the diffused topside metal layer observed by Höffner and Friedman [10, 11], the density perturbations of the high-altitude TIFe layers show clear downward phase propagation with tidal or gravity-wave period. The thermospheric K layer reported from Arecibo descended with a similar speed as the downward phase speed of the modeled semidiurnal tide over Arecibo [6], so does the thermospheric Na layers reported from Lijiang, China [7] and Cerro Pachón, Chile [8]. The phase speed consistency strongly implies the essential roles played by the tidal-wave-generated winds in the formation of the thermospheric metal layers in the mid and low latitudes. However, the diurnal variations of the thermospheric Fe layers at Antarctica showed in Lübken et al. [2011] and Chu et al. [2016] exhibits more complex features rather than just downward phase speed of tides.

In this paper, we adopt the 1-D TIFe model to investigate the roles of the tidal winds and convective electric field in the formation of the diurnal features of the TIFe layers in Antarctica.

2. OBSERVATIONS

The Fe Boltzmann temperature lidar installed at Arrival Heights Observatory (77.8° S, 166.7° E) by the University of Colorado Boulder has been described in Chu et al. [2011, 2016]. The unique location of the Arrival Heights enables that the lidar observations produce outstanding datasets, leading to several scientific breakthroughs [13]. The nearly-zero artificial light pollution reduce the background noise to the lowest level so that the lidar system is sensitive to very low Fe density under relatively short integration time. The extended darkness during the polar nights enabled the continuous observations of TIFe layers over two days. Such data runs have benefitted greatly from the hard work of winter-over lidar students.

A diurnal TIFe layer case is the 50-h continuous Fe observations taken by Dr. Cao Chen in June 2014 at McMurdo (see Figure 1). Variations of TIFe layers clearly present diurnal features associated with the highest layer around 8 and 32 UT at 150–160 km. Before 8 UT, the peak altitudes of the three thermospheric Fe layers gradually increase from 115 to 150 km and then descend with relatively slower speed. Eventually the envelope of the thermospheric Fe layers forms a shape similar to a skewed Gaussian profile. The upward extension of the thermospheric Fe layers is ascending at the rate of ~1.4 m/s (the solid black line in Figure 1), and descending rate is ~0.7 m/s (the dashed black line). Within a single envelope of the thermospheric Fe layer, the Fe densities are perturbed by several cycles of short-period gravity waves, which clearly show downward phase progression.

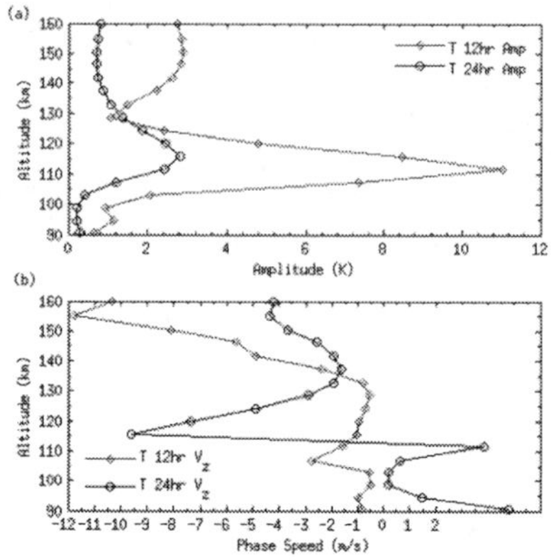

Fig. 2. (a) Amplitudes of diurnal temperature tides (black) and semidiurnal temperature tides (red), (b) the corresponding vertical phase speeds computed by GSWM09 in June at McMurdo.

The descending rate (the dashed black line in Figure 1) is much slower than the gravity-wave phase speed in the thermospheric Fe layers observed on 28 May 2011 at McMurdo [3], and it is comparable with the semidiurnal tide phase speed around 1 m/s calculated by the Global Scale Wave mode, GSWM09 [14], as showed in Figure 2(b). The semidiurnal tide computed by GSWM09 includes both migrating and nonmigrating components with wavenumber from -6 to +6. Like the tidal amplitudes in the thermospheric K layer case at Arecibo [6], the semidiurnal amplitude dominates over that of the diurnal tide in the thermosphere. Compared to the thermal tides at Arecibo, the amplitudes of the diurnal and semidiurnal tides in June at McMurdo are smaller by a scale factor of roughly 0.2, because of lack of thermal forcing [15]. It suggests less influences imposed by the thermal tides on the formation of the thermospheric Fe layers.

The second diurnal thermospheric Fe layer case is from Lübken et al. [2011]. By taking a 24-h Fe density composite 171 hours observations within a period of 12 days, Lübken et al. [2011] presented a contour plot of daily relative Fe density

https://doi.org/10.1051/epjconf/202023704002

perturbations from 75 to 130 km. Diurnal tidal perturbation is found in the main Fe layer from ~75 km to ~108 km, with an estimated tidal phase speed of ~5.4 m/s and a vertical wave length of ~36 km. Above 108 km, the positive phase of the Fe density perturbations dramatically changed and the shape becomes virtually vertical, indicating large descent rate of the Fe layer center. The descent rate of the positive perturbations is faster than 10 km/h, which obviously is not consistent with the classical tidal phase speed in altitudes at polar region.

Unlike those thermospheric metal layers with tidal features in mid and low latitudes [6, 7, 8], the two diurnal cases discussed here in polar region show complex layer envelope rather than just simple tidal shapes. As pointed out by Chu et al. [2016], the diurnal variations of the convection electric field probably play essential roles in the formation of the diurnal shape of thermospheric Fe layers. In the Section 3, we adopt the 1-D TIFe model to investigate the relative contribution of polar electric fields and tidal waves to the formation of thermospheric Fe layers in polar region.

3. SIMULATIONS OF DIURNAL CYCLES

3.1 Setup of Numerical Simulations

The numerical TIFe model is developed from first principles by Yu and Chu at the University of Colorado Boulder [5]. It is one-dimension model used to explore the formation and evolution of TIFe layers and related transport and convergence of Fe^+ in the E-F regions. The model includes the Fe and molecular ion chemistry in the E-F regions with newest reaction rate coefficients. A high-order flux corrected transport scheme is adopted to more accurately simulate the layer shapes.

In this paper, we try to reproducing the In this paper, we try to reproduce the TIFe layers observed on 18-20 June 2014 at McMurdo. Our major purpose is to investigate the roles of tidal winds and convection fields; therefore, we utilize the same initial Fe/Fe^+ profiles and auroral input as in Chu and Yu [2017]. The neutral winds including zonal and meridional winds are generated by the GSWM09, as shown in Figure 3. The interplanetary magnetic field (IMF) and solar winds data are taken from ACE satellites, and the electric field computed by Weimer-2005 is shown in Figure 3(a). As argued in Chu and Yu [2017], we reduce the electric field strength to 25% of the original Weimer field for simulations shown in Figure (3b) and 5-6. The neutral and ionic parameters are input from MSIS and IRI models for the specific date and geomagnetic conditions.

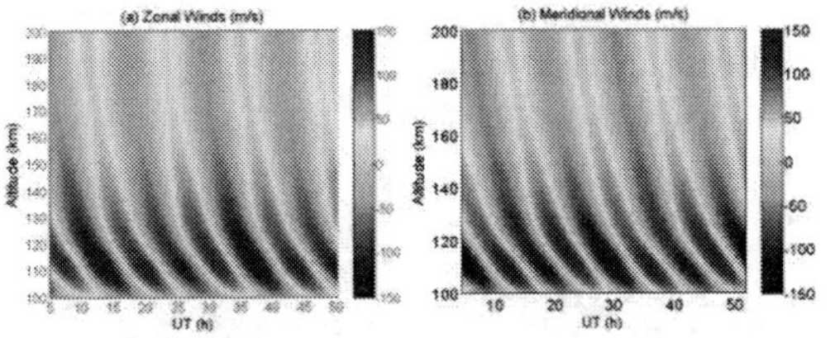

Fig. 3. Zonal (a) and meridional winds generated by the GSWM09 with monthly mean amplitude of the diurnal and semidiurnal tides in June over McMurdo.

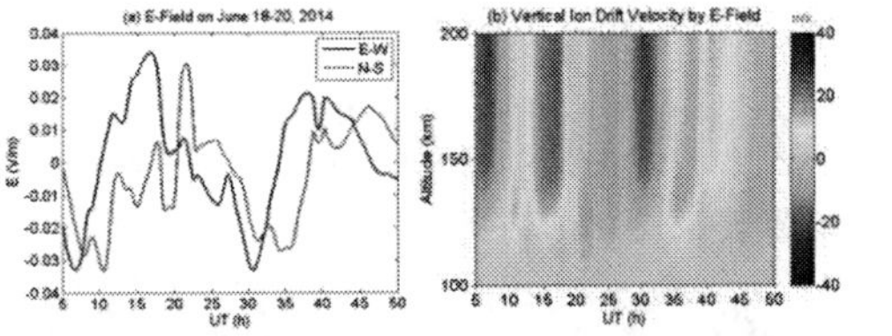

Fig. 4. The electric field (a) on June 18-20 2014 calculated from Weimei-2005 model at McMurdo, and the vertical ion drift velocity driven by electric field.

3.2 Reproducing the case on June 18-20 2014

The simulation starts at 4 UT on June 18 and runs for 48 hours. The chemistry and GLOW model are turned on for the entire period of simulations. Two scenario simulations have been tested: (1) Fe^+ and Fe transported only by the electric field, (2) Fe^+ and Fe transported by the GSWM09 winds and the electric field.

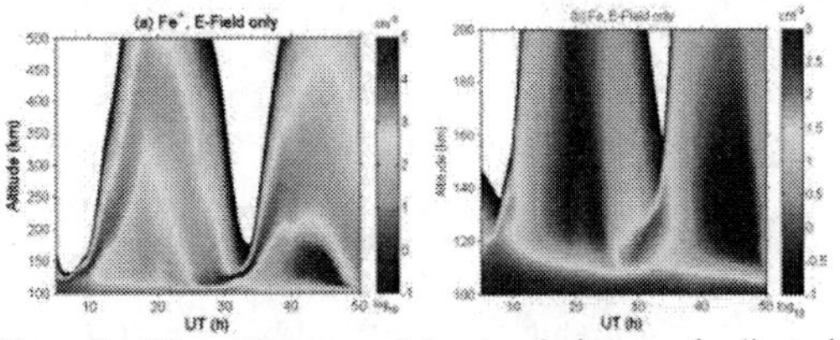

Fig. 5. The TIFe model simulations of diurnal variations of thermospheric Fe^+ (a) and Fe (b) layers transported only by the electric field.

The patterns of Fe^+ and Fe layers on June 18-20 2011 are presented in Figure 5 when the electric field is turned on but neutral winds are turned off. As shown in Figure 4(b), the ion drift velocity is downward above ~140 km, and upward below it from 5 UT to 10 UT, leading a converged layer around ~140 km. The upward transport from 10 UT to 20 UT lift the entire layer up to F region, and then move downward by the downward transport from 20 UT to 24 UT. Similar Fe^+ layer movement is found for the next 24 hours. Eventually the Fe^+ layers exhibit periodical patterns, producing the Fe layers with upward and downward movement, as shown in Figure 5(b).

Figure 6 shows the patterns of Fe^+ and Fe layers under the same conditions as that of Figure 5 except the GSWM09 winds are turned on. Similarity between the overall envelope of Fe^+ layer in Figures 5(a) and 6(a) is found, but below 150 km, several newly converged Fe^+ layers around 12 UT, 24 UT, 36 UT and 48 UT are formed by the tidal winds, which do not occur in Figure 5(a). The Fe layers are formed from the tidal-wind-converged Fe^+ correspondingly as in Figure 6(b), although Fe layers in Figure 6(b) show similar envelope as those in Figure 5(b).

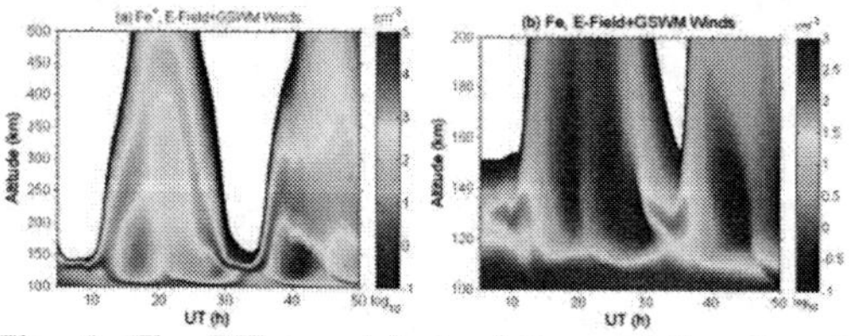

Fig. 6. The TIFe model simulations of the diurnal variations of thermospheric Fe^+ (a) and Fe (b) densities transported by the GSWM09 winds and electric fields simultaneously. The other setup parameters are the same as that of the Figure 5.

4. DISCUSSION AND OUTLOOK

The simulations presented in Figure 5 and Figure 6 suggest that the electric field is the dominant factor determining the periodic occurrence and envelopes of the thermospheric Fe^+ and Fe layers, and the diurnal and semidiurnal thermal tidal winds have minor contributions. However, the large amplitude of the semidiurnal tides below 120 km is still able to significantly shape the Fe^+ layers via relatively strong neutral-ion coupling, eventually produces tides wave-shaped Fe layers.

In this study, the winds generated by the gravity waves are not included. The simplified tidal wind inputs probably lead to the large discrepancies between the simulated and observed Fe layers. In addition, the simulated Fe^+ and Fe layers are under the ideal assumptions of constant auroral input, zero meteor ablation input and horizontal homogeneity, which can be hardly satisfied, especially for the long simulation time. In the future, a sophisticated wind model including the gravity wave perturbations will be embedded to the TIFe model. The time varying auroral particle precipitations and meteor inputs will be taken into account for further numerical simulations.

ACKNOWLEDGEMENTS

We appreciate Jeffrey Forbes, Art Richmond and John Plane for their contributions to the TIFe model development. We thank Xiaoli Zhang for her assistance in the GSWM09 setup. This work was supported by NSF grants OPP 1246405 and 1443726.

REFERENCES

[1] F. T. Djuth, L. D. Zhang, D. J. Livneh, I. Seker, S. M. Smith, M. P. Sulzer, J. D. Mathews, and R. L. Walterscheid, JGR., 115, A08305 (2010).
[2] S. L. Vadas, JGR, 112, A06305 (2007).
[3] X. Chu, et al., Lidar observations of neutral Fe layers and fast gravity waves in the thermosphere at McMurdo, Antarctica, Geophys. Res. Lett., 38, L23807 (2011).
[4] X. Chu and G. Papen, Chapter 5 in "Laser Remote Sensing", ISBN: 0-8247-4256-7, pp. 179-432 (2005).
[5] X. Chu and Z. Yu, Formation mechanisms of neutral Fe layers in the thermosphere at Antarctica studied with a TIFe model, J. Geophys. Res., 122, 6812-6848 (2017).
[6] J. S. Friedman, X. Chu, C. Brum and X. Lu, J. Atmos. Sol. Terr. Phys. (2013).
[7] Q. Gao, X. Chu, X. Xue, X. Dou, T. Chen, and J. Chen, JGR Space Physics, 120(10), 9213–9220 (2015).
[8] A. Z. Liu, Y. Guo, F. Vargas, and G. R. Swenson, GRL, 43, 2374–2380 (2016).
[9] Lübken, F.-J., J. Höffner, T. P. Viehl, B. Kaifler, and R. J. Morris, Geophys. Res. Lett., 38, L24806, (2011).
[10] Höffner, J., and J. S. Friedman, Atmos. Chem. Phys., 4, 801–808 (2004).
[11] Höffner, J., and J. S. Friedman, J. Atmos. Sol. Terr. Phys., 67, 1226–1237 (2005).
[12] X. Chu, Z. Yu, W. Fong, C. Chen, J. Zhao, I. F. Barry, J. A. Smith, X. Lu, W. Huang, and C. S. Gardner, invited paper in Proceedings of the 27th International Laser Radar Conference, New York, July 2015 (2016).
[13] X. Chu, et al., Antarctic Fe and Na lidars: laser spectroscopy in space for exploring cosmic dust and gravity waves, Proceedings of the 29th International Laser Radar Conference, Hefei, July 2019 (to be published).
[13] X. Zhang, J. M. Forbes, and M. E. Hagan, JGR, Vol 115, A06317 (2010).
[15] M. E. Hagen and J. M. Forbes, JGR, Vol. 108, NO. A2, 1062 (2003).
[16] T. T. Tsuda, X. Chu, T. Nakamura, M. K. Ejiri, T. D. Kawahara, A. S. Yukimatu, and K. Hosokawa, GRL, 42, 3647-3653 (2015).

LIDAR OBSERVATION OF MESOSPHERIC CLOUDS ABOVE BEIJING: A CASE STUDY

Shaohua Gong[1, 2, *], Guotao Yang[2], Jihong WANG[2], Yuchang Xun[2], Foju Wu[2], Qinzeng Li[2], Xiao Liu[3], Chunxia Chen[1], Dawei Tian[2]

[1] *School of Physics and Electronics Engineering, Hainan Normal University, Haikou, 571158, China;*
[2] *State Key Laboratory of Space Weather, Chinese Academy of Sciences, Beijing, 100190, China*
[3] *College of Mathematics and Information Science, Henan Normal University, Xinxiang, 453007, China*
* *shhgong@hainnu.edu.cn*

ABSTRACT

We report on the first observation of mesospheric cloud (MC) at the altitude range of 50-65 km with lidars at Yanqing (40.5°N, 116°E) and Pingquan (41°N, 118.7°E) on 30 October 2018. The MC occurred at the 51-56 km altitude range during the hours of dawn. It had an obvious double-layer structure, but the cloud layer was sparse. The MC was re-observed at the 56-62 km altitude range in the twilight, and the double-layer structure was still obvious. However, the cloud layer became thicker with a maximum volume backscatter coefficient (BSC) $3.1\times10^{-10}\mathrm{m}^{-1}\mathrm{sr}^{-1}$. Atmospheric temperature structure was derived according to the lidar observations, and it was found that, several hours before the MC occurrence, a temperature anomaly with coldest temperature ~185 K was propagating downward at the altitudes of 50-65 km. This MC layer was simultaneously observed with lidar at Pingquan in the twilight. It could be a regional MC event and possibly formed locally by a transient cooling due to small-scale disturbances in the mesosphere.

1. INTRODUCTION

Mesospheric clouds (MCs) observed at mid-latitudes were accepted as being very unusual atmospheric phenomena. They are possibly related to exceptional dynamical forcing or perhaps are an early indicator of significant long-term changes in the Earth's climate. For a long time, two kinds of typical MCs, i.e., noctilucent clouds and polar mesospheric clouds, have been extensively studied by imaging or lidar observations [1-3]. It was generally agreed that MCs are composed of ice crystals and often occur in summer near polar latitudes in the upper mesosphere between 76 and 86 km [4]. Meanwhile, it was reported that, due to the global climate change, the occurrence frequency of MC has increased and it has been occasionally observed at some mid-latitudes in the last several decades [5-7]. However, for the MC event under the mesopause, it has been seldom reported so far. In this paper, the lidar-observed features of an MC layer in the 50-65 altitude range above Beijing will be introduced. On the basis of the simultaneous lidar observations at different locations, the generation mechanism of this rare atmospheric phenomenon will be discussed.

2.INSTRUMENTATION &METHODOLOGY

The Yanqing (40.5°N, 116°E) lidar station was built in early 2010. Its lidar system works with two laser beams (@532 nm & 589 nm), and their emitted energies are approximately 320 mJ and 60 mJ per shot, respectively [8]. For an individual lidar profile, the photon counts accumulate for every 1000 laser shots in 34 s, and the spatial resolution and temporal resolution are 96 m and 34 s, respectively. The Pingquan (41°N, 118.7°E) lidar station was established in the summer of 2016, and its lidar system currently works with one laser beam at 589 nm. All the parameters of sodium lidar are the same as those at Yanqing. At altitudes between approximately 35 km and 110 km the lidar records the photons backscattered from air molecules, the sodium layer and aerosols.

An MC event is detected as an enhanced signal relative to the background noise and the air molecule signal at the altitude range of 50-80 km. Such enhanced signal can last for more than 30 min and not vary when the laser frequency is detuning. The MC is quantified by determining the volume backscatter coefficient, β_{MC}, which is defined as [9]:

$$\beta_{MC}(z,\lambda) = n(z)\cdot\frac{d\sigma(180^\circ)}{d\Omega} \quad (1)$$

where n is the number density of MC particles and $\frac{d\sigma(180^\circ)}{d\Omega}$ is the effective cross section for backscatter by an individual MC particle for a specific particle size distribution and the applied wavelength of 589 nm or 532 nm. In practice, β_{MC} was calculated with the following two equations:

$$R(z) = \frac{S_{total}(z)}{S_{mol}(z)} \quad (2)$$

$$\beta_{MC}(z) = (R\,(z) - 1)\cdot\beta_{mol}(z) \quad (3)$$

where R is the backscatter ratio of MC, $\beta_{mol}(z)$ is the BSC of atmospheric molecules, S_{mol} is the backscatter signal from molecules, and S_{total} is the total signal after subtraction of the background. At those altitudes where there was no MC, the backscatter signal was normalized to the atmospheric molecular backscatter signal,

$$S_{total} = S_{mol} = c\cdot\beta_{mol} \quad (4)$$

and it was calculated from the atmospheric densities taken from NRLSISE-00 model. Then, the constant c was used to constrain $\beta_{mol}(z)$ at MC altitudes.

3. RESULTS & DISCUSSION

3.1 Mesospheric cloud observed in the dawn on 30 October 2018.

During the regular lidar observation at Yanqing (40.5°N, 116°E) in the dawn on 30 October 2018, it was found that some obvious enhancement began to occurred in the photon-count profiles at the altitude range of 51-56 km due to the occurrence of MCs. The time variation of the volume backscatter coefficient of this MC was analyzed and plotted in Fig.1. It was shown that, around 04:30 LT, the cloud began to occur near the altitude of 51 km, and it was very sparse. About 15 min later, another cloud layer occurred at the altitude range of 53.5-55.5 km, and a double layer structure was observed during the rest of lidar observations. Unfortunately, this MC observation has been carried out for only ~90 min, and it was terminated around the sunrise (~06:00 LT) because the lidar observation can be performed just in nighttime.

According to the definition in Eq.(1), the volume BSC of cloud layer is directly related to the number density of MC particles. It is presented in Fig.1 that the maximum β_{MC} was only $1.2\times10^{-10}\ m^{-1}sr^{-1}$ although the density of cloud particles increased gradually with time. That means the MC layer was very sparse during the hours of dawn, and the number density of MC particles was extremely tiny.

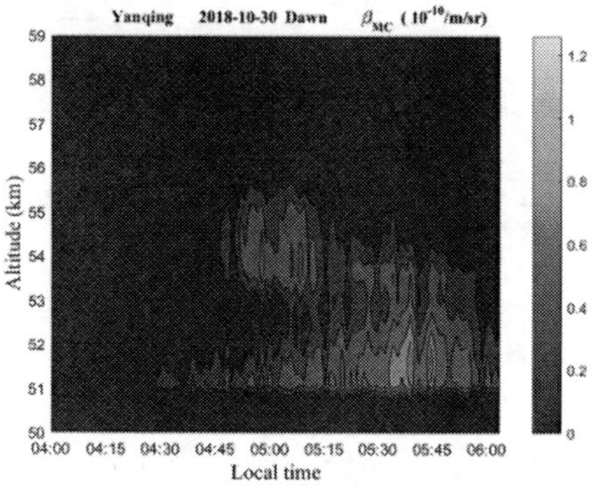

Fig. 1. Contour plot of the volume backscatter coefficient showing the time evolution of MC layer observed at Yanqing (40.5°N, 116°E) in the dawn on 30 October 2018.

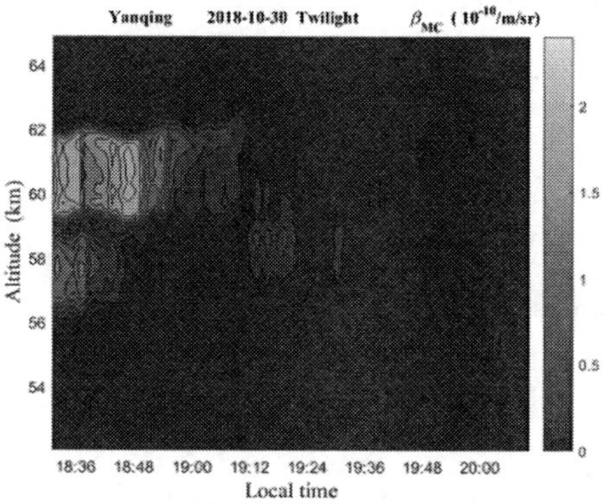

Fig.2. Contour plot of the volume backscatter coefficient showing the time evolution of MC layer observed at Yanqing (40.5°N, 116°E) in the twilight on 30 October 2018.

3.2 Mesospheric cloud observed in the twilight on 30 October 2018.

The lidar observations restarted around 18:30 LT in the twilight, and the MC layers were observed again at the very beginning of the observation. The time-altitude variation of the volume BSC of cloud layer was plotted in Fig.2. It was shown that the MC layer occurred at the altitude range of 56.2-62.3 km, and the double-layer structure was still obvious. However, the number density of MC particles in the upper layer became greater than that in the lower layer, and the maximum β_{MC} occurred in the upper layer around 18:48 LT. The value was $2.8\times10^{-10}\ m^{-1}sr^{-1}$, and it is about twice of the maximum observed in the dawn. Moreover, Fig.2 presents that the volume BSC decreased gradually in 19:00-20:00 LT. It means the number density of MC particles decreased gradually in this time period, and the MC almost totally faded away around 20:00 LT.

At Pingquan (41°N, 118.7°E), this MC event has been simultaneously observed with lidar, and the analysis result about the time variation of the volume BSC was plotted in Fig.3. It is found that the MC was firstly observed by lidar at the altitude range of 52-55 km around 18:45, and a very sparse MC layer existing for only several minutes was observed at the altitude range of 59-62 km. For the layer observed at the altitude range of 52-56 km, the volume BSC firstly increased gradually and then decreased, and the maximum $3.1\times 10^{-10}\ m^{-1}sr^{-1}$ occurred around 19:30 LT. This observation result indicates that the MC was very sparse at the very beginning and gradually became thicker till 19:30 LT, and then it gradually diffused and totally fade away around 20:45 LT.

Fig. 3. Contour plot of the volume backscatter coefficient showing the time evolution of MC layer observed at Pingquan (41°N, 118.7°E) in the twilight on 30 October 2018.

3.3 Comparison and discussion

To compare the two simultaneous observation results from lidars at Yanqing (40.5°N, 116°E) and Pingquan (41°N, 118.7°E), although an MC event has been simultaneously observed at both the two locations, the double-layer structure of MC over Pingquan was not such obvious as that observed at Yanqing. In Fig.3, the upper layer was very tiny, and it was visible to lidar for just several minutes around 19:15 LT. Meanwhile, the lower layer mainly located in the altitude range of 53-56 km, and it was ~3 km lower in altitude than that shown in Fig.2. In terms of time, at Pingquan, the maximum concentration of cloud occurred around 19:30 LT, and it was ~40 min earlier than that occurred over Yanqing. Moreover, the MC event disappeared around 20:45 LT at Pingquan, and it was around 20:00 LT at Yanqing. That is, different features of the cloud layer have been observed by lidars at the two locations which are ~230 km far from to each other. This suggests that the MC layer was probably inhomogeneous in horizontal direction.

In additional, at Yanqing, the same MC event was simultaneously observed by the lidar working at 532 nm, and identical features of the MC layer were found. However, at Pingquan, the MC event was observed by lidar in the twilight only, and no clue of MC has been found when the lidar observations in the dawn were checked. This may indicate that a small-scale MC event was observed above Beijing on 30 October 2018.

According to lidar-observed features above, this MC observed above Beijing is likely not a noctilucent cloud or a polar mesospheric cloud which are often occur in summer near polar latitudes in the upper mesosphere between 76 and 86 km. It occurred in the late autumn of Beijing, and the main altitude located near stratopause. However, this lidar-observed MC possibly had similar generation mechanism as them.

By employing the method introduced by Chanin and Hauchecorne [10], the atmospheric temperature structure before the MC occurrence has been retrieved according to the lidar observations at Yanqing. It was found in Fig.4 that, in the time period of 01:00-04:00 LT (i.e., 1~3 hours before the occurrence of MC), a lower temperature structure was transporting downward, and the atmospheric temperature decreased gradually from 210 to 185 K in the 65-50 km altitude range. Moreover, an extremely cold temperature (~140 K) was measured at the altitude of 80 km by SABER/TIMED (footprint@(41.8°N,

119.7°E)) at 23:09 LT on 29 October 2018 (i.e., ~6 hours before the occurrence of MC).

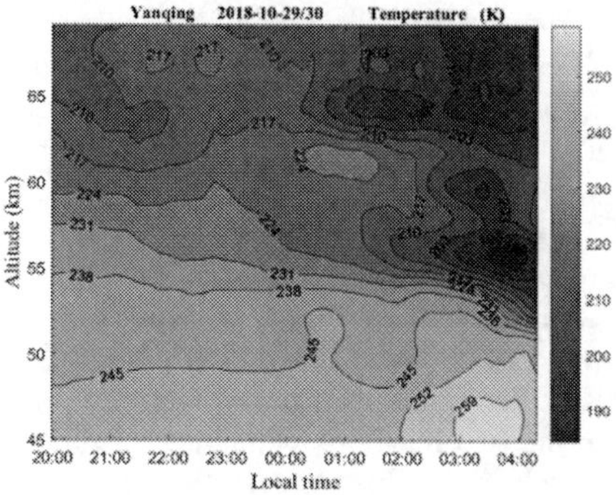

Fig. 4. Time-altitude variation of atmospheric temperature (40-65km) derived from lidar observations at Yanqing (40.5°N, 116°E) before the occurrence of MC.

It is well known that, in the past few decades, greenhouse-gas concentrations in the atmosphere have unequivocally risen, and weather extremes (e.g., extreme temperatures and precipitations) frequently occurred in the Norther Hemisphere mid-latitudes due to the global warming [11]. The mesosphere probably became cooler because of the increasing levels of CO_2 in the atmosphere. Meanwhile, more water vapor occurred in the mesosphere due to the oxidation of elevated levels of methane in the atmosphere.

Therefore, for this lidar-observed MC with a small scale at middle latitudes, it was possibly related to an exceptional dynamical forcing (e.g., gravity wave activities) which was leading to greater adiabatic cooling in a small region and forming a temperature minimum near the altitude range of 50-65 km. When the atmospheric temperature was sufficiently cold, water vapor formed those mesospheric clouds at these altitudes.

4. CONCLUSIONS

At the dawn and twilight on 30 October 2018, an MC event in the 50-65 km altitude range was observed with lidars at Yanqing (40.5°N, 116°E) and Pingquan (41°N, 118.7°E). It is an extremely rare atmospheric phenomenon over Beijing and has not been observed and reported in China so far. Analysis results presented that, this MC layer had a double structure with a small scale, and it was probably inhomogeneous in horizontal direction. Although the composition and the formation mechanism of this MC are still not such clear for us, but this mid-latitude detection is significant because of what it may signify about the global changes of the Earth's climate.

ACKNOWLEDGEMENTS

This work was supported by the High-level Talents Project of Hainan Province (No. 2019RC202), the National Natural Science Foundation of China (No. 41864005), and the Project Supported by the Specialized Research Fund for State Key Laboratories. We acknowledge the use of data from the Chinese Meridian Project.

REFERENCES

[1] T.W. Backhouse, The luminous cirrus cloud of June and July. Meteorol. Mag., 20,133 (1885).

[2] L. Thomas, et al., Lidar observations of ice crystals associated with noctilucent clouds at middle latitudes, Geophys. Res. Lett., 21 (5), 385-388 (1994).

[3] G. von Cossart, et al., Size distributions of NLC particles as determined from 3-color observations of NLC by ground-based lidar. Geophys. Res. Lett., 26 (11), 1513-1516 (1999).

[4] M. Gadsden, The North-West Europe data on noctilucent clouds: a survey. J. Atmos. Sol. Terr. Phys., 60, 1163-1174 (1998).

[5] V.B. Wickwar, et al., Visual and lidar observations of noctilucent clouds above Logan, Utah, at 41.7°N, J. Geophys. Res. 107 (D7), 4054 (2002).

[6] M.J. Taylor, et al., Mesospheric cloud observations at unusually low latitudes. J. Atmos. Sol. Terr. Phys., 64, 991-999 (2002).

[7] H. Suzuki, et al., First imaging and identification of a noctilucent cloud from multiple sites in Hokkaido (43.2-44.4°N), Japan, Earth, Planets and Space, 68, 182 (2016).

[8] S. Gong, et al., Statistical characteristics of atmospheric gravity wave in the mesopause region observed with a sodium lidar at Beijing, China. J. Atmos. Sol. Terr. Phys., 97, 143-151 (2013).

[9] J. Höffner, et al., First observations of noctilucent clouds by lidar at Svalbard, 78° N, Atmos. Chem. Phys., 3, 1101-1111 (2003).

[10] M.L. Chanin, et al., Lidar observation of gravity and tidal waves in the stratosphere and mesosphere. J. Geophys. Res., 86 (C10), 9715–9721 (1981).

[11] J. Cohen, et al., Recent Arctic amplification and extreme mid-latitude weather, Nature Geosci., 2014, doi: 10.1038/NGEO2234.

https://doi.org/10.1051/epjconf/202023704004

OBSERVATIONS OF THE NICKEL LAYER IN THE MESOPAUSE REGION AT MID-LATITUDES

Michael Gerding[1], Kathrin Baumgarten[1], John M. C. Plane[2]

[1]Leibniz-Institude of Atmospheric Physics at the Rostock University, 18225 Kuehlungsborn, Germany
[2]School of Chemistry, University of Leeds, Leeds LS2 9JT, United Kingdom
**Email: gerding@iap-kborn.de*

ABSTRACT

Observations of the mesospheric Ni layer have been performed by lidar in January-March 2018 at Kuehlungsborn/Germany (54°N, 12°E). These soundings provide only the second Ni data set after initial observations by Collins et al. at Chatanika/Alaska (65°N, 147°W)[1]. We utilized for the first time a transition from the low-lying excited Ni(^{3}D) state at 341 nm. For all soundings, nightly mean peak densities varied between ~280 cm^{-3} and 450 cm^{-3}, which is a factor of ~40 less than previously reported for Chatanika [1]. The observed Ni abundance is especially important if compared with the abundance of other metals like Fe, and with their respective abundances in evaporating meteoroids, which form the source of the metal layer in the upper mesosphere. Here, we present exemplarily a sounding from January 8, 2018. Beside the Ni raw data and density profiles we show a temperature profile as measured simultaneously be the co-located RMR lidar and the temperature variation due to gravity waves and tides.

1. INTRODUCTION

Ablation of meteoroids entering the Earth's atmosphere forms a metal layer between ~80 and 110 km. This layer has been detected first 90 years ago by Na airglow observations [2]. Nowadays, ground-based lidars are used for quantitative studies, and more species have been detected subsequently (e.g., Fe, K, Ca, Ca$^+$, Li) [3-7]. It turned out quickly that the abundances of the different species do not resemble the ratios found in typical meteoroids [8,9]. Lidar observations of the metal concentrations as well as measurements of chemical reaction rates and modelling of meteoroid heating revealed a complex interplay of different factors determining the height-dependent concentrations of the particular metal atoms [10]. Hence, any additional metal species provides new independent information to deepen our knowledge about the middle atmospheric dynamics and chemistry.

Collins et al. described first Ni observations from Chatanika/Alaska [1]. During two lidar soundings, they observed peak densities of ~16,000 cm^{-3} and a total abundance of 2.7 x 10^{-10} cm^{-2}, which is about 0.8 times the typical Fe abundance at this site. This Ni abundance is surprisingly large compared to the Fe/Ni ratio of ~18 in CI-Chondrites [8]. The observation of Ni by lidar and other methods is generally technically challenging because of the small resonance backscatter cross section and the short UV wavelength. We have been able to observe the mesospheric Ni layer by lidar during six nights between January and March 2018 at Kuehlungsborn/Germany (54.1°N, 11.7°E). First results are already published in Geophys. Res. Lett [11], hereafter cited as GDP2019. Here, we present additional technical information about the soundings. Additionally we show two examples of one-hour mean profiles from a single night. Simultaneous soundings with a co-located RMR lidar provided independent temperature observations and allowed to deduce information about activity of gravity and tidal waves for that night.

2. METHODOLOGY

The Ni lidar observations at Kuehlungsborn have been made with a dye laser system pumped by an excimer laser at 308 nm [6]. The dye laser has been tuned to the resonance wavelength of the Ni(^{3}D) state at λ_{air} = 341.4764 nm. (For additional soundings we also tuned the laser to the Ni(^{3}F) resonance λ_{air} = 336.9563 nm, but for clarity we concentrate here on the description of the first case. More information on advantages and disadvantages of the different resonance transitions is provided in GDP2019.) With two amplifier stages, we got a laser pulse energy of ~14 mJ. The bandwidth of the laser was limited to 0.4 pm by an intra-cavity etalon. The backscattered photons are collected by a 78-cm telescope and detected by a Hamamatsu R7600-200 photomultiplier. For background suppression we used an interference filter with 29 nm bandwidth (full width at half maximum) and ~85% transmission. The comparatively large bandwidth allowed to detect all relevant backscatter wavelengths, but limited the observations to moonless nights.

Fig. 1 shows the differential backscatter cross section σ_{diff} of the Ni(^{3}D) resonance line, the laser spectrum and the correspondent effective differential backscatter cross section σ_{eff}. Ni densities at altitude *z* are calculated by the equation

$$\rho_{Ni}(z) = \rho_{air}(z_R) \cdot \sigma_{Ray}/\sigma_{eff} \cdot N_{res}^{corr}(z)/N_{Ray}^{corr}(z_R) \cdot P$$

with $\sigma_{eff} = 1.08 \times 10^{-17}$ m^2/sr and $\sigma_{Ray} = 3.57 \times 10^{-31}$ m^2/sr, $N_{c,res}$ and $N_{c,\,Ray}$ range-corrected resonance and Rayleigh count rate and $\rho_{air}(z_R)$ air density at reference altitude z_R, taken from NRLMSISE-00 [12]. An additional factor P accounts for the thermal population of the Ni(^{3}D) state of ~15% at typical temperatures of 200 K around 90 km altitude.

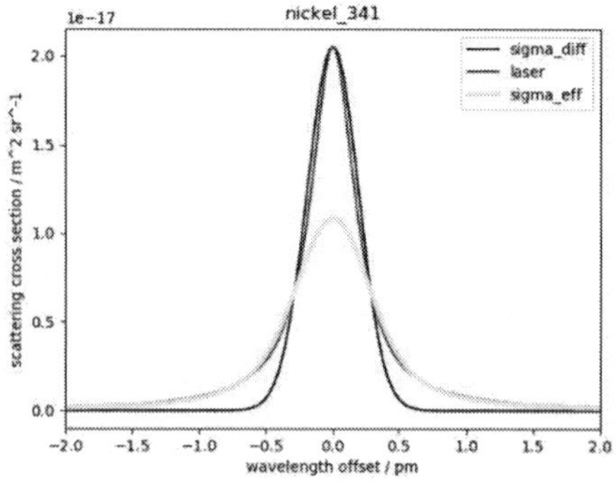

Fig. 1: Blue: Differential backscatter cross section for excitation of the low-lying excited Ni(3D) state. Red: Approximated laser spectrum (Lorentz-shape). Yellow: Effective differential backscatter cross section (convolution of laser and σ_{diff}).

Simultaneous soundings with the co-located RMR lidar [13] can be used to check the focusing of the beam widening optics and the overlap with the telescope. Fig. 2 shows the background-corrected backscatter at 341 nm together with the count rate of the RMR lidar 532 nm high-altitude channel. At altitudes without Ni resonance backscatter, both profiles perfectly agree, confirming good adjustment of beam widening optics and telescope.

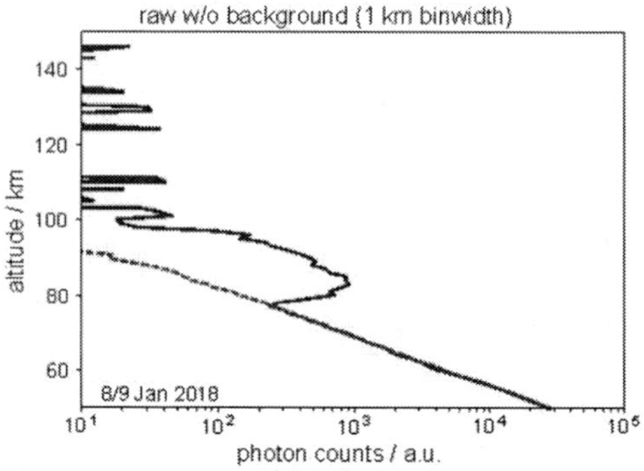

Fig. 2: Background-corrected raw data at 341 nm (blue), integrated 21-22 UTC on Jan. 8, 2018 with 1 km bin-width. The green dashed line is the 532 nm backscatter of the RMR lidar for the same period.

The main purpose of the RMR lidar are temperature soundings in the middle atmosphere. Taking advantage of different optical filtering techniques, we are able to suppress the daylight background efficiently and continue the temperature soundings during day and night [13]. This allows detailed studies of gravity waves as well as tidal waves [14,15].

3. RESULTS

We examined the Ni layer during six moonless nights between January and March 2018. Each sounding lasted between 40 min and 2.5 h. Fig. 3 shows exemplarily two one-hour mean profiles from January 8, 2018 (18:30 – 19:30 UTC, blue, and 21:00 – 22:00 UTC, red). In the beginning of the sounding the Ni layer appeared between 77 km and ~100 km with a clear maximum at 84 km. A peak density of ~290 atoms/cm^3 was observed. The layer shape has changed in the later profile. The layer was getting narrower but showed a broader maximum (or maybe even a double maximum). Peak density decreased to ~240 atoms/cm^3. Also the column abundance decreased by about 20%, from 3.48×10^{-8} cm^{-2} to 2.74×10^{-8} cm^{-2}. We will discuss potential reasons for the density change below.

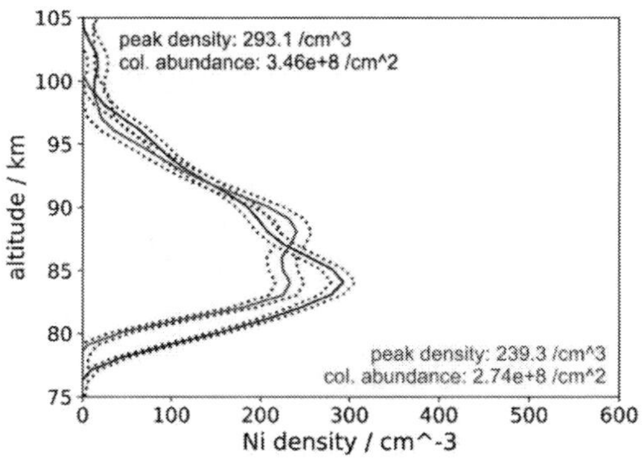

Fig. 3: Ni density profiles for 18:30-19:30 UTC (blue) and 21:00 – 22:00 UTC (red) on Jan. 8, 2018. Bin width is 1 km. The plot is smoothed by a 7-point Hann-filter for visualization only.

Comparing all different nights, we detected Ni mean peak densities between 280 and 450 cm^{-3}. Peak altitudes varied between 83 km and 87 km.

The Ni backscatter cross section is calculated for an ambient temperature of 200 K. Fig. 4 shows that this assumption is valid on average, but that the temperature above 75 km is quite variable due to the presence of gravity waves and tides. The lower mesosphere on Jan 8/9 is slightly warmer than usual and shows prominent large-scale waves. We would like to note that the polar vortex was slightly shifted towards the North-Atlantic and Kuehlungsborn was on its outer edge. Nevertheless, conditions for a Sudden Stratospheric Warming were not fulfilled.

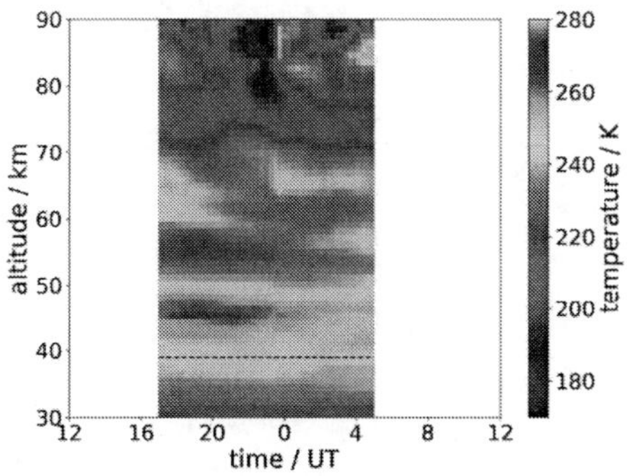

Fig. 4: Temperature observations with the RMR lidar at Kuehlungsborn. Data are limited to nighttime to cover also the upper mesosphere. Integration time is 2 h. The dashed line denotes transition of height channels.

Fig. 5 shows the temperature variability after filtering with a long-pass Butterworth filter (15 km cut-off). The wavelike structures are now most probable due to tides. Please note the amplitudes of at least ±10 K above 80 km, which add up to the temperature variation by gravity waves. Of course, the short data set, even if extending up to at least 90 km, limits the detectability of tides. Below 73 km this soundings covers nearly 48 h continuously. These results are in general agreement with the nighttime data set (not shown).

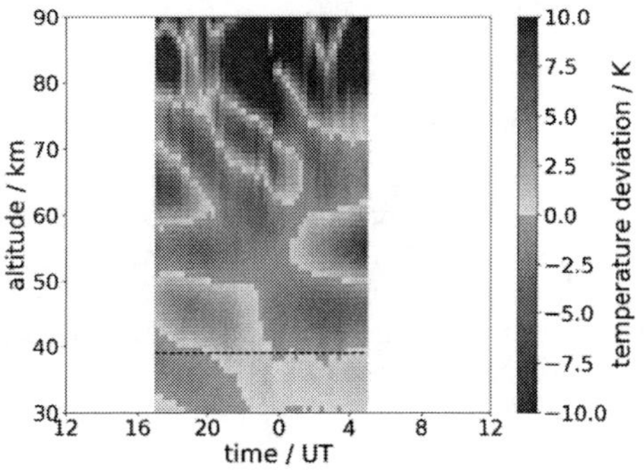

Fig. 5: Temperature variation after filtering with 15 km low-pass Butterworth filter. Waves with these large vertical wavelength are typically attributed to tides.

The temperature variability above 80 km implies some uncertainty on the absolute Ni densities. As mentioned above, the backscatter cross section is calculated for a constant temperature of 200 K, which is a reasonable assumption. We checked the population of the lower Ni state also for 180 and 220 K and found a variation by ~20 % (cf. Supporting Information of GDP2019). This is in the order of magnitude of the Ni variation shown above. Therefore, we attribute a large fraction of the observed density change to the systematic uncertainty in the backscatter cross section. We plan to take the temperature-dependance of the backscatter cross section into account in a future study. On the other hand, the large night-to-night variability of the Ni density is hardly an artifact but due to dynamical and chemical effects. Similar variability has been found also with other metals [5-7].

In summary, we were able to observe the second set of Ni observations after initial studies by Collins et al. [1]. We found a similar layer shape, but a strongly reduces Ni density. Comparing with mid-latitude Fe observations [16] we got a Fe/Ni ratio of 38, which is within a factor of ~2 of the ratio found in CI-Chondrites (Fe/Ni ≈ 18). These data provide an important benchmark for future studies including chemical transport models etc. So far, we do not have an explanation for the different results obtained at Chatanika/Alaska and Kuehlungsborn/Germany. Finally we would like to note that the observations of the $Ni(^3F)$ ground-state gave similar results, but a strongly reduced lidar performance. Therefore, we suggest to use the $Ni(^3D)$ transition in future.

ACKNOWLEDGEMENTS

We thank our colleagues Josef Höffner, Torsten Köpnick, Reik Ostermann and Michael Priester (all IAP) as well as Shane Daly (U Leeds) for their help with installation of the Ni lidar. JMCP acknowledges funding by the Natural Environment Research Council (grant NE/P001815/1).

REFERENCES

[1] R. Collins et al., Geophys. Res. Lett., 42, 665–671, doi: 10.1002/2014GL062716 (2015).
[2] V. M. Slipher, Publ. Astro. Soc. Pac., 41, 262-265 (1929)
[3] J. P. Jegou et al., Geophys. Res. Lett., 7, 995-998 (1980
[4] J. Qian and C. S. Gardner, J. Geophys. Res., 100, 7453-7461 (1995)
[5] V. Eska et al., J. Geophys. Res. 103, 29,207-29,214 (1998)
[6] M. Gerding et al., J. Geophys. Res., 105, 27,131-27,146 (2000)
[7] X. Chu et al., Geophys. Res. Lett., 38, L23807 (2011)
[8] J. M. C. Plane, Int. Rev. Phys. Chem., 10, 55-106 (1991)
[9] S. Raizada et al., J. Atmos. Solar-Terr. Phys., 66, 596-606 (2004)
[10] J. M. C. Plane et al., Chem. Rev., 115, 4497-4541 (2015)
[11] M. Gerding et al., Geophys. Res. Lett., 46, 408-415, doi: 10.1029/2018GL080701 (2019)
[12] J. M. Picone et al., J. Geophys. Res., 107, 1468 (2002)
[13] M. Gerding et al., Atmos. Meas. Tech., 9, 3707-3715 (2016)
[14] M. Kopp et al., J. Atmos. Solar-Terr. Phys., 127, 37-50 (2015)
[15] K. Baumgarten et al., Atmos. Chem. Phys., 18, 371–384 (2018)
[16] T. J. Kane and C. S. Gardner, J. Geophys. Res., 98, 16,875-16,886 (1993)

SPECIAL STRUCTURES OF SODIUM LAYER OBSERVED IN THE DAYTIME OVER BEIJING, CHINA

Yuan Xia[1], Guotao Yang[2]*, Jihong Wang[2], Xuewu Cheng[3], Faquan Li[3]

[1]School of Electronic Engineering, Nanjing Xiaozhuang University, Nanjing, Jiangsu 211171, China
[2]State Key Laboratory of Space Weather, National Space Science Center, Chinese Academy of Sciences, Beijing, 100190, China
[3]State Key Laboratory of Magnetic Resonance and Atomic and Molecular Physics, Wuhan Institute of Physics and Mathematics, Chinese Academy of Sciences, Wuhan, Hubei 430071, China
**Email: gtyang@nssc.ac.cn*

ABSTRACT

In this paper the observation of sodium (Na) layer in mesosphere and lower thermosphere (MLT) region over complete diurnal cycles based on broadband Na lidar at Yanqing Station, Beijing, China (40.5°N,116°E) was reported. Faraday filters with dual-channel design were used in the lidar receiving unit to suppress the strong background light in the daytime, which allow observation of Na layer with an acceptable signal-to-noise ratio (SNR) under sunlit condition. Several special structures of Na layer observed in the daytime was discussed. The simultaneous continuous observation of zonal wind by meteor radar was presented for comparison. These observation results can provide direct and reliable supports for the study of mesopause dynamics and solar effect on Na layer.

1. INTRODUCTION

Metallic Na layer in the mesopause region is a good tracer for studying atmospheric dynamics and photochemistry. Based on resonance fluorescence scattering of Na atoms, Na lidar can achieve the measurement of Na layer parameters with high temporal and vertical resolution, which effectively promotes the development of the scientific research of the middle and upper atmosphere [Krueger et al., 2015]. However, due to the severe interference of the strong solar background light on the received signal in the daytime, there are far fewer lidar observations of the Na layer in daytime than at night, which limit our understanding of the diurnal variations of Na layer, especially on its underside where neutral chemistry plays a major role [Clemesha et al., 2002; States and Gardner, 1999].

In this paper, the observation of Na layer over complete diurnal cycles based on the lidar of Chinese Meridian Project in Beijing was reported. The special structures of Na layer observed in the daytime, including sporadic Na layer (Nas) and the downward extension of the layer bottom side, were discussed.

2. METHODOLOGY

The resonant fluorescence lidar of Chinese Meridian Project in Beijing was upgraded to a double-laser beam system for Na layer and K layer measurements [Jiao et al., 2015; Wang et al., 2012]. The 589 nm laser for the detection of Na layer was generated by a pulsed dye laser which was pumped by a pulsed Nd:YAG laser with wavelength of 532 nm. The output energy of 589 nm pulses was about 40 mJ. Backscattered resonant fluorescence photons from Na layer were received by a Cassegrain telescope with a 1000 mm diameter primary mirror. Narrowband Faraday anomalous dispersion optical filters (FADOF) were used for Na lidar receiving to suppress sky background light [Chen et al., 1996], so that reliable Na echo signal can be obtained under sunlit condition, allowing diurnal continuous observation of Na layer. In order to obtain higher effective transmission of echo fluorescence signal in condition of guaranteeing background suppression, the filtering unit was designed with dual channels. The echo signal was splitted to horizontal and vertical polarization beams with a polarization beam splitter (PBS), and then combined with a polarization beam combiner (PBC) after each beam pass through a Faraday filter respectively, which allow echo fluorescence photons in both polarizations to be

https://doi.org/10.1051/epjconf/202023704005

received at the same time, effectively improving the SNR of Na layer echo signal during the day.

The spatial and temporal resolution of raw data were 96 m and 33.3 s (corresponding to 1000 laser pulses integrated to produce a profile), respectively. The data was generally integrated 15 min . The main parameters of broadband Na lidar in YanQing, Beijing are presented in table 1.

Table 1. The main parameters of broadband Na lidar in YanQing, Beijing

Parameters	Value
Laser wavelength	589 nm
Laser energy	~40 mJ
Repetition rate	30 Hz
Telescope aperture	Φ1000 mm
Receiving field of view	0.2~2 mrad
Bandwidth of filter	~3 GHz
Effective aperture of filter	Φ15-18 mm
Range resolution (raw data)	96 m
Time resolution (raw data)	33.3 s

3. RESULTS

The diurnal operations of Na lidar have been conducted from May, 2014 and more than 4500 hours of observation data covering all months in a year over complete diurnal cycles were collected, which provides effective support for the study of the diurnal variation of Na layer and the formation mechanism of special phenomena occurred in Na layer. Figure 1 shows the distribution of observation hours of Na lidar in each month in the period from May, 2014 to December, 2016. According to the statistics, there are more than 30 campaigns with continuous data acquisition of longer than 48 hours, and more than 5 campaigns longer than 96 hours. Some examples of Na layer observed in daytime are as follows.

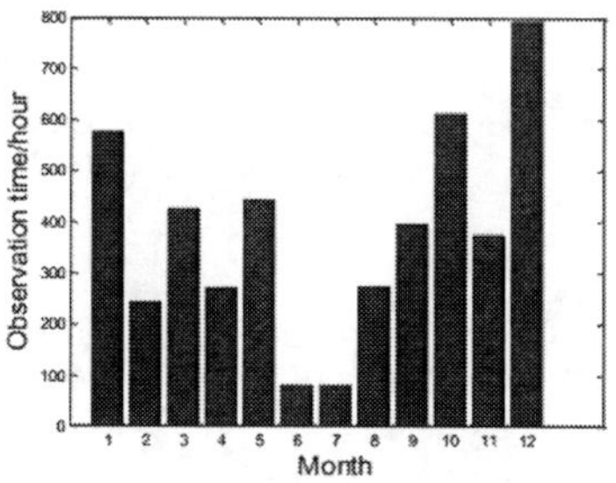

Fig. 1 The distribution of cumulative observation hours in each month of the broadband Na lidar of Yanqing Station, Beijing during the period of May, 2014 – December, 2016

3.1 Sporadic sodium layers occurred in daytime

Nas event is one of the research hotspots in the field of atmospheric and space physics, the diurnal observation of Na layer allows for a specialized analysis of Nas occurred in the daytime. An example of Nas events observed in the daytime of three consecutive days in August, 2016 is shown in Figure 2. As can be seen from the data sequences of the Na profiles in all three plots, the Nas events started in the morning or around noon, and the total process lasted for about 4-6 h (the Na profile from about 13:15 LT to 13:30 LT in 10th August was removed due to the low SNR). The Nas occurred around the altitude between 95 km and 105 km with very strong density enhancements, especially during the day of August 8th. According to the statistical analysis of nocturnal Nas in Beijing by Jiao et al. [Jiao et al., 2015], the Nas descending times were generally longer than the ascending times. In contrast to this, as shown in Figure 2, both the Nas events observed in the daytime of 8th and 9th August have longer ascending times (more than 3 h) than descending times (less than 1 h).

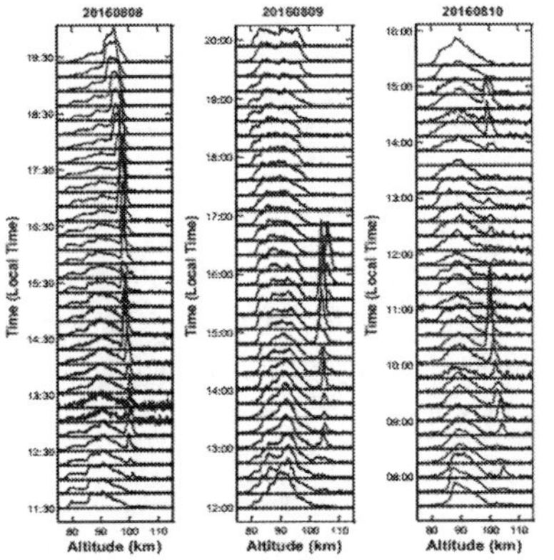

Fig. 2 Nas observed in the daytime for three consecutive days (each profile was integrated with 15 min, and the spatial resolution is 96 m)

3.2 The variations on the underside of Na layer in the daytime

The altitude of Na layer is usually between 80 km and 105 km. Lots of observation results have shown that Na layer generally rises steeply on the bottom side around 80 km and falls slowly on the topside around 105 km. In recent years, the special phenomena on the topside of the sodium layer, such as the high-altitude Nas [Collins et al., 1996; Yuan et al., 2014] and the thermosphere Na layers[Dou et al., 2013; Xue et al., 2013; Tsuda et al., 2015; Gao et al., 2015], have been observed and studied by several groups. Besides, Gong et al. and Wang et al. reported the double sodium layers (DSL), which have secondary sodium layers (SeSL) above the normal layer [Gong et al., 2003; Wang et al., 2012]. More recently, Xun et al. reported neutral Na atoms layer at higher altitudes which extended to around 200 km over Beijing, China [Xun et al., 2019]. However, the distinctive structures on the underside of the sodium layer, have been rarely reported. Figure 3 is a contour of relative sodium density in time-altitude, which shows an obvious downward extension of the bottom of the sodium layer to around 72 km in the morning. The x-axis in Figure 3 represents local solar time. As shown in Figure 3, the bottom side gradually decreased with time from about 4:30 LT, and lasted about 5 h. By around 9:30 LT, the bottom of Na layer showed upward contraction. It is worth mentioning that the contraction process is faster than the extension process.

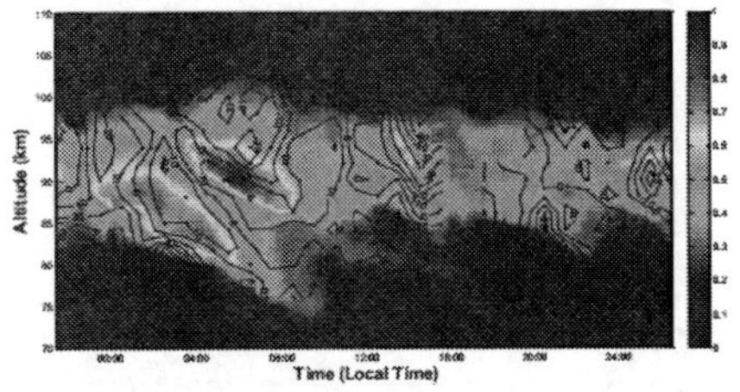

Fig. 3 Relative sodium density and simultaneous zonal wind measured by Radar. Sodium density are shown with color mapping scaled from 0 to 1, winds are shown as contours with black lines.

The formation of metal atomic layers is considered to be controlled by the interaction of meteoric ablation injection, dynamical transport and chemical processes. This large-scale extension of the bottom side of Na layer started in the early morning of December in winter, which was quite different from the seasonal characteristics of the diurnal variation of Fe layer bottom side reported by Yu et al [Yu et al., 2012]. Moreover, the downward extension process of the Na layer did not last for the whole day, and it showed a fast upward contraction before noon, which was not linearly related to the variation of solar radiation. We may infer that there were some differences in the response of Na atoms and Fe atoms in the MLT region to solar radiation.

We can also find that the peak density of the Na layer in Figure 3 exhibits obvious downward tidal fluctuations propagation. We plotted the simultaneous zonal wind results measured by meteor radar, which was shown as contours with black lines in Figure 3 for comparison. The meteor radar was deployed in Beijing observation station of Institute of Geology and Geophysics Chinese Academy of Sciences (IGGCAS) (116°E, 40°N), which was very close to the YanQing Na lidar station. The radar wind data has a spatial resolution of 2 km and temporal resolution of 1 h. As can be seen from Figure 3, from about 00:00 to 08:00 LT, the null point of the zonal wind with a westward wind above an eastward wind descended from around 90 km to 75 km, and another null point with an eastward wind above an westward wind descended from around 95 km to 80 km, which were similar to the evolution trends of the two peak altitudes of the Na layer in Figure 3. It is obvious that there is a close relationship between the zonal wind and the vertical distribution of Na layer.

These observation results can provide direct evidence for the study of the solar effects on the Na chemistry and tides in the mesosphere. However, the possible formation mechanisms may be complicated, and more case and statistical analyses and comparative studies combined with the joint observation data of other equipments such as ionosonde, meteor radar and satellite are needed.

ACKNOWLEDGEMENTS

This work was supported by the National Natural Science Foundation of China (No. 41627804), the Natural Science Foundation of the Jiangsu Higher Education Institutions of China (18KJB170009), Scientific Research Project of Nanjing Xiaozhuang University (4177022, 2018NXY46). We acknowledge the use of the data from the Chinese Meridian Project and World Data Centre for Geophysics, Beijing(http://wdc.geophys.ac.cn).

REFERENCES

[1] Collins R., Hallinan T., Smith R., et al. Geophys. Res. Lett., 23(24): 3655-3658 (1996)
[2] Chen H., White M.A., Krueger D.A. et al. Opt. Lett., 21:1003-1005 (1996)
[3] Clemesha B.R., Batista P.P., and Simonich D.M., J. Atmos. Sol. Terr. Phys., 64:1321-1325 (2002)
[4] Dou X.K., Qiu S.C., Xue X.H., et al. J. Geophys. Res., 118(10): 6627-6643 (2013)
[5] Gao Q., Chu X.Z., Xue X.H., et al. J. Geophys. Res., 120, 9213-9220 (2015)
[6] Jiao J., Yang G., Wang J., et al. J. Geophys. Res. Space Physics, 120: 5214-5225 (2015)
[7] Krueger D. A., She C. Y., and Yuan T. Appl. Opt., 54(32): 9469-9489 (2015)
[8] States R.J., and Gardner C.S. J. Geophys. Res., 104: 11783-11798 (1999)
[9] Tsuda T., Chu X, Nakamura T, et al. Geophys. Res. Lett., 42(10):3647-3653 (2015)
[10] Wang J., Yang Y., Cheng X., et al. Geophys. Res. Lett., 39, L15801 (2012)
[11] Xue X.H., Dou X.K., Lei J.H., et al. J. Geophys. Res. Space Physics, 118(5): 2409-2418 (2013)
[12] Xun Y., Yang G., She C., et al. Geophys. Res. Lett., DOI: 10.1029/2018GL081645 (2019)
[13] Yu, Z., X. Chu, W. Huang ,et al. J. Geophys. Res., 117, D22303 (2012)
[14] Yuan T., Wang J., Cai X., et al. J. Geophys. Res. Space Physics, 119(7): 5985-5999 (2014)

STUDYING THE VARIATIONS IN AEROSOL LOADING AND THERMAL REGIME OF THE STRATOSPHERE OVER TOMSK ON THE BASIS OF LIDAR MONITORING

G.G. Matvienko, V.N. Marichev, D.A. Bochkovsky

V.E. Zuev Institute of Atmospheric Optics, Siberian Branch of the Russian Academy of Sciences,

Academician Zuev Square, Tomsk 634021, Russia, E-mail: mgg@iao.ru

ABSTRACT

One of the important applications of lidar techniques is the study of aerosol content and thermal regime of the stratosphere. Such investigations in monitoring mode were initiated at the Institute of Atmospheric Optics since 1986 (for aerosol) and 1994 (for temperature), and are continued to the present. The main attention is paid to studying the annual variations in aerosol content in the stratosphere and sudden disturbances caused by winter stratospheric warming. In this paper we present the results of studying the aerosol content and its vertical stratification and vertical distribution of temperature in the stratosphere over Tomsk for the last three years.

1 INTRODUCTION

The period of 2016-18, which presents the results was characterized by the absence of volcanic eruptions which could affect the disturbance of the aerosol component of the stratosphere of the northern hemisphere including western Siberian region. So it was a convenient possibility to observe the peculiarities of the temporal variability of the background aerosol loading of the stratosphere over Western Siberia during quite long time interval. In this period, as in previous years, winter stratospheric warming's were observed. This gave the opportunity to continue to study the peculiarities of their manifestations.

The represented data array of 222 total signals collected in some nights was used as initial data for analysis in 2016-18.The height interval was from 15 till 50-60 km, the spatial resolution was 192 m. Lidar signals were received in the photopulses counting mode with accumulation of по 12×10^4 laser pulses (accumulation time was about 2 hours during a night).

2 METHODOLOGY

Lidar methods of the Raman and elastic aerosol and molecular scattering of light (Raman, Mie and Rayleigh scattering) were used to determine the vertical profile of aerosol characteristics and temperature in the stratosphere. The atmospheric sounding was carried out by laser radiation at a wavelength of 532nm, the reception of the lidar signals at wavelengths 532 and 607 nm. The optical characteristics of the scattering ratio R(H) and the integral aerosol backscattering coefficient B(H) were used to estimate the aerosol filling of the stratosphere.

R(H) (H is the current height) describes the aerosol vertical stratification. By definition, R(H) is the ratio of the sum of aerosol and molecular backscattering coefficient to the molecular backscattering coefficient. For example, fulfillment of the condition R(H) = 1 means the absence of aerosol at the certain height, and contrary, aerosol appears where R(H) > 1. B (H) describes the temporal dynamics of the aerosol filling of the stratosphere according to the formula

$$B = \int_{H_1}^{H_2} \beta(h)dh$$

where *β(h)*- aerosol backscattering coefficient, H_1 is the height of the tropopause, and H_2 is chosen about 30km.

The vertical profiles of temperature T(H) were retrieved from the Rayleigh and Raman signals using the formula

$$T(H) = \frac{P_1(H)P_2(H)}{N(H)H^2}\left[\frac{N(H_m)}{P_1(H_m)P_2(H_m)}T(H_m) + \frac{1}{R^*}\int_{H_m}^{H}\frac{N(h)h^2 g(h)dh}{P_1(h)P_2(h)}\right]$$

Here *N(H)* - lidar signals, *P1(H)* and *P2(H)* are the transparencies of the atmosphere from the level of arrangement of the lidar up to the height H at the wavelength of 532 nm (Rayleigh signals) and 532 and 607 nm (Raman signals), respectively; R^* is the universal gas constant, $g(h)$ is the gravity acceleration, *Hm* is the maximum height, from which quite reliable for processing signals are detected (so-called calibration height, at which the temperature values *T(Hm)* are set).

3 RESULTS

3.1. Aerosol

The summary results of lidar observations of the stratospheric aerosol layer for the period 2016-18 are presented in Fig.1. In this picture measurements of the monthly mean dynamics of the aerosol vertical stratification are shown.

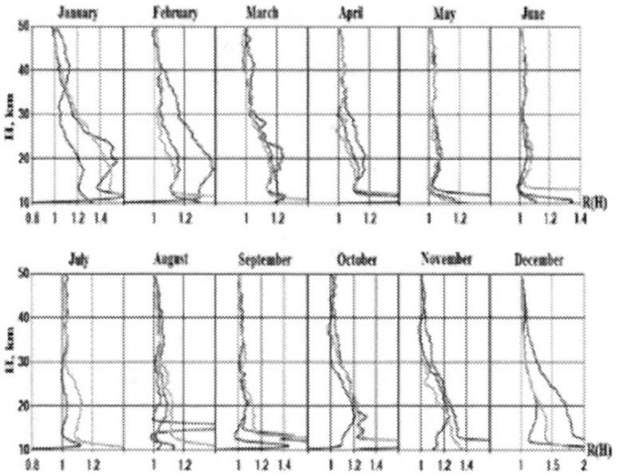

Fig. 1. Monthly mean profiles of the aerosol vertical stratification in 2016-2018. Red lines show the mean profiles in 2016, , green lines show the mean profiles in 2017, orange lines show the mean profiles in 2018.

The observations for this period of time indicate that the common feature for the region of the Western Siberia is the maximal aerosol loading of the stratosphere in winter months, almost no aerosol loading for the entire stratospheric layer in summer months, and no loading for the upper stratospheric layer from 30 to 50 km, except in winter months. Also, there are substantial differences in aerosol stratification. For instance, the aerosol loading was stronger in December 2017 as compared to 2016 and 2018, it was stronger in January 2016 and 2018, and stronger in February 2016.

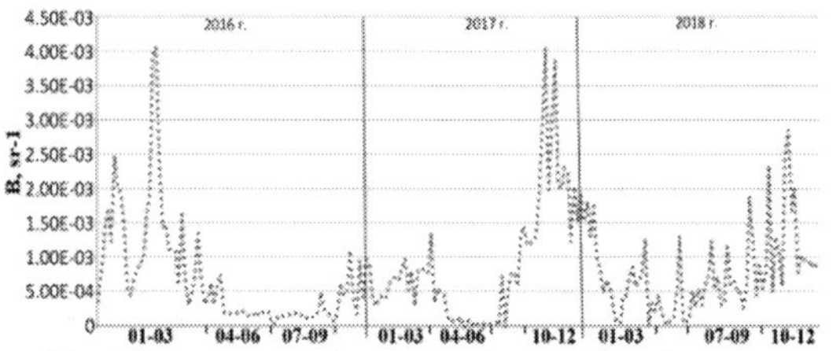

Fig.2 Time series of integrated aerosol backscattering coefficient for the period of 2016 – 2018.

Time dynamics of the total stratospheric loading with background aerosol in 2016-2018, expressed via a parameter of integrated aerosol backscattering coefficient, is presented in Fig. 2.

The observation time series, presented in Fig. 2, indicate a weak aerosol loading for cold period in November 2016 – April 2017 ($B(H) \leq 10^{-3} sr^{-1}$); quite a marked aerosol loading, strongly varying in time, for the period of January – April 2016; and a large aerosol loading for the period of October 2017 – February 2018 (up to $B(H) \approx 4 \cdot 10^{-3} sr^{-1}$). The aerosol loading was minimal in May – September 2016 and in May – August 2017. Abrupt bursts ($B(H) \geq 10^{-3} sr^{-1}$), and not only minima in aerosol content, were observed in spring, summer, and fall of 2018. The tendency of aerosol loading of the lower stratosphere, peaking in January, decreasing in spring, and almost vanishing in May – September, is confirmed by the comparison with observations in previous years [1,2].

3.2. Temperature.

Observations for January 2016 are presented in Fig. 1. In this figure, like in figures below, the vertical temperature profiles (VTP), obtained from lidar measurements, are compared with those, measured on Aura satellite [3] and at aerological sensing station in Novosibirsk [4], as well as those from CIRA-86 model [5].

From Fig. 3 it can be seen that the stratospheric warming (SW) began in the second decade of January, and continued on the next days of observations in a classical form of half-waves with a positive deviation from monthly average temperature in the upper half of the stratosphere and a negative deviation in its lower half.

The maximal positive deviation reached 40 K (January 12, altitude of 45 km), and maximal negative deviation was 35 K (January 5 and 7, altitude of about 30 km). From January 19 to 26, the vertical temperature distribution stabilized somewhat (approached the model distribution) in the upper stratosphere. However, in the entire layer of the lower atmosphere, the temperature profiles obtained from lidar, balloon, and satellite data remained shifted toward negative values relative to the model profile.

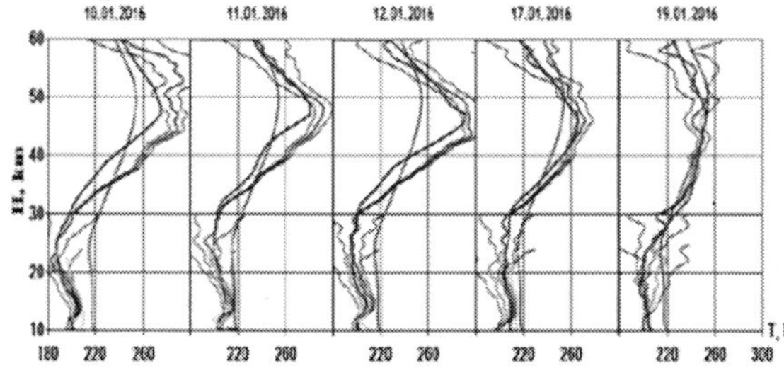

Fig. 3. Lidar and satellite observations of manifestation of stratospheric warming in January 2016. Temperature profiles, calculated from lidar measurements of Rayleigh and Raman signals (red and yellow lines) and their standard deviations (black lines), observations from Aura satellite and meteorological sondes (blue and light blue lines), and monthly average profile according to CIRA-86 model (green line).

The last overshoot of warming was observed on February 1 of 2016 in the height range of 40-60 km (see Fig.4). The analogous warming, but in the narrower height range of 45-60 km, was also observed from the data of satellite measurements. According to the data of the following observations, the final stage of stabilization of the vertical temperature distribution occurred. From Fig. 4, we can see that the temperature profiles obtained in the entire height range of 10-60 km from lidar, satellite, and balloon measurements become quantitatively and qualitatively closer to the model profile.

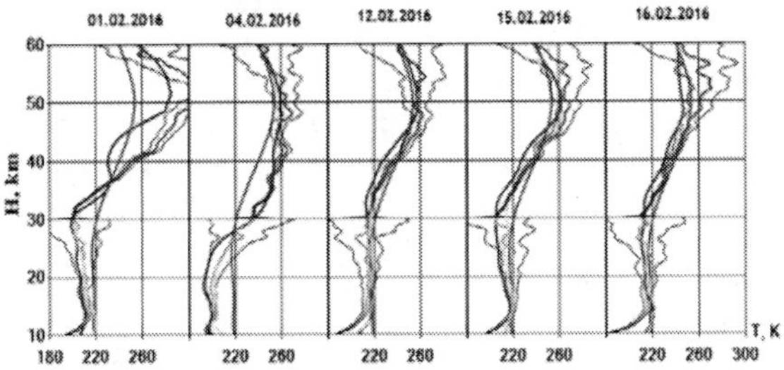

Fig. 4. Vertical distribution of temperature in some nights of February 2016.

The observed SW in 2016 was of the minor type, as indicated by the data taken from the website of the European center for medium-term weather forecasts [9].

Temperature profiles for the quiescent period of 2016 (from April to September) are shown in Fig. 5. In this long period of time, the stratosphere evolved into a stable thermal regime, in which the vertical temperature distribution well agrees with the monthly average model distribution.

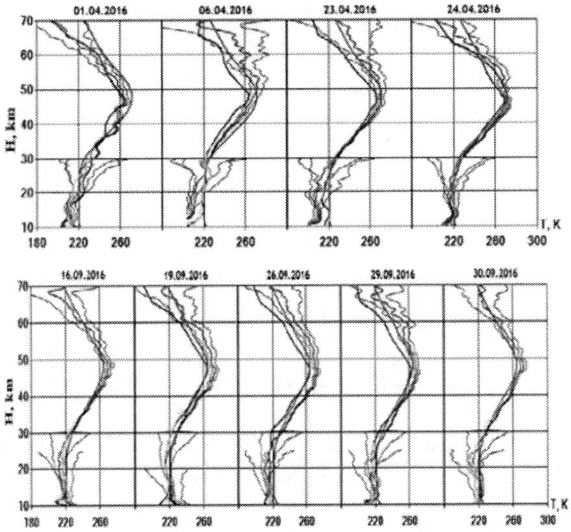

Fig. 5. Vertical temperature distribution on separate nights in April – September 2016.

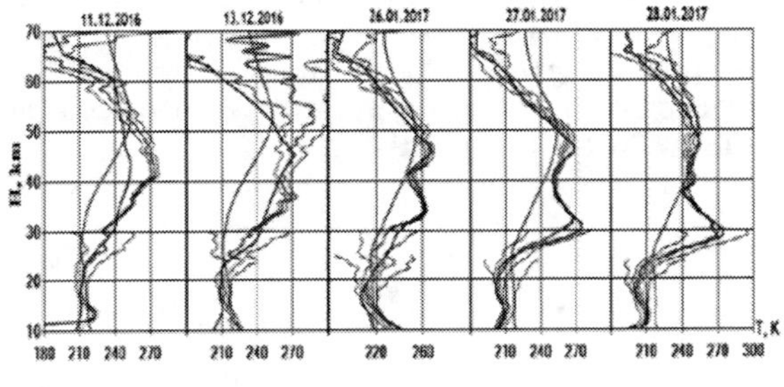

Fig. 6. Lidar and satellite observations of manifestation of stratospheric warming in winter 2016/2017.

The next stratospheric warming began in December (Fig. 6). For instance, on December 11 at the altitude of about 50 km there had been a positive deviation of temperature from monthly average by about 30 K. Then, the weather conditions were unfavorable, so that observations had been resumed as late as the end of January 2017. From Fig. 6 it can be seen that, as indicated by both lidar and satellite observations, the stratospheric warming continued for the third decade of January. It is noteworthy that the lidar and satellite measurements of temperature well agree. Based on these independent measurements, the warming pattern shows two peaks at altitudes of 30-32 and 47-50 km.

Then, in the next period of observations from February to December, the thermal regime of the stratosphere evolved into the stable state, indicated by a good correspondence between lidar/satellite and model-based temperature profiles (not shown).

Occurrence of stratospheric warming in winter 2017/2018 was recorded both by lidar and satellite measurements in the last decade of January 2018 (Fig. 7). The warming peaked in the stratopause region (January 21, H=50 km). On next days of observations, the stratopause altitude decreased to 40 km, and the maximal positive deviation reached 50 K.

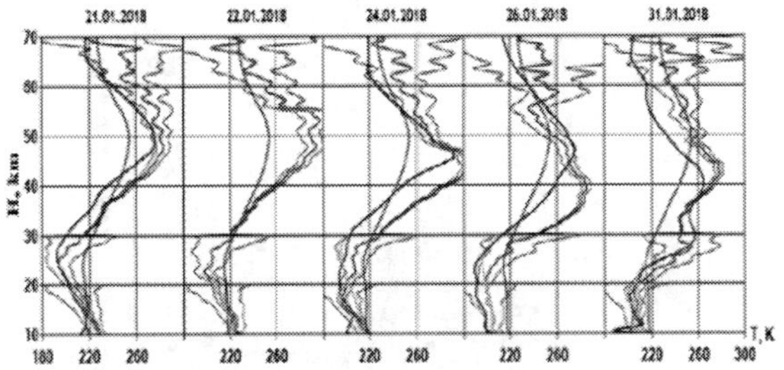

Fig. 7. Lidar and satellite observations of manifestation of stratospheric warming in winter 2017/2018.

The vertical temperature distribution during the SW episode in December 2018 – January 2019 is shown in Fig. 8. A pronounced warming was localized at altitudes from 20 to 50 km, and was accompanied by a cooling at altitudes from 10 to 20 km. We note that the stratospheric warming was as long as the whole month and had a large (up to 60 K) amplitude of the positive deviation.

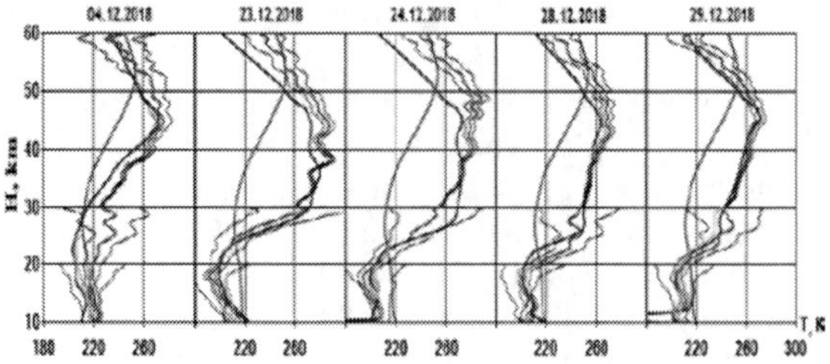

Fig. 8. Lidar and satellite observations of the dynamics of stratospheric warming during winter 2018/2019.

The SW in 2018-2019 was major in type, i.e., westerly air mass transport in the stratosphere reversed to easterly transport (see Fig. 9).

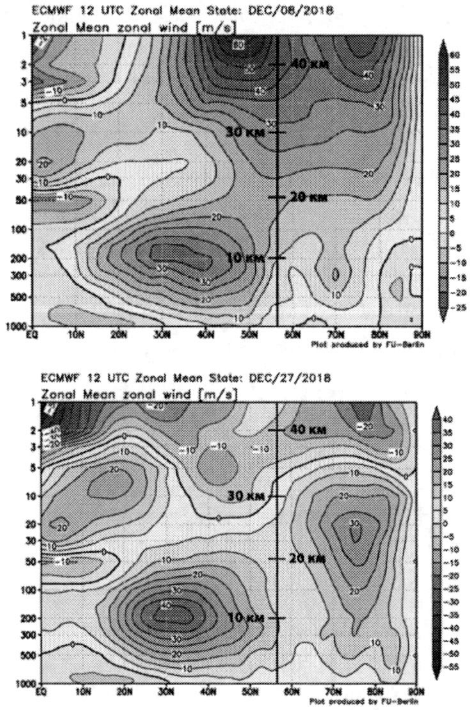

Fig. 9. Altitudinal distribution of zonal wind speed and direction in the Northern Hemisphere, recorded on December 8 and 27, 2018. Orange areas are for westerlies and blue areas for easterlies. The vertical thick lines indicate the altitudinal transects for the latitude of Tomsk.

CONCLUSION

The lidar studies of the stratosphere over Tomsk in the period of 2016-2018, as well as earlier years, revealed the following:

1. The loading of the stratosphere with background aerosol occurs in the cold period of a year: start in October, maximum in January, and end in April. In the warm period, there is practically no aerosol in the stratosphere.
2. Every year the winter warming takes place in the stratosphere. It begins in December, is most pronounced in January, and is sometimes extended to February. Weak and, more rarely, strong warmings are observed. The amplitude of positive temperature deviations from the monthly average value can achieve 60 K, and the height of the stratopause can go down to 30 km. In the period of March till November, the vertical temperature distribution is inasatisfactory agreement with the CIRA-86 model distribution.

This work was supported by the Russian Science Foundation (grant no. 19-7720067) in carrying out the monitoring measurements, and by the Russian Fund for Basic Research (project no. 19-45-700008) in the analysis of the lidar sensing data and the grant of President of RF MK-4592.2018.8

REFERENCES

[1] Marichev V.N. Investigation of variability of the background aerosol vertical structure above Tomsk based on lidar observations in 2010-2011, Atmos. Ocean Opt., 2012, 25. No. 11. P. 976-984.
[2] Marichev V.N., Bochkosvkii D.A. Studying the variations in background aerosol loading of the stratosphere in 2014, Proceedings of SPIE. 2015. V.9680. CID: 9680 6U. [9680-209]
[3] http://mirador.gsfc.nasa.gov
[4] http://weather.uwyo.edu/upperair/sounding.html
[5] Rees D., Barnett J.J., Labitske K. COSPAR International Reference Atmosphere: 1986. Part II, Middle Atmosphere Models // Adv. Space Res. 1990. V. 10, N 12. 525 p.
[6]http://users.met.fu-berlin.de/~Aktuell/strat-www/wdiag/eczm.php?alert=8&forecast=all&var=u&lng= eng#fig1
[7] Marichev V.N. Lidar investigations of stratospheric warming events above Tomsk in 2008-2010. // Atmos. Ocean Opt. 2011. V. 24. No. 05. pp. 386-391.
[8] Marichev V.N. Investigation into features of manifestation of winter stratospheric warming events over Tomsk from the data of lidar temperature measurements in 2010-2011. // Atmos. Ocean Opt. 2011. V. 24. No. 12. P. 1041-1046.
[9] Marichev V.N. The analysis of the air density and temperature behaviour in the stratosphere above Tomsk in periods of perturbed and quiet states performed according to the results of lidar measurements. // Atmos. Ocean Opt. 2013. V. 26. No. 09. pp. 783-792.

THE MAJOR SUDDEN STRATOSPHERIC WARMING IMPACT ON MID-LATITUDE SURFACE WEATHER

Yuke Wang[1]*, Oleksandr Evtushevsky[2], Gennadi Milinevsky[1,2,3], Valery Shulga[1,4], Yuliia Yukhymchuk[3], Wei Han[1], Dmitri Shulga[4], Asen Grytsai[2]

[1]*College of Physics, International Center of Future Science, Jilin University, Changchun 130012, China*
[2]*Taras Shevchenko National University of Kyiv, Kyiv 01601, Ukraine*
[3]*Main Astronomical Observatory of Ukraine, Kyiv 03143, Ukraine*
[4]*Institute of Radio Astronomy, National Academy of Sciences of Ukraine, Kharkiv 61002, Ukraine*

**Email: wangyk16@mails.jlu.edu.cn*

ABSTRACT

The possible relation of a major sudden stratospheric warming (SSW) with the mid-latitude surface weather conditions was investigated using data from the ERA-Interim and NCEP–NCAR reanalyzes. An important feature of the SSW event is the impact on lower altitudes, when temperature and wind anomalies descend downward into the high- and mid-latitude troposphere during the weeks or even month and influence the surface weather [1, 2]. Owing to known SSW impacts on the surface weather [2], we consider the possible relation of the SSW event in winter 2018 to cold weather anomaly in the Northern Ukraine and North-East China in February 2018.

1. INTRODUCTION

The event of major sudden stratospheric warming (SSW) which happened roughly each two years in the North Polar region is produced by strong planetary wave activity. The major SSW is accompanied by dramatic increase of the stratosphere temperature up to 50 K and zonal wind reverse from climatologically westerlies to easterlies during several days [3, 4, 5]. The SSW event can impact on lower altitudes and mid-latitude troposphere changing the surface weather significantly. The SSW event in February 2018 shows enhanced warming in the stratosphere and cooling in the mesosphere. These processes trigger the chain of effects that propagate down in altitudes and moving from high to lower latitudes. In the paper we consider the possible relation of the SSW event in winter 2018 with cold weather anomaly in the Northern Ukraine and North-East China regions in February 2018.

2. METHODOLOGY

2.1 DATA

The data of air temperature, zonal wind, and geopotential height were used from the NCEP–NCAR reanalysis data ([6], https://www.esrl.noaa.gov/psd/cgi-bin/data/). The NCEP–NCAR reanalysis data on time–latitude air temperature anomaly sequence allow to retrieve the delayed response of the near-surface temperature to the stratospheric anomalies in February–March 2018 (Fig. 1). The warm stratospheric anomaly at 10–200 hPa expands to lower latitudes and cold anomaly to –8°C appears at the surface at 45–70°N with about one week delay after the SSW beginning and it was observed from February 20 till March 5 (Fig. 1d). The air surface temperature negative anomaly was observed overall at the end of February and beginning of March 2018 in the North-East Ukraine including region around Kharkiv city, where the microwave radiometer site for detecting mesosphere parameters is located [5]. The air surface temperature negative anomaly was observed also over North-East China in the same time period.

The latitude–height and longitude–height sections of daily air temperature anomalies from the NCEP–NCAR reanalysis data (https://www.esrl.noaa.gov/psd/data/composites/day/) were used to analyze the possible downward influence of the SSW event (Fig. 2). The temperature anomalies have been averaged in segment 30–40°E centered at Kharkiv (Fig. 2a–2e)

https://doi.org/10.1051/epjconf/202023704007

and zone 45–60°N (Fig. 2f–2j). Surface negative temperature anomalies are observed mainly in the zone 45–60°N in late February–March (Fig. 1d), therefore the vertical profiles along the eastern longitudes 0–180°E (Fig. 2, lower panel) are averaged for this zone.

3. RESULTS

3.1 Time–latitude air temperature anomaly

The longest cold period in time–latitude plot of surface temperature anomaly is seen on February 20–28 (Fig. 1d).

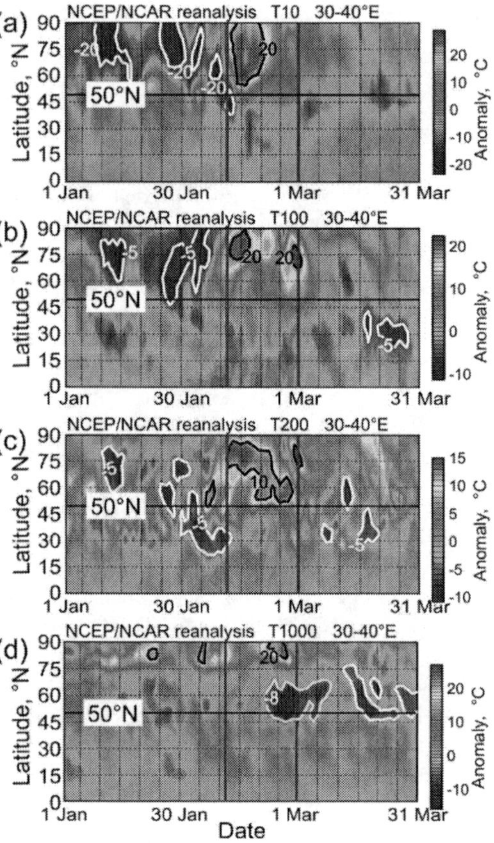

Figure 1. Time–latitude air temperature anomaly plots at the pressure levels 10, 100, 200 and 1000 hPa (T10, T100, T200 and T1000, respectively) during SSW event. Latitude 50°N is marked by black solid line. Climatology 1981–2010 is used for the NCEP–NCAR reanalysis anomalies.

This cold temperature anomaly period is also confirmed by altitudinal distributions in Fig. 2. Initially, the negative regional anomaly in the tropospheric temperature is formed at the north of the Kharkiv latitude 50°N (vertical line in Fig. 2). It locates in middle troposphere (white contours in Fig. 2a and 2b) and is vertically aligned with the southern edge of the strong positive stratospheric anomaly (> 10°C, thick black contour). The cold anomaly gradually descends between the tropopause (black dashed curve) and the surface, reaches the level < –10°C and moves a little south of latitude 50°N synchronously with the positive stratospheric anomaly (Fig. 2a–2e). In zonal direction, both anomalies shift westward, however descent is observed only in the troposphere (Fig. 2f–2j) similarly to the meridional section (upper panel in Fig. 2). The similar but weaker processes took place in the atmosphere at North-East China region in the longitudinal sector 110–130°E and averaged in latitudinal zone 45–60°N (see Fig. 2f–2j).

So, appearance of the cold surface anomaly in late February is associated with the strong warm anomaly in the stratosphere, which meridionally extends between the middle and polar latitudes and zonally covers the eastern longitudes (upper and lower panels, respectively, in Fig. 2). The downward stratosphere–troposphere coupling involves vertical dipole-like structure of the temperature anomalies appeared near 20 February, with a 10-day time delay relative to the starting date of the SSW (10 February).

It should be noted that the cold anomaly intensity is sensitive to the thermal tropopause, which declines in the North-East Ukraine (near Kharkiv) region to 400 hPa (~7 km) on 24 and 26 February (black dashed curve in Fig. 2c, 2d and 2h). The similar smaller tropopause altitude decline over the cold temperature anomaly is seen at the 110–130°E latitudinal sector in the atmosphere at North-East China (see Fig. 2f–2j). This regional tropopause decline resembles the tropopause folding and suggests that the processes associated with stratosphere–troposphere exchange of air masses that can play a critical role in the cold weather formation [7]. Since the coldest air masses concentrated near the tropopause, their downward penetration could cause a cold anomaly occurrence in the troposphere. It should be noted that the warm tropospheric anomalies are sometimes observed under the elevated tropopause (Fig. 2b–2g).

4. DISCUSSION

As known, cold surface anomalies can be produced by downwelling of the stratospheric anomalies, they are associated with the negative phase of annular mode and the direct downward influence involves the whole polar region [8, 9, 10]. The Arctic Oscillation (AO) index that describes annular mode phase had slightly positive value 0.1 in February 2018 (https://www.esrl.noaa.gov/psd/data/climateindices/). This does not suggest significant anomaly in zonal mean circulation and surface weather. The strongest surface anomaly up to –10 K in late February covered mid- to high-latitude Europe between about 45°N and 70°N (Fig. 1d and Fig. 2c–2e) in the longitude sector 0–50°E (Fig. 2h–2j). This shows that anomaly was relatively small zonally to be seen in global scale AO index.

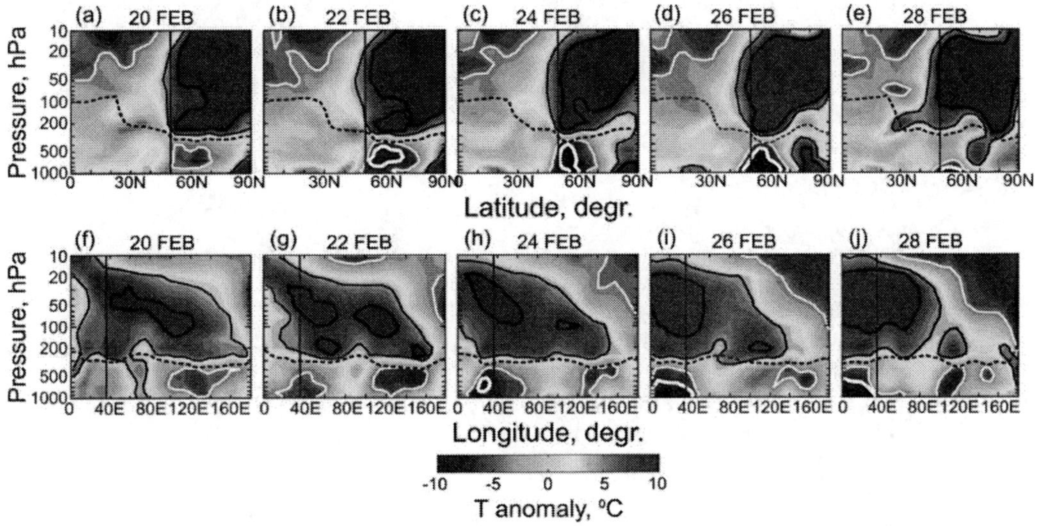

Figure 2. Daily vertical profiles of the air temperature anomalies in the troposphere–stratosphere in (a–e) latitude–height section, 0–90°N, and (f–j) longitude–height section, 0–180°E, averaged in segment 30–40°E centred at longitude of Kharkiv and in zone 45–60°N, respectively. Latitude and longitude of the Kharkiv site is marked by vertical line. Longitudinal sector of North-East China region is 110–130°E (not marked). Thin (thick) contours outline temperature anomalies of ±5 (±10) °C. Black dashed curve shows thermal tropopause pressure. From the NCEP–NCAR reanalysis data.

Cool surface anomaly in late February–early March 2018 lasted about two weeks (Fig. 1d) and it started 10 days after the SSW onset (10 and 20 February, respectively). This time delay is consistent with the time scale of stratospheric anomaly downwelling to the surface due to the SSW influences [8]. Downwelling of cold anomaly through the troposphere is also clearly observed and it occurs in coupling with warm anomaly in the lower stratosphere (Fig. 2). Therefore, generally, cold surface anomaly in Fig. 1d and Fig. 2 could be interpreted as the regional midlatitude effect of the large-scale SSW event concentrated mainly in the polar region.

It is worthy of note that the surface cooling peak was reached 26 February (Fig. 2d and 2i) and it was preceded by a sharp weakening of upward wave flux. It may be suspected that wave reflection cold be involved in stratosphere–troposphere coupling [11, 12] during this event. Whether or not the wave reflection took place in surface cooling in February 2018 should be additionally studied considering the longitudinal dependence of upward and downward wave propagation [11, 12] not represented in zonal means.

The stratosphere–troposphere coupling in late February 2018 was accompanied by significant decrease of the thermal tropopause height above the cold tropospheric anomaly in the North-East Ukraine including Kharkiv region (down to 400 hPa, or ~7 km, on 24 and 26 February; black curve in Fig. 2c, 2d, 2h and 2i). The similar but weaker processes took place in the atmosphere at North-East China region. It is possible that this tropopause anomaly evolves similarly to the tropopause folding with stratosphere–troposphere exchange of air masses, which can also contribute to the cold weather formation [7, 13]. These processes also need more detailed analysis.

ACKNOWLEDGEMENTS

This work was supported in part by Taras Shevchenko National University of Kyiv, project 19BF051-08; by the Institute of Radio Astronomy of the National Academy of Sciences of Ukraine; by the College of Physics, International Center of Future Science, Jilin University, China. NCEP–NCAR reanalysis data were provided by the NOAA/OAR/ESRL PSD, Boulder, Colorado, USA, from their Web site at http://www.esrl.noaa.gov/psd/.

REFERENCES

[1] A. H. Butler, et al. Bull. Amer. Meteor. Soc., 96, 1913–1928, doi:10.1175/bams-d-13-00173.1 (2015)

[2] Y. Yu, et al. Mon. Weather Rev., 146, 2717–2739, doi:10.1175/MWR-D-18-0110.1 (2018)

[3] A. Chandran and R. L. Collins, Ann. Geophys., 32, 859–874, doi:10.5194/angeo-32-859-2014 (2014)

[4] Butler, A. H. et al., Earth Syst. Sci. Data, 9, 63–76, doi:10.5194/essd-9-63-2017 (2017)

[5] Y. Wang, et al. Atmos. Chem. Phys. Discuss. https://doi.org/10.5194/acp-2018-1361 (2019)

[6] E. Kalnay, et al. B. Am. Meteorol. Soc., 77, 1057–1072 (1996)

[7] B. Škerlak, et al. J. Geophys. Res. Atmos., 120, 4860–4877, doi:10.1002/2014JD022787 (2015)

[8] M. P. Baldwin and T. J. Dunkerton, Science, 294(5542), 581–584, doi:10.1126/science.1063315 (2001)

[9] D. W. J. Thompson, et al. J. Clim., 15, 1421–1428 (2002)

[10] J. Kidston, et al. Nature Geoscience, 8, 433–440, doi:10.1038/ngeo2424 (2015)

[11] K. Kodera, et al. J. Geophys. Res. Atmos., 121, 80–94, doi:10.1002/2015JD023359 (2016)

[12] D. Nath, et al. Scientific Reports, 6, Article ID 24174, doi:10.1038 (2016)

[13] J. R. Holton, et al. Rev. Geophys., 33(4), 403–440, doi:10.1029/95RG02097 (1995)

The RELEVANCE OF HANLE EFFECT ON NA AND FE LIDARS

Chiao-Yao (Joe) She[1*]

[1] *Physics Department, Colorado State University, Fort Collins, CO, 80523 U.S.A.*
**Email: joeshe@colostate.edu*

ABSTRACT

A laser resonant scattering process involves two steps, excitation and emission. That emission occurs spontaneously is well accepted. That the atoms involved in the emission are excited coherently by a laser beam leading to a non-isotropic angular distribution of emission (an antenna pattern) is not well known. The difference between coherent and incoherent excitation leads to the Hanle effect. In this paper, I discuss the physics of Hanle effect, and its influences on the backward scattering intensity of Na, K, and Fe atomic transitions and the associated Na and Fe resonant fluorescence lidar systems.

1. INTRODUCTION

It is well known that the dipole moment responsible for non-resonant Rayleigh scattering is coherently excited by a laser beam. In a resonant scattering process, the scattered photons are spontaneous emissions from atoms excited during the process. That such atoms, quite different from those involved in a fluorescence light, are also coherently excited is not as well known. Unlike incoherent excitation in the case of a fluorescence light, which leads to isotropic emission, the coherent excitation by a laser beam leads to angular distribution of emitted (or scattered) light, an antenna pattern if you wish. The difference between the two leads to the Hanle effect. We discuss the physics and formulation of the Hanle effect, and apply it to the backscattering of Na, K, and Fe atoms, and in turn treat the associated consequences on the associated resonance fluorescence lidar signals.

2. METHODOLOGY

A semiclassical/quantum mechanical treatment aimed at lidar applications was recently presented [1]. The resonant scattering cross-section from initial state $|f>$ via excited state $|F>$ to the final state $|f'>$ is given in (16) of the paper. By recognizing $\Gamma_{mm'} = A_{Ff'} / \left(2\alpha_{Ff'}\right)$, we have Eq. (1):

$$\frac{d\sigma_{fFf'}^{RS}}{d\Omega} = \frac{1}{4\pi}\frac{\lambda_{Ff}^2}{4}\frac{3}{2}\frac{g_F}{g_f}A_{Ff}g(\omega-\omega_{Ff})\alpha_{Ff'}C_{fFf'}(\hat{e};\hat{e}')$$

Here, $\lambda_{Ff}, g_F, g_f, A_{Ff}, \alpha_{Ff'}, and\ g(\omega-\omega_{Ff})$ are respectively, transition wavelength, excited state degeneracy, ground state degeneracy, Einstein A coefficient from excited to initial state, fractional transition rate, and absorption line-shape. The backscattering strength factor $C_{fFf'}(\hat{e};\hat{e}')$ is responsible for the angular distribution in question and it depends on the incident and scattering polarizations, $\hat{e}\ and\ \hat{e}'$. Referring to the basic scattering coordinate and geometry shown in Fig. 3 of She et al. [1], there are 4 cases denoting as: $C_{fFf'}(\hat{e}_a;\hat{e}_a')$, $C_{fFf'}(\hat{e}_a;\hat{e}_b')$, $C_{fFf'}(\hat{e}_b;\hat{e}_a')$, and $C_{fFf'}(\hat{e}_b;\hat{e}_a')$, each of which may be realized in laboratory experiments by the use of a polarizer and analyzer. However, for atmospheric applications, we consider only the cases where the radiated power of both $\hat{e}'_a$ and $\hat{e}'_b$ polarizations are received. Thus, we only evaluate $C_{fFf'}(\hat{e}_a;\hat{e}') \equiv C_{fFf'}(\hat{e}_a;\hat{e}'_a) + C_{fFf'}(\hat{e}_a;\hat{e}'_b)$ and $C_{fFf'}(\hat{e}_b;\hat{e}') \equiv C_{fFf'}(\hat{e}_b;\hat{e}'_a) + C_{fFf'}(\hat{e}_b;\hat{e}'_b)$ as given in (B8a) and (B8c) of [1]. To allow for the possibility of more than one emission final state, we define the differential cross section, summing all allowed emission pathways, with $\alpha_{Ff'} = A_{Ff'} / \sum_{f'} A_{Ff'}$, we have the desired differential cross-section as Eq.(2):

$$\frac{d\sigma_{fF\Sigma}^{RS}}{d\Omega} = \sum_{f'}\frac{d\sigma_{fFf'}^{RS}}{d\Omega} = \frac{1}{4\pi}\frac{\lambda_{Ff}^2}{4}\frac{3}{2}\frac{g_F}{g_f}A_{Ff}g(\omega-\omega_{Ff})\left[3\sum_{f'}\alpha_{Ff'}C_{fFf'}(\hat{e};\hat{e}')\right]$$

We denote the quantity in the square bracket as the distribution function of the fluorescence intensity, $I \equiv \left[3\sum_{f'}\alpha_{Ff'}C_{fFf'}(\hat{e};\hat{e}')\right]$. To make a complex discussion manageable, we simplify our discussion to a judiciously selected case with incident polarization as $\hat{e} = \hat{e}_b = -\hat{z}$, as in the textbook discussion of classical scattering, giving the fluorescence intensity as Eq. (3):

$$I \equiv \left[3\sum_{f'}\alpha_{Ff'}C_{fFf'}(\hat{e}_b;\hat{e}')\right] = \frac{2}{3}\left\{1+\frac{9}{2}\left(g_F\sum_{f'}\alpha_{Ff'}B_{fFf'}^{(2)}\right)\left[\frac{-2\left(2-3\sin^2\theta\right)}{30}\right]\right\}$$

Here the coefficient is given in terms of the 6-j coefficient in (16) of [1].

With the above simplification, we now consider two cases for pathedagogy reason and lidar application.

2.1 Physics demonstrated in a simple system

We now apply these formulae to the simplest quantum structure without spin-orbit or hyperfine interaction, and substitute $f = f' = 0$ and $F = 1$ into Eq. (3) the 6-j coefficients for $B^{(2)}_{fFf'}$. As a result, we obtain $g_1 = 3$, $\alpha_{10} = 1$ and $B^{(2)}_{101} = 5/9$, leading to the fluorescence intensity functions, $I = 3C_{010}(\widehat{-z;e'}) = \sin^2\theta$, an angular distribution agreeing with the classical results in textbook [2]. This clearly demonstrated the difference from the spontaneous emission distribution that the Hanle effect is not of quantum origin. Rather, it is because of the excitation of the atomic system by a laser-like source, which coherently prepares the mixed excited sub-states before fluorescence emission. Indeed, the Hanle effect has a classical explanation and is most prominent when the excited sub-states are degenerate, i.e., when there is no Zeeman splitting at B = 0; thereby, it is an effective tool for measuring the lifetime of atomic excited states [3].

2.1 Hanle effect of Na, K and Fe transitions

In order to apply the formulation to atoms of practical interest, we need to evaluate this factor $g_F \sum_{f'} \alpha_{Ff'} B^{(2)}_{fFf'}$ for the transitions of Na, K, and Fe atoms. Though tedious, this can be down in a straightforward manner. The results for Na and K are given in Table 2.1.a and for Fe in Table 2.1.b below. Substitute these values into Eq. (2), we can calculate the fluorescence intensity distribution function $I(\theta)$ as well as that in the backward direction, $I(\theta) = \left[3\sum_{f'} \alpha_{Ff'} C^{\pi}_{fFf'}(-\hat{z};\hat{e}')\right]$ for realistic atoms for lidar application.

Table 2.1.a $g_F \sum_{f'} \alpha_{Ff'} B^{(2)}_{fFf'}$ for Na D2 transitions

$g_F \sum_{f'} \alpha_{Ff'} B^{(2)}_{fFf'}$	F=3	F=2	F=1	F=0
$\left(2, \sum_{f'}\right)$	2/5	0	-1/15	
$\left(1, \sum_{f'}\right)$		0	1/3	0

Table 2.1.b $g_J \sum_{j'} \alpha_{Jj'} B^{(2)}_{jJj'}$ for Fe transitions

$g_J \sum_{j'} \alpha_{Jj'} B^{(2)}_{jJj'}$	$z^5F^0_4$	$z^5F^0_5$	$z^5D^0_4$
$a\,^2D_3, \left(\sum_{j'}\right)$	0.257		- 0.380
$\left(a\,^2D_4, \sum_{j'}\right)$	-0.400	0.289	0.532

Although the Hanle effect for Na D transitions in connection with lidar applications has been discussed in the literature, see for example Fricke and von Zahn [4] and Krueger et al. [5], its implications on Fe lidar considered [Höffner, private communication] but not yet evaluated. We present the radiation patterns for both for comparison here.

3. RESULTS

3.1 Radiation (antenna) patterns of laser induced fluorescence from Na and Fe atoms

Using the recipe presented for the simple case of z-polarized $\left(\hat{e} = -\hat{z}\right)$ incident beam, we compute and plot the radiation patterns for Na atom in Fig. 3.1.a and that for Fe atom in Fig. 3.1.b below.

The backward values, $\left[3\sum_{f'} \alpha_{Ff'} C^{\pi}_{fFf'}(-\hat{z};\hat{e}')\right]$, for Na excitations (2,1), (1,1) and (2,3), shown respectively in solid (blue, green and red) are 0.653, 0.733, and 0.747, corresponding to, respectively, 98%, 110% and 112% of its isotropic average. The backscattering values for Fe transitions at 386 nm (black), 372 nm (red), 374 nm (blue), 392 nm (light green) and 368 nm (dark green), respectively, are 1.160, 1.087, 1.077, 0.886 and 0.880 times their isotropic average of 2/3.

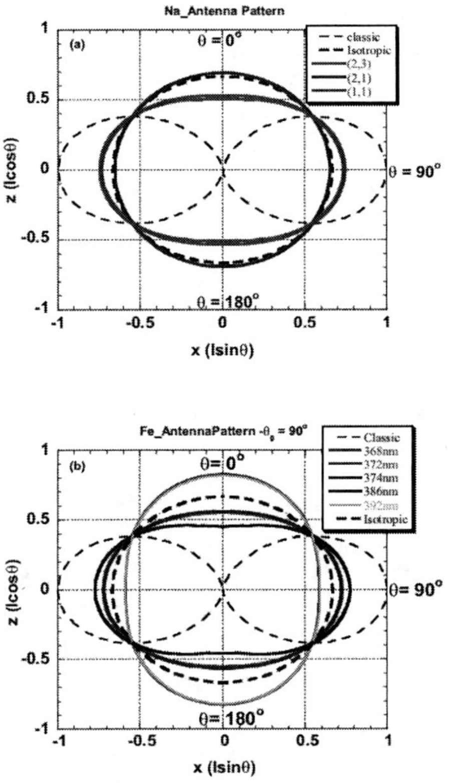

Fig. 3.1. Radiation patterns of the z-polarized, $\hat{e}=-\hat{z}$, dipole with fluorescence intensity $I=\left[3\sum_{f'}\alpha_{Ff'}C_{fFf'}(-\hat{z};\hat{e}')\right]$ as a function of scattering angles for (a) Na D_2 and (b) ^{56}Fe transitions. In (a), the Na (2,2), (1,2) and (1,0) excitations exhibit spontaneous emission behavior (bold-black dashed) with a backward scattering value of interest equal to the isotropic value, $3C^{\pi}_{fFf'}(-\hat{z};\hat{e}')=<3C_{fFf'}(-\hat{z};\hat{e}')>=2/3$. In (b), none of the Fe transitions exhibits spontaneous emission behavior (shown in bold-black dashed for reference). All transitions, shown in solid lines, exhibit the Hanle Effect. In both figures, for comparison the classic dipole pattern is shown in the thin-black dashed line.

3.2 Relevance to resonance fluorescence lidars

We consider the relevance of Hanle effect and apply the results outlined in Section 3.1 to Na and Fe fluorescence lidars. We do not discuss K lidar in part because its similarity with Na lidar and smaller atmospheric abundance. The simplest lidar (often broadband) for measuring metal density uses a transmitting laser beam. In this case, the Hanle effect could amount to up top 12 % density error. For atmospheric temperature and wind measurements, typically, intensity ratios are invoked. If the two or three frequencies in the ratio(s) are from the same atomic transition, as in the 3-frequency Na lidar [5], then the correction of the Hanle effect does not affect the temperature or wind retrieval. However, the accompanying Na density determination will still be affected. A different situation exists in the temperature determination with iron Boltzmann lidar, in which, the intensity ratio of 374 nm and 372 nm transitions is used. In this case, unless the multiplicative factors on the intensity of the two lines due to the Hanle effect are the same, they should be included. Fortunately, in this case, the multiplicative factors are nearly the same, respectively 1.077 and 1.087, ignoring the Hanle effect is justifiable.

We had only discussed the Hanle effect at B = 0. This is also justifiable, not only because the earth magnetic field is small compare to the internal magnetic field in the atom [5], also the Hanle effect is largest at B=0.

Note added in proof: A more extensive version of this paper has since been published [7].

REFERENCES

[1] C.-Y. She, H. Chen, and D. A. Krueger, "Optical Processes for middle atmospheric doppler lidars: cabannes scattering and laser-induced resonance fluorescence," *J. Opt. Soc. Am.* **B 32**, 1575–1592 (2015).

[2] D. J. Griffiths, *Introduction to Electrodynamics*, 2nd ed. (Prentice-Hall, 1981), p. 381.

[3] A. Corney, *Atomic and Laser Spectroscopy* (Oxford, 1977).

[4] K. H. Fricke and U. von Zahn, "Mesopause temperatures derived from probing the hyperfine structure of the D_2 resonance line of sodium by lidar", *J. Atmos. Terr. Phys.* **47**, 499–512 (1985).

[5] D. A. Krueger, C.-Y. She, and T. Yuan, "Retrieving mesopause temperature and line-of-sight wind from full-diurnal-cycle Na lidar observations," *Appl. Opt.*, **54**, 9469 – 9489 (2015).

[6] X. Chu, W. Pan, G. Papen, C. S. Gardner, and J. Gelbwachs, "Fe Boltzmann temperature lidar: Design, error analysis, and initial results at the North and South Poles", *Appl. Opt.*, **41**, 4400 – 4410 (2002).

[7] C.-Y. She, "Hanle effect in laser-induced fluorescence and Na and Fe resonance scattering lidars", **Appl. Opt., 58**, 8354 – 8361 (2019).

A MICRO-PULSE DIFFERENTIAL ABSORPTION LIDAR TEST NETWORK

Scott Spuler*[1], Todd Bernatsky[1], Catharine Bunn[3], Joshua Carnes[1], Matthew Hayman[1], Kevin Repasky[4], Robert Stillwell[2], Tammy Weckwerth[1]

[1]*National Center for Atmospheric Research, Earth Observing Lab, Boulder, CO 80301, USA, *spuler@ucar.edu*
[2]*National Center for Atmospheric ResearchAdvanced Study Program, Boulder, CO 80301, USA*
[3]*Montana State University, Physics Department, MT 59717, USA*
[4]*Montana State University, Electrical and Computer Engineering, Bozeman, MT 59717, USA*

ABSTRACT

The National Center for Atmospheric Research (NCAR) and Montana State University (MSU) have developed a test network of five micro-pulse Differential Absorption Lidar (DIAL) instruments to continuously measure high-vertical-resolution water vapor in the lower atmosphere. The instruments are accurate, low-cost, operate unattended, and eye-safe – all key features to enable larger 'national-scale' networks needed to characterize atmospheric moisture variability which influences important processes related to weather and climate.

1. INTRODUCTION

Water vapor is one of the fundamental thermodynamic variables that define the state of the atmosphere. It is highly variable in space and time and influences many important processes related to weather and climate. Several National Research Council (NRC) reports and a recent review paper, detailing the state of the art for thermodynamic profiling,- all highlight the need for improved water vapor measurements as a necessary step toward improving mesoscale numerical weather prediction and quantitative precipitation forecasting skills [1, 2, 3, 4]. To help address this observational need, we have developed a testbed of five diode-laser-based, micro-pulse DIAL (MPD) instruments for measuring the spatial and temporal distribution of water vapor in the lower atmosphere.

Developments in semiconductor laser and amplifier technology have enabled compact DIAL transmitters with narrow linewidths, and high spectral purity at low cost and with long lifetimes [5, 6, 7]. Semiconductor lasers are well suited to lidar systems designed for network deployment where instruments must be capable of unattended operation for long periods without maintenance and should not pose a risk to aircraft or bystanders. The network testbed described herein, was built from a field-capable semiconductor-based lidar prototype first demonstrated in 2015, which was shown to deliver accurate measurements of water vapor in the lower troposphere and produce scientifically significant data [8, 9].

2. INSTRUMENT DESIGN

The instrument, shown schematically in Figure 1, utilizes a diode-laser-based master oscillator power amplifier (MOPA) configured transmitter, and a shared telescope transmitter and receiver to achieve opto-mechanically stable, and eye-safe, operation. A multistage filtered receiver is implemented to allow continuous measurement in high background conditions (e.g., bright daytime clouds).

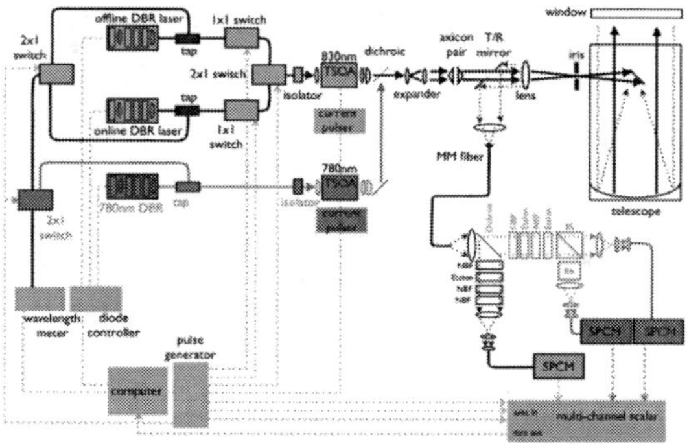

Figure 1. Schematic of the MPD system. One of the five units has been integrated with an HSRL channel (shown in red). BS = beam-splitter; T/R = transmit receive; Rb = Rubidium vapor cell; SPCM = Single Photon Counting Module, NBF = narrow-band filter.

A pair of distributed Bragg reflector (DBR) semiconductor lasers nominally centered at 828 nm are interleaved at 100 Hz and used to seed a tapered semiconductor optical amplifier (TSOA). Current pulses to the TSOA amplify the laser and produce approximately 5μJ over 900 ns duration. The beam is shaped into an annular ring with an axicon pair and then expanded to fill the inner part of the telescope with a radius of 10 cm. The laser repetition rate, set to 7 kHz, limits the maximum altitude of the lidar system to 21 km. Direct exposure to the laser exiting the telescope is within eye-safe limits. The backscattered light received is directed into a 105μm core multimode fiber in the receiver system. The light from the fiber is collimated and passes through a pair of interference filters and a solid etalon, with a combined passband of 0.014 nm for daylight rejection. The filtered light is then directed into a fiber coupled single photon counting module (SCPM), which utilizes a silicon avalanche photodiode operated in 'Geiger-mode'.

A diode-laser-based high spectral resolution lidar (HSRL) channel – based on the same low-cost lidar ar-

https://doi.org/10.1051/epjconf/202023705001

chitecture [10] – was incorporated into one of the MPD units. A schematic of the combined system is shown in Figure 1. The upgraded instrument combines water vapor with quantitative measurements of aerosols and could significantly contribute to studies of aerosol-cloud interaction. Since HSRL do not require a correction for the geometric overlap function to obtain the aerosol backscatter coefficient, they are well suited to quantitative retrievals at comparatively low altitudes. As part of the network deployment, the diode-laser-based HSRL offers opportunities for higher resolution measurements of boundary layer height for model and satellite comparison studies as well as for continuous monitoring of this key observational parameter for weather and climate. Furthermore, to reduce alignment time at the start of field deployments, a fiber coupled transmitter package is being tested for these systems. This effort builds on a NASA funded SBIR which developed fiber-coupled isolated DBR lasers sources [11].

Following the completion of each water vapor MPD unit, they were operated continuously to test basic system functionality. For MPD unit #3, this initial test period was approximately 6 months starting in the summer of 2018 as shown in Fig. 2. This initial test, done for all of the MPD units, demonstrated a high continuity time series in all weather and provided profiles of water vapor – including periods of bright daytime clouds – from 300 m above ground level to 4-6 km (or cloud base, whichever is lower) with 150 m nominal vertical resolution and 1-5 minute temporal resolution. It is important to note, that this 6 month dataset was autonomously obtained – without human intervention – unless a subsystem was being upgraded as part of scheduled development and testing.

2.1. Multi-Channel Scaler (MCS)

A new MCS was developed for the testbed units to increase overall functionality, simplify the system (i.e., reduce part count), and improve the system integration reliability. The MCS used in the original prototype was a commercially available product from Sigma Space Inc., which required a proprietary software package that was then wrapped by another software package (in our case, Labview). The new MCS uses no proprietary software, and has advanced features while reducing overall system part counts. First, the MCS uses software defined switches local to the system's field-programmable gate array (FPGA). This feature allowed removal of the radio-frequency (RF) hardware switches, and associated wiring, used in the prototype to route the signals from the photon counting module to the MCS. Second, the MCS has power monitoring channels, which allows for independent monitoring of rapidly switched laser pulses (e.g., both online and offline channels) via the FPGA software switches. This was not possible with the original hardware configuration and simplifies the system design by allowing removal of the third party power monitoring components and associated software. The gigabit Ethernet speed and software integration simplicity is a major advantage over the original system design.

The MCS includes eight input channels each capable of 90×10^6 pulse counts per second. Pulses are accumulated into range histograms that support 10 nanosecond to 20 microsecond bin sizes ($\approx$ 1.5 meters to 3 kilometers in two-way distance), up to 3,000 bins, and as many as 30,000 profile accumulations. Complete histograms are output at rates as fast as 10 Hz. Pulses received on a given input channel may be digitally and synchronously route-switched to different histogram accumulation channels, enabling the separation of online and offline wavelength accumulations without the need for external RF switches. Three separate synchronization inputs may be used to synchronize the start of histogram profile accumulations. Built-in data and synchronization test pattern options aid in system integration and development.

The MCS electronics are composed of a PicoZed processing module card and custom carrier card that includes pulse power detector channels, input buffering, power conditioning, and connectorization. The PicoZed processing card – a low-cost, off-the-shelf component developed by AVNET – includes a Xilinx Zynq System-on-Chip (SoC), 1GB memory, Ethernet and USB physical layer interfaces. The application utilizes less than 10%

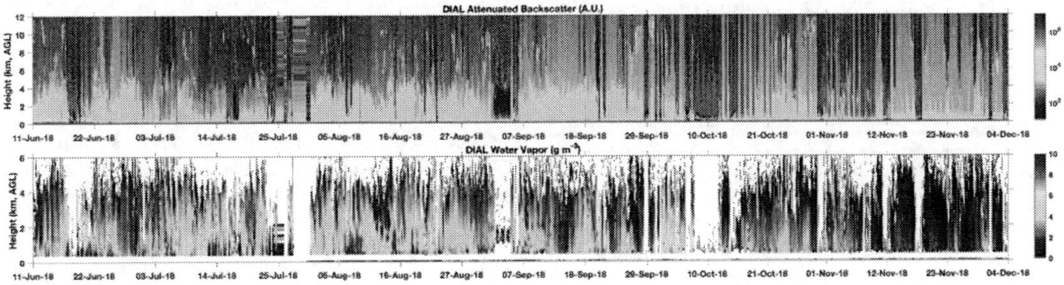

Figure 2. Time series of decimated 5-min 150-m MPD unit 3 profiles showing (top) range-corrected relative backscatter and (bottom) absolute humidity (gm^{-3}). Vertical white bars signify missing data due to MPD downtime from instrument testing periods (i.e., not hardware related) or attenuated signal above optically thick clouds.

of available SoC's FPGA resources while operating at a 200 MHz core clock rate, and the core processor readily handles the histogram data and metadata flow with its 600 MHz CPU clock and direct memory access (DMA) routing to either the Gigabit Ethernet or USB2.0 interfaces. The MCS case occupies only the front half of a US-standard 19 inch rack 1U slot. Pulse power detector channels utilize a filtering plus threshold detection technique to translate the amplitude of a pulse – with known pulse duration – into a time difference value that is measured via an FPGA input and simple timer counter. The time-based power measurement concept was adopted from the technique used by the University of Wisconsin lidar group.

2.2. Data Acquisition and Processing Software

The acquisition software of the MPD systems was completed with the following main goals: increasing fault tolerance, allowing flexibility to facilitate future system expansion, improved data quality and overall system duty cycle, and increased runtime stability. Most of these goals were achieved by altering the prototype operational structure of the acquisition software to run in parallel instead of a serial manner. Fault tolerance was achieved by isolating system critical components (e.g., the MCS control and laser control) from less critical components (e.g, housekeeping and real time data display). More monitoring resources are directed towards system critical components. Flexibility is further achieved using parallel control by isolating components that are expected to change with instrument modifications. An example of this is the number of lasers controlled. The water vapor DIAL profiling technique requires two laser transmitters (an online and an offline wavelength) but a third transmitter (for an HSRL) was added to unit two. Planned future upgrades to all of these systems (e.g., to measure temperature) will require additional lasers transmitters. Therefore, for the code to smoothly handle these hardware differences, the new architecture allows for additional controllers without changing any source code, just software controlled options. Data acquisition duty cycle is also improved by operating the control components in parallel, more completely utilizing computer resources. The combination of all of these changes allows for improved runtime stability without any loss of data quality. Furthermore, this combination of changes enhanced flexibility to accommodate software or hardware failures in the field or further development. Further modifications to the acquisition software beyond the overall structure includes the addition of more metadata for operational monitoring while deployed to field sites and the simplification of the metadata available. This adds redundancy for field operations. Calibration data and metadata such as that provided by the uninterruptible power supply were also added.

The post processing software used to verify system operation and performance was also heavily modified from the original prototype. The processing software was modified to accept changes to the acquisition software described above as well as adding necessary abstraction to facilitate future hardware changes. As a bonus of enhanced abstraction, simplification of the post processing software was achieved. Finally, minor changes were made to the archiving system hosted on NCAR servers to accommodate expansion from a single lidar unit to five.

2.3. Environmental Housing

The MPD testbed units were built using an aluminum frame that integrates the transceiver, custom telescope, and instrument rack. This aluminum frame is mounted to a shock and vibration isolation system, which helps minimize misalignment and potential damage during shipping. The housing optical window for each unit has a heating unit to help keep it clear during most operating conditions. Modifications from the initial prototype design are listed in Table 1 and bullet items below. The interior volume of the new system is 83% of the previous design, with reduced height (to makes transportation easier and simplify setup), and utilizes the same HVAC system to provide temperature stability over a wider range of environmental conditions.

Table 1. Interior dimensions of environmental enclosures

	Unit 1	Units 2-5
Width	1.88 m (74 in)	1.80 m (71 in)
Depth	0.86 m (34 in)	0.86 m (34 in)
Height	1.73 m (68 in)	1.49 m (58.5 in)
Volume	2.80 m^3 (99 ft^3)	2.32 m^3 (82 ft^3)

- Telescope redesigned for improved stability, ease of alignment, and reduced weight (e.g., 93 kg telescope assembly weight used in unit 1 was reduced to 54 kg in units 2-5.)
- Shock and vibration isolation system redesigned
- Heated window cell redesigned for lower fabrication costs
- Volume of container decreased (detailed in Table 1 above)
- Quick Pin mounted caster/wheels for easy removal
- Quick Pin mounted mast for weather station and cell modem
- Changed HVAC thermostat for tighter temperature control (tightening control band by 90%)

3. FUTURE WORK

As part of its 6-month commissioning phase, the five-unit testbed network is scheduled to be field tested from Apr to Jul 2019 at the Atmospheric Radiation Measurement (ARM) Southern Great Plains (SGP) atmospheric observatory in Oklahoma, USA. The test – conducted during a convectively active period, and performed alongside radiosondes, Raman lidar, and passive remote sensors for validation – is expected to provide valuable data to the science community. It is hoped that this initial demonstration of the technology will lead to fundamental advances within both the climate and weather forecasting communities, as it will allow observations of the evolution of the vertically-resolved thermodynamic environment at high temporal resolution for the first time. Furthermore, the evaluation of a test MPD network will be valuable in providing understanding if deployment of MPD type devices at a national scale could provide enough additional information to improve weather forecasts.

The complete five-unit MPD testbed is anticipated to be ready for service in late 2019, and will be made available to the larger research community for field campaigns and deployment through NCAR. As the units were built with modular hardware and software – we are planning to integrate the testbed units with upgrades; such as HSRL channels and O_2:DIAL channels for temperature retrievals.

4. SUMMARY

A small testbed network of validated micro-pulse water vapor DIAL instruments has been completed. The diode-laser-based transmitter is inherently low-maintenance, low-cost, and designed to be eye-safe at the exit port. Each node provides continuous profiles of water vapor with complete daytime coverage – including periods of bright clouds – from 300 m above ground level to 4-6 km (or cloud base, whichever is lower) with 150 m nominal vertical resolution and 1 to 5 minute temporal resolution. A low-cost HSRL channel has been integrated into one of the instruments. The testbed is currently scheduled to be fielded at the ARM SGP site from Apr to Jul 2019. The test will be a significant step towards improving our understanding of the distribution of atmospheric water vapor, and provide the opportunity to evaluate the impact potential of a national scale network.

ACKNOWLEDGMENTS

NCAR is sponsored by the National Science Foundation. The authors would like to acknowledge the support of the National Science Foundation grant number 1624736.

REFERENCES

[1] National Research Council, 2009: *Observing Weather and Climate from the Ground Up: A Nationwide Network of Networks*. The National Academies Press, Washington, DC.

[2] National Research Council, 2010: *When Weather Matters: Science and Service to Meet Critical Societal Needs*. The National Academies Press, Washington, DC.

[3] National Research Council, 2012: *Weather Services for the Nation: Becoming Second to None*. The National Academies Press, Washington, DC.

[4] Wulfmeyer, V., R. M. Hardesty, D. D. Turner, A. Behrendt, M. P. Cadeddu, P. Di Girolamo, P. Schlssel, J. Van Baelen, and F. Zus, 2015: A review of the remote sensing of lower tropospheric thermodynamic profiles and its indispensable role for the understanding and the simulation of water and energy cycles *Rev. Geophys.*, **53**, 819–895.

[5] Nehrir, A. R., Repasky, K. S., Carlsten, J. L., Obland, M. D., and Shaw, J. A., 2009: Water Vapor Profiling Using a Widely Tunable, Amplified Diode-Laser-Based Differential Absorption Lidar (DIAL). *Journal of Atmospheric and Oceanic Technology*, **26(4)**, 733–745.

[6] Nehrir, A. R., Repasky, K. S., and Carlsten, J. L., 2011: Eye-Safe Diode-Laser-Based Micropulse Differential Absorption Lidar (DIAL) for Water Vapor Profiling in the Lower Troposphere. *Journal of Atmospheric and Oceanic Technology*, **28(2)**, 131–147.

[7] Nehrir, A. R., Repasky, K. S., and Carlsten, J. L., 2012: Micropulse water vapor differential absorption lidar: transmitter design and performance. *Optics express*, **20(22)**, 25137–51.

[8] Spuler, S. M., Repasky, K. S., Morley, B., Moen, D., Hayman, M., and Nehrir, A. R., 2015: Field deployable diode-laser-based differential absorption lidar (DIAL) for profiling water vapor. *Atmospheric Measurement Techniques*, **8**, 1073–1087.

[9] Weckwerth, T. M., K. Weber, D. D. Turner, and S. M. Spuler, 2016: Validation of a Water Vapor Micropulse Differential Absorption Lidar (DIAL). *J. Atmospheric and Oceanic Technology*, **33**, 2353–2372.

[10] Hayman, M. and S. Spuler, 2017: Demonstration of a diode-laser-based high spectral resolution lidar (HSRL) for quantitative profiling of clouds and aerosols. *Optics Express*, **24**, A1096.

[11] Overton G, 2016: Lidar: Differential absorption lidar demands high-power, narrowband DBR sources *Laser Focus World*, **52(9)**, 14–15.

A RECIPE TO OBTAIN LIDAR POLARISATION CALIBRATION PARAMETERS G, H AND K

Joelle Buxmann[1*], Martin Osborne[1], George Georgoussis[2] and Volker Freudenthaler [3]

[1]Met Office, FitzRoy Road, Exeter, Devon, EX1 3PB, United Kingdom
[2]Raymetrics S.A., Spartis 32, Metamorfosis, Greece.
[3]Ludwig-Maximilians-University, Munich, Germany

**Email: Joelle.c.buxmann@metoffice.gov.uk*

ABSTRACT

The accuracy of the polarisation calibration is of prime importance for aerosol classification using lidars. We present a detailed description how to obtain the calibration parameters introduced in 2016 [1] accounting for various effects of non-ideal optics, lasers and atmospheric conditions. We find that crucial parameters such as the rotation angle of the plane of polarisation of the Laser (*RotL*) as well as the degree of linear polarisation (*DOLP*) influence the volume linear depolarisation ratio significantly.

1. INTRODUCTION

In a paper on the effects of polarising optics on lidar signals and the polarisation calibration [1] a set of general analytical equations has been introduced; a major change to the way of calibration used before [2]. This can be applied to a large variety of lidar systems, and will be discussed for one specific lidar within the Met Office Volcanic Ash (VA) network [3, 4] in the following. Using this new approach, the linear volume depolarisation ratio δV can be expressed as:

$$\delta V = \frac{\delta^*(G_T + H_T) - (G_R + H_R)}{(G_R - H_R) - \delta^*(G_T - H_T)} \quad (1)$$

The parameters G and H account for effects without and with atmospheric polarisation in the reflected ($_R$) and transmitted ($_T$) channel (explicit definitions can be found in [1]), respectively. The measured range dependent signal ratio is here defined as:

$$\delta V = \frac{P_R(r)}{\eta P_T(r)} \quad (2)$$

The calibration factor η can be determined from the gain ratio η*, which we get with a Δ90-calibration, and the theoretical correction K of the gain ratio.

$$\eta = \frac{1}{K}\eta^* = \frac{1 P_R}{K P_T}(\pm 45) \quad (3)$$

The correction K can be retrieved from an analytical expression (see page 4203, 2016 [1]). Combining equation (1) -(3) we get:

$$\delta V = \frac{\frac{P_T(\pm 45) P_R}{P_R(\pm 45) P_T} K (G_T + H_T) - (G_R + H_R)}{(G_R - H_R) - \frac{P_T(\pm 45) P_R}{P_R(\pm 45) P_T} K (G_T - H_T)} \quad (4)$$

2. METHODOLOGY

In order to calculate the G_R, G_T, H_R, H_T we need to describe the optical set up of the lidar using the Müller-Stokes formalism. The multiplication of Müller matrices is in direction of the propagation of light. In essence our lidar can be represented by the set up as shown in *Figure 1.*

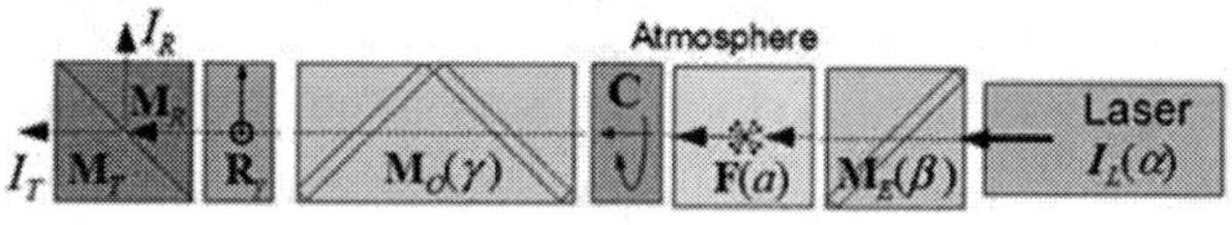

*Figure 1: Müller-Stokes schematic of a two-channel, polarisation-sensitive lidar setup (adapted from[1].). Müller matrix block elements are indicated by **M** and the calibrator **C** is a rotator in our set-up. The elements are as follows IL: laser Stokes vector, ME: emitter optics; **F**: atmospheric backscatter matrix with polarisation parameter a; **MO**: receiver*

*optics; **Ry**: rotation matrix for the 0° (y=+1) and 90° (y=-1) detection setup (see text); **MT,R**: transmitted and reflected part of the polarising beam splitter; **IT,R**: transmitted and reflected detection signals. Angles α, β, and γ are rotations around the optical axis.*

1 Application and calculation of the G, H, K parameters with an example

We will now describe how to obtain the needed parameters on the example of the datasheets provided for the Camborne lidar of the VA-network [3, 4]. The same approach can be applied to other lidar systems.

The correction parameters G, H and K can be easily calculated using a Python code developed by V. Freudenthaler. The latest official version of the open source code is available from https://bitbucket.org/iannis_b/atmospheric_lidar_ghk. The code works with Python 2.x and 3.x. The code reads the lidar system parameters from an input file (as downloaded from bitbucket and edited for a specific lidar). Therefore, we will use the same nomenclature as in the input parameter file for easy application to future studies. The input parameters *X* are indicated in italic, their uncertainties as *dX*, and the number of steps in the error calculation as *nX*. In order to adapt the optical input parameters for our own lidar we need the optical performance values for each element. We define the plane of linear polarisation of the laser as the reference plane of our optical system.

1.1 Polarizing beam splitter and cleaning polarizing sheet filters

We start from the left in figure 1 with the PBS transmission T_S and T_P in the planes perpendicular ($_S$) and parallel ($_P$) to the incidence plane, respectively, which are given by the manufacturer for 355nm as:

TP, dTP, nTP = 0.9837, 0.01, 1

TS, dTS, nTS = 0.00035, 0.0000035, 1

The retardance RetT is assumed to be 0°.

RetT, dRetT, nRetT = 0.0, 0., 0

All values of the reflected path of the PBS, i.e. *RP* and *RS* and their uncertainties, are directly calculated from the transmission values: *RP* = 1 – *TP* and *RS* = 1 – *TS*.

We use high quality polarisation filters behind the transmitted and reflected path of the PBS, to minimize the cross talk. Their extinction ratio *ERaT* and *ERaR* are given by the manufacturer:

ERaT, dERaT, nERaT = 0.00014, 0.00002, 1

ERaR, dERaR, nERaR= 0.00014, 0.00002, 1

The Rotation *RotaT* of the pol.-filter behind the transmitted path is 0°+/- 2° and *RotaR* behind the reflected path is 90°+/- 2° with respect to the incidence plane of the PBS.

RotaT, dRotaT, nRotaT = 0., 2., 1

RotaR, dRotaR, nRotaR = 90., 2., 1

1.2 Rotation matrix parameter

As the laser polarisation rotation *Rootle* = 0° and the parallel laser polarisation is detected in the reflected channel, the polarising beam splitter orientation parameter *y = -1*. This leads to the rotation matrix **R**y:

$$R(y) = \begin{matrix} 1 & 0 & 0 & 0 \\ 0 & -1 & 0 & 0 \\ 0 & 0 & -1 & 0 \\ 0 & 0 & 0 & 1 \end{matrix}$$

1.3 Receiver optics including telescope

The diattenuation parameter is defined as *D* = (I – I) / (I + I) (see [1] p. 4190, Eq.(15)) of the 355/387nm beamsplitter as given by the manufacturer (Raymetrics, pers. Communication):

DiO, dDiO, nDiO = 0.0075567, 0.0008, 1

TiO = 0.9925

RetO, dRetO, nRetO = *0., 0., 0*

RotO, dRotO, nRotO = *90., 0.5, 0*

1.4 Calibrator

We use a mechanical rotator before the receiver optics, which is described by the calibrator and location parameters:

LocC = 3 ; TypeC = 1

1.5 Emitter optics

For two rotated retarding linear diattenuators (in our case two steering mirrors, one before and one after the beam expander) at 0° and 90° with respect to incident plane, we get for the combined diattenuation (*DiE*), transmittance (*TiE*) and retardance (*RetE*) (adapted from [1], eq. 10.10.10):

$$DiE = \frac{Di3 - Di4}{1 - Di3Di4};$$

$$TiE = Ti1Ti2(1 - Di1Di2);$$

$$RetE = Di3 - Di4$$

Both highly reflective mirrors were custom designed (*Laseroptik GmbH*) so that the phase shift is zero at 355 nm and the reflection (here referred to as *TiE* > 0.95). The polarisation state, when the laser beam hits the incident plane, is considered as s-polarisation. Because the mirrors are cross aligned, their effects for *DiE* and *RetE* cancel each other (*DiE=RetE=0*). The intensity transmittance is Tmax~0.97 for s-polarisation and Tmin~0.93 for p-pol. (see Figure 2 & 3).

This leads to the following parameters:

DiE, dDiE, nDiE = *0.0, 0.0, 0*

TiE = *0.9025*

RetE, dRetE, nRetE = *0., 0., 0*

RotE, dRotE, nRotE = *0., 0.0, 0*

RotE (rotation of optical element in degrees) only matters if *RetE* and *DiE* do not cancel out (e.g. if angle between mirrors is ≠90°).

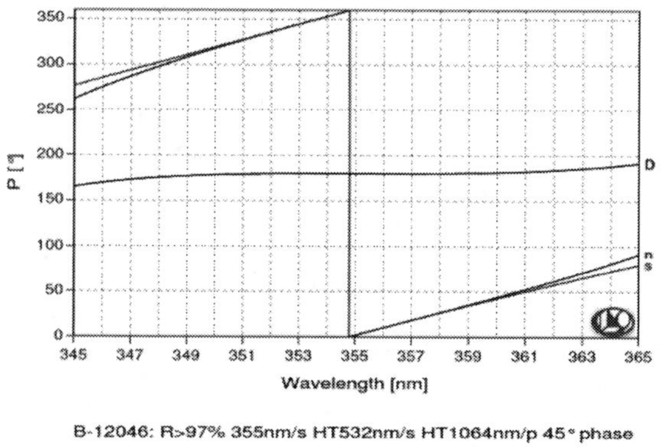

Figure 2: Retardation of the two mirrors in the emitting optics.

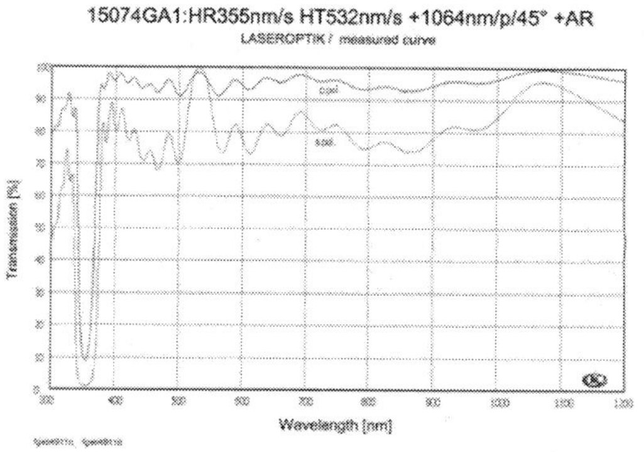

Figure 3:Transmission of the two mirrors in the emitting path

Laser

Specifications for the *DOLP* or the rotation of laser polarisation (*RotL*) in degrees have not been provided or measured. The polarisation purity of the laser was given to 99.7% by the manufacturer; so we assume as best case:

DOLP, dDOLP, nDOLP = 1.0, 0.0 0

RotL, dRotL, nRotL = 0.0, 0.0 1

The effects of other *RotL* and *DOLP* on δV are evaluated further down in the results section and in an experimental study (see contribution to ILRC29 by M. Osborne).

3. RESULTS for G, H and K

We can now add the values from above in the input file for the Python script ver. 0.96. This results in:

GR, GT, HR, HT, K (0.004), K (0.2), K (0.45)

1.95277, 2.01411, 1.94690, -2.00807, 0.984654, 0.984654, 0.984654

K is evaluated for different δV in the calibration measurements.

We further asses the effect of the laser's *DOLP* and *RotL* on δV shown in figure 4. We find a δV of 0.0298 for the best case scenario (*RotL*=0, *DOLP*=1), which also is in agreement with the former way of calibration [2]. We find that δV ranges between 0.0181-0.0298 for *RotL* between 4°-0° and *DOLP* between 0.99-1, respectively and δV =0.0283 assuming *DOLP*=0.997, *RotL*=0°, as given by the manufacturer. This means an added uncertainty of up to -0.0102 to +0.0015 of the δV in this case.

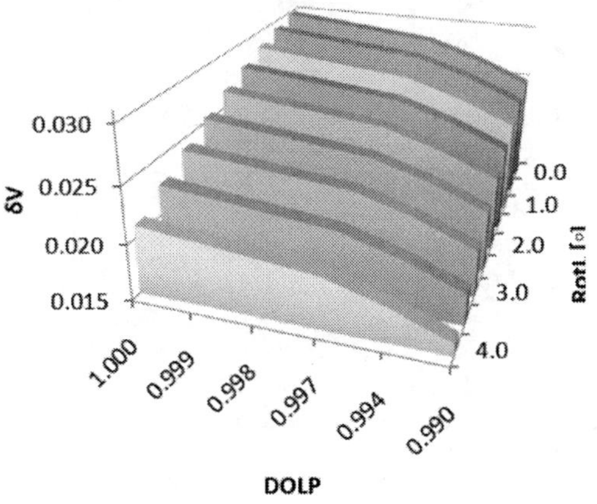

Figure 4 δV depending on variation of DOLP and RolL.

4. CONCLUSION

We show how to practically apply the calibration method as described in [1]. We find, that additional assessment of values such as the rotation angle of the plane of polarisation of the Laser RotL as well as the *DOLP* need to be measured or additional uncertainty needs to be added to the calculated δV.

ACKNOWLEDGEMENTS

The VA Lidar network is funded by UK Civil Aviation Authority and Department for Transport.

We thank T.Allott for managing the network and all the engineers for their hard work.

REFERENCES

[1] Freudenthaler , V., Atmos. Meas. Tech., 9, 4181–4255. (2016).

[2] V. Freudenthaler et al. Tellus, Ser. B Chem. Phys. Meteorol. 61(1) 165–179 (2009)

[3] M. Osborne et al. Atmos. Chem. Phys. 19(6) 3557-3578. (2019).

[4] Adam, M et al., Proceedings of the 28th International Laser Radar Conference. (2017).

 https://doi.org/10.1051/epjconf/202023705003

ACTRIS AND ITS AEROSOL REMOTE SENSING COMPONENT

Ulla Wandinger*[,1], Doina Nicolae[2], Gelsomina Pappalardo[3], Lucia Mona[3], Adolfo Comerón[4]

[1]Leibniz Institute for Tropospheric Research, Leipzig, Germany
[2] National Institute of R&D for Optoelectronics, Magurele, Romania
[3]Istituto di Metodologie per l'Analisi Ambientale, Tito Scalo, Potenza, Italy
[4]Universitat Politècnica de Catalunya, Barcelona, Spain
**Email:* ulla@tropos.de

ABSTRACT

The Aerosol, Clouds and Trace Gases Research Infrastructure ACTRIS is currently being developed with support from more than 20 countries and more than 100 research-performing organizations in Europe. The pan-European distributed research infrastructure shall provide data and services related to short-lived atmospheric constituents to facilitate high-quality Earth system research in the long term (over at least 20 years). While some of the activities are already in place, ACTRIS functionality will be further ramped up until full operation in 2025. The observation of aerosol, clouds and reactive trace gases with in-situ and remote-sensing techniques in ACTRIS is supported by six Topical Centres, which are responsible for common standards and quality assurance. Free and open virtual access to ACTRIS data is provided by the Data Centre. International users will also have physical access to ACTRIS observatories, atmospheric simulation chambers and mobile platforms as well as remote or physical access to calibration services, digital services and training. Access provision is organized through a single-entry point by the Head Office. In this contribution, the general principles and structure of ACTRIS are introduced, and the observational component related to aerosol remote sensing, which builds on the heritage of the European Aerosol Research Lidar Network (EARLINET) and the European part of the Aerosol Robotic Network (AERONET-Europe), is explained in more detail.

1. ACTRIS OBJECTIVE

ACTRIS intends to provide data and services related to the 4D-variability and the physical and chemical properties of short-lived atmospheric constituents and the processes leading to their formation, transformation and removal. ACTRIS data and services shall facilitate and enhance research in atmospheric science, with particular focus on climate change and air quality, and support the worldwide user community working on models, satellite retrievals and analysis and forecast systems. An overview on the current status of ACTRIS can be found in the ACTRIS Stakeholder Handbook 2018 [1].

2. ACTRIS STRUCTURE

ACTRIS is organized around Central Facilities and National Facilities as shown in Figure 1.

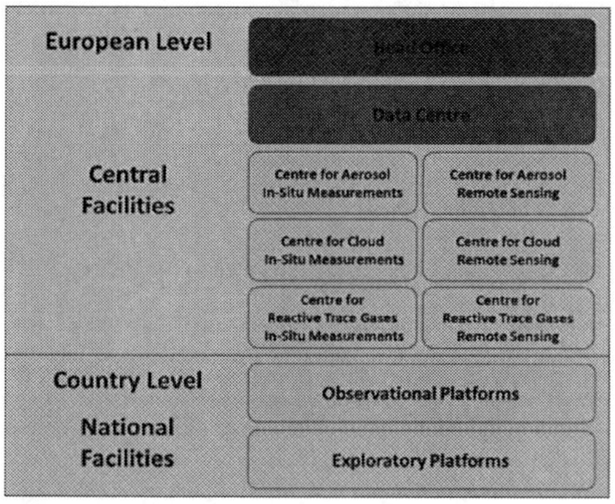

Figure 1: ACTRIS structure consisting of eight Central Facilities operated at European level and National Facilities comprising Observational and Exploratory Platforms operated at the national level.

ACTRIS Central Facilities represent the key operative entities of the research infrastructure and have a fundamental role as they provide data and services to users according to the ACTRIS policies [2, 3] as well as operation support to the National Facilities. ACTRIS Central Facilities are the Head Office, the Data Centre, and six Topical

Centres. The Central Facilities will be established at the European level and will be part of the governance and decision-making structure of ACTRIS. Each Central Facility has several operational Units, which are typically located at national research-performing organizations with respective expertise in different countries.

ACTRIS Topical Centres are organized around the main scientific themes of ACTRIS: aerosol, clouds and reactive trace gases, each with a particular focus on either remote sensing or in-situ measurement techniques (see Figure 1). The key services and operation support provided by the Topical Centres will be

a) To develop procedures and tools for quality assurance and quality control (QA/QC) of ACTRIS measurements and data,
b) To perform knowledge transfer and training to ACTRIS operators and users,
c) To foster the implementation of validated new techniques in ACTRIS,
d) To stimulate new techniques and methodologies and to sustain a high level of performance,
e) To contribute to expert collaboration networks.

ACTRIS National Facilities consist of Observational and Exploratory Platforms, which are developed, managed and operated by research-performing organizations at the national level. Observational Platforms are fixed ground-based stations that deliver long-term data based on a regular measurement schedule and common operation standards. They are widely distributed over Europe and also located at selected sites outside of Europe. Observational Platforms must contribute to at least one of the six observational components, i.e., remote sensing or in-situ measurement of aerosol, clouds and reactive trace gases, following the principles and technical requirements of ACTRIS [4].

Exploratory Platforms are atmospheric simulation chambers, laboratory platforms and mobile platforms that perform dedicated experiments and contribute data on atmospheric compounds, processes, events or regions by following common ACTRIS standards [5]. The major task of the National Facilities is the acquisition and delivery of quality-controlled data. National Facilities, in particular the exploratory platforms and large observatories, shall also provide physical access to users, which will be centrally managed by the Service and Access Management Unit within the ACTRIS Head Office.

3. LEGAL IMPLEMENTATION

ACTRIS will be implemented as a European Research Infrastructure Consortium (ERIC), a legal entity based on an agreement of EU Member States and associated countries. The task of ACTRIS ERIC is to establish and operate the distributed research infrastructure, including liaisons and establishment of long-term, sustainable and formal agreements with ACTRIS Central Facilities and management of the service provision for a wide user community. ACTRIS ERIC will provide the coordination, integration, development, monitoring and governance of ACTRIS and is in charge of the strategic and financial development and eventual long-term operation and sustainability of ACTRIS. The tasks and activities of ACTRIS ERIC are defined in the ACTRIS ERIC Statutes [6].

4. TECHNICAL REQUIREMENTS FOR AEROSOL REMOTE SENSING IN ACTRIS

Aerosol remote sensing in ACTRIS builds on the heritage of EARLINET and AERONET-Europe and applies standards developed for these observation networks over the past two decades. ACTRIS requires lidar-photometer synergy, i.e., the equipment at the Observational Platforms always consists of at least a high-power aerosol lidar and an automatic sun/sky (and lunar) photometer [4]. The distribution of aerosol remote-sensing sites over Europe as foreseen by 2025 is shown in Figure 2.

ACTRIS high-power aerosol lidars provide quantitative measurements of aerosol extinction and backscattering profiles throughout the troposphere and lower stratosphere with high spatial and temporal resolution. Spectral and polarization information (extinction-related and backscatter-related Ångström exponents and particle linear depolarization ratio) is retrieved in order to classify aerosol types and to determine natural and anthropogenic contributions to the aerosol load. ACTRIS variables also include information on geometrical boundaries of aerosol layers and layer-mean optical properties.

Figure 2: Distribution of ACTRIS aerosol remote-sensing sites over Europe as foreseen by 2025. Several additional sites are located outside of Europe (Arctic, Tropics, Central Asia).

Automatic sun/sky/lunar photometers measure the columnar spectral aerosol optical depth and sky radiance, from which information on particle microphysical and radiative properties, e.g., particle size distribution, absorption, complex refractive index, spherical/non-spherical fraction, phase matrix, is retrieved. The synergistic use of photometer and lidar observations aims at the retrieval of higher-level variables such as fine and coarse particle mass concentration and spectral extinction, backscatter and absorption profiles.

In general, ACTRIS recommends an optimum measurement setup for its Observational Platforms and requires at least a minimum configuration in order to qualify as a National Facility and receive the respective label. The optimum setup of an ACTRIS aerosol remote-sensing platform consists of a three-wavelength Raman or high-spectral-resolution (HSR) lidar with polarization discrimination capability and an automatic sun/sky/lunar photometer (optionally also polarized), both operating continuously. Instrument Level 0 data must be uploaded automatically for central processing in real real-time (RRT, data available within 3 hours after acquisition), with the aim to provide products for assimilation purposes, fast model evaluation and immediate information of the public about special events and hazardous situations.

As a minimum setup, a one-wavelength Raman or HSR lidar with polarization discrimination capability and a sun/sky photometer are required. The photometer is operated automatically, and the Level 0 data are uploaded for central processing in RRT or near real-time (NRT, data available within 3 days after acquisition). The lidar must be operated following the measurement schedule for climatological observations and shall in addition contribute to the observation of special events and satellite validation activities. Lidar data have to be uploaded in RRT or NRT in order to provide immediate public Level 1 images (quicklooks).

All higher-level variables are centrally produced, quality controlled and provided to the users through the Centre for Aerosol Remote Sensing and the ACTRIS Data Centre. Calibration and quality assurance tools for photometer and lidar instruments have to be applied following the guidelines of the Centre for Aerosol Remote Sensing.

5. CENTRE FOR AEROSOL REMOTE SENSING (CARS)

The main duty of the Centre for Aerosol Remote Sensing (CARS) is to enable the functionality of the National Facilities operating high-power aerosol lidars, automatic low-power aerosol lidars/ ceilometers (used for cloud remote sensing in ACTRIS) and sun/sky/lunar photometers and to assure the compliance of their methodologies and products with the requirements of full traceability, high quality and full documentation. The operation support provided by CARS to National Facilities consists of: a) procedures and tools for quality assurance and quality control of ACTRIS measurements and data; b) transfer of knowledge and training to ACTRIS operators; c) improvement of measurement methodologies. In addition, CARS will provide services for users.

CARS consists of eight Units operated by seven institutions in Romania, Germany, Italy, France and Spain, out of which three are dedicated to high-power lidars, two to low-power lidars and ceilometers and three to sun/sky/lunar photometers. Each Unit is specialized in providing certain services, however crucial services are covered simultaneously by at least two Units in order to minimize the risks and reduce the costs. The three Units responsible for the quality assurance of high-power lidars have been operating the Lidar Calibration Centre (LiCal) during the ACTRIS-2 project. During the ACTRIS Implementation Phase (2020-2024), their task will be to document and help in implementing, at all ACTRIS National Facilities for aerosol remote sensing, the standards and calibration techniques developed previously in the framework of EARLINET and ACTRIS, e.g., calculation of Rayleigh scattering coefficients, Rayleigh fit, trigger-delay determination, telecover test, dark-signal subtraction and polarization calibration. These tests, when used regularly and properly, are the basis for the QA/QC program of the ACTRIS high-power aerosol lidars, assuring a controlled quality of the Level 0 datasets, which are then processed with the Single Calculus Chain (SCC) in the Data Centre to produce Level 1 to Level 3 ACTRIS data. CARS support will impact on data quality mainly through avoiding signal disturbances, which may not be completely compensated by the SCC. Through training, consultancy, direct comparisons with reference systems and other specific activities, CARS will facilitate the labelling of the National Facilities.

In the following years, when ACTRIS is fully operational, CARS will work on new developments with the scope of: a) facilitating fast and correct use of the QA program by the ACTRIS operators and users on a regular basis; b) establishing new tools to quantify signal disturbances; c) producing complete error analysis of the final products, which includes and combines systematic and random errors. Also, CARS will consider future lidar developments as, e.g., the HSR lidar and measurements of the circular depolarization ratio.

6. SUMMARY AND OUTLOOK

ACTRIS shall establish, operate and develop a pan-European distributed research infrastructure for short-lived atmospheric constituents. Among other European observation and exploration capacities, it fosters the integration of EARLINET and AERONET-Europe in a synergistic aerosol remote sensing component, supported by the Centre for Aerosol Remote Sensing and the Data Centre. In the near future, ACTRIS shall provide effective access for a wide user community, including virtual and remote access to data and services as well as physical access to Topical Centres and National Facilities.

Currently, ACTRIS is in the Preparation Phase, which is supported by the EC-funded ACTRIS Preparatory Phase Project (PPP, 2017-2019). ACTRIS plans to enter the Implementation Phase in 2020 and to be fully operational in 2025. The ACTRIS ERIC shall be in place in the beginning of 2021. This legal entity with statutory seat in Finland will govern the long-term sustainable operation of the research infrastructure and facilitate the interaction with the worldwide user community over the next decades.

ACKNOWLEDGEMENTS

This work has received funding by the European Commission under the Horizon 2020 – Research and Innovation Framework Programme, Grant Agreement no. 654109 (ACTRIS-2) and Grant Agreement no. 739530 (ACTRIS-PPP).

REFERENCES

[1] ACTRIS Stakeholder Handbook 2018, http://www.actris.eu/Outreach/Publications/ACTRISStakeholderHandbook.aspx

[2] ACTRIS Data Policy, https://www.actris.eu/Portals/46/Documentation/ACTRIS PPP/Deliverables/Public/WP2_D2.3_M30.pdf

[3] ACTRIS Access and Service Policy, https://www.actris.eu/Portals/46/Documentation/ACTRIS PPP/Deliverables/Public/WP2_D2.6_M32.pdf

[4] Documentation on technical concepts and requirements for ACTRIS observational platforms, https://www.actris.eu/Portals/46/Documentation/ACTRIS%20PPP/Deliverables/Public/WP5_D5.1_M18.pdf

[5] Documentation on technical concepts and requirements for ACTRIS exploratory platforms, https://www.actris.eu/Portals/46/Documentation/ACTRIS PPP/Deliverables/Public/WP5_D5.2_M18_.pdf

[6] Document defining the founding articles for the legal entity, https://www.actris.eu/Portals/46/Documentation/ACTRIS PPP/Deliverables/Public/WP2_D2.4_M32.pdf

EPJ Web Conferences **237**, 05004 (2020)
ILRC 29
https://doi.org/10.1051/epjconf/202023705004

ADVANCES IN LIDAR OBSERVATIONS FOR AIRBORNE HAZARDS FOR AVIATION IN THE FRAMEWORK OF THE EUNADICS-AV PROJECT

Arnoud Apituley[1,*], Lucia Mona[2], Nikolas Papagiannopoulos[2], Rolf Rufenacht[3], Saskia Wagenaar[1], Piet Stammes[1], Jos de Laat[1], Wim Som de Cerff[1], Marijn de Haij[1], Francesco Marchese[2], Alfredo Falconieri[2], Alexander Haefele[3] and Maxime Hervo[3]

*[1]Royal Netherlands Meteorological Institute (KNMI), The Netherlands, *apituley@knmi.nl*

[2]IMAA-CNR, Potenza, Italy

[3]MeteoSwiss, Payerne, Switzerland

ABSTRACT

The vulnerability of the (European) aviation system to the airborne hazards was evident during the Eyjafjallajökull volcanic eruption in 2010. For support of Air Traffic Control (ATC) many observations of the event were available from satellites, ground based instruments and airborne platforms, at pan-European scale. However, efficient use of the data for ATC proved difficult mainly due to sub-optimal aggregation and integrated assessment of the available information in near-real-time.

The project EUNADICS-AV (European Natural Disaster Coordination and Information System for Aviation) aims to close this gap. The observational component of the project will make existing data products for airborne hazards more accessible, visible and used, and to foster tailored product development. Once assimilated into models and integrated in the EUNADICS-AV Data Portal these data can be the base to efficiently improve European airspace resilience to airborne hazards.

Since 2010 new data products have become available specifically for for airborne hazard alerting and monitoring together with specific tailored products designed for replying to user needs and recommendations.

This paper describes the EUNADICS-AV approach and will focus on the role of operational and research grade ground based lidars.

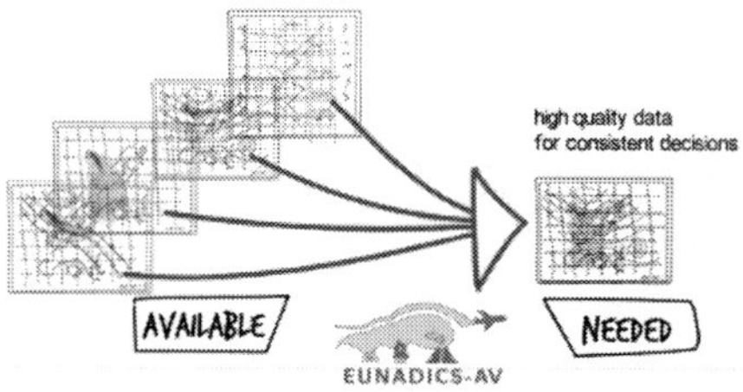

Figure 1 *The EUNADICS-AV project aims to make existing data products for airborne hazards more accessible, visible and used.*

1. INTRODUCTION

Aviation is one of the most critical transport systems of our time. Even short interruptions in flight schedules will cause major economic damage. The main objective of the H2020 EUNADICS-AV project (2016–2019) is closing the significant gap in data and information availability during airborne hazards resulting from environmental emergency scenarios and nuclear incidents, that affect air traffic. The focus is on atmospheric aerosols with trace gases used as indicators for: desert dust storms, forest fires, volcanic eruptions, and nuclear incidents (6).

The environmental emergencies the project focusses on are considered rare, but with an extremely high impact such as volcanic eruptions. The observational component of the project aims to provide integrated and advanced data sets of atmospheric constituents relevant to the project goal. The observational datasets from satellite, ground based remote sensing and air- borne in-situ measurements, are the base for the

https://doi.org/10.1051/epjconf/202023705004

development of an Early Warning System and for the assimilation into dispersion models. EUNADICS-AV activities in this sense aim to foster the use of the existing data in a more efficient way and develop new products specific tailored for this kind of applications. This is illustrated in Fig.1.

In this paper we focus on the role of ground based lidar network data in the project.

2. OBSERVATION INFRASTRUCTURE

To support the EUNADICS-AV goals, observations are needed that will be assimilated in models that subsequently provide high-quality 3D-maps of the affected hazard areas. The main priorities for observations are: 1) tailored products that can improve situational awareness (e.g. high resolution, NRT availability, global coverage, low and characterized uncertainty; 2) satellite / airborne visualisation of hazard identification, 3) visualisation of lidar imagery for volcanic ash identification.

Among the novel observations available within EUNADICS-AV are the aerosol and trace gas observations at much improved spatial resolution of 3.5x7 km^2 at nadir provided by TROPOMI on board of the Copernicus Sentinel-5p polar orbiting satellite. TROPOMI has already shown its high potential in detecting and monitoring with high definition desert dust, volcanic plumes and forest fires (3).

In EUNADICS-AV data from space-borne, ground based remote sensing and in-situ air- borne sensors are used. Here we focus on the ground based lidar data and refer to the link to satellite data and models.

2.1 Ground based observations

2.1.1 ACTRIS-EARLINET

ACTRIS (Aerosol, Clouds and Trace Gases Research Infrastructure) is a pan-European research infrastructure

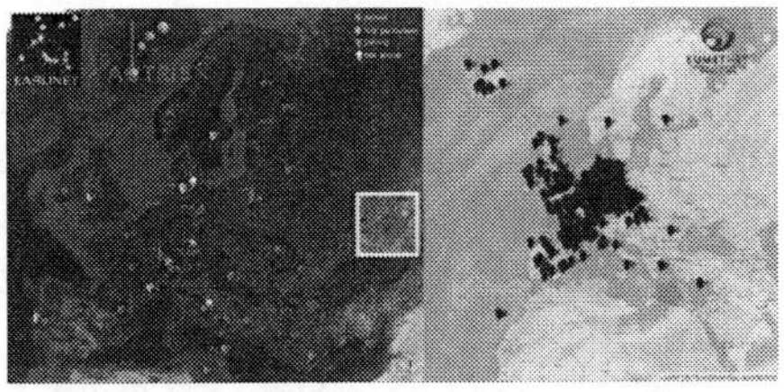

Figure *2 Maps of ACTRIS-EARLINET stations (left) and of the EUMETNET automated lidar ceilometer network (right).*

consolidating actions amongst European partners producing high- quality observations of aerosols, clouds and trace gases (1). To address many societal and environmental challenges, such as air quality, health, sustainability and climate change, detailed knowlede of atmospheric processes is needed. ACTRIS contributes to resolving these challenges by offerring a platform for re- searchers to combine their efforts effectively, and by providing open and free observational data of aerosols, clouds and trace gases.

Part of ACTRIS is EARLINET (Fig.2), the European Aerosol Research Lidar Network, with the goal of creating a quantitative, comprehensive, and statistically significant database for the horizontal, vertical, and temporal distribution of aerosols on a continental scale. Aerosol Raman lidars are the backbone of the network. At most stations the lidars measure at multiple wavelenths (1064, 532 and 355 nm) and include depolarisation channels (10).

These properties of the EARLINET lidars enable retrieval of aerosol optical properties (backscatter, extinction, lidar ratio), as well as microphysical aerosol properties, such as effective size (9), and distinction of particle shapes through the use of polarisation data. These abilities have proven invaluable in analysis of volcanic ash plumes (11), desert dust plumes as well as forest fires. In addition, it should be stressed that from well-calibrated depolarisation lidar data, an estimate of the volcanic ash mass

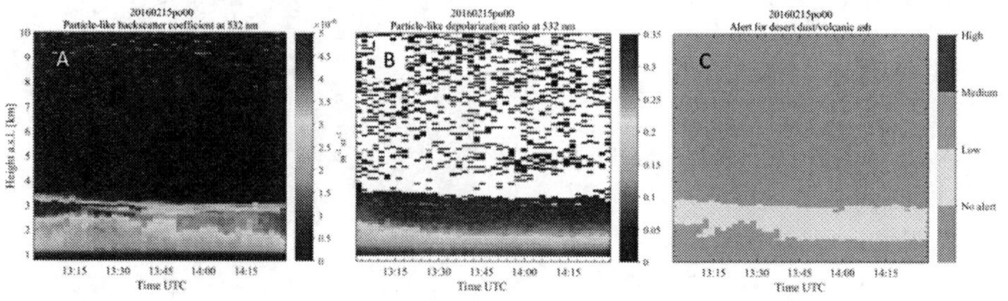

***Figure 3** Example of a tailored data product applied for Saharan dust over Potenza (It). A. Backscatter profiles at 532 nm, B. Particle depolarisation ratio at 532 nm, C. the resulting altitude dependent alert levels.*

concentration, which is the quintessential parameter needed to decide on safe versus hazardous airspace, can be made (7), even in near- real-time (8).

2.1.2 E-Profile ALC Network

EUMETNET is a grouping of 31 European National Meteorological Services that provides a framework to organise co-operative programmes between its members in the various fields of basic meteorological activities. These activities include observing systems, data processing, basic forecasting products, research and development and training.

The E-PROFILE Programme manages the European network of radar wind profilers and automatic lidars and ceilometers (ALC) for the monitoring of vertical profiles of wind and aerosols including volcanic ash (2).

Due to technical advances of the last years, ceilometers, originally designed to provide cloud base height, now also measure the vertical distribution of aerosols derived from the backscatter profile. To make this new observation capacity available E-PROFILE has developed a framework to produce and exchange profiles of attenuated backscatter profiles. ALCs of stations across Europe are added to the operational network (Fig.2). One of the goals for the ALC network is to liaise with International Programmes. Use of the data within EUNADICS-AV is an excellent example.

2.2 Lidar generated alert masks

The EARLINET Single Calculus Chain (SCC) delivers high resolution data: the calibrated total attenuated backscatter and the calibrated volume depolarization ratio time series. From these profiles, new high resolution parameters can be derived, such as particle-like intensive properties, and provide the basis of a NRT particle classification mask. A decision tree based on threshold criteria identifies dust/ash layers and associated risk levels for aviation (Fig.3).

2.3 Space Borne Observations and Models

The lidar information can now be merged with satellite information (such as from existing mechanisms like SACS (Support to Aviation Control Service) and models into a color-coded map that according to the ash/dust hazard level.

3. DATA PORTAL INTEGRATION

Based on the VAAC (Volcanic Ash Advisory Center) needs (4; 5), quicklooks of the lidar data from across European from EARLINET and products from E-PROFILE, will be made disseminated through the interactive EUNADICS-AV portal (Fig.4). The portal also displays satellite data and output from models that have assimilated the lidar and satellite data in near-real-time. Additionally, a new tailored ACTRIS/EARLINET products is under investigation for the NRT identification of dust/ash cloud, to be made available through EUNADICS-AV and ingested in the Early Warning System.

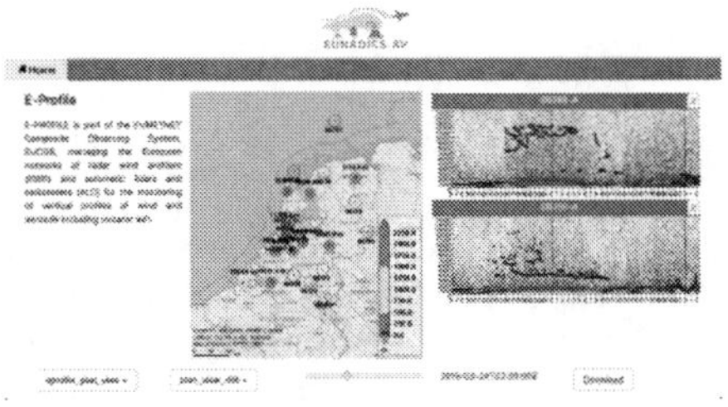

Figure 4 *Example of display of e-profile ALC data in the interactive EUNADICS-AV data portal. The interactive interface enables, amongst others, selection of multiple stations and data sets and allows continuous through time with live updates.*

4. CONCLUSIONS

In EUNADICS-AV lidar data plays a major role for detection of aerosol plumes during events that impact aviation. In case of emergencies like these, availability of reliable data in near- real-time is essential for efficient air traffic control. The operational automated lidar ceilometers (ALC) in E-PROFILE provide quality controlled data in a relatively dense network. However, ALCs are limited in capabilities for quantitative aerosol profiling. Therefore, the ALC network and the much more advanced capabilities of the quantitative aerosol lidars in EARLINET provide a unique combination of information for the tasks defined in EUNADICS-AV. Further combination with space-borne data provides a solid basis for the information needed to make airspace safer in the presence of airborne disasters.

Within the EUNADICS-AV project a frame- work for the integration of observations, mod els and data visualisation and dissemination has been set up and is available, for further optimisation, and eventually operationalisation.

ACKNOWLEDGEMENTS

This work received funding from the EC Horizon 2020 Research and Innovation Programme, Grant Agreement no.723986.

REFERENCES

[1] actris.eu.

[2] ceilometer.e-profile.eu.

[3] https://apps.sentinel-hub.com/eo-browser.

[4] https://www.metoffice.gov.uk/aviation/vaac/.

[5] http://www.meteo.fr/vaac.

[6] www.eunadics.eu.

[7] A. Ansmann, et al. Ash and fine-mode particle mass profiles from EARLINET-AERONET observations over central Europe after the eruptions of the Eyjafjallajökull volcano in 2010. *J. Geophys. Res.*, 116, 06 2011.

[8] D. P. Donovan and A. Apituley. Practi- cal depolarization-ratio-based inversion procedure: lidar measurements of the Eyjafjallajökull ash cloud over the Netherlands. *Appl. Opt.*, 52(11):2394–2415, Apr 2013.

[9] D. Müller, et al. Microphysical parti- cle properties derived from inversion algo- rithms developed in the framework of ear- linet. *Atmospheric Measurement Techniques*, 9(10):5007–5035, 2016.

[10] G. Pappalardo, et al. Earlinet: towards an ad- vanced sustainable european aerosol lidar net- work. *Atmospheric Measurement Techniques*, 7(8):2389–2409, 2014.

[11] G. Pappalardo, et al. Four-dimensional dis- tribution of the 2010 eyjafjallajökull volcanic cloud over europe observed by earlinet. *Atmo- spheric Chemistry and Physics*, 13(8):4429– 4450, 2013.

EPJ Web Conferences **237**, 05005 (2020) https://doi.org/10.1051/epjconf/202023705005
ILRC 29

BIOMASS BURNING MEASUREMENTS IN EARLINET

Mariana Adam[1*], Doina Nicolae[1], Livio Belegante[1], Iwona S. Stachlewska[2], Dominika Szczepanik[2], Maria Mylonaki[3], Christiana Anna Papanikolaou[3], Nikolaos Siomos[4], Kalliopi A. Voudouri[4], Arnoud Apituley[5], Luca Alados-Arboledas[6], Juan Antonio Bravo-Aranda[6], Aleksander Pietruczuk[7], Anatoli Chaikovski[8], Michaël Sicard[9,10], Constantino Muñoz-Porcar[9], Ina Mattis[11], Nikolaos Papagiannopoulos[12], Lucia Mona[12], Holger Baars[13], Ulla Wandinger[13], Daniele Bortoli[14], Ivan Grigorov[15], Zahary Peshev[15], Bogdan Antonescu[1]

[1]*National Institute of Research and Development for Optoelectronics - INOE 2000, Magurele, 077225, Romania*
[2]*University of Warsaw, Faculty of Physics, 02093, Poland*
[3]*National Technical University of Athens, Department of Physics, Athens, 15780, Greece*
[4]*Laboratory of Atmospheric Physics, Aristotle University of Thessaloniki, Thessaloniki, 54124, Greece*
[5]*KNMI – Royal Netherlands Meteorological Institute, De Bilt, 3731, the Netherlands*
[6]*Andalusian Institute for Earth System Research, Department of Applied Physics, University of Granada, Granada, 18071, Spain*
[7]*Institute of Geophysics, Polish Academy of Sciences, Warsaw, 01-452, Poland*
[8]*Institute of Physics, NAS of Belarus, Minsk, 220072, Belarus*
[9]*RSLab / CommSensLab, Universitat Politècnica de Catalunya, Barcelona, 08034, Spain*
[10] *CTE-CRAE / IEEC, Universitat Politècnica de Catalunya, Barcelona, 08034, Spain*
[11]*Deutscher Wetterdienst, Meteorologisches Observatorium Hohenpeißenberg, Hohenpeißenberg, 82383 Germany*
[12]*CNR-IMAA, C.da S. Loja, Tito Scalo, 85050, Italy*
[13]*Leibniz Institute for Tropospheric Research, Leipzig, 04318, Germany*
[14]*Earth Sciences Institute, Physics Department, University of Évora, Évora, 7000, Portugal*
[15]*Institute of Electronics, Bulgarian Academy of Sciences, 1784, Sofia*
**Email: mariana.adam@inoe.ro*

ABSTRACT

The biomass burning events are analyzed using the EARLINET-ACTRIS atmospheric profiling of aerosols using lidars. The period of 2008-2017 was chosen to analyze all the events assigned in the EARLINET database under Forest Fire category. A number of fourteen stations were considered. The data provided, ranged from complete data sets (backscatter, extinction and particle linear depolarization ratio) to single profiles (backscatter coefficient). A thorough quality control was performed. Smoke layers geometry was evaluated and the mean properties within each layer were computed. The Hysplit backward-trajectory technique and the FIRMS fire database were used to double check the source of each layer. Discussions were made under the following scenarios: fire events seen by two stations, long range transport from North America, and geographical clusters.

1. INTRODUCTION

Biomass burning (BB) represents an important source of atmospheric particles (aerosols) which influences the radiative transfer in the atmosphere with in general, a negative effect over the globe [1]. At the ground level or within *PBL* (Planetary Boundary Layer), strong BB episodes can cause large reduction of the visibility over various regions [2], which can affect the traffic and more importantly can cause serious health issues for humans through the low air quality [3-4]. The monitoring of the wild forest fires events occurring each year in Europe are reported since 2000 in the annual reports from European Commission [5]. Lidars can provide smoke measurements being able to deliver the boundaries of the smoke layers and the optical properties therein. Both the layer geometry and optical properties can be used for validation of the transport model output [6]. The optical properties can be also used in closure studies to model the microphysical properties of the aerosol [7]. Regarding the impact of BB aerosol on weather forecast, the study by Zhang et al. [8] suggests that the BB' effect is seen for *AOD* (aerosol optical depth) at 550nm larger than 1.

Within this context, the use of ACTRIS-EARLINET database will bring more insights about the BB aerosol on both spatial and temporal evolution. Selection of results shown, focuses on particular events and one example of geographical clusters.

2. METHODOLOGY

2.1 The EARLINET database

The EARLINET database consists currently of 31 lidar stations covering most of Europe and one location outside Europe. A review of the EARLINET network is given by Pappalardo et al. [9]. There are several categories of submitted data file, one of them being Forest Fire (FF), which are built based on voluntary basis, as opposed to Climatology and Calipso categories. For the present study, we chose the 2008-2017 interval, as most of the data are submitted after 2008, while Hysplit model [10] is available from late 2007. We considered only the data with emission wavelength at 355, 532 and 1064 nm. The initial data set contained ~3700 files at 20 stations.

2.1 Data quality control

Several criteria were implemented for data quality control (QC). Firstly, the EARLINET tools on QC were used to eliminate the b (backscatter) and e (extinction) files which did not pass the checks on technical characteristics or the optical properties values [11]. Secondly, additional data checks were performed in-house. After preliminary QC, ~ 2300 files from 14 stations were considered.

2.2 Evaluation Procedure

- Firstly, an automatic algorithm was developed to estimate the boundaries of the aerosol pollution layers visible in the lidar profiles. Secondly, a manual check was performed over the automatic boundaries' retrievals. When the automatic retrievals were not accurate, the boundaries were corrected manually.
- Mean values of optical properties in the layers were computed
- NOAA Hysplit backtrajectory and fires location from NASA FIRMS [12] were used to select the pollution layers of BB origin, whereby quite often, several layers with different sources were recorded at once. Only the fires on a 100 km radius along backtrajectory and within ± 1h of the airmass pass were selected. The backtrajectories for layers without fires found in this range were not considered in analyses.
- profiles with layers of BB origin are selected
- The selected events were discussed based on the intensive parameters (lidar ratio *LR*, backscatter Ångström exponent *BAE*, extinction Ångström exponent *EAE*, particles linear depolarization ratio *PDR*).
- The measurements were quantified as "pure" (with single fire contributing) or "mixed" (more fires contributing)

Please note that all the criteria involved in various steps of the procedure will be discussed in detail during conference.

3. RESULTS

3.1 Fire events seen at two stations

Fig. 1 shows the BB detection at Thessaloniki ("th") and Athens ("at") stations on 29-31 May 2017. The square in the upper plot shows the common fire. The second plot shows the histogram of the number of fires' detections for the two stations. The third plot shows the common fire' coordinates versus time of occurrence. The bottom plot shows the common fire' coordinates as a function of measurement time at the two stations. Thus, the smoke from the fire occurring at 00:00 on 26th (48.171N, 30.622E) arrived in "th" at 13:36 on 29th at 1260m altitude and in "at" at 08:40 on 31st at 1742m altitude. The intensive parameters (Fig. 2) were: *BAE*@532/1064 = 0.52 ("th") and *BAE*@355/532 = 1.69, *BAE*@532/1064 = 1.32 ("at"). Smoke was considered "pure" (i.e. with one source) for both stations. Figures 3-4 depict backtrajectories and the fires' location for "th" and "at". As better seen on bottom plots, for each station there was only one fire contributing to the measurement. The smoke was observed after 85h30m in "th" and 128h40m in "at" from the time of occurrence. The common variable was *BAE*@532/1064. The value in "th" was smaller than the value in "at". The difference can be explained by the slightly different path and travel time.

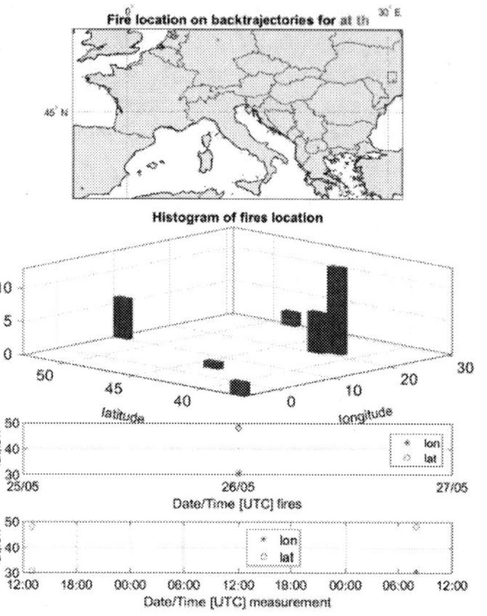

Fig. 1. Upper: location of the fires captured by "at" and "th" stations on 29 May - 2 June 2017. Square shows the common fire. Second: histogram of the number of fires' detection. Third: common fire' coordinates and time of fires' observation. Bottom: fire' coordinates versus lidar measurement time.

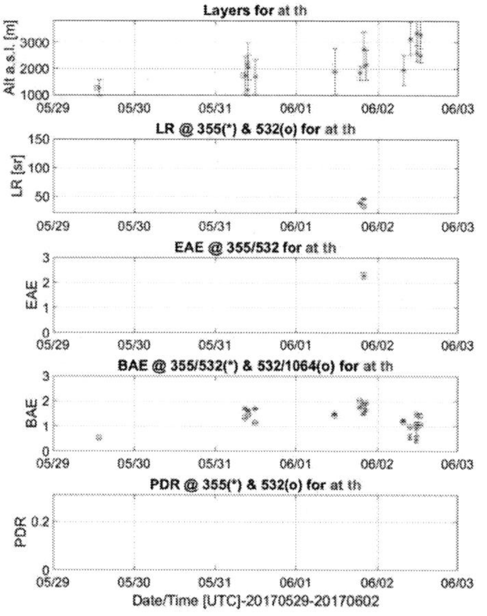

Fig. 2. Intensive variables for "at" and "th", event 29 May - 2 June 2017. Squares show the layers with BB origin.

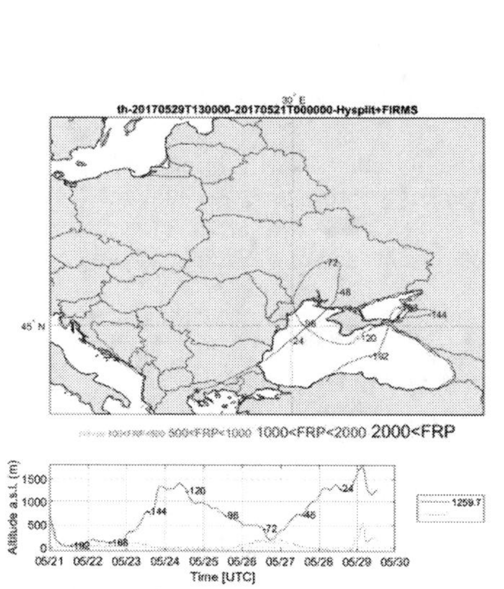

Fig. 3. Hysplit backtrajectory and fires' location along for "th" station. Backtrajectory starting time: 13:00 20170529, altitude 1260m asl.

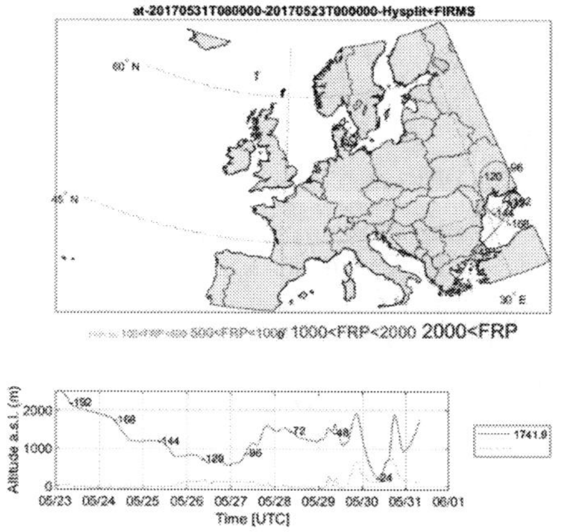

Fig. 4. Hysplit backtrajectory and fires' location along for "at" station. Backtrajectory sarting time: 08:00 20170531, altitude 2123m and 1742m asl.

3.2 Long range transport from N America

We identified a number of 24 BB events with long range transport (LRT) from N America. The strongest event was recorded by three stations (Belsk -be, Cabauw-ca and Warsaw -wa) during 8-10 July 2013 with the main fires occurring between 30 June to 7 July. Some of those measurements were discussed in literature (e.g. [13-14]). This event will be discussed from a larger perspective during the conference. The statistics of the intensive parameters of LRT from

N America (over all 24 events) is shown in Table 1. We divided the events into two categories: "pure N America" (NA only) and "mixed" (N America + European). The values from literature on N America LRT (Lit res) were added for comparison. Note that we do not have information if the LRT values reported were pure or mixed fires (in the way we defined them). Values for "pure" cases are closer to literature except PDR.

Table 1. Mean intensive parameters for long range transport from N America.

Intensive parameter	LR355 mean	LR355 std	LR532 mean	LR532 std	EAE mean	EAE std	BAE 355/532 mean	BAE 355/532 std	BAE 532/1064 mean	BAE 532/1064 std	PDR 355 mean	PDR 355 std	PDR 532 mean	PDR 532 std
All	46.2	15.6	70.7	29.8	1.18	0.83	1.44	0.42	1.24	0.52	2.44	0.51	3.10	2.24
NAonly[1]	**42.2**	13.9	**62.0**	28.1	**1.06**	0.77	**1.37**	0.47	**1.31**	0.58	2.48	0.49	3.59	2.89
Mixed[2]	52.44	17.2	77.6	32.3	1.34	0.99	1.53	0.31	1.10	0.33	2.38	0.61	2.38	0.16
Lit res[3]	47.6	14.7	59.2	12.4	1.02	0.79	1.47	0.36	1.28	0.66	17.17	10.29	9.53	6.30
	(23)		(30)		(0.2)		(0.5)		(0.2)		(5)		(2)	
	(100)		(129)		(3)		(2.1)		(2.85)		(28)		(20)	

* In parenthesis, minimum and maximum values.

3.1 Geographical clusters

The EARLINET stations considered in BB study were split into four geographical clusters: SE Europe (Athens-at, Bucharest-bu, Potenza-po, Sofia-sf and Thessaloniki-th), SW Europe (Barcelona-ba, Evora-ev, Granada-gr), NE Europe (Belsk-be, Minsk-mi and Warsaw-wa), Central Europe (Cabauw-ca, Leipzig-le, Observatory Hohenpeissenberg-oh).

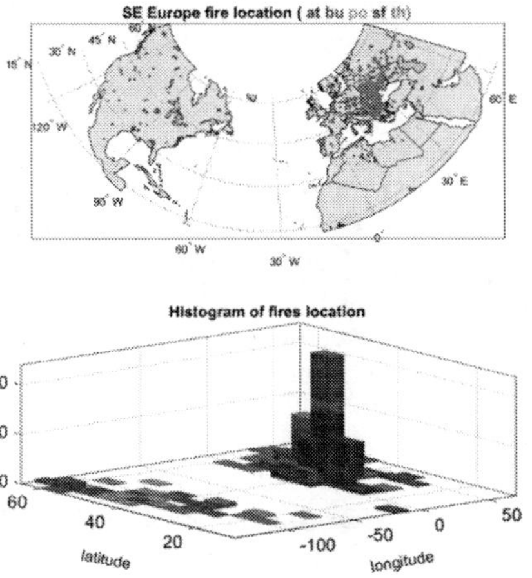

Fig. 5. SE Europe cluster (Athens, Bucharest, Potenza, Sofia, Thessaloniki). Upper plot represents the location of the fires. Lower plot represents the histogram of the number of fires' detection.

As it will be shown during the conference, each of those clusters measures predominantly smoke from fires occurring in a specific region. As an example, SE Europe cluster is shown in Fig. 5. The stations from this region record BB events with the major source in SE Europe. Each cluster will be discussed in connection with the main meteorological features of each region.

3. CONCLUSIONS

The Forest Fire category of the EARLINET - ACTRIS database allows to perform a variety of studies on biomass burning aerosols. One can exploit fires' smoke as i) recorded by several stations, ii) long range transport (e.g. N America or W Asia), on the basis of iii) statistics on geographical clusters, and iv) the cases with complicated meteorology when one station can measure mixed BB aerosols, from different sources.

ACKNOWLEDGEMENTS

This work was supported by the H2020 ACTRIS-2 No 654109 and the Romanian National contracts 18N/08.02.2019 and 19PFE/17.10.2018.

REFERENCES

[1] IPCC, https://www.ipcc.ch/ipccreports/tar/wg1/235.htm (2001)
[2] M. Adam, et al. J. Geophys. Res., 109, D16S02, doi:10.1029/2003JD004047 (2004)
[3] A. Sapkota, et al. Environ. Sci. Technol., 39, 24-32 (2005)
[4] E. Alonso-Blanco, et al. Atmos. Chem. Phys., 18, 10615–10643 (2018)
[5] http://effis.jrc.ec.europa.eu/reports-and-publications/annual-fire-reports/ (2000-2017)
[6] B. L. van Drooge, et al. Atmos. Poll. Res., 7, 494-502 (2016)
[7] M. Fiebig, et al. Atmos. Chem. Phys., 3, 881–891 (2003)
[8] J. Zhang, et al. Atmos. Chem. Phys., 16, 6475–6494 (2016)
[9] G. Pappalardo, et al. Atmos. Meas. Tech., 7, 2389–2409, 2014
[10] A. F. Stein, et al. Bull. Amer. Meteor. Soc., **96**, 2059-2077 (2015)
[11] https://www.earlinet.org/index.php?id=125
[12] https://firms.modaps.eosdis.nasa.gov/
[13] L. Janicka, et al. Atmos. Environ., 169, 162-174 (2017)
[14] P. Ortiz-Amezcua, et al. Atmos. Chem. Phys., 17, 5931–5946 (2017)

FULLY AUTOMATED LIGHT PRECIPITATION DETECTION FROM MPLNET AND EARLINET NETWORK LIDAR MEASUREMENTS.

Simone Lolli[1,*], Gemine Vivone[4], Ellsworth J. Welton[2], Jasper R. Lewis[3], James R. Campbell[5], Michael Sïcard[6], Adolfo Comeron[6], Gelsomina Pappalardo[1]

[1] CNR, Institute of Methodologies for Environmental Analysis, 85050 Tito Scalo (PZ), Italy.
[2]NASA Goddard Flight Space Center, CODE 612, Greenbelt, MD 20771, USA
[3]JCET-UMBC, 1000 Hilltop Cir, Baltimore, MD 21250, USA
[4] Univ. of Salerno, Dept. of Inform. Engin., Electr. Engin. and App. Maths., 84084 Fisciano (SA), Italy
[5] Naval Research Laboratory, 7 Grace Hopper Ave, Monterey, CA 93943, USA
[6] RSLAB, Univ. Politec de Catalunya, Carrer de Jordi Girona, 1, 3, 08034 Barcelona, Spain

**Email: simone.lolli@imaa.cnr.it*

ABSTRACT

The water cycle strongly influence life on Earth and precipitation especially modifies the atmospheric column thermodynamics through the evaporation process and serving as a proxy for latent heat modulation. For this reason, a correct light precipitation parameterization at global scale, it is of fundamental importance, bedsides improving our understanding of the hydrological cycle, to reduce the associated uncertainty of the global climate models to correctly forecast future scenarios. In this context we developed a full automatic algorithm based on morphological filters that, once operational, will make available a new rain product for the NASA Micropulse Lidar Network (MPLNET) and the European Aerosol Research Lidar Network (EARLINET) in the frame of WMO GALION Project

1. INTRODUCTION

In the last decades, thanks to internet, ground-based network of instruments started to develop to measure important climate-related variables, among which the most important the aerosol optical and microphysical properties measured by lidar instruments, developed after CO_2 laser invention [1]. The NASA Micropulse Lidar Network (MPLNET, [2], [3], [4]) is a global network of homogeneous elastic lidar instruments that monitor the atmosphere 24/24 under any meteorological condition. Version 3 Level 1 and 2 MPLNET products provide, among others, aerosol and cloud backscattering and extinction atmospheric profiles [5], [6] and the boundary layer height, used to characterize aerosol sources and transport [7]. Elastic, multi-wavelength and Doppler [8] lidar measurements containing raining events are unjustifiably disregard. Precipitations, which modifies the atmospheric column thermodynamics, are crucial to reduce global model uncertainty to forecast future scenarios [9]. [10]. In this work we present the first tests of the automatic light rain detection algorithm developed to be operative on NASA MPLNET network to characterize light precipitation characteristics at global scale.

2. METHODOLOGY

Previous studies demonstrated that it is possible to quantitatively assess light rain microphysical characteristics using a multi-wavelength lidar [11], [12] and an elastic lidar and a disdrometer [13]. However, for the latter applications, the rain episodes were eye detected. For this reason, the need to develop an automatic algorithm to be implemented operationally for future product release [14]. The implemented methodology employs vertical and horizontal morphological filters to detect the rain episodes, i.e rain start, rain stop, and relative cloud height. The algorithm uses as input the 24hr depolarization channel range corrected signal.

3. RESULTS

An example on how the rain looks on lidar data is

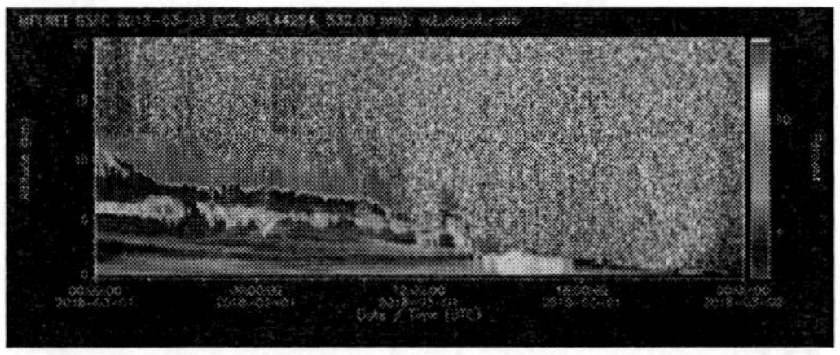

Figure. 1 Depolarization Channel from MPLNET GSFC station on March, 1st 2018. A rain event is clearly detected from about 1400 to 1800 UTC.

showed in Figure 1 representing the range corrected depolarization channel from NASA GSFC permanent observation station on March, 1st 2018. Multi-layers clouds are clearly visible around 10 km and 6 km. The cloud base is lowering and starts to rain at 2.5 km at around 1400 UTC and finishing at around 1745 UTC. A second rain episode is visible at around 2330 UTC.

After testing several edge detectors, the morphological filters produce the best results and were chosen for final implementation. With respect to Figure 1, the algorithm detected the following rain episodes:

Rain Starts	Rain ends
16:58	18:08
18:39	19:13
23:49	23:54

Table 1 Rain episodes on March 1st, 2018 at NASA GSFC permanent observational site

The results were verified data from ground-based rain-gauge measurements. The algorithm will be exhaustively tested on MPLNET and EARLINET data, to be further implemented operationally on both networks. The result is encouraging in view of the new constituting network of networks GALION, of the World Meteorological Organization.

ACKNOWLEDGEMENTS

This study and the NASA Micropulse Lidar Network (MPLNET) are supported by the NASA Radiation Sciences Program (H. Maring)

REFERENCES

[1] Ciofini, et al.,. "Diffractive optical components for high power laser beam sampling." *Journal of Optics A: Pure and Applied Optics* 5.3 (2003): 186.

[2] Welton, E.J., J. R. Campbell, J. D. Spinhirne, and V. S. Scott, Global monitoring of clouds and aerosols using a network of micro-pulse lidar systems, Proc. SPIE, 4153, 151-158 (2001)

[3] Campbell, J.R. J.R., D.L. Hlavka, E.J. Welton, C.J. Flynn, D.D. Turner, J.D. Spinhirne, V.S. Scott, and I.H. Hwang. Full-time, Eye-Safe Cloud and Aerosol Lidar Observation at Atmospheric Radiation Measurement Program Sites: Instrument and Data Processing, *J. Atmos. Oceanic Technol.*, 19, 431-442 (2002)

[4] Lolli, S., P. Di Girolamo, Principal Component Analysis Approach to Evaluate Instrument Performances in Developing a Cost-Effective Reliable Instrument Network for Atmospheric Measurements. *J. Atmos. Oceanic Technol.*, **32**, 1642–164 (2015)

[5] Lolli, S., Madonna, F., Rosoldi, M., Campbell, J. R., Welton, E. J., Lewis, J. R., Gu, Y., and Pappalardo, G.: Impact of varying lidar measurement and data processing techniques in evaluating cirrus cloud and aerosol direct radiative effects, *Atmos. Meas. Tech.*, 11, 1639-1651 (2018)

[6] Lolli, S., J.R. Campbell, J.R. Lewis, Y. Gu, J.W. Marquis, B.N. Chew, S. Liew, S.V. Salinas, and E.J. Welton: Daytime Top-of-the-Atmosphere Cirrus Cloud Radiative Forcing Properties at Singapore. *J. Appl. Meteor. Climatol.*, **56**, 1249–1257 (2017)

[7] Milroy, C, et al. "An assessment of pseudo-operational ground-based light detection and ranging sensors to determine the boundary-layer structure in the coastal atmosphere." *Advances in Meteorology* 2012 (2012).

[8] Lolli, S., Delaval, A., Loth, C., Garnier, A., and Flamant, P. H.: 0.355-micrometer direct detection wind lidar under testing during a field campaign in consideration of ESA's ADM-Aeolus mission, Atmos. Meas. Tech., 6, 3349-3358 (2013)

[9] Lolli, S., Khor, W. Y., Matjafri, M. Z., & Lim, H. S. (2019). Monsoon Season Quantitative Assessment of Biomass Burning Clear-Sky Aerosol Radiative Effect at Surface by Ground-Based Lidar Observations in Pulau Pinang, Malaysia in 2014. *Remote Sensing, 11*(22), 2660.

[10] Campbell, J.R., C. Ge, J. Wang, E.J. Welton, A. Bucholtz, E.J. Hyer, E.A. Reid, B.N. Chew, S. Liew,

S.V. Salinas, S. Lolli, K.C. Kaku, P. Lynch, M. Mahmud, M. Mohamad, and B.N. Holben,: Applying Advanced Ground-Based Remote Sensing in the Southeast Asian Maritime Continent to Characterize Regional Proficiencies in Smoke Transport Modeling. *J. Appl. Meteor. Climatol.,* **55**, 3–22 (2016)

[11] Lolli, S., E.J. Welton, and J.R. Campbell, Evaluating Light Rain Drop Size Estimates from Multiwavelength Micropulse Lidar Network Profiling, *J. Atmos. Oceanic Tech.*, 30, 2798-2807 (2013)

[12] Lolli, S., P.D. Girolamo, B. Demoz, X. Li, and E.J. Welton, 2016. Rain evaporation rate estimates from dual-wavelength lidar measurements and intercomparison against a model analytical solution. J. Atmos. Oceanic Tech., DOI: 10.1175/JTECH-D-16-0146.1.

[13] Lolli, S.; D'Adderio, L.P.; Campbell, J.R.; Sicard, M.; Welton, E.J.; Binci, A.; Rea, A.; Tokay, A.; Comerón, A.; Barragan, R.; Baldasano, J.M.; Gonzalez, S.; Bech, J.; Afflitto, N.; Lewis, J.R.; Madonna, F. Vertically Resolved Precipitation Intensity Retrieved through a Synergy between the Ground-Based NASA MPLNET Lidar Network Measurements, Surface Disdrometer Datasets and an Analytical Model Solution. *Remote Sens.*, *10*, 1102 (2018

[14] Lolli, S., Vivone, G., Lewis, J. R., Sicard, M., Welton, E. J., Campbell, J. R., ... & Pappalardo, G. (2020). Overview of the New Version 3 NASA Micro-Pulse Lidar Network (MPLNET) Automatic Precipitation Detection Algorithm. *Remote Sensing*, *12*(1), 71.

IAOOS OBSERVATIONS OF AEROSOLS AND CLOUDS IN THE HIGH ARCTIC BY AUTONOMOUS DRIFTING LIDAR PLATFORMS

J. Pelon[1], C. Di Biagio[2], V. Mariage[1], P. Genau[1], L. Loyer[1], J.-C. Raut[1], G. Ancellet[1]

[1]LATMOS, Sorbonne Université-CNRS-UVSQ Paris, France, [2]LISA, Université Paris-Créteil, France

ABSTRACT

New drifting platforms have been deployed within the French project IAOOS (Ice-Atmosphere-Ocean Observing System) in the Arctic since 2014. Radiation and meteorological parameters are measured at the surface and profiles of aerosol and cloud properties are obtained with autonomous backscatter lidar systems. These platforms are indeed equipped for ocean-ice-atmosphere studies over the Arctic to better understand processes and interactions controlling sea-ice changes [1]. As stations in the Arctic are sparse, they can also be used as reference measurements for satellite observations. They are deployed in the Arctic almost every year and allow to perform regular measurements of the vertical structure and optical properties of the atmosphere in complement to satellite observations. Other data on snow, ice and ocean are simultaneously measured. Comparisons were made with CALIPSO/CALIOP observations. Measurements on the atmosphere are presented and results are discussed.

1. IAOOS OBSERVATIONS

The upper part of the IAOOS platforms is dedicated to atmospheric measurements. It includes an integrated meteorological station, radiometric sensors, a backscatter lidar (possibly including polarization separation) and a camera. Figure 1 presents the mounting of the platforms. Additional temperature measurements are performed in ice, and upper ocean from a multiple temperature sensor chain [2].

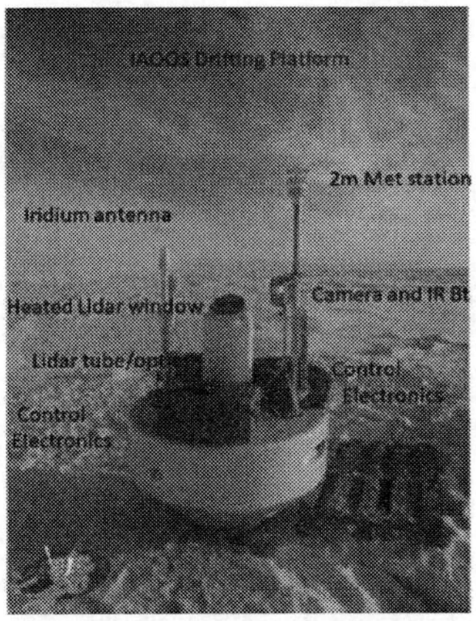

Fig. 1: Implementation of the instrumentation on the IAOOS platform

Other sensors are adapted for temperature, salinity, gases and nutrients for measurements in the ocean [1]. Data are taken at regular intervals (up to 8 times per day) over a given period of time (minutes to an hour) Several platforms can be deployed each year, depending on the opportunities by the collaborations with groups performing ocean campaigns in Arctic, or the possibility to deploy from the Russian Barneo camp in spring at the North Pole. Field measurements are starting in spring or summer and may last up to two years as the platforms are drifting over the Arctic, depending on where they are placed and problems that may occur.

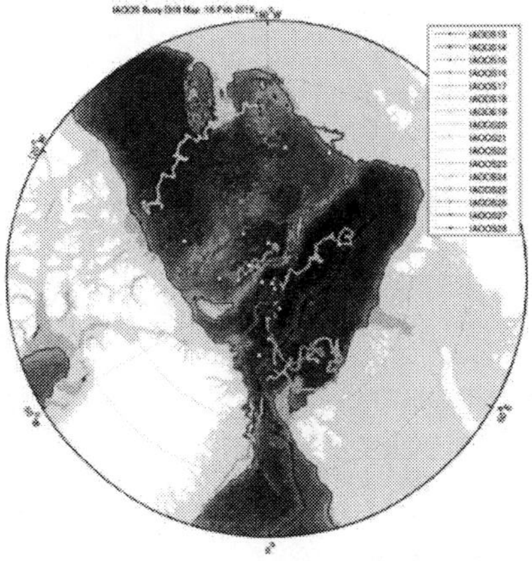

Figure 2 : Trajectories of all deployed IAOOS platforms over high Arctic since 2013, showing the transpolar drift of sea-ice. Depth of Artic ocean correspond to shades of gray. Land is reported as white areas (courtesy N. Sennéchael, LOCEAN/SU).

The sea-ice drift is such as most of the ice exits through the Fram straight between Greenland and Svalbard Islands. Some moves to the Beaufort gyre north of Canada in western Arctic. This can be seen in Fig. 2 presenting the trajectories of all the buoys that have been deployed. Platforms are usually destroyed when they reach the limit of ice

https://doi.org/10.1051/epjconf/202023705007

and ocean, in summer. Some have been crashed in ridges formed by ice floes, some by bears playing with them. Table 2 lists the deployments of platforms equipped with lidars since the beginning of the project, their operation and their duration of operation. Observations are taken over several months and some platforms have provided more than 1,5 year of data, some much less. The deployment is done during ocean cruises, from various ice-breakers in the frame of collaborative international efforts, most particularly with Korea (Araon), Norway (Lance), Germany (Polarstern) and from the North pole station (Barneo).

Platform name	Lidar type	Year	Deploy	Location	Operation
IAOOS 4	back	2014	Barneo	N. Pole	4/2014-11/2014
IAOOS 7	back	2015	N-ICE	Svalbard	1/2015-4/2015
IAOOS 9	Polar	2015	N-ICE	Svalbard	1/2015-4/2015
IAOOS10	back	2015	N-ICE	N. Pole	4/2015/6-2015
IAOOS 12	Polar	2015	Pstern	Siberia	-
IAOOS 16	back	2015	Araon	Siberia	9/2015-7/2016
IAOOS 20	back	2016	Barneo	N. Pole	10 days*
IAOOS 21	Polar	2016	Barneo	N. Pole	10 days*
IAOOS 23	back	2017	Barneo	N. Pole	4/2017-12/2017
IAOOS 24	back	2017	Araon	Siberia	4/2017-11/2017
IAOOS 25	back	2017	Barneo	N. Pole	8/2017-11/2018
IAOOS 26	back	2018	Barneo	N. Pole	4/2018-
IAOOS 27	back	2018	Barneo	N. Pole	4/2018-
IAOOS 28	back	2018	Oden	N. Pole	7/2018-8/2018*

Table 1. IAOOS Lidar Platforms deployed since 2014

Icing of the window remains an issue, due to the discontinuous operation, although heating is applied at the end of summer.

2. IAOOS ATMOSPHERIC SYSTEM

2.1 Lidar

The laser systems were designed to be eye-safe (regarding the NF EN 60825-1 norm). Energy was therefore limited, and high operation frequency (up to 10 kHz) has been selected. The time duration of the sequence was chosen as a compromise between the optimization of the signal to noise ratio, the atmospheric variability, and the sampling need. Vertical sampling resolution is 15m, but signals are further averaged before transmission (see Table 2). A typical 10 minutes averaging sequence for each profile was adopted. Four sequences a day were aimed at to allow a sampling of cloud properties compatible with inputs/outputs of meteorological model analyses.

The optical head was developed in France at INSU by Division Technique, who also designed the control system and interface. Integration of the lidar system was performed by the French company Cimel, that also designed the photon counting sampling and averaging electronics. Backscattering lidar systems can be implemented with and without polarization discrimination. A similar overall design is used for both systems, including a polarizing device and a polarization splitter at the emission and at the reception of the polarized system. A two-channel reception system is used for dual polarization signal analysis. Detectors are avalanche photodiode from Excelitas used in Geiger mode. Ancillary data are acquired such as those from accelerometers to inform about the tilt angles of the buoy linked to ridge formation and sea-ice breaking. All data are transferred to IPEV (http://iaoos.ipev.fr) using Iridium satellite link. To reduce data transmission costs, vertical averaging is performed in progressively increasing altitude bins (Table 2).

Parameters	Value
frequency / pulse width	~ 5 kHz / 200 ns
wavelength / bandwidth (FWHM)	~ 800 nm / < 0.6 nm
emitted energy / pulse	~ 2 µJ
diameter emission / reception lens	~ 70 mm
emission / reception full FOV	~ 650 µrad
detection filter bandwidth (FWHM)	~ 0.6 nm
overlap range (90%)	~ 300 m
Sampling vertical resolution Vertical resolution (transmitted data)	15m
	15 m [0-1 km] 30 m [1-3 km] 60 m [3-15 km] 120 m [15 -25 km]
background noise (average and standard deviation)	25 to 30 km

Table 2: Typical IAOOS lidar system parameters

Data processing is performed according to [3]. Saturation (or dead time limitation) is corrected for high backscattering signals (as due to clouds and background). Measurements have been made in laboratory using a specific procedure (varying the background level for a given pulse energy received) to check the dead time value. Results were close to that provided by the manufacturer and manufacturer's values are used. In case of dense cloud occurrence at a short distance oversaturation of the APD can be observed, with a strong decrease in signal (undershoot), biasing the cloud return signal. A polynomial fit has been used to correct this undershoot, allowing the recovery of tens of meters of useful signal in the cloud.

The average background signal B is then calculated in the upper channels (between 25 and 30 km). It is subtracted to the signal S(r) in each bin. Overlap function O(r) and calibration constant Cs can then be determined from clear air signal analysis (previously performed during tests

https://doi.org/10.1051/epjconf/202023705007

before expedition and from very clear air profiles observed in Arctic [3]). The attenuated backscatter is derived from the range-corrected signal as

$$\beta_{att}(r) = [S(r) - B].r^2 / [C_s.O(r)] \quad (1)$$

Altitude of cloud and aerosol layers can be determined directly, whereas forward lidar signal inversion is used to determine optical depths and scattering ratio or extinction profiles.

2.2 Met and radiation

As seen in Fig. 1 IAOOS platforms are equipped with several meteorological sensors (pressure, temperature, moisture and wind (from sonic anemometer)), and infrared brightness temperature measurement in the 8-12 µm region. Lidar is also used as a radiometer to measure the solar radiance scattered at nadir by the sky above. A large bandwidth solar radiometer is also mounted on some platforms. A low cost camera is used to take views of the platform environment during sunlight seasons.

3. IAOOS ATM. OBSERVATIONS

An example of time series obtained on the atmosphere, results obtained by the B24 platform set up at the North Pole in 2017 are shown in Figure 3. The variation of the lidar signal background is reported in the upper panel. It is a standard measure of IAOOS, used as a proxy for the downward diffuse solar flux. The sky's brightness temperature T_B (which indicates the presence of cloud when T_B is high) and the altitude-time cross-section of the attenuated backscattering signal. Areas of strong positive correlations (high particle scattering, high T_B, high diffuse flux) are seen, marking the presence of dense clouds (red rectangles). Areas of small scattering, low TB and small diffuse fluxes marking clear-air atmospheric cases (blue rectangles). Between the two, all types of cloud structures can be observed. This information is used for the determination of radiative flux at the surface and the analysis of the surface energy balance which leads to the melting of snow. The T_B measurements provide an additional constraint on the optical depth and type of the clouds. Lidar measurements show that the occurrence of super-cooled boundary layer clouds can be significantly high. They are predominating in spring, as observed for all deployment years. Aerosols are frequently observed in winter [2, 3].

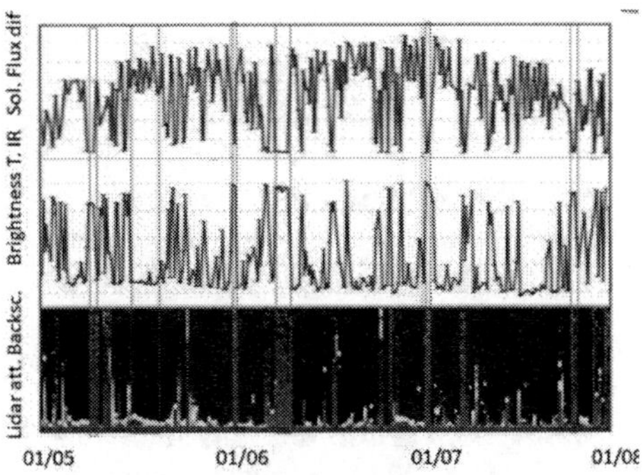

Fig. 3 : Time series of lidar background radiance (top), temperature brightness (middle) and lidar signal between 0 and 5 km (low) measured by platform B24 from the beginning of May to the end of July 2017, showing cloud and aerosol occurrence observed with the lidar and impact on radiation (see text).

4. AEROSOL TRANSPORT

Properties of aerosol have been analyzed over the high Arctic north of Svalbard between October 2014 and June 2015 using the new version 4 (V4)

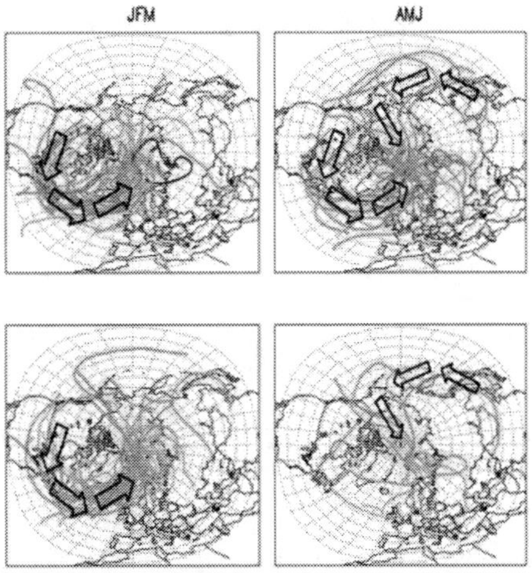

Fig. 4: Backward trajectories for desert dust (upper panel), and polluted dust (lower panel) subtypes in the periods of January to March 2015 (JFM, left) and April to June 2015 (AMJ, right). The color of the trajectories indicates the arrival altitude in the high Arctic: orange 2–5 km, and green 5–10 km). The arrows show principal transport patterns.

CALIPSO (Cloud and Aerosol Lidar and Infrared Pathfinder Satellite Observations) data [see AMT special issue]. Space–borne observations indicate a maximum in aerosol occurrence at the end of winter attributed to low–level (0-2km). Mid–tropospheric (2-5 km) particles are mostly identified by CALIOP (Cloud and Aerosol Lidar with Orthogonal Polarization) as highly

https://doi.org/10.1051/epjconf/202023705007

depolarizing. Systematic trajectories have been performed for all aerosol layers identified by CALIOP. The analysis of backward trajectories indicates that below 2 km aerosols originate mostly from Russia and Europe, while above 2 km significant sources are also North America and the Pacific areas. Fig. 4 reports trajectories for dust and polluted dust.

5. IAOOS AND CALIPSO OBSERVATIONS

Results from IAOOS drifting platforms have been compared to CALIOP observations for a few cases of CALIOP identified dust types. Fig. 5 shows an example of the attenuated scattering ratio (attenuated backscatter divided by the molecular backscattering derived from a model) profile obtained by IAOOS on February 14th, 2015. The aerosol content measured by IAOOS at 800nm was rather large

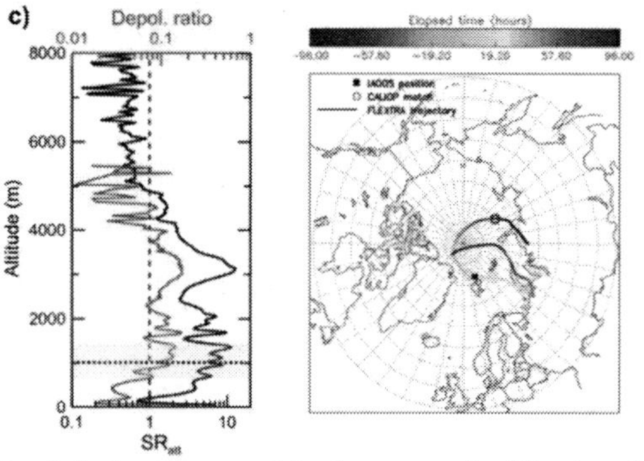

Fig. 5: Left:: Attenuated backscatter ratio (SRatt) and pseudo-depolarization ratio vertical profiles measured by IAOOS lidar at 800 nm on February 14th, 2015. The horizontal dotted line indicates the altitude at which the FLEXTRA backward and forward trajectories were performed, while the yellow band shows the vertical interval (±0.5 km) used for CALIOP data analysis. Right: Map showing the 4 days backward and forward FLEXTRA trajectories.

(IAB~7.3×10^{-3} sr^{-1}), and CALIOP identified the presence of aerosols both along backward and forward trajectories from the IAOOS position. On February 14th, the dominant subtypes identified by CALIPSO were dusty marine at 1 km and polluted dust at 3 km. Rather high-depolarization ratios (larger than 10%) in both layers are indicative of the presence of non-spherical particles. Assuming that clear air is observed on this IAOOS profile above 5 km, as suggested by an extended range of altitude where SR_{att} is low and almost constant, one can determine that the two-way transmission due to particle scattering and absorption is about 60% (see Fig. 5). The IAB retrieved for the 14 February IAOOS profile combined with the measured transmission allows to estimate an average IAOOS lidar ratio (LR) of 27(±6) sr at 800 nm, for the aerosol layer between 0 and 5 km, considering measurement uncertainties of ±10% for both transmission and IAB and multiple scattering correction [2]. It corresponds to a 532 nm lidar ratio of 29 (±8) sr assuming the spectral dependence proposed by [4]. This value better corresponds to ice clouds (LR~25–30 sr at 532 nm) possibly including clean continental particles (LR~35 sr). Pure desert dust, polluted dust or polluted particles/smoke, or Arctic haze having much larger LR values [4].

6. CONCLUSION

Observations provided by IAOOS platforms deployed in the Arctic allow to characterize super-cooled boundary layer clouds frequently observed in spring and elevated aerosol layers in winter. Comparisons with space-borne CALIOP observations lead to suggest that the aerosol classification in Arctic may be biased by the occurrence of diamond dust.

Acknowledgements: The IAOOS program has been developed in partnership with LOCEAN at SU and funded by the Agence Nationale de la Recherche (ANR-10-EQPX-32-01). This work was supported by funding from the ICE-ARC program from the EU FP7, grant 603887, ICE-ARC 072. We thank the DT/INSU/CNRS for the preparation of the buoys, deployment, and survey during N-ICE. The N-ICE2015 campaign was supported by the Norwegian Polar Institute' s Centre for Ice, Climate and Ecosystems (ICE)

REFERENCES

[1] Provost, C., et al. (2015), IAOOS (Ice-Atmosphere-Arctic Ocean Observing System, 2011–2019), Mercator Ocean Quart. Newsl., 51, 13–15. [http://www.mercator-ocean.fr/]

[2] Mariage, V., J. Pelon, F. Blouzon, S. Victori, N. Geyskens, N. Amarouche, C. Drezen, A. Guillot, M. Calzas, M. Garracio, N. Wegmüller, N. Sennéchael, and C. Provost (2017), IAOOS microlidar-on-buoy development and first atmospheric observations obtained during 2014 and 2015 arctic drifts, Optics Express, 25 (4), A73-A84.

[3] Di Biagio, C., Pelon, J., Ancellet, G., Bazureau, A., & Mariage, V. (2018), Sources, load, vertical distribution, and fate of wintertime aerosols north of Svalbard from combined V4 CALIOP data, ground-based IAOOS lidar observations and trajectory analysis.J. Geophys. Res., 123.

[4] Cattrall, C., Reagan, J., Thome, K., & Dubovik, O. (2005), Variability of aerosol and spectral lidar and backscatter and extinction ratios of key aerosol types derived from selected Aerosol Robotic Network locations. J. Geophys. Res, 110, D10S11.

LALINET NETWORK STATUS

Eduardo Landulfo[1*], Albeht Rodríguez Vega[2], Alexandre Calzavara Yoshida[1,3], Alvaro Bastidas[4] Amanda Vieira dos Santos[5], Antonieta Silva[6], Antonio Arleques Gomes[1], Boris Barja Gonzalez[7] Carlos Andrés Melo-Luna[9], Carlos D. Hoyos[9], Cristina Tobler de Souza Rae[10], Daniel Camilo[10], David Vivas[9], Diego Alves Gouveia[5], Elian Wolfram[11], Estela Collini[12], Fábio J. S. Lopes[1], Gregori de Arruda Moreira[1], Henrique de Melo Jorge Barbosa[5], Hernan Ciminari[12], John H. Reina[9], Jonatan João da Silva[1], Jonnathan Céspedes[9], Juan Carlos Antuña Marrero[2], Juan Luis Guerrero Rascado[14] , Juan Pallotta[11], Judith Hoelzemann[10], Lucas Alados Arboledas[14], M.Alejandra Salles[12], Marcos Paulo Araújo da Silva[10], Renata Sammara[10], Ricardo N. Forno[13], Sebastian Papandrea[12]

[1]Center for Lasers and Applications, Nuclear and Energy Research Institute, Av. Prof. Lineu Prestes, 2242, São Paulo, Brazil
[2]GOAC (INSMET/CUBA) - Cuba
[3]Instituto de Ciências Exatas e Naturais do Pontal, Universidade Federal de Uberlândia, ICENP/UFU - Ituiutaba, MG, 38304-402 – Brazil
[4]GLEO-UNALMED - Colombia
[5]Instituto de Física, Universidade de São Paulo, Rua do Matão 1371, São Paulo, SP, 05508-090, Brazil
[6]Departamento de Ciencias Físicas, Universidad de La Frontera, Temuco - Chile
[7]Atmospheric Research Laboratory (LIA), University of Magallanes (UMAG). Av. Bulnes 01855, 6210427, Punta Arenas - Chile
[8]Universidad Nacional de Colombia - SIATA (AMVA) - Colombia
[9]Centre for Bioinformatics and Photonics (CIBioFi), Universidad del Valle, Calle 13 No. 100-00, Edificio E20 No. 1069, 760032 Cali - Colombia
[10]Departamento de Ciências Atmosféricas e Climáticas - Universidade Federal do Rio Grande do Norte (DCAC/UFRN) – Brazil
[11]Unidad de Investigación y Desarrollo Estratégico para la Defensa. Juan Bautista de la Salle 4397, Villa Martelli, Buenos Aires, Argentina CP:B1603ALO
[12]National Meteorological Service (SMN), Av. Dorrego 4019, CABA, Argentina CP: C1425
[13]Laboratorio de Física de la Atmósfera, Universidad Mayor de San Andrés (UMSA) , Campus Universitario Calle 27 de Cota-Cota, La Paz Bolivia
[14]Andalusian Institute for Earth System Research (IISTA-CEAMA), University of Granada, Autonomous Government of Andalusia, Granada, 18006, Spain
[15]Department of Applied Physics, University of Granada, 18071, Granada, Spain
**Email: elandulf@ipen.br*

ABSTRACT

LALINET (Latin American Lidar Network) follows its goal to consolidation as a federative lidar network to provide regional coverage over Latin America in providing aerosol and greenhouse gas profiles following QA/QC protocols and promoting the development of researchers and students in atmopheric science field. We show recent results on different approaches for studying the optical properties of the atmosphere regarding aerosols at tropospheric and stratospheric level and greenhouse gas mixing ratio profiles followed by our recent support and validation efforts towards present and future satellite missions.

1. INTRODUCTION

LALINET became operational as a scientific network in 2013[1], as a member of GALION (GAW Lidar Observation Network) and covers mostly of South America and Central America. The scientific drives rely on observation of aerosol related events such as biomass burnings [2], volcanic eruptions [3], dust transport [4], aerosol-cloud interaction [5], as well GHG mixing ratio profiles [6] and related activities [7]. The majority of observations are at tropospheric level but strospheric observations are also available during special operation cycles. There are also efforts towards an algorithm consolidation following the strategy and goals of a robust database creation. As part of network organization a biannual workshop has been strictly followed in order to improve the network operational ability, to establish new scientific goals, to add more member stations to the network and foster education in atmospheric science.

2. METHODOLOGY

2.1 Network Stations and lidar systems

At the present there are 19 stations in LALINET, and 3 more proposed. They operate mostly with Nd:YAG laser typical wavelengths (1064, 532 and 355 nm) and some with Raman and or polarization detection capability, there is also one station for O_3 detection (RGL). Colocated with the lidar systems are sunphotometers, nearby radiossounding stations and other instruments.

Fig. 1 Lidar Station sites in LALINET. There are centres involved in data analyis in the network.

The schedule for operation is on Mondays and Thursdays but given the spread of the station along the continent this is not a tight contraint for measurements and preferably one run under coordinate condition only during extreme events such larg biomass burning activities, dust outbreaks and volcanic eruptions.

3. RESULTS

3.1. Tropospheric aerosols

Tropospheric aerosols are measured in all stations of LALINET, showing different aerosol type contributions to the sites. Long term monitoring has been conducted in PAR site, located in pristine environment of the Southern - hemisphere midlatitude. Usually clean and pristine conditions prevail at this site, but occasionally dust and biomass burning aerosol layers were observed [xx]. Figure 2 shows the attenuated backscattered coefficient for 1064 nm channel (upper panel) and the volume depolarization ratio for 532 nm (bottom panel) during two days of measurements at PAR during November 27 to 29, 2018.

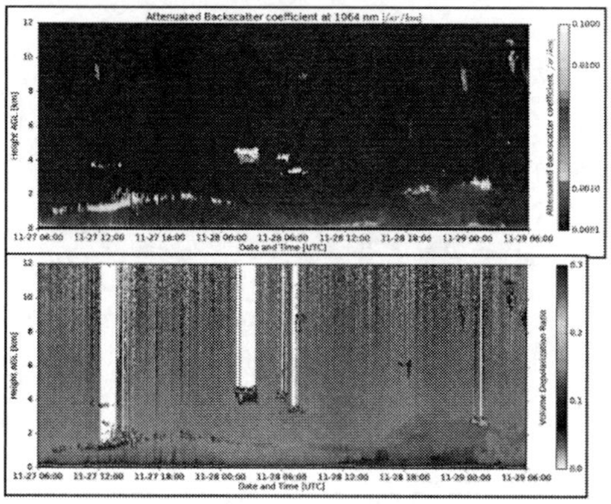

Fig 2. Attenuated Backscatter Coefficient for 1064 nm and Volume depolarization ratio for 532 nm at PAR station.

3.2 Stratospheric aerosols

A recent paper reports a multi instrument observational study of the aerosols from the April 2015 Calbuco volcanic eruption in Chile (41.33°S, 72.62°W). Combining the lidar and AERONET sun-photometer local at São Paulo, with MODIS, CALIPSO and OMPS_LP the mean features of the UTLS aerosols transported over São Paulo were characterized. It was relevant the good agreement between the measurements from different instrument. In the case of coincident profiles of São Paulo lidar and OMPS_LP, the disagreement found was explained by the inhomogeneity of the UTLS aerosol layer [3]. LALINET is also coordinating its contribution to VOLRES (Volcanic Response), an initiative from SSiRC (Stratospheric Sulfur and its Role in Climate) a SPARC activity. Locations hosting LALINET lidars were evaluated to be fitted for balloon and sounding launching following technical requirements. Six potential LALINET

https://doi.org/10.1051/epjconf/202023705008

sites has been identified: NQN, BRC, VMA, SMN, SPU and PAR, thus preparing Latin America to respond to large volcanic eruptions [8].

3.2 Clouds

Intensive measurement campaign in different stations allowed to identify cirrus clouds over a delimited height range between 7.0 to 18.0 km where they generally ocurred. Geometrical and optical properties allow a primary classification based on the average base and top of cirrus clouds and accounted their morphologial variety. Collocated data from radiosounding lauchings were used to observe relative humidity profiles.

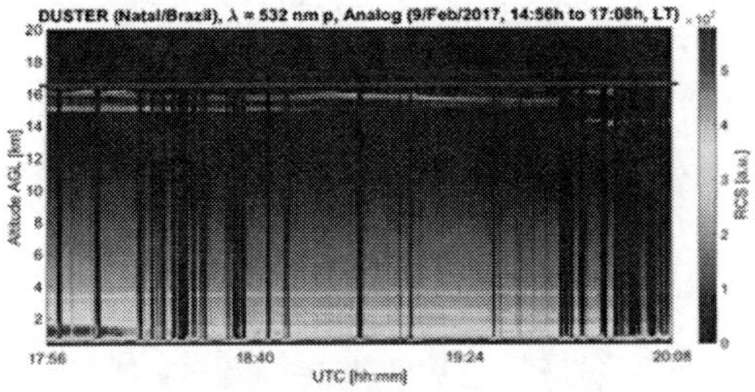

Fig 3. Cirrus cloud observation at NAT station. The presence of low clouds made a special operation schedule to be performed.

It has been derived also the first comprehensive statistics of optical and geometrical properties of cirrus clouds in Amazonia and estimates of its radiative forcing. For that, 1 year of ground-based lidar observations in the central Amazon forest, Brazil. The retrieved optical and geometrical properties (published by [5]) were then considered to calculate the cirrus radiative forcing (CRF) and heating rates profiles computed in detail by the radiative transfer models. The MAO cirrus frequency of occurrence was found to be as high as 88% during the wet season and not lower than 50 % during the dry season, with mean values of cirrus cloud top and base heights, cloud thickness and cloud optical depth were 14.3 ± 1.9 (std) km, 12.9 ± 2.2 km, 1.4 ± 1.1 km, and 0.25 ± 0.46, respectively. With such high frequency of occurrence and altitude over the dark-pristine Amazon forest (albedo $\alpha \approx 0.12$), these clouds produced a net CRF at the top and bottom of the atmosphere (TOA and BOA) of +14.7 and -3.6 Wm-2, respectively. Together, the vertical profiles with total optical depth (column integrated COD) > 0.3 was responsible for about 72%(62%) of the TOA(BOA) net CRF.

3.3 Greehouse gas profiles – O_3

Height-resolved ozone measurements over RG station Gallegos were measured with a differential absorption lidar (DIAL) instrument. Llidar profiles are carried out in November 2009 have a duration that ranges from 3 –5 h. Ozone densities as a function of altitude for selected days are shown within a dynamic range from 14.5 to 45 km. An austral O_3 climatology in mixing ratios was possible together with satellite and ozonesonde observations during the 1980–1991 period and is available for 17 latitudinal bands from 80 to 80 degrees spacing at 17 pressure levels between 100 and 1 hPa. Complete description a this atmospheric event intrusion of polar airmass with depleted ozone content over RG station can be found in Wolfram el al [9].

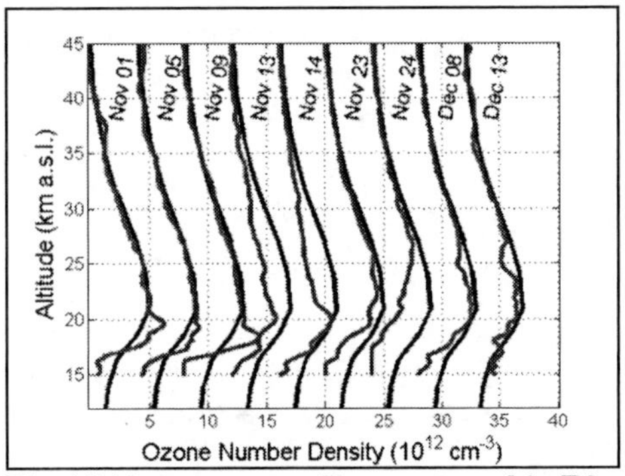

Fig 4 Lidar ozone profiles measured in Rio Gallegos sites (blue lines) from 1 November to 13 December 2009.

3.4 Planetary Boundary Layer

The PBL Height estimation methods in the network database are performed from several algorithmsIn the last five years many campaigns about validations of proposed methodologies have been performed. Among them, the MCity Campaign stands out, due to comparison between radiosonde and lidar data performed during 20 days consecutively (figure The micrometeorological studies are concentrated in the analysis of high-order moments (until fourth order) from elastic lidar data. A good example is presented from Cali Station [20] in Colombia where GM and WCT methods are employed.

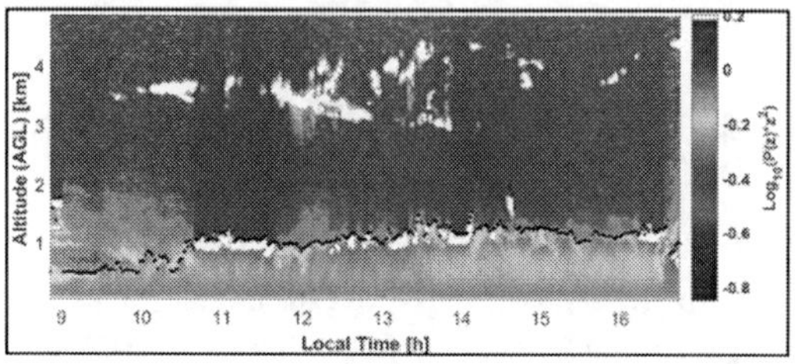

Fig 4. Cali Station temporal evolution of range corrected signal in the period 08:30 – 17:00 December 13th, 2018. The scatter plot represent PBL top.

3.5 Satellite validation

In order to support present and future satellite mission LALINET is working on increase the accuracy of aerosol measurements and improve the network observation capability by the implementation of more rigorous and systematic protocols. The ultimate goal is to develop a robust database of the vertical distribution of aerosols over South America. The more recent effort is LALINET's paticipation in the ESA AEOLUS mission validation by performing correlative measurements as closa as possible in time to Aeolus overpasses and establish a timeframe to make observations based on the station-overpass distance, e.g., a 2h-period for 40 km distances and a 4 h-period for 100 km.

ACKNOWLEDGEMENTS

The authors would like to thank for the support given by ILRC 2019 committee. A.V.S., H.B. and E.L. would like to thank FAPESP for the research grants: 17/23810-9, 13/50510-5, 17/17047-0 and 15/12793-0, respectively. F.S., D.G. and G.A. would like to thank for CNPq, F.S. was under grant numbers: 152156/18-6, 432515/18-6 and 1507116/17-6. B.B would like to thank FONDCYT for research grant 11181335. C.R. would like to thank Area Metropolitana del Valle de Aburrá for the Science and Technology grant number 511/2017.

REFERENCES

[1] Antuña-Marrero, J.C., E. Landulfo, R. Estevan, B. Barja, A. Robock, E. Wolfram, P. Ristori, B. Clemesha, F. Zaratti, R. Forno, E. Armandillo, Á.E. Bastidas, Á.M. de Frutos Baraja, D.N. Whiteman, E. Quel, H.M. Barbosa, F. Lopes, E. Montilla-Rosero, and J.L. Guerrero-Rascado, LALINET: The First Latin American–Born Regional Atmospheric Observational Network. *Bull. Amer. Meteor. Soc.*, **98**, 1255–1275, (2017)

[2] A. G. Guedes, E. Landulfo, E. Montilla-Rosero, F. J.S. Lopes, J. J. Hoelzemann, J. H. Fernandez, M. P.A. Silva, R.S.S. Santos, J.L. Guerrero-Rascado, L. Alados-Arboledas, Detection of saharan mineral dust aerosol transport over brazilian northeast through a depolarization lidar, EPJ Web of Conferences, 2018

[3] J. S. Lopes, F.; Silva, J.J.; Antuña Marrero, J.C.; Taha, G.; Landulfo, E. Synergetic Aerosol Layer Observation After the 2015 Calbuco Volcanic Eruption Event. Remote Sens. 2019, 11, 195.

[4] F.J.S. Lopes, J. L. Guerrero-Rascado, J.A. Benavent-Oltra, R. Román, G.A. Moreira, M.T.A. Marques, J. J. da Silva, L.Alados-Arboledas, P. Artaxo, E. Landulfo, Rehearsal for Assessment of atmospheric optical Properties during biomass burning Events and Long-range transportation episodes at Metropolitan Area of São Paulo-Brazil (RAPEL), , EPJ Web of Conferences, 2018

[5] Gouveia, D. A.; Barja, B.; Barbosa, H. M.; Seifert, P.; Baars, H.; Pauliquevis, T.; Artaxo, P. Optical and geometrical properties of cirrus clouds in Amazonia derived from 1 year of ground-based lidar measurements. Atmospheric Chemistry and Physics, v. 17, n. 5, p. 3619–3636, 2017.

[6] E. A. Wolfram, J. Salvador, F. Orte, R. D'Elia, S. Godin-Beekmann, J. Kuttippurath, A. Pazmino , F. Goutail, C. Casiccia, F. Zamorano, N. Paes Leme, and E. J. Quel, The unusual persistence of an ozone hole over a southern mid-latitude station during the Antarctic spring 2009: a multi-instrument study, Ann. Geophys., 30, 1435–1449, 2012

[7] G. A. Moreira, F.J.S. Lopes, J.L. Guerrero-Rascado, J.J. Silva, A. A. Gomes, E. Landulfo, and L. Alados-Arboledas. Analyzing the Atmospheric Boundary Layer by high-order moments obtained from multiwavelenght lidar data: impact of wavelength choice. *Submitted, 2019.*

[8] Thomason, L. and Kremser, S., 2017, SSiRC Meeting at ISSI, SPARC Newsletter, No. 49,
July 2017, pp. 26 – 29.

[9] Wolfram, E. A., Salvador, J., Orte, F., D'Elia, R., Godin-Beekmann, S., Kuttippurath, J., Pazmiño, A., Goutail, F., Casiccia, C., Zamorano, F., Paes Leme, N., and Quel, E. J.The unusual persistence of an ozone hole over a southern mid-latitude station during the Antarctic spring 2009: a multi-instrument study,: Ann. Geophys., 30, 1435-1449, doi:10.5194/angeo-30-1435-2012, 2012.

[10] J. Céspedes, C. A. Melo-Luna, J. H. Reina, A first measurement of the planetary boundary layer top in santiago de cali-colombia: elastic lidar application, in: Remote Sensing of Clouds and the Atmosphere XXIII, Vol. 10786, International Society for Optics and Photonics, 2018, p.107860L

LONG-RANGE-TRANSPORTED MINERAL DUST FROM AFRICA AND MIDDLE EAST TO EAST ASIA OBSERVED WITH THE ASIAN DUST AND AEROSOL LIDAR OBSERVATION NETWORK (AD-NET)

Nobuo Sugimoto[1*], Atsushi Shimizu[1], Tomoaki Nishizawa[1], Yoshitaka Jin[1], and Keiya Yumimoto[2]

[1]National Institute for Environmental Studies, 16-2 Onogawa, Tsukuba, Ibaraki 305-8506 Japan
[2]Research Institute for Applied Mechanics, Kyushu University, Kasuga, Fukuoka, 816-8580 Japan
**Email: nsugimot@nies.go.jp*

ABSTRACT

Mineral dust generated in Africa and Middle East is sometimes transported to East Asia. Some cases were observed with the Asian Dust and aerosol lidar observation Network (AD-Net). In the large Sahara dust event in March 2018, which was reported by mass media as that snow in Sochi, Russia was stained into orange, the dust was transported to Sapporo, Japan in 4 days from Sochi and observed with the AD-Net lidar. Sahara dust events were also observed in April 2017 and April 2018 with AD-Net. The source areas and transport paths were studied with chemical transport models and trajectory analysis and also confirmed with CALIPSO data. This study showed that long-range transport from Africa and Middle East to East Asia is not rare in springtime.

1. INTRODUCTION

Continuous lidar network observations have been conducted since 2001 in AD-Net. Currently, lidars are continuously operated at 20 locations in East Asia (Japan, Korea, China, Mongolia and Thailand). The major targets of the observations are Asian dust and regional air pollution aerosols in the lower atmosphere [1].

The source areas of major dust phenomena in East Asia are the Gobi desert in Mongolia and Chinese Inner Mongolia, and the Taklamakan desert in China. Dust from the Gobi is transported near the surface and significantly affect the environment of the lower atmosphere [2, 3]. It is often mixed with air pollution [4, 5]. Dust from Taklamakan is often uplifted and transported long distance in the free troposphere [6]

We reported a Sahara dust case observed with the AD-Net lidars in 2005 [7]. We observed lofted dust layer in Suwon and Gosan in Korea and Tsukuba in Japan which was not reproduced by a regional dust transport model. We compared with global aerosol transport models and analyzed with back trajectory analysis, and we found the dust was transported from Africa. We finally confirmed the source region by retrospective simulation separately masking Sahara and Taklimakan. The extinction coefficient estimated at the dust plumes in Gosan and Tsukuba was ~0.05 km^{-1} at the maximum. After that, we did not pay much attention to long-range-transported dust cases, because our major targets were heavy dust events and regional air pollution.

In 2015-2016, Leibniz Institute for Tropospheric Research (TROPOS) conducted continuous observation with a multi-wavelength Raman lidar in Dushanbe, Tajikistan in Central Asia [8], and they recently started long-term observation. That motivated us to study long-range-transported dust cases again using recent AD-Net data. It would be interesting if we could observe the same air mass along the transport path to study the change in optical characteristics of dust. Also, recent studies of bioaerosols (microbes attached to dust particles) suggest such long-range transport might be important even if the density of the transported dust is low.

2. METHODOLOGY

We used the AD-Net two-wavelength (1064 nm and 532 nm) and polarization sensitive (532 nm) lidar data (http://www-lidar.nies.go.jp/AD-Net/). We firstly used archived results of NAAPS (Navy Aerosol Analysis and Prediction System) of the U.S. Naval Research Laboratory (https://www.nrlmry.navy.mil/aerosol/) to find possible long-range-transported dust cases to East Asia. We then searched corresponding plumes in AD-Net data. When we found corresponding dust plumes, we performed backward trajectory analysis using HYSPLIT (Hybrid Single-Particle Lagrangian Integrated Trajectory model) of U.S. National Oceanic and Atmospheric

 https://doi.org/10.1051/epjconf/202023705009

Administration (https://ready.arl.noaa.gov/HYSPLIT.php). We also confirmed dust transport using the CALIPSO browse data when available. (https://www-calipso.larc.nasa.gov)

3. RESULTS

3.1 Sahara dust case in March 2018

It was reported by mass medias (BBC, CNN, etc.) that dust from Sahara stained snow in Sochi, Russia into orange on March 22, 2018. We observed the transported dust plume by the lidar in Sapporo on March 26 as shown in Fig.1.

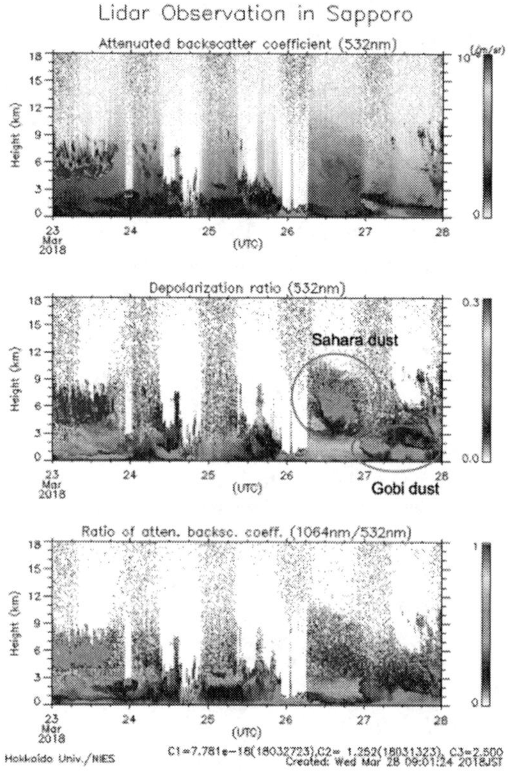

Fig.1 Time-height indications of attenuated backscatter coefficient at 532 nm (top), volume depolarization ratio at 532 nm (middle), and color ratio of attenuated backscatter coefficients at 1064 nm to 532 nm (bottom), observed in Sapporo, Japan in March 2018.

Figure 2 show the HYSPLIT backward trajectory of the plume. It shows the trajectory exactly passed over Sochi on March 22. The plume was transported through Kazakhstan and Russia.

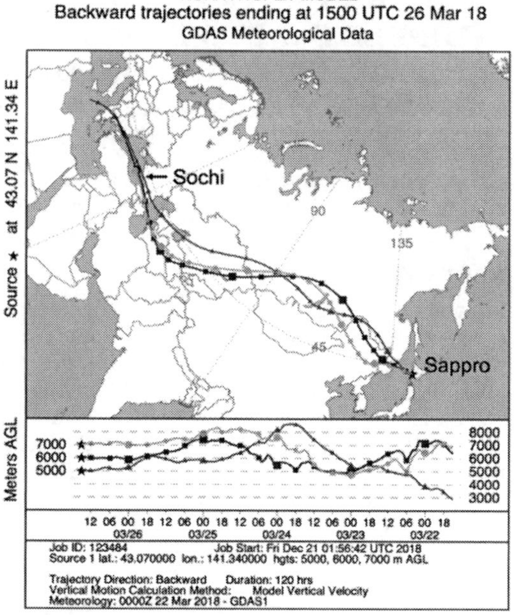

Fig.2 HYSPLIT backward trajectory of the Sahara dust plume indicated in Fig. 1.

The extinction coefficient of the plume derived with the Fernald method with an assumption of S1=50 sr was ~0.06 km^{-1} at the maximum. The particle depolarization ratio was ~0.28. Optical depth of the plume estimated from the lidar extinction coefficient was ~0.3 at the maximum. Backscattering color ratio ($\beta_{1064nm}/\beta_{532nm}$) of the plume was about 0.4. It is very low compared to that in strong Gobi dust cases (~1.0). This suggests the particle size of the long-range-transported Sahara dust was small, or the plume was mixed with other aerosols such as sulfate. The result of NAAPS showed however the mixing was not likely. Though it is not possible to estimate the mass concentration quantitatively because the size distribution is not known, in a very rough estimation it was several tens microgram per cubic meter. It was high as a long-range-transported dust. Also, the dust layer stayed above Sapporo for a half day.

3.2 Sahara dust case in April 2017

Figure 3 shows another example of Sahara dust observed in Sainshand, Mongolia. A lofted dust plume is seen on April 27-28, 2017, and it looks a part of the dust plume was taken into the boundary layer and reached the ground. Estimated extinction coefficient of the dust plume was ~0.05 km^{-1}, and particle depolarization ratio was ~0.30. Backward trajectory of the dust plume is shown in Fig.4. Judging from the NAAPS archived results, the emission source of the dust was in Africa. Dust was transported through Kazakhstan and Russia and entered into Mongolia from the northwest. The transport pass was similar to that of March 2005 case [7].

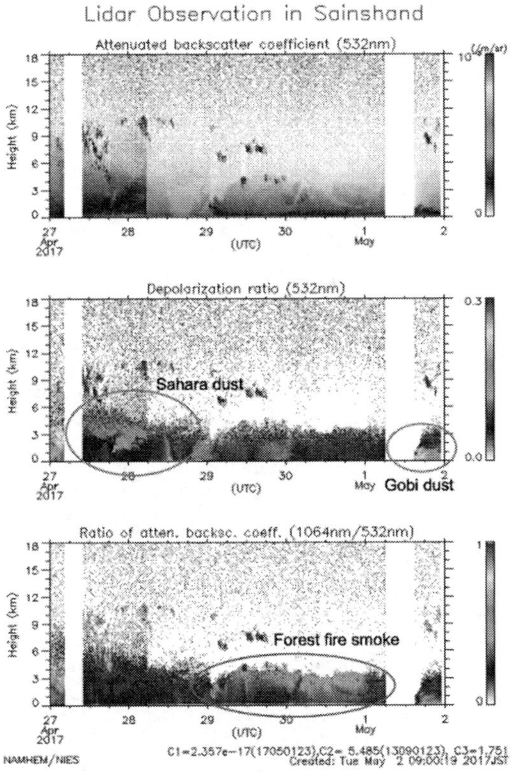

Fig.3 Time-height indications of attenuated backscatter coefficient at 532 nm (top), volume depolarization ratio at 532 nm (middle), and color ratio of attenuated backscatter coefficients at 1064 nm to 532 nm (bottom), observed in Sainshand, Mongolia in April 2017.

An interesting structure is seen in the color ratio of attenuated backscattering coefficient on April 29 and 30 in Fig.3. The NAAPS results indicate there was forest fire smoke in the boundary layer. The color ratio increased with height because the attenuation at 532 nm was larger than at 1064 nm.

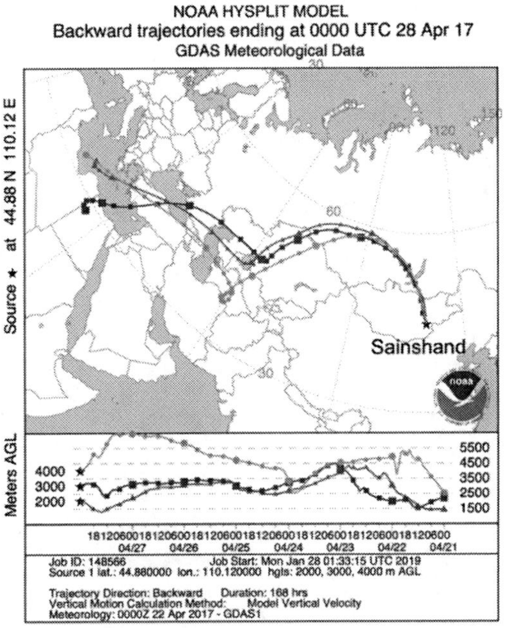

Fig.4 HYSPLIT backward trajectory of the Sahara dust plume indicated in Fig. 3.

3.3 Middle East dust cases

We also identified several cases of long-range transport of dust from Middle East. A good example is the case reported by Hofer et al. [8]. They observed a lofted dust layer over Dushanbe, Tajikistan on April 12, 2015. We observed probably the same dust plume in Nagasaki on April 17.

Figure 5 shows backward trajectory arriving at Nagasaki on April 17. The trajectory shows the plume passed near Dushanbe on April 12. The dust was transported across the high mountain in the west of the Taklamakan desert and across the Taklamakan desert.

Another example of Middle East dust was the plume observed in Sendai, Japan on April 27, 2015. In this case the dust was transported through the north side of Mongolia similar to the transport path in Fig.2.

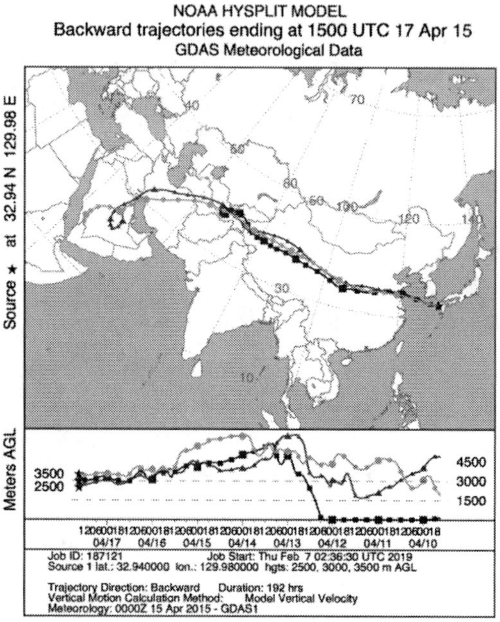

Fig.5 HYSPLIT backward trajectory of the dust observed in Nagasaki on April 17, 2015.

4. Discussion

Dust from Africa and Middle East are transported to East Asia every year in the springtime (mostly in March and April). The transport path arriving in East Asia often passes through Kazakhstan and Russia. Dust plumes transported across the high mountain in the west of the Taklamakan desert were also observed. In such cases, however, long-range-transported dust is mixed with the dust from the Taklamakan and the Gobi.

To understand dust emission source and mixing of dust from different sources during transport, we require a global dust transport model having source identification tags.**

In seasons other than spring, no significant long-range-transported dust cases from Africa and Middle East to East Asia have been observed, so far, after 2015. The heavy Central Asia dust events observed in Dushanbe in summer reported by Hofer et al. [8] were not transported to East Asia. Interestingly, those events themselves were not well reproduced in the global aerosol transport models (NAAPS and SPRINTARS).

We realized through this work that Mongolia is a good location also for studying characteristics of long-range-transported dust from Africa and Middle East. We think it is reasonable to improve one of the AD-Net lidars in Mongolia to a multi-wavelength Raman lidar or HSRL. Also, it will be very useful if we have an observation site in Kazakhstan.

**Comparison with the global chemical transport model MASINGAR separately calculated for different dust source regions was recently published in Ref.(9).

ACKNOWLEDGEMENTS

We would like to thank Teppei J. Yasunari, Dashdondog Batdorj, and all other people involved in AD-Net for their contributions in the lidar observations.

REFERENCES

[1] A. Shimizu, T. Nishizawa, Y. Jin, S.-W. Kim, Z. Wang, D. Batdorj, N. Sugimoto, Opt. Eng. 56 (3), 031219 (2017)
[2] N. Sugimoto, I. Uno, M. Nishikawa, A. Shimizu, I. Matsui, X. Dong, Y. Chen, H. Quan, Geophys. Res. Lett. 30 (12), 1640 (2003)
[3] A. Shimizu, N. Sugimoto, I. Matsui, K. Arao, I. Uno, T. Murayama, N. Kagawa, K. Aoki, A. Uchiyama, A. Yamazaki, J. Geophys. Res. 109, D19S17 (2004)
[4] N. Sugimoto, T. Nishizawa, A. Shimizu, I. Matsui, H. Kobayashi, J. Quant. Spectrosc. Radat. Trans. 150 , 107 (2015)
[5] X. Pan, I. Uno, Z. Wang, T. Nishizawa, N. Sugimoto, S. Yamamoto, H. Kobayashi, Y. Sun, P. Fu, X. Tang, Z. Wang, Sci. Rep. 7, 335 (2017)
[6] I. Uno, I., K. Eguchi, K. Yumimoto, T. Takemura, A. Shimizu, M. Uematsu, Z. Liu, Z. Wang, Y. Hara, N. Sugimoto, Nature Geosci. 2, 557 (2009)
[7] C.-B. Park, N. Sugimoto, I. Matsui, A. Shimizu, B. Tatarov, A. Kamei, C.-H. Lee, I. Uno, T. Takemura, D. L. Westphal, SOLA 1, 121 (2005)
[8] J. Hofer, D. Althausen, S. Abdullaev, et al., Atmos. Chem. Phys., 17 14559 (2017)
[9] N. Sugimoto, Y. Jin, A. Shimizu, T. Nishizawa, K. Yumimoto, SOLA 15, 257 (2019)

SAHARAN DUST EVENTS OVER THE NORTHERN MEDITERRANEAN: 4 YEARS OF MEASUREMENTS OVER 4 EARLINET STATIONS

Ourania Soupiona[1*], Alex Papayannis[1], Maria Mylonaki[1], Nikolaos Papagiannopoulos[2], Pablo Ortiz-Amezcua[3,4], Rodanthi E. Mamouri[5], Silke Groß[6], Christina A. Papanikolaou[1], Panagiotis Kokkalis[7]

[1]*Applied Mathematics and Physical Sciences, National Technical University of Athens, 15780, Greece*
[2]*Consiglio Nazionale delle Ricerche, Istituto di Metodologie per l'Analisi Ambientale, Tito Scalo, 85050, Italy.*
[3]*Andalusian Institute for Earth System Research, 18006, Granada, Spain*
[4]*Department of Applied Physics, University of Granada, 18071, Granada, Spain*
[5]*Department of Civil Engineering and Geomatics, ERATOSTHENES Research Centre, Cyprus University of Technology, 3036, Limassol, Cyprus*
[6]*Institute of Atmospheric Physics, Deutsches Zentrum für Luft- und Raumfahrt (DLR), Oberpfaffenhofen, 82234, Germany*
[7]*Department of Physics, Kuwait University, P.O. Box 5969, Safat 13060, Kuwait*
**Email: soupiona.rania@gmail.com*

ABSTRACT

Four years (2014-2017) of observations of depolarization Raman Lidar systems of four EARLINET (European Aerosol research Lidar Network) stations [from West to East: Granada (Spain), Potenza (Italy), Athens (Greece) and Limassol (Cyprus)] were collected and used to a statistical analysis of Saharan dust events over Mediterranean basin.

In this study, emphasis is given to the consistency of the particle linear depolarization ratio (δ_{p532}), the extinction-to-backscatter ratio mentioned as Lidar Ratio (LR_{532}) and the Aerosol Optical Thickness (AOT_{532}) within the observed Saharan dust layers, corresponding to the visible range (532 nm). Geometrical properties and clusters of aerosol mixtures are also presented. Our clustering was based on previous classification by airborne High Spectral Resolution Lidar (HSRL) observations and was further supported by backward trajectory analysis. We found mean δ_{p532} values of 0.24±0.05, 0.26±0.06, 0.28±0.05 and 0.28±0.04, mean LR_{532} values of 52±8 sr, 51±9 sr, 52±9 sr and 49±6 sr, mean AOT_{532} values of 0.40±0.31, 0.11±0.07, 0.12±0.10 and 0.32±0.17 and mean layer thicknesses of 3392±1458 m, 2150±1082 m, 1872±816 m and 1716±567 m for Granada, Potenza, Athens and Limassol respectively.

This work could assist in bridging the existing gaps related to the extensive and intensive dust aerosol properties over the Mediterranean and enriching the bibliography about mixed aerosol layers from different sources (e.g. dust and biomass burning (BB) aerosols, dust and urban/ industrial aerosols).

1. INTRODUCTION

The Mediterranean basin plays a crucial role in different origin air mass cross paths and transitions since it covers portions of three continents: Europe, Asia and Africa. Depending on the season and the area of the observation, dust can originate from different places of the Saharan and Asian region. Desert dust is one of the major natural aerosol sources while Sahara is one of the biggest deserts all over the world.

Saharan dust advections to Mediterranean countries are modulated by meteorology along rather regular seasonal patterns. In the Western Mediterranean (Granada), the African dust occurrence is higher in summer [1]. In Central Mediterranean (Potenza), spring and summer are associated to dust aerosol loads extending up to altitudes of 3–4 km [2]. In Eastern Mediterranean (Athens and Limassol), the main dust transport occurs from spring to autumn [3, 4, 5] as a result of the high cyclonic activity over northern Africa during this period [6].

Mineral dust may highly influence the atmospheric radiation balance as well as the cloud nucleation, formation and lifetime, through scattering and absorption processes [7]. Considerable uncertainties in quantifying the global direct radiative effects of aerosols arise from the variability of aerosols' spatial distribution and aging/mixing processes that can affect their optical and microphysical properties.

Light detection and ranging (Lidar) instruments are among the most powerful and suitable tools for retrieving vertically the aerosol optical properties with high temporal and spatial resolution [8, 9, 10]. Up to now, there are a lot of studies about Saharan dust optical properties [3, 4, 10, 11, 12], however many Lidar stations did not have depolarization sensitive channels until a few years ago, for long-term statistical analysis.

In this study, we introduce the four EARLINET stations and present the data selection and methodology in Section 2. In section 3, we present our results, while in Section 4 our conclusions are shortly presented.

2. METHODOLOGY

2.1 The four EARLINET Lidar systems

Four EARLINET stations have been considered in this study, as representative of mineral dust events over the Northern Mediterranean basin, due to their location close to the African continent. The corresponding institutes, along with their location and Lidar setup, are presented in Table 1. In order to provide homogeneous Lidar data and results we used a common study period (2014-2017) and the vertical profiles of the 532 nm backscatter (β), extinction (α) and depolarization (δ) signals, only. Raman-Lidar (nighttime) observations were used in order to provide the LR_{532}, and depolarization (daytime or nighttime) observations were used to provide the δ_{p532}.

2.2. Data selection and analysis

The total number of dust cases per station for the period 2014-2017 used here are 15 (Granada), 18 (Potenza), 12 (Athens) and 6 (Limassol). For the region of Cyprus the situation is more complex since Middle East dust outbreaks occur frequently [13, 14]. Taking that into account, the selection of the Saharan dust events over Limassol should be handled with care.

Since pure dust layers (δ_{p532}=0.31±0.02, [15]) are rare over urban Mediterranean cities due to urban, polluted or even marine aerosols and their mixtures, a sufficiently lower δ_{p532} threshold value should be considered for long-range transported dust aerosol layers in our data set. The LR_{532} values are expected to range between 40 and 60 sr with some exceptions/deviations.

Consequently, for the total set of dust events over each station, we considered for further analysis only cases that meet two basic criteria: i) the δ_{p532} to exceed 0.15 in the free troposphere and ii) the thickness of the detected layer to be 500 m, at least. The critical height (in meters asl) in which the first criteria was met, was considered to be the base of the layer. This assumption was deemed necessary to be made since usually, the lofted dust layers cannot be distinguished from the top of the Atmospheric Boundary layer (ABL), while the presence of urban haze takes low δ_p values (0.03-0.10). The top of the dust layer was estimated as the height where the signals were similar to the molecular scattering in the free troposphere. For some cases of Athens station, where only Raman-Lidar measurements were available, the values of the base and top were calculated from the Lidar signals, following the procedure proposed by [11]. Thus, the number of the total cases of each station that fulfills the aforementioned criteria per optical property may differ (see Fig.2). For each case, mean δ_{p532}, mean LR_{532} and mean AOT_{532} values were calculated inside the layers. The corresponding standard deviation (sd) values give an indication of the variability of these parameters from base to top within the layer.

Moreover, we performed a careful investigation of the air mass origin and long-range dust aerosol transport by means of backward trajectory analysis. This analysis was carried out using the HYbrid Single-Particle Lagrangian Integrated Trajectory (HYSPLIT) model (http://www.arl.noaa.gov /HYSPLIT.php).

3. RESULTS

Statistics of the dust aerosol optical properties

The backward trajectories from HYSPLIT per station are shown in Fig.1 revealing the origin of each observed layer. In the majority of the cases, air masses originate from the west and northwest Africa (Morocco, Mauritania, Algeria and Tunisia).

Fig.2 shows the mean values of the optical (δ_{p532}, LR_{532}, AOT_{532}) and geometrical (base, top, thickness) properties per station and per event. For all stations no large variations in the mean values of each parameter are observed with Granada to hold the minimum values for base and the maximum for top and layer's thickness. The mean δ_{p532} for Athens and Limassol is shifted a bit to higher values, while no remarkable variations in LR_{532} values were observed (51±1 sr) among the four stations.

Based on the HSRL classification by [16] we plotted the intensive aerosol quantities, LR_{532} versus δ_{p532} (Fig. 3a), and we identified three of the six existing clusters in our data. Our first cluster (green marks/errorbars, green trajectories, Fig.3b) represents a mixing state of Saharan dust and BB aerosols.

Table 1: Institute, location and Lidar setup of the selected EARLINET stations providing data for this study.

Station	Abbreviation	Location	Lidar setup	References
Andalusian Institute for Earth System Research, University of Granada, Spain	IISTA-CEAMA, Gr	37.16° N, -3.61° E, elev. 680 m	MULHACEN 3β+2α+1δ(532 nm)	[17]
Consiglio Nazionale delle Ricerche – Istituto di Metodologie per l'Analisi Ambientale, Potenza, Italy	CNR-IMAA, Po	40.60° N, 15.72° E, elev. 760 m	MUSA 3β+2α+1δ(532 nm)	[18]
Laser remote Sensing Unit, National Technical University of Athens, Greece	LRSU-NTUA, At	37.96° N, 23.78° E, elev. 212 m	EOLE/AIAS 3β+2α+1δ(532 nm)	[19]
Cyprus University of Technology, Limassol, Cyprus	CUT, Lm	34.67° N, 33.04° E, elev. 10 m	Polarisation Raman Lidar, 1β+1α+1δ, (532 nm)	[20]

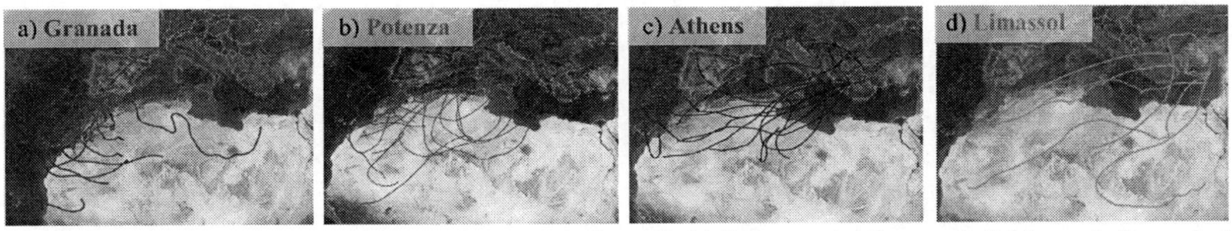

Figure 1: 96-120 hour backward trajectories over a) Granada, b) Potenza, c) Athens and d) Limassol, for arrival heights of approximately the center of each layer of all selected cases from the period 2014-2017.

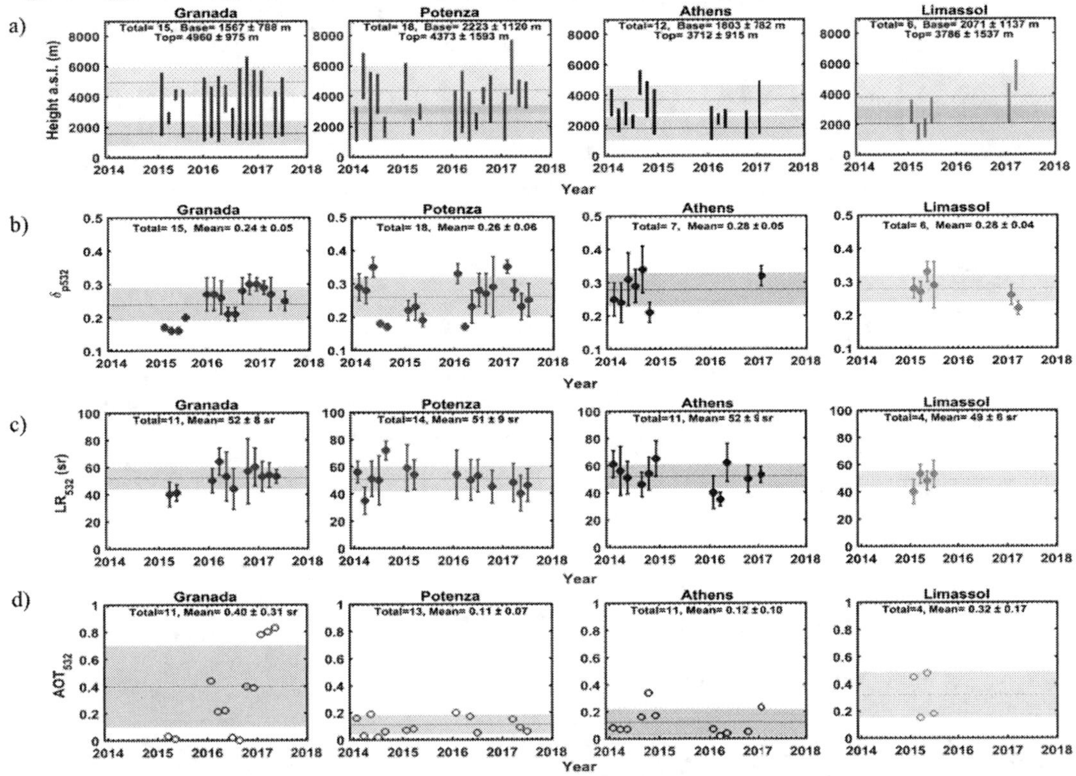

Figure 2: Mean values of a) base and top, b) δ_{p532}, c) LR_{532} and d) AOT_{532}, per station (text and banded lines) and per case (symbols), along with their sd (error bars) inside the observed dust layers.

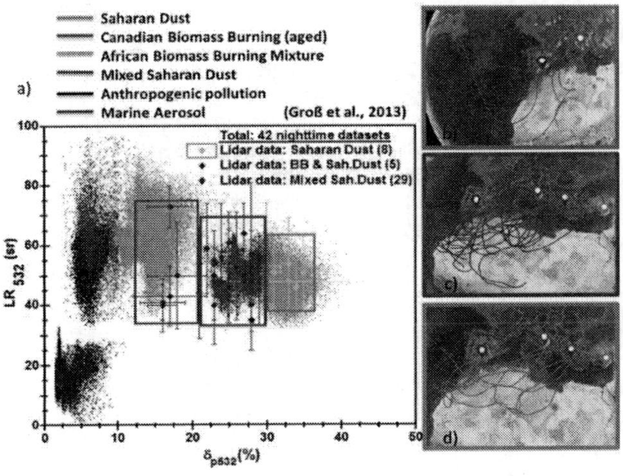

Figure 3: a) LR_{532} vs. δ_{p532} from HSRL observations (colored dots) along with the selected datasets from the four EARLINET stations (symbols and error bars) b) backward trajectories for BB & Saharan dust (green) cluster c) backward trajectories for mixed Saharan dust (red) cluster and d) backward trajectories for Saharan dust (orange) cluster.

The second cluster (red marks/errorbars, red trajectories, Fig.3c) is attributed to mixed Saharan dust, where dust aerosols are dominant, but BB, urban or marine aerosols are also possibly present. The third cluster (orange marks/errorbars, orange trajectories, Fig.3d) is in good agreement with former findings of pure Saharan dust layers [15, 16].

The 29 of the 42 datasets fulfill the specifications of the mixed dust cluster, as air masses are usually circulating over Mediterranean and Europe (and

Turkey for Limassol cases) (Fig.3c), while the 8 cases corresponding to the Saharan dust cluster seem to travel shorter paths, in time and distance, before reaching the observation stations. The rest 5 seem to have a clear influence of BB aerosols not only from Africa but also from Canada (Fig.3b, with inserted fire information from MODIS, https://firms.modaps.eosdis.nasa.gov). As expected, the air masses of the Saharan dust cluster have higher mean δ_{p532} and AOT_{532} values, larger layers' vertical thickness (m) and less mixing hours (26±13 h) after leaving African continent in comparison with the mixed Saharan dust cluster (66±41 h) and the BB & mixtures cluster (41±26 h). No clear conclusions can be extracted from the parameter LR_{532} (see Fig.4).

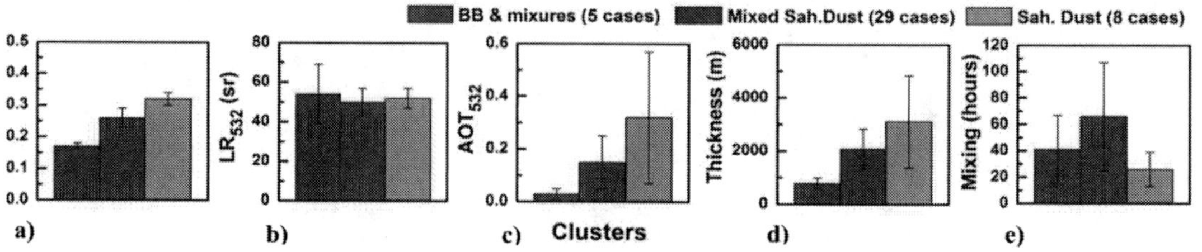

Figure 4: Mean values of a) δ_{p532}, b) LR_{532}, c) AOT_{532}, d) thickness and e) mixing hours (after leaving African continent) for the three identified clusters.

4. CONCLUSIONS

A total of 51 cases of Saharan dust events observed over 4 southern Europe cities were studied with regard to their optical and geometrical properties. Backward trajectory analysis allowed for source identification of the elevated Saharan dust layers over the Mediterranean region. The dust layers are usually observed between ~1.6 and ~5 km. δ_{p532} and LR_{532} values range from 0.16 to 0.35 and from 35 sr to 73 sr respectively, depending on the mixing state. Significantly high AOT_{532} values were found for Granada indicating that the dust outbreaks occurred over this area are more intense. Results of LR_{532} vs. δ_{p532} are presented in order to elucidate the difference of dust and dust with mixtures. Layers with lower δ_{p532} (0.17±0.01), AOT_{532} (0.03±0.02) and thicknesses (786±212 m) indicate high mixing ratio, while the properties of the least or no mixed dust layers (δ_{p532}=0.32±0.02, AOT_{532}=0.31±0.23 and thickness=3158±1606 m) are in a good agreement with literature findings for pure Saharan dust. LR seems to be less effective in extracting conclusions about the mixing state of a layer, possibly due to the same absorption efficiency of BB and dust.

ACKNOWLEDGMENTS

O. S.'s research project has been financed through a scholarship from the General Secretariat for Research and Technology (GSRT) and the Hellenic Foundation for Research and Innovation (HFRI).

REFERENCES

[1] P. Salvador, et al. Atmos. Chem. Phys. 14, 6759–6775. doi.org/10.5194/acp-14-6759-2014 (2014)

[2] F. Barnaba and G. P. Gobbi, Atmos. Chem. Phys., 4, 2367–2391. www.atmos-chem-phys.net/4/2367/2004/ (2004)

[3] A. Papayannis, et al. Journal of Geophysical Research, VOL. 113, D10204. doi:10.1029/2007JD009028 (2008)

[4] O. Soupiona, et al. Atmos. Environ. 183, 165–174. doi.org/10.1016/j.atmosenv.2018.04.011 (2018)

[5] A. Nisantzi, et al. Proc.Volume 8182, SPIE Remote Sensing, 818217. doi:10.1117/12.898357 (2011)

[6] E. Flaounas, et al. 2015. Atmos. Sci. Lett. 16, 473–478. doi.org/10.1002/asl.584 (2015)

[7] IPCC. Cambridge University Press, pp. 571–658. doi.org/10.1017/CBO9781107415324.016 (2014)

[8] D. Balis, et al. Geophys. Res. Lett. 31, 10–13. doi.org/10.1029/2004GL019881 (2004)

[9] I. Mattis, et al. Geophys. Res. Lett. 29, 20–21. doi.org/10.1029/2002GL014721 (2002)

[10] L. Mona, et al. Adv. Meteorol. Vol. 2012, Article ID 356265, 36 pages. doi.org/10.1155/2012/356265 (2012)

[11] L. Mona, et al. J. Geophys. Res. Atmos. 111, 1–13. doi.org/10.1029/2005JD006569 (2006)

[12] M. Tesche, et al. Tellus B Chem. Phys. Meteorol. 63, 649–676. doi.org/10.1111/j.1600-0889.2011.00548.x (2011)

[13] S. Solomos, et al. Atmos. Chem. Phys. 17, 4063–4079. doi.org/10.5194/acp-17-4063-2017 (2017)

[14] P. Kokkalis, et al. MDPI, Remote Sensing, 10, 1807, doi.org/10.3390/RS10111807 (2018)

[15] V. Freudenthaler, et al. Tellus, Ser. B Chem. Phys. Meteorol. 61, 165–179. doi.org/10.1111/j.1600-0889.2008.00396.x (2009a)

[16] S. Groß, et al. Atmos. Chem. Phys., 13, 2487-2505. doi.org/10.5194/acp-13-2487-2013 (2013)

[17] J.L. Guevrrero-Rascado, et al. Atmos. Environ. 42, 2668-2681. doi:10.1016/j.atmosenv.2007.12.062 (2008)

[18] F. Madonna, et al. Atmos. Meas. Tech., 4, 1191–1208 doi:10.5194/amt-4-1191-2011 (2011)

[19] P. Kokkalis, et al. 26th ILRC, June 25-29, 2012, Porto Heli, Greece. (2012)

[20] A. Nisantzi, et al. Atmos. Chem. Phys., 15, 7071–7084. doi:10.5194/acp-15-7071-2015 (2015)

VARIOUS APPLICATIONS OF AD-NET PRODUCTS FOR ENVIRONMENTAL IMPACT STUDIES RELATED TO ASIAN DUST

Atsushi Shimizu, Tomoaki Nishizawa, Yoshitaka Jin, Nobuo Sugimoto

National Institute for Environmental Studies, Tsukuba 305-8506 Japan
**Email: shimizua@nies.go.jp*

ABSTRACT

AD-Net, the Asian dust and aerosol lidar observation network, is operated in East Asian counties for creating useful dataset for environmental impact studies. The products from AD-Net have been utilized as the source of data assimilation in a chemical transport model, and as the index of exposure of Asian dust to human in the area. Long term observation is indispensable to estimate climatorogical change of dust density, and sustainable operation strategies are needed for further development of regional lidar network.

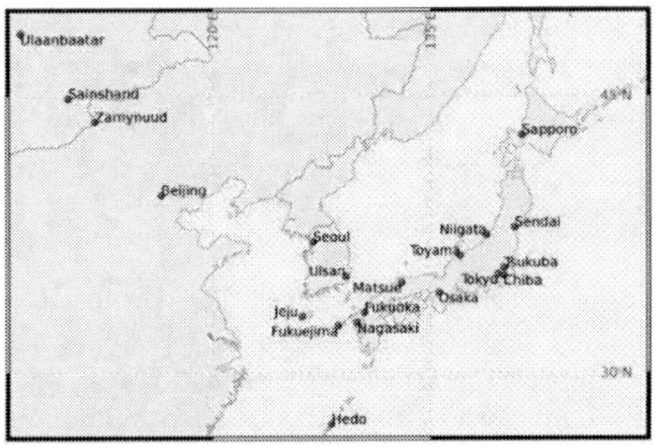

Fig.1 Locations of lidars in AD-Net.

1. INTRODUCTION

In East Asia, mineral dust particles have been one of major target of lidar observations. Iwasaka et al.[1] reported the existence of aerosol layer at around 4-8 km over Nagoya, Japan using a lidar. Since then there were a lot of observational evidence of floating mineral dust with polarization lidars because the mineral dust particles have non-spherical shape which causes depolarized backscatter if lidars were employed to identify density and altitude of aerosol layer. The National Institute for Environmental Studies (NIES) has been operating dual wavelength polarization lidars in east Asian countries including Japan, Korea, China and Mongolia (Fig.1), and they are known as AD-Net (Asian dust and aerosol lidar network) which is a member of GALION (Gaw aerosol lidar observation network). For more than 15 years AD-Net has obtained vertical profiles of dust and spherical particle backscatter, and the product were utilized in various field of environmental impact studies. This paper exhibits the methodology of data production, examples of environmental impact studies, and future outlook of application of AD-Net.

2. PRODUCTS FROM AD-NET

Each AD-Net lidar is operated automatically in 24 hours regardless of whether conditions. Every 15 minutes the vertical profiles of three backscatter intensities, parallel / perpendicular components at 532 nm and total at 1064 nm, are recorded with vertical resolution of 6 meters. Based on assumptions that lidar ratio (S1) is constant and two-components of particles (mineral dust and spherical particles) are externally mixed, 'dust extinction coefficient' and 'spherical particle coefficient' are calculated in semi real time[2]. Products in netCDF format and visualized figures are available online (http://www-lidar.nies.go.jp/AD-Net/) except for data from stations in China. Two components of extinction coefficients are available with vertical resolution of 30 m below 9 km (in clear case) or below cloud base height, and the information related to cloud base / (apparent) top is also supplied. An example of visualized product is shown in Fig.2.

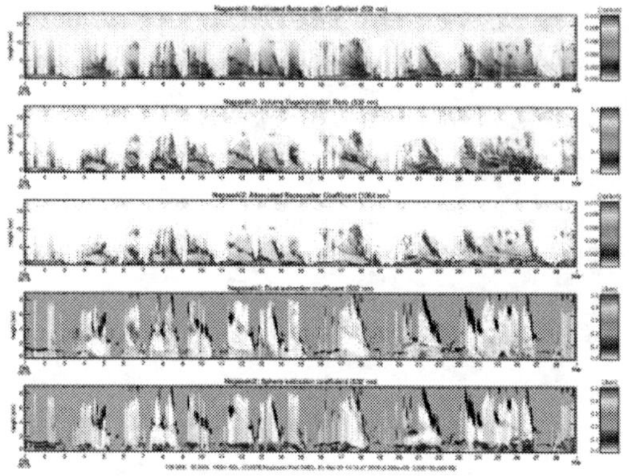

Fig.2 Time-height sections of [from top] attenuated backscatter coefficient (ABC) at 532 nm, volume depolarization ration at 532 nm, ABC at 1064 nm, dust extinction coefficient with cloud mask, spherical particle extinction coefficient with cloud mask for February 2019 in Nagasaki, Japan.

3. APPLICATIONS

3.1 Validation and data assimilation with numerical chemical transport models

Validation of vertical distribution of particles estimated in chemical transport model (CTM) is one of traditional application of lidar observation, especially in Asian dust seasons[3]. Yumimoto et al.[4] firstly employed dust extinction coefficient as a source of data assimilation in 4D-VAR method. The effect of assimilation was confirmed by independent surface PM10 measurements in Asian dust event on March 2007. Re-analysis of aerosol distribution is under calculation by MRI [5] and dust extinction coefficient will be incorporated into it soon.

Recently Goto et al.[6] compared modeled vertical distribution of extinction coefficient by sulfate, carbonaceous and sea-salt with spherical particle extinction coefficient by lidars in AD-Net. Interpretation of spherical particle extinction coefficient is not straight like dust because it is affected by relative humidity of the ambient atmosphere that is not directly measured. Also, variations of lidar ratio (S1) in anthropogenic particles introduce the uncertainty of extinction coefficient. Thus, effective method to utilize spherical particle extinction coefficient with CTM is one big challenge of AD-Net.

3.2 Epidemiology related to health impact of Asian dust

Dust extinction coefficient near the surface is widely utilized as the index of mineral dust density because there is no other method to estimate the concentration of dust except for chemical analysis with filter sampling. Dust density is required in daily basis because health related data is usually daily, however simple averaging of dust extinction coefficient in 24 hours is sometimes contaminated by miss classification of cloud/rain. To avoid the problems, averaging of segmented medians, or correlation between dust extinction and SPM (suspended particulate matter, which is measured with beta-ray method at the surface) hourly data are examined. Several studies exhibited correlation between symptoms and Asian dust events in Japan [7-10].

3.3 Climatology of Asian dust

The occurrence of Asian dust at meteorological observatories in Japan made a peak around A.D.2000 and recently it decreases gradually. However the quantitative assessment based on daily SYNOP report is not easy. Thus, 'Dust AOD'(tau_d) is defined from dust extinction coefficient and annual cycle of tau_d was fitted to the combination of linear function and sinusoidal curve with a period of one year [11]. As a result, tau_d decreased -2.5% per year over Japan (Fig.3), and this trend was more significant in upper region (5-6km). This trend was not clear in Mongolian region, which suggests that the negative trend of Asian dust over Japan is not directly related to the change of generation of Asian dust in source region but to the change of transportation route to Japan/northwestern Pacific region. Another challenge is discrimination of long-range transported Asian dust and locally generated dust in Japan. In early spring dust particles are blown from crop fields in Japan, and it causes highly depolarized backscatter. Although current algorithm of AD-Net cannot distinguish such local dust and Asian dust, optical properties must be different among these two kinds of dust. More sophisticated algorithm is required to

https://doi.org/10.1051/epjconf/202023705011

identify the climatological change of Asian dust transport.

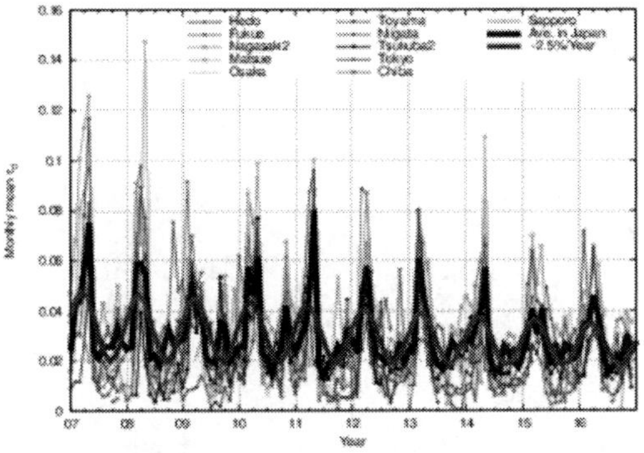

Fig.3 Monthly variations of Tau_d since 2008 observed by independent lidars in Japan (thin lines), average in Japan (thick line) and fitted curve to average (solid purple). Reproduced from [11].

3.4 Contribution to International Activities

Products of AD-Net are also utilized in the international collaborative activities. In Dust and Sand Storm Working Group 1 (DSS-WG1) under Tripartite Environmental Ministry Meeting (TEMM), Japan, Korea and China exchange observational results during specified Asian dust events, and data of AD-Net is important resource to identify such events. AD-Net is also expected to supply products to WMO Sand and Dust Storm Warning Advisory and Assessment System (SDS-WAS).

4. FUTURE OUTLOOK

Although rapid economic growth in East Asia caused significant enhancement of anthropogenic aerosols in past few decades, the emission of precursors of aerosols are now decreasing. Thus the natural particles like Asian dust becomes relatively more important to evaluate atmospheric conditions. To identify the shift of dust emission area, or the main transport path of Asian dust to western Pacific region, observational evidence by AD-Net and CTM must be combined more closer. Once reanalysis data set made by CTM becomes available, health impact studies are possible even in the region where lidar is not operated. Additionally hemispheric transport of mineral dust, namely Saharan/Arabian dust transport to East Asia is a new topic in observational studies. To catch them, lidars must be located in wider area in the region. To maintain the existing lidar stations and to identify the effective locations of (possible) new stations are important to sustain AD-Net activities.

ACKNOWLEDGEMENTS

Operation of AD-Net is supported by The Ministry of Environment Japan, and Environmental Research and Technology Development Fund (E-1502) by Environmental Restoration and Conservation Agency, Japan.

REFERENCES

[1]Iwasaka et al., Tellus 35B:189-196 (1983)
[2]Shimizu et al., Opt. Eng. 56(3):031219 (2016)
[3]Shimizu et al., J. Geohpys. Res. 109:D19S17 (2004)
[4]Yumimoto et al., Atmos. Chem. Phys. 8(11):2869-2884 (2008)
[5]Yumimoto et al., Geosci. Model Dev. 10:3225-3253 (2017)
[6]Goto et al., J. Geophys. Res. 120:6247-6270 (2015)
[7]Ueda et al., Toxicology 24(12):858-867 (2012)
[8]Higashi et al., J. Atmos. Env. 97:537-543 (2014)
[9]Kanatani et al., Annals. Allergy Asthma Immunology 116(5):425-430 (2016)
[10]Nakamura et al., J. Epidemiology 26(11):593-601 (2016)
[11]Shimizu et al., Sci. Online Let. Atmos. 13:205-208 (2017)

WATER VAPOUR AND TEMPERATURE MEASUREMENTS BY RAMAN LIDAR IN THE FRAME OF THE NDACC

Benedetto De Rosa[1*], Paolo Di Girolamo[1], Donato Summa[1]

[1] *Scuola di Ingegneria, Università degli Studi della Basilicata, Potenza, 85100, Italy*
**Email: bundit@hotmail.it*

ABSTRACT

In November 2012, the University of BASILicata Raman Lidar system (BASIL) was approved to enter the International Network for the Detection of Atmospheric Composition Change (NDACC). Since then measurements were routinely carried out on a once per week basis. This paper illustrates specific measurement examples from this effort, with a dedicated focus on temperature and water vapour measurements, with the ultimate goal to provide a characterization of the system performance. Case studies illustrated in this paper demonstrate the ability of BASIL to perform measurements of the temperature profile up to 50 km and of the water vapour mixing ratio profile up to 15 km, based on an integration time of 2 hours and a vertical resolution of 150 m, with measurement bias not exceeding 0.1 K and 0.1 g kg^{-1}, respectively. Raman lidar measurements are compared with measurements from additional instruments, such as radiosondings and satellite sensors (IASI and AIRS), and with model re-analyses data (ECMWF and ECMWF-ERA). Comparisons in this paper cover the altitude interval up to 15 km for water vapour mixing ratio and up to 50 km for the temperature. Comparisons between BASIL and the different sensor/model data in terms of water vapour mixing ratio indicate a mean absolute/relative bias of -0.024 g kg^{-1}(or -3.9 %), 0.342 g kg^{-1}(or 36.8 %), 0.346 g kg^{-1} (or 37.5 %), -0.297 g kg^{-1} (or -25 %), -0.381 g kg^{-1} (or -31 %), when compared with radisondings, AIRS, IASI, ECMWF, ECMWF-ERA, respectively. For what concerns the comparisons in terms of temperature measurements, these indicate a mean absolute bias between BASIL and the radisondings, AIRS, IASI, ECMWF, ECMWF-ERA of -0.04, 1.99, 0.48, 0.14, 0.62 K, respectively. Based on the available dataset and benefiting from the circumstance that the Raman lidar BASIL could be compared with all other sensor/model data, it has been possible to estimate the absolute bias of all sensors/datasets, this being 0.004 g kg^{-1}/0.30 K, 0.021 g kg^{-1}/-0.34 K, -0.35 g kg^{-1}/0.18 K, -0.346 g kg^{-1}/-1.63 K, 0.293 g kg^{-1}/-0.16 K and 0.377 g kg^{-1}/0.32 K in terms of water vapour mixing ratio/temperature for BASIL, the radisondings, IASI, AIRS, ECMWF, ECMWF-ERA, respectively.

1. INTRODUCTION

Water vapour is the most important atmospheric greenhouse gas and its increasing concentration, though indirectly, is primarily driven by human activity. The increasing concentration of CO_2 and CH_4, primarily associated with fossil fuel combustion, leads to warmer tropospheric temperatures, which determine an increased atmospheric humidity content and ultimately lead to a warmer climate. Moreover, water vapour in the upper troposphere/lower stratosphere (UTLS) region plays a crucial role in the Earth radiative balance, and consequently in the climate system, its presence being primarily associated with two main sources: transport from the troposphere, taking place mainly in the tropics, and the in-situ oxidation of methane. More specifically, changes in temperature and water vapour concentration in the UTLS result in radiative forcing alterations [1], which trigger climate changes (Intergovernmental Panel on Climate Change, 2007). Observations demonstrate that stratospheric water vapour increases with increasing tropospheric temperature, implying the existence of a stratospheric water vapour feedback [2]. The strength of this feedback has been estimated to be ~ 0.3 W m^{-2} K^{-1}. Stratospheric water vapour has also an important role in stratospheric clouds formation, which are a key element in stratospheric ozone depletion mechanisms [3, 4]. Furthermore, stratospheric water vapour is important in the formation of hydrogen radicals, which play a role in stratospheric chemistry and stratospheric ozone depletion. Despite the major importance of having accurate water vapour and temperature measurements in the troposphere and stratosphere, few measurements of these variables and of their long-term variability are available, especially in the UTLS region. Water vapour, measurements in the UTLS region have been traditionally guaranteed by the use of operational radiosondes or balloon-borne frost-point hygrometers, the latter sensor remaining the best source of high quality water vapour measurements in this region. However, these are too expensive to be used on an operational basis and thus do not allow a sufficiently intense launch schedule to guarantee the high temporal resolution needed for the above mentioned scientific scopes. Similar considerations apply to temperature, the main source of measurements covering the troposphere and stratosphere being represented by microwave and infrared satellite sounders. All the above weather and climate-related issues call for highly accurate measurements of both the water vapour and temperature profiles throughout the atmosphere, with a specific focus on the UTLS region. These motivations pushed the international Network for the Detection of Stratospheric Change (NCSC), now Network for the Detection of Atmospheric Composition Change (NDACC), to include in the early 2000s water vapour Raman lidars and temperature lidars among its instruments.

2. METHODOLOGY

2.1 The lidar system

The Raman Lidar system BASIL joined NDACC in November 2012, with the primary goal of providing accurate routine measurements of the vertical profiles of both water vapour mixing ratio and temperature. The setup of BASIL has been described in a variety of previous papers [5-7]. A mobile version of the system has been deployed in several international field campaigns [8-18]. BASIL measurements of the temperature profile cover the full altitude interval from the surface up to the stratopause. Measurements over such a wide altitude interval are possible based on the combined application of the pure rotational Raman technique [19, 20] which is used to cover the lowest 20 km, and the integration technique, covering the altitude region from 20 km to typically 50-55 km. The combined application of these two techniques is possible because of the presence of an overlap region (20-25 km) where both techniques properly work. Recently, the aerosol backscattering coefficient at 354.7 nm has been added to the set of atmospheric variables measured by BASIL and uploaded to NDACC repository.

3. RESULTS

3.1 Measurements

For the purpose of this paper, we focused our attention on four selected case studies collected during the first 2 years of operation of the system, namely 7 November 2013, 9 October 2014 and 2 and 9 April 2015. While a larger data-set could have been chosen, we decided to focus our attention to a limited number of case studies, which have been carefully analyzed with a customized approach, instead of considering a larger dataset analyzed with a standard routine analysis approach. This approach was considered with the aim of minimizing the effects associated with the application of the data analysis approach. The four selected case studies cover a two year time interval from the early stage of BASIL operation in the frame of NDACC (7 November 2013 to April 2015). Figure 1a illustrates the mean water vapour mixing ratio profile measured by BASIL on 9 October 2014 over the time interval 16:00-18:00 UTC, together with the closest in time vertical profiles the water vapour mixing ratio profiles from IASI (at 19:20 UTC) and AIRS (at 14:35 UTC) and the model re-analysis ECMWF ERA-15 and ECMWF ERA-40 (at 18:00 UTC). The vertical resolution of the Raman lidar data is 150 m up to 6 km, 300 m between 6 and 8 km and 600 m above 8 km, with the water vapour mixing ratio profile extending up to an altitude of ~ 14 km, with values down to 0.02 g kg^{-1}. Similar performances are found in all two-hour profiles collected in the frame of NDACC. The present comparison also includes the radiosonde profile launched at 18:00 UTC from the nearby station IMAA-CNR (~ 7 km W). Deviations between BASIL and the radiosonde are not exceeding 0.1 g kg^{-1} up to ~ 9 km. Figure 1b illustrates the mean atmospheric temperature profile measured by BASIL over the same time interval considered in figure 1a, together with the corresponding profiles from IASI, AIRS and the model re-analysis (ECMWF ERA-15 and ECMWF ERA-40). The lidar measurement is based on the use of the rotational technique up to 20 km and the integration technique above 20 km.

The deviations observed above 35 km are possibly associated with the effects of gravity waves propagation, whose presence is revealed by the Raman lidar, but missed by the other sensors/models. The temperature profile from the radiosonde extends up to 30 km. Water vapour mixing ratio and temperature deviations between BASIL and the radiosonde are smaller than those characterizing any other sensors/models pair.

3.1 Mean Bias

The performance of the different profiling sensors and models considered in the study are assessed through a statistical analysis. Figures 2a and b illustrate the water vapour mixing ratio absolute and percentage BIAS for all sensor/model pairs. As expected, the absolute BIAS shows larger values in the ABL (in the range -3/+5 g kg^{-1}), with values being typically smaller than ±1 g kg^{-1} above 2 km. More specifically, in the ABL the absolute BIAS of BASIL vs. the radiosondes is in the range ±0.3 g kg^{-1}.For all sensor/model pairs, the absolute BIAS shows values smaller than ±0.1 g kg^{-1} above 8 km and smaller than ±0.02 g kg^{-1} above 10 km. In the altitude region 8-16 km the mutual bias of BASIL vs. the radiosondes or ECMWF is smaller than ±0.01 g kg^{-1}, while the mutual bias of BASIL vs. ECMWF-ERA40 and AIRS vs. IASI is smaller than ±0.07 g kg^{-1} above 10 km. For all sensor/model pairs, the relative or percentage BIAS shows values in the range ±60 % all the way up to 16 km. The smallest percentage bias values are found in the comparison of BASIL vs. the radiosondes, with values not exceeding ±18 % all the way up to 12 km and values in the range ±13 % above the ABL up to 4 km. Small percentage bias values are also found in the comparison of BASIL vs. ECMWF, with values not exceeding ±30 %. Figure 2c illustrates the temperature absolute BIAS profiles for all sensor/model pairs. The BIAS of BASIL vs. the radiosondes is in the range ±1 K above the ABL up to 12 km, with deviations in the ABL not exceeding 2 K. With the exception of a few data points, BIAS values are within ±2 K up to 30 km, which is the maximum altitude reached by the radiosonde. The BIAS of BASIL vs. ECMWF-ERA40 is in the range ±0.8 K up to 12.5 km. For all sensor/model pairs, the BIAS shows values in the range ±5 K all the way up to 50 km.

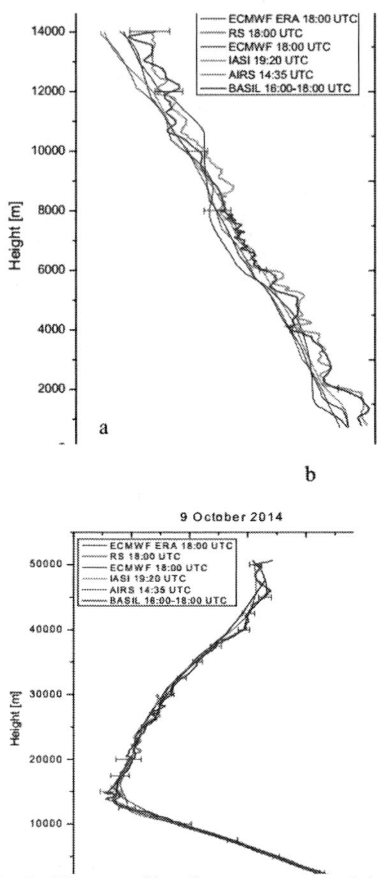

Fig. 1: Mean profile of water vapour mixing ratio (a) and temperature (b) as measured by BASIL over the time interval 16:00-18:00 UTC on 9 October 2014, together with the closest in time profiles from IASI (at 19:20 UTC), AIRS (at 14:35 UTC) and the model re-analysis ECMWF (ERA-15 & ERA-40, at 18:00 UTC).

RMS deviation values for both water vapour mixing ratio and temperature for all sensor/model pairs (not shown here) are slightly larger than BIAS values, which testifies the limited contribution of statistical uncertainties and geophysical changes in the deviations between the different measured/modeled profile pairs. The vertically-averaged mean bias over the whole inter-comparison range is obtained through the application of a weighted mean. Results from the present inter-comparison effort provide evidence of the high quality of water vapour and temperature profile measurements from BASIL and on the possibility to use these measurements for the long-term monitoring of atmospheric composition and thermal structure changes.

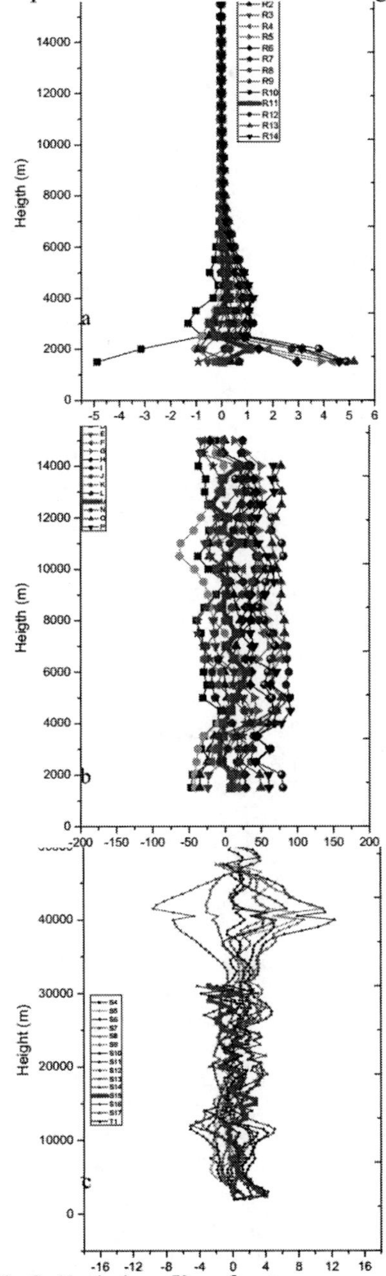

Fig 8: Vertical profiles of water vapour mixing ratio mean absolute BIAS (a) percentage BIAS (b) and vertical profiles of temperature mean absolute BIAS (c).

REFERENCES

[1] M. Riese, et al. Geophys. Res. 117, D16305, doi: 10.1029/2012JD017751 (2012).
[2] A. Dessler, et al. Natl. Acad. Sci. USA 110, 18087-18091, doi: 10.1073/pnas.1310344110 (2013).
[3] A. Di Sarra, et al. Geophys. Res. Lett. 19, 1823-1826, doi: 10.1029/92GL01887 (1992).
[4] P. Di Girolamo, et al. Geophys. Res. Lett. 21, 1295-1298, doi: 10.1029/93GL02892 (1994).
[5] P. Di Girolamo, et al. Multiparameter Raman Lidar Measurements for the Characterization of a Dry Stratospheric Intrusion Event, J. Atmos. Oceanic Technol. 26, 1742-1762, doi: 10.1175/2009JTECHA1253.1 (2009).
[6] P. Di Girolamo, et al., UV Raman lidar measurements of relative humidity for the characterization of cirrus cloud microphysical properties, Atmos. Chem. Phys. 9, 8799-8811, doi: 10.5194/acp-9-8799-2009 (2009).
[7] R. Bhawar, et al. Q. J. Roy. Meteor. Soc. 137, 325-348, doi: 10.1002/qj.697 (2011).
[8] R. Bhawar, et al. Geophysical Research Letters 35, L04812, doi: 10.1029/2007GL032207 (2008).
[9] C. Serio, et al. Optics Express 16/20, 15816-15833 doi: 10.1364/OE.16.015816 (2008).
[10] V. Wulfmeyer, et al. Bulletin of the American Meteorological Society 89, 1477-1486, doi: 10.1175/2008BAMS2367.1 (2008).
[11] L. J. Bennett et al., Q. J. Roy. Meteor. Soc. 137, 176-189, doi: 10.1002/qj.760 (2011).
[12] V. Ducrocq, et al. Bull. Amer. Meteor. Soc. 95, 1083-1100, doi: 10.1175/BAMS-D-12-00244.1 (2014).
[13] A. Macke, et al. Atmos. Chem. Phys. 17, 4887-4914, doi:10.5194/acp-17-4887-2017 (2017).
[14] Di Girolamo, et al. Atmos. Environ. 50, 66-78, 10.1016/j.atmosenv.2011.12.061 (2012).
[15] P. Di Girolamo, et al. Atmos. Chem. Phys. 12, 4143-4157, doi: 10.5194/acp-12-4143-2012 (2012).
[16] P. Di Girolamo, et al. Q. J. Roy. Meteor. Soc., 142, 153-172, doi: 10.1002/qj.2767 (2016).
[17] P. Di Girolamo, et al., Characterisation of Boundary Layer Turbulent Processes by the Raman lidar BASIL In the frame of HD(CP)2 Observational Prototype Experiment, Atmos. Chem. Phys., 17, 745-767, doi:10.5194/acp-17-745-2017 (2017).
[18] P. Di Girolamo, et al. Atmos. Chem. Phys., 18, 4885-4896, doi: 10.5194/acp-18-4885-2018 (2018).
[19] A. Behrendt and J. Reichardt, Appl. Opt. 39, 1372–1378, doi: 10.1364/AO.39.001372 (2000).
[20] P. Di Girolamo, et al., Rotational Raman Lidar measurements of atmospheric temperature in the UV, Geophys. Res. Lett. 31, doi: 10.1029/2003GL018342 (2004).

A NEW AIRBORNE DOPPLER LIDAR FOR BOUNDARY LAYER RESEARCH & LES-BASED SIMULATION

P. Gasch[1*], A. Wieser[1], N. Kalthoff[1], U. Corsmeier[1], T. Feuerle[2] and C. Kottmeier[1]

[1]Institute of Meteorology and Climate Research, Karlsruhe Institute of Technology, Karlsruhe, Germany
[2]Insitute of Flight Guidance, Technical University Brunswick, Brunswick, Germany
**Email: philipp.gasch@kit.edu*

ABSTRACT

Airborne Doppler lidar measurements are a valuable tool in meteorological research. A new airborne Doppler lidar system has been developed jointly by the Karlsruhe Institute of Technology (KIT) and the Technical University of Brunswick. The system is designed for use aboard a medium-range turboprop aircraft, thereby targeting meso-scale flow phenomena. In order to optimize system setup and retrieval strategy, as well as evaluate measurement quality under turbulent conditions, a first LES-based airborne Doppler lidar simulator has been developed. The results provide detailed insight into airborne Doppler lidar wind profiling characteristics.

1. INTRODUCTION

Airborne Doppler lidar measurements can provide added value compared to ground-based measurements due to their flexibility allowing for targeted measurements at high spatial resolution [1]. Due to their benefits, various airborne Doppler lidar systems have been developed in the past. Most of the existing systems are based on long-range aircraft flying high and fast, thereby they are destined for sensing the free troposphere due to the coarser spatial resolution [2,3]. Some deployments of airborne Doppler lidar have been reported based on medium-range aircraft flying low and slow for wind profiling purposes [4,5].

In a joint project by the Karlsruhe Institute of Technology (KIT) and the Technical University of Brunswick a new airborne Doppler lidar system for boundary layer research has been designed. The new system has a flexible scanning geometry, allowing for both wind profiling and Nadir measurements of the wind field.

Wind profiling at highest resolution in the boundary layer suffers from errors due to a violation of the homogeneous conditions assumed for retrieval, e.g. due to turbulence in the scanned volume. In order to investigate the sensitivity of the error characteristics to system setup and retrieval strategy, a first LES-based airborne Doppler lidar simulator (ADLS) has been developed. The ADLS has the advantage that the atmospheric input data during the measurement process are known exactly. Thereby, representativeness problems due to differing measuring geometries, complicating airborne Doppler lidar validation for commonly used tools (e.g. dropsondes, towers, other profilers), can be avoided [6,7]. Further, a wide range of system setups and retrieval strategies can be evaluated.

2. THE NEW AIRBORNE DOPPLER LIDAR SYSTEM

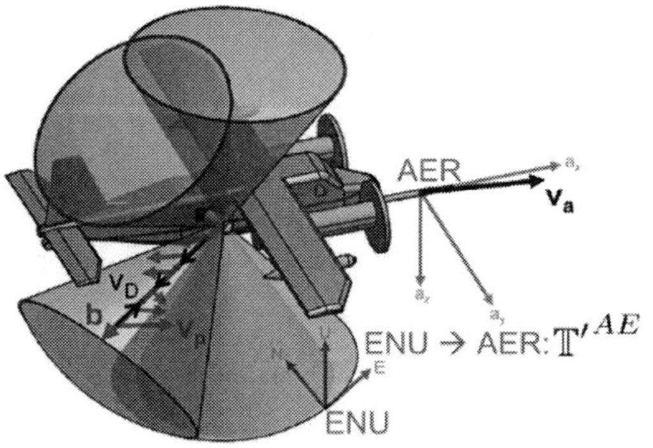

Figure 1: Schematic depiction of the airborne Doppler lidar system and idealized scan geometries. The coordinate systems used in the ADLS are indicated (ground based ENU, aircraft based AER). Further, the particle velocity $\mathbf{v_p}$, beam pointing direction **b**, measured radial Doppler velocity v_D and aircraft velocity $\mathbf{v_a}$ used in the simulations are also sketched.

A new airborne Doppler lidar (ADL) system is designed for use aboard the Dornier 128-6 research aircraft [8]. A Lockheed Martin WindTracer WTX Doppler lidar owned by KIT, measuring at 1.6 μm wavelength, with a pulse duration of 300 ns and 5.4 kW peak power, is operated at a pulse repetition frequency of 750 Hz and 100 ms accumulation time. It is coupled with a new airborne scanner built by Arges GmbH. The 2-axis scanner can point both up and downward (±180° around the aircraft pitch axis), as well as ±30° sideway (swath) in all positions, at scan rates of 20° s^{-1}. The system allows for flexible scanning geometries and automatic correction of aircraft movement. It yields high spatial resolution due to the slow flying aircraft (65 m s^{-1}) and high measurement frequency.

In order to characterize the expected accuracy of wind profiling under inhomogeneous flow conditions, as well as to optimize system setup and retrieval strategies, an ADLS has been developed.

Table 1: System setup and retrieval parameters as used for this study.

Parameter	Setting
Aircraft altitude	1700 m
Flight directions	16, 22.5° apart
Scan elevation	30° – 80°
Scan rotation	20 s
Data acquisition	1 Hz
Components retrieved	u,v,w
Averaging distance	1300 m
Vertical resolution	62 m
Radial velocity noise σ v_D	0 – 1 m s^{-1}
Beam pointing error σ **b**	0° – 1°

3. LES-BASED SIMULATION

The ADLS allows for analysis of varying system setups as well as retrieval strategies by simulating the new ADL system numerically.

3.1 Operating scheme

The ADLS uses PArallelized LES Model (PALM) data as LES input [9,10], which provides wind fields with 10 m grid-spacing at geostrophic background wind speeds of u_G = 0, 5, 10 and 15 m s^{-1}. The surface kinematic heat flux is set to 0.03 K m s^{-1} (0.23 K m s^{-1} for u_G = 0 m s^{-1}). The setup results in an unstably stratified flow with the presence of organized convective structures. The turbulent boundary layer, wherein profiling is conducted, extends up to a height of 800 m (1400 m). For wind profiling, the aircraft transects the LES 25 times in 16 directions in order to create a sufficient number of profiles. The scanner and lidar system are emulated geometrically, without simulation of the backscattering process or laser photonics. For the measurement, all geometric transformations applicable in real-world aircraft measurements are included in the ADLS. Two system simulations are performed: 1. Ideal measurement system, no error added on any component in order to focus on errors due to flow inhomogeneity. 2. Noisy measurement system, to emulate lidar and scanner error, random deviations can be added to the radial velocity (σ v_D) and beam pointing direction (σ **b**). The specific setup used for all components is shown in Tab. 1.

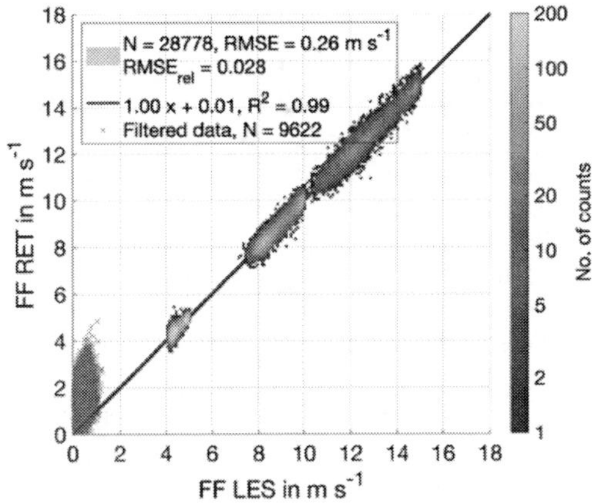

Figure 2: LES input wind speed (FF LES) and ADL retrieved wind speed (FF RET) for an ideal measurement system with specifications according to Tab. 1.

After the measurement, an airborne velocity azimuth display (AVAD) retrieval is performed on the measured data [11]. For error analysis, the originally supplied LES wind field data can then be evaluated against the retrieved wind data. For system setup and retrieval characterization, as well as optimization, a number of parameters adjustable in real-world measurements are varied to evaluate their effect on retrieval quality. Last, using the ADLS, it is possible to evaluate the performance of commonly used quality filtering criteria in Doppler lidar wind profiling.

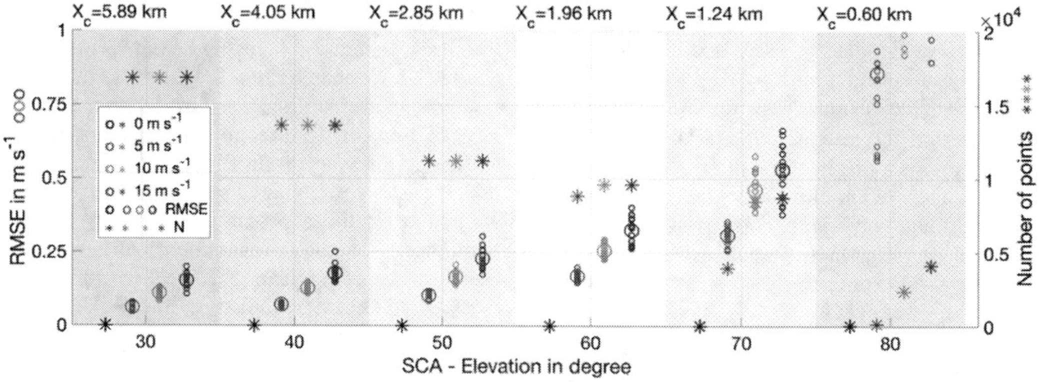

Figure 3: Root-mean-squared error (RMSE) and number of retrieved profiles for varying scan elevation angles and different background wind cases. All other parameters according to Tab. 1. For the 0 m s^{-1} background wind speed case no retrievals pass quality filtering.

4. RESULTS

4.1 Error in wind profiling under turbulent conditions

The comparison between LES input and ADL retrieved wind speeds for an ideal system is shown in Fig. 2. The overall retrieval quality is high for all wind speed cases, except the lowest (RMSE = 0.26 m s^{-1}, R^2 = 0.99). For the settings investigated, turbulence can cause considerable profiling error above 1 m s^{-1}. A striking feature is that no useful retrieval is possible for the 0 m s^{-1} background wind speed case, where the retrieval is biased. However, all biased points are successfully eliminated by applying commonly used quality filtering criteria in the least-square fit analysis (R^2 > 0.9, condition number < 9). The vertical distribution of profile retrieval error mirrors the underlying boundary layer turbulence (not shown). Profile error increases towards the middle of the boundary layer where amplitude and size of the inhomogeneities are largest. Towards the surface, a decrease in profiling error can be observed.

4.2 Influence of system setup variation on retrieval error

Using the ADLS, the parameters in Tab. 1 can be varied to investigate their influence on retrieval quality. A parameter that can be easily varied is the scan elevation angle. Steeper scan elevation angles (closer to nadir) allow for a smaller footprint of the measurement, but at the cost of decreasing the contribution of the horizontal wind to the measured radial velocity. Retrieval characteristics for varying scan elevation angles, using an idealized system (the new ADL allows for scans with elevation < 60° only at forward and backward azimuth angles), can be evaluated from Fig. 3.

The results show that scanning at elevation angles steeper than 70° is problematic for the conditions and setup investigated, as the retrieval error becomes large. When using more shallow elevation angles retrieval quality is improved, however, at the cost of having a much wider measurement footprint, denoted by the across-track measurement diameter x_c. From this analysis, the usage of a 60° scan elevation angle provides a reasonable tradeoff between profiling error and measurement footprint.

4.3 Influence of system errors on retrieval quality

Using the ADLS, it is possible to evaluate the influence of various error sources on wind profiling quality. The two system errors considered here are random radial velocity noise σ_{v_D} (e.g. due to detector noise, speckle effects and turbulence within the measurement volume) and beam pointing error $\sigma_{\mathbf{b}}$ (due to scanner or aircraft positioning error). Both are varied using worst-case assumptions, the results are shown in Fig. 4.

Results show that turbulence is the main driver of profiling error for the higher wind speed cases. Even the worst-case error assumptions do not lead to retrieval error of comparable magnitude to that due to turbulence (Fig. 4). For the 5 m s^{-1}

background wind speed case, system accuracy becomes important. Here an increase of error above 0.25 m s^{-1} or 0.25° does not allow for retrieval of meaningful wind profiles anymore. Further, quality filtering seriously degrades the number of retrieved profiles for these cases. However, the error levels of 0.25 m s^{-1} and 0.25° already present levels which are not expected to occur in a real system [4].

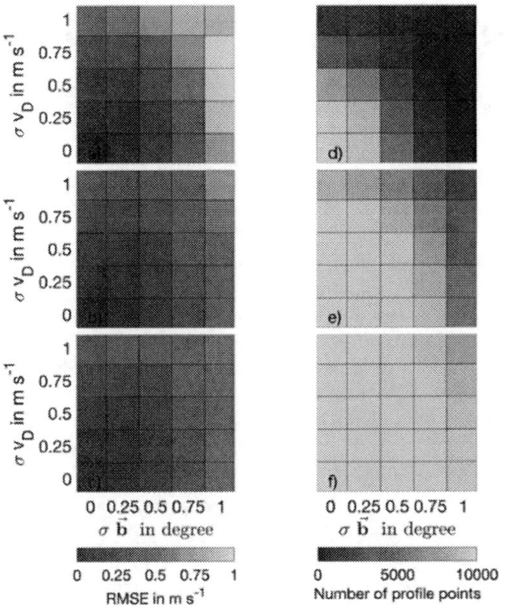

Figure 4: Retrieval error due to emulated noisy measurement system. a), d) 5 m s^{-1}; b), e) 10 m s^{-1}; c), f) 15 m s^{-1} background wind speed case. Retrieval error due to turbulence (determined using ideal system) has been subtracted from all entries.

5. CONCLUSIONS

A new airborne Doppler lidar system is designed for use aboard a medium-range twin-engine unpressurized turboprop aircraft. The new system is expected to provide insight into meso-scale flow phenomena. In preparation, for evaluation of wind profiling accuracy under inhomogeneous flow conditions, a first LES-based airborne Doppler lidar simulator has been developed. Results show that wind profiling is possible with high resolution in the boundary layer, except for situations where there is no background wind speed. Turbulence is the main driver of profiling error for higher wind speeds, at lower wind speeds system accuracy becomes important. Profiling error can be minimized through appropriate system setup and retrieval strategy. For optimization, the ADLS provides a valuable link between theory and measurements.

ACKNOWLEDGEMENTS

The first author acknowledges helpful remarks and comments on the LES-based simulator work by Prof. J. K. Lundquist.

REFERENCES

[1] Baker, W. et al.: Lidar-measured wind profiles: The missing link in the global observing system, Bull. Am. Meteorol. Soc., 95, 543–564, 2014.
[2] Chouza, F. et al.: Saharan dust long-range transport across the Atlantic studied by an airborne Doppler wind lidar and the MACC model, Atmos. Chem. Phys., 16, 11581–11600, 2016.
[3] Bucci, L. R., et al.: Validation of an Airborne Doppler Wind Lidar in Tropical Cyclones, Sensors, 18, 4288, 2018.
[4] De Wekker, S. F., et al.: Airborne Doppler lidar measurements of valley flows in complex coastal terrain, J. Appl. Meteorol. Climatol., 51, 1558–1574, 2012.
[5] Koch, G. J., et al.: Three-dimensional wind profiling of offshore wind energy areas with airborne Doppler lidar, J. Appl. Remote Sens., 8, 083662, 2014.
[6] Weissmann, M., et al.: Targeted observations with an airborne wind lidar, J. Atmos. Ocean. Technol., 22, 1706–1719, 2005.
[7] Wainwright, C. E., et al.: A time series sodar simulator based on large-eddy simulation, J. Atmos. Ocean. Technol., 31, 876–889, 2014.
[8] Corsmeier, U., et al.: Airborne turbulence measurements in the lower troposphere onboard the research aircraft Dornier 128-6, D-IBUF, Meteorol. Zeitschrift, 10, 315–329, 2001.
[9] Raasch, S. and Schröter, M.: PALM - A large-eddy simulation model performing on massively parallel computers, Meteorol. Zeitschrift, 10, 363–372, 2001.
[10] Stawiarski, C., et al.: Assessment of Surface-Layer Coherent Structure Detection in Dual-Doppler Lidar Data Based on Virtual Measurements, Boundary-Layer Meteorol., 156, 371–393, 2015.
[11] Leon, D. and Vali, G.: Retrieval of three-dimensional particle velocity from airborne doppler radar data, J. Atmos. Ocean. Technol., 15, 860–870, 1998.

https://doi.org/10.1051/epjconf/202023706002

CHARACTERISATION OF SMALL-SCALE ATMOSPHERIC WIND-FIELD STRUCTURES USING COHERENT WIND LIDAR WITH SHORT PULSES

D. T. MICHEL[1*], M. VALLA[1], D. GOULAR[1], L. LOMBARD[1], A. DOLFI-BOUTEYRE[1], C. BESSON[1]
[1]ONERA/ DOTA Université Paris Saclay
F 91123 Palaiseau - France
**Email: david-tomline.michel@onera.fr*

ABSTRACT

A lidar design has been developed at ONERA that uses short square pulses (75 ns) to have a small spatial resolution (22.5 m) and be able to measure small-scale atmospheric wind-field structures. Results show that the system is able to resolve the small-scale structures of vortices and to measure wind field structures of a turbulent wind field down to ~20 m.

1. INTRODUCTION

Measuring meter-scale wind field structures are important for a wide variety of physics: in meteorology, to measure atmospheric turbulence [1-6], its effect on ocean elevation [6] and diffusion of pollutants in cities [7], in airport to detect wind shear [8], vortices and measure their time of life [9-11], for airship which are very sensitive to small-scale structures [12], for wind turbine which can be optimized according to the wind structures [13], to perform formation flight [14], or for drone transport. These measurements are generally done remotely using a Coherent Doppler light detection and ranging (lidar) sensor in the Mie regime. This instrument uses a laser to measure the speed of the particles (which is similar to the wind speed) projected along its axis. The backscattered signal is frequency shifted due to particle speed (Doppler Effect) that is determined using heterodyne detection. Improving the spatial resolution of the lidar will improve the restitution of the wind field spatial structure.

To perform this measurement, a first kind of lidar uses pulsed laser [1, 10, 11] where the length of the pulses (ΔT) determines the spatial resolution ($\Delta R = c\Delta T$ where c is the speed of light). If the pulses are long, the spectral resolution, given by the Fourier transform of the pulse (and so the accuracy in velocity measurement), is high but the spatial resolution is low. Up to now, pulses of 400 ns-800 ns [1, 10] down to 150-250 ns [11, 15] have been used which limit the resolution to ~50 m. In addition, pulsed laser have a delay before any measurement can be done because of the reflection of the laser on the optics of the system that mixes with the backscattered signal. This blind zone corresponds to a distance of $c\Delta T/2$ from the output optic. So the longer the pulse, the longer the distance of the first detectable point is.

Another technique consists in using a lidar with continuous wave laser [11, 16] that is focused at the region of interest. In this case, the spatial resolution is given by the Rayleigh length of the system over which the backscattered signal is much larger than the signal coming from the other regions. However, such lidar only allows performing the measurement at one location each time which makes the measurement of wind field structures more difficult. In addition, it requires a large output optic to obtain a small Rayleigh length and minimize the spatial resolution. Such optics are expensive and less practical in particular for airborne applications. Furthermore, it is sensitive to large reflections (due to solid target or clouds behind the focus region) that produce a backscattered signal that mixes with the useful signal and corrupts it.

In this article, a pulsed lidar that has been set up at the ONERA is presented. It emits short pulses (~75 ns) to enable measuring meter-scale wind field structures. The pulse length results in a spatial resolution of 22.5 m and limits the blind distance to 15 m ahead of the exit lens.

2. DESIGN OF THE LIDAR

The lidar was designed to measure the wind field velocity of wake vortices perpendicular to the path of a plane [Fig. 1]. A short pulse lidar was chosen because the amount of reflecting particles was large and the vortices induced short wind field structures.

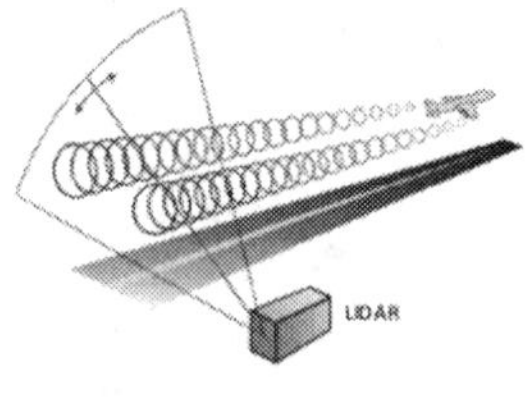

Figure 1: Schematic of the wake vortex characterization by the lidar.

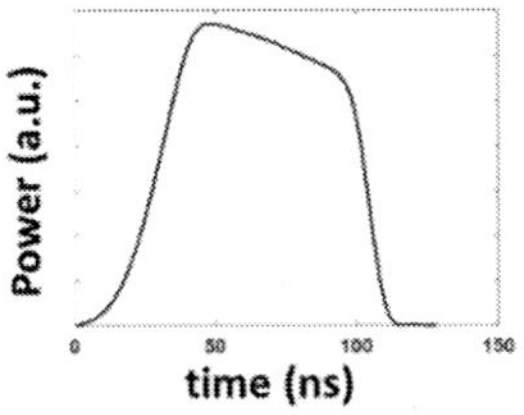

Figure 2: Measured lidar pulse time profile.

The lidar uses a master oscillator power fiber amplifier configuration that emit at 1545-nm [17]. It delivers short pulses with narrow linewidth (close to transform limited) appropriate for heterodyne detection. Their temporal profiles –so called "square" pulse- were shaped with sharp edges (see Fig. 2) to minimize the blind zone, optimize the spatial resolution and maximize the energy per pulse. The pulse shape has intensity at FWHM of 75 ns (~100 ns at -30 dB) which results in a spatial resolution of 22.5 m and a blind distance of 15 m [that correspond to the strong signal at the left of Fig. 2].

3. SIGNAL PROCESSING ON VORTEX

The lidar has been used on the field to collect experimental data in order to localize vortex center and characterize the induced wind field

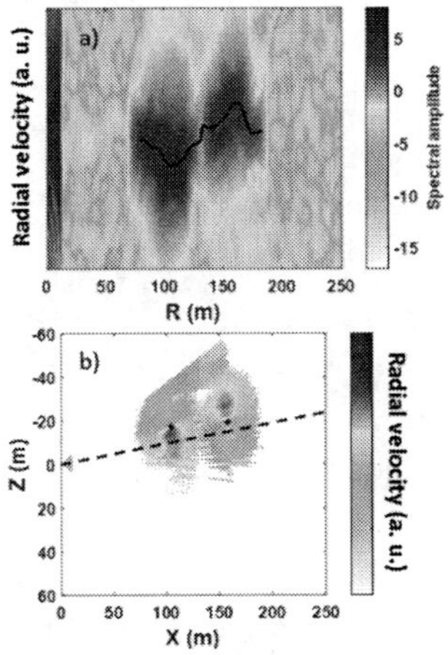

Figure 3: a) frequency versus time plot: Comparison of the lidar spectra measured at θ=-5.5° [dashed line in b)] as a function of distance (R). The wind velocity corresponds to the maximum spectral amplitude (black curve). In the figure, the frequencies are converted into velocity. b) Wind field velocities measured during a lidar scan.

Figure 3 a) shows the spectra of the lidar heterodyne current, calculated at each time bin. In the figure, the time bins are converted into distance (R=ct/2 where t is the time bin) and the frequencies are converted into velocity [$V=-\lambda/2\ (f-f_{IF})$ where λ is the laser wavelength, f is the frequency and f_{IF} is the frequency shift of the pulses]. For each bin, the velocity is obtained by determining the maximum of the spectrum.

The wind field velocity is measured using a scan of the lidar between -20° and 20°. Figure 2 b) shows an example of modelled wind field measurement for a vortex structure.

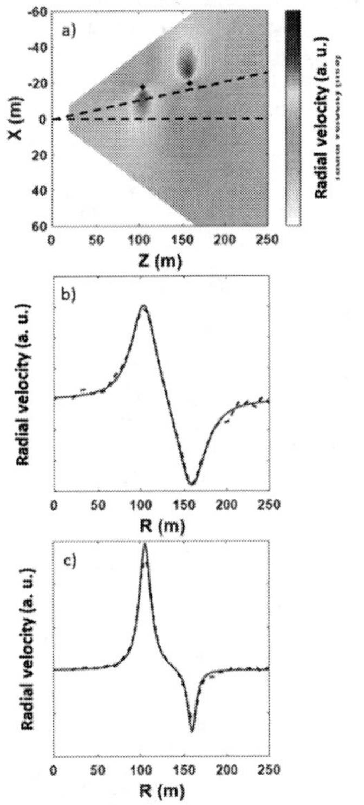

Figure 4: a) Simulated lidar wind field measurement for a two-vortex wind field structure. Comparison of the simulated wind velocity (red curves) and the wind velocity determined from the maximum of the synthetic lidar spectra (dashed blue curves) at an angle θ=0° (b) and θ=-5.5° (c).

The spatial resolution for wake vortex measurement was estimated using simulations. A vortex wind field was determined using the Hallock-Burham model [18] that best fit the data shown in Fig. 3 b). The lidar was simulated to determine the spectra for angles between - 20° and 20°. Figure 4 a) shows the retrieved wind field obtained from the maximum of the synthetic spectra.

The vortex wind field structures are clearly observables with the positive and negative wind around the vortex core. Experimentally this structure is also well shown which allows determining precisely the vortex core. Post-treatments show that the core position can be determined within less than 1 m.

Figure 4 b) and 4 c) shows the effect of the lidar on the wind profile. At θ=0°, the lidar axis is at ~15 m from the first vortex core position and 20 m from the second vortex core position. In this case, the lidar allows measuring precisely the velocity profile. In the second case (θ=-5.5°), the lidar axis is at 5 m (respectively 3 m) from the first (respectively second) vortex. In this case, the maximum velocity is under estimated by 8 % and 10% respectively. In conclusion, the lidar allows measuring vortex wind velocities with an excellent precision affecting only the maximum velocity between 3 m and 8 m from the vortex core with a maximum difference of 10 %.

4. SIGNAL PROCESSING ON TURBULENT WIND FIELD

A second application of the lidar was tested to characterize turbulent wind field.

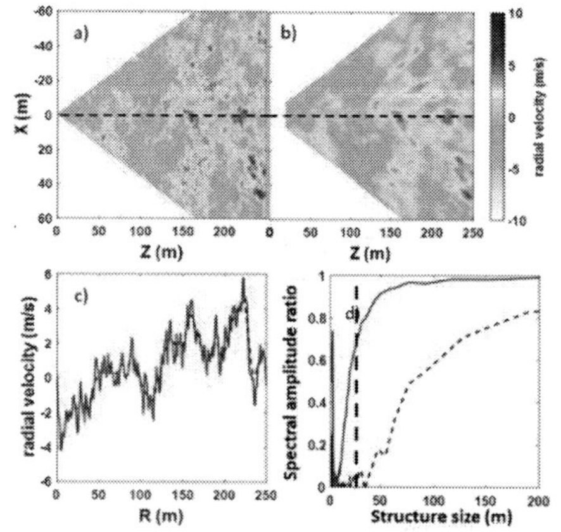

Figure 5: a) Simulated turbulent wind field for a moderate turbulence. b) Velocities determined after processing of the simulated lidar spectra using the wind field in a). c) Comparison of the simulated wind profile at θ=0° with the wind profile determined from the synthetic lidar spectra. Comparison of the spectral amplitude of the velocity profiles as a function of the size of the velocity structure for the 75-ns square pulse (solid red curve) and for an energy equivalent 200-ns-FWHM Gaussian pulse (dashed blue curve). For the 75-ns pulse lidar, the spatial resolution of the lidar of about 25 m is plotted (dashed black line).

Figure 5 a) shows a turbulent wind field simulated using a Von Karman model [19, 20] with a characteristic scale length of 200 m and a velocity variance of 3.2 m/s (typical parameters for moderate atmospheric turbulence). The lidar was simulated to determine the spectra.

In Fig. 5 b), we see that after processing of the synthetic lidar spectra, most of the wind structure can be distinguished. This is more apparent when the velocity profiles obtained for one line of sight are compared [Fig. 5 c)]. It shows that the amplitude of the high spatial frequencies of the velocity has a low magnitude (according to the model) so that the error in velocity is lower than 1 m/s throughout the profile.

To estimate more accurately the lidar filtering on the amplitude of the spatial frequencies of the velocity, we have calculated the Fourier transform of the velocity profiles for each angle for the simulated wind field and for the reconstructed wind field. The spectral amplitude ratio was obtained by dividing the spectral amplitudes of the reconstructed profiles by ones of the simulated profiles [Fig. 5 d)]. This ratio was determined using a χ^2 analysis over all the lineouts. It was plotted as a function of the structure sizes that correspond to the inverse of the spatial frequency. Figure 4 d) shows that the amplitude ratio remains large until the size of the structures gets lower than the spatial resolution of the lidar (22.5 m). In addition, using this curve, the statistic of measured frequencies can be reconstructed up to ~10 m by dividing the measured amplitude by this factor.

In Fig. 5 d), this curve was compared with the curve that corresponds to the reduction of the spectral amplitude ratio induced by a lidar that uses a 200-ns Gaussian pulse shape with equivalent energy per pulse. It shows that, for the 200-ns pulse shape, this ratio gets very small (<0.1) below 60 m -70 m (which roughly corresponds to the spatial resolution for this lidar) and is significantly lower than for the 75-ns pulse shape for all the structure sizes shown here. This shows that this new pulse shape design results in a strong improvement of the ability of the lidar to measure small scale structures.

5. CONCLUSION

We have designed and tested a special fiber lidar to retrieve wind field velocity with a high spatial resolution. It has been tested experimentally on wake vortices. Its performances on turbulent wind field have been modelled and a strong improvement compared to 200-ns Gaussian pulse is expected.

ACKNOWLEDGEMENTS

The authors wish to acknowledge Airbus for the support of this study.

REFERENCES

[1] W. L. Eberhard and R. E. Cupp J. Atmos. Oceanic Technol 6 809-819 (1989)
[2] R. Frehlich J. Atmos. Oceanic Technol 14 54-75 (1997)
[3] R. Frehlich and R. Cornman Appl. Opt. 38 7456-7466 (1999)
[4] R. Frehlich and R. Cornman J. Atmos. Oceanic Technol 19 355-366 (2006)
[5] S. Florian et al. J. Phys.: Conf. Ser. 318 072019 (2011)
[6] M. Shinozuka and C. M. Jan J. of Sound and Vibration 25(1) 111-128 (1972)
[7] R. Frehlich et al. J. of appli. Meteo. And clim. 45 821-837 (2006)
[8] R. Targ, et al. Appl. Opt., 30 2013–2026 (1991)
[9] M. Harris et al. Aerospace Sci. and Tech. 6 325–331 (2002)
[10] F. Köpp et al. J. Atmos. Oceanic Technol 21 194-206 (2004)
[11] A. Dolfi-Bouteyre et al. AerospaceLab Journal AL01-07 (2009)
[12] K. El Omari et al., Monografías del Seminario Matemático García de Galdeano 31, 545–554 (2004)
[13] T. Mikkelsen et al. Proceedings online European Wind Energy Association (2010)
[14] S. Ning, A., Flanzer, T. C., & Kroo, I. M. (2011). Aerodynamic performance of extended formation flight. Journal of aircraft, 48(3), 855-865
[15] S. W. Henderson et al. Opt. Lett., 16, 773–775 (1991)
[16] J.-P. Cariou et al. 13th Coherent Laser Radar conference (CLRC), Kamakura (2005).
[17] A. Dofli-Bouteyre et al. 19th Coherent Laser Radar conference (CLRC), Okinawa (2018).
[18] D. C. Burnham and J. N. Hallock Chicago monostatic acoustic vortex sensor system. Report No. DOT-TSC-FAA-79-103.IV 206pp (1982)
[19] V. Karman Proc. Nat. Acad. Sci. 34 530-539
[20] D. K. Wilson U.S. Army Research Laboratory, ARL-TR-1677 (1998)

COMPARISON BETWEEN SINGLE-APERTURE AND MULTI-APERTURE APPROACHES TO COMPUTER CORRECTION OF IMAGES

Vadim Dudorov, Anna Eremina

V.E. Zuev Institute of Atmospheric Optics, Rus. Acad. Sci., Tomsk 634021, Russia

ABSTRACT

The effectiveness of the reconstruction of images formed by single- and multi-aperture systems and distorted by an atmospheric turbulence is analyzed in this work. Based on numerical simulation, we show that the use of multi-aperture observation systems for the computer correction of atmospheric distortions under anisoplanar conditions can significantly reduce the exposure time. Main distortions are well corrected during image synthesis for a short exposure time, which corresponds to a "frozen" turbulence. The time required for the correction of residual small-scale distortions is an order of magnitude shorter than in the case of synthesis of long-exposure images with the use of traditional single-aperture observation systems.

1. INTRODUCTION

Methods for post-detection (computer) correction of images are often used in observations of distant objects through the atmosphere. One of the most effective approaches to computer correction is based on the solution of the inverse problem of retrieval of the structure of an object from its blurred image [1, 2]. For the correct solution of the inverse problem (deconvolution) under conditions of strong turbulence, the accumulation of images is commonly used [2–5]. A disadvantage of this approach to the formation of isoplanar distorted images is impossibility of analyzing dynamically changing scenes of observation.

An alternative to the long-exposure method for the formation of isoplanar distorted images is an approach based on the division of the receiving aperture [1, 7–8]. This approach can be implemented in multi-aperture systems when several identical subimages are simultaneously recorded: the number of subimages is determined by the number of receiving subapertures. To solve the inverse problem in this case, an image is synthesized as the sum (superposition) of simultaneously recorded subimages. In addition to averaging the inhomogeneity of short-exposure distortions, which is required for correct solution of the inverse problem, this approach allows compensation of local (within each subaperture) tilts of the wavefront arrived from an object and distorted by turbulent air inhomogeneities.

In this work, we compare the effectiveness of the reconstruction of turbulence-distorted images when using single-aperture and multi-aperture observation systems.

2. LONG-EXPOSURE IMAGING BASED ON SINGLE-APERTURE APPROACH

If an object is observed through the atmosphere along horizontal paths, short-exposure images formed during a time substantially shorter than the time of correlation of turbulent air inhomogeneities are characterized by strongly anisoplanar distortions, which is a significant obstacle in computer reconstruction of images using deconvolution algorithms. In the case of long exposure, anisoplanar distortions are averaged for a time much longer than the time of correlation of turbulent air inhomogeneities, which results in uniform blurring of the entire image. Although the width of the long-exposure point spread function (PSF) in this case is larger than the effective width of the short-exposure PSF, the isoplanatism of a long-exposure image allows the deconvolution to be performed the most accurately.

Figure 1 shows how the quality of an image synthesized by a single-aperture system depends on the exposure time, which was determined in our numerical experiment by the number of realizations of turbulent air inhomogeneities over which short-exposure images were averaged. It is seen that the anisoplanar distortions are averaged as the number of pseudo-random realizations of a "frozen" turbulent medium increases. The sum of five non-correlated short-exposure images already allows much better identification of an object than any short-exposure realization. An increase in the number of realizations of turbulence to 100 during long-exposure imaging allows us to speak about almost isoplanar distortions even under sufficiently strong turbulence along an observation path. It should be noted that an increase in the receiving aperture increases the rate of averaging of anisoplanar distortions.

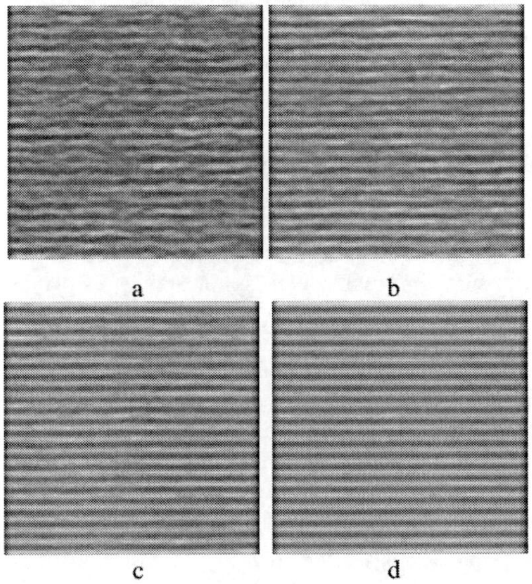

Figure 1. Effect of the exposure time on image parameters in anisoplanar turbulence. The observation range $z = 0.05ka_t^2$; $D/r_0 = 9.6$, where $D = 2a_t$. $N_{pix} = 512$; averaging over (a) 1, (b) 5, (c) 20, and (d) 100 realizations of short-exposure images.

The operation of reconstruction of the object structure from its image based on fast Fourier transform *FT*, used in the numerical simulation, is written as

$$I_{obj}^{rec}(\mathbf{r}) = FT^{-1}\left[\frac{FT[I_{im}(\mathbf{r})]}{FT[PSF(\mathbf{r})]}\right], \quad (1)$$

where $I_{im}(\mathbf{r})$ is the distorted image and *PSF* is a selected point spread function.

Figure 2 shows that the reconstruction of a short-exposure image by means of deconvolution increases the image contrast, but almost does not improve the object identification. As the exposure time increases (the number of pseudo-random short-exposure images), the inhomogeneity of distortions decreases. In addition to the object contrast, a possibility of object identification (resolution) significantly enhances.

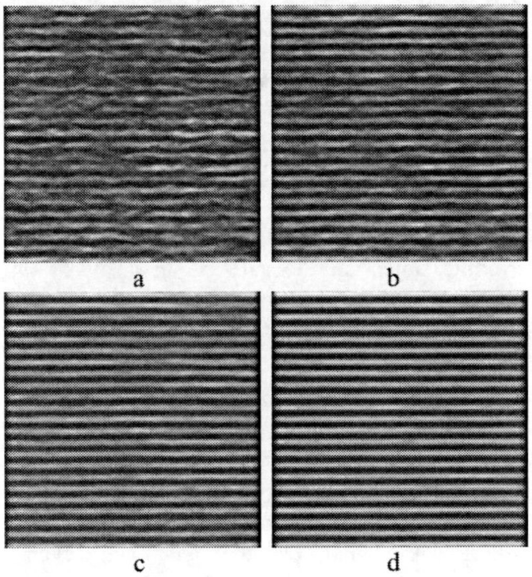

Figure 2. Effect of the exposure time on the efficiency of reconstruction of the object structure in anisoplanar turbulence. The conditions are similar to those in Fig. 1.

3. LONG-EXPOSURE IMAGING BASED ON MULTI-APERTURE APPROACH

Traditional systems with a single focusing lens (system of lenses along the optical axis) form a single image of an object observed on a CCD array. A multi-aperture system, considered in this work, forms a number of images, where their number is determined by the number of receiving subapertures. By analogy with corrected long-exposure images, an image synthesized by this system is understood as a sum (superposition) of the subimages with corrected random turbulent shifts.

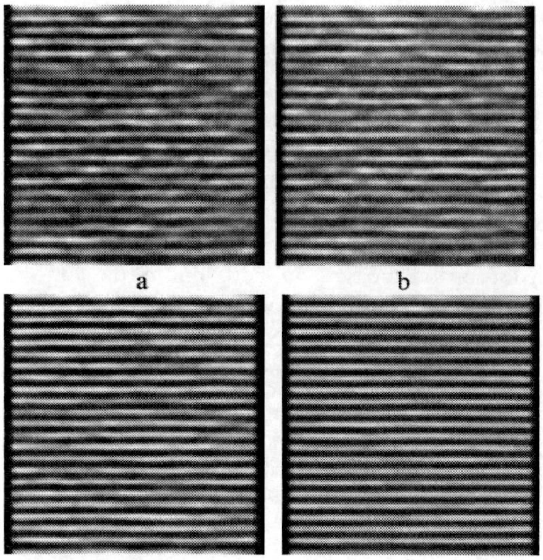

Figure 3. Reconstructed image of a periodic harmonic object in the case of an array of 64 subapertures for the number of short-exposure realizations N_{SE} = 1 (a), 2 (b), 5 (c), and 20 (d) used for the image synthesis.

It can be seen from Fig. 3 that the deconvolution method allows much better identification of an object even with the use of a short-exposure image synthesized in the multi-aperture system than in the case of a single-aperture system. When averaging the image synthesized over several short exposures, the quality of the image reconstructed noticeably improves. An increase in the number of short-exposure images synthesized in the multi-aperture system up to 10–20 can significantly improve the quality of correction.

4. RESULTS AND DISCUSSION

The quality of a reconstructed image of a single-aperture observation system synthesized of long-exposure images noticeably improves when the number of short exposures $N_{SE} > 100$. A significant decrease in the number of short-exposure images when using multi-aperture observation systems (see Fig. 3) can be explained by the fact that the strongest large-scale distortions can be corrected by superposition of subimages for each short exposure in this case. And accumulation of short exposures allows one to get rid of residual small-scale distortions.

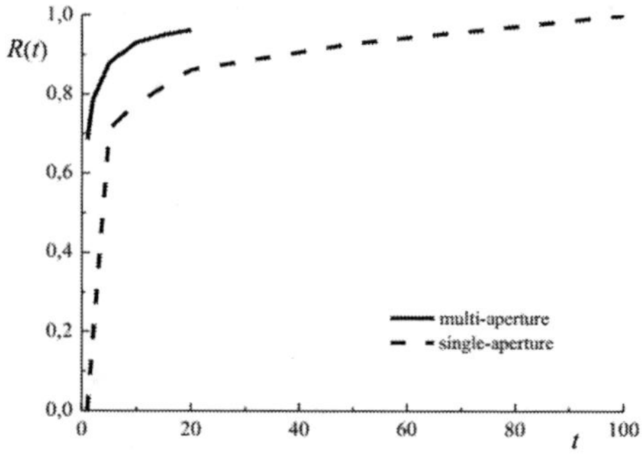

Figure 4. The coefficient of correlation between reconstructed and non-distorted images as a function of the number of short-exposure frames for multi- and single-aperture systems.

Thus, we can conclude that the use of multi-aperture observation systems for computer correction of atmospheric distortions in an anisoplanar turbulence can significantly reduce the exposure time (accumulation of short-exposure images) when synthesizing isoplanar distorted images. Main distortions are well corrected when synthesizing images for a short exposure time, which corresponds to a "frozen" turbulent medium. The time required for the correction of residual small-scale distortions is an order of magnitude shorter than in the case of synthesizing long-exposure images using a single-aperture observation system.

ACKNOWLEDGEMENT

The work was supported by the Russian Science Foundation (project No. 18-79-00179).

REFERENCES

[1] Invanov, M. and McGaughey, D., "Image reconstruction by aperture diversity blind deconvolution," Proc of the AMOS Technical Conference, E78 (2007).

[2] Huebner, C. S. and Greco, M., "Blind deconvolution algorithms for the restoration of atmospherically degraded imagery: a comparative analysis," Proc. SPIE 7108, 71080M-1-12 (2008).

[3] Schulz, T., "Multiframe blind deconvolution of astronomical images," J. Opt. Soc. Am. A 10(5), 1064–1073 (1993).

[4] Averin, A. P., Morozov, Yu. B., Pryanichnikov, V. S. and Tyapin, V. V., "Computer correction of turbulent distortions of image of extended objects on near-Earth paths," Quant. Electron. 41(5), 475–478 (2011).

[5] Zhu, X. and Milanfar, P., "Image reconstruction from videos distorted by atmospheric turbulence," Proc. SPIE 7543, 75430S-1-8 (2010).

[6] Hope, D. A., Jefferies, S. M., Hart, M. and Nagy, J. G., "High-resolution speckle imaging through strong atmospheric turbulence," Opt Express 24(11), 12116-12129 (2016).

[7] Dudorov, V. V. and Eremina, A. S., "Digital compensation of turbulent distortions of incoherent optical images using multiaperture systems," Opt. Atmos. Okeana 31 (02), 95–102 (2018) [in Russian].

[8] Dudorov, V. V. and Eremina, A. S., "Computer correction of turbulent distortions of incoherent optical images using multi-aperture systems," Proc. SPIE 10787, 107870E -1-9 (2018).

https://doi.org/10.1051/epjconf/202023706004

DETECTION OF COMPLEX TERRAIN-INDUCED WIND SHEAR BY DOPPLER LIDAR AT BEIJING CAPITAL INTERNATIONAL AIRPORT

Xiaoying Liu[1*], Songhua Wu[1, 2], Hongwei Zhang[1], Jianjun Zhang[3], Zhiqiang He[3], Xi Zhang[3]

[1] *Ocean Remote Sensing Institute, College of Information Science and Engineering, Ocean University of China, Qingdao 266100, China.*

[2] *Laboratory for Regional Oceanography and Numerical Modeling, Pilot National Laboratory for Marine Science and Technology (Qingdao), Qingdao 266237, China.*

[3] *North China Regional Air Traffic Management Bureau of CAAC, Beijing 100621, China.*

**Email: lxy6302@stu.ouc.edu.cn*

ABSTRACT

In November 2018, the lidar-based wind shear synchronous experiment was performed at Beijing Capital International Airport (BCIA). In this experiment, aiming at the measurement of the terrain-induced wind shear and the wind field around the runway, the glide path scanning mode, and the RHI strategy were conducted alternately. Radial velocity retrieved from the glide path scanning can obviously present the wakes caused by complex terrain (e.g., hills, tall trees, residential and terminal buildings). The Pulse Coherent Doppler Lidar (PCDL) warned the terrain-induced wind shear, which was verified by the pilot report. The wind field structure around the runway under the wake effect and the building shielding effect is also analyzed.

1. INTRODUCTION

Wind shear has a great threat to aviation safety according to statistical reports [1]. Wind speed or wind direction changes rapidly during wind shear, which alters the lift of aircraft and causes deviations from the predetermined flight path. On account of the limitations of conventional wind shear detection instruments (such as anemometer networks, weather radars, and wind profilers), lidar has gradually become the popular technology to observe and alert wind shear, with its high spatial-temporal resolutions and flexible scanning modes [2]. Hong Kong International Airport (HKIA), which is often plagued by wind shear due to its special geographical location, introduced lidar to the airport for the first time for routine observations throughout the day and developed an alerting system for extensive researches [2-3].

Recently, wind shear studying has made great progress based on lidars in the design of effective wind shear scanning strategies [4-7] and the accuracy of alerting methods [8-10]. And the distribution law of wind shear has been further explored through wind shear statistical characteristics and turbulence feature analysis along the glide path [7-8, 11-12]. Wind shear can occur in a wide range of weather conditions, like terrain effects under stably stratified boundary layer, sea breeze with sunshine, and gust or microburst in severe convection weather. The terrain-induced wind shear becomes difficult for warning due to its high spatial and temporal variability wind field, which plays a dominant role in many factors with accounting for 70% in HKIA [2].

This paper explores the complex terrain-induced wind shear by using Pulse Coherent Doppler Lidar (PCDL) and the wind field structure around the runway under the wake effect and the building shielding effect at Beijing Capital International Airport (BCIA).

2. LIDAR EXPERIMENT AND METHODOLOGY

As the largest passenger throughput airport in China, BCIA is located in the center of Beijing. Although there are no high mountains around it, the presence of hills, trees, and buildings makes the wind field more complicated. The weather conditions in Beijing are controlled by a typical monsoon climate. Therefore, the northerly wind and northwesterly wind become prevailing in the winter and early springtime at BCIA, accompanied by more frequent windy weather.

BCIA contains three runways, namely 36L/18R, 36R/18L and 01/19, as shown in Fig.1. Statistics data reported by the pilots (from 2015, 2016 and 2018) showed that the wind shear mainly occurred

https://doi.org/10.1051/epjconf/202023706004

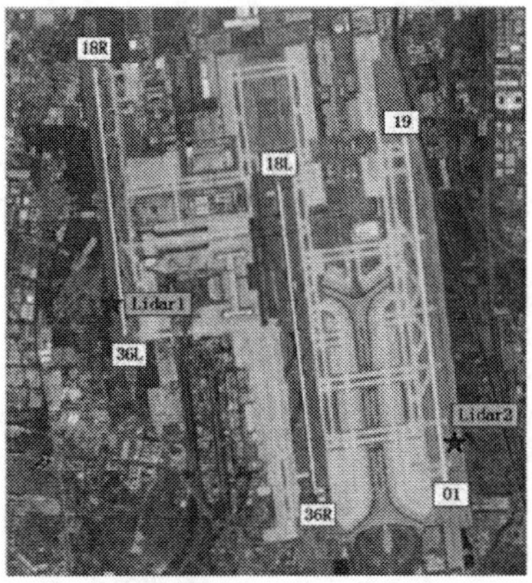

Fig.1. The geographical environment of BCIA.

in the 36L and 01 runway at BCIA, accounting for 43% and 31% respectively, as shown in Fig.2. Subsequently, the wind direction when the pilots reported the occurrence of wind shear in 36L and 01 runway corridor (from 2018) was analyzed in Fig.3, and the results reveal that the northwester, norther, and wester prevailed. In consideration of the geographical environment around BCIA, complex terrains under the above-mentioned prevailing wind should be responsible for this. The residential buildings, tall trees and small mounds on the west of 36L/18R runway and the terminal buildings which are adjacent west side to 01/19 will cause disturbance to the wind field along the glide path in 36L and 01.

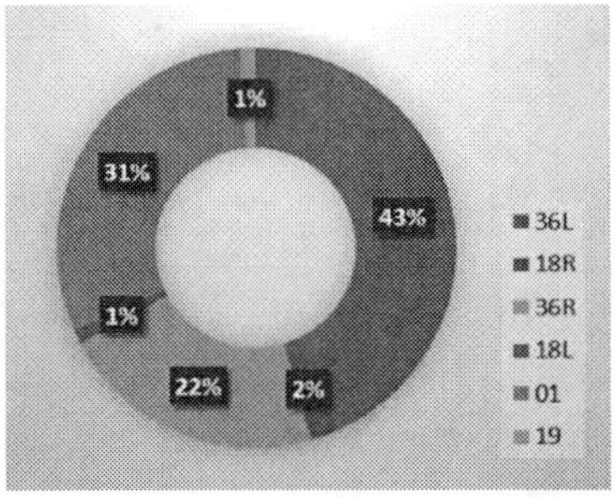

Fig.2. Wind Shear statistics reported by the pilots in different runways at BCIA (from 2015, 2016 and 2018).

Lidar has an advantage in exploring the complex terrain-induced wind shear with a rapid change of wind field in the clear sky. Since January 2018, BCIA has introduced two lidars located at 36L (Lidar1) and 18R (Lidar 2) runway for scientific

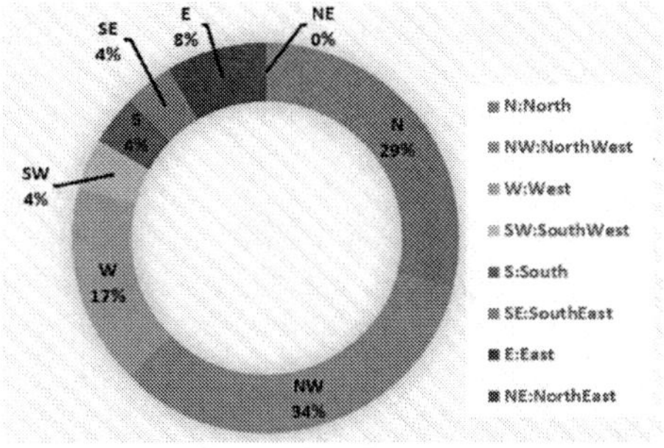

Fig.3. Wind direction statistics during wind shear at 36L and 01 runway corridor at BCIA (from 2018).

research. With the arrival of the winter monsoon, in November 2018, Lidar2 moved to 01 runway (as shown in Fig.1) to observe the terrain-induced wind shear under prevailing wind. During the experiment, Lidar1 and Lidar2 both adopted the glide path scanning strategies [2, 4] with the coordinate change of azimuth and pitch angles to carry out the synchronous observation. Table 1 lists the detailed specifications of the PCDL. In the experiment, the distance between the two lidar locations and the centerline of 36L and 01 runway are all 70m, so that the velocity along the laser beam can be as nearly equal to the real velocity along the glide path. And then, the radial velocities of lidars were used to construct the headwind profiles experienced by the aircraft along the glide path.

Table 1. Specifications of the PCDL system at BCIA.

Parameters	Specification
Radial Spatial Resolution	30m
Measurement Range*	60 to 4000 m
Wavelength	1.5 μm
Pulse Energy	150 μJ
Pulse Repetition Frequency	10 kHz
Scanner Positioning Accuracy	0.1°
Resolution of Velocity	0.1 m/s

* Weather condition related

3. RESULTS

3.1 Complex terrain wakes and wind shear

In December each year, northerly and northwest winds are prevalent at BCIA. The runway corridor 01 and 36L are located downwind of complex terrain (as shown in Fig.1), which is susceptible to the wakes of the hills (or tall trees) and buildings. Fig.4 shows the obvious influence of wakes formed by the terminal under prevailing wind in 01 runway. The wind strips will disturb the velocity along the glide path and increase the bumpiness of the aircraft with potential hazards. The pilot reported the presence of wind shear in 01 runway at 19:03 (local time) on December 27, 2018, at this time, the background wind field is shown in the right half of Fig.4. Meanwhile, the PCDL also issued a wind shear alert, as illustrates in Fig.5. The velocity dropped from 17.7m/s to 8.9m/s during landing from far afield to the touchdown points, with changing about 8.8m/s (17kt) which is much more than 7.7m/s (15kt) adopted by the International Civil Aviation Organization (ICAO).

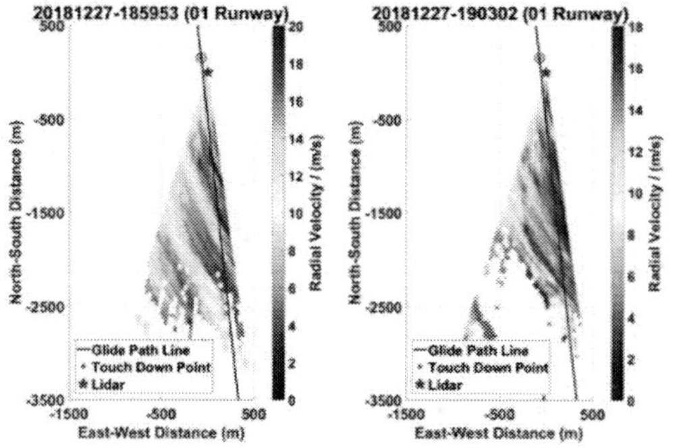

Fig.4. Glide path radial velocity of 01 runway on December 27, 2018.

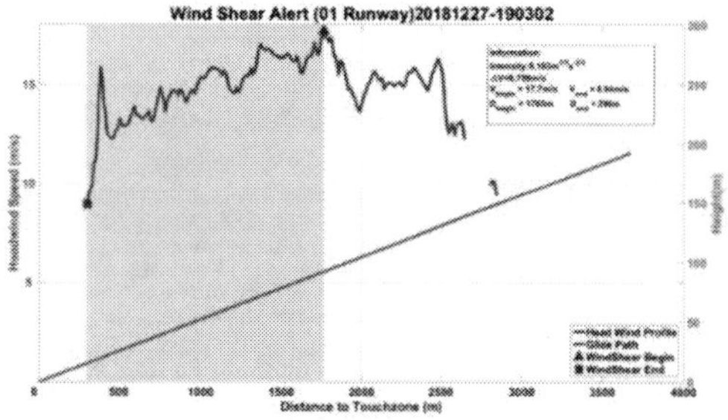

Fig.5. Headwind profile of 01 runway on December 27, 2018.

During this period, under the influence of strong northwest wind, the velocity in the 36L runway was affected by the wakes of tall trees, hills, and residential buildings, as the stripes formed in Fig.6. Lidar also issued a wind shear signal that the difference of velocity was up to 10.6m/s (20kt) which depicted the situation is more serious than 01 runway at the same time (If the time difference is less than 1.5min, it is considered to be synchronous between 36L and 01 runways in this paper.) as shown in Fig.5 and Fig.7. In addition to the wake effect, the smaller velocity at the lower altitude (the blue area near the lidar, as shown in Fig.6) should contribute to this serious situation. Fig.7 presents that the higher-altitude velocity was maintained above 15m/s, thus forming a sharp velocity change along the glide path.

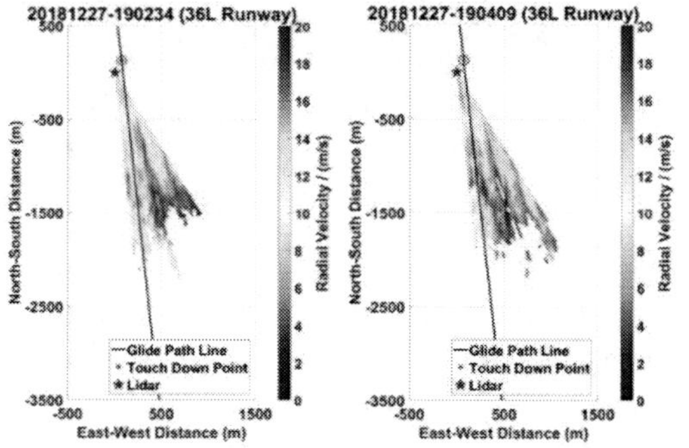

Fig.6. Glide path radial velocity of 36L runway on December 27, 2018.

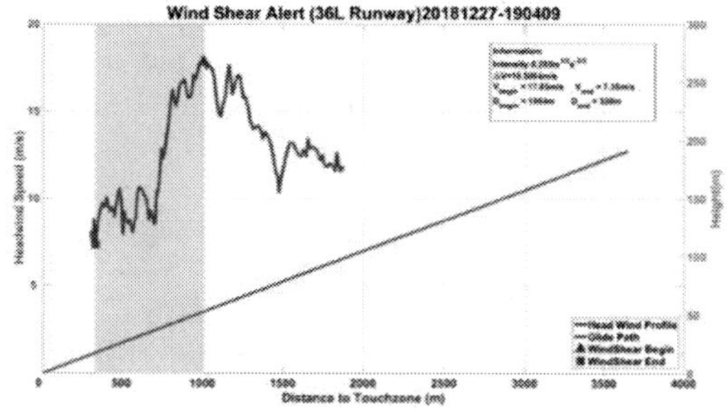

Fig.7. Headwind profile of 36L runway on December 27, 2018.

3.2 Complex topography and wind field along the runway

In this experiment, the RHI scanning strategy and the glide path scanning mode were designed alternately. For the RHI scanning, the azimuth of PCDL was consistent with the 01/19 runway's azimuth, and the pitch angle varied between 0 and 45 degrees. Wind shear alerts were issued by the PCDL at 03:40 and 03:47 (local time) on January 04, 2019, respectively. The glide path radial velocity figures showed evident stripes that are the terminal building wakes, which was not shown in this paper. Fig.8 shows the radial velocity retrieved by RHI scanning for the above time period. In this graph, with zero representing the location of PCDL, the positive distances represent the positions at the south of lidar along the runway and vice versa for the northern positions. In the upper graph of the Fig.8, there are two lower-velocity zones in the wind field below the height of 100m at 600m and 1700m account of wake effect. And the flight glide path just passed through the two zones, which were accompanied by the wind shear or turbulence. In the lower part of Fig.8, a wind area of lower-velocity was distributed at the distance of 60m to 1600m below the altitude of 200m, mainly because this area is located on the east side of the terminal building (referring to Fig.1) with strong shielding effect.

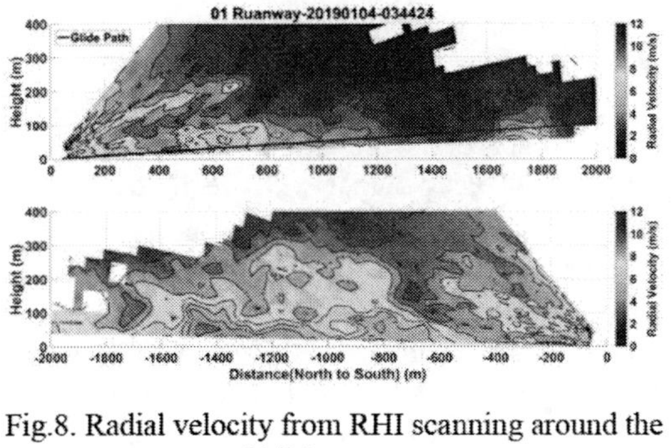

Fig.8. Radial velocity from RHI scanning around the 01/19 runway.

4. CONCLUSIONS

Glide path mode designed to detect the wind shear and RHI scanning strategy used to explore the wind field around the runway observed alternately in this synchronous experiment between 36L runway and 01 runway. The stripes appearing in the glide path scanning radial velocity indicate the wake effects produced by complex terrain. The PCDL warned the terrain-induced wind shear, and which was verified by the pilot report. The results show that the difference between higher-velocity and lower-velocity at different altitudes in 36L runway exacerbates the variation of the headwind profile and brings a more severe wind shear situation. The structure around the runway can be obtained by RHI scanning. The complex terrain wake will form lower-velocity areas along the glide path, which will bring the wind shear and turbulence. In addition, under the shielding effect of the terminal building, a wide range of lower-velocity was formed at low altitudes.

ACKNOWLEDGEMENTS

This work was supported by the National Key Research and Development Program of China, grant number 2016YFC1400904 and North China Regional Air Traffic Management Bureau of CAAC. We thank our colleagues including Xiaoye Wang and Xiaomin Chen from Ocean University of China for preparing and conducting the experiment.

REFERENCES

[1] Hallowell, R.G., et al. Lincoln Laboratory Journal 18 (2) (2010)
[2] Shun, C. M., et al. Journal of Atmospheric and Oceanic Technology 25(5): 637-655 (2008)
[3] Shun, C M, et al. 10th Conference on Aviation, Range, and Aerospace Meteorology (2002)
[4] Zhang, H, et al. 19th Coherent Laser Radar Conference (2018)
[5] CHAN, P. W., et al. 12th Conference on Aviation, Range, and Aerospace Meteorology (2006)
[6] Chan, P. W, et al. Journal of Atmospheric and Oceanic Technology 29(2): 207-220 (2012)
[7] Zhang, H, et al. Infrared Physics & Technology 96:113-122 (2019)
[8] Chan, P. W, et al. Meteorologische Zeitschrift, 20(6): 661-670 (2011)
[9] Chan, P. W. Meteorologische Zeitschrift 21(2): 193-204 (2012)
[10] Lee, Y. F, et al. Meteorological Applications 21(1): 86-93 (2014)
[11] Chan, P. W. 12th Conference on Aviation, Range, and Aerospace Meteorology (2006)
[12] Chan, P W. Meteorologische Zeitschrift 19(6): 549-563 (2010)

https://doi.org/10.1051/epjconf/202023706005

DEVELOPMENT OF AN ALL-FIBER COHERENT DOPPLER LIDAR IN THE IAO SB RAS

Artem Sherstobitov[1,2]*, Viktor Banakh[1], Alexander Nadeev[1], Igor Razenkov[1], Igor Smalikho[1], Andrey Falits[1]

[1]*Institute of Atmospheric Optics of Siberian Branch of the Russian Academy of Sciences, Russia,*
[2]*National Research Tomsk State University, Russia, *shrarm@iao.ru*

ABSTRACT

Paper presents a model of the all-fiber pulsed coherent Doppler lidar (IAO-lidar) build in the IAO SB RAS. Here is described lidar design, the algorithm for processing of lidar signals and the software-hardware system that implements signal processing in real time, created with the use of open source software. The results of joint measurements of the radial velocity by the IAO-lidar and the HALO Photonics (Stream Line) lidar are given.

1. INTRODUCTION

For experimental study of turbulent wind fields of the atmospheric boundary layer and coherent structures, including aircraft wake vortex, a pulsed coherent Doppler lidar (IAO-lidar) was recently build at the Institute of Atmospheric Optics (IAO) of the SB RAS. This paper presents design features of the IAO-lidar, algorithm for processing lidar signals, its implementation in real time, and comparative measurements of the radial velocity by IAO-lidar and by HALO Photonics (Stream Line) lidar.

2. LIDAR DESIGN

A construction of the developed IAO-lidar is shown in Figure 1. CW laser radiation (20 mW, 1550 nm) with a frequency f_0 passes through coupler 75/25, enters the isolator (75%), and then goes to the circulator. After the circulator there is an acoustic optical modulator that generates the sounding pulses. At the other end of the modulator we have the Bragg grating (FBG), which acts as a selective mirror. Radiation passes through the modulator twice. Double passage of radiation allows suppress the signal leakage after the sounding pulse. This is the basic difference between our design and the well-known [1]. The frequency of the radiation after the modulator is increased by $f_i = 80$ MHz. The optical pulse after double passage through the modulator returned to the circulator and redirected to an erbium amplifier (EDFA). Second circulator is combined with EDFA in one module.

Sounding pulse with the frequency $f_0 + f_i$ goes to the atmosphere via telescope. The backscattered signal coming from the atmosphere due to Doppler effect acquires an additional frequency shift f_d, depending from the radial component of the wind speed. Two inputs of the coupler 50/50 receive CW laser signal with frequency f_0 and continuous echo from the atmosphere with frequency $f_0 + f_i + f_d$. Coupler divides CW signal and return in half, and then both go to the two inputs of the balanced PIN detector. Balanced detector extracts $f_i + f_d$ frequency component. The pulse repetition frequency is 10 kHz, the average outgoing power is 400 mW. Aperture of the transceiver telescope is 50 mm. The system is equipped with the scanner.

The fiber junctions in Figure 1 are shown with an inclined short lines. All fibers maintaining radiation polarization (PM). To prevent reflections at the joints, the type of connectors FC/APC is used.

3. LIDAR SINGNAL PROCESSING ALGORITHM

The radial velocity is calculated from the Doppler frequency shift of the signal $X(t)$ from the balanced PIN detector. Signals $X(t)$ are digitized by ADC, then from the signal array $X_l(nt_s)$ the signal power spectra $S_m(f_k)$ is calculated by the use of FFT with the averaging over L sounding pulses:

$$S_m(f_k) = \frac{t_S}{LNT_W}\sum_{l=1}^{L}\left|S_{ml}(f_k)\right|^2, \qquad (1)$$

where

$$S_{ml}(f_k) = \sum_{n=-N/2}^{N/2-1} X_l(nt_s + 2R_m/c)\exp(-\frac{2\pi jnk}{N})W(n),$$

t_s is the ADC sampling rate, $f_k = k\Delta f$, $\Delta f = (Nt_s)^{-1}$, $k = 0,1,2,...,N/2$, $S_m(f_k)$ is the spectrum at the range $R_m = R_0 + m\Delta R$, $m = 0,1,2,...,M$ is the number of the power spectrum, ΔR is the range gate length, $W(n)$ is the window function, T_W is the effective width of the window function , c is light speed.

The following parameters were used: t_s = 4 ns; the dimension of the Fourier transform N = 64, which corresponds to the spectral resolution Δf = 3.9 MHz; M = 390; a rectangular time window was used, that is, $W(n)$ = 1; for more accurate calculation of the radial velocity profile from the power spectra Fourier interpolation was used.

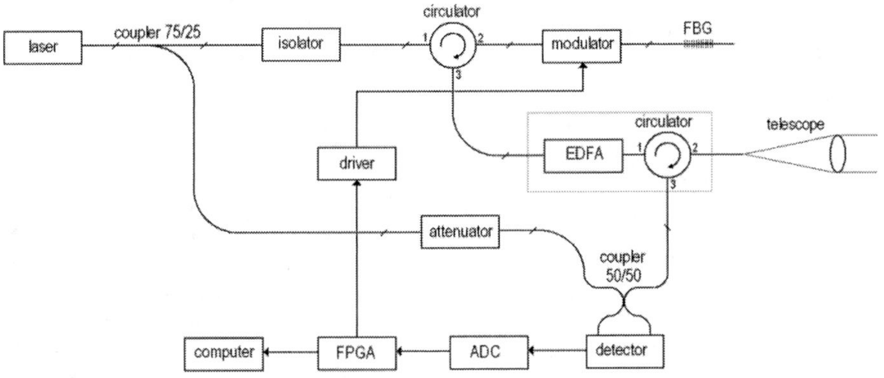

Fig. 1. Diagram of the all-fiber Doppler IAO-Lidar.

In the general case, the power spectrum $S(f)$ of a lidar signal can be represented as the sum of the signal $S_S(f)$ and noise $S_N(f)$ components [2]. To eliminate the influence of the noise component, two methods were applied: a) heuristic calibration method: the power spectra sequence obtained by subtraction of the blocked beam sequence from spectrum $S(f)$;

b) noise subtraction: this method is described in [2], that is, the spectrum $S_m(f)$ starting from the index M_1 is considered as the noise, thus, the average value of the noise spectrum is calculated by

$$< S_N(f) > = \sum_{m=M_1}^{M-1} S_m(f) \Big/ (M - M_1) \qquad (2)$$

Signal-to-noise ratio (SNR) was estimated as follows:

$$\mathrm{SNR}_m = \frac{\sum_{k=b_1}^{b_2}(S_m(f_k) - < S_N(f_k) >)}{\sum_{k=b_1}^{b_2} < S_N(f_k) >}, \qquad (3)$$

where $b_1 = (f_i - B/2)/\Delta f$, $b_2 = (f_i + B/2)/\Delta f$ and B = 50 MHz is the bandwidth for estimation of the radial velocity.

4. HARDWARE-SOFTWARE IMPLEMENTETION OF THE REAL TIME SIGNAL PROCESSING

The hardware-software module consists of analog-to-digital converter ADS4249EVM, TSW1400EVM board with field-programmable gate array (FPGA) STRATIX IV on it, and portable PC.

https://doi.org/10.1051/epjconf/202023706005

The FPGA firmware calculates averaged power spectra by the formula (1) in real time, and controls the sounding pulse (Figure 2). The calculation of averaged power spectra is a conveyor algorithm using the open Fourier conveyor transform module [3] (Figure 3). The firmware is written in the Verilog HDL in the Intel Altera Quartus II development environment.

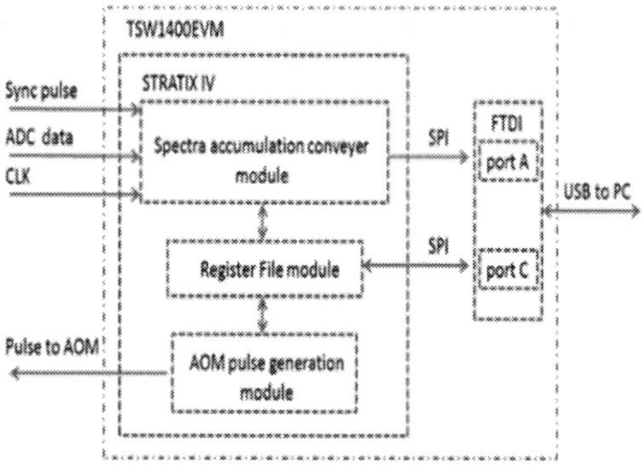

Fig.2. Software-hardware data processing system block diagram.

The PC program communicates with the FPGA via USB cable, connected to the FPGA board by using the 2 channels of FTDI chip, available on the TSW1400EVM. One channel are used for reading accumulated spectra (1) via SPI mode 3, another channel is also configured in SPI mode 3 and is used to read and write the block of registers to the FPGA.

The registers block is implemented as Verilog HDL module and is used for storing parameters (M, N in eq. 2, the time between the sounding pulses, etc.) Information about the ending current averaging by L sounding pulses is also recorded to the register file.

Real-time spectra-accumulating conveyer shown at figure 3 uses floating point (single precision float) based on open source Fourier conveyor transform module [3]. Floating point computations are quite slow, so we process four numbers of coming ADC samples per time. Parallel computations allow to reduce conveyer frequency from 250 MHz (ADC data-capturing frequency) to 62.5 MHz.

The PC program receives the average Doppler spectra as the input, from which the radial wind speed is calculated using the Doppler equation after Fourier interpolation of the spectra $S_m(f_k)$. The signal-to-noise ratio is calculated by Eq. (3).

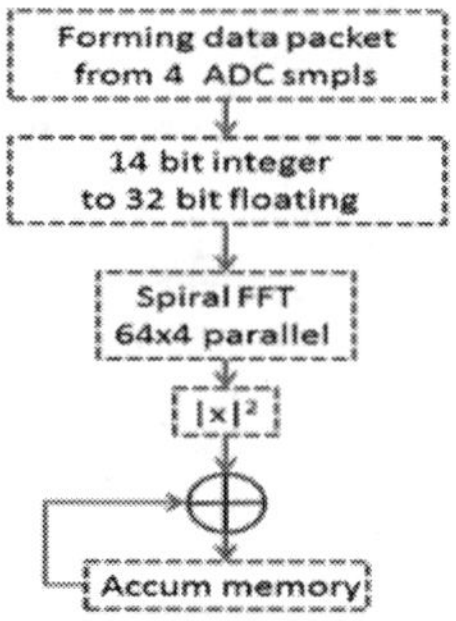

Fig.3. Spectra accumulating conveyer block, which implements equation (1) in real time.

5. EXPERIMENT

We carried out joint measurements of SNR and wind velocity by the IAO-lidar and the commercial lidar “Stream Line” made by HALO Photonics. Sounding of the radial velocity was held on fixed horizontal path (Fig. 4). Both paths were located in parallel at a distance of ~ 3 meters from each other. The results of sounding are presented at figures 5, 6. Figure 5 shows the range profiles of the signal-to-noise ratio. SNR curves point out that lidars characteristics are the similar. The sensing range is limited by 2 km.

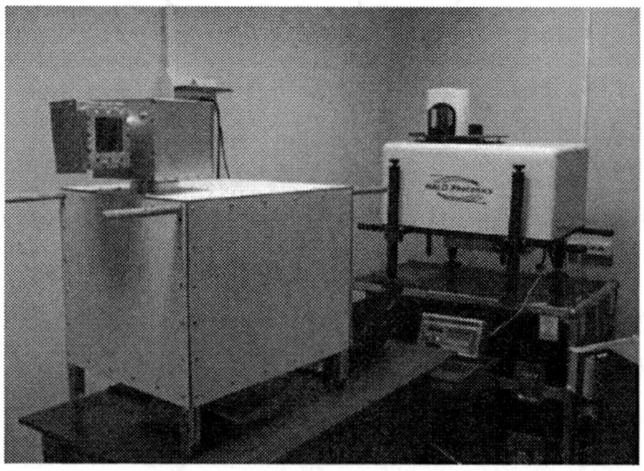

Fig. 4. Lidars: IAO -lidar (left) and "Stream Line" (right) during joint experiment.

 https://doi.org/10.1051/epjconf/202023706005

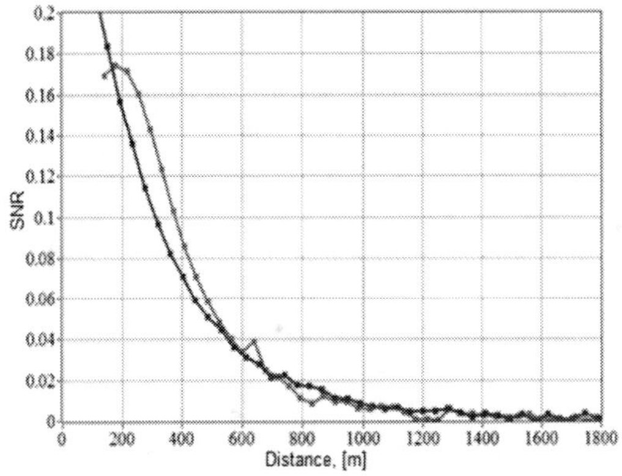

Fig.5. SNR measured by Stream Line lidar (blue line) and by IAO-lidar (red line) 04.02.2019.

The time dependence of the radial velocity measured by the IAO-lidar (Fig. 6) is similar to the results obtained from the Stream Line lidar data. The correlation coefficient between two curves is 0.84.

CONCLUSIONS

Laboratory coherent Doppler all-fibered lidar was built at the IAO SB RAS for field measurements during atmospheric turbulence research experiments. The transmitter design is differ from the known ones by the additional fiber circulator and FBG, which provide a double pass of radiation through the modulator. This allows one to suppress the signal leakage after the sounding pulse and increase the signal-to-noise ratio.

Software-hardware module performs signal processing in real time. Moreover, the module was created using open source software, which made the module more portable and flexible. The field experiment was carried out together with the Stream Line lidar. In a result we have good correlation between measured lidar's wind radial velocities.

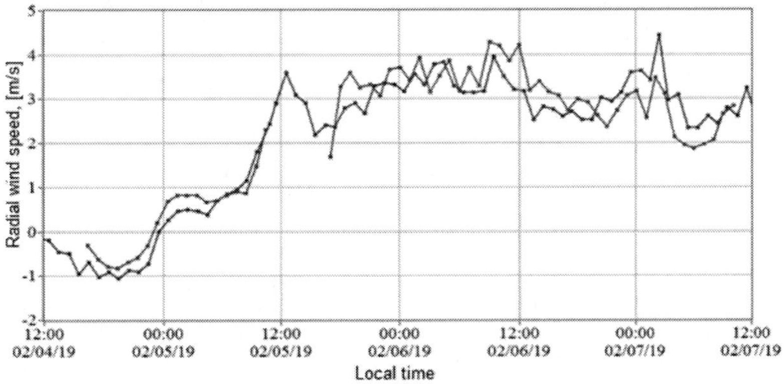

Fig.6. Time dependencies of radial wind speed at the distance of 442 m from lidars. Red line – IAO-lidar, blue line – "Stream Line" lidar. Average time is one hour.

REFERENCES

[1] S. Kameyama, T. Ando, K. Asaka, Y. Hirano, and S. Wadaka "Compact all-fiber pulsed coherent Doppler lidar system for wind sensing", Applied Optics, 46, 1953-1962 (2007)

[2] Banakh V.A., Smalikho I.N. Coherent Doppler wind lidars in a turbulent atmosphere. Boston – London: Artech House, 2013. 248 p.

[3] Verilog HDL generator of conveyer Fast Fourier transform: [link]. URL http://www.spiral.net/hardware/dftgen.html

DEVELOPMENT OF DIRECT-DETECTION DOPPLER WIND LIDAR FOR VERTICAL ATMOSPHERIC MOTION

Shoken Ishii[1,2*], Makoto Aoki[1], Kanna Tominaga[2], Tomoaki Nishizawa[3], Yoshitaka Jin[3], Nobuo Sugimoto[3], Yasukuni Shibata[2], Atsushi Sato[4, 1], Kaori Sato[5], and Hajime Okamoto[5]

[1]*National Institute of Information and Communications Technology, Koganei Tokyo 184-8795, Japan*
[2]*Tokyo Metropolitan University, Hino Tokyo 191-0065, Japan*
[3]*National Institute of Environmental Studies, Tsukuba Ibaraki 305-8506, Japan*
[4]*Tohoku Institute of Technology, Sendai Miyagi 982-8577, Japan*
[5]*Kyushu University, Kasuga Fukuoka 816-8580, Japan*
**Email: sishii@nict.go.jp*

ABSTRACT

Wind is fundamental in many atmospheric phenomena. Global wind profile observation is important to improve numerical weather prediction (NWP) and various meteorological studies. Wind profile observations are measured mainly by radiosonde networks. A Doppler Wind Lidar (DWL) is a useful remote sensing technique for wind measurement. DWL would provide us with a wind profile having high vertical resolution, low bias, and good precision. The National Institute of Information and Communications Technology (NICT) studies DWL has been developing various DWL. In the paper, we report development of a 355-nm direct-detection DWL and describe recent results of a 2-μm coherent DWL at NICT.

1. INTRODUCTION

Current space-based passive observing systems for wind measurement is limited [1]. The current space-based passive observing systems have a large coverage area and high temporal and horizontal resolutions but has a low vertical resolution. The World Meteorological Organization (WMO) wants to develop space-based wind profiling systems [2]. A Doppler Wind Lidar (DWL) can provide us with a wind profile with high vertical resolution, low bias, and good precision. A space-based DWL can fill the gap for the current space-based passive observing systems. The European Space Agency (ESA) launched a first space-based DWL called 'Aeolus' for the global wind profile observation in 2018 [3]. The National Institute of Information and Communications Technology (NICT) is studying the feasibility to realize the future space-based DWL [4-6]. NICT developed a ground-based 2-μm coherent DWL for CO_2 and wind measurements [7]. We have a plan to make wind measurement around and inside clouds with 355-nm direct-detection and 2-μm coherent DWL measurements. We are developing a ground-based direct-detection DWL. In the paper, we describe development and progress of the two DWLs.

2. DEVELOPMENT of 355-NM DIRECT-DETECTION DOPPLER WIND LIDAR

2.1 355-nm direct-detection lidar system

A direct-detection DWL has the advantage that wind measurement can be made any aerosol conditions because it uses the light backscattered by molecules (Rayleigh scattering). This feature is one of the characteristics of the direct-detection DWL. NICT is developing a direct-detection DWL with a double-edge technique. Figure 1 and Table 1 show a block diagram and specifications of the direct-detection DWL. The direct-detection DWL uses a flash-lump-pumped Q-switched Nd:YAG laser (Contuinum: Surelite I), an o Schmidt-Cassegrain type (Celestron: C14), a Fabry-Perot etalon, a knife-edge right-angle prism mirror, and two photomultiplier tubes (Hamamatsu: R9880U-113), The single-frequency a flash-lump-pumped Q-switched Nd:YAG operating at a wavelength of 0.355 μm can emit an output energy of 100 mJ with a pulse width of 10 ns (FWHM) at a pulse repetition frequency of 10 Hz. The wavelength of backscattered light is Doppler-shifted from that of the incident laser light by the moving molecules and aerosols. The backscattered light is collected by a telescope and split into two paths using a wedge plate. Each split light beam passes through a Fabry-Perot étalon (which functions as an edge filter) and is focused on each detector. The number

https://doi.org/10.1051/epjconf/202023706006

of photons, $N_{i=1,2}$ detected at each detector for a line-of-sight (LOS) wind speed of v_{LOS} (m/s) is given as Eq. (1),

$$N_{s,i}(v_{LOS}) = T_{BS} \cdot \frac{E_T \cdot Freq \cdot \Delta t}{h v_L} \cdot A_T \cdot T_O \cdot O_A(r) \cdot Q_e \cdot \frac{\Delta r}{r^2} \cdot T_A^2 \times \left[\beta_M(r) \cdot \int_{-\infty}^{\infty} T_{F,i}(v) \cdot \{G_L * G_M(v, \Delta v)\} dv + \beta_A(r) \cdot \int_{-\infty}^{\infty} T_{F,i}(v) \cdot \{G_L * L_A(v, \Delta v)\} dv \right], \quad (1)$$

where T_{BS} is the beam-splitter transmittance, E_T is the transmitted laser energy, Freq is a pulse repetition frequency, Δt is integration time, v_L is the laser frequency, A_T is the telescope area, T_o is optical efficiency of the system without etalon transmission, $O_A(r)$ is an overlap function, Q_e is quantum efficiency of the detector, Δr is range bin length, r is a distance to the target atmosphere, T_A is atmospheric transmission, $T_{F,i=1,2}(v)$ is transmission at a frequency of v for an edge filter centered at v_L, β_m and β_a are the molecule and aerosol backscattering coefficients, G_L and G_m are Gaussian line-broadening functions for the laser-line spectrum and the Rayleigh-Brillouin spectrum, L_a is a Lorentzian function for the aerosol-line spectrum, Δv is the Doppler-shifted frequency, the asterisk denotes a convolution operation, 'h' is Planck's constant, and 'c' is the light speed.

Since the photon noise is the dominant noise source, we assume that the standard deviation of N_i is given by a Poisson distribution. SNR_i is given as Eq. (2),

$$SNR_i = \sqrt{Freq \cdot \Delta t} \cdot \frac{N_i / Freq \cdot \Delta t}{\sqrt{N_i / Freq \cdot \Delta t + N_{DC} + N_{BG}}}, \quad (2)$$

where N_{DC} and N_{BG} are the dark current noise of the detector and the background noise. The location of the *i*th edge filter is determined by the following condition described by Flesia and Korb [8]. the LOS wind speed error σ_{LOS} is given by

$$\sigma_{LOS} = \frac{c}{\frac{2}{v} \cdot \left(\frac{1}{N_1(0)} \frac{d}{dv} N_{s,M,1}(0) + \frac{-1}{N_2(0)} \frac{d}{dv} N_{s,M,2}(0) \right)} \cdot \sqrt{\frac{1}{SNR_1^2} + \frac{1}{SNR_2^2}} \quad (3)$$

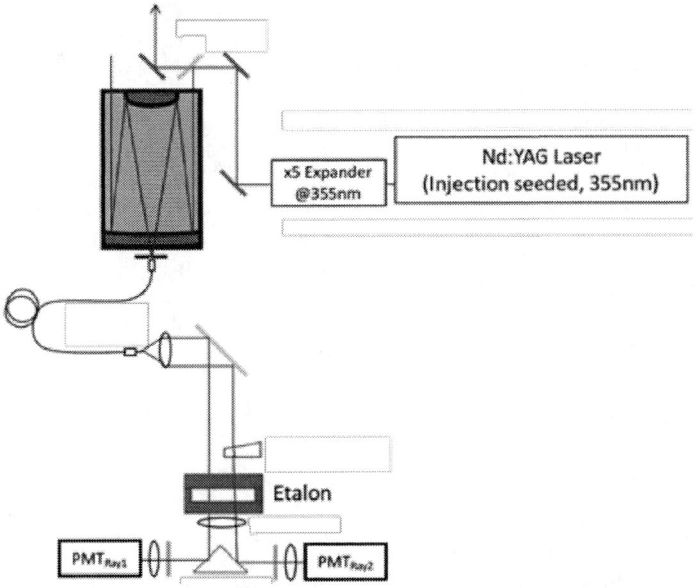

Figure 1. Block diagram of 355-nm direct-detection Doppler wind lidar.

Table 1. Specifications of 355-nm direct-detection Doppler wind lidar.

Transmitter	
Laser	Nd:YAG,
Wavelength	355 nm
Pulse energy	100 mJ
Pulse width	10 nsec
PRF	10 Hz
Beam divergence	0.2 mrad
Receiver	
Telescope	Schmidt-Cassegrain
Diameter	280 / 350 mm
FOV	0.1 mrad
Separators	Fabry-Perot etalon
Detector	PMT
Data acquisition	
Signal processing	11 Bit A/D
Sampling frequency	40 MHz
Sampling points	65536

2.2 Wind Measurement performance of 355nm direct-detection DWL

Figure 2 shows the profile of the LOS wind error calculated from Eqs (1) through (3) for a LOS wind speed of 1 m/s for the various accumulation laser pulse: 100, 600, 3000, and 6000. The figure shows that altitudes for the LOS wind speed errors of 0.5 m/s for the four accumulation laser pulses are 2, 5, 10, and 15 km. Our coherent lidar developed for and wind measurements can make vertical wind speed measurement at altitudes up to about 10 km for the accumulation laser pulse of 30 corresponding to 1 sec, where depends on the load of aerosol or clouds. When SNR is high, the 2-μm coherent lidar can make wind measurement with random error of 0.12 m/sec and vert low bias of <0.1 m/sec under the real atmospheric conditions [9]. The appropriate accumulation laser pulses for the dual DWL measurements are 100 and 600 shots.

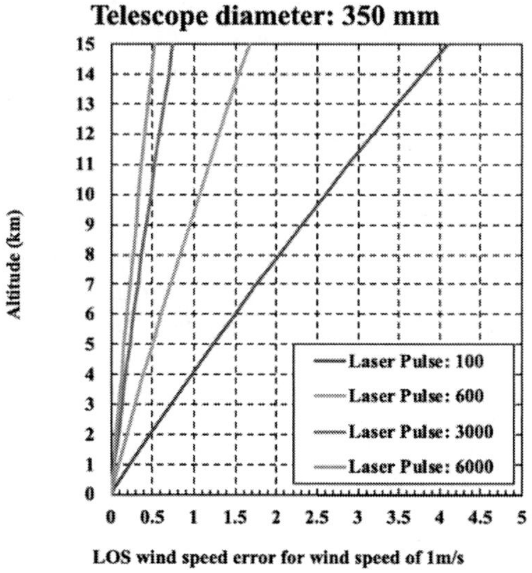

Figure 2. Profiles of the LOS wind speed error for various accumulation laser pulse:100, 600, 3000, and 6000.

3. VERTICAL WIND MEASUREMENT PERFORMANCE OF 2-μM COHERENT DOPPLER WIND LIDAR

A ground-based 2-μm coherent DWL was developed at NICT in 2013. The block diagram specifications of the ground-based coherent DWL are shown in Figure 3 and Table 2, respectively. The ground-based DWL uses a conductively-cooled laser-diode-pumped Q-switched Tm,Ho:YLF laser, an off-axis telescope, two heterodyne detectors, and signal processing devices. The single-frequency Q-switched Tm,Ho:YLF laser an emit an output energy of 90 mJ with a pulse width of 150 ns (FWHM) at a pulse repetition frequency of 40 Hz. The bias and error in the LOS wind speed measurement were also investigated by returns backscattered from a building (hard target) located at about 25 km. The ground-based coherent DWL can make the LOS wind speed with a low bias of <0.0 m/sec and random error of about 0.1 m/sec estimated from 10-shot averaged zero LOS wind speeds. Figure 4 shows the vertical wind speed measurements of atmosphere, aerosols and clouds by the ground-based coherent DWL at NICT headquarters on 27 March 2018. The ground-based coherent DWL detected details of thermal convections near in the atmospheric boundary layer.

4. SUMMARY

A DWL is a useful remote sensing technique for wind measurement. DWL would provide us with a wind profile with high vertical resolution, low bias, and good precision. We have plan to make vertical wind measurement around and in side clouds with

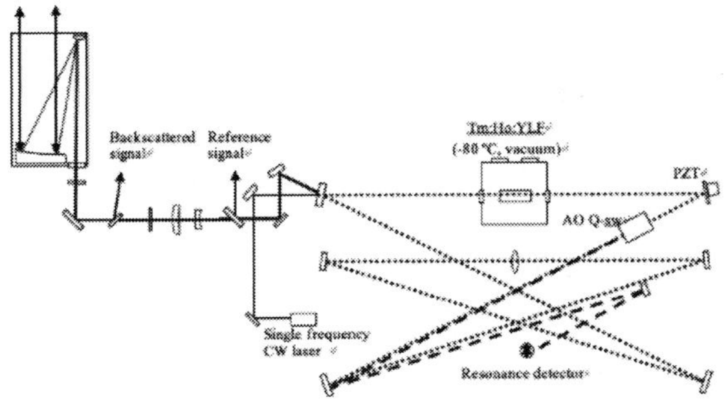

Figure 3. Block diagrams of 2-μm coherent Doppler wind lidar.

Table 2. Specifications of 2-μm coherent Doppler wind lidar.

Transmitter	
Laser	Tm,Ho:YLF
Wavelength	2051.002–2051.058 nm (On) 2051.250 nm (Off)
Pulse energy	50–80 mJ/pulse (Target)
Pulse width	150 ns (FWHM)
Pulse repetition	> 30 Hz
Polarization	Circular
Receiver	
Telescope type	Mersenne off-axis
Diameter	10 cm
Magnification	10 x
Detector	InGaAs-PIN photodiode
DET_1	Balanced InGaAs-PIN photodiode
Resolution	14-bit A/D conversion
Sampling frequency	400 MHz

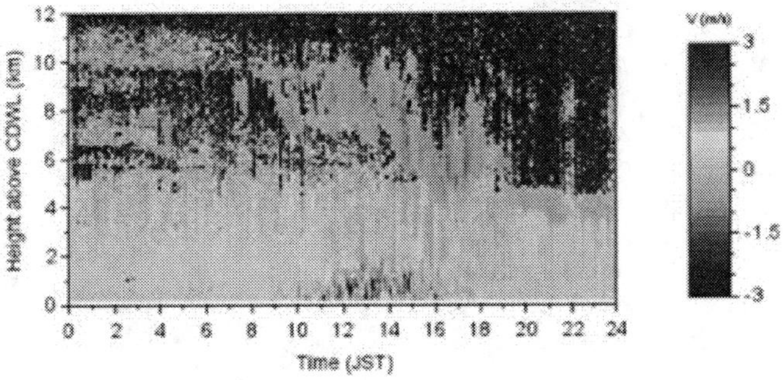

Figure 4. Temporal variation of vertical wind speed and SNR observed by 2-μm coherent DWL on 27 Mach, 2018.

direct-detection and coherent DWL wind measurements. NICT is a developing a direct-detection DWL with the double-edge technique. Our result shows that the appropriate accumulation laser pulses of 100 and 600 shots is appropriate for the direct-detection DWL wind measurement. The 2-μm coherent DWL can describe temporal variation of atmosphere. The combination of direct-detection and coherent DWL wind measurement would be useful for investigating details of various motions in the atmosphere. The dual DWL wind measurements with a cloud radar measurement. would be suitable to evaluate and characterize vertical wind motion around and inside clouds.

ACKNOWLEDGEMENTS

This research is supported by JSPS KAKENHI under Grant Numbers JP17H06139. A part of this research was supported by the Collaborated Research Program of Research Institute for Applied Mechanics, Kyushu University.

REFERENCES

[1] W. E. Baker et al. Bull. Am. Meteorol. Soc., 95, 543-564 (2014).
[2] WMO. WMO Tech. Rep. 2012-1, 6 (2012).
[3] ESA. SP-1233(4), (1999).
[4] S. Ishii et al. J. Meteorol. Soc. Jpn.95, 301-317 (2017).
[5] P. Baron et al. J. Meteorol. Soc. Jpn 95, 319-342 (2017).
[6] K. Okamoto et al. J. Meteorol. Soc. Jpn. 96, 179-199 (2018).
[7] S. Ishii et al. Appl. Opt., 49, 1809-1817 (2010).
[8] Flesia and Korb, Appl. Opt. **38**, 432-440 (1999).
[9] Iwai et al. J. Atmos. Ocean. Technol., 30(3), 429-449 (2013).

LIDAR OBSERVATIONS OF MOUNTAIN WAVES DURING BORA EPISODES

Longlong Wang[1*], Marija Bervida[1], Samo Stanič[1], Klemen Bergant[1,2], William Eichinger[3] and Benedikt Strajnar[2]

[1]*Center for Atmospheric Research, University of Nova Gorica, Nova Gorica, 5270, Slovenia*
[2]*Slovenian Environment Agency, Ljubljana, 1000, Slovenia*
[3]*Department of Civil and Environmental Engineering, University of Iowa, Iowa City, IA 52242, USA*
**Email: longlong.wang@ung.si*

ABSTRACT

Airflows over mountain barriers in the Alpine region may give rise to strong, gusty downslope winds, called Bora. Oscillations, caused by the flow over an orographic barrier, lead to formation of mountain waves. These waves can only rarely be observed visually and can, in general, not be reliably reproduced by numerical models. Using aerosols as tracers for airmass motion, mountain waves were experimentally observed during Bora outbreak in the Vipava valley, Slovenia, on 24-25 January 2019 by two lidar systems: a vertical scanning lidar positioned just below the peak of the lee side of the mountain range and a fixed direction lidar at valley floor, which were set up to retrieve two-dimensional structure of the airflow over the orographic barrier into the valley. Based on the lidar data, we determined the thickness of airmass layer exhibiting downslope motion, observed hydraulic jump phenomena that gave rise to mountain waves and characterized their properties.

1. INTRODUCTION

Orographically-generated mountain waves occur when stably stratified airflow passes over a mountain ridge. These waves play an important role in redistributing momentum and energy, thus making a significant impact on the atmospheric dynamics both on the local and regional scale [1]. In the Alpine region, in particular in the Dinaric Alps along the Adriatic coast, mountain waves are, under certain synoptic conditions, related to the strong, cold and gusty downslope wind, called Bora [2,3]. Occurrences of mountain waves at heights considerably lower than the orographic barrier have been ascribed to hydraulic jumps in the Bora flow [4]. Although the role of mountain waves has been theoretically explained, their modeling remains a challenge due to the need for high temporal and spatial resolution as well as the non-periodic variability of Bora flow [5,6]. Systematic observations of airflow properties became possible with the advance of active remote sensing devices and techniques, for example by the use of a Doppler wind lidar or a scanning Mie lidar that exploits aerosols as tracers for air mass motion [7].

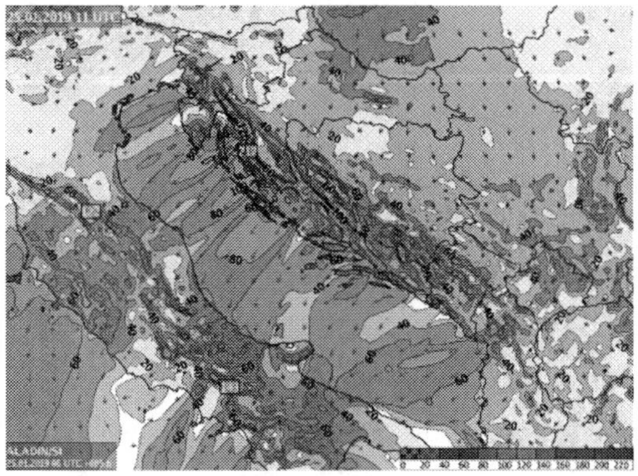

Fig. 1. ALADIN forecast of peak wind gust speeds during a major windstorm across the western Mediterranean on 25 January 2019 between 9:00 and 11:00 UTC. The location of the Vipava valley, where gust speeds exceeded 20 m/s, is denoted by a red arrow.

The aim of this study is to characterize the properties of mountain waves during a Bora outbreak based on lidar remote sensing and in-situ wind measurements. The Vipava valley in the SW Slovenia as our area of interest is ideal for this purpose being frequently exposed to strong Bora outbreaks with peak wind gust speeds exceeding 30 m/s.

2. METHODOLOGY

The Vipava valley is a deep mountain valley in southwestern Slovenia. Towards the Adriatic coast in the south-west, the valley (125 m above the sea level) is closed by the Karst plateau (up to 300 m a.s.l.), while to the north-east the terrain rises steeply to the Trnovski gozd plateau (up to 1200 m a.s.l.). Being at the uppermost part of the Dinaric Alps barrier (Fig. 1), it is frequently exposed to strong Bora outbreaks and can serve as a representative location for the study of Bora and mountain waves. Airflow structures on the lee side of the Trnovski gozd were investigated using simultaneous lidar and in-situ wind measurements at two locations (Fig. 2), one just below the peak of the ridge (Otlica, 951 m a.s.l.) and the other at the valley floor (Ajdovščina, 125 m a.s.l.). The horizontal distance between the two stations was 5 km. The retrieved data allowed us to observe both the structure of the Bora airflows along the lee mountain side as well as temporal dynamics of mountain waves above the valley.

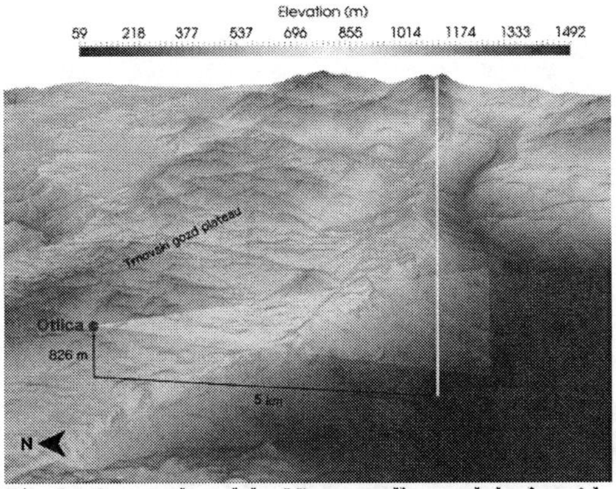

Fig. 2. Orography of the Vipava valley and the lee side of the Trnovski gozd plateu. Monitoring stations (Otlica and Ajdovščina) are marked by red points. The lidar at Otlica was oriented with the wind towards Ajdovščina and was scanning elevations between 12° and -5° (shaded area). The lidar at Ajdovščina was stationary and pointed towards the zenith (white line).

2.1 Instrumentation

The self-developed scanning Mie-scattering lidar at Otlica, operating at 1064 nm with 10 Hz repetition rate [8] was oriented towards Ajdovščina (azimuth 225°) and retrieved two-dimensional scans of aerosol loading distributions for elevation angles between 12° and -5° with 0.1° step. Each scan took about 4 minutes. It was also used in fixed mode at 0° to provide 10 minute long time series of horizontal aerosol loading distribution between Otlica and Ajdovščina. The self-developed vertical lidar at Ajdovščina, operating at 1064 nm with 10 Hz repetition rate [9] was used to retrieve temporal dynamics of vertical atmospheric structure above the valley. Both lidars were deployed indoors and were not directly mechanically affected by Bora. Backscattering signals were averaged over 10 consecutive laser shots, digitized with range resolution of 3.75 m and range square corrected. In addition to remote sensing, wind speed and directions were monitored at both sites by two-dimensional Vaisala WMT700 ultrasonic wind sensors. At Ajdovščina, wind data was available with a 1 second temporal resolution, while the sensor at Otlica provided average wind speed and direction as well as maximum gust speed and direction for 10 minute intervals.

2.2 Experimental details

The measurements took place on 24-25 January 2019 in neutral atmosphere conditions. Synoptic situation with a deep cyclone in central Mediterranean resulted in a typical Bora situation causing moderate wind gusts also in Vipava valley (Fig. 1). The predominant wind direction at Ajdovščina was NE (N at Otlica) with average wind speeds of 10 m/s and peaks of about 20 m/s. Thirty lidar scans were performed at Otlica with the aim to retrieve airflow characteristics. In order to estimate the effects of scan duration on the observed structures, the measurements were performed with alternating scanning directions. Properties of mountain waves arising from a hydraulic jump about 150 m below the observatory were extracted from five horizontal time series measurements. Periods and amplitudes of mountain waves were extracted from vertical measurements at Ajdovščina, which are available for the entire two-day period.

3. Results

Stratified atmospheric layers with periodic atmospheric structures were observed in vertical lidar data throughout the Bora outbreak at the height of the Trnovski gozd ridge and below (Fig. 4, (a)) with typical amplitudes of about 100 m. The most significant period of the mountain waves above Ajdovščina was found to be 180 s, with the presence of considerably shorter periods (Fig. 4, (b)).

In order to investigate if the origin of mountain waves at heights considerably lower than the orographic barrier may be attributed to hydraulic jumps in the Bora flow, we investigated the presence of such jumps in two dimensional vertical lidar scans (Fig. 3). A hydraulic jump was observed at horizontal distance of about 1.5 km from the lidar site, where the about 100 m thick aerosol rich downslope Bora flow layer of cold air

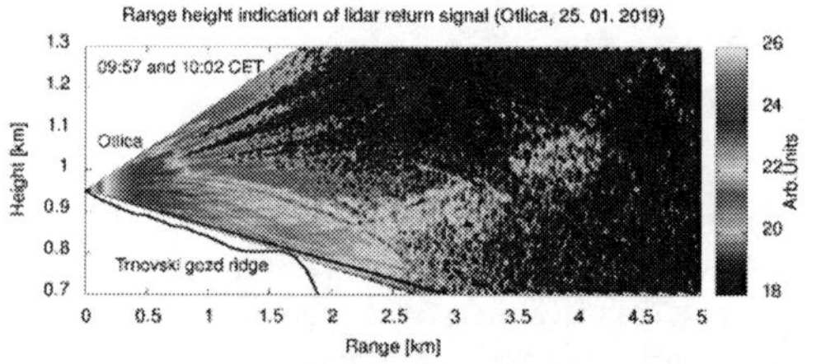

Fig. 3. Range-height indication of aerosol loading at Otlica on 25 January 2019. Two subsequent lidar scans were merged in this plot, an upward scan from -5° to 12° at 09:57 and a downward scan from 12° to -5° at 10:02 CET to minimize scanning direction effects.

rose higher above the surface due to a secondary ridge about 150 m below Otlica. In order to confirm that the observed jump of cold air masses gave rise to mountain waves, we searched for periodic structures in the horizontal profiles from Otlica, which captured the dynamics of aerosol loading 150 m above the jump. The lidar trace was found to reach through a number of waves

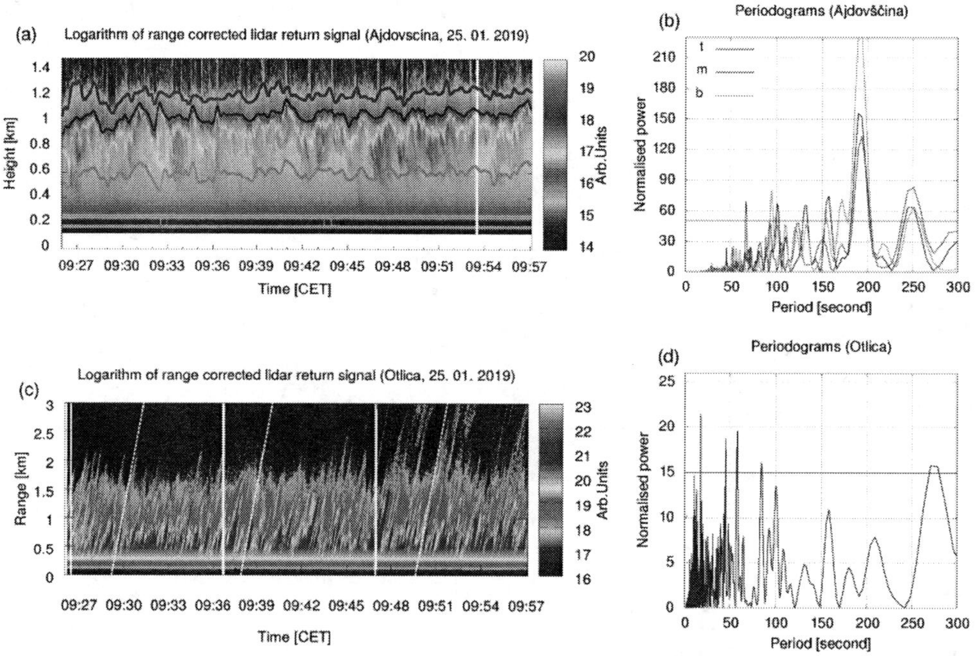

Fig. 4. a) Temporal evolution of vertical aerosol loading profiles at Ajdovščina. Air mass streamlines at different heights are denoted by solid lines; c) Temporal evolution of horizontal aerosol loading profiles at Otlica. The observed structures are inclined at 23 m/s (denoted by dashed white line); b) periodogram of the extracted streamlines, obtained using least square spectral analysis method; d) periodogram of the observed peaks in horizontal profiles. Black horizontal lines represent 0.001 significance level.

https://doi.org/10.1051/epjconf/202023706007

propagating away from the mountain side (Fig. 4, c). The most significant period of the observed structures was found to be as short as 18 s, which may be attributed to local micro-orographical terrain features of this particular jump. They were all found to be inclined at the same angle of 23 m/s, which represents horizontal wave propagation speed. Constant wave propagation speed is an indication that Taylor's frozen turbulence hypothesis is an appropriate assumption for the description of mountain wave propagation during Bora episodes. Furthermore, spatial periodic structure of aerosol loading was revealed by sampling of horizontal profiles with the same phase (with 18 s intervals between them). A delay of the starting time of profile sampling caused the wave to move away from the jump. Relative delay of 9 s between start times of two samplings yielded waves with opposite (180° shifted) phases (Fig. 5). In both cases, the wavelength was about 400 m, which is in agreement with the retrieved propagation speed and period of the waves.

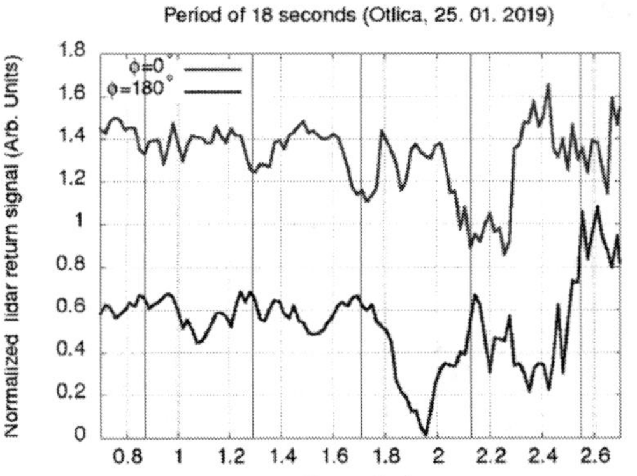

Fig. 5. Using the same data as presented in Fig. 4c, range square corrected horizontal lidar returns with the same mountain wave phase (occurring every 18 s) were added, corrected for the attenuation and normalized. Blue and red line represent waves shifted for half a period (9 s), which are manually vertically offset by 0.8 for easier viewing. Vertical lines denote peak-to-peak distance (wavelength) of 400 m.

3. CONCLUSIONS

Two coordinated lidar measurements providing simultaneous vertical and horizontal airflow dynamics allowed us to directly observe and quantify the parameters of mountain waves, which appeared during a Bora episode above and at the lee side of the Trnovski gozd plateau. We identified a hydraulic jump of cold, narrow layer of Bora flow descending downslope the lee side of the mountain as a source of mountain waves at a hight well below the top of the orographic barrier. The propagation speed of these waves was found to be constant at 23 m/s (at a higher speed than wind speed at Otlica) and their wavelength was approximately 400 m with a period of 18 s. The constant propagation speed allows us to use Taylor's frozen turbulence hypothesis for estimating horizontal propagation of mountain waves. Based on vertical lidar data, which yielded the most significant periodicity of 180 s at the height of the orographic barrier, and under the assumption that wave propagation speed is the same as for waves observed at lower heights, the wavelength for these waves was estimated to be about 4 km. As a next step, mountain waves will be investigated using a combination of stereo lidar measurements, large-eddy CFD simulations and high resolution numerical weather prediction modeling to interpret the observations.

ACKNOWLEDGEMENTS

This work was supported by the Slovenian Research Agency (grant number P1-0385) and Fulbright Scholar program.

REFERENCES

[1] L. Strauss, *et al.*, *J. Atmos. Sci.* 73.4, 1481-1506 (2016).
[2] M. Yoshino, *Local Wind Bora*. University of Tokyo Press, 1976.
[3] T.N. Palmer, *et al.*, Q. J. R. Meteorol. Soc. 112, 1001–1039 (1986)
[4] J. Kuettner, *et al.*, Bull. Am. Meteorol. Soc. 62, 793–805 (1981).
[5] A.Tsiringakis, *et al.*, *Q. J. Meteorol. Soc.* 143.704, 1504-1516 (2017).
[6] R.B. Smith, *et al.*, J. Atmos. Sci. 42, 2597–2603 (1985)
[7] M. Mole, *et al.*, *J. Quant. Spect. Radiat. Trans.* 188, 39-45 (2017).
[8] T-Y. He, *et al.*, Proc. SPIE, 7832, 7832U (2010).
[9] L.Wang, *et al.*, Atmosphere, 10, 128 (2019).

EPJ Web Conferences **237**, 06008 (2020) https://doi.org/10.1051/epjconf/202023706008
ILRC 29

LIDAR OBSERVATIONS VERSUS FORECAST OF WATER VAPOR TRANSPORT

Brian Carroll[1, 2,*], Belay Demoz[1, 2], Ruben Delgado[1, 2]
[1]Atmospheric Lidar Group, Physics Department, University of Maryland Baltimore County, Baltimore, Maryland, U.S.
[2]Joint Center for Earth Systems Technology, Baltimore, Maryland, U.S.
**Email: brian.carroll@umbc.edu*

ABSTRACT

Water vapor advection is an important diagnostic parameter for initiation and sustenance of convective systems. This variable has traditionally been explored with balloon soundings or models, but lidar provides a unique opportunity to continuously resolve spatial and temporal evolution of water vapor transport with relatively high resolution. In this study we utilize co-located Doppler wind lidars and water vapor lidars (DIAL and Raman) at multiple sites to investigate water vapor advection. A case of nocturnal convection in the Great Plains is analyzed with these observations and compared to the North American Mesoscale Forecast System (NAM) operational forecast model.

1. INTRODUCTION

Mesoscale convective systems (MCSs) are large, intense convective systems that bring substantial rainfall and potentially damaging winds or hail. MCSs are particularly impactful in the Great Plains region of the United States, where warm-season MCSs can form and persist overnight. Forecasting these events and related phenomena such as low-level jets (LLJs) and bore waves is challenging. In 2015 a large field campaign was executed to address these issues, called Plains Elevated Convection at Night (PECAN) [1].

PECAN had several fixed and mobile sites equipped with in situ and remote sensing instrumentation. Three of these sites, spread across Oklahoma and Kansas with significant latitudinal and longitudinal separation, featured both a Doppler wind lidar and a water vapor lidar. This setup is conducive to observing the mesoscale phenomena of interest, especially water vapor transport caused by LLJs. LLJs are strong wind events within the lowest 2 km of the atmosphere and are known to advect moisture and warm temperatures northward, critical for initiating or fueling convection [2]. Because these events are often poorly forecasted, this study will address a case of a LLJ feeding into an MCS through the lens of lidar observations to understand details of the event as well as compare the observations to an operational forecast of the domain. The chosen forecast model is the North American Mesoscale Forecast System (NAM) model, run and archived by NCEP.

2. METHODOLOGY

2.1 Lidar systems

The data shown here comes from four lidar systems; two sites each featuring wind and water vapor systems. The northern site, closer to the MCS of interest, is denoted as "FP3" and featured a water vapor DIAL developed by NCAR [3]. The southern site, "FP1", featured the ARM SGP Raman lidar for water vapor profiling [4]. Nocturnal measurements from these lidars compare very well to balloon soundings, as shown in Figure 1. The lidar sites are marked in Figure 2.

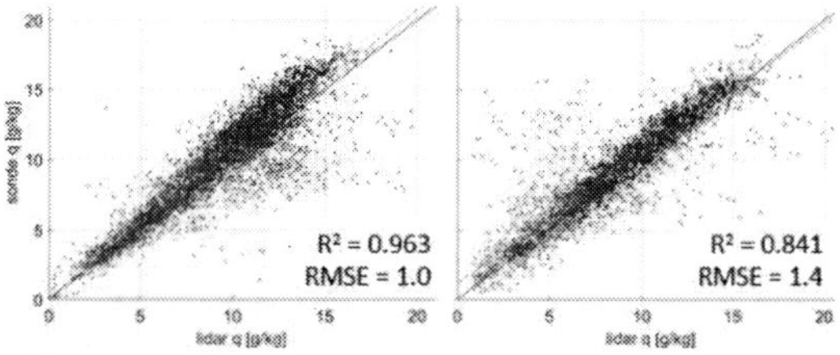

Figure 1. Comparison of water vapor mixing ratio (g/kg) from sonde and lidar at FP1 (left) and FP3 (right), with dashed line of best fit and solid 1-to-1 line. Color indicates measurement height, fading black to red from 0 km to 3 km.

The Doppler lidars are both commercial heterodyne systems operating at 1540 nm. The lidar at FP3, a Leosphere WINDCUBE®70, operated with the common four-azimuth Doppler beam swinging (DBS) pattern to allow wind profile retrieval. FP1's Halo Stream Line Pro

https://doi.org/10.1051/epjconf/202023706008

performed conical scans with beams at 8 azimuths. Wind profiles from these Doppler systems compared very well with balloon soundings (figures not shown).

2.2 Water vapor advection calculation

Water vapor advection is calculated as the product of the wind speed and the water vapor mixing ratio. These fields are retrieved by the lidars at different spatial and temporal resolutions, so the higher-resolution profile at a given site is interpolated to the lower resolution. The same approach is used in comparing the observed advection fields to the relatively coarse NAM model output.

2.3 Case of interest: 11 July 2015

On 11 July 2015, the PECAN experiment focused on an MCS on the Kansas-Nebraska border (overlapping with FP4 in Figure 2). MCSs also developed in north-central Nebraska and northern Missouri. The main system of interest began shortly before sunset and grew overnight into a multi-celled MCS, slowly propagating eastward and dying out at sunrise. It produced over 35mm of rain along much of its track. NAM predicted the convection initiation but did not capture the correct motion, intensification, and rainfall of this event, nor did it accurately develop the two nearby MCSs.

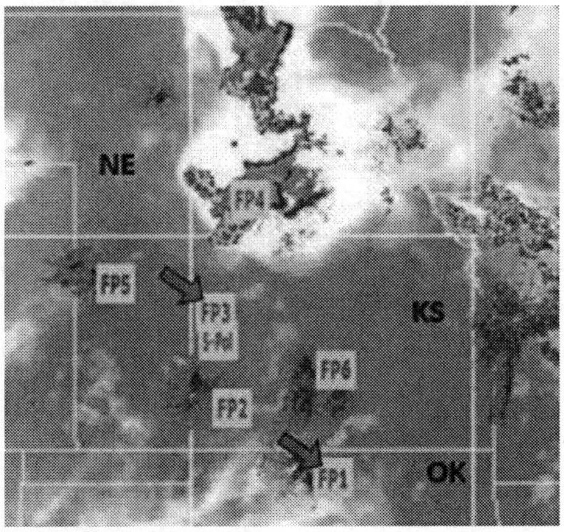

Figure 2. Radar mosaic at 0615 UTC. Ground sites, including FP1 and FP3, are identified. Shading is GOES-13 near-IR.

The 12km-resolution NAM model run used in this study is the 0000 UTC run from 10 July 2015, so convective events of interest are happening at forecast hours 20-36.

3. RESULTS

The lidars were able to capture features of the LLJ and boundary layer water vapor in great detail for this study. The product water vapor advection profiles are shown in the top row of Figure 3. NAM shows qualitatively similar evolution of the water vapor advection profile (Figure 3, middle row) but critical shortcomings are revealed in the forecast error plots (Figure 3, bottom row). Plots of the separate lidar fields of winds and water vapor also aided in analysis (not shown here).

Wind profiles from the lidars detected a southerly direction for the strong LLJ, thus the latitudinal separation of FP1 and FP3 gives spatial context to the LLJ's along-track transport. FP1, the southern site, had a very moist afternoon boundary layer that dried out overnight. FP3 experienced the opposite; a drier afternoon boundary layer became moister overnight, eluding to northward transport of water vapor by the LLJ. Water vapor advection shown in Figure 3 depicts fairly constant transport throughout the night at FP1, as the site grew drier and the LLJ simultaneously strengthened.

The maximum in water vapor advection observed on this night was in the core of the LLJ around 0600 UTC at FP3. This timing corresponds with the appearance and rapid intensification of a new cell in the MCS along the Kansas-Nebraska border, highlighting the importance of water vapor advection to MCS evolution. This same water vapor advection feature was the largest instance of underprediction by NAM. Investigation of forecast errors in the separated wind and water vapor fields shows the NAM underpredicting the FP3 peak LLJ wind speed by 6 m/s, with less consistent or large bias in the FP3 water vapor profile. While this failure of NAM is critical towards MCS development, other model dynamics and parameterizations are presumably also important to NAM's misplaced location and development of the MCSs in this time frame.

4. SUMMARY

Co-located lidar observations of winds and water vapor during PECAN provided a rare opportunity to investigate water vapor advection, an important parameter for convective initiation and intensification. A case study of 11 July 2015 was presented here. The lidar profiles revealed a maximum in water vapor advection at the northern site, closest to the MCS of interest, at the same time that a new cell developed and intensified. The NAM model severely underpredicted this advection feature and failed to capture the rainfall location, timing, and amount over the course of the night within the PECAN region.

ACKNOWLEDGEMENTS

This work was supported by the National Science Foundation award number 1503563, and NASA MUREP IRO NNX15AQ03A. The authors also thank all fellow PECAN participants.

REFERENCES

[1] L. K. Berg, et al. Journal of Climate, 28(17), 6682–6706. (2015)
[2] B. Geerts, et al. Bulletin of the American Meteorological Society (April), BAMS-D-15-00257.1. (2017)
[3] S. M. Spuler, et al. Atmospheric Measurement Techniques, 8(3), 1073–1087. (2015)
[4] R. A. Ferrare, et al. Journal of Geophysical Research: Atmospheres, 106(D17), 20333-20347. (2001)

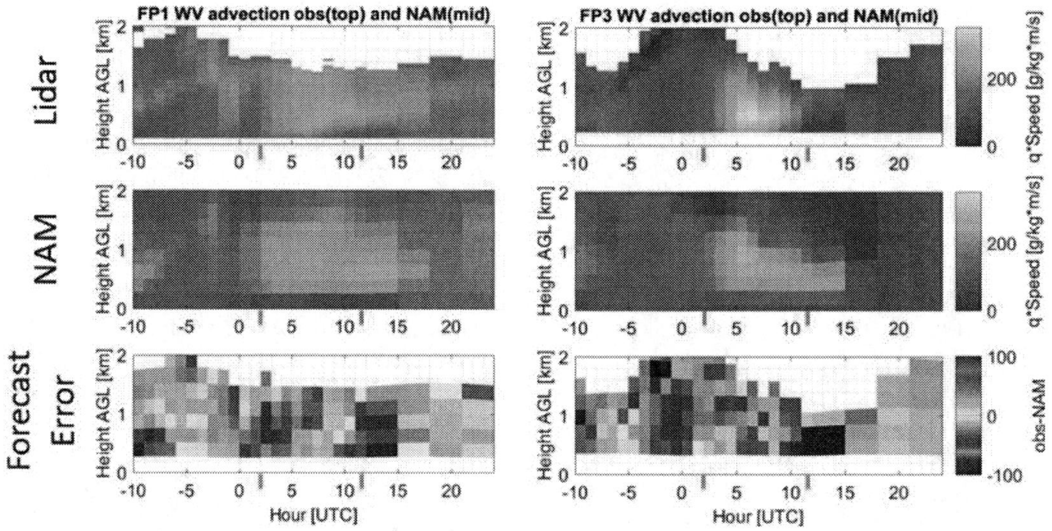

Figure 3. Water vapor (*WV*) mixing ratio (*q*) and advection profile timeseries from observations and NAM, along with forecast error (obs-NAM). Left column is at site FP1, right column is FP3. 0 on the x-axis is 0000 UTC, 11 July. Magenta lines indicate sunset (~0200 UTC) and sunrise (~11 UTC).

https://doi.org/10.1051/epjconf/202023706009

OBSERVATION OF THE URBAN WIND ISLAND EFFECT

Sunil Baidar[1, 2*], Tim Bonin[1,2], Aditya Choukulkar[1,2], Alan Brewer[2], Mike Hardesty[1,2],

[1]CIRES, University of Boulder, Boulder, CO 80309, USA
[2]Chemical Sciences Division, ESRL, NOAA, Boulder, CO 80305, USA
**Email: sunil.baidar@noaa.gov*

ABSTRACT

Urban wind island effect (UWI) is defined as a phenomenon in which boundary layer mean wind speeds in an urban area are noticeably higher than its neighboring rural areas. Unlike urban heat island effect which has been extensively studied, the UWI was only recently observed in a modeling study. Here we study existence of the UWI over Indianapolis, Indiana using wind profile measurements from two Doppler wind lidars (DWL) that were deployed in climatologically upwind and downwind of the city. Under certain atmospheric conditions higher wind speeds and turbulence were observed at the downwind site over the entire urban boundary layer outside the urban canopy layer.

1. INTRODUCTION

More than half of the world population now lives in urban areas and increasingly in denser cities. By 2050, it is projected that two-third of the world population will be dwelling in urban areas [1]. Landscape changes with development of urban areas creates micro climate over these urban areas. With more people living in urban areas these micro climates disproportionately affect higher number of people. The urban heat island effect, where a higher temperature is observed over urban areas compared to their rural surroundings [2], is one such example. This effect has been widely studied and it is known to be caused by change is surface properties, such as heat capacity, roughness etc., over urban areas.

Similarly, under certain conditions wind speeds can also be higher over urban areas compared to their neighboring rural surroundings, a phenomenon named the urban wind island effect (UWI) [3]. Using a conceptual bulk model of atmospheric boundary layer, Droste et al. (2018) showed that for certain atmospheric conditions, boundary layer mean wind speeds over a city can be higher than its surrounding rural areas. They found a combination of differences in atmospheric boundary layer growth, surface roughness and ageostrophic wind between city and rural areas causes the UWI in the afternoons. Considering that increase in surface roughness over urban areas is expected to reduce wind speeds, occurrence of the UWI in the model is a surprising finding.

Understanding the UWI is very important for urban air quality and transport of pollutants from cities. In this study we investigate existence of the UWI over Indianapolis, Indiana by comparing wind speeds measured by two Doppler wind lidars (DWL) located upwind and downwind of the city. We compare mean wind speeds in the urban canopy layer (UCL) and in the urban boundary layer (UBL) between the two sites under difference wind direction regimes. We explore individual case studies to identify conditions that might favor development of the UWI.

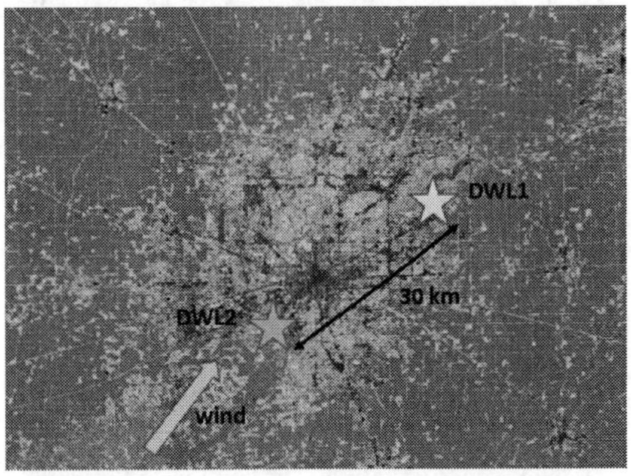

Figure 1: Land usage map from the 2011 National Land Cover Database [4] showing locations of the two Doppler wind lidars in Indianapolis, IN. Cyan arrow shows the predominant wind direction. Major land uses are developed areas (red, darker indicating denser development), and cultivated crops (light brown).

2. OBSERVATIONS

A coherent DWL (DWL1; HALO Photonics Stream Line XR) has been making measurements of mean horizontal wind profiles and mixing layer heights in the suburban area of Indianapolis, IN as part of the Indianapolis Flux Experiment (INFLUX) [5] for quantification of greenhouse gas emissions from the urban area. Location of this DWL is shown as yellow star in Figure 1.

The lidar has been operating with a repeating 20 min scan cycle which consists of conical scans at 3, 10, 35.3, and 60° elevation angles (EA); vertical slice scans to the south and east; a zenith stare (lasting 10 min during the day, 4 min at night); and quasi-horizontal stares at 20° EA to the south and east at night. The scan cycle was optimized for retrieval of horizontal wind profiles, atmospheric turbulence properties and mixing layer height.

Horizontal wind profiles are retrieved using the velocity-azimuthal display (VAD) technique [6]. Mixing layer height is retrieved using an automated composite fuzzy logic technique which combines information from various scan types to determine a unified measurement of the mixing height and its uncertainty [7].

A second coherent DWL (DWL2; Leosphere Windcube 200S) was deployed climatologically upwind of the city for a three-month period in Fall 2017 to study the urban influence on winds and boundary layer. Location of DWL2 is shown as green star in Figure 1. The DWL2 was operated with the same scan settings, and horizontal wind profiles, atmospheric turbulence properties and mixing layer height was retrieved using the same methods as the DWL1.

3. RESULTS

3.1 Composite Analysis

We segregated individual measurement days into 4 groups based on the predominant wind direction that day (sector 1: 0-89°, sector 2: 90-179°, sector 3: 180-269° and sector 4: 270-359°). The mean horizontal wind speed inside the UCL (lowest 100 m layer) and inside the UBL (400-700 m layer) measured every 20 minutes at the two sites each day were then compared. The tallest building in Indianapolis is 247 m [8]. Thus, the lowest 100 m layer is very likely within the UCL and the layer between 400-700 m is outside the UCL but very likely within the urban atmospheric boundary layer (UBL here on).

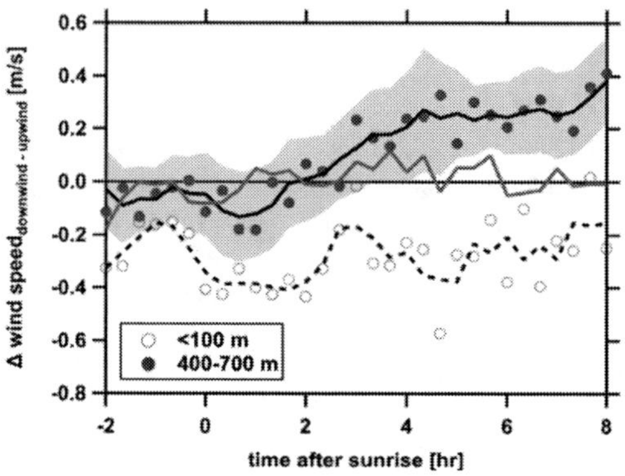

Figure 2: Mean difference in horizontal wind speed inside the UCL (red open circles), and inside the UBL (red closed circles) for wind direction = 180-269° (N = 41). The black dashed and solid lines are hourly averaged (3 points) running mean to the red open and closed circles respectively and included for visual aid. The grey shading represents the standard error of the mean. The green line is same as the black solid line for wind direction = 90-179° (N = 16).

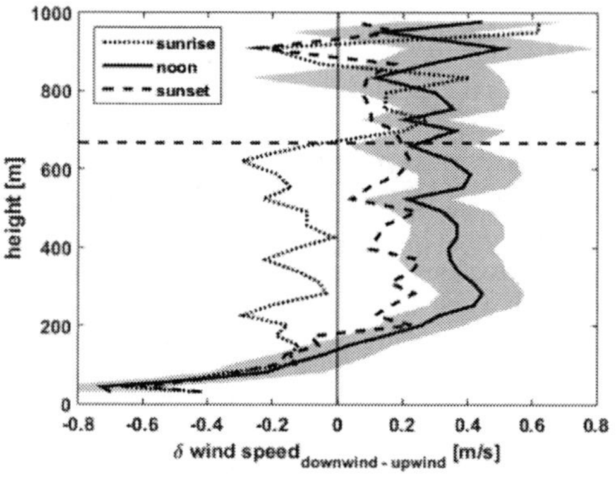

Figure 3: Vertical profile of mean difference in wind speed between the two DWL measurements at sunrise (dotted), solar noon (solid) and sunset (dashed). The grey shading represents the standard error of the mean. The blue dashed line shows the mean mixing layer height for the days included in the analysis.

https://doi.org/10.1051/epjconf/202023706009

The mean horizontal wind speed difference between the two lidar measurements (DWL1-DWL2) in the UCL and UBL as a function of time after the sunrise for sector 2 and 3 are shown in Figure 2. When the wind was blowing from sector 3, DWL2 was located upwind of the city and was not affected by the city whereas DWL1 was located downwind of the city and was affected by the city. Thus, sector 3 provided cases for studying the UWI. When wind was blowing from sector 2, both DWLs were not affected by the city and thus, provided a null hypothesis cases. For sector 3, within the UCL, the mean wind speeds were lower downwind of the city throughout the course of the day. This was very likely due to increased surface roughness of the urban area slowing down the wind. Inside the UBL, wind speeds were comparable between the two sites up to the first two hours after the sunrise. Two hours after the sunrise, the mean wind speeds became higher in the downwind site and stayed higher for the rest of the day. This is consistent with the observation by Droste et al. (2018). In contrast, for sector 2, when both the sites were not affected by the city, there was no difference in wind speeds in the UBL between the two sites (Figure 2 green line).

Figure 3 shows the vertical profiles of mean difference in wind speeds at sunrise, solar noon and sunset for sector 3. The higher wind speeds at the downwind site were observed through the entire UBL outside the UCL at solar noon and sunset but not at sunrise. In contrast, wind speed is higher upwind of the city over the entire UCL.

3.2 Case Study

Individual case studies provide an opportunity to further explore the causes for UWI. Figure 4 shows an example of time-height cross section of horizontal wind speed profiles measured by the two DWLs on 26 September 2017 when wind was blowing from sector 3. The differences between wind speed profiles at the two sites are also shown in Figure 4. Both the sites showed very similar overall wind speed indicating both the sites experienced the same general wind flow. Although individual profile differences were noisy, we can see that early in the morning the wind speed differences were negative inside the mixed layer i.e. the downwind site experienced lower wind speeds. This could be due to wind speed was being used to generate turbulence for eroding night time inversion layer and mixing layer height growth at the downwind site (see Figure 5). Indeed, the mixing layer height started to grow earlier at the downwind site.

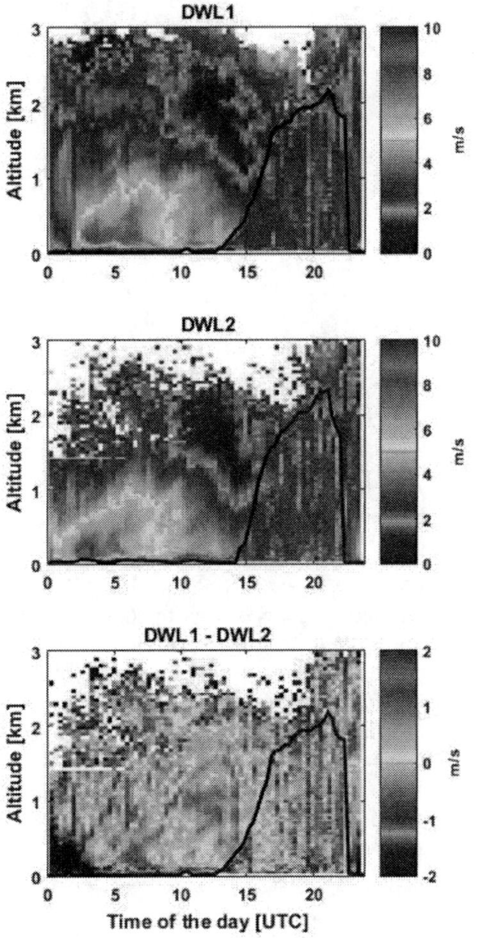

Figure 4: Time-height cross section showing horizontal wind speed profiles measured by DWL1 (top), and DWL2 (middle) on 26 September 2017. The black solid lines represent the measured mixing layer height at the two sites. The bottom panel shows the difference between the top two panels.

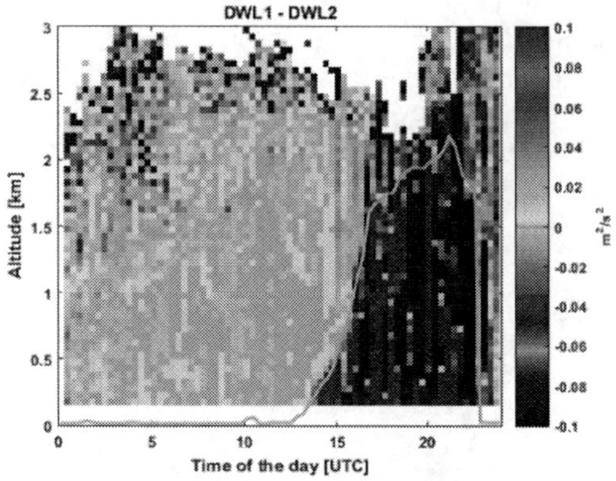

Figure 5: Time-height cross-section showing the difference in vertical velocity variance between two sites on 26 September 2017. The cyan solid like represent mixing layer height measured by DWL1.

As seen in the Figure 2, around 2 hours after the sunrise, wind speeds were generally higher in the downwind location. This higher wind speed coincided with higher turbulence inside the UBL during the mixing layer height growth phase and lower turbulence during the decay phase. Figure 5 shows difference in vertical velocity variance between the two sites on 26 September 2017. This enhanced mixing in the UBL at the downwind site during the mixing layer growth phase facilitates entrainment of free tropospheric air and this very likely resulted in higher wind speed. Droste et al. (2018) also identified enhanced turbulence in the urban area as one of the causes for UWI.

4. CONCLUSIONS

Horizontal wind profile measurements by DWLs at an upwind and a downwind location of Indianapolis, Indiana showed existence of an UWI effect i.e. mean horizontal wind speed inside the boundary layer is higher in the urban areas than the surrounding rural areas. This higher wind speeds at the downwind site was observed over the entire UBL and it was found to occur with increased turbulence. Enhanced turbulence aids mixing of free tropospheric air and this very likely results in UWI. Inside the UCL, wind speeds were lower in the downwind site. This is consistent with our conventional understanding of winds over urban areas i.e. increase surface roughness over urban areas results in lower wind speeds. In a control experiment where neither of the DWLs were affected by the city, there was no difference in mean horizontal wind speeds between the two sites.

Future work to understand UWI would involve measurements of horizontal wind profiles, atmospheric turbulence properties, and mixing layer height over rural-urban areas using airborne and/or mobile Doppler wind lidars to continuously monitor the progression of UWI effect across urban areas.

ACKNOWLEDGEMENTS

This work was supported by the NIST Greenhouse Gas Measurements Research Program via Interagency Agreement (IAA) M121271. We are grateful to Ivy Tech and their staff for hosting the instrument and providing assistance when needed. We thank Scott Sandberg and Ann Weickmann for developing, deploying, and maintaining the instruments.

REFERENCES

[1] United Nations, Department of Economic and Social Affairs, Population Division, World Urbanization Prospects: The 2018 Revision, Online Edition. (2018)
[2] T. R. Oke, Quarterly J. of Royal Meteo. Soc. 108, 1-24 (1982)
[3] A. M. Droste, G.J. Steeneveld, and A. A. M. Holtslag, Environ. Res. Lett. 13, 094007 (2018)
[4] B. C. Homer et al. Photogramm. Eng. Remote Sens., 81, 345–354 (2015)
[5] K. Davis et al. Elem. Sci. Anth., 5, p21 (2017)
[6] K. A. Browning, and R. Wexler, J. Appl. Meteor. 7, 105-113 (1968)
[7] T. A. Bonin et al. J. Atmos. Oceanic Technol. 35, 473-490 (2018)
[8] http://skyscraperpage.com/diagrams/?stateID=70, Accessed 12 March 2019

EPJ Web Conferences **237**, 06010 (2020) https://doi.org/10.1051/epjconf/202023706010
ILRC 29

OBSERVATIONS AND MODELS OF TURBULENCE DURING HEAT EVENTS IN COMPLEX URBAN COASTAL REGIONS USING MULTIPLE DOPPLER WIND LIDARS

MARK AREND[1], DAVE A. LIGON[2], DERYCK JAMES[2], MARK J. CAMPMIRE[3], DAVID MELECIO-VAZQUEZ[4], JORGE GONZALEZ[4], FRED MOSHARY[1]
[1]*NOAA-CREST and Electrical Engineering Department, CUNY CCNY, NY, NY, 10031*
[2]*U. S. Army Research Laboratory, Adelphi, MD, 20783*
[3]*Civil, Architectural, and Environmental Engineering Dept. UT Austin, TX*
[4]*NOAA-CREST and Mechanical Engineering Department, CUNY CCNY, NY, NY, 10031*
**Email: marend@ccny.cuny.edu*

ABSTRACT

Multiple Doppler Lidars have been co-operated in the NYC region during the summer of 2018 to provide detailed observations of the turbulent atmosphere especially during heat wave events. The co-operating Doppler Lidar observations allow for mean flows to be distinguished from complex flows so that a better understanding of the transport of air masses can be provided to investigate the fidelity of high resolution numerical weather prediction models that are being developed to interpret and model turbulence during such events.

1. INTRODUCTION

Understanding the dynamics of ground and upper level pollutant transport in urban coastal areas during heat wave events is becoming a major concern. Even though there have been many advancements on the modeling capabilities to predict the occurrences of such episodes[1], there is a need to improve the observational capabilities. Multiple Doppler Lidars can be operated in a collaborative mode to provide detailed observations of the flow of air mass[2-3]. The analysis of the Doppler Lidar return signals using the volume velocity processing (VVP) retrieval technique[4-5] is an approach that is analogous to the types of inverse modeling techniques that are becoming popular in Data Science. By using the VVP technique in conjunction with forward dynamical models allows both inverse and forward modeling techniques to be implemented that can bring insight to the complexity of turbulent flows. This will advance previous observational studies focused on urban coastal regions where the forcings encompass complex thermodynamic and geographic effects[6] during major heat events[7].

2. METHODOLOGY

A heat event in NYC that occurred on July 2nd, 2018 was observed and modeled. Multiple Doppler Lidars were operated in unison and scan strategies were chosen so that the complexity of the urban coastal flows could be revealed. Fig. 1 shows the three Doppler Lidars (A Halo Streamline, a Halo Pro and a Vaisala/Leosphere 200S) located on top of a roof in upper Manhattan in NYC and the location of these instruments (N 40.82187, W 73.94789) on a map.

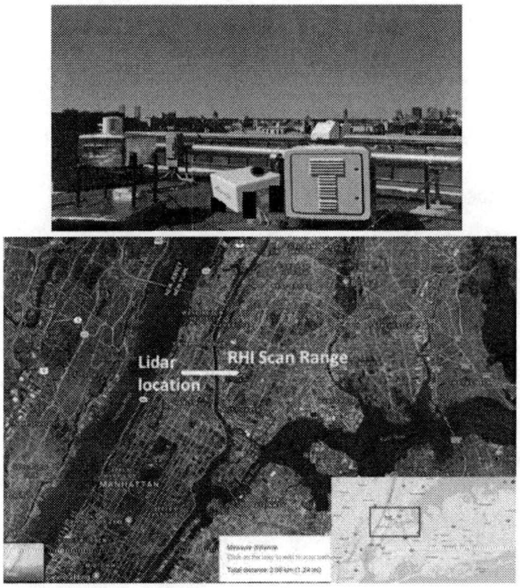

Fig. 1 Siting of Doppler Lidars in NY City

A number of different scans were selected from the various Doppler Lidars to observe mesoscale and microscale flows. Fig. 2 represents the types of scan strategies.

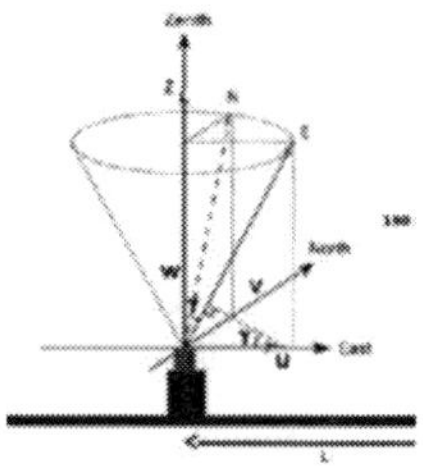

Conical Scan: Profiles mean wind fields.

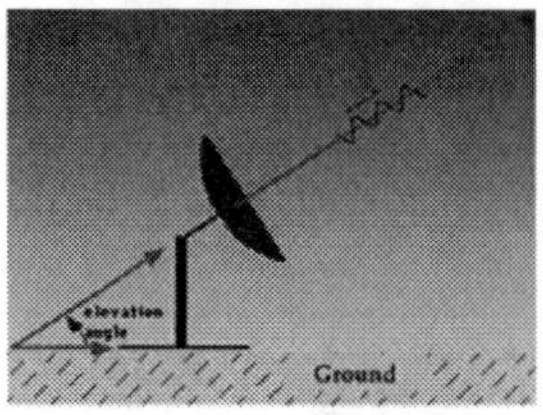

Range Height Indicator (RHI) scan: Lidar is scanned in Elevation at a fixed Azimuth (East in this case)

Fig. 2 Conception of Lidar Scan Strategies

The Halo Pro Doppler Lidar was programmed to perform Conical Scans to observe the vertical profile of the mean flows. Range Height Indicator (RHI) Scans were implemented on the Halo Streamline so that the VVP processing method can be used to invert the lidar returns. The Azimuth angle for the RHI Scans were fixed (pointed due East) and the Elevation angel was varied between 0 and 90 degrees. The scans implemented on the 200S Doppler Lidar were switched between Vertical Staring for 2 minutes and Doppler Beam Swinging (DBS) for 5 minutes. By comparing these observations with the outputs from a numerical weather prediction model that takes in to account urban-coastal effects (uWRF)[8], the dynamics of the vertical structure of the atmosphere becomes apparent.

3. RESULTS

Plots of the vertical profiles of the horizontal wind velocity a function of time (on July 2nd, 2018) are plotted in the Figures 3, 4 and 5. Fig. 3 is the Vaisala/Leosphere 200S Doppler Lidar, Fig. 4 is the Halo Pro Doppler Lidar and Fig. 5 is the output from the uWRF model. These plots show some similarities of the turbulent activities on this heat wave event. The horizontal wind speeds are reasonably low until the evening. At around 15:00 EDT both the observations and the uWRF plots show a significant low level jet that forms. To portray the formation of turbulent structures throughout the day and into the evening, the results from the RHI scans of the Halo Streamline are shown in Figures 6-8.

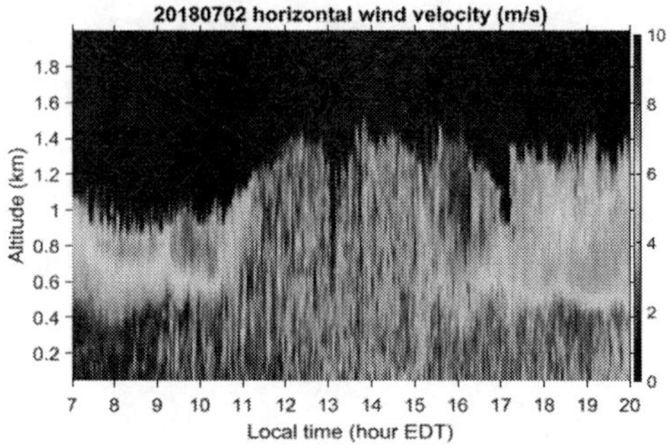

Fig. 3 200S Horizontal Wind Speed

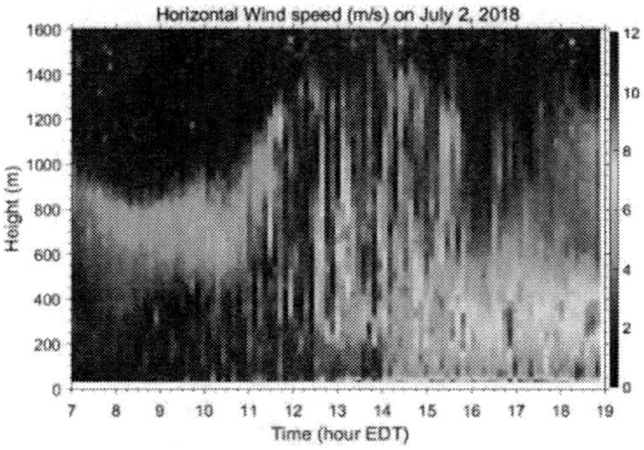

Fig. 4 Halo Pro Horizontal Wind Speed

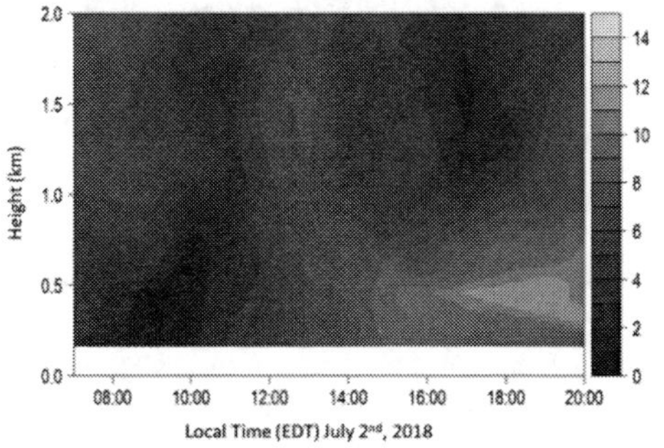

Fig. 5 uWRF Horizontal Wind Speed

Plots of the Halo Pro and Streamline Lidars results are shown in Figures 6-8 to show how the turbulent eddies evolve throughout the day. The Pro plots (6a, 7a, 8a) show how the the evolution of the low level jets and how horizontal direction has a slight amount of vertical shear. For the RHI plots (6b,7b,8b), the x-axis of these plots is the horizontal distance from the lidar, the y axis is the height above ground level, the color scales represent the line of sight wind speed and the arrows represent the wind direction only along the direction that the RHI scans are performed (East/West and vertical). Figure 7 and Figure 8 clearly demonstrate how the turbulent structures become more apparent later in the day.

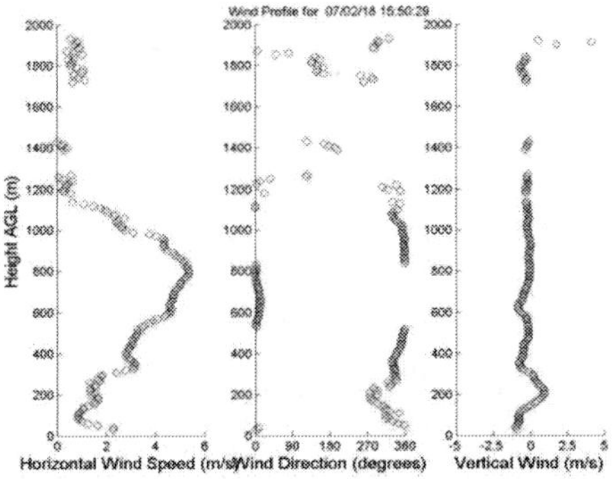

Fig. 6a Halo Pro profiles at 11:30 EDT

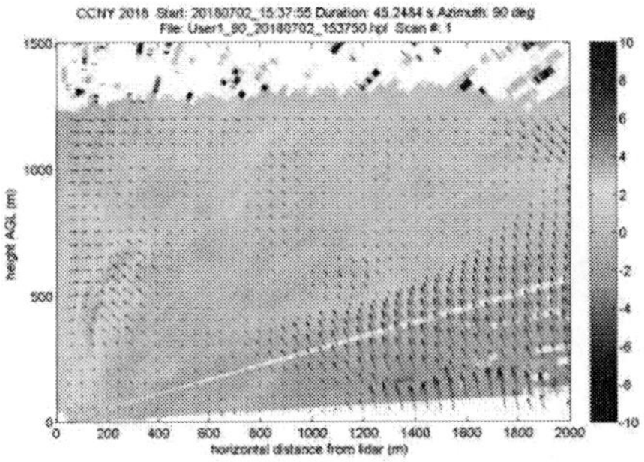

Fig. 6b Halo Streamline RHI output 11:30 EDT

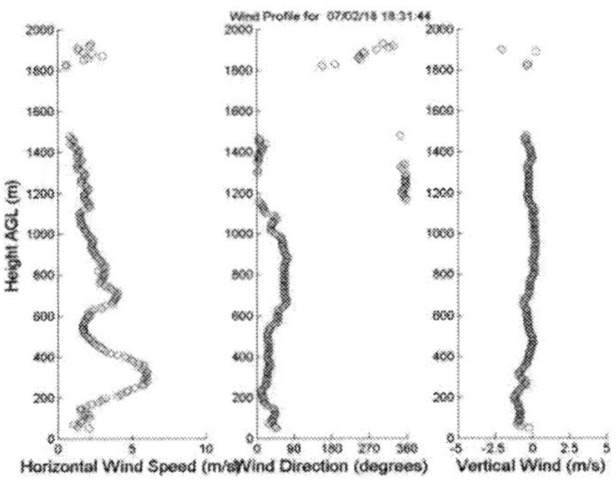

Fig. 7a Halo Pro profiles at 13:30 EDT

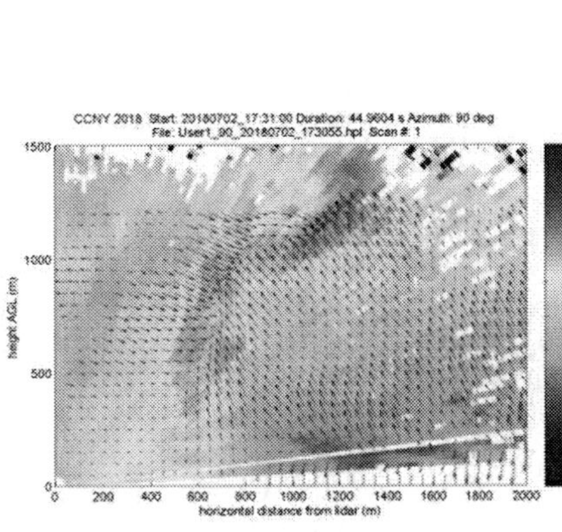

Fig. 7 Halo Streamline RHI output 13:30 EDT

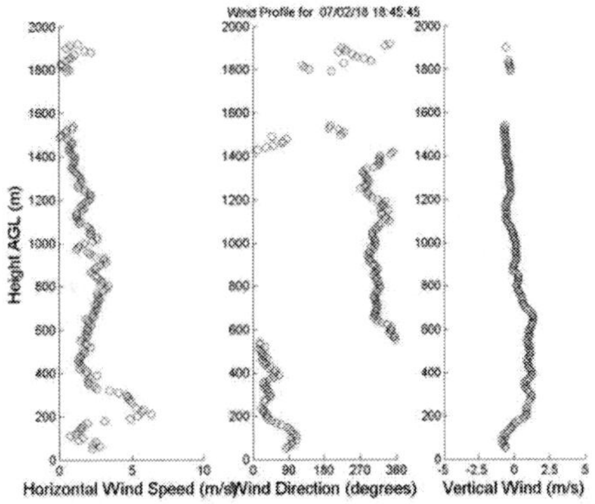

Fig. 8a Halo Pro profiles at 13:45 EDT

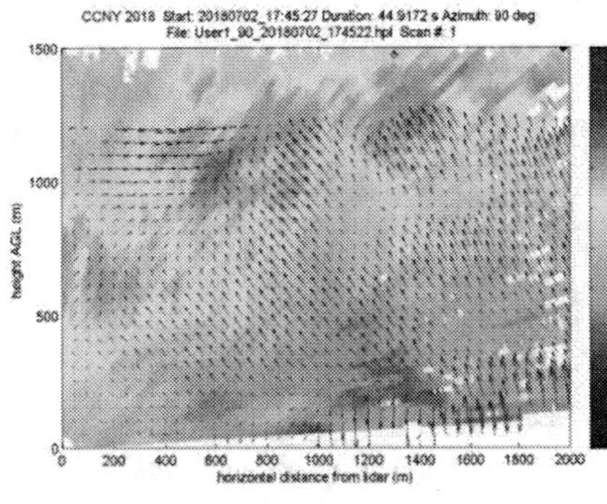

Fig. 8b Halo Streamline RHI output 13:45 EDT

Analysis of these events was made to explore the turbulent structures that evolved over the period of the event and the low level jets. The analysis performed in this study when combined with evolving Doppler Lidar inversion techniques[7-9], will be able to advance the methods described in this study.

ACKNOWLEDGEMENTS

This study is partly supported by the NESCAUM project #2411, New York State Energy Resources Development Authority (agreement # 100415) and the NOAA under the CREST Cooperative Agreement Grant # NA16SEC4810008.

REFERENCES

[1] Kaihui Zhao, Yunxuan Bao, Jianping Huang, Yonghua Wu, Fred Moshary, Mark Arend, Yongwei Wang, Xuhui Lee, "A high-resolution modeling study of a heat wave-driven ozone exceedance event in New York City and surrounding regions", Atmos. Environ., 2019, 199, 368–379

[2] Newsom, Rob & Calhoun, Ron & Ligon, David & Allwine, Jerry. (2008), "Linearly Organized Turbulence Structures Observed Over a Suburban Area by Dual-Doppler Lidar. Boundary-Layer Meteorology", 127. 111-130. 10.1007/s10546-007-9243-0.

[3] Newsom RK, Ligon D, Calhoun R, Heap R, Cregan E, Princevac M. "Retrieval of microscale wind and temperature fields from single- and dual-Doppler lidar data", Journal of Applied Meteorology. 2005 Sep;44(9):1324-1344. https://doi.org/10.1175/JAM2280.1

[4] Boccippio, D. J., 1995, "A diagnostic analysis of the VVP single-Doppler retrieval technique", *J. Atmos. Oceanic Technol.*, **12,** 230–248.

[5] Waldteufel, P., and H. Corbin, 1979,"On the analysis of single-Doppler radar data", *J. Appl. Meteor.*, **18,**532–542.

[6] Melecio-Vázquez, D., Ramamurthy, P., Arend, M. et al., "Thermal Structure of a Coastal–Urban Boundary Layer", Boundary-Layer Meteorol (2018) 169: 151. https://doi.org/10.1007/s10546-018-0361-7

[7] Ramamurthy P, González J, Ortiz L, Arend M, Moshary F (2017) Impact of heatwave on a megacity: an observational analysis of New York City during July 2016. Environ Res Lett 12:54011. https://doi.org/10.1088/1748-9326/aa6e59

[8] Gutiérrez E, González JE, Martilli A, Bornstein R, Arend M (2015) Simulations of a heat-wave event in New York City using a multilayer urban parameterization. J Appl Meteorol Climatol 54:283–301. https://doi.org/10.1175/JAMC-D-14-0028.1

[7] A. Choukulkar, R. Calhoun, B. Billings and J. D. Doyle, "A Modified Optimal Interpolation Technique for Vector Retrieval for Coherent Doppler LIDAR," in *IEEE Geoscience and Remote Sensing Letters*, vol. 9, no. 6, pp. 1132-1136, Nov. 2012. doi: 10.1109/LGRS.2012.2191762

[8] Abdelazim, Sameh; Santoro, David; Arend, Mark; Moshary, Fred and Ahmed, S., "Development and Operational Analysis of an All-Fiber Coherent Doppler Lidar System for Wind Sensing and Aerosol Profiling," in *IEEE Transactions on Geoscience and Remote Sensing*, vol. 53, no. 12, pp. 6495-6506, Dec. 2015. doi: 10.1109/TGRS.2015.2442955

[9] Abdelazim, Sameh; Santoro, David; Arend, Mark; Moshary, Fred; Ahmed, Sam. 2018. "A Hardware Implemented Autocorrelation Technique for Estimating Power Spectral Density for Processing Signals from a Doppler Wind Lidar System." Sensors 18, no. 12: 4170

OPERATION OF THE AIRBORNE 355-NM HIGH SPECTRAL RESOLUTION AND DOPPLER LIDAR LNG

D. Bruneau (1), J. Pelon (1),

F. Blouzon (2), Q. Cazenave (1), H. Collomb (1), A. Irbah (1), J-F Mariscal (1), J. Delanoë (1)

1 – LATMOS, Sorbonne Université-CNRS-UVSQ Paris, 2DT-INSU, CNRS, Meudon, France

ABSTRACT

High spectral resolution lidar (HSRL) are known to offer capabilities of separating attenuated aerosol and molecular backscattering so that particle extinction and backscattering can be separately retrieved. UV operation provides high energy in eye-safety conditions. Further to that, it could be important for most meteorological or environmental studies to get wind measurements at the same time. LNG is now the only HSR Doppler Lidar (HSRDL) system capable of this. Results obtained during ground-based and airborne measurements show that the backscatter and extinction coefficients at 355 nm can be measured with a relative precision better than 10% (adjusting altitude and time resolution from 60 m to 240 m and 30s to 2mn, respectively) in aerosol layers of 0.5 10^{-6} m^{-1} sr^{-1} backscatter coefficient from ground and aircraft. The same relative precision is obtained in cirrus clouds of a 10^{-5} m^{-1} sr^{-1} backscatter coefficient. The capacity of the system to perform wind velocity measurements has also been demonstrated with precisions in the range of 1 to 2 ms^{-1} in same conditions. We present the main characteristics and illustrate observational capabilities from ground-based and airborne measurements.

I. Instrument

The High Spectral Resolution (HSR) lidar technique allows to differentiate molecular and particular scattering thus improving aerosol or cloud extinction and backscatter coefficients retrieval which are critical parameters for environmental and meteorological studies. The first airborne HSR lidar systems were developed at DLR in Europe [1] and NASA in the USA [2]. The LNG system is based on the use of a Mach–Zehnder interferometer (MZI) as a spectral discriminator [3,4]. This presents the advantage to not require any frequency locking between the emitter and the interferometer and is moreover compatible with a multimode laser emission [5].
Measurement method, system description and first results can be found in [3,6]. We focus here on new results from ground-based and airborne field experiments presenting both scattering and wind measurements.

LNG is a three-wavelength (1064 nm, 532 nm and 355 nm) backscatter lidar with polarization analysis at 355 nm. The HSR capability, based on a MZI, has been added at 355 nm following a previous concept analysis [3]. The lidar is designed to be installed in the Falcon 20 or ATR 42 aircraft of SAFIRE, the French research aircraft service.

A. Emitter

The emitter is based on a Quantel Nd:YAG Q-Switched oscillator, injection seeded with a CW narrow-line emission. It provides single longitudinal mode pulses with 7 ns duration and a 70 MHz linewidth. The fundamental infrared emission is directed though second and third harmonic generators and then through a beam expander. The emitted energy at 355 nm is about 50 mJ@20 Hz, with a divergence smaller than 0.2 mrd. at the beam expander output. The divergence at 532 and 1064 is larger (4 and 5.5 mrd, respectively, to reduce eye-safety distance to a few hundred meters.

B. Receiver

The backscattered light is collected by a Cassegrain telescope with a 300 mm aperture. Behind the telescope focus, an optical package (receiving optics, RO) ensures the separation of the wavelengths, the background light spectral and spatial filtering and the parallel and cross polarization separation at 355 nm. The receiver field-of view is increased at 532 and 1064nm, accordingly to emission. This difference in field-of-views allows particle size estimates from multiple scattering analyses.

C. MZI design

The actual MZI design, detailed in [6], has an optical path difference (OPD) Δ= 20 cm and is field compensated in order to accept the large fiber diameter input without reduction of interference contrast. The MZI optical components and the detectors are mounted in a temperature stabilized box (+/- 0.1°C).

II. Measurement principle

According to (3), the signal delivered by each channel, for i = 1 to 4, can be written

$$S_i = \frac{S_t}{4} a_i \left[1 + M_i M_a \sin\left(\varphi + (i-1)\frac{\pi}{2}\right)\right] \quad (1)$$

with S_t the total signal, and a_i, M_i the relative photometric sensitivity and instrumental interference contrast of channel i, respectively. M_a is the interference contrast given by the atmospheric backscattered signal and φ the interference phase. The four signals are combined two-by-two in order to produce a complex signal Q (with in-phase and quadrature components), whose modulus is the contrast given by the atmospheric signal:

Always according to [3], the OPD of 20 cm has been chosen so that the interference contrast given by the molecular scattering is null. As the spectral broadening of the particulate scattering is negligible, the resulting contrast is equal to the instrumental contrast produced by the laser itself. Thus, the retrieved contrast M_a is equal to the particulate to total backscatter ratio (PBR) :

$$M_a = |Q| = PBR = \frac{\beta_p}{\beta_m + \beta_p} = \frac{R-1}{R} \quad (2)$$

where R is the usual lidar backscatter ratio (R=1+βp/βm) and βm, βp are the molecular and particulate backscatter coefficient, respectively. This result is independent of the exact molecular linewidth and line-shape provided it is sufficiently large to yield a null contrast. This avoids the necessity of precise line-shape modeling based otherwise needed [3,4]. Note also that the PBR retrieval is not affected by a Doppler shift of the signal caused by the platform motion and pointing, which is of importance for an airborne instrument.

In addition to the contrast, the interference phase φ, is obtained by the argument of the complex signal Q. Subtracting the reference interference phase φ_r, obtained on a highly attenuated pick-up of the laser emission one can obtain the Line-of-sight particles velocity V_{LOS} with:

$$\varphi - \varphi_r = \frac{2\pi\Delta}{\lambda_0}\left(1 + \frac{2V_{LOS}}{c}\right) \quad (3)$$

This ensures the Doppler measurement capability with the same MZI device. Of course the phase can only be measured when a significant interference contrast occurs, i.e. on particulate backscattering, but in this case, it is independent of the backscatter ratio.

B. Calibration and signal processing

Before processing actual atmospheric signals, it is necessary to perform a calibration of the MZI for determining the a_i, M_i parameters in Eq. (1). For this purpose we record the reference signal on the 4 channels on long time series. A fit of the recorded signals with the model described by Eq. (1) is obtained with a correlation better than 99.9% that allows determining the a_i and M_i parameters. With an atmospheric density model, we can calculate β_m (only dependent of molecular density) as a function of altitude. The attenuated particulate signal S_{par} and attenuated molecular signal S_{mol} can then be obtained as

$$S_{par} = M_a S_{tot} \quad and \quad S_{mol} = (1 - M_a) S_{tot} \quad (4)$$

with

$$S_{tot} = \frac{4}{M_1^{-1} + M_2^{-1} + M_3^{-1} + M_4^{-1}}\left(\frac{S_1}{a_1 M_1} + \frac{S_2}{a_2 M_2} + \frac{S_3}{a_3 M_3} + \frac{S_4}{a_4 M_4}\right) \quad (5)$$

IV. Scattering measurements

We first present here the measurements performed during tests performed on May 11, 2016 from ground.

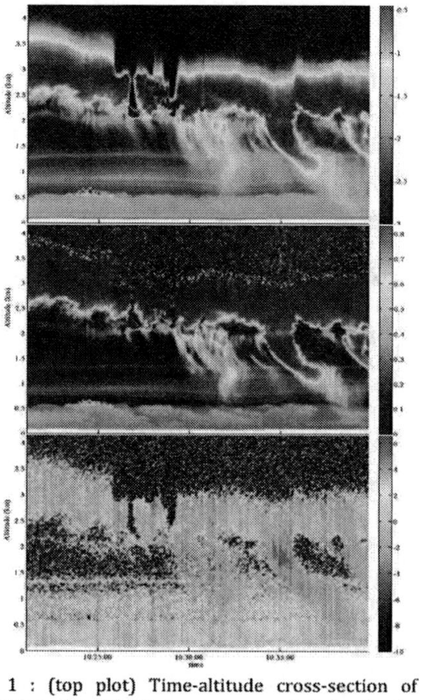

Fig. 1 : (top plot) Time-altitude cross-section of the logarithm of the 1064 nm range-corrected signals showing precipitating altostratus cloud above the 500m high boundary layer ; (middle plot) UV contrast M determined from MZI and (bottom plot) vertical air speed retrieved in the cloud and boundary layer. A 20 mn sequence of G/B measurements is shown for 11 May 2016.

The cross-section of the 1064 nm backscatter signals over the whole test duration is presented in Fig 1 (top panel). The aerosol layers can be clearly identified below 1.5 km from the 1064 nm signal and on the 355 nm MZI contrast. One can see that the boundary layer top is detected from the signal gradient at about 500 m above surface. The contrast M of MZI measurements shows a good dynamics from less than 0.02 to about 1 near cloud top. It allows retrieving the particle backscatter coefficient using Eq. (3). Vertical upward motions are seen in the boundary layer and above, the Doppler shift associated with the precipitations can also be clearly observed.

The 355 nm HSR analysis shows a weak PBR value above the boundary layer, increasing in the cloud up to 0.9, corresponding to an average backscatter ratio = 10 and to a particulate backscatter coefficient of 10^{-1} $km^{-1}.sr^{-1}$. The HSR vertical profiles of particulate backscatter (β_p) and extinction (α_p) coefficients at 355 nm and the related errors (standard deviation) profiles are presented in Fig. 2 for an averaging over 1 minute. The vertical resolutions are 60 m and 240 m for β_p and α_p, respectively. Random errors caused by the detection noise are calculated according to the equations developed in [3], from the total sensitivity of the detection channel and the number of photo-electrons derived from the recorded analog signals.
The particular depolarization ratio varies from 20% to 30% in the cloud. Results show that for long measurement series, the contrast and speed exhibit negligible biases even under variable environmental conditions.

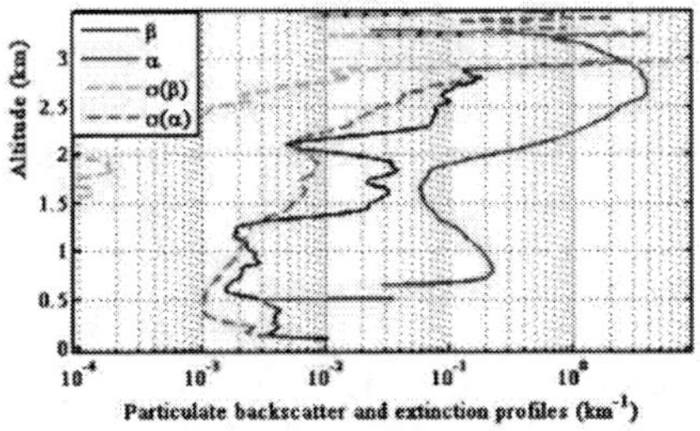

Fig. 2 : Profiles of HSR measurements over the aerosol layer a) particulate backscatter coefficient ($km^{-1}sr^{-1}$) and extinction coefficient (km^{-1}) with related errors (standard deviation).

IV. Airborne measurements

Pointing can be done in nadir or slant viewing (37° inclination angle) modes. We present here the measurements performed in nadir viewing mode during test flights in 2014-2015 and during the NAWDEX/EPATAN field experiment in 2016 [7].

A. Verification of ground echo speed

As a preliminary check of the HSR Doppler performance, measurements of the apparent ground speed Vs during flights over land have shown that the mean difference between the apparent ground speed calculated from the aircraft navigation data and measured by the lidar is 0.1-0.15 ms^{-1} with a standard deviation of about 1.5 ms^{-1}.

B. Cloud microphysics

Flight of 16 October 2016 has been used to perform a multiple field of view analysis. Two ratios are calculated from measured signals: the 532/355nm and the 1064/532 nm. The first one is sensitive to multiple scattering, whereas the second is not due to their similar field of views, and can be used as a reference.

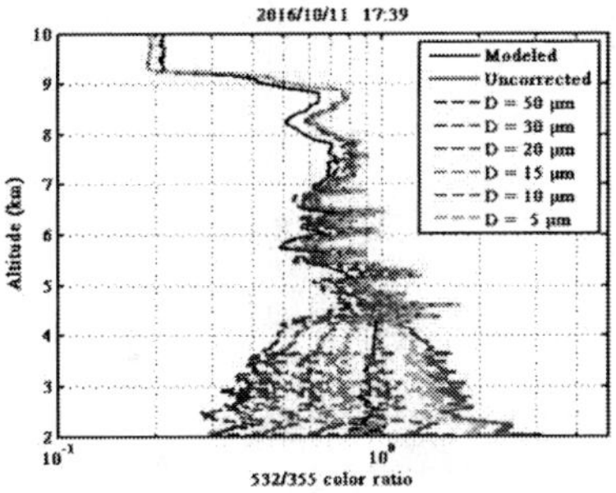

Fig3.: Ratio of the 532 to 355 nm in a mid-level altostratus (below 4.5 km) and calculation using the forward scattering model.

Fig. 3 shows that the correction of the 355/532 nm ratio using the Hogan model [8] allows to bring values close to the 532/1064 nm, indicative of particle size of about 20 µm.

C. Wind speed measurements on VAD

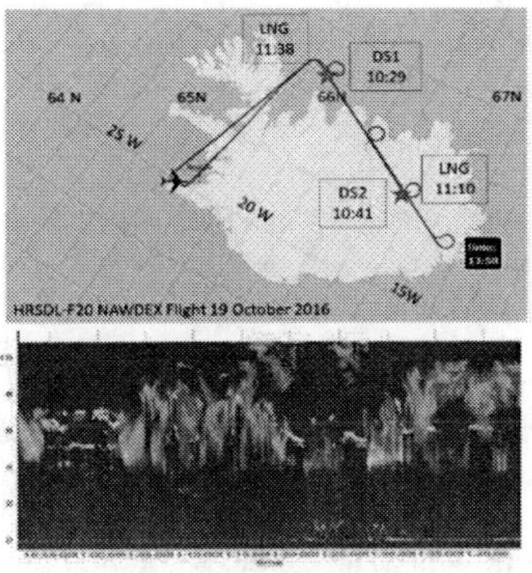

Fig. 3 : Flight plan and circle (top) made by the F20 to perform VAD measurements with LNG in vertical pointing.

Dropsonde launches have been performed on the way to the south-east part of Iceland, whereas VADs have been made on the way back. Backscatter signal in observed cirrus derived from contrast measurements using MZI after roll angle corrections in the four VADs.

During the flight of October 30th, 2016 (track is reported in Fig. 3 (top panel)), the F20 was flying over cirrus clouds over Iceland. Four wind profiles were performed during this flight using the lidar VAD technique on 360° aircraft turns together with two dropsonde launches. Results are presented in Fig. 4.

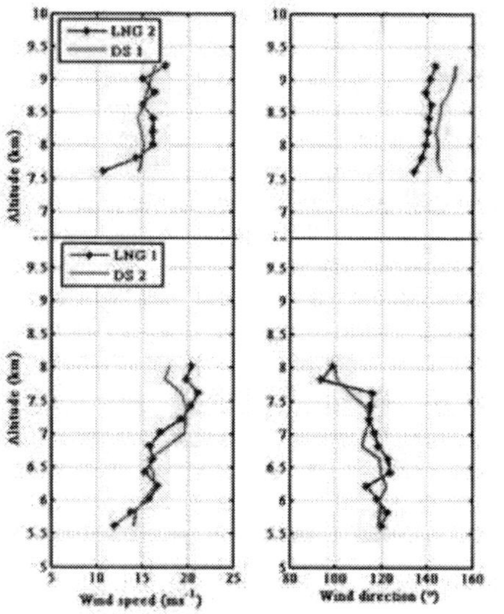

Fig. 4: Vertical profiles of wind speed and direction retrieved from VADs 1 and 3. Note that VAD 3 is closer in time and space to the dropsonde launch.

Comparison between VAD lidar measurement and dropsonde profiles show an excellent agreement, in both wind speed and direction for DS2 closer in time with lidar measurements.

V. Conclusion

The measurement capabilities of the 3-wavelength 2-polarisation HSR Doppler airborne lidar LNG have been successfully demonstrated during field campaigns. Results show that the backscatter and extinction coefficients at 355 nm can be measured with a relative precision of 10% for 60 m and 240 m vertical resolution respectively in aerosol layers of 10-6 $m^{-1}.sr^{-1}$ backscatter coefficient with a 30 km horizontal resolution. The same precision with the same vertical resolution is obtained in cirrus clouds 2 10-5 $m^{-1}.sr^{-1}$ backscatter coefficient with a horizontal resolution of 5 km. The Doppler measurement capability has also been demonstrated. The velocity measurement is performed on the particle backscattering and its precision is strongly dependent on the backscattering ratio. In favorable conditions, LOS wind velocity measurements can be made with a precision in the range of 1 to 2 ms^{-1}. VADs provide accurate profiles of wind components. LOS speed measurements performed on ground echoes show averaged biases less than 0.15 ms^{-1}. This instrument is particularly matched for the cloud studies in combination with cloud radar as planned in the ongoing RALI project [8]. It is planned to be used for aerosol-cloud process studies and for the validation of the ESA space-borne ADM-Aeolus and Earthcare-ATLID lidars. In this frame, further comparisons with ground-based and other airborne instruments are planned to ensure the validation process.

Acknowledgements

Development and test flight of LNG have been supported by INSU and CNES which are acknowledged. The authors also wish to thank the technical support of the Division Technique for the development and flights of the HRSD-LNG and the SAFIRE group for its assistance during the airborne campaign. NAWDEX-EPATAN flights were supported by ESA (Contract No. 4000119015/NL/CT/gp), we particularly acknowledge.

References

[1] M. Esselborn, M. Wirth, A. Fix, M. Tesche, and G. Ehret, "Airborne high spectral resolution lidar for measuring aerosol extinction and backscatter coefficients," Appl. Opt. 47, 346–358 (2008).
[2] J. W. Hair, C. A. Hostetler, A. L. Cook, D. B. Harper, R. A. Ferrare, T. L. Mack, W. Welch, L. R. Izquierdo, and F. E. Hovis, "Airborne high spectral resolution lidar for profiling aerosol optical properties," Appl. Opt. **47**, 6734–6752, (2008).
[3] D.Bruneau and J. Pelon, "Simultaneous measurements of particle backscattering and extinction coefficients and wind velocity by lidar with a Mach–Zehnder interferometer: principle of operation and performance assessment", Appl. Opt., 42, 1101–1114 (2003).
[4] D. Bruneau, A. Garnier, A. Hertzog, and J. Porteneuve, "Wind-velocity lidar measurements by use of a Mach–Zehnder interferometer, comparison with a Fabry–Perot interferometer", Appl. Opt., 43, 1, 173-182, (2004)
[5] D. Bruneau, F. Blouzon, J. Spatazza, F. Montmessin, J.Pelon and B. Faure, "Direct-detection wind lidar operating with a multimode laser", Appl. Opt., 52, 4941-4949, (2013).
[6] D.Bruneau, J. Pelon, F. Blouzon, J. Spatazzan , P. Genau, G. Buchholtz, N. Amarouche, A. Abchiche and O.Aouji, "355 nm high spectral resolution airborne lidar LNG: system description and first results", Appl. Opt. 54, 8776-8785, (2015)
[7] Schäfler and co-authors, The North Atlantic Waveguide and Downstream Impact Experiment, Bull. Amer. Met. Soc., 1607-1637, doi.org/10.1175/, 2018.
[8] Hogan, R. J., 2006: Fast approximate calculation of multiply scattered lidar returns. Appl. Opt., 45, 5984– 5992.
[9] Protat A.; Pelon J., Testud J.; Grand N., Delville P.; Laborie P., Vinson J.-P., Bouniol D.;, Bruneau D.; Chepfer H.; Delanoe J.; Haeffelin M., Noel V. and Tinel C., 2004 : Le projet RALI : Combinaison d'un radar nuage et d'un lidar pour l'étude des nuages faiblement précipitants, La Météorologie, n° 47, 23-33.

POISSON TOTAL VARIATION DENOISING FOR MICROPULSE WATER VAPOR DIAL

Matthew Hayman[1], Willem Marais[2], Robert Stillwell[1], Scott Spuler[1]

[1]*National Center for Atmospheric Research, Earth Observing Lab, Boulder, CO 80303, USA*
[2]*Cooperative Institute for Meteorological Satellite Studies, University of Wisconsin-Madison, USA*
**Email: mhayman@ucar.edu*

ABSTRACT

We have adapted the Poisson Total Variation lidar signal processing technique for Micro-Pulse DIAL water vapor estimates. This processing technique ingests data at 10 second-37.5 meter resolution where it adaptively adjusts the retrieval resolution based on signal-to-noise and performs range deconvolution. The result is high resolution data in regions with ample signal, preservation of sharp discontinuities where they exist, and a data product that better represents of the acquired photon counting data than the standard processing technique.

1. INTRODUCTION

Water vapor DIAL uses two closely spaced wavelength, one tuned to a water absorption feature (online) and a second tuned off the feature (offline) to estimate range resolved water vapor concentration in the atmosphere [1]. Through knowledge of the water vapor absorption cross section, the two observations can be directly related to the water vapor number concentration. The power of this approach is that it requires no ancillary observations of atmospheric state for processing or calibration, making it a truly independent of other water vapor observations.

In a collaboration between the National Center for Atmospheric Research (NCAR) and Montana State University, a Micro-Pulse DIAL (MPD) has been developed based on diode laser technology for remote unattended operations over long periods of time [2,3,4]. The instrument lasers are invisible and eye-safe and the receiver employs narrowband filtering and photon counting to observe weak signals day and night in clear and cloudy conditions. In order to obtain useful water vapor estimates, the standard MPD signal processing employs fixed temporal smoothing between 1 and 5 minutes and range smoothing between 150 and 750 m (depending on signal-to-noise).

Poisson Total Variation is a processing technique recently developed at University of Wisconsin for high spectral resolution lidar (HSRL) retrievals and has been shown to significantly improve extinction and lidar ratio estimates from the UW HSRL instrument [5]. The technique leverages numerical methods developed for Poisson image denoising and deblurring [6,7]. PTV assumes that the estimated lidar data products can be approximated by a linear piecewise function and effectively adapts the region of estimation based on signal-to-noise. It leverages correlations in the observed image (in the case of lidar a time/range curtain) to infer the data product from the random noise without forcing the underlying signal to be band limited.

In a collaboration between NCAR and University of Wisconsin, we have demonstrated the PTV technique can be adapted for MPD water vapor estimates. While the approach is still under development, it shows promise for improving the instrument's capabilities by providing flexibility to capture variations at higher resolution and increasing retrieval accuracy.

2. METHODOLOGY

The PTV method for HSRL is detailed in [5], and the technique was adapted for processing DIAL signals without sacrificing the advantages of the conventional DIAL technique. Specifically, it allows estimation of water vapor without direct knowledge of the atmospheric aerosol loading, instrument effects common to both wavelengths (e.g. geometric overlap) or scalar differences between the two wavelength channels.

PTV requires a forward model to evaluate retrieved parameters against raw photon counts. To implement this we add a second parameter retrieval. The forward model consists of a profile that is dependent on the wavelength (λ), the water vapor

https://doi.org/10.1051/epjconf/202023706012

density (n_j) and a term that represents all terms that are common multipliers between the two wavelength channels (ϕ_i). The observed photon counts at bin i and operational wavelength λ is given by

$$u_i(\lambda) = C(\lambda)\phi_i exp\left(-2\Delta r \sum_{j=1}^{i} \sigma_j(\lambda)n_j\right) + b$$

Where $C(\lambda)$ is a constant multiplier specific to the laser wavelength of operation, $\sigma_j(\lambda)$ is the water vapor absorption cross section at that wavelength, Δr is the range between data points and b is the background.

In principle, the common term can be estimated using a conventional DIAL processing approach, but it is not necessary and provides limited scientific value. For this analysis, we require estimating ϕ_i to forward model the retrieval and evaluate its accuracy.

In our retrieval we allow the estimator to adjust the scalar gain of each channel (C). Like ϕ_i, this falls out in standard DIAL processing but is needed for forward modeling. It has no impact on the actual water vapor estimate.

The forward modeled profiles are evaluated using the inverse log-likelihood for a Poission distribution. Lower values of this function indicate a better fit. We also include a regularizer term to penalize variation in the retrieved profile. The overall minimization is performed on

$$\epsilon(\phi, n) = \sum_{\lambda}\sum_{j=1}^{i} u_i(\lambda) - U_i(\lambda)\ln u_i(\lambda) + \alpha_n \|n\| + \alpha_\phi \|\phi\|$$

such that the solution is

$min_{\phi,n}\ \epsilon(\phi, n)$,

where the forward model $u_i(\lambda)$ is a function of the water vapor density (n_j) and multiplier term (ϕ_i). The term $\|n\|$ evaluates the sum of the absolute value of all changes in n in both time and range and α_n and α_ϕ are regularization coefficients that determine the amount of penalty imposed for variation in the estimated water vapor and common terms.

3. RESULTS

Figure 1 shows an example of MPD retrievals applied to data from August 16, 2018. The top panel shows our standard water vapor processing routine. The middle panel shows the water vapor estimate from the PTV processing method described here and the bottom panel shows the common terms obtained through the PTV method.

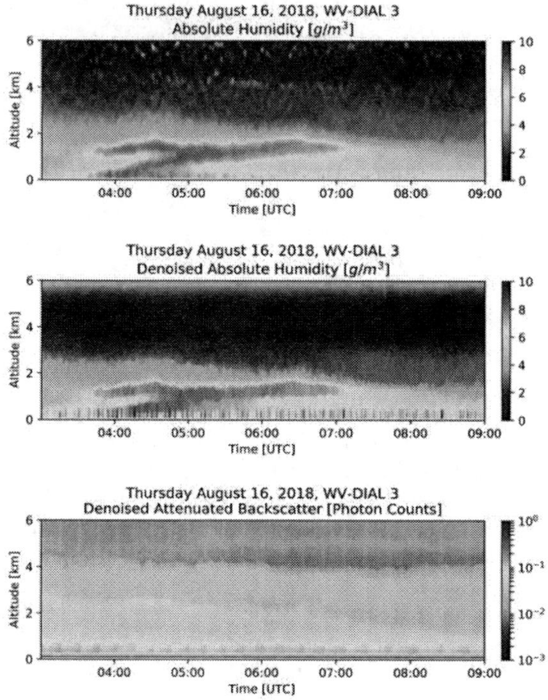

Figure 1: Water vapor estimates from an MPD using the standard inversion (top) and PTV inversion (middle). For context, the PTV estimated common terms are shown (bottom).

We note that there are some key differences between the standard and PTV processing techniques. First the PTV method captures sharper discontinuities in the water vapor profile because PTV approximates the water vapor field as a linear piecewise function. This is in contrast to the our standard Savitzky-Golay filtering and Gaussian smoothing approach used in the top panel, which effectively limits the sharpness of the profile.

In the region above 3 km, the PTV better captures the dryness of the air by using larger pieces, while the standard processing approach has substantial noise that deceptively looks like enhanced water (a product of the color scale having a minimum value of zero).

The PTV estimate shows an artificially high water vapor content at the top of the profile above 5.5 km. We suspect this artificially high region is the result of low signal, where the optimizer did not adjust the water vapor estimate because it had little impact on the forward model error.

The PTV method also appears to be having problems estimating water vapor in the lowest range bins below 500 m where there is significant striping throughout the time series. This may be the result of the outgoing laser shot contaminating the return signal. We note that the standard method also appears to have problems in this region, though they are less severe.

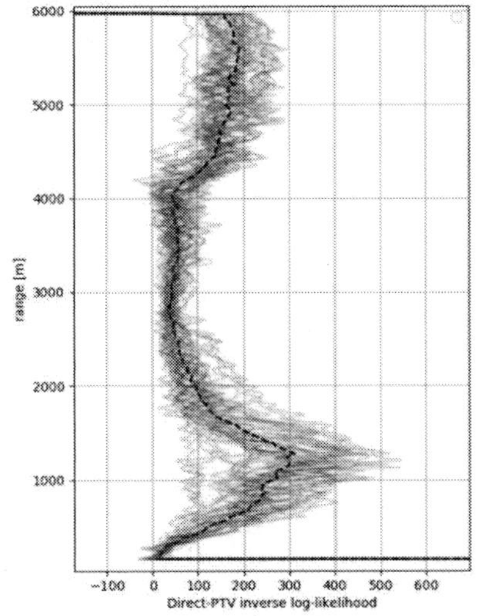

Figure 2: Difference in inverse log-likelihood between direct water vapor inversion and PTV inversion for the vertical profiles in Figure 1. The median is indicated by the dashed black line. Positive values indicate better validation in the PTV retrieval.

To analytically assess the two retrieval methods we use the inverse log-likelihood $\epsilon(\phi, n)$,without the regularizer term where we compare the forward modeled retrievals against a verification observation that is statistically independent from the original retrieval data. These separate retrieval and verification datasets are obtained from a single photon counting MPD profile using Poisson thinning, where the underlying signals are common to both datasets, but their Poisson noise is statistically independent. We forward model the two water vapor estimation approaches using the common terms from the PTV method in both cases. We found using the standard approach to estimate ϕ_i made its forward model evaluate worse and we do not want to penalize the approach for a secondary, typically unused, parameter. The difference in the inverse log-likelihoods are shown in Figure 2, where positive values indicate better agreement from PTV water vapor estimates and negative indicates better agreement from the standard approach.

From Figure 2, it is clear that the PTV method provides a better overall reflection of the lidar profiles. The preservation of higher frequency structure seems to improve the forward signal reconstruction below 2 km and above 4 km. Notably the artificially high water vapor does not seem to have a significant impact on the verification process, again suggesting that there is not enough information content in this region (i.e. there is high signal uncertainty) to estimate water vapor concentration.

3. CONCLUSIONS

The PTV signal processing method appears to be promising, for developing high accuracy water vapor estimates from photon counting MPD. However there are practical challenges to the technique. Signal processing is computationally expensive. The data set shown here was processed on a desktop computer and took approximately one day to complete. At present, there is no fast way to estimate uncertainty in the retrievals. Current techniques employed for HSRL retrievals employ ensemble approaches with various initial conditions making the computationally expensive though parallelizable.

Never-the-less these issues are not necessarily major limitations. Advances in and availability of high speed computing resources continue to improve. There is currently work to implement

PTV on GPUs and we continue to pursue opportunities to develop faster uncertainty estimates.

ACKNOWLEDGEMENTS

The National Center for Atmospheric Research is sponsored by the National Science Foundation.

REFERENCES

[1] Schotland, R. M. "Some observations of the vertical profile of water vapor by means of a laser optical radar," in *4th Symposium on Remote Sensing of the Environment*, (Michigan, 1964).

[2] Nehrir, A. R., et al., "Water vapor profiling using a widely tunable, amplified diode-laser-based differential absorption lidar (DIAL)," J. Atmos. Ocean. Technol., 26, pp. 733–745, doi:10.1175/2008JTECHA1201.1 (2009).

[3] Spuler, S. M., et al., "Field-deployable diode-laserbased differential absorption lidar (DIAL) for profiling water vapor," Atmps. Meas. Tech, 8, pp. 1073–1087, doi:10.5194/amt-8-1073-2015 (2015).

[4] Weckwerth, T. M., et al., "Validation of a Water Vapor Micropulse Differential Absorption Lidar (DIAL)," J. Atmos. Ocean. Technol., 33, pp. 2353–2372, doi: 10.1175/JTECH-D-16-0119.1 (2016).

[5] Marais, W. J., et al., "Approach to simultaneously denoise and invert backscatter and extinction from photon-limited atmospheric lidar observations," Appl. Opt., 55, pp. 8316-8334, doi:10.1364/AO.55.008316 (2016)

[6] Wright, S., et al., "Sparse reconstruction by separable approximation," IEEE Trans. On Signal Proc., 57, pp. 2479-2493 (2009), doi:10.1109/TSP.2009.2016892.

[7] Harmany, Z. and R. M. Willett, "This is SPIRAL-TAP: Sparse poisson intensity reconstruction algorithms-theory and practice," IEEE Trans. On Image Proc., 21, pp. 1084-1096, doi:10.1109/TIP.2011.2168410 (2012).

RAMAN LIDAR FOR SYNCHRONOUS WATER VAPOR, LIQUID WATER AND ICE WATER PROFILINGS

Wang Yufeng, Zhang Jing, Wang Qing, Gao Fei, Di Huige, He Tingyao, Yan Qing, Liu Jingjing, Hua Dengxin

School of Mechanical and Precision Instrument Engineering, Xi'an University of Technology, Xi'an 710048, China

Email: wangyufeng@xaut.edu.cn

ABSTRACT

Water is the only atmospheric parameter with three-phase state. An ultraviolet Raman lidar was developed for synchronous measurements for water vapor, liquid water and ice water in Xi'an University of Technology, Xi'an, China (34.233°N, 108.911°E). An accurate retrieval method on the basis of interference degree is proposed for synchronous three-phase water mixing ratio profiles. Preliminary measurements are carried out in the Laser Radar Center of Remote Sensing of Atmosphere (LRCRSA). Several representative examples are obtained and validated the performance of Raman system. Combined with atmospheric temperature profiles, the synchronous water vapor, liquid water and ice water profiling are retrieved and revealed the variation characteristics in three-phase water. The effective detection can reach up to a height of 5 km under cloudy weather, and synchronized growth in water vapor and liquid water content was obtained in cloud layers. Continuous observations are also made under haze weather condition, and the temporal and spatial evolution trend of three-phase water in clouds at 2 km altitude are successfully realized.

Keywords: Raman lidar, water vapor, liquid water.

1. INTRODUCTION

Water vapor is the only atmospheric parameter with three-phase state, which plays an important role in the global water cycle. Water vapor plays an important role in cloud evolution, precipitation and climate change. Liquid water in clouds is not only an important component in the balance of atmospheric water budget, but also an important microphysical parameter to understanding cloud physical processes. Super-cooled water in cloud (which remains liquid below 0° C), is also one of the most important parameters indicative to the potential of artificial precipitation enhancement. Therefore, the study of water three-phase distribution with high precision and high-spatial-temporal resolution is of great significance in understanding the formation of cloud and precipitation, fine analysis and prediction of precipitation, and artificial precipitation enhancement.

Lidar is a powerful tool in the remote sensing of atmospheric components and atmospheric properties. Many ground-based Raman lidar systems have achieved currently exist around the world, and significant achievements have been made in the measurement of atmospheric water vapor and aerosol profiles [1-4]. Recent years, several institutes have carried out research work on three-phase water. In 2000, Veselovski et.al developed a Raman lidar for detecting water vapor and liquid water in tropospheric layer, and the relative lidar power under different atmospheric conditions was mainly discussed [5]. In 2004, Zhien Wang et.al developed a Raman lidar to detect the solid-water content in cirrus clouds, and proposed the inversion method of solid-water mixing ratio [6]. In 2012, Yi Fan et.al obtained complete Raman spectra of three-phase water by a 32-channel grating spectrometer, and achieved lidar returned signals of atmospheric water vapor and liquid water in the range of 0-6 km [7]. It is required to further studies on synchronous detection, especially on synchronous retrievals of accurate mixing ratio profiling.

In this paper, an ultraviolet Raman lidar system has been established for synchronous measurement of atmospheric three-phase water. We also carried out preliminary experiments to validate the performance of Raman lidar under different weather condition. Furthermore, an accurate retrieval method on the basis of interference degree is proposed for synchronous three-phase water mixing ratio profiles. The

https://doi.org/10.1051/epjconf/202023706013

results show that the synchronous detection of atmospheric three-phase water below 5 km can be realized, and the synchronous growth of liquid water and water vapor in clouds is achieved.

2. METHODOLOGY

2.1 The Raman lidar system

The ultraviolet Raman lidar system used in this study was developed at Xi'an University of Technology. A schematic diagram of the system is presented in Fig. 1. A set of dichroic mirrors (DMs) and narrow-band interference filters (IFs) is used to construct a high-efficiency polychromator, which divides the returned signals into five channels: channel 1 is used for the detection of elastic Mie-Rayleigh signals, at central-wavelength of 354.7nm; channel 2-5 are used for the detection of different vibrational Raman scattering signals; channel 2 is for vibrational Raman lidar returns with a central-wavelengths of 386.7 nm; and channel 3-5 are used to detect Raman scattering signals of ice water, liquid water and water vapor molecules, with central-wavelengths of 397.0nm, 403.0nm and 407.6 nm, respectively.

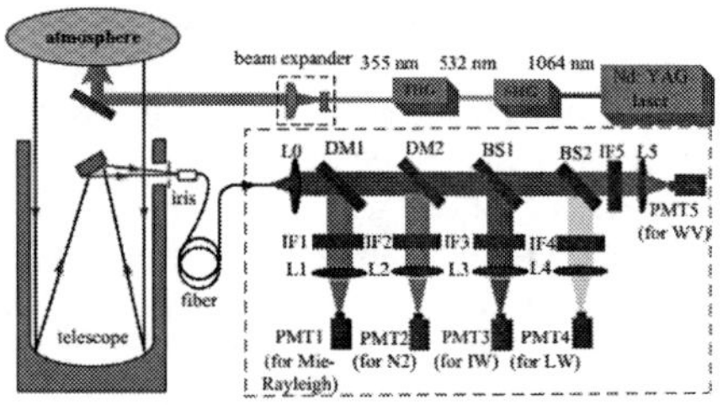

Fig.1 Diagram of Raman lidar for synchronous three-phase water detection. DM: Dichroic Mirror, IF: Interference Filters, BS: Beam Splitter.

2.2 Retrieval methods

According to its definition, three-phase water mixing ratio can be retrieved by using the ratio of Raman scattering signals of three-phase water and nitrogen molecules, and can be given as,

$$\begin{cases} W_{WV}(z) = 0.485 \times \frac{P_{WV}(z)}{P_N(z)} \cdot \frac{k_N}{k_{WV}} \cdot \frac{\sigma_N(\pi)}{\sigma_{WV}(\pi)} \cdot \exp\left\{ \int_0^z [\alpha_{\lambda_{WV}}(z') - \alpha_{\lambda_N}(z')] dz' \right\} \\ W_{LW}(z) = \frac{P_{LW}(z) - C_{LW} P_{WV}(z) - C_{LI} P_{IW}(z)}{P_N(z)} \cdot \frac{k_N \sigma_N(\pi)}{k_{LW} \sigma_{LW}(\pi)} \cdot \exp\left\{ \int_{z_0}^z [\alpha_{LW}(z') - \alpha_N(z')] dz' \right\} \\ -f_{\Delta k} \frac{1}{\Delta Z_{CB}} \times \int_{z_{CB}-\Delta z}^{z_{CB}} \frac{P_{LW}(z') - C_{LW} P_{WV}(z')}{P_N(z')} \exp\left\{ \int_{z_{CB}-\Delta z}^{z_{CB}} [\alpha_{LW}(z') - \alpha_N(z')] dz' \right\} dz' \\ W_{IW}(z) = \frac{P_{IW}(z) - C_{IL} P_{LW}(z)}{P_N(z)} \cdot \frac{k_N \sigma_N(\pi)}{k_{IW} \sigma_{IW}(\pi)} \cdot \exp\left\{ \int_{z_0}^z [\alpha_{IW}(z') - \alpha_N(z')] dz' \right\} \end{cases}$$

where the subscripts LW, WV, IW and N denote liquid water, water vapor, ice water and N2 molecules, respectively. P is the output power of the channel, σ is the cross section of Raman scattering, α_λ is the extinction coefficient at the wavelength λ, k is the system efficiency factor, including the optical efficiency and electrical efficiency of system. C represents the interference degree from different water phase Raman channels. Thus, due to spectral overlap characteristics of three-phase water, it is essential to take into account mutual interference degree of different phase water, to achieve synchronous retrieval of three-phase water.

3. MEASUREMENTS EXAMPLES AND RESULTS

The Raman lidar system was built and operated at Xi'an University of Technology, Xi'an (34.233°N, 108.911°E), China. Some tests were conducted to validate the system performance. Figure 2 presented two representative measurement examples taken at under cloudy condition. The left figure corresponds to the results taken at 21:39 CST on 29 Sep, 2018. Two obvious peaks appear at heights of 4.5 km and 5.5 km from the elastic scattering signals. The effective range can reach 6km and 5km for water vapor and liquid water Raman channel, and synchronous grow of water vapor and liquid water in the clouds can be observed. Another measurement example was taken at 22:35 CST on March 4, 2019, as shown in right figure. Two obvious peaks appear at heights of 1-1.2 km (cloud 1) and 2-2.2 km (cloud 2) from the elastic scattering signals, which correspond to the information of thin cloud layers. Fortunately, we can obviously synchronous increase in the RSCS of water vapor, liquid water and ice water in cloud layer 1 of ~1.2km. At the heights of 2-2.2km of cloud layer 2, synchronous growth can also be observed in water vapor and liquid water, however, it has failed to capture the information in ice water because of the weak signal itself, indicating that the Raman lidar system has successfully realized the simultaneous detection of atmospheric water vapor, liquid water and ice water in the clouds.

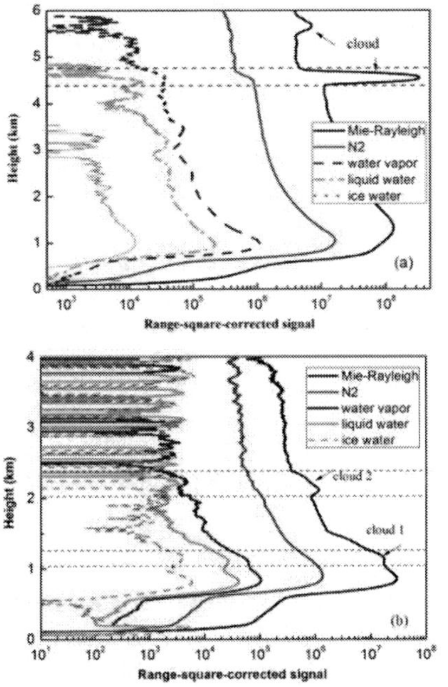

Fig.2 Representative measurement examples under cloudy weather condition. (a) Range-square-corrected signal taken at 21:39 CST on 29 Sep, 2018. (b) Range-square-corrected signal taken at 22:35 CST on March 4, 2019.

Another measurement example was taken at 23:54 CST on December 24, 2018, as shown in Fig.3. The example is also used to validate the performance of the Raman lidar under haze weather condition. The local ground temperature is 4 °C, and the visibility is 5 km, and the concentration of PM 2.5 and PM10 are 75 ug/m^3 and 145 ug/m^3 by , respectively, AQI is 99, indicating mild haze weather. From the RSCS of Mie-Rayleigh scattering signals, there exists a thick cloud layer at heights of 1.8-2.1 km, and thus the effective detection range can only reach 2 km. Meanwhile, the cloud information does not appear in the attenuated nitrogen vibrational Raman signals, which verifies the high rejection rate to elastic scattering signals. The retrieval mixing ratio of three-phase water profiles are also shown at the right. Evident enhancement of water vapor, liquid water and ice water can be obtained at the cloud layer.

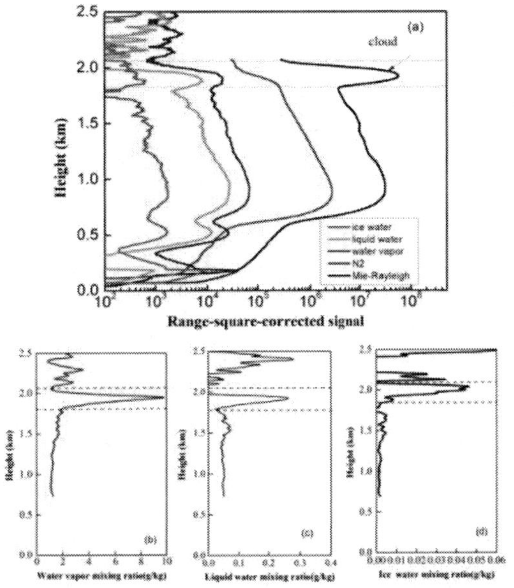

Fig. 3 Measurement results taken at 23:54 CST on December 24, 2018 under mild haze and cloud condition. (a)Range-square-corrected signal, (b)water vapor mixing ratio, (c)liquid water mixing ratio and (d)ice water mixing ratio.

ACKNOWLEDGEMENTS

This work was supported by the National Natural Science Foundation of China (NSFC) (grant No. U1733202, 41627807, 41575027 and 41027004).

REFERENCES

[1] M. Froidevaux., et al. Advances in Water Resources, 51, 345-356 (2013)

[2] T. Leblanc., et al. Atmos. Meas. Tech. 5, 17-36 (2012)

[3] Wang Yufeng, et al. Journal of Quantitative Spectroscopy and Radiative Transfer, 205, 114-126 (2018)

[4] A. Behrendt., et al. Appl. Opt., 41(36), 7657-7666 (2002)

[5]Veselovskii I A, et al. Applied Physics B, 71(1):113-117(2000)

[6]Wang Z, et al. Geophysical Research Letters, 31(311):121-141(2001).

REMOTE SENSING OF ATMOSPHERIC TURBULENCE PROFILES BY LASER GUIDE STARS

Xiwen Qiang

The State Key Laboratory of Laser Interaction with Matter, Xi'an 710024, P. R. China

ABSTRACT Ranged-resolved profiles of atmospheric turbulence are necessary and important for many applications in astronomical and adaptive optics communities. In order to characterize the vertical atmospheric structure in field, a technique is put forward to remote sensing ranged-resolved profiles of atmospheric turbulence by combined with laser guide stars and differential image motion method. Laser guide stars are formed at several successive altitudes by projecting pulsed laser, returned signals of images are received by a optical system with two receiving telescopes, and variance of centroids' distance is derived from the images with two spots at the same altitude. So, based on a inversion algorithm, atmospheric turbulence profiles are retrieved from differential image motion variance of distance of centroids at various altitudes. The structure constants of refractive index of atmosphere range from $10^{-14}m^{-2/3}$ at lower altitudes to $10^{-16}m^{-2/3}$ at higher altitudes are remote sensed experimentally. The results show it is a effective method that combined laser guide stars with differential image motion method and could sense atmospheric turbulence profiles remotely in real time.

KEYWORDS atmospheric turbulence profiles, remote sensing, atmospheric optics, laser guide stars, differential image motion method

1. INTRODUCTION

Characteristics of atmospheric turbulence varies temporally and spatially, so remote sensing of atmospheric turbulence profiles in real time is important and necessary for applications such as performance analysis of astronomical adaptive optics systems, astronomical sites surveys and selection, validation of atmospheric turbulence prediction model, free space laser communication, and laser beam propagation in the atmosphere. The astronomical and adaptive optics communities have established several optical techniques for remote sensing Fried parameter, a parameter of path-integrated structure constant of refractive index of atmosphere, but mote sensing ranged-resolved information of atmospheric turbulence. Laser guide stars technique is used commonly in adaptive optics community, and differential image motion method is a matured technique in astronomical community. So, a new technique is put forward for remote sensing atmospheric turbulence profiles. Combined with laser guide stars and differential image motion method, remote sensing of atmospheric turbulence profiles in real time is realized in experiments effectively.

The configure of experiment consist of a powerful pulsed laser, a beam-expanding lens and a negative lens for beam-focusing, a reflective mirror for beam-projecting, a optical device for receiving and imaging of returned signals, a ICCD camera for recording images of laser guide stars, and a computer for controlling and data processing. By expanding and focusing laser beam projected from a powerful pulsed lasers, laser guide stars are formed at several successive altitudes. A optical system with two receiving telescopes is developed based on differential image motion method. The laser guide stars are observed with the optical system developed, and images of laser guide stars with two spots at a receiving ICCD camera. The images with two spots at the same altitude are processed and centroids of two spots at every images are given. Differential image motion variance of distance of centroids could be derived from the images at the same altitude. So, the motion of the images of the laser guide stars from each altitude is characterized as the differential image motion variance.

Based on a inversion algorithm, atmospheric turbulence profiles could be retrieve from differential image motion variance of distance of centroids at various altitudes. Experiments of remote sensing have been performed, and the profiles of atmospheric turbulence are obtained. The structure constants of refractive index of atmosphere range from $10^{-14}m^{-2/3}$ at lower altitudes to $10^{-16}m^{-2/3}$ at higher altitudes. The results show it is a effective method that combined laser guide stars with differential image motion method and could sense atmospheric turbulence profiles remotely in real time and the profiles data remote sensed are reasonable.

2. BACKGROUND AND PRINCIPLE of REMOTE SENSING

2.1 Background

The atmospheric turbulence profiles had been measured in field by various methods, including radio-sonde with microthermal sensors, MASS(Multi-Aperture Scintillation Sensors)[1], SCIDAR(SCIntillation Detection And Ranging)[2,3], SODAR(SOund Detection And Ranging)[4], and other techniques. The turbulence profiles measured by radio-sonde are not in real time and do not along vertical direction, the spatial resolution of turbulence

https://doi.org/10.1051/epjconf/202023706014

profiles sensed by MASS are lower in vertical direction which are only seven layers, SCIDAR need binary stars within a certain angle, and SODAR only sense lower altitudes within boundary layers and altitude sensed is limited. In order to characterize the vertical structure of atmospheric turbulence, it is necessary to remote sense the whole turbulence profiles in real time.

It is a new technique to remote sense turbulence profiles by combined laser guide stars with differential image motion and its concept was put forward by M. S. Belen'kii et al[5]. In recent years, Georgia Tech Research Institute have developed brassboard turbulence LIDAR[6] in 2007 and remote sensed turbulence profiles at horizontal path, in boundary layer measured and compared with microthermal sensors simultaneously, and along vertical paths in field. The layout of the brassboard turbulence LIDAR[7] is illustrated in Fig. 1, the example guide stars image[7] is shown in Fig. 2, example of Differential Image Motion versus camera frame number is shown in Fig. 3, and the time-height diagram of the log of the strength of turbulence is shown in Fig. 4.

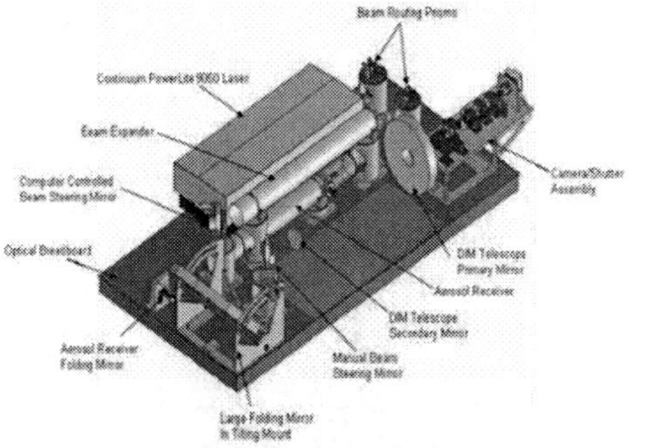

Fig. 1 Brassboard Turbulence LIDAR Layout

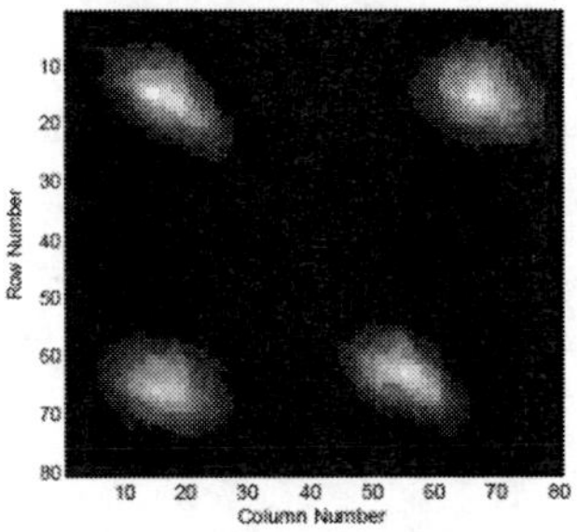

Fig. 2 Example Guide stars image

2.2 Principle of remote sensing

The Fried parameter could be sensed by differential image motion method(DIMM)[7] utilizing a smaller telescope by place a mask with two holes in front of the telescope to remote sense a star. By this DIMM technique, the path-integrated parameter, Fried parameter, not a range-resolved profile, is sensed. If a star could be formed at any height and its brightness is enough, the range-resolved profiles of atmospheric turbulence could be sensed remotely in real time. So, a technique for remote sensing turbulence profiles of atmosphere is put forwards by combined differential image motion method and laser guide stars(LGS). Here differential image motion method is frequently-used and mature in astronomical community, and the method to form laser guide star is also frequently-used and mature in adaptive optics community. So, combining these two mature methods, a new technique "DIMM + LGS" is put forward for sensing turbulence profiles remotely in real time. The principle of remote sensing of atmospheric turbulence profiles is illustrated in Fig. 5[8].

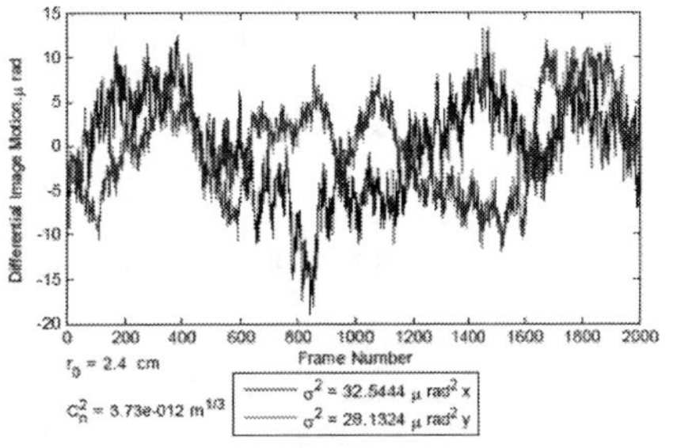

Fig. 3 example of Differential Image Motion versus camera frame number

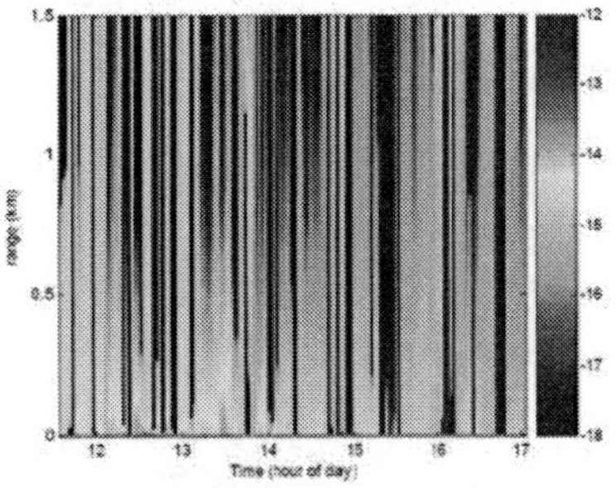

Fig. 4 Time-height diagram of the log of the strength of turbulence

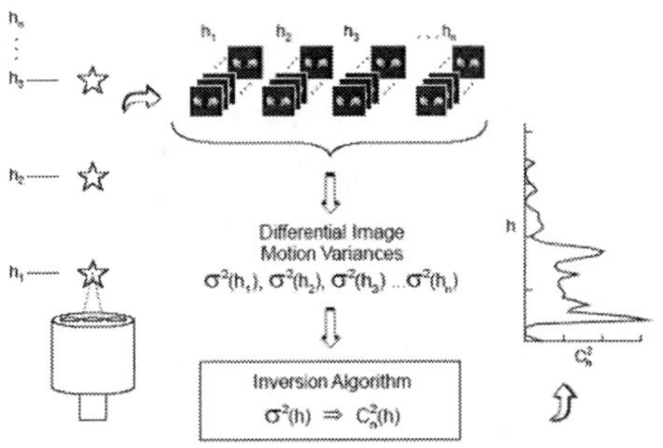

Fig. 5 The principle for remote sensing of atmospheric turbulence

As shown in Fig. 5, a powerful pulsed laser is focused and projected and laser guide stars are formed at various selected heights. The light from the scattering volume is received by a telescope with two spatially separated subapertures and the image of a laser guide star is separated to two images in the focal plane of ICCD camera by a wedge. Variance of centroids distance of images at various heights is obtained by the DIMM introduced above[7], and a turbulence profile could be retrieved.

3. REMOTE SENSING of TURBULENCE PROFILES

3.1 Experimental configuration

Bases on "DIMM + LGS" technique, a experiment was carried out and the experimental configuration is illustrated in Fig. 6. The experiment system comprises four major parts: (1) the powerful pulsed laser which generates and transmites intense pulses of 532nm, duration about 6ns, and energy 200mJ per pulse; (2) lens assembly for expanding, focusing, and projecting laser pulses, to form laser guide stars at selected heights; (3) a DIMM system for receiving signals of laser guide stars and forming images in focal plane of ICCD camera; (4) a computer for data processing to retrieve the profile of the atmospheric turbulence. The experimental configuration is off-axis, laser and lens assembly are placed on a platform in the top storey of a building, and laser pulses are transmitted from a circular hole in the ceiling of the building top and the aperture diameter 500The DIMM system placed above ceiling of the building top and could receive the returned signals of laser guide stars. After differential images and variance of centroids obtained, the computer processes the data, and retrieve profile of the atmospheric turbulence.

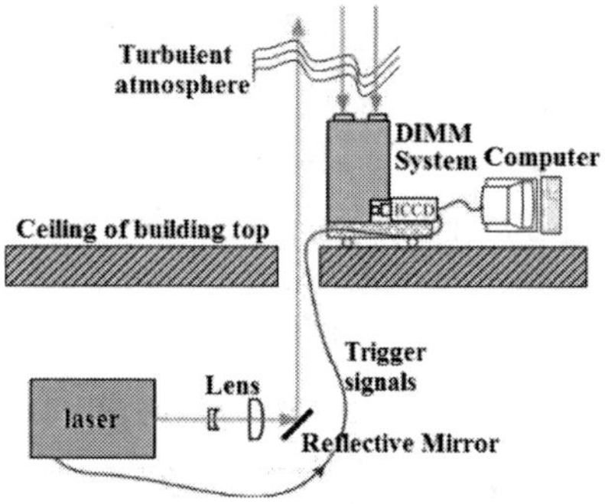

Fig. 6 Experimental configuration for remote sensing of turbulence profiles of atmosphere

3.2 Experiment of remote sensing

In the experiment for remote sensing turbulence profiles of atmosphere, firstly the pulsed laser is focused and transmitted, and then the laser guide stars are formed at pre-setting heights. After the returned signals of laser guide stars are received by the DIMM system, the differential motion images are formed in the focal plane of the ICCD camera, and the computer acquires the images, processes the data, and retrieves the turbulence profiles of atmosphere. At every heights, 2000 frames images should be acquired.

3.3 Data processing and a typical result

After images of laser guide stars are acquired, the locations of two centroids of spots in every images are calculated and distance between centroids are presented. Variance of differential image motion, σ_{DIM}^2, is calculated utilizing N(here $N = 2000$) data at every heights. The Fried parameter could be calculated based on variance of differential image motion, and turbulence profile is retrieved based on variance of differential image motion at various heights. A typical data is presented in table 1. The values of turbulence strength are descending along with increasing heights in negative exponential mode.

Table 1 Data of Remote Sensing and Turbulence Profile Retrieved

Height/m	Number of Frames	σ^2_{DIM}/rad^2	C_n^2/m$^{-2/3}$
200	2000	6.53e-12	2.75e-15
350	1999	3.27e-11	1.47e-14
500	1966	1.31e-10	5.69e-14
650	1982	1.36e-10	1.27e-15
800	1983	2.02e-10	3.68e-14
950	1988	2.12e-10	5.90e-15
1100	1982	2.85e-10	4.06e-14
1250	1976	3.00e-10	8.87e-15
1400	1960	4.23e-10	6.89e-14
1550	1934	5.40e-10	6.56e-14

4. CONCLUSIONS

By analyzing the technique and remote sensing of turbulence profiles experimentally, it shows that turbulence profiles of atmosphere could be by "DIMM + LGS", and the results shows that the turbulence profile obtained is reasonable. By this technique, turbulence profiles could be sensed remotely in real time.

The validity of the data sensed remotely should be verification, and the data should be compared with the data measured with other method, such as microthermal sensors at certain heights.

In order to get $C_n^2(h)$ at higher altitudes, the energy of per pulse of pulsed lasers should be powerful enough.

REFERENCES

[1] Kornilov, V., Tokovin, A., Vozyakova. O., "MASS: a monitor of the vertical turbulence distribution," . Proc. of SPIE 4839, 837-845(2003).

[2] Fuchs, A., Tallon, M., Vernin, J., "Focusing on a turbulence layer: principles of the generalized SCIDAR," Astronomical Society of the Pacific 110, 86-91(1998).

[3] Rodríguez-Hernández, M. A. C., Delgado, J. M., Fuensalida, J. J., et al., "The hybrid Shark-

Hartmann/G-SCIDAR instrument," Proc. of SPIE 6747, 646712(2007).
[4] Travouillon, T., "SODAR calibration for turbulence profiles in TMT site testing," Proc. of SPIE 6267, 626720(2006).
[5] Belen'kii, M. S., Roberts, D. W., Stewart, J. M., Gimmestad, G. G., "Experimental validation of the differential image motion lidar concept," Optics Letters 25, 518-520(2000).
[6] Gimmestad, Gary G., Roberts, David A., Stewart, John M., et al., "LIDAR system for monitoring turbulence profiles," Proc. of SPIE 6551, 655110H(2006).
[7] Sarazin, M., Roddier, F., "The ESO differential image motion monitor," Astron. and Astrophys. 227, 294-300(1990).
[8] Gimmestad, Gary G., Roberts, David A., Stewart, John M., et al., "Testing of lidar system for turbulence profiles," Proc. of SPIE 6951, 6951109(2008).

REMOTE SENSING OF STABLE BOUNDARY LAYER OF ATMOSPHERE

V.A. Banakh[1*], I.N. Smalikho, [1], A.V. Falits[1]

[1] *V.E. Zuev Institute of Atmospheric Optics SB RAS 1 Akademik Zuev Square, Tomsk, 634050, Russia*
**Email: banakh@iao.ru*

ABSTRACT

The stability of the atmospheric boundary layer (ABL) has been studied experimentally with a Stream Line coherent Doppler wind lidar and an MTP-5 temperature profilometer. Two-dimensional height-time distributions of the parameters characterizing the ABL temperature regime and wind turbulence, namely, the Richardson number, the dissipation rate of the kinetic energy of turbulence, the variance of fluctuations of the radial velocity, and the outer scale of turbulence, have been obtained. The amplitudes and period of oscillations of wind velocity components during atmospheric waves arising in the stable ABL have been determined.

1. INTRODUCTION

The understanding of turbulence processes in stably stratified media remains an unsolved problem in many fields of geophysics. Just this problem is a challenge to fundamental results of the Kolmogorov—Obukhov—Monin theory of isotropic turbulence, which has found wide use not only in geophysics, but also in other research fields [1, 2]. Stable stratification usually arises when the underlying surface is cooler than the upper-lying air layers. In summer time in the atmospheric boundary layer, this occurs at night over land and in daytime over large cold water bodies. The stable stratification in the boundary layer is characterized by formation of low-level jet streams (LLJS) and atmospheric waves.

To study turbulence and temperature-wind regimes in the stable atmospheric boundary layer, we have carried out an experiment in 2018 at the west coast of Lake Baikal near the village of Listvyanka with the use of the Stream Line coherent Doppler wind lidar and the MTP-5 temperature profilometer. The experiment was conducted on August 6 till 23. Cold water of Lake Baikal provides a stability of ABL during the experiment. The results of this experiment are reported in the paper.

2. METHODOLOGY

The Stream Line wind lidar was installed a few meters far from the building of the Big Solar Vacuum Telescope (BSVT) at the territory of the Baikal Astrophysical Observatory of the Institute of Solar-Terrestrial Physics (ISTP) SB RAS at a distance of 340 m from Baikal at a height of 180 m. The site coordinates are 52° 50' 47" N and 104° 53' 31"E). Wind measurements with the use of conical scanning by a probing beam around the vertical axis were carried out continuously during the experiment. The elevation angle of the lidar ray was $\varphi = 60°$, and the time for one complete scan (the azimuth angle θ varying from 0° to 360°) was T_{scan} = 36 s. In the most cases, the probing beam was focused to a distance of 300 m during the measurements. The heights from 280 to 1180 m above the Lake Baikal level were covered. Rough data of measurements by the wind lidar are arrays of estimates of the signal-to-noise ratio $\text{SNR}(R_k, \theta_m; n)$ and the radial velocity (projection of the wind velocity vector to the optical axis of the probing beam) $V_L(R_k, \theta_m; n)$, where $R_k = R_0 + k\Delta$. is the distance between the lidar and the center of the sensed volume, $k = 0, 1, 2, ..., K-1$, ΔR = 30 m is the range step, $\theta_m = m\Delta\theta$ is the azimuth angle, $\Delta\theta = 2°$ is the azimuth angle resolution, $m = 0, 1, 2, ... M-1$, $M = 180$ is the number of rays for one conical scan, and $n = 1, 2, 3, ...$ is the scan number. The arrays of estimates of the radial velocities $V_L(R_k, \theta_m; n)$ allow retrieval of vertical profiles of the wind velocity vector $\mathbf{V} = \{V_z, V_x, V_y\}$, where V_z is the vertical component and V_x, V_y are the horizontal components of the wind. Thus space-time distributions of the wind (distributions of $\mathbf{V}$ in height $h_k = R_k \sin\varphi = h_0 + k\Delta h$ and in time $t_n = t_0 + (n-1)\Delta t$) with the rather high vertical $\Delta h = \Delta R \sin\varphi$ = 26 m and time

$\Delta t \approx T_{scan} = 36$ s resolution are obtained. To determine the wind velocity vector, the filtered sine-wave fitting (FSWF) was used [3, 4].

As a result, we have visualized the spatiotemporal distributions $U(h_k,t_n)$, $\theta_V(h_k,t_n)$ and $V_z(h_k,t_n)$ for $K = 30$ height levels from $h_0 = 91$ m to $h_{K-1} = 844$ m. The obtained estimates of the wind velocity vector are averaged in the scan time $T_{scan} = 36$ s and in the space along the circle of the scan cone base having the length $L_k = (2\pi / \mathrm{tg}\varphi)h_k$, which is equal to 330 m at the height $h_0 = 91$ m and 3061 m at the height $h_{K-1} = 844$ m.

The parameters of wind turbulence (dissipation rate of turbulent energy ε, variance of the radial wind velocity σ_r^2, and integral scale of turbulence L_V) were estimated from the array of estimates of the radial velocity $V_L(R_k,\theta_m;n)$ by the method of azimuth structure function (ASF) [5,6]. Acceptable estimates of the wind turbulent parameters by the ASF method can be obtained if the percentage of bad estimates in the array of $V_L(R_k,\theta_m;n)$ is nearly zero. This restricts the height of retrieval of the wind turbulence parameters in comparison with the height of retrieval of the wind velocity vector.

To retrieve vertical profiles of wind turbulence parameters, we used the arrays of estimates of radial velocities $V_L(R_k,\theta_m;n)$ obtained for 15 scans, so that the averaging time was approximately 9 minutes. As a result, we have obtained two-dimensional height-time distributions $\varepsilon(h_k,t_n)$, $\sigma_r^2(h_k,t_n)$, and $L_V(h_k,t_n)$ for $K = 17$ height levels from $h_0 = 91$ m to $h_{K-1} = 506$ m.

The MTP-5 temperature profilometer was installed at the top of a hill near the foot of BSVT. The profilometer allows us to obtain the vertical temperature profiles every three minutes with a resolution of 25 m for heights from 0 to 100 m and 50 m for heights from 100 m to 1000 m. The profilometer data were used to obtain the spatiotemporal distributions of the mean temperature of air $T(h,t)$, its derivative $\partial T(h,t)/\partial h$, and potential temperature $T_p(h,t) = T(h,t) + \gamma_a h + C_1$, where $\gamma_a = 0.01$ deg/m is the dry-adiabatic gradient $(\partial T_p(h,t)/\partial h = \partial T(h,t)/\partial h + \gamma_a)$. For temperature averaging, we used the data measured for 24 min. The averaged data were used to calculate the parameter

$$N^2 = \frac{g}{T_p}\frac{\partial T_p}{\partial h}, \tag{1}$$

which characterizes the thermal stratification of the atmosphere, where g is the free-fall acceleration. The stratification is unstable at $N^2 < 0$, neutral at $N^2 = 0$, and stable at $N^2 > 0$. In the case of stable stratification, N is the Brunt—Vaisala frequency.

The Richardson number Ri was calculated as

$$\mathrm{Ri} = N^2\left(\frac{\partial U}{\partial h}\right)^{-2}, \tag{2}$$

where lidar data for the horizontal components of the wind vector V_x and V_y averaged over 31 scans (20 min) were used to obtain the average wind velocity U and its derivative $\partial U/\partial h$.

3. RESULTS

It follows from the obtained temperature data that the conditions for stable stratification occurred in the boundary layer during the entire period of measurements in August 6 to 23. The vertical temperature gradient always exceeded the dry-adiabatic one. Figure 1 shows the spatiotemporal distributions of the wind velocity and the Richardson number for August 21 to 23 as obtained from measurements with the wind lidar and the temperature profilometer.

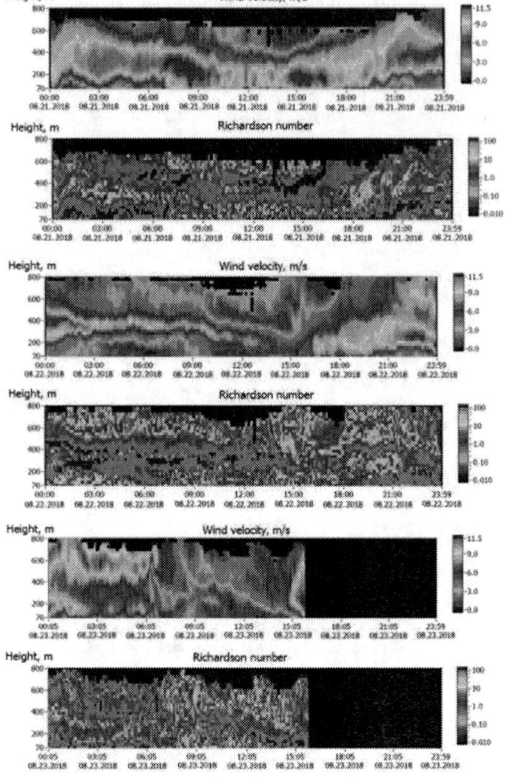

Figure 1 – Spatiotemporal distributions of the wind velocity and the Richardson number for August 21-23 of 2018.

It can be seen from the figure that jet streams were observed in the boundary layer of the Baikal coastal zone during this period. Two jet streams were observed at the night of August 22 to 23 and in the daytime of August 23. A characteristic feature of the data shown in Figure 1 is that beyond streams the Richardson number takes large positive values far exceeding the critical value $Ri_{cr} = 0.25$, which is indicative of the high stability of the boundary layer. At the heights of location of jet streams, the Richardson number takes much smaller values. It still keeps positive, but its value becomes smaller than Ri_{cr}. That is, in the region of jet streams, the stability of thermal stratification of the boundary layer decreases, and channels arise, in which the conditions are close to neural stratification.

Atmospheric waves may appear in the stable boundary layer simultaneously with jet streams. Among all the data measured by the Stream Line lidar in August 6 to 23 of 2018, we revealed 14 cases of atmospheric wavy formation. From the obtained distributions $U(h_k,t_n)$, $\theta_V(h_k,t_n)$, and $V_z(h_k,t_n)$, we first determined visually the heights and time of appearance of periodic variations in the wind velocity components. Then we determined the parameters of atmospheric waves by the method described in [7]. In all the cases of observation of atmospheric waves, the amplitude of the wave component of the horizontal wind velocity A_v was no smaller than 0.6 m/s. The amplitude of oscillations of the vertical velocity was no smaller than 0.2 m/s. The maximal amplitudes of wind oscillations were seen in the wave observed from 02:30 to 03:30 LT on August 23 of 2018. They amounted to $A_v \approx 3$ m/s for the horizontal wind velocity component and $A_v \approx 1$ m/s for the vertical one.

The lifetime of atmospheric waves observed in the region under study in August 6 to 23 of 2018 varied from 40 min to 5 hours. The oscillation period ranged from 5 to 20 min. This oscillation period far exceeds the Brunt-Vaisala frequency (1) calculated from the data of temperature profilometer. This means that the mountain relief of the Baikal coastal zone is critical for formation of atmospheric waves observed in our experiment.

Figure 2 exemplifies the height-time distribution of wind turbulence parameters retrieved from lidar measurements of the radial velocity on August 21 of 2018. At that day, jet streams were observed from 00:00 to 08:30 LT and from 21:00 to 24:00 LT at heights of 200–400 m. It can be seen from Figs. 2 (a, b) that turbulence is quite weak at the heights of location of jet streams. To the contrary, turbulence is very strong beyond jet streams, at heights of 100–200 m. At these heights, the dissipation rate ε achieves 0.09 m^2/s^3, while the variance σ_r^2 is 2.5 $(m/s)^2$. An atmospheric wave formed approximately at 7 a.m. for the short 40-min time interval. An increase in the turbulence intensity from 07:00 to 09:00 LT at heights exceeding 350 m is likely connected with the destruction of this atmospheric wave.

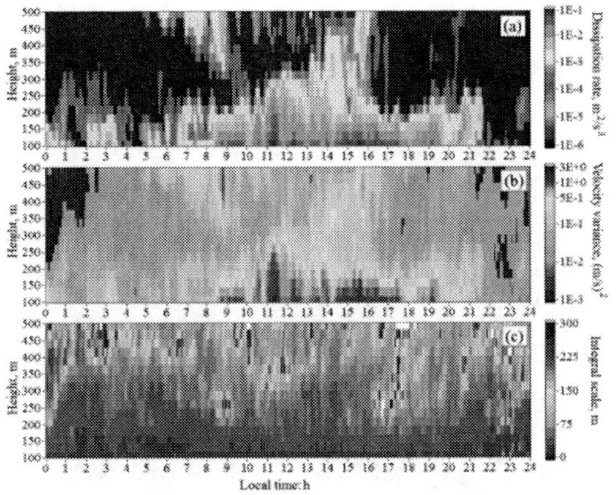

Figure 2 – Height-time distribution of the dissipation rate of turbulence energy (a), variance of the radial wind velocity (b) and the integral scale of turbulence (c) obtained from measurements by the Stream Line lidar on August 21 of 2018.

The analogous distributions $\varepsilon(h_k,t_n)$, $\sigma_r^2(h_k,t_n)$, and $L_V(h_k,t_n)$ obtained on August 23 of 2018 demonstrate that during the atmospheric waves observed at that day from 02:30 to 03:30, from 05:30 to 06:30, and from 08:30 to 09:30 LT, the intensity of wind turbulence was different at different heights in ABL. The atmospheric wave from 02:30 to 03:30 formed in the presence of two jet streams (see Fig. 1), the angle between the directions of wind flows in which was approximately 90°. In this wave, the amplitudes of oscillation of wind velocity components were maximal for the entire period of measurements.

The oscillation period was identical at all the heights of lidar measurements, in particular, at the heights of location of the both jet streams. The dissipation rate of the turbulence energy in the lower jet stream appeared to be an order of magnitude smaller than that at the heights of location of the upper jet stream. As the wave disappeared, the turbulence became moderate at all the heights. Thus, during atmospheric waves, the wind turbulence may be both moderate and even weak and strong.

ACKNOWLEDGEMENTS

This study was performed under project №AAAA-A17-117021310149-4 and supported, in part, by the Russian Foundation for Basic Research, Project No. 18-05-00220.

REFERENCES

[1] G.S. Golitsyn, *Statistics and Dynamics of Natural Processes and Phenomena: Methods, Instruments, and Results* (Krasand, Moscow, 2013), 400 pp.
[2] G.S. Golitsyn, Gidrol. Meteorol. 3, 5-15 (2018)
[3] V.A. Banakh, et al. Atmospheric and Oceanic Optics 23(5): 381–388 (2010)
[4] V.A. Banakh, et al. Atmospheric and Oceanic Optics 28(2): 185–191 (2015)
[5] I.N. Smalikho, V.A. Banakh, Atmospheric Measurement Techniques. 10(11): 4191–4208 (2017)
[6] V.A. Banakh, I.N. Smalikho, A.V. Falits, Optics Express 25(19): 22679-22692 (2017)
[7] V.A. Banakh, I.N. Smalikho, Atmospheric Measurement Techniques 9(10): 5239-5248 (2016)

RESEARCH ON A WIND LIDAR WITHOUT BLIND ZONE BASED ON THE TECHNOLOGIES OF PSEUDO RANDOM CODE PHASE MODULATION AND HETERODYNE DETECTION

FU YANG[1]*, ZHENG LIU[1], XIAOPENG ZHU[2]

1: College of Science, Donghua University,, Shanghai 201620, China
2: Shanghai Institute of Optics and Fine Mechanics, Chinese Academy of Science, 201800, China
**Email: fuyang@dhu.edu.cn*

ABSTRACT

In this measurement method, a wind lidar system with no blind zone based on the technologies of heterodyne detection and pseudorandom code phase modulation is proposed. The simulation results show the feasibility of the method. 15m range resolution and 7.75cm/s line of sight (LOS) wind velocity resolution can be achieved from 0 m to 300m, when the laser transmitted power is 2mW, the transmitted pulse length is 100us, the receiving telescope aperture is 2cm, and an accumulation times of 10. The simulation results also show the positive effect of the transmitted pulse length.

Key words: wind lidar, heterodyne detection, blind zone

1. INTRODUCTION

An accurate measurement of the wind velocity plays a key role in the fields of weather forecast, aviation safety, air pollution control, wind energy utilization and so on[1]. The direct wind detection system extracts the wind LOS velocity through an optical frequency discriminator [2] while the coherent wind detection system get the wind LOS velocity through Doppler frequency [3]. No matter which detection mechanism, the system exists an area of blind zone due to the end reflection signal. In addition, in order to obtain better range resolution, the pulse width of the laser should be short, so a huge peak power is required. In this paper, a wind lidar measurement system with no blind zone is proposed. The high range resolution is realized by random phase modulation, the high velocity resolution is realized by transmitted long pulse length, and high sensitivity is realized by adopting heterodyne detection.

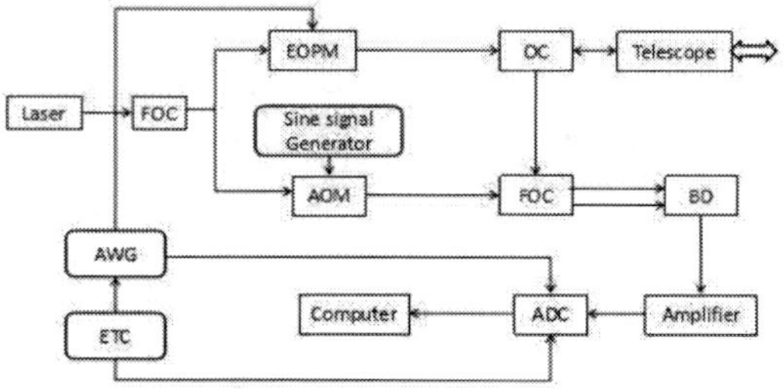

Figure 1 *A schematic view of the system*

2. METHODOLOGY

2.1 The lidar system

The schematic of the system is shown in Fig.1. [4]

FOC: fiber optic coupler;
EOPM: electro-optic phase modulator;
AOM: acousto-optic modulator;
OC: optical circulator;
BD: Balance detector;
ADC: dual channel AD capture card;
AWG: Arbitrary waveform generator;
ETC: external trigger circuit

The system uses an eye-safe 1550nm narrow linewidth fiber laser as the light source. The output laser is split into two beams by a 1 x 2 FOC. One part serves as reference light and the other as signal light, and the other serves as local oscillator (LO). The EOPM produces phase modulation of the signal light according to the PRC sequence, and its drive signal can be provided by the AWG. The output signal after PRC phase modulation is shown in Eq. 1

$$S = A_T \cos[\omega_c + P'_{RC}(t)\pi + \emptyset_1] \qquad (1)$$

Where A_T is the amplitude of the laser, ω_c is the carrier frequency, $\emptyset_1$ is an unchanged random

https://doi.org/10.1051/epjconf/202023706016

phase, and the PRC sequence $P'_{RC}(t)$ is shown in Eq. 2:

$$P'_{RC}(t) = \sum_{n=-\infty}^{n=+\infty} b_n p(t - nT_0) \qquad (2)$$

Where b_n is a pseudo-random code, and the value is 0 or 1 with the same probability. The system uses the M-sequence PRC, p(t) is the gate function shown in Eq. 3:

$$\mathrm{p(t)} = \begin{cases} 1, & |\mathrm{t}| \le T_0/2 \\ 0, & |\mathrm{t}| > T_0/2 \end{cases} \qquad (3)$$

Where T_0=1/F, is the PRC code width. It is the inverse of modulation rate. The local oscillator (LO) signal is

$$\mathrm{L} = A_{LO}\cos[(\omega_c + \omega_m)t + \emptyset_2] \qquad (4)$$

Where A_{LO} is the amplitude of the LO, ω_m is the frequency shift of the AOM, and $\emptyset_2$ is an unchanged random phase. The drive signal of the AOM is provided by a signal generator. The modulated signal light is transmitted by a telescope through port 1 of the OC. The return signal light is received by the same telescope, and coherent with the LO through port 3 of OC in a FOC. The return signal carries the LOS wind velocity. The coherent signal is detected by a BD. After amplification, the signal is sampled by channel 1 of the dual-channel ADC. The other channel of the ADC samples the output signal of the AWG which is used to offer variable delay pseudo random signals needed in the signal process. The ADC and the AWG generator are triggered by the same external trigger circuit (ETC) to ensure time synchronization.

The coherent signal is a phase-discontinuous triangular wave signal, whose frequency is decided by the AOM and the Doppler frequency shift.

The output of BD can be expressed as:

$$i_s(\mathrm{t}) = A_S A_{LO}\cos\left[(\omega_m + \omega_d)\left(t - \frac{2R}{c}\right) + P'_{RC}\left(t - \frac{2R}{c}\right)\pi + \varphi_3\right] \qquad (5)$$

Where R is the distance between the detected wind and the system, c is the speed of light, and φ_3 is an unchanged random phase. A_S is the amplitude of the echo signal, and its relationship with the echo signal power P_{sig} is $A_S = \sqrt{P_{sig}}$. $P_{sig} = \eta_{atm}\eta_{sys}P_T c T_0 \beta \frac{\pi D^2}{4R^2}$, η_{atm} and η_{sys} are the optical transmittance of the atmosphere and radar system, respectively, P_T is the transmitted laser peak power, c is the speed of light, T_0 is the length of each PRC code, β is backscatter coefficient. D is the telescope diameter, and R is the range of the wind. φ_3 is an unchanged random phase.

According to the character of cosine function, the Eq. 5 can be expressed as Eq. 6:

$$i_s(\mathrm{t}) = A_S A_{LO} P_{RC}\left(t - \frac{2R}{c}\right)\cos[(\omega_m + \omega_d)t + \varphi_4] \qquad (6)$$

Where φ_4 is an unchanged random phase.

The acquired signals from the 2 channels of the ADC are simultaneously multiplied and FFT transformed, the final signal is shown in Eq. (7)

$$S_{IG} = \mathcal{F}(\mathrm{Y}) = \mathcal{F}[i_s(\mathrm{t}) \times P_{RC}(\mathrm{t} - \mathrm{n}T_0)]$$
$$= A_S A_{LO}\mathcal{F}[P_{RC}(\mathrm{t} - 2\mathrm{R/c}) \times P_{RC}(\mathrm{t} - \mathrm{n}T_0) \times \cos((\omega_m + \omega_d)t + \emptyset_4)]$$
$$\mathrm{n} = 1,2,3 \ldots \ldots \mathrm{N} \qquad (7)$$

For each time delay amount nT_0, the wind information located at around $cnT_0/2$ can be extracted. So the range resolution is $cT_0/2$, which is 15m in our simulation. The corresponding spectral peak f can be obtained. The difference of f with respect to ω_m is the Doppler frequency shift caused by the wind LOS velocity. The relation between the Doppler frequency shift and LOS wind velocity v is $v = \frac{\lambda\omega_d}{2}$. If v >0, it means the wind moves towards the system. Otherwise, it means the opposite direction. All of the system signal processing can be done in the computer.

For the end face reflection signal, we know the distance of it, and the Doppler frequency is 0. So we remove the peak, which is contributed by the end face reflection, in the FFT spectrum when the delay range $cnT_0/2$ matches the end face.

3. RESULTS

The system is simulated by using Matlab. The simulation parameters are shown in Tab. 1.

Table 1

Parameter	Value
P_T	2mW
P_{LO}	0.2mW
D (Telescope diameter)	0.1m
Br (End reflection coefficient)	10^{-6}
Backscatter coefficient	$3.5 \times 10^{-6} m^{-1} \cdot S_r^{-1}$
F (Phase Modulation frequency)	10MHz
Line of sight (LOS) wind velocity	3.875m/s
AOM	50MHz
T (The transmitted pulse length)	50us

The relation between the intensity of the ADC channel 1 (which is in proportional to the output of BD) signal and the range is shown in Fig. 2

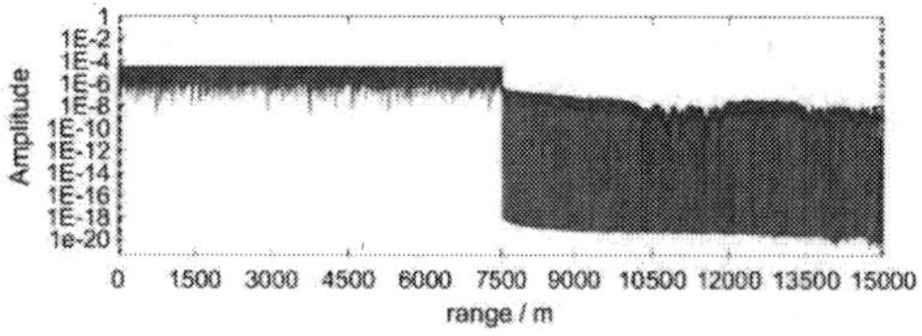

***Figure 2** The amplitude of the ADC channel 1 versus the wind range*

We use logarithmic coordinate for the Y-axis. It can be seen that the intensity of the first half section in Fig. 2 is much stronger than the second half section. Because we use the long transmitting pulse length which is 500 us in the simulation, the end face reflection effect lasts 50 us (corresponding to 7500 m for the range in air). The intensity of the first half is mostly contributed by the end face reflection. In the second half section, we can see the attenuation of the return wind signal due to the increase of range.

After an accumulation times of 25, the final signal for the range of 1m and 150 m are shown in FIG.3

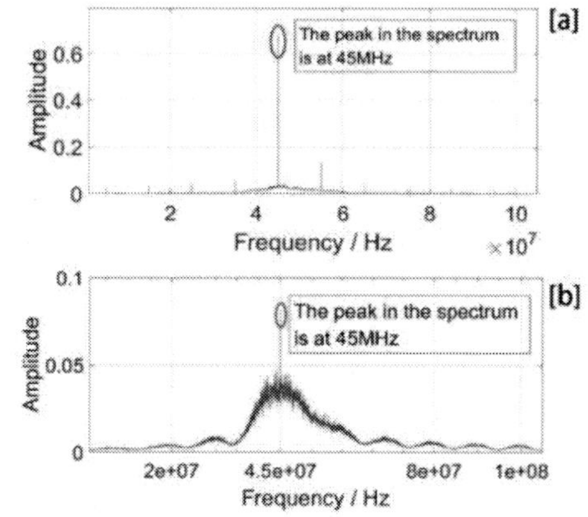

***Figure 3** The final signal spectrum for the range of (a) 1m (b)150m*

From Fig. 3, it is observed that the wind information at 1m and 150m can both be detected clearly. Fig. 3(a) shows clearly that the end face signal effect can be removed. The peak frequency deviates from the 50MHz(AOM frequency shift) is due to doppler effect. The LOS wind velocity is 3.875m/s.

The LOS wind velocity resolution is decided by the transmitted pulse length.

$$dv = \frac{\lambda}{4T} \qquad (8)$$

Where dv is The LOS wind velocity resolution, T is the transmitted pulse length. According to the simulation parameters in Tab.1, the LOS wind velocity resolution is 0.775 cm/s.

The frequency shift of the AOM determines the wind detection range according to $V = \frac{\lambda AOM}{2}$.Based on Tab. 1, the LOS wind velocity range is 0~38.75m/s, which is adequate for wind detection.

The relationship between the maximum detection range and the accumulation times under

 https://doi.org/10.1051/epjconf/202023706016

variable sequence length is shown in Fig.4 In the simulation, the effects of noise have been considered.

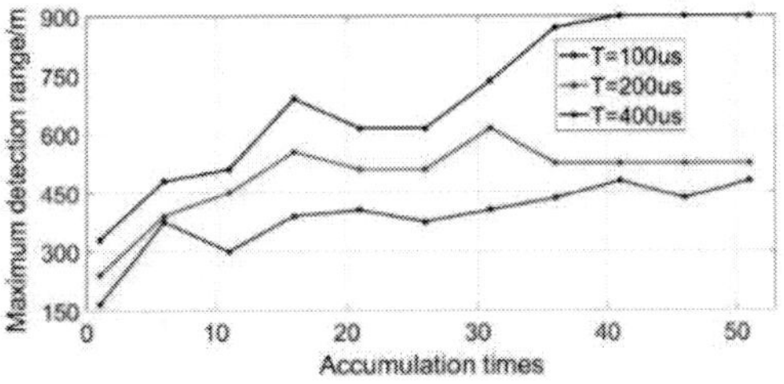

Figure 4 *The relationship between the maximum detection range and the accumulation times under variable sequence length*

From Fig. 4, it's observed that the transmitted pulse length plays a positive effect on the maximum detection range. And the effect of accumulative times will tend to saturation.

The relationship between the maximum detection range and the transmitted laser peak power under variable sequence length is shown in Fig.5. And the accumulative times is set at 50.

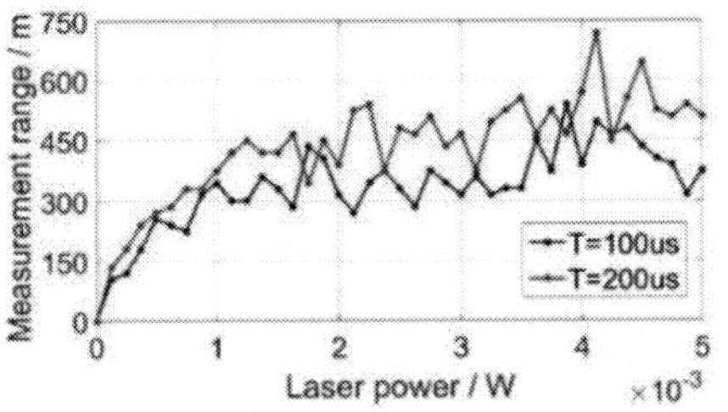

Figure 5 *The relationship between the maximum detection range and the transmitted laser peak power under variable sequence length*

From Fig. 5, it is observed that the transmitted pulse length plays a positive effect on the maximum detection range. And the effect of laser transmitted peak power will tend to saturation.

At the same time, this system does not require too much laser power. 2~3 mW transmitted laser peak power can detect the wind peak power of several hundred meters.

4. Conclusion

A wind lidar without blind zone based on the technologies of pseudo random code phase modulation and heterodyne detection is demonstrated in the simulation. The high range resolution is realized by fast modulation rate, the high velocity resolution is realized by transmitting long pulse length, and high sensitivity is realized by adopting heterodyne detection. 0.775cm/s line of sight (LOS) wind velocity resolution and range resolution can be achieved from 0 m to 300 m, when the laser transmitted power is 2 mW, the transmitted pulse length is 100 us, the receiving telescope aperture is 2cm, and an accumulative times of 10. Unlike conventional systems, the system has neither close-range blind spots nor excessive peak power requirements.

ACKNOWLEDGE

The research is sponsored by the opening foundation of Chinese Academy of Science Key Laboratory of Space Laser Information Transportation and Detection Technology.

REFERENCES

[1] Zhou Yanzong, Wang Chong, Liu Yanping, Xia Haiyun. Research Progress and Application of Coherent Wind Lidar[J]. Progress in Laser and Optoelectronics, 2019, (02): 9-26.

[2] Sun Dongsong, Yang Zhao, Xue Guogang. Performance Analysis of Direct Detection of Wind Lidar[J].Infrared and Laser Engineering,2003(02):115-117+126.

[3] Pan Jingyan, Yan Shuangyang, Liu Guo, Dong Guangyan, Zhang Pengfei, Chen Jing. Wind field measurement technology of coherent laser wind radar[J].Infrared and Laser Engineering,2013,42(07):1720-1724.

[4] Fu Yang, Yan He, Weibiao Chen, and Yage Zhan IEEE PHOTONICS TECHNOLOGY LETTERS, VOL. 26, NO. 23, DECEMBER 1, 2014

https://doi.org/10.1051/epjconf/202023706017

SPECTRALLY-RESOLVED RAMAN LIDAR TO MEASURE ATMOSPHERIC THREE-PHASE WATER SIMULTANEOUSLY

Fuchao Liu[1, 2,*], Fan Yi[1, 2]

[1]*School of Electronic Information, Wuhan University, Wuhan 430072, China*
[2]*State Key Laboratory of Information Engineering in Surveying, Mapping and Remote Sensing, Wuhan University, Wuhan 430079, China*
**Email: lfc@whu.edu.cn*

ABSTRACT

We report on a spectrally-resolved Raman lidar that can simultaneously profile backscattered Raman spectrum signals from water vapor, water droplets and ice crystals as well as aerosol fluorescence in the atmosphere. The lidar emits a 354.8-nm ultraviolet laser radiation and samples echo signals in the 393.0-424.0 nm wavelength range with a 1.0-nm spectral resolution. A spectra decomposition method is developed to retrieve fluorescence spectra, water vapor Raman spectra and condensed (liquid and/or ice) water Raman spectra successively. Based on 8 different clear-sky nighttime measurement results, the entire atmospheric water vapor Raman spectra are for the first time obtained by lidar. The measured normalized water vapor Raman spectra are nearly invariant and can serve as background reference for atmospheric water phase state identification under various weather conditions. For an ice virga event, it's found the extracted condensed water Raman spectra are highly similar in shape to theoretical ice water Raman spectra reported by *Slusher* and *Derr* (1975). In conclusion, the lidar provides an effective way to measure three-phase water simultaneously in the atmosphere and to study of cloud microphysics as well as interaction between aerosols and clouds.

1. INTRODUCTION

The traditional water vapor Raman lidar that extracts the Q-branch vi-rotational Raman signals from atmospheric water vapor using narrowband optical devices works well under clear weather conditions. However, it fails in cloud area: since liquid and ice water Raman spectra overlap in spectrum with that of water vapor, Raman signals from cloud condensed (liquid and/or ice) water also contributes to the water vapor Raman echo, yielding unreasonable measurement results (e.g., relative humidity>100%). Over the years, there have been two different approaches to measure simultaneously vaporous and condensed water. One approach utilizes two discrete detection channels to extract the backscattered Raman signals respectively from condensed water and water vapor. When the lidar emits an ultraviolet radiation (e.g., 355 nm), atmospheric (biogenic) aerosol-induced fluorescence or possible inner-instrument fluorescence might span a rather wide spectrum range which crams the entire Raman band of three-phase water. In this case, the discrete-channel Raman lidar would often produce false measurement results. Another approach introduces a spectrometer to record the full Raman spectra of water [1-4]. To convince the presence of (aerosol) fluorescence, *Reichardt* [4] proposed using spectral signals between 392 and 395 nm as an indicator, as no Raman scattering is observed in this spectrum area. However, the selected spectra are too narrow to effectively determine the fluorescence spectra, making it difficult to remove the fluorescence contamination from the Raman signals of water. Based on our former work [3], we present here a spectrally-resolved Raman lidar with well-arranged spectra coverage so that it can simultaneously record backscattered Raman spectrum signals from three-phase water as well as aerosol fluorescence in the atmosphere. A spectra decomposition method is also developed to retrieve fluorescence spectra, water vapor Raman spectra and condensed (liquid and/or ice) water Raman spectra successively. Typical measurement results are provided to verify the lidar performance.

2. LIDAR SETUP

Fig. 1 shows a schematic layout of the spectrally-resolved Raman lidar. Compared to the former version [3], the main changes are summarized as following: (1) a beam splitter (BS) is placed after the first bandpass filter (BP1) to reflect ~10% of incident light to be filtered as a reference N_2 Raman signal. The BP1 has >94% transmission in

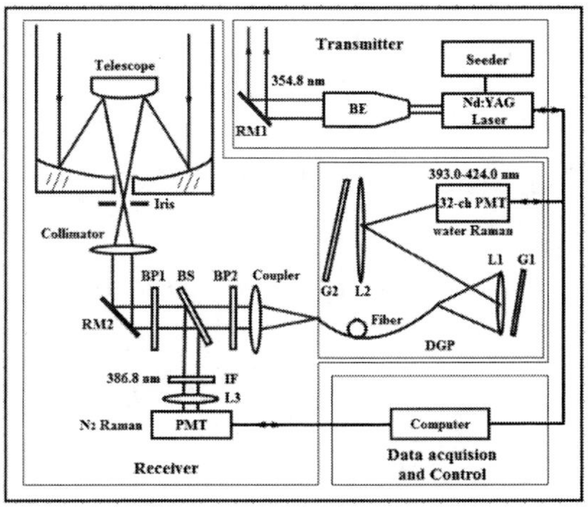

Fig. 1 *Schematic layout of the spectrally-resolved Raman lidar. BE, beam expander; BP, bandpass filter; RM, reflecting mirror; BS, beam splitter; IF, interference filter; L, lens; G, grating; PMT, photomultiplier tub.*

the 387-447 nm range and a suppression of >6 orders of magnitude to elastic signal around 354.8 nm; (2) a second bandpass filter (BP2) is added before the fiber coupler to transmit >80% of signal light in the 393.0-424.0 nm range, but reject elastic signal around 354.8 nm as well as the O_2 and N_2 Raman signals around 375.5 and 386.8 nm respectively with a suppression of more than 3 orders of magnitude; (3) the DGP (dual-grating polychromator) is updated to have an overall linear dispersion of 1.0 mm nm^{-1} in the 393-424 nm wavelength range (fiber: core diameter=0.6 mm, numerical aperture=0.12; L1: diameter=100 mm, focal length=300 mm; G1: groove density=600 gr mm^{-1}, blazing angle=6.89°, working angle=9.27°; L2: diameter=100 mm, focal length=400 mm; G2: groove density=600 gr mm^{-1}, blazing angle=21.10°, working angle= 21.72°). A 32-channel linear-array photomultiplier tubes (H7260-200, Licel, DE) is placed on the focal plane of the DGP. The spectrally-resolved Raman lidar records simultaneously the spectral signal in the 393-424 nm wavelength range with a 1.0-nm resolution. For further understanding of the lidar system can refer to [3].

3. SPECTRA DECOMPOSITION METHOD

The lidar equation for each detection channel can be written as (background subtracted):

$$\begin{cases} N_k(z) = \dfrac{C_k \cdot O_k(z)}{z^2} \cdot \beta_{k,t}(z) \cdot \tau(\lambda_0) \cdot \tau(\lambda_k) \\ \beta_{k,t} = \beta_{k,F} + \beta_{k,V} + \beta_{k,L} + \beta_{k,I} \\ \lambda_k = (k-16) + 407.57 \approx k + 392 \ nm \\ k = 1, 2, \ldots 32 \end{cases} \quad (1).$$

Here $N_k(z)$ denotes recorded photon counts at height z for the kth detection channel, C the system constant, O the overlap function (become unity when z >1 km), τ the atmospheric transmission, λ_0 the laser wavelength and λ_k the signal wavelength. The total (t) volume backscatter coefficient β is contributed by that of vaporous (V), liquid (L) and ice (I) water plus fluorescent (F) aerosol.

Defining normalized spectra $\{S_k\}$:

$$S_k = \frac{N_k}{N_{16}} = \frac{C_k \cdot O_k \cdot \tau(\lambda_k)}{C_{16} \cdot O_{16} \cdot \tau(\lambda_{16})} \cdot \frac{\beta_{k,t}}{\beta_{16,t}} \quad (2).$$

Considering the maximum wavelength deviation between λ_k and λ_{16} is <17 nm, the return-trip atmospheric transmission difference can well be neglected. As for the system overlap function O_k, they can be regarded as nearly identical to O_{16} due to the optics of the DGP. Moreover, the ratios of C_k to C_{16} is calibrated by a white light source with known spectral characteristics (similar to [3]) and found to have values close to 1. Thus for simplification here, the normalized spectra are approximated by:

$$S_k \approx \frac{\beta_{k,F} + \beta_{k,V} + \beta_{k,L} + \beta_{k,I}}{\beta_{16,t}} \quad (3).$$

For a 354.8-nm laser radiation, the water vapor, bulk water and ice generate vi-rotational Raman spectra in the spectrum ranges of 401-418 nm, 396-410 nm and 395-409 nm, respectively. The spectral signals in the ranges of 393-394 and 419-424 nm are regarded as (aerosol) fluorescence, yielding normalized fluorescence spectra:

$$S_k^F \approx \frac{\beta_{k,F}}{\beta_{16,t}}, \quad k = 1, 2, 27, 28, \ldots 32 \quad (4).$$

Since the lidar has a limited wavelength detection range of only 32 nm, an assumption is made that the fluorescence spectra vary slowly in the 393-424 nm range. Then the full fluorescence spectra in the 393-424 nm spectrum range are estimated (interpolated) using a linear fitting method:

$$\begin{cases} (\lambda_k, S_k^F),\ k=1,2,27,28,...32 \\ \quad \Downarrow\ linear\ \ fitting \\ \{S_k^F\},\ k=1,2,...32 \end{cases} \qquad (5).$$

Simultaneous Equations (3) and (5), we get the normalized three-phase water Raman spectra in the 395-418 nm range:

$$S_k^w = S_k - S_k^F = S_k^V + S_k^L + S_k^I, k=3,4,...26 \quad (6).$$

Under clear-sky conditions no cloud liquid and/or ice water are present, so Equation (6) regresses to:

$$S_k^w = S_k - S_k^F = S_k^V, k=3,4,...26 \qquad (7).$$

Equation (7) indicates that the normalized water vapor Raman spectra can be obtained under clear-sky conditions.

After the retrieval of fluorescence spectra and water vapor Raman spectra, the condensed liquid and/or ice water Raman spectra in the 395-410 nm range are obtained:

$$S_k^{L,I} = S_k - S_k^F - S_k^V, k=3,4,...18 \qquad (8).$$

Thus, a spectra decomposition method has been developed to retrieve fluorescence spectra, water vapor Raman spectra and condensed water Raman spectra successively. The liquid and ice water Raman spectra can also be separated, but to economize words, we don't make any further discussion here. Moreover, the separated true water vapor Raman signals, e.g.:

$$N_{16}^V = N_{16} \cdot S_{16}^V \qquad (9),$$

can be directly used to retrieve the water vapor content. Similarly, the liquid and/or ice water content in cloud area can also be determined.

4. TIPICAL MEASUREMENT RESULTS

4.1 Clear-sky case

Fig. 2 presents measurement result performed on the night of July 28, 2015. The weather was clear as a co-located polarization lidar convinced that there were no clouds or obvious aerosol load present in laser sight. The data between 21:00 and 22:00 LT and in the height range of 1.5-2.0 km were accumulated to generate the normalized spectra (Fig. 2a). The spectra only show a single peak around 407.57 nm, corresponding to the Q-branch Raman spectra of water vapor. The fluorescence spectra are first separated (Fig. 2b, magenta) using the method introduced above. The residual spectra (Fig. 2b, blue) are regarded as lidar-measured water vapor Raman spectra.

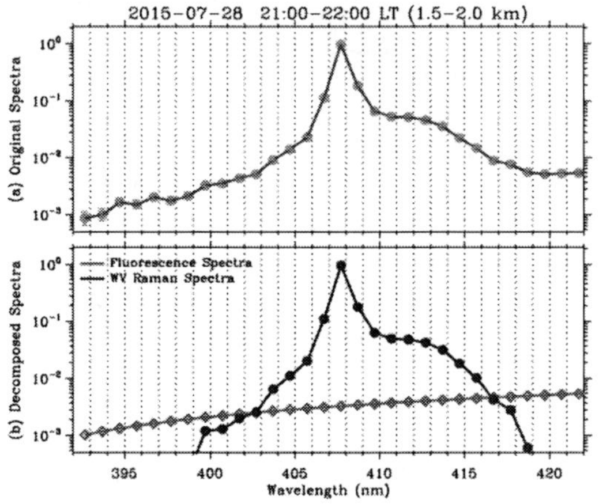

Fig. 2 *(a) original normalized spectra by the spectrally-resolved Raman lidar. Error bars indicate statistical uncertainty; (b) separated spectral components after applying the spectra decomposition method: fluorescence spectra (magenta) and lidar-measured water vapor Raman spectra (blue).*

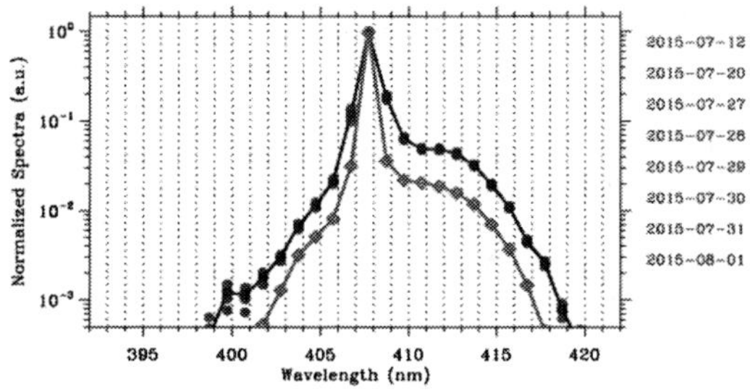

Fig. 3 *Lidar-measured water vapor Raman spectra: red, nightly-mean spectra for 8 different clear-sky nights; blue, their average spectra; magenta, theoretical water vapor Raman spectra (according to [5]) smoothed with a 1.0-nm resolution. Note the lidar-measured water vapor Raman spectra are broadened by the lidar receiving optics, but nearly invariant and quite similar to the theoretical spectra in shape.*

Fig. 3 shows the results of lidar-measured water vapor Raman spectra on 8 different clear-sky nights. For each night, water vapor Raman spectra from different 1-h integration periods are retrieved and averaged to yield nightly-mean spectra (red). The blue curve plots the average spectra of the 8 nights. For comparison, the theoretical water vapor Raman spectra calculated according to [5] and smoothed with a 1.0-nm resolution (Fig. 3, magenta) are also added. It can be seen that the lidar-measured spectra are broadened by the lidar receiving optics, but quite similar in shape to the theoretical spectra. Note the lidar-measured water vapor Raman spectra are nearly invariant, thus can serve as background reference for separation of mixed-phased water Raman spectra.

4.2 An ice virga case

The co-located and concurrent polarization lidar indicated that there were clouds in the height range of 2.5-3.5 km and between 03:00 and 04:00 LT on the night of Jan 01, 2016 with volume depolarization ratio values of >0.3 (Figures not provided here). It is regarded as an ice virga event. Fig. 4a shows the original normalized spectra obtained by the spectrally-resolved Raman lidar and a second peak is visible near 400 nm. After applying the above spectra decomposition method, Fig. 4b plots the successively separated spectra components: fluorescence spectra (magenta), total mixed-phased water Raman spectra (dash black), vaporous (solid blue) and condensed (solid red) water Raman spectra. For comparison, theoretical water vapor (dash blue) and ice (dash red) water Raman spectra are also added. The theoretical ice Raman spectra data are taken from *Slusher* and *Derr* (1975). It is found the extracted condensed water Raman spectra are highly similar to the theoretical ice water Raman spectra, convincing that the cloud is dominated by ice water which is consistent with the polarization lidar measurement results. The cloud ice water content can further be determined using the separated ice water Raman signals.

In conclusion, the presented spectrally-resolved Raman lidar is capable of measuring atmospheric three-phase water simultaneously despite presence of fluorescent aerosols. It's regarded as an effective tool for studying of cloud microphysics and interaction between aerosols and clouds.

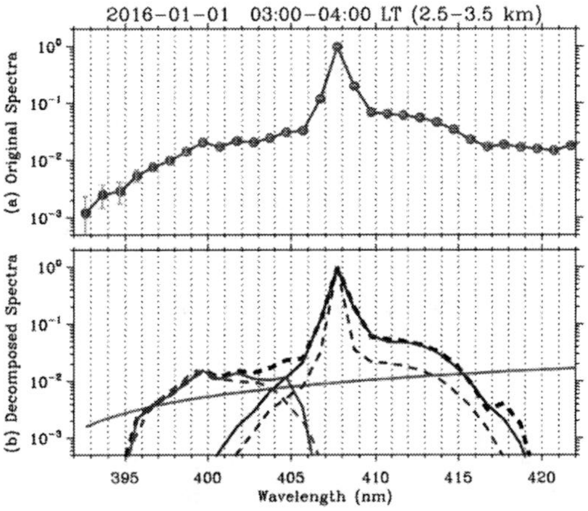

Fig. 4 *Spectra measurement results for an ice virga event: (a) original normalized spectra with a second peak near 400 nm; (b) separated spectra components: fluorescence spectra (magenta), total mixed-phased water Raman spectra (dash black), vaporous (solid blue) and condensed (solid red) water Raman spectra. Theoretical water vapor (dash blue) and ice (dash red) water Raman spectra are added for comparison. The theoretical ice Raman spectra are taken from [6].*

ACKNOWLEDGEMENTS

This work was supported by the National Natural Science Foundation of China (No. 61505148) and National Key Scientific Instrument and Equipment Development Project (No. 41327801).

REFERENCES

[1] D. Kim, et al. Lidar measurement of a full Raman spectrum of water by using a multichannel detector, *J. Korean Phys. Soc.*, 54, 38–43 (2009) .

[2] T. Sakai, et al. Liquid water cloud measurements using the Raman lidar technique: Current understand-ding and future research needs, *J. Atmos. Oceanic Technol.*, 30, 1337–1353 (2013) .

[3] F. Liu and F. Yi. Spectrally resolved Raman lidar measurements of gaseous and liquid water in the atmosphere, *Appl. Opt.*, 52(28), 6884–6895 (2013) .

[4] J. Reichardt. Cloud and Aerosol Spectroscopy with Raman Lidar, *J. Atmos. Oceanic Technol.*, 31, 1946–1963 (2014) .

[5] G. Avila, et al. Ro-vibrational Raman cross sections of water vapor in the OH stretching region, *J. Mol. Spectrosc.*, 196, 77–92 (1999).

[6] R. B. Slusher, V. E. Derr. Temperature dependence and cross sections of some Stokes and anti-Stokes lines in ice Ih, *Appl. Opt.*, 14, 2116–2120 (1975).

TOWARDS DEVELOPING A MICROPULSE DIFFERENTIAL ABSORPTION LIDAR TO MEASURE ATMOSPHERIC TEMPERATURE

Robert A. Stillwell[1,2,*], Scott M. Spuler[2], Matthew Hayman[2], Catharine E. Bunn[3], and Kevin S. Repasky[2,3,4]

[1]*Advanced Study Program, National Center for Atmospheric Research, Boulder, CO 80301, USA.*
[2]*Earth Observing Laboratory, National Center for Atmospheric Research, Boulder, CO 80301, USA.*
[3]*Physics Department, Montana State University, Bozeman, MT 59717, USA.*
[4]*Electrical and Computer Engineering, Montana State University, Bozeman, MT 59717, USA.*
**Email: stillwel@ucar.edu*

ABSTRACT

It has generally been assumed that differential absorption lidar (DIAL) systems are incapable of measuring atmospheric temperature with useful accuracy. This assumption is a direct result of errors that arise in standard DIAL retrievals due to differential Rayleigh-Doppler broadening from aerosols and molecules. We present here, a combined high spectral resolution (HSRL) and DIAL system that addresses this identified source of uncertainty by measuring quantitative aerosol parameters as well as oxygen absorption parameters. This system, in combination with a perturbative retrieval method, accounts for the Rayleigh-Doppler broadening effects on the oxygen absorption. We describe this combined DIAL/HSRL system and retrieval to evaluate the first retrieval parameters exploring the likelihood that it is possible to measure atmospheric temperature using a DIAL system.

1. INTRODUCTION

Measurements of thermodynamic variables, such as temperature and water vapor mixing-ratio, are critical to understanding and predicting the state of the atmosphere. Much of the predictive skill of numerical weather forecasting is related to the knowledge of these variables in both space and time. Twice daily, hundreds of radiosondes are released all over the globe, which are assimilated into operational weather forecasting systems, to specify the thermodynamic state of the atmosphere and guide weather prediction. This system has proven effective but is limited in application. The infrastructure needed to launch hundreds of sondes simultaneously is fundamentally limited, especially regarding temporal resolution. Many reports have called for an improvement in this atmospheric measurement system [1-5]. Lidar systems show promise to dramatically increase the temporal resolution of thermodynamic profiles, but current temperature lidar systems are not well suited for wide unattended network deployment.

An ongoing collaboration between Montana State University (MSU) and the National Center for Atmospheric Research (NCAR) has resulted in the advancement of diode-laser-based (DLB) lidar systems that are used to perform eye-safe, continuous, and unattended measurements of the atmosphere over the time period of several weeks to months [6-9]. This DLB architecture is adaptable to a number of relevant lidar system types. A prototype water vapor differential absorption lidar (DIAL) has been described and validated [9,10]. Additionally, a high spectral resolution lidar (HSRL) system based on a rubidium vapor cell has been demonstrated [11]. Together, a combination of DIAL and HSRL along with a perturbative retrieval shows promise to measure atmospheric temperature [12].

Previous attempts to measure atmospheric temperature using DIAL leveraging absorption in the oxygen A-band have highlighted major error sources due to differential Rayleigh-Doppler broadening between aerosol and molecular scatterers [13]. The absorption of light after scattering is fundamentally a function of the ratio of total scattering to molecular scattering, i.e. the backscatter ratio. It was proposed, therefore, that a Klett type inversion might suffice using the DIAL offline wavelength to retrieve the backscatter ratio, but this failed to yield useful accuracy - errors on the order of 10°C were observed [13].

We have demonstrated a DIAL system that uses the oxygen A-band and retrieves aerosol information from an integrated HSRL. This data is then used with a perturbative retrieval to retrieve temperature. This system uses a potassium vapor cell with the D_1 absorption line acting as a narrow band (aerosol blocking) filter. This presentation will describe the system and the initial measurement results and progress towards making an autonomous DLB system capable of measuring full atmospheric thermodynamic profiles.

2. LIDAR HARDWARE DESIGN

The measurement of temperature using DIAL relies on the availability of three high quality measurements: oxygen differential absorption, water vapor number density, and the backscatter ratio. Using absorption measurements of a well-mixed species like oxygen forms the basis of the measurement technique. If not accounted for, the exact mixing-ratio of oxygen can be sufficiently altered by the water vapor mixing-ratio to bias the retrieval. Additionally, the backscatter ratio must be known at all altitudes in common volume with the DIAL measurements to correct for the differential Rayleigh-Doppler broadening of the backscattered light in the temperature retrieval. Figure 1 shows a block diagram of a combined DIAL/HSRL system capable of making all three measurements.

Figure 1 shows a combined two-species DIAL system. The first, highlighted in gray, is the WV DIAL system described previously [9]. The red boxes show the needed additions to measure temperature. The transmitters are, by design, identical except their output wavelength. The receivers are similar as well, with an addition of a potassium gas vapor cell in the oxygen receiver channel.

Several design features are of note that allow for the measurement of atmospheric temperature in an unattended, field-deployable, and eye-safe manner. At its core, the system described in Figure 1 is an oxygen DIAL system leveraging the oxygen A-band. A pair of Distributed Bragg Reflectors (DBRs) are used, which are fiber coupled. A set of opto-electric switches interleaves the outgoing pulses at a rate of 100 Hz. Both lasers use a common amplifier, which, in combination with the receiver design, allows for the removal of all overlap corrections. The water vapor and oxygen DIAL beams are combined with a dichroic beam splitter and use a shared telescope design to transmit the beam to the environment. This design provides needed passive opto-mechanical stability for long term unattended operation.

The receiver for the combined system is fiber coupled by a first-stage receiver, providing long-term stability and simplifying alignment. The second-stage receiver for both water vapor and oxygen consist of interference filters, with full widths at half maximum (FWHM) of approximately 0.5 nm and 1 nm, respectively. The filters provide high out of band blocking, on the order of OD 12. For the WV system, the filters are 2 identical custom filters made by Alluxa. For the oxygen system, a single 13 nm FWHM filter from Semrock (part number FF01-766/13-25) and a single 1 nm FWHM filter from Aluxa (part number C11-3268-7270-769BP # 102) are used. Both channels also have a custom narrowband solid etalon with FWHM of 1 GHz and 10 GHz, respectively. Both designs also pass both on- and offline wavelengths in adjacent longitudinal modes of the etalon to use a common detector for a DIAL wavelength pair.

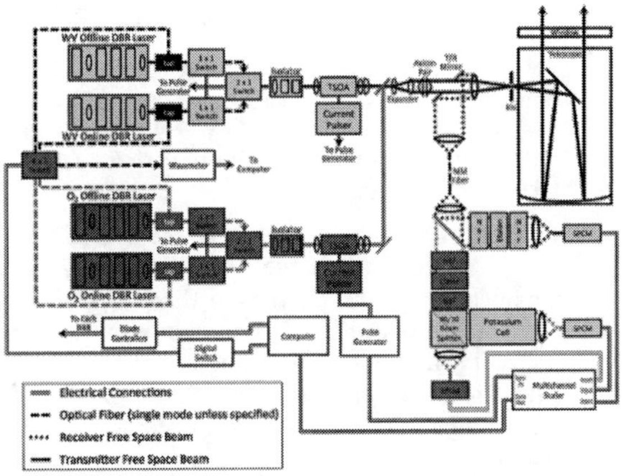

Figure 1: Block diagram of a combined DIAL/HSRL system capable of measuring atmospheric thermodynamic profiles.

Specifications of the combined DIAL/HSRL system are given in Table 1. The wavelengths for water vapor measurements can be changed to increase or decrease the absorption as governed by the prevailing environmental conditions. The wavelengths for oxygen DIAL are not as flexible as in the water vapor system, as the offline must be tuned to the potassium D_1 line and the online wavelength tuned to the 769.8 nm oxygen line.

Table 1: Transmitter and Receiver specifications.

Transmitter	Spec	Receiver	Spec
Locking stability [MHz]	25.3	Telescope Diameter [cm]	40
Online λ [nm] (WV, O_2)	$\approx$ 828.195, 769.7958	Bandpass [nm] (WV, O_2)	0.012, 0.080
Offline λ [nm] (WV, O_2)	$\approx$ 828.295, 770.1085	Field of View [μrad]	100
Power [mW] (WV, O_2)	$\approx$ 25, $\approx$ 17.5	Range Resolution [m]	150

As a result, the oxygen channel etalon's free spectral range is tightly constrained as its transmission lines must align with both the oxygen and potassium lines. A scan of the oxygen receiver etalon with the modeled oxygen spectrum and potassium D_1 line are shown in Figure 2.

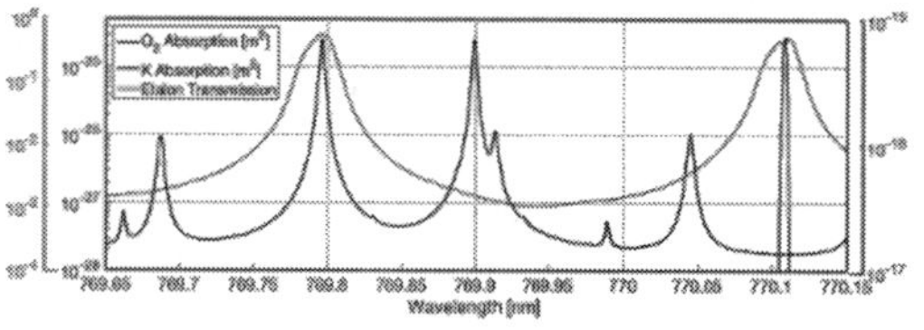

Figure 2: Spectral characteristics of the oxygen A-band, potassium D_1 line, and etalon. The etalon that passes both DIAL wavelengths was designed with a FSR of 157.90 GHz, with measured FSR of 158.13 GHz.

Finally, this etalon design provides a method of easily calculating the HSRL calibration parameters. As the on- and offline wavelengths have the same overlap function and use the same detectors, the online wavelength can be used to calculate the range dependent calibration information. The ratio of the oxygen online raw photons measured on the combined and molecular detectors (so named from the offline HSRL signals observed) yields the calibration constants directly without having to remove optical elements or alter the receiver in any way.

3. FIRST MEASUREMENTS

Temperature measurements are made using the above-described system using a retrieval based on a perturbative solution to the two-wavelength DIAL equation [12]. This retrieval takes an initial estimate of the temperature profile to retrieve water vapor and backscatter ratio. We use ground-based weather station temperature and an assumed constant lapse rate. These values are, by design, relatively insensitive to temperature. Using these retrieved water vapor mixing-ratio and backscatter ratio, the temperature-sensitive oxygen absorption coefficient is retrieved. Temperature is retrieved from the oxygen absorption coefficient by using an iterative version of the perturbative retrieval.

First temperature measurements are shown in Figure 3. For context, surface based weather station temperature values are also provided. These data are taken over a clear 38-hour period without the etalon described in Section 2. The surface temperature and first range bin show qualitative agreement, especially highlighting the day-to-night temperature change.

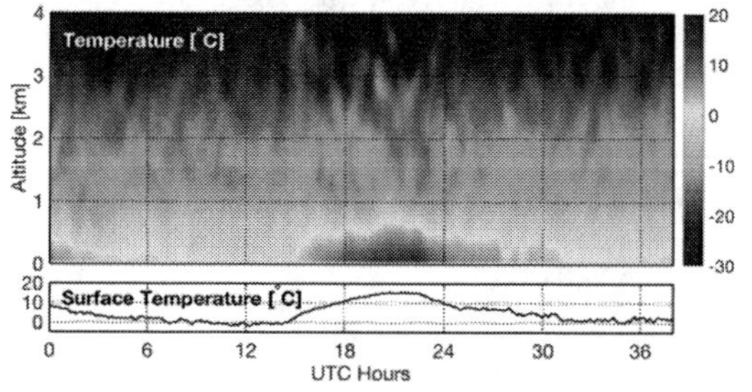

Figure 3: First temperature measurements taken with the described combined DIAL/HSRL system. Range resolution is 300 meters and temporal resolution is 20 minutes. First range observed is 300 meters. The first range bin here is filled with weather station data and does not come from the DIAL system.

For more qualitative comparisons, the radiosonde launched from Denver International Airport (DIA) is used to assess the vertical temperature structure. Temperature differences between the two radiosondes, launched at 12 and 36 UTC from Figure 3, and the DIAL system are shown in Figure 4. This comparison is used to assess vertical structure but is not an ideal comparison as the radiosonde launch site and DIAL are separated by 50 km. As a result, the ground temperatures at the sites are not equal on this day. To address this issue, the vertical structure determined from the radiosonde is used with a bias applied to the whole profile to force ground temperatures to align.

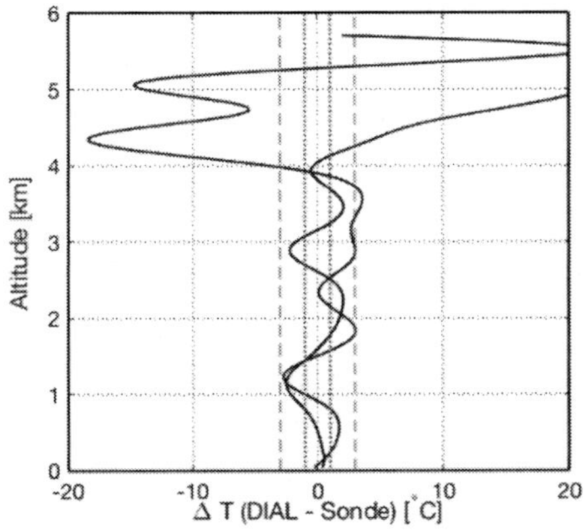

Figure 4: Comparisons of DIAL temperature data with DIA radiosonde launces. The red dashed lines are 3 °C bounds and the dotted lines are 1 °C.

While the temperature overlap is not perfect, some promising features can be noted from this data. First, the WMO OSCAR database identifies an accuracy of 3°C as the beginning of useful temperature data for numerical weather prediction in the lower troposphere [14]. These data suggest that such accuracies should be possible. Second, the maximum range of the measurements shown is 4 km. This is a direct result of low SNR measurements. Receiver optimization has increased the measured signals by a factor of approximately 5 since these data were taken.

More detailed comparisons with co-located radiosondes are planned to occur starting in April 2019. These results show promise, but need to be further investigated to understand the absolute accuracy of the designed system.

4. CONCLUSION

The work presented here demonstrates that DIAL systems hold promise for providing continuous monitoring of temperature in the lower troposphere. We present a design and retrieval technique that addresses previously identified issues with DIAL temperature measurements caused by molecular broadening of the backscatter spectrum. This design is a combined DIAL/HSRL system capable of measuring ancillary parameters such as water vapor mixing ratio and backscatter ratio in common volume with differential absorption measurements from oxygen. This combination along with a perturbative retrieval technique shows promise to measure atmospheric temperatures. Furthermore, given that DLB lidar systems are relatively inexpensive, and stable over periods on the order of months, this system also shows promise to be developed into a field deployable network platform. This presentation will address the development of this system. It will also demonstrate ongoing testing with this system to describe operational performance.

ACKNOWLEDGEMENTS

The National Center for Atmospheric Research is sponsored by the National Science Foundation.

REFERENCES

[1] NRC (2009): Observing Weather and Climate from the Ground Up: A Nationwide Network of Networks

[2] NRC (2010): When Weather Matters: Science and Services to Meet Critical Societal Needs

[3] NSF (2014): Dynamic Earth: GEO Imperatives & Frontiers 2015-2020

[4] Wulfmeyer, V. et al. (2015): Rev. Geophys., 53, 819-895

[5] NASEM (2018): The Future of Atmospheric Boundary Layer Observing, Understanding, and Modeling: Proceedings of a Workshop

[6] Nehrir, A. R., et al. (2009): J. Atmos. Oceanic Technol., 20, 733-745

[7] Nehrir, A. R., et al. (2011): J. Atmos. Oceanic Technol., 28, 131-147

[8] Repasky, K. S., et al., (2013): Remote Sens., 5, 6241-6259

[9] Spuler, S. M. et al., (2015): Atmos. Meas. Tech., 8, 1073-1087

[10] Weckwerth, T. M., et al. (2016): J. Atmos. Oceanic Technol., 33, 2353-2373

[11] Hayman, M. and S. M. Spuler, (2017): Opt. Express, 25, A1096-A1110

[12] Bunn, C. E., et al. (2018): Appl. Opt., 57, 4440-4450

[13] Boesenberg, (1998): Appl. Opt., 37, 3845-3860

[14] WMO (2019): OSCAR: Observing Systems Capability Analysis and Review Tool. https://www.wmo-sat.info/oscar/requirements

EPJ Web Conferences **237**, 06019 (2020) https://doi.org/10.1051/epjconf/202023706019
ILRC 29

TURBULENT UV-LIDAR AND FIELD EXPERIMENT IN AIRPORT

Igor A. Razenkov*, Victor A. Banakh

*V.E. Zuev Institute of Atmospheric Optics, Tomsk, 634055, RUSSIA, *email: lidaroff@iao.ru*

ABSTRACT

Built the special UV (355 nm) eye safe lidar for remote detection of the atmospheric turbulence. Lidar works on the backscatter enhancement effect. Testing of lidar was carried out on the airfield at the airport Tolmachevo (Novosibirsk) in July, 2018. Sounding path was crossing the place of take-off and landing. Lidar surely fixed the position, size and intensity of turbulent zones.

1. INTRODUCTION

In 1972, the effect of backscatter enhancement (BSE) in the propagation of radiation in a turbulent atmosphere was discovered [1]. The effect is appeared in double propagation of radiation through the atmospheric heterogeneities of the refractive index. As a result, the *average intensity* of backscattered radiation on the beam's axis is increased by rays from the beam's periphery. The area in which there is an increase in intensity is small and approximately equal to the double Fresnel zone (it is equal 66 mm at 355 nm wavelength and for 3 km distance). Therefore, the lidar aperture should be small.

To register an enhancement in the average radiation intensity on the beam's axis, a complete coincident of the probing beam with the incoming beam is required. Herewith, the echo signal must be accumulated within a certain time interval. Practically, this is possible when probing with weak light pulses to ensure the operation of the detector and not to damage it by radiation of parasitic stray light on optics.

It is obvious that turbulent lidar is a kind of micro pulse lidar operating in the photon counting mode. Also, the turbulent lidar must have a second receiving channel (off-axis), which is not affected by turbulence. So, the normalization of the signal of the main receiving channel (on-axis) on the off-axis will eliminate the influence of a possible inhomogeneous distribution of the aerosol scattering coefficient along the sounding path [2].

For lidar to be used in special places, for example, at the airport, its radiation must be invisible and absolutely safe for the human eye. Therefore, we chose the wavelength of 355 nm.

2. LIDAR DESIGN

The lidar diagram is shown in Figure 1. The receiving-transmitting telescope is located on side 2 of the optical bench (the upper part of the figure). The laser, optics and receiving box are placed on the other side 1 of the bench (the lower part of the picture). The radiation of the pulsed laser 3 by mirrors 4-9 is directed to the collimator 10-11. Next are a half-wave plate 12 and a thin-film polarizer 13. Mirrors 14 and 15 send the beam to the afocal telescope 17-18. In front of the telescope is a quarter wave plate 16. Behind the telescope is a screen 19 with a pair of round holes. In Fig. 1 the on-axis channel is green and the off-axis is blue. Lidar characteristics are shown in Table. 1. Calibration procedures for turbulent lidar are detailed in [2].

3. THEORY AND INVERSE PROBLEM

The theory for the proposed scheme of turbulent lidar [2] was developed by Gurvich and Vorobyov [3]. Lidar equation taking into account the influence of turbulence (backscatter enhancement) can be represented as

$$\frac{\langle P(x,R,\rho_0)\rangle}{\langle P_0(x,R,\rho_0)\rangle} = 1 + q(x,R,\rho_0), \qquad (1)$$

where P_0 is the power at the receiver in the absence of turbulence; x is the distance; ρ_0 is the distance between the axes of the receivers; R is the receiving aperture. The value of $q(x,R,\rho_0)$ is called the influence factor of turbulence on the average power of the scattered light at the receiver.

https://doi.org/10.1051/epjconf/202023706019

For the case of weak and isotropic turbulence [3]

$$q(x,R,\rho_0) = \frac{4\pi}{R}\int_0^\infty F_I(\kappa,x)J_1(\kappa R)J_0(\kappa\,\rho_0)d\kappa, \quad (2)$$

where both J_0 and J_1 are Bessel functions of zero and first order; k_0 – wave number; $\Phi_n(\kappa,x) = 0{,}033C_n^2(x)|\kappa|^{-11/3}$ - spectrum of refractive index fluctuations; C_n^2 - structural characteristic of refractive index; and

$$F_I(\kappa,x) = 8\pi k_0^2 x\int_0^\infty \Phi_n\left[\kappa(1+\eta),\frac{x}{1+\eta}\right]\times$$
$$\times\left[1-\cos\left(\frac{\kappa^2 x}{k_0}\eta\right)\right]d\eta\,. \quad (3)$$

For statistically homogeneous turbulence (C_n^2=const), a good approximation of these formulas is

$$q(x,R,0) = \frac{0{,}49C_n^2 k_0^{7/6}x^{11/6}}{\left[1+0{,}4\frac{R^2k_0}{x}\right]^{7/6}}. \quad (4)$$

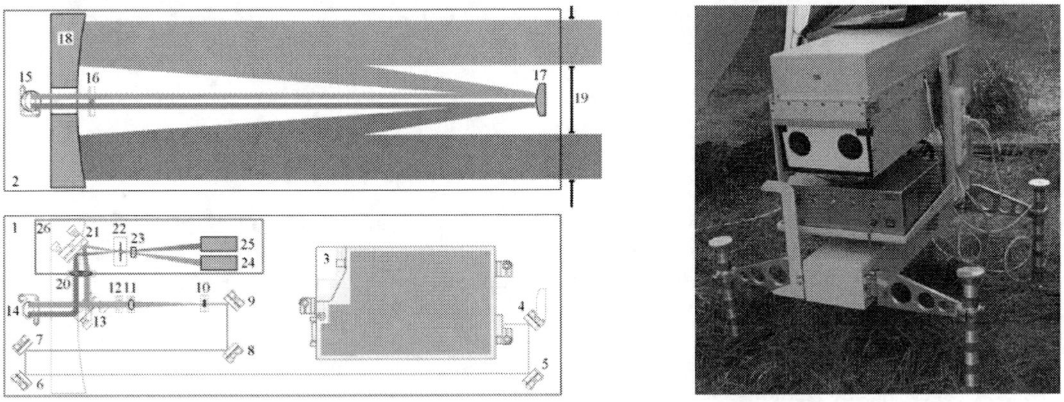

Fig. 1. Optical diagram of turbulent UV lidar (left) and real lidar on the field experiment (right).

Table 1.

Transmitter		**Receiver**	
Laser	Solar JX 330	Afocal telescope	Mersen
Wavelength	355 nm	Telescope diameter	250 mm
Pulse duration	30 ns	Telescope magnification	12,5$^{\times}$
Pulse repetition	10 KHz	Number of channels	2
Laser output	1 W	Receiver aperture	70 mm
Pulse energy	100 μJ	Field of view	300 μRad
Beam quality	TEM_{00}, M^2<1,2	Space resolution	15 m
Beam size (~$1/e^2$)	70 mm	Time resolution	> 15 s
Beam divergence	<30 μRad	PMT (H10682-210)	QE=33%

4. EXPERIMENT IN AIRPORT

To test the UV turbulent lidar, the experiment was organized on the airfield of Tolmachevo airport (Novosibirsk). Lidar was located to the left of the glide path at a distance of 300 m. The sounding path was approximately horizontal and crossed the runway at the landing point (Fig. 2).

Figure 3 schematically shows the approximate location of the (a) landing, braking and taxiing zones. And also, (b) the location of turbulent zones along the direction of sounding. The first turbulent zone appears after landing as a result of braking, the second zone occurs as a result of taxiing aircraft in the parking area.

 https://doi.org/10.1051/epjconf/202023706019

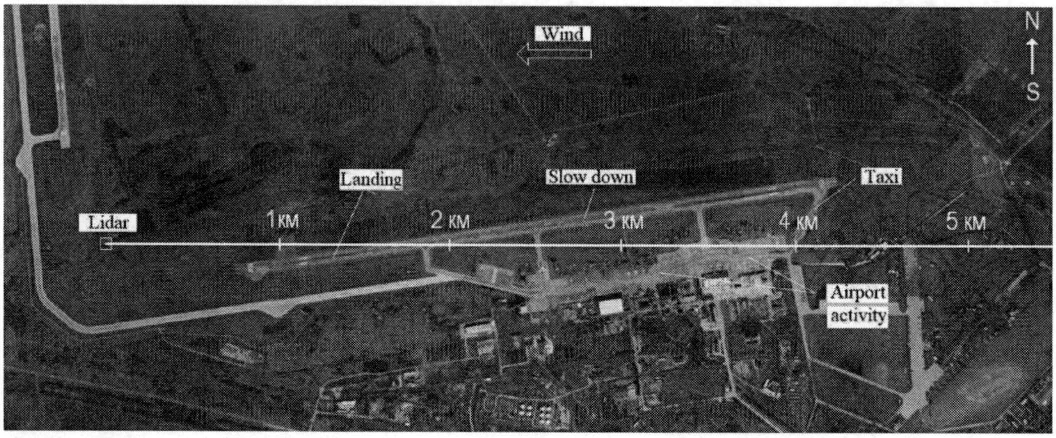

Fig. 2. Tolmachevo airport. The lidar sounding path crossing the run way at 1.5 km. Maximum range is 5 km.

Figure 4 shows (a) the photo of MD-11 landing and (b) touchdown of MD-11, when as a result of the wheels burning rubber an aerosol cloud appears. During the measurements the wind was from East.

Figure 5 presents examples of lidar returns. The collecting time was 1 min. In Fig. 5(a) the echo signals of the on-axis 1 and off-axis 2 receiving channels coincide, so from formula (1) it follows that there is no turbulence and $q(x)=0$. On Fig. 5(b) there is a second turbulent zone at the distance of 3.5 km...4 km from the lidar. Fig. 5(c) shows the situation 2 minutes after landing MD-11 (Fig. 4). At the distance of 1300 m from the lidar is an aerosol cloud. The first turbulent zone is at the distance of 1500 m…1800 m. Note, intense turbulence took place in a limited area above runway, but its influence affects the echo signal of the axial receiving channel for some time (aftereffect).

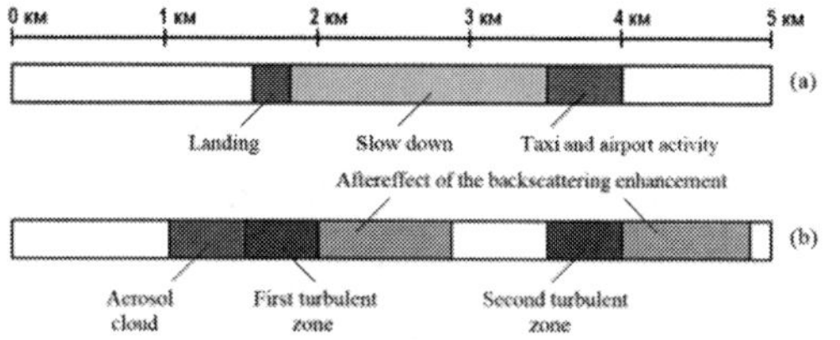

Fig. 3 Scheme of zones location: (a) landing, braking and taxiing; (b) turbulent zones and aerosol cloud.

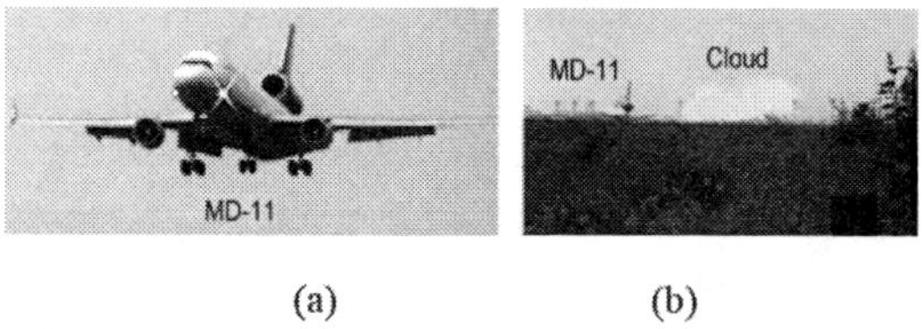

Fig. 4. Landing of the aircraft MD-11: (a) aircraft descend; (b) aircraft on the ground, aerosol cloud appearance at the moment of touching the runway.

Figure 6 presents examples of q-factor behavior. The first two graphs (a, b) are obtained under conditions of homogeneous turbulence ($C_n^2=const$) at different levels of its intensity, since the values of q-factors differ significantly. To estimate the value of the structural characteristic of C_n^2, we used the formula (4), fitting the theoretical curve (dotted line) to the experimental data (circles). For the data in Fig. 6(a) and 6(b), respectively, were obtained values $C_n^2 = 2{,}4\times10^{-15}$ m$^{-2/3}$ and $C_n^2 = 7{,}0\times10^{-14}$ m$^{-2/3}$.

Figure 6(c) presents an example of an inhomogeneous distribution of the turbulence intensity along the sounding path. On Fig. 6(d) an example of superposition of the first turbulent zone on the daily turbulence intensity profile is shown. Note, the solution of the inverse problem of reconstructing the inhomogeneous profile of the structural characteristic $C_n^2(x)$ is a nontrivial problem. After landing or take-off of the aircraft, the lifetime of the turbulent zones above the runway was 2-3 minutes with size about 1 km.

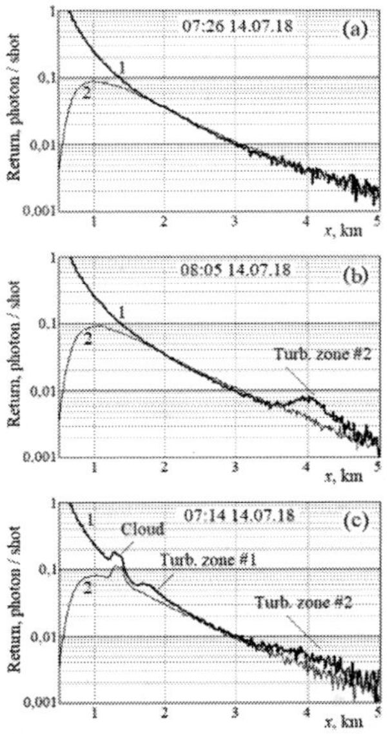

Fig. 5. Examples of returns: (a) no activity in airport; (b) the second turbulent zone is clearly recorded; (c) after landing MD-11 (see Fig. 4), there is present an aerosol cloud and both turbulent zones.

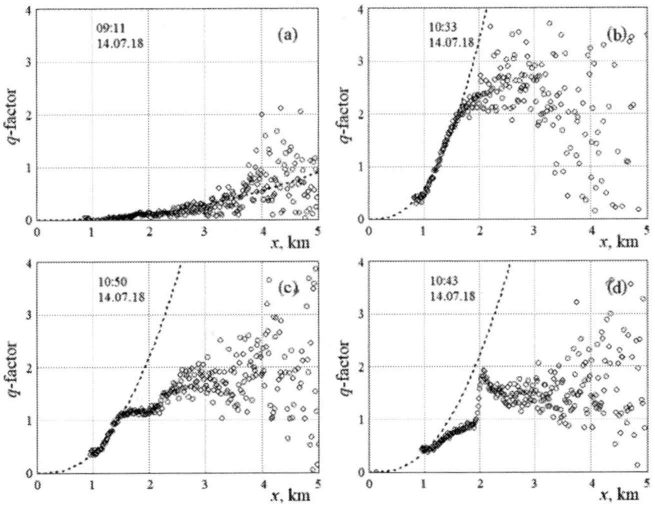

Fig. 6. Examples of the behavior of q-factor: (a) weak homogeneous turbulence intensity in the morning (C_n^2 =2.4×10^{-15} m$^{-2/3}$); (b) moderate homogeneous intensity of turbulence afternoon (C_n^2 =7,0×10^{-14} m$^{-2/3}$); (c) a non-uniform intensity profile (at the beginning of path C_n^2 =4,6×10^{-14} m$^{-2/3}$); (d) overlay the first turbulent zone at distance 2 km after aircraft landing on the daily profile of turbulence intensity.

CONCLUSIONS

In Institute of Atmospheric Optics was built a turbulent UV-lidar working due to the backscatter enhancement effect. The lidar has a micro-pulse laser and an afocal telescope with a receiving-transmitting channel and a second receiving channel designed for normalization of the return of first one. Lidar implements the principle of relative measurement, which provides stability and reliability.

For the solution of the inverse problem used the Gurvich-Vorobyov theory. In conditions of homogeneous turbulence, the device allows to estimate the average value of the structural characteristic C_n^2, proportional to the intensity of atmospheric turbulence.

Lidar was tested on the airfield at the airport Tolmachevo (Novosibirsk). Lidar allowed to register turbulent zones that occurred during landing and take-off, as well as during the taxiing, and working engines in the parking.

ACKNOWLEDGMENTS

The authors thank E.V. Gordeev, O.A. Rynkov and K.A. Rynkov for their technical support. This study was carried out according to the project of the Program for Basic Research of the Russian Academy of Sciences AAAA-A17-117021310149-4.

REFERENCES

[1] A. G. Vinogradov, Yu. A. Kravtsov, and V. I. Tatarskii, “The effect of intensification of back scattering by bodies that are situated in a medium having random inhomogeneities,” Radiophys. Quantum Electron. **16** (7), 818–823 (1973).

[2] I. A. Razenkov, “Turbulent lidar. I. Design,” Atmos. Ocean. Opt. **31** (3), 273–280 (2018).

[3] V. V. Vorob’ev, “On the applicability of asymptotic formulas of retrieving "optical" turbulence parameters from pulse lidar sounding data: I—Equations,” Atmos. Ocean. Opt. **30** (2), 156–161 (2017).

Water vapor mixing ratio distribution inversion by Raman lidar in Beijing

SiQi Yu[1, 2], Dong Liu[2*], JiWei Xu[2], ZhenZhu Wang[2], DeCheng Wu[2], Yingjian Wang [2,1]

[1]School of Environmental Science and Optoelectronic Technology, University of Science and Technology of China, Hefei, Anhui 230026, China

[2]Key Laboratory of Atmospheric optics, Anhui Institute of Optics and Fine Mechanics, Chinese Academy of Sciences, Hefei, Anhui 230031, China

**Email: dliu@aiofm.ac.cn*

ABSTRACT

Water Aerosol Raman Lidar-II is an active detection instrument with high temporal and spatial resolution at Nanjiao observation station, and that could continuous water vapor mixing ratio (WVMR) measurements. WVMR profiles inversion from lidar data and water ratio retrieved from radiosonde data are in good agreement. The statistical results of the vertical distribution of WVMR indicate that WVMR seasonal mean distribution is consistent with precipitation. In addition, WVMR in Nanjiao station is related to total cloud cover.

1. INTRODUCTION

Water vapor plays a very important role at climate change. Global water vapor amounts increase in a warmer climate, leading to a positive feedback via its enhanced greenhouse effect[1]. Lidars (DIAL and Raman) are the only instruments available for high temporal and vertical resolution of continuous WVMR measurements[2]. What's more, obtaining a dataset of atmospheric water vapor content with high spatial and temporal coverage conduce to understand the vertical structure[3].

In the current study, the water vapor mixing ratio profiles obtained by the means of Water Aerosol Raman Lidar-II in Nanjiao station. Observation period from September 2017 to August 2018. The lidar and the data set, as well as the inversion approach to get water vapor mixing ratio and water ratio are described in Section 2. The results are presented in Section 3, including the relationship between water vapor mixing ratio and water ratio, as well as the statistical results of water vapor mixing ratio.

2. METHODOLOGY

2.1 Experimental site and instrumentation

The instrument used to detect atmosphere is Water Aerosol Raman Lidar-II (named as WARL-II), which has been installed at the Nanjiao observation station, Beijing, China. Nanjiao station (39.81N,116.48E) located in North China Plain, which experiences distinct seasons, namely, winter (December, January and February), spring (March, April and May), summer (June, July and August) and autumn (September, October and November). Figure1 gives the schematic diagram of WARL-II system. WARL-II emits at 532nm (output energies per pulse of 150 mJ) and receive elastic channels at parallel and perpendicular polarization channels at 532 nm as well as Raman-shifted channels at 607nm (from N_2) and 660nm (from H_2O). The spatial and temporal resolution of WARL-II is 7.5m and 5minutes, respectively. The ground-based passive device continuously measured aerosol and cloud vertical distribution and humidity profiles during the studied period.

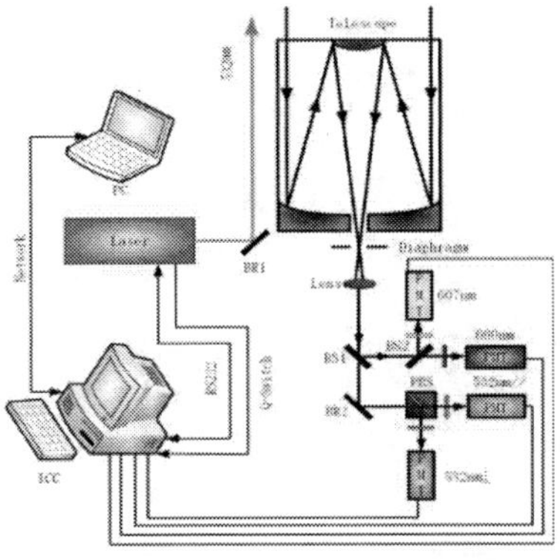

Figure 1 The schematic diagram and photo of WARL-Ⅱ

Table 1 Specification of WARL-Ⅱ

Characteristic	Value
Transmitter	
Wavelength	532nm
Pulse Energy	150mJ
Pulse Width	<6nm
Pulse frequency	10Hz
Divergence	<0.5mrad
Receiver	
Type of telescope	Cassegrain
Diameter	~300mm
FOV	0.5-2mrad
receive wavelength	532nm,607nm,660nm
Filter bandwidth	~1nm
Optical sensor	PMT
Data Acquisition	
Resolution	16bit

The department of atmospheric sciences of University of Wyoming has a radiosonde database for the whole world. The Beijing observation (39.93N,116.28E) is about 21km away from Nanjiao station. The radiosonde data obtained twice a day, at local standard time 8:00 and 20:00 respectively. The soundings are available on http://weather.uwyo. edu/upperair/sounding.html.

2.2 inversion method

2.2.1 Ground-based LIDAR

The Raman lidar is an efficient measurement tool of the mixing ratio of water vapor to dry air. A classical retrieval approach of water vapor mixing ratio(WVMR) provided by Ansmamn et al.[4] at 1992,which can be defined as follows[5]:

$$m(z) = K_m \frac{P_{\lambda 660}}{P_{\lambda 607}} \times \frac{\exp\left\{-\int_0^z [\alpha_{\lambda 607}^{aer}(z) + \alpha_{\lambda 607}^{mol}(z)]dz'\right\}}{\exp\left\{-\int_0^z [\alpha_{\lambda 660}^{aer}(z) + \alpha_{\lambda 660}^{mol}(z)]dz'\right\}} \quad (1)$$

Where, the P_λ denotes the Raman-shifted return lidar signals from distance z at the laser wavelength λ for the water vapor (660 nm) and the nitrogen(607 nm). The $\boldsymbol{\alpha}$ denotes the light extinction at wavelength λ, as well as 'aer' and 'mol' is represent for aerosol extinction and molecular extinction respectively. The water vapor mixing ratio calibration constant $K_m = \frac{K_{\lambda 607}}{K_{\lambda 660}}$ could determine from a comparison of the lidar measurement with critically evaluated data from a radiosonde ascent.

Generally, the Raman signal are operated more conveniently during night time. This is because, the lidar return signal is dominated by noise during day-time.

2.2.2 Radiosondes

The Radiosonde water ratio(WR) profile retrieval is according to equation(2,3), which are used by Wu et al.2016[6].

$$\mathrm{WR} = \mathrm{RH} * \mathrm{S} = \mathrm{RH} * \frac{0.622 * P_S(T)}{P - 0.378 * P_S(T)} \quad (2)$$

$$P_S(T) = 6.1121 * \exp\left(\left(18.678 - \frac{T}{234.5}\right) * \left(\frac{T}{257.14+T}\right)\right) \quad (3)$$

Where, RH denotes relative humidity, S denotes specific humidity. P is the atmospheric pressure, and P_S is the saturated vapor pressure (mb) at temperature T (℃) and can be calculated by equation(3).

3. RESULTS

3.1 COMPARISON RESULTS

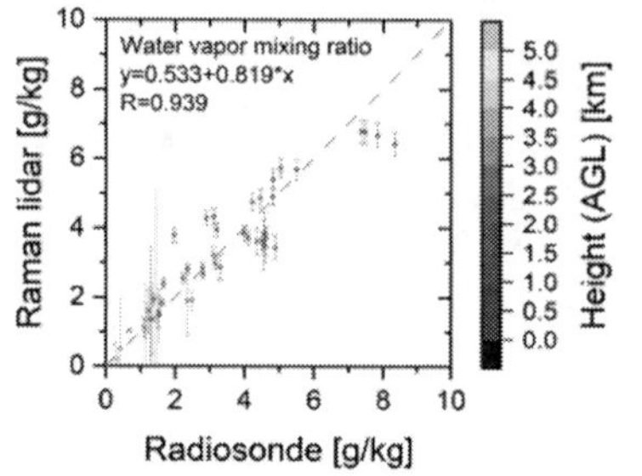

Figure 2 regression of water vapor mixing ratio profile to radiosonde measurement

Figure 2 is the comparison results between WARL-II and radiosonde. It is obtained for profile by the linear regression method. The regression analysis results used the data of three days of October ,2017, 13,19 and 23 respectively. The correlation coefficient is 0.939 and that means the calibration of WARL-II measurement can give a reasonably accurate estimate of water vapor profile for the routine observation.

3.2 statistical results of water vapor vertical distribution

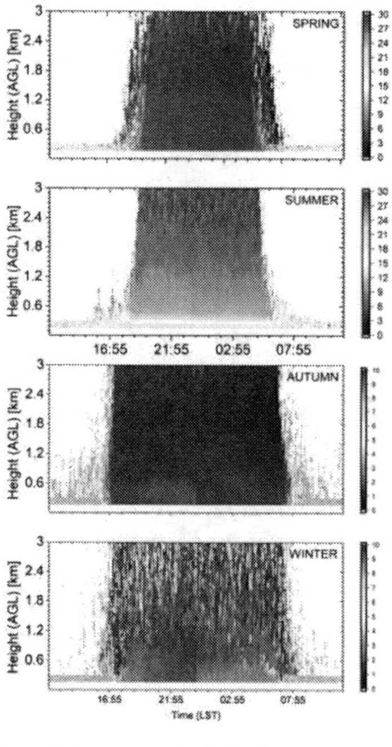

Figure 3 WVMR vertical distribution at difference seasons

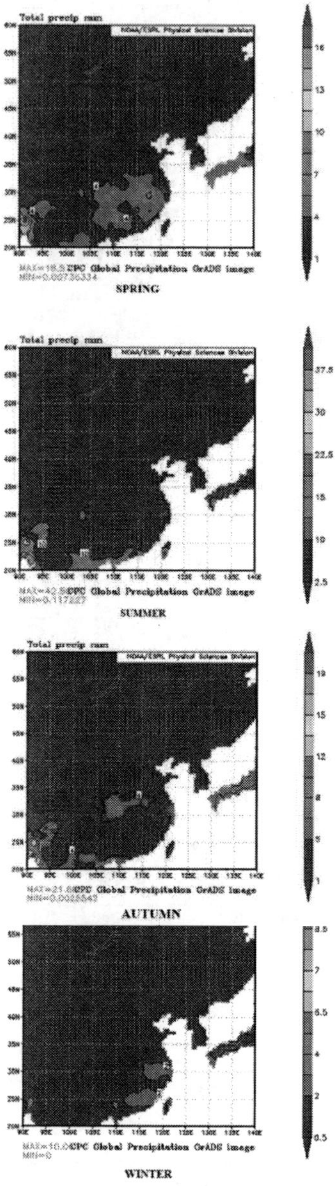

Figure 4 seasonal mean precipitation

As shown in Figure 3, statistical analysis results of Water Vapor Mixing Ratio (WVMR) measured by lidar among spring, summer, autumn and winter. Atmospheric observations of autumn were performed from 1 September to 30 November 2017. The period of winter is between 00:00 LST on 1 December 2017 and 24:00 LST on 28

https://doi.org/10.1051/epjconf/202023706020

February 2018. Spring is from 00:00 LST on 1 March 2018 to 24:00 LST on 31 May 2018. Summer, the last season of observation, ending at 24:00 LST on 31 August 2018. The seasonal mean values of WVMR in spring, summer, autumn, winter are range from 0 to 9 g/kg, 8 to 18 g/Kg, 1 to 8 g/Kg, 0 to 3 g/Kg, respectively. These observation results are consistent with the consequence of Wang et al.[7]. Nanjiao station is at the monsoon zone, which with seasonal rainfall variability. In order to explain the relationship of water vapor mixing ratio and precipitation, seasonal mean precipitation at 116.48E,39.81N from the National Center for Environmental Prediction/National Center for Atmospheric Research (NCEP/NCAR) Reanalyzed data are examined in Fig. 4. The four seasons of Figure 4 cover the same time period of lidar observation. According to Figure 3 and figure 4, water vapor mixing ratio measured by lidar have a good consistency with precipitation. The nocturnal value of WVMR in summer is the largest, while in winter is minimum. The reason for the difference obvious for the monsoon.

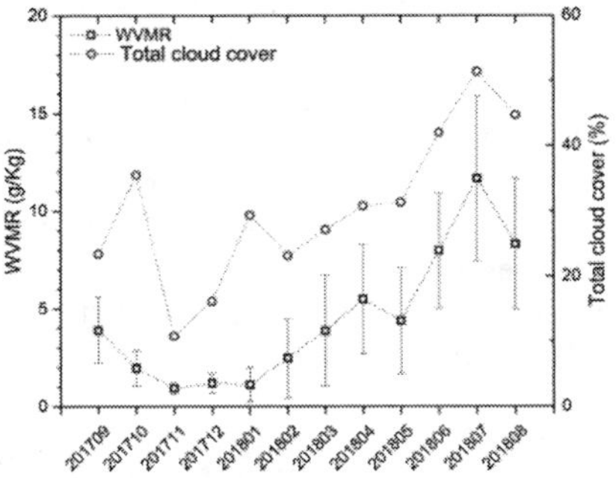

Figure 5 water vapor mixing ratio (WVMR) and total cloud cover

Figure 5 reveals the correlation between monthly mean WVMR and monthly mean total cloud cover. To ensure the accuracy of the measurement, the measured the water vapor profiles measured by WARL-II statistic from 00:00 LST to 06:00 LST and 18:00 LST to 24:00LST. The data of monthly mean total cloud cover is from (NCEP/NCAR) website. Exclude October 2017 and January 2018, red line and blue line have consistent trend. Frequency that occurs during the day of Clouds may contribute to the result.

ACKNOWLEDGEMENTS

This work was s supported by the Science and Technology Service Network Initiative of CAS (KFJ-STS-QYZD-022).

Thanks to the NCEP/NCAR for the Reanalyzed data. As well as thanks to University of Wyoming for radiosonde data.

REFERENCES

1. Boucher, O., et al., *Clouds and aerosols*, in *Climate change 2013: the physical science basis. Contribution of Working Group I to the Fifth Assessment Report of the Intergovernmental Panel on Climate Change*. 2013, Cambridge University Press. p. 571-657.
2. Filioglou, M., et al., *Profiling water vapor mixing ratios in Finland by means of a Raman lidar, a satellite and a model*. Atmospheric Measurement Techniques, 2017. **10**(11): p. 1-23.
3. Fujita, M. and T. Sato, *Observed behaviours of precipitable water vapour and precipitation intensity in response to upper air profiles estimated from surface air temperature*. Sci Rep, 2017. **7**(1): p. 4233.
4. ANSMANN, et al., *Combined Raman elastic-backscatter LIDAR for vertical profiling of moisture, aerosol extinction, backscatter, and LIDAR ratio*. Applied Physics B, 1992. **55**(1): p. 18-28.
5. Stachlewska, I.S., M. Costa-Surós, and D. Althausen, *Raman lidar water vapor profiling over Warsaw, Poland*. Atmospheric Research, 2017. **194**: p. 258-267.
6. Wu, S., et al., *Observations of water vapor mixing ratio profile and flux in the Tibetan Plateau based on the lidar technique*. Atmospheric Measurement Techniques,9,3(2016-04-01), 2016. **8**(11): p. 11925-11952.
7. Wei, W., et al., *Long-Term Measurement for Low-Tropospheric Water Vapor and Aerosol by Raman Lidar in Wuhan*. Atmosphere, 2015. **6**(4): p. 521-533.

https://doi.org/10.1051/epjconf/202023706021

WIND PROFILING BY PASSIVE OPTICAL METHOD

Vadim Dudorov, Anna Eremina

V.E. Zuev Institute of Atmospheric Optics, Rus. Acad. Sci., Tomsk 634021, Russia

ABSTRACT

Possibilities of the wind speed profiling along an observation path of a distant object from the analysis of a video sequence of images of the object are studied in this work. The method is based on the analysis of two neighbor frames of a video sequence of incoherent images. The wind velocity retrieved is compared with data of acoustic anemometers.

1. INTRODUCTION

There are several optical approaches to the remote measurements of crosswind speed [1-6]. An evident advantage of passive techniques is that they do not need in proper radiation sources. This allows measurement instruments to be simplified and their cost to be reduced, as well as the domain of applicability of contactless wind speed meters to be expanded to the situations where it is impossible to mount a radiation transmitter/receiver at one of the ends of an atmospheric path under study, e.g., in problems of line-of-sight correction.

We suggest an approach which is based on the analysis of the dynamics of anisoplanar (inhomogeneous) turbulent distortions in object images formed by wide-field cameras [7–10]. In this case, the shift of distortions in an object image, which characterizes the speed of displacement of turbulent inhomogeneities, can be determined from the analysis of two neighbor frames of a video sequence of images. The advantages of this approach are a high speed due to the lack of a need in accumulation of a series of observations to construct temporal correlation functions and a capability of determining the instantaneous wind speed.

Two key problems are to be solved for the development of the method for wind profiling: first, it is necessary to derive a relationship between turbulent inhomogeneities and image distortions caused by them, and, second, it is required to identify (visualize) turbulent distortions in images of distant objects.

2. INFLUENCE OF LOCATION OF TURBULENT INHOMOGENEITIES ON THE CHARACTER OF DISTORTIONS CAUSED BY THEM

To determine the relation between the character of distortions in the image of an object and the location of atmospheric inhomogeneities in the region between the object and an observer, the process of imaging in the presence of some bounded region (thin layer) of turbulent inhomogeneities along the observation path was numerically simulated. To suppress the effect of the structure of complex objects, we choose a test object with the rough (Lambertian) surface and periodic variations in the reflection coefficient like

$$T(\mathbf{r}) = \exp\left\{-\left[1+\sin\left(x/b\right)\right]^{10}\right\}, \qquad (1)$$

Figure 1 shows the scheme of formation of the incoherent image of an object in the presence of a thin layer of turbulent inhomogeneities (one phase screen) located at a different distance from the observer z_{ts}. In this case, the displacement of the phase screen along the optical axis allows us to separate the effect of air inhomogeneities located at different distances from the receiving system along the observation path.

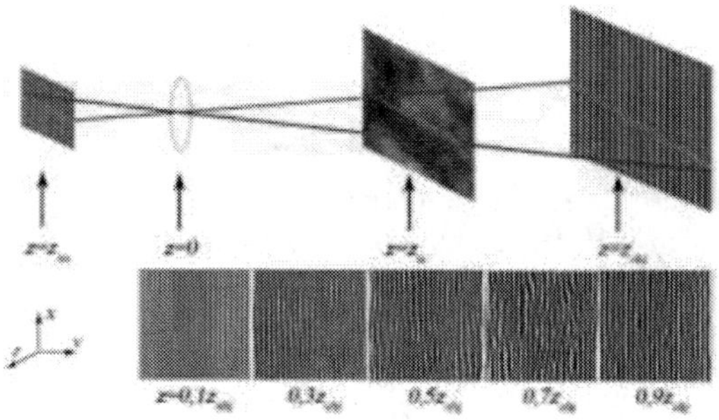

Fig.1. Scheme of formation of an incoherent image of an object in the presence of a thin layer of turbulent inhomogeneities (single phase screen).

The numerical simulation results allowed us to ascertain that the characteristic size of anisoplanar distortions produced by the layer of a turbulent medium (phase screen) located at the distance z_{ts} from the observer inversely relates to the distance:

$$a_{anisopl}(z_{ts}) = \frac{A}{z_{ts} / z_{obj}}, \qquad (2)$$

where A is the dimensional parameter and is a function of the atmospheric conditions and receiving system.

3. TURBULENT INHOMOGENEITIES VISUALISATION

During the processing of images analyzed, the information about the object structure is to be filtered, i.e., the so-called quality map of an image should be constructed, which includes only data on atmospheric distortions. Based on the sharpness functional, the image quality map can be defined in the form:

$$Q(\mathbf{r}, a_k, t) = \left|\nabla_{\mathbf{r}} I(\mathbf{r}, t)\right|^2 \otimes K(\mathbf{r}, a_k), \qquad (2)$$

where $\nabla_{\mathbf{r}} = \vec{i}\frac{\partial}{\partial x} + \vec{j}\frac{\partial}{\partial y}$ is the gradient vector, $\left|\nabla_{\mathbf{r}} I(\mathbf{r}, t)\right|^2$ is the image sharpness functional; a_k is the smoothing coefficient; $K(\mathbf{r}, a_k)$ is the certain smoothing function defined as Gaussian:

$$K(\mathbf{r}, a_k) = \frac{1}{\pi a_k^2} \exp\left(-r^2 / a_k^2\right) \qquad (3)$$

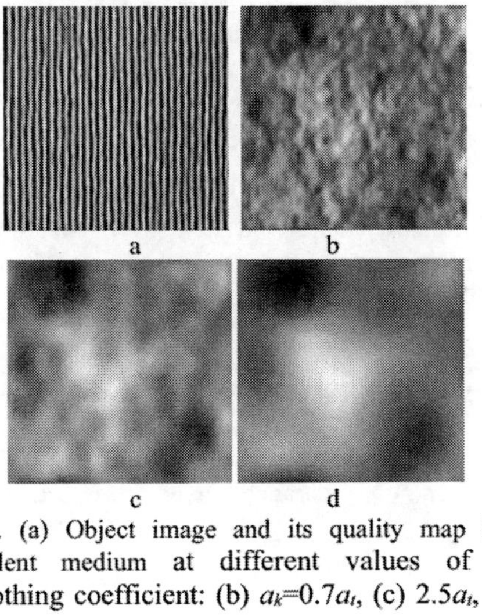

Fig.2. (a) Object image and its quality map in a turbulent medium at different values of the smoothing coefficient: (b) a_k=0.7a_t, (c) 2.5a_t, and (d) 6.0a_t. The distance to the object $z_{obj} = 0.1 k a_t^2$; b=0.2a_t; N_z=5; R_{obj}=40a_t; N_R=512; the ratio of the receiving aperture diameter to the Fried length $2a_t/r_0 = 4$.

Figure 2 shows an object and the quality maps for different values of the smoothing coefficient a_k. It is seen that the characteristic size of distortions $a_{anisopl}$ can be controlled by variations in a_k when analyzing the dynamics of the distortions in the video sequence of images of the object under study with the aim of distinguishing air regions at different distances from the recording equipment.

4. WIND PROFILING METHOD

To profile the wind speed in an arbitrary section of the atmosphere, the observation path between the object and receiving optical system is divided into N segments; the contribution of every segment is considered similar to the contribution of an infinitely thin screen. The algorithm for wind speed profiling at the object observation path is the following:

- the number N of layers of turbulent inhomogeneities, for which the drift velocity is to be calculated, is determined;

- the smoothing parameters $a_{filtr}^{(i)} = \frac{A}{z_{ts}^{(i)} / z_{obj}}$ are calculated for each ith inhomogeneous layer centered in the plane $z_{ts}^{(i)}$,

- the image quality map $Q(\mathbf{r}, a_{\min}, t) = \nabla_{\mathbf{r}}^2 I(\mathbf{r}, t) \otimes K(\mathbf{r}, a_{\min})$ is calculated, which allows filtering only the object structure and remaining the information about distortions. The convolution kernel can be defined as $K(\mathbf{r}, a) = \exp\left(-r^2 / a^2\right)$;

- the function $Q(\mathbf{r}, a_{filtr}^{(1)}, t) = \nabla_{\mathbf{r}}^2 I(\mathbf{r}, t) \otimes K(\mathbf{r}, a_{filtr}^{(1)})$ is calculated, which provides for filtering small-scale inhomogeneities introduced by turbulent inhomogeneities located farther than the first layer;

- the drift velocity of the first layer $\mathbf{V}_1$ is calculated by the equation $\mathbf{V}_i(t) = \frac{\Delta \mathbf{r}(t)}{\Delta t} \cdot \frac{z_i}{z_{im}}$, where $\Delta \mathbf{r}(t) = \arg\max\left[R(\mathbf{r}, t)\right]$,

$R(\mathbf{r},t)=Q(\mathbf{r},a_{filtr}^{(i)},t)\otimes Q(-\mathbf{r},a_{filtr}^{(i)},t+\Delta t)$ is the correlation function of the image quality maps at different time points;

- the difference $\Delta Q^{(1)}(\mathbf{r},t)=Q(\mathbf{r},a_{\min},t)-Q(\mathbf{r},a_{filtr}^{(1)},t)$ is calculated, which allows filtering distortions introduced by the first layer of turbulent inhomogeneities;

- the function $Q(\mathbf{r},a_{filtr}^{(2)},t)=\Delta Q^{(1)}(\mathbf{r},t)\otimes K(\mathbf{r},a_{filtr}^{(2)})$ is calculated to filter the small-scale distortions introduced by turbulent inhomogeneities located farther than the second layer;

- the drift velocity of the second layer $\mathbf{V}_2$ is calculated;

- the difference $\Delta Q^{(2)}(\mathbf{r},t)=\Delta Q^{(1)}(\mathbf{r},t)-Q(\mathbf{r},a_{filtr}^{(2)},t)$ is calculated;

- the last steps are repeated until $\mathbf{V}_N$ is calculated.

This algorithm allows serial filtration of distortions introduces by each layer of turbulent inhomogeneities (phase screen), beginning from the layer the closest to the observer toward the object. It should be noted that there are inhomogeneities on different scales in each turbulent layer. It is evident that imaging through a real atmosphere can be subjected to distortions on a certain scale induced by air inhomogeneities located at different distances from the observer. In this case, when using the method suggested, distortions introduced by different turbulent layers are filtered only partially.

For more rigorous filtration of the distortions of near turbulent layers, an additional parameter can be used, i.e., the angular size of the image region to be analyzed. We suggest reducing this region to sizes of about the characteristic size of the distortions which are to be filtered. This reduction also allows the spatial averaging of the measured drift velocity of distortions over the whole area of the object observed and, thus, a decrease in the random error of wind speed calculated along the observation path.

5. COMPARISON OF THE RESULTS OF PROCESSING OF VIDEO SEQUENCE WITH CONTACT MEASUREMENTS OF WIND VELOCITY

The efficiency of the method suggested was estimated during processing of a video image of a real object. Shooting was conducted with a frequency of 100 Hz. The accuracy of wind speed retrieval by a passive optical method is determined from the comparison with measurements of ten acoustic anemometers, equispaced along a 500-m atmospheric path. Figure 3 shows the object image and its sharpness functional. The upper part of the image (a fragment of deciduous forest) was used during the processing.

The wind velocity strongly changed along the observation path during imaging. Therefore, to weakens the effect of the random measurement error, the path was divided into three segments, within which the wind velocity value averaged over three (or four) anemometers was found.

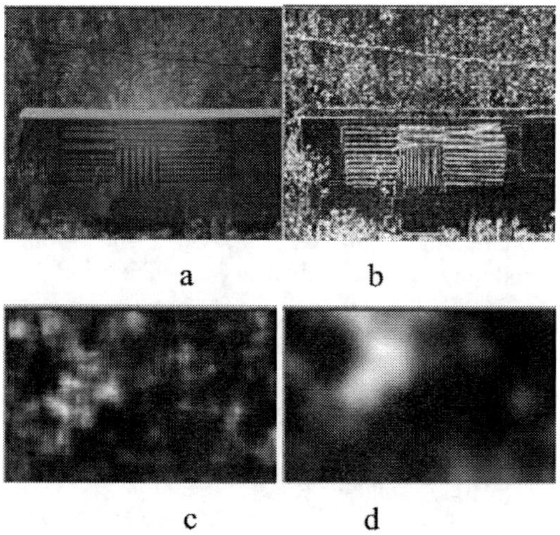

Fig. 3. Object image (a), its sharpness functional (b) and quality maps (c-d). The horizontal size of the region observed is 5 m; a_f=5 cm, the distance z_{obj} =500 m, and Δt = 10 ms.

Figure 4 shows that the wind speed retrieved in the nearby layers of turbulent inhomogeneities is in a good agreement with the wind speed contact measurements averaged over three acoustic anemometers.

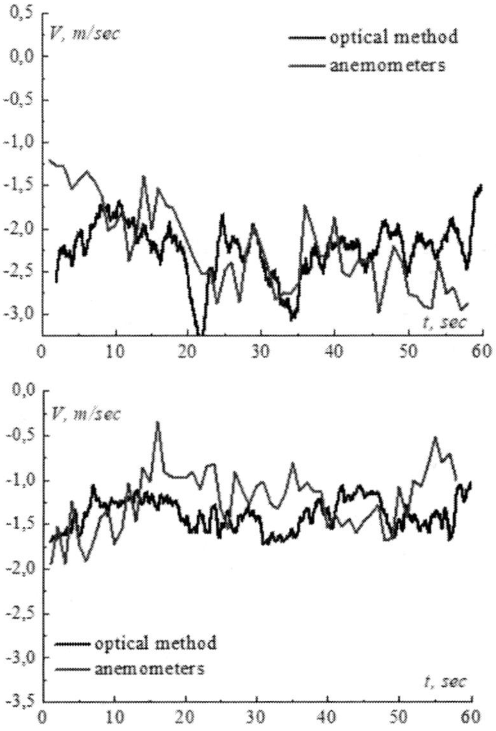

Fig. 4. Dynamics of the (a) retrieved wind speed near the observer (black curve) and (b) wind speed averaged over the central path segment in comparison with contact measurements by 10 acoustic anemometers equispaced along the path (red curve).

4. CONCLUSIONS

The comparison of the results of the algorithm developed by us showed a good agreement with contact measurements. The wind velocity is retrieved with the minimal error near the receiving optical system. Retrieval of wind speed in the atmospheric region near the object is problematic due to the small scale of the distortions introduced by it. To increase the range of crosswind determination by the method suggested, it is necessary to increase the camera resolution.

ACKNOWLEDGEMENTS

The work was supported by the Russian Science Foundation (project No. 18-79-00179).

REFERENCES

[1] Holmes, R.B., "Passive optical wind profilometer,". US patent US005469250A, (1995).
[2] Belenkii M., "Passive crosswind profiler." US patent 0128136 A1, (2010).
[3] Lukin, V. P., Lavrinov, V. V., Botugina, N. N.,Emaleev, O. N. and Nosov, V. V., "Differential turbulence and wind velocity meters," Proc. SPIE 6733, 67330N (2007)
[4] Banakh, V. A. and Marakasov, D. A., "Wind velocity profile reconstruction from intensity fluctuations of a plane wave propagating in a turbulent atmosphere," Opt. Lett. 32(15), 2236–2238 (2007).
[5] Varshneya, D., Griggs, S. and Jeffers, L., "Electro-optic system for crosswind measurement," US patent US9157701B2, (2013).
[6] Porat, O., Shapira, J., "Crosswind sensing from optical-turbulence-induced fluctuations measured by a video camera," Appl. Opt. 49(28), 5236-5244 (2010).
[7] Dudorov, V. V., Eremina, A. S., "Determination of atmospheric turbulent inhomogeneity wind drift from sequence of incoherent images," Proc. SPIE 9292, 92921F (2014).
[8] Dudorov, V. V., Eremina A.S., "Filtration of optical image distortions for retrieving the drift velocity of atmospheric turbulence inhomogeneities," Proc. SPIE 9680, 96802E (2015).
[9] Dudorov, V. V., Eremina A.S., "Possibilities of crosswind profiling based on incoherent imaging," Proc. SPIE 10035, 100351Q (2016).
[10] Dudorov, V. V., Eremina, A. S., "Retrieval of crosswind velocity based on the analysis of remote object images: Part 1 — drift of a thin layer of turbulent inhomogeneities," Atmospheric and Oceanic Optics 30(5), 422-428 (2017).
[11] Dudorov, V. V., Eremina, A. S., "Retrieval of crosswind velocity based on the analysis of remote object images: Part 2—Drift of turbulent," Atmospheric and Oceanic Optics 30(6), 596-603 (2017).
[12] Dudorov, V. V., Eremina A.S., "Visualization of the wind drift of turbulent inhomogeneities," Proc. SPIE 10787, 1078708-1-9 (2018).

A METHOD OF MEASURING FABRY-PAROT ETALON TRANSMITTANCE BY USING FREQUENCY COMB LASER

Yufei Chu[1, 2], Dong Liu[1], Decheng Wu[1], Zhenzhu Wang[1], Qian Deng[1, 2], Zhiqiang Kuang[1, 2], Lu Li[1, 2], Peng Zhuang[1, 2] Zhiyuan Fang[1, 2], Chenbo Xie[1, 2], Yingjian Wang[1, 2*]

[1]*Key Laboratory of Atmospheric optics, Anhui Institute of Optics and Fine Mechanics, Chinese Academy of Sciences, Hefei, Anhui 230031, China*
[2]*University of Science and Technology of China, Hefei, Anhui 230031, China*
**Email: wyj @aiofm.ac.cn*

ABSTRACT

This paper presents a method for instantaneous measurement of Fabry-Parot(F-P) transmittance curve. Compared with the previous method, this method has many advantages, such as scanning-free, simple in structure and accuracy.

1. INTRODUCTION

The F-P etalon is an interferometer which mainly consisting by two flat glass or quartz plates. It can be used in high-resolution spectroscopy as a high-resolution optical filter or a precision wavelength meter. In laser systems, it is often used to narrow the linewidth in the cavity or to make the laser system operate in a single mode. It can be used as the medium bandwidth control and tuning device in broadband picosecond laser. It also be widely used in the lidar system for Doppler wind measurement and aerosol detection. Due to the very wide application of the F-P etalon, the standard transmission curve of the F-P is a very important parameter. The general F-P etalon measures the transmittance curve by tuning the wavelength of the laser source or tuning the angle of incident light. The F-P etalon generally measures the transmittance curve of the etalon by tuning the wavelength of the laser source. Depending on the bandwidth of the etalon, the laser source with wavelength tuning function is very expensive, and the FP etalon of wide free spectral range is difficult to find a corresponding laser source with a wide range of wavelength tuning. Moreover, both the tuning of the wavelength and the angle of incident light are nonlinear, and the measurement accuracy is insufficient. This paper proposes a new method for detecting the transmittance curve of a F-P etalon using a frequency comb laser source. Due to the equal spacing of the frequency of the frequency comb laser, the transmittance curve of the F-P etalon can be measured at one time. And the frequency interval and spectral range are tunable, which can avoid the above defects. This new method greatly reduces the cost of measuring transmission curve, improve measurement accuracy and effectiveness and has great theoretical and practical value.

This paper introduces the working principle of F-P etalon and common transmittance measurement method, then introduce the characteristics of the frequency comb laser source, finally introduce the method of measuring F-P transmittance based on frequency comb laser and its simulation results.

2. METHODOLOGY

2.1 Working principle of F-P etalon and common transmittance measurement method

A traditional etalon consists of two reflective surfaces that are parallel to each other. As shown in Figure 1, the amplitude reflectance of the two reflective surfaces are R_1 and R_2, respectively, and the refractive index of the medium between the two reflective surfaces is n, the thickness of the medium is h.

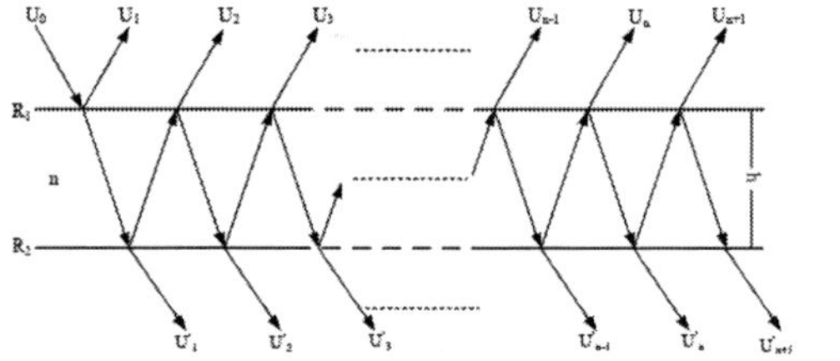

Figure 1 Principle of etalon transmittance

Transmittance of F-P etalon is defined as[1]:

$$T = \frac{U'U'^*}{U_0U_0^*} = \frac{(1-R_1)(1-R_2)}{(1-R_1R_2)^2 + 4\sqrt{R_1R_2}\sin^2(\frac{\delta}{2})}$$

When the amplitude reflectance of the two surfaces of the etalon is equal, that is, $R_1 = R_2$, and the intensity reflectance of each surface is $R = R_1^2$. The transmittance of the etalon can be simplified as follows

$$T = \frac{1}{1 + \frac{4R}{(1-R)^2}\sin^2(\frac{\delta}{2})}$$

Which, $\delta = \frac{4\pi nh\cos\theta}{\lambda}$, from the expression of T and δ, we can get that the transmittance is related to the length of the cavity, the wavelength and the incident angle, etc. The transmittance curve can be obtained directly by changing the wavelength, or the transmittance change caused by the change of the corresponding wavelength can also be deduced by changing the length of the cavity or changing the incident angle[2]. However, no matter the angle or the length of the cavity or the wavelength of the laser is changed, there is no guarantee that the change is linear, which will introduce errors. Next section we will introduce a frequency comb laser source that can provide multiple frequency components at a time.

2.2 The characteristic of the frequency comb laser source

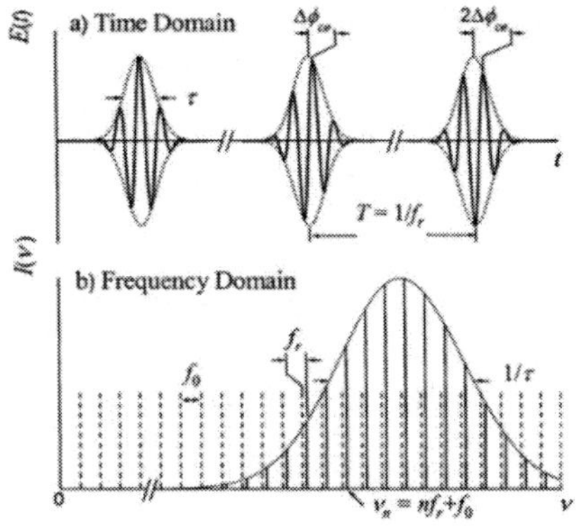

Figure 2 Standard comb: time (upper) and frequency (down) domain

In optics, a frequency comb is a laser source whose spectrum consists of a series of discrete, equally spaced frequency lines. Frequency combs can be generated by a number of mechanisms, including periodic modulation (in amplitude and/or phase) of a continuous-wave laser, four-wave mixing in nonlinear media, or stabilization of the pulse train generated by a mode-locked laser. Much work has been devoted to the latter mechanism, which was developed around the turn of the 21st century and ultimately led to one half of the Nobel Prize in Physics being shared by John L. Hall and Theodor W. Hänsch in 2005. [3、4]

The time and frequency characteristics of ultrashort laser pulses are shown in the figure 2. Because the spectrum is very similar to the comb in daily life, it is called "optical frequency comb" [5]. The frequency comb light source is more expensive, so we can replace it by a similar frequency comb laser source[6].

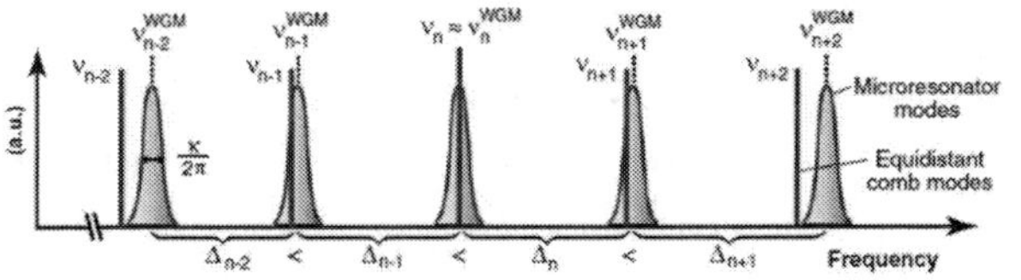

Figure 3 Whispering-gallery-mode Laser

The whispering gallery mode[7] is a typical similar frequency comb laser source which has multiple frequency components, but not equal intervals, as shown in Figure 3.

2.3 Measuring F-P transmittance with frequency comb laser

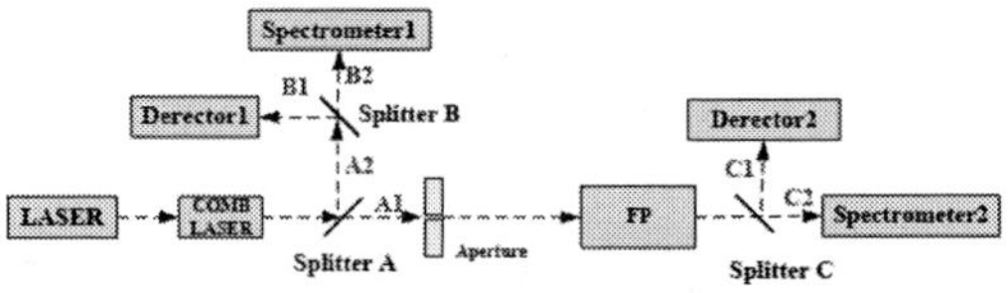

Figure 4. Schematic of measuring FP transmittance curve using frequency comb light source

The seed laser generates a variety of frequency components through a frequency comb laser, and passes through splitter A, then beam would be divided into beam A1 and A2. The beam A2 is divided into beam B1 and B2 by splitter B. The beam B2 is detected by the spectrometer1 and the beam B1 is received by the detector1. The beam A1 pass through the F-P etalon, and is divided into beam C1 and C2 by splitter C. Beam C2 is detected by the spectrometer2 and beam C1 is received by the detector2. By comparing the

https://doi.org/10.1051/epjconf/202023707001

relative changes of the corresponding frequency components before and after the F-P, the transmittance of the corresponding frequency can be obtained, and the transmission curve can be obtained by connecting all the frequency components.

3. RESULTS

3.1 Parameters of the F-P etalon

The main technical parameters of solid state etalon for aerosol detection are given in the table 1.

Table 1: Main parameters of FP etalons

Parameters	Index
Wavelength(nm)	532
Refractive index	1.461
Beam divergence full angle(mrad)	8
Thickness(mm)	0.1311
Optical aperture(mm)	40
Effective aperture (mm)	30
Reflectivity	*86%*
Surface accuracy @633nm)	λ/100
Fine number	17.9
Peak transmittance	0.860
FWHM(pm)	41.3
Center wavelength (nm)	532.12
Free spectral range (pm)	739

From the parameters in the table, we can get Numerical simulation results of etalon transmittance curves:

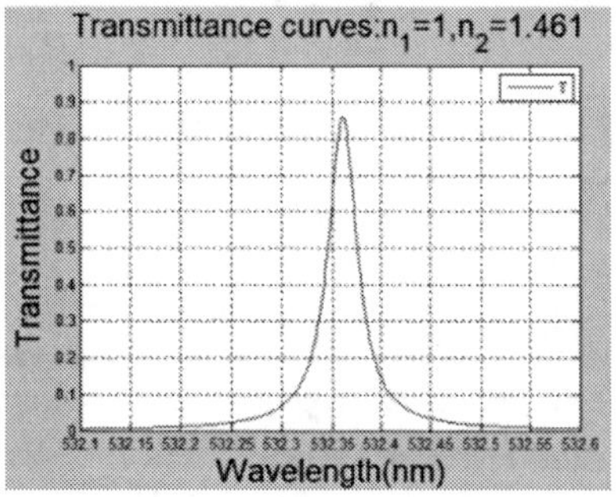

Figure 5 Numerical simulation results of solid state etalon transmittance curves

From the above numerical simulation results, when the thickness of the etalon is 131um and the two optical surfaces are plated with 86% of the reflective film, the center wavelength of the etalon is 532.37nm, and the peak transmittance is 86%.

3.2 Simulation results

We use the method introduced in section 2.3 to test the passing rate curve of F-P.

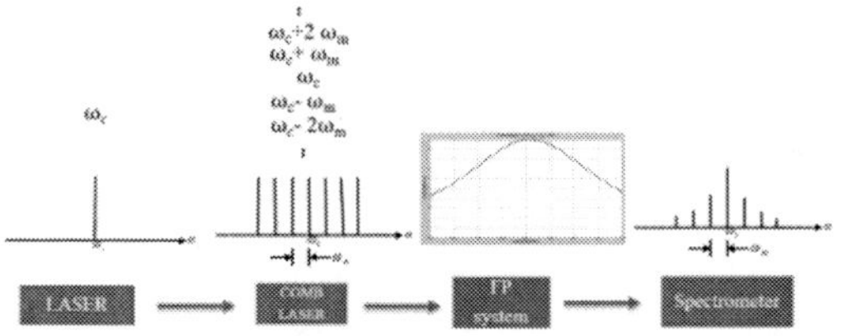

Figure 6 Numerical simulation results of solid state etalon transmittance curves

As we know in section 2.1, the different frequency components have different transmission rates for F-P etalon. According to the different frequency components in Figure 4, When the different frequency components pass through the F-P etalon, the rate of change of frequency will be consistent with the rate of transmission.

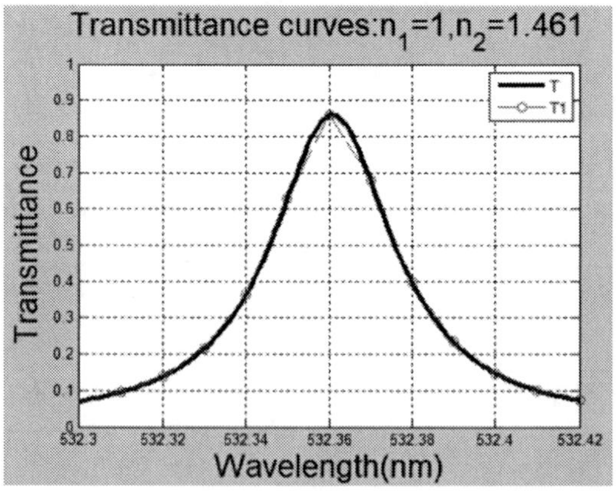

Figure 7 Theoretical transmittance curve comparison

T represents the theoretical pass rate curve, and T1 represents the transmittance curve obtained by the method described in this paper. As can be seen from the figure, the match is very good.

ACKNOWLEDGEMENTS

This work was supported by the Youth Innovation Promotion Association CAS (2017482) and by the general propgram of National Natural Science Foundation of China (41875033). Key Program for External Cooperation of Bureau of International Cooperation, Chinese Academy of Sciences (GJHZ201976) is also acknowledged.

REFERENCES

[1] Shen F H., et al. Wind lidar based on Fizeau interferometer[J]. Infrared and Laser Engineering, 2006, 35(6): 691-695 .
[2] Sun D S., et al. Performance analysis of direct detection Doppler lidar for wind measurement[J]. Infrared and Laser Engineering, 2003, 32(2): 115-117
[3] Hall J L. Nobel Lecture: Defining and measuring optical frequencies[J]. Review of Modern Physics, 2006, 78(4):1279-1295.
[4] Hänsch T W. Nobel Lecture: Passion for precision[J]. Review of Modern Physics, 2006, 78(78):1297-1309.
[5] Ye J, Cundiff S T. Femtosecond Optical Frequency Comb: Principle, Operation, and Applications[M]. 2005.
[6]Chu YF., etal. Fluorescence radiation characteristics based on evanescent wave pumping in a microfluidic chip [J]. Acta Physica Sinica, 2017(10):156-162.
[7] Zhang Y X., et al. Threshold characteristics of evanescent-wave pumped whispering-gallery-mode fiber laser [[J]. Acta Physica Sinica, 2009, 58(5).

ATMOSPHERE ACTIVITY MEASUREMENT BY LED RAMAN MINI LIDAR

Tatsuo Shiina

Graduate School of Engineering, Chiba University, Chiba-shi, Chiba 263-8522, Japan
**Email: shiina@faculty.chiba-u.jp*

ABSTRACT

The LED mini lidar was improved to monitor Raman scattering echoes. The Raman scattering signal indicates a certain gas concentration and it can distinguish the target from the other materials. It is so weak, 1/1000 of Mie scattering echoes, but even enough to be stimulated from the target gas with LED pulsed beam. At first, we developed a compact Raman lidar with micro pulse DPSS laser to detect hydrogen gas quantitatively. We replaced it with the LED pulse module, which was calculated enough potential to stimulate Raman scattering and detect the target gas. The next task is the activity measurement of such a target gas. To validate the potential of the LED Raman lidar, we conducted the sea surface atmosphere measurement. As a result, the unique relationship between the surface atmosphere and sea water echoes was observed. In this report, we state the concrete specification of the LED mini Raman lidar and some results of the activity observations.

1. INTRODUCTION

The Raman scattering light has an information of the target gas, which is characteristic Raman shift with the target itself, and we can distinguish the target from the other gases [1]-[2]. It is effective measurement for a certain gas which has no absorption wavelength in visible and near infrared light region [3]-[4]. We have focused on the hydrogen gas detection with a compact Raman lidar with a DPSS laser [5]-[7].

The leakage detection of hydrogen gas is important for the hydrogen gas station and fuel cell manufacture scene. To detect it on site, the lidar should be compact and bring it everywhere. The system has to act with a buttery and its transmitting beam intensity is lowered enough for safety to human eyes. Sometimes, the laser beam is not accepted in certain facilities, for example, in a nuclear power facility. Under such backgrounds, we developed a LED mini Raman lidar, which was replaced the light source of the DPSS laser to a LED pulse module. We calculated the adequate optical power to detect the low percentage hydrogen gas and fixed the concrete lidar design. It could have successfully accomplished the quantitative measurement of the hydrogen gas concentration within 1 – 100% [8]-[9].

The next task is the activity monitoring of such a target gas. To examine the precise concentration, the measurement has to follow the target gas flow. That is, the lidar should evaluated its echo within a higher sampling rate. To validate the potential of the LED Raman lidar, we conducted model experiments with a moving target; the sea surface atmosphere measurement. In this report, we state the concrete specification of the LED mini Raman lidar and some results of the atmospheric activity observations.

2. LED MINI RAMAN LIDAR

At first, we had calculated the minimum detectable values of Raman scattering intensity referring to the DPSS laser based compact Raman lidar. The verification test was conducted by detecting the actually lowered Raman echo of hydrogen gas. The laser beam is entered into the quartz cell filled with hydrogen gas. The concentration was 100%. As the laser power was gradually decreased up to the detection limit, which is estimated by the numerical analysis. The estimated detection limit was obtained by the transmitting pulse power of 1W with the pulse width of 10ns. The experimental result coincided with the calculation.

Considering the above results, we fixed the design of the LED based mini Raman lidar. The lidar optics was biaxial type and its optical paths to distinguish the gases were as same as the laser based

Table 1 Raman scattering from each molecule.

Target	Raman Shift(cm^{-1})	Wavelength (nm)	Scattering Cross section ratio
N_2	2331	400	1
Water	3440	420	-
Water Vapor	3652	415	2.18
H_2	4160	430	3

 https://doi.org/10.1051/epjconf/202023707002

compact Raman lidar. The detecting targets here was the atmospheric nitrogen, water, water vapor and hydrogen gas. Their Raman shifts and its wavelengths against the incident LED beam of 365nm were summarized in Table 1 [10]. The LED lamp module had the center wavelength of 365nm and its spectrum width was 10nm. And the wavelengths of Raman scattering echoes become 400nm for atmospheric nitrogen, 415nm for water vapor, 420nm for water, and 430nm for hydrogen gas. Water Raman channel was used to calibrate the water vapor signal in lab-experiment. LED beam is not easy to collimate in comparison with the laser beam, while it never focuses after the transmission. It becomes benefit in the closed space for the safety [11]. The diameter of the transmitting beam was 50mmϕ and its divergence was about 10mrad. The LED pulsed power was <1W (=10nJ/10ns). The pulse repetition frequency was improved to >500kHz. The receiving telescope was Cassegrain type and its aperture is 127mmϕ. The receiver's FOV was restricted to 3mrad.The system was so compact and handy that one can use it in any observation scenes. The system power consumption was about 2W and could be act with a small PC battery in 10 hours. The specification of the LED mini Raman lidar optics was summarized in Table 2 and its optics was illustrated in Fig.1.

Table 2 LED mini Raman lidar specifications.

Transmitter	
LED Pulse power	750mW(7.5nJ/10ns) (NCSU033C Nichia Corp.)
Wavelength	365nm
Beam Diameter	50mmϕ
Beam Divergence	10mrad.
Repetition Frequency	>500kHz
Receiver	
Telescope	Cassegrain type
Diameter	127mmϕ
Field of View	<5mrad.
Interference Filters	$\Delta\lambda = 10nm$
Detector	PMTs (R6350P, Hamamatsu Photonics K.K.)

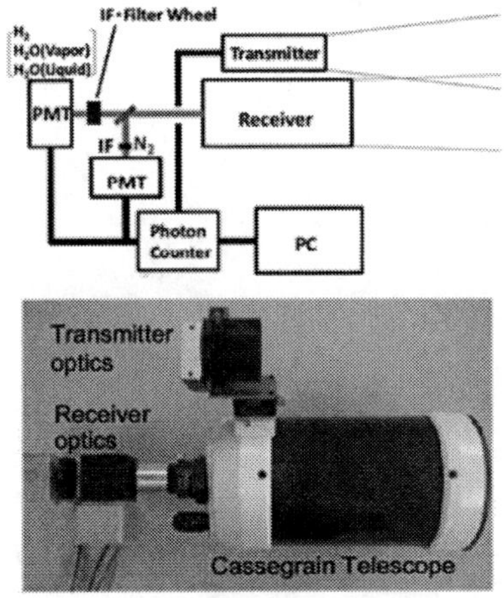

Fig. 1 Optical setup of LED mini Raman lidar.

3. FUNDAMENTAL OBSERVATION

To catch the target gas flow activity, the lidar should evaluated its echo within a higher sampling rate. The high-speed and high-resolution photon counter was developed for this purpose. It can follow the pulse repetition frequency of >500 kHz. Its time resolution is >5ns. It is equivalent to the spatial (distance) resolution of >0.75m. It has 4 channels, and their simultaneous measurements are possible with a single input or output trigger signal [12].

Figure 2 shows the results of the atmospheric nitrogen gas and the water vapor measurement. It was conducted in nearly horizontal direction in night time. Accumulation was 3.5 minutes. Nitrogen echo was obtained up to 70m (Fig.2(a)). After that the fluorescence echo of the tree was also detected at 80m.

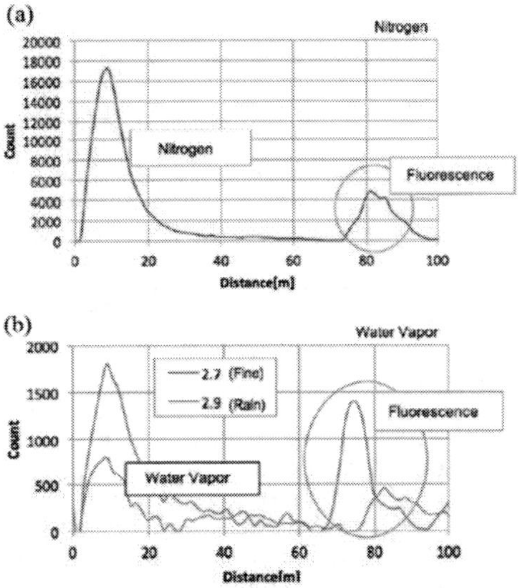

Fig.2 Atmosphere monitoring with nitrogen and water vapor echoes. (a)Nitrogen Raman echo (b)Water vapor Raman echo.

As the tree leaves scattered the LED beam partially, their echoes were observed over about 20m. The atmospheric nitrogen was stable and its echo waveform became smooth. On the other hand, the water vapor echo fluctuated because of low counts (Fig.2(b)). It was drastically changed by depending on the humidity. In the figure, the humidity was 25% on Feb. 7 in fine day, while it was 90% on Feb. 9 in rain day. The observation range of the water vapor was 70m, which is equal to that of the nitrogen echo. The fluorescence echoes of trees were detected, too. With the enough summation time, the signal-to-noise ratio improves, while the spatial and temporal activity of the target gas flow are compromised.

4. SEA SURFACE OBSERVATION

We conducted the sea surface measurement to verify its activity visualization. Figure 3 shows the concept. The lidar fixes on the top of sea wall or the bow of fisher boat. And the lidar is trained on the surface atmosphere with a shallow angle. The beam divergence is wider than the laser beam, and echo can detect from a few meters to 300m. Thanks to the shallow angle measurement, the interaction between the surface atmosphere and sea water can be captured. With a certain shorter summation time, their activity is visualized, too. The mini Raman lidar was brought to the coast and set with shallow angle. The location was Inage coast, Chiba, Japan on 8th Dec. 2018. At that time, the weather condition was calm and wind speed was 1.5m/s.

The Figure 4 shows the observation results of the atmospheric nitrogen and water vapor echoes. Every echo was stored at 0.5s interval. The horizontal axis represents the time and the vertical axis was distance. The echo counts represented with color. One can see that the echoes appeared on 15 – 25m on nitrogen echo and 20 -30m on water vapor echo, respectively. That is, the nitrogen echo is nearer than the water vapor echo. The hair line like echoes indicates the individual sea waves. From the analysis of them, it is confirmed that the wavelength and speed of sea wave were 8m and 2.2m/s, respectively, and its period was 3.67s. The height of the wave was calculated as 2.7m by considering the shallow angle orientation. As the statistical consideration, the captured waves on the atmospheric nitrogen echo and the water vapor echo were related each other. Figure 5 indicates the one echoes' A-scope graphes on the images in Fig. 4. It is of interest that

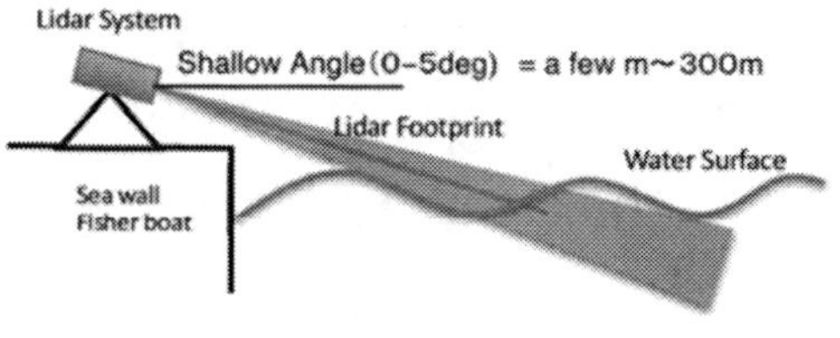

Fig.3 Shallow angle observation with LED Raman lidar.

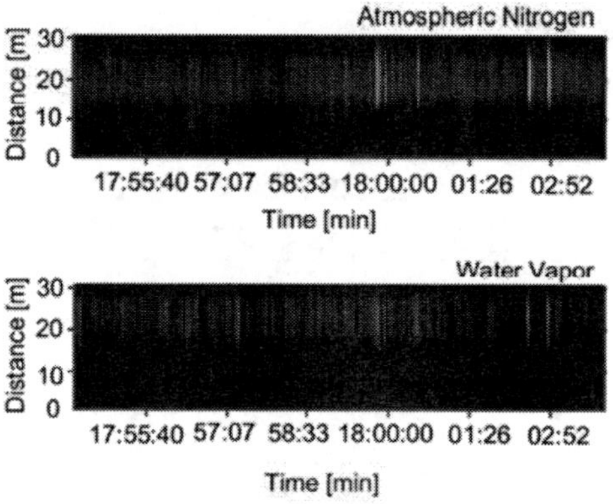

Fig.4 Sea surface atmosphere and water echoes at Inage coast, Chiba, Japan on 28th Dec. 2018.

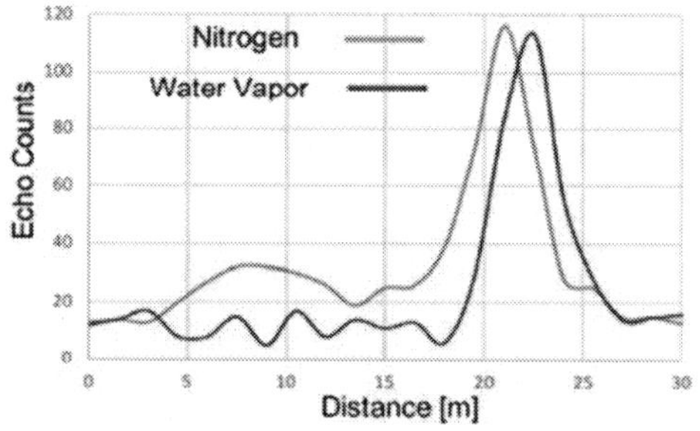

Fig.5 Interaction between sea surface atmosphere and water echoes at 17:59:57 on 28th Dec. 2018.

https://doi.org/10.1051/epjconf/202023707002

Both of waveforms on Nitrogen echo and Water vapor echo had the same width. It indicates their synchronized motions, too. The stronger intensity echoes were synchronized each other as shown in Fig.4, while the weak intensity echoes of the atmospheric nitrogen had a certain delay against the motion of the water vapor echoes. Figure 6 shows such peak motions on the atmospheric nitrogen and the water vapor echoes in Fig.4. It seems that the nitrogen echoes fluctuate on their delay. It suggests that the sea wave moves with the atmosphere shuffle. The sea wave motion reflects the wind condition, water flow and tide. The bottom feature of the sea also affects the sea wave especially at a shallow. It is confirmed that they were represented on the Raman lidar echoes due to the sea and weather conditions.

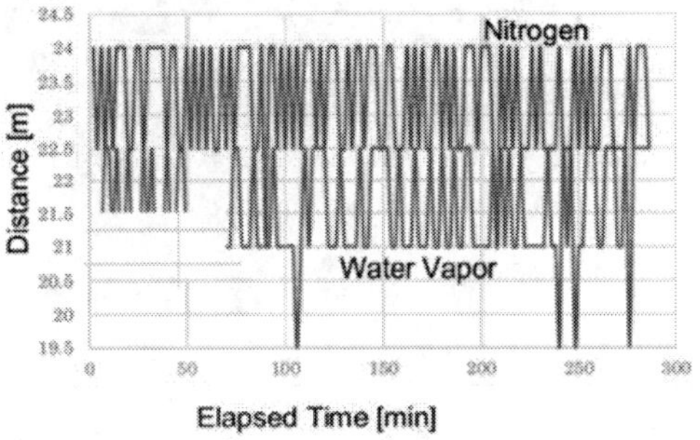

Fig. 6 Peak motions on the atmospheric nitrogen and the water vapor echoes in Fig.4.

5. SUMMARY

The LED based mini Raman lidar was proposed and improved. In this study, we focused on the target gas activity. To visualized it, the observation was conducted with the shorter summation time and higher resolution. The atmospheric nitrogen and water vapor were aimed as the targets. The unique behavior of the relationship between the atmospheric nitrogen and water vapor on the sea surface could be successfully visualized.

We have already spread the target to a certain gas as hydrogen and methane gas. Furthermore, the artificial twister was utilized to monitor its activity, too. These results analyzed in view point of quantitative measurement for concentration calculation and activity visualization. It will lead to the true concentration measurement on the temporal and spatial change.

ACKNOWLEDGEMENTS

This work was supported by JSPS KAKENHI Grant Number JP16H03129.

REFERENCES

[1] D. A. Leonard, *Nature* **216**, 142-143, 1967.
[2] S. H. Melfi, , *Appl. Opt.* **11**, 1605-1610, 1972.
[3] H. Gu, Z. Wang and Y. Hu, Sensors, **12**, 5517 (2012).
[4] X. Q. Zeng, M. L. Latimer, Z. L. Xiao, S. Panuganti, U. Welp, W. K. Kwok, and T. Xu, Nano letters, **11**, 262 (2011).
[5] H. Ninomiya, S. Yaeshima, K. Ichikawa, and T. Fukuchi, Opt. Eng. **46**, 094301 (2007).
[6] H. Miya, T. Shiina, T. Kato, K. Noguchi, T. Fukuchi, I. Asahi, S. Sugimoto, H. Ninomiya, Y. Shimamoto, CLEO/Pacific Rim, (2009).
[7] T. Fukuchi and T. Shiina Eds., *Industrial Applications of Laser Remote Sensing*, Bentham Science Publishers Ltd (2012).
[8] T. Shiina, K. Noguchi, and K. Tsuji, *Proceedings of 23th international conference on nuclear engineering (ICONE23)*, 17 (2015).
[9] Djezzar, B. (Ed.) *Ionizing Radiation Effects and Applications*; IntechOepn: Chapter 3, London, UK (2018).
[10] R. M. Measures, *Laser Remote Sensing; Fundamentals and Applications*, John Wiley & Sons, New York (1984).
[11] T. Shiina, Sensors, **19**, 569(2019)
[12] Britun, N.; Nikiforov, A. (Eds). *Photon Counting - Fundamentals and Applications*, IntechOpen: Chapter 11, London, UK (2018).

BECOOL: A BALLOON-BORNE MICROLIDAR SYSTEM DESIGNED FOR CIRRUS AND CONVECTIVE OVERSHOOT MONITORING

François Ravetta[1*], Vincent Mariage[1], Emmanuel Brousse[1], Eric d'Almeida[1], Frédéric Ferreira[1], Jacques Pelon[1] , Stéphane Victori[2]

[1] *Laboratoire Atmosphère Milieux, Observations Spatiales (LATMOS), CNRS-INSU, Sorbonne Université,Université Versailles St Quentin, Paris, France*
[2] *Cimel Electronique, Paris, France*
**Email: francois.ravetta@latmos.ipsl.fr*

ABSTRACT

A balloon-borne microlidar has been built at LATMOS to monitor cirrus optical properties and convective overshoot topography during long duration flights in the lower tropical stratosphere. Weighting less than 7 kg in a reduced volume and consuming less than 10 W, it will be involved in the CNES-Strateole2 campaign. This instrument paves the way to the use of microlidar technology for planetary missions.

1. Scientific context and objectives

The CNES project Strateole2 aims at flying different configurations of multi instrumented open stratospheric balloons (OSB) in the lower tropical stratosphere to study the tropical tropopause layer (TTL). Long duration (several months) flights are planned to improve our knowledge of air-masses transport and dehydration processes. Relationship between temperature anomaly and cirrus occurrence has already been highlighted thanks to the Airborne Tropical TRopopause EXperiment (ATTREX) NASA campaign [1], with the Cloud Physics Lidar [2] onboard an unmaned aircraft system. The BeCOOL (Balloonborne Cirrus and convective overshOOt Lidar) system was conceived to be one of the payloads of the OSB configurations (see Fig.1). This profiler will measure optical properties of cirrus in nadir viewing mode and characterize the upper topography of convective system. The payload of the flight configuration also involves a GPS radio-occultation [3] to reveal temperature anomalies and a bolometer [4] to measure upward longwave and shortwave radiative fluxes. The goal is to better understand modulation of cirrus properties by gravity waves [5], and to better assess the radiative heating [6] and forcing [7] of the lower stratosphere by cirrus and water vapor anomalies.

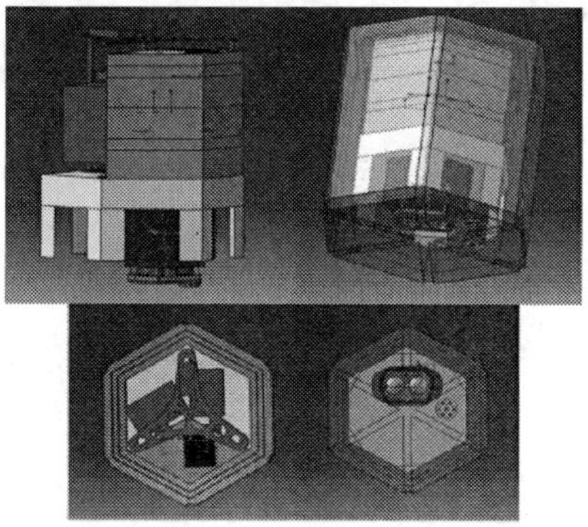

Fig. 1 3D drawing of the BeCOOL microlidar in the STRAT configuration gondola. The insulated hexagonal shaped boxes used to contain parts of the instrument are shown in brown. They are located in the same volume as the batteries and other instruments. The telescope with its two lenses can be seen in the lower "cold" volume of the gondola.

2. Instrumental development

The BeCOOL system is based on the biaxial microlidar architecture developed for the French Ice Atmosphere Arctic Ocean Observing System (IAOOS project) which aimed at deploying a network of multi-instrumented buoys in the Arctic region [8]. Its main characteristics (see Tab.1) are comparable to those of balloon-borne microjoule lidars first developed [9]. The use of optical fibers between the components of the microlidar eases the integration of the system, particularly to separate the telescope (located in the "cold" area of the gondola) and the thermally sensitive parts. A specific laser driver board and an onboard computer were developed by Cimel Electonique, respectively to handle the supply of the laser diode as well as its thermal regulation, and to deal with data acquirement and communication with the gondola. Given the balloon-borne constraints and stratospheric environment, a solution had to be found to put all the components in a very small volume. An appropriate insulation is made for the thermally sensitive parts like the laser diode and the interferential filter. The telescope is placed apart and operated in external air conditions. The weight was also decreased by removing some parts of the IAOOS telescope and using lighter materials. Numerous tests were performed in a test chamber at very low temperatures and low pressure to ensure a proper behaviour of all the components. The thermal regulation of the laser diode is particularly difficult to handle because of the lack of convection in the OSB.

Repetition rate/impulsion length	4.8 kHz/150 ns
Wavelength/FWHM	~802 nm/~0.3 nm
Impulse energy	10 µJ
Lens diameter (emission/reception)	7 cm
FOV	650 µrad
Filter FWHM	0.6 nm
Optical transmission (receiver)	10-15 %
Total mass	< 7 kg
Average power supply	< 10 W

Tab.1 Main characteristics of the balloon-borne lidar BeCOOL

3. First balloon-borne measurements

A first flight was successfully achieved within the STRATOSCIENCE campaign in August 2018, at Timmins, Canada. Flying at an altitude of 22 kilometers, the lidar has demonstrated its ability to observe cirrus at the tropopause and aerosol in the boundary layer during nighttime (see Fig. 2). The SNR is seen to decrease with the sunrise at the end of the flight, but cirrus are still well captured. The increase of SNR during the first 30 minutes is due to the self-heating of the laser diode emitting continuously.

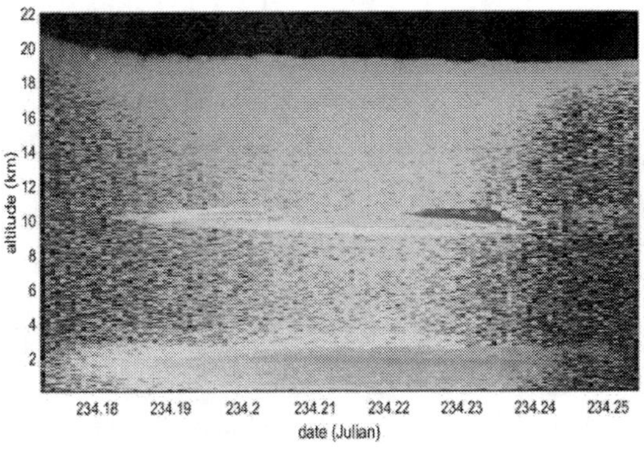

Fig. 2 Vertical cross section of the range-corrected attenuated backscattered signal (arbitrary unit) averaged over one minute for a leg of 2 hours LT on August 23 2018.

Conclusion and Perspectives.

The first Strateole2 field campaign will take place in winter 2019-2020. Beyond the scientific objectives of this CNES program, the development of this balloon-borne system is also an opportunity to increase the technology readiness level (TRL) of the microlidar technology so that, with further instrumental improvements, it may be implemented onboard a planetary rover or a small satellite.

Acknowledgments
We thank the French Centre National d'Etudes Spatiales (CNES) for its support to the Strateole2 program, the Canadian Space Agency for the flight at Timmins, Sylvain Meunier and Joseph Leclercq from Cimel for their contribution.

REFERENCES

[1] E. J. Jensen et al, Bull. AMS, 129-143 (2017)
[2] M. McGill et al., Appl. Opt.,(41) 3725-3734 (2002)
[3] J. S. Haase et al., Geophys. Res.L., 39, (2012)
[4] P. Zhu et al., Geos. Inst. Meth. and Data Systems, 89-98, (2015)
[5] J.-E. Kim et al, Geophys. Res. L. (43), 5895-5901 (2016)
[6] S. Davis et al., J.Geophys Res, 115 (2010)
[7] S. Solomon et al., Science (327), 1219-1223 (2010)
[8] V. Mariage et al., Optics Express 25 (4), A73-A84 (2017)
[9] G. Di Donfrancesco et al., Appl. Opt. (45) 5701-5708 (2006)

CHARACTERIZATION OF A NOVEL TEMPERATURE LIDAR RECEIVER BY MEANS OF LABORATORY RAYLEIGH-BRILLOUIN SCATTERING MEASUREMENTS

Jiaqi Xu[1, 2, *], Benjamin Witschas[1], Kun Liang[2], Yuanqing Wang[3], Wim Ubachs[3]

[1] *German Aerospace Center (DLR), Institute of Atmospheric Physics, 82234 Oberpfaffenhofen, Germany*
[2] *School of Electronic Information and Communications, Huazhong University of Science and Technology, Wuhan, Hubei, 430074, China*
[3]*Department of Physics and Astronomy, LaserLaB, VU University, 1081 HV Amsterdam, Netherlands*
**Email: jiaqi.xu@dlr.de*

ABSTRACT

A new receiver setup based on an imaging Fizeau interferometer measuring the Rayleigh Brillouin scattering (RBS) spectra to retrieve tropospheric temperature profiles is introduced. The optical layout of the proposed receiver design is discussed and its functionality is demonstrated by means of accurate and controlled laboratory RBS measurements. The proposed temperature receiver is based on a Fizeau interferometer and a 32 channel PMT array which can resolve the RBS spectrum without applying any scanning procedures. In order to avoid any incidence angle fluctuations on the Fizeau interferometer caused by for instance atmospheric turbulence, the application of a fiver coupled telescope was investigated. It is shown that a 50 µm fiber (NA=0.12) provides both, sufficient coupling efficiency and sufficiently low divergence in order to keep the Finesse of the interferometer. The performance of the novel lidar receiver is verified by means of accurate laboratory RBS measurements.

1. INTRODUCTION

Temperature is a key parameter in the atmosphere and plays a major role in many atmospheric processes. The global observation of temperature is important in predicting the atmospheric state, as well as retrievals of other atmospheric properties such as wind, pressure or other trace gas concentrations.

Currently, various in-situ and remote sensing technologies are available to obtain global temperature data. To get temperature profiles with both, high horizontal and vertical resolution and high accuracy, it would be beneficial to utilize the lidar technology. For instance, Raman lidar can be used to acquire temperature profiles with high vertical and temporal resolutions [1-3], and are discussed to be designable for airborne [4] and space-borne application [5]. However, due to the low Raman differential backscattering cross section, high-power lasers and large telescopes would be needed for this purpose

As the differential backscattering cross section is more than a factor of 50 larger than the one of rotational Raman scattering, it is beneficial to use Rayleigh scattering in order to derive temperature [6-8]. In particular, atmospheric temperature can be derived by resolving the RBS spectrum with high spectral resolution filters, such as atomic vapor cells [8] or Fabry–Pérot interferometers [6], and subsequently relating the measured spectrum to an appropriate line shape model. Hua et al. introduced a Rayleigh-Mie lidar for daytime-temperature profiling of the troposphere using a frequency tripled Nd: YAG laser (λ = 355 nm) and, for the first time, a Fabry Perot interferometer (FPI) [6]. They showed temperature profiles which fit well to radiosonde temperatures from ground to 4 km and calculated a temperature uncertainty of less than 1 K within this range. Witschas et al based on a Doppler wind lidar system to obtain the RBS spectrum. In order to do so, the frequency of the used laser was changed in defined steps over a range of 12 GHz (at 354.89 nm) and a Fabry-Perot interferometer was used as frequency discriminator. The sampling of the entire RBS line shape required 28 min, and the results show temperature profiles from 2 km to 15.3 km [9].

To further reduce the measurement time and stabilize the measurement procedure, it is suggested to sample the RBS spectrum without applying any scanning procedures. This can be realized by means of an imaging Fizeau interferometer and a multichannel photomultiplier

(PMT) array. By using such a setup, even a varying aerosol load would not negatively influence the temperature retrieval.

2 The improved new receiver system

The suggested temperature receiver optical design is sketched in Fig. 1.

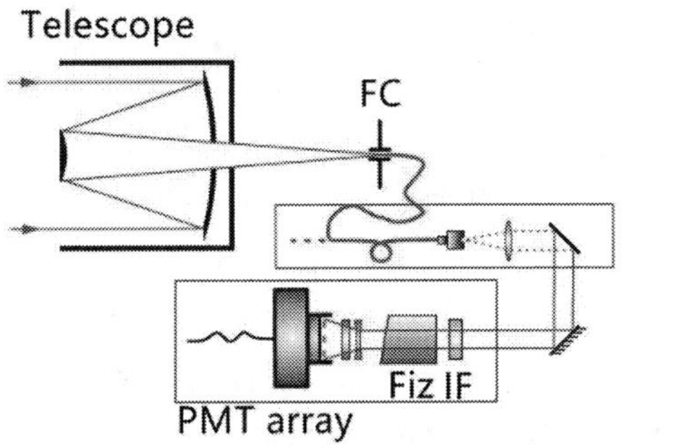

Fig. 1. Setup principle of an imaging high spectral resolution temperature receiver by using a Fizeau interferometer.

The specifications of the suggested temperature lidar receiver are listed in Table 1.

Table 1. Parameters of the Fizeau interferometer

Interferometer	
Type	Fizeau
Nominal plate distance/(mm)	19
FSR/(GHz)	8
Wedge angle/(μrad)	8.3
Plate reflectivity	0.80
Detector	
Type	32-channel PMT

The backscattered light from the atmosphere is collected by a telescope. A fiber coupler (Fig. 1, FC) is used to couple the backscattered light and direct it to the Fizeau interferometer. By using the fiber, any incidence angle fluctuations in the atmosphere are not transported to the Fizeau interferometer and thus, do not lead to any systematic error. After the fiber coupler, the beam diameter is adapted to match the aperture size (32 mm) of the Fizeau interferometer (Fig. 1, Fiz), then laser beam is sent to a narrowband interference filter (Fig. 1, IF) (bandwidth is 0.72 nm) to additionally remove background light and signal from Raman scattering. Finally, the RBS spectrum or rather the convolution of the RBS spectrum with the Fizeau transmission function is adapted by two cylindrical lenses in order to match the detector size of 1mm x 32 mm (32 channel PMT array). The PMT device is able to detect the very weak backscatter signal. With this setup. entire RBS spectrum can be detected at once without applying any scanning procedures.

2.2 Characterization of the receiver system by means of laboratory measurements

The basic idea of the suggested temperature receiver is to retrieve temperature from measured RBS spectra. This is possible as the RBS line shape depends among other quantities also on temperature. Basically, the RBS spectrum is an image of the velocity distribution of the molecules in the scattering medium, and therefore, it contains information on gas transport properties like heat capacity, thermal conductivity, shear and bulk viscosity as well as on temperature, density (pressure) and bulk velocity [13-15]. Laboratory measurements give one the possibility to perform the RBS measurements under controlled conditions regarding the aforementioned quantities.

3. Laboratory measurements for measuring RBS spectra with the lidar receiver

In order to investigate the feasibility of the designed receiving system, the controlled laboratory measurements are currently performed at the LaserLaB of VU University Amsterdam. A block diagram of the Laboratory experimental setup used to measure RBS spectrum in air is shown in Fig. 2.

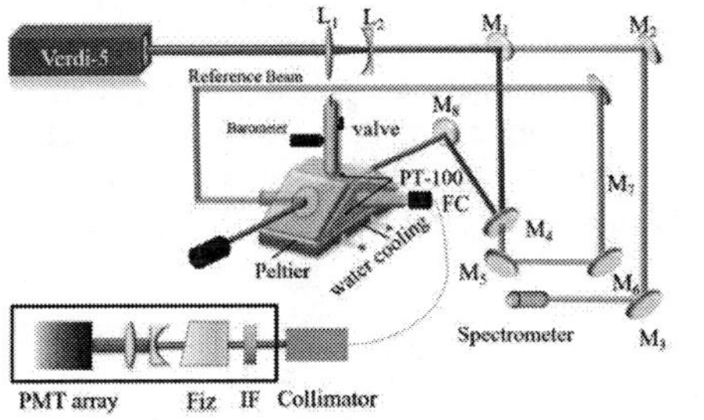

Fig. 2. Schematic diagram of the experimental setup. Dark green: laser beam (532nm), light green: reference beam (532nm), SC: scattering cell with Peltier and

water-cooling to heat & cool, Barometer: monitor the pressure, PT-100: monitor the temperature, FC: fiber coupling, Collimator: collimating laser beam, IF: narrowband filter, Fiz: Fizeau interferometer.

A Verdi-V5 laser provides continuous wave light at 532.22 nm at a power of 5 Watt and bandwidth less than 5 MHz. The laser light is split into two beams: The pump beam crosses the RB-scattering gas cell whereas the temperature and pressure are controled and measured with an accuracy of 0.15% (pressure) and ± 0.2 K (temperature). The light scattered with a angle of $\theta = 124.2°$ is collected an directed to an optical fiber. The small fraction reference beam transmitted through M4 is used to align the beam path after the gas cell towards the detector. The scattered light is coupled to a fiber then after the narrowband interference filter (bandwidth is 0.72 nm) (Fig. 2, IF) to remove background light and signal from Raman scattering, then it is analyzed the Fizeau Interferometer (Fig.2 Fiz and directed to the PMT array.

During the measurements, we found that the coupling efficiency of single mode fiber is too low for receiving scattered light, and thus, a multimode (MM) fiber was applied. However, MM fibers suffer from Speckle and reduce the Finesse of the used interferometer. Thus, investigations were performed in order to find the best compromise between light throughout and divergence. In particular we used the following fibers: 50μm 0.12 NA fiber (a), 105μm 0.12 NA fiber (b), 50μm 0.22 NA fiber (c). First, we use the reference beam and the power meter to measure the coupling efficiency. It shows that (a) is 18.7 %, (b) is 28.1%, (c) is 84.3%.

The number of transmitted modes is also an important parameter for MM fiber selection, which will affect the finesse of the interferometer and the quality spectrum effectively obtained by The PMT. The number of modes can be calculated according to:

$$M \approx \frac{V^2}{2} = \frac{\left(\frac{2\pi\alpha}{\lambda} \cdot NA\right)^2}{2} \tag{1}$$

where V is the normalized frequency (V-number), α is the fiber core radius, and λ is the free space light wavelength. The mode number of fiber (a), (b), (c) is 2510, 11072, 8438, respectively. The finesse of Fizeau Interferometer is determined by the ratio of FSR and the FWHM shown from different fibers, and is correlated with the fiber mode. Therefore, less modes of fiber will lead to the higher finesse of the interferometer. Secondly, less modes will result in a lower number of speckles pattern, and thus to a better image of the RBS spectrum, Considering that, the 50μm 0.12 NA fiber has the lowest in the mode number, which is the best choice for our experiment,

Another important point to consider for using the fiber to collected scattering light and inject it into the interferometer is the area of the source as seen by the interferometer, which determines the étendue of the Fizeau Interferometer [16]. It must be sufficiently large to achieve the light to make the required measurements. Therefore, in order to have enough area of light source and a good collimation, we choose the Thorlabs Large-Beam Fiber Collimators (C40FC-A). By measuring the spot diameter of the fiber after the collimator, it is found that 0.12 NA multi-mode fibers have the diameter 9.6 mm which is matched to the aperture size of the Fizeau interferometer.

After the optical fiber is determined, first we use a broadband LED through to measure the illumination function of the system shown in Fig. 3. After that, we can obtain the instrument function of the system using reference beam shown in Fig.4.

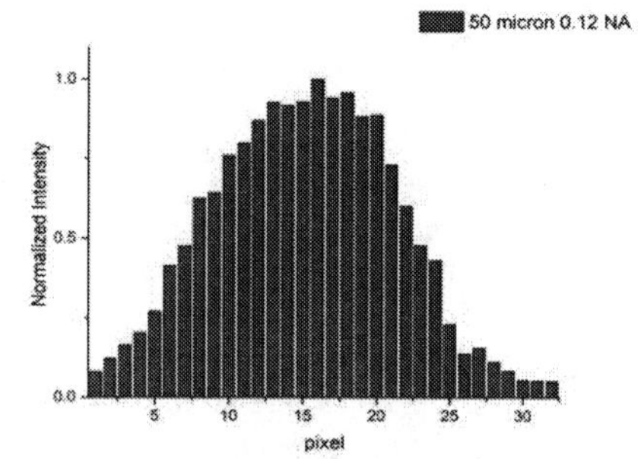

Fig. 3. The illumination function of system using the uniformly circular aperture from 50 μm 0.12 NA.

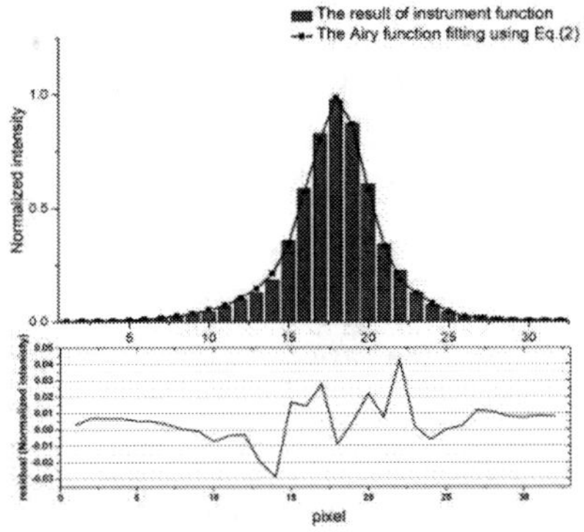

Fig. 4. top: The instrument function measured with the PMT array (red bar) and the best fitting using Eq. (2) of an Airy function (black dots), bottom: the residual of the fitting.

The transmission function of an Fizeau Interferometer can be described by an Airy function *Airy(f)* because of the lower reflectivity and lower asymmetry. Then the instrument function *I(f)* shown in Fig. 4. can be fitted by the Airy function multiply by illumination curve:

$$I(f) = ill(f) * Airy(f) = ill(f) * \frac{1}{FSR}\left(1 + 2\sum_{k=1}^{\infty} R^k \cos(\frac{2k\pi f}{FSR})\right) \quad (2)$$

I(f) is the instrument function of system, *ill(f)* is the illumination curve from PMT array, and the *airy(f)* is the airy function which can be calculated by the Free spectrum range (*FSR*) of interferometer and reflectivity (*R*). From the fitting results, the FWHM calculate from FSR and R of our Fizeau interferometer system is 4.66 pixel, The FSR (30.76 pixel) and the FWHM of the fringe give a finesse of 6.60.

In order to verify the feasibility of the receiving system to obtain the scattered light, we will first measure the gas CO2 which has a high scattering cross section. The averaging spectrum shown in Fig.5 is sum of 10000 shots from PMT, which only need for 100s (one shot is 10ms), and average for a more accurate experiment results and removing the influence of random noise.

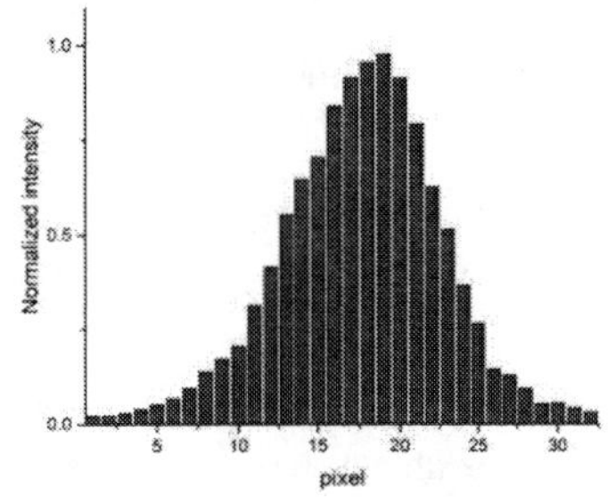

Fig. 5. The RBS spectrum of CO2 at pressure 3.00 bar, temperature 293.0 K.

Further work is to acquire the RBS spectrum of air from different and the pressure with the new receiver system in the lab, then the explained retrieval algorithms (both analytical model and Tenti S6 model) are applied to the measured RBS line shapes. Latest results of this retrieval will be discussed during the conference.

ACKNOWLEDGEMENTS

Jiaqi Xu is supported by the China Scholarship Council (CSC).

REFERENCES

[1] A. Behrendt, et al. Appl. Opt. 43, 2930 (2004)
[2] M. Radlach, et al. Atmos. Chem. Phys. 8, 159 (2008)
[3] A. Behrendt et al. Appl. Opt. 39, 1372 (2000)
[4] M. Fraczek, et al. Appl. Opt. 51, 148 (2012)
[5] P. Di Girolamo, et al. Appl. Opt. 45, 2474 (2006)
[6] Hua D., et al. Appl. Opt., 44(7): 1315-1322 (2005)
[7] Hair, J., et al. Appl. Opt., 40(30): 5280-5294 (2001)
[8] Liu, et al. Appl. Opt., 48(27): 5143-5154 (2009)
[9] B. Witschas, et al. The 27th International Laser Radar Conference (ILRC 27). EDP Sciences 119, 27004 (2016)
[10] B. Witschas, in Atmospheric Physics: Background Methods Trends, U. Schumann, ed. (Springer), 69–83 (2012)
[11] Paffrath, U. PhD thesis, Technical university of Munich. (2006)
[12] O. Reitebuch, et al, J. Atmos. Ocean. Technol. 26, 2501 (2009)
[13] B. Witschas, et al, Appl. Opt. 51, 6207 (2012)
[14] B. Witschas, et al. Opt. Lett. 39(7): 1972-1975 (2014)
[15] K. Liang, et al. sensors, 17, 1503 (2017)
[16] O. Novák, et al. Review of sci. instrument 82, 023105 (2011)

CONTINUOS MONITORING OF LIQUID WATER CLOUDS AND AEROSOLS WITH DUAL-FOV LIDAR POLARIZATION TECHNIQUE

Cristofer Jimenez[1*], Albert Ansmann[1], Ronny Engelmann[1], Patric Seifert[1], Robert Wiesen [1], Martin Radenz[1], and Ulla Wandinger[1]

[1]Leibniz Institute for Tropospheric Research, Permoserstr. 15, 04318 Leipzig, Germany
**Email: cristofer.jimenez@tropos.de*

ABSTRACT

In this work we evaluate the possibilities to assess aerosol-cloud interactions in short time scales (2 min.) on an observational base. Retrievals of the cloud effective radius and number concentration in a liquid-water cloud by using the multiple scattering technique Dual-FOV Polarization lidar, together with the aerosol extinction coefficient in the boundary layer has shown a correspondence between the aerosol and cloud properties in the 6 hours case presented, obtaining a value of $ACI_N = 0.76 \pm 0.29$, which corroborates the potential of lidar observations to study the relation between aerosols and clouds on small scales.

1. INTRODUCTION

Atmospheric aerosols have been on the hotspot of climate research in the last decades due to their influence on the transfer of radiant energy and distribution of latent heat through the atmosphere. Directly through the absorption, emission and scattering of solar and terrestrial radiation and indirectly, acting as cloud condensation nuclei (CCN) and as ice nucleating particle (INP) in water clouds. To better understand the influence of these particles in weather and a changing climate, observational evidence of the aerosols playing his role in clouds is needed. For this reason, several techniques for the monitoring and characterization of aerosols and clouds, have emerged in the last years.

In the case of liquid-phase clouds, a metric to aerosols and clouds is the so called aerosol-cloud interaction index, defined as the variation of the cloud droplet number concentration (CDNC), in the cloud base region at a constant liquid water content, divided the variation of the aerosol extinction σ_{aer} (or alternatively the CCN concentration) [1].

$$ACI_N = \frac{\Delta \ln CDNC}{\Delta \ln \sigma_{aer}}, \qquad (1)$$

Lidar systems have proved to have great potential for the assessment of this index, allowing estimations of CCN concentrations, from measurements of the aerosol extinction coefficient and depolarization in the non-saturated region below cloud base [2]. Lidar systems can also provide information about the cloud optical and microphysical properties, which is contained in the lidar returns affected by multiple scattering (MS) in water clouds [3]. The Dual Field-of-View Raman (DFOV-Raman) technique, implemented on the MARTHA system (Multiwavelength Tropospheric Raman lidar for Temperature, Humidity, and Aerosol profiling) at TROPOS, permits non-ambiguous retrievals of droplet extinction and effective size profiles in the cloud base region by means of the MS-affected Raman return at two FOVs [4]. The technique needs however stable cloud base layers and can only be applied during nighttime. In order to decrease the time resolution and perform observations also during daytime, a Dual-FOV Polarization method (DFOV-Pol) was developed and proposed, which by measuring elastic cross- and co-polarized backscattering can resolve the liquid-cloud properties around the clock and with narrow temporal resolutions (1-3 min) [5].

2. METHODOLOGY

2.1 Dual-FOV technique for retrievals of liquid-cloud microphysics.

This lidar-based method makes use of the volume depolarization ratio measured at two FOVs (similarly as the Raman method).

When the light is scattered by cloud droplets, it will be mostly in the forward direction, having a more sharped lobe for large droplets as for small droplets. Forward scattered light may be scattered several further times by different droplets, one wide-angle backscattering event must occur so that it is detected by the receiver telescope. Cross polarized

https://doi.org/10.1051/epjconf/202023707005

light has shown to be sensitive to the droplet size, since in liquid water clouds, depolarization occurs exclusively in multiple scattering regime [5], when the backscattering angle differs slightly from 180°, increasing rapidly as this angle decrease. In the other hand, the larger the FOV, the larger the possible backscattering angles that they detect, so it can be expected that for large droplets, characterized by a sharp forward scattering lobe, the measured depolarization at the two FOVs will be much similar as for small droplets (Figure 1). How much sensitive are the measurements will depend on the size of the FOVs sizes and also the altitude of the cloud base [6].

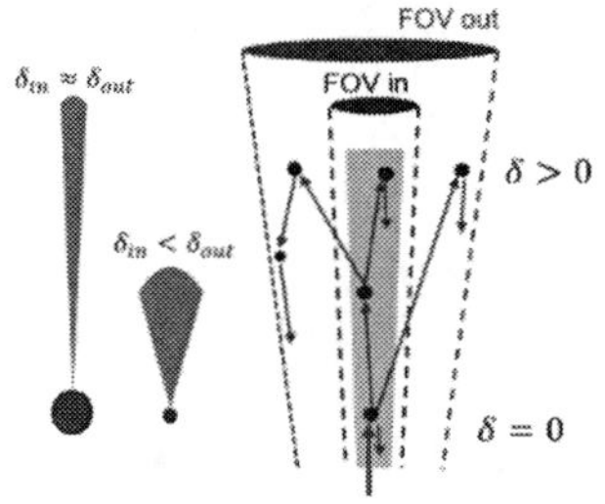

Figure 1: Schematic view of the multiple scattering in cloud droplets.

To quantify this dependence, Simulations of the MS-affected co- and cross-polarized lidar returns at two FOVs were perform. The cloud has been assumed as adiabatic, having a constant increase in the liquid water content with height from the cloud base, while the cloud droplet number concentration remains constant. In this study the FOVs used are $FOV_{in} = 1.0$ and $FOV_{out} = 2.0$ mrads, for a laser with 0.2 mrads divergence. Computations of the volume depolarization ratio at the FOVs "in" and "out" for a wide range of cloud scenes were carried out [7]. The cloud scenes are the vertical profiles of extinction and effective radius, for different reference values and cloud base altitude.

For the retrieval of the cloud parameters, the mean depolarization ratios are calculated ($\bar{\delta}_{in}, \bar{\delta}_{out}$), and then the ratio between them, we define this term as the relative depolarization: $\delta_r = \bar{\delta}_{in},/\bar{\delta}_{out},$

With the relative depolarization the profile of effective radius can be estimated by using pre-computed lookup tables. The penetration depth until which the volume depolarization ratios are averaged was set to 75 meters, which offers the best compromise between measurement uncertainty and sensitivity to the droplet size. For the retrieval of the droplet extinction coefficient we make use of the single-FOV depolarization ratio $\bar{\delta}_{in}$, and the retrieved effective radius profile. The larger δ_r larger the droplet size, and the larger $\bar{\delta}_{in}$, larger extinction values. The conversion curves are defined for each cloud base altitude from 1 to 6 km. More details about the computed lookup tables will be available in [8].

2.2 Implementation of DFOV-Pol technique into PollyXT system for continuous observations.

In the last two decades, at TROPOS one of the goals has been to develop sophisticated, portable and quality-assured lidar systems (Polly), dedicated to stand-alone operation at remote places [9]. The current setup of the polly system permits continuous observation of the aerosol backscattering at three wavelengths, extinction and depolarization at two wavelengths, and of water vapor mixing ratio profiles.

Recently the Polly system has received the instrumental upgrade to integrate the dual-FOV Polarization retrieval technique into the automated lidar for continuous observations of clouds. The current setup consist of a far-range telescope, with a FOV of 1.0 mrads, equipped with polarization channels (total and cross), calibrated with the Δ90° method [10]. For the near range a 50mm-diameter telescope is located next to the beam expander, having a FOV of 2.0 mrads. This telescope is coupled by an optical fiber, therefore the polarization components cannot be separated to retrieve the volume depolarization ratio. To achieve this goal, a second 50 mm telescope was located concentric to the main Telescope, the telescope measures the cross polarized backscatter component and together with the near-range telescope, measuring the total backscatter, they provide the volume depolarization ratio at a second FOV 2.0 mrads [8].

Currently, the polly system, part of the mobile station LACROS (Leipzig Aerosol and Cloud Remote Observations System) is deployed in the city of Punta Arenas, Chile (53° S, 71° W), for a one-year measurement campaign. The goal is to study the differences between the cloud properties in different extreme environments. Measurements of aerosols and clouds in Leipzig and Punta Arenas,

https://doi.org/10.1051/epjconf/202023707005

offers a unique scenario to study the interaction between aerosols and clouds, contrasting the entanglement in a polluted and in a pristine environment.

3. RESULTS

A measurement case to evaluate the interaction between aerosols and clouds in the small scale can visualized in the color map of the Attenuated backscatter at 532 nm (Fig. 2). The cloud case, with cloud mean cloud base at 1.7 km, has been classified as a liquid layer by the CLOUDNET classification scheme [11]. Figure 3 shows the volume depolarization ratio of the FOV_{in} in the cloud layer. An increase in the depolarization as MS take place can be observed. Figure 3b presents the so-called relative depolarization ratio. These two observables permits the retrieval of the effective radius and extinction coefficient up to 75 meters in the cloud [8].

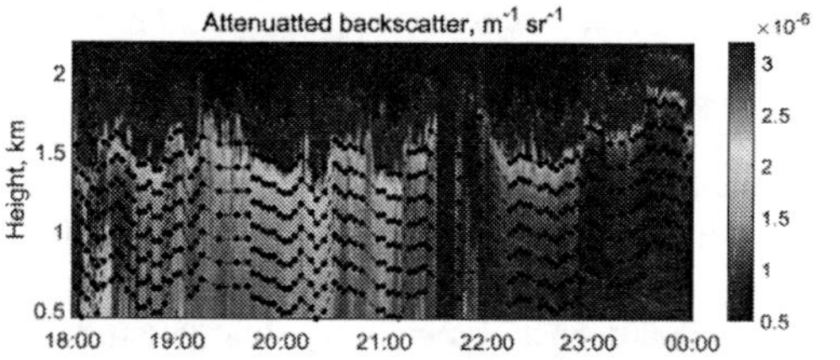

Figure 2: Attenuated backscatter, the black dashed lines describe the aerosol regions considered to evaluate ACI.

The cloud droplet number concentration is then estimated by assuming a modified Gamma distribution with quasy-shape parameter k=0.8 (marine air masses) [Miles 2000]. In order to find the coupled aerosol region, we estimated the so called aerosol cloud interaction index (Eq. 1) diving the region below cloud base in height steps of 150 meters (Figure 2).

As aerosol proxy, we use the aerosol extinction coefficient. Since the measurements were performed at day-time conditions, the retrieval of the backscattering coefficient by using two channels (elastic, Raman N_2) cannot be afforded. To retrieve this parameter we fixed a constant value for the lidar ratio LR=25, the single-channel lidar constant was previously determined in the cloud-free period after the presented case, between 2:00 and 3:00 (not showed). The lidar constant is then used to calculate the attenuated backscatter and the backscattering coefficient is obtained by evaluating the lidar equation so that $\sigma_{aer} = LR\ \beta_{aer}$.

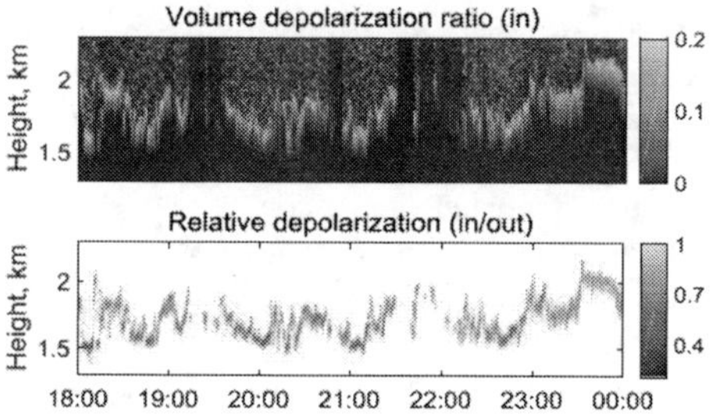

Figure 3: Values of the depolarization δ_{in} and ratio of depolarization δ_r used for the retrieval of cloud droplet size and extinction coefficient.

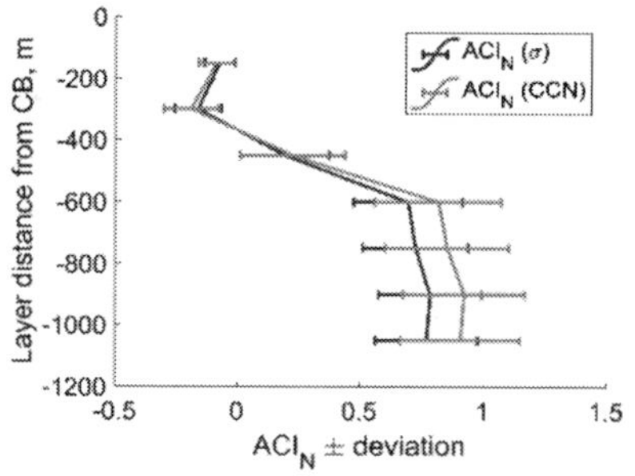

Figure 4: Vertical structure of the Aerosol Cloud Interaction index for each 150m layer below cloud base.

To describe the aerosol-cloud interaction based on he observed parameters, the cloud droplet number concentration $CDNC(t)$ and the aerosol extinction coefficient $\sigma(z,t)$ below cloud base, the ACI_N was estimated for each layer equidistant to the cloud base during the whole period (Fig. 4). Here it can be noted how the CDNC correlates with the aerosol layers between 600 and 1050 meters, while no correlation is seen in the region right below cloud base, where enhanced aerosol optical properties due to hygroscopic growth mask any correspondence between the measured properties and the CDNC. Then mean values of the extinction in the highly correlated region were used to evaluate the overall cloud response to the aerosol.

Figure 5 present the time series of the CDNC and the extinction coefficient of aerosols.

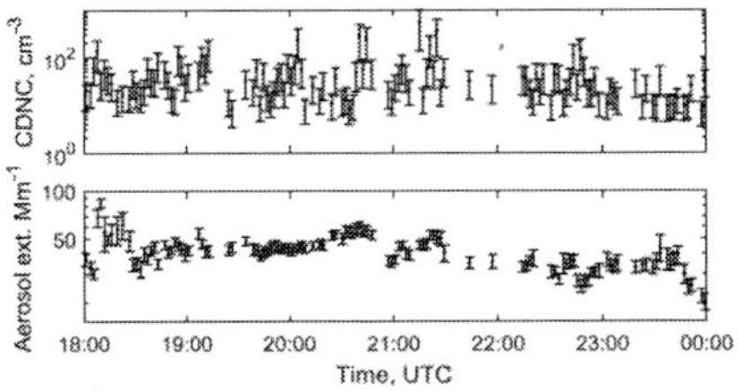

Figure 5: Time series for the CDNC and aerosol extinction with temporal resolution of 2 min.

Figure 6 present the average extinction profile of the 6 hours period. Here, the dashed line is just an indication of the cloud, and does not represent the values of the extinction, since the elastic signal is affected by multiple scattering and the lidar ratio in the cloud (~18) would differ from that of marine aerosols (25). The mean value of the CDNC is 48 cm^{-3}, while the mean extinction value is 41 Mm^{-1}, from which we estimated a CCN concentration of 150 cm^{-3}. This results suggests that 1/3 of the aerosols in the layer 600 meters below cloud base has been activated into a cloud droplet.

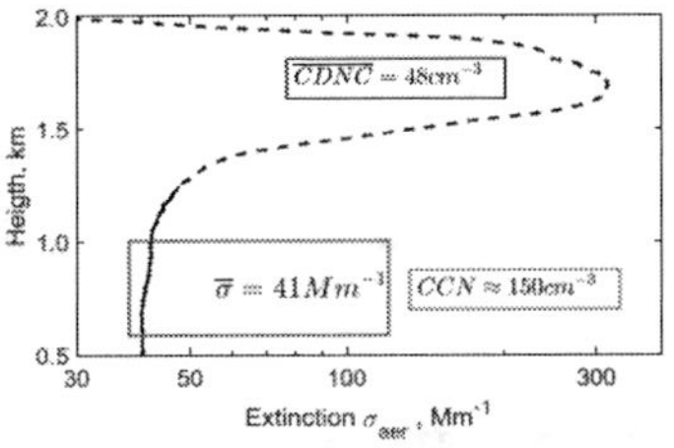

Figure 6: Mean profile of the aerosol extinction coefficient.

The ACI$_N$, associated to the 2-min observations of the cloud and aerosol parameters (extinction and corresponding CCN concentrations) are presented in Figures 7a and 7b. A linear regression of the data yields to $ACI_N(\sigma) = 0.76 \pm 0.29$ and $ACI_N(CCN) = 0.89 \pm 0.37$. The pristine ambient of the region surrounding the city of Punta Arenas, and the aerosol type, identified as marine from backward trajectories presents favorable conditions for aerosol-cloud interactions given the large numbers obtained for its index. These results are also in agreement with [12], where the relation between extinction and CCN constrains the maximum ACI$_N$ from extinction observations to 0.8.

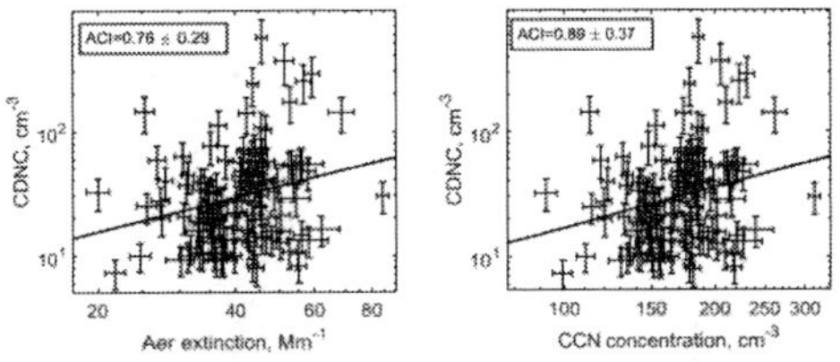

Figure 7: (a) CDNC vs aerosol extinction for the 6 hours period. (b) CDNC vs CCN concentration.

4. CONCLUSIONS

This work presents an application of the observation capabilities of a Dual-FOV Polarization system. The 2-min. retrievals of the cloud droplet number concentration and of aerosol extinction coefficient at different layers below cloud base shows how the cloud responds to the aerosol presence in the boundary layer, obtaining large correspondence in the region between 600 to 1105 meters below cloud base $ACI_N = 0.76 \pm 0.29$. Further analysis of more cases will allow us to study aerosol-cloud interactions in different environments (polluted and not-polluted). Vertical movement is another important parameter to consider. Long-term measurements of aerosol and cloud properties, combined with vertical wind information (Doppler Lidar) will permit a deep statistical analysis of the interaction.

REFERENCES

[1] G. Feingold, J. Geo. Res., 106, 22907-22922 (2001).
[2] R. Mamouri, A. Ansmann, Atmos. Chem. Phys., 16, 5905–5931 (2016).
[3] L. Bissonnette, et al., Appl. Phys. B 60, 355-362 (1995).
[4] J. Schmidt, et al., Appl. Opt., 52, 2235–2247 (2013).
[5] I. Veselovskii, et al., Appl. Opt. 45, 6839-6848 (2006)
[6] C. Jimenez, et al., Proc. SPIE, 10429, 1042907 (2017)
[7] E. Zege, et al., Appl. Phys. B 60, 345-353 (1995).
[8] C. Jimenez, et al., Atnos, Meas. Tech, (in preparation, 2020).
[9] R. Engelmann, et al., Atmos. Meas. Tech., 9, 1767-1784 (2016)
[10] V. Freudenthaler, Atmos. Meas. Tech., 9, 4181–4255 (2016).
[11] H. Baars, et al., Atmos. Meas. Tech., 10, 3175–3201 (2017).
[12] Y. Shinozuka, et al., Atmos. Chem. Phys., 15, 7585-7604 (2015)

DESIGN AND DEVELOPMENT OF A RAMAN LIDAR FOR CHERENKOV GAMMA ARRAY EXPERIMENTS

George Vasileiadis[1*], Patrick Brun[1], Omar Gabella[1], Stephane Rivoire[1], George Avdikos[2], Alexandros Louridas[2], George Georgoussis[2] and Alexandros Papayannis[3]

[1]*University of Montpellier 2, LUPM, Montpellier, FR34095, France*
[2]*Raymetrics S.A., Metamorfossis, GR14452, Greece*
[3]*National Technical University of Athens, Physics Department, Zografos, GR15780, Greece*
**Email: george.vasileiadis@lupm.in2p3.fr*

ABSTRACT

Future Cherenkov Gamma Ray (CGR) experiments will reach a sensitivity and energy resolution never obtained until now by any other high energy gamma–ray experiment. It is well known that atmospheric conditions contribute particularly in this aspect. Raman lidars can help reduce the systematic uncertainties of the molecular and aerosol components of the atmosphere so these performances can be reached. The motivation for Raman lidars and the design and development of the Montpellier Raman lidar system is described. It provides both multiple elastic and Raman readout channels and custom-made optics design. Preliminary lidar tests and signals show the actual performance of the lidar in consistency with the desired goals.

1. INTRODUCTION

Future Cherenkov Gamma Ray (CGR) experiments will reach a sensitivity and energy resolution never obtained until now by any other high energy gamma–ray experiment [1]. It is well known that atmospheric conditions contribute particularly in this aspect.

Current Imaging Atmospheric Cherenkov Telescope (IACT) experiments and the Pierre Auger Observatory have pioneered several ways to introduce atmospheric calibration devices, among them the use of Lidars [2,3]. Their experience has led to a coherent atmospheric calibration strategy for the future CGR experiments. One of its key components consists of the assessment of the atmospheric extinction profile throughout the entire path that Cherenkov photons may take before getting imaged, at two wavelengths within the sensitive regime of the light sensors used in CGR telescope cameras.

To achieve this, one or several Raman lidars, operating at various wavelengths will provide extinction profiles, to continuously assess the atmospheric extinction across the observed science targets. Such an approach is sufficiently precise when either a single or no cloud layer are present in the observed field–of–view, a situation fulfilled practically all the time when the CGR experiments will be operative.

To precisely characterize extinction profiles up to 25 km distance, the Raman lidars should use powerful lasers and large mirrors. These choices guarantees the highest precision at the expense of strong interference with the light sensors making it impossible to simultaneously operate both. Since the telescopes need to re-point frequently, the Raman lidars must therefore be able to measure the extinction profiles within short time scales (a few minutes).

The Laboratoire Univers et Particules de Montpellier (LUPM) is using a recycled 1.8 m mirror from the CLUE (Cherenkov Light Ultraviolet Experiment) experiment [4], to achieve the required sensitivity to backscattered light (Figure 1). A specific spectrometer (wavelength separation unit) has been designed and manufactured for the detection of signals at 355, 387, 532 and 607nm. Zemax [5] and Solidworks [6] simulations have been applied, leading to a compact and robust spectrometer. Preliminary lidar tests and signals show the actual performance of the lidar in consistency with the desired goals.

2. METHODOLOGY

2.1 The LUPM lidar system

The LUPM Raman lidar is optically and mechanically based on a telescope and mechanical assembly from the CLUE former experiment. Due to the small cross section of the Raman lines, about 2 to 3 orders of magnitude smaller than the elastic cross section, a large reflecting area is needed to

https://doi.org/10.1051/epjconf/202023707006

collect a sufficient amount of light. For this reason, the telescope reflector is 1.8m in diameter and coated to optimize UV reflection. It is installed on an altazimuth mount. Measurements of the point spread function and reflectivity show that the mirror has maintained a good optical quality, although the reflectivity has decreased to 64% at 350nm [7].

Figure 1. The refurbished CLUE container and UV coated mirror used for the Raman LIDAR.

Pointing the telescope to a specific direction is achieved by a pair of stepper motors with a precision of 0.02 degrees. The motors are piloted by a dedicated automation system built around a Panasonic PF7 PLC. A Quantel CFR 400 Nd:YAG laser is mounted at a fix point on the telescope's structure. It is a military-grade III 9 ns pulsed laser, with a repetition rate of 20 Hz, providing three different wavelengths at 1064 nm, 532 nm and 355 nm in a coaxial configuration, in order to achieve a good overlap between the laser beam light cone and the telescope field of view.

A good overlapping factor is needed to measure the optical depth starting from a few tens of meters height. This is important since measuring the extinction of the atmosphere as low as possible permits a better estimation of the photon absorption.

The laser beam exits the telescope using a double UV-coated mirror guiding system. The laser head is firmly mounted on the telescope dish. A single 1-inch fixed mirror will guide the laser beam to the center of the dish at the level of the focal point. To achieve optimal performance, a precise alignment system is needed to assure that at any moment the optical axis of the telescope and the laser beam are co–linear. We have opted for a computerized industrial system by ThorLabs, with a step-size precision of less than 1 mm.

A support for the light guide has also been installed in the structure of the telescope. The pieces fix it to the structure in such a way that movements are restricted and a minimum bending radius for the liquid fiber light guide is assured.

2.2 The spectrometer system

The light output from the focal point of the telescope is fed to the entrance of the Raman polychromator via a Lumatec 300 series liquid fiber of 8 mm diameter with a numerical aperture of 0.59. We use this kind of fiber to avoid stress effects due to bending during the pointing phase of the telescope, while keeping an excellent UV transmission (better than 70%). Other solutions based on custom rigid optical components were studied but did not have good enough transmission. Figure 2 presents the polychromator built by the Raymetrics company.

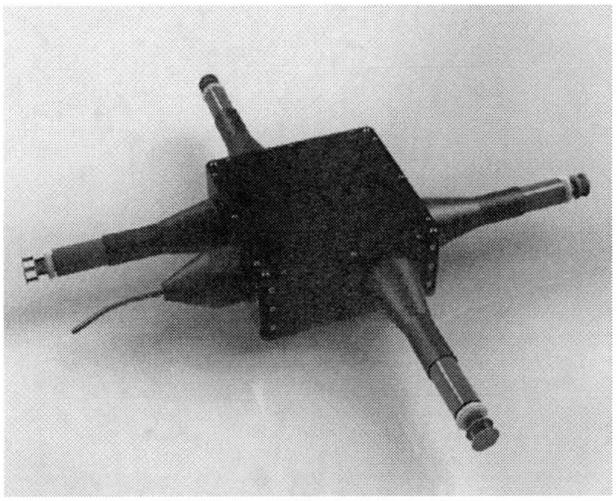

Figure 2. Image of the Raymetrics Raman spectrometer. A four channel Elastic/Raman design with a liquid light guide input option.

The polychromator is designed to be lightweight, mechanically modular, and optically efficient. Dichroic mirrors are used to separate the different Raman and elastic lines, while a dedicated two-lens eye piece is used in front of every PMT to focus uniformly the incident light. Custom 2-inch optics are used throughout the spectrometer to match efficiently the PMT entrance window size.

3. RESULTS

3.1 Optical design simulations

In order to efficiently simulate the image formation from near and far end achieving 80% coverage at the PMT photocathode level (considering the specifications of the big telescope), it was necessary to use 2-inch optical components. A set of 2-inch collimating lenses at the output of the fiber was used for collimating the beam and inserting this into the spectrometer. Then, specific custom dichroic beam splitters (CVI) were used for the separation of the wavelengths into different directions. Also 2-inch interference filters with FWHM equal to 10 nm were used in front of every 2-inch Hamamatsu R329P Photomultiplier Tubes (PMT). Simulations with Zemax software shown that there is a very homogeneous distribution of the image for both near and far field into the PMT cathode, almost eliminating all redundant effects caused by PMT inhomogeneities (Figure 3).

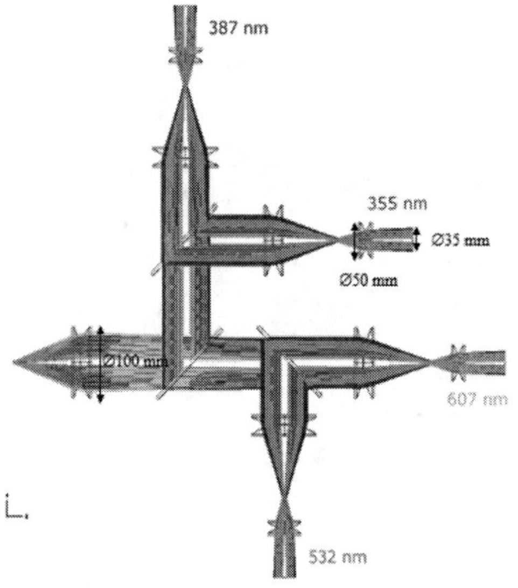

Figure 3. Zemax simulations of the spectrometer.

3.2 Expected performance of the Raman Lidar

Studies of the PSF and reflectivity of the telescope mirror, show that it has maintained a good optical quality, keeping basically the same PSF as when it was produced (6.5 mm), although the reflectivity has decreased to 64 % at 350 nm. The mirror spot size has been characterized through its PSF by different methods. Different energy containments at different radii are calculated. The results obtained show that 80% of the light is contained in an area of 6mm diameter while the 90% falls in a circle with a diameter of about 6.4mm (Figure 4).

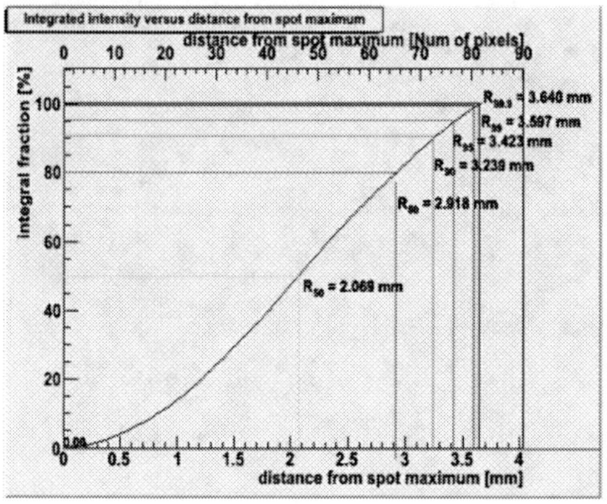

Figure 4: Measurement of the primary mirror PSF. All light is enclosed in the liquid light guide perfectly. In this case, D90 around 6.4mm.

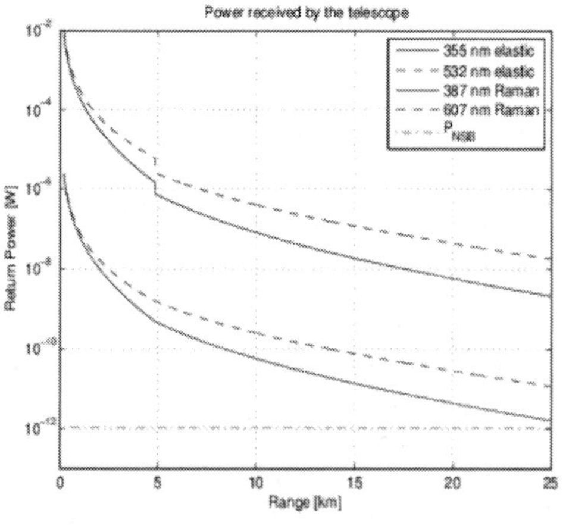

Figure 5: Estimated return power from the link-budget simulation of the Raman LIDAR. The horizontal solid red line is the background power calculated for typical NSB levels.

Finally, dedicated link-budget simulations [8] have revealed that this system can characterize the atmosphere up to 25 km with the elastic channel,

and up to 18 km with the Raman channels in less than a few minutes (see Fig. 5).
Preliminary return signal profiles are also shown at Figure 6. Only results obtained using the 532nm line are presented here. Even if we are still in the testing phase these first results are very much promising for the future. A loss of signal is observed at lower altitudes due to misalignment issues, under investigation, while the performance at high altitudes is as expected.

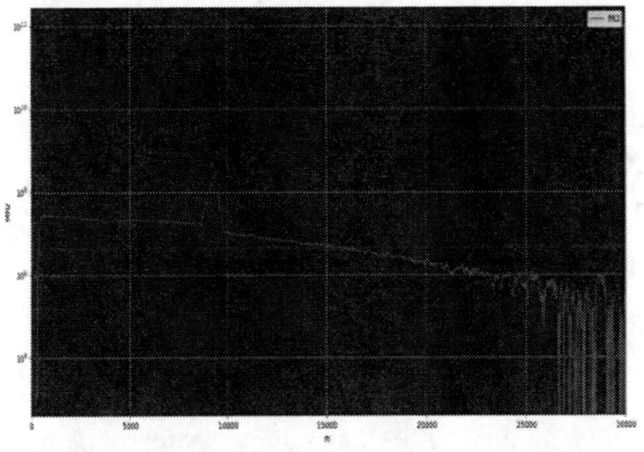

Figure 6: Preliminary results obtained using the 532nm laser line. The log of the return signal, range corrected, is plotted against altitude.

Conclusions

For future CGR experiments a proper characterization of the atmosphere has to be done, in order to estimate the extinction that the UV photons undergo from the top of the atmosphere to the ground and correctly reconstruct the energy of the primary gamma ray. A suitable device to achieve this goal is a Raman LIDAR, an instrument that can provide a height resolved characterization of the atmosphere. The Raman LIDAR would allow to correct systematic biases on the energy scale and flux, reducing the systematic uncertainties from a current 20 % down to better than 5 %. This will help not only to reach the desired energy resolution but also to increase the duty cycle.

Preliminary results both on simulations and actual expected performance are promising. The optical system is currently under assembly and first test bench confirms system requirements. As soon as the behaviour of the complete chain is completely understood the LIDAR will be operational as a full Raman one.

References

1. Actis, M. et al. Design concepts for the Cherenkov Telescope Array CTA: an advanced facility for ground-based high-energy gamma–ray astronomy, Exp. Astron. 32 193 (2011)
2. Will M. *Atmospheric Monitoring at the Site of the MAGIC Telescopes,* Proc. AtmoHEAD 2016, EPJ 144 01002 (2017)
3. G.Vasileiadis et al. An elastic lidar system for the H.E.S.S. NIM A819(2016)60-66
4. Alexandreas D. et al. Status report of CLUE NIM A 360 385 (1995)
5. www.zemax.com
6. www.solidworks.com
7. A.Oramas et al. ICRC 2013 (1517).
8. M. Eizmendi, Master Thesis (2011)

EPJ Web Conferences **237**, 07007 (2020) https://doi.org/10.1051/epjconf/202023707007
ILRC 29

DESIGN OF THE SHIP-BORNE MULTI-WAVELENGTH POLARIZATION OCEAN LIDAR SYSTEM AND MEASUREMENT OF SEAWATER OPTICAL PROPERTIES

Qi Liu[1], Bingyi Liu[1, 2], Songhua Wu[1, 2 *], Jintao Liu[1], Kailin Zhang[1], Xiaoquan Song[1,2], Xiangcheng Chen[1], Peizhi Zhu[1]

[1] *Ocean Remote Sensing Institute, College of Information Science and Engineering, Ocean University of China, Qingdao 266100, China*

[2]*Laboratory for Regional Oceanography and Numerical Modeling, Pilot National Laboratory for Marine Science and Technology (Qingdao), Qingdao 266237, China*

**Email: wush@ouc.edu.cn*

ABSTRACT

A ship-borne multi-wavelength polarization ocean lidar system LOOP (Lidar for Ocean Optics Profiler) is introduced in detail, aiming to obtain high-precision vertical profiles of seawater optical characteristics. Based on Monte-Carlo simulation, the receiving telescope is designed with a variable field of view, producing system attenuation coefficient (K_{lidar}) approximating the optical parameters of seawater under a different field of view and water body conditions. At first, a sea trial was conducted in Jiaozhou Bay, and the measured diffuse attenuation coefficient (K_d) of seawater was 0.3m^{-1}, being in good agreement compared with the results measured by field instrument TriOS. Then a field campaign was organized in the South China Sea. The measurement of the seawater diffuse attenuation (K_d) was 0.035m^{-1}. These results support the prospects that lidar, as an effective tool supplement to traditional passive ocean color remote sensing, can provide the vertical distributions of optical properties in the upper ocean.

1. INTRODUCTION

Ocean Color Sciences is an important part of Earth observation. The results of measurement and analysis have revolutionized the field of biological oceanography and have made important contributions to biogeochemistry, physical oceanography, ocean systems models, marine fisheries and coastal zone management [1][2]. The optical properties of solutes and particles in seawater vary with the type and concentration, which decides the optical properties of seawater showing great spatial and temporal changes. The optical properties of seawater have a strong relationship with biological, chemical, geographic and physical environments, which determine that the Ocean optics is the basis and key role of the Ocean Color Sciences and the Remote Sensing [3]. At present, the measurement of the inherent optical properties of seawater mainly uses the in-situ optical cage and passive ocean color satellite [4]. The in-situ optical cage can only measure several profiles per day, which is difficult to perform large-scale measurements in a short period of time and cannot fully characterize the temporal and spatial variation of ocean optical properties. Passive ocean color satellites can only observe during the daytime, and the observed results are affected by atmospheric correction, having large errors. Besides , the inversion algorithm is based on the assumption that the seawater is vertically uniform, and the vertical profile information of the ocean optical properties cannot be given [5].

Lidar using the blue-green band laser can easily penetrate the water body, which can directly and accurately obtain the high-precision vertical profile of seawater optical properties [6], thus making up for the limitations of in-situ optical cages and passive ocean color satellites, achieving breakthrough from 2D plane to 3D stereoscopic observation.

For the anticipation, this paper describes in detail of the self-developed ship-borne multi-wavelength polarized lidar system, LOOP (Lidar for Ocean Optics Profiler), in Sect2. The results in Jiaozhou Bay and the South China Sea and case studies are provided in Sect. 3.

2. METHODOLOGY

2.1 The lidar system

Utilizing three-wavelength Nd:YAG narrow pulse laser with high energy, the LOOP system has five regular channels including 355 nm parallel-cross polarization, 532 nm parallel-cross polarization and 1064nm. The wavelength of 355nm is used for fluorescence. The backscattered light at the wavelength of 532nm is used for ocean optical properties. Both wavelengths of 532nm and 355nm with parallel-cross polarization are used for particle recognition obtained by the depolarization ratio. The water surface distance is measured at the 1064-nm wavelength. The LOOP system specifications are shown in Table 1.

Laser	wavelength(nm)	355&532&1064
	Pulse energy (mj)	1064nm@400 532nm@200 355nm@120
	Repetition rate(Hz)	20
	Pulse duration (ns)	≤5
	Divergence(mrad)	≤0.5
Receiver	Aperture(mm)	60
	Field(mrad)	5~100
	Filter_FWHM(nm)	2
	Sampling frequency	AD@250MHz PC@1GHz

Table 1. LOOP system hardware configuration.

Based on the Monte Carlo simulation, the lidar system attenuation coefficient K_{lidar}, is close to different ocean water optical properties in the different fields of view [7]. As shown in Fig.1, the telescope is designed with a set of lens and a variable stop located behind the aspherical lens is used to control the system's field of view from 5mrad to 100mrad. The return signal travels to the five regular channels through beam splitters and narrow-band interference filters positioned at the front of photomultiplier tubes (PMT). Besides, the 420nm-700nm spectrum channel is reserved for the future detection of hyper fluorescence signals in seawater.

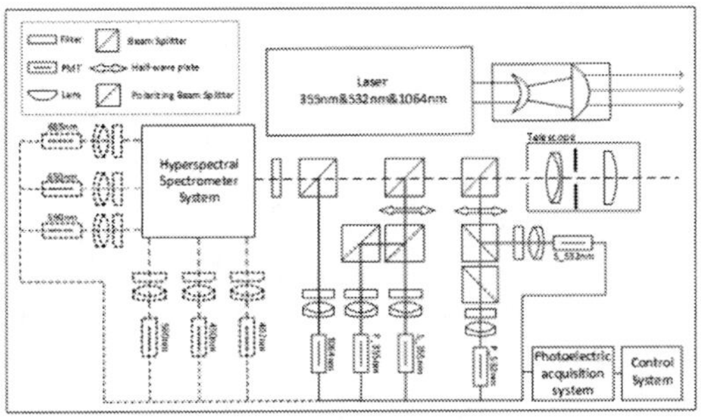

Fig 1. The LOOP system.

Notice that at front of the polarizing beam splitters (PBS) of 355nm and 532nm, a half-wave plate was used for the method called " Δ 45°-calibration" to calibrate the volume linear depolarization ratio [8]. And at the cross polarization channels of 355nm and 532nm, the PBS equipped to ensure T_p/T_s extinction ratio cube reaches 10^6:1. Finally, the return signals were recorded by independent research and development of digitizer at a sampling rate of 1 GHz for photon counting (PC) and 250MHz for analog with 14-bit digitally resolution.

2.2 Method

The LOOP was initially field tested aboard a boat in Jiaozhou Bay on July 30, 2018. A field instrument TriOS measured the seawater diffuse attenuation (K_d) simultaneously. Then the system was deployed aboard the No.9 Marine Geology research ship in the South China Sea for four days trial (March 7 to 10, 2019, Fig.2).

Fig 2. LOOP field experiment. The left is in Jiaozhou Bay and the right is in the South China Sea.

For the ship-borne ocean lidar, the return signal can be expressed as [9]:

$$P(z) = \frac{A\beta(z)}{n(nH+z)^2} exp\left[-2\int_0^z K_{lidar}(y)dy\right] \quad (1)$$

Where z is depth, A is a calibration factor including geometric losses , incidence of the sea surface and two-way transmission through sea surface, n is the refractive index of sea water, H is the distance from the lidar to the surface, z is the path length in water, β is the volume scattering coefficient at a scattering angle of π radians and K_{lidar} is the lidar attenuation coefficient.

The system attenuation coefficient (K_{lidar}) was calculated using the slope method [10]. This method derives both β and K_{lidar} from the signal. But considering the ocean optical conditions, there is an approximation algorithm. If the water column is well mixed in the upper ocean [11], both K_{lidar} and β will be constant over the profile. In this case, K_{lidar} can be derived from the slope of the logarithm of the signal.

3. RESULTS

3.1 Jiaozhou Bay (near-shore water)

The prototype of the LOOP system was first compared with the field instrument TriOS (field instrument) to obtain the optical properties profile of the coastal water column on July 30, 2018.

Due to the turbidity of the coastal water column and large dynamic range of the return signal, it is difficult to improve laser effective penetration depth. As the system was deployed above water on the boat, when the laser emitted with high energy, the specular return off the sea surface cause digitizer saturation that cannot obtain the signal of the water column even to 20m below the surface. Otherwise, the laser energy rapidly with depth, making the effective penetration only 5m. Finally, the lidar measurement of Klidar in the range from 3.6m to 7.8m was selected for comparison with the TriOS measurement of Kd in the range from 0.8m to 5.8m.

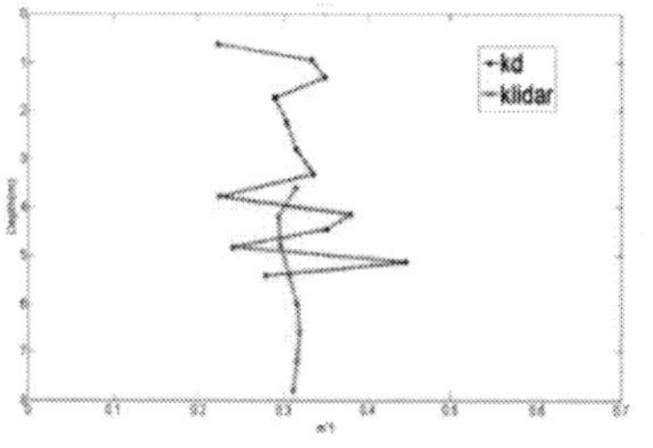

Fig 3. K_{lidar} compared with K_d measured by field instrument TriOS

3.2 The South China Sea (off-shore water)

A field campaign was organized on March 10, 2019, in the South China Sea. Fig 4. shows the lidar attenuated backscatter for two different fields of view. A strong saturation from the water surface to 18m underwater was found when the field of view is 100mrad. It can easily be observed that the signal decayed more rapidly in the narrow field of view than in wide one.

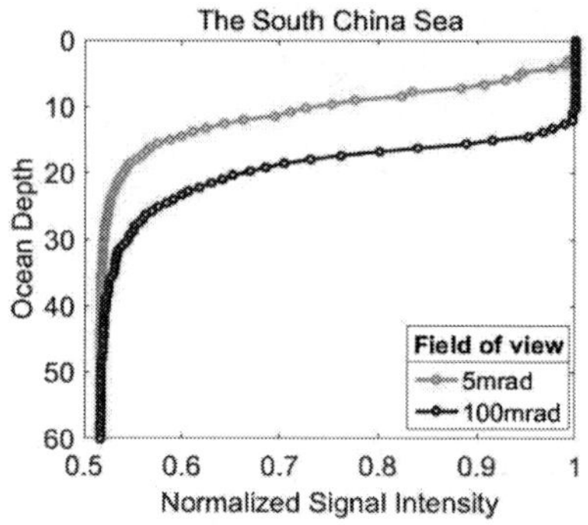

Fig 4. Lidar attenuated backscatter in the narrow field of view (5mrad) and wide field of view (100mrad).

Fig 5. Shows the K_{lidar} derived from above lidar attenuated backscattering. In the narrow field of view, K_{lidar} has a large value initially then decayed to the same value as the wide field of view. It is considered that K_{lidar} in both fields of view tends to be diffuse attenuation coefficient K_d around 0.035 m^{-1} below 40m depth of water column. The perturbation which did not do any smoothing may be caused by strong winds and ocean waves on the day. And the backscatter response is expressed

as the combination of convolution in terms of the water column and strong reflection of the sea surface, which caused K_{lidar} a large error above the water depth of 18m.

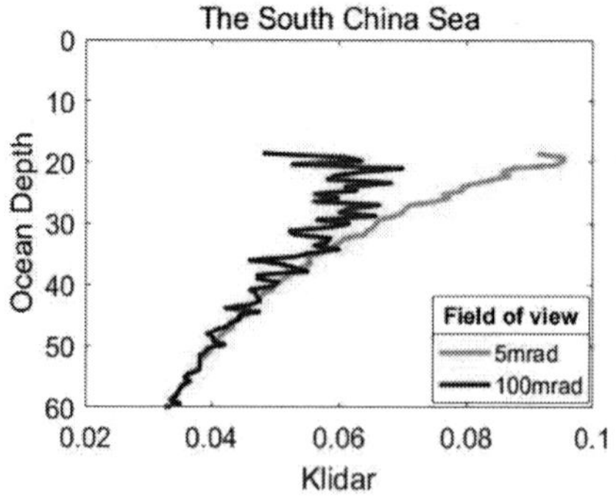

https://doi.org/10.1051/epjconf/202023707007

Fig 5. Measurement of K_{lidar} in the different field of view.

4. Discussion

These results support the proposition that the ship-borne ocean lidar can provide an effective way to obtain the vertical distributions of optical properties and geochemical constituents for remotely measuring in the upper ocean. However, in order to improve the detection capability of ocean lidar, there are many challenges. Most important, for example, the up to 10^{10} dynamic range of the lidar signal transmission to 70m needs a high-speed digitizer with sufficient dynamic range. Even though there are several approaches to maximizing dynamic range, all of them have advantages or disadvantages. In addition, the Gaussian laser beam transmits in the water column with a larger divergence, limiting the laser penetration depth of the seawater column. A new approach such as the Bessel beam may be an ideal way to solve this problem. Furthermore, the Optical Parametric Oscillator (OPO) laser which can emit tunable laser wavelength with higher energy and repetition rate should develop because of different types of water columns having a different optimal optical penetration window. Farther into the longer term with the development of lidar, global ocean coverage from a satellite oceanographic lidar is likely. The measurement of the absorption coefficient a, backscatter coefficient b, plankton layers and particulates discrimination related to depolarization will be investigated in the future.

ACKNOWLEDGEMENTS

This research was funded by National Key Research and Development Program of China, grant number 2016YFC1400904, supported by Pilot National Laboratory for Marine Science and Technology (Qingdao).

REFERENCES

[1] Behrenfeld M J, et al., Nature Climate Change, 2016, 6(3): 323-330.
[2] Dutkiewicz S, et al., Nature Communications, 2019, 10(1).
[3] Mcclain C R, et al., Experimental Methods in the Physical Sciences, 2014, 47:73-119.
[4] Lee Z, et al., Applied Optics. 57, 3463-3473 (2018).
[5] Lu X, et al., Journal of Geophysical Research: Oceans, 2014, 119(7): 4305-4317.
[6] Churnside J H, et al., Applied Optics, 1998, 37(15): 3105-3112.
[7] Gordon H R, Applied Optics, 1982, 21(16): 2996-3001.
[8] Dai G, et al., Remote Sensing. 2018;10(3).
[9] Churnside J H, Optical Engineering, 2013, 53(5):051405.
[10]Churnside J H, et al., Applied Optics. 2017;56 (18):5228-33.
[11]C de Boyer, et al., Journal of Geophysical Research, 2004, 109(C12).

DEVELOPMENT AND TEST OF A FRINGE-IMAGING DIRECT-DETECTION DOPPLER WIND LIDAR FOR AERONAUTICS

Patrick Vrancken*, Jonas Herbst*

* *Institute of Atmospheric Physics, German Aerospace Center (DLR), 82234 Weßling, Germany*
Email: patrick.vrancken@dlr.de

ABSTRACT

DLR currently investigates the use of Doppler wind lidar as sensor within feedforward gust alleviation control loops on fast-flying fixed-wing aircraft. Such a scheme imposes strong requirements on the lidar system such as sub-m/s precision, high rate, high spatial resolution, close measurement ranges and sensitivity to mixed and pure molecular backscatter.

We report on the development of a novel direct-detection Doppler wind lidar (DD-DWL) within these requirements. This DD-DWL is based on fringe-imaging of the Doppler-shifted backscatter of UV laser pulses in a field-widened Michelson interferometer using a fast linear photodetector.

A prototype for airborne operation has been ground-tested in early 2018 against a commercial coherent DWL, demonstrating its ability of measuring close-range wind speeds with a precision of 0.5 m/s, independent of the actual wind speed.

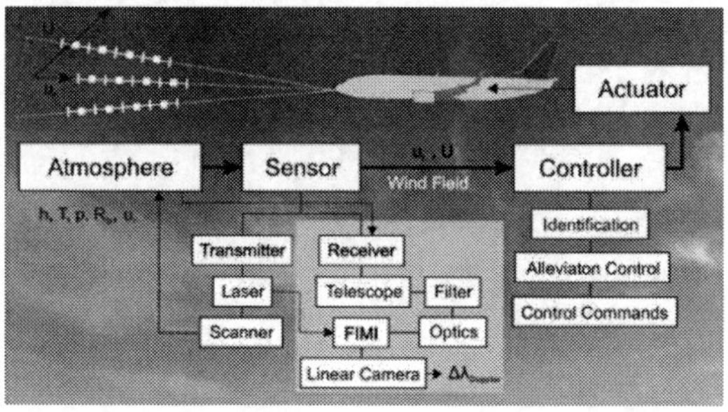

Figure 1: Lidar-based feed-forward gust alleviation system (the grey-shaded box is treated herein)

1. INTRODUCTION

Applications of Doppler wind lidars (DWL) in aeronautics have been under discussion since their invention [1], and test flights of actually all generations of coherent DWLs (from CO_2 to solid-state and fiber laser technology) and also direct-detection DWLs to detect turbulence and wake vortices have been undertaken since the early 70ies. Despite these efforts and despite turbulence being a major cause for (non-fatal) incidents in civil aeronautics, with increasing trend, none of these setups has found its way into an application or product so far. Apart from the objective complexity of lidars, possible reasons might be identified in the increasing structural resilience of aircraft towards the effects of turbulence. Today's urge for efficiency within the air transport sector ultimately implying more lightweight, and thus vulnerable structures, will meet higher turbulence abundance in projected flight altitudes [2]. This shall lead to a turning point of the above described situation.

In this context, DLR-FT (Institute of Flight Systems) is studying feed-forward flight control systems for mitigation of atmospheric effects. Key ingredient for such feed-forward alleviation (for turbulent gusts or wake vortices) systems is a number of measured wind speed values in the flightpath ahead of the aircraft. Today's generation of these controllers may handle partial or comparatively low data yield (i.e. realistic in terms of lidar delivery) due to the use of a-priori information on the phenomenon to be encountered [3,4]. This dramatically decreases the requirements on DWLs for delivery of these wind data, i.e. pushing these into technically satisfiable bounds. These requirements are being iterated within the present studies, together with feasible lidar parameters, delivered by the LIDAR group of DLR-PA (Institute of Atmospheric Physics).

Briefly, owed to the high speed of high-flying aircraft (30 kft to 40 kft), such a feed-forward flight controller would impose following rough parameters on a DWL system: a full-field (several measurement directions, as within a cone, e.g.) update rate of 10 to 20 Hz; a wind speed precision of around or less 1 m/s, mainly due to the angular penalty (line-of-sight / LOS measurement versus vertical required) in a cone-like scanning setup;

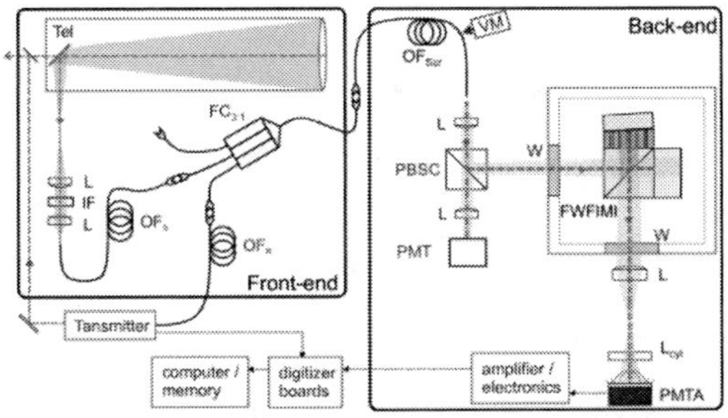

Figure 2: Simplified synopsis of FWFIMI-based Doppler Wind Lidar, technical details in the text

a spatial resolution of 20 m to 50 m, depending on the aircraft mass, wingspan and flight speed (and thus its frequency response to turbulence); close ranges 50 m to 300 m ahead of the aircraft being comfortable with the SNR but actually representing challenges of optical nature; and last, full availability of the measurement in cruise flight conditions, dictated by certifying authorities, and eventually negating the use of aerosol-dependent coherent DWLs.

DLR-PA theoretically studied and compared several possible implementation forms of direct-detection (DD) DWLs [5]. We deliberately discarded coherent DWL for the risk of non-availability due to possible virtual aerosol absence within the targeted clear air volumes of cruise flight airspace, following the same argumentation as the AWIATOR project before [6,7].

The needed close measurement range (< 300 m) infers a maximized overlap within this region and thus a large field-of-view (FOV) of several mrad. Even for reasonable (i.e. aircraft-compatible) telescope sizes and ~ back-end optics diameters, the magnification translates this FOV to important angular distributions. The here [5] studied interferometers within DD-DWLs thus have to be field-widened in order to accept such angular distributions without considerable loss of contrast. This field-widening (FW) may be realized with two-path interferometers such as Mach-Zehnder (MZ) and Michelson (MI), as opposed to multiple-path interferometers such as Fizeau and Fabry-Perot interferometers. Several groups have successfully adopted the MZ technique, as within LNG of French LATMOS [8] or the different generations of Ball Aerospace's OAWL lidar [9,10], both using quadrature phase retrieval. Our group has implemented the fringe-imaging (FI) option of MI [5], for several reasons: First, a FIMI may be implemented with slanted mirrors, producing linear fringe patterns, easily to be imaged on fast linear detector arrays. This grants both optimum photon efficiency and possible high range resolution, in contrast to two-dimensional CCD-type detectors, as used in AWIATOR [6,7], e.g. Second, an airborne application calls for resilience towards vibrations and temperature variations, what may be responded to by a monolithic interferometer architecture, comparatively easily to be achieved with a MI. Third, a monolithic FIMI can be constructed to be both field widened (FWFIMI) and temperature compensated. Finally, such a FW design may additionally be arranged to be tilted to the incident light, enabling two-channel (transmit and reflected channels) operation.

Once the FW option established, this allows for implementing large-core multi-mode optical fibers (OF) for the transport of the backscattered optical flux from telescope collection to the interferometer, representing additional advantages: First, using OF eases the construction and implementation of such lidar systems, eventually for the operation within the confined space of an aircraft. They also allow for easy combination, splitting etc. via typical fiber modules such as couplers. Second, large-core OF provide scrambling [11, e.g.], i.e. annihilating of angular information, thus (at least theoretically) overcoming the need for range-resolved calibration. Further, OF allow for shaping the optical flux to nearly arbitrary shapes (like rectangular-top hat) better suited for detector illumination than "natural" circular-Gaussian ones. And last, while at the same time generating speckles, large-core OF facilitate the spatial averaging of atmospheric (molecular and aerosol backscatter-generated) speckles.

Our theoretical studies and simulations [5] regarding the possible performance of such an FWFIMI-based DD-DWL also addressed different types of pulsed laser types (regarding power and pulse repetition frequency – PRF), all in UV for its higher molecular backscatter and less eye-safety constraints. While these studies revealed the superiority of medium-high PRFs (around 1 kHz), in our demonstration experiments we used the airborne-proven WALES/DELICAT laser [12] with 100 Hz PRF due to its immediate availability. We realized the described FWFIMI option as an experimental DWL receiver system. In the next section we give a simplified synopsis of the developed FWFIMI-DD-DWL, while in Sect. 3 we shortly address the first ground-based validation experiments and their results.

2. THE FWFIMI-DWL SYSTEM

Fig. 2 shows a fairly simplified schematic of the DW lidar setup. The frequency-locked laser transmitter emits short UV laser pulses into the atmosphere in a mono-static, co-axial arrangement with the receiver telescope. The lidar receiver may be broken down into a front-end (RFE) and back-end (RBE) part. The RFE has the simple function of collecting the backscattered light, collimating it (by lens L, see Fig. 2) for background noise suppression by a narrow interference filter (IF) and subsequent concentration (L) in a large-core (600 μm) optical fiber (OF_S, with subscript S for signal) which actually defines the FOV. The transmitter is also equipped with a fiber launch and thus delivers its spectral reference pulses in an OF_R (subscript R for reference) that further serves as delay line in order to arrange the reference laser pulses temporally after the Doppler-shifted signal. Both paths are combined within a 3:1 custom-built 600 μm fiber coupler ($FC_{3:1}$). The third port may be used for analysis illumination by a UV LED, e.g.

For transport of the light to the RBE, scrambling and beamforming, we implemented, after performance tests of various large-core multi-mode OF, a square-core OF_{Scr} of 600 μm edge length, including a vibrating device (VM) for temporal fiber speckle averaging.

The RBE comprises the collimation of the light from the OF_{Scr} (by lens L), and the polarization separation by a polarizing beam splitter cube (PBSC). The loss of polarization (and thus 50% of the signal power here) is the major disadvantage of using MM OF since the beamsplitter coating of the MI requires s-polarization for delivering adequate fringe contrast. The p-part is focused onto a single PMT used for calibration and alignment (overlap) purposes. The s-part is then

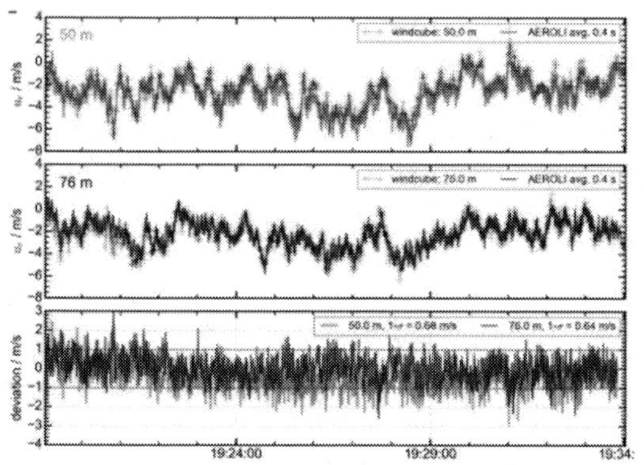

Figure 3: Exemplary horizontal wind measurement time series of the FWFIMI-DWL, bias removed, compared to Windcube® 200S data, both averaged to 2.5 Hz rate at center ranges 50 and 76 m. Lower panel: Respective deviations between the DWLs

fed into the MI which is mounted in an Invar bracket within a steel temperature-controlled housing (with optical windows W). The FWFIMI consists of a cubic beamsplitter of 30 mm edge length, an air arm of 11 mm with composite spacers (for temperature tuning optimization) and skewed mirror, and a 16 mm glass arm; these elements are all optically-contacted one to another resulting in a compact monolithic device.

The fringe localization plane is imaged by a lens L onto 16 elements of a linear photomultiplier array (PMTA), the vertical dimension (orthogonal to sketch plane in Fig. 2) being compressed by a cylindrical lens (L_{cyl}). The signal current of each element of the PMTA is converted to a voltage by a transimpedance amplifier and is then digitized by two eight-channel 16 bit, 30 MHz-bandwidth analog-to-digital converter boards, yielding a sampling rate of 30 MS/s. The atmospheric signal, the laser reference signal and a pre-trigger background signal are stored on a computer for each shot.

Currently, the retrieval of wind speed is performed offline since this routine is subject of ongoing research. The imaged compressed fringe amplitude has the approximate form of a cosine with skewness and kurtosis, resulting from imperfections of the MI optical surfaces, illumination etc. This fringe has to be corrected by an illumination function which is determined by a laser frequency sweep over the MI free spectral range (FSR). After a certain range averaging (six bins in the following example) and temporal averaging (40 pulses here) the temporally separated signal and reference fringes are each approximated by a four-parameter function (including the phase ϕ) employing a downhill search method. The phase comparison directly yields the wind speed, a shift of 5.6 MHz ($\approx$ 1/1900 of the FSR of 10.7 GHz) translating to 1 m/s.

3. GROUND-BASED VALIDATION

For validation of the concept and of our physical implementation we performed ground-based tests at DLR Oberpfaffenhofen. Here, our FWFIMI-DWL was arranged with horizontal LOS, alongside a commercial coherent DWL (Leosphere Windcube® 200S), the latter set at most to similar settings in terms of focusing, range and time averaging. Fig. 3 shows such comparative measurements for synthetic range gates of 30 m, centered at 50 m and 76 m distance to the lidar over a 15 min interval. An important bias (see below) has been removed but the respective short-term fluctuations of the very moderate wind fit well one to another. The lower panel shows the (bias-removed) difference between the two DWL series. Here, a joint standard deviation of 0.7 m/s may be determined. Assuming an upper bound for the Windcube® precision of 0.5 m/s (specified by the manufacturer), our DWL also yields this level of precision. Pearson correlation coefficients of 0.89 and 0.86, respectively, have been determined for these two series.

The above mentioned illumination function currently is a key point of investigation since its spurious fluctuations ultimately lead to time-variable bias of the wind measurement.
The shown analysis does not yet include investigations regarding signal-to-noise ratio and many other intricacies but merely demonstrates the potential of such a FWFIMI-DWL setup.

4. CONCLUSION AND OUTLOOK

We demonstrated the to our knowledge first direct-detection Doppler wind lidar based on fringe-imaging within a field-widened Michelson interferometer. We could show theoretically, by simulation and experimentally that this architecture is fundamentally appropriate for delivering wind speed measurements at high spatial and temporal resolution while yielding sub-m/s precision. In the near future we strive to evolve and enhance our experimental system based on the deficiencies identified so far. Part of the activities ahead are atmospheric trials in high altitude (i.e. molecular backscatter conditions), ultimately aiming airborne tests.

REFERENCES

[1] P. Vrancken, in *Aviation Turbulence*, edited by R. Sharman and T. Lane (Springer Int. 2016).
[2] P. D. Williams, Advances in Atmospheric Sciences **34**, 576 (2017).
[3] J. Ehlers et al., in *14th AIAA Aviation Technology, Integration, and Operations Conference* (AIAA 2014).
[4] N. Fezans and H.-D. Joos, in *AIAA Atmospheric Flight Mechanics Conference* (AIAA 2017).
[5] J. Herbst and P. Vrancken, Applied Optics **55**, 6910 (2016).
[6] N. P. Schmitt et al., Aerospace Science and Technology **11**, 546 (2007).
[7] G. J. Rabadan et al., Journal of Aircraft **47**, 392 (2010).
[8] D. Bruneau et al., Appl. Opt., AO **54**, 8776 (2015).
[9] S. C. Tucker et al., in *Lidar Remote Sensing for Environmental Monitoring XV* (SPIE 2015), p. 96120E.
[10] S. C. Tucker et al., JTECH **35**, 2079 (2018).
[11] G. Avila et al., in Proc. SPIE 7735, Ground-based and Airborne Instrumentation for Astronomy III, (2010), p. 773588.
[12] P. Vrancken et al., Applied Optics **55**, 9314 (2016).

EPJ Web Conferences **237**, 07009 (2020) https://doi.org/10.1051/epjconf/202023707009
ILRC 29

Development of a multiple-field-of-view multiple-scattering polarization lidar system at 355nm for cloud measurements

Tomoaki Nishizawa[1]*, Yoshitaka Jin[1], Nobuo Sugimoto[1], Hajime Okamoto[2], Masahiro Fujikawa[2], and Shoken Ishii[3]

[1]National Institute for Environmental Studies, 16-2 Onogawa, Tsukuba, 305-8506, Japan,
[2]Research Institute for Applied Mechanics, Kyushu University, 6-1 Kasugakoen, Kasuga 816-8580, Japan
[3]National Institute of Information and Communication Technology, 4-2-1 Nukiikitamachi, Koganei, 184-8795, Japan
**Email: nisizawa@nies.go.jp*

ABSTRACT
We developed a multiple-filed-of view multiple-scattering polarization lidar (MFMSPL) at 355nm to study microphysics of optically thick clouds and understand multiple scattering effects on space lidar measurements (e.g., EarthCARE). This system is based on depolarization Mie lidar technique; it uses a compact receiver module including a small telescope, polarizer, and detectors to measure co-polar and cross-polar backscatter signals, separately. The MFMSPL system has all the 10 measurement channels. The MFMSPL uses the five receiver modules with FOV of 10mrad tilted with different zenith angles (θ); it measures on-beam signals (θ=0mrad) as well as four off-beam signals (θ=10,20,30,40mrad). This MFMSPL system was based on previously developed 532nm MFMSPL system, however, we made the system more stable, robust, and compact by integrating co-polar and cross-polar channels in the receiver module than the previous system. In addition, we improved the dynamic range of the measurement in nighttime by synergy use of analog and photon counting measurements. In the conference, we present the MFMSPL system, data analysis method including calibration method, and observation results.

1. INTRODUCTION

Generally, lidar cannot penetrate optically thick clouds (or aerosols) due to strong attenuation; it generally detects only the bottom of the layer [1]. This limitation is different for space-borne lidar observations. The two-wavelength depolarization Mie lidar CALIOP [2] has a much larger footprint size (~90m) compared to that of conventional ground-based lidar (~1m at 1km altitude). This larger foot print enables the CALIOP to collect more multiple scattered signals and detect the inner part of the cloud layer. Ground-based lidars to measure multiple scattering signals have been developed. For example, Polonsky [3] developed a wide-angle imaging lidar by means of off-beam lidar returns. Cahalan [4] developed a multiple FOV lidar system using a fiber bandle. Though several "multiple scattering lidars" have been developed, there has been few multiple scattering lidar to measure depolarization ratio.

To obtain more information in the inner part of cloud layer and overcome the limitations of conventional ground-based lidars, we developed a multiple-filed-of view multiple-scattering polarization lidar (MFMSPL) at 532nm that can detect backscatter coefficients and depolarization ratios as similar as those observed by CALIOP [5]. This system was constructed by combining four channels with different zenith angles, such that each receiver channel has a 10mrad FOV and a polarization function. Thus, the total FOV of this system is 70mrad, comparable to the footprint size of the CALIOP at 1km altitude.

EarthCARE is a joint Japanese (JAXA)-European (ESA) satellite observation mission to understand the interaction between cloud, aerosol, and radiation processes in the earth climate [6]. In the EarthCARE mission, cloud and aerosol observation will be conducted by 355nm high-spectral resolution lidar ATLID.

In this study, we developed a MFMSPL system at 355nm by improving the previously developed MFMSPL, to develop and improve algorithms to retrieve optical and microphysical properties of clouds and aerosols from ATLID

https://doi.org/10.1051/epjconf/202023707009

measurements and validate the ATLID measurements. The total FOV of the developed system is 90mrad, covering the footprint of the ATLID (~30m). This paper describes the developed MFMSPL system (Section2), calibration method (Section3), observation results (Section 4), and summary (section 5).

2. MFMSPL SYSTEM

The configuration and specifications of the MFMSPL developed in this study are depicted in Fig. 1 and Table 1.

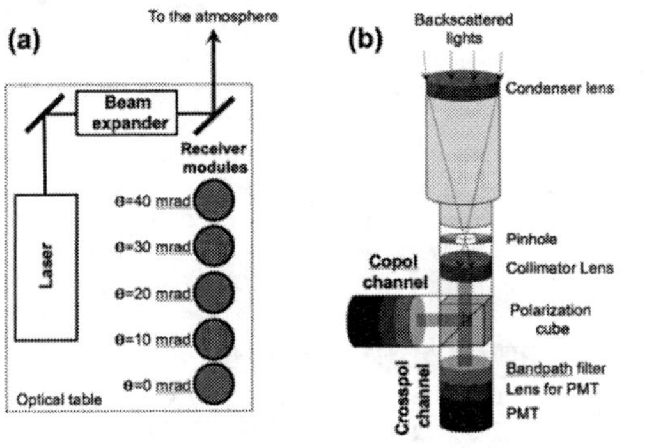

Fig.1 The MFMSPL system developed in this study. (a) Configuration of the entire system. (b) Configuration of the receiver module.

Table 1. Specifications of the MFMSPL specifications.

Transmitter	
Laser type	Nd:YAG, Q-switched, linearly polarized
Wavelength	355nm
Pulse energy	80mJ
Repetition rate	10Hz
Divergence	0.1mrad (using a 5x expander)
Receiver module (5 modules used in this system)	
Condenser lens	Φ=50mm, F~380mm
FOV	10mrad
Detector	PMT
Band-path filter	1nm (FWHM)
Data acquisition	
Transient recorder for analog measurement (40MHz, 16bit) and photon counting measurement (800MHz counting rate)	

The MFMSPL employs a commercial Nd:YAG laser with second- and third-harmonics generators (Surelite I, Continuum) and transmit the laser beam at 355nm to the atmosphere upward in the vertical direction. Backscattered lights are received by five receiver modules.

Each receiver module corrects the backscattered light with a 50mm-diameter lens and measures signal intensities of the co-polar and cross-polar components with two PMTs; the FOV of the receiver module is 10mrad. The five receiver modules are mounted and tilted from the vertical direction with different zenith angles θ; thus, they measure on-beam signals (θ=0mrad) and four off-beam signals (θ=10, 20, 30, and 40mrad). This MFMSPL system is based on previously developed 532nm MFMSPL system [5], however, we made the system more stable, robust, and compact by integrating co-polar and cross-polar channels in the receiver module. In addition, we improved dynamic range of the measurement in nighttime by synergy use of analog and photon counting measurements (TR40-16bit, Licel).

3. CALIBRATION

The receiver modules are tilted with different zenith angles in usual observation to measure on-beam signals and off-beam signals (Fig. 1), however, to calibrate the 10 channels of the MFMSPL relatively, we make all the five receiver modules upward in the vertical direction (i.e., θ=0mrad) and set a polarization sheet on each receiver module rotating in angle of 45° to the direction of the linear polarization of the laser as conducted in calibration of depolarization Mie lidars [7]. Using the measured signals, we obtain relative calibration constants CR. CR is obtained as the ratio of the measured signal at a channel P_{ch} to the signal for the co-polar component channel at θ=0mrad in the usual observation (P_1) (i.e., $CR_{ch} = P_1/P_{ch}$). This relative calibration measurement is periodically conducted.

To obtain attenuated backscatter coefficients for each channel (β_{atn}), we periodically calibrate the on-beam co-polar channel (i.e., co-polar channel at θ=0mrad) using the data measured in usual observation and Fernald method [8] as conducted in calibration of depolarization Mie lidar measurements [7]. In this calibration, we use data measured in clear-sky condition without cloud layers. Using the obtained calibration constant for the on-beam co-polar channel C_1 and relative calibration constant CR_{ch}, β_{atn} for each channel are obtained (i.e., $\beta_{atn,ch}= Z_2C_1CR_{ch}P_{ch}$, here Z is an altitude).

4. OBSERVATION AND ANALYSIS

The MFMSPL system is constructed at National Institute of Information and

Communication Technology (NICT) in Koganei, Tokyo, Japan. The signals are measured with 10s temporal resolution and with 3.75m range resolution. The measured data are averaged every 5min and 30m range; we used the averaged data in the calibration and data analysis.

Fig. 1 depicts an example of the observed and analyzed data. The data with less than 10 in signal-to-noise ratio are removed. We started usual observation from 10UTC after we conducted the relative calibration measurements. The on-beam channels indicate that cloud layers appeared from 4km to 8km on the day and it rained from 19UTC to 20UTC. The off-beam channels, which measure only multiple scattering signals and do not include single scattering signals, succeed in grasping cloud layers and rain drops appropriately. The figure also shows that the on-beam signals are fully attenuated in the lower layer of the cloud due to strong cloud attenuation (e.g., cloud layers appeared at 4km from 18UTC to 21UTC), however, the off-beam channels could detect the signals scattering from the higher layers of the cloud by multiple-scattering effect.

5. SUMMARY AND PLAN

More stable, robust, and compact the MFMSPL system was developed. We constructed the 355nm MFMSPL system in Japan and started continuous measurement in 2018. The measured data were calibrated periodically. A 355nm HSRL system is being constructed at NICT; this HSRL data will be used for the absolute calibration of the 355 MFMSPL in the near future. An improved 532nm MFMSPL system based on the system developed in this study are being constructed at National Institute for Environmental Studies in Tsukuba, Japan. The 355nm MFMSPL as well as the 532nm MFMSPL and the 355 HSRL are used for developments of the algorithms to retrieve optical and microphysical properties of clouds and aerosols and validations for EarthCARE and ADM-Aeolus [9].

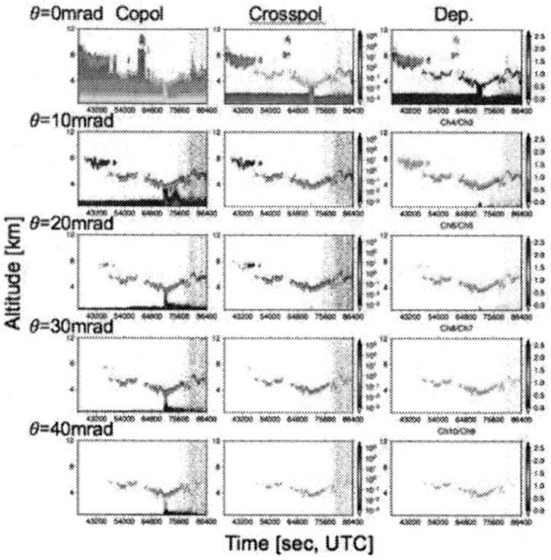

Fig. 2 Time-height cross sections of copol and crosspol components of the attenuated backscatter coefficients [/m/sr] and depolarization ratio for on-beam (θ=0mrad) and off-beam (θ=10, 20, 30, and 40 mrad) signals observed at Koganei, Tokyo, Japan on June 27, 2018.

ACKNOWLEDGEMENTS

This research was supported by a Grant-in-Aid for Scientific Research (JP17H06139) from the Japan Society for the Promotion of Science (JSPS) and research grant for the EarthCARE algorithm and validation research from JAXA.

REFERENCES

[1] Sassen K., et al., 1995: *Opt. Rev.*, **2**, 394-400.
[2] Winker D. M., et al., 2007: *Geophys. Res. Lett.*, **34**, L19803.
[3] Polonsky N. I., et al., 2005: *J. Atmos. Ocean. Tech.*, **22**, 628-648.
[4] Caharan R. F., et al.,2005: *J. Atmos. Ocean. Tech.*, **22**, 605-627.
[5] Okamoto et al., 2016: *Opt. Exp.*, **24**, 30053-30067.
[6] Illingworth et al., 2015: *Bull. Am. Meteor. Soc.*, **96**, 1311-1332.
[7] Shimizu, A., et al., 2016: *Opt. Eng.*, **56**, 031219, doi:10.1117/1.OE.56.3.031219.
[8] Fernald F. G., 1984: *Appl. Opt.*, **23**, 652.
[9]https://www.esa.int/Our_Activities/Operations/Aeolus_operations

Development of multifunction micro-pulse lidar at 1.5 micrometer

Mingjia Shangguan[1], Haiyun Xia [1,*], Xiankang Dou [1,2], Jiawei Qiu [1], Chao Yu [1]

[1]School of Earth and Space Science, University of Science and Technology of China, Hefei, 230026, China

[2]School of Electronic Information, Wuhan University, Wuhan, 430072, China

**Email: hsia@ustc.edu.cn*

ABSTRACT

Taking advantage of the 1.5 μm lidar, a series of 1.5 μm micro-pulse lidars have been developed at the University of Science and Technology of China, in Hefei, China. According to the different characteristics of three kinds of single-photon detectors at 1.5 μm, namely superconducting nanowire single-photon detector, up-conversion SPDs and InGaAs/InP single-photon avalanche diodes, different kinds of lidar systems have been constructed to realize the detection of atmospheric visibility, cloud, depolarization, wind field at the atmospheric boundary layer.

1. INTRODUCTION

Aerosol lidars play an important role in many fields, such as atmospheric environment management, air chemistry, hydrological cycle, radiation budget and public health. Compared with ultraviolet (UV) and visible systems, 1.5 μm lidars offer several advantages, including higher maximum permissible exposure to human eyes, lower atmospheric attenuation, weaker sky radiance and minor disturbance from Rayleigh scattering.

In order to guarantee a detection range of about 10 km, the pulse energy of the aerosol lidar should be higher than 100 mJ during clear weather conditions [1]. However, the peak power of such a laser pulse is far beyond the stimulated Brillouin scattering (SBS) threshold of any erbium-doped fiber amplifier (EDFA). Thus, complex and expensive solid-state techniques including stimulated Raman scattering (SRS), optical parametric oscillator, and resonator with Cr4_: YAG crystal are developed to generate high-power lasers at 1.5 μm for aerosol lidars [2]. In principle, the SNR of lidar is dominated by the product of laser power and the telescope area. Although the detector used in a lidar finally decides the quality of the raw data, its contribution is often neglected. Because the quantum efficiency is usually treated as a constant, the cathode material of a detector is selected according to its sensitivity versus the working wavelength. For example, InGaAs avalanche photodiode (APD) is used for 1.5-μm detection commonly.

Generally, traditional micro-pulse lidar is thought quite unsuitable for long-range aerosol detection with high temporal/spatial resolution [1]. In our work, by optimizing different single-photon detectors at 1.5 μm, different micro-pulse lidars incorporating a fiber laser and a small optical antenna are proposed and demonstrated for long-range atmospheric parameters detection during day and night.

2. METHODOLOGY

Using single-photon detectors (SPDs) at 1.5 μm can greatly improve the performance of aerosol lidars. Currently, the main techniques for single-photon detection at the telecom wavelength [2] include superconducting nanowire single-photon detectors (SNSPDs), up-conversion SPDs and InGaAs/InP single-photon avalanche diodes, as listed in Table I.

SNSPDs exhibit the excellent performance of high photon detection efficiency (PDE), low dark count rate (DCR) and low timing jitter [3]. Thus, by using the SNSPD to detect the weak backscattering with a high signal-to-noise ratio, the performance of the lidar systems can be enhanced.

The up-conversion SPDs exhibit moderate performance. It upconverts photons at communication band to visible photons when quasi-phase matching (QPM) is achieved in a periodically poled lithium niobate waveguide. Then single photons at 1.5 μm can be counted by using a Si: APD with high efficiency, low noise, and less afterpulse.

Compared with the above detectors, InGaAs/InP SPADs have relatively poor performance. However, such detectors are widely used in practical applications due to their advantages of small size, multi-mode fiber (MMF) coupling,

https://doi.org/10.1051/epjconf/202023707010

being polarization-independent, and low cost [4]. MMF coupling highly enhances the collection efficiency of backscattering signals distorted by atmospheric turbulence compared with SMF coupling, which, therefore, significantly reduces the volume and weight of telescopes.

Table 1 Comparison of different free-running SPDs

	SNSPD [3]	UCSPD [2]	InGaAs [4]
PDE	93 %	20 %	10 %
DCR	1cps	60 cps	950 cps
Afterpulse	0.0%	0.9%	18.0%
MCR	25 Mcps	37 Mcps	1.6 Mcps
Temperature	1 K	300 K	223 K
Fiber mode	SMF/MMF	SMF	SMF/MMF
Spectral response	Wide	Narrow	Wide
Polarization dependence	Yes/NO	Yes	NO

3. RESULTS

3.1 Visibility

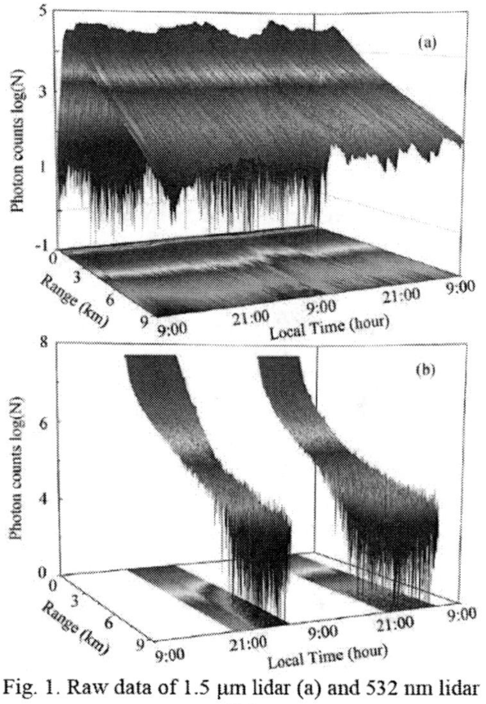

Fig. 1. Raw data of 1.5 μm lidar (a) and 532 nm lidar (b).

An adaptive inversion algorithm for 1.5 μm visibility lidar is proposed by using the particle size distribution of atmospheric aerosol [5]. The visibility with a temporal resolution of 5 min is detected from 9:00 on April 21 to 9:00 on April 23, 2017, in Hefei (31.83 °N, 117.25 °E). As shown in Fig. 1, 1.5 μm visibility lidar can work during day and night with lower noise than 532 nm lidar. In the two-day experiment, the extinction coefficient at 1.5 μm is always less than the extinction coefficient at 532 nm, as shown in Fig. 2. As we can see from Fig. 3, the average visibility error between the new method and a visibility sensor (Vaisala, PWD52) is 5.2% with the R-square value of 0.96, while the relative error between another reference visibility lidar at 532 nm and the visibility sensor is 6.7% with the R-square value of 0.91.

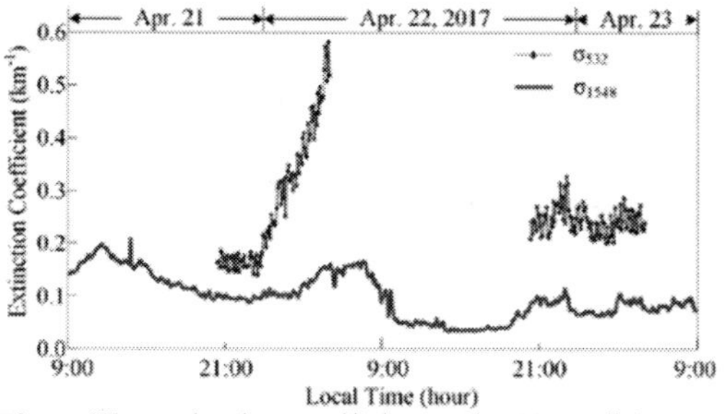

Fig. 2. The extinction coefficients of 1.55 μm lidar and 532 nm lidar.

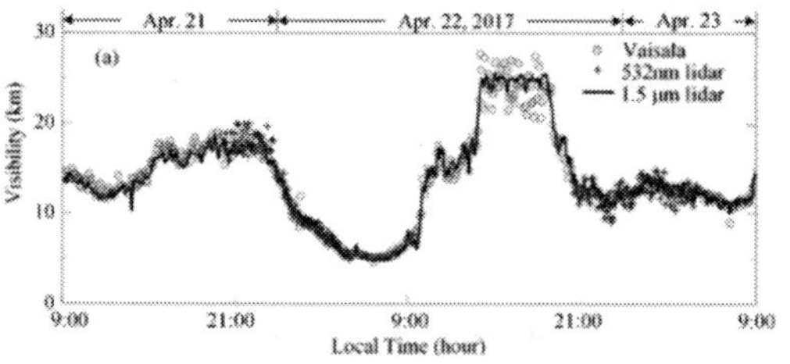

Fig. 3. Visibilities derived from different methods

3.2 Cloud

Since 1.5 μm laser has advantages of low atmospheric attenuation and low solar background, it's an ideal wavelength for cloud detection. To collect a weak cloud signal within 15 km, a multi-mode fiber (MMF) coupling InGaAs/InP single-photon detector is developed [6]. MMF matches larger receiving telescope than single-mode fiber (SMF) due to its large diameter and high numerical aperture. Using the MMF coupling receiver can enhance the collection efficiency by about 5 times, as shown in Fig.4. A specific

https://doi.org/10.1051/epjconf/202023707010

algorithm is also developed to obtain accurate cloud signals by correcting InGaAs/InP SPD noises and amplified spontaneous emission (ASE) noises of EDFA. Finally, we achieve the multi-layer cloud base height and thickness detection with 15 m range resolution and 1 s time resolution. The range corrected signal of a continuous observation is shown in Fig.5.

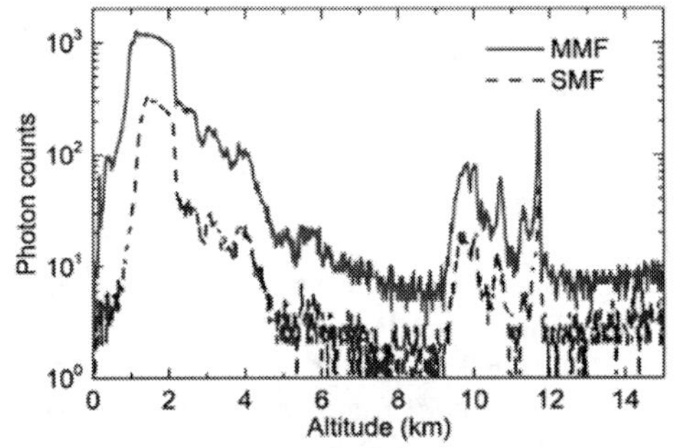

Fig. 4. Original photon counts for a vertical cloud detection observation using SMF and MMF receivers with a measurement time of 1 s.

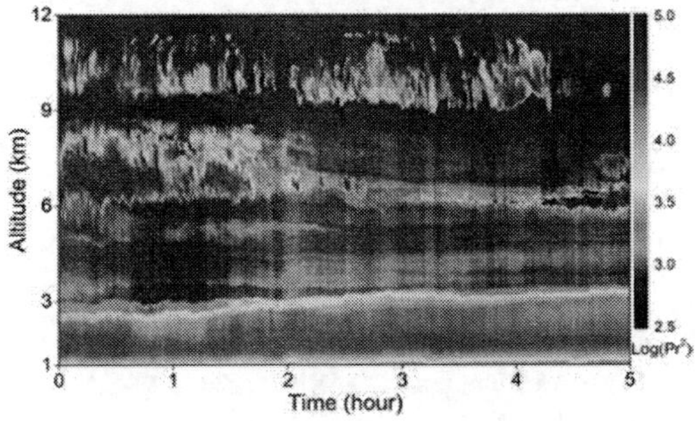

Fig. 5. Range corrected signal (log Pr2) of multi-layer clouds using the MMF receiver, observed from 00:00 to 05:00 on 24 March 2018.

3.3 Polarization

The polarization lidar has shown the capability of identifying different types of aerosols and clouds by analyzing the linear depolarization ratio (LDR). Taking advantage of the lowest attenuation (<1 dB/km) in the fiber and the maturity of optical devices at optical communication band, an all-fiber eye-safe polarization lidar incorporating a time-division multiplexing (TDM) technique is developed [7]. Only a single piece of the detector is used to detect the backscatter signals at two orthogonal states in an alternative sequence. Thus, regular calibration of the two detectors in traditional polarization lidars is avoided. The signal-to-noise ratio of the lidar is guaranteed by using an SNSPD, providing high detection efficiency and low dark count noise. The LDR of the urban aerosol is observed horizontally over 48 h in Hefei as shown in Fig. 6.

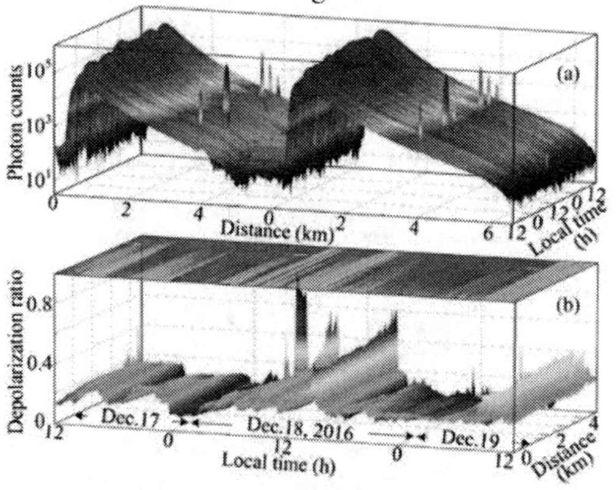

Fig. 6. (a) Raw signal of the polarization lidar over 48 h and (b) 48 h continuous measurement results of the LDR.

3.4 Wind filed

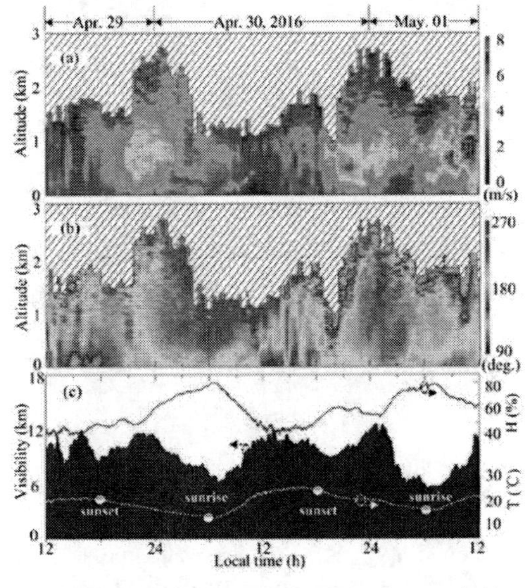

Fig. 7. Forty-eight-hour observation of atmospheric wind and visibility. (a) Wind speed, (b) direction, and (c) visibility, temperature, and humidity.

Wind field measurement is not only essential for flight safety and airport capacity, but also crucial importance for understanding the transport and dispersal processes of atmospheric pollutants [8]. High accuracy and real-time wind measurement will also address some vital needs such as Air Force operations, hurricane tracking, epidemic prevention and potential chemical biological release trajectory prediction.

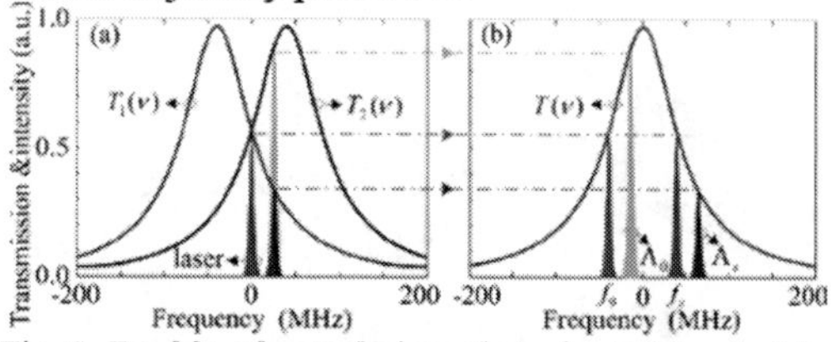

Fig. 8. Double-edge technique that adopts (a) a double-channel FPI and (b) dual-frequency laser pulses.

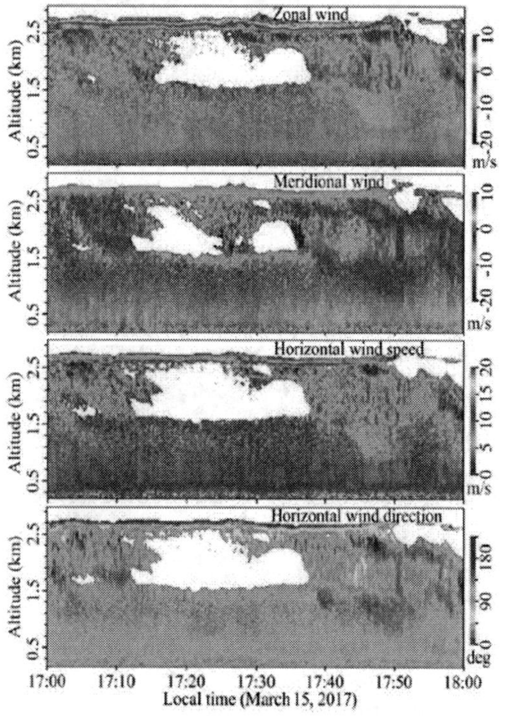

Fig. 9. One-hour observation results

By using an up-conversion single-photon detector, a versatile direct detection Doppler lidar is developed [9]. An all-fiber and polarization maintaining architecture is realized to guarantee the high optical coupling efficiency. Using integrated-optic components, the conservation of etendue of the optical receiver is achieved by manufacturing a fiber-coupled periodically poled lithium niobate waveguide and an all-fiber Fabry–Perot interferometer (FPI). The double-edge technique is implemented by using a convert single-channel FPI. The backscatter photons at 1548.1 nm are converted into 863 nm via mixing with a pump laser at 1950 nm. In experiments, atmospheric wind and visibility over 48 h are detected in the boundary layer, as shown in Fig. 7. In order to enhance the spatial resolution, a dual-frequency direct detection Doppler lidar is developed using an SNSPD at 1.5 μm [10]. The so-called double-edge technique is implemented by using dual-frequency laser pulses, rather than using a double-channel FPI, as shown in Fig. 8. Such a modification enhances the frequency stability at the system level. Two telescopes that point westward and northward at a zenith angle of 30° are used to detect the line-of-sight wind components, which are used to synthesize the horizontal wind profile. Horizontal wind profiles up to an altitude of about 2.7 km are calculated with vertical spatial/temporal resolution of 10 m/10 s, as shown in Fig. 9. The dynamic evolution of wind and vertical wind shears are observed clearly.

REFERENCES

[1] S. M. Spuler and S. D. Mayor, Proc. SPIE 6681, 668102 (2007).

[2] H. Xia, G.-L. Shentu, M. Shangguan, X.-X. Xia, X. Jia, C.Wang, J. Zhang, J. S. Pelc, M. M. Fejer, Q. Zhang, X. Dou, and J.-W. Pan, Opt. Lett. 40, 1579–1582 (2015).

[3] F. Marsili, V. B. Verma, J. A. Stern, S. Harrington, A. E. Lita, T. Gerrits, I. Vayshenker, B. Baek, M. D. Shaw, R. P. Mirin, and S. W. Nam, Nat. Photonics 7, 210–214 (2013).

[4] C.Yu, M. Shangguan, H. Xia, J. Zhang, X. Dou, and J.-W. Pan, Opt. Express 25, 14611–14620 (2017).

[5] X. Shang, H. Xia, X. Dou, M. Shangguan, M. Li. and C.Wang, Opt. Commn. 418, 129–134 (2018).

[6] Chao Yu, Jiawei Qiu, Haiyun Xia, Xiankang Dou, Jun Zhang, and Jian-Wei Pan, Rev. Sci. Instrum. 89, 103106 (2018).

[7] J. Qiu, H. Xia, M. Shangguan, X. Dou, M. Li, C. Wang, X. Shang, S. Lin, and J. Liu, Opt. Lett. 42, 4454–4457 (2017).

[8] M. Shangguan, H. Xia, C. Wang, J. Qiu, G.-L. Shentu, Q. Zhang, X. Dou, and J.-W. Pan, Opt. Express 24, 19322–19336 (2016).

[9] H. Xia, M. Shangguan, C. Wang, G.-L. Shentu, J. Qiu, Q. Zhang, X. Dou, and J.-W. Pan, Opt. Lett. 41, 5218–5221 (2016).

[10] M. Shangguan, H. Xia, C. Wang, J. Qiu, S. Lin, X. Dou, Q. Zhang, and J.-W. Pan, Opt. Lett. 42, 3541–3544 (2017).

DEVELOPMENT OF SYNERGETIC-ACTIVE SENSOR-SYSTEM FOR EVALUATION OF OBSERVATIONS BY EARTHCARE

Hajime Okamoto[1*], Kaori Sato[1], Masahiro Fujikawa[1], Eiji Oikawa[1], Tomoaki Nishizawa[2], Shoken Ishii[3], Yoshitaka Jin[2], Makoto Aoki[3], Nobuo Sugimoto[2]

[1]*Research Institute for Applied Mechanics, Kyushu University, Kasuga Fukuoka 816-8580, Japan*
[2]*National Institute for Environmental Studies, Tsukuba Ibaraki 305-8506, Japan*
[3]*National Institute of Information and Communications Technology, Koganei Tokyo 184-8795, Japan*

**Email: okamoto @riam.kyushu-u.ac.jp*

ABSTRACT

We develop the synergetic ground-based active-sensor-system for the evaluation of observations by space-borne lidars. The system consists of second version of multi-field-view multiple-scattering polarization lidar (MFMSPL-2), multiple-field-of-view high spectral resolution polarization lidar, direct-detection Doppler wind lidar, coherent Doppler wind lidar and 94GHz cloud profiling radar. The system can simulate observed signals from sensors onboard the joint Japanese/European mission Earth Clouds, Aerosols and Radiation Explorer (EarthCARE). The observation system can provide unique opportunity to study interaction of cloud microphysics, aerosol microphysics, vertical air motion and vertical distribution of horizontal wind and it will lead to evaluate cloud-convective parameterization and to reduce uncertainties in climate change predictions.

1. INTRODUCTION

Recently, we developed the first version of the multi-field-view multiple-scattering polarization lidar system (MFMSPL) by combining a 10 mrad-FOV and a polarization function with a total FOV >35 mrad [1]. The MFMSPL-1 has eight channels (four parallel and four perpendicular). The FOV of each telescope was chosen to be 10mrad. A wavelength 532nm was detected. There were four tilt angles, 0, 10, 20, and 30 mrad used to detect signals. The lidar was designed to detect lidar signals from optically thicker parts of clouds that a conventional lidar could not penetrate. It is also designed to detect similar magnitude of lidar signals influenced by multiple scattering detected by space-borne lidars such as Cloud-Aerosol Lidar with orthogonal Polarization (CALIPSO). Calibration procedure was described in [1]. Depolarization ratio values of low-level optically thick water clouds detected by the MFMSPL were as large as those detected by CALIPSO.

Recent theoretical achievement to treat lidar multiple scatterings was as follows. A practical model (called Physical Model: PM) for accounting lidar attenuated backscattering coefficient (ß) was developed for global space-borne lidar data [2]. Higher-order phase function was analytically derived to drastically reduce computational cost for simulating the angular distribution of the multiple scattering irradiance. The decay rate of the multiple scattering backscattered irradiance was given based on the path integral approach. The estimated ß by the PM was in good agreement with Monte Carlo simulations. The PM was extended with a polarization function to create a vectorized physical model (VPM) to analyze vertical profile of the observed depolarization ratio due to multiple scattering from water clouds by space-borne lidar [3]. The VPM incorporates a high-order scattering matrix and accommodates mechanics that modify the polarization state during multiple scattering events. The estimated depolarization ratio by the VPM showed good agreement with vectorized Monte Carlo simulations. VPM can simulate polarization properties under multiple scattering conditions effectively for the first time.

Japanese/European joint mission EarthCARE is scheduled to be launched in 2021. It will provide global profiles of microphysics of clouds, aerosols, rain and snow with their radiative fluxes and heating rates [4]. It will carry four instruments; (1) Atmospheric lidar (ATLID), (2) Cloud profiling radar (CPR), (3) Multispectral imager (MSI) and

(4) Broad band radiometer (BBR). ATLID is a 355-nm high spectral resolution lidar so that extinction coefficient and backscattering coefficient will be directly derived. Depolarization ratio will be also observed. Its foot print size is about 30m that is three times smaller than that of CALIPSO. Thus less degree of multiple scattering for ATLID is expected compared with CALIPSO for the same clouds. A two-dimensional diagram of lidar ratio and depolarization ratio was proved to be effective for discriminating among ice particle types and orientations [5].

In a frame of the MFMSPL-concept, we develop the 355-nm-MFMSPL-2 and HSRL in order to quantify the EarthCARE and Direct detection Doppler lidar at 355nm. Evaluation of EarthCARE and Aeolus algorithms is possible by the synergetic active sensor systems.

2. Synergetic active sensor system

MFMSPL-2 has ten channels, five parallel and five perpendicular. Five telescopes are used instead of ten telescopes by using polarization cube. The wavelength of 355nm is detected.

High spectral resolution lidar is developed. It has narrow (1mrad) and wide (10mrad) FOV to study multiple scattering effect in the backscattering and extinction retrievals in cloud observations. Wavelengths of 355nm and 532nm are panned to be detected. 532nm-HSRL system was completed. 355-nm-HSRL system and 355-nm-direct-detection Doppler lidar are under development.

3. RESULTS

Observation of low and high clouds by MFMSPL-2 was conducted on 23 October 2018 in Koganei, Tokyo, Japan. Attenuated backscattering coefficient detected by vertical point channel (CH.1) showed that the values of low-level clouds located at 5km at 17, 18.5 and 20 UTC exceeded 10^{-4} [1/m/sr] (Fig.1a).

Attenuated backscattering coefficient of upper clouds located between 8 to 10 km, were around 10-5 [1/m/sr]. Depolarization ratio of low-level clouds was 0.2-0.3 at vertical channel (Fig.1b). Upper clouds show depolarization ratio > 0.4, indicating ice clouds. Same clouds detected by off-beam channel (CH3), whose tilt angle is 10 mrad, showed smaller backscattering than for vertical channel (Fig,2a). Attenuated backscattering coefficient of low-level water clouds detected by off-beam channel showed larger values than those of ice clouds due to the large optical thickness and large contribution of multiple scattering component .

Depolarization ratio of water clouds detected by CH.3 and CH.4 ranged from 0.3 to 0.6 (Fig,2b). These values were larger than those by CH.1 and CH.2 and the increase was attributed to multiple scattering in optically thick water clouds. Depolarization of ice clouds revealed two different features. Thick ice clouds observed between 17 and 17.5 UTC showed significantly large depolarization ratio (>0.5) for off-beam channels. The value is larger than the value detected by vertical channels. Ice clouds detected after 18 UTC in off-mean channel produced small backscattering and depolarization ratio was not detected. These differences can be explained by the difference in optical thickness.

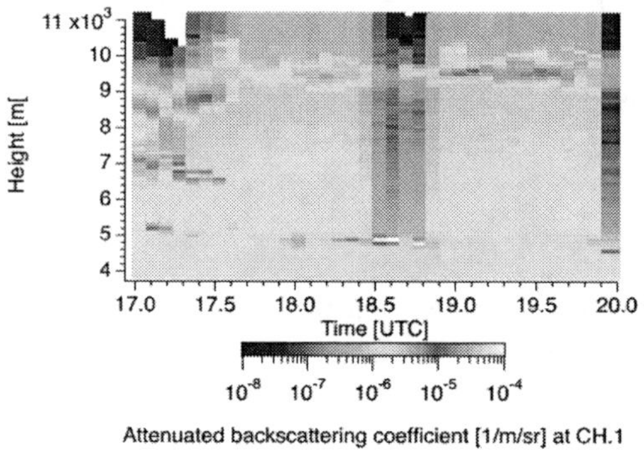

(a)

https://doi.org/10.1051/epjconf/202023707011

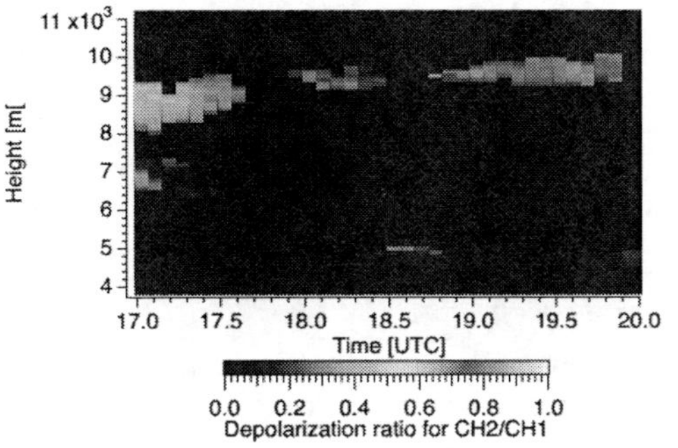

(b)

Fig. 1 (a) Time-height cross section of total attenuated backscattering coefficient of clouds and aerosols by on-beam channel, CH.1. Wavelength is 355nm in 23 October 2018. (b) Same as (a) but for depolarization ratio by CH.1 and CH2.

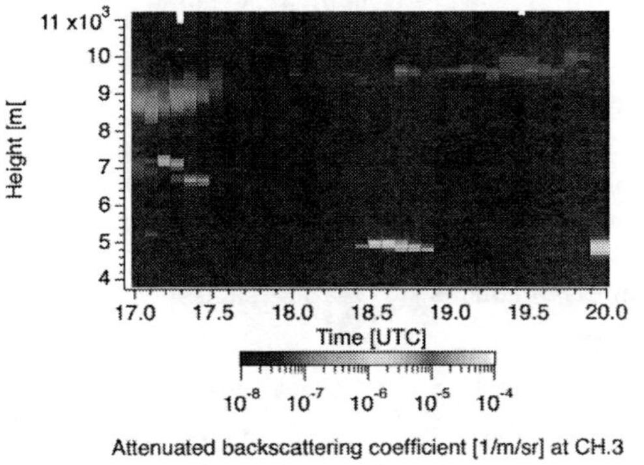

(a)

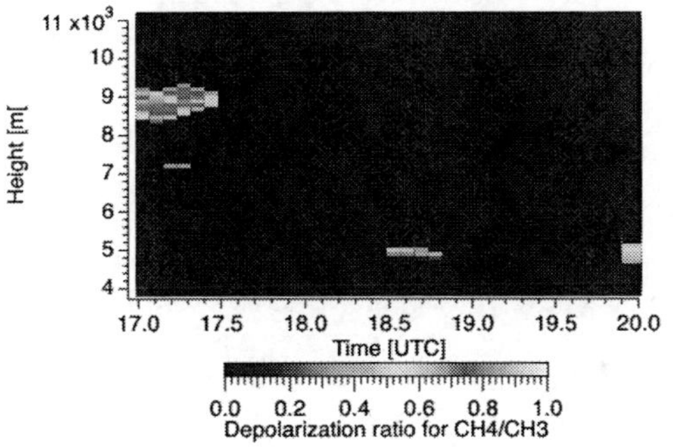

(b)

Fig.2 Same as Fig.1 but for off-beam channel. (a) attenuated backscattering coefficient for CH.3 and (b) depolarization ratio for tilt angle of 10 mrad by CH.3 and CH.4.

3. SUMMARRY

Ice and water clouds were observed by the MFMSPL-2 at 355nm. Results showed increase of depolarization for both of water and ice clouds when optical thickness in off-beam channel compared with on-beam values. This indicated backscattering coefficient as well as depolarization ratio are affected by multiple scattering in space-borne lidar such as CALIPSO lidar, ATLID and Aeolus. It is important to take into account FOV dependent multiple scattering contribution in the observed values to derive cloud microphysics. Multiple field view-HSRL will possibly provide further information of the effect and synergetic use of these two new lidars will help to develop algorithms for EarthCARE and Aeolus satellites.

ACKNOWLEDGEMENTS

This work was supported by the JSPS KAKENHI JP17H06139 and JP18K03745; by the Collaborated Research Program of Research Institute for Applied Mechanics, Kyushu University; the Japan Aerospace Exploration Agency for EarthCARE Research Announcement; The Arctic Challenge for Sustainability (ArCS).

REFERENCES

[1] H. Okamoto, et al. Opt. Express 24(26), 30053-30067 (2016)
[2] K. Sato, et al. Opt. Express 26(6), A301-A319 (2018)
[3] K. Sato, et al. Opt. Express 27(4), A117-A132 (2019)
[4] A. Illingworth, et al. BAMS 96(8), 1311-1332 (2015)
[5] H. Okamoto, et al. Opt. Express 27(25), 36587-36600 (2019)

DISTRIBUTION MAP OF PLANT FLUORESCENCE SPECTRUM IN THREE-DIMENSIONS CREATED BY A LASER-INDUCED FLUORESCENCE SPECTRUM (LIFS) LIDAR OBSERVATIONS

S. Utsunomiya, Y Saito*, Y. Kumagai, and T.Tomida.

Institute of Engineering, Shinshu University, 4-17-1 Wakasato, Nagano City, Nagano, 380-8553, Japan
**Email: saitoh@cs.shinshu-u.ac.jp*

ABSTRACT

We have developed a vegetation monitoring lidar called a laser-induced fluorescence spectrum (LIFS) lidar which is able to get the living status of plants by observing their fluorescence remotely. The features of its operation are; daytime observations possible even outdoors, mobility and self-sufficiency, capability of mapping plant living information, and a user-friendly operation by unifying the controls of different equipment of the lidar using software. These features make observations by our LIFS lidar possible at any time and any place. In forest observations, we could depict three-dimensional structures of fluorescence spectrums. We also discuss in this work the possibility of monitoring other plant physiological information such as the concentration of chlorophyll and photosynthesis secondary metabolites in this work.

1. INTRODUCTION

Photosynthesis, the most well-known activity of plants, supplies food and oxygen that is indispensable to our survival. Furthermore, various effects of the biological, chemical, and physical processes of plants influence the formation of the earth's natural environmental system. For example, gaseous biogenic volatile organic compounds (BVOCs) like aromatic compounds generate new compounds or exclude some through chemical reaction with various kinds of molecules in the Earth.[1] Physical reactions in a forest such as the evaporation of a water droplet splashing on a leaf surface[2] or a water micro-burst on the ground surface[3] could be a great source of forest aerosol. These activities occur continuously, so information about the living status of a plant itself or vegetation such as forests or mountainsides is considerably important in maintaining the continuity of not only the ear-th's environment, but also human beings' society.

In this paper, we discuss the potential of the LIFS lidar, which we developed for such purposes, based on the results of field investigations.

2. METHODOLOGY

2.1 Fluorescence spectroscopy

Plants as living organisms need to be monitored using non-destructive techniques. Chemical analyses which involve picking, sampling, crushing plant parts or applying chemical liquid treatments should be avoided.

On the other hand, in plant science, many optical methods have been applied to understand the living status of plants. In particular, fluorescence spectroscopy is widely used and there are many reports on plant physiological monitoring using the fluorescence method for studying chlorophyll photosynthesis[4], photochemical reactions[5], environmental-stress[6] and others. In this work, we study the changes of fluorescence spectrum shapes of plants to gather such physiological information.

2.2 Lidar remote sensing techniques

Monitoring the living status of plants should be done under natural conditions where the plants grow, but these locations/places are often out of reach for humans. Lidar remote sensing techniques provide important possibilities in such situations.

2.3 Laser-induced fluorescence spectrum (LIFS) lidar

A fluorescence lidar that combines fluorescence spectroscopy and lidar remote sensing techniques can be a powerful apparatus in tree/forest surveys. We named this type of fluorescence lidar as a laser-induced fluorescence spectrum (LIFS) lidar and we can use it to investigate the living status of plants by monitoring the changes of their fluorescence spectrum shapes remotely.

https://doi.org/10.1051/epjconf/202023707012

3. CONSTRUCTION OF LIFS LIDAR SYSTEM

3.1 Requirements

In the development of the LIFS lidar, we required the following functions.

1) Daytime and outdoor operation: As described above, photosynthesis is one of the key characteristics of plants and we require the LIFS lidar to be able to operate in daylight. This requires a technique for reducing the background sun which is a major source of noise in lidar observations.
2) Mapping: To survey the forest living status in which many plants and trees are mixed and grow at different locations, the system should have a function to create a three-dimensional structure of the distribution.

3) User-friendly operation: Because the system consists of several devices and equipment having different types of control mechanisms, it is necessary to unite the differences.

3.2 Construction and performance check

The configuration and specification of the LIFS lidar we developed based on the requirements in 3.1 are shown in Fig. 1 and in Table 1.

1) To enable daytime and outdoor operations, we adapted the synchronous detection technique[7] and prepared a photonic multi-channel analyzer system (PMA-12 (C10029-01), Hamamatsu Photonics, Japan) whose detector was a gated intensified CCD. The CCD gate used as a lidar detector opens to coincide only with the arrival of fluorescence emissions corresponding to laser irradiations.
The actual effect of the synchronous detection technique on reducing the solar background was checked. In Fig. 2, we can see that the back-ground intensity constantly decreases in proportion to the CCD gate time-width. The noise that was still left at 10 ns gate time-width appears to be the noise of the detection system itself which includes the thermal noise of the CCD device and the electrical noise of the system.
The size of the LIFS lidar was 2.5 m x 0.5m x 0.8 m and it operates on a stretcher using a small dynamo with power of 900 VI, so that it can be loaded in a small commercial van-type car which

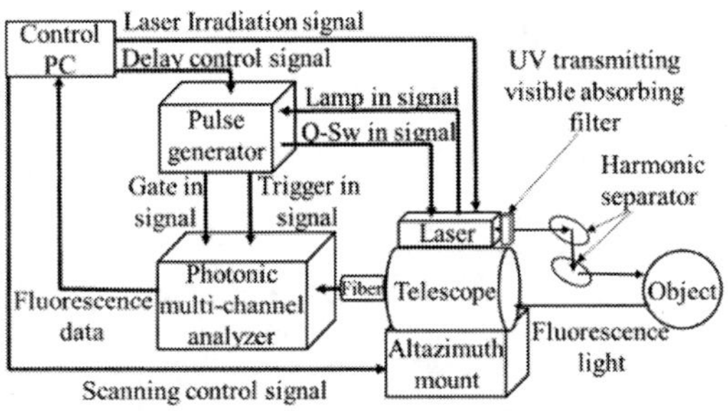

Fig. 1 Block diagram of the LIFS lidar. Operational parameter of each device and the flow of signals and data are shown.

Table 1 Specifications of the LIFS lidar

Transmitter		
Laser	Wavelength	355nm
	Pulse energy	20mJ
	Pulse width	6ns
	Reception	10Hz
	Filter	Short pass < 400nm
Receiver		
Telescope	Type	Schmidt-Cassegrainian
	Effective area	425cm²
	Focal length	2000mm
Fiber	Inlet	1mm
	Outlet	Rectangle-shape
	Filter	Long-pass >370nm
Detector		
Spectrometer	Range	200-860nm
	Resolution	<3nm
Sensor	CCD	Electronically cooled type
	Image Intensifier	10ns gate
Control/Processing		
Delay/pulse generator	Resolution	5ns
PC	OS	Windows10

is indispensable for outdoor observations such as forest investigations where the actual roadways to approach the place may be narrow and winding. This mobile and self-sufficient LIFS lidar arrangement made observations at any time and any place possible.

2) Mapping: A scanning function was added to the telescope operation. A laser beam is passed into the scanning area to get information at various depth. Horizontal and vertical scanning offers two-dimensional data at various distances and a three-dimensional structure of plant fluorescence can be made.

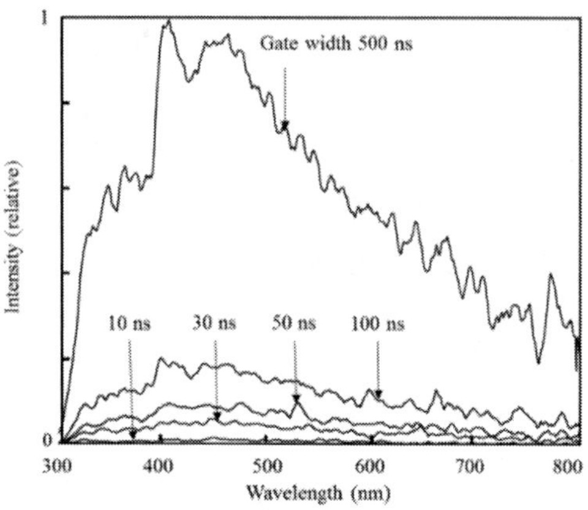

Fig. 2 Dependency of solar spectrum intensity on gate time-width.

3) User-friendly operation: We developed new software that could unite the operations of different devices/equipment. Each device had its own specifications for communication methods, software language, and operating systems. The flow of control of signals and data is shown in Fig. 1 and the structure of the new software for the LIFS lidar is shown in Fig. 3(a). The PMA was controlled only by GUI software with the AutoIt language that was provided by the manufacturer. Other pieces of equipment were unified in C++ for maintainability. To be able to use these languages on a Microsoft Windows PC, a cygwin environment that allows programs of Unix-like systems to run was prepared. Each program was again integrated with the Ruby language and We were able to operate the LIFS lidar operation on the command line.

4. OBSERVATIONS

4.1 Three-dimensional structures of chlorophyll concentrations of a single tree

Before conducting field observations, we checked the performance of the LIFS lidar. The lidar was positioned on a third floor veranda to point to the upper part of a zelkova tree.

To capture information at different depths, the distance to the object was changed by varying the delay-time of the CCD gate-open time in 10 ns increments which corresponds to a range resolution changes of 1.5m. The lidar scanned over an area of 13 m (horizontal distance, H) with

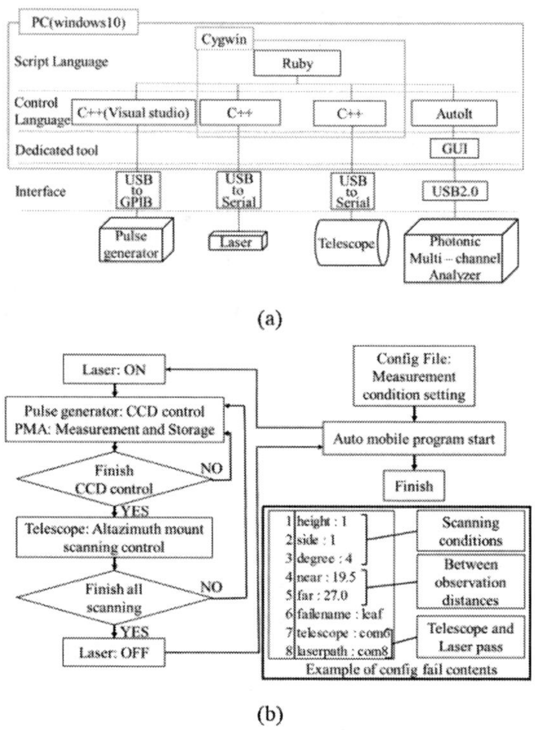

Fig. 3 Structure of the LIFS lidar software (a) and flow chart (b).

a horizontal angle resolution of 1 degree x 13m (vertical distance, V) with a vertical angle resolution of 1 degree.

It is known that the intensity ratio of chlorophyll fluorescence at 685 nm and 740 nm is an index of chlorophyll concentration[8]. The relationship between the ratio and chlorophyll concentrations measured by a high performance liquid chromatograph (HPLC) was investigated. The relationship was applied to the lidar observation result which offered the ratio. The distribution maps of chlorophyll concentrations of the zelkova tree are depicted in Fig. 4.

4.2 Large scale structures of forest fluorescence spectrums

Forest observations were done in locations where different kinds of trees grew naturally. Fig.5 shows the structure of fluorescence spectrum of vegetation. The covered area was 30 m (H) with a horizontal angle resolution of 3 degrees x 30 m (V) with a vertical resolution of 3 degrees x 20 m

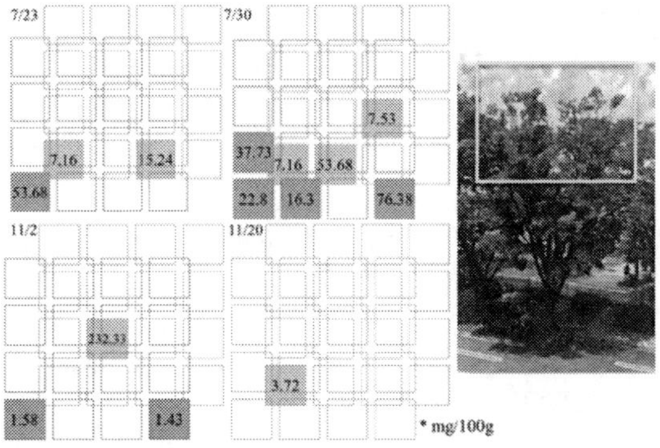

Fig. 4 Distribution map of chlorophyll concentration of a single zelkova tree.

(depth) with depth range resolution of 1.5 m. The total number of pixels of the structure was 192. The range resolution in the depth direction was constant, but those in the horizontal and vertical directions became larger as the laser moved away from the system as shown in the figure.

The spectral shape appears to change depending on the tree location and type. Several papers report that plant fluorescence spectrums at around 400-650 nm involve information about photosynthetic secondary metabolites which are one of the key molecules to understand the status of plant growth.[5] A detailed analysis of the lidar data still remains, but the usefulness of the LIFS lidar and the time effectiveness of constructing distribution maps in three-dimension was fully shown.

5. CONCLUSIONS

A mobile, self-efficient, automated, scanning LIFS lidar was developed and its performance was tested in the construction of three-dimensional structures of plant fluorescence in daytime observations.

The distribution maps of chlorophyll concentrations of a single tree clearly showed differences in leaf location and in season. In forest observations, the fluorescence spectrum of different kinds of trees distributed over a large area could be obtained.

The growth status of vegetation may complexly depend on environmental factors such as soil, water, air, and others. We have already checked that our LIFS lidar has the potential to detect fluorescence for such parameters.[9] Our interest is a fusion experiment of monitoring plant living information together with plant-growth environment conditions using the LIFS lidar.

Fig. 5 Distribution map of the fluorescence spectrum of a forest.

ACKNOWLEDGEMENTS

This work was supported by the 27th Botanical Research Grant of The New Technology Development Foundation and the joint research program of CEReS at Chiba University.

REFERENCES

[1] Y. Saito et al., Applied Optics, Vol. 40, pp. 3572-3574, 2001.

[2] S. Murakami, J. Hydrol., Vol. 319, pp.72-82, 2006.

[3] Y. S. Joung and C. R. Buie, Nat. Commun, Vol. 6, 6083, 2015.

[4] *Chlorophyll a Fluorescence* (G. C. Papageorgiou and Govindjee Eds.), Springer, 2004.

[5] Z. G. Cerovic et al., Agronomie 19, pp.543-578, 1999.

[6] Y. Saito, in *Recent Progress of Bio/Chemiluminescence an Fluorescence Analysis in Photosynthesis* (N. Wada an M. Mimura Eds.), Chapter 11, Research Signpost, 2005.

[7] Y. Saito et al., Applied Optics, Vol.#55, pp.6727-6734, 2016.

[8] H. K. Lichtenthaler, J. Plant Physiol. 131, pp.101-110, 1987.

[9] Y. Saito et al., Proceeding of SPIE, Vol. 10779 107790M; doi: 10.1117/12.2324428.

 https://doi.org/10.1051/epjconf/202023707013

DRONE-BASED FLUORESCENCE LIDAR SYSTEMS FOR VEGETATION AND MARINE ENVIRONMENT MONITORING

Zheng Duan[1], Ying Li[1], Xun Wang[1], Jinlei Wang[1], Mikkel Brydegaard[2], Guangyu Zhao[1] and Sune Svanberg[1,2*]

[1]Center for Optical and Electromagnetic Research, South China Academy of Advanced Optoelectronics, South China Normal University, University City Campus, Guangzhou 510006, China
[2]Department of Physics, Lund University, SE 221 00 Lund, Sweden
**Email: sune.svanberg@fysik.lth.se*

ABSTRACT

We have developed two different types of drone-based fluorescence lidar systems for vegetation and marine environment monitoring, both based on violet CW diode lasers. An inelastic hyperspectral Scheimpflug lidar system was used for vegetation profiling combined with fluorescence spectral recordings. A light-weight fluorosensor set for fixed-height recordings was employed for monitoring of marine environments, featuring water Raman signals, algal chlorophyll and strong oil spill fluorescence.

1. INTRODUCTION

In the study of vegetation in agriculture and forestry, chlorophyll, responsible for photo-synthesis, is an essential indicator to reflect the growing conditions of vegetation. Chemical methods to monitor chlorophyll inside vegetation, e.g., extraction by acetone [1] are time-consuming and only available for laboratory quantification. Optical methods to monitor vegetation have merits of fast-response, non-invasive detection and high-sensitivity and are adapted to meet the need for large-scale monitoring and remote sensing. Laser-induced fluorescence spectroscopy (LIF) techniques are highly suitable for real-time, in-situ and non-invasive monitoring of vegetation status. When a short-wavelength excitation laser beam is impinging on vegetation, chlorophyll will efficiently absorb the radiation and re-emit fluorescence at longer wavelengths, which provides a finger-print characteristic of chlorophyll [2]. Combined with remote sensing techniques, the laser-induced fluorescence technique was applied to early remote monitoring studies of vegetation, e.g., spruce, maize and maple [3]; sugar beets [4] and corn [5]. In the past years we constructed a flexible mobile pulsed laser lidar system for environmental monitoring [6] and we reported fluorescence lidar applications for hybrid maize and hybrid rice [7,8]. However, while perfect for monitoring of, e.g., atmospheric atomic mercury [9], such a system is complex and expensive to operate when it comes to fluorescence monitoring, which is much less demanding [10]. The introduction of a compact inelastic hyperspectral Scheimpflug lidar system [11], makes it possible to realize simultaneous spatial and spectral information of distributed targets. Based on a commercial drone platform, we have developed a miniaturized hyperspectral lidar system for profiling the vegetation structure [12].

2. METHODOLOGY

2.1 Drone-based inelastic hyperspectral Scheimpflug lidar for vegetation monitoring

The setup used in our vegetation studies can be seen in Fig. 1. A commercial drone (DJI M600 Pro) with a maximum lifting capacity of 6.2 kg is utilized to carry the hyperspectral Scheimpflug lidar system. The excitation laser is a 1.5 W 445 nm Nichia diode laser with an adjustable-focus objective from Computar. A common industrial control computer is used for data capturing and storage. When the backscattered light from the objects is captured by the receiving telescope, the light, after beam redirection by a folding mirror, will pass through a slit with a width of 100 µm. A prism-grating-prism structure is used to disperse the light. After collimation by lenses, the light will be distributed on the imaging plane of an area CCD. A combination of the spatial and spectral information will be obtained in every frame.

https://doi.org/10.1051/epjconf/202023707013

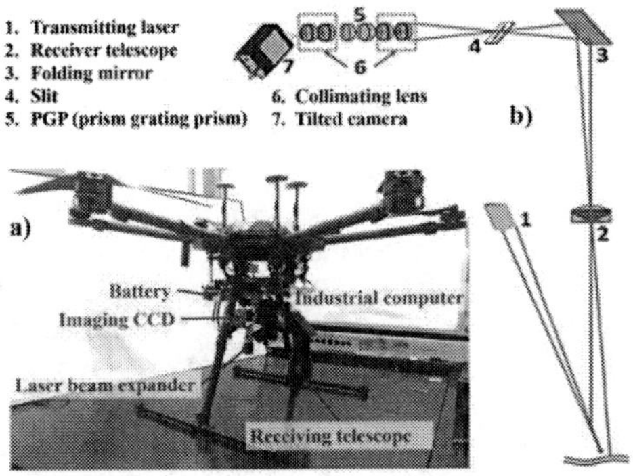

Fig. 1. a) Drone-based inelastic hyperspectral Scheimpflug lidar for vegetation monitoring; b) Schematic of the lidar system, modified from [12].

2.2 Drone-based fluorosensor for marine environment monitoring

The fluorosensor is equipped with a 1W, 412 nm diode laser (Xinrui, Model 412) and uses a compact spectrometer (USB4000 Ocean Optics) as the detector as seen in Fig. 2. A 425 nm long-pass filter (Edmund Optics #84-742) is used to cut off the excitation laser radiation. By tilting the laser, which is mounted non-coaxially next to the 50 mm diameter telescope with 200 mm focus, the measurement range is set to about 10 m. The laser beam is transmitted from the diode laser using a small collimator, and a telescope with a filter mounted after the objective is used to collect the fluorescence from the target. The laser can be directed at a proper angle to overlap with the telescope field-of-view as tuned according to the distance to the target. After the deflection from the mirror, the fluorescence is focused into the spectrometer and the data will be stored in an industrial control computer.

3. RESULTS

3.1 Vegetation monitoring

As a demonstration, we have monitored two trees behind our laboratory on the University City campus of South China Normal University with

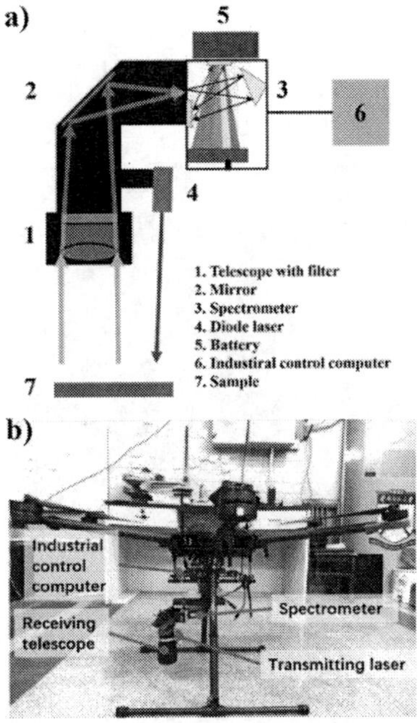

Fig. 2. a) Schematic of the fluorosensor; b) Drone-based fluorosensor for marine environment monitoring.

our inelastic hyperspectral Scheimpflug lidar based on a drone. As shown in Fig. 3, through integrating GPS (global positioning information) and INS (inertia navigation system) data with recorded frames, we can reconstruct these two trees with 3D point clouds with color coding of the spectral information. The fluorescence ratio between 690 nm and 730 nm was calculated as a relative measure of the concentration of chlorophyll in the vegetation. The yellow spots indicate the elastic signals without spectral information. Light from the road lamp constituting background noise could be suppressed by modulating the laser output.

3.2 Marine environment monitoring

The drone-based fluorosensor system is utilized to perform marine environment monitoring on the Zhujiang (Pearl) river, flowing with many branches through the city of Guangzhou. Two 90-cm-diameter floating containers were positioned on the surface of the river as artificial targets. One was filled with 1 litre of pure engine oil, resulting in layer thickness of typically 1-2 mm, and the other one was filled with 10 litres of a dilute solution of Rhodamine 610 dye with the

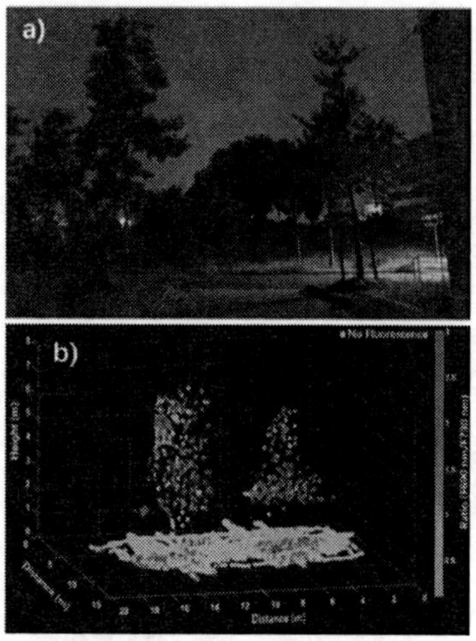

Fig. 3. a) Photo of two trees as targets; b) Point cloud figure of two trees with spectral information captured by the drone lidar; Color map coding corresponding to the ratio of chlorophyll fluorescence (F690/F730) is employed; modified from [12].

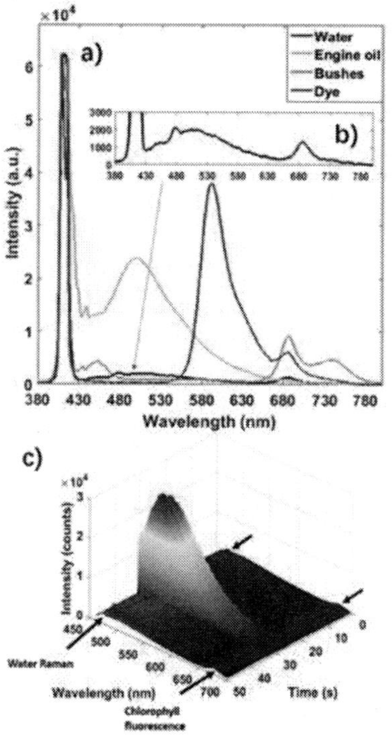

Fig. 4. a) Spectra captured by the drone-based flurosensor over the Zhujiang river; b) Zoom-in spectrum of natural Zhujiang river; c) Result of a linear scan of an oil-filled floating container on Zhujiang river. The exposure time of the spectrometer was set to 0.5 s; modified from [17].

concentration of 0.03g/l, resulting in a 1-2 cm layer thickness. The dye is dissolved in natural Zhujiang river water. With the drone scanning along a setup flight route covering the artificial targets, the measurements are illustrated in Fig. 4 a) and b). Four typical spectra recorded by the flurosensor are presented. The characteristic fluorescence peak of oil at 500 nm and dye fluorescence peak at 580 nm can be observed. Vegetation on the riverbank was also covered, and the strong chlorophyll characteristic peaks at 685 nm and 745 nm are recorded. In the zoom-in spectrum of the natural Zhujiang water, strong dissolved organic matter (DOM) fluorescence and algal chlorophyll signals are shown, as well as the water Raman peak at 480 nm. A linear scan of the oil-filled container with a continuous recording of 50 s is presented in Fig. 4 c). High sensitivity of the fluorosensor towards oil spills can be noticed, as contrasted to the background water Raman, algal chlorophyll and DOM signals.

ACKNOWLEDGEMENTS

This work was supported by the National Science Foundation of China under Grant No. 61705069 and the Chinese Ministry of Science and Technology through the National Key Research and Development Program of China under Grant No. 2018YFC1407503 and the Special Funds Program for the Cultivation of Guangdong College Students´ Scientific and Technologicaol Innovation Climbing Project under Grant No. pdjha0127.

REFERENCES

[1] H. K. Lichtenthaler, Methods in Enzymology 148: 350-382 (1987)
[2] H. K. Lichtenthaler, et al. CRC Critical Reviews in Analytical Chemistry 19: S29-S58 (1988)
[3] S. Svanberg, Physica. Scripta T58: 79-85 (1995)
[4] Z. G. Cerovic, et al. Biochimica et Biophysica Acta (BBA) - Bioenergetics 1188: 58-68 (1994)
[5] B. Xu, et al. Journal of Remote Sensing 7(S1): 51-59 (1992)
[6] G. Y. Zhao, et al. Applied Optics 56(5): 1506-1516 (2017)
[7] G. Y. Zhao, et al. Applied Optics 55(19): 5273-5279 (2016)
[8] Z. Duan, et al. Applied Optics 57(13): 3481-3487 (2018)
[9] M. Lian, et al. Environmental Pollution 240: 353-358 (2018)

[10] S. Svanberg, in Laser Remote Sensing (Eds: T. Fujii, T. Fukuchi), CRC Press, Boca Raton, FL 2005, pp 433-467
[11] G. Y. Zhao, et al. Lasers & Photonics Reviews 10(5): 807-813 (2016)
[12] X. Wang, et al. Applied Physics B 124: 207 (2018)
[13] S. Svanberg, in Optoelectronics for Environmental Science (Eds: S. Martellucci, A. N. Chester), Springer, Berlin Heidelberg 1990, pp 15-27
[14] F. E. Hoge, et al. Applied Optics 19(19): 3269-3281 (1980)
[15] M. Lenon, et al. European Oceans 2005(2): 848-853 (2005)
[16] A. Pisao, et al. Remote Sensing 7(1): 1112-1134 (2015)
[17] Z. Duan, et al. Applied Physics B 125(11): 108 (2019)

EXPERIMENTAL DETERMINATION OF LIDAR OVERLAP PROFILE BASED ON DUAL FIELD-OF-VIEW HIGH SPECTRAL RESOLUTION LIDAR

Nanchao Wang[1], Xue Shen[1], Yudi Zhou[1], Chong Liu[1], Yupeng Zhang[1], Ju Ke[1], Da Xiao[1], Tianfen Zhong[1], Yuhang Rong[1], Binyu Wang[1] , Jie Chen[1] and Dong Liu[1,*]

[1]State key Laboratory of Modern Optical Instrumentation, College of Optical and Engineering, Zhejiang University, Hangzhou, Zhejiang 310027, China
**Email: liudongopt@zju.edu.cn*

ABSTRACT

This paper presents two approaches to calibrate the overlap factor under inhomogeneous atmospheric condition without critical assumption and delivers detailed analysis about the retrieval errors of overlap profile in High-Spectral-Resolution-Lidar (HSRL). The first method employs an additional optical subsystem with different field-of-view, that is dual field-of-view HSRL, for the retrieval of overlap profile. The second method takes advantage of the difference of the result between the HSRL and Klett method, that is about the retrieval of backscatter coefficient for uncorrected lidar signal, to correct overlap profile. Surprisingly, two methods show very high-level consistency and stability of the result. It is potential that this technique would be an excellent solution for experimental determination of lidar overlap in ground-based HSRL.

1. INTRODUCTION

In general, literature has reported various methods to correct the overlap profile, which could be grouped into two categories: theoretical and experimental approaches. The calculations of theoretical approaches are based on the specifications of the Lidar system as the divergence angle of laser beam, the direction of laser beam and the field-of-view of the receiver, etc. The theoretical approaches could indicate what effect the configuration of Lidar system change will have on the overlap profile, and have directive meaning for designing Lidar system. The inconvenient of such approaches is that they require a good knowledge of specification to derive an overlap profile with enough accuracy, which is quite difficult to measure in the general case. The experimental determination of the overlap profile is required to derive an accurate correction with few priori knowledge. Sasano et al. [1] proposed a simple method to derive overlap profile assuming homogeneous atmospheric condition. Also, Tomine et al.[2] presented a method under light fog and mist condition. An approach was reported by Dho et al.[3], that could be applied in inhomogeneous condition, based on a polynomial fitting. An enlightening approach widely used in Raman lidar about experimental determination of lidar overlap profile was reported by U. Wandinger[4], and without critical assumption except lidar ratio. And on this basis, J. Luis et al.[5] compared this method with the measurement of overlap profile with additional 910nm ceiliometer in Ramen lidar, which is similar to our idea.

As a summary of previous work, either system specifications, atmospheric conditions, or lidar ratios are required for estimation of the overlap profile. Additional assumption inevitably leads to uncertainties in the results. And for all we know, there is limited research for experimental determination of lidar overlap without assumption in the HSRL. In this paper, we present two simple method for experimental determination of overlap profile for dual-field-of-view HSRL with no prior assumption.

2. INSTRUMENTATION

A highly robust dual-field-of-view HSRL that provides measurement of aerosol optical properties at 532nm has been developed, tested, and deployed on field experiments by Zhejiang University. And it provides the ability to measure aerosol backscatter, extinction coefficient and depolarization.

The main system is based on Q-switched Nd:YAG laser seeded by DBR laser with fundamental emission at 1064nm, and additional emissions at 532nm by using second harmonic generators is employed. The receiving system mainly consists of a 280mm diameter Cassegrain telescope with a full field-of-view of 0.36mrad, an interferential filter, a polarization cube and a beam splitter, that discriminates three detecting channels corresponding to parallel channel, perpendicular channel and molecular channel. The separation distance between transmitting and telescope is 180mm. The backscatter signal of molecular channel is directed to the iodine filter module which contains the iodine vapor filter and temperature controller. And it could suppress the aerosol backscatter signal and transmit most of molecular backscatter signal, that is why it is called molecular channel. The lidar backscatter signal is registered with a vertical resolution of 7.5m and temporal resolution of 2-minute within 0.5-8km detecting range.

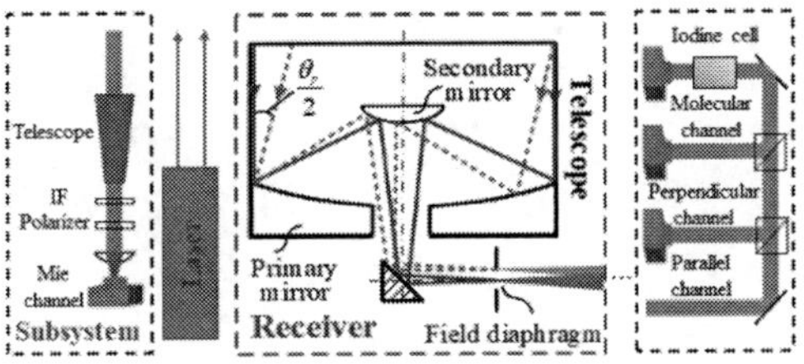

Fig.1 The system layout of dual field-of-view HSRL system.

An additional subsystem is designed for experimental determination of overlap profile. It receives backscatter signal with different field-of-view, which is compared to main Lidar in order to retrieve overlap profile. Lidar subsystem mainly consists of a small telescope with large field-of-view, a polarizer, an interferential filter, and a photomultiplier detector which is the fourth channel called Mie channel. The detailed technique would be discussed in Section 3.

3. METHODOLOGY

The fundamental idea for experimental determination of overlap profile that motivates our approach is to compare two backscatter signals generated from same area with different field-of-view, that is, the deviation between two signals affected by incomplete overlap allows for retrieving the lidar overlap function. The most of experimental approaches require specific assumptions such as homogeneous atmospheric conditions or lidar ratio. Whereas our method does not need any assumption and could retrieve overlap profile real-time by taking advantage of a compact low-cost optical sub-system or an iterative approach. In section.4, we could see that the results of two method are in accordance with each other. The following paragraphs describe in detail the proposed procedure.

For the purpose of increasing the signal-to-ratio, it is convenient to let the polarization of subsystem parallel to the parallel channel of main system. The backscatter signal in all four channels is listed below: P_{Mie} is the channel of subsystem through the polarizer, P_m is the molecular channel filtered by iodine cell, $P_{//}$ and $P_{\perp}$ are the parallel and perpendicular channels, respectively.

3.1 Dual field-of-view method

We consider the backscatter signal of subsystem as the reference profile. The measurement of subsystem is unaffected by incomplete overlap already from above the blind zone ($O_1(\mathrm{r})=0$). Note here and reflected in Eq. (1) and (4), the only difference between P_{Mie} and $P_{\parallel}$ is the overlap terms due to different field-of-view of two system. It could be seen as an approximation that $O_2(\mathrm{r})\approx 1$ in near-filed if the filed-of-view of subsystem receiver is large enough. Then uniform $O_1(\mathrm{r})$ could be easily derived from the Eq. (4) by applying Eq. (6) and used to correct other channels as follow

$$O_1(\mathrm{r}) \propto {P_{\parallel}(\mathrm{r})}\Big/{P_{Mie}(\mathrm{r})} \qquad (1)$$

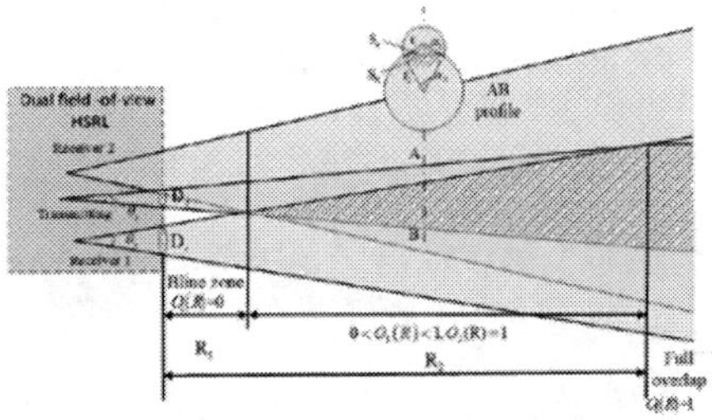

Fig.2 The schematic view of dual field-of-view.

3.2 Iterative method

The basic idea underlying the iterative method is that HSRL could retrieve backscatter coefficient

https://doi.org/10.1051/epjconf/202023707014

without knowing the overlap profile and the backscatter coefficient retrieving by Klett method is affected by the incomplete overlap.

The difference could be used to correct overlap profile as following:

$$\frac{\beta_{HSRL}-\beta_{Klett}}{\beta_{HSRL}+\beta_{m}} \propto \frac{P(\mathrm{r})\cdot O^{-1}(\mathrm{z})-P(\mathrm{r})}{P(\mathrm{r})\cdot O^{-1}(\mathrm{z})}=1-O(\mathrm{z}) \tag{2}$$

However, Klett method can often be unstable to retrieve overlap profile by simply applying Eq. (14). We will try to use the relative difference to reduce iteratively the incomplete overlap by correcting the backscatter signal $P(\mathrm{r})$. In next step (i+1), the corrected backscatter signal is used to retrieve the aerosol backscatter profile again.

$$\Delta O^{(i)}(\mathrm{r})=\frac{\beta_{HSRL}(\mathrm{r})-\beta_{Klett}^{(i)}(\mathrm{r})}{\beta_{HSRL}(\mathrm{r})+\beta_{m}(\mathrm{r})} \tag{3}$$

$$P^{(i+1)}(\mathrm{r})=P^{(i)}(\mathrm{r})\cdot[1+\Delta O^{(i)}(\mathrm{r})] \tag{4}$$

After few iterations (about 10-20 times), this procedure could correct the effect of incomplete overlap completely. The overlap profile could be calculated as:

$$O(\mathrm{r})=\mathrm{P}^{(1)}(\mathrm{r})\Big/\mathrm{P}^{(\mathrm{end})}(\mathrm{r}) \tag{5}$$

4. RESULTS

Data used in this paper were measured at Yuquan Campus, Zhejiang University, Hangzhou, Zhejiang, China (N30°16', E120°07'07") during the CALIOL (Cloud-Aerosol Lidar Observation Laboratory) field campaign. The measurement of 11th Mar 2019 took place from 00:44 until 03:29 UTC+8.

Fig.3 the field campaign of CALIOL.

Figure 4(a) presents an example of the backscatter signals obtained from parallel channel and Mie channel. The comparison of two profile shows that they are very similar above the area of full overlap. And we could easily to calculate the overlap profile.

The result of iterative method used to correct backscatter signal is also shown in Figure 4(b) for the measurement of same time. The consequence of corrected backscatter signal for each step is also illustrated. The value of lidar ratio is assumed as 40 according to the above result in the retrieval of Klett method, which is typical value of urban aerosols.

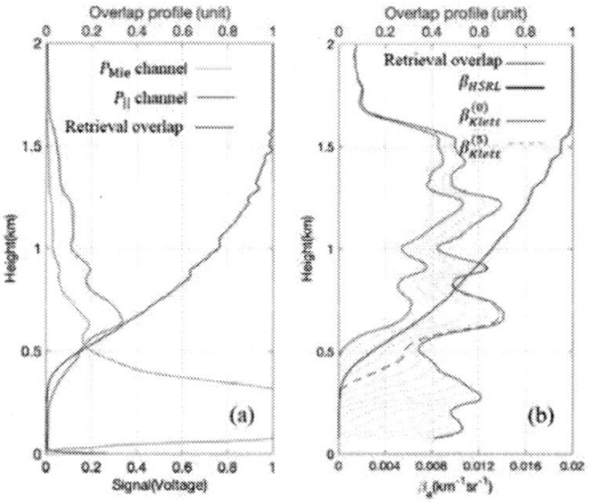

Fig.4 Overlap profile derived from dual field-of-view (a) and iterative method (b).

We take into account all 1211 profiles measured from 0:44 until 3:29 to calculate the overlap profile with two method, and the results is shown in Figure 5. The error bars (three standard deviation) show the variability of the overlap profile. It seems like that iterative method is more stable than dual field-of-view method. But the average overlap profiles of two method are almost entirely the same. Then two average overlap profiles fitting by exponential model[5], both correlation index R^2>0.99, indicating that the high-level consistency of two methods had proved the certain dependability of the results.

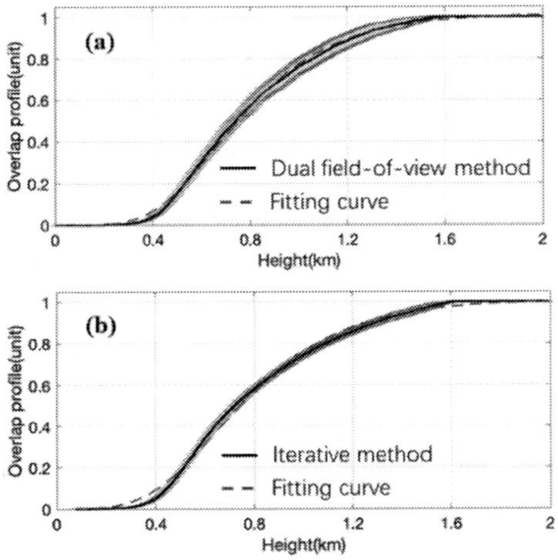

Fig.5 The results of dual field-of-view method (a) and iterative method (b).

ACKNOWLEDGEMENTS

This work was supported by the National key Research and Development Program of China (2016YFC0200704).

REFERENCES

[1] SASANO Y, SHIMIZU H, TAKEUCHI N, et al. Geometrical form factor in the laser radar equation: an experimental determination [J]. Applied Optics, 1979, 18(23): 3908-10.
[2] TOMINE K, HIRAYAMA C, MICHIMOTO K, et al. Experimental determination of the crossover function in the laser radar equation for days with a light mist [J]. Applied Optics, 1989, 28(12): 2194-5.
[3] DHO S W, PARK Y J, KONG H J. Experimental determination of a geometric form factor in a lidar equation for an inhomogeneous atmosphere [J]. Applied Optics, 1997, 36(24): 6009-10.
[4] WANDINGER U, ANSMANN A. Experimental determination of the lidar overlap profile with Raman lidar [J]. Applied Optics, 2002, 41(3): 511.
[5] JUAN LUIS G R, MARIA JO?O C, DANIELE B, et al. Infrared lidar overlap function: an experimental determination [J]. Optics Express, 2010, 18(19): 20350-9

EPJ Web Conferences **237**, 07015 (2020) https://doi.org/10.1051/epjconf/202023707015
ILRC 29

EXPLORING THE TWILIGHT ZONE: A MULTI-SENSOR APPROACH

JONATAN DA SILVA[1,2*], FERNANDO G. MORAIS[1,3], MARCO A. FRANCO[3], FÁBIO J. S. LOPES[1], GREGORI DE A. ARRUDA[1], ALEXANDRE C. YOSHIDA[1,4], ALEXANDRE CORREIA[3] AND EDUARDO LANDULFO[1]

[1]*Centro de Lasers e Aplicações (CLA), Instituto de Pesquisas Energéticas e Nucleares - IPEN, Brasil.*
[2]*Centro de Ciências Exatas e das Tecnologias, Universidade Federal do Oeste da Bahia - UFOB, Brasil*
[3]*Instituto de Física - Universdiade de São Paulo - USP- Brasil*
[4]*Instituo de Ciências Exatas e Naturais do Pontal, Universdidade Federal de Uberlândia - UFU, Brasil*
**Email: jonatansilva@usp.br*

ABSTRACT

This study shows a set of analysis of measurements from ground-based and satellite instruments to characterize the twilight zone (TLZ) between clouds and aerosols in São Paulo, Brazil. In the vicinity of clouds turbulence measurements showed an intense upward movement of aerosol layers, while sunphotometer results showed an increase in aerosol optical depth, and lidar measurements showed an increase in the backscatter vertical profile signal.

1. INTRODUCTION

The Twilight Zone concept was introduced more than 10 years ago as a mean to describe the transition from cloud-free regions in the atmosphere to cloudy environments [1]. The TLZ represents the region where clouds and aerosols interact with the presence of forming and evaporating clouds fragments and hydrated aerosol particles [1]. Radiation fluxes through the TLZ can extend from meters to tens of kilometers, are highly sensitive local conditions and still have uncertainties when observed from space or ground-based instruments.

Several studies demonstrate an increase of cloud cover and cloud optical depth correlated with an increase of aerosol optical depth (AOD) in areas near clouds. The city of São Paulo in Brazil is among the largest and most polluted cities in the world, where we examined columnar aerosol optical properties both on areas nearby and far away from clouds, using a ground-based lidar system operating in synergy with an Aerosol Robotic Network (AERONET) sunphotometer [3], satellite measurements, an all-sky camera, and a pyranometer. We have identified a subgroup of measurements carried out between March and November of 2018, which presents cloud field structures compatible with the TLZ definition. Here we present preliminary results for the 10th April 2018 using a multi-sensor approach analysis.

2. INSTRUMETATION AND METHODOLOGY

2.1 The lidar system

The lidar system is located at Center for Laser and Applications (CLA) of the Nuclear and Energy Research Institute (IPEN) (23.560 S, 46.740 W, 740 m above sea level) [4]. The lidar system is a multiwavelength Raman $(3\beta + 2\alpha + H_2O)$ configured in a monostatic biaxial alignment pointing vertically to the zenith, operating with a laser energy per pulse of about 600, 400 and 230 mJ at 1064, 532 and 355 nm, respectively, at a repetition rate of 10 Hz and pulse duration of approximately 5 ns.

2.2 AERONET sunphotometer, satellite retrievals pyranometer and all-sky camera

AERONET is a network of automated sun and sky radiometers that has been working for more than 25 years continuously, maintaining a readily accessible public domain database of aerosol optical, microphysical and radiative properties for aerosol research and characterization, validation of satellite retrievals, and synergism with other databases. This network provides observations of spectral AOD, inversion products, and columnar precipitable water in diverse aerosol regimes [3]. The Advanced Baseline Imager (ABI) aboard the Geostationary Operational Environmental Satellite (GOES-16) retrieves the AOD at 550nm over land during daytime every 15 minutes.The MS-80 pyranometer measures the global solar radiation incident on the Earths surface, in a spectral range of 285 to 3000 nm, with a sensitivity of 7 to 14 $\mu V/W/m^2$. The Automated Sky Imaging System (ASI) model J1006 [5] is installed within five meters of the lidar system. This system is made up of a fisheye lenses, software for data acquisition and analysis. The measurements are taken every five minutes with a resolution of 1600 by 1200 pixels.

2.3 Methodology

The lidar measurements allowed us to identify from a 2018 dataset occurrences of cloud field structures and free-cloud zones compatible with the TLZ definition. These days were filtered by correlative measurements retrieved by the AERONET sunphotometer, GOES-16 AOD retrievals, and the all-sky camera images. We selected for this case study the lidar measurements for 10th April 2018 which were performed with at an acquisition rate of 2 s (so-called turbulence measurement). Using the turbulence measurement approach [7] it was possible to analyze high order moments of scattering profiles, such as the skewness, that describes descending and ascending movements of the aerosol parcels in the atmospheric boundary layer (ABL).

Retrieved data from GOES-16 satellite with a footprint of 24 km centered at the lidar station allowed the investigation of the AOD product. AOD variation has an impact in the all-sky camera red-green-blue (RGB) channel magnitudes, in this sense, some analysis was conducted by comparison of mean red-blue ratio (RBR) [5] with interpolated AERONET AOD at 532 nm. The RBR image analysis allows classify cloud cover. The all-sky camera and sunphotometer are about 350 meters apart. All-sky images for 10th of April were retrieved together with sunphotometer and lidar data. During this day the cloud covering was moderate (30-40%), allowing mixed lidar measurements with clear and cloudy conditions identified by RGB image components [5].

In this context we apply the lidar system and the ASI to map and select possible TLZ scenarios, while the AERONET, GOES-16 and lidar products are used to study the aerosol optical properties in atmospheric regions far away and near the clouds (timewise and spacewise).

3. RESULTS

Cloud free and the cloudy regions, together with the temporal evolution of the boundary layer is presented in figure 1, that shows the range corrected signal (RCS) retrieved at 532 nm by the lidar station.

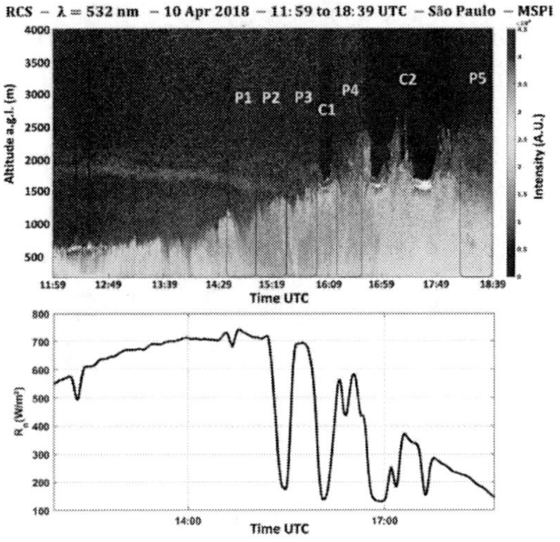

Figure 1: RCS at 532 nm retrieved by lidar station at 10th of April of 2018 showing the clouds altitude, and below, the pyranometer profile showing radiation blockage by clouds.

Seven different scenarios are selected, P_{1-5} are the cloud free conditions to be used to investigate TLZ conditions while the C_1 and C_2 are scenarios containing clouds around 1500 meters of altitude, represented in the figure by pink open circles.

The values of net radiation (R_n) are estimated from the solar global irradiance (R_{Sd}) measured by a pyranometer using the following seasonal model: $R_n = a_1 R_{Sd} + b_1$, where $a_1 = 0.716$ and b_1=-46.5 W/m^2 [6]. Figure 1 also presents a comparison between R_n and the lidar RCS. The sharp reductions in the R_n signal along the day are in agreement with the presence of clouds detected by the lidar system. This information can help us to identify the optical thickness of different clouds, since the thicker the cloud, the deeper the valley in pyranometer profile.

The values of Skewness (S) obtained from RCSs are related with vertical aerosol movements, so that $S > 0$ represents case of updrafts and $S < 0$ indicates downdrafts. From the skewness profile in figure 2 it is possible to infer vertical movements of aerosol parcels. The predominance of atmospheric updrafts in periods P_3 and P_4 are an indication of the onset of shallow convection with aerosol contribution for cloud formation or the increasing of aerosol

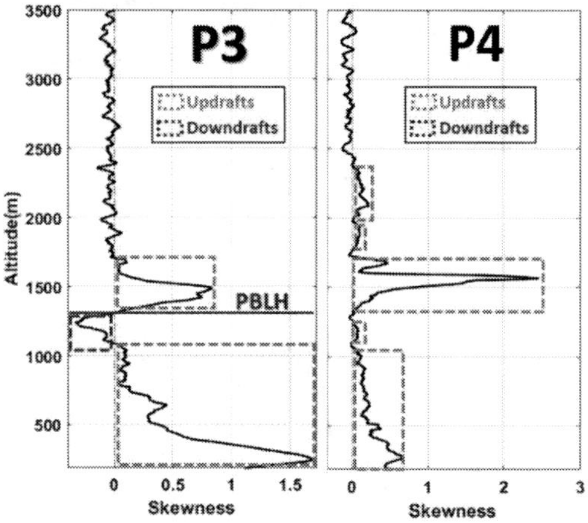

Figure 2: Skewness profile for two periods without clouds, P3 and P4 (fig.1).

quantity in the TLZ. In order to derive aerosol optical properties from the lidar measurements we retrieved the backscatter profile at 532 nm for each cloud-free conditions presented at figure 1.

As can be seen in figure 3, the backscatter profile values, at the altitude region between 1.5 to 2.5 km, are increasing as the profiles P_i are approaching, in temporal scale, to the cloud profile C_1, which is an indication of the increased aerosol backscatter near clouds.

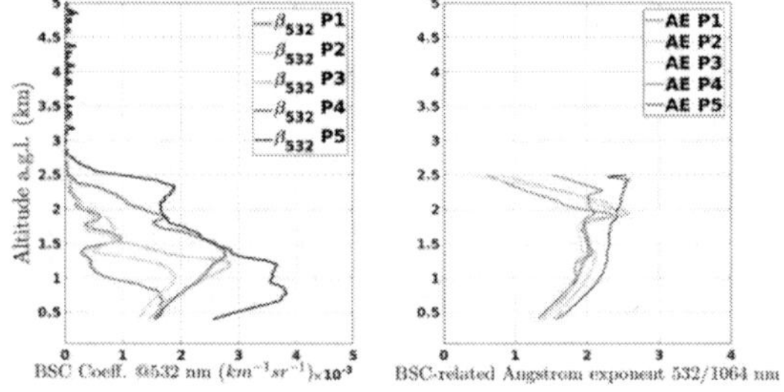

Figure 3: Lidar backscatter profile at 532 nm and backscatter-related Ångström exponent-532/1064nm.

The Backscattering-related Ångström exponent-532/1064 nm also suggests that the amount of aerosols increases near cloud profile C_1, and stay constant in the profiles P_4 and P_5, which is an expected behavior according to [2].

Further exploration of this feature should be undertaken with more cases where similar cloud and aerosol conditions are present. The AOD retrieved from AERONET (figure 4) are always below 0.50 indicating that the atmosphere is not heavily polluted, BSC-related Ångström exponent values about 2 indicate size distributions dominated by fine mode aerosols.

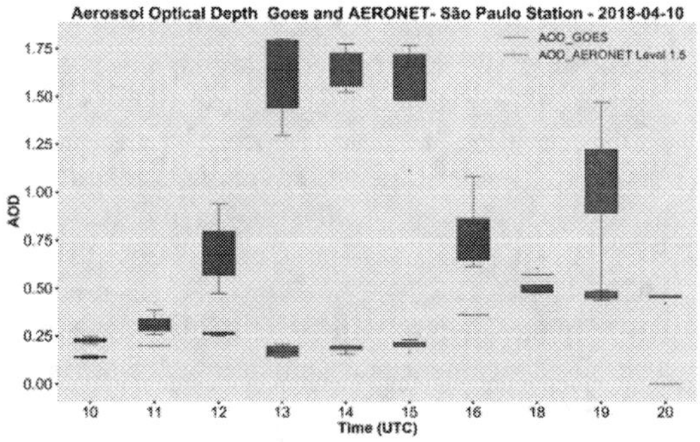

Figure 4: AOD retrievals from AERONET sunphotometer and Goes-16 Satellite.

Comparation between AOD from GOES-16 and AERONET reveals the GOES (Beta) product overestimates the "ground truth" AOD represented by AERONET, which can be due to enhanced solar radiation 3-D scattering in the TLZ cloud field [1].

The interpolated AOD values retrieved at 500 nm by the AERONET sunphotometer were correlated with the mean all-sky image RBR values for coincident measurement (≈ 5 min) of both instrument.

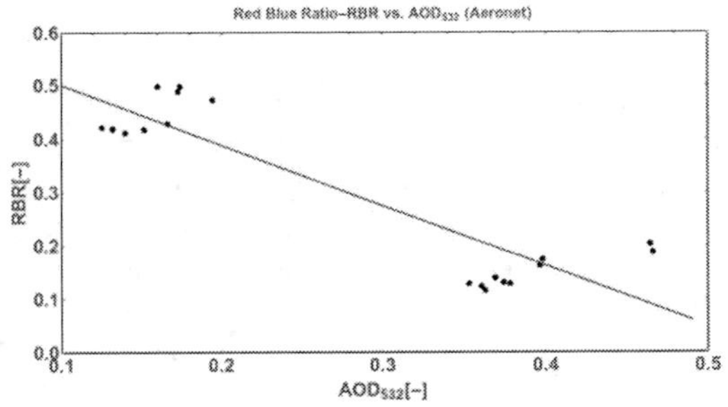

Figure 5: RBR-AOD at 532 nm correlation for 10th of April for Sun zenith angles between 30^0 and 70^0.

The RBR-AOD correlation obtained was: $RBR = 0.61 - 1.13\ \tau_{532}$. The sky conditions allow the correlation between RBR and AOD value retrieval, which might be a trend to be inspected when other cases will be analyzed and reveal a stronger correlation.

4. CONCLUSIONS

We employed a set of ground-based and satellite measures to map twilight zone features. The AOD results retrieved by the AERONET revealed an aerosol increasing in the vicinity of clouds, whereas retrieved GOES-16 AOD showed significant biases. These AERONET results are corroborated by the skewness analysis, which presented an updraft movement of aerosol layers and an increase of aerosol load at in the TLZ. The same indication is shown by the backscatter profile at 532 nm and the BSC-related Ångström exponent retrieved by the lidar system.

ACKNOWLEDGEMENTS

The authors would like to acknowledge the FAPESP (Projects 15/12793-0, 18/03413-8, 152156/2018-6, and 10/15959-3) and CNPQ (Projects 152156/2018-6, 432515/2018-6, and 429112/2018-1).

REFERENCES

[1] Ilan Koren et al., Geophys. Res. Letters. 34, L08805 (2007), doi:10.1029/2007GL029253.
[2] W. Su et al., J. Geophys. Res., 113, D24202 (2008), doi:10.1029/2008JD010588.
[3] Brent Holben et al., Remote Sens. of envir., 66, 1-16 (1998),doi:10.1016/S0034-4257(98)00031-5
[4] Fábio J. S. Lopes et al., Remote Sens., 11, 195 (2019), doi:10.3390/rs11020195.
[5] M. S. Ghonima et al., Atmospheric Measurements Techniques, 11, 2881-2892 (2012), doi:10.5194/amt-5-2881-2012.
[6] Inmaculada Alados et al., Agricultural and Forest Meteorology, 116, 221-227 (2003), doi:10.1016/S0168-1923(03)00038-8.
[7] Gregori de A. Arruda et al., Atmos. Chem. Phys., 19, 1263-1280 (2019). doi:10.5194/acp-19-1263-2019.

FLUORESCENCE DATABASE OF AEROSOL-CANDIDATE-SUBSTANCES FOR FLUORESCENCE LIDAR APPLICATION

Y. Saito[1*], K. Shiraishi[2], A. Nishimura[1], T. Kirinaka[1], Y. Sakurai[1] and T. Tomida[1]
[1]*Institute of Engineering, Shinshu University, 4-17-1 Wakasato, Nagano City, Nagano, 380-8553 Japan*
[2]*Department of Earth System Science, Fukuoka University, 8-19-1 Nakamura, Jonan-ku, Fukuoka, 814-0180 Japan*
**Email: saitoh@cs.shinshu-u.ac.jp*

ABSTRACT

A database containing spectrum and cross-section of the fluorescence of substances has been made. In test of forest environment monitoring by our Laser-Induced Fluorescence Spectrum (LIFS) lidar, the database showed that the origin substance of the aerosol observed by the lidar was cedar pollen, and the concentration was calculated using the cross-section. In the urban atmosphere monitoring, three substances stored in the database were proposed to be the origins of the aerosol. Based on these experiments, we discuss the usefulness of the fluorescence database in lidar observations.

1. INTRODUCTION

Until recently, using fluorescence lidars had been limited to hard targets such as vegetation, water, living things, and buildings because low cross-section of aerosol fluorescence makes dear observations difficult. However recent spectrum detection technologies using an intensified CCD or a multi anode photomultiplier single-photon-counting have been expanding the fluorescence lidar observation capabilities to atmospheric aerosol monitoring where the target concentrations are much less than that of hard targets.

Tatarov et a.l[1)] and Saito et al.[2)] first reported detections of the fluorescence spectrum from the atmosphere in KOSA events at the same time but independently using their own different fluorescence lidars. Sugimoto et al. showed that fluorescence in the lower atmosphere could be of air-pollution aerosols transported from urban and industrial areas[3)]. The co-existence of clouds and fluorescing aerosols was discussed by Reichadt et al.[4)]

The problem in lidar atmospheric observations is how to identify the species of aerosol observed by the lidar. Even if we use a fluorescence lidar, the fluorescence information of substances is needed.

In this paper, we describe a fluorescence database developed for this purpose. It contains fluorescence information about the spectrum and cross-section of substances which we can use in the discussion of fluorescence lidar observations.

2. FLUORESCENCE LIDAR OBSERVATIONS

Two types of LIFS lidars were developed (Fig.1). Both systems use a 355-nm laser for the lidar transmitter and a multichannel spectrometer for the lidar spectroscopic receiver.

(a) (b)

Figure 1. (a) LIFS-field lidar and (b) LIFS-lab lidar.

2.1 Pollen fluorescence monitoring by LIFS-field lidar[5)]

Many people suffer from pollen allergies, so we decided to test the possibility of pollen monitoring by a fluorescence lidar.

A fluorescence lidar capable of operating outdoors was developed at Shinshu University and named the LIFS-field lidar. We used a multi-channel spectrometer (PMA-12; Hamamatsu, Japan) for spectral detection which consisted of a spectrometer with a gated-intensified CCD array detector. Its short gating time of several tens of ns made it possible to detect weak fluorescence

https://doi.org/10.1051/epjconf/202023707016

signals even in daytime. The LIFS-field lidar was a mobile and self-sufficient system operating with a small dynamo (900 VI).

Spring-time observations were made at the mountain-side of a town (Matsushiro, Nagano City, Japan) where there was a forest with several different species of trees. The lidar was placed on the ground so that it was about 3 m higher than the ground of the forest. The distance to the forest was about 20 m. The detected fluorescence spectrum is shown in Fig. 2a. The signal reached the peak at 13:33. Time depended signal intensity variation was caused by the wind.

Autumn observations were made at a riverbed of the town where the ragweed had spread over large areas. The laser beam passed horizontally 1 m above the ragweed area. Fig. 2b shows the detected fluorescence spectrum which has a different spectral shape from the cedar and the intensity was slightly lower than that of cedar pollen.

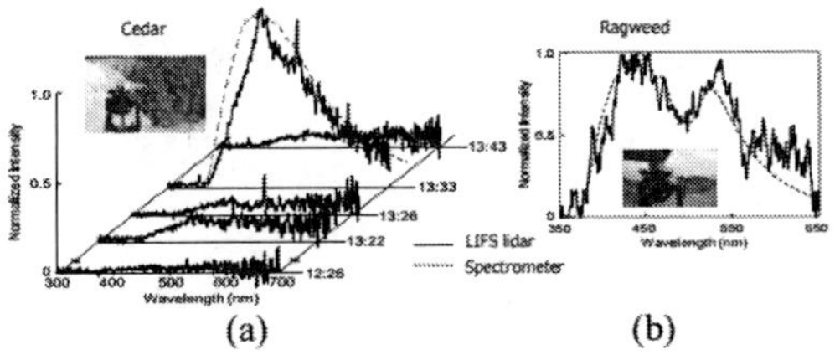

Figure 2. (a) Cedar and (b) ragweed pollen monitoring by the LIFS-field lidar. Each of the spectra stored in the database are comparatively shown in the observation.

2.2 Atmospheric aerosol fluorescence monitoring by LIFS-lab lidar

Fukuoka has a population of over one million. Above the troposphere, the westerly wind constantly flows carrying KOSA particles, and the area faces the Japan Sea. Investigation of urban atmosphere is one of our research interests, since they can be affected by various interactions among the atmosphere, sea, and human activity.

A LIFS-lab lidar was developed at Fukuoka University for observing urban atmospheric environments. The detection system was consisted of a spectrometer (SP-2758, Acton, USA) and a 32-channel multi-spectral lidar detector (MS-125, Licel, Germany). The LIFS-lab lidar had functions of Raman and depolarization measurement. Multiple wavelength lidar using 532 nm and 1064 nm also worked together with the LIFS-lab lidar.

A laser beam was transmitted vertically and observation was done at night automatically and routinely. A fluorescence signal was observed in almost every observation. One of the examples is shown in Fig. 3 in the state where urban air pollution and KOSA particles were mixed. The signal intensity was saturated in the atmosphere lower than 1 km and doubled compared with only KOSA particle inflow. Many other observations showed that the fluorescence signal was higher in KOSA events and decreased if the depolarization was high, and it increased if the particles were liquid-oriented ones.

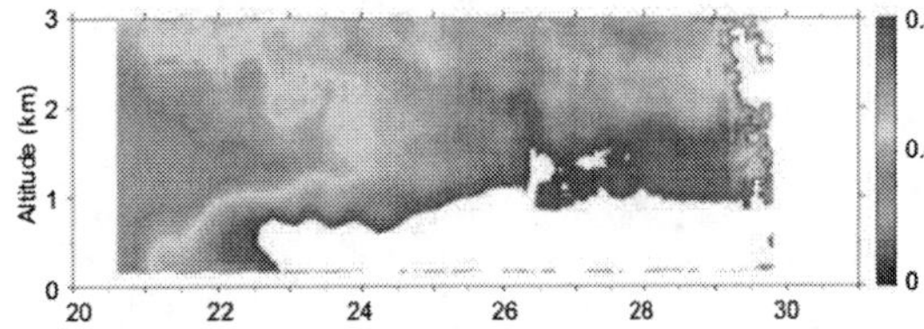

Figure 3. An example of the urban atmosphere in a KOSA event monitored by the LIFS-lab lidar. Total fluorescence signal intensity was normalized to N2 Raman signal at 420-510 nm.

3. FLUORESCENCE DATABASE

3.1 Preparation of fluorescence substance

Traditional lidar analysis seems to have mainly dealt with inorganic substances. Our scenario of aerosol generation is conceptually illustrated in Fig. 4, in which we examine every kind of

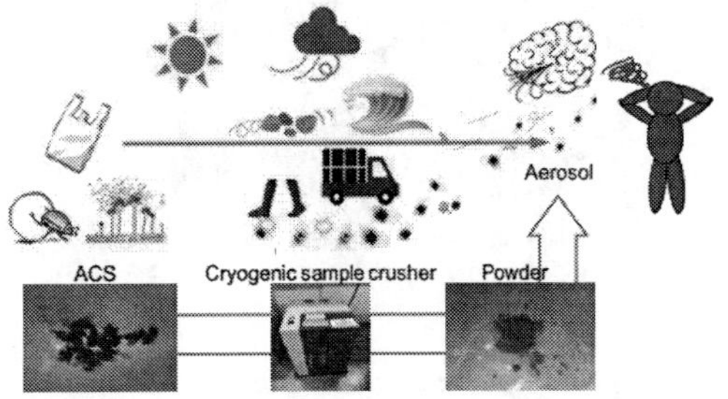

Figure 4. Scenario of aerosol generation (upper) and procedure of making powder of ACS (lower).

substance without limiting the type. Plastic garbage, plants, living things, and minerals are included. Here we named this Aerosol-Candidate-Substance (ACS) which can be expected to be aerosols.

ACSs were collected from the surroundings and made into powder by a cryogenic sample crusher (JFC-3000, Japan Analytical Industry, Japan) to resemble aerosols in the atmosphere. The size after crushing was around 50 μm. ACSs such as pollen, which is a powder naturally (particle), was used as it is without crushing.

3.2 ACSs Fluorescence spectrum

Excitation-Emission-Matrix (EEM) characteristics of each powder were measured by a fluorescence spectrometer (F-7000, Hitachi High-Tech Science, Japan).

Several examples of EEM are shown in Fig. 5. The emission (fluorescence) spectrum was measured by varying the excitation wavelength from 250 or 300 nm to 540 nm with 10 nm separation.

We developed software which is useful in identifying aerosol species in lidar observations and installed it into the fluorescence database. The database stores fluorescence spectra of 50 species so far and they are roughly divided into plant-, biological-, and artificial-related ones.

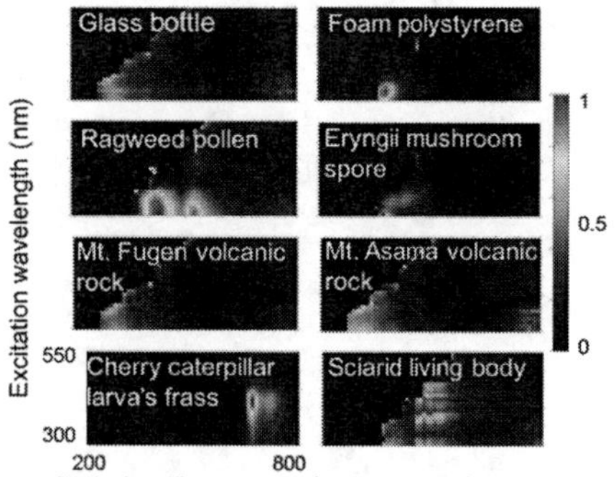

Figure 5. Examples of EEM of ACS powders.

3.3 ACSs fluorescence cross-section

Information about cross-section (or quantum efficiency) of aerosol is needed to calculate aerosol concentrations from lidar observations, but there are very few reports on the data. In this work, we developed a fluorescence cross-section measurement system. A 355-nm pulse laser irradiated ACS powder set on a non-fluorescent quartz plate and a photomultiplier measured the voltage of the fluorescence. A particle counter (KC-20, RION, Tokyo, Japan) determined the number of powdered particles of the ACS.

Cedar pollen was investigated. Figure 6 shows the fluorescence intensity (voltage) depending on the number of particles. From this we could derive that the fluorescence cross-section of cedar pollen was 1.5×10^{-13} cm^2 sr^{-1} nm^{-1} $particle^{-1}$ at 460 nm. This is compatible with that of pine pollen reported by Stephen[6] and those of Betula alba and other pollens by Weichert[7]. It should be added that their excitation was at 280 nm.

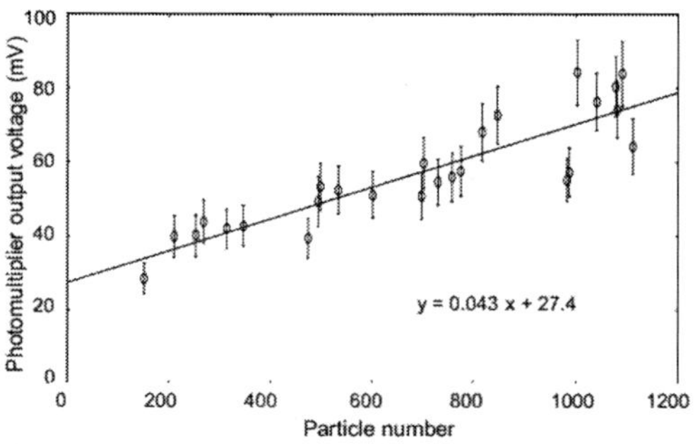

Figure 6. Fluorescence intensity of cedar pollen depending on the numbers of particles.

4. ADAPTATION OF DATABASE

4.1 Pollen observation[5]

There were several species of trees growing in the area where pollen observations were made as described in 2.1. They included locust tree, cedar, pine and cypress. Fluorescence spectrum of each stored in the database helped us identify the species, so that we knew that the spectrum observed by the LIFS-lab lidar was of cedar pollen. In the autumn observation, it was ragweed pollen.

We calculated the concentration of the cedar pollen by applying the cross-section data obtained in 3.3 in the lidar result. In this case, the sensitivity of the PMA detection of the LIFS-field lidar was checked by a comparative experiment with a photomultiplier detection. Then the calculation indicated that the number of cedar pollens observed by the LIFS lidar was $1.5x10^7$ particle m^{-3}.

4.2 Atmospheric aerosol observation[8)]

We investigated ACSs which might be the origin of aerosols observed by the LIFS-lab lidar.

As the Environmental Scanning Electron Microscope Energy Dispersive X-ray (ESEM-EDX: Quanta FEG 200, FEK and XL-30, EDAX, U.S.A) analysis of the substances sampled on a building roof suggested that aerosols included mostly Si with Al, Mg, and smaller concentrations of Fe, S, Ca, K, we selected three fluorescence spectra from the database which might contain such minerals. These included the fluorescence of glass bottles, Mt. Asama volcanic rock, and concrete blocks.

Figure 7 compares the fluorescence spectra of atmospheric aerosols and a composition of three ACSs and their tendencies of increasing and decreasing were similar. Among the three ACSs, the spectrum of a glass bottle, whose main component is SiO_2, seemed to fit well.

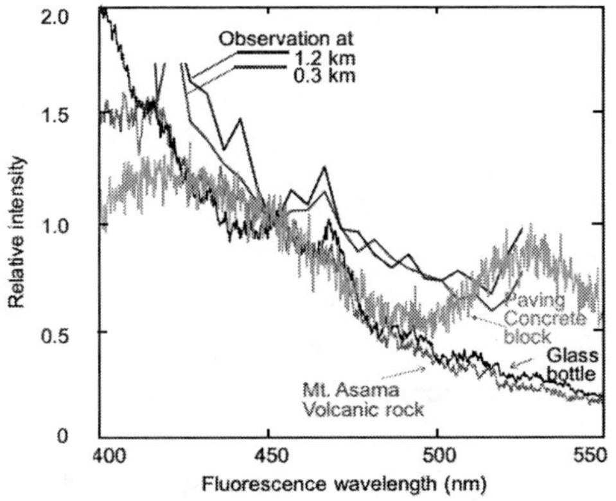

Figure 7. Adaptation of fluorescence spectrum of three ACSs into lidar atmospheric fluorescence data. The values of each spectrum were compared relatively with their own intensities at 452.4 nm.

5. CONCLUSION

We developed two types of LIFS lidars that made it possible to observe atmospheric fluorescence. Fluorescence spectrum database was useful in estimating the origins of atmospheric aerosols. Cedar pollen concentrations could be estimated by applying its cross-section value into the computation.

These experimental results show that the LIFS lidar with a fluorescence database is a new apparatus for conducting research of atmospheric environments.

ACKNOWLEDGEMENTS

Authors thank to the students of the Saito-Tomida Lab. of Shinshu Univ. and of the Shiraishi Lab. of Fukuoka Univ. for their helpful assistance with experiments. This research has been supported by Japan Society for the Promotion of Science (Grant No. 17H01478, PI: Professor Y. Igarashi of Kyoto Univ.) and Takahashi Industrial and Economic Research Foundation.

REFERENCES

[1] B. Tatarov, N. Sugimoto, I. Matsui, D.-H. Shin, D. Muller, The 8th Inernational Symposium on. Advanced Environmental Monitoring (Sapporo, Japan; June 29-July 2, 2010), I-AO3, pp. 18-23.
[2] Y. Saito, K. Kobayashi, F. Kobayashi, ibid., II-EO1, pp. 54-57.
[3] N. Sugimoto, Z. Huang, T. Nishizawa, I. Matsui, B. Tatarov, Optics Express 20, pp.20800-20807, 2012..
[4] J. Reichardt, R. Leinweber, A. Schwebe, 28th International Laser Radar Conference (Bucharest, Romania, 25-30 June 2017), 050-189.
[5] Y. Saito, K. Ichihara,K. Morishita, K. Uchiyama, F. kobayashi, T. Tomida, Remote Sensing, 10, 1533, 2018; doi:10.3390/rs10101533
[6] J. R. Stepens, Proceedings of the 1996 Edgewood RDEC Scientific Conference on Obscuration and Aerosol Research (Aberdeen, MD, USA, 19-22 November, 1996).
[7] R. Weichert, in *Landbauforschung Volkenrode Sonderheft 235* (T. Hinz, B. Ronnpagek and S. Linke, Eds), pp. 83-87, 2002.
[8] Y. Saito, T. Tomida, K. Shiraishi, Proceedings of SPIE Vol. 10779, 107790M (Hawaii, USA, 24 October, 2018); doi: 10.1117/12.2324428

LIDAR INNOVATIONS FOR TECHNOLOGIES AND ENVIRONMENTAL SCIENCES (LITES) – AN REMOTE SENSING INFRASTRUCTURE FACILITY: SETUP AND MEASUREMENTS EXAMPLES

Boyan Tatarov[1*], Detlef Müller[1], Matthias Tesche[1,2], Sung-Kyun Shin[1]

[1]School of Physics, Astronomy and Mathematics, University of Hertfordshire, College Lane, Hatfield, United Kingdom

[2]now at: Leipzig Institute for Meteorology (LIM), Leipzig University, Leipzig, Germany

**Email: b.tatarov@herts.ac.uk*

ABSTRACT

At the University of Hertfordshire, we have been developing a new remote sensing facility (LITES) to explore the feasibility of using Raman and/or fluorescence backscattering for chemical aerosol profiling. This paper provides an overview of the instruments of the facility and measurement examples. LITES includes a ultra-high-energy Nd:YAG/OPO setup, spectroscopic equipment with high spectral resolution, several imaging and single detectors that allow for time-resolved (lidar) signal detection, a Raman/fluorescence microscope, and a suite of gas and aerosol chambers. We present examples of elastic, rotational and vibrational spectroscopic lidar signals, as well as in-situ microscopic spectrums of dust and bio-aerosol compounds.

1. INTRODUCTION

Lidar is the only remote measurement method that allows for spatially (vertically) and temporally highly-resolved observations of atmospheric state parameters, trace gases and particulate pollution [1,2]. The combination of various lidar techniques (nitrogen Raman and high spectral resolution lidar) with mathematical inversion algorithms nowadays allows for temporally and vertically resolved observations of aerosol optical and microphysical properties [3,4,5]. Differential absorption lidar is another concept that allows for quantitative range-resolved measurements of the concentration of chemical components in the atmosphere [6]. However, lidar and passive remote sensing instruments currently cannot provide us with vertically-resolved information on the chemical composition of aerosols under ambient atmospheric conditions. We are exploring the potential of a multi-channel high-resolution spectrometric lidar for the vertically-resolved chemical characterization of aerosols and reactive gases by measuring range-resolved Raman and fluorescence/photoluminescence spectra. For that purpose, we have been developing a new remote sensing facility, called **L**idar **I**nnovations for **T**echnologies and **E**nvironmental **S**ciences (LITES). The facility combines a unique **Li**dar **S**pectroscopy **I**nstrument (LiSsI) with in-situ instruments such as Raman and fluorescence microscope, gas and aerosol chambers.

In this work we present the setup of LITES and show measurement examples. LiSsI allows for the profiling of trace gases, chemical components in particles, and bio-aerosols in atmospheric aerosol pollution in the troposphere through combining different non-linear spectroscopy techniques (photoluminescence, fluorescence, Raman and coherent anti-Stokes Raman spectroscopy) on a single measurement platform.

2. INSTRUMENTS

The instruments of LITES were designed to operate as a single platform that allows for an easy switch between different experimental setups. Most of the equipment parts can be used for more than one instrument component of LITES. Table 1 summarizes the specifications of the system.

2.1 The lidar system

The main components of the lidar subsystems include:

- an ultra-high-energy seeded Nd:YAG laser (Continuum Powerlite Furie LD) with SHG and THG. This system emits laser light with energies up to 8 J per pulse (at the fundamental 1064-nm wavelength). This output corresponds to a pulse power of more 500 GW and an average power of more than 50 W;
- an Optical Parametric Oscillator (Horizon OPO) that is optically pumped by the main laser. The OPO emits laser light from the deep UV (192 nm) to the near infrared (2750 nm);

Table 1. Technical specification of the lidar systems

Continuum Powerlite Furie LD	
Laser type Pulse energy Beam divergence Repetition rate Linewidth Pulse duration	Nd:YAG, injection seeded 8000 mJ@1064 nm 5000 mJ@532 nm 2500 mJ@355 nm 0.5 mrad 10 Hz <0.003 cm^{-1} <15 ns
Horizon Optical Parametric Oscillator	
Pump wavelength Wavelengths Resolution/Scan step Pulse energy Beam divergence Repetition rate	355 nm from 192 to 2750 nm 0.01 nm 120 mJ@400 nm 60 mJ@600 nm 25mJ@300 nm <2 mrad (both axes) 10 Hz
HORIBA 1250M Research Spectrometer	
Focal length Spectral range Grating size Dispersion @500 nm Accuracy/Repeatability	1.25 m, F/9 0-1500 nm @ 1200 g/mm 110 mm x 110 mm 0.65 nm/mm ±0.15 nm / ±0.005 nm
Gratings and blaze 2400 gr/mm @250 nm 1800 gr/mm @ 400 nm 1200 gr/mm @ 330 nm 600 gr/mm @ 500 nm	max resolutions at 313.183 nm 0.003 nm 0.004 nm 0.006 nm 0.012 nm
Detection	
Mie and Rayleigh scattering	355 nm, PMT HV-R9880U-20 532 nm, PMT HV-R9880U-20 1064 nm, APD InGaAs50, Si
Spectroscopic 1 Licel SP32-20	Hamamatsu H7260-20, 0.8 mm x 7 mm x 32 anodes spectral response 300-920 nm
Spectroscopic. 2 Princeton Instruments PI-MAX4 ICCD camera	Number of pixels 1024x1024 Pixels size 16 x 16 μm Gen III filmless intensifier Sensitive range 290-710 nm
Spectroscopic 3 ANDOR iXon 3 EMCCD camera	Number of pixels 512x512 Pixels size 12.8 x 12.8 μm Sensitive range 300-710 nm
Schmidt-Cassegrain Telescope	
Primary Second	f/10, FL=3910 mm, D=391mm f/10, FL=2000 mm, D=200mm
Data acquisition system	
Mie and Rayleigh scattering	Licel transient recorders, 16-bit, 20 MHz A/D converters and photon-
Multi-anode PMT	Single-photon counting, maximum count rate 100 MHz, 50 ns
ICCD	Digitization 16 bit, 32 MHz, minimum gate width 2 ns
EMCCD	Digitization 16 bit, 17 MHz

- a customized motorized beam combiner for selecting the wavelengths. The beam combiner allows for switching between seven modes, namely emission of 1) 1064 nm only, 2) 532 nm only, 3) 355nm only, 4) 1064 nm and 532 nm (simultaneously), 5) 1064 nm and 532 nm and 355 nm (simultaneously), as well as 6) optimized OPO pumping, and 7) side-port emission of laser light for non-lidar applications;
- a Horiba 1250M imaging spectrometer with four exchangeable gratings, dual entrance ports/slits and two exits ports – one imaging (front) and one high-spectral resolution (side) port/slit;
- three imaging detectors that can be used with the spectrometer;
- an intensified CCD camera (ICCD, Princeton Instruments PI-MAX4 1024i-HBf);
- a 32-channel Licel PMT;
- an EMCCD (ANDOR iXon 3);
- three kinds of single detectors – Photomultiplier tubes, InGaAs ADP, and silicon diode.

The system allows for using three types of acquisition approaches by which we can detect lidar signal on the basis of a) the use of single detectors, b) simultaneous detection of multi-channel spectroscopically-resolved lidar signals, c) determination of Raman/fluorescence spectrums by point-to point swapping of wavelengths, d) swapping of the range of wavelengths for the imaging detectors we use for LITES.

Depending on the setup of the experiment, the lidar instrument can be used as a a) multiwavelength elastic backscatter lidar for measurements of aerosols and temperature from the troposphere to the mesosphere, b) multi-channel spectroscopic Raman lidar (using Stokes and anti-Stokes, rotational, and rotational-vibrational Raman scattering), c) multi-channel spectroscopic photoluminescence/fluorescence lidar, d) high spectral resolution lidar (HSRL), e) polarization lidar, or f) infrared absorption and differential absorption lidar (DIAL).

2.2 The Raman and fluorescence microscope

This instrument allows for investigations of Raman and fluorescence spectrums of solid-state aerosol samples taken by in-situ particle-collection methods. The instrument is based on an

https://doi.org/10.1051/epjconf/202023707017

Olympus BX51TRF-6 microscope. The microscope can use the OPO laser emissions or light from a 532-nm CW DPSS laser that is used as pumping source for delivering Raman and fluorescence spectrums. The microscope is equipped with objective lenses with magnification factors x5, x10, x20, x50, and x100. In that way we obtain a spatial resolution better that 1μm. An additional CMOS imaging camera (xiQ, MQ013CG-) is available for acquiring high spatial resolution reflection or transmission images of the samples in the visible range of radiation. The spectrums can be obtained with the main spectrometer HORIBA 1250M and all detection systems available in the LITES facility.

2.3 Customized gas and aerosols gas chambers

LITES also includes a suite of gas and aerosol chambers for the investigation of general optical properties of gases and particles, namely ATR (Attenuation, Transmission, Reactance) and Raman and fluorescence spectrums of gases and aerosols particles. The chambers can simulate the conditions of gas mixtures of different types of air-pollutions, reactive trace gases and aerosol particles under real atmospheric conditions up to tropopause heights.

3. RESULTS

3.1 Elastic analog and photon counting signals

Figure 1 shows an example of lidar signals measured by the elastic configuration of LITES on 7 March 2018. The red curves present signals taken from 1930 - 2130 UTC with 100 mJ laser emission power. The black curves present the analog and photon counting signals obtained from 2312 - 2334 UTC on the same date with 500 mJ laser emission power

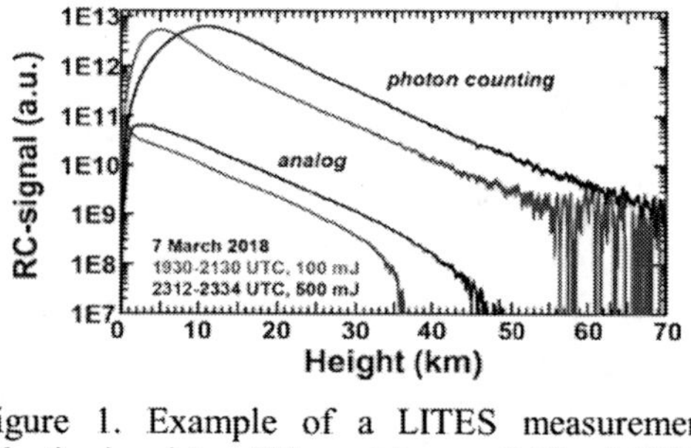

Figure 1. Example of a LITES measurement (elastic signals) at 532 nm taken on 7 March 2018.

3.2 Pure rotational and vibrational-rotational signals

Results of the first pure-rotational Raman scattering measurements that were obtained with the ICCD camera and the Horiba 1250M spectrometer are presented in Figure 2. The emitted laser intensity was 300 mJ at 354.9 nm. The spectrum was averaged over 10 laser shots and the background has been subtracted. The time gate of the camera was set to 0.03-2.00 μs which corresponds to 4.5-300.0 m distance (black curve) and 2-4 μs which corresponds to 300-600 m distance (red curve). A 355-nm RazorEdge ultrasteep long-pass edge filter (Semrock LP02-355RE-25) was used to suppress elastic scattering. The spectrometer grating was set to 1200 g/mm with a 50 μm entrance slit that allows for a spectral resolution of less than 5 cm^{-1}.

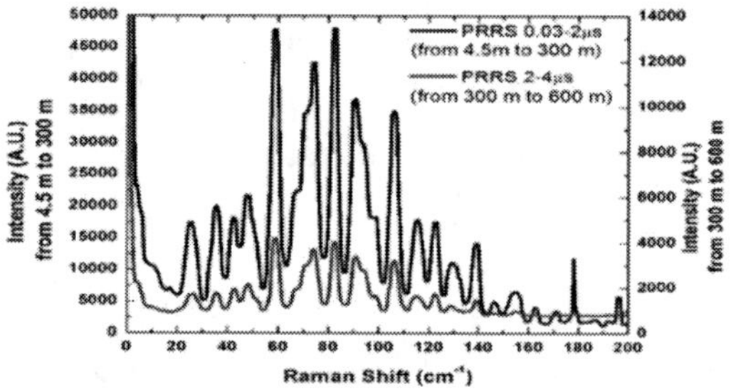

Fig. 2. Pure rotational Raman spectrum of air at 2017 UTC on 4 November 2017 averaged over 10 shots (1 s).

An example of spectra taken with the rotational-vibrational lidar Raman set-up is shown in Fig. 3.

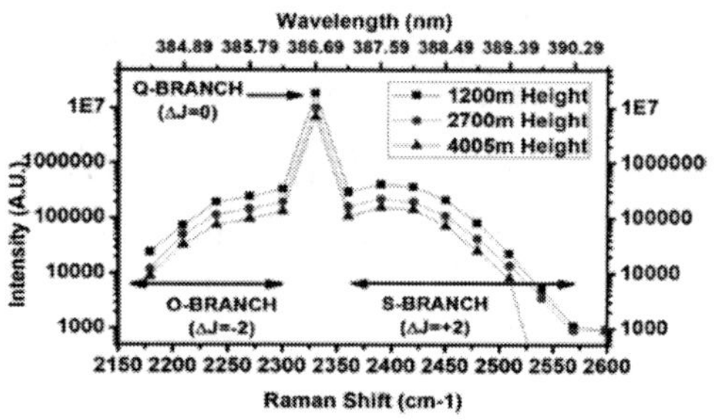

Fig. 3. Rotational-vibrational Raman spectrum of air nitrogen molecules taken from 2030 to 2359 UTC on 19 October 2018 averaged over 125576 shots (209 minutes). Laser power: 1100 mJ at 354.9 nm. Grating: 1200 g/mm. Entrance slit: 50 μm. Resolution ≈ 5 cm^{-1}.

3.3 Nitrogen dioxide measurements

Figure 4 shows one of the first LiSsI Raman measurements performed between 1726 and 1735 UTC on 3 November 2017 using the laser at 354.9 nm with a power of approximately 300 mJ. The signals were taken with a SP32-20 multi-channel PMT in photon counting mode and averaged over 5400 laser shots.

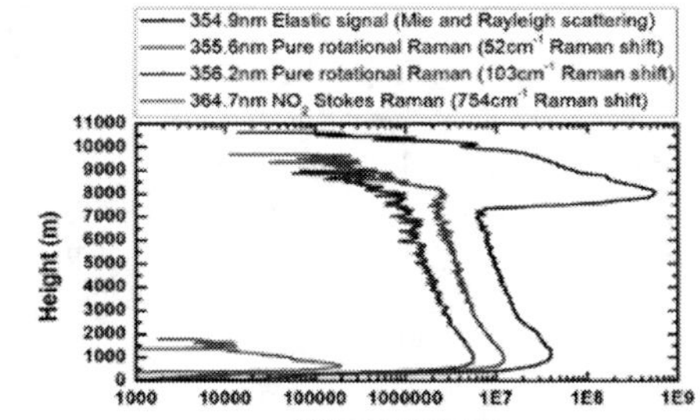

Figure 4. Elastic and Raman lidar signals at 1726 UTC on 3 November 2017.

The spectrometer grating was set to 1200 g/mm with a 50-µm entrance slit. Aerosols can be seen in the elastic signal (354.9nm) up to 3-km height above ground. Cirrus clouds are clearly visible between 7.0 and 10.5 km height. Profiles from pure rotational Raman lidar signals from nitrogen molecules were detected at two wavelengths up to approximately 9 km height. Raman signals were also observed in the 364.7-nm channel (754 cm^{-1} Raman shift) up to 1.4 km height. This signal is most probably caused by nitrogen dioxide (NO_2) which has a strong Raman peak at 754 cm^{-1}.

3.4 Microscopic Raman spectra of standard samples and mineral dust

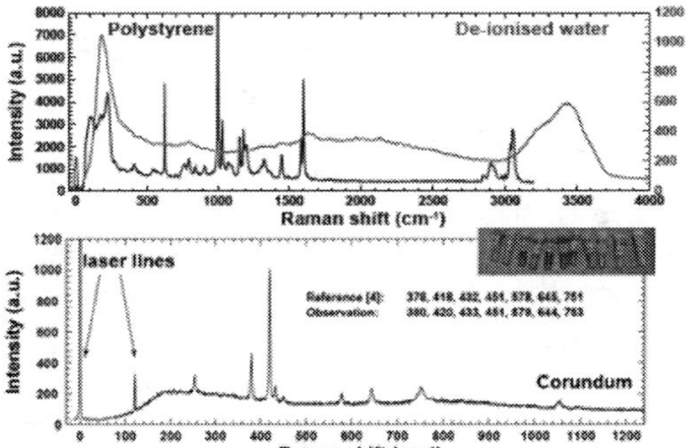

Figure.5. Raman spectrums of: a) polystyrene and de-ionized water and b) corundum (Al_2O_3)

Figure 5 presents examples of Raman spectrums obtained by the Raman microscope of calibration standard samples (polystyrene and de-ionized water) and corundum (Al_2O_3). The spectral resolution is 0.02nm, we used 150mW@532nm laser power and a grating with 2400gr/mm and slit width of 200µm.

3.5 Microscopic fluorescence spectra of bioaerosols.

Figure 6 presents examples of measurements of fluorescence spectra of different types of fresh (Cupressus) and aged (Platanus, Cupressus,) pollen taken from different urban locations at the city of Granada (Spain, 680 m a.s.l.). The spectra were obtained using the laser wavelength 532 nm, a grating with 2400 grooves mm^{-1}, and 200 µm slit width. Gold-coated slides (up to 50-nm thick coating) were used to improve signal intensity.

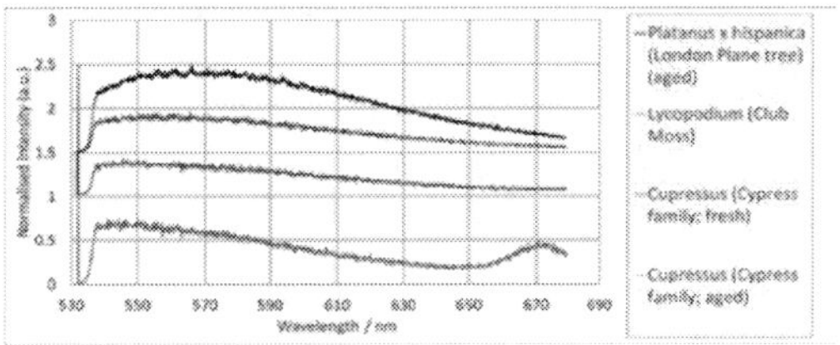

Figure 6. Fluorescence spectra of fresh and aged pollen.

ACKNOWLEDGEMENTS

This work was supported by the University of Hertfordshire through capital investment and the proof-of-concept grant MARBLE. Boyan Tatarov is supported by a Marie-Skłodowska-Curie Action Fellowship of the European Commission (CAPABLE, H2020-MSCA-IF-2015).

REFERENCES

[1] C. Weitkamp: Springer, New York 2005, ISBN 0-387-40075-3.
[2] R. M. Measures, Laser Remote Sensing, 510 pp., John Wiley, New York, 1984
[3] A. Ansmann, M. Riebesell, and C. Weitkamp, Optics Letters, 15 (1990) 746.
[4] P. Piironen and E. W. Eloranta, Optics Letters, 19 (1994) 234.
[5] D. Müller, et al., Applied Optics, 38 (1999) 2346.
[6] Bösenberg, J., Springer Series in Optical Sciences, Springer-Verlag, New York, 2005.

LIDAR OVERLAP FUNCTION DETERMINATION USING THE RAMAN LIDAR SIGNALS

Jaswant[1,2], Shishir Kumar Singh[1,2], Radhakrishnan S.R.[1,2], Devesh Shukla[1], Chhemendra Sharma[1,2]

[1]Environmental Sciences and Biomedical Metrology Division, CSIR-National Physical Laboratory, Dr K.S. Krishnan Marg, New Delhi 110012, India

[2]Academy of Scientific and Innovative Research (AcSIR), CSIR Campus, CSIR Road, Taramani, Chennai 600113, India

**Email: rkrishnan@nplindia.org*

ABSTRACT

The determination of vertical distribution of optical properties of clouds and aerosols using the lidar system is affected by the incomplete overlap between the field of view of transmitter i.e. laser beam & the receiver in the near-field range. Thus, the study of vertical profiles of aerosol optical properties in the lower atmosphere is erroneous without the correction of lidar overlap function. Here we have analysed the effect of overlap using a simple technique proposed by Ansmann and Wandinger to determine overlap function. We have determined the overlap factor for 5 different days of June 2016 and then calculated the mean overlap profile and determined the relative deviation of each day with respect to mean overlap factor. Results reveal that the complete overlap was achieved beyond 300 meters.

1. INTRODUCTION

In recent years, the lidar system has drawn greater attention for the measurement of optical aerosol properties at near altitudes due to increasing air pollution episodes. The lidar system can provide the range resolved measurements of optical properties of aerosols and clouds. However, the lidar observation is affected by the incomplete overlap between the transmitter and receiver field of view (FOV) in particularly at lower heights. This incomplete overlap makes a significant impact on the vertical profiles up to a few kilometres in the systems having a FOV as narrow as 0.5 mrad or below. Without correcting the lidar backscattered signal for the overlap effects, it is not possible to study the boundary layer pollution and its sources using the lidar system.

There are several methods to analytically determine the overlap factor [2], by applying raytracing model [1], and also experimentally [3]. However, there are several limitations to these techniques since it requires us to know the technical parameters and we have to make an assumption of homogeneous aerosol environments. In this work we are using a simple technique proposed by Ansmann and Wandinger to determine the overlap function [4]. This method relies on the combination of elastic and Raman backscattered signals under clear atmospheric conditions.

2. METHODOLOGY

A Raman Lidar system has been deployed at the CSIR-Institute of Himalayan Bioresource Technology (IHBT), Palampur, Himachal Pradesh, India. The station is situated at an altitude of 1347 m (amsl). This lidar system is currently being used to obtain the vertical distribution of clouds and aerosol over Palampur.

2.1 Experimental Setup

Table 1 specifies the technical specifications of the Raman lidar system installed at Palampur. This lidar system has a space resolution of 7.5m and operated at 355nm (elastic channel) with Raman channel at 387 nm. The return signal is detected in the binary format after which it is converted to ASCII format using the software provided by the manufacturer for further processing. The co-polarized and cross polarized Mie channels of the lidar system provides the linear depolarization ratio which is an indicator of the shape of the aerosol/cloud particle. The nitrogen Raman channel at 387 nm of the lidar system can be used to derive the lidar ratio, which is an important parameter for the identification of different types of aerosols in the troposphere.

Table 1 Lidar System Details

Emitter	
Laser Source	Nd:YAG (Quantel

https://doi.org/10.1051/epjconf/202023707018

	ULTRA100 Series)
Wavelength	355 nm (frequency tripled)
Pulse Energy	32.8 mJ
Repetition Rate	20 Hz
Near Field Beam Diameter	3.59 mm
Laser beam divergence	<0.8 mrad
Receiver	
Telescope type	Cassegrain
Primary Diameter	200 mm
Secondary Diameter	48 mm
Field of View	2.3 mrad (adjustable from 0.5 to 3mrad)
Detection Unit	
Transient Recorder	Licel (20-12bit) with 7.5m Raw Spatial Resolution
Detectors	3 PMTs

2.2 Sampling Site

The sampling site is in Palampur located at the southern region of Western Himalayas and lies within the latitude 32.093° N and longitude 76.543° E. Figure 1 shows the topographic map of Palampur. The experimental site is situated at Chandpur village which is at the altitude of 1347 m (amsl).

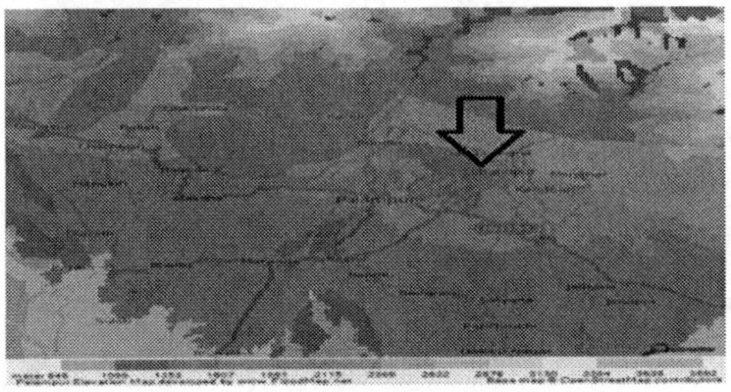

Figure 1: Topographic map of Palampur

2.3 The Overlap Correction

The lidar equations for both elastic and Raman channel are given as follows:

$$P_o(z) = C_o O_o(z) z^{-2}[\beta_{o,P}(z) + \beta_{o,M}(z)] T_o^2(z) \quad (1)$$

$$P_R(z) = C_R O_R(z) z^{-2} \beta_R(z) T_o(z) T_R(z) \quad (2)$$

Here P denotes received power and 0 & R denote the wavelength of the laser at λ_0 and λ_R is the Raman wavelength; system constants are denoted by C_0 and C_R for both the elastic and Raman channels respectively; and $O(z)$ represents the overlap factor. The $O(z)$ reaches to zero near the lidar system (without overlap) and reaches to 1 in the time of complete overlap at large distances. The basic assumption made here in this approach is that $O_0(z) = O_R(z)$ [1].

From Eq. (1), $\beta_{0,P}$ and $\beta_{0,M}$ denote the backscatter coefficients for elastic channel at λ_0 for particles and molecules respectively. The nitrogen Raman backscatter coefficient at the wavelength λ_R is denoted by β_R in Eq.(2). Atmospheric transmission at λ_0 is represented by T_0 i.e. two way between the lidar and the backscatter region whereas T_R at λ_R is atmospheric transmission after the Raman scattering process along the way back to the lidar.

The iterative approach used here is based on the fact that the deviation between the Klett solution [5], [6] for the backscatter coefficient i.e. $\beta_{Klett}(z)$, determined from the elastically backscattered signal and the Raman solution [7] $\beta_{Raman}(z)$, contains the important information about the incomplete overlap [5].

This iterative approach makes use of the fact that the aerosol backscatter signal, after overlap and range corrections is directly proportional to the total backscatter coefficient (see Eq. (1)).

This iterative procedure to reduce the overlap effect on elastically backscattered signal is based on the aforementioned principle. As a first step (i=1), we apply the Klett inversion method to the elastically backscattered uncorrected signal. The solution thus obtained is used to solve the expression:

$$\Delta O_i(z) = \frac{\beta_{Raman}(z) - \beta_{Klett,i}(z)}{\beta_{Raman}(z) + \beta_{o,M}(z)} \quad (3)$$

The overlap correction to elastically backscattered signals is applied as follows:

$$P_{o,i+1}(z) = P_{o,i}(z)[1 + \Delta O_i(z)] \qquad (4)$$

This signal profile $P_{0,2}(z)$ is used as an input (step i=2) in the Klett method to derive backscatter coefficient $\beta_{Klett,2}(z)$ and complete the next iteration. By repeating the same procedure again and putting $\beta_{Klett,2}(z)$ into Eq. (3) we can derive a new $\Delta O_2(z)$ which can be used in Eq. 4 again to correct the signal for overlap effect. This procedure is repeated till 10-12 iterations so as to completely remove overlap effect. In each step, the new difference between $\beta_{Klett}(z)$ and $\beta_{Raman}(z)$ decrease. The final overlap profile is derived by comparing the improved signal with the measured signal.

3. RESULTS

We have used here the iterative method to successfully determine the overlap function at Palampur station. The results are shown in Figure 2(a) and 2(b) obtained on five days of June 2016 with the help of Raymetrics Raman lidar system. The Palampur has pristine environment and rainy conditions during this time of the year so a lidar ratio of 35sr at 355nm is assumed for Klett inversion.

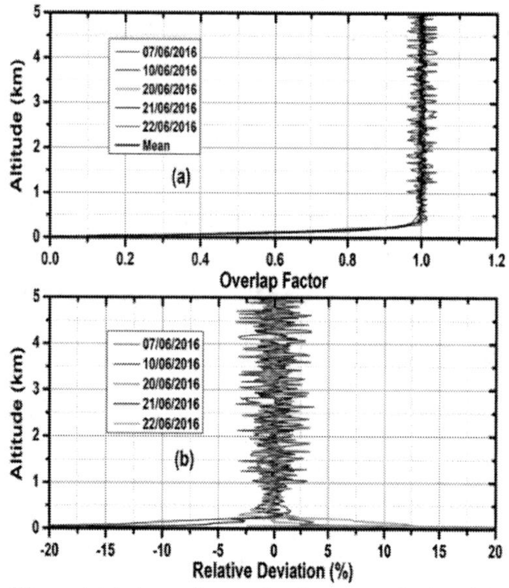

Figure 2(a) shows the overlap function calculated for the five days in the month of June 2016 with the mean overlap factor (black line) calculated using iterative method. Figure 2(b) shows the relative deviation of the overlap factor for 5 days with respect to mean overlap factor.

The results of Klett inversion and Raman inversion were derived and the solutions for several different iterations were derived using the method discussed in section 2.3. Here we have performed 12 iterations to determine the complete overlap height for each day. We have taken 5 different days of June 2016 and determined the overlap factor for each day and then computed the mean overlap profile with the relative deviation of each day with respect to mean. The results are shown in Figure 2(a) and 2(b). We can see from figure 2(a) that there was incomplete overlap below 300 meters and the complete overlap was achieved beyond 300 meters. Below 300 meters, the value of overlap function was less than 1 and after 300 meters, the value was close to 1. Figure 2(b) shows the relative deviation of overlap factor calculated for each day with respect to mean overlap factor. The deviation above 300 meters was less than 5%, however for lower heights i.e. below 300 meters the deviation was high.

4. CONCLUSIONS

It is important for the lidar users to have a proper knowledge about the geometrical form factor to remove or correct the lidar backscattered signals for the overlap effect which occurs at lower altitude due to the incomplete overlap between laser and telescope field of view. Here we have used an iterative approach to determine the overlap function. Results reveal that the lidar system was in complete overlap above 300m and the lidar data has been applied with this correction. The overlap correction has helped us to determine the optical properties with improved results especially in planetary boundary layer.

ACKNOWLEDGEMENTS

Authors are grateful to Director, NPL for necessary support. Authors are thankful to CSIR network project PSC-0112 for necessary financial support. Mr. Jaswant is thankful to Council of Scientific and Industrial Research (CSIR) for providing research fellowship and also to Academy of Scientific and Innovative Research (AcSIR) for facilitating as its PhD student.

REFERENCES

[1] R. Velotta, B. Bartoli, R. Capobianco, L. Fiorani, and N. Spinelli, *Appl. Opt.*, 37, p. 6999 (1998)

[2] G. M. Ancellet, M. J. Kavaya, R. T. Menzies, and A. M. Brothers, *Appl. Opt.*, 25, p. 2886 (1986).

[3] K. Tomine, C. Hirayama, K. Michimoto, and N. Takeuchi, *Appl. Opt.*, 28, p. 2194 (1989)

[4] U. Wandinger and A. Ansmann, *Appl. Opt.*, 41, p. 511 (2002)

[5] F. G. Fernald, *Appl. Opt.*, 23, p. 652 (1984)

[6] J. D. Klett, *Appl. Opt.*, 24, p. 1638 (1985)

[7] A. Ansmann, U. Wandinger, M. Riebesell, C. Weitkamp, and W. Michaelis, *Appl. Opt.*, 31, p. 7113 (1992)

LIDAR TECHNOLOGY BASED ON FIBER SYSTEM AND ITS APPLICATION

Xuewu Cheng[1], Yuan Xia[2]*, Yulian Yang[1], Xin Lin[1], Yong Yang[1], Linmei Liu[1], Faquan Li[1]

[1]*State Key Laboratory of Magnetic Resonance and Atomic and Molecular Physics, Wuhan Institute of Physics and Mathematics, Chinese Academy of Sciences, Wuhan, Hubei 430071, China*
[2]*School of Electronic Engineering, Nanjing Xiaozhuang University, Nanjing, Jiangsu 211171, China*
**Email: xiayuanxxyy@163.com*

ABSTRACT

The sodium atom existed in the metal layer of the earth's atmosphere has a high atomic number density and a large scattering cross section. Sodium layer can act as a good tracer for atmospheric detection in the middle and lower-thermosphere (MLT) region. The sodium fluorescence lidar uses ultrashort pulsed laser to excite sodium atoms, which enabling simultaneous detection of wind and temperature in the middle and upper atmosphere. This paper reports on the development of sodium fluorescence laser radar in recent years, especially the integration of fiber-coupled optical switches and fiber-coupled acousto-optic frequency modulation technologies, which greatly improved the stability and reliability of lidar system and reduced the maintenance of lidar operation, laying a good foundation for the application of lidar observations under harsh environments. This technology has been applied to the sodium wind /temperature lidar in Yangbajing, Tibet and has been running stably for a long time.

1. INTRODUCTION

A layer of metal atoms exists at a height of 80-120 km of the earth's atmosphere. The metal layer contains several kinds of metal atoms and ions, such as iron, sodium, potassium, calcium, and calcium ions [1]. Due to the high column density of sodium atoms and the relatively large scattering cross section, is the sodium layer is relatively easier to be detected [2,3]. Based on several decades of research and development, the lidar technique for sodium layer detection based on sodium fluorescence scattering mechanism became more and more mature. In 1990s, the simultaneous detection of wind and temperature of the sodium layer was carried out using narrow linewidth laser and multi-wavelength transmitter [5]. In the late 1990s, ultra-narrow bandwidth optical filtering technology was applied to the sodium lidar receiver, which enabling the full diurnal observation capability of sodium lidar thus permitting sodium layer wind and temperature measurement under sunlit conditions [6]. However, in the early studies of sodium wind temperature lidar, in order to realize the three-wavelength laser emission for simultaneous wind and temperature measurement, a ring-cavity dye laser, a free-space acousto-optic frequency modulation unit are usually used in the laser transmitter, so the optical path of the laser system is complex, and the adjustment and maintenance are also difficult, greatly limiting the lidar application. In recent years, optical fiber technology has been developed rapidly. Fiber-coupled fast optical switching technology and fiber-coupled acousto-optic frequency modulation technology can improve the configuration of laser unit from free-space light transport to fiber-coupled mode, greatly simplifying the optical path adjustment and reducing the maintenance difficulty and providing a good means for mobile and field observations. This article reports the latest development of narrowband sodium lidar and its application.

2. METHODOLOGY

In the early laser radar emitting system, a ring cavity dye laser is usually used. Since the ring cavity laser has a long optical path, it is particularly sensitive to environmental factors such as vibration, pressure, temperature, etc., and frequency locking and adjusting are quite difficult. In addition, in the three-frequency switching unit, two free-space acousto-optic modulator (AOM) with very small aperture of about 0.3 mm are used. The frequency switching unit has an optical path of several meters, requiring a very high adjustment accuracy, resulting in that the operation and maintenance are extremely difficult.

The lidar system introduced in this paper adopts a commercial solid-state diode laser (Toptica Company), which can output 1178 nm single-mode continuous-wave (CW) laser with power of more than 1W. After fiber splitting, its frequency is multiplied by a periodically poled lithium niobate (PPLN) waveguide and input into a sodium atom cell to obtain Doppler-free saturation absorption spectrum for laser frequency stabilization [7]. The other part of the laser light is input into the one-in-three-out fiber-coupled switcher for AOM+, AOM- and F0. The output of the optical switcher is then sent to a Raman fiber amplifier [8], and the amplified 1178nm CW laser light is multiplied by PPLN crystal to obtain laser at wavelength of 589 nm. Finally, the 589 nm CW laser is injected into a pulsed dye amplifier. This system replaced the conventional ring cavity laser and free-space AOM unit [9]. Since the entire optical path adopts the optical fiber mode, less optical path adjustment is required, and the conventional observation operation and daily maintenance became simpler, which can be used in unmanned lidar system.

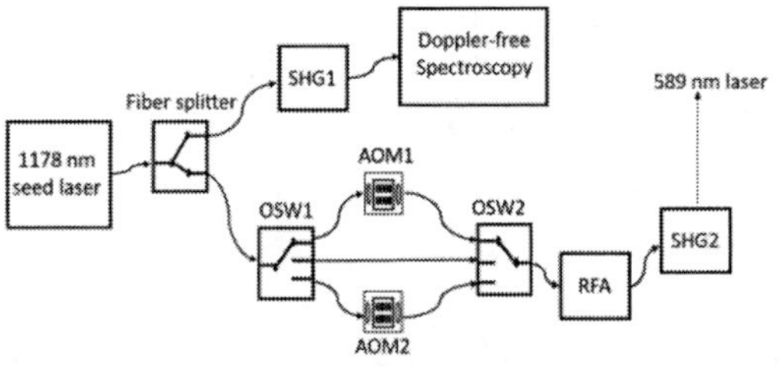

Fig. 1 Principle of fiber-coupled laser system.OSW1-3: fiber-coupled optical switcher; SHG1-2: second harmonic generation; RFA: Raman fiber amplifier; AOM1-2: acousto-optic modulator.

3. RESULTS

The design of laser transmitter has been successfully applied to the Yangbajing sodium layer wind and temperature lidar system, and the lidar operation is simple and the working state is stable. Typical wind and temperature observational results in MLT region are shown in Figure 2.

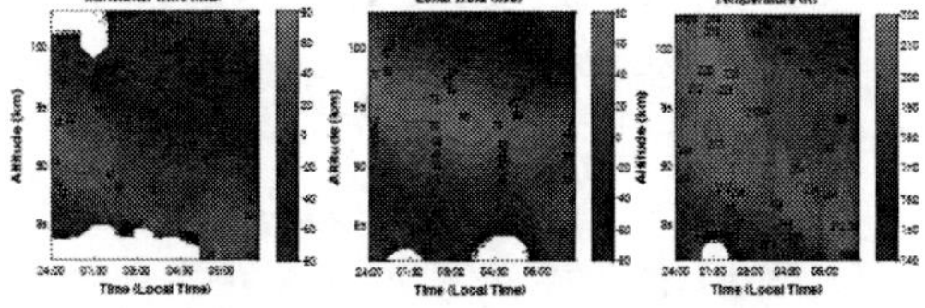

Fig. 2 The observational results of sodium lidar on Oct 17th 2017. (a) meridional wind. (b) zonal wind. (c) temperature.

ACKNOWLEDGEMENTS

This work was supported by the National Natural Science Foundation of China (No. 41127901, 41505027, 41627804, 2016YFC1400301), the Natural Science Foundation of the Jiangsu Higher Education Institutions of China (18KJB170009).

REFERENCES

[1] M. Gerding, M. Alpers, J. Hoffner, et al.. Sporadic Ca and Ca+ layers at mid-latitudes: Simultaneous observations and implications for their formation, Annales Geophysics, 19: 47-58 (2001)

[2] John A. Smith, Xinzhao Chu. High-efficiency receiver architecture for resonance fluorescence and Doppler lidars, Appl. Opt. 54(11): 3173-3184 (2015)

[3] Li, T., Fang, X., Liu, W., et al. Narrowband sodium lidar for the measurements of mesopause region temperature and wind. Appl. Opt. 51, 5401-5411(2012)

[4] Bowman, M.R., Gibson, A.J., Sandford, M.C.W., Atmospheric Sodium measured by a Tuned Laser Radar. Nature, 221, 456-457 (1969)

[5] She, C.Y., Yu, J.R., Simultaneous three-frequency Na lidar measurements of radial wind and temperature in the mesopause region. Geophys. Res. Lett, 21, 1771-1774 (1994)

[6] T. Yuan, C.-Y. She, T. D. Kawahara, et.al. Seasonal variations of midlatitude mesospheric Na layer and their tidal period perturbations based on full diurnal cycle Na lidar observations of 2002-2008. J. Geophys. Res. 117, D11304 (2012)

[7] Yue, J., She, C.Y., Williams, et.al. Continuous-wave sodium D2 resonance radiation generated in single-pass sum-frequency generation with periodically poled lithium niobate. Opt. Lett. 34, 1093–5 (2009)

[8] Feng, Y., Taylor, L.R., Calia, D. B.. 25 W Raman-fiber-amplifier-based 589 nm laser for laser guide star. Opt. Express 17, 19021–19026 (2009)

[9] Yulian Yang, Yong Yang, Yuan Xia, Xin Lin, Lei Zhang, Huawei Jiang, Xuewu Cheng, Linmei Liu, Kaijun Ji and Faquan Li. Solid-state 589 nm seed laser based on Raman fiber amplifier for sodium wind/temperature lidar in Tibet, China. Opt. Express 26(13):16226-16235 (2018)

 https://doi.org/10.1051/epjconf/202023707020

MULTICHANNEL POLARIZATION LIDAR MEASUREMENTS OF AEROSOLS AND CIRRUS CLOUDS

Jan Bai Nee[1, 2*]

[1]*National Central University, Chungli, Taiwan 320*
[2]*Aerosol Sciences Research Center, National Sun Yat-sen University, Kaohsiung, Taiwan, 700*
**Email:jbnee@phy,ncu.edu.tw*

ABSTRACT

In this paper we report using a 6-channel polarization detector to measure optical properties of aerosols and clouds. The polarization lidar system is designed to measure Stokes vectors and Mueller matrices from back-scatterings of air, aerosols and clouds by using several polarizers of setting at different angles, and a retarder to measure circular polarization. The 4 components Stokes vectors of the scattering media are constructed and a case of tropopause cirrus cloud and stratospheric aerosols are measured with the Mueller matrix derived.

1. INTRODUCTION

Polarization lidar system is an important tool to measure aerosols and clouds by deriving the depolarization ratios based on two component parallel and perpendicular polarization states. However, as demonstrated a century ago, the 4 component Stokes vector is a more complete description of the polarization state which can be used to derive the Mueller matrix to provide information about optical properties of various subjects. Stokes vector consists of four parameters $S=[I, Q, U, V]^T$ defined in terms of the components of the electric fields of the light waves. Here a vector A^T indicates the transpose of the A.

The four components of the Stokes vector are defined as follows:

$$I= Ex\ Ex^*+ EyEy^* \quad (1a)$$

$$Q= Ex\ Ex^*-EyEy^* \quad (1b)$$

$$U= Ex\ Ey^*+ Ex^*Ey \quad (1c)$$

$$V = Ex\ Ey^*- Ex^*Ey \quad (1d)$$

In the above equations, Ex and Ey are the two orthogonal components of the electric fields of the light wave which propagates along the z direction. Taking cloud as an example, if S_{in} is the Stoke vector of the incident light and S_{out} is the Stoke vector of scattering light from the cloud, we can use the following relationship to describe the interaction:

$$S_{out}= M_c\ S_{in}.. \quad (2)$$

where M_c, is a 4x4 Mueller matrix for the cloud as shown in Fig.1. $S_{in}=[1,1,0,0]^T$ is the Stokes vector for light of linear polarization along x direction.

S_{out} is measured by the lidar receiver which includes the telescope, filters, and detector system. The detector system has an instrument Mueller matrix M_x described as $T=M_x\ S_{out}$ where T is the Stokes vector of the signals measured by the lidar detector system with $T=[T_0,T_1,T_2,T_3]^T$.

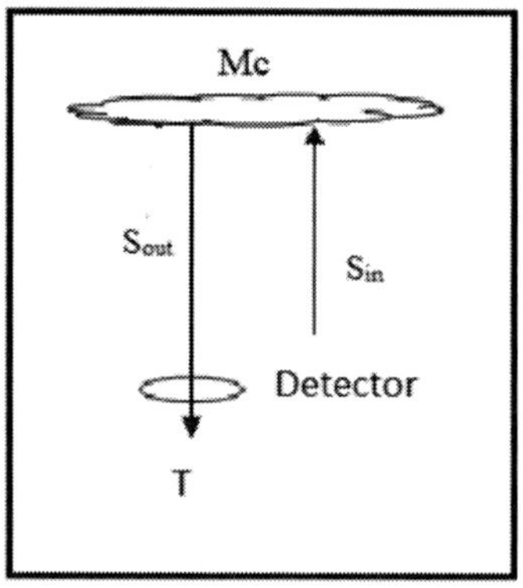

Fig.1 Stokes vectors S_{in},S_{out}, and T

The components of T are determined by the polarization filters and the retarder of the lidar system as will be describe later.

The Mueller matrix Mc describes the state of the cloud which consists of randomly oriented particles, As shown in previous works [3-5] the Mueller matrix for randomly oriented particles can be simply shown as

$$M = \begin{pmatrix} a_1 & b_1 & 0 & 0 \\ b_1 & a_2 & 0 & 0 \\ 0 & 0 & a_3 & b_2 \\ 0 & 0 & -b_2 & a_4 \end{pmatrix} \quad (3a)$$

which can be further simplified as:

$$M = \begin{pmatrix} 1 & 0 & 0 & 0 \\ 0 & 1-d & 0 & 0 \\ 0 & 0 & d-1 & 0 \\ 0 & 0 & 0 & 2d-1 \end{pmatrix} \quad (3b)$$

where d is the depolarization ratio [3-5].

2. METHODOLOGY

2.1 The lidar system

The lidar system include a 532 nm laser and a Cassegrain telescope of 20 cm diameter. Signals are measured by the detector system which consists of a photomultiplier tube (Hamamatsu R9880) and filter systems which include polarizers and interference filters (0.3 nm FWHM at 532 nm). Signals are treated by a transient recorder (Licel system) with a spatial resolution of 7.5 m. Each profile is accumulated for 30-60 sec. The lidar system has been described in previous papers [1-2]. The signals are accumulated 1 min for each channel and takes about 5 min for a complete measurements of 5 channels.

The 5 polarization channels are defined by using linear polarizers at orientations of 0^o(#6), $\pm 45^o$(#3,4), 90^o(#5), which are arranged by comparing their intensities with the laser emission whose polarization direction exiting to the sky is set by using a half-wave plate. The 532 nm filter is put in front of the polarizing filter system to receive the laser backscattering light. The sixth channel is a dark channel to check the background. Signals are recorded by a multichannel analyzer LICEL system. The telescope and photomultiplier system are considered as mainly constant in response to any polarizer.

We assume $S_{out}=[I_0,Q_0,U_0,V_0]$ as the Stokes vector of the back-scattered signals from the cirrus clouds. As shown previously, the measurements produce another Stoke vector T:

$$T=M_xS_{out} \quad (4)$$

The combined Mueller matrix for the detector as $M_x=M_1(\theta)\ M_2(\phi)$,with M_2 and M_1 the Mueller matrices for the retarder and linear polarizer at a specific angle (θ=0,90,45). When the measurement involves only linear polarizer without retarder (such as channels 1 and 2) we have $\phi=0$.

$$T=M_1M_2S_{out} \quad (3)$$

$$\text{For example, } M_1= \begin{pmatrix} 1 & 0 & 0 & 0 \\ 0 & 1 & 0 & 0 \\ 0 & 0 & 0 & 1 \\ 0 & 0 & -1 & 0 \end{pmatrix} \quad (5)$$

The four component of $T=[T_o,T_1,T_2,T_3]$ are measured by the lidar system as shown in Fig. 1. The 1st term, T_o is the most interesting since it is the intensity term [6]. We can derive T_0 for a liner polarizers at angle θ and a retarder of phase $\phi=\pi/2$. After the expansion, we get:

$$T_o(\theta,\phi)= (1/2)(I_0+Q_0 \cos 2\theta + U_0 \sin 2\theta \cos\phi + V_0 \sin 2\theta \sin\phi) \quad (6)$$

where θ (0,45,90) is the orientation of the linear polarizer. Again, ϕ =0 means without the retarder. From Eq. (6), we can derive

$$T_o(0,0) = (1/2)(Io+Qo) \quad (7a)$$

$$T_o(90,0) = (1/2)(Io - Qo) \quad (7b)$$

$$T_o(45,0) = (1/2)(Io + Uo) \quad (7c)$$

$$T_0(-45,0) = (1/2)(Io - Uo) \quad (7d)$$

From the above equations, the Stokes vectors S_{out} can be derived as follows

$$I_o= T_o(0,0)+ T_o(90,0) \quad (8a)$$

$$Q_o =T_o(0,0) - T_o(90,0) \quad (8b)$$

$$U_o=2T_o(45,0) - I_o \quad (8c)$$

$$V_o=2T_o(45,\pi/2) - I_o \quad (8d)$$

 https://doi.org/10.1051/epjconf/202023707020

$T_o(\theta,\phi)$ is the signals read from the transient recorder shown in Fig. 2. Therefore, a complete determining of the Stoke Vector of S_{out} can be made by a few measurements with a quarter wave plate and a linear polarizer setting at 0°, 90°, ±45°. Normally, we will normalize these quantities to set I_o=1. So the measurements are relative.

3. RESULTS

3.1 Cirrus cloud observations

Fig.2 shows the backscattering signals measured by the 5 filters. Below14 km, signals are from the Rayleigh scattering of air. At 14-17 km, a cirrus cloud was measured.

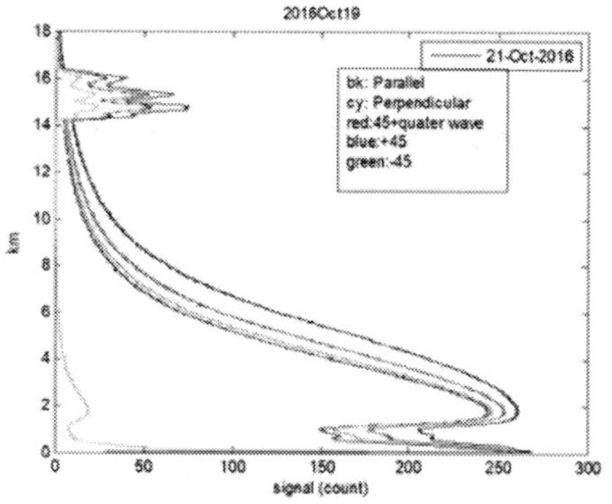

Fig. 2 Backscattering signals measured by various linear polarizers.

The cirrus clouds consists of roughly three layers at 14.8,15.4 and 16 km. For example, the Stokes vector of the 16 km layer is found to be:

(1.00, 0.498, 0.048, 0.097)

from which we found a2~ 0.5. For input light of linear polarization, the backscattering light is S_{out}= [1, δ, p, q] as shown in the cases of cirrus clouds. In order to determine a3, b4, we have to use circular polarization as the light source.

a2-a3+a4=1

For atmospheric application, the circular polarization is very small as shown in Hansen and Travis [7]. In practice, the 4th row and column can be ignored with M_c left as 3x3 matrix. Under this condition, we find a3=a2~0.5. So the Mueller matrix of the 16 km cloud is:

$$M = \begin{pmatrix} 1 & 0 & 0 & 0 \\ 0 & 0.5 & 0 & 0 \\ 0 & 0 & 0.5 & 0 \\ 0 & 0 & 0 & 0 \end{pmatrix}$$

4. Conclusion

REFERENCES

[1] J. B. Nee et al., J. Geophys. Res. *112*, D15202, doi:10.1029/2007JD008476.

[2] Z. W. Huang et al. Remote Sensing (MDPI), 10:1017 (2018)

[3] J. W. Hovenier and D. W. Mackowski, J. Quant. Spectrosc. Radiat. Transfer 60: 483-492, (1998)

[4] A. J. Brown and Y. Xie, J. Quant. Spectrosc. Radiat. Transfer 113:644–651 (2012)

[5] G. Gimmestad, Appl. Opt. 47: 3795–3802 (2008).

[6] Goldstein D. Polarized Light., 2nd ed. Marcel Dekker Inc. (2003)

[7] J. E. Hansen and L. D. Travis, Space Sci. Rev. 16:527-610 (19740

OCEANIC LIDAR: THEORY AND EXPERIMENT

Dong Liu[1, *], Yudi Zhou[1, #], Qun Liu[1], Weibiao Chen[2], Aleksey Malinka[3], Bing Han[4], Zhihua Mao[5], Peituo Xu[1], Zhipeng Liu[1], Xiaoyu Cui[1], Xiaobin Wang[1], Haochi Che[6], Peng Chen[5], Qingjun Song[7], Xiaolei Zhu[2], Chengfeng Le[8], Chong Liu[1]

[1]*State Key Laboratory of Modern Optical Instrumentation, College of Optical Science and Engineering, Zhejiang University, Hangzhou 310027, China*

[2]*Shanghai Institute of Optics and Fine Mechanics, Chinese Academy of Sciences, Shanghai 201800, China*

[3]*Institute of Physics, National Academy of Sciences of Belarus, Pr. Nezavisimosti 68-2, Minsk 220072, Belarus*

[4]*National Ocean Technology Center, Tianjin 300112, China*

[5]*The Second Institute of Oceanography, State Oceanic Administration, Hangzhou 310012, China*

[6]*Department of Physics, University of Oxford, Oxford OX1 3PU, UK*

[7]*National Satellite Ocean Application Service, Beijing 100081, China*

[8]*College of Ocean, Zhejiang University, Hangzhou 310027, China*

*Email: *liudongopt@zju.edu.cn #zhouyudi@zju.edu.cn*

ABSTRACT

Study on the upper ocean is of great significance to the global climate change and carbon cycle. Lidar can be used to effectively detect depth-resolved optical properties of the ocean. However, both theory and experiment of oceanic lidar are limited by complex multiple scattering. Several progresses by Zhejiang University will be illustrated in this paper: 1) a polarized lidar system was developed, and a Monte Carlo model and a radiative transfer model were established (Zhou, et al. *remote sensing*, 2019; Zhou, et al. *Journal of remote sensing*, 2019; Xu, et al. and Liu, et al. *Journal of remote sensing*, 2019); 2) Cross validations are demonstrated to verify the availability of the lidar system and models (Liu, et al. *IEEE TGRS*, 2019); 3) phase function effects on backscatter and attenuation are studied considering multiple scattering, respectively (Liu, et al. *Optics Express*, 2019). Oceanic lidar is proven to have great potential in marine studies.

1. INTRODUCTION

Study on the upper ocean is of great significance to the global climate change and carbon cycle. The ocean color remote sensing based on the airplane or satellite, like SeaWiFS, is able to collect the global data over a long term efficiently. Nevertheless, the limited information about the depth and the dependence on the natural light partly restrict its applications. So far, the oceanographic lidar, one of the active remote sensing methods, has been employed in detecting fisheries, phytoplankton layers and internal waves, etc., in the upper ocean [1-3].

In general, interpretation of lidar signals is not difficult under the well-known single-scattering approximation. However, in dense media, such as clouds, seawater, etc., the preceding and following events are accompanied by strong multiple scattering [3]. A portion of photons that are lost in the scattering events could eventually contribute to the lidar signals through multiple scattering. As a result, large errors could be introduced into the single-scattering approximation. Therefore, it is important to quantify and analyze lidar signals with multiple scattering.

Several progresses by Zhejiang University will be illustrated in this paper: 1) a polarized lidar system was developed, and a Monte Carlo model and a radiative transfer model were established; 2) Cross validations are demonstrated to verify the availability of the lidar system and models; 3) phase function effects on backscatter and attenuation are studied considering multiple scattering, respectively.

2. METHODOLOGY

2.1 The lidar system

The developed shipborne oceanic lidar transmitted a laser pulse into the seawater and detected the lidar returns for the retrieval of the seawater optical properties, as shown in Fig. 1. The transmitter included a frequency-doubling Q-switched Nd:YAG pulsed laser at 532 nm, with a single pulse energy of 5 mJ, a pulse width of 10 ns and a repetition frequency of 10 Hz.

Fig. 1. The lidar field experiment in the Yellow Sea. Lidar system mounted on the fore deck of the ship and *in situ* instruments going into the water.

2.2 Monte Carlo models

The standard MC and semianalytic MC models are employed to provide MC-simulated results. The basic principle of the MC simulation is to treat photons as classical particles [4] and simulate the trajectories of a number of photons to measure the relevant information. The standard MC algorithm refers to the method described in [4]. The semianalytic MC algorithm refers to the method illustrated in [5], where an analytical estimate is calculated for the possibility of the collection of scattered photons at certain points. The semianalytic MC algorithm greatly improves the calculation efficiency compared with the standard MC algorithm [6].

2.3 Analytical model

The analytical model employed in this paper follows the work of Katsev [7] and Malinka [8]. Seawater has very sharp forward peaks in its phase function that make the probability of scattering into the near-forward directions much larger than that of scattering into the backward hemisphere. If the optical thickness of the media is not too large, it is assumed that the trajectories that contribute to lidar signals primarily consist of single backscattering and small-angle forward multiple scattering on the outgoing and returning legs [7], namely the QSA approximation, which forms the foundation of the analytical model [6].

3. RESULTS

3.1 Cross validation

The lidar signals calculated by the analytical model, semianalytic MC algorithm and standard MC algorithm are shown in Fig. 2. The effects of height, FOV and water type are shown in Fig. 2 (a)-(c), respectively. The semianalytic MC and standard MC algorithms are simplified as "Semi MC" and "Stan MC", respectively, in each legend. To give a more explicit picture, the lidar signals are normalized by their maximums, and the dynamic ranges are set to 4 orders of magnitude. In Fig. 2, the results of the three algorithms agree very well at different conditions [6].

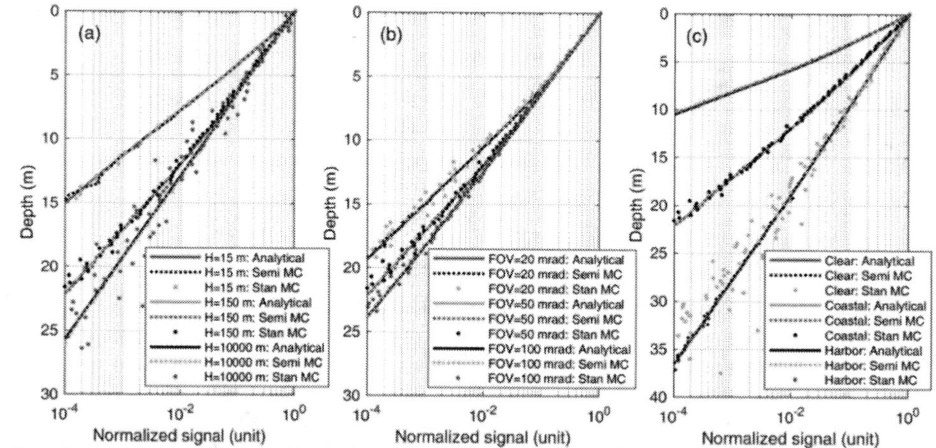

Fig. 2. Normalized lidar signals are calculated by the analytical model, semianalytic MC algorithm and standard MC algorithm at different: (a) heights; (b) FOVs; and (c) water types.

https://doi.org/10.1051/epjconf/202023707021

Figure 3 shows lidar signals calculated by the analytical model (orange solid lines) and measured by the lidar (blue solid lines with error bars) with a large, full FOV of 200 mrad. The average values in blue solid lines and standard deviations in blue error bars of 10 lidar-measured signals during the *in situ* operation period are plotted. Dynamic ranges of 3-4 orders of magnitude can then be realized in most cases. As shown in Fig. 3, the analytical model can perfectly match the lidar-measured results at different stations [6, 9].

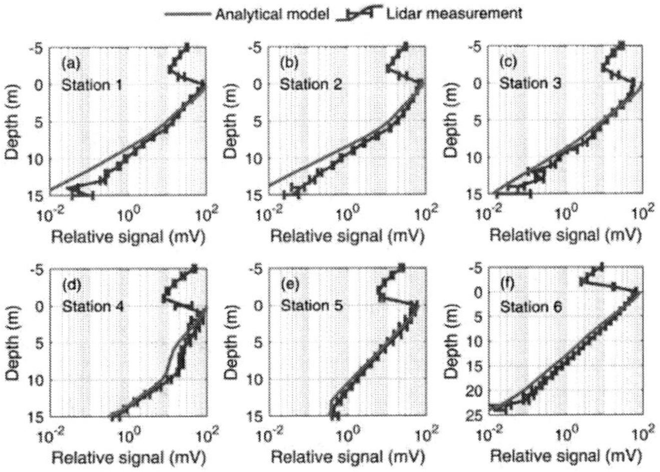

Fig. 3. (a)-(f) are the lidar signals calculated by the analytical model (orange solid lines) and measured by the lidar (blue solid lines with error bars) at Stations 1-6, respectively.

3.2 Phase function effects

The effective 180° VSF $\beta_p'^{\pi}$ can be obtained from the simulated signals. $\beta_m'^{\pi}$ is assumed to be equal to true molecular 180° VSF β_m^{π} because the molecular VSF is smooth in the backward direction. Figures 4(a)-4(c) show the simulation results of Cases A-C, respectively. The true 180° particulate VSFs β_p^{π} (dashed lines) that are the products of the scattering coefficients and phase functions at 180° remain constant with depth. The effective 180° particulate VSFs $\beta_p'^{\pi}$ (solid lines) typically remain consistent with β_p^{π} at the water surface but deviate from β_p^{π} with increasing depth. The degrees of deviations are closely related to the simulation conditions and phase function [10].

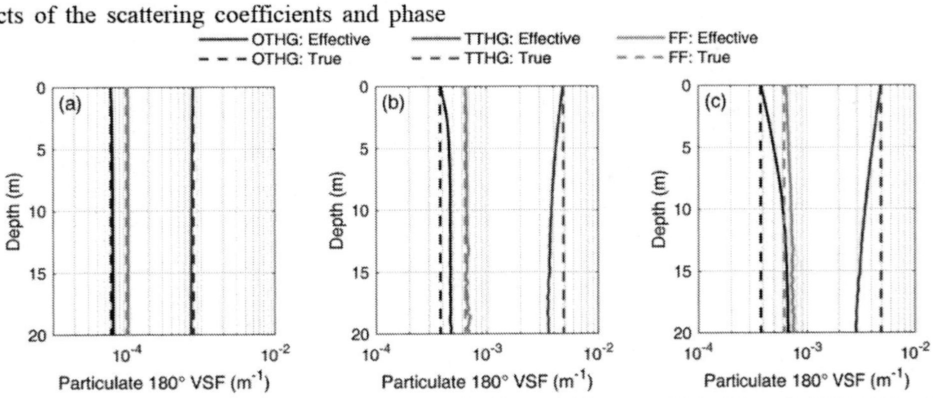

Fig. 4. Phase function effects on the effective particulate 180° VSF under (a) Case A (clear ocean, FOV of 50 mrad), (b) Case B (coastal ocean, FOV of 50 mrad) and (c) Case C (coastal ocean, FOV of 200 mrad).

The lidar attenuation coefficient α can be retrieved, as shown in Fig. 5. The sum of the absorption and backscattering coefficients $(a+b_b)$ (black dashed lines) is used as a reference because α should be always greater than $(a+b_b)$ under QSA approximation [3, 11]. The term α is close to $(a+b_b)$ for all phase functions at the water surface and increases with depth because of loss during multiple scattering.

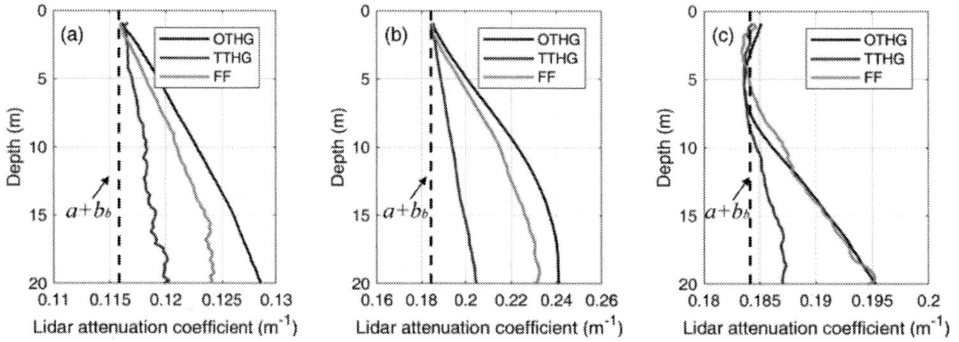

Fig. 5. Phase function effects on the lidar attenuation coefficient under (a) Case A (clear ocean, FOV of 50 mrad), (b) Case B (coastal ocean, FOV of 50 mrad) and (c) Case C (coastal ocean, FOV of 200 mrad).

ACKNOWLEDGEMENTS

This work was supported in part by the National Key Research and Development Program of China under Grant 2016YFC1400900 and Grant 2016YFC0200700, in part by the National Natural Science Foundation of China (NSFC) under Grant 41775023 and Grant 61475141, in part by Zhejiang Provincial Natural Science of China under Grant LR19D050001.

REFERENCES

1. Churnside, J.H., *Review of profiling oceanographic lidar*. Optical Engineering, 2014. **53**(5): p. 051405.
2. Hostetler, C., et al., *Spaceborne Lidar in the Study of Marine Systems*. Annual review of marine science, 2017: p. 13.1-13.27.
3. Bissonnette, L.R., *Lidar and Multiple Scattering*. 2005: Springer. 43-103.
4. Wang, L., S.L. Jacques, and L. Zheng, *MCML—Monte Carlo modeling of light transport in multi-layered tissues*. Computer methods and programs in biomedicine, 1995. **47**(2): p. 131-146.
5. Poole, L.R., D.D. Venable, and J.W. Campbell, *Semianalytic Monte Carlo radiative transfer model for oceanographic lidar systems*. Applied Optics, 1981. **20**(20): p. 3653-6.
6. Zhou, Y., et al., *Validation of the Analytical Model of Oceanic Lidar Returns: Comparisons with Monte Carlo Simulations and Experimental Results*. Remote Sensing, 2019. **11**(16): p. 1870.
7. Katsev, I.L., et al., *Efficient technique to determine backscattered light power for various atmospheric and oceanic sounding and imaging systems*. Journal of the Optical Society of America A, 1997. **14**(6): p. 1338-1346.
8. Malinka, A.V. and E.P. Zege, *Analytical modeling of Raman lidar return, including multiple scattering*. Applied optics, 2003. **42**(6): p. 1075-1081.
9. Liu, D., et al., *Lidar Remote Sensing of Seawater Optical Properties: Experiment and Monte Carlo Simulation*. IEEE Transactions on Geoscience and Remote Sensing, 2019. **57**(11): p. 9489-9498.
10. Liu, D., et al., *Phase function effects on the retrieval of oceanic high-spectral-resolution lidar*. Optics Express, 2019. **27**(12): p. A654-A668.
11. Kopilevich, Y.I. and A. Surkov, *Mathematical modeling of the input signals of oceanological lidars*. Journal of Optical Technology, 2008. **75**(5): p. 321-326.

https://doi.org/10.1051/epjconf/202023707022

RECENT ADVANCES IN AEROSOL AND TRACE GAS MONITORING BY EMPLOYING THE SCHEIMPFLUG LIDAR TECHNIQUES

Liang Mei[1], Zheng Kong[1], Teng Ma[1], Zhi Liu[1], Zhenfeng Gong[1], and Kun Liu[1,*]
[1]School of Optoelectronic Engineering and Instrumentation Science, Dalian University of Technology, Dalian 116024, China
**Email: liukun@dlut.edu.cn*

ABSTRACT

Lidar techniques, based on the time-of-flight principle, have been widely employed in atmospheric remote sensing since decades. The Scheimpflug lidar (SLidar) technique, which employing robust high-power laser diodes as light sources and highly integrated CCD/CMOS image sensor as detectors, has been recently developed for various atmospheric applications. Range-resolved atmospheric backscattering signal is obtained by capturing the backscattering imaging of transmitted continuous-wave laser beam based on the Scheimpflug principle. This paper reported recent advances in aerosol and trace gas monitoring by employing the SLidar techniques.

Keywords: Scheimpflug lidar, laser diode, aerosol, pollution, remote sensing, NO_2.

1. INTRODUCTION

Lidar techniques have been widely employed for atmospheric aerosol [1-4], trace gases [5,6], temperature [7,8] and wind speed [9,10] monitoring for decades. The Scheimpflug lidar (SLidar) technique has been recently demonstrated for atmospheric remote sensing [11,12]. The laser beam that is transmitted into atmosphere can be clearly focused on a tilted image sensor, if the optical layout of the lidar system satisfies the Scheimpflug principle. Infinite depth-of-focus is achieved while employing large aperture optics. Thus, the SLidar technique can measure range-resolved atmospheric backscattering signal from recorded pixel intensities. The SLidar technique significantly reduces the complexity of laser sources and detection units, the requirement on system maintenance and the overall cost of an atmospheric lidar system by utilizing high-power continuous-wave (CW) laser diodes as light sources and highly integrated CMOS/CCD image sensors as detectors. This paper reports our recent advances of the SLidar technique in the applications of atmospheric aerosol and trace gas monitoring.

2. AEROSOL MONITORING

Single-band Mie scattering Scheimpflug lidar (SLidar) systems, operating at several visible and near-infrared wavelengths, e.g., 407 nm, 450 nm, 520 nm and 808 nm, have been implemented and employed for real-time continuous atmospheric aerosol remote sensing [13-16], as shown in Fig. 1. A polarization 808-nm SLidar system has also been developed and evaluated by employing two linearly polarized 808-nm laser diodes and a single CMOS image sensor based on the time-division multiplexing detection scheme [17].

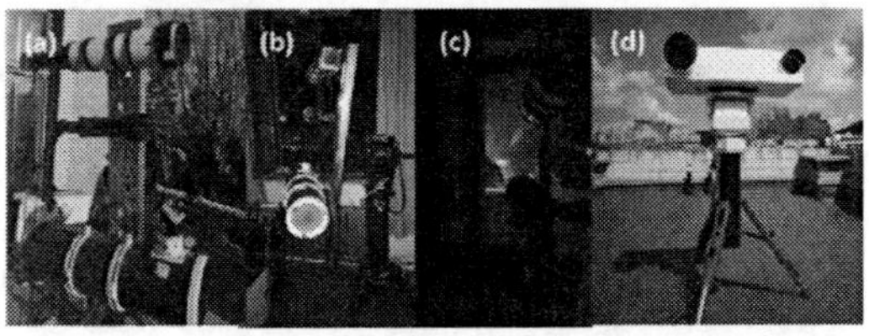

Fig. 1. Atmospheric remote measurements by employing (a) 808 nm SLidar system, (b) 450 nm SLidar system, (c) 520 nm SLidar system and (d) a portable unmanned 808-nm SLidar system.

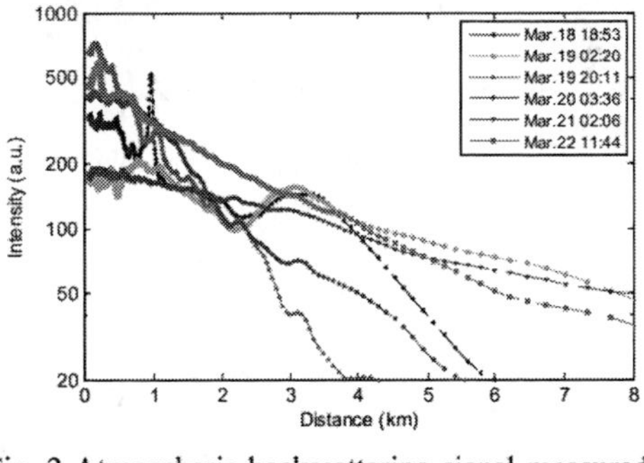

Fig. 2 Atmospheric backscattering signal measured on a near horizontal path by employing an 808-nm SLidar system in 2017.

All-time operation is feasible for the 407-nm and 808-nm SLidar systems due to relatively lower sunlight background and the narrow bandwidth of interference filters (less than 3 nm). However, the 450-nm and 520-nm Scheimpflug lidar systems, where the FWHMs of interference filters were about 10 nm, could be saturated under strong sunlight conditions. It has been found out that the sunlight shot noise dominates during daytime operation particular during a sunny weather condition, while the photon-response non-uniformity (PRNU) noise is the primary noise source during nighttime operation after signal averaging, e.g., 45 s [18]. Nevertheless, the maximum measurement distance of the SLidar systems could reach up to 7 km under clean weather conditions, as shown in Fig. 2. In several case studies, the temporal evolution of the extinction coefficients retrieved from the SLidar measurements were in good agreement with the PM10 or PM2.5 concentrations and the relative humidity (RH) measured by conventional air pollution monitoring stations, showing a great potential of employing the single-band SLidar technique in the applications of atmospheric aerosol monitoring.

The promising results achieved by single-band SLidar systems promoted the development of the dual-band SLidar system, which can be employed for the studies of aerosol extinction coefficients, Ångström exponent and particulate sizes. An 808-nm laser diode and a 407 nm laser diode were employed as the light sources. Backscattering signals at two wavelengths were detected by two CMOS cameras, respectively. Two-week continuous atmospheric remote measurements have been performed in May 2018. Hygroscopic studies of the aerosol extinction coefficient revealed that the shorter wavelength (407 nm) had a larger enhancement factor of water absorption for the aerosol particles off the Yellow Sea coast during the measurement period. The Ångström exponent, which varied between 0 and 2 during the measurement season, was influenced by the mass concentration fraction of fine model particles (PM2.5) and the RH. Moreover, the Ångström exponent was positive correlated with the RH, implying a bimodal or multimodal size distribution of aerosol particles. Nevertheless, further studies on the performances and applications of the dual-band as well as the three-band polarization SLidar system will be performed in the future.

3. ATMOSPHERIC POLLUTION TRACKING

Real-time monitoring of atmospheric pollution emissions as well as their transportations in a large area is also feasible by performing scanning lidar measurements. A portable Mie-scattering 808-nm Scheimpflug lidar system has been designed and implemented for atmospheric pollution monitoring, as shown in Fig. 1(d). The portable SLidar system employed a continuous-wave high-power 808-nm laser diode as the light source and the emitted laser beam is collimated by an achromatic lens (F6) with a 100-mm aperture. Atmospheric backscattering light is collected by a F5 achromatic lens with a 150-mm aperture and then detected by a 45° tilted CMOS image sensor. The separation between the transmitting and the receiving optics is about 756 mm to satisfy the Scheimpflug principle.

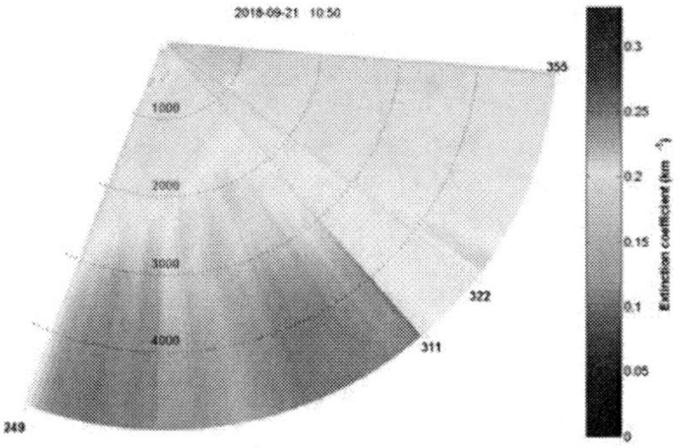

Fig. 3. Scanning map of the aerosol extinction coefficient measured at 10:50 on September 21st, 2019. Region A: 249°-311°, elevation angle 5.5°; region B: 312°-322°, elevation angle 8.5°; region c: 323°-355°, elevation angle 4.2°.

Outdoor atmospheric measurements were performed in urban area to investigate the system performance. Localized emissions can be identified by performing scanning measurements over the urban atmosphere for 107° in approximately every 17 minutes. Fig. 3 shows the scanning maps of the atmospheric extinction coefficient retrieved according to the Klett method. Clear boundaries due to different elevation angles can be observed between different scanning regions, which also implied a vertically layered structure of the atmosphere during the measurement period. As can be seen, localized high values of the aerosol extinction coefficient can be found at many places. The promising result shows a great potential of the portable lidar system for unmanned atmospheric pollution monitoring in urban areas, and more field measurement campaigns are planned in the near future.

4. ATMOSPHERIC NO_2 MONITORING

Differential absorption lidar (DIAL) technique provides unique capability of remotely monitoring urban/rural area localized NO_2 emissions and profiling tropospheric vertical NO_2 distribution. By measuring and evaluating atmospheric backscattering echoes of transmitted laser pulses at on-line absorption (λ_{on}) wavelength and off-line absorption (λ_{off}) wavelength of NO_2, range-resolved NO_2 concentration can be retrieved. However, the prevailing pulsed DIAL techniques mainly employ Nd:YAG laser pumped dye lasers with coumarin dye [19-21], which has a very short lifetime and low-conversion efficiency when pumping by the ultraviolet light (355 nm). Thus, frequent change of dye and system alignment are required during measurements, which prevents practical applications of the NO_2 DIAL technique.

Atmospheric NO_2 has a relatively broadband absorption spectrum in the visible region around 450 nm, which provides the possibility of employing high-power multimode laser diodes for differential absorption measurements. Recently, we have developed a DIAL technique based on the Scheimpflug lidar, by introducing the tunable diode laser absorption spectroscopy (TDLAS) into the lidar fields [22]. The λ_{on} wavelength at around 448.6 nm and the λ_{off} wavelength at around 452.1 nm can be alternatively generated by modulating the driving current and tuning the case temperature of the high-power laser diode. The emitted laser beam of the laser diode is collimated and transmitted into atmosphere through a refractor. Atmospheric backscattering signal is first collected by a Newtonian telescope and then detected by a 45-degree tilted high-sensitivity CCD image sensor, when the optical layout satisfies the Scheimpflug principle. Range-resolved NO_2 concentration is then retrieved from the recorded DIAL curve according to the lidar equations. By employing compact and robust laser diodes as light sources, the present lidar technique can continuously operate without interruption, while conventional pulsed lidar technique employing sophisticated pulsed tunable dye lasers often requires extensive maintenance during measurements. Remote measurements have been performed during nighttime on a near horizontal path and shown a good agreement with the NO_2 concentration measured by a local national pollution monitoring station. A detection sensitivity up to ppbv level has been achieved in the region between 0.3-3 km. The low-cost and robust DIAL system opens up many possibilities for field NO_2 remote sensing applications in the near future. However, the spectral stabilities of the multimode laser diode and the SNR of the retrieved lidar curves should be further improved in the near future by performing real-time spectral calibration and reducing the PRNU noise, etc.

5. CONCLUSION

The SLidar technique has been successfully demonstrated for atmospheric extinction coefficient retrieval, depolarization and Ångström exponent studies, atmospheric pollution tracking, as well as NO_2 monitoring during recent years. It has shown a great potential for practical atmospheric applications with the advantages of being low cost, less complexity, and low maintenance. Future work involves updating the portable single-band 808-nm scanning SLidar system, development and applications of multi-band polarization SLidar system, and improving the detection sensitivity and stability of the NO_2-DIAL technique for field deployment.

ACKNOWLEDGEMENTS

This work was supported by National key research and development program of China (2016YFC0200600); National Natural Science Foundation of China (61705030); Natural Science Foundation of Liaoning Province, China (201602163); Fundamental Research Funds for the Central Universities (DUT18JC22).

REFERENCES

1. H. Baars, P. Seifert, R. Engelmann, and U. Wandinger, Atmos. Meas. Tech. **10**, 3175-3201 (2017).
2. D. Althausen, D. Muller, A. Ansmann, U. Wandinger, K. Franke, and F. Wagner, Laser Radar Technology and Applications Viii **5086**, 139-144 (2003).
3. A. Comeron, C. Munoz-Porcar, F. Rocadenbosch, A. Rodriguez-Gomez, and M. Sicard, Sensors-Basel **17**, 1450 (2017).
4. A. Behrendt and T. Nakamura, Opt. Express **10**, 805-817 (2002).
5. C. Weitkamp, R. Baumgart, A. Behrendt, P. Bisling, J. Glauer, U. B. Goers, S. Kohler, W. Lahmann, and J. Reichardt, Igarss 2001: Scanning the Present and Resolving the Future, Vols 1-7, Proceedings, 1023-1025 (2001).
6. U. N. Singh, J. Yu, M. Petros, T. F. Refaat, R. G. Remus, and K. Reithmaier, Proceedings of SPIE, Toulouse, 9612, 961204, (2015).
7. J. Hildebrand, G. Baumgarten, J. Fiedler, and F. J. Lubken, Atmos. Chem. Phys. **17**, 13345-13359 (2017).
8. A. Behrendt, T. Nakamura, and T. Tsuda, Appl. Opt. **43**, 2930-2939 (2004).
9. S. H. Wu, J. P. Yin, B. Y. Liu, J. T. Liu, H. W. Zhang, X. Q. Song, and K. L. Zhang, EPJ Web Conf. **119**, (2016).
10. R. M. Banta, Y. L. Pichugina, W. A. Brewer, J. K. Lundquist, N. D. Kelley, S. P. Sandberg, R. J. Alvarez, R. M. Hardesty, and A. M. Weickmann, J. Atmos. Ocean Tech. **32**, 904-914 (2015).
11. L. Mei and M. Brydegaard, Opt. Express **23**, 247841 (2015).
12. M. Brydegaard, A. Gebru, and S. Svanberg, PIER **147**, 141-151 (2014).
13. L. Mei, Z. Kong, and P. Guan, Opt. Express **26**, A620-A274 (2018).
14. L. Mei, P. Guan, Y. Yang, and Z. Kong, Opt. Express **25**, A628-A638 (2017).
15. L. Mei, Z. Liu, Z. Kong, and P. Guan, presented at the Proc. SPIE 10846, Optical Sensing and Imaging Technologies and Applications, 2018.
16. L. Mei, Z. Kong, Z. Liu, L. Zhang, and P. Guan, Sensors-Basel **18**, 1880 (2018).
17. L. Mei and P. Guan, Opt. Lett. **42**, 3562-3565 (2017).
18. L. Mei, L. Zhang, Z. Kong, and H. Li, Opt. Commun. **426**, 463-470 (2018).
19. Q. Liu, Y. Chen, J. Wang, J. Huang, and S. Hu, Proc. SPIE **10605**, 106053L (2017).
20. K. F. Boersma, D. J. Jacob, M. Trainic, Y. Rudich, I. DeSmedt, R. Dirksen, and H. J. Eskes, Atmos. Chem. Phys. **9**, 3867-3879 (2009).
21. H. J. Kolsch, P. Rairoux, J. P. Wolf, and L. Wöste, Appl. Opt. **28**, 2052-2056 (1989).
22. L. Mei, P. Guan, and Z. Kong, Opt. Express **25**, A953-A962 (2017).

RESEARCH ON NON-INVASIVE LASER DOPPLER CURRENT MEASUREMENT FOR HYDROTHERMAL VENTS FLOW RATES PROFILE

Hongwei Zhang[1,*], Songhua Wu[1,2], Jintao Liu[1], Yuanshuai Zhang[1], Xiaomin Chen[1]

[1] *Ocean Remote Sensing Institute, College of Information Science and Engineering, Ocean University of China, Qingdao 266100, China*

[2] *Laboratory for Regional Oceanography and Numerical Modeling, Pilot National Laboratory for Marine Science and Technology (Qingdao), Qingdao 266237, China*

**Email: zhanghongwei8944@ouc.edu.cn*

ABSTRACT

Hydrothermal vents play important roles in the dynamic process of the stratosphere and the circulation of energy materials. Nowadays, invasive devices such as vane-type flow meter and turbine flow meter are commonly utilized to estimate the flow rates of hydrothermal vents. In-situ observation data obtained by LDCP system and CFD methods will be combined to solve the blind area of LDCP system during *in-situ* observation measurements. The flow rates profiles obtained in this project will improve the research on the circulation of energy materials and mineralization.

1. INTRODUCTION

Generally, hydrothermal vents form in volcanically active areas, especially on mid-ocean ridges. Ocean water percolates into the crust through cracks and porous rocks and is heated by underlying magma. The heat helps drive chemical reactions that remove oxygen, magnesium, sulfates and other chemicals from the water that entered the ocean through rain, rivers, and groundwater. In the process, the fluids also become hotter and more acidic, causing them to leach metals such as iron, zinc, copper, lead, and cobalt from the surrounding rocks. Hydrothermal vents act as natural plumbing systems that transport heat and chemicals from the interior of the Earth and that help regulate global ocean chemistry. In the process, they accumulate vast amounts of potentially valuable minerals on the seafloor [1].

A number of different methods have been used to estimate the flow rates of hydrothermal vents since the discovery of vents, starting with visual observations using a vane-type flow meter from the Alvin submersible [2, 3]. Lately, turbine flow meters combined with temperature-probe data allowed the measurement of fluid flow rates exiting from high-temperature vents. The turbine flow meter was held 3cm above the fluid exit to limit jamming of the turbine by small particles and to prevent melting of the plastic insulation [4]. However, it was shown that the sensor was significantly affected by fouling during long-term deployment [5]. An alternative method, visual tracking of particles and eddies in the ascending fluids, was used by Rona and Trivett and Ramondenc [6, 7]. In contrast, non-invasive methods that use image processing techniques to track refractive index anomalies can estimate diffuse velocities or matter fluxes [8, 9]. Fluid velocity, however, estimated through video imagery, which requires the presence of eddies and particles and does not allow for autonomous long-term measurements [10].

Nowadays, there are no non-invasive instruments designed for high-resolution flow rates measurements of hydrothermal vents, making the estimation of heat flux with great error [11, 12].

In this paper, *in-situ* observation data obtained by LDCP system and CFD methods will be combined to solve the blind area of LDCP system during *in-situ* observation measurements. The detailed information of hydrothermal vents obtained by in-situ devices will be analyzed in this project to confirm the specification of Laser Doppler Current Probe for hydrothermal vent in-situ observation. Measurement scheme to *in-situ* observation of hydrothermal will be designed based on a LDCP system with features of high-resolution and non-invasive.

2. The Laser Doppler Current Profiler system

The LDCP system is non-invasive flow measurement tool based on LDA (laser Doppler anemometry) principles which has been developed to be a high-level standard method for flow measurement. A standard one-component LDA system consists of two laser beams. For simplicity two laser beams of equal frequencies (v_0) are considered to intersect at an angle 2α (Fig. 1). The cross area of these two laser beams in the flow is called the measurement volume. A particle that is suspended in the flow is assumed to pass through the measurement volume and so scatters the lights of two laser beams simultaneously. Because of the different spatial layout of two laser beams, the moving particle of velocity up perceives the different light frequencies resulted from different Doppler effects. A detector is spatially located along ℓ_2 for receiving the light that is scattered from the measurement volume. The velocity of the moving particles are considered to be the speed of flow.

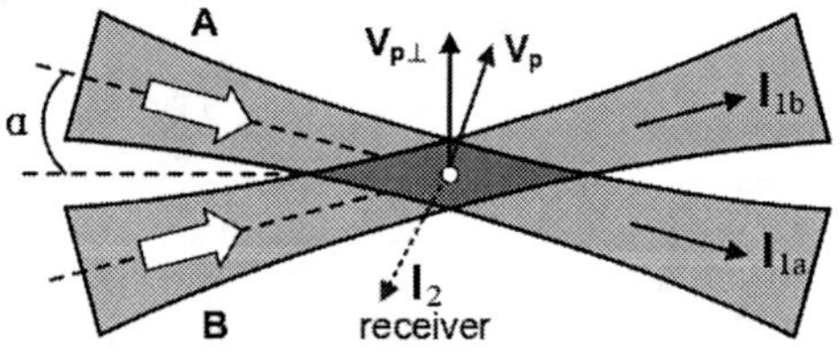

Fig.1 The scheme of particle passing through the measurement volume formed by LDCP system

A LDCP system based on LDA technique has been designed for the measurements of ocean current as Fig.2 shows. The LDCP system (including: optical transmitting and receiving parts, electronical instruments and data processing devices) is assemabled into a 1cm Alumium tube to protect it from the high pressure of sea water.

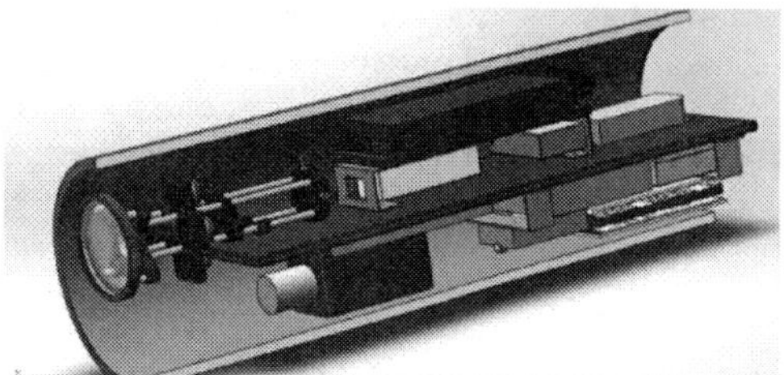

Fig.2 LDCP system assembly diagram generated by *Solidworks* software

The continuum laser source is a DPSS laser with excellent beam quality (M2=1.1), whose average energy is 50mW at each wavelength. The Doppler shift of acoustic optical modulator (AOM) is 40MHz, while the sampling rate of Field Programmable Gate Array (FPGA) is 250 MHz to make the signal accurately. The detailed specifications of the LDCP system are listed in the Table 1.

Table 1 the specification of the LDCP system

Qualification	Specification
Wavelength	491 nm & 532 nm
Power	50mW@each wavelength
Accuracy	1%
Velocity Range	0.001-10 m/s
Measurement Range	0.2 m

3. Simulation of Flow Rates

Computational fluid dynamics (CFD) is a branch of fluid mechanics that uses numerical analysis and data structures to analyze and solve problems that involve fluid flow. Generally, high-speed computers are used to perform the calculations required to simulate the free-stream flow of the fluid, and the interaction of the fluid (liquids and gases) with surfaces defined by boundary conditions.

A model of glass trough has been constructed by *Rahino* software, and the flow field has been calculated by *Fluent* software as Fig. 3 shows. The characteristic of the glass trough's central plane is illustrated in Fig. 3. The colors reveal the flow speed while the arrow mean the flow direction.

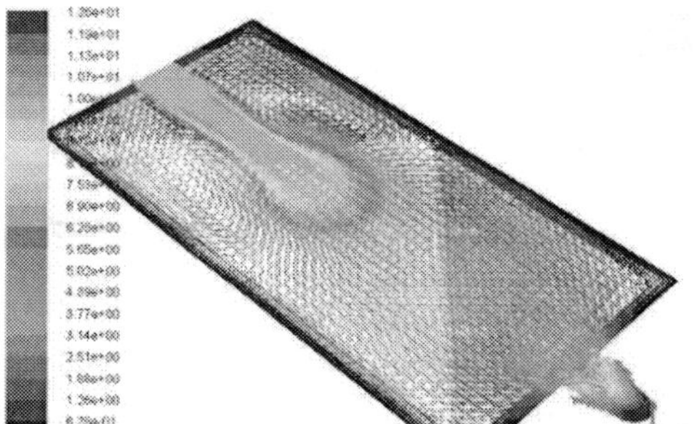

Fig.3 Flow field of glass trough's central plane simulated by *Fluent* software

To verify the accuracy of simulated results, the flow information (include flow rates and flow direction) of selected position in the model can be picked out. The selected positions are illustrated in Fig.4, while the simulated results are listed in Table 2.

Table 2 simulation of flow rates under different input speed conditions.

	5m/s	10m/s	15m/s	20m/s
Point-1	1.35	5.68	11.16	5.40
Point-2	0.60	9.05	14.48	2.40
Point-3	0.97	2.07	3.63	3.88
Point-4	0.92	1.71	2.83	3.68
Point-5	1.90	2.51	5.04	7.60
Point-6	1.14	2.40	3.65	4.56
Point-7	1.12	1.92	2.83	4.48
Point-8	0.80	1.76	2.64	3.20
Point-9	1.87	0.74	0.95	7.50
Point-10	5.54	10.00	15.05	22.16
Point-11	0.60	0.60	1.12	2.40

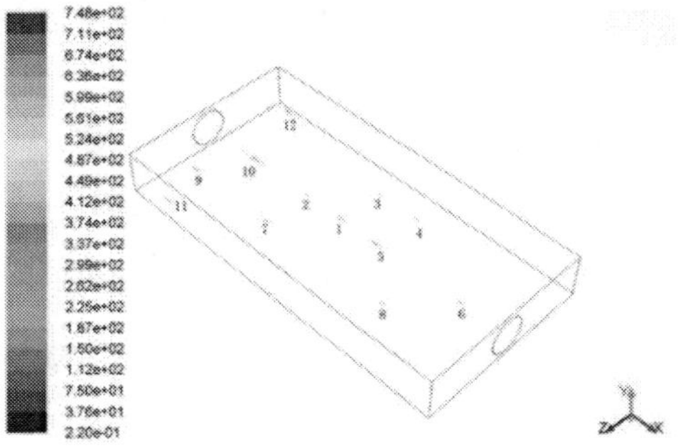

Fig.4 Selected position in flow field simulated by *Fluent* software

4. METHODOLOGY

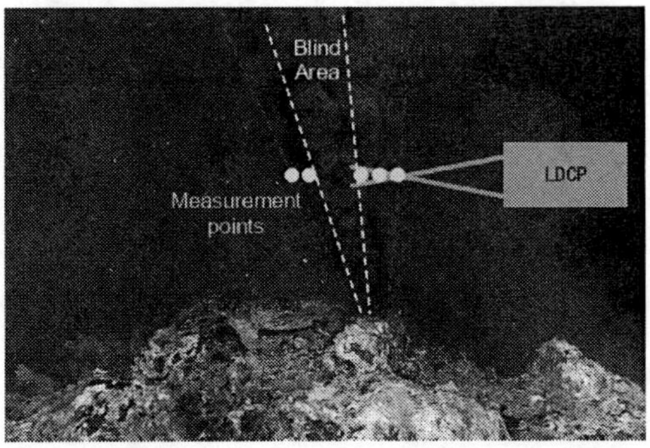

Fig.5 The scheme of hydrothermal vent flow rates profile measurements using the LDCP system

For measurements of flow such as hydrothermal vents flow, non-invasive techniques with high spatial resolution should be applied. Conventional laser Doppler anemometry (LDA) is not capable of obtaining high-spatial-resolution flow rate profiles because of existing of blind area of LDCP system as Fig.5 shows.

To solve the existing of blind area during the *in-situ* measurement, this paper intends to combine the LDA technology and CFD technology to illustrate the whole flow field characteristics of hydrothermal vents. The Schematic diagram of retrieving velocity profile from the fusion of simulated data and measured data can be depicted from Fig.6. The modified laser Doppler current probe was used to measure flow rates in the penetrable area of the hydrothermal vents, and the measured data will be used as the input data of the fluid simulation software to simulate the flow velocity profile of the hydrothermal vents. The input parameters of the simulation software (such as boundary conditions, fluid viscosity, number of iterations, etc.) will be adjusted to fit the measurement curves. Then the whole flow rates profile will be calculated by *Fluent* software.

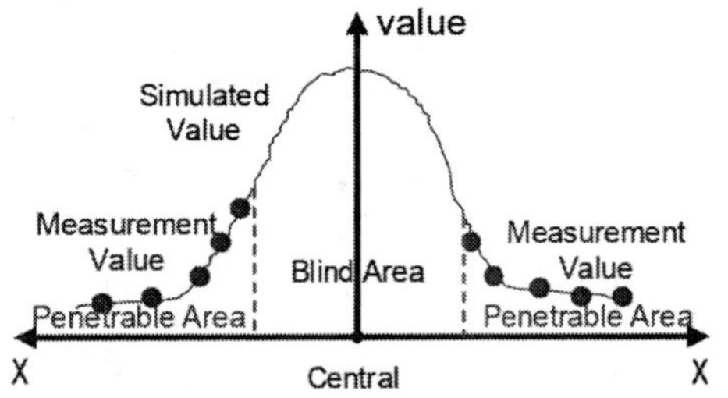

Fig.6 Schematic diagram of retrieving velocity profile from the fusion of simulated data and measured data

5. CONCLUSION

For the deep sea *in-situ* measurements, no technology has enough advantages to replace other measurement techniques, and various measurement methods play a role in the field in which it is good. The traditional devices are invasive instruments with the advantages of low cost and easy operation. Their body shell, however, disturb the flow field and cause large measurement error especially under low speed conditions. Comparing to the point measuring device, the instruments based on the acoustic Doppler technology have large measuring range and high-time resolution. Due to low spatial resolution, the acoustic instruments perform not well in high resolution flow rate measurements of hydrothermal vents. Methods based on the optical video image sequences to retrieve the flow information of vertical plane can make full use of the existing seabed hydrothermal video data. The measurement result, however, focus on the surface of the hydrothermal vent and cannot provide the flow rates information inside the hydrothermal spray column.

In this paper, we focus on the combination of LDA technology and CFD technology for flow field measurements based on an optical non-invasive instrument. However, the flow rates profiles vary greatly with different exit speed or the shape of hydrothermal vent models. Table of flow rates under different exit speed or different hydrothermal vent shape conditions will be established to estimate the heat flux of entire hydrothermal vents field.

6. FUTURE WORK

The method for measuring the flow rate profiles of the hydrothermal vent models mentioned in this paper needs to be verified by experimental work and simulations. Table of flow rates under different input speed or different hydrothermal vent shape conditions should be constructed to retrieve the detailed information of the flow field.

The temperature and velocity information perform important roles in estimating the heat flux. However, the LDCP system is designed for flow rates measurement. The temperature sensor and conductivity sensor should be combined with the LDCP system during the field experiments for the estimation of heat flux.

Although the method mentioned in this paper can estimate flow rates, measurements encompass small areas, still making estimation of fluid fluxes across an entire vent field difficult without other constraints.

ACKNOWLEDGEMENTS

We thank our colleagues for their kindly support during the laboratory experiments, including Kailin Zhang and Xiangcheng Chen with Ocean University of China for performing the LCDP system, Lei Liu and Fanghan Wang for CFD simulation work. This research was funded by National Key Research and Development Program of China under Grant No. 2016YFC1400904.

REFERENCES

[1] https://www.whoi.edu/main/topic/hydrothermal-vents

[2] Corliss, J. B., Dymond, J., et al. Science, 203(4385), 1073-1083 (1979).

[3] Schultz, W., Apicella, P., Scarnati, E., et al. Journal of neuroscience, 12(12), 4595-4610 (1992).

[4] Rona, P. A., et al. Eos, Transactions American Geophysical Union, 83(44), 497-502 (2002).

[5] Smart, C. J., Roman, C., et al. Geochemistry, Geophysics, Geosystems, 14(11), 4743-4757 (2013).

[6] Xu, G., Jackson, D. R., et al. Geochemistry, Geophysics, Geosystems, 14(7), 2369-2382 (2013).

[7] Zhang, X., et al. Deep Sea Research Part I: Oceanographic Research Papers, 57(2), 297-306 (2010).

[8] Mittelstaedt, E., et al. Deep Sea Research Part II: Topical Studies in Oceanography, 150, 67-81 (2018).

[9] Dumke, I., Ludvigsen, et al. IEEE Transactions on Geoscience and Remote Sensing (2018).

[10] Sarrazin, J. et al. Deep Sea Research Part I: Oceanographic Research Papers, 56(11), 2065-2074 (2009).

[11] Xu, G., Jackson, D. R., et al. Earth and Planetary Science Letters, 404, 220-231 (2014).

[12] Thomson, R. E., et al. Journal of Geophysical Research: Oceans, 94(C4), 4709-4720 (1989).

[13] Michel, A. P., et al. Deep Sea Research Part II: Topical Studies in Oceanography, 150, 57-66 (2018).

The design of focal plane splitting unit in a hyperspectral lidar system

Qian Liyong[1 2], Wu Decheng[1], Zhou Xiaojun[1], Zhong Liujun[1], Wei Wei[1], Deng Qian[1 2], Chu Yufei[1 2], Liu Dong[1 2*], Wang Yingjian[1 2]

1. Key Laboratory of Atmospheric Optics, Anhui Institute of Optics and Fine Mechanics, Chinese Academy of Sciences, Hefei, Anhui 230031

2.University of Science and Technology of China, Hefei, Anhui 230026

**Email: dliu@aiofm.ac.cn*

ABSTRACT

A hyperspectral lidar system with focal plane splitting unit is proposed, including an off-axis receiving telescope, a grating spectrometer, and a single-tube detector array. The spectrum of the system covers 380-930nm, and is separated by grating spectrometer. The microlens-fiber coupling system guides the echo signal of 50 channels into each detector. The system solves the data processing problem of the bandwidth and gain the line array and area array detector in traditional hyperspectral lidar. And it also meets the requirement of the high efficiency splitting coupling and weak signal acquisition and detection.

1. INTRODUCTION

Airborne/space-borne lidar usually works in a single-wavelength mode to acquire the earth surface data which contained single-wavelength echo intensity information[1, 2]. In order to make better use of these observation capability in the existing lidar technology and scene classification, the intensity information and three-dimensional information of single-wavelength laser echo signals are combined to use[3]. Based on the backscattered signal reflected from the laser footprint, traditional method classifies the features such as road, grassland, crops, etc. The measurement based on the intensity information of the laser footprint to detect large rocks and cliffs in the forest area needs to get more accuracy of the mountain terrain data[4]. Seldom studies fundamentally improve the lack of physical detection capability of airborne lidar due to a single wavelength[5]. Although it has outstanding advantages in the acquisition of information in three-dimensional space, it is limited by the single-wavelength detection capability in terms of the physical property detection classification and state of the ground object. Compared with traditional passive remote sensing methods such as multi-hyperspectral imaging, it cannot reflect the essential advantages[6, 7].

According to the fast application of lidar and its technical characteristics, multi-spectral and multi-wavelength lidar remote sensing is carried out. This research with high spatial resolution detection capability will have important value and application prospects in the detect physical properties[8, 9]. Just as passive optical remote sensing as a trend from a full-color camera to a multi-spectral camera-hyperspectral camera, the lidar geodetic observation technology is also shown a trend from single wavelength to double wavelength to multiple wavelengths.

This paper proposes a new type of Earth observation high-spectral lidar technology. The spectrum of the lidar system covers from 380 to 930nm at wavelength. Due to the transmitted high-spectral laser, it can received the spectral and the collocated elevation information which improves the system's ability to distinguish vegetation, soil, rock and other ground objects which can be applied to forest survey, agricultural monitoring, land monitoring, resource exploration and other fields in a rapid, economical and accurate manner.

2 System Introduction

As shown in Figure 1, by using the focal plane efficient spectroscopic technology based on microlens and fiber array, the wideband echo signals are efficiently divided into different channels detected by the high-speed, high-sensitivity detectors through optical fibers. Amplitude and dynamics of wide-spectrum laser echo signals adopt APD and PMT hybrid array detection technology to solve the problem of hyperspectral fast imaging under weak light conditions, realizing the full coverage detection of wide-spectrum bands of detection targets. Research on optical noise reduction technology should be used background light suppression technology to improve the detection performance of hyperspectral lidar. The design goal requires that the laser spectral band covers 380-930nm, and the hyperspectral lidar imaging detection with a band number of not less than 50 channels is realized.

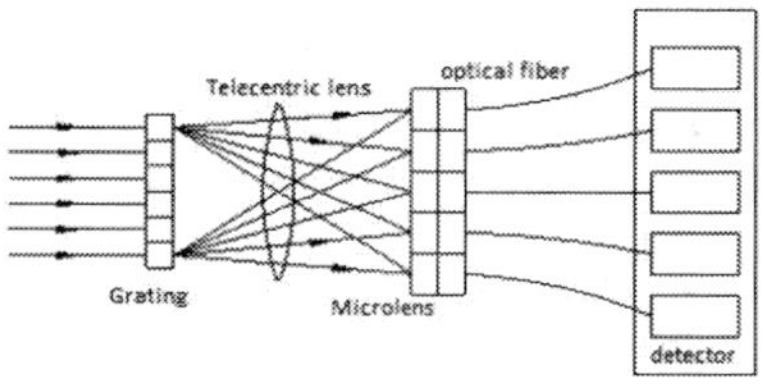

Figure 1 overall system design

3 KEY TECHNOLOGY

3.1 Grating spectrometer

As shown in Fig.2, the grating spectrometer is a spectroscopic device in the hyperspectral lidar, which mainly includes a small aperture diaphragm, a collimated off-axis parabolic mirror, a grating, and a telecentric spectral focusing lens. In order to improve the quality of remote sensing imaging, 380-930nm laser wide-spectrum coverage is used. The scattered echo light signal passing through the aperture pupil is collimated by a collimated off-axis parabolic mirror and then sent to the grating for splitting. And the telecentric lens put the spectrally separated spectrum refocuses.

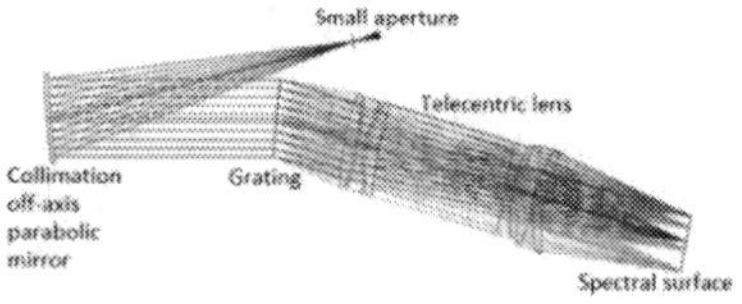

Figure 2 Optical design of grating spectrometer

The aperture is used to limit the receiving field of view in the receiving telescope, and the apertures are blackened on both sides to reduce the scattered light generated by the apertures of the apertures. An off-axis parabolic mirror is used to collimate the full-band echo optical signal through the aperture stop. Compared with the transmissive collimation method, the off-axis parabolic mirror can well collimate the full-band echo signal, which can reduce the dispersion of the focal plane of the grating focal plane because of the insufficient collimation parallelism. Thereby the spectral resolution could be improved.

Under the premise of satisfying the spectral resolution, the system is made more compact as much as possible, and the transmission grating is selected as the spectroscopic device. The focusing mirror behind the grating adopts a transmission focusing mirror system, which can effectively reduce the focusing spot aberration of each spectral channel after grating splitting to improve the spectral resolution[10]. In addition, the focusing mirror system adopts the structural form of the image telecentric system, ensuring the main rays of each spectral channel in perpendicular to the surface of the microlens-fiber detection array, effectively matching the receiving numerical aperture and improving the optical coupling efficiency.

According to the existing parameters, the spectral range is from 380 to 930nm at wavelength, and the grating transmitted light is centered on the wavelength of 632nm. The diffraction effect of the grating is the effect of single-slit diffraction and multi-slit interference superposition. For single-slot interference, phase

 https://doi.org/10.1051/epjconf/202023707024

there is an optical path difference $\Delta L = 2d\sin\theta$ between the adjacent groove faces in the transmission direction,Satisfy $2d\sin\theta = \lambda$, λ is a class 1 blaze wavelength, and the grating equation is:

$$\sin\alpha + \sin\beta = m\lambda / d$$

Where α and β represent the angle of incidence and diffraction angle of the grating, respectively; m is the diffraction order; λ is the wavelength; d is the grating constant. The relationship between the incident light wavelength λ and the grating blaze angle α of the transmission grating is:

$$\sin\alpha = \frac{m\lambda}{2d}$$

Therefore, the incident angle θ of the grating spectrometer is:

$$\theta = \arcsin\left(\frac{m\lambda}{2d}\right) = \arcsin\left(\frac{632\times10^{-6}\times360}{2}\right) = 6.$$

Substituting the grating spectrometer line dispersion is:

$$\frac{1}{D_l} = \frac{\cos\theta}{line\cdot f_c}\cdot10^6 = 10nm/mm$$

3.2 Focus splitting system

As shown in Fig. 3, each fiber corresponds to each microlens, and the grating splits the broad spectral signal with the spectral segment $N \sim M$ into $n+1$ spectral segments such as $N \sim N_1$ 、 $N_1 \sim N_2$ 、 ... $N_n \sim M$. Spectral resolution is $\frac{M-N}{n+1}$. In this paper, the spectrum covers from 380 to 930nm at wavelength, achieving 50 channels with a spectral resolution of 10nm.

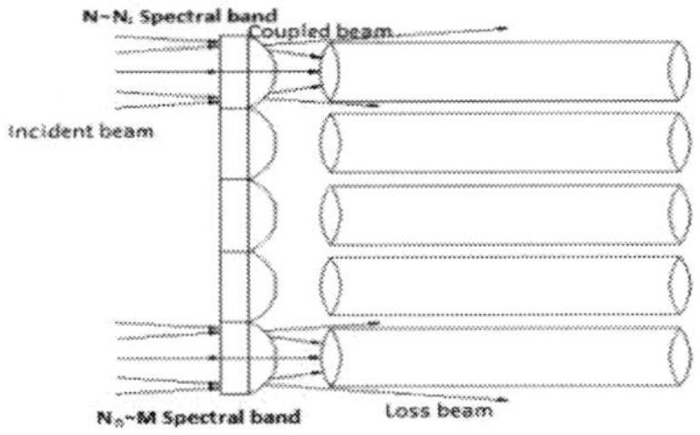

Figure 3 Microlens-fiber coupling system

The optical design of the microlens array is shown in Figure 4. It can be seen from the figure that after the microlens, the spot which concentrated on the end face of the fiber is equivalent to increasing the filling rate of the fiber array. Figure 5 shows the simulation results of the effect of the microlens on the spot energy distribution on the fiber's end face. It can be seen from Fig. 5 (a) and (b) that the energy of the spot is uniformly distributed without the addition of the microlens, and only about 50% of the energy is included in the effective clear aperture of the fiber, coupling the efficiency is 50%. After adding the microlens, it can be seen from (c) and (d) that the energy of the spot is mainly concentrated in the center of the fiber, and contains about 90% of the spot energy in the range of about 100um in the center of the fiber. Therefore, after using the microlens array, the coupling efficiency of the fiber array can be increased from 50% to more than about 95%.

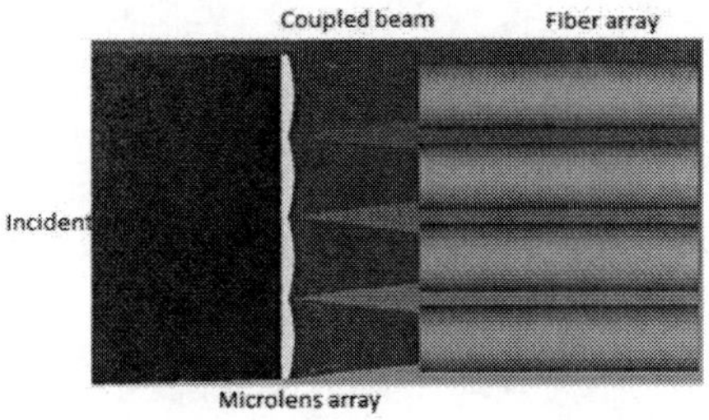

Figure 4 Microlens-fiber coupling design

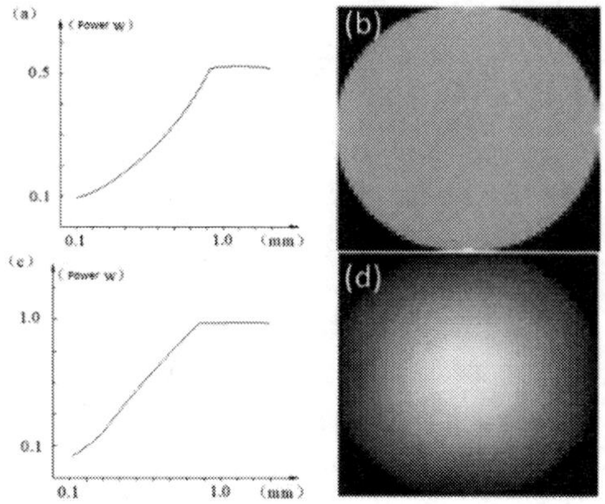

Figure 5 Spot energy distribution

4 CONCLUSION

In this paper, a new design of Spectroscopic system in hyperspectral lidar is proposed. The spectral coverage is 380-930nm, the spectral resolution is better than 10nm, and 50 channels of high-speed synchronous data acquisition are realized. Further work needs to complete the development of the system prototype, and realize the airborne flight test. This work could also obtain the hyperspectral laser detection data and simultaneously realize the cross-domain of the high-spectrum lidar successfully developed by demonstration prototype theory. It would truly promote the hyperspectral lidar to practical development.

ACKNOWLEDGEMENTS

Supported by National Key R&D Program of China(2018YFB0504500).

REFERENCES

1. Song.S,et al, Isprs Journal of Photogrammetry & Remote Sensing, 66(5): p. 672-682 (2011)
2. Elaksher, A.F, Optics and Lasers in Engineering, 46(7): p. 493-498 (2008)
3. Dalponte, M,et al, IEEE Transactions on Geoscience & Remote Sensing, 46(5): p. 1416-1427 (2008)
4. Hakala, T, et al, Optics Express, 20(7): p. 7119-7127 (2012)
5. Nichol, C. J, et al, in Agu Fall Meeting. 2009.
6. Harding, D. J, et al, Remote Sensing of Environment, 76(3): p. 283-297 (2001)
7. Lovell, J.L, et al, Canadian Journal of Remote Sensing, 29(5): p. 16 (2003)
8. Bourliaguet, B, et al, GIS Applications, and Geology IX. 2009.
9. Lin, D, et al, International Journal of Applied Earth Observation & Geoinformation,44: p. 136-143 (2016)
10. Helmut, R, et al, Optics Express, 17(6): p. 4268 (2009)

THE ESA-EVE POLARIZATION LIDAR FOR ASSESSING THE AEOLUS AEROSOL PRODUCT PERFOMANCE

Peristera Paschou[*1,2], Emmanouil Proestakis[1], Alexandra Tsekeri[1], Nikos Siomos[2], Antonis Gkikas[1], Anna Gialitaki[1], Eleni Marinou[1,3], Ioannis Binietoglou[4], Charikleia Meleti[2], Volker Freudenthaler[5], George Georgoussis[6], George Doxastakis[6], Alexandros Louridas[6], Jonas Von Bismarck[7] and Vassilis Amiridis[1]

[1]*IAASARS, National Observatory of Athens, Athens, 15236, Greece*
[2]*Aristotle University of Thessaloniki, Greece*
[3]*Institut für Physik der Atmosphäre, Deutsches Zentrum für Luft und Raumfahrt, Germany*
[4]*National Institute of R&D for Optoelectronics, Romania*
[5]*Ludwig-Maximilians-Universitat, Germany*
[6]*Raymetrics S.A., Greece*
[7]*Directorate of Earth Observation Programmes, ESA-ESRIN, Italy*

[*]*Email: pepaschou@noa.gr*

ABSTRACT

We present the EVE lidar concept, a combined linear/circular polarization system, tailored to evaluate the spaceborne ALADIN Doppler lidar system aerosol retrievals. EVE, currently under development, aims to provide the ESA-Aeolus mission with a flexible, mobile reference ground-based lidar system capable of providing well-characterized fiducial reference measurements of aerosol optical properties. Since ALADIN detects only the copolar component of the backscattered circularly polarized radiation, a portion of the received radiation gets lost, leading to an underestimation of the backscatter coefficient and the circular depolarization ratio in strongly depolarizing scenes with non-spherical particles. The main focus of the new EVE lidar is to quantify these uncertainties and to evaluate aerosol backscatter/extinction retrievals for Aeolus, and later also for EarthCARE product validation, quality assessment and improvement.

1. INTRODUCTION

The validation and calibration procedures of advanced spaceborne earth observation instruments have traditionally relied on well-calibrated ground-based instruments to provide fiducial reference measurements, capable for assessing the performance of the space sensors [1]–[3]. The atmospheric missions of the European Space Agency (ESA), and among them the Earth Explorers Aeolus [4] and EarthCARE [5], are no exceptions, particularly with respect to the validation of profile products related to aerosols and clouds.

The Atmospheric LAser Doppler INstrument (ALADIN) on board Aeolus, is designed to provide global profiles of the Horizontal Line-Of-Sight (HLOS) wind component in the troposphere and the lower stratosphere [4], [6]. ALADIN is a sophisticated Doppler Wind Lidar (DWL) operating in the ultraviolet region of the spectrum (355 nm), implemented in a transceiver configuration and tilted 35° from nadir [7]. It is also a High Spectral Resolution Lidar (HSRL), capable of deriving aerosol optical parameters such as particle backscatter and extinction coefficient profiles. Furthermore, the EarthCARE future mission of ESA is designed to utilize an array of scientific instruments, including the ATmospheric LIDar, ATLID. ATLID is an HSRL, operating at the wavelength of 355 nm with a high-spectral resolution receiver and circular depolarization channel, that will provide vertically resolved optical products for aerosol and thin clouds.

Both ALADIN and ATLID have an identified need for co-located measurements with ground-based lidars operating at the same wavelength (355 nm) for the data product's assessment and validation. While a growing fraction of the ground-based lidar systems operate a depolarization channel at 355 nm, the common lidar types featured for example in the European Aerosol Research Lidar Network (EARLINET), do not perform circular depolarization measurements alongside linear depolarization measurements [8].

The main aim of EVE (Enhancement and Validation of ESA products) lidar is to provide the upcoming ESA missions of Aeolus and EarthCARE, with a well-characterized, mobile ground-based lidar system, capable of providing fiducial reference measurements for ALADIN and ATLID.

2. AEOLUS AEROSOL PRODUCT

The aerosol and cloud spin-off product of Aeolus comprises of profiles of extinction coefficient, backscatter coefficient, and extinction-to-backscatter (lidar) ratio at 355 nm. An inherent weakness of ALADIN is the lack of detection of the cross-polarized return of the emitted circularly-polarized signal. This limitation has almost negligible impact when near-spherical aerosol particles are probed. However, when non-spherical, depolarizing particles, such as desert dust, volcanic ash, and ice crystals are encountered along the Line-of-Sight (LoS) of ALADIN, the retrievals will deviate from those obtained by an instrument that would measure both the co-polar and the cross-polar components. The deviation of the measured backscatter coefficient by ALADIN (middle panel), with respect to the corresponding amount, which would have been measured if cross- and co-polar components were taken into account (left panel), is depicted in Figure 1 for a case of non-spherical dust particles.

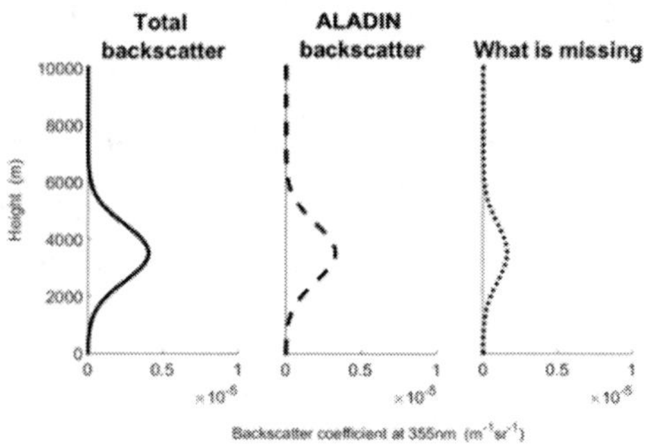

Figure 1: Total backscatter coefficient of dust at 355 nm as a function of altitude (left), the corresponding co-polar component detected by ALADIN (middle) and the undetected-missing cross-polar component (right)

The underestimation of the total backscatter coefficient retrieved by ALADIN, due to the detection only of the co-polar component, compared to the total backscatter when both cross- ($b_\perp$) and co-polar ($b_\parallel$) components would be measured, is depicted in Figure 2. For this estimation, we used CALIPSO (Cloud-Aerosol Lidar and Infrared Pathfinder Satellite Observation;[9]), based on the LIVAS (LIdar climatology of Vertical Aerosol Structure for space-based lidar simulation studies) three-dimensional aerosol and cloud database [10]. More specifically, based on CALIPSO Level 2 version 3 aerosol products we calculated the profiles of linear depolarization ratio at 532 nm, and accordingly, converted them to 355 nm. Considering the direct relationship between linear and circular depolarization for randomly oriented particles ($\delta_C=2\delta_L/(1-\delta_L)$) [11], we calculated the circular depolarization profiles at 355 nm ($\delta_C=b_\perp/b_\parallel$) and computed the average underestimation of the total aerosol backscatter due to the detection of only the co-polar component ($-100\cdot b_\perp/(b_\parallel+b_\perp)$). On a global scale, the underestimation of the average backscatter coefficient could be up to 65%. Similarly, Wandinger et al. [12] in their presentation in the "ADM-Aeolus Science and Cal/Val Workshop" showed that the underestimation of the backscatter coefficient can be up to 50% for desert dust and up to 70% for cirrus.

https://doi.org/10.1051/epjconf/202023707025

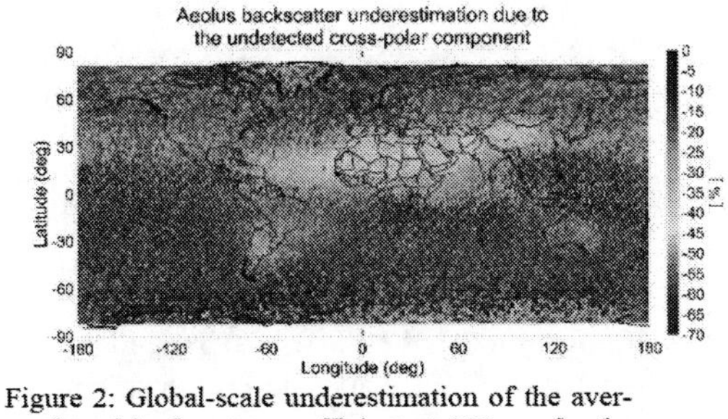

Figure 2: Global-scale underestimation of the averaged total backscatter coefficient at 355 nm for the year 2007, due to detection of only the co-polar component, based on LIVAS database results.

Furthermore, polarization-dependent multiple scattering effects, especially in the case of cirrus clouds, can significantly bias the retrievals for both extinction and backscatter coefficients. Since the EVE lidar will measure both linearly and circularly polarized signals, the validation of the relation between circular and linear depolarization ratios for randomly oriented particles could be used as indicator of the multiple scattering effects [13].

Orientation of the particles may also bias the retrievals [14] mainly for the backscatter coefficient and less for the extinction coefficient as shown in the simulations in Figure 3. The Figure *3* shows that the extinction and backscatter coefficients of randomly oriented particles (black line) are different than the extinction and backscatter coefficients of oriented particles (blue and red lines). Moreover, the latter depend on the viewing angle, thus, in case of oriented particles a zenith-looking ground-based lidar will measure different values (blue line) than ALADIN, looking at 35° (red line). EVE is optimally designed to reproduce the ALADIN measurements at 35°. It will also have the capability to perform measurements at multiple angles and provide indicators of particle orientation.

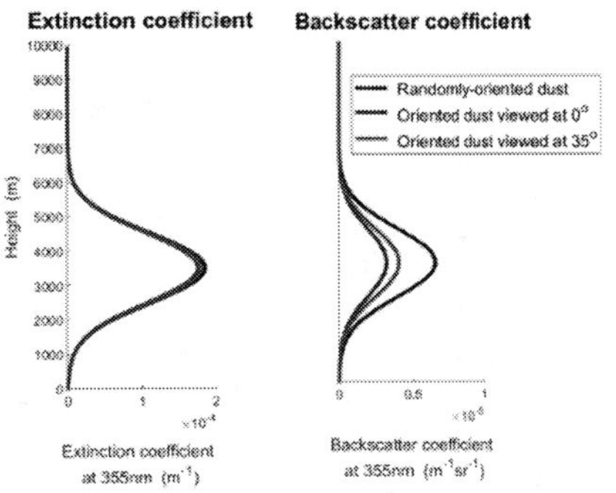

Figure 3: Simulations of the extinction coefficient (left) and backscatter coefficient at 355 nm (right) as a function of altitude for a layer of randomly oriented (black line) and vertically oriented particles (blue and red lines). For oriented particles, the extinction and backscatter coefficients viewed with a lidar at 0° are denoted with blue and viewed at 35° with red.

3. EVE CONCEPTUAL DESIGN

The system design will utilize a dual-telescope/dual-laser configuration, operating in the UV at 355 nm, and will allow the adjustment of the zenith angle from 3 to 35 degrees and full adjustment of the azimuth angle to fulfil the requirements of both Aeolus and EarthCARE. The overall design will also allow for an easy integration and operation of additional emission/receiver channels (for example at 532 nm and 1064 nm for aerosol characterization) at a later stage, and it enables the transportation to and participation in Aeolus (and EarthCARE) CalVal campaign activities, such as the upcoming ASKOS tropical campaign for the evaluation of Aeolus aerosol products. The conceptual design for the EVE mobile lidar is presented in Figure 4.

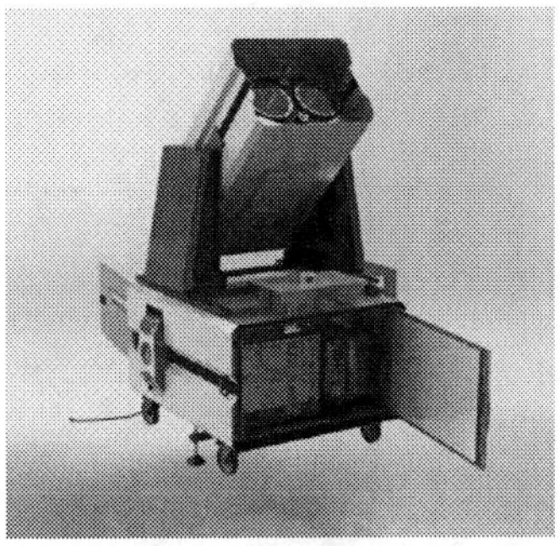

Figure 4: Conceptual design for the EVE mobile lidar. The system measurements are based on two lasers emitting circularly and linearly polarized radiation at 355nm and two receiving telescopes. It will be

easily transportable to facilitate CalVal operations at a variety of sites and aerosol/cloud conditions.

First, we aim to develop a lidar simulator to demonstrate the optimal measurement strategy of EVE for the Aeolus CalVal applications. In the test phase, EVE will be operated under different atmospheric environments and conditions in order to refine the prototype design, to demonstrate its fitness for purpose and conformity with the design specifications. As such, the testing and validation of the early prototype version of the EVE lidar includes the operation alongside a quality assured local lidar system for a period of at least one month, for intercomparison and performance evaluation purposes. This intercomparison will be performed at the PANhellenic GEophysical observatory of Antikythera (PANGEA), a fixed observational platform of NOA (National Observatory of Athens) at the island of Antikythera-Greece. PANGEA hosts the multi-wavelength lidar system PollyXT-NOA (portable aerosol Raman lidar system with eXTended capabilities; [15]) and the AERONET's photometer. The observatory location is considered representative of the broader region in the Mediterranean as it is a crossroad of different aerosol pathways, significantly impacted by mineral dust from Africa, smoke from frequent forest fires and agricultural biomass burning, marine particles, as well as anthropogenic background pollution from megacities such as Athens and Istanbul [16].

In parallel, we will develop the processing algorithms for the retrieval of the lidar optical products. The processing algorithms will provide the attenuated backscatter, circular and linear depolarization ratios, backscatter coefficient and extinction coefficient at 355 nm, including the associated uncertainties.

ACKNOWLEDGEMENTS

EVE is implemented by NOA and Raymetrics S.A. under an ESA contract. We acknowledge the support of (a) the project "PANhellenic infrastructure for Atmospheric Composition and climatE change" (MIS 5021516) which is implemented under the Action "Reinforcement of the Research and Innovation Infrastructure", funded by the Operational Programme "Competitiveness, Entrepreneurship and Innovation" (NSRF 2014-2020) and co-financed by Greece and the European Union (European Regional Development Fund); (b) the European Research Council under the European Community's Horizon 2020 research and innovation framework program/ERC Grant Agreement 725698 (D-TECT) and (c) the Stavros Niarchos Foundation.

REFERENCES

[1] L. Mona *et al.*, *Atmos. Chem. Phys.*, 2009.
[2] G. Pappalardo *et al.*, *J. Geophys. Res. Atmos.*, 2010.
[3] E. Proestakis *et al.*, *Atmos. Chem. Phys. Discuss.*, pp. 1–30, Feb. 2019.
[4] A. Stoffelen *et al.*, *Q. J. R. Meteorol. Soc.*, vol. 132, no. 619, pp. 1927–1947, Jul. 2006.
[5] A. J. Illingworth *et al.*, *Bull. Am. Meteorol. Soc.*, 2015.
[6] A. Dabas, *Comptes Rendus Geosci.*, vol. 342, no. 4–5, pp. 370–379, Apr. 2010.
[7] S. Lolli *et al.*, *Atmos. Meas. Tech.*, vol. 6, no. 12, pp. 3349–3358, Dec. 2013.
[8] G. Pappalardo *et al.*, *Atmos. Meas. Tech.*, vol. 7, no. 8, pp. 2389–2409, Aug. 2014.
[9] D. M. Winker *et al.*, *Bull. Am. Meteorol. Soc.*, 2010.
[10] V. Amiridis *et al.*, *Atmos. Chem. Phys.*, vol. 15, no. 13, pp. 7127–7153, Jul. 2015.
[11] M. I. Mishchenko and J. W. Hovenier, *Opt. Lett.*, vol. 20, no. 12, p. 1356, Jun. 1995.
[12] U. Wandinger *et al.*, in *ADM-Aeolus Science and CAL/VAL Workshop, ESA-ESRIN*, 2015.
[13] G. Roy and N. Roy, *Appl. Opt.*, 2008.
[14] B. V. Kaul *et al.*, *Appl. Opt.*, vol. 43, no. 36, p. 6620, Dec. 2004.
[15] R. Engelmann *et al.*, *Atmos. Meas. Tech.*, 2016.
[16] E. Marinou *et al.*, *Atmos. Chem. Phys.*, 2017.

THREE-DIMENSIONAL LASER IMAGING BASED ON A PHOTON-COUNTING AVALANCHE PHOTODIODE ARRAY

Peng Ge[1,2*], Cong Chen[1,2], Zhen Shang[1,2], Yanen Fan[1,2], Jingjing Guo[1,2], Zibo Zhuang[3], Jialong Ge[1,2]

[1]East China Research Institute of Electronic Engineering, Hefei 230088, Anhui, China
[2] Key Laboratory of Aperture Array and Space Application, Hefei 230088, Anhui, China
[3] Civil Aviation University of China, Tianjin 300300, China
**Email: gepeng_cas@qq.com*

ABSTRACT

Photon-counting detector array is very desired for high-resolution laser imaging based on direct time-of-flight measurement. Such systems have potential applications in remote sensing with long distances. We will perform three-dimensional imaging using a large InGaAs Geiger-mode avalanche photodiode array which has single-photon sensitivity. To improve the image quality with only a few photon detections, the photon counting imaging process is analyzed and a regularization method based on pixel spatial correlation is employed for image reconstruction. The performance of the method is compared with that of conventional maximum likelihood estimation on intensity and range image reconstructions of a building about six hundred meters away.

1. INTRODUCTION

Laser radar can achieve much higher angular resolution and range precision than microwave radar due to the short wavelength of laser light[1]. For imaging at a long stand-off distance with laser radar, the received reflected light is usually very low, and single-photon detectors are widely used for the weak signal detection with its unparalleled sensitivity. The transverse resolution of the imaging is obtained either by point-by-point scanning[2] or by large detector array[3]. As the scanning way is time consuming for large imaging, detector array is much preferred which can acquire scene information in a parallel way. Currently Geiger-mode avalanche photodiode (APD) based array is the most mature and widely used single-photon detector array, and plenty of effort has been done on the development of large Geiger-mode APD array[4][5] and its applications on three-dimensional laser imaging[6][7].

With the help of flood illumination of short laser pulses, Geiger-mode APD array is mainly used for direct-detect three-dimensional imaging. Benefit from the single-photon sensitivity, it is always operated at a low incident photon rate per laser pulse that leads to a low detection probability for a laser pulse. Multiple pulse collection is commonly used to achieve a high detection probability[8][9]. Geiger-mode APD records a number of discrete photon detections which is different with energy measurement with traditional light detectors[10]. The discrete photon detection data cannot be deal with conventional Gaussian noise model for image reconstruction. Alternatively, Poisson noise model shall be applied. Therefore, besides the performance of Geiger-mode APD array, the image reconstruction algorithm is also important to achieve high quality imaging[11][12].

In this paper, we will demonstrate the three-dimensional laser imaging based on a single-photon camera which is composed of a large InGaAs Geiger-mode APD array. Compared with that made of silicon, the Geiger-mode APD made of InGaAs absorption material has good response to longer wavelengths beyond 900nm that can be used for eye-safe laser imaging. We will discuss the photon counting imaging process on the Geiger-mode APD for the intensity and range estimation. Instead of using conventional pixel by pixel maximum likelihood method, a regularization method based on the spatial correlation is employed to improve the image reconstruction quality with a few photon detections.

2. EXPERIMENTAL SETUP

The schematic diagram of our imaging setup is illustrated in Fig. 1 It is a bistatic system using separated transmitting and receiving apertures so that the reflections from the transmitting lens will not blind the single-photon camera. For imaging at kilometer distances, a Nd:YAG microchip laser is employed to generate 500ps laser pulses at

1064nm with high peak power. The laser is passively Q-switched internally and operated at a repetition rate of 1 kHz with output pulse energy of 80uJ via external trigger current. For the laser cavity building time is about 150us with a fluctuation about 200ns, the emitting time of each pulse is precisely measured via an auxiliary detector to synchronize the single-photon camera to start to work.

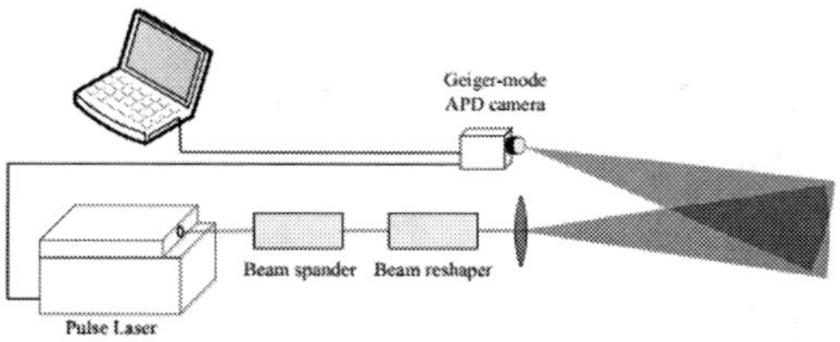

Figure 1 Schematic diagram of the imaging setup.

The laser pulses are spread to illuminate a scene of interest. However the light intensity on the laser beam cross-section is not a uniform distribution with high intensity in the center and low intensity in the edge following Gaussian distribution. As a result, the illumination efficiency is low and the received light intensities on detectors have large spatial difference. A beam shaper, pishaper_1064_6_6, is employed to transform the beam spatial intensity from Gaussian distribution to uniform distribution. As the beam shaper requires a collimated input beam spot with a diameter of 6mm, a beam expander is placed between the laser and the shaper. The shaper has nearly 100% transforming efficiency at 1064nm. The output beam is then spread to flood illuminated the scene with an optical lens.

The reflected light is collected by a small telescope with 25mm receiving aperture and imaged on an focal plant array (FPA) developed by CETC44. The array is fabricated with 64×64 InGaAs Geiger-mode APDs. Each APD is back-illuminated with 1μm pixel diameter and 50μm pixel pitch. A microlens array is aligned and attached on the surface of the FPA to increase the fill factor to about 60%. The device is cooled to 258K and operated with a reverse bias voltage of 58.5V for photon-counting detection. The quantum efficiency at 1064nm is about 15% and the normalized dark count rate is 8kHz - 20kHz. A readout integrated circuit (ROIC) of 64×64 array reads out APD signal with 2ns time resolution through a 500MHz clock. The device can be operated at a maximum frame rate of 20kHz through a CameraLink interface.

3. IMAGE RECONSTRUCTION

3.1 Photon-counting Imaging Process

With the receiving telescope, detectors on the two-dimensional array collect reflected light from the corresponding scene patches. According to the quantum nature of light, the photons measured on a detector follow a Poisson process[13]. Apart from the reflected laser light, solar background and dark counts also contribute to the process. The received photon rate from the reflected laser light is time-varied determined by laser pulse shape and scene reflection, while solar background and dark counts have constant rates. The combined process is a time inhomogeneous Poisson process[14]. The received photon rate at detector i can be written as

$$r_i(t) = a_i s(t - 2z_i / c) + b_i .$$

Where $s(t)$ is the temporal intensity distribution of laser pulses illuminating on each scene patch, a_i is the fraction of light collected at detector i, including the effects of transmitting and receiving attenuation and object reflection, z_i is the distance between the system and scene patch i, c is the speed of light and b_i is background light rate.

In a fix dwell time, the average received photon number v_i can be obtained by integration on $r_i(t)$. The actual received photon number is an integer obeying Poisson statistics with mean value of v_i. It is a probability process that the received photons create primary photoelectron on the detector, which is donated by quantum efficiency η. The detector is then fired from the primary photoelectron. In fact the firing of the detector is also a Poisson process. Considering dark count rate d the firing rate can be written as

$$\lambda_i(t) = \eta r_i(t) + d = \eta a_i s(t - 2z_i / c) + \eta b_i + d$$

Once the detector is fired, subsequent photoelectron within a short time interval will be blocked due to dead time effect and inability of counting multiple pulses. For the array's ROIC doesn't have the capacity of counting multiply pulses during a frame, only the first photoelectron can be recorded. In other word, if background photons or dark counts fired the detector before signal photons, the signal event will not be recorded.

3.2 Noise Censoring

The noise censoring is performed on the time-of-flight (TOF) histogram on each detector according to their temporal difference with that of signal. As the signal has an impulsive shape, the measured TOF shall span only a few time bins in the histogram. The signal's TOF distribution is mainly determined by laser pulse width and time resolution of synchronizing. In our experiment, the laser pulse width is 500ps and time resolution of synchronizing is 4ns, so the convolution of the two terms is about 4.03ns. We take three sigma confidence interval of Gaussian distribution as signal interval. As the histogram has a time bin width of 2ns, the signal shall span about 6 bins. To locate the signal interval, we calculate the sum value of any six adjacent time bins and take the interval with maximum value as signal interval

$$f = \max \sum_{i=k}^{k+5} n_i$$

In case of no object present in the scene path, all photon detections come from noise and the signal interval is also wrong due to noise fluctuation. To solve this problem, we calculated the average photon number in each bin, and if f exceed noise fluctuation with 5 sigma we then take it as a signal. Once the signal interval is located, the other events in the histogram will be censored.

3.3 Regularization Method

At very low light level, the collected number of photon detections per pixel is always a few in a limited dwell time[15][16]. As in Poisson statistics the variance is equal to the observed variable, the imaging will suffer much from low statistics. In addition, it has to take into account the impact of background light and dark counts. So it is a challenge to carry out precise image reconstruction, and one effective way is using regularization method[17][18]. The regularization method takes advantages of spatial correlations in natural images, or sparse in their spatial frequencies, for smoothness. With regularization method, a cost function is created incorporating a log-likelihood term and regularization term as

$$C(x) = -\ln L(x;n) + \lambda R(x).$$

Where x is the image signal to be reconstructed, $L(x;n)$ is the likelihood term based on measured dataset n, $R(n)$ is the regularization term based on spatially correlation and τ is the regularization factor to balance the two terms. The image reconstruction is then converted to the minimization of the cost function $C(x)$.

For the regularization term, a number of functions can used to calculate the sparsity. We choose the total-variation(TV) norm of image signal in wavelet-domain for smoothness. The discrete wavelet transform (DWT) derived from Daubechies's 4-tap filters is used. So the cost function in our image reconstruction is

$$C(x) = -\ln L(x;n) + \lambda \left\| W^T x \right\|_{TV}$$

Where, W is the wavelet basis. Based on the cost function, we use Poisson negative likelihood for intensity imaging reconstruction, and use Gaussian negative likelihood for range imaging reconstruction. As the problem is strictly convex, the minimizations are carries out with matlab toolkit SPRIAL-TAP which uses quadratic separable approximations[11].

4. RESULTS

With the experimental setup, we attempt to obtain the imaging of a building that is about six hundred meters away as illustrated in Fig. 2(a) The captured region of the building is marked with yellow lines in the figure. The camera is operated with a frame rate of 1kHz that synchronized by the microchip laser. In order to reduce ambient noise, a delay of 2us after each pulse emitting is applied on the time gate of the camera. The detector is exposure for 2us for the measurement of back-scattered laser pulses. Fig. 2(b) shows the building's intensity image after noise censoring with photon detection data of 500 pulses. The image resolution is 64×64 as the same as the single-photon detector array size, and the average photon number per pixel per pulse is about 0.09. Fig. 2(c) shows the reconstructed intensity image with regularization method from which we can see the image quality is improved with smoothness. Fig. 2(d) shows the building's range image with censored photon detection data of 500 pulses through maximum likelihood estimation. Fig. 2(e) shows the reconstructed range image with regularization method. Compared with the conventional method, the regularization method gives a much better result of three-dimensional imaging with spare photon detection data.

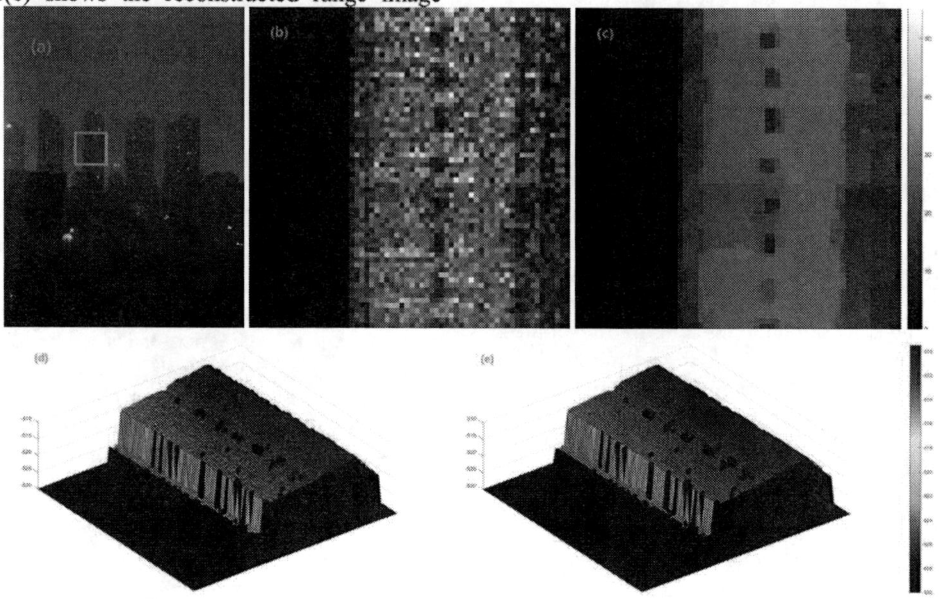

Figure 2 Photograph and photon-counting images of a building about six hundred meters away. (a) Building photograph. (b) Intensity image after noise censoring. (c) Intensity image reconstructed with regularization method. (d) Range image with maximum likelihood estimation. (e) Range image reconstructed with regularization method.

ACKNOWLEDGEMENTS

This work was supported by the CETC38 Innovation Fund and the National Natural Science Foundation of China under Grant No. U1433202.

REFERENCES

[1] P. F. McManamon, Opt. Eng. 51(6), 060901 (2012)
[2] A. McCarthy, et al. Appl. Opt. 48, 6241-6251 (2009)
[3] R. Heinrichs et al. Proc. SPIE 4377, 106-117 (2001)
[4] B. F. Aull, et al. Lincoln Lab. J. 13, 335-349 (2002)
[5] M. A. Itzler, et al. J. Mod. Opt. 58(3): 174-200 (2011)
[6] R. M. Marino, et al. Lincoln Lab. J. 15(1): 23-36 (2005)
[7] R. M. Marino, et al. Proc. SPIE 5791, 138-152 (2005)
[8] W. Becker, Springer, (2005)
[9] G. S. Buller, et al. IEEE 13, 1006-1015 (2007)
[10] P. F. McManamon, et al. Opt. Eng. 56(3): 031223 (2017)
[11] Z. T. Harmany, et al. IEEE Trans. 21, 1084-1096 (2012)
[12] L. Mertens, et al. Scientific reports, 7, 42164 (2017)
[13] D. G. Fouche, Appl. Opt. 42(27), 5388-5398 (2003)
[14] S. Kim, et al. Sensors 13(7), 8461-8489 (2013)
[15] A. C. Watts, et al. Remote Sens. 4, 1671-1692 (2012)
[16] M A. Itzlera et al. J. Mod. Opt. 58(3–4), 174-200 (2011)
[17] A. Kirmani, et al. Science 343, 58-61 (2014)
[18] D. Shin, et al. Nat. Commun. 7, 12046 (2016)

VEHICULAR MOTION EXPERIMENT AND DATA RETRIEVAL OF A COMPACT FLOATING LIDAR SYSTEM

Qichao Wang[1]*, Shengguang Qin[3], Jiaping Yin[3], Hongwei Zhang[1], Tong Cui[1], Songhua Wu[1,2]

[1] *Ocean Remote Sensing Institute, College of Information Science and Engineering, Ocean University of China, Qingdao 266100, China*
[2] *Laboratory for Regional Oceanography and Numerical Modeling, Pilot National Laboratory for Marine Science and Technology (Qingdao), Qingdao 266237, China*
[3] *Leice Transient Technology Co. LTD, Qingdao 266100, China*

**Email: wqc@stu.ouc.edu.cn*

ABSTRACT

Accurate and rapid observation of sea surface wind is important for the research of ocean dynamic prediction model, offshore wind resource assessment, air-sea interaction and flux. A compact floating coherent Doppler lidar system named WindMast 350-M was developed by Ocean University of China (OUC) and Leice Transient Technology Co. LTD (LEICE) for the observations of sea surface wind profiles. As an observation device installed on buoy platforms, the first vehicular motion experiment was conducted at Laoshan campus of Ocean University of China(120.49° E , 36.16° N) on 06 and 12 March, 2019. During the first experiment, the wind profiles measured by the WindMast 350-M were compared with the results from a well calibrated Ground-based Coherent Doppler lidar WindMast WP350. In this contribution, the systematic design and the specifications of 350-M are presented in detail. The preliminary results of the vehicular motion experiment are discussed as well.

Keywords: Floating Coherent Doppler lidar, Motion Correction, Data Retrieval and Validation.

1. INTRODUCTION

As an effective method for wind detection， Coherent Doppler lidar has the advantages of high accuracy, high spatial and temporal resolution. It is capable of low-altitude wind measurement. By combining buoy and wind lidar, floating lidar systems offer a great potential to offshore wind resources assessment, ocean dynamic prediction model research, air-sea interaction and flux study [1]. Compared with traditional fixed offshore meteorological masts, floating lidar system is cheaper and flexible. It also helps to improve the ability of sea surface wind detection under bad conditions, which compensates for the shortcomings of shipborne observation equipments.

Therefore, a compact floating lidar system WindMast 350-M was developed by the Ocean University of China (OUC) and Leice Transient Technology Co. LTD (LEICE). In this paper, the system design, the details and preliminary results of the first experiment are presented.

2. METHODOLOGY

2.1 System design

WindMast 350-M system consists of laser transmitter, scanning servo, optical receiver, signal collector and processor. All fiber and modular design are adopted to achieve the anti-vibration ability, thus improves the stability and reliability of whole system. To obtain the attitude and velocity information, a compact GPS/MEMS (Micro Electro Mechanical System) was equipped inside the lidar system.

Table 1. The specifications of the WindMast 350-M

Qualification	Specification
Wavelength	1.5 μm
Pulse Energy	12 μJ
Pulse Duration	100/200 ns
Pulse Repetition Frequency	10 kHz
Pitch Angle	60°
Radial Spatial Resolution	10 m
Measurement Range	40 to 350 m
Wind Profile Update Rate	1 min
Velocity Measurement Range	± 60 m/s
Accuracy of Velocity	0.3 m/s

https://doi.org/10.1051/epjconf/202023707027

Power Consumption	< 80w
Weight	< 30kg
Size	452*351*430 mm
Eye Safety Standards	EN60825-1:2007

2.2 Experiments design

During the first vehicular motion experiment, an undulating area was selected at the Laoshan campus of OUC (Figure 1) in order to simulate/demonstrate the undulate motion of buoy. The WindMast 350-M system was carried by a pickup truck (Figure 2), moving on a circle route in clockwise direction constantly. For purpose of intercomparison, a ground-based Coherent Doppler lidar of the type WindMast WP350 was deployed nearby simultaneously (green star in Figure 1).

Figure 1. The Route of First Experiment (Yellow Line) and the Placement Location of WindMast WP350 (Green Star)

Figure 2. Floating Lidar System - WindMast 350-M

The total length of the circular route for vehicular motion experiment is about 570 meters. The height of zenith point (red star in Figure 1) is 75m and the height of nadir point is 58m (yellow star in Figure 1) in this route. The pickup truck was running at different speed between 15 to 30km/h and the attitude information (Yaw angle, Pitch angle and Roll angle). The motion velocity information (Eastward speed, Northward speed and Upward speed) and the height information of lidar system during the experiment could be acquired by GPS/MEMS with temporal resolution of 0.05s, which was used for motion correction and wind data retrieval.

Figure 3. Comparison Experiment between WindMast 350-M and WindMast WP350 Systems

2.3 Algorithm

The recorded velocity corresponds to the relative velocity along the laser beam direction between the lidar and the atmospheric target. Considering the mobility of lidar system, the motion will add to the recorded radial velocity [2]. Therefore, a necessary step in data retrieval is the calibration of the radial velocity caused by the motion of lidar [3]. For this purpose, the real point angle (azimuth and pitch angle) of each laser radial need to be calculated first by combining the roll angle, pitch angle, yaw angle which are recorded by the GPS/MEMS and the recorded point angle of each laser radial [4].With the angle information, the projection of lidar motion velocity in each radial direction could be removed.

Afterwards, the real height of each range gate in all laser radial could be calculated with GPS height information. The radial velocity at selected altitude could be obtained by interpolation, which consists of the horizontal wind speed projection and the vertical speed projection in each radial direction, as shown in the equation:

$$V_{los} = u \cdot \cos\alpha \cdot \cos\theta + v \cdot \sin\alpha \cdot \cos\theta + w \cdot \sin\theta$$

V_{los} is the LOS velocity in each direction, α is the azimuth angle and θ is the pitch angle of laser radial after correction. u , v , w represent the north-south speed, east-west speed and vertical speed of wind. u , v could be used to retrieve the wind horizontal speed and the wind direction.

3. Results

During the first vehicular motion experiment, different measuring range setting were used, more than 5 hours wind information was observed by WindMast 350-M and WindMast WP350. After the motion correction and the wind retrieval, the THI (Time-Height-Intensity) diagrams of the wind profiles measured by these two systems are shown as below.

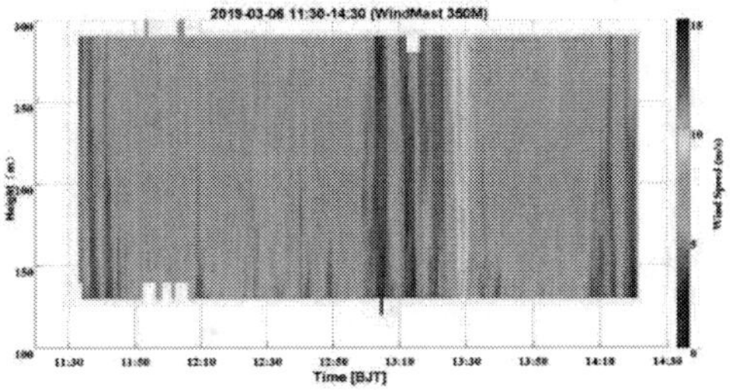

Figure 4. Wind THI diagram from 11:30 to 14:30BJT on 6 March 2019 observed by 350-M

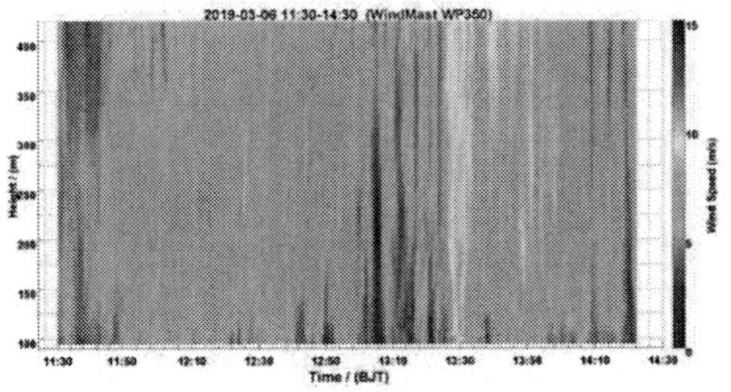

Figure 5. Wind THI diagram from 11:30 to 14:30BJT on 6 March 2019 observed by WP350

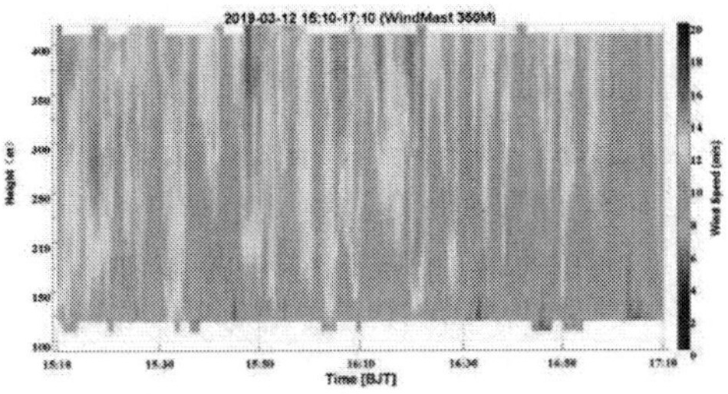

Figure 6. Wind THI diagram from 15:10 to 17:10BJT on 12 March 2019 observed by 350-M

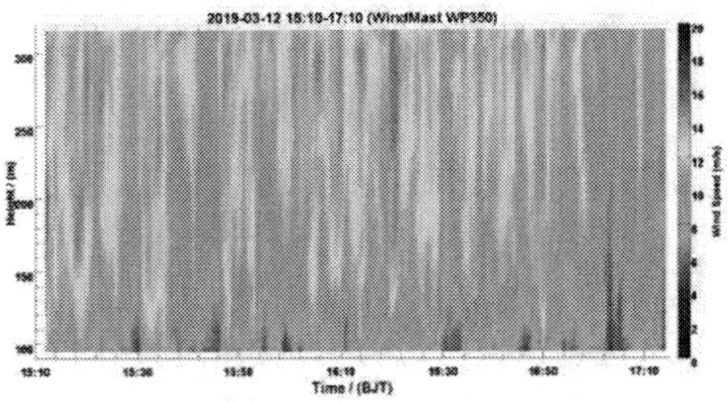

Figure 7. Wind THI diagram from 15:10 to 17:10BJT on 12 March 2019 observed by WP350

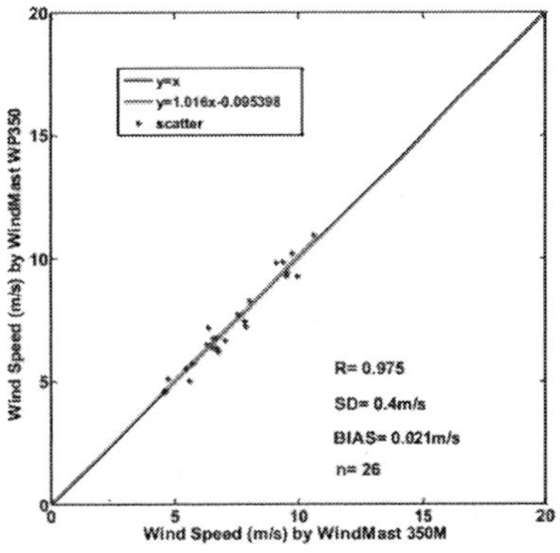

Figure 8. Scatter diagram of 150m wind speed (10 mins) of 350-M and WP350

As presented in the THI diagrams, the wind profiles observed by 350-M system fits well with the results from WP350 system after data correction and retrieval. 10-minute average horizontal wind speed at the height of 150 meters were selected for the statistical comparison in order to make more analysis of the wind speed details, as presented in Figure 8. The number of samples(n), standard deviation (SD), bias and correlation coefficient (R) are provided. The wind speed linear regression shows a correlation coefficient of 0.975, SD of 0.4m/s, demonstrating the reliability of WindMast 350-M system and the feasibility of the data correction and wind reversal method.

4. Discussion and future works

During the first vehicular motion experiment, carried by a pickup truck, the WindMast 350-M system was operating around the WindMast WP350 system. The longest distance between two systems is about 170 meters. There are some measurement deviations between these two systems, which were caused by the differences of location and laser pitch angle (WindMast 350-M: 60°, WindMast WP350: 71.38°). Due to the influence of the surrounding hill and buildings, the wind field of low-altitude area became inhomogeneous. Additionally, result from the limitation of the existing GPS/MEMS, the height error and the attitude error were induced into the wind retrieval.

Consequently, as an outlook, the further error analysis should be finalized in the following schedule.

ACKNOWLEDGEMENTS

We thank our colleagues for their great support during the field experiments, including Guangyao Dai, Yuanshuai Zhang from Ocean University of China and Xitao Wang, Caili Zhu, Tongtong Wang from QINGDAO Leice Transient Technology Co. LTD.

This research was supported by National Key Research and Development Program of China, grant number 2016YFC1400904 and the Fundamental Research Funds for the Central Universities, grant number 201822004.

REFERENCES

[1] Gottschall, J., et al. Energy Procedia 53: 156-161 (2014).

[2] X. Zhai, et al., Atmospheric Measurement Techniques, 11(3): 1313-1331 (2018)

[3] S. Wu, et al. EDP Sciences. EPJ Web of Conferences Vol. 176, p. 02012 (2017).

[4] Li Z. G., et al. Proceeding 'Dragon 3 Mid-Term Results Symposium', ESA SP-724, November (2014).

A NEW METHOD FOR A SIDE-SCATTERING LIDAR TO RETRIEVE THE AEROSOL BACKSCATTERING COEFFICIENT

Zhenzhu Wang[*,1], Zongming Tao[1], Dong Liu[1], Chenbo Xie[1], Yingjian Wang[1,2]

[1]*Key Laboratory of Atmospheric optics, Anhui Institute of Optics and Fine Mechanics, Chinese Academy of Sciences, Hefei, Anhui 230031, China*
[2]*University of Science and Technology of China, Hefei, Anhui 230031, China*
**Email: zzwang@aiofm.ac.cn*

ABSTRACT

The side-scattering lidar based on the CCD camera is a powerful tool to measure the near-ground aerosol, which is most interesting for the environmental and meteorological departments. The inversion method for the side-scattering lidar is different from the conventional Fernald method due to the differences in two lidar equations. A new inversion method for the side-scattering lidar to retrieve the aerosol backscattering coefficient is proposed for the first time, which is based on the aerosol backscattering coefficient at the ground as a restriction condition. Numerical simulation is used to analyze the convergence of this new method. Case studies indicate that this new method is reasonable and feasible.

1. INTRODUCTION

Aerosols are colloids of fine solid particles or liquid droplets suspended in the atmosphere. Atmospheric aerosols have substantial influences on human health [1-2], air quality [3-4] and climate changes [5-6] and their loading has significantly increased. Thus backscattering lidar takes an important place in the studies of the atmospheric environment, air pollution and diffusion with its capable of providing range-resolved atmospheric profiles continuously. But the backscattering lidar system has a shortcoming in the lower hundreds of meters because of the overlap factor caused by the configuration of the transmitter divergence and the receiver's field of view(FOV) at ranges close to the instrument [7]. This limits the backscattering lidar to be applied to the near-range measurement, especially for the fixed vertical-pointing lidar [8]. Therefore, to study the aerosol properties in haze day a new technique without the overlap factor near the ground would be quite useful. And a new inversion method for the side-scattering lidar to retrieve the aerosol backscattering coefficient is also proposed for the first time, which is based on a prior reference value measured by itself at the ground as a constrained condition.

2. METHODOLOGY

In the side-scattering lidar, we receive side-scattering light instead of backscattering light and use a CCD detector instead of a telescope [9]. The side-scattering lidar equation can be written as

$$P(z,\theta)=\frac{P_0KA}{D}\beta(z,\theta)T_tT_r\Delta\theta \qquad (1)$$

Where P(z,θ) is the received photo electron number for the altitude z and the scattering angle θ in pixels, P_0 is the laser photo electron number, K is the calibration constant representing the system optical efficiency and A is the effective collecting area of the optics, D is the distance from the CCD to a laser beam, T_t and T_r and is the total atmospheric (aerosol and molecular) transmittance from the laser to the altitude z and from the altitude z along the slant path to the CCD receiver, respectively. B(z,θ) is the total atmospheric side-scattering coefficient, Δθ is the FOV of a pixel. For the lidar equation, there are two different aspects compared to the backscattering lidar: (I) received signals without range square dependence, (II) no overlap factor. Without the overlap problem, the side-scattering lidar is suitable for detecting the aerosol vertical profile in the near range, especially [8, 10].

As to the backscattering lidar, there are two variables -- the aerosol extinction coefficient and the backscattering coefficient. Fernald [11] proposed an iterative inversion solution for the backscattering lidar, which is not suitable for side-scattering lidar due to six unknown variables in the side-scattering lidar equation, i.e. the relative phase functions, the backscattering and extinction coefficients for aerosol and molecules.

Using the relative aerosol phase function, our group has developed a numerical inversion method for the side-scattering lidar [12]. In order to use the backward integral for reducing inversion errors, the reference point is selected at about 1 km altitude in the method. However, there is a shortcoming lying in the fact that the aerosol backscattering coefficient value at about 1 km altitude is not easy to get, especially in the foggy and hazy weather condition, which is usually obtained simultaneously from the backscattering lidar. Fortunately, the aerosol backscattering coefficient value at the ground is found easily even in foggy and hazy weather. So in this letter, using the aerosol backscattering coefficient value at the ground as a restriction condition, we propose an iterative inversion method for the side-scattering lidar to retrieve the aerosol backscattering coefficient based on the backward integral. Based on the previous method in the literature [12], our new inversion procedure can be obtained to get the solution for the aerosol backscattering coefficient profile.

3. RESULTS

In order to check the convergence of this new inversion method, numerical simulation has been investigated from the backward and forward integral directions. Supposing there is a simulation aerosol backscatter coefficient profile shown in Fig. 1, then the simulated side-scatter lidar signal can be derived from the side-backscattering lidar equation (1).

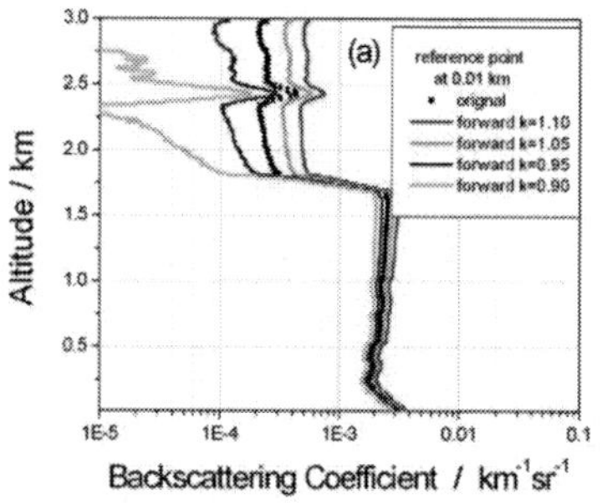

Fig. 1 Retrieval of the aerosol backscattering coefficient profile using the forward integral

As to the forward integral, when we select the reference point at 0.01 km altitude, the original values of the aerosol backscattering coefficient at the reference point are 0.0033 $km^{-1}sr^{-1}$ in Fig.1. Supposing the aerosol backscattering coefficient is times of the original value at the reference point, and =1.10, 1.05, 0.95, 0.90 , respectively, the retrieval aerosol backscattering coefficient profiles are calculated in Fig. 1.

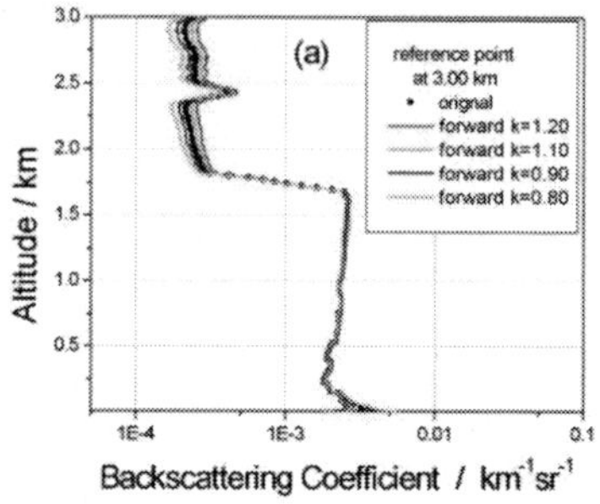

Fig. 2 Retrieval of the aerosol backscattering coefficient profiles using the backward integral

As to the backward integral, the reference point is selected at 3.00 km altitude, the original values of the aerosol backscattering coefficient at the reference point are 0.00024 $km^{-1}sr^{-1}$ in Fig. 2. Supposing the aerosol backscattering coefficient is times the original values at the reference point, and =1.20, 1.10, 0.90, 0.80 , respectively, the retrieval aerosol backscattering coefficient profiles are calculated in Fig. 2.

It should be noted that the backward integral will be convergent, and the forward integral cannot be always convergent as shown in Fig. 1 and Fig. 2. As to the forward integral, the bigger the aerosol coefficient and the larger the bias at the reference point, the worse the convergence. The simulation calculations indicate that the backward integral is reasonable direction for the side-scattering lidar inversion.

One measurement by the side-scattering lidar system are presented in the following. In the evening on February 13, 2015, the sky was overcast. The side-scattering lidar and the visibility instrument were operated together at the same place. The visibility instrument was on top of a building at 16 m altitude. At 19:00, the visibility was 3.87 km, and the corresponding aerosol extinction coefficient was 1.01 km-1 and the corresponding aerosol backscattering coefficient was 0.024 km-1sr-1. Taking the aerosol backscattering coefficient 0.024 km-1sr-1 at 16 m altitude as the restriction condition, the aerosol backscattering coefficient profile was calculated from the side-scattering lidar equation using the new method, which is shown in Fig. 3 (a)(black line). The red line in Fig. 3(a) was the aerosol backscattering coefficient profile from the previous method [12].

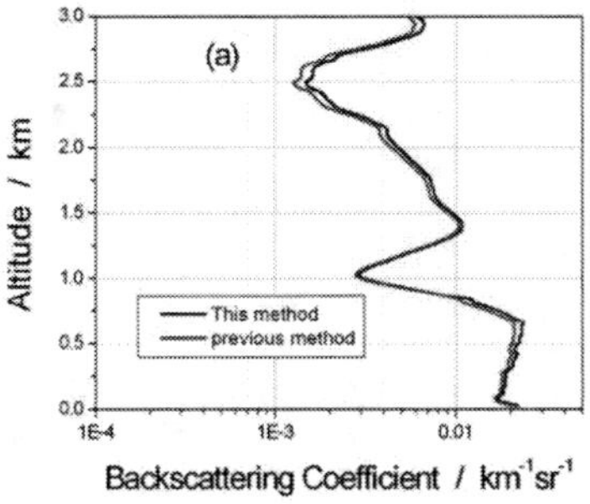

Fig. 3 Retrieved profiles of the aerosol backscattering coefficient from the two methods, at 19:00 LST on February 13, 2015

In summary, a new method for the side-scattering lidar to retrieve the aerosol backscattering coefficient is proposed. This new method solves the important problem: retrieving the aerosol backscattering coefficient without using an expensive backscattering lidar. Numerical simulation and compared experiments indicate that this method is feasible and very suitable to foggy and hazy weather conditions, especially.

ACKNOWLEDGEMENTS

This work was supported by the National Natural Science Foundation of China under Grant No. 41975038, 41590871, 41811530088 and the Youth Innovation Promotion Association CAS under Grant No. 2017482 and the CAS President's International Fellowship Initiative under Grant No. 2016VEA044.

REFERENCES

[1] Nel, A. Science, 308(5723), 804-806,(2005).
[2] Zheng, S., et al. Atmos. Chem. Phys., 15(10), 5715-5725, (2015).
[3] Wang, X., et al. J. Geophys. Res., 115, D00K27, (2010).
[4] Huang, R. J., et al. Nature, 514, 218 - 222,(2014).
[5] Ramanathan, V., et al. J. Geophys. Res., 106(D22), 28371-28398, (2001).
[6] Che, H. Z., et al., Atmos. Chem. Phys., 15, 7619 - 7652, (2015).
[7] F. Mao, et al. Opt. Laser Technol. 44(4), 907 - 912 (2012).
[8] Z. Wang, et al. Opt. Lett. 40, 1749-1752 ,(2015).
[9] J. E. Barnes, et al. Appl. Opt. 42(15), 2647-2652, (2003).
[10] Z. Tao, et al. Opt. Express 1127, (2014) .
[11] F. G. Fernald, Appl. Opt. 23(5), 652-653, (1984).
[12] Z. Tao, In Proceeding of International conference on remote senseing, environment and transportation engineering, (Nanjing, China), 250-252 ,(2013).

 https://doi.org/10.1051/epjconf/202023708002

AEROSOL - CLOUD TARGET CLASSIFICATION IN HALO LIDAR/RADAR COLLOCATED MEASUREMENTS

Eleni Marinou[1*], Florian Ewald[1], Silke Gross[1], Martin Wirth[1], Andreas Schaefler[1], Quitterie Cazenave[2] and Julien Delanoë[2]

[1] *Deutsches Zentrum für Luft- und Raumfahrt (DLR), Institut für Physik der Atmosphäre, Oberpfaffenhofen 82234, Germany.*

[2]*Laboratoire Atmosphères, Milieux, Observations Spatiales, Guyancourt, France*

**Email: Eleni.Marinou@dlr.de*

ABSTRACT

Particle attenuated backscatter and depolarization ratio at 532 nm from the WALES (Water Vapor Lidar Experiment in Space) instrument are used in combination with the radar reflectivity from the 35 GHz MIRA35 cloud radar in order to perform an aerosol-cloud target classification on the lidar/radar collocated observations carried out with the High Altitude and LOng-range research aircraft (HALO), in high temporal and spatial resolution.

The methodology is applied to the measurements conducted during the North Atlantic Waveguide and Downstream Impact Experiment (NAWDEX) in fall 2016. Here, we present one case study to show the feasibility and information context of the aerosol-cloud discrimination, which can serve as a good complement to the Cloud target categorization applied already in the measurements. Our results demonstrate that the developed mask is capable to identify complex stratifications with different aerosol and cloud types and even aerosol layers of low signal.

1. INTRODUCTION

Aerosol, clouds and their interactions with each other, but also with radiation and precipitation, influence weather and climate. Aerosols act as cloud condensation nuclei and ice nuclei influencing the cloud macro- and micro-physical properties and precipitation initiation. Despite of the significant influence aerosols and clouds have on climate, the quantification of their effect presents a large uncertainty, which can be partly attributed to their spatiotemporal variability and the diversity of the occurring processes [1].

In order to better quantify the aerosols and clouds effect on climate, and to improve the determination of their interactions, high spatiotemporal and vertically resolved observations of these features are of great assistance. Lidar/radar collocated measurements of aerosols and clouds is of unprecedented significance towards this goal. Several synergistic observations are currently utilized under this concept, applied at the framework of ground-based Research Infrastructures (e.g. ACTRIS, ARM) and at airborne and space platforms (e.g. HALO, RALI, CALIPSO, CLOUDSAT).

For an in-depth study of the aerosol and cloud properties observed with the aforementioned synergistic datasets, the atmospheric features detected are primarily classified as aerosols or clouds, based on lidar-only or synergistic lidar/radar categorization methods [e.g. 2, 3, 4]. For the HALO observations, a sophisticated aerosol classification has been developed [5] and a cloud classification scheme [6] has been implemented. However, there is no aerosol-cloud target classification scheme which will be used in combination to the other two categorizations to further assess aerosol and clouds features and study their interactions. Motivated by this need, we aim to develop a basic aerosol-cloud categorization of the observed features, utilizing the collocated airborne lidar and radar measurements. To this end, we use key parameters estimated by the two systems, which are independent of calibration sensitivities that may vary from flight-to-flight operations. The basic quantities used are the particle attenuated backscatter coefficient (β_{par}) and particle depolarization ratio (δ_{par}) at 532 nm from the lidar, and the radar reflectivity (Z_e) from the cloud radar.

To develop the classification mask and demonstrate its' performance, we use the datasets obtained during the NAWDEX experiment,

https://doi.org/10.1051/epjconf/202023708002

including collocated lidar-radar aerosol and cloud observations from overflights of the North Atlantic Ocean for a period of one month. From this dataset we derive the β_{par}, δ_{par} and Z_e parameters and perform the aerosol-cloud classification of the atmospheric features, in high temporal resolution. In the future, this basic typing will be used for detecting the necessary features, as for example on the aerosol-cloud-interaction processes and on the retrieval of aerosols and clouds optical and microphysical properties [e.g. 4, 7].

The work is structured as follows: first, the NAWDEX experiment is briefly introduced in section 2.1, then the methodology to derive the aerosol-cloud categorization is described in section 2.2. The methodology is applied and analyzed in section 3, and conclusions and future work are drawn in Section 4.

2. METHODOLOGY

2.1 NAWDEX

The North Atlantic Waveguide and Downstream Impact Experiment (NAWDEX) was conducted above the domain [70W – 10E; 45N – 80N] during 17/9 to 22/10 of 2016 [8]. In this campaign, four aircrafts were deployed with a sophisticated remote-sensing and in-situ payload and ground-based measurements. For this study, the dataset collected with the HALO are used, and particularly the lidar observations from the WALES instrument [9] and the cloud radar observations of the MIRA-35 instrument of HAMP (HALO Microwave Package) [10]. During the campaign, HALO performed 16 flights with a total of 96 flight hours and the observations include low, medium and high level clouds (ice, water and mixed phase) along with several marine and elevated aerosol layers in between the clouds.

2.2 Aerosol Classification

In order to demonstrate the quantities used for the aerosol - cloud classification, the time-height cross sections of these quantities are shown in Figure 1 for the day of 27/9/2016. The atmospheric features are well seen in the radar reflectivity (1-tob) the particulate attenuated backscatter coefficient (1-center) and the particle depolarization ratio (1-bottom). Several ice clouds are visible (up to 8 km) along with water clouds (bellow ~4 km). Between the clouds, two distinctive aerosol layers are observed, one between 15:00 and 16:00 UTC with $\beta_{par} > 10^{-6} m^{-1} sr^{-1}$ and enhanced δ_{par} (30 ± 3%) and one between 16:30 to 17:30 UTC with low attenuation ($< 10^{-6} m^{-1} sr^{-1}$) and depolarization (between ~3% and 18%).

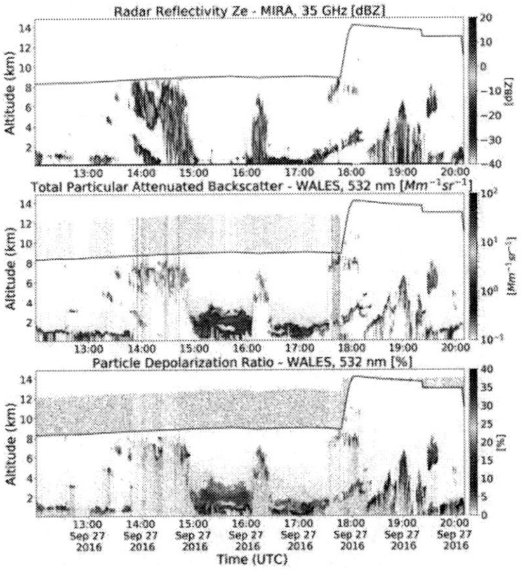

Figure 1. HALO observations at NAWDEX on 27/9/2016. From top to bottom: MIRA35 cloud radar reflectivity, WALES particle attenuated backscatter coefficient at 532 nm and WALES particle depolarization ratio at 532 nm.

For the typing of atmospheric features, i.e. the optical dominant scattered type, Z_e, β_{par} and δ_{par} observations are used for the classification between aerosols and clouds. Auxiliary to the observations, the temperature, pressure and water vapor fields from the ECMWF ensemble (European Center from Medium Range Weather Forecast) is used. The complete typing procedure is listed in Table 1 and is described below. Overall, the classification is based on 60 m vertical and 5 sec (~ 1 km) averaged profiles of lidar and radar observations.

In the first step, the classification utilizes the cloud radar observations and where a feature is detected from the radar (Z_e not equal to the fill value), this feature is consider to be cloud, surface or insects and its specific classification is handled by the methodology described in [6]. For the bits where both lidar and radar signals are totally attenuated, the cell is categorized as present of liquid unknown (same as in [6]). The atmosphere is classified as clear sky if $\beta_{par} < 1 \times 10^{-8} m^{-1} sr^{-1}$, a threshold which is well below the given range of aerosols of the CALIPSO classification [11]. The observations with $\beta_{par} > 2 \times 10^{-5} m^{-1} sr^{-1}$ are classified as clouds, a well suited threshold for the discrimination between aerosols and clouds, as the largest overlap between the two features is found in the β_{par} region between $4 \times 10^{-6} m^{-1} sr^{-1}$ and $1 \times 10^{-5} m^{-1} sr^{-1}$ [12]. This threshold value in the β_{par} is also used in the Cloudnet for the discrimination between aerosols and clouds [3]. Utilizing the δ_{par}, the lidar observations with $\delta_{par} < 1\%$ are considered to be water clouds and with $\delta_{par} > 38\%$ ice clouds.

The aforementioned decisions are based on each lidar/radar observational bin. Furthermore, some filteres are applied to account for the Cirrus fringes and the features coherence. The filters are described hereon.

Cirrus fringes for tenuous ice clouds: If in a target bin the temperature is less than 0°C and the bin is adjusted to an ice cloud (within the ± 180 m vertical and ± 2 km horizontal window), then the target bin is categorized as cirrus (similar as "Fringe amelioration" in CALIPSO [4] and in Cloudnet [3] categorizations).

Coherent filter for clear sky: if in the 3x3 cell around a target bin there are more than 5 bins (out of the total 9) that are categorized as clear sky (from the previous filters), the bin of interest is considered to be clear sky. This filter deals with the random molecular nose.

Coherent filter for clouds: if in a 3x3 cell around a target bin there are more that 5 bins (out of the total 9) that are categorized as clouds, the bin of interest is considered to be cloud.

Coherent filter for aerosols: if in a 3x3 cell around a target bin there are less than 4 bins (out of the total 9) that are categorized as clouds, clear sky, totally attenuated, surface or insects, the bin of interest is considered to be aerosols.

The remaining observations that are not categorized as clouds, clear sky, totally attenuated, surface or insects are considered to be aerosols.

Table 1. Overview of feature typing.

Aerosol – Cloud target classification
If detected Z_e: Clouds, surface and/or insects
If lidar and radar signals are totally attenuated: presence of liquid unknown
If $\beta_{par} > 2 \times 10^{-5} m^{-1} sr^{-1}$: Clouds
If $\beta_{par} < 1 \times 10^{-8} m^{-1} sr^{-1}$: Clear sky
If $\delta_{par} < 1\%\ or > 38\%$: Clouds
If T < 0 and there is an adjusted cloud: Cirrus fringe
If in the 3x3 cell around a target bin clear sky prevail: Clear sky
If in the 3x3 cell around a target bin clouds prevail: Clouds
If in the 3x3 cell around a target bin aerosols prevail: Aerosols
All the remaining uncategorized targets: Aerosols

* Further cloud sub-classification is applied with [6]

3. RESULTS

3.1 Aerosol cloud separation

In this section, we demonstrate the aerosol – cloud separation acquired with the discussed classification, for the flight of 27/9/2016. Figure 3 shows the distributions of occurrence frequency of the WALES measurements on the 2D space of $log_{10}(\beta_{par})$ vs δ_{par} for all features (top), for the features categorized as aerosols (bottom-left) and for the features categorized as clouds (bottom-right). The aerosol layers in this case occupied mainly the β_{par} area between $10^{-6.5}$ – $10^{-5.5}$with δ_{par} as high as 36%, indicating a dust layer. Although the aerosol measurements with $\beta_{par} < 10^{-6.5}$ are acquired with low confidence, they are still included in the aerosol class. In the clouds 2D β_{par} vs δ_{par} space, we can distinguish the water clouds region with $\delta_{par} < 5\%$ and the cirrus region with $\delta_{par} > 20\%$.

3.2 The new target classification

In Figure 3 we demonstrate the aerosol – cloud separation acquired with the discussed target categorization, along with its' combination with the cloud sub-categorization of [6], for the discussed case of 27/9/2016. We see that the aerosol and cloud features are successfully detected and separated. With the combination of the cloud sub-categorization mask, we observe the several liquid clouds (purple color) above the domain up to 4km, and the aerosol layers (orange color) which are located right on top of the water clouds between 13:45 to 17:15 UTC. Exactly before and after the dust observations, ice clouds are detected (blue color), elevated up to 7 km.

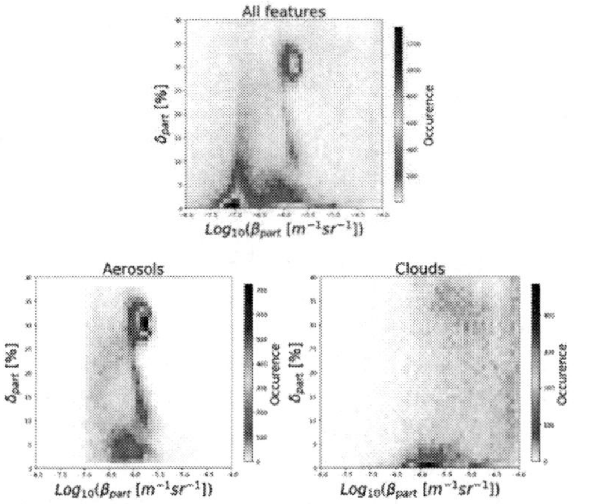

Figure 2. Distribution of occurrence frequency as a function of $log_{10}(\beta_{par})$ and δ_{par} for all the observations (top), the aerosols (bottom-left) and the clouds (bottom-right), for the WALES observations on 27/9/2018.

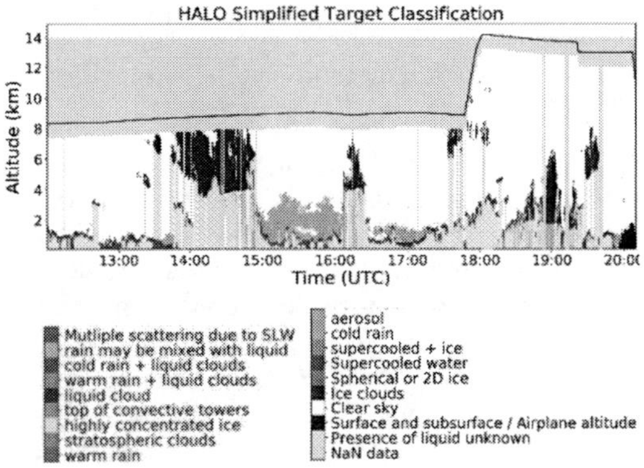

Figure 3. Target classification for 27/9/2016.

4. CONCLUSIONS

In this work, we combine lidar and radar observations from the HALO platform with high temporal and spatial resolution, with the scope to categorize the detected atmospheric features in aerosols and clouds. For the demonstration of the classification, we apply our methodology on data collected during the NAWDEX experiment. We conclude that the variables of the radar reflectivity, the particulate attenuated backscatter coefficient and the particle depolarization ratio at 532 nm are sufficient to discriminate aerosol and cloud features. The new target categorization works well and is in agreement with Cloudnet aerosol-cloud categorization. More cases will be analyzed, in a variety of atmospheric conditions and aerosol loads so as to further evaluate the performance of the new mask. Additionally, the mask will be combined with the HALO aerosol mask presented in [5] for a complete aerosol and cloud characterization of the scenes.

ACKNOWLEDGEMENTS

The HALO campaign received funding from DLR (project "Klimarelevanz von atmosphärischen Spurengasen, Aerosolen und Wolken" KliSAW), ETH Zurich, NRL Monterrey, the German Science Foundation (DFG; within SPP1294 HALO and SFB/TRR165 Waves to Weather), the European Facility for Airborne Research (EUFAR, project NAWDEX Influence) and the European Space Agency (ESA, EPATAN project, contract No. 4000119015/16/NL/CT/gp). Special thanks go to the Max Planck Institute Hamburg, University Hamburg, University Cologne and DLR for sharing the HAMP payload during the campaign. This work has been done with funding of a DLR VO-R young investigator group.

REFERENCES

[1] Seinfeld, et al., PNAS,113 (21) 5781-5790 (2016)
[2] Baars et al., Atmos. Meas. Tech., 10, 3175-3201 (2017)
[3] Hogan and O'Connor, Dept. of Meteorol. Univ. of Reading, UK, available at: http://www.met.reading.ac.uk/~swrhgnrj/publications/categorization.pdf (last access: March 2019) (2004)
[4] Liu et al., Atmos. Meas. Tech., 12, 703-734 (2019)
[5] Gross et al., Atmos. Chem. Phys., 13, 2487–2505 (2013)
[6] Ceccaldi et al., J. Geophys. Res. Atmos.,118, 7962–7981 (2013)
[7] Cazenave et al., Atmos. Meas. Tech. Discuss., in review (2018)
[8] Schäfler, et al., Bull. Amer. Meteor. Soc., 99, 1607–1637 (2018)
[9] Wirth et al., Appl. Phys. B, 96, 201-213 (2009)
[10] Ewald, et al., Atmos. Meas. Tech. Discuss., in review (2018)
[11] Winker et al., Atm. Ocean. Tech., 26, 2310-2323 (2009)
[12] Liu et al., J. Atmos. Oceanic Technol., 26, 1198–1213 (2009)

AEROSOL TYPING BASED ON MULTIWAVELENGTH LIDAR OBSERVATIONS AND METEOROLOGICAL MODEL DATA

Maria Mylonaki[1*], Elina Giannakaki[2,3], Alexandros Papayannis[1], Elena Floca[2], Mika Komppula[3]

[1]Laser Remote Sensing Unit, Physics Department, School of Applied Mathematics and Physical Sciences, National Technical University of Athens, 15780 Zografou, Greece
[2]Department of Environmental Physics and Meteorology, Faculty of Physics, National and Kapodistrian University of Athens, 15780 Zografou, Greece
[3]Finnish Meteorological Institute, P.O.Box 1627, 70211 Kuopio, Finland
**Email: mylonaki.mari@gmail.com*

ABSTRACT

Three different aerosol classification methods have been used to characterize lidar observations: Mahalanobis distance automatic aerosol type classification, Neural Network Aerosol Typing Algorithm (NATALI) and Source and Analysis (SCAN) aerosol classification. The data selection has been made through the EARLINET database depending on the 3b+2a+1δ optical property availability. One hundred aerosol layers from four EARLINET stations (Bucharest, Kuopio, Leipzig and Potenza) have been classified. We present a typical case study of aerosol characterization observed by the MUSA system over Potenza on the 11th of April 2016 (20:30-21:30 UTC).

1. INTRODUCTION

Aerosol typing is essential for understanding the impact of the different aerosol sources (desserts, wildfires, sea spray, anthropogenic emissions, volcanoes) on various atmospheric research topics such as air quality, meteorology, cloud microphysics and radiative forcing [1].

Lidars have proven to be an appropriate tool in detecting aerosol layers with high spatial and temporal resolution [2,3]. Furthermore, multi-wavelength Raman/depolarization lidars can be used on aerosol characterization (ie. dust, smoke, continental, etc.) as they provide vertically-resolved information of extensive (particle backscatter and extinction coefficient, volume depolarization ratio) and intensive (lidar ratio, Ångström exponent, linear particle depolarization ratio) optical properties.

2. METHODOLOGY

2.1 Classification methods

Automatic aerosol type classification - Mahalanobis distance

This method is a lidar stand-alone classification algorithm which uses four (or three) lidar intensive properties (Lidar Ratio (LR), ratio of Lidar Ratio ($LR_{\lambda1}/LR_{\lambda2}$), Ångström Exponent ($AE_{\lambda\alpha}/_{\lambda\beta}$) and Linear Particle Depolarization Ratio (LPDR) (if provided)) in order to classify the measured aerosol data into 8 types: Volcanic (V), Smoke (S), Marine (M), Dust (D), Continental (C), Polluted Dust (PD), Polluted Continental (PC) and Mixed Dust (MD) or C, D, M, MD, PC, PD if LPDR is not provided (Table 1). Basically, this technique assigns any given N-dimensional point (x_1, x_2, …, x_N) (observation) to the cluster (created with EARLINET network-wide typing results already published in literature) that has minimum Mahalanobis distance from that point [4].

Neural Network Aerosol Typing Algorithm - NATALI

NATALI is based on artificial neural networks (ANNs) trained (using data from OPAQ components + relative humidity (RH), GADS microphysics and Mischenko spheroidal model) to estimate the most probable aerosol type from, solely, a set of multispectral lidar data (Color Index (CI), Color Ratio (CR), LR, $AE_{\lambda\alpha}/_{\lambda\beta}$ and LPDR (if provided)). The 14 aerosol types (high - resolution typing) are: Continental, Continental Polluted, Smoke, Dust, Marine, Volcanic, Continental Dust, Marine Mineral, Continental Smoke, Dust Polluted, Coastal, Coastal Polluted, Mixed Dust and Mixed Smoke (Table 1) [5].

Source and Analysis aerosol classification - SCAN

SCAN aerosol classification uses the backward-trajectories for certain height levels (ie. from 0.5 km to 6.0 km with a step of 250 m) created with Hybrid Single-Particle Lagrangian Integrated Trajectory (HYSPLIT) [6] in combination with satellite data (e.g fire/hotspot information based on FIRMS/MODIS [7]). Each aerosol layer is associated to one trajectory level regarding the middle height of the layer. To identify the source region of particles SCAN, firstly, assumes certain regions above Earth: West – Northwest, North – Northeast, Maritime and Local areas for Europe and long-range transport for the other continents. Then, it calculates the hours that air-mass is above of each region, the mean height in each region as well as the minimum of the height. In order to associate the observed layers to certain aerosol source, it is required that the mean height of each region is lower than 3 km. If the mean height is above 3 km, the minimum height of each region is checked and if it is lower than 2 km then the certain region is chosen as the source of the observed layer. If these two criteria are not indulged to none of the regions, then the layer is characterized as unknown. If these two criteria are indulged into more than 2 regions a mixture of more than one aerosol types is assumed. The 12 aerosol types of SCAN aerosol classification are: Smoke, Marine, Dust, Continental, Dust Polluted, Marine Mineral, Mixed Dust, Continental Smoke, Coastal and Mixed Smoke. The variable of the height of the air parcel above the source region is also used.

Table 1: Aerosol types according to Mahalanobis, NATALI and SCAN classification algorithms.

	Aerosol Type	Mahalanobis (with LPDR)	NATALI (with LPDR, high resolution)	SCAN
1	Smoke	Smoke (S)	Smoke (S)	Smoke (S)
2	Marine	Marine (M)	Marine (M)	Marine (M)
3	Dust	Dust (D)	Dust (D)	Dust (D)
4	Clean Continental	Clean Continental (CC)	Clean Continental (CC)	Continental (C)
5	Continental Polluted	Continental Polluted (CP)	Continental Polluted (CP)	Continental (C)
6	Dust with Smoke/Continental	Dust Polluted (DP)	Dust Polluted (DP) Continental Dust (CD)	Dust Polluted (DP)
7	Dust/Volcanic with Marine	Mixed Dust (MD)	Marine Mineral (MM)	Marine Mineral (MM)
8	Continental with Dust and Marine	Unknown	Mixed Dust (MD)	Mixed Dust (MD)
9	Continental with Smoke	Unknown	Continental Smoke (CS)	Continental Smoke (CS)
10	Continental with Marine	Unknown	Coastal (Coas)	Coastal (Coas)
11	Continental Polluted with Marine	Unknown	Coastal Polluted (CoasP)	Coastal (Coas)
12	Continental with Smoke and Marine	Unknown	Mixed Smoke (MS)	Mixed Smoke (MS)
13	Volcanic	Volcanic (V)	Volcanic (V)	-

2.2 Data and lidar stations

Data and lidar station selection depended on the availability of the vertical profiles of the full set of aerosol optical properties: backscatter coefficient (b_{355}, b_{532}, b_{1064}), extinction coefficient (e_{355}, e_{532}), Lidar Ratio (LR_{355}, LR_{532}), Ångström exponent ($AE_{355}/_{532}$, $CI_{355/532}$, $CI_{532/1064}$) and Linear Particle Depolarization Ratio ($LPDR_{532}$) at the EARLINET database (https://www.earlinet.org) within the years 2014-2018. The 4 lidar stations performed the used measurements are: Kuopio, Finland; Bucharest, Romania; Leipzig, Germany; and Potenza, Italy (Table 2).

Table 2: EARLINET lidar station information

Location	ACTRIS Code	Institute	Coordinates (lat, long, altitude amsl)	Ref.	No of cases	Year
Bucharest	INO	National Institute of R&D for Optoelectronics (INOE) Link	44.35 N, 26.03 E, 93 m	[8]	7	2017
Kuopio	KUO	Finnish Meteorological Institute (FMI), Atmospheric Research Centre of Eastern Finland, Kuopio Link	62.74 N, 27.54 E, 190 m	[9, 10]	9	2015, 2016
Leipzig	LEI	Leibniz Institute for Tropospheric Research, Leipzig Link	51.35 N, 12.43 E, 90 m	[9, 10]	17	2018
Potenza	POT	Consiglio Nazionale delle Ricerche - Istituto di Metodologie per l'Analisi Ambientale (CNR-IMAA), Potenza Link	40.60 N, 15.72 E, 760 m	[11]	64	2015-2016

https://doi.org/10.1051/epjconf/202023708003

3. RESULTS

3.1 Case study

In the following we present a typical case study of aerosol characterization. On the 11th of April 2016 (20:30-21:30 UTC) an aerosol layer was observed by the MUSA system over Potenza (Fig. 1a, b).

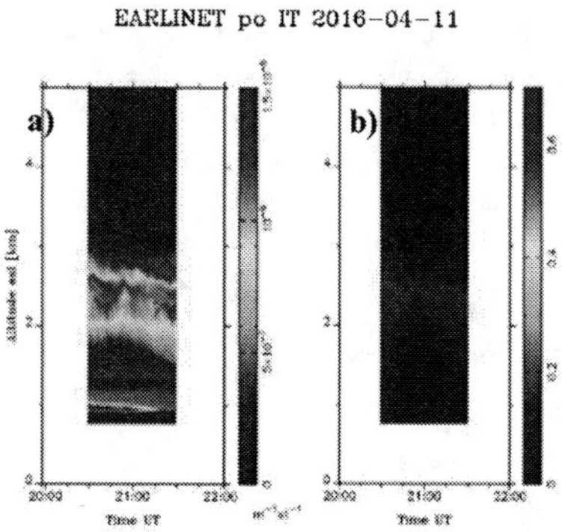

Figure 1: Temporal evolution of the a) 1064 nm range-corrected lidar signal and b) 532 nm volume linear depolarization ratio obtained by the MUSA system over Potenza on the 11th of April 2016 (20:30-21:30 UTC).

The aerosol layer boundaries were calculated by applying the gradient method on the 1064 nm backscatter coefficient profile [12] and found 1.9 km (cf. Fig.2-bottom lower horizontal red line) and 2.7 km (cf. Fig.2-top upper horizontal red line) above mean sea level (amsl).

Optical profiles of the extensive and intensive aerosol properties were retrieved. The mean values and standard deviation of the properties of the aerosol layer (1.9-2.7 km amsl) are: LPDR 15 ± 4 [%], LR_{355} 37 ± 4 [sr], LR_{532} 49 ± 5 [sr], $AE_{355/532}$ -0.7 ± 0.2, $CI_{355/532}$ -0.1 ± 0.3 and $CI_{532/1064}$ 0.3 ± 0.1 (Fig. 2).

Mahalanobis distance automatic aerosol type classification classified the aerosol layer observed over Potenza on the 11th of April 2016 (20:30–21:30 UTC) between 1.9-2.7 km amsl as "Mixed Dust". The maximum weighted probability of the classification was 59.1% for Mixed Dust, followed by Dust with 20.1% probability and Clean Continental with 11.1% probability. Lower values of probability were found for the aerosol types of Smoke, 6.3%; Polluted Dust, 1.8%; Marine, 1.3% and Volcanic, 0.3% (Fig. 3).

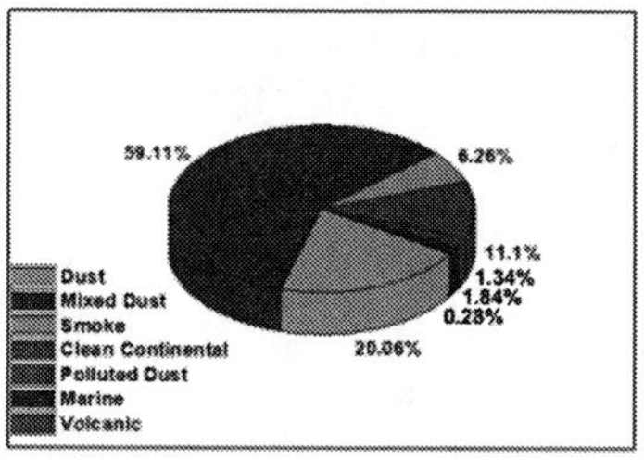

Figure 3: Mahalanobis distance automatic aerosol type weighted probabilities for the case study measured over Potenza on the 11th of April 2016 (20:30–21:30 UTC) between 1.9-2.7 km amsl.

NATALI classified the aerosol layer observed over Potenza on the 11th of April 2016 (20:30–21:30 UTC) between 1.9-2.7 km amsl as "Mixed Dust" with 93% confidence and 11347 Agreements (Table 3).

Finally, the 96-hour HYSPLIT backward trajectories arriving at 1.5 km above ground level (agl) (2.2 km amsl) (Fig. 4, red line) and 2.0 km agl (2.7 km amsl) (Fig. 4, blue line) over Potenza on the 11th of April 2016 (20:30–21:30 UTC) used by SCAN are shown in Figure 4.

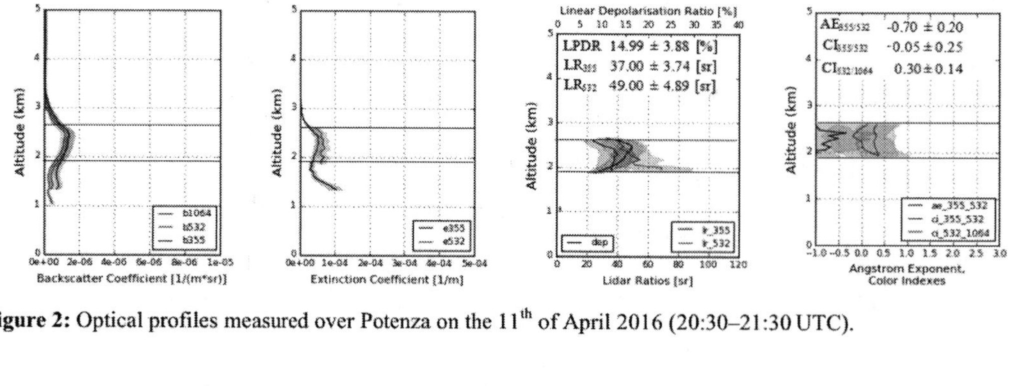

Figure 2: Optical profiles measured over Potenza on the 11th of April 2016 (20:30–21:30 UTC).

Table 3: NATALI classification of the aerosol layer measured over Potenza on the 11th of April 2016 (20:30–21:30 UTC) between 1.9- 2.7 km amsl.

Predominant Component	Dust
Aerosol Type	Mixed Dust/CC
A1H Answer	Mixed Dust/CC
A1H Confidence	0.88
A1H Agreements	13200
A2H Answer	Unknown
A2H Confidence	0
A2H Agreements	0
A3H Answer	Mixed Dust/CC
A3H Confidence	0.93
A3H Agreements	11347

The maximum probability of the first classification was 78.1% for Marine, followed by 13.4% for Dust and 9.3% for Continental. However, the minimum height that the air parcel reached during its 96-hour travel was 2.5 km amsl, so the marine contribution assumed to be negligible. As a result, SCAN classified the aerosol layer observed over Potenza on the 11th of April 2016 (20:30–21:30 UTC) between 2.0-2.5 km amsl as "Dust Polluted".

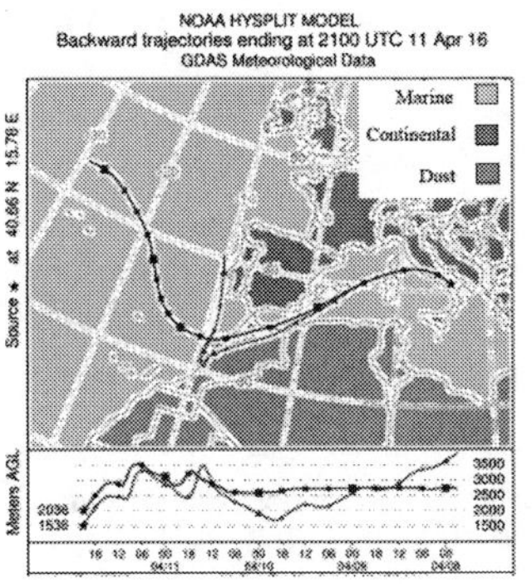

Figure 5: 96-hour HYSPLIT backward trajectories arriving at 1.5 km a.g.l. (2.2 km amsl) (red line) and 2.0 km agl (2.7 km amsl) (blue line) over Potenza on the 11th of April 2016 (20:30–21:30 UTC).

For the case study of the 11th of April 2016 (20:30–21:30 UTC) the 3 different classification algorithms classified the aerosol layer observed between 1.9 and 2.7 km as mixture of Dust with Marine and/or Continental.

Future work targets at the investigation of the discrepancies of such cases and the statistical analysis of the comparison of the 3 aerosol layer classification methods.

ACKNOWLEDGEMENTS

We acknowledge support of this work by the project "PANhellenic infrastructure for Atmospheric Composition and climatE change" (MIS 5021516) which is implemented under the Action "Reinforcement of the Research and Innovation Infrastructure", funded by the Operational Programme "Competitiveness, Entrepreneurship and Innovation" (NSRF 2014-2020) and co-financed by Greece and the European Union (European Regional Development Fund). The financial support for EARLINET by the European Union under grant RICA 025991 in the FP6 is gratefully acknowledged. Part of this research has been co-financed by additional funding from the European Union FP7 (/2007-2013) under grant agreement no 262254.The authors gratefully acknowledge the NOAA Air Resources Laboratory for the provision of the HYSPLIT transport and dispersion model and READY website (http://www.ready.noaa.gov) used in this publication. MODIS-FIRMS data were provided by EARTHDATA (EOSDIS-NASA).

Με τη συγχρηματοδότηση της Ελλάδας και της Ευρωπαϊκής Ένωσης

REFERENCES

[1] IPCC. Cambridge University Press, 571–658 (2014)

[2] E. Giannakaki, et al., Atmos. Chem. Phys., 16, 8109–8123 (2016)

[3] O. Soupiona, et al., Atmos. Environ., 183, 165–174 (2018)

[4] N. Papagiannopoulos, et al., Atmos. Chem. Phys., 18, 15879-15901 (2018)

[5] D. Nicolae, et al., Atmos. Chem. Phys., 18, 14511–14537 (2018)

[6] R. R. Draxler, et al., Aust. Meteor. Mag., 47, 295–308 (1998)

[7] D. K. Davies, IEEE J-STARS, 47, 72-79 (2009)

[8] A. Nemuc, et al., Atmos. Meas. Tech, 6, 3243-3255 (2013)

[9] D. Althausen, et al., J. Atmos. Oceanic Technol., 26, 2366-2378 (2009)

[10] R. Engelmann, et al., Atmos. Meas. Tech., 9, 1767-1784 (2016)

[11] F. Madonna, et al. Atmos. Meas. Tech., 4, 1191–1208 (2011)

[12] L. Belegante, et a;., Acta Geophys., 62, 276-289 (2014)

ALGORITHM FOR SOLVING THE LIGHT SCATTERING PROBLEM FOR ARBITRARY SHAPED ICE PARTICLES WITH ABSORPTION

Timofeev D.N.*, Konoshonkin A.V., Kustova N.V., Borovoi A.G.

V.E. Zuev Institute of Atmospheric Optics, Rus. Acad. Sci., Tomsk, 1 Academician Zuev Sq. 634055, Russia

**Email: tdn@iao.ru*

ABSTRACT

The work presents the estimation of the absorption influence effect on the light scattering problem for atmospheric ice particles. The calculation of light scattering matrices was performed for two types of particles: the solid hexagonal columns and the arbitrary shaped particles. The range of the size for both types of particles varies from 10 μm to 1000 μm. The results of the research show an insignificant decrease of the value of the M11 element of the light scattering matrix for small-sized particles for all chosen wavelengths (0.355 μm, 0.532 μm, 1.064 μm и 1.6 μm), and a significant decrease for big-sized particles for a 1.6 μm wavelength.

1. INTRODUCTION

The light scattering problem of ice crystal particles in cirrus clouds is crucial for investigation of the solar energy transfer process. Cirrus clouds are an essential part of all clouds that covers Earth's surface because of this the recovering of their optical and microphysical properties is required to refine global climate models [1].

Cirrus clouds consist of ice crystals particles with various shape and size. For solving the light scattering problem for ice particles it is difficult to apply exact numerical methods such as the finite difference time domain method (FTDT) [2, 3] or the discrete dipole approximation method (DDA) [4, 5] due to their high requirements for computational resources for big-sized particles [6]. For this reason the problem is often solved with the geometrical and the physical optics approximation [7].

During the numerical calculation of light scattering matrices for ice particles it is important to take the specific factors that could influence on scattering into account. One of these factors is the absorption of light in ice particles.

The one of the important properties of an absorptive medium is refractive index that is complex and consists of two parts:

$$n = N_r + N_i \quad (1)$$

, where N_r is the real part and N_i is the imaginary part. When electromagnetic wave propagates in absorptive medium the amplitude of the electromagnetic field decreases. This decrease is defined by following factor:

$$\exp\left(-\frac{2\pi N_i}{\lambda}\cdot l\right) \quad (2)$$

, where λ is the wavelength, and l is the distance that optical wave passed in absorptive medium [8]. When the value of the wavelength is deep in the near infrared (IR) zone of the spectrum the value of the imaginary part of the effective refractive index increases, therefore the influence of the absorption should decrease [9].

For taking the absorption into account the new version of the modified beam-splitting (MBS-1) algorithm [10] was developed. It calculates light scattering matrices within geometrical and physical optics approximation for particles with flat facets. In addition, there is the feature that allows defining an arbitrary shaped particle in the new version of the algorithm.

2. ESTIMATION OF ABSORPTION

For estimation of the absorption influence to the light scattering process on ice particles with different shapes it was performed the calculation of light scattering matrices for the particles with two shapes: solid hexagonal column and arbitrary shaped particle (Figure 1). The calculation was performed with following parameters: random spatial orientation of particle, three different wavelengths with corresponded complex refractive indices (Table 1), and the maximal dimension of the particle varies from 10 μm to 1000 μm. The aspect ratio of hexagonal column is corresponded to D. Mitchel dependency [11]. The size of the arbitrary formed particle varies proportionally. As the maximal dimension it means the distance between two the most distant vertices of the particle.

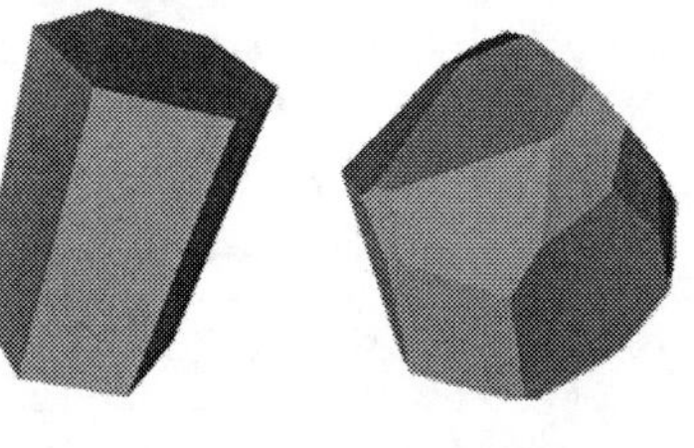

Figure 1 Particles for the calculation: the solid hexagonal column (left) and the arbitrary formed particle (right).

Table 1 Refractive indices (n) for chosen wavelengths (λ)

λ, μm	*n*
0.355	$1.3243 + i3.6595\cdot10^{-9}$
0.532	$1.3116 + i2.6180\cdot10^{-9}$
1.064	$1.3004 + i1.9000\cdot10^{-6}$
1.600	$1.2893 + i3.5365\cdot10^{-4}$

It is convenient to define the estimation quantity of the absorption influence in the backscattering direction as χ and for the whole diagram as ψ:

$$\chi = \frac{M_{11}(180°) - \bar{M}_{11}(180°)}{M_{11}(180°)}\cdot 100\% \quad (3)$$

$$\psi = \frac{\int_0^{\pi} \bar{M}_{11}(\theta)sin(\theta)d\theta}{\int_0^{\pi} M_{11}(\theta)sin(\theta)d\theta} \quad (4)$$

, where M_{11} is the intensity of the scattering light without absorption, $\bar{M}_{11}$ is the same quantity with absorption.

3. CALCULATION

For laser sounding problems only backscattering point (180°) is interesting, but the whole scattering diagram is important for radiation transfer problems. The dependency of the maximal dimension of the particle (D_{max}) on χ in the backscattering direction for the solid hexagonal column and the arbitrary formed particle is presented in Figure 2 and the dependency of D_{max} on ψ is in Figire 3a and 3b.

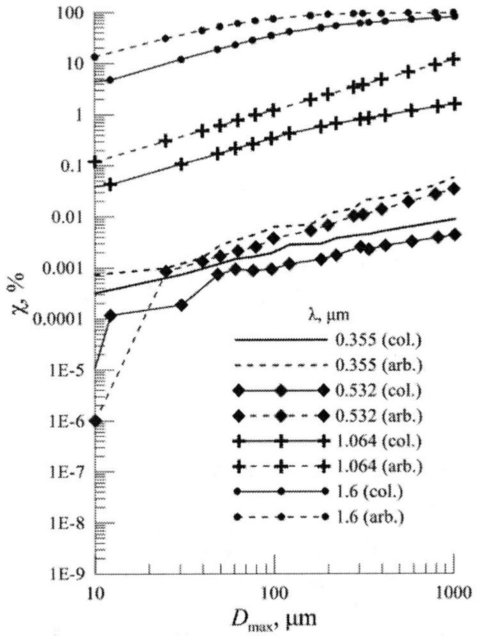

Figure 2 Dependence of D_{max} on χ for the solid hexagonal column and the arbitrary formed particle

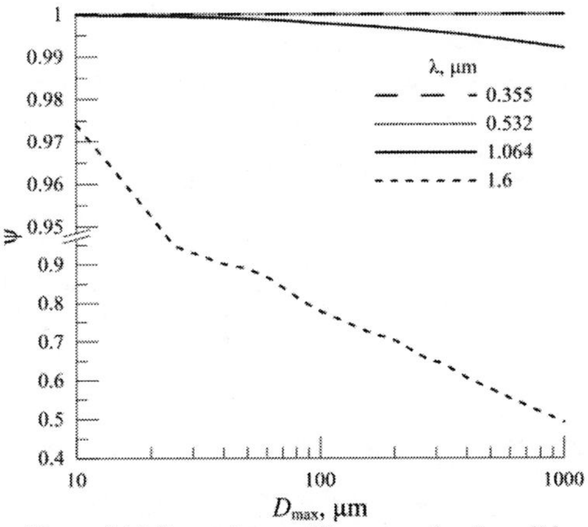

Figure 3(a) Dependence of D_{max} on ψ for the solid hexagonal column

https://doi.org/10.1051/epjconf/202023708004

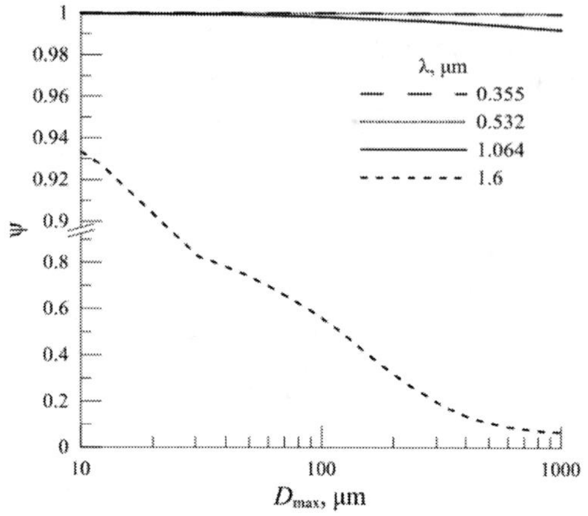

Figure 3(b) Dependence of D_{max} on ψ for the arbitrary formed particle

4. CONCLUSIONS

The analysis of results show that the absorption influence is insignificant for the ultraviolet and visible zone of the spectrum (0.355 μm and 0.532 μm wavelengths) for solving of the light scattering problem for ice crystal particles of cirrus clouds, because it's value is less than basis points of percent regardless to the shape, orientation, and size of particle. However in the near IR zone the absorption influence could reach several dozen percent almost in the whole range of the shape and size of particles. It is convenient to choose 1.064 μm with the value of imaginary part of the refraction index (N_i) $1.9{\cdot}10^{-6}$ as the threshold value of the wavelength where the absorption is significant. Thus for the wavelengths with N_i less than $1.9{\cdot}10^{-6}$ (near IR zone) it is obligatory to take the absorption into account for solving the direct light scattering problem. Moreover the absorption influence significantly depends on shape and size of particle for the 1.064 μm wavelength. In conclusion, the absorption influence is less than 2% for the chosen range of size for hexagonal ice particles.

ACKNOWLEDGEMENTS

This work is supported by the Russian Foundation for Basic Research under Grants no. 18-05-00568, no. 18-55-53046, and no. 19-01-00351 and supported in part by Russian Science Foundation (18-77-10035).

REFERENCES

[1] Liou K.N. Influence of cirrus clouds on weather and climate processes: A global perspective // Mon. Wea. Rev. 1986. V. 114, N 6. P. 1167–1199.
[2] Kunz K.S., Luebbers R.J. Finite Difference Time Domain Method for Electromagnetics. — Boca Raton, FL: CRC Press, 1993
[3] Taflove A. Advances in Computational Electrodynamics: The Finite-Difference Time-Domain Method. — Boston: Artech House, 1998
[4] Purcell E.M., Pennypacker C.R. Scattering and absorption of light by nonspherical dielectric grains // Astrophys. J. — 1973. — T. 186. — C. 705–714.
[5] Yurkin M.A., Maltsev V.P., Hoekstra A.G. The discrete dipole approximation for simulation of light scattering by particles much larger than the wavelength // J. Quant. Spectrosc. Radiat. Transfer. — 2007. — T. 106. — P. 546–557.
[6] Yurkin M.A., Hoekstra A.G. The discrete-dipole-approximation code ADDA: Capabilities and known limitations // J. Quant. Spectrosc. Radiat. Transfer. — 2011. — T. 112. — P. 2234–2247.
[7] Volkovitcky O.A., Pavlova L.N., Petrushin A.G. Optical properties of the Ice Clouds. — Leningrad: Gidrometeoizdat, 1984. — P. 198.
[8] Bohren C.F., Huffman D.R. Absorption and Scattering of Light by Small Particles. – John Wiley & Sons, 1986. – P. 660.
[9] Timofeev D.N., Konoshonkin A.V., Kustova N.V., Borovoi A.G. Backscattering matrices calculation for atmospheric ice crystals within the physical optics approximation with absorption effect // Proc. SPIE. 2018. V. 10833. 10833-174.
[10] Konoshonkin, A., Kustova, N., and Borovoi, A., "Beam splitting algorithm for the problem of light scattering by atmospheric ice crystals. Part 2. comparison with the ray tracing algorithm," Atmospheric and Oceanic Optics, 28, 448-454 (2015).
[11] Mitchell D.L., Arnott W.P. A model predicting the evolution of ice particle size spectra and radiative properties of cirrus clouds. Part II. Radiation. // J. Atmos. Sci. 1994. V. 51. P. 817–832.

ANALYZING THE IMPACT OF DIFFERENT FILTERING METHODS ON SATELLITE ALTIMETRY FULL WAVEFORM DECOMPOSITION

Zhijie Zhang[1, 2], Huan Xie[2, *], Xiaohua Tong[2], Binbin Li[2], Yunwen Li[2], Hanwei Zhang[1]

[1]*School of Surveying and Land Information Engineering, Henan Polytechnic University, Jiaozuo 454000, China*

[2]*College of Surveying and Geo-Informatics, Tongji University, Shanghai 200092, China*

**Email: huanxie@tongji.edu.cn*

ABSTRACT

Filtering is an essential step in the denoising of satellite altimetry full waveform data, since any deformation and distortion in the shape of the waveform can cause errors in range estimation and further waveform decomposition will also be adversely affected. This paper evaluated comprehensive performance of the popular filtering approaches like Gaussian filter, Taubin filter, Wavelet filter, and EMD based filter by simulated waveform data and ICESat/GLAS waveform. Firstly, according to the principle of each filter, the optimal parameters of filtering algorithm by ergodic tests were selected, then the Gaussian function using Levenberg-Marquardt method was used to fit full waveform to exact waveform parameters (i.e. peak amplitude, position, and half-width). Thirdly, through comparing SNR, RMSE of the pre and post filtering simulation waveform, and the consistency ratio, the average error of peak amplitude, position, and half-width in each Gaussian components of the fitted simulation waveform, verified the effectiveness of these filters and analyzed their influence on decomposition accuracy. Both the simulation experiments and ICESat/GLAS experimental results suggested that the Taubin filter had superior performance with the lowest peak position error, which turns out it has advantage in full waveform denoising and contributes to better full waveform decomposition. However, it introduces more parameters needed to be selected. The self-adaptive EMD based approach has the highest consistency, which shows EMD-based method is more suitable for the denoising of satellite altimetry full waveform decomposition.

Key words: satellite altimetry, full waveform, denoising, filtering, decomposition

1. INTRODUCTION

Space-borne laser altimetry is a powerful remote sensing technology, which can provide high resolution information on surface elevations and ground cover to the Earth science community by digitally recording the return laser pulse. NASA's Ice, Cloud, and land Elevation Satellite (ICESat/GLAS) was the first Earth-orbiting laser altimeter mission that operated from 2003–2009[1], its product data have been widely used in monitoring glaciers[2], ice shelves[3], vegetation, and land elevation[4], etc. Building on this heritage, the next generation of space-borne Advanced Topographic Laser Altimeter System (ICESat-2/ATLAS) is seeking to monitor a diverse range of geophysical phenomena at unprecedented levels of precision and accuracy by photon-counting technolgy[5]. Comparing with America, though China started later in this aspect, currently ZY3-02 as the first experimental altimeter acquired 44 tracks effective data, and the combined adjustment of its laser and image data can actually improve the accuracy of uncontrolled elevation[6]. GaoFen7 (GF-7) is the first civil laser three-dimensional mapping satellite in China, and will launch this year whose main objective is to aid high precision mapping.

Critical to the success of full waveform satellites is the ability to interpret the shape of the digitally-recorded return laser pulse and extract accurate time for specific reflecting surfaces within the footprint (e.g., first and last), and thus to geolocate the desired reflecting surface directly. Brenner proposed Gaussian function to decompose the waveforms[7]. Specifically, from the real waveform signal, initial parameters need to be identified first for the fitting step, more precisely the number of peaks or modes together with width, amplitude and location for each mode. Due to the noisy nature of many waveforms, estimation of initial values from the raw signal results in a large number of modes with a low amplitude and a narrow width.

Therefore, it is necessary to smooth the waveforms in order to avoid the noise-induced Gaussian components and help increase the fitting accuracy.

Many methods have been proposed for both noise removal and improvement of the signal-to-noise ratio (SNR) in full waveform decomposition. But some methods are not appropriate in all cases and need specific assumptions. Typically, in Low-Pass filters, the high frequency segments are assumed to contain noise. Wiener filter requires to pre-estimate a cross-correlation vector between input signal and expected signal, which is rather difficult to carry out because the desired signal is unknown before filtering. To Gaussian filter, the selection of an appropriate kernel width is a key but difficult work, and the denoising effect comparing with other filter like Taubin filter, Wavelet filter and EMD filter lack explicit comparison. In this paper, we report our experience of using those different kinds of filters. In order to keep the results objective, the principle of the filters are first described and their smoothing parameters are investigated and evaluated. And following the results of these methods applied to simulation and ICESat/GLAS data are presented and compared. At last, experimental results are discussed and concluding remarks are offered, with expectation to provide reference for the upcoming GF-7 data process.

2. METHODOLOGY

2.1 Gaussian Filter

Gaussian filtering is realized by convolution between the Gaussian kernel function with the original signal[8]. The smoothing performance of it could be adjusted by Gaussian kernel width σ. The width of flat plane pulse is derived as the best Gaussian kernel width for a fixed width Gaussian filtering[9]. In the paper, we chose width of GLAS transmit pulse 6ns as the kernel width to determine the weight of each bin. When the bin length is 9, the smoothing effect is better through many tests.

2.2 Taubin Filter

On the basis of Gaussian filter, Taubin filter adds two scale factors μ and λ to adjust the smoothing degree[10], which effectively avoid the defects of edge data shrink led by Gaussian filter. It uses formula eq.(1) for iterative processing of waveform data.

$$\begin{cases} x_i' = x_i + \lambda \sum_{j\epsilon i^*} w_{ij}(x_j - x_i) \\ x_i'' = x_i' + \mu \sum_{j\epsilon i^*} w_{ij}' (x_j' - x_i') \end{cases} \qquad (1)$$

In the equation, x_i denotes the raw waveform data before filtering; x_i' means the waveform after the first smoothing; i^* is the adjacent point set of i, and j is the point of the set; x_i'' stands for the waveform after the second smoothing; λ and μ mean the scale, and $\lambda > 0$, $\mu < 0$, $\mu < -\lambda$. $w_{ij}(w_{ij}')$ denotes the weight of $x_j(x_j')$ impact on $x_i(x_i')$, which could adopt the reciprocal of point number in the set[11]. In the paper, we chose λ equals 0.9057, μ equals -0.9072 by many experiments.

2.3 Wavelet filter

While the wavelet filter is quite different from the above two methods, the basic idea of it is to start with a basis wavelet function with regularity, locality and oscillation, and to get a function family through scaling and translation transformations[12]. In this paper, we adopted "Daubechies" basis (db4) with the level 3, and the soft threshold method for estimating wavelet coefficients.

2.4 EMD Filter

Empirical mode decomposition (EMD), introduced by Huang et al[13] is a method for decomposing complex, multi-component signal into several elementary Intrinsic Mode Functions (IMFs) as eq.(2) shows. Differing from Fourier and wavelet analysis which have predefined basis, EMD uses only scale and frequency characters of the original signal. EMD is a local, fully data-driven and self-adaptive analysis approach. In order to let its instantaneous frequency have a meaningful interpretation, the IMF has to satisfy two conditions: (1) in the whole data set, the number of extrema and the number of zero-crossings must either equal or differ at most by one; (2) at any point, the local average of upper and lower envelop is zero.

$$S_0(t) = \sum_{i=j+1}^{n} IMF_i(t) + r_n(t) \qquad (2)$$

Where $S_0(t)$ means the smoothing signal after deleting the first j IMFs that are considered as noise. In this paper, if the Hurst exponent of a certain IMF is less than 0.5, then it is regarded as noise.

3. RUSULTS AND DISCUSSIONS

In this section, we used simulated waveform data and ICESat/GLAS waveform to test each filter performance. Due to the true Gaussian components are known in simulation waveform data, therefore, we compared average absolute error of peak value, peak position, half-width error between the fitted waveform and real waveform. While, the ICESat/GLAS data we just compared these waveform with same Gaussian number and their fitting accuracy.

3.1 Simulation tests and results

In order to analyze fitting correct rate and fitting accuracy, a group of the full-waveform echoes 5000 in total were simulated using addition of Gaussian functions and Gaussian white noise. The number of Gaussian function from fitting result was compared with the number of Gaussian function given in simulation signal to verify correctness of fitting result. If these two number are equal and the location error is less than 1ns, the fitting result is considered as consistency. The specific parameters are as follows:

Table1. Parameters set of simulation full waveform signal

SNR_set = [10, 15, 20, 25, 30, 35, 40]
Amp_set = [0.1, 0.15, 0.2, 0.25, 0.3,0. 4, 0.5, 0.6, 0.7, 0.8, 0.9, 1.0]
Loc_set = [100, 130, 160, 190, 220, 250, 280, 310, 340, 370,400,430, 460, 500]
sigma_set = [6, 7, 8, 9, 10, 15, 20, 25, 30, 35, 40]

Figure 1 presents one of the typical waveforms denoising effect of different filters, we can see that Wavelet filter and EMD filter and keep the shape of waveform better, while the Gaussian filter and Taubin filter slightly weaken peak value. Figure 2 reveals SNR gain[14] (the SNR after waveform smoothing/ raw waveform SNR) all increased with smaller RMSE. When the maximum peak number is three and random SNR, fitting difference of consistency ratio (CR), and absolute error difference of peak value (A), peak location (T), peak half width (σ), and fit deviation (FitDev) of the total waveforms are shown in figure 3. It shows EMD filter has superior performance with biggest CR, and Taubin filter has the highest T precision. (G W T E are abbreviation of Gaussian filter,Wavelet filter,Taubin filter and EMD filter respectively).

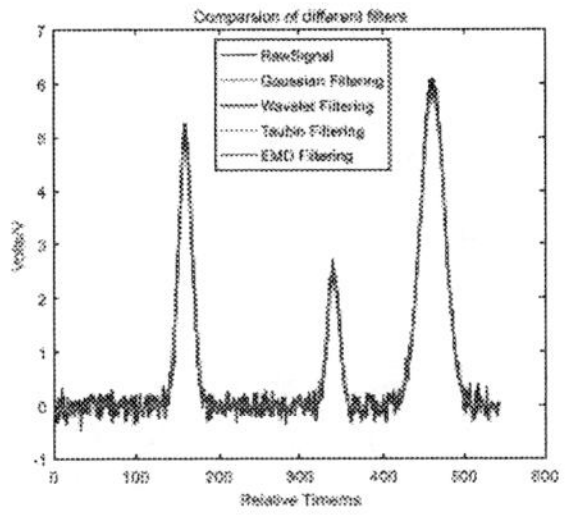

Fig.1 Denoising effect comparision of different filiters

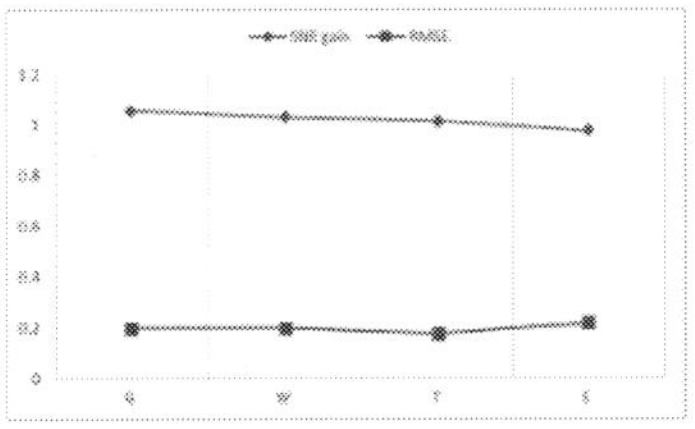

Fig.2 SNR gain and RMSE of different filiters

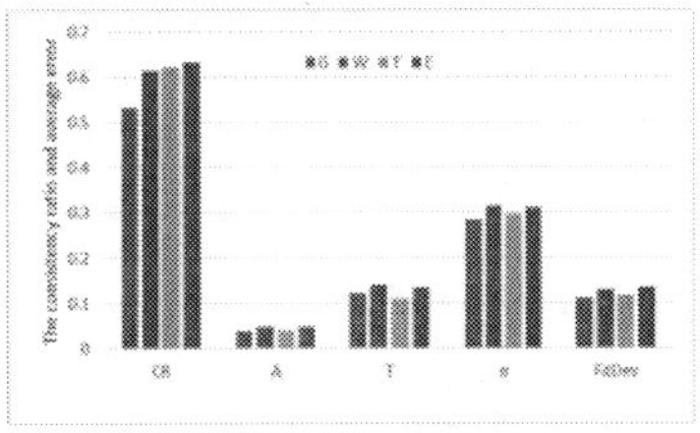

Fig.3 Fitting effect comparartion of different filiters

3.2 ICESat/GLAS data experiments

In this article we selected GLAS01 waveform data on March 9, 2008, and the corresponding GLAS05 peak number, peak value, peak location, and sigma of each Gaussian component as validation parameters. First we chose the available data through ancillary science packet APID availability flag (apid_ADSci_flg), then the land waveform data by waveform type flag (i_waveformType), and sample padding with zero filling by sample padding flag (i_samp_pad) to acquire the 1000 pieces of input waveform data. During decomposition, some constraint conditions were set to fitted Gaussian parameters, like minimum peak value (minPeak>noiseAverage+3*Devnoise), and the peak width (2.5<σ<300). Figure 4 shows the number waveform ratio (number of waveforms could be decomposed successfully/1000, NumberWf Ratio), consistency ratio, and absolute ratio (consistency ratio/number ratio, Absolute Ratio). It depicts some waveforms cannot be decomposed after using Wavelet filter and EMD filter. The reason is that after waveform smoothing, the sigma is too narrow which is less than the width of transmit pulse and thus the peak is deleted during Levenberg-Marquardt optimization. Figure 5 illustrates the root mean squared error of fitting reduced near one time comparing than GLAS05 results, and Taubin filter has the smallest fit deviation.

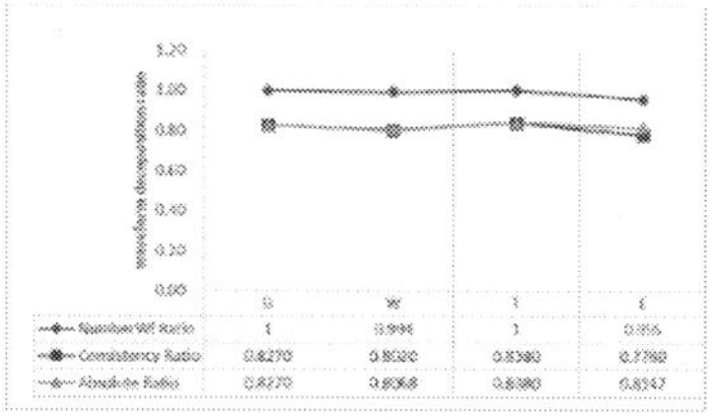

Fig.4 Decomposition comparartion of different filiters with GLAS05

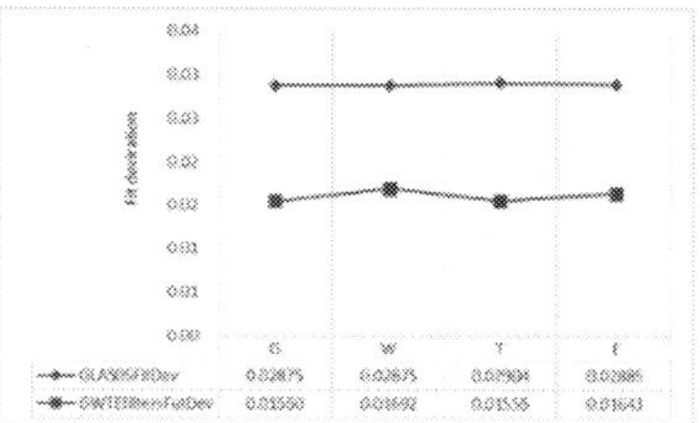

Fig.5 Fit deviation comparartion of different filiters and GLAS05

4. CONCLUSIONS

The paper evaluated the influence of popular filtering methods on satellite full waveform decomposition. Besides Gaussian filter, the other three filters are all effective denoising methods for full waveform decomposition. And each method has its advantage over others, for example, smoothing parameters in Gaussian are relatively easy to set, while others introduce scale and thresholds coefficients, etc. The simulation experiments turn out that Taubin filter can acquire higher time accuracy, and EMD filter can get higher consistency ratio in waveform decomposition. In ICESat/GLAS waveform data tests, the GLAS05 validation results also show that Taubin filter has better performance. And the fit deviations all excel the results of GLAS05.

ACKNOWLEDGEMENTS

This work was supported by the National Natural Science Foundation of China under Grant No. 41822106 and 41571407, National High Resolution Ground Observation System of China (GF-7) under Grant No. 11-Y20A12-9001-17/18, Science and Technology Innovation Action Plan Program of Shanghai under Grant No. 18511102100, Dawn Scholar of Shanghai under Grant No. 18SG22, and the Fundamental Research Funds for the Central Universities of China.

REFERENCES

[1] B. E. Schutz, H. J. Zwally, C. A. Shuman, D. Hancock, and J. P. DiMarzio, "Overview of the ICESat Mission," Geophysical Research Letters, 32(21) (2005)

[2] A. S. Gardner et al., "A reconciled estimate of glacier contributions to sea level rise: 2003 to 2009," Science, 340(6134): 852-857 (2013)

[3] H. D. Pritchard, S. R. M. Ligtenberg, H. A. Fricker, D. G. Vaughan, M. R. V. D. Broeke, and L. Padman, "Antarctic ice-sheet loss driven by basal melting of ice shelves," Nature, 484 (7395): 502 (2012)

[4] C. C. Carabajal and D. J. Harding, "ICESat Validation of SRTM C-Band Digital Elevation Models," Agu Fall Meeting, (2004)

[5] W. Abdalati et al., "The ICESat-2 Laser Altimetry Mission," Proceedings of the IEEE, 98(5): 735-751 (2010)

[6] G. Li et al., "improve the zy-3 height accuracy using icesat/glas laser altimeter data," International Archives of the Photogrammetry Remote Sensing & S, XLI-B1: 37-42 (2016)

[7] A. C. Brenner et al., "The Algorithm Theoretical Basis Document for the Derivation of Range and Range Distributions from Laser Pulse Waveform Analysis for Surface Elevations, Roughness, Slope, and Vegetation Heights," (2012)

[8] V. H. Duong, "Processing and Application of ICESat Large Footprint Full Waveform Laser Range Data," None (EN), Delft University of Technology, (2010)

[9] H. Zhou. "Laser footprints information inversion and positioning of satellite laser altimeter (dissertation)". (in Chinese). Wuhan: Wuhan University, (2007)

[10] G. Taubin, "A Signal Processing Approach To Fair Surface Design," in Conference on Computer Graphics & Interactive Techniques, (1995)

[11] X. D. Lai, et al. Iterative decomposition method for small foot-print LiDAR waveform. Journal of Infrared & Millimeter Waves 32.4: 319-324 (2013)

[12] C. E. Parrish, "Exploiting full-waveform lidar data and multiresolution wavelet analysis for vertical object detection and recognition," in IEEE International Geoscience & Remote Sensing Symposium, (2008)

[13] N. E. Huang, et al. "The empirical mode decomposition and the Hilbert spectrum for nonlinear and non-stationary time series analysis." Proceedings A 454.1971: 903-995: (1998)

[14] D. Yong. "Research on modern methods for landslide deformation analysis and prediction (dissertation)". (in Chinese) Wuhan: Wuhan University, (2008)

ASSESSING THE LONG TERM STABILITY OF THE DEPOLARIZATION CONSTANT FOR THE CLUJ STATION

H. Stefanie[1*], N. Ajtai[1], A. Radovici[1], D. Ene[2], L. Belegante[2]

[1] *Faculty of Environmental Science and Engineering, Babeş-Bolyai University, Cluj-Napoca, Romania,*

[2] *National Institute for Research and Development in Optoelectronics INOE 2000 Magurele, Romania*

**Email: horatiu.stefanie@ubbcluj.ro*

ABSTRACT

The study aims to assess the additional systematic error caused by the assessment of the depolarization constant. Based on the long time assessment of this constant, a mean value and a standard deviation can be retrieved. A sensitivity study could assess the deviation of the depolarization value in high and low depolarizing atmospheric layers. Based on these deviations, the systematic error of the calibration is retrieved and included in the final depolarization products.

1. INTRODUCTION

At the present state, the LIDAR systems have different configurations and geometries, according to their applications. The multi-wavelength Raman and depolarization LIDAR systems used in the European Aerosol Research Lidar Network – EARLINET [2] are suitable to provide an extended set of optical parameters for aerosol characterization which can be used in climatology studies, models and different events like Saharan dust outbreaks, transport of smoke plumes or volcanic ash [3, 4, 6].

The study aims to assess the long term stability of the depolarization constant for the Cluj lidar station and retrieve the additional systematic error from the assessment of this constant [1, 5]. For this study, daily depolarization calibrations were performed over an extended time period.

2. METHODOLOGY

2.1 The lidar system

The Cluj-Napoca lidar (CLOP) emission system is based on a Nd:YAG laser (Continuum INLITE II -30) with a repetition rate of 30 Hz, equipped with second and third harmonics. The laser beams at 1064 nm, 532 nm and 355 nm are simultaneously emitted into atmosphere. Currently no beam expander is mounted on the instrument's emission unit.

The backscattered radiation is collected by a Raymetrics D300 Cassegrain type telescope with a focal length of 1500 mm. The signal detection unit has 4 detection channels for the elastically backscattered radiation at 1064, 532 (cross and parallel) and 355 nm and 2 detection channels for the Raman radiation backscattered from the N2 molecules at 607 and 386 nm.

2.2. Calibration of the depolarization channels

The calibration of the depolarization channels is performed using a mechanical rotator in front of the wavelength separation module. The calibration consists in performing two consecutive Δ90° according to Freudenthaler (2016).

3. RESULTS AND DISCUSSIONS

Figure 1 shows the variation of the depolarization constant over a period of more than one month. The data shows large deviations during the first calibration sets, a 20% deviation over the next 20 days and a decreased deviation for the remaining 8 days (<10%). The mean value for the depolarization constant is 0.4 with an overall deviation of 0.1 (25%).

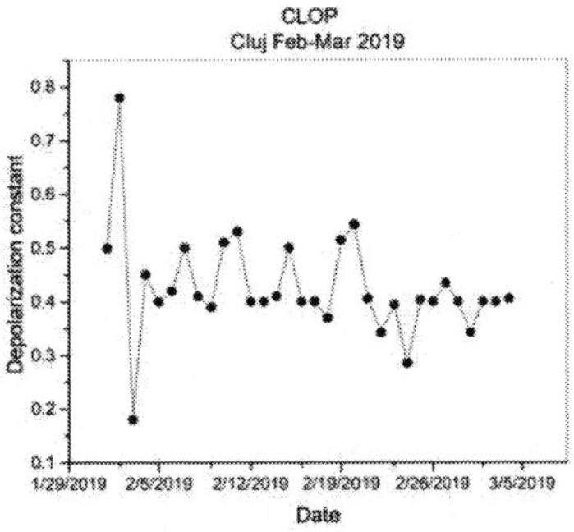

Figure 1 Depolarization constant values over extended time period

Different relative deviations over different time periods could indicate either external factors like instrument alignments or could simply indicate changes to the measurement setup (signal intensity or integration times).

Further on for the study, a mean value of 0.4 for the depolarization constant and a 0.1 error were considered.

Fig 2 a, b shows the error of the depolarization profile for 24.02.2019. The sensitivity study shows that the error of the high depolarizing layers is more affected by the accuracy of the retrieval than the error of the low depolarizing regions.

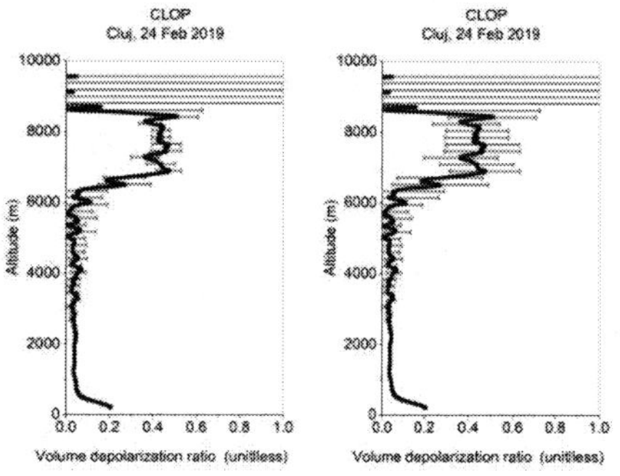

a b

Figure 1 Depolarization profile for 24 of February 2019
a) systematic retrieval error not included
b) systematic retrieval error included

3. CONCLUSIONS

The retrieval of the depolarization calibration constant is an important procedure since the accuracy of the constant can have a great influence on the depolarization value of the high depolarizing layers (either linear volume or particle depolarization products).

The time series of the depolarization calibration constant could be used as a stability indicator. Changes to the lidar instrument could easily be seen in the time series.

ACKNOWLEDGEMENTS

The research for this paper was financially supported by the ROSA STAR C3 contract no. 152/20.07.2017 - Satellite products validation using ground-based LIDAR stations - STRATUS.

The research for this paper was financially supported by the ROSA STAR C3 contract no. 183/05.09.2017 - Intelligent Software for Aerosol typing Based on Earth observation products, remote sensing measurements and transport modelling – ISABEL.

REFERENCES

[1] Bravo-Aranda, J. A., Belegante, L., Freudenthaler, V., Alados-Arboledas, L., Nicolae, D., Granados-Muñoz, et al.,: Assessment of lidar depolarization uncertainty by means of a polarimetric lidar simulator, Atmos. Meas. Tech., 9, 4935–4953, https://doi.org/10.5194/amt-9-4935-2016, 2016.

[2] G. Pappalardo, A. Amodeo, A. Apituley, A. Comeron, V. Freudenthaler, H. Linné, A. Ansmann, J. Bösenberg, G. D'Amico, I. Mattis, L. Mona, U. Wandinger, V. Amiridis, L. Alados-Arboledas, D. Nicolae, M. Wiegner, EARLINET: towards an advanced sustainable 20 European aerosol lidar network, Atmos. Meas. Tech. Discuss., 7(3) 2929–2980, (2014).

[3] L. Mona, Z. Liu, D. Muller, Lidar Measurements for Desert Dust Characterization: An Overview, Advances in Meteorology, (2012), ID356265.

[4] D. Nicolae, A. Nemuc, D. Muller, C. Talianu, J. Vasilescu, L. Belegante, A. Kolgotin, Characterization of fresh and aged biomass burning events using multiwavelength Raman lidar and mass spectrometry, J. Geophys. Res. Atmos., 118, (2013)

[5] Freudenthaler, V.: About the effects of polarizing optics on lidar signals and the Δ90 calibration, Atmos. Meas. Tech., 9(9), 4181–4255, doi:10.5194/amt-9-4181-2016, 2016.

[6] G. Pappalardo, L. Mona, G. D'Amico, U. Wandinger, M. Adam, A. Amodeo, A. Ansmann, A. Apituley, L. Alados Arboledas, D. Balis, A. Boselli, J. A. Bravo-Aranda, A. Chaikovsky, A. Comeron, J. Cuesta, F. De Tomasi, V. Freudenthaler, M. Gausa, E. Giannakaki, H. Giehl, A. Giunta, I. Grigorov, S. Groß, M. Haeffelin, A. Hiebsch, M. Iarlori, D. Lange, H. Linné, F. Madonna, I. Mattis, R.-E. Mamouri, M.A.P. McAuliffe, V. Mitev, F. Molero, F. Navas-Guzman, D. Nicolae, A. Papayannis, M.R. Perrone, C. Pietras, G. Pietruczuk, Pisani, J. Preißler, M. Pujadas, V. Rizi, A. Ruth, J. Schmidt, F. Schnell, P. Seifert, I. Serikov, M. Sicard, V. Simeonov, N. Spinelli, K. Stebel, M. Tesche, T. Trickl, X. Wang, F. Wagner, M. Wiegner, K.M. Wilson, Four-dimensional distribution of the 2010 Eyjafjallajökull volcanic cloud over Europe observed by EARLINET, Atmos. Chem. Phys. Discuss., 12, 30203- 30257, (2012).

EPJ Web Conferences **237**, 08007 (2020) https://doi.org/10.1051/epjconf/202023708007
ILRC 29

AUTOMATIC LIDAR CALIBRATION AND PROCESSING PROGRAM FOR MULTIWAVELENGTH RAMAN POLARIZATION LIDAR

Zhenping Yin[1, 2, 3,*], Holger Baars[1], Patric Seifert[1], Ronny Engelmann[1]

[1]*Leibniz Institute for Tropospheric Research, Permoserstraße 15, 04318 Leipzig, Germany*
[2]*School of Electronic Information, Wuhan University, Wuhan, China*
[3]*Key Laboratory of Geospace Environment and Geodesy, Ministry of Education, Wuhan, China*
**Email: zhenping@tropos.de*

ABSTRACT

A new version of automatic lidar calibration and processing program was developed to process the data from multiwavelength Raman polarization lidar. The absolute lidar calibration and water vapor calibration algorithms were applied. The program can provide plentiful products, like aerosol backscatter and extinction coefficients, lidar ratio, Ångström exponent, volume and particle depolarization ratios, water vapor mixing ratio and aerosol target classification. Good agreement was found in the comparison with manual quality-assured profiles or radiosonde measurement. Lidar calibration based on the aerosol optical properties retrieved with Raman method, Klett method and AOD-Constrained method were implemented. Good consistency was found.

1. INTRODUCTION

Aerosol can influence the Earth radiation budget by absorption and scattering solar radiation, and modulate cloud formation and cloud microphysical properties. Therefore, obtaining the vertically resolved aerosol optical and microphysical properties is very important for climate modeling. Lidar, as a very efficient detection tool for aerosol measurement, can provide highly resolved vertical and temporal information. In order to observe the aerosol distribution in a large scale, several ground-based lidar networks were also established, e.g. EARLINET [1], the ceilometer network of the German Meteorological Service (DWD, [2]), MPLNET [3] and multiwavelength Raman polarization lidar network PollyNET [4]. However, more systems bring more data, which can easily exceed our limit of manual analysis capacity. Although there are some automatic algorithms and programs for lidar data processing, most of them are only suitable for single wavelength or elastic lidar systems ([5], [6]). In order to realize automatic analysis for multiwavelength Raman polarization lidar measurements, the first version of the automatic lidar processing program was developed to obtain aerosol backscatter and extinction coefficients, lidar ratio and volume/particle depolarization ratio [4]. As the emerging of water vapor calibration and aerosol target classification algorithms, we designed a more powerful version to include new products, like water vapor mixing ratio, relative humidity, lidar calibration constants and aerosol target classifications.

2. POLLYXT

Polly^{XT} is a multiwavelength Raman polarization lidar system [7]. It has 12 channels with 355 nm, 387 nm, 532 nm and 607 nm at near- and far-range channels, 355 nm and 532 nm at cross channels, 407 nm and 1064 nm at far-range channels. The incomplete overlap zone for far- and near-range channels have been suppressed to 500 and 120 m. The system is operated 24/7 with a raw resolution of 7.5 m and 30 s and only water vapor channel needs to be turned off at daytime.

3. METHODOLOGY

The flowchart of the new processing program is presented in Fig. 1. The entire structure is very straightforward and consists of well-defined modules and algorithms ([8], [9], [10], [11]). In the first step, the raw signal is pre-processed with deadtime correction, background correction and first bin adjustment. Then, the 355 and 532 nm cross channels are calibrated with ±45° calibration method [8], see Fig. 2c. This calibration process is implemented three times a day to ensure reliable results. After that, the profiles with cloud contamination are screened out based on the derivative of the range corrected signal at both near- and far-range channels, see

https://doi.org/10.1051/epjconf/202023708007

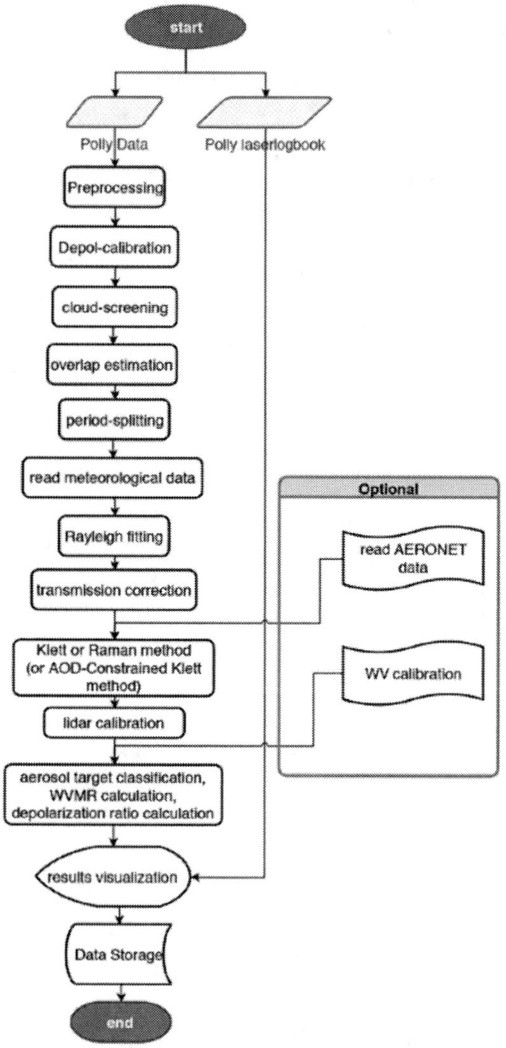

Figure 1. Flowchart of the PollyNET processing chain. (Optional parts mean the modules are dependent on the available AERONET sites nearby).

Fig. 2d. The remained cloud-free profiles are divided into several 1-hour continuous subgroups and integrated for further analysis. Then the Rayleigh fitting algorithm [11] is applied to search for the reference height (nearly aerosol-free region). After that, the Klett method [12], Raman method [13] and AOD-Constrained method [14] are implemented according to the SNR of Raman signal and the availability of sunphotometer measurements. The lidar ratios of 55 sr, 55 sr and 55 sr at 355, 532 and 1064 nm are used in Klett method. The elastic far-range channels are absolutely calibrated based on the retrieved aerosol profiles. The calibration range is determined with the smallest variability of the calibration results within a sliding window. After the lidar calibration, the aerosol classification algorithm is implemented to differentiate the aerosol types according to the size and shape of the particles [10]. This is the entire structure of the calibration and processing program. The source code is based on MATLAB and can be freely accessed via GitHub [15].

4. RESULTS

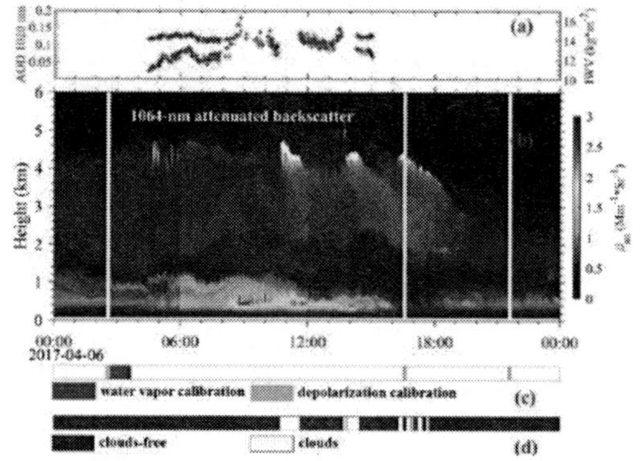

Figure 2. Dust case on 6 April 2017 at Limassol, Cyprus. (a) Level 1.5 data from AERONET (red: 1020 nm AOD, blue: IWV), (b) height-time plot of 1064 nm attenuated backscatter with Polly^{XT} and (c) calibration and cloud status for each profile.

4.1 Vertical profiles

In Fig. 3, we presented the aerosol vertical profiles from the automatic processing program and manual standard EARLINET analysis during the period of 00:00-00:59 UTC on 6 April 2017. The detailed aerosol evolution can be checked in Fig. 2b. For the comparison in Fig. 3, we can find very good agreement for the backscatter and extinction coefficients and volume/particle depolarization ratio.

4.2 Water vapor calibration results

The water vapor channel can be calibrated with the integrated water vapor (IWV) either from

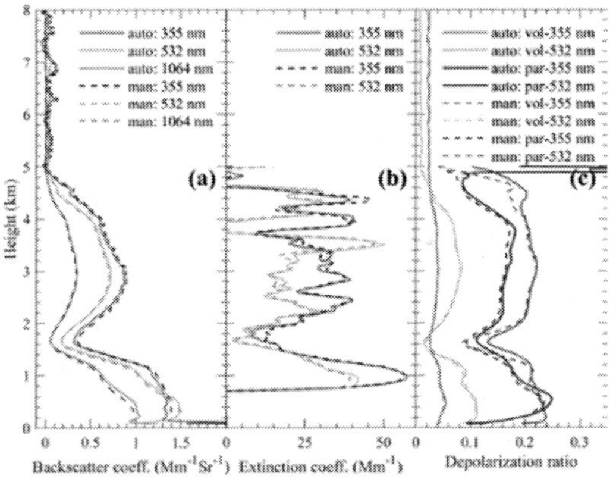

Figure 3. Comparison of manually (man) and automatically (auto) derived aerosol optical properties for 6 April 2017 in Limassol, Cyprus, averaged from 00:00 to 00:59 UTC. (a) aerosol backscatter coefficient at 355 nm, 532 nm and 1064 nm, (b) aerosol extinction coefficient at 355 nm and 532 nm and (c) volume and particle depolarization ratio at 355 nm and 532 nm.

collocated sunphotometer [9] or from microwave radiometer measurements [16]. For the case on 06 April 2017 at Limassol, Cyprus, we used the sunphotometer measurement at 04:30 UTC for the calibration and the water vapor calibration constant was 6.50 g kg^{-1}. This constant was used to calculate the water vapor mixing ratio with lidar at 02:41-03:40 UTC to compare with radiosonde measurement (see Fig. 4). We can find very good agreement between them.

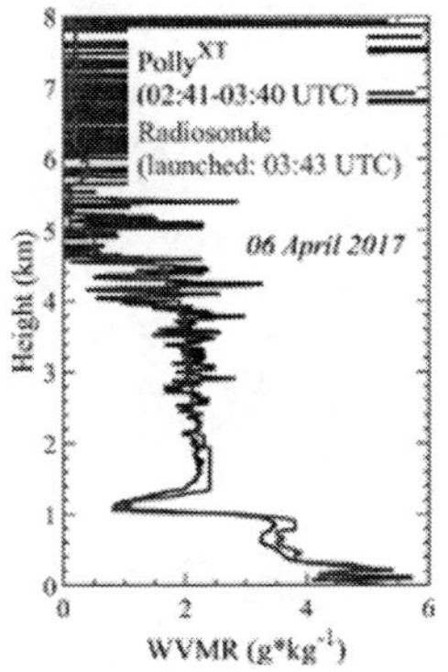

Figure 4. Comparison of water vapor mixing ratio profile with Polly^{XT} and radiosonde.

4.3 Lidar calibration results

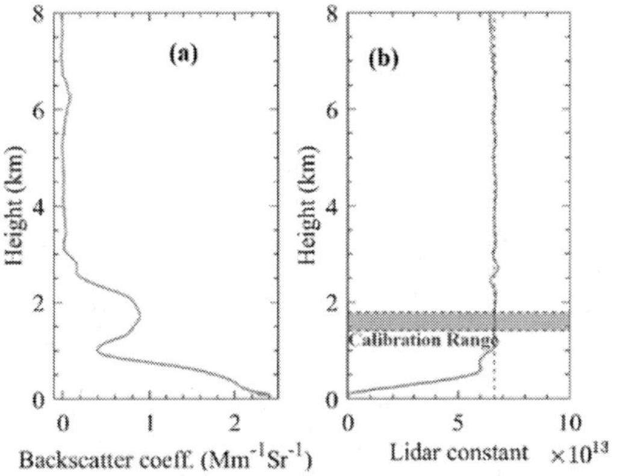

Figure 5. Lidar calibration results on 06 April 2017, at Limassol, Cyprus. (a) Aerosol backscatter coefficient at 532 nm with Raman method, averaged from 22:41 to 23:59 UTC, (b) lidar calibration constant profile.

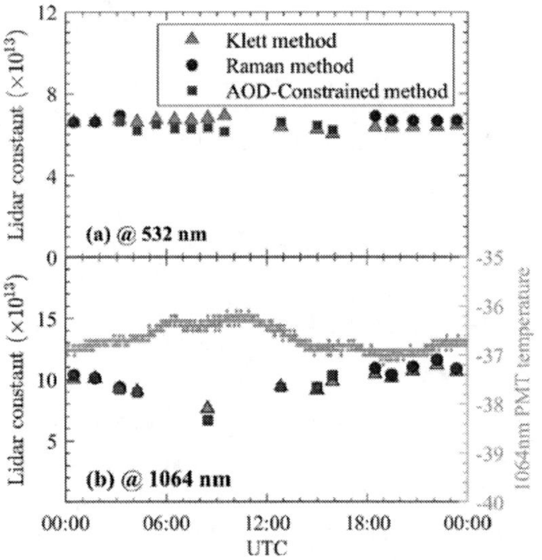

Figure 6. Time series of lidar constants at 532 nm (a) and 1064 nm (b) with Klett method, Raman method and AOD-Constrained method on 06 April 2017 at Limassol, Cyprus.

According to single-scattering elastic lidar equation [12], the lidar system can be absolutely calibrated with accurately obtained aerosol backscatter and extinction profiles [17]. Normally, the extinction coefficient profile with Raman method is very noisy and can hardly be used to calibrate the lidar system in an automatic way. Instead, we used the backscatter coefficient multiplied with a fixed lidar ratio of 55 sr. With the constrain of AOD to be less than 0.3, this simplification introduces an error less than 10 % for the lidar calibration under the conditions of clean and polluted continental aerosol, dust and smoke [10]. At the height with incomplete overlap, the backscatter coefficient with Klett method or AOD-Constrained method is considered as height-independent. This leads to an error less than 10 % for Polly^{XT} lidar and can be further reduced with overlap correction. In Fig. 5, we showed the lidar calibration results at 532 nm on 06 April 2017 22:41-23:59 at Limassol, Cyprus. The backscatter coefficient profile in Fig. 5a was retrieved with Raman method. The lidar constant profile was displayed in Fig. 5b and the calibration program calculated the lidar constant within the most stable region, which located between 1.4 and 1.8 km in this case. In Fig. 6, the whole day lidar calibration results on 06 April 2017 were presented. In Fig. 6a, we can see the calibration results with different retrieving methods are very stable and in good agreement. In Fig. 6b, we can see some variations at daytime. And the variations could be explained by the change of 1064 nm PMT temperature at daytime.

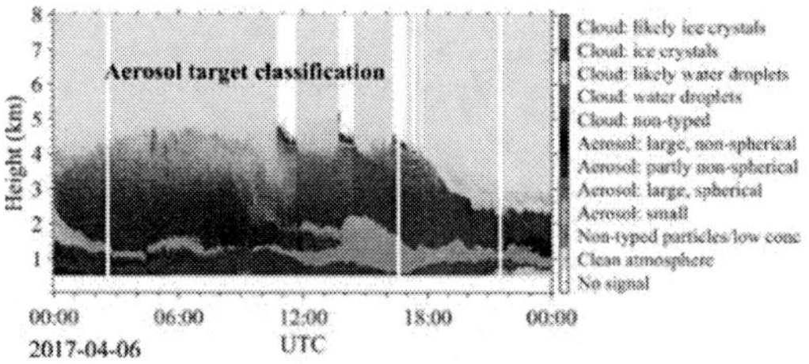

Figure 7. Aerosol target classifications on 6 April 2017 for Limassol, Cyprus.

4.4 Aerosol target classification

With the lidar calibration constants at 532 nm and 1064 nm, the aerosol target classification algorithm can be implemented to obtain the aerosol morphology information [10].

In Fig. 7, the aerosol classification result on 6 April 2017 at Limassol, Cyprus was displayed. Partly non spherical and large non-spherical aerosol were dominant in the whole day. These categories normally stand for mixed dust or dust particles. This finding was corroborated by the backward trajectories, which showed that partly of the trajectories can be traced to European continent and partly can be traced to Saharan desert (not shown here).

ACKNOWLEDGEMENTS

This work was funded by the European Union's Horizon 2020 research and innovation programme under grant agreement nos. 654109 (ACTRIS-2) and 763643 (EXCELSIOR). The authors appreciated the support from Cyprus University of Technology and Remote Sensing and Geo-Environment Lab (ERATOSTHENES Research Centre) during CyCARE. Zhenping Yin thanks the support from the Chinese Scholarship Council (CSC) to conduct this research under the CSC no. 201706270117.

REFERENCES

[1] G. Pappalardo, et al. Atmospheric Measurement Techniques 8: 2389-2409 (2014)

[2] H. Flentje, et al. Atmospheric Measurement Techniques Discussions 3: 3643-3673 (2010)

[3] E.J. Welton, et al. Proc. SPIE, 4153, 151-158 (2001)

[4] H. Baars, et al. Atmospheric Measurement Techniques 16.8: 5111-5137 (2016)

[5] Y. Morille, et al. Journal of Atmospheric and Oceanic Technology 24.5: 761-775 (2007)

[6] A. Omar, et al. Journal of Atmospheric and Oceanic Technology 26.10: 1994-2014 (2009)

[7] R. Engelmann, et al. Atmospheric Measurement Techniques 9.4: 1767-1784 (2016)

[8] V. Freudenthaler, et al. Tellus B 61.1: 165-179 (2009)

[9] G. Dai, et al. Atmospheric Measurement Techniques 11.5: 2735-2748 (2018)

[10] H. Baars, et al. Atmospheric Measurement Techniques 10.9: 3175 (2017)

[11] V. Freudenthaler, Lidar Rayleigh-fit criteria, in: EARLINET-ASOS 7th Workshop, available at: https://epub.ub.uni-muenchen.de/12970/ (2009)

[12] J.D. Klett, Applied Optics 20.2: 211-220 (1981)

[13] A. Ansmann, et al. Applied Optics 31.33: 7113-7131 (1992)

[14] B. Heese, et al. Atmospheric Measurement Techniques 3.6: 1763-1770 (2010)

[15] https://github.com/PollyNET/Pollynet_Processing_Chain

[16] A. Foth, et al. Atmospheric Chemistry and Physics 15.14: 7753-7763 (2015)

[17] Y. Wu, et al. Proc. Of SPIE Vol.7479 (2009)

EPJ Web Conferences **237**, 08008 (2020) https://doi.org/10.1051/epjconf/202023708008
ILRC 29

CALCULATIONS OF LIGHT SCATTERING MATRIX OF DUST AEROSOLS FOR PROBLEMS OF LIDAR REMOTE SENSING

Natalia Kustova*[,1], Alexander Konoshonkin[1], Anatoli Borovoi[1], Zhenzhu Wang[2], Dong Liu[2], Chenbo Xie[2], Alexandra Tsekeri[3], Josef Gasteiger[4]

[1]*V.E. Zuev Institute of Atmospheric Optics SB RAS, Tomsk 634055, Russia*
[2]*Anhui Institute of Optics and Fine Mechanics, Chinese Academy of Sciences, Hefei 230031, China*
[3]*Institute for Astronomy Astrophysics Space Applications and Remote Sensing, Athens 15236, Greece*
[4]*Faculty of Physics, University of Vienna, Vienna 1010, Austria*
**Email: kustova@iao.ru*

ABSTRACT

The light scattering matrix is calculated for large dust particles with irregular shape and refractive index of 1.3116+*i*0.0. The scattering matrix in the backward direction needed for lidar studies is separately discussed. In this case, the obtained results for the lidar and depolarization ratios are in good agreement with experimental data. It is shown that for randomly oriented particles the number of particle orientations needed for numerical calculations by exact methods like DDA becomes a crucial parameter. In particular, for particles with size parameters larger than 40 the number of orientations should be more than 1000.

1. INTRODUCTION

Dust aerosols are of great importance for the Earth radiative balance. The optical characteristics of dust particles with realistic, irregular shapes can be calculated with good accuracy by exact methods like the finite-difference time-domain FDTD [1], the discrete dipole approximation (DDA) [2,3], and so on. However in literature there are rather few data for the scattering properties of irregular dust particles with size parameters larger than 20 [4,5]. The aim of this contribution is to fill up this gap using the physical-optics method (PO) [6]. Although the scattering properties shown herein do not represent the scattering properties of realistic dust particles considering the low refractive index, this is a first step towards the application of PO for dust particles.

2. NUMERICAL RESULTS

The optical properties of a particle at a given wavelength λ depend on its microphysical characteristics such as shape, size, and the refractive index m. Size of nonspherical particles can be determined in different ways. We define the size parameter x at a given wavelength λ for a convex randomly oriented particle as

$$x = 2\pi r / \lambda,\ \ r = \sqrt{C_{geo} / \pi}\ , \qquad (1)$$

where r is the radius of a sphere with equal geometric cross section and C_{geo} is the orientation-averaged geometrical cross-sectional area.

For numerical solutions of light scattering by randomly oriented particles of complex shapes, it is important to choose the necessary number of orientations [7]. Usually a criterion of the necessary number of orientations is some concurrence of the calculated and theoretical data for the extinction cross-section. Of course, such a criterion is not applicable to the lidar investigations where only the backscatter cross-section is measured. In this contribution, we state that the number of necessary orientations for the backscatter should be much more than for the concurrence of the extinction cross-section.

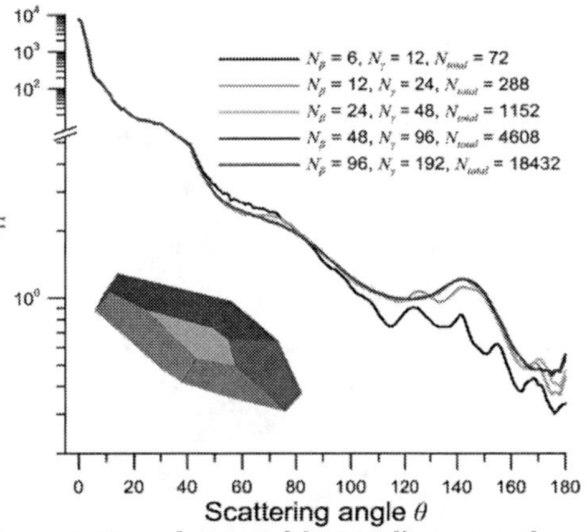

Figure 1. First element of the Mueller matrix for a randomly oriented dust particle calculated with the physical-optics approximation for different number of particle orientations.

Figure 1 supports this statement. Here the first element of the Mueller matrix is presented for a typical shape of dust particles at $x = 44.21$, $\lambda = 0.532$ μm, and $m = 1.3116+i0.0$.

Figure 1 shows that the numerical data obtained for different orientation numbers oscillate more strongly in the backward scattering direction semi-sphere than in the forward counterpart. In particular, the numerical data converge more or less satisfactorily at 18432 orientations in the back hemisphere and at 72 orientations in the forward hemisphere. Since the forward scattering direction determines the extinction cross section according to the optical theorem, we guess that 72 orientations would be enough for the corresponding convergence of the extinction cross-section while this is not applicable to the backscatter. The same convergence is illustrated in Fig. 2 for other elements of the Mueller matrix.

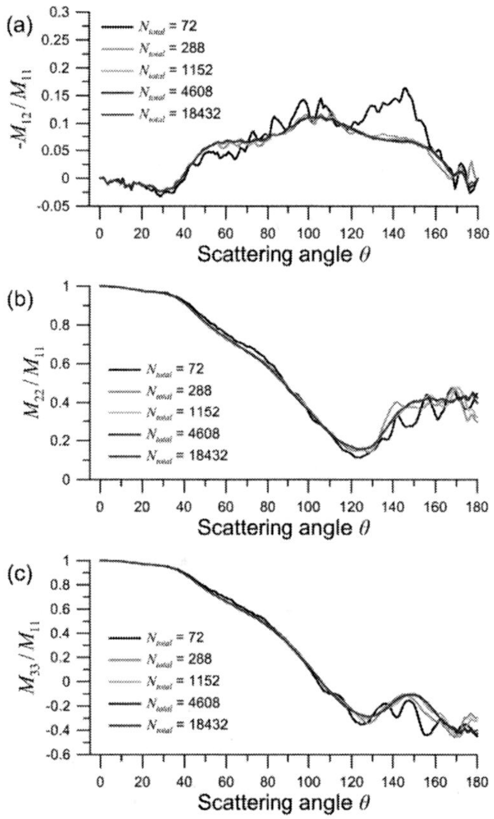

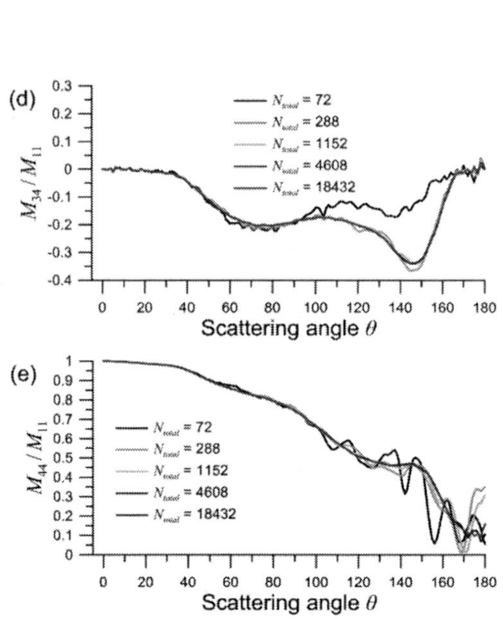

Figure 2. The same as in Fig. 1 for other elements of the Mueller matrix

This conclusion is easily explained within the framework of the physical-optics approximation as following. Let us determine a particle orientation by three Euler angles α, β, and γ, where α describes particle rotation around the light incident direction, γ describes rotation about an axis chosen inside a particle and β corresponds to a tilt between the incident direction and the chosen axis. It is instructive to exclude the averaging over the angle α and to represent the Mueller matrix averaged only over the angles β and γ. In this case, unlike Figs. 1 and 2, the Mueller matrix is represented as function of two variables: azimuth φ and zenith θ scattering angles. Such two-dimensional functions are shown in Figs. 3 and 4 for different numbers of particle orientations.

Let us note that the averaging over angle α in our numerical algorithm is worked out analytically and this averaging is not included in the number of orientations. Figures 3 and 4 demonstrate that the interrupted view of the two-dimensional functions is created by the discrete sets of the orientations used at the numerical averaging. These interrupted views of the functions prove that the number of orientations was not sufficient for the reliable averaging over the random orientations. And only at larger number of orientations the Mueller matrixes become smooth functions demonstrating that the orientation number was more or less satisfactory.

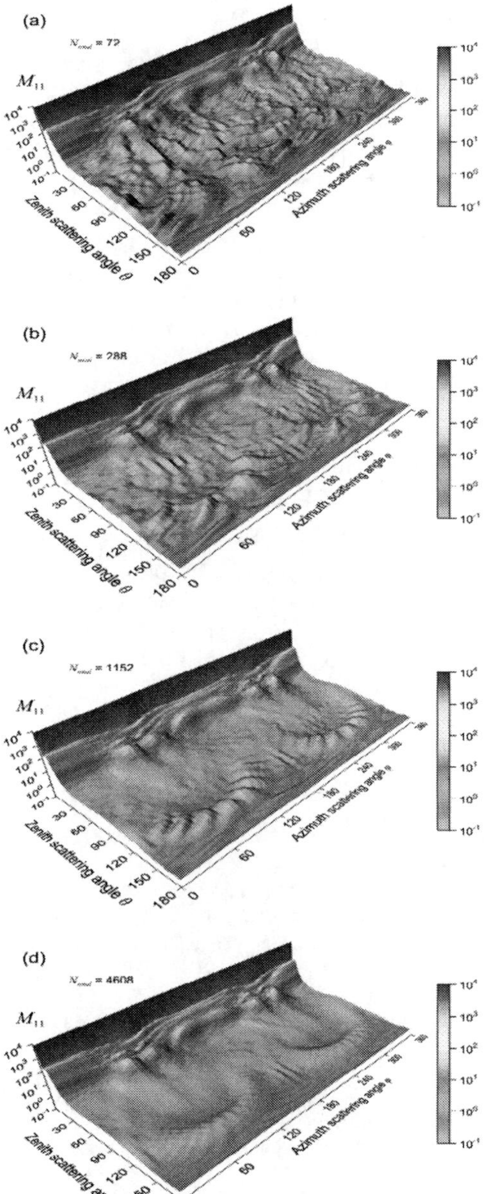

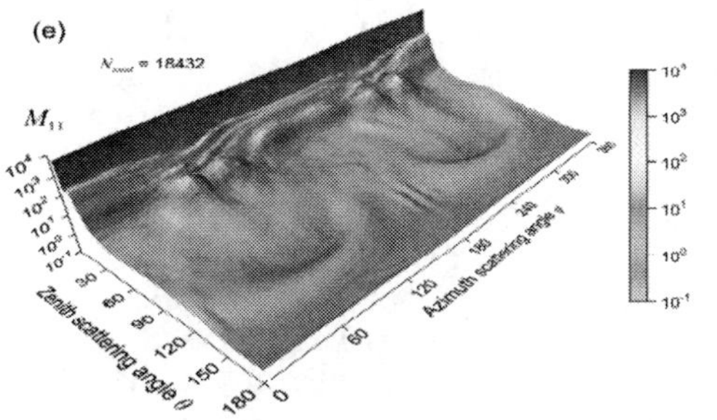

Figure 3. The first element of the Mueller matrix at the partial averaging over particle orientations for different numbers of orientation.

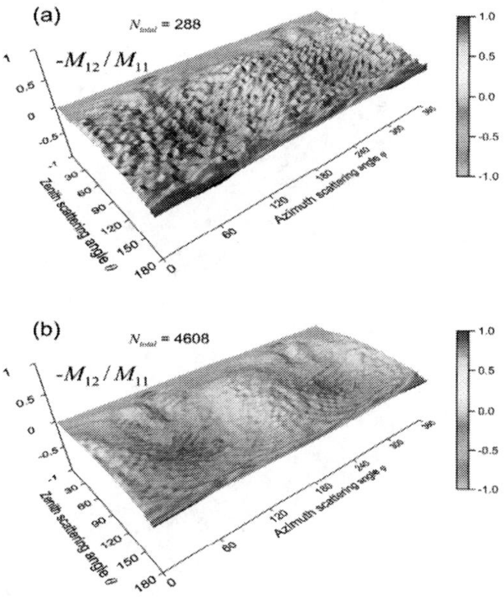

Figure 4. The same as in Fig. 3 for the element M_{12}

In the forward scattering direction, the diffracted field that is weakly changing with particle orientations becomes predominant. Similarly, the light beams scattered near the forward direction undergo a little number of reflections inside the faceted particles.

Therefore the scattering to the forward hemisphere is less sensitive to the orientation numbers as compared with the backward hemisphere.

The simple estimation of the necessary orientation numbers can be described by the equation

$$N_{total} = N_\beta \cdot N_\gamma \cdot N_\alpha = \frac{\pi}{0.3\xi} \cdot \frac{2\pi}{0.3\xi} \cdot \frac{2\pi}{0.3\xi}, \quad (2)$$

where $\xi = \lambda/a$ is the diffraction angle, a – the maximum size of a side of a faceted particle. In Eq. (2), the rotation angle of a particle is taken as 1/3 of the diffraction angle. This is the upper estimate. In practice, anybody may decrease the estimate in 2-3 times. Such an estimate corresponds to our calculations presented in Fig. 1. Indeed, the convergence should be at 167·335=55945 orientations according to Eq. (2) while the convergence takes place already at 18432 orientations.

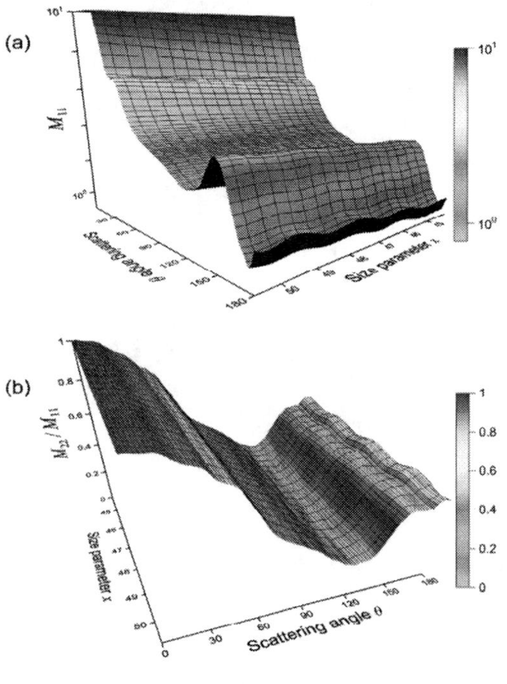

Figure 5. Interference oscillations for the backscattering Mueller matrix

It is worthwhile to note that a change of the particle size leads to interference oscillations at the backward scattering direction which are important in lidar studies. Such oscillations are seen both for M_{11} and M_{22} elements (see Fig. 5). Here the maximum quantity is transformed to the minimum quantity if a change of the size parameter is equal to one. Consequently, correct calculations of the backscattering Mueller matrix should be done with the mesh less than one for the size parameter.

3. CONCLUSION

Thus, in calculations of the light scattering matrixes for large particles with irregular shape resembling dust particles, but having low refractive index of 1.3+i0 (size parameter more than 40) it is necessary to take more than 1000 particle orientations.

Number of the necessary orientations increases essentially with particle size. Moreover, the step of calculations relative to the size parameter should be less than one. This circumstance also increases the calculation cost. Such facts are usually underestimated in the literature. As a result, the exact methods like DDA should meet a lot of difficulties while our physical-optics approximation looks as a promising method for solving this problem.

ACKNOWLEDGEMENTS

This work is supported by the RFBR under Grants no. 18-05-00568, no. 18-55-53046, and budget project AAAA-A17-117021310145-6. Calculation of Mueller matrixes is supported only by Russian Science Foundation (18-77-10035).

REFERENCES

[1] H. Ishimoto, Y. Zaizen, A. Uchiyama, K. Masuda, Y. Mano, J. Quant. Spectrosc. Radiat. Transfer 111, 2434-2443 (2010)
[2]T. Mehri, O. Kemppinen, G. David, H. Lindqvist, J. Tyynelä, T. Nousiainen, P. Rairoux, A. Miffre, J. Atmos. Res. 203, 44-61 (2018)
[3] M. A. Yurkin, V. P. Maltsev, A. G. Hoekstra, J. Quant. Spectrosc. Radiat. Transfer 106, 546-557 (2007)
[4] J. Gasteiger, M. Wiegner, S. Groß, V. Freudenthaler, C. Toledano, M. Tesche, K. Kandler, Tellus B: Chemical and Physical Meteorology 63(4), 725-741 (2011)
[5] Y. Grynko, Y. Shkuratov, J. Förstner Optics Letters 41, 3491-3494 (2016).
[6]A. Borovoi, A. Konoshonkin, N. Kustova, J. Quant. Spectrosc. Radiat. Transfer 146, 181-189 (2014)
[7] A. V. Konoshonkin, A. G. Borovoi, N. V. Kustova, H. Okamoto, H. Ishimoto, Y. Grynko, J. Förstner, J. Quant. Spectrosc. Radiat. Transfer 195, 132-140 (2017)

EPJ Web Conferences **237**, 08009 (2020) https://doi.org/10.1051/epjconf/202023708009
ILRC 29

FIELD PERFORMANCE OF ALL-FIBER PULSED COHERENT DOPPLER LIDAR

Heng Liu[1, 2], Xiaopeng Zhu[*1], Chunhui Fan[3], Decang Bi[1], Jiqiao Liu[*1], Xin Zhang[1], Xiaolei Zhu[1], Weibiao Chen[1]

[1] *Key Laboratory of Space Laser Communication and Detection Technology, Shanghai Institute of Optics and Fine Mechanics, Chinese Academy of Science, Shanghai 201800, China*
[2] *Center of Materials Science and Optoelectronics Engineering, University of Chinese Academy of Sciences, Beijing 100049, China*
[3] *College of Meteorology and Oceanography, National university of defense technology, Nanjing 211101, China*
**Email: xp_zhu@siom.ac.cn, liujiqiao@siom.ac.cn*

ABSTRACT

Coherent Doppler lidar (CDL) has been a powerful instrument to measure atmospheric wind velocity. In this work, an 1.5 μm all-fiber pulsed CDL has been developed and deployed to measure wind profiles in campaign experiment in 2018. The CDL has 0.1-5 km detection range with range resolution of 30 m, temporal resolution of 16 s at Velocity-Azimuth-Display (VAD) mode. Field experiments were implemented and the wind vector profiles were retrieved. As compared with sounding balloon, the discrepancy of wind speed and direction are nearly 0.7 m/s and 5 degrees, respectively. It shows that the CDL is a powerful tool for wind speeds measurement in lower atmospheric troposphere.

1. INTRODUCTION

1.5 μm all-fiber pulsed CDL has attracted much attentions due to its eye safety, compact size, flexible deployment and mature fiber components technology from telecommunication industry.[1] Mitsubishi Electric Corporation has developed an airborne 1.5 μm pulsed CDL which can detect as far as 9.3 km with changeable range resolution of 30, 75 and 150 m.[2] Leosphere displays a long range lidar (Windcube 200S) which can measure 3D wind profiles up to 7 km with 70 m range resolution, 0.5 m/s velocity resolution and 1 s time resolution.[3] Prasad et al. have presented a 1.54 μm all-fiber pulsed CDL which can measure wind velocity greater than 120 m/s over ranges greater than 10 km and its range resolution is less than 15 m.[4] Wu et al. have designed a 1.5 μm pulsed CDL with high updating rate of 4 Hz and variable physical spatial resolution from 15-60 m.[5] In this study, an all-fiber single mode laser with pulse energy of 300 μJ, pulse width of 400 ns and repetition rate of 10 kHz is used to measure wind velocity in range of 0.1-5 km. The CDL shows good stability and reliability in field campaign.

2. METHODOLOGY

2.1 The lidar system

The pulsed CDL for wind velocity measurement is operated by transmitting high energy narrow bandwidth laser beam into the atmosphere, which can acquire LOS velocity by analyzing the Doppler shift between objective backscattered signal and local oscillator. The key technologies are composed by: High-power single-frequency pulsed laser technology, optical antenna system, high sensitivity heterodyne detection technology, signal acquisition and real-time processing technology, as is shown in Fig. 1.

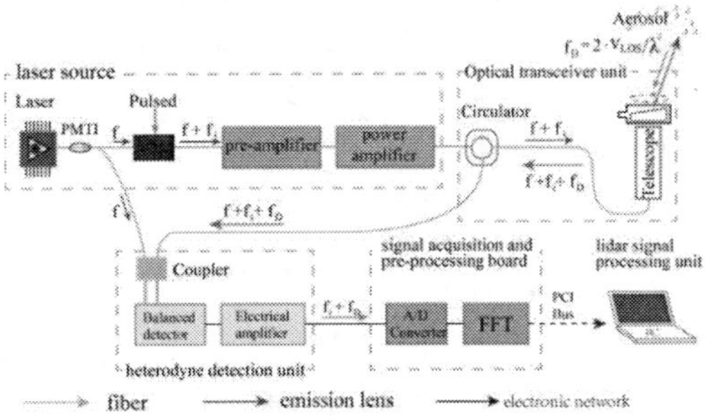

Figure 1. System diagram of the ground-based lidar.

The main system parameters are shown in Tab. 1.

Table 1. Main parameters of the pulsed coherent Doppler lidar system.

https://doi.org/10.1051/epjconf/202023708009

Component	Qualification	Specification
Transmitter	Operating wavelength	1540nm
	Pulse energy	300 μJ
	Pulse repetition rate	10 kHz
	Pulse width	400 ns
Transceiver	telescope diameter	100 mm
	scan mode	Conical
	zenith angle	20 °
	azimuth	8
	scanning time	16 s
Data-Acquisition	sampling frequency	1 GHz
	sampling length	40 μs
	range resolution	30 m

2.2 Comparison of CDL and sounding balloon

Sounding balloon detects wind speed through calculating its azimuth angle and elevation angle in unit time. The dataset from sounding balloon record 30 s average wind speed, updating time of 1 s, vertical range resolution of near 5 m. Horizontal space range between sounding balloon and CDL is near 50 m. The comparison scheme is given based on the quality of sounding balloon and CDL.

Table 1: comparison scheme between sounding balloon and CDL.

	Temporal resolution	Dataset updating time	Range resolution
Sounding balloon	30 s	1 s	5 m
CDL	16 s	2 s	30 m
Comparison	30 s	~1 s	30 m

3. RESULTS

3.1 Continuous measurement of CDL

On Aug. 10th ,2018, the comparison experiments are implemented in Jingbian city, Shanxi province, China, with local altitude of 1332 m. The ground-based CDL obtained a continuous 3D wind field dataset, as shown in Fig. 2.

There is a frontal surface (the interface between cold and warm wind) in range of 1500 to 2000 m. The continuous wind-profiles measurement displays the detailed characteristic wind zone which is meaningful in wind field monitoring.

Another point is that CDL signal shows the change of cloud base height from 22:00 to 23:00. The cloud base is higher than 5000 m (Altitude value) in that time. This feature captured by CDL can express its operational capacity below the boundary layer.

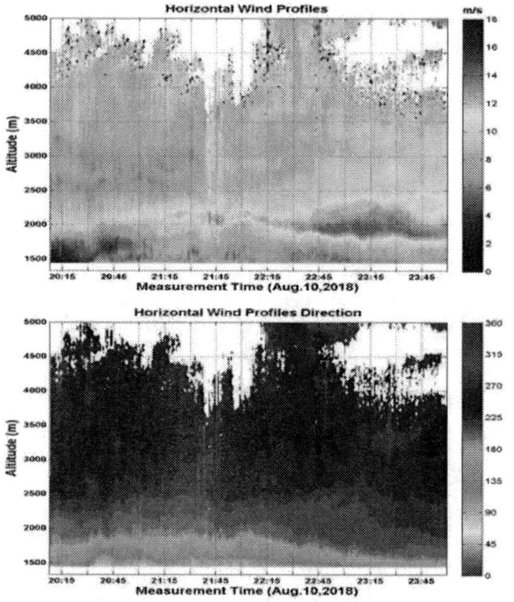

Figure 2: Horizontal wind profiles measured by CDL on Aug. 10th, 2018.

3.2 CDL and sounding balloon measurement

Measurements experiment of CDL and sounding balloon almost on the same time is implemented in August, 2018. Relative position of the two instruments are marked in Fig. 3. The horizontal space distance between CDL and sounding balloon on the ground is less than 50 m.

https://doi.org/10.1051/epjconf/202023708009

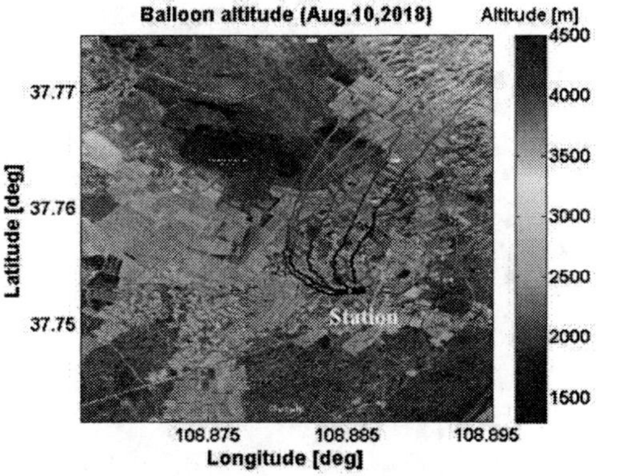

Figure 3: Relative position of sounding baloon and CDL. The color lines show tracks of the altitude of sounding baloon.

Samples of the comparison of wind speed are shown on the left side below. Wind speed of two instruments shows similar tendency. There are differences in 30 s average wind speed in range of 1~2 m/s, this discrepancy might become decrease after longer temporal average. On the right side, comparison of wind direction shows accordance in most situations.

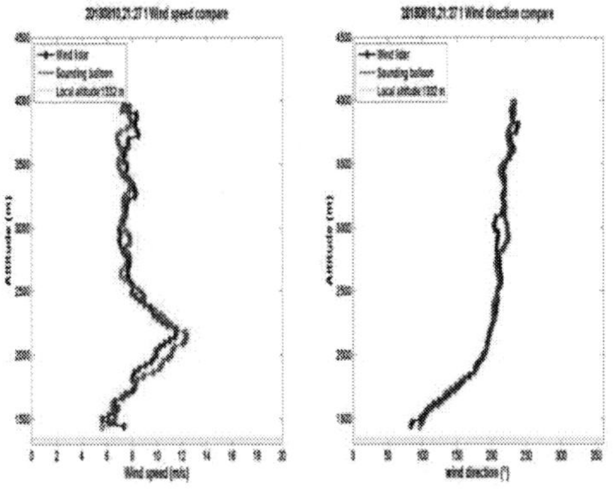

Figure 4: Comparison of wind speed and wind direction measured by CDL and sounding balloon.

Numerous analysis of the differences is shown on Fig. 5. Take the dataset on Aug. 10th, 2018 as an example, the discrepancy of wind speed and wind direction is less than 1 m/s and 5 degrees between CDL and sounding balloon, respectively. It shows good agreement of wind detection in two different methods, which prove the accuracy of CDL.

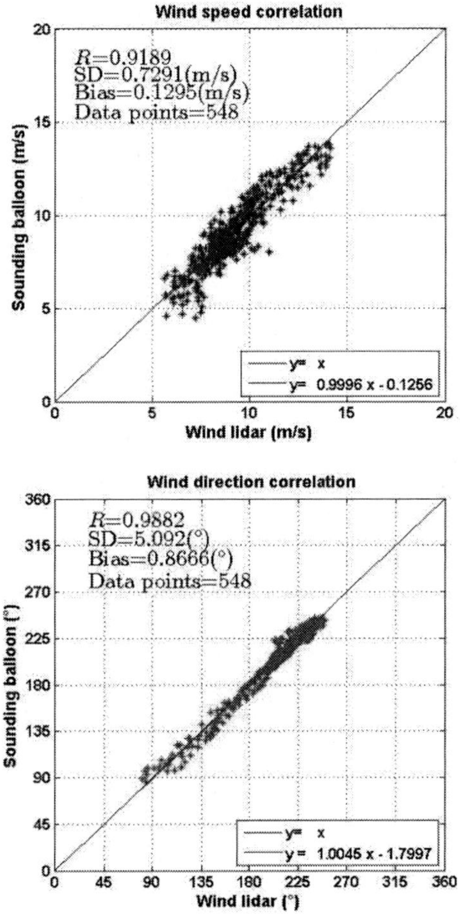

Figure 5: Linear correlation of wind speed and wind direction detected by CDL and sounding balloon.

The 1.5 μm all-fiber pulsed CDL is used to measure wind profiles in the campaign experiments in Aug. 2018. The practical detection altitude is more than 4 km in hazy weather. The comparison with sounding balloon shows good agreement. This 1.5 μm all-fiber pulsed CDL shows good performance in field experiments and potential application on ground-based or airborne platform in future.

ACKNOWLEDGEMENTS

This work was supported by the National Key R&D Program of China under Grant No. 2017YFF0104600.

REFERENCES

[1] Liu, J., et al. *All-Fiber Airborne Coherent Doppler Lidar to Measure Wind Profiles*. in *EPJ Web of Conferences*. 2016. EDP Sciences.

[2] Inokuchi, H., H. Tanaka, and T. Ando, *Development of an onboard doppler lidar for flight safety*. Journal of aircraft, 2009. **46**(4): p. 1411.

[3] Cariou, J. P., et al. *Long range scanning pulsed Coherent Lidar for real time wind monitoring in the Planetary Boundary Layer*. in *Proceedings of 16th Conference on Coherent Laser Radar*. 2011.

[4] Prasad, N.S., et al. *Innovative fiber-laser architecture-based compact wind lidar*. in *SPIE OPTO*. 2016. International Society for Optics and Photonics.

[5] Wu, S., et al., *Wind turbine wake visualization and characteristics analysis by Doppler lidar*. Optics express, 2016. **24**(10): p. A762-A780.

EPJ Web Conferences **237**, 08010 (2020) https://doi.org/10.1051/epjconf/202023708010
ILRC 29

IMPACTS OF SINKHOLES ON SALINITY LEVEL OF GROUNDWATER IN FINEGAYAN AREA, GUAM, USA

Yuming Wen[1,*], John Jenson[1]

[1] *Water and Environmental Research Institute, University of Guam, Mangilao, GU 96923, USA*
**Email: ywen@triton.uog.edu*

ABSTRACT

Guam is a strategic location for the United States of America, which is located in the western Pacific. Guam's economy is mainly supported by tourism and military related activities. Because of buildup activities and increasing population of tourists in the following few years, much pressure has been imposed on water resources and water quality in Guam. With about 5,000 marines relocated to Guam from Okinawa in the next few years, the Finegayan area will be a concerned area for buildup activities. Therefore, the study area for the project is located in the Finegayan area, and the research concentrates on application of LiDAR data to figure out sinkholes in the Finegayan area, and application of GIS in evaluating salinity levels from drinking wells, and locating wells with problems with salinity in the study area. More concerns are focused on relationship between sinkholes and salinity, and whether sinkholes affect salinity levels in the study area.

1. INTRODUCTION

Guam, an unincorporated U.S. territory in the western Pacific, is the largest (about 541.3 km^2) and southernmost island in the Mariana Islands chain. Ground water supplies about 90% of the drinking water for the island's about 170,000 residents and over 1.5 million visitors annually, and aims to welcome 2 million of visitors in 2020 [1][2]. However, there is a serious concern that the projected population increase will severely compromise the island's drinking water supplies and have an unprecedented impact on water quality and quantity. In Guam, water is pumped from over 121 wells, most of which are located in the northern Guam [3]. The high porosity and rapid recharge characteristics of the northern Guam lens aquifer make it especially susceptible to contamination from urban runoff, chemical spills, effluents from septic tanks and sewage overflows. Because of run-off and human induced activities, and unreliable drinking water delivery systems and management [4], the potential for contamination is high. The relocation of about 5000 marines, their dependents and foreign labors to support buildup activities for the next few years impose pressure on availability and quality of water resources. Among the concerns of water contaminants and water quantity, baseline study on determination of contaminants and water quantity in Guam's groundwater monitoring wells was identified as one of the highest priority research needs for Guam. Based on the above statement, groundwater quality and quantity is a main concern in Guam. Since around 5,000 marines will be relocated to Guam, particularly the Finegayan area in the next few years, it's very important to evaluate whether relocation of the marines and their dependents, and related activities will affect the quantity and quality of the water resources, especially groundwater resources. Based on the concerns mentioned above, it is crucial to conduct research on water quantity and quality, and other environmental and ecosystem concerns affected by buildup activities in the Finegayan area. Some recent research on impacts of human induced activities on water quality and historical evaluation of the factors contributing to changes of groundwater salinity in the Northern Guam Lens Aquifer are discussed by Wen [5] and Simard et. al [6] respectively. However, this is no profound research on relationship between sinkholes and salinity or impacts of sinkholes on salinity in Guam. This research mainly focuses on how to use LiDAR data to figure out sinkholes in the Finegayan area, and application of GIS in evaluating salinity levels from drinking/monitoring wells, and locating wells with problems with salinity in the study area. Emphasis is imposed on evaluation of relationship between sinkholes and salinity, and whether sinkholes affect salinity levels in the study area. The Government of Guam agencies, particularly Guam Waterworks Authority (GWA) and Guam Environmental Protection Agency (GEPA), and U.S. federal government agencies are interested in groundwater quality and quantity in Guam, including the Finegayan area. Findings from the project will help local and federal government agencies, and other interested parties in better understanding and management of the Finegayan Area, and therefore will facilitate buildup activities in the area.

2. METHODOLOGY

2.1 Study Area and Data Sources

Guam is a U.S. territory in the western Pacific. It is close to some major countries in the western Pacific region, including China, Australia, Japan, Korea, and the Philippines. It takes about 4 or 5 hours by air from Guam to some major cities such as Manila, Hong Kong, Shanghai, Seoul, and Tokyo in the region. The study area for this research is located in the northwestern part of Guam (Figure 1).

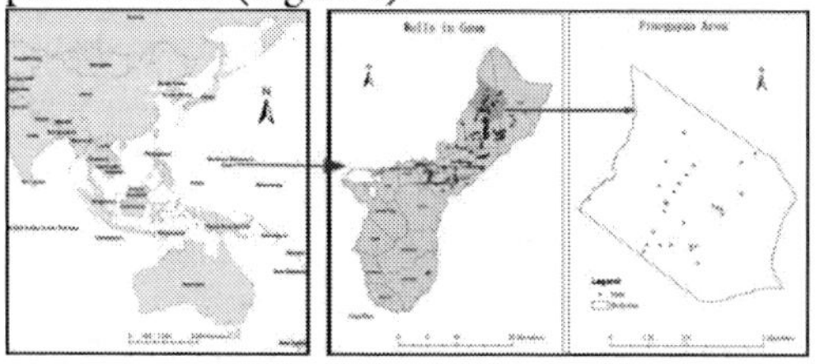

Figure 1. Location of Study Area

Available data for the research include boundary data for basins and Guam municipalities, bare earth data derived from LiDAR data of 2007, wells, chloride data collected quarterly from 2001 to 2009.

2.2 Methods

In order to evaluate the impacts of sinkholes on salinity of groundwater in the Finegayan area, salinity data and sinkholes need to be collected. GWA samples water from wells and tests the samples for contaminants including chlorides quarterly. Chloride data used for this research was collected between 2001 and 2009. Since the data of chlorides available from GWA is stored in the format of MS Excel spreadsheet, it needs to be geocoded and saved in a GIS format (In this case, it is saved in a shapefile) so that it can be further processed and analyzed in GIS. Spatial analysis is applied to locate wells with problems with high levels of salinity. As for sinkholes, LiDAR data of 2007 is employed. The bare earth data processed from the LiDAR point cloud data of 2007 is utilized to analyze the locations of sinkholes in the study area. GIS is the main tool applied to figure out wells with problems with salinity and sinkholes. Geospatial analysis and statistics are used to evaluate the relationship between salinity levels of groundwater from wells and locations of sinkholes.

3. RESULTS

3.1 Salinity Problems in the Finegayan Area

There are 27 wells considered in the Finegayan area, and data of chlorides for this research is available from GWA, collected from 2001 to 2009. Based on U.S. EPA's regulations, the suggested maximum contaminant level (MCL) for chloride, an indicator for salinity, is 250 mg/L. Based on spatial analysis of the chloride data, three wells are significant in salinity levels, three wells with a few occurrences with salinity levels over MCL and there are no salinity problems in all other wells. During the period of sampling, 23 samples for Well F-10, 21 samples for Well F-6 and 19 samples for Well F-13 out of 24 samples exceed the MCL of 250 mg/L. Those three wells with non-significant problems with salinity are Wells F-19, F-4 and F-12.

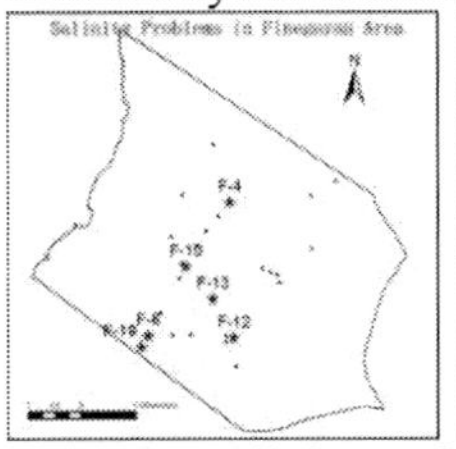

Well_ID	Frequency
F-10	23
F-6	21
F-13	19
F-19	6
F-4	4
F-12	1

Figure 2. Salinity Problems in Finegayan Area

3.2 Sinkholes in the Finegayan Area

The LiDAR data of 2007 includes two pieces of information, i.e. first return and last return data. The first return point cloud data is applied to achieve bare earth data, which can be employed to locate sinkholes and other information related to elevations. The spatial resolution of the digital elevation model (DEM) data for the bare earth derived from the LiDAR data is 1 meter by 1 meter. ArcGIS Spatial Analyst Tools such as hydrology and Map Algebra are utilized to obtain sinkholes in the study area (Figure 3). First of all, spatial analysis is applied to locate land depressions in the study area. The left map in Figure 3 indicates all land depressions based on the bare earth LiDAR data, the center map shows all land depressions with areas more than 10 square meters (m^2), and the left map manifests the preliminary result of delineated sinkholes with areas greater than 10 m^2, and average depths at least about one foot.

Figure 3. Sinkholes in Finegayan Area. Top: All land depressions; Center: Land depressions more than 10 m^2; Bottom: Preliminary Sinkholes.

3.3 Impacts of Sinkholes on Salinity

The sinkholes may affect salinity levels in different ways. This research discusses whether areas and numbers of sinkholes located in buffer zones of specific distances to wells affect salinity levels in the study area. Table 1 demonstrates the relationship between some characteristics of sinkholes and wells. Buffer 50 means using buffer distance of 50 meters to create buffer zones for concerned wells. Same explanation exists for Buffer 100 and Buffer 150. It indicates that there are no sinkholes within 50 meters to Wells F-10, F-13 and F-19. There are no sinkholes within 100 meters to Well F-13. It seems that sinkholes have impacts on salinity levels for Wells F-6, F-4 and F-12, though they may not affect Wells F-4 and F-12 significantly. Sinkholes can be located within 150 meters to all problematic wells with salinity.

	Buffer 50		Buffer 100		Buffer 150	
Well_ID	Area(m^2)	# of Sinkholes	Area(m^2)	# of Sinkholes	Area(m^2)	# of Sinkholes
F-10	0	0	2280	1	8600	2
F-6	5299	2	23058	2	34455	3
F-13	0	0	0	0	134	1
F-19	0	0	5014	1	18677	2
F-4	3103	2	5314	4	16374	6
F-12	3	1	3089	2	8600	3

Table 1. Relationship between Sinkholes and Wells with Salinity Problems

3.4 Discussions

Sinkholes have impacts on salinity levels to some degrees in the study area. More efforts will be invested to evaluate whether sizes, depths, distances to wells, fragmentation and distribution of sinkholes affect salinity levels. Other factors may also affect salinity levels. Those factors may include hydrological features such as slope and slope length, vegetation types, geological features, well pumping rates, human induced activities including ponding basins and other man-made structures, and climate factors such as El Niño and La Niña.

ACKNOWLEDGEMENTS

This project is partly funded by USGS 104b Program, and Guam Hydrological Survey.

REFERENCES

[1] Guam Visitors Bureau (GVB), Tourism 2020, Tumon, Guam. (2014)
[2] GVB, 2018 Annual Summary. (2019)
[3] Guam Waterworks Authority, 2017 Annual Water Quality Report, Upper Tumon, Guam. (2018)
[4] EPA Pacific Southwest/Region 9, EPA Progress Report 2006, San Francisco, California (2006)
[5] Y. Wen, Impacts of Human Activities on Groundwater Quality in Guam, Mariana Islands, International Journal of Environmental, Cultural, Economic and Social Sustainability, Vol. 7, Issue 5, pp 243 – 256 (2011)
[6] C. Simard, et al. Salinity in the Northern Guam Lens Aquifer, *Technical Report 143*, Water and Environmental Research Institute of the Western Pacific, University of Guam (2015)

https://doi.org/10.1051/epjconf/202023708011

LIGHT BACKSCATTERING BY ICE CRYSTALS OF CIRRUS CLOUDS WITHIN THE PHYSICAL OPTICS APPROXIMATION

Alexander Konoshonkin*[1], Anatoli Borovoi[1], Natalia Kustova[1], Zhenzhu Wang[2], Dong Liu[2], Chenbo Xie[2]

[1]*V.E. Zuev Institute of Atmospheric Optics SB RAS, Tomsk 634055, Russia*
[2]*Key Laboratory of Atmospheric Optics, Anhui Institute of Optics and Fine Mechanics, Chinese Academy of Sciences, Hefei 230031, China*
**Email: sasha_tvo@iao.ru*

ABSTRACT

Lidar technologies are widely used for retrieving microphysics of cirrus clouds, i.e. sizes, shapes and spatial orientation of ice crystals constituting the clouds. Interpretation of the lidar signals are based on the backscattered light. However, properties of the light backscattered by the ice crystals have not been well known and understood yet. The reason of this is that the problem of light scattering by the ice crystals much larger that incident wavelengths has not been solved satisfactorily yet because of great demands to computer resources. In this contribution we review the physical optics approximation as a prospective method to solve the light scattering problem on large nonspherical particles in lidar application.

1. INTRODUCTION

There are several recently developed numerical exact methods based on the Maxwell equations like the PSDT (pseudo-spectral time domain method) and II-TM (invariant imbedding T-matrix method) [1,2]. At present, these methods are capable to solve the problem numerically for rather small ice crystals while the larger crystals can only be calculated within the physical-optics approximation [3]. The scattering problem is characterized by the size parameter $x = \pi a/\lambda$, where a is the characteristic particle dimension and λ is the incident wavelength. The maximum size parameter achieved within the numerical exact methods is reported as large as 100 in [1] and 130 in [2] while in natural cirrus clouds this parameter runs up to several thousands.

In this contribution, a review of the backscattered data obtained by the physical-optics approximation is proposed. One of the discussed topics is existence of a sharp peak around the backward scattering direction in the phase function of cirrus clouds. Earlier, such a peak was predicted within the physical-optics approximation [3]. It was explained as the corner-reflection effect caused by dihedral angles of 90 degrees between some crystal facets. Later this peak was obtained for smaller crystals as well by means of the numerical exact methods [1,2]. Moreover, the peak took place for the roughed crystals and irregular crystal shapes. Nevertheless, this problem has not completely clarified yet. In this paper, the new data obtained within the physical-optics approximation are presented that clarify whether the backscattering peak appears in cirrus clouds.

Lidar community usually does not operate with the phase function. The microphysics is inferred from the backscatter ratios: the depolarization, color, and lidar (extinction-to-backscatter) ratios. These quantities are main sources of information about size, shape and spatial orientation of the crystals [4]. In this contribution, these ratios are reviewed in application to the scanning and tilted lidars.

2. CORNER REFLECTION EFFECT

The corner reflection effect predominates in backscattering only for pristine hexagonal ice crystals. In real cirrus clouds the crystals often have a complicated shape, such as hollow columns, rough crystals and aggregates. These types of particles are often randomly oriented. This is the reason why lidars cannot observe the corner reflection effect in their everyday observations.

It is worthwhile noting that pristine hexagonal ice plates are frequently observed with their quasi-horizontal special orientation [5-8]. In this case the vertically pointing lidar cannot detect the corner reflection effect because the backscatter mainly consists of the specular reflection. Only recently Veselovskii et al. [9] using the tilted lidar have observed the corner reflections effect of the quasi-horizontally oriented pristine ice plates. The observation showed that the backscattering Angström exponent (BAE) changed from zero to 0.55 and two depolarization ratios at two wavelengths diverged.

The information about the fraction of the quasi-horizontally oriented pristine ice plates is important for radiative transfer models. Cirrus clouds consisted of either quasi-horizontally or randomly oriented particles with the same optical thickness produce essentially different radiative forcing.

The physical optics method allowing us to estimate the fraction of the quasi-horizontally oriented pristine ice plates by using the data from tilted lidar sensing linear depolarization ratio at two wavelengths. As an example, Fig. 1 shows the result of retrieving the fraction of quasi-horizontally oriented pristine ice plates from data presented by Veselovskii et al. [9]. The retrieving can be made separately by using only the BAE (red line) or by using only two depolarization ratios (green and blue lines). The results agree well [10].

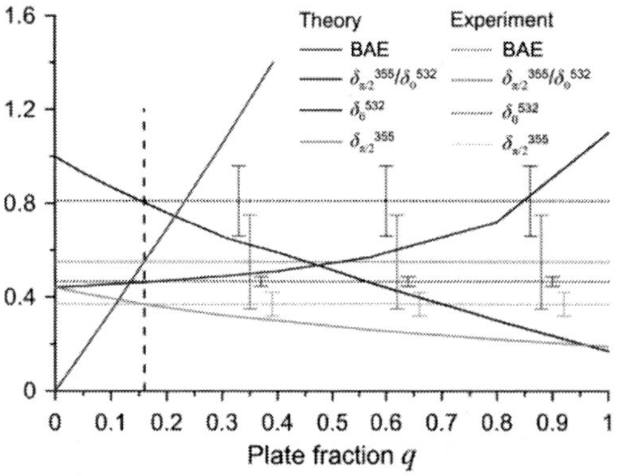

Figure 1. Experimental and theoretical data for the depolarization ratios and the backscattering Angström exponent (BAE) at the height of 8 km.

3. THE DATA BANK OF THE MUELLER MATRICES

Within the physical optics method we have found the solution of the light scattering problem by ice crystal particles of cirrus clouds for particles with size varied from 10 to 1000 μm at wavelengths from 0.355 to 1.6 μm.

The solution of the light scattering by randomly oriented hexagonal ice plates and columns in the backscattering direction is presented as a data bank of Mueller matrices. As an example, Fig. 2 shows the lidar and depolarization ratios calculated with help of this data bank for the hexagonal plates.

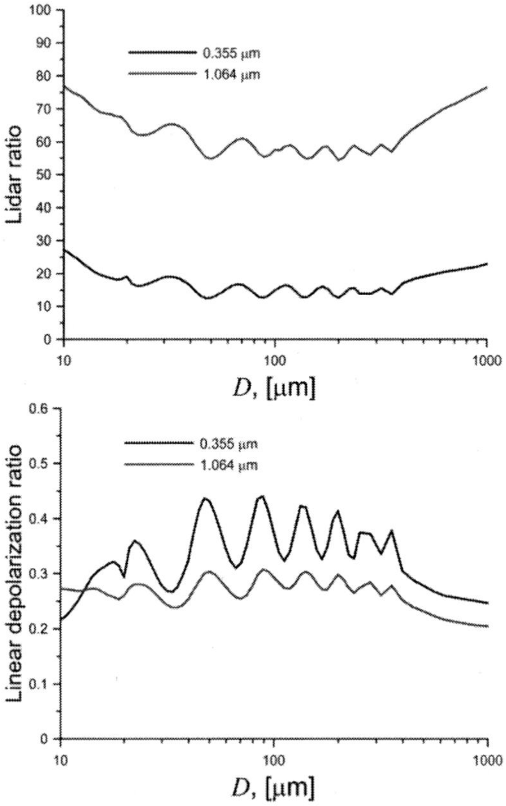

Figure 2. The lidar ratio and linear depolarization ratio for randomly oriented hexagonal ice plates as a function of the diameter (D) of the hexagonal facet for two wavelengths of 0.355 and 1.064 microns.

Figure 2 shows that the lidar ratio has a strong dependence on the wavelength of the incident light but it has a weak dependence on the particle size. The linear depolarization ratios, on the contrary, have a weak dependence on the wavelength but it significantly oscillates with particle size.

It is of great interest to study the dependence of the optical characteristics of an ensemble of ice particles as a function of the effective slope angle

 https://doi.org/10.1051/epjconf/202023708011

of particles relative to the horizontal plane, since the particles in clouds often have predominant orientation. The optical characteristics of predominantly oriented crystal particles obtained by us using the physical optics approximation are also presented in the data bank of the Mueller matrices. As an illustration, Fig. 3 shows the lidar and depolarization ratios for hexagonal columns and plates, as well as for the droxtals at a wavelength of 0.355 μm, depending on the effective slope angle β_{eff} of the particle.

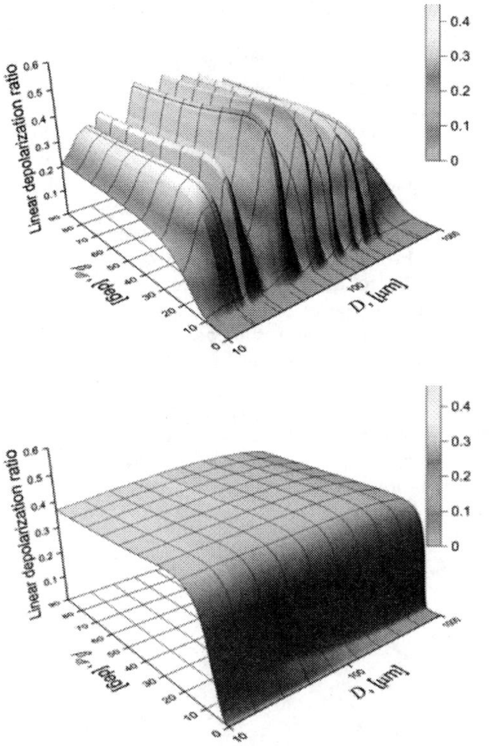

Figure 3. The linear depolarization ratio for quasi-horizontally oriented hexagonal ice plates and droxtals.

4. SCANNING LIDAR

Based on the data bank we have already studied the signal of the scanning lidar. The example of this important result is shown in Fig. 4. Here we can see that there is the sharp peak of the lidar ratio and of the first element of the Mueller matrix (M_{11}) in the vicinity of small lidar angles. It is important to analyze this vicinity for interpreting the scanning lidar data.

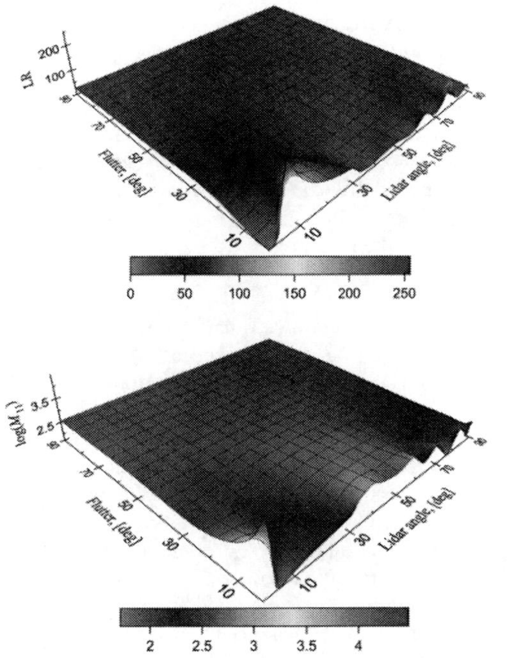

Figure 4. The lidar ratio (LR) and first element of the Mueller matrix for the cloud of ice plates with size of 100 μm depending on the flutter and the lidar scanning angle.

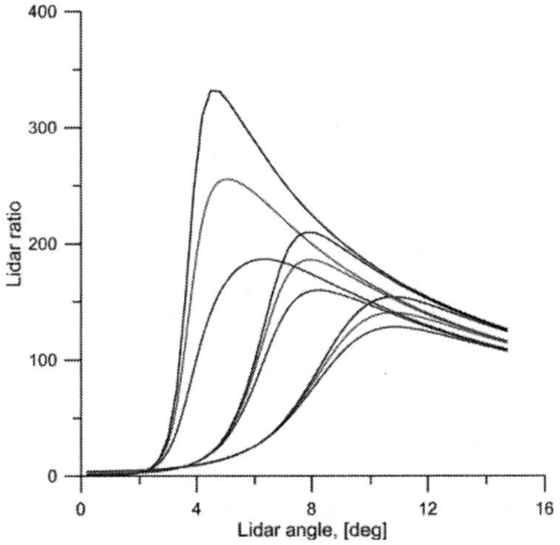

Figure 5. Comparison of the lidar ratio depending on the lidar angle for the plate with different size: 30 μm (blue), 100 μm (red) and 300 μm (black). The flutter varies from 1 to 3 degrees (from left to right slopes).

We have calculated the lidar ration for different sizes of particle and different effective tilt of the quasi-horizontally oriented pristine ice plates. The results are shown in Fig. 5. It is clearly seen that the size of the particle influence less than the particle orientation. Changing the size in 10 times (from 30 to 300 micron) leave the localization of the slope in the same place, while changing the flutter from 1° to 2° shifts the slope from 3 to 5 degrees of the lidar angle.

3. CONCLUSION

In this study we showed that the physical optics approximation can be successfully used to calculated the data bank of solutions of light scattering problem to interpret the lidar signals. Also scanning lidars are prospective tools to detecting the main parameter of horizontally oriented particles: the flutter, i.e. the degree of particle orientation. It is very important parameter that has to be added to the numerical models of the planet's radiation balance. It has been shown that the degree of particle orientation can be found directly from the lidar ratio profile of the scanning lidar. In this profile the lidar ratio has a strong slope and the localization of this slope gives the information of the flutter. It is also shown that the wavelength has little effect on the localization of the slope. The size of the particle, as far as the mixture of the randomly oriented particles, influence only to the size of the slope but not on the localization, so gives no difficulties to interpretation algorithms.

ACKNOWLEDGEMENTS

This work is supported by the RFBR under Grants no. 18-05-00568, no. 18-55-53046. Calculation of Mueller matrixes within the physical optics method is supported only by Russian Science Foundation (18-77-10035).

REFERENCES

[1] C. Zhou, P. Yang, "Backscattering peak of ice cloud particles", Opt. Exp. 23(9), 11995, 2015.

[2] J. Ding, P. Yang, et al., "Ice cloud backscatter study and comparison with CALIPSO and MODIS satellite data", Opt. Exp. 26(1), 620, 2016.

[3] A. Borovoi, A. Konoshonkin, N. Kustova, "Backscattering by hexagonal ice crystals of cirrus clouds," Opt. Lett. 38(15), 2881(2013).

[4] A. Konoshonkin, A. Borovoi, N. Kustova, J. Reichardts, "Power laws for backscattering by ice crystals of cirrus clouds," Opt. Exp. 25(19), 22341, 2017.

[5] J. Gasteiger, M. Wiegner, S. Groß, V. Freudenthaler, C. Toledano, M. Tesche, K. Kandler, "Modelling lidar-relevant optical properties of complex mineral dust aerosols", Tellus B: Chemical and Physical Meteorology 63(4), 725-741, 2011

[6] M. A. Yurkin, V. P. Maltsev, A. G. Hoekstra "The discrete dipole approximation for simulation of light scattering by particles much larger than the wavelength", J. Quant. Spectrosc. Radiat. Transfer 106, 546-557 (2007)

[7] A. Borovoi, A. Konoshonkin, N. Kustova "The physical-optics approximation and its application to light backscattering by hexagonal ice crystals", J. Quant. Spectrosc. Radiat. Transfer 146, 181-189 (2014)

[8] A. V. Konoshonkin, A. G. Borovoi, N. V. Kustova, H. Okamoto, H. Ishimoto, Y. Grynko, J. Förstner "Light scattering by ice crystals of cirrus clouds: From exact numerical methods to physical-optics approximation", J. Quant. Spectrosc. Radiat. Transfer 195, 132-140 (2017)

[9] A. V. Konoshonkin, N. V. Kustova, A. G. Borovoi "Beam-splitting code for light scattering by ice crystal particles within geometric-optics approximation", J. Quant. Spectrosc. Radiat. Transfer 164, 175-183 (2015)

[10] A. G. Borovoi, A.V. Konoshonkin, N.V. Kustova, I.A.Veselovskii "Contribution of corner reflections from oriented ice crystals to backscattering and depolarization characteristics for off-zenith lidar profiling", J. Quant. Spectrosc. Radiat. Transfer 212, 88-96 (2018)

LIGHT SCATTERING BY PARTICLES WITH ARBITRARY SHAPE IN THE VICINITY OF THE BACKWARD SCATTERING DIRECTION WITHIN GEOMETRICAL OPTICS APPROXIMATION

Victor Shishko*, Alexander Konoshonkin, Natalia Kustova, Anatoli Borovoi, Dmitry Timofeev

V.E. Zuev Institute of Atmospheric Optics SB RAS, Tomsk, 634055, Russia
**Email: sva@iao.ru*

ABSTRACT

The work presents the solution for the light scattering problem by arbitrarily-shaped particles in the vicinity of the backward scattering direction. The solution was obtained within the framework of the geometrical optics approximation. The refractive index was equal to 1.3116. It was shown that the general contribution of scattering light for arbitrarily-shaped particles in the vicinity of the backscattering direction consists of the specular reflection of the particles and two types of non-specular optical beams. It is shown that the optical characteristics of the ice particles with arbitrary shapes correspond to experimental data.

1. INTRODUCTION

Despite the fact that attempts to solve the inverse problem of laser polarization sensing of ice clouds have been made for a long time [1], this problem has not been solved yet. The main difficulty in interpreting experimentally obtained lidar data is the absence of an effective method for solving the direct problem of light scattering on ice crystal particles typical for real cirrus clouds.

2. METHODOLOGY

To build such a solution, the physical optics approximation was developed in the V.E. Zuev Institute of Atmospheric Optics SB RAS [2]. This method is effective for solving the direct problem of light scattering by crystal particles of cirrus clouds ranging in size from 10 to 1000 μm [3]. Using this method, the Mueller matrices database for the main types of atmospheric ice crystals (hexagonal plates and columns, droxtals, bullets, deformed particles, and aggregates) was calculated. The work on the theoretical calculation of Mueller matrices for all this variety of forms and sizes of atmospheric ice crystals has been going on for many years. The database is available at the IOA SB RAS ftp server [4]. However, the most traditional shape of ice crystals reveal some spectral dependence in their optical characteristics, which is not observed in experiments.

To overcome this discrepancy, some more complicated shapes of the ice crystals should be considered. In this contribution, the differential cross sections for arbitrarily shaped particles have been calculated within the geometrical optics approximation.

3. RESULTS

The full Mueller matrices have been calculated for a set of 50 particle shapes with random orientations. The Maximal dimensions of all particles were approximately equal to 20 μm. As an example, Fig. 1 presents the results of the calculation of the differential scattering cross section (DSCS) of these shapes. The refractive index was taken equal to 1.3116.

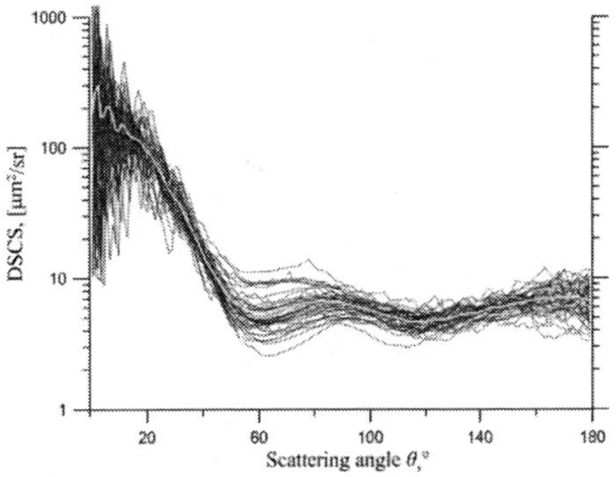

Figure 1. Differential scattering cross section obtained for a set of 50 ice particles of arbitrary shape. The yellow line – the average value over the entire set of particles

For a preliminary analysis, five particles those DSCS scarcely deviated from the average values in the vicinity of backward scattering direction ($\theta = 180°$) in Fig. 1 were chosen. These particles and their optical properties are shown in Fig. 2 and table 1. It can be seen that the optical characteristics of these five ice particles are in the range of experimentally observed values [5, 6], despite the fact that such particles are not typical for cirrus clouds.

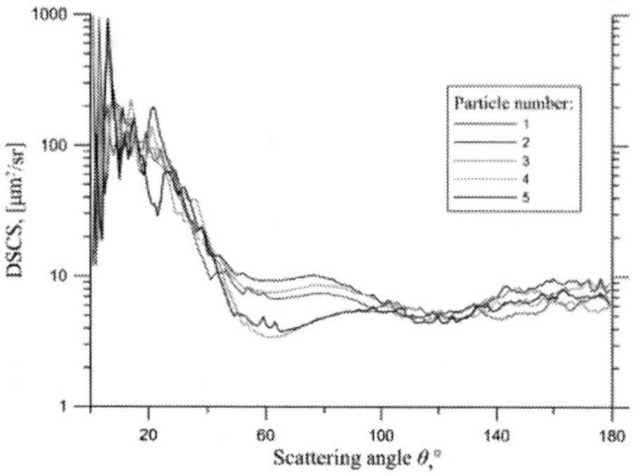

Figure 2. Differential scattering cross section for chosen five ice particles of arbitrary shape

Despite the fact that the selected particles differ in the values of their optical characteristics over almost the entire range of scattering angles, in the vicinity of the backward scattering direction, these particles have similar values of the optical characteristics. Therefore, it is sufficient to consider in more detail only one particle, for example, the particle of number 5.

Table 1. Five particles of arbitrary shape, their appearance and optical characteristics in the backscattering direction

Particle number	Appearance	LR, sr	DR
1		51.26	0.51
2	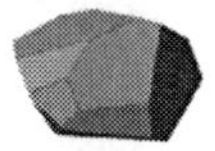	55.72	0.52
3		62.1	0.47
4	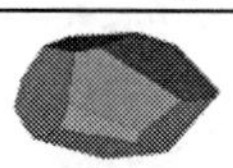	38.12	0.54
5		51.57	0.51
Average value		50.58	0.51

For the particle of number 5, the DSCS in the vicinity of the backward direction is formed due to specular reflection from the particle and about 2200 non-specular beams that occurred in the backward scattering direction, Fig. 3. Each non-specular beam is characterized by its own trajectory.

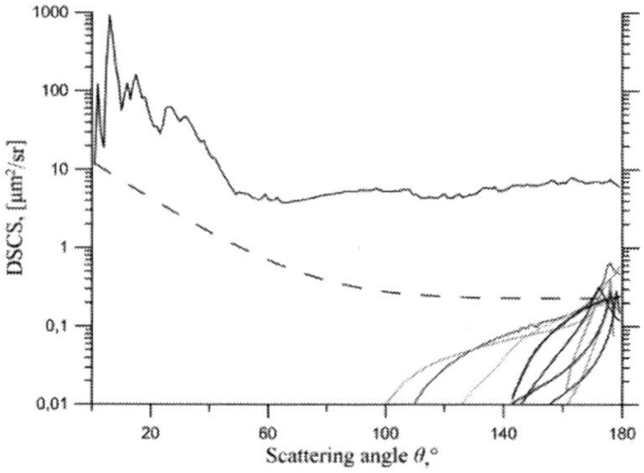

Figure 3. Differential scattering cross section for the particle of number 5 and the contributions of some typical non-specular beams (dotted line is the contribution of specular reflection, color lines were the contribution of various non-specular beams)

To differentiate the non-specular beams, distributions of light intensities over the backward scattering semi-sphere have been calculated for any beams (Figs. 4 and 6). The intensities are drawn as the dots of different brightness at a fixed particle orientation. A set of the dots shown in Figs. 4 and 6 gives 2D distributions of scattered intensities over the backward semi-sphere for a randomly oriented particle where rotation around the incident direction is excluded.

As a result, all non-specular trajectories occurred in the backward scattering directions can be divided into two types: the trajectories having a uniform distribution in the vicinity of the backward scattering direction (Fig. 4) and trajectories with uneven distribution (Fig. 6), concentrated in an area close to the backscattering direction.

Consider the non-specular beams with a uniform distribution in the vicinity of the backward scattering direction on the example of the non-specular beam highlighted by blue in Fig. 3. The

https://doi.org/10.1051/epjconf/202023708012

ray trajectories for this non-specular beam are shown in Figure 5.

The non-specular beam that corresponds to Figs. 4 and 5 gives a high depolarization ratio close to 0.5 in backscattering. Exactly these beams form the main contribution in backscattering when calculating within geometrical optics approximation.

Further, it is worth to consider the non-specular beams with uneven distribution concentrated in an area close to the backscattering direction, for example, the non-specular beam highlighted by red in Fig. 3. The trajectory of rays for this non-specular beam is shown in Fig. 7.

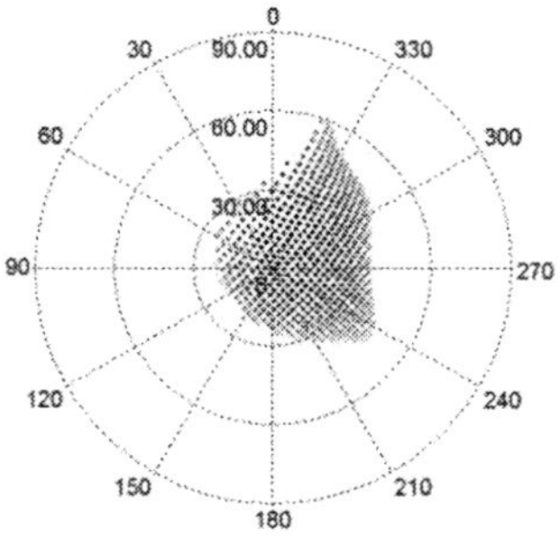

a)

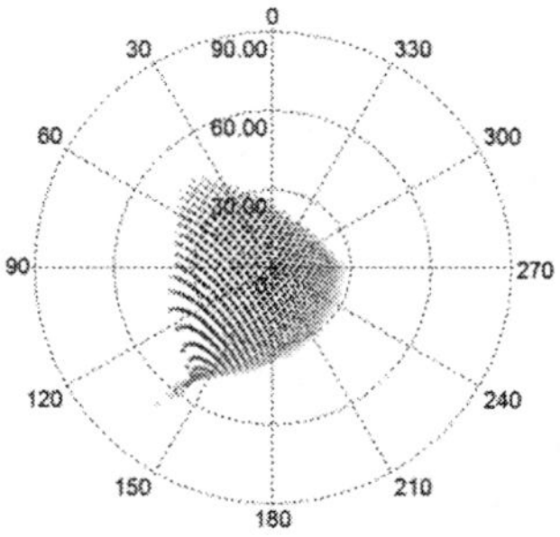

b)

Figure 4. Backscattering semi-sphere, the point in the center of semi-sphere corresponds to the backscattering point: a) – the distribution of the non-specular beam in backscattering direction (highlighted by blue in Fig. 3); b) – the distribution of the reciprocal non-specular beam

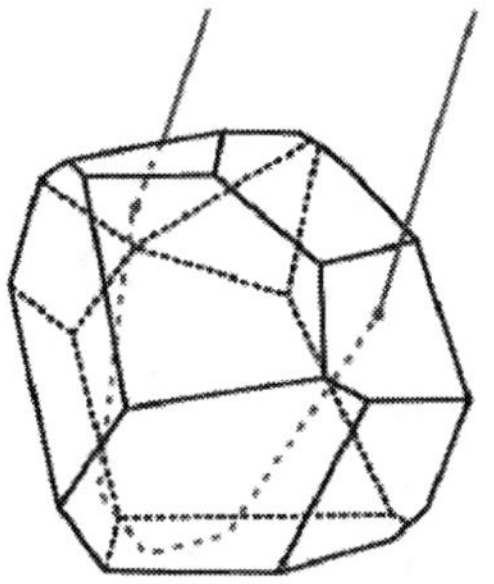

Figure 5. Visualization of non-specular trajectory presented in Fig. 4

However the non-specular beam presented in Fig. 6 and Fig. 7 is noteworthy because of the fact that it gives an insignificant contribution in backscattering in any calculations within geometrical optics approximation. Consequently, it is necessary to take these trajectories into account within the physical optics approximation because these beams would give a noticeable contribution by diffraction rings in backscattering.

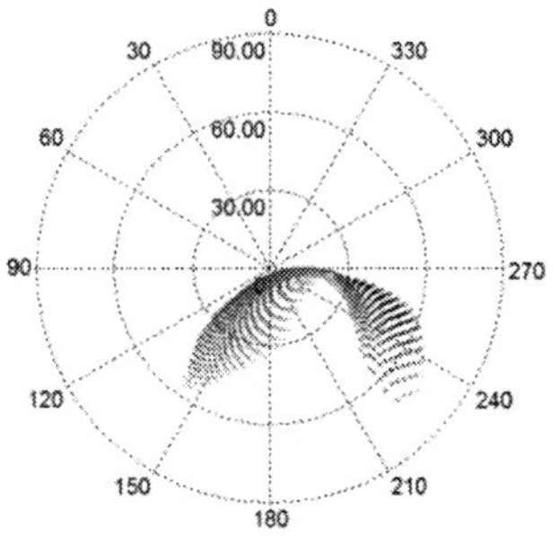

a)

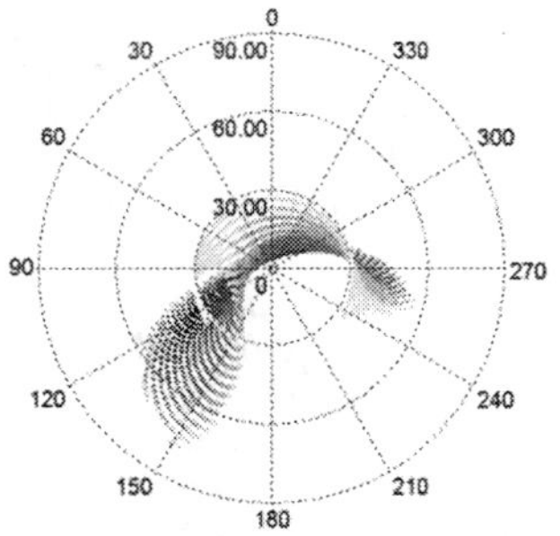

b)

Figure 6. a) – The same as Fig. 4, but for the optical beam highlighted by red in Fig. 3; b) – the distribution of the reciprocal non-specular beam

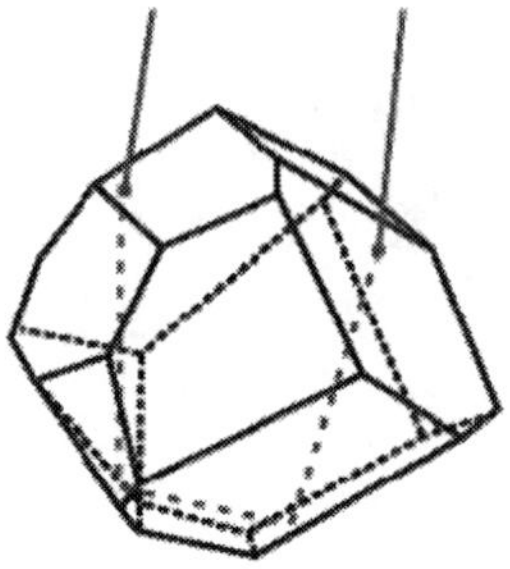

Figure 7. Visualization of non-specular trajectory presented in Fig. 6

4. CONCLUSION

The arbitrarily shaped particles without particularity in backward scattering direction were analyzed. It was shown that the main contribution in the vicinity of backscattering direction for such kind of particles is formed by both specular reflection and two types of beams described in Section 3. Also it was shown that their optical characteristics approximately correspond to the experimental data.

For further calculating the optical characteristics of this kind of particles it is necessary to take into account the fact that a part of optical beams would give a contribution by diffraction in backscattering.

ACKNOWLEDGEMENTS

This work is supported by the Russian Foundation for Basic Research under Grants no. 18-05-00568, no. 18-55-53046, and no. 19-01-00351, in part by the President of the Russian Federation MD-3306.2019.5. The calculation of the irregular particle is supported only by Russian Science Foundation (18-77-10035).

REFERENCES

[1] D. N. Romashov, B. V. Kaul, I. V. Samokhvalov, Atmospheric and oceanic optics. 13(09), 794-800 (2000)
[2] A. V. Konoshonkin, N. V. Kustova, A. G. Borovoi, D. N. Timofeev, V. A. Shishko, Proc. SPIE. 10466, 104662C (2017)
[3] Z. Wang, V. A. Shishko, A. V. Konoshonkin, N. V. Kustova, A. G. Borovoi, G. G. Matvienko, C. Xie, D. Liu, Y. Wang, Atmospheric and Oceanic Optics. 30(3), 234-235 (2017)
[4] URL: ftp://ftp.iao.ru/pub/GWDT/Physical_optics/Backscattering/
[5] W.-N. Chen, C.-W. Chiang, J.-B. Nee, Appl. Opt. 41, 6470-6476 (2002)
[6] R. Kuehn, R. Holz, E. Eloranta, M. Vaughan, J. Hair, EPJ Web of Conferences. 119 (2016) doi:10.1051/epjconf/201611916009

MONITORING ICE CRYSTALS CLOUDS: INVESTIGATION OF THE LIDAR DEPOLARIZATION RATIOS

Kalliopi – Artemis Voudouri[1*], Elina Giannakaki[3,4], Mika Komppula[3], Anna Gialitaki[5], Athanasios Natsis[1] and Dimitris Balis[1].

[1]Laboratory of Atmospheric Physics, Physics Department, Aristotle University of Thessaloniki, Greece

[3]Finnish Meteorological Institute, P.O.Box 1627, FI-70211, Kuopio, Finland

[4]Department of Environmental Physics and Meteorology, Faculty of Physics, University of Athens, Greece

[5]National Observatory of Athens, NOA, Institute for Astronomy, Astrophysics, Space Applications and Remote Sensing, Greece

***Email: kavoudou@physics.auth.gr**

ABSTRACT

A cirrus cloud dataset from a groundbased lidar in a sub-arctic station is analyzed in terms of the particle depolarization ratio. The depolarization values -δ- showed a broad distribution, ranging between 0.25 and 0.7, with a mean value of 0.38 ± 0.07. This variability is examined in correlation with temperature dependencies and the cirrus optical depth. Depolarization values greater than 0.45 are observed from temperatures between -60°C and -40°C, where horizontally oriented planar ice crystals are to be expected. Two case studies are also, discussed and analyzed in terms of depolarization height dependence.

1. INTRODUCTION

Cirrus clouds are present all seasons globally with a mean global coverage of about 20–30 % in both hemispheres. This type of clouds are composed almost exclusively of nonspherical ice crystals and the crystal shape is found to be a hexagonal column/plate [1]. As the meteorological conditions and the evolution processes differ in different parts, a broad diversity of macro-physical properties of cirrus layers, is reported, given that ice particles can vary in size and number, crystal shapes, and orientation. Indeed, cirrus layers can be found in a large variety of shapes and sizes [2]. A quantity representative of the particle shape, the lidar depolarization ratio, can be regarded as an indicator of the shape of particles in the probed cirrus layer. Higher depolarization ratios are linked to higher shape ratios (columns) and low depolarization ratios to lower shape ratios (plates), according to literature [1]. Thus, differences in the measured depolarization ratio values can be linked to different ice crystals shape.

Given that, classification of cirrus clouds has important cloud microphysical and radiative consequences, their parameterization in circulation models, require a detailed understanding of their microphysical properties [3]. As it is still unclear how these properties drive ice cloud radiative properties, the variations of shape inside the cirrus layers could lead to important influence on the cloud radiative properties.

The high sensitivity of the lidar instrument to optically thin clouds makes it one of the instruments best suited to the study of ice clouds, especially when equipped to monitor the state of light polarization.

In this study, a dataset of depolarization values of cirrus layers monitored by a ground-based lidar is analyzed. Measurements were performed in a specific environment in continental Finland. The variability of the measured depolarization values with height, cloud optical depth, temperature, etc., is analyzed.

2. METHODOLOGY

2.1 Data and Methodology

Data set from a ground-based multi-wavelength-polarization-Raman lidar (Polly^{XT}) system is analyzed in this study. The system was automatically and continuously operational 24/7- providing observations of clouds and aerosol particles in Kuopio (62.74°N, 27.54°E, 190m a.s.l) Finland during period 2012-2016. Detailed description of the system, can be found [4] and observations can be found online at http://polly.tropos.de.

The detection of the cirrus geometrical boundaries was made by applying an automatic algorithm to the lidar measurements. The method was based on a wavelet covariance transform (WCT), a method that detects discontinuities in lidar signal profile, such as the cloud boundaries [5]. The recognised loud layers were classified as cirrus layers only when the following criteria were met: i) the depolarization value was higher than 0.25, ii) the altitude of the layers was higher than 7km and iii) the temperature base was below -27°C.

The depolarization ratio is computed by using the the cross signal of the 532nm channel and the total signal of 532. The calibration of the depolarization measurements was made by using the ±45 method, as proposed by [6].

3. RESULTS

3.1 Measurement cases

Two case studies from different days of observations are presented and discussed. These days showed almost persistent cirrus layers. Temperature observations used in the present study come from radio soundings, launched daily at 6 and 18 UTC at the Jyvaskyla Airport, located to the southwest of the lidar station.

The 16th September case developed around 00:00 UTC and was observed for its entire lifetime until 04:30 UTC. This case has the cloud base average altitude around 9560m and the cloud top average around 10964m (Fig. 1a). Lidar observations of backscatter and extinction coefficient profiles and depolarization ratio d are shown in Figure 1b,c,d. The mean Lidar Ratio was 30 sr and the cirrus optical depth was calculated 0.36. Temperatures between -36°C (cloud base) and -51°C (cloud top), confirm the that the cloud is mainly composed of ice crystals. In this case, a height dependent depolarization ratio is obvious (Fig. 1d). We can see a gradual decrease of the depol ratio inside cirrus layer, at heights between 10-11.5 km. We can link this height dependence with variety of the ice particle size and orientation inside the layer.

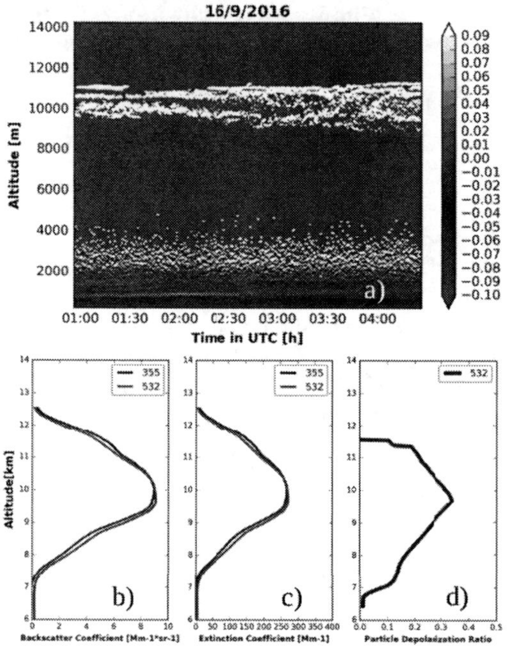

Figure 1: Cirrus cloud evolution observed at Kuopio on 16 September 2016 (a) and the calculated backscatter (b), extinction coefficient profile (c) and depolarization ratio (d).

The second case of the 20th May is presented in Figure 2. For this case, the cloud base average altitude calculated 8800m and cloud top average around 10334m. Lidar observations of backscatter and extinction coefficient profile and depolarization ratio δ are shown in Figure 2b,c,d. This case reveals that the depolarization ratio is almost constant inside the cirrus layer (9-10.5 km) and equal to 0.3. The temperature values were equal to -45.0 °C and -58.9°C for the cloud base and cloud top, corresponding to effective radius of 40 and 30μm accordingly [7]. The stability of the depolarization ratio could be linked to the stability of the temperature and as a consequence to the harmonised crystals shape inside the layer.

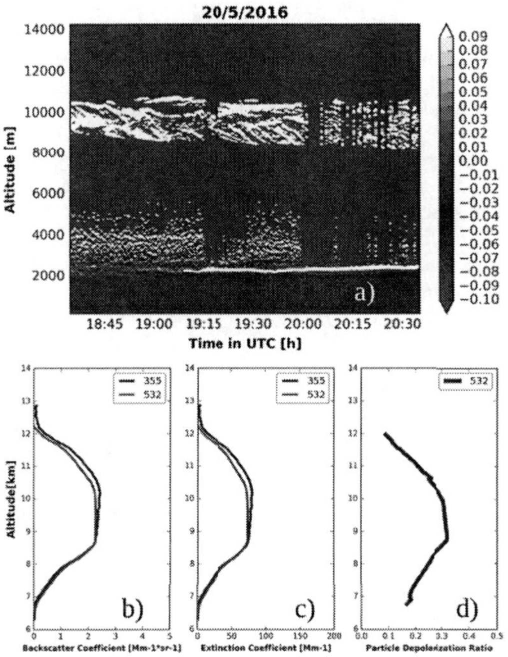

Figure 2: Cirrus cloud evolution observed at Kuopio on 20 May 2016 (a) and the calculated backscatter (b), extinction coefficient profile (c) and depolarization ratio (d).

3.2 Results of the entire dataset

The mean depolarization value was calculated for the whole period, with a large variability of observed values. Concerning the shape of crystals, higher depolarization values are linked to higher shape ratios (columns) and low depolarization values are linked to lower shape ratios (plates), according to literature [8]. A histogram of the occurrence of the depolarization values derived from our lidar dataset is presented in Figure 3. The figure shows that the most frequently observed depolarization values are found between 0.3 and 0.4, corresponding to irregular crystall shapes.

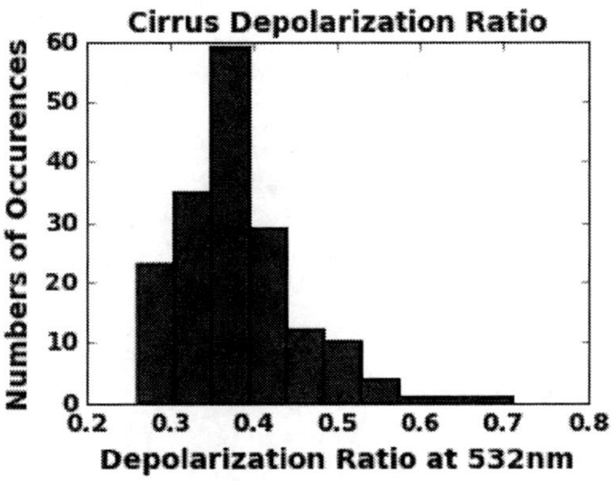

Figure 3: Depolarization ratio distribution of cirrus clouds.

In Fig.4 we present the mean δ values in 5°C intervals of cirrus base temperature. The number of cases per temperature bin are also labeled. The error bars show the standard deviation of the observed mean depolarization values in each temperature interval. No clear tendency is found, as the variability of depolarization is relatively constant, with a slightly increase of the particle depolarization with the increasing mid temperature.

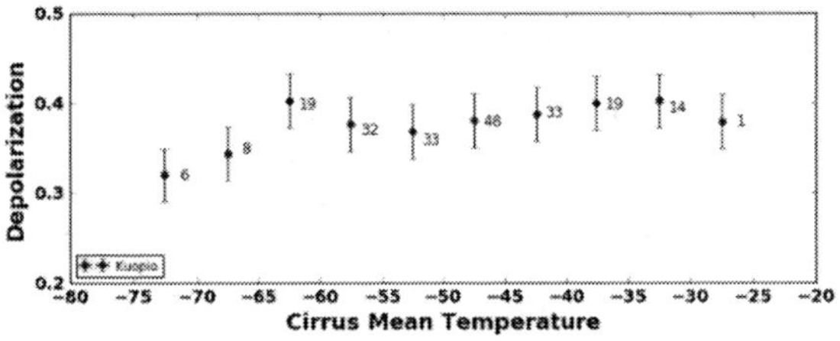

Figure 4: Mean d values in 5°C temperature intervals. The number of cases per temperature bin are also labeled.

4. CONCLUSIONS

A dataset of mean depolarization ration values of cirrus clouds measured between 2012 and 2016 with a ground-based lidar was analyzed. The correlations with height, temperature and optical depth were also investigated and presented.

In the future we intend to apply the investigation to a larger-scale climatological study in measurements from spaceborne instruments and examine the variation of ice crystals microphysical properties with geographical location. Furthermore, we aim to investigate the influence of the different ice crystal shapes to the radiative properties, so as to improve the ability to model the effects of cirrus clouds on climate.

ACKNOWLEDGEMENTS

We acknowledge support of this work by the project "PANhellenic infrastructure for Atmospheric Composition and climatE change" (MIS 5021516) which is implemented under the Action "Reinforcement of the Research and Innovation Infrastructure", funded by the Operational Programme "Competitiveness, Entrepreneurship and Innovation" (NSRF 2014-2020). Voudouri K.A acknowledges the support of the General Secretariat for Research and Technology (GSRT) and Hellenic Foundation for Research and Innovation (HFRI). (Scholarship Code: 95041).

REFERENCES

[1] Korolev, A. , Isaac, G. A. and Hallett, J., Ice particle habits in stratiform clouds. Q.J.R. Meteorol. Soc., 126: 2873-2902. doi:10.1002/qj.49712656913, 2000

[2] Noel, V., D. M. Winker, M. McGill, and P. Lawson, Classification of particle shapes from lidar depolarization ratio in convective ice clouds compared to in situ observations during CRYSTAL-FACE, J. Geophys. Res., 109, D24213, doi:10.1029/2004JD004883, 2004

[3] Stephens, G. L., S.-C. Tsay, P. W. Stackhouse, and P. J. Flatau, The relevance of the microphysical and radiative properties of cirrus clouds to climate and climate feedback, J. Atmos. Sci., 47, 1742 – 1753, 1990

[4] Engelmann R., Kanitz T., Baars H., Heese B., Althausen D., Skupin A., Wandinger U., Komppula M., Stachlewska IS, Amiridis V., Marinou E., Mattis I., Linne H., Ansmann A., 2015: EARLINET Raman lidar PollyXT: the next generation. Atmos Meas Tech Discuss 8:7737-7780. doi10.5194/amtd-8-7737-2015, 2015

[5] Baars, H., Ansmann, A., Engelmann, R., and Althausen, D.: Continuous monitoring of the boundary-layer top with lidar, Atmos. Chem. Phys., 8, 7281-7296, https://doi.org/10.5194/acp-8-7281-2008, 2008.

[6] Freudenthaler, V., Esselborn, M., Wiegner, M., Heese, B., Tesche, M., Ansmann, A., Muller, D., Althausen, D., Wirth, M., Andreas, F. I. X., Ehret, G., Knippertz, P., Toledano, C., Gasteiger, J., Garhammer, M., and Seefeldner, M.: Depolarization ratio profiling at several wavelengths in pure saharan dust during SAMUM 2006, Tellus B, 61, 165–179, 2009.

[7] Sassen, K. , P. W. Arnott, D. O.' C. Starr, G. G. Mace, Z. Wang, and M. R. Poellot: Midlatitude cirrus clouds derived from Hurricane Nora: A case study with implications for ice crystal nucleation and shape. J. Atmos. Sci., 60, 873–891, 2003a.

[8] Korolev, A. , Isaac, G. A. and Hallett, J., Ice particle habits in stratiform clouds. Q.J.R. Meteorol. Soc., 126: 2873-2902. doi:10.1002/qj.49712656913, 2000.

OBSERVATIONS OF THE HORIZONTALLY ORIENTED CRYSTALLINE PARTICLES WITH A SCANNING POLARIZATION LIDAR

Grigorii Kokhanenko*, Yurii Balin, Anatolii Borovoi, Marina Klemasheva, Sergei Nasonov, Mikhail Novoselov, Ioganes Penner, Svetlana Samoilova

Institute of Atmospheric Optics, Siberian Branch of Russian Academy of Sciences, Tomsk, Russia
**Email: kokh@iao.ru*

ABSTRACT

Scanning lidar LOSA-M3 makes it possible to measure the polarization characteristics of backscattering signals from aerosol and clouds at wavelength 532 and 1064 nm. The lidar transceiver is placed on a scanning column, which allows changing the direction of sounding within the upper hemisphere at a speed of 1 degree per second. The polarization characteristics of the transmitter (linear or circular polarization) can be changed by rotating the phase plates synchronously with the laser pulses. Conical scanning of the lidar allows identifying cloud areas with preferential zenith or azimuthal orientation of the crystal particles. The article presents the results of observations of the cloud polarization structure carried out in Tomsk. Methods of the calibrations of lidar polarization channels are described.

1. INTRODUCTION

Cirrus clouds cover a significant part of the earth's surface. They have a significant impact on the radiation balance and climate. Radiation reflection by and transmission through these clouds depends essentially on particle shapes and orientations[1-3]. The use of the scalar model, even assuming a random orientation of the particles, gives a noticeable error in the calculation of the radiation passage[4]. Traditionally, most radiative models of cirrus suppose a random orientation of crystals[5]. According to Ref. 6, in the presence of a preferential orientation of cloud particles, lidar depolarization measurements can yield ambiguous results. Numerous work on sounding cirrus clouds by polarization lidars[6-10] have shown that ice crystals in cirrus clouds often show not only a predominant orientation relative to the horizon (zenith orientation), but they also have a predominant orientation in the horizontal plane (azimuthal orientation) connected with the direction of the wind. The lidar polarization measurements for which the lidar was directed or strictly vertical, or deviated from the vertical direction by some fixed angle,[11-13] made it difficult to interpret the obtained data. The advantage of scanning lidars is the ability to diagnose the preferential orientation of crystals in the clouds. Two wavelengths of the radiation source of 532 nm and 1064 nm provide additional information on the microphysical parameters of the clouds

2. METHODOLOGY

2.1 The lidar system

Lidar LOZA-M3 is a continuation of a series of small-sized lidars LOZA-MS and LOZA-M2[14,15], developed and operated in the IAO SB RAS. All these lidars are intended primarily for use in expeditionary conditions, which imposes certain restrictions on the weight-dimensional characteristics and the energy potential of the device. The main features of LOZA-M3 lidar are: 1) - automatic scanning drive, which allows to change the direction of sounding within the upper hemisphere at a speed of up to 1 degree per second with the accuracy of the set angle measurement no worse than 1 minute, and 2) polarization channel that allows to determine some elements of the scattering matrix of aerosol particles at two wavelengths simultaneously.

The optical scheme of the lidar is shown in Fig. 1. The Q-Smart 850 (Quantel) laser *1* with a fundamental harmonic energy of 850 mJ is used, the repetition frequency is 10 Hz. The radiation is linearly polarized, but the plane of polarization of the second harmonic (532 nm) is rotated by 45 ° relative to the first. To reconcile the polarization planes, a phase plate *2* having a phase shift of 20 wavelengths for λ = 532 nm and 9.5 wavelengths for λ = 1064 nm is used. Turning the plate achieves the coincidence of the planes of polarization for two harmonics. The Glan prism *3* improves the polarization contrast of the radiation. The phase plate *4*, optionally a quarter wave λ/4 or half-wave λ/2, serves to rotate the plane of polarization of the radiation (λ/2) or to transform the polarization state (λ/4, linear-circular).

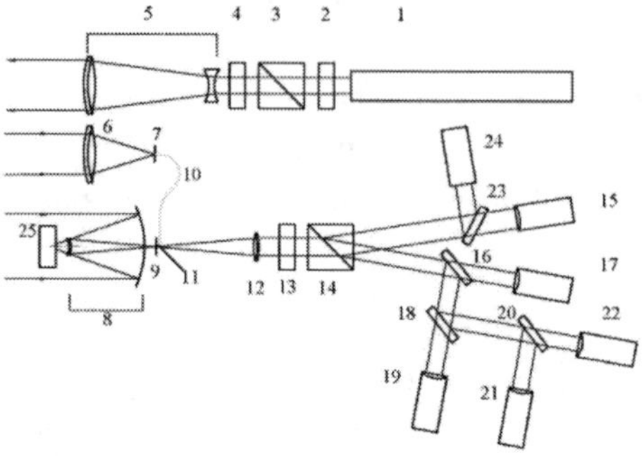

Fig. 1. Optical circuit of the lidar LOZA-M3

The Wollaston prism *14* divides the beam into components with orthogonal polarization, which are further divided along the wavelengths by dichroic beamsplitters. Unlike schemes in which wavelength division is carried out before the separation of polarization components, there is no distortion of the polarization state when reflected from dichroic elements and there is no need to apply laborious calculations of the instrument vector and correction of the measured polarization[16].

2.2 Calibration of the polarization channels

To measure the depolarization ratio of the backscattered radiation, the relative sensitivity of detectors must be known. We use the technique of continuous rotation of λ/2 phase plate (*13* at Fig. 1), installed in receiver module. Let us assume that laser radiation is linearly polarized (the initial vector is $\mathbf{S}_0^L = [1,1,0,0]^T$). If we set the λ/2 phase plate before the polarization prism with matrix **L**(φ), φ is the rotation angle, the scattered Stokes vector will be

$$\mathbf{S} = \mathbf{LMS}_0^L = \begin{pmatrix} m_{11}+m_{12} \\ (m_{12}+m_{22})C_4-(m_{13}+m_{23})S_4 \\ (m_{12}+m_{22})S_4+(m_{13}+m_{23})C_4 \\ m_{14}+m_{24} \end{pmatrix}, \quad (1)$$

where $C_4 = \cos 4\varphi$, $S_4 = \sin 4\varphi$. Polarization components $I_\parallel, I_\perp$ are different, but the integral over angle φ: $\int_{2\pi} I_{\parallel,\perp} d\varphi = \pi(m_{11}+m_{12})$ is the same for each polarization component regardless of the values of matrix elements. Using the continuous rotation of the λ/2 phase plate, we can measure the relative sensitivity of the detectors and calibrate polarization channels.

3. RESULTS

3.1 Zenith scanning at 1064 nm wavelength

Below we present the results of the crystal cloud observations at the zenith scanning. Scanning was performed at a rate of 0.5 degrees per second. The angle is measured from the vertical position of the radiation beam, which was set with an error of ±5 minutes. In this case, the scan was carried out with the lidar axis passing through the zenith (ranging from -1° to+4°), which allowed controlling the accuracy of the lidar vertical setting.

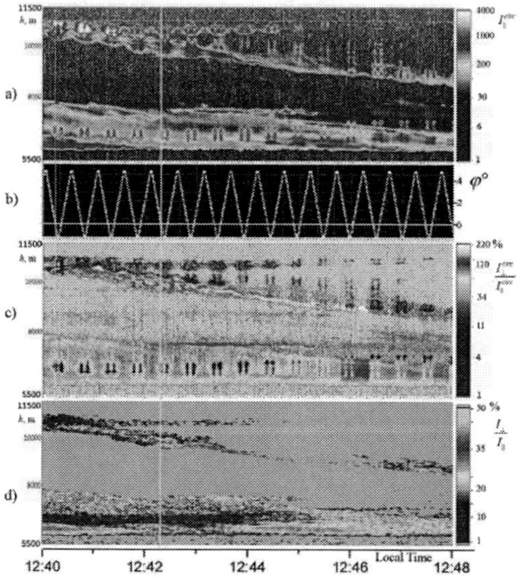

Fig. 2. Cloud structure, observed with zenith scanning

Figure 2 shows the lidar data for 01.10.2018. The tilte angle is shown in Fig.2.b. Upper frame (2.a) indicates the parallel component $I_\parallel^{circ}$ (circular initial polarization) in ADC codes (0-4096). Fig. 2c shows the depolarization ratio $\delta = I_\perp / I_\parallel$. Positions of vertical orientation α=0 (one of them is indicated with yellow line) are characterized by maximal intensity $I_\parallel$ and minimal δ. The bottom frame 2d (depolarization ratio for linear polarization) shows the record from lidar LOSA-S, placed near the LOSA-M3 lidar and permanently oriented vertically.

Fig. 3 indicates the dependence of the signal intensity on the tilt angle α, obtained in measurements on 02.06.2018. The abscissa shows the tilte angle α, ordinate shows the intensity in ADC codes. The record in Fig. 3a was registered at 12^{00} LST with averaging of the characteristics over the layer from 11470m to

11600m. The record 3b was made at 10^{00} LST with averaging of the characteristics over the layer from 10980m to 11270m. These records show the opposite cases of the degree of the specular reflection. For 3a depolarization ratio $\delta = I_{\perp}/I_{\parallel}$ δ=0.006 at vertical position (α=0) and δ=0.11 at α=4°. For 3b δ(0)=0.24 and δ(4°)=1.42.

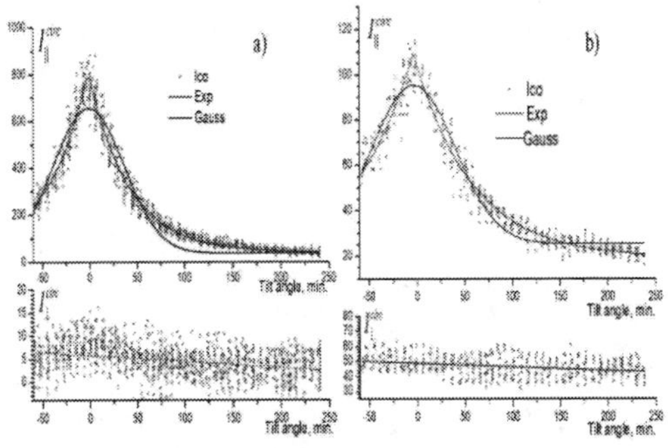

Fig. 3. Dependence of the signal intensity on the tilt angle

All data measured on June 02, 2018 well described by the exponential distribution

$$I(\alpha) = I_0 + A\exp(-|\alpha - \alpha_0|/w), \qquad (2)$$

(red line in Fig. 3), where w determines the width of the distribution, and α_0 indicates the error of the lidar targeting to vertical. For Fig. 3a w=42′, for Fig. 3b w=82′. The shift of the curve maximum is less than 2 minutes, which indicates a good orientation of the lidar to the zenith.

As opposed to Ref. 10 the Gaussian function poorly describes the observed distribution (blue line in Fig. 3). The main difference is determined by the underestimation of the sharp peak at α=0°. The values $I_{\perp}$ in most cases shows a weak decline with the angle, but its change are smaller than the measurement errors. We can conclude that $I_{\perp}$ are practically independent of the inclination angle. It should be noted that the intensities $I_{\perp}$ for linear polarization in the mirror reflecting zone at some cases do not exceed the noise level of the receiver and are difficult to process. That's why we present here the date of measurements with circular polarization.

Selected cases of observed on June 02 distributions are shown in Fig. 4. These cases cover the entire range of observed widths w of the distribution $I_{\parallel}(\alpha)$.

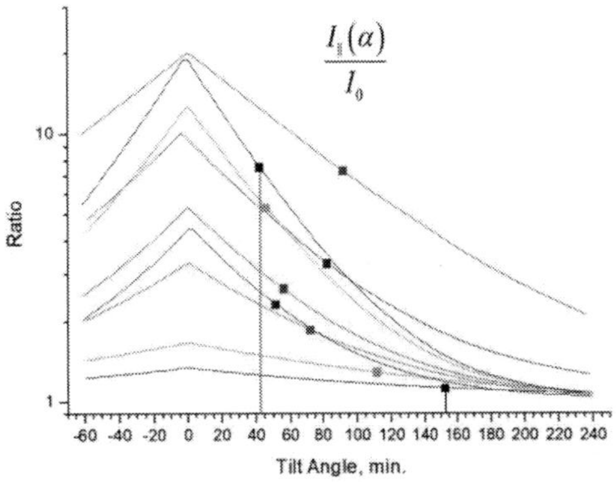

Fig. 4. Different cases of observed distributions $I(\alpha)$

The value of intensity is normalized to the value I_0 (Eq. 2). If we assume that the signal intensity for α>>4° is determined by randomly oriented particles, this value represents the contribution of specularly oriented particles to the total backscattering intensity. We can assume that this corresponds to the distribution of the inclination angles of crystalline particles relative to the horizontal plane.

3.2 Green and infrared dependencies

Simultaneous recording of signals allows comparing the angular dependences for wavelengths 532 and 1064 nm. Above, we assumed that the angular dependence of the specular component reflects the angular distribution of particles. But the relative intensity of this component may differ for two wavelength because i) backscattering coefficient for α>>4° as rule is different, and ii) the intensity of specular reflection may be differ because of the contour diffraction. Unfortunately, we have not an absolute calibration of recorders for two wavelength, so in the Fig. 5 we present the relative angle dependencies.

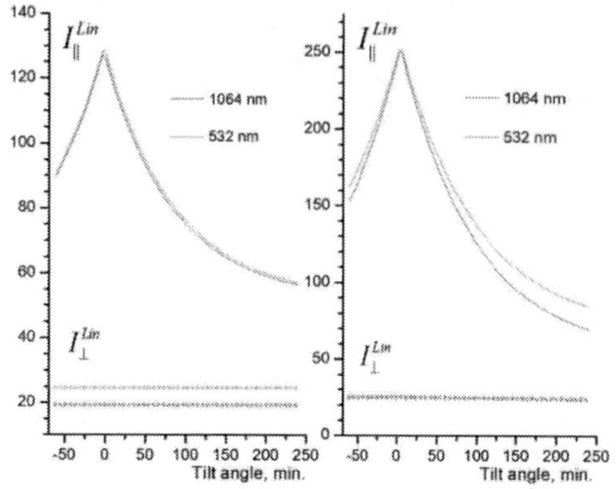
Fig. 5. Angle dependencies of polarization components for 532 nm (green) and 1064 nm (red)

The curves are aligned so that the intensity maxima at α=0° are equal. We show two cases with maximum angular differences. In the left frame the angle distributions are the same, in the right frame the distribution for 532 nm is wider ($I_{532}(4°)/I_{1064}(4°)=1.5$). In all other cases the distribution width w for 532 nm is equal or more then for 1064 nm.

ACKNOWLEDGEMENTS

The work was supported in part by Ministry of Science of Education of RF (Agreement No. 14.616.21.0104, unique identifier RFMEFI61618X0104).

REFERENCES

[1] Liou, K. N., "Influence of cirrus clouds on weather and climate processes: a global perspective," J. Geophys. Res. 103, 1799-1805 (1986).

[2] Sassen, K., Griffin, M. K., and Dodd, G. C., "Optical scattering and microphysical properties of subvisual cirrus clouds, and climatic implications," J. Appl. Meteorol. 28, 91-98 (1989).

[3] Masuda, K., and Ishimoto, H., "Influence of particle orientation on retrieving cirrus cloud properties by use of total and polarized reflectances from satellite measurements," J. Quant. Spectrosc. Radiat. Transfer 85, 183–193 (2004).

[4] Lawless, R., Xie, Yu, Yang, P., Kattawar, G. W., Laszlo, I., "Polarization and effective Mueller matrix for multiple scattering of light by nonspherical ice crystals," Opt. Express 14, 6381-6393 (2006).

[5] Takano, Y., Liou, K.-N, and Minnis, P., "The effect of small ice crystals on cirrus infrared radiative properties," J. Atmos. Sci. 49, 1487–1493 (1992).

[6] Kaul, B.V., Samokhvalov, I.V., and Volkov, S.N., "Investigating of particle orientation in cirrus clouds by measuring backscattering phase matrices with lidar," Appl. Opt. 43, 6620 –6628 (2004).

[7] Balin, Yu. S., Kaul, B. V., Kokhanenko, G. P., and Penner, I. E., "Observations of specular reflective particles and layers in crystal clouds," Opt. Express 19, 6209-6214 (2011).

[9] Borovoi, A., Balin, Y., Kokhanenko, G., Penner, I.; Konoshonkin, A., and Kustova, N., "Layers of quasi-horizontally oriented ice crystals in cirrus clouds observed by a two-wavelength polarization lidar," Opt. Express 22, 24566-24573 (2014).

[9] Chepfer, H., Brogniez, G., Goloub, P., Breon, F. M., and Flamant, P. H., "Observations of horizontally oriented ice crystals in cirrus clouds with POLDER-1/ADEOS-1," J. Quant. Spectrosc. Radiat. Transfer 63, 521–543 (1999).

[10] Noel, V., and Sassen, K., "Study of Planar Ice Crystal Orientationsin Ice Clouds from Scanning Polarization Lidar Observations," J. Appl. Meteor. 44, 653-664 (2005).

[11] del Guasta, M., Vallar, E., Riviere, O., Castagnoli, F., Venturi, V., and Morandi, M., "Use of polarimetric lidar for the study of oriented ice plates in clouds," Appl. Opt. 45, 4878-4887 (2006).

[12] Hayman, M., Spuler, S., Morley, B., and Van Andel, J., "Polarization lidar operation for measuring backscatter phase matrices of oriented scatterers," Opt. Express 20,29553–29567 (2012).

[13] Veselovskii, I , Goloub, P., Podvin, T., Tanre, D., Ansmann, A., Korenskiy, M., Borovoi, A., Hu, Q., and Whiteman, D. N., "Spectral dependence of backscattering coefficient of mixed phase clouds over West Africa measured with two-wavelength Raman polarization lidar: features attributed to ice-crystals corner reflection," J Quant Spectrosc Radiat Transfer 202, 74–80 (2017).

[14] Bairashin, G.S., Balin, Yu.S., Ershov, A.D., Kokhanenko, G.P., and Penner, I.E., "Lidar "LOZA-MS" for investigation of aerosol fields in troposphere," Optical Engineering, 44(7), 071209-1-071209-7 (2005).

[15] Bairashin, G.S., Balin, Yu.S., Klemasheva, M.G., Kokhanenko, G.P., Penner, I.E., and Samoilova, S.V. "Aerosol-Raman lidar "LOSA-M2," Quantum electronics, 41(10), 945-949 (2011).

[16] Di, H.G., Hua, H.B., Cui, Y., Hua, D.X., Li, B., and Song, Y.H., "Correction technology of a polarization lidar with a complex optical system," J. Optical Soc. Amer. A., 33(8), 1488-1494 (2016).

ON THE SHAPES AND SIZES OF CIRRUS PARTICLES CAUSING SPECULAR REFLECTIONS IN LIDAR SIGNALS

Jens Reichardt[1,*]

[1] *Richard-Aßmann-Observatorium, Deutscher Wetterdienst, Am Observatorium 12, 15848 Lindenberg, Germany*

**Email: jens.reichardt@dwd.de*

ABSTRACT

A new method to determine the projection-area-equivalent radius of cirrus particles is presented. Together with direct lidar measurements of ice water content and a simple particle model, the technique facilitates the retrieval of ice particle shape and size. As a first application, cirrus clouds are studied that feature layers with lidar and depolarization ratios highly depressed by mirror reflections from horizontally aligned particles. It is found that in most cases not thin hexagonal plates dominate the ensemble of scatterers but rather compact particles.

1. INTRODUCTION

Specular reflections of horizontally aligned (2D) cirrus particles cause a decrease in the measured depolarization and lidar ratios. It is widely accepted that these particles are of thin planar shape. However, the measurements with the RAMSES lidar [1] at the Lindenberg Meteorological Observatory in northern Germany, which is equipped with a water spectrometer [2] that facilitates direct measurements of ice water content (IWC), suggest otherwise, because IWC values measured in 2D cirrus layers are in most cases too high to be explained by the ice volume of large, thin hexagonal plates.

To study this issue further, a new technique was developed to derive the projection-area-equivalent radii in cirrus clouds from the extinction measurements with RAMSES. This information is used as input for a particle model to assess the plausible ranges of particle shape and size of the ice clouds, and thus IWC. By comparing the modeled IWC to the direct RAMSES IWC measurements, one can draw conclusions on the particles that most likely populated the 2D layers. This conference contribution presents first results of the analysis.

2. METHODOLOGY

2.1 Determination of cirrus-particle projection-area-equivalent radius

RAMSES is equipped with two receivers of different field-of-view (FOV) (0.33 mrad vs. 3.23 mrad) [1] which makes the cirrus extinction measurements with these receivers susceptible to multiple scattering (MS) at varying degrees. Once properly corrected for MS, the obtained single-scattering extinction profile should be the same for both. This identity opens up an elegant method to estimate the projection-area-equivalent radius (r_{pae}) of cirrus clouds: For a prescribed cirrus profile of r_{pae} the MS contributions to the N_2 vibrational-rotational Raman signal are calculated with a fast MS code [3] for both FOVs, and the MS-corrected extinction profiles are calculated and compared. The r_{pae} profile is adjusted and the procedure repeated until satisfying agreement is reached.

As a side note, a similar 2-FOV Raman lidar technique was applied previously to estimate the effective droplet radius of warm clouds [4]. However, the experimental setup differed from RAMSES', in that only a single telescope was used. By drilling a hole in one of the receiver's deflecting mirrors, a narrow-FOV measurement could be obtained in transmission, and a wide-FOV measurement in reflection. This approach would probably not be applicable to ice clouds, because the scattering phase function of the comparatively large ice particles is narrowly peaked in the forward direction leaving insufficient signal for the wide-FOV observation.

2.2 Measurement of ice water content

Cloud-water Raman backscattering coefficients are measured with the RAMSES water spectrometer [2]. Spectroscopic data [5, 6] are used to convert these coefficients into IWC. Differences in Raman efficiency between cloud droplets and ice particles are accounted for. However, IWC measurements are only possible at night, under favorable atmospheric conditions and often only in the lower cirrus ranges because of the extremely weak inelastic return signals from clouds.

3.3 Retrieval of particle shape and size

Modeled IWC profile depends on r_{pae}, the particle number concentration, and the assumed shape, orientation and length-to-width (aspect) ratio of the particles. Overfitting is avoided by assuming that all ice particles are identical. The number concentration is obtained directly from the single-scattering extinction-coefficient profile and the particles' geometrical cross section $G = \pi\ r_{pae}^2$. Random (3D) orientation can be expected when cirrus depolarization and lidar ratios are high (> ~0.2 and > ~10 sr, respectively), at least partial 2D orientation otherwise. The basic particle shape is assumed to be hexagonal. Selection of aspect ratios follows loosely Ref. [7]. The set of model parameters is varied until reasonable agreement with the measured IWC profile is reached.

3. RESULTS

3.1 Cirrus with randomly oriented particles

First, examples of 3D cirrus are discussed. Figure 1 presents two measurement cases on 05 Dec 2015 where the particle optical properties clearly indicate the lack of 2D effects. At 19:00 UTC, MS-corrected extinction profiles show very good agreement for r_{pae} = 50 μm, and measured and modeled IWC are alike when 3D columns with an aspect ratio of 2.5 are assumed. Half an hour later, at 19:30 UTC, the cirrus cloud has changed. The center of the cloud is lower, retrieved r_{pae} is larger (r_{pae} = 80 μm), and the particle ensemble is probably a mixture of 3D columns and plates.

3.2 Cirrus with layers containing horizontally oriented particles

Next, two cases are examined where the 2D cirrus layers consisted predominantly of thin planar particles. As a matter of fact, it was difficult to find measurements that would be suited as convincing examples. The best candidates are presented in Fig. 2. Lidar and depolarization ratios are decisively smaller than in the flanking 3D

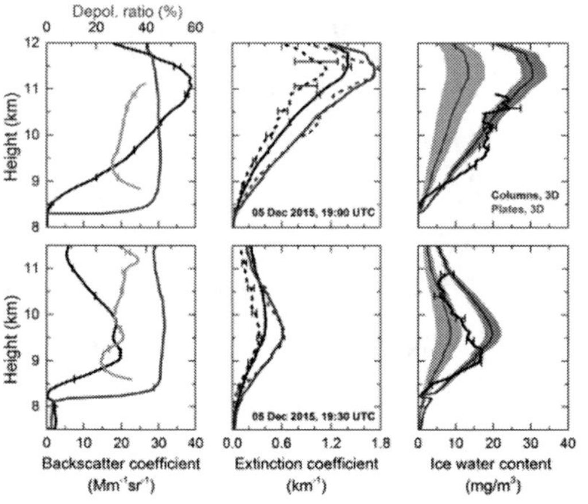

Figure 1. Retrieval results for two observations of 3D cirrus. (left) Particle backscatter coefficient, lidar ratio and particle depolarization ratio. (middle) Particle extinction coefficient as measured with the wide-FOV receiver (black dashed curve) and with the narrow-FOV receiver (black solid curve), and after the correction for multiple scattering (colored curves). (right) Measured (black curve) and retrieved ice water content (colored curves). Either 3D plates or columns are assumed, aspect ratios are 0.08 and 2.5, respectively. Shaded areas are supposed to illustrate natural variability. Lidar data integration time is 20 min, bars indicate statistical errors. The wide-FOV receiver has a telescope diameter four times smaller than the narrow-FOV receiver, which explains the different noise levels of the extinction profiles.

layers, albeit not extremely low (cf. Fig. 3). Therefore, the effect of mirror reflections is not dominant – perhaps because the fraction of 2D particles was small, or the oriented particles were poor reflectors (rough surfaces).

In the case of the measurement on 9 Nov 2015, good agreement in the single-scattering extinction profiles could only be achieved below 5.1 km (r_{pae} = 80 μm). Still, above 4.5 km it seems justified to conclude that only a mix of thin 2D and 3D planar crystals can explain the measured IWC values. At lower altitudes, however, one has to assume rather compact columns instead.

On 5 Dec 2015, both MS-corrected extinction profiles fit together at all altitudes (r_{pae} = 130 μm). Again, in the upper part of the 2D layer only thin

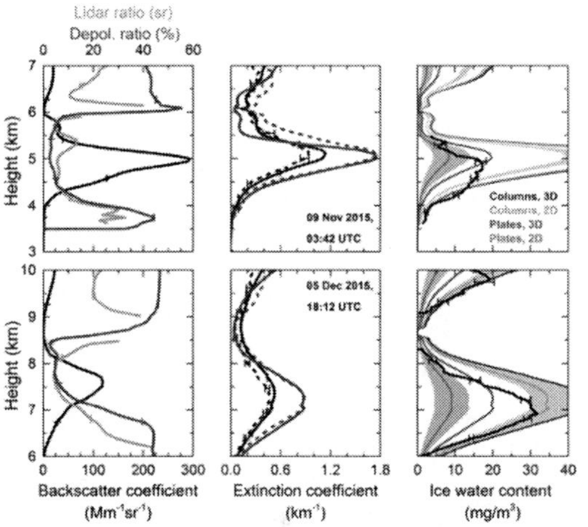

Figure 2. Same as in Fig. 1, but for two observations of 3D cirrus with embedded 2D layer of low ice water content. Either 3D, or 2D, plates or columns are assumed, aspect ratios are 0.06 and 1.5 (top), and 0.08 and 3.0 (bottom), respectively.

plates reproduce the measured IWC values well while elongated columnar particles are required below. Note that the latter are a good particle model for the 3D cirrus above 9 km as well.

Finally, measurement examples are presented that can be considered characteristic of most 2D cirrus clouds observed with RAMSES (Fig. 3). In all cases, single-scattering extinction profiles agree quite well, derived r_{pae} values are about 90, 120, and 90 μm (Fig. 3, top to bottom). Note that on 21 Dec 2015 the particle backscatter coefficient had rather extreme peak values of more than 1 $km^{-1}sr^{-1}$ which, together with miniscule particle depolarization ratios (~0.02) and lidar ratios (~2 sr), demonstrate an overwhelming scattering contribution of 2D particles. Only thick plates or compact hexagons are voluminous enough to explain the significant IWC of the layer. Incidentally, for hexagonal particles with aspect ratios around one the geometrical cross section does not depend strongly on orientation which is why 2D and 3D IWC profiles are close to each other.

The same conclusion can be drawn from the other two examples depicted in Fig. 3: Thick planar or compact columnar particles are required to reproduce the IWC profiles directly measured

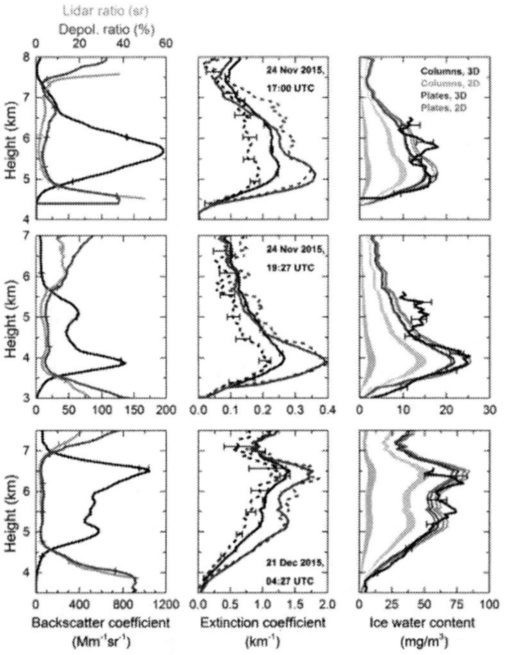

Figure 3. Same as in Fig. 1, but for three observations of 3D cirrus with embedded 2D layer of substantial ice water content. Either 3D, or 2D, plates or columns are assumed, aspect ratios are 0.55 and 1.0 (top, center), and 0.50 and 1.0 (bottom), respectively. Shaded areas illustrate results obtained for thin 2D plates (aspect-ratio range of 0.04-0.08) and elongated 2D columns (2.5-3.0).

with RAMSES while thin plates or elongated columns cannot.

4. CONCLUSIONS

The new 2-FOV Raman lidar technique offers a straightforward way to estimate the effective projection-area-equivalent radii of cirrus cloud particles. The method yields robust and reasonable results, for instance retrieved r_{pae} tends to decrease with height as one would expect. However, because the MS code used employs the small-scattering-angle approximation it is limited to ice clouds with extinction coefficients smaller than ~2 km^{-1} so that light scattering does not become too diffusive.

From the retrieved values of r_{pae} and aspect ratio one can calculate the maximum dimension of the

https://doi.org/10.1051/epjconf/202023708015

ice hexagons. For the 2D cirrus layers shown in Fig. 3, maximum dimensions of thick plates range from 200 to 260 μm, while those of compact columns are somewhat smaller (165 – 210 μm). Because of the poor agreement between retrieved and measured IWC, the results for the probably thin 2D plates of Fig. 2 should be considered inconclusive, but still seem to indicate plate diameters slightly larger than those of the thick plates (Fig. 3). Finally, the elongated 3D columns of the cold cirrus cloud in Fig. 1a have lengths of about 150 μm, the particle maximum dimension retrieved for the ice cloud of Fig. 1b is around 235 μm, irrespectively of basic shape.

The main result of this study is that in cirrus layers with the optical characteristics of 2D particles IWC is often considerable, which is inconsistent with the notion that thin planar particles dominate the particle ensemble here. Instead, compact hexagons with aspect ratios around one explain the observations best. This outcome suggests that 2D cirrus clouds may be better described as being composed of ensembles of compact scatterers that are horizontally aligned, a fraction of which such that they function as mirrors.

Depending on the perfectness of the particle surfaces (and interior), varying fractions may be required to noticeably affect the optical properties. The relation between the flawlessness of the reflecting particle, fraction of 2D particles, and lidar and depolarization ratio is illustrated in Fig. 4, for the derivation of the underlying equations see Ref. [8]. Cloud optical properties always reach their minima for a 2D fraction of 100%, and the minimum values themselves can be assumed to be connected with the irregularity of the particles [8]. So the better the optical quality of ice particles the smaller the fraction of 2D particles acting like mirrors needs to be to obtain a drop to a certain lidar ratio or depolarization ratio. In summary, the encouraging results presented here warrant future studies of 2D cirrus layers. In particular, the quest for convincing measurement examples of ice clouds with horizontally aligned *thin* planar particles will continue.

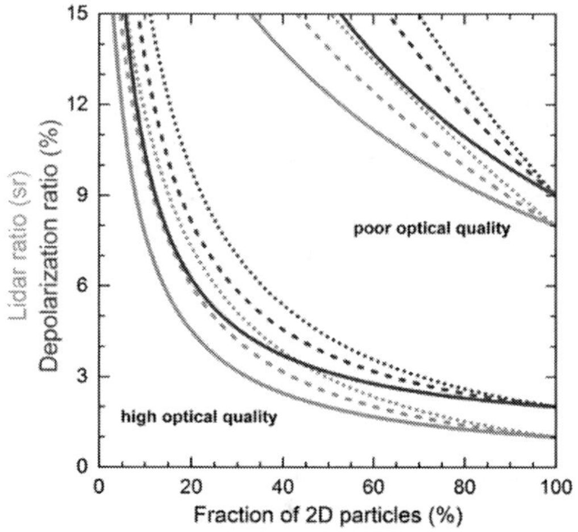

Figure 4. Cirrus lidar ratio and depolarization ratio as functions of the fraction of horizontally aligned ice particles. For the calculations, common 3D values of 25 sr and 40%, respectively, are selected. Assumed 2D values fall in the ranges of the lowest 2D lidar and depolarization ratios ever measured with RAMSES (high optical quality, 1 – 2 sr, 1% – 2%) or are chosen to illustrate the effect of poor particle optical quality (8 sr, 9%). Results are presented for aspect ratios of 0.5 (solid curves), 1.0 (dashed curves), and 1.5 (dotted curves).

REFERENCES

[1] J. Reichardt, et al. Appl. Opt. 51, 8111-8131 (2012)
[2] J. Reichardt, J. Atmos. Oceanic Technol. 31, 1946-1963 (2014)
[3] R. Hogan, J. Atmos. Sci. 65, 3621-3635 (2008)
[4] J. Schmidt, et al. Appl. Opt. 52, 2235-2247 (2013)
[5] T. Plakhotnik, et al. J. Quant. Spectrosc. Radiat. Transf. 194, 58-64 (2017)
[6] T. Plakhotnik, et al. J. Quant. Spectrosc. Radiat. Transf. 208, 172-178 (2018)
[7] A. H. Auer, Jr., et al. J. Atmos. Sci. 27, 919-926 (1970)
[8] J. Reichardt, et al. J. Geophys. Res. 107, doi: 10. 1029/2002JD002589 (2002)

https://doi.org/10.1051/epjconf/202023708016

OPTICAL PROPERTIES OF CANADIAN BIOMASS BURNING PARTICLES OVER EUROPE OBSERVED WITH CALIPSO AND GROUND-BASED LIDAR SYSTEMS

Christina-Anna Papanikolaou[1, 2*], Elina Giannakaki[1,3], Alex Papayannis[2], Maria Tombrou[1], Maria Mylonaki[1, 2] and Ourania Soupiona[2]

[1]Department of Environmental Physics and Meteorology, Faculty of Physics, National and Kapodistrian University of Athens, 157 72 Zografou, Greece
[2]Laser Remote Sensing Unit, Physics Department, School of Applied Mathematics and Physical Sciences, National Technical University of Athens, 157 80 Zografou, Greece
[3]Finnish Meteorological Institute, Kuopio, Finland
**Email: papanikolaou.christiann@gmail.com*

ABSTRACT

A long-lasting biomass burning event affected Europe from 27 August to 3 September 2018. The biomass burning aerosol layers were observed with ground- and space-based lidars in heights ranged between 2-7 km (a.s.l.). The mean backscatter coefficient for the ground-based stations ranged between 0.29 and 1.51 $Mm^{-1}sr^{-1}$, while the CALIPSO retrieved values ranged between 0.43 and 1.83 $Mm^{-1}sr^{-1}$. Moreover, the mean Ångström exponent (AE_b) values, relevant to backscatter, ranged from 0.83 to 1.04 for the aforementioned lidar stations. At the same time, the mean AE_b values obtained from CALIPSO ranged between 0.17 and 1.89. The mean particle depolarization ratio ranged between 0.037 and 0.080.

1. INTRODUCTION

Biomass burning particles (BBPs) through interactions with radiation and clouds play an important role in the Earth's climate system as they scatter and absorb solar radiation (so-called direct effect) while influencing the cloud processes as cloud condensation nuclei (CCN) (so called indirect effect) [1-3].

Forest fires are a significant source of BBPs, especially at northern latitudes especially in spring and summer period [4]. It has been shown that large smoke plumes from large forest fires can be injected into the free troposphere, and then easily transported by air masses along the Earth [1,5-6].

In this context, the CALIPSO (Cloud-Aerosol Lidar and Infrared Pathfinder Satellite Observation) mission provides the opportunity to represent the four-dimensional distribution of aerosols and clouds, offering global coverage with high spatial resolution [8]. Nevertheless, CALIPSO observations offer only snapshots and therefore only the synergic use with terrestrial systems, passive remote sensing measurements and models allows a detailed description of the spatial and temporal distribution of biomass burning long transport incidents and its evolution on a global scale.

In order to increase and validate the accuracy of aerosol optical properties retrieved from CALIPSO's products, a comparison with ground-based lidar observations is required. Thanks to its geographic coverage and the deployment of advanced Raman aerosol lidars, EARLINET offers a unique opportunity for the validation of the CALIPSO [7, 8].

In this contribution, we study an extreme biomass burning aerosol event, in terms of geometrical and optical properties of the biomass burning plume. Wildfires started in August 2018 from the Canadian Province of British Columbia, causing enormous forest areas to burn up. The BBPs travelled across the Atlantic Ocean and affected Europe between 26th of August and early September. The event was studied with the synergy of data from the CALIOP (Cloud-Aerosol Lidar with Orthogonal Polarization) lidar system on board the CALIPSO satellite, the EARLINET network, the NASA's tool Giovanni and the HYSPLIT (Hybrid Single-Particle Lagrangian Integrated Trajectory) model.

2. METHODOLOGY

2.1 CALIOP lidar System

CALIPSO is an Earth Science observation mission that was launched on 28 April 2006. It flies at a nominal orbital altitude of 705 km. The CALIPSO mission includes active lidar and passive sensors to obtain unique data on aerosol and cloud vertical structure, cloud particle phase and classification of the aerosol size.

CALIOP is an elastically backscattered lidar operating at 532 and 1064 nm, equipped with a depolarization channel at 532 nm that provides high-resolution vertical profiles of aerosols and clouds [9,10]. The Level 2 CALIPSO products include a cloud layer product, an aerosol layer product and an aerosol profile product (backscatter, extinction, and depolarization ratio) [11]. In this study, the aerosol optical properties reported in the Level 2 products (version 3.40) are used.

2.2 Passive satellite products

The Moderate Resolution Imaging Spectroradiometer (MODIS), a major NASA EOS instrument, was launched aboard the Terra satellite in 1999 for global monitoring of the atmosphere, terrestrial ecosystems, and oceans, while in 2002, a similar instrument was launched on the EOS-Aqua. MODIS, flying on two satellites in orbits that complement each other, provide the diurnal variation of the rapidly varying parameters for the study of climate and global change studies [12].

Giovanni is an online tool for visualization, exploration, and analysis of NASA Earth Science data. Atmospheric Optical Depth (AOD) observations from several instruments can be accessed through Giovanni. Total and fine mode AOD from MODIS, as well as from other instruments are available [13]. The Giovanni tool was used to verify the path of smoke's motion and represent the AOD at 550 nm, according to MODIS (Aqua & Terra) data for temporal resolution equal to one day and spatial resolution of 1° (not shown here).

2.3 Ground-based lidar stations

Four ground-based lidar stations, within EARLINET, were selected based on their data availability at time periods that CALIOP detected smoke over Europe. The stations, along with their geophysical coordinates and the available data products, are presented in Table 1 [8]. The corresponding profiles of the aerosol backscatter (β) and extinction (α) coefficients, $AE_{b,a}$ and depolarization ratio (δ) of ground-based lidar measurements were used to validate the space-borne products for this biomass burning event.

Table 1: Location and available data products for 4 EARLINET stations used in this study.

EARLINET stations	
Location	Data Products Available
Athens, Greece 37.96°N, 23.78°E (212 m a.s.l.)	β (532, 1064 nm) δ (355 nm)
Leipzig, Germany 51.35° N, 12.43° E (90 m a.s.l.)	β (532, 1064 nm) α (532 nm)
Madrid, Spain 40.456°N, 3.729°W (669 m a.s.l.)	β (532, 1064nm) α (532, 1064nm)
Thessaloniki, Greece 40.6300°N, 22.9500°E (50 m a.s.l.)	β (532, 1064 nm) α (532 nm)

2.4 Trajectory Analysis

The analysis of the forward trajectories was performed using the ensemble calculation of HYSPLIT model [14,15]. Multiple trajectories start from the selected starting point, and each member of the trajectory ensemble is calculated by offsetting the meteorological data by a fixed grid factor, resulting in 27 members for all-possible offsets in longitude, latitude and altitude. The 12-day-forward analysis was performed for each one of the days that active fires and smoke were observed upon Canada. As long as the aerosol layers over Canada identified as smoke by CALIPSO, their coordinates and altitude were used as initial values for the model.

3. RESULTS

3.1 CALIPSO smoke observation

Active fires in the Canadian region were detected by CALIOP and MODIS, between 16 and 22.08.2018 as presented in Figures 1 and 2.

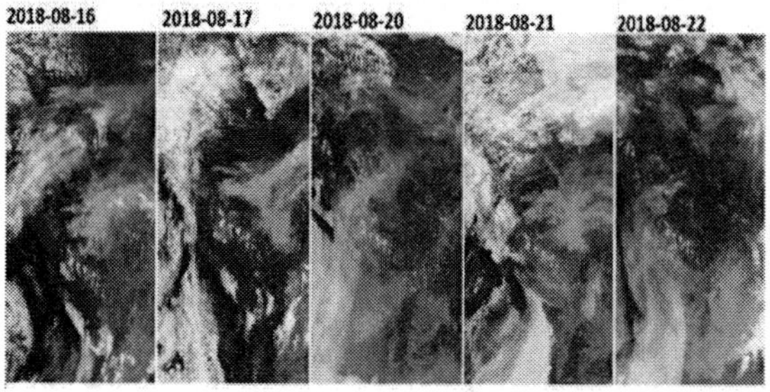

Figure 1. Active fires and smoke detected in British Columbia, Canada during 16-22 August 2018 (Terra and Aqua/ MODIS).

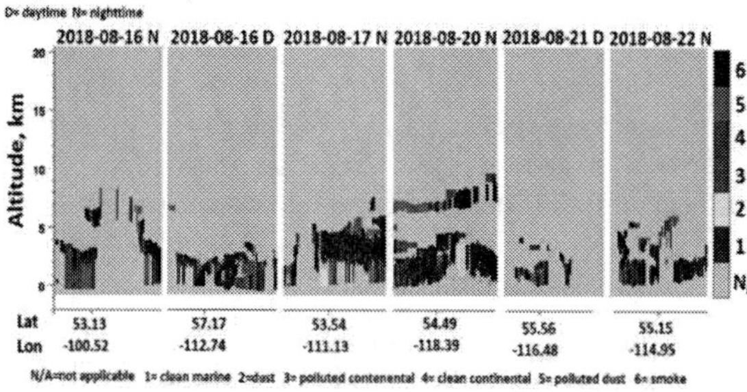

Figure 2. Different types of the aerosol layers detected by CALIPSO for each of the overpasses upon British Columbia, Canada, during 16-22 August 2018. Black color indicates smoke aerosol particles.

The Canadian wildfire plume has traveled in the free troposphere and reached Europe several days later. The aerosol layers were detected by CALIOP and four EARLINET stations [Athens (Greece), Leipzig (Germany), Madrid (Spain) and Thessaloniki Greece)]. These ground-based stations have been used for comparison with CALIPSO Level 2 version 3.40 aerosol products on 27, 29, 30 August and 03 September 2018 (Figs. 3-4).

As shown in Fig. 4, the CALIOP observations show aerosol layers in the 2–7 km altitude range above

https://doi.org/10.1051/epjconf/202023708016

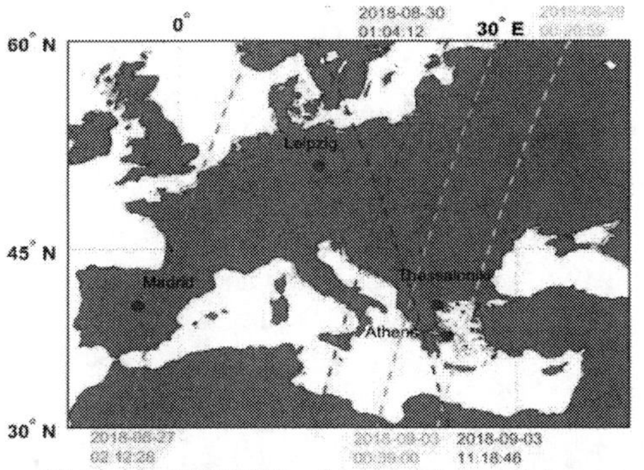

Figure 3. CALIPSO orbits on 27, 29, 30 August and 03 September 2018. Red dot correspond to Athens, Leipzig, Madrid and Thessaloniki lidar stations.

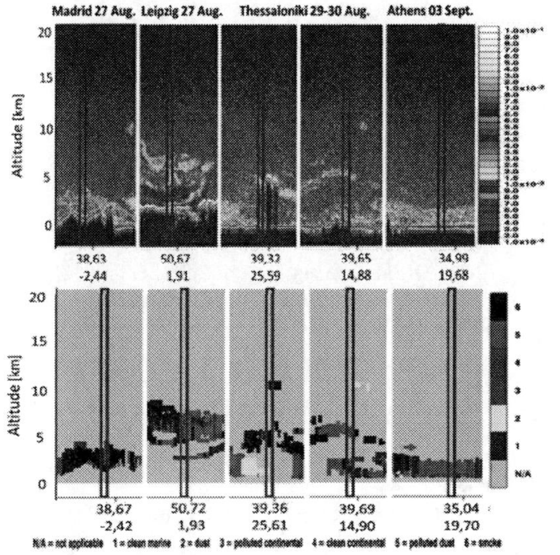

Figure 4. CALIPSO total attenuated backscatter at 532 nm and aerosol subtyping, for each of the overpasses tracking the masses arriving at Madrid, Leipzig, Thessaloniki and Athens. Red lines correspond to the nearest overpass over each lidar station.

Europe. According to subtyping, aerosol layers correspond to smoke particles and particles of polluted dust (dust mixed with smoke) [16].

Moreover, Satellite observations from MODIS instrument, NASA's tool GIOVANNI, as well as modeling tools such as HYSPLIT, have been used to estimate transport paths and the source of those forest fire smoke particles (not shown here).

3.2 Comparison of CALIPSO and EARLINET optical profiles

In principle, all CALIPSO level 2 data can be directly compared with EARLINET data for nearby overpasses and stations providing data at 532 and 1064 nm [7]. For the period from 27 August to 03 September for every station with available data through CALIPSO overpasses, the vertical profiles of aerosol optical properties compared to the ones of CALIPSO are plotted. The CALIPSO aerosol retrievals are performed at 5, 25, 45 and 105 km spatial resolution. Indicatevely, the profiles for Madrid and Athens lidar stations are presented in Figs. 5-6.

CALIPSO overpassed Madrid during the night of 27th of August, the minimum distance from the orbit equals to 123.878 km. The aerosol layer observed both from CALIOP and ground-based station, ranged between 2 and 3.5 km, was attributed to smoke particles, with a value of 0.95±0.05 $Mm^{-1}sr^{-1}$ for the backscatter coefficient while the mean value measured by Madrid station was equal to 1.03±0.02 $Mm^{-1}sr^{-1}$. The agreement between the two coefficients, calculated in (%) (CALIPSO-EARLINET/EARLINET), gives a 7.8% difference. The extinction coefficient at 532 nm was equal to 62.1±4.6 Mm^{-1} and 41.2±0.7 Mm^{-1} measured by CALIPSO and ground-based station respectively. The mean backscatter-related Ångström

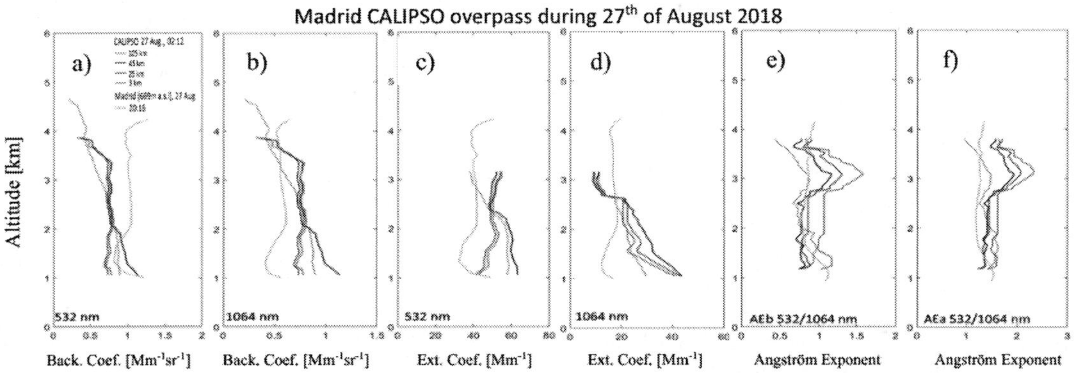

Figure 5: Aerosol backscatter coefficient profiles at 532 and 1064 nm (a,b), extinction coefficient profiles at 532 and 1064 nm (c,d), backscatter and extinction Ångström exponent at 532/1064 nm (e,f), measured over Madrid station on 27 August (20.16 UTC) and the corresponding profiles at 532 and 1064 nm derived from CALIPSO on 27 August (02.12 UTC).

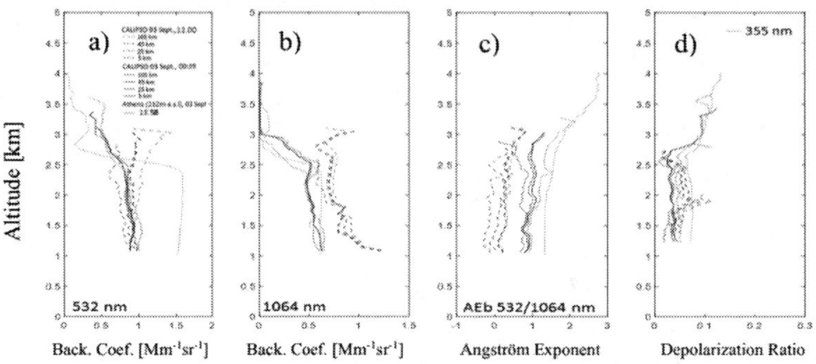

Figure 6: Aerosol backscatter coefficient profiles at 532 and 1064 nm (a,b), Ångström exponent at 532/1064 nm (c) and particle depolarization ratio at 355 nm (d), measured at Athens station on 03 September (11.58 UTC). The dashed lines correspond to profiles at 532 and 1064 nm derived from daytime CALIPSO orbit on 03 September (12.00 UTC), while the continuous lines from nighttime orbit (00.45 UTC).

exponent value from CALIPSO was 0.86±0.15 while in Madrid station had a mean value of 0.83±0.04. The mean extinction-related Ångström exponent value from CALIPSO was 1.45±0.13 while in ground-based station had a mean value of 1.24±0.04.

During the CALIPSO overpass over Athens on 3 September, the minimum distance from the daytime and nighttime orbit was 173.253 and 275.387 km, respectively. Two mixed biomass burning aerosol layers were observed between 2 and 2.5 km and between 3 and 3.5 km height. The mean value of the aerosol backscatter coefficient for the 2km-layer, obtained from both orbits was found equal to 0.88±0.11 $Mm^{-1}sr^{-1}$ and 0.43±0.03 $Mm^{-1}sr^{-1}$ for the 3 km-layer, while the corresponding value measured by Athens station was 1.51±0.08 $Mm^{-1}sr^{-1}$ and 0.29±0.10 $Mm^{-1}sr^{-1}$, respectively. The agreement between these coefficients, calculated in (%), gives a 42% for the lower layer and a 48% difference between CALIPSO and the ground-based station. The mean value of the AE_b obtained by CALIPSO ranged from 0.17 to 1.89, while the same parameter for the ground based system was 1.54±0.37. The mean particle depolarization ratio of 0.075 measured at Athens station, agrees with the CALIPSO value of 0.080.

ACKNOWLEDGMENTS

We acknowledge support of this work by the project "PANhellenic infrastructure for Atmospheric Composition and climatE change" (MIS 5021516) which is implemented under the Action "Reinforcement of the Research and Innovation Infrastructure", funded by the Operational Programme & Competitiveness, Entrepreneurship and Innovation & (NSRF 2014-2020) and co-financed by Greece and the European Union (European Regional Development Fund). Part of this research has been co-financed by additional funding from the European Union 7th Framework Programme (FP7/2007-2013) under grant agreement No 262254.

Με τη συγχρηματοδότηση της Ελλάδας και της Ευρωπαϊκής Ένωσης

REFERENCES

[1] IPCC AR5, Summary for Policymakers (2013)
[2] P. Ortiz-Amezcua,et al., Atmos. Chem. Phys., 17, 5931-5946 (2017)
[3] I. McKendry, et al., Atmos. Chem. Phys., 11, 465–477 (2011)
[4] S. Generoso, et al., Atmos. Chem. Phys., 3, 1211-1222 (2003)
[5] G. Ancellet, et al., Atmos. Chem. Phys., 16, 4725–4742 (2016)
[6] J. L. Guerrero-Rascado, et al., Proc. 26th ILRC, July 5-9, 2010, St.-Petersburg, Russia (2010)
[7] G. Pappalardo, et al., J. Geophys. Res., 115, 148-227 (2010)
[8] G. Pappalardo, et al., Atmos. Meas. Tech., 7, 2389–24092014 (2014)
[9] D. M. Winker, et al, Geophys. Res. Lett., 34 (2007)
[10] W.H Hunt, et al., J. Atmos. Oceanic Technol., 26, 1214–1228 (2009)
[11] R. E. Mamouri, et al., Atmos. Meas. Tech., 2, 513-522 (2009)
[12] A. Savtchenko, et al., Advances in Space Research 34, 710–714 (2004)
[13] A. Prados, et al., IEEE J. Sel. Top. Appl. Earth Obs. Remote Sens., 3, 1939-1404 (2010)
[14] A. F. Stein, et al., Bull. Amer. Meteor. Soc., 96, 2059–2077 (2015)
[15] G. Rolph, et al, Env. Mod. & Soft., 95, 210-228 (2017)
[16] A. H. Omar, et al., Jour. of Atm. and Ocean. Tech., 26, 1994-2014 (2009)

POLARIZATION OF RADIATION BY THE AEROSOL-GAS COMPONENT OF THE ATMOSPHERE FOR LIDAR WAVELENGTHS.

ANNA ZIMOVAYA[1] , ALEXANDER KONOSHONKIN[2*]

[1]*Laboratory of Optical Signals Propagation, V.E. Zuev Institute of Atmospheric Optics SB RAS, Tomsk, 634055, Russia*

[2]*Laboratory of Atmospheric Radiation, V.E. Zuev Institute of Atmospheric Optics SB RAS, Tomsk, 634055, Russia*

**Email: avk@iao.ru*

ABSTRACT

The report presents an original algorithm for solving the radiation transfer equation based on the Monte Carlo method for problems of atmospheric laser sensing, taking into account the polarization of radiation. A number of test calculations were performed to analyze the data of the polarization scanning lidar of the V.E. Zuev Institute of Atmospheric Optics SB RAS. It is shown that the proposed algorithm, taking into account the polarization, makes it possible to reduce the discrepancy in interpreting lidar data in comparison with the algorithm without taking into account polarization.

1. INTRODUCTION

In problems of atmospheric laser sensing, the interaction of radiation with aerosol is paid close attention. As a rule, to take into account the influence of the aerosol component, it is necessary to solve the radiation transfer equation. The most suitable method for solving the radiation transfer equation is the Monte Carlo method, since it allows us to perform calculations with controlled accuracy. For problems of polarization laser sensing, it is necessary to take into account the influence of polarization. The paper presents the original algorithm developed by the authors, which allows to solve the radiation transfer equation taking into account the polarization.

2. METHODOLOGY

The simulation of energy flow with allowance for radiation polarization in the considered formulation of the problem was performed by the Monte Carlo method with local estimates at points of collision by the backward scheme. In contrast to the kernel of the scalar integral radiation transfer equation (RTE), the kernel of vector integral RTE for the Stokes vector parameters is not self-conjugated, which introduces some specificity in comparison with the scalar case. For simulation of the radiation transfer with allowance for polarization by the backward scheme with the use of local estimates at every point of collision, it is necessary to store the product of scattering matrices at the previous collisions. That is, at every new collision the following matrix should be formed:

$$M_i = M_{i-1} \cdot P(\boldsymbol{\omega}', \boldsymbol{\omega}, \mathbf{r})$$

$$P(\boldsymbol{\omega}', \boldsymbol{\omega}, \mathbf{r}) = L(\pi - i_2) R(\mu, \mathbf{r}) L(-i_1)$$

$$R(\mathbf{r}, \mu) = \frac{\sigma_{s,a}(\mathbf{r})}{\sigma_{t,a}(\mathbf{r}) + \sigma_{t,m}(\mathbf{r})} \times \times \frac{3}{4} \begin{pmatrix} \frac{(1+\mu^2)}{2} & -\frac{(1-\mu^2)}{2} & 0 & 0 \\ -\frac{(1-\mu^2)}{2} & \frac{(1+\mu^2)}{2} & 0 & 0 \\ 0 & 0 & \mu & 0 \\ 0 & 0 & 0 & \mu \end{pmatrix} + + \frac{\sigma_{s,m}(\mathbf{r})}{\sigma_{t,a}(\mathbf{r}) + \sigma_{t,m}(\mathbf{r})} \cdot \begin{pmatrix} r_{11} & r_{12} & 0 & 0 \\ r_{12} & r_{22} & 0 & 0 \\ 0 & 0 & r_{33} & r_{34} \\ 0 & 0 & -r_{34} & r_{44} \end{pmatrix},$$

where $\boldsymbol{r}$ is the radius-vector of the scattering point; L is the rotation matrix; R is the total scattering matrix; P is the phase scattering matrix; μ is the cosine of the scattering angle, r_{11} , r_{12} , r_{33} , r_{34} are elements of the aerosol scattering matrix on the assumption of homogeneity and sphericity of aerosol particles [1], i_1 and i_2 are the angles between the scattering plane and the planes passing through the vertical at the point of

 https://doi.org/10.1051/epjconf/202023708017

collision and the vectors of direction before and after scattering, respectively.

Then, in local estimates, the received radiation is determined as a product of the obtained matrix M_i and initial values of the Stokes vector-parameter. The values of elements of the aerosol scattering matrix depend on the state of aerosol and should be determined for every particular situation.

2. MODEL OF THE ATMOSPHERE WITH ALLOWANCE FOR THE RADIATION POLARIZATION

For development of efficient models of the atmosphere, NASA organized the AERONET network of stations for ground-based observation of the state of the atmosphere [2]. The data of measurement of the aerosol state are available at [2]. These data include the aerosol optical depth (AOD) of the atmosphere, complex refractive index of aerosol particles, particle size distribution, and scattering phase functions. The scattering phase functions are given only for wavelengths of 0.34, 0.38, 0.44, 0.5, 0.675, 0.87, 1.02, and 1.64 µm. However, for operation with actual lidar data, it is necessary to have the scattering phase functions for the lidar wavelengths. Toward this end, the elements of the aerosol scattering matrix were calculated from the data on the complex refractive index and the particle size distribution by the program based on [3]. An example of calculation, the measurement date is shown in Fig. 1.

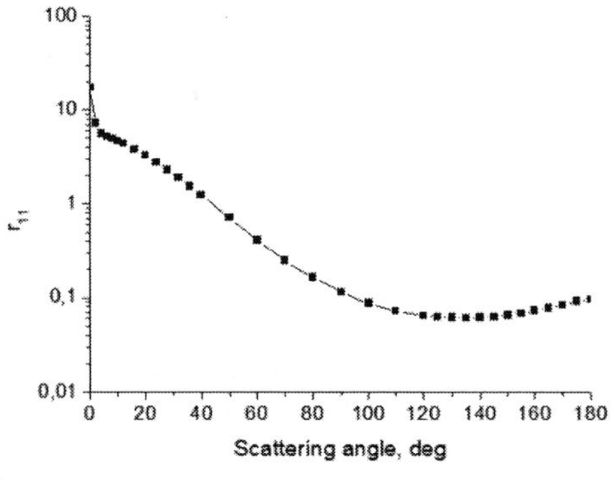

a)

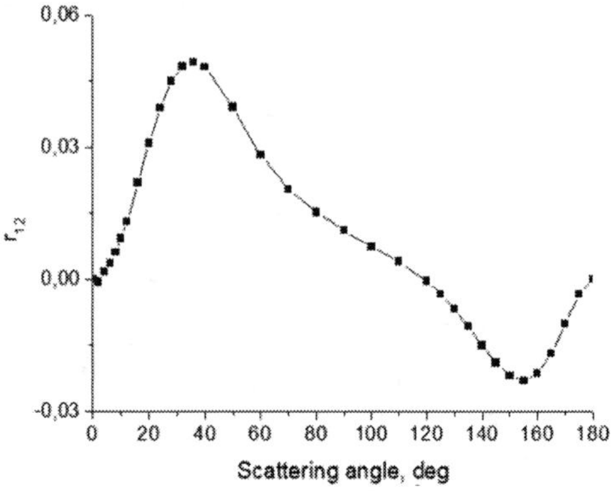

b)

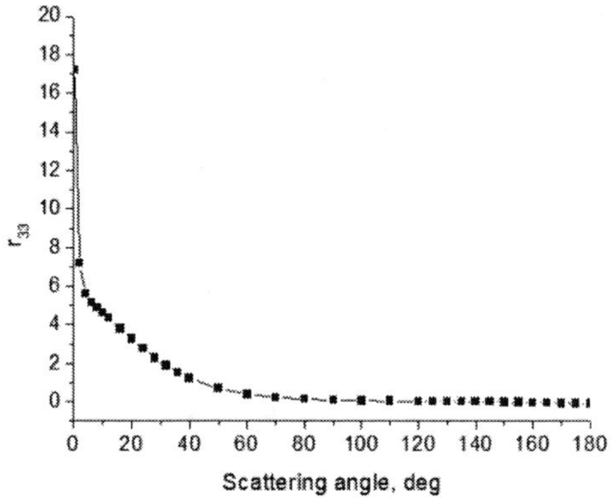

c)

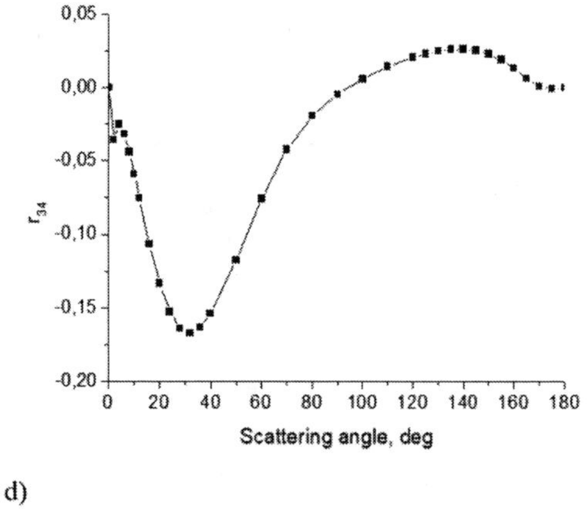

d)

https://doi.org/10.1051/epjconf/202023708017

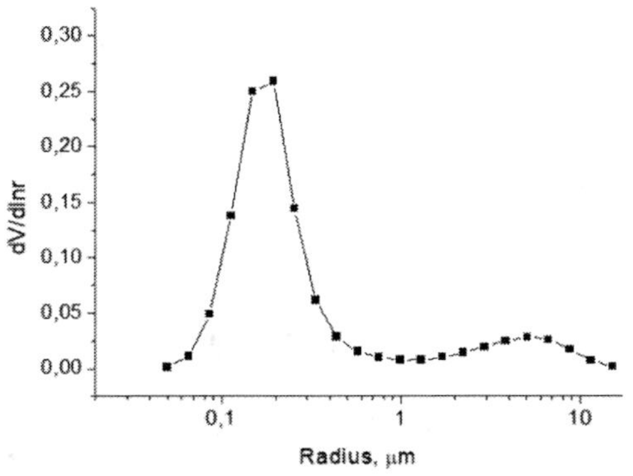

e)

Figure 1. a-d) elements of aerosol scattering matrix r_{11}, b) r_{12}, c) r_{13}, d) r_{14}, e) Particle size distribution function

3. RESULTS OF NUMERICAL CALCULATIONS

The proposed algorithm supplemented with the program for calculation of light scattered in atmosphere with polarization taken into account was tested for a southern area of the Tomsk Region with coordinates of 55.95–56.85^0 N and 84.05-84.95^0 E for scanning lidar of V.E. Zuev Institute of Atmospheric Optics. The correction was made for λ=0.535 μm, 1.064 μm. In the near-surface layer (up to 5 km), the aerosol scattering matrix was calculated from measurements of the AERONET station for the corresponding days. For the upper layers, the changing of polarization was neglected. The vertical profile of the optical coefficients of aerosol was set by the LOWTRAN-7 [4] model closest in AOD to the AERONET measurements. The point with the coordinate 56.5°N, 84.3°E, was selected as the test point. The results of comparison of average errors of retrieval are shown in Figure 2.

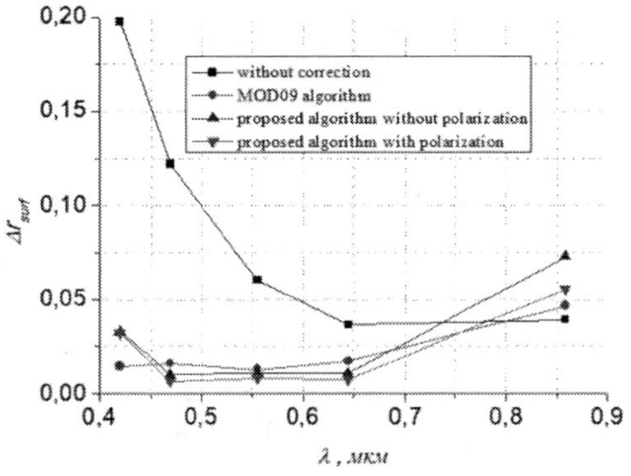

Figure 2. Comparison of average differences between algorithm without polarization and proposed algorithm with polarization.

The figure shows that the allowance for the radiation polarization and the use of AERONET measurements allow us to significantly decrease the error of the proposed algorithm. It can be important for lidar applications.

REFERENCES

[1] H. C. van de Hulst., Light scattering by small particles. By New York, London, 1957. Pp. 470;

[2] Aerosol Robotic Network (AERONET), http://aeronet.gsfc.nasa.gov.

[3] Bohren C.F., Huffman D.R., “Absorption and Scattering of Light by Small Particles,” Wiley, 530 (1983).

[4] Kneizys F.X., Shettle E.P., Anderson G.P., Abreu L.W., Chetwynd J.H., Selby J.E.A., Clough S.A., Gallery W.O. User Guide to LOWTRAN-7. – ARGL-TR-86-0177. ERP 1010. Hanscom AFB. 1988. MA 01731. 137 p.

RETRIEVAL OF AEROSOL OPTICAL PROPERTIES BASED ON HIGH SPECTRAL RESOLUTION LIDAR

Da Xiao[1], Tianfen Zhong[1], Xue Shen[1], Nanchao Wang[1], Yuhang Rong[1], Chong Liu[1], Yupeng Zhang[1], Dong Liu[1*]

[1] *State Key Laboratory of Modern Optical Instrumentation, College of Optical Science and Engineering, Zhejiang University, Hangzhou 310027, China*

*Email: *liudongopt@zju.edu.cn*

ABSTRACT

The detection of clouds and aerosols is important for climate research. Lidar has been widely used in atmospheric remote sensing research because of its high spatial and temporal resolution and ability to detect profiles. High spectral resolution lidar (HSRL) accurately calculates the optical properties of aerosols and clouds without relying on any assumptions. Based on the 532nm iodine HSRL system, the lidar ratio of the urban aerosol in Hangzhou is 40-50sr, and the average lidar ratio of the cirrus is 24.79sr, demonstrating that the HSRL system and retrieval algorithms accurately obtain the optical properties of clouds and aerosols.

1. INTRODUCTION

Aerosols and clouds play an important role in global and local climate and becoming two of the largest uncertainty sources in climate sensitivity research [1, 2]. Aerosols directly affect the solar and terrestrial radiation in the atmosphere by scattering and absorbing light, influencing the Earth's radiation budget, while clouds play a major role in determining The Earth's climate [3]. The acquisition of cloud and aerosol optical parameters and vertical profile information is of great significance for atmospheric research.

Lidar, regarded as the only means to obtain the vertical profile distribution characteristics of clouds and aerosols, has been widely used. The traditional lidar cannot directly calculate the extinction coefficient. The extinction coefficient can only be calculated with the assumption of the ratio of the extinction coefficient to the backscatter coefficient [4]. However, the aerosol lidar ratio is a highly variable quantity, so that large errors in the retrieval must be expected if the lidar ratio is not known exactly [5]. Raman lidar and high spectral resolution lidar (HSRL) can accurately invert the extinction coefficient without any assumptions. Raman lidars are limited by low signal-to-noise ratios and require long time averaging, which is not suitable for daytime measurements [6]. The HSRL system cleverly uses the spectral difference of the lidar echo signals to separate the aerosol return from the molecular return through a spectral discriminator with high spectral resolution, so the extinction coefficient and lidar ratio can be accurately measured.

This paper briefly introduces the principle of retrieval and describes the process of data retrieval to obtain the optical properties of atmospheric aerosols in detail.

2. METHODOLOGY

2.1 The principle of retrieval

A general 532nm iodine HSRL system block diagram with abstract structure is shown in Fig.1. A bundle of parallel light received by the receiving system is split by the PBS, and the reflected beam is received by a PMT as a mixed vertical channel. The PBS transmitted light consists of parallel components, which are split by a splitting prism (BS), and the reflected light is received by the detector as a mixed parallel channel. The transmitted light of the BS passes through an iodine pool and is received by the detector as the molecular channel.

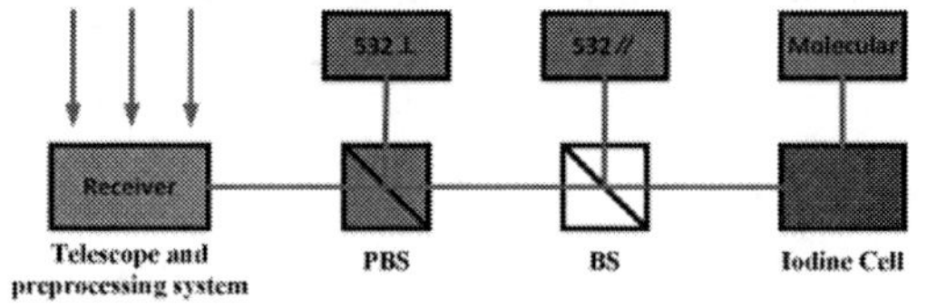

Fig.1 A general 532nm iodine HSRL system block diagram with abstract structure.

The echo signals received by the three detectors can be represented by Equation 1.1.

$$B_c^{\perp}=(\beta_a^{\perp}+\beta_m^{\perp})e^{(-2\tau)}$$
$$B_c^{//}=(\beta_a^{//}+\beta_m^{//})e^{(-2\tau)} \quad (1.1)$$
$$B_m^{//}=(\beta_a^{//}T_a+\beta_m^{//}T_m)e^{(-2\tau)}$$

where $B_c^{\perp}$, $B_c^{//}$ and $B_m^{//}$ are the attenuation scatter signals obtained by the mixed vertical channels, mixing the parallel channels and the molecular channels, respectively. β_a and β_m are the backscattering coefficients of aerosol and atmospheric molecules, respectively. The superscripts "$\perp$" and "//" indicate their vertical polarization and parallel polarization components, respectively. τ is the atmospheric optical thickness, T_m and T_a are the transmittances of the spectral discriminator for the molecular scattering signal and the aerosol scattering signal, respectively.

Combining the three signals, we can get the atmospheric backscatter coefficient

$$\beta=\beta_m\frac{(1+\delta)(T_m-T_a)K}{(1+\delta_m)(1-T_aK)} \quad (1.2)$$

Where δ_m is the depolarization ratio of atmospheric molecules, which can be accurately obtained with atmospheric model. δ is the total depolarization ratio of atmospheric molecules and aerosols, which can be obtained by the ratio of the mixed vertical channel to the mixed parallel channel, as in Equation 1. 3.

$$\delta=\frac{B_c^{\perp}}{B_c^{//}} \quad (1.3)$$

The definition of K is similar to the depolarization ratio, which can be obtained by the ratio of mixed parallel channels to molecular channels, as in Equation 1.4.

$$K=\frac{B_c^{//}}{B_m^{//}} \quad (1.4)$$

Meanwhile, we can get the expression of atmospheric optical thickness in Equation 1.5.

$$\tau=-\frac{1}{2}\ln\left(\frac{(1-T_aK)(1+\delta_m)B_m^{//}}{(T_m-T_a)\beta_m}\right) \quad (1.5)$$

After obtaining the basic quantities β and τ, we can perform the retrieval of the aerosol extinction coefficient α_a, as in Equation 1.6.

$$\alpha_a=\frac{\partial\tau}{\partial r}-\alpha_m \quad (1.6)$$

Where α_m is the extinction coefficient of atmospheric molecules.

2.2 Retrieval algorithm and process

A HSRL system based on 532nm iodine cell has been set up by Zhejiang University. Based on this system, we developed a retrieval algorithm for calculating the optical properties of atmospheric aerosols, as shown in Fig.2.

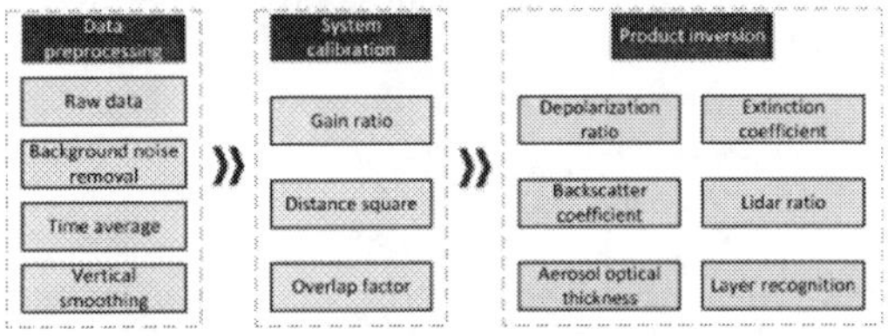

Fig.2 HSRL retrieval algorithm block diagram

The retrieval algorithm is divided into three modules, data pre-processing, system calibration, and product retrieval. The purpose of the data pre-processing module is to remove background noise and improve signal to noise ratio. The original lidar echo signal can be represented by Equation 1.7.

$$S(z,\lambda)=S_a(z,\lambda)+S_m(z,\lambda)+S_{atm}(\lambda)+S_{el} \quad (1.7)$$

Where $S_a(z,\lambda)$ and $S_m(z,\lambda)$ are the signal contributions backscattered by aerosol particles (a) and molecules (m) at altitude z and at wavelength λ, $S_{atm}(\lambda)$ is the optical signal background from the atmosphere, which is independent of range, S_{el} represents circuit background noise, depending on the electronic effects of signal detection and data acquisition. The constant background components $S_{atm}(\lambda)$ and S_{el} can be determined by the long range of lidar signals, far enough to make the expected contribution of atmospheric backscatter negligible, or to utilize the pre-trigger range before the laser pulse [7].

Time integration and vertical smoothing are used to reduce noise and improve signal to noise ratio. The time average is limited to around 1000 profiles to ensure adequate high time resolution. For vertical smoothing, there are common methods such as wavelet denoising, EMD noise reduction, S-G filtering and Butterworth filtering.

System calibration is used to calibrate the effects of system parameters on the retrieval results, which includes the gain of each channel, overlap factor and the distance squared correction. The gain ratio correction for each channel eliminates the effects of different detectors and acquisition amplifier circuits on signal gain. Distance correction primarily considers the trigger delay effect due to the delay in the timing of the laser output pulse and the timing at which the digital acquisition system begins recording the laser profile. The correction of the overlap factor helps to correctly invert the near-field aerosol. In the HSRL system, accurate measurement of the overlap factor is a problem that is significant for the retrieval of optical thickness and extinction coefficient. For the HSRL system developed by Zhejiang University, to correct the overlap factor effectively, we specially add an auxiliary channel with a telescope in small size, which is shown in Fig.4.

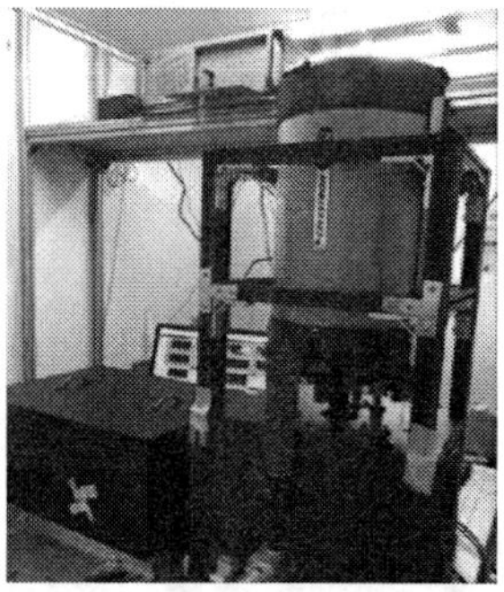

Fig 4 A receive module schematic diagram of HSRL

As mentioned before, HSRL has the advantage of inverting aerosol optical parameters without any assumptions. A polarized HSRL system can not only calculate the optical characteristic parameters of the particles, such as backscattering coefficient and extinction coefficient, but also invert the depolarization ratio and extinction-to-backscatter ratio that reflect the microscopic physical properties of the particles. Extinction coefficient, obtained by the derivative of the optical thickness, is the main cause of the fluctuation of lidar ratio. A Savitzky–Golay filter of first order is used for the numerical derivative of the AOT profile [5]. This is equivalent to a linear fit of the specified number of points in the moving window, with the slope of the fitted line for each point as the derivative value.

With the obtained optical properties, we can get more accurate other atmospheric information, such as boundary layer and atmospheric layer information.

3. RESULTS

With long-term observation HSRL data, we verify the feasibility of the retrieval algorithm. Figure 3 shows the data retrieval result during the measurement on March 12, 2019 at the Yuquan Campus of Zhejiang University. Figure 3(a)-(d) are the depolarization ratio, the backscattering coefficient, the extinction coefficient, and the extinction-to-backscatter ratio, respectively. There is a strong aerosol load below 3km, and there is a cloud at an altitude of about 6km. The average backscatter coefficient and depolarization ratio is $0.004 km^{-1}sr^{-1}$ and 0.1 in aerosol layer, respectively. The average value of the extinction coefficient is $0.18\ km^{-1}$, and the lidar ratio of the urban aerosol in Hangzhou is 40-50sr. For the cloud layer, the values are 0.37, $0.04 km^{-1}sr^{-1}$, $1.02\ km^{-1}$, 24.80sr, respectively..

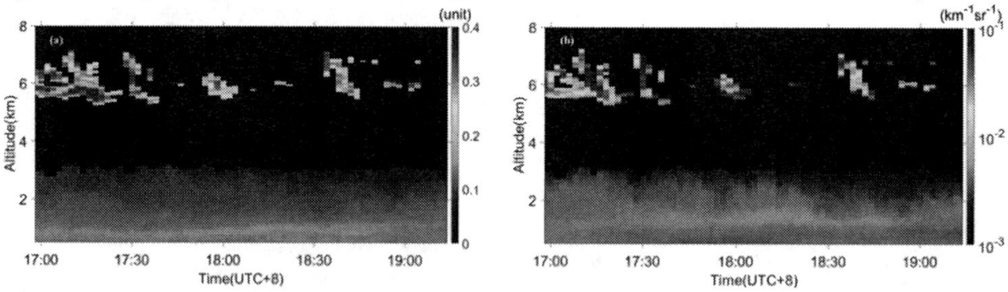

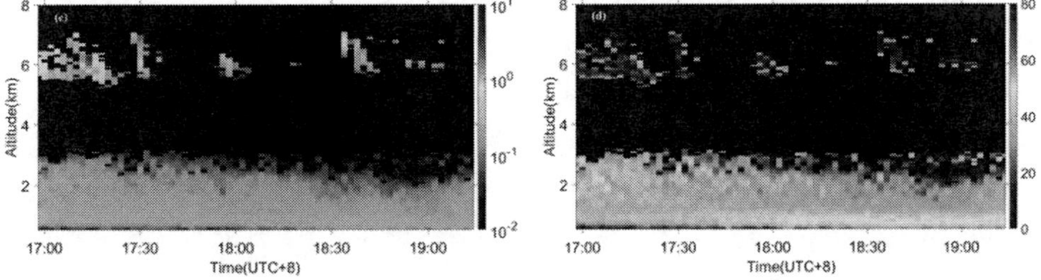

Fig 3 The data retrieval result during the measurement on March 12, 2019 at the Yuquan Campus of Zhejiang University. These four pictures are: (a) the depolarization ratio; (b) the backscatter coefficients; (c) the extinction coefficients; (d) the extinction-to-backscatter ratio.

The results are in good agreement with the data range of urban aerosols and cirrus clouds, which demonstrates that the HSRL system and retrieval algorithms accurately obtain the optical properties of clouds and aerosols.

ACKNOWLEDGEMENTS

This work was supported by the National Key Research and Development Program of China (2016YFC0200700, 2016YFC1400905), National Natural Science Foundation of China (41775023, 41305023, 61475141), Natural Science Foundation of Zhejiang Province (LR19D050001).

REFERENCES

[1] Liou, K. N. . "An Introduction to Atmospheric Radiation." Physics Today 34.7(1981):66-67.

[2] Mao, Feiyue , W. Gong , and Z. Zhu . "Simple multiscale algorithm for layer detection with lidar." Applied Optics 50.36(2011):6591-8.

[3] Hair, J. W. , et al. "Airborne high spectral resolution lidar for profiling aerosol optical properties. " Applied Optics 47.36(2008):6734-52.

[4] Klett, James D. . "Stable analytical retrieval solution for processing lidar returns." Applied Optics 20.2(1981):211-20.

[5] Esselborn, Michael , et al. "Airborne high spectral resolution lidar for measuring aerosol extinction and backscatter coefficients." Applied Optics 47.3(2008):346-358.

[6] Liu, Dong , et al. "Retrieval and analysis of a polarized high-spectral-resolution lidar for profiling aerosol optical properties." Optics Express 21.11(2013):13084.

[7] D'Amico, Giuseppe , et al. "EARLINET Single Calculus Chain - technical - Part 1: Pre-processing of raw lidar data." Atmospheric Measurement Techniques,9,2(2016-02-12) 8.10(2016):10387-10428.

RETRIEVAL OF DUST MICROPHYSICAL PROPERTIES

Böckmann, Christine[1,*], Samaras, Stefanos[2] and Haarig, Moritz[3]

[1]*University of Potsdam, Institute of Mathematics, Potsdam, Germany,*

[2]*German Aerospace Center, Oberpfaffenhofen, Germany,*

[3]*Leibnitz Institute of Tropospheric Research, TROPOS, Leipzig, Germany*
**bockmann@uni-potsdam.de*

ABSTRACT

An intense mineral dust event from the Saharan desert was observed over the Island of Barbados after a long-range transport over the Atlantic Ocean during SALTRACE Campaign in June 2014. We analyze data from a multi-wavelength Raman-lidar system of Leibniz Institute for Tropospheric Research (TROPOS) called BERTHA (Back-scatter Extinction lidar Ratio Temperature Humidity profiling Apparatus) to derive the aerosol microphysical properties of the inspected layer via regularization using the software SphInX (Spheroidal Inversion eXperiments). These parameters were found to be within credible ranges.

1. INTRODUCTION

The data was obtained from the multi-wavelength Raman-lidar system of TROPOS located at the ground-based remote sensing station in Husbands, north of the capital Bridgetown at the west coast of Barbados (13.15°N, 59.62°W, and 110 m above sea level) on 20 June 2014. The BERTHA system is a container-based, multi-wavelength polarization Raman lidar. It is able to take measurements of the depolarization ratio at three wavelengths (355, 532 and 1064 nm) simultaneously, which is crucial to characterize the dust after long-range transport, see [1, 2, 3] for more details. It operates as a 3+2+3 lidar system (3 backscatter coefficients β, 2 extinction coefficients α and 3 depolarization ratios). The signals are detected with a range resolution of 7.5 m and a time resolution of 10s. Additionally, an AErosol RObotic NETwork (AERONET) sun photometer was operated at the field site. The layers of interest can be seen in Fig. 1. Here we selected an altitude range from 1.5(2.0)-2.75 km and 3.25-4 km for the microphysical inversion. The optical profiles are shown in Fig. 2.

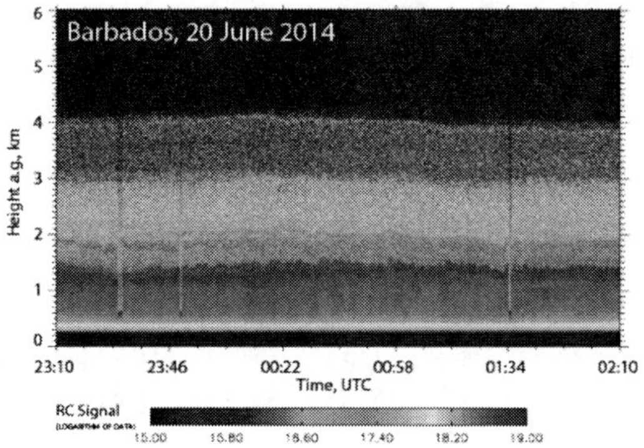

Figure 1. Lidar range-corrected signal at 1064 nm.

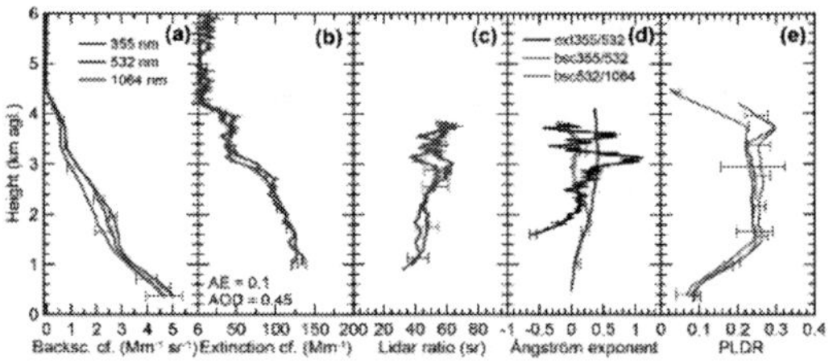

Figure 2. Optical properties captured by BERTHA (20 June 2014 23:10-02:10 UTC, 750 m smoothed). PLDR-particle linear depolarization ratio.

2. METHODOLOGY

We consider a spheroid-particle approximation, which needs only one additional parameter, i.e. the aspect ratio a, to be fully defined. The model relating the optical parameters $\mathcal{L}(\lambda)$ with the volume size distribution $v(r,a)$ is described by the action of a 2D Fredholm integral operator of the 1st kind

$$\mathcal{L}(\lambda) = \int_{a_{\min}}^{a_{\max}} \int_{r_{\min}}^{r_{\max}} \frac{3A}{16\pi r^3} Q_{\alpha/\beta}(r,\lambda,a;m) v(r,a) dr da, \quad (1)$$

https://doi.org/10.1051/epjconf/202023708019

where A is the particle surface area, m is the complex refractive index (RI), λ is the wavelength, r is the volume equivalent radius, $[r_{\min}, r_{\max}]$ and $[a_{\min}, a_{\max}]$ are sensible radius and aspect ratio ranges. $\mathcal{L}(\lambda)$ denotes either the 2 extinction or 6 backscatter (cross and parallel) coefficients, and Q stands for either the extinction or the backscatter (dimensionless) efficiencies, respectively. Identifying $\mathcal{L}(\lambda)$ as our measurement data and $v(r,a)$ as the unknown volume distribution, the problem reduces to the inversion of Eq. (1). Knowing the volume distribution, we can then extract the following microphysical parameters

- Total surface-area concentration ($\mu\text{m}^2\text{cm}^{-3}$)

$$s_t = \int_{a_{\min}}^{a_{\max}} \int_{r_{\min}}^{r_{\max}} \frac{3}{\pi r^3} G(r,a) v(r,a) dr da$$

- Total volume concentration ($\mu\text{m}^3\text{cm}^{-3}$)

$$v_t = \int_{a_{\min}}^{a_{\max}} \int_{r_{\min}}^{r_{\max}} v\,(r,a) dr da,$$

- Effective radius (μm) $\quad r_{\text{eff}} = 3\frac{v_t}{s_t}$.

The function $G(r,a)$ denotes here the spheroidal geometrical cross section. In addition, the complex RI and the single scattering albedo (SSA) in 355 nm and 532 nm are retrieved. Note, that in this work the common assumption of wavelength-independent RI is made, as an element of the predefined grid introduced.

A precalculated data-base will be used, created by the software tool Mieschka [4] which is able to perform scattering-related T-matrix calculations for spherical particles and rotationally symmetric non-spherical particles. Mieschka's look-up tables include scattering efficiencies for a 6×7 ($\Re(m) \times \Im(m)$) refractive index grid (a total of 42 RI values), 7 different aspect ratios and a size parameter range $[0.02, 40\ \mu\text{m}]$ with a resolution of 0.2μm. While the maximum size parameter is reasonably large for other applications, its potential cannot be fully exploited here in terms of the radius extent, since the lidar wavelengths restricts the maximum radius ($r_{\max}$) to appr. 2.2 μm. The resolution gap in the aspect ratio needed for the integrations is handled by interpolation to the nearest neighbor. Solving Eq. (1) requires discretization, regularization and a parameter choice rule handled by the software SphInX, see e.g. [5, 6, 7].

3. RESULTS

A reasonable layer selection where both lidar ratio and the Ångström exponent are relatively constant,

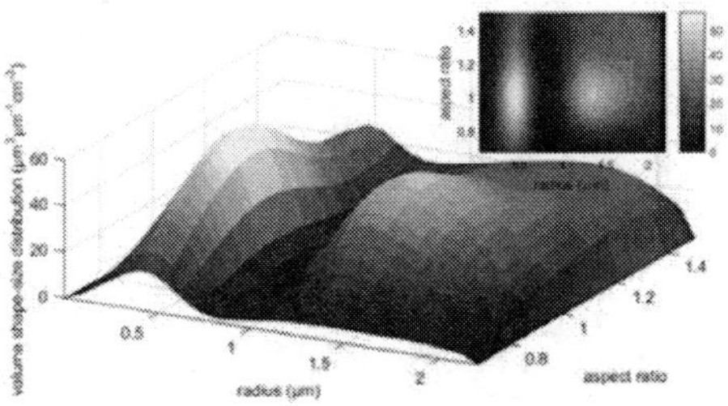

Figure 3. Retrieved shape-size distribution shown in 3D and in ar-plane for 1.5-2.75 km.

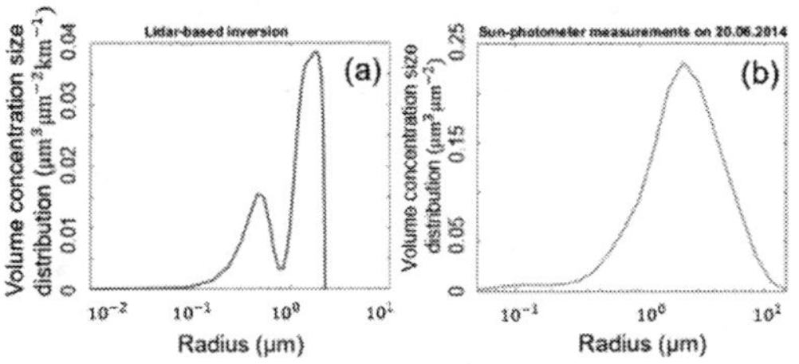

Figure 4. Retrieved lidar-reduced VCSD for 1.5-2.75 km (left) and AERONET VCSD (right).

the aerosol properties are more representative of the total behavior and where the optical profiles (α and β) are still intense is the one in 2-2.75 km, see Table 1. For the inversion we used 100 iterations of Padé regularization [6, 7], 9-14 spline points and spline degree within the range 2-5. The refractive index grid (RIG) was defined in the broad form of RRI × IRI $= [1.4, 1.5, 1.6, 1.7, 1.8] \times [0, 0.001, 0.005, 0.01, 0.05]$ (RIG 1). The upper integration boundary was set to the (maximum available) $r_{\max} = 2.2$. The results reveal a pattern

Table 1. Dust layer 2-2.75 km (averaged).

Optical properties				
α355 (Mm⁻¹)	$\beta355^{\parallel/\perp}$ (Mm⁻¹ sr⁻¹)	α532 (Mm⁻¹)	$\beta532^{\parallel/\perp}$ (Mm⁻¹ sr⁻¹)	$\beta1064^{\parallel/\perp}$ (Mm⁻¹ sr⁻¹)
98.50	1.69/0.41	92.51	1.56/0.45	1.17/0.27
Intensive properties				
δ355/532/1064 (%)	LR355 (sr)	LR532 (sr)	AE^{α} 355/532	AE^{β} 355/1064
24±0.0/29±0.0/23±0.0	47.71±5.2	46.34±2.94	0.16±0.09	0.48±0.02

which allows to categorize them in three (second not shown here) altitude ranges, namely 1.5-2.75km, 2.75-3.25km and 3.25-4km, only by observing the largest parameter variation, see Table 2. In this table, the variability (Var %) refer to the 5 best solutions for a single dataset corresponding to a specific altitude range, and the mean variability pertains to more than one datasets are involved. The uncertainty (Unc %) of the mean parameter value, found for every altitude range, is also given; here the input-datasets correspond to different altitude ranges.

Especially for the complex refractive index and the single scattering albedo retrieved by AERONET, the given values represent spectral mean values in the four almucantar wavelengths. Hence, Var (%) corresponds to a mean spectral variability, and the given Unc (%) corresponds to the uncertainty of the mean retrieved parameters (spectrally) between the consecutive measurements. We can see clearly from Table 2 that our algorithm demonstrates very good stability. There is virtually no difference in the microphysical retrieval between considering the whole range $2-2.75$ km and splitting the even larger range $1.5-2.75$ km to 5 smaller ranges of 250 m. The uncertainty of all parameters is very low and thus we could consider any of these height ranges to arrive to similar results. We should note that this efficiency could be largely the result of good quality data and the fact that we use the maximum number of depolarization data available at the present time.

Table 2. Retrieved microphysical parameters.

RIG 1, RRI=1.4 (Var, Unc: 0.0%, IRI=0.05 (Var, Unc: 0.0%)						
2.0-2.75 km / SSA532=0.685 (Var: 1.40%)						
Parameter Unit	a_t $\mu m^2 cm^{-3}$	v_t $\mu m^3 cm^{-3}$	r_{eff} (total/fine /coarse μm	svcf -	a_{eff} .	a_{var} .
Average	160.68	36.44	0.68/0.29 /1.45	0.31	1.10	0.046
Variability %	3.71	5.36	4.62/3.09 /1.54	4.01	0.38	3.72
1.5-2.75 km, step: 250m / SSA532=0.684 (Var: 3.20%, Unc: 1.11%)						
Average	163.07	38.63	0.71/0.35 /1.45	0.31	1.10	0.046
Variability %	2.88	7.00	5.54/4.67 /1.68	3.79	0.53	2.87
Uncertainty %	2.84	6.75	5.00/27.17 /0.58	1.33	0.21	1.55
3.25-4.0 km, step: 250m / SSA532=0.677 (Var: 2.44%, Unc: 0.81%)						
Average	65.24	16.28	0.753/0.398 /1.46	0.32	1.10	0.046
Variability %	2.65	8.04	6.71/7.26 /1.36	3.75	0.54	2.89
Uncertainty %	10.05	3.22	10.12/32.83 /0.37	3.89	0.57	2.42

Our algorithm predicts large particles, as expected for an intense dust event, with a total effective radius of 0.71 μm ($1.5-2.75$ km) which is constituted by fully separated modes (Fig. 3), i.e. with r_{eff} fine 0.35 μm and r_{eff} coarse 1.45 μm. Focusing on higher altitudes, i.e. ($3.25-4$km), where AE$^{\alpha}$ 355/532 is close to smaller values we also see in $1.5-2.75$ km (or even tinier), the effective radii (total, fine or coarse) follows the expected upscaling relatively to values for 2.75 - 3.25 km (not shown). Another notable characteristic is that the effective radius in ($3.25-4$km) is still well-retrieved despite the dramatic attenuation of signal in these heights. The attenuation is already visible above 2.75 km, see Fig. 2, and in terms of the microphysical retrieval it is translated into a diminished surface-area and total volume concentration see Table 2, as it was intuitively expected.

The effective shape behavior is attributed to prolate particle ensembles with $a_{\text{eff}} = 1.10$ and oblate particle ensembles with $a_{\text{eff}} = 0.046$. The sphericity rate 0.20% (Var: 17.18%) predicted by AERONET (Table 3) shows no involvement of spherical and (to some extent) fine particles. By comparison, the spherical volume concentration fraction (lidar) svcf, found 0.31 (Var: 3.79%, Unc: 1.33) in $1.5-2.75$ km, shows that there could be spherical-particle contributions. We note that AERONET's sphericity parameter refers to the percentage of the spherical- vs the non-spherical component (assumed by AERONET's forward model), and thus there is only a rough correspondence of this quantity with svcf.

We get a more detailed visual of our retrieval by looking at the (mean) retrieved shape-size distribution for the altitude range $1.5-2.75$ km in Fig. 3. There we see two very well separated modes, namely a narrower fine-to-medium-coarse mode and a much broader coarse mode, with volume concentration peaks about 58 and 48 $\mu m^3 cm^{-3}$at about 0.43 and 1.45 μm respectively. There are contributions of all kinds of spheroidal particle geometries (also spherical), and for higher aspect ratios (a ≥ 1.20) the fine-coarse peak difference fades, so that both modes maxima equalize at about 30 $\mu m^3 cm^{-3}$. Fig. 4 shows also the reduced volume concentration size distribution (left) and the volume concentration size distribution retrieved by AERONET (right). The latter shows merely a very prominent coarse mode predicting very large particles. The complete absence of a fine mode is often a mathematical artifact when one of the expected modes is much more dominant than the other, and the smoothing process suppresses or eliminates the smaller one.

Although the lidar-based (reduced) VCSD is not directly comparable quantitatively to the latter, as mentioned earlier, we can see that the volume concentration is of the same order in both subfigures by specifying the altitude. In this sense and by considering the thickness of the whole dust layer about 3.5 km, the estimated maximum lidar-retrieved concentration is about $0.14\, \mu m^3 \mu m^{-2}$ which is comparable with AERONET's maximum of about $0.24\, \mu m^3 \mu m^{-2}$. Since the volume concentration is obviously not constant along the aerosol layer and also the sampled volume is different in the two cases, this calculation has no practical purpose other than the order comparison.

The refractive index is found $1.4 + 0.05i$ through all layers $(1.5 - 4)$ and with 0% uncertainties (Var, Unc), and $\mathrm{SSA532} = 0.684$

Table 3. AERONET sun-photometer-based inversion results.

Para-meter Unit	u_t $\mu m^3 \mu m^{-2}$	r_{eff} (total/fine /coarse) µm	speri-city %	RRI -	IRI -	SSA -
Average	0.40	1.30/0.16 /1.87	0.20	1.52	0.0017	0.963
Varia-bility %	6.21	7.25/6.35 /7.27	17.18	1.02	21.13	0.77
Uncer-tainty %	-	-	-	1.54	64.24	2.27

(Var: 3.20%, Unc: 1.11%) which points to highly absorbing particles. This might not be as common for such an event but it is also not incredible. Experimental findings have shown that such a high absorption can mainly be found near the sources of dust events. For instance, a soil sample from Burkina Faso (rich in hematite and kaolinite) was found in [8] using also a spheroid-particle-based inversion scheme to have $\mathrm{IRI} = 0.0495 \pm 0.0206$ at 305 nm. Moreover, [9] found that the IRI increases a lot (up to 0.03) with decreasing particle size ($< 0.5\, \mu m$), again due to the predominant hematite / soot component in smaller particles. AERONET's retrieval suggests, on the contrary, a refractive index with $\mathrm{RRI} = 1.52$ (Var: 1.02%, Unc: 1.54%) and $\mathrm{IRI} = 0.0017$ (Var: 21.13%, Unc: 64.24%), and $\mathrm{SSA} = 0.963$ (Var: 0.77%, Unc: 2.27%), i.e. weakly absorbing particles. However, this IRI is even lower than the usual values considered for Saharan dust particles, see e.g. [10]. The latter study also finds a dramatic increase in IRI (up to five times higher at 637 nm) when the dust concentration is lower so that a soot-type absorber prevails.

4 CONCLUSIONS

The results generally conform to usual dust characteristics, i.e. relatively large size (r_{eff}) and reduced sphericity (svcf) and they further show a volume concentration order in accordance to the one of AERONET independent retrieval, although the high absorption found requires further investigation. Finally, multiple inversions in sublayers of 250 m demonstrated the stability of the algorithm along with its sensitivity to particle size variation.

ACKNOWLEDGEMENTS

The work has been supported partially by the EU under grant agreement No.289923 - ITaRS and EU's Horizon 2020 under No.654109 (ACTRIS-2).

REFERENCES

[1] M. Haarig, et al. Atmospheric Chemistry and Physics, 17(17): 10767–10794, (2017).
[2] M. Haarig, et al. Atmospheric Chemistry and Physics, 17(23): 14199–14217, (2017).
[3] R.-E. Mamouri, and A. Ansmann, AMT, 10(10): 3403–3427, (2017).
[4] T. Rother and M. Kahnert, Springer, NY,(2014).
[5] S. Samaras, PhD thesis, Potsdam University, (2017).
[6] C. Böckmann, and A. Kirsche, Computer Physics Communications, 174(8): 607–615, (2006).
[7] C. Böckmann, and L. Osterloh, Inverse Problems in Science and Engineering, 22(1): 150–165, (2014).
[8] R. Wagner, et al. Atmospheric Chemistry and Physics, 12(5): 2491–2512, (2012)
[9] K. Kandler, et al. Atmospheric Environment, 41(37): 8058 – 8074, (2007).
[10] A. Schladitz, et al. Tellus B, 61(1): 64–78, (2009).

SIMULATION EFFECTS OF SURFACE GEOMETRY AND WATER OPTICAL PROPERTIES ON HYDROGRAPHIC LIDAR RETURNS

Song Yang*[1, 2,3,4] , Qian Sun[1,2,3,4] ,Yongchao Zheng[1,2]

[1]*Beijing Institute of Space Mechanics & Electricity, Beijing 100094, China*
[2]*Key Key Laboratory for Space Laser Information Perception Technology of CAST,*
[3]*Beijing Key Laboratory of Advanced Optical Remote Sensing Technology, Beijing 100094, China*
[4]*Key Laboratory for Optical Remote Sensor Technology of CAST, China*
**Email: kellyyszx@163.com*

ABSTRACT

. Water LiDAR model was applied to simulate the returned waveforms of hydrographic LiDAR considering the effects of surface geometry and water optical properties. The signal to noise ratio(SNR) of bottom returned peak was considered as a criterion for performance of hydrographic LiDAR. The behavior of LiDAR was sensitive to water optical properties and it was insensitive to water surface roughness.

1. INTRODUCTION

Water surface altimetry and bathymetry are important factors for geomorphology, hydrology[1], and meteorology. Water surface altimetry and bathymetry can be measured with several methods by using sonar, radar, and optical imaging. However, these techniques have some difficulties when used on swallow water, such as: 1) cannot achieve full spatial coverage on coastal water (sonar), 2) the lack of high depth accuracy due to the low penetration of radar and optical imaging.

Airborne hydrographic LiDAR has been considered as a suitable sensor for bathymetry because of its low latency, high accuracy and high density data. Take advantages of the blue-green visible wavelength window, blue-green laser has a good performance in long distance underwater sensing, especially the subsurface sensing. Thus the airborne hydrographic LiDAR can get the returned signals from the water surface[1], water column[2] and water bottom[3,4] for altimetry and bathymetry.

Due to the scattering and absorption property of water, the effects of energy loss and direction changing can be generated for the optical beams. These effects may introduce the pulse spreading and pulse stretching, which play important roles in beam propagation. Several Mathematical models[1,3,4] are developed to simulate the returned signals from the hydrographic Lidar in order to optimize parameters of lidar systems under development or to estimate a system efficiency in a given conditions.

In this paper, we take water LiDAR simulation model to simulate the returned waveforms for airborne LiDAR with 532nm frequency. The SNR of simulated waveforms are taken to evaluate the behavior of LiDAR in different conditions. The effects of water surface slopes and water optical properties are also discussed.

.

2. METHODOLOGY

2.1 Water LiDAR Model

The Water LiDAR model[5] presented by H. Abdallah et.al., is a generalized model for LiDAR waveforms with wavelength in the optical spectrum domain from 300nm to 1500nm. This model has good features including: 1) description of the water surface geometry based on bidirectional reflectance distribution function (BRDF) [6], 2) description of the optical property of water column for different wave length based on the laws of radiative transfer, 3) calculation of SNR which considers detector noise and background solar noise.

The returned waveforms from hydrographic LiDAR system can be written as[4,5]

$$P_T(t) = P_s(t) + P_c(t) + P_b(t) + P_{bg}(t) + P_N(t) \quad (1)$$

where $P_T(t)$ is the total power received, $P_s(t)$ is the power returned from the surface, $P_c(t)$ is the power returned from the column, $P_b(t)$ is the power returned

https://doi.org/10.1051/epjconf/202023708020

from water bottom, $P_{bg}(t)$ is the background solar radiation power, $P_N(t)$ is the detector noise power, and t is the time scale.

2.2 Signal to Noise Ratio

The SNR of the hydrographic LiDAR is shown as:

$$SNR = \frac{P_s(t)+P_c(t)+P_b(t)}{\sigma_{bg}(t)+\sigma_N(t)} \quad (2)$$

$\sigma_{bg}(t)$ is the standard deviation of the background solar radiation power, and $\sigma_N(t)$ is the detector noise power.

2.3 Effect of Surface Geometry

The returned power from the surface is given by

$$P_s(t) = \omega_s(t) * P_s, \quad (3)$$

with $\omega_s(t)$ is the returned waveform of the pulse propagation in a perfect and lossless channel from the water surface, and P_s is the instant echoes from the water surface[3,5],

$$P_s = P_e T_{atm}^2 A_R \eta_e \eta_R \frac{f}{(H/\cos\theta)^2}. \quad (4)$$

Here P_e is the emitted power of the LiDAR system, T_{atm} is the loss due to the atmosphere, A_R is the receiver area, H is the sensor altitude, θ is the local incidence angle, η_e and η_R are the optical efficiency of the emitter and receiver. f is the loss factor from the surface and is based on BRDF of the water surface. Here we use the geometric model of Cook and Torrance to describe the behavior of surface geometry. The equation of loss factor f is[6]

$$f = \frac{DOF_r}{4\cos^2\theta}, \quad (5)$$

where

$$D = \pi^{-1} r^{-2} \cos^{-4}\theta \exp(-(\tan\theta / r)^2) \quad (6)$$

is the microfact distribution function[6,7] , O is the geometric attenuation factor of BRDF, and F_r is the Fresnel reflection of light on the water surface.

2.4 Effect of Optical properties of water

The optical properties of water can be described by absorption coefficient $a(\lambda)$, scattering coefficient $b(\lambda)$, attenuation coefficient $c(\lambda)$, scattering function $\beta(\varphi)$, and diffuse attenuation coefficient k. The empirical relationship between diffuse attenuation and optical properties[1] is established as

$$k = c(\lambda)[0.19\times(1-b(\lambda)/c(\lambda))]^{\frac{b(\lambda)}{2c(\lambda)}} \quad (7)$$

where[8] $c(\lambda) = a(\lambda) + b(\lambda)$,

$a(\lambda) = a_w(\lambda) + a_y(\lambda) + a_{ph}(\lambda) + a_s(\lambda)$,

$b(\lambda) = b_w(\lambda) + b_{ph}(\lambda) + b_s(\lambda)$.

Here the subscripts *'w', 'y', 'ph', 's'* represent pure water, yellow substances, phytoplankton and sediment, respectively.

The returned waveform from the water column and bottom is calculated as

$$P_{c/b} = P_e T_{atm}^2 A_R \eta_e \eta_R \frac{(1-f)^2 \exp(-2kz/\cos\theta)}{(nH + z/\cos\theta)^2} \cdot (\mathrm{sgn}(\frac{z-Z}{Z})\beta(\varphi) + (1-\mathrm{sgn}(\frac{z-Z}{Z}))\frac{R_b}{\pi}) \quad ,(8)$$

where n is the water refractive index, R_b is the bottom albedo, z is the depth underwater, Z is the depth of the bottom, and sgn(•) is the step function.

3. RESULTS

3.1 Simulation results for hydrographic LiDAR

According to water LiDAR model, the returned waveforms from traditional hydrographic LiDAR are shown in Fig.1. The relevant parameters are shown in Table. 1.

Table.1 Parameters used in simulation

Parameters	LIDAR	Parameters	LIDAR
λ (nm)	532	η_R	0.5
H(m)	400	T_{atm}	0.9
P_e(MW)	1	$\beta(\varphi)\,(m^{-1}sr^{-1})$	0.0014
θ (°)	20	r	0.1

 https://doi.org/10.1051/epjconf/202023708020

A_R(m^2)	0.025	n	1.33
η_e	0.9	R_b	0.1

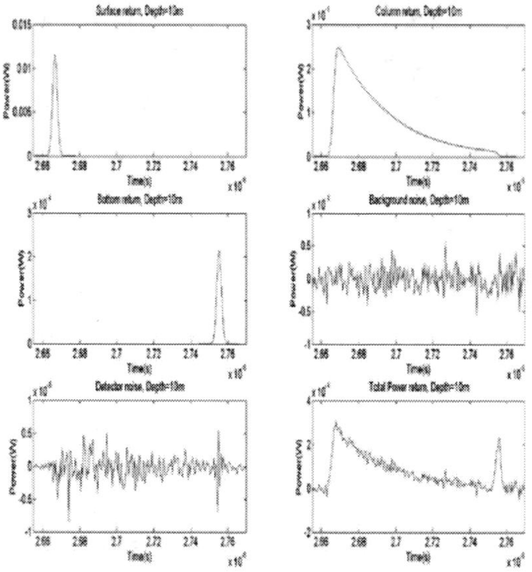

Fig.1 Simulated returned waveforms of hydrographic LiDAR with wavelength 532nm. The water is of Case II quality (the average concentration of phytoplankton $C_{ph} = 4\text{mg}/\text{m}^3$, sediment $S = 2.6\text{mg}/1$)

3.2 SNR for different depth

The SNR of bottom returned waveforms at different depth is discussed in this section. Fig.2 shows SNR distribution of peaks of the bottom returned waveforms. The results shows that SNR is linearly decreases with water depth.

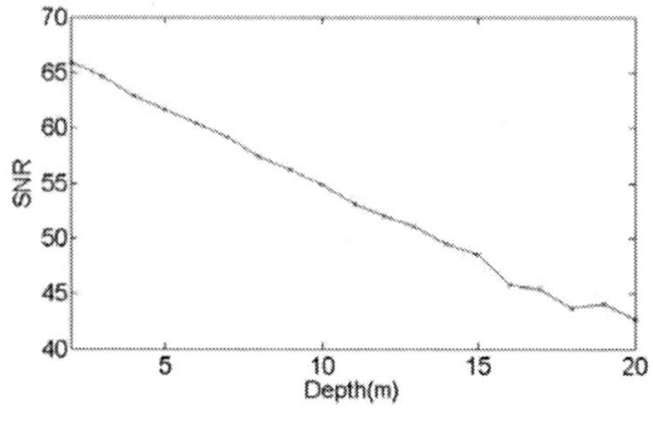

Fig.2 SNR of the bottom returned waveform vs. water depth

3.3 Effect of Surface Geometry

The effect of surface geometry is mainly relevant with the surface slope parameter r. Small values of r represent gentle facet slopes and behaves as a specular facet, while large values of r represent steep facet slopes and behaves as a diffuse facet. The performance of SNR of bottom returned waveforms versus surface slope parameter r is exhibited in Fig.3. When r was small(r<0.1), SNR of bottom waveform was low since the returned light is highly directional and has low overlap with receiver area. When r was large (r>0.1), SNR became stable since the returned light is unidirectional and get insensitive with incident angle.

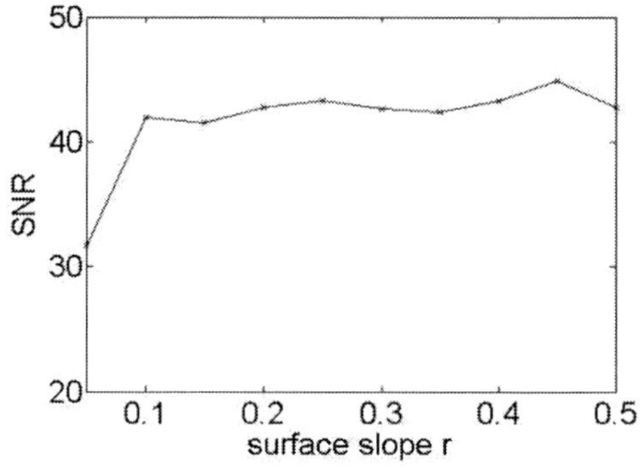

Fig.3 SNR of the bottom returned waveform vs. surface slope

3.4 Effect of Optical properties of water

In Section 2.4, the relationship between optical properties of water and diffuse attenuation coefficient k is shown in Eq.(7). We compare the total power return of Case I water and Case II water. The scattering and attenuation effects for Case II water are more serious than that for Case I water.

Table 2. Comparisons of two sites with different optical properties

Parameters	Case I	Case II
k(m^{-1})	0.0449	0.1572
c (m^{-1})	0.0731	0.4912
a(m^{-1})	0.0412	0.1233
b (m^{-1})	0.0319	0.3679

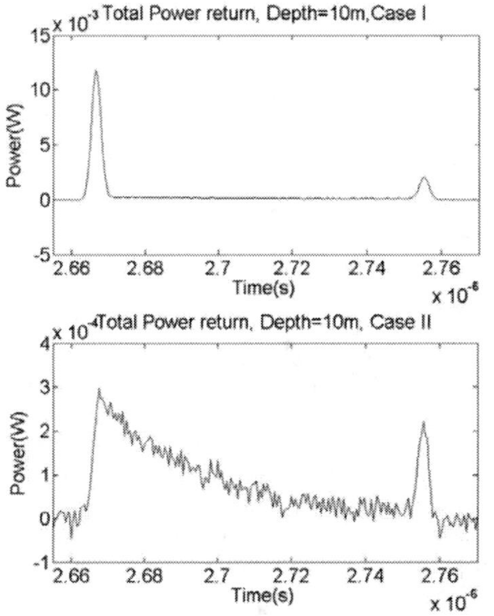

Fig.4 Comparison of total power return in Case I water and Case II water. The water depth is 10m.

4. CONCLUSION

Water LiDAR simulation model was applied to describe the returned waveforms of airborne hydrographic Lidar. The SNR of peaks of bottom returned waveforms are analyzed and SNR is linearly decreases with depth. The Effect of Surface Geometry is also considered in this paper, and the SNR is stable with the water surface slope $r \in [0.1, 0.5]$. The performance of hydrographic Lidar is very sensitive to optical properties of water, and prior information of water optical properties is significant for estimation the maximum depth of hydrographic Lidar.

ACKNOWLEDGEMENTS

This work was supported by National Natural Science Foundation of China(11574400 , 11204379, 11304013, 61705007), GuangXi innovative Development Grand Grant under Grant No. GuiKe AA 18118038, CAST Foundation for Distinguished Young Scholars(2019).

REFERENCES

[1] G. C. Guenther, presented at the NOAA Professional Paper Ser. NOS1, Rockville, MD, 1985.
[2]A. Collin, P. Archambault, and B. Long, IEEE Trans. Geosci. Remote Sens., vol. 46, no. 10, pp. 2947–2955, Oct. 2010.
[3] J. Feigels, Ocean Opt. XI, vol. 1750, pp. 473–484, 1992.
[4] H. M. Tulldahl and K. O. Steinvall, Appl. Opt., vol. 42, no. 12, pp. 2462–2483, Apr. 2004.
[5] Hani Abdallah, Nicolas Baghdadi, et.al. IEEE Trans. Geosci. Remote Sens., Vol. 9, no. 4, pp.744-748, July.2012.
[6]R. L. Cook and K. E. Torrance, ACM Trans. Graph., vol. 1, no. 1, pp. 7–24, Jan. 1982.
[7]P. Beckmann and A. Spizzochino, *The Scatter of Electromagnetic Waves From Rough Surfaces.* Norwood, MA: Artech House, 1987.
[8] C. D. Mobley, *Light and Water: Radiative Transfer in Natural Waters.*San Diego, CA: Academic, 1994.

THE DEPOLARIZATION PROPERTIES OF CIRRUS CLOUD BY LIDAR: OBSERVATION AND MODEL MATCH

Zhenzhu Wang*[,1], Anatoli Borovoi[2], Gennady Matvienko[2], Dong Liu[1], Chenbo Xie[1], Yingjian Wang[1,3], Alexander Konoshonkin[2], Natalia Kustova[2]

[1]*Key Laboratory of Atmospheric optics, Anhui Institute of Optics and Fine Mechanics, Chinese Academy of Sciences, Hefei, Anhui 230031, China*
[2]*V. E. Zuev Institute of Atmospheric Optics, Rus. Acad. Sci , Tomsk, 634055, Russia*
[3]*University of Science and Technology of China, Hefei, Anhui 230031, China*
**Email: zzwang@aiofm.ac.cn*

ABSTRACT

The Polarization Lidar has been developed for cirrus cloud optical properties measurement. The Nd-YAG laser is employed as emitting linear source and two channels are used for detecting back-scattering signals from cirrus cloud including two 532 nm Orthogonal Polarization channels. Measurements were carried out from December 2010, to February 2013. In this work, we consider the only linear depolarization ratio measured at a wavelength of 532 nm. The dependences of depolarization ratios on both the size of ice crystals and the distribution of their distortion angles were also calculated. These results were compared with each other and used for retrieving the microphysical parameters of cirrus clouds observed in Hefei.

1. INTRODUCTION

Cirrus cloud is one of the most commonly occurring cloud types globally. It forms in the upper part of troposphere, due to synoptic scale lifting, or as a result of moisture transport by deep convection [1-3]. Cirrus clouds cover about 30% of the Earth surface and they essentially impact on the radiative budget of the Earth and, consequently, on the climate [4-5]. The radiative characteristics of cirrus clouds are poorly studied yet because of the strong spatial and temporal variability of the clouds. And it is also a significant challenge to accurately measure and model their optical and microphysical properties in order to assimilate them into global climate models. On the other hand, lidar has been extensively demonstrated for observing cloud properties. The previous work with lidar techniques concentrate on cirrus clouds in experiment [6-12] and theory [13-14].

2. METHODOLOGY

2.1 The lidar system

The Three-wavelength Raman Mie Polarization Lidar (named as TRMPL) has been developed for cirrus cloud properties measurement. This idar system has built in Hefei and passed the performance assessment in the end of 2010, and then improved to be a movable lidar system to carry out the continuous measurements of vertical distribution of regional cloud.

The TRMPL is a moveable and continuous working lidar installed in a standard container with a glass window on the roof. Fig.1 gave the system diagram and the photo of the TRMPL.For this paper, only polarization signals were used for cirrus clouds.

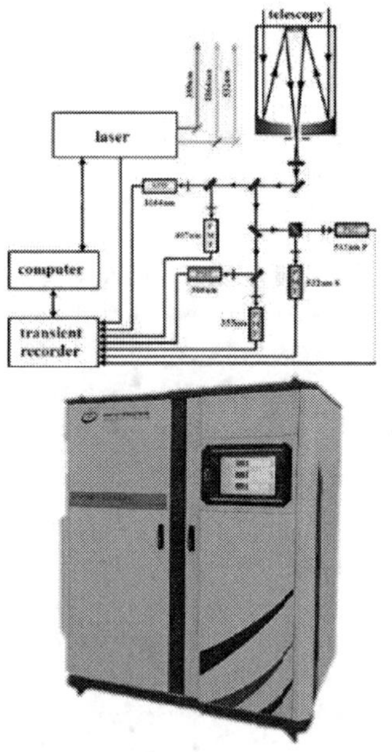

Figure 1. The diagram and photo of TRMPL

2.2 The inversion methods

In general, the depolarization ratio (namely) is determined from parallel (P_p) and perpendicular (P_s) polarization signals at 532 nm together with the k value [7-8]:

$$\delta = \sigma_{\perp}/\sigma_{//} = \kappa \cdot P_s/P_p \qquad (1)$$

The theoretical problem of light backscattering by ice crystals of cirrus clouds has been solved only for the simplest crystal shapes—hexagonal columns and plates of an ideal shape [15]. In this work, we present the results of calculations for a more realistic model, that is, randomly oriented distorted hexagonal columns, where hexagonal faces of a crystal are skewed relative to rectangular faces at a small angle ξ(distortion angle), similarly to [16]. At such a distortion of the crystal shape, the right angle between crystal faces results in the strong dependence of the backscattering on the crystal sizes.

However, averaging over the distortion angles ξ, inherent in real clouds, levels the dependence of the depolarization ratio on the crystal sizes. Thus, as our calculations show, the measured values of the depolarization ratio can be meant to weakly depend on crystal sizes. Consequently, the depolarization ratio can be considered as an indicator of only the crystal shape and its spatial orientation in a cloud.

3. RESULTS

The statistical characteristics of depolarization ratio for cirrus cloud have been investigated using measured data from Dec. of 2010 to Feb. of 2013, shown in figure 2. The averaged depolarization ratio is 0.36 at 532 nm varying almost from 0.2 to 0.5. In this work, we calculate the depolarization ratio as a function of crystal sizes at different distributions of crystal distortion angles for the given model of crystal shape. This allows us to retrieve, with a known uncertainty, microphysical parameters of cirrus clouds, observed at Heifei in this period.

Sizes of hexagonal columns are defined by one parameter L (the column length). The diameter D of a hexagonal column in this case is found from the empirical relation [17]. Figure 3 shows depolarization ratios calculated in the physical optics approximation for randomly oriented hexagonal columns, with the distortion angle ξ distributed by the Gauss law. As seen in Figure 3, the sufficiently strong distortion of crystal shape (ξ_{eff} > 3°) levels the dependence of the depolarization ratio on the crystal sizes and increases it from δ ≈ 0.2, which is typical for crystals of regular shape, to δ ≈ 0.35.

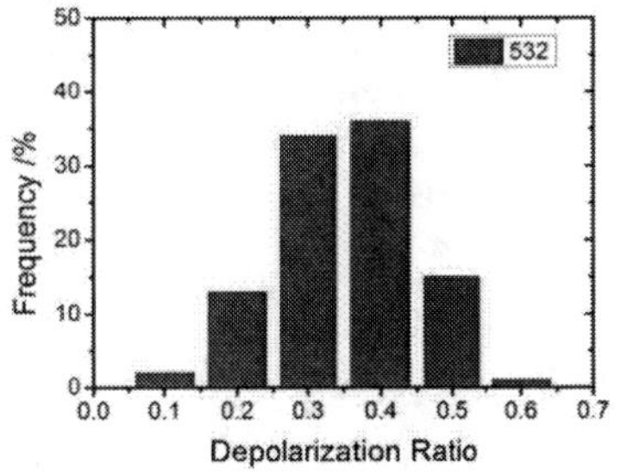

Figure 2. The frequency of depolarization ratio for cirrus observed at Heifei

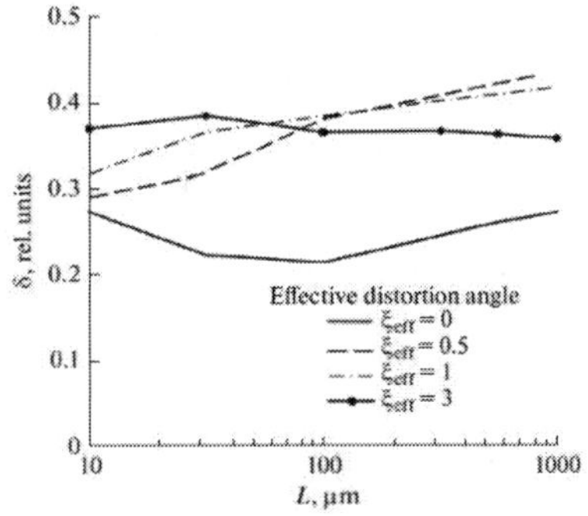

Figure 3. Linear depolarization ratio calculated for randomly oriented hexagonal columns with effective distortion angles of 0, 0.5, 1, and 3 as a unction of the crystal length at a wavelength of 532 nm

Thus, all the above written and Figs. 2 and 3 allow us to draw a conclusion that in 70% of 85% cases of observation of chaotically oriented particles in the period from December 2010, to February 2013, they had rather simple shapes of the hexagonal column type (bullet or bullet-rosette), and only in 15% of the cases the shape was of aggregate type.

4. SUMMARY

A moveable TRMPL lidar system is developed for cirrus detecting at Hefei, China. The depolarization ratios for cirrus cloud have been investigated from statistical characteristics. The parameters are matched with the developed model well and This allows us to retrieve, with some uncertainty, microphysical parameters of cirrus clouds.

ACKNOWLEDGEMENTS

This work was supported by the National Natural Science Foundation of China under Grant No. 41975038, 41590871, 41811530088 and the Youth Innovation Promotion Association CAS under Grant No. 2017482 and the CAS President's International Fellowship Initiative under Grant No. 2016VEA044.

REFERENCES

[1] Goldfarb, L., et al. J. Geophys. Res. Lett. 9(28), 1687-1690 (2001).
[2]Heymsfield, A. J., et al. J. Atmos. Sci. 41, 846-855 (1984).
[3]Jensen, E. J., et al. J. Geophys. Res. 101, 21361-21375 (1996).
[4]Chen, T., et al. J. Climate 13, 264-286 (2000).
[5]Nazaryan, H., et al. J. Geophys. Res. 113, D16211 (2008).
[6]Young, S. A., Appl. Opt. 34, 7019-7030 (1995).
[7]Wang, Z., et al. Optical Review 16, 566-570 (2009).
[8]Wang, Z., et al. Chinese Optics Letters 6, 235-237 (2008).
[9]Tao, Z., et al. Appl. Opt. 47, 1478-1485 (2008).
[10]Sassen, K., Bull. Am. Meteorol. Soc., 72, 1848-1866 (1991).
[11]Borovoi, A., et al. Optics Express 22(20), 24566-24573 (2014).
[12]Sassen, K., et al. J. Atmos. Sci. 58(15), 2103-2112 (2001).
[13]Borovoi, A.,et al. J. Quant. Spectrosc. Radiat. Transfer. 146, 181-189 (2014).
[14]Yang, P., et al. Appl. Opt. 35(33), 6568-6584 (1996).
[15] A. Borovoi, et al. Opt. Lett. 39 (19), 5788–5791 (2014).
[16] A. Borovoi, et al. Opt. Express. 23 (19), 24557–24571 (2015).
[17] D. L. Mitchell et al. J. Atmos. Sci. 51, 817–832 (1994).

THE NONLINEAR EFFECTIVE ATTENUATION COEFFICIENT OF SPACEBORNE OCEANIC LIDAR SIGNAL

Qun Liu, Dong Liu*, Jian Bai, Xiaoyu Cui, Yudi Zhou, Peituo Xu, Zhipeng Liu, Xiaobin Wang

State Key Laboratory of Modern Optical Instrumentation, College of Optical Science and Engineering, Zhejiang University, Hangzhou 310027, China

*Email: *liudongopt@zju.edu.cn*

ABSTRACT

Multiple scattering is an inevitable effect in spaceborne oceanic lidar because of the large footprint size and the high optical density of seawater. The effective attenuation coefficient k_{lidar} in oceanic lidar equation, which indicates the influence of the multiple scattering effect on the formation of lidar returns, is an important parameter in the retrieval of inherent optical properties (IOPs) of seawater. In this paper, the nonlinearity of k_{lidar} and the relationships between k_{lidar} and the IOPs of seawater are investigated by solving the radiative transfer equation with an improved semianalytic Monte Carlo model. k_{lidar} is found to decrease exponentially with the increase of depth in homogeneous waters. k_{lidar} is given as an exponential function of depth and IOPs of seawater. The results in this paper can help to have a better understanding of the multiple scattering effect of spaceborne lidar and improve the retrieval accuracy of the IOPs of seawater using spaceborne lidar.

1. INTRODUCTION

Lidars obtain depth-resolved return profiles by collecting backscattered signals from molecules and particles. It is essential to establish an accurate model of the radiative transfer process. The return signal of spaceborne oceanic lidar is strongly affected by multiple scattering due to the large footprint size and the high optical density of seawater [1]. To characterize the multiple-scattering radiative transfer process, an effective attenuation coefficient k_{lidar} is introduced into the oceanic lidar equation. Gordon [2] found that the ratio of the radius of the spot on the sea surface viewed by the lidar receiver optics to the mean free path of photons in water is a significant parameter which is related to k_{lidar}. When this ratio is near zero, k_{lidar} is given by the beam attenuation coefficient. If the ratio is greater than 5, k_{lidar} is given by the diffuse attenuation coefficient K_d. Phillips et al. [3] proposed a slightly different assumption that the upper and lower limits of k_{lidar} is given by the beam attenuation coefficient and the absorption coefficient, respectively, depending on field of view (FOV). Collister et al. [4] exhibited the similarity between k_{lidar} and the absorption coefficient under wide FOV using shipborne lidar measurements. The depth characteristics of k_{lidar} are ignored in aforementioned studies. the values of k_{lidar} at different depths in homogenous water are not constant, due to the multiple scattering effect [5, 6], especially near the ocean surface. This phenomenon leads to errors in the retrieval of optical properties, which further affects the assessment accuracy of the phytoplankton biomass and primary productivity. Thus, the depth characteristics of k_{lidar} and the relationship between k_{lidar} and the optical properties of seawater need to be studied in detail.

In this paper, the potential of using an improved semianalytic Monte Carlo radiative transfer simulation model (SALMON) to accurately identify the characteristics of both homogeneous and inhomogeneous waters is investigated. The use of SALMON is proposed by Poole et al. [5] and widely applied in oceanic and atmospheric lidars, therefore, it would be beneficial to develop an improved SALMON that can better simulate the inhomogeneous waters in the real oceans. We use this improved model to investigate the characteristics of lidar return profiles and the corresponding k_{lidar}. Analysis of the influence of the lidar system parameters on k_{lidar} and the depth characteristic of k_{lidar} for homogenous water are also inclueded. In addition, an exact relationship between the depth-resolved k_{lidar} and IOPs of seawater is derived based on the statistical properties of case 1 waters.

2. METHODOLOGY

2.1 Lidar return signals

The schematic diagram of the theoretical model of laser backscatter from water is illustrated in Fig. 1. The laser pulse transmits downward towards the flat horizontal water surface at the height of H. The photons are scattered and absorbed by the molecules and particles in seawater, and the backscattered photons are received by the detector forming the lidar return signal, which can be described by the lidar equation [7, 8]:

$$P(z) = P_0 M \frac{v\tau}{2n} T_{atm}^2 T_{surf}^2 \frac{A}{(nH+z)^2} \times \beta_\pi(z)\exp\left(-2\int_0^z k_{lidar}(z')dz'\right) \quad (1)$$

where $P(z)$ is the power received from a depth of z, P_0 is the transmitted pulse power, M is the lidar instrument constant, v is the speed of light in vacuum, τ is the temporal pulse length, $v\tau/(2n)$ is the depth resolution in water, T_{atm} is the transmission through the atmosphere, T_{surf} is the Fresnel transmittance of the air-sea surface, A is the receiver area, n is the refractive index of sea water and β_π is the volume scattering coefficient at a scattering angle of π rad.

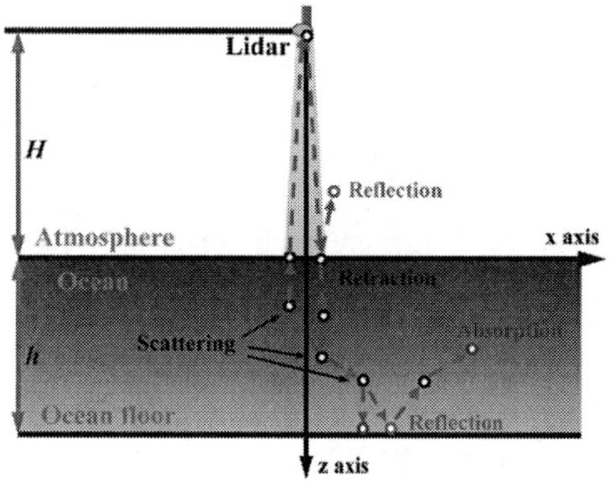

Figure 1. The schematic diagram of spaceborne lidar system.

Rewriting Eq.(1), we have the $P_{norm}(z)$ expressed by

$$P_{norm}(z) = \frac{P(z)*(nH+z)^2}{P_0 M\left(v\tau/2n\right)T_{atm}^2 T_{surf}^2 A\beta_\pi(z)} = \exp\left(-2\int_0^z k_{lidar}(z')dz'\right) \quad (2)$$

Thus, the effective attenuation coefficient can be written as

$$k_{lidar}(z) = -\frac{1}{2}\frac{d}{dz}\ln\left(P(z)_{norm}\right) \quad (3)$$

k_{lidar} is an important parameter for retrieving optical properties of seawater. The IOPs considered here include the scattering coefficient $b(\lambda)$ (m^{-1}), the absorption coefficient $a(\lambda)$ (m^{-1}), and the beam attenuation coefficient $c(\lambda)$ (m^{-1}).

2.2 An improved semianalytic Monte Carlo radiative transfer simulation model

Monte Carlo method is a numerical method of solving the radiative transfer. In 1981, Poole et al. [5] proposed a semianalytic Monte Carlo radiative transfer simulation model (SALMON) for atmosphere lidar, which is widely used in the current oceanic lidars. The improved SALMON proposed here is applicable for both homogeneous and inhomogeneous waters.

The SALMON model simulates the random collision and propagation in water of individual photons from the transmitted beam through a series of absorption and scattering until the photons move outside of the field of view of the lidar receiver or reach the boundaries of medium. The distance travelled before collision is the step size (s) in dimensionless unit and defined as the integration of the attenuation coefficient over the photon pathway [9].. In the SALMON model, s is calculated by $-\ln(R)/c$, where R is a random value between 0 and 1. The larger the attenuation coefficient c, the shorter distance the photon traveled before it is scattered or absorbed. For the stratified inhomogeneous waters, the IOPs are assigned separately for each layer. If the photon passes through more than one layer, c will change along the pathway. Therefore, the actual step size in layer $i \pm 1$ could be calculated as $s_{i\pm1} = (s_i - d_i)c_{i\pm1}/c_i$, where d is the geometrical distance along the ray in layer i where the scattering took place. The sign indicates the photon propagates up (-) or down (+). The photon travels until the initial step size has been consumed ($s_t < d_t$). All the calculations in this paper are executed with more than 10^{11} photon packets to ensure a low level of standard deviation of the statistical estimates.

3. RESULTS

3.1 Characteristics of lidar return signals for homogeneous case 1 waters

The effective attenuation coefficient k_{lidar} in lidar equation is the reflection of the IOPs of seawater. Determining the value of k_{lidar} is beneficial to improve the retrieve accuracy of IOPs from lidar equation. In terms of the orbit altitude of 400 km, the characteristics of lidar return signals from homogeneous case 1 water with various chlorophyll concentrations are studied. In these simulations, the divergence angle of the laser beam is assumed to be 0.1 mrad and the diameter of the telescope is 1.5 m [10]. The ocean is assumed to have a depth of 50 m with a black bottom and the depth resolution is set to be 0.3 m. The normalized lidar return signal power and the associated effective attenuation coefficients of seven groups of seawaters with [Chl] between 0 to 2 mg/m^3, which covers most open ocean and productive coastal environment, are provided in Fig. 2. The normalized lidar return profiles in Fig. 2(a) shows that the linearity of the lidar return signal decreases with the increase of [Chl]. For seawaters with high [Chl], k_{lidar} is not constant even if the IOPs are uniform in the vertical direction [Fig. 2(b)]. k_{lidar} decreases with the increase of depth at first, then it becomes more stable. The full range of the nonlinearity in k_{lidar} depends on the sensitivity of the receiving system. If this nonlinearity in k_{lidar} is ignored in homogeneous waters, it may cause error in the retrieval of lidar signal.

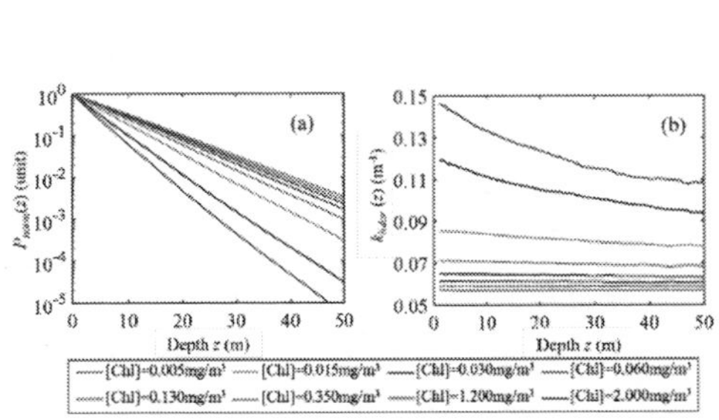

Figure 2. (a) Depth profiles of the normalized lidar return signals $P_{norm}(z)$ and (b) the corresponding effective attenuation coefficients $k_{lidar}(z)$ for [Chl] between 0 to 2 mg/m^3.

The nonuniformity of k_{lidar} caused by multiple scattering is inevitable and it increases the uncertainty of the retrieval.

3.2. The relationships between the klidar(z) and IOPs of water

Based on the results shown in Fig. 3(b), an exponential formula is proposed to describe the depth characteristic of k_{lidar}, which is given by

$$k_{lidar}(z) = m \times \exp(-n \times z) + p \tag{4}$$

The slope and the asymptotic value of k_{lidar} in Eq. (4) are determined by m, n and p, respectively. Moreover, k_{lidar} involves the IOPs of seawater, indicating these three parameters are related to the IOPs of seawater.

The influence of the IOPs on the qualitative behavior of k_{lidar} is analyzed using a series of calculations with one variable at a time (Fig. 3). Fig. 3(a) shows that the curves of k_{lidar} for different ω_0 (with b=0.2 m^{-1}) are parallel to each other, which means that the absorption coefficient a only affects the parameter p. Fig. 3(b) shows that the scattering coefficient b influences both the slope and the asymptotic value of k_{lidar}. Fig. 3(c) shows that the phase function mainly affects the slope of the k_{lidar} but has little influence on the asymptotic value of k_{lidar}. The dependence of k_{lidar} on the phase function and b can be summarized by a backscattering coefficient b_b, which is obtained by integrating the phase function in the range from $\pi/2$ to π.

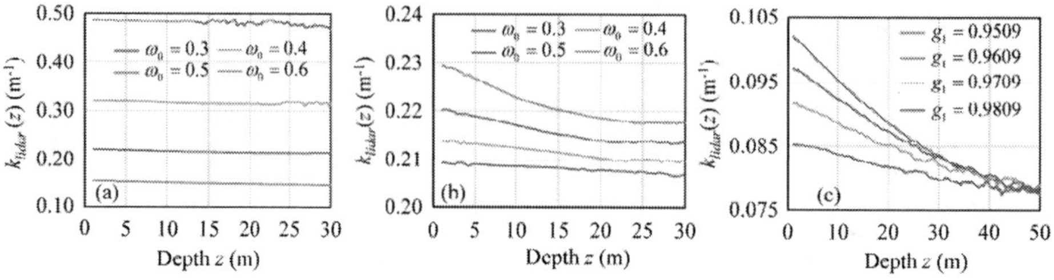

Figure 3. The influence of IOPs on k_{lidar} for (a) $b = 0.2\text{m}^{-1}$ and $\omega_0 = 0.3(0.1)0.6$, (b) $a = 0.2\text{m}^{-1}$ and $\omega_0 = 0.3(0.1)0.6$, and (c) [Chl]=0.35 mg/m^3 and g_1= 0.9509(0.01)0.9809.

In summary, the relationships among the parameters m, n, p and IOPs can be derived. Both m and n show a linear relation with b_b, and p is a linear function of a and b_b. The linear functions can be expressed by

$$\begin{aligned} m &= 4.8907b_b - 0.0004 \\ n &= 4.2506b_b - 0.0055 \\ p &= a + 0.3582b_b - 0.0042 \end{aligned} \tag{5}$$

4. CONCLUSIONS

Multiple scattering effect on the spaceborne oceanic lidar returns is investigated with an improved SALMON, which includes a Gaussian beam emission model and is applicable for both homogeneous and inhomogeneous waters. The effective attenuation coefficient decays exponentially with the increase of depth in water according to the simulations of lidar return signal from case 1 waters, because high order scattering signals increase with depth and predominate the total signal at shallow depth. An exponential function is proposed to describe the depth characteristic of k_{lidar} and the parameters in this exponential function is related to IOPs. The exact function of k_{lidar}, depth and the IOPs of seawater is given and examined using real ocean waters simulated by a model of inhomogeneous water.

ACKNOWLEDGEMENTS

National Key Research and Development Program of China (2016YFC1400900, 2016YFC0200700); National Natural Science Foundation of China (41775023, 61475141).

REFERENCES

[1] G. M. Krekov, M. M. Krekova, and V. S. Shamanaev, "Laser sensing of a subsurface oceanic layer. I. Effect of the atmosphere andwind-driven sea waves," Appl Opt **37**, 1589-1595 (1998).

[2] H. R. Gordon, "Interpretation of airborne oceanic lidar: effects of multiple scattering," Applied Optics **21**, 2996-3001 (1982).

[3] D. Phillips and B. Koerber, "A Theoretical Study of an Airborne Laser Technique for determining Sea Water Turbidity," Australian Journal of Physics **37**, 75 (1984).

[4] B. L. Collister, R. C. Zimmerman, C. I. Sukenik, V. J. Hill, and W. M. Balch, "Remote sensing of optical characteristics and particle distributions of the upper ocean using shipboard lidar," Remote Sensing of Environment **215**, 85-96 (2018).

[5] R. E. Walker and J. W. Mclean, "Lidar equations for turbid media with pulse stretching," Applied Optics **38**, 2384-2397 (1999).

[6] V. I. Feygels, "Mathematical modeling of input signals for oceanographic lidar systems," Proc Spie **5155**, 30-39 (2003).

[7] C. A. Hostetler, M. J. Behrenfeld, Y. Hu, J. W. Hair, and J. A. Schulien, "Spaceborne Lidar in the Study of Marine Systems," Annual Review of Marine Science **10**(2017).

[8] J. H. Churnside, "Review of profiling oceanographic lidar," Optical Engineering **53**, 051405 (2014).

[9] L. Wang, S. L. Jacques, and L. Zheng, *MCML—Monte Carlo modeling of light transport in multi-layered tissues* (Springer US, 1995), pp. 131-146.

[10] D. M. Winker, M. A. Vaughan, A. Omar, Y. Hu, K. A. Powell, Z. Liu, W. H. Hunt, and S. A. Young, "Overview of the CALIPSO Mission and CALIOP Data Processing Algorithms," Journal of Atmospheric & Oceanic Technology **26**, 2310-2323 (2009).

TOWARDS AN ALGORITHM FOR NEAR REAL TIME PROFILING OF AEROSOL SPECIES, TRACE GASES, AND CLOUDS BASED ON THE SYNERGY OF REMOTE SENSING INSTRUMENTS

Nikolaos Siomos[1*], Dimitrios Balis[1], Alkiviadis Bais[1], Mariliza Koukouli[1], Katerina Garane[1], Kalliopi A. Voudouri[1], Fani Gkertsi[1], Athanasios Natsis[1], Dimitrios Karagkiozidis[1], Ilias Fountoulakis[1,2]

[1]*Laboratory of Atmospheric optics, Laboratory of Atmospheric Physics, Physics Department, Aristotle University of Thessaloniki, Greece*
[2]*Aosta Valley Regional Environmental Protection Agency (ARPA), Saint-Christophe, Italy*
**Email: nsiomos@physics.auth.gr*

ABSTRACT

In this manuscript we present the concept of a novel algorithmic chain that aims to a dataset of unprecedented detail in the vertical distribution of multiple atmospheric components in near real time conditions. The analysis will be based on the following remote sensing instruments: a depolarization Raman lidar, a visible and a thermal all-sky camera, a Brewer spectrophotometer, and up to three mini DOAS/MAX-DOAS systems. Based on both individual and synergistic processing of the data collected, novel products will be made available in near real time conditions to the end users. Columnar aerosol information from the spectrophotometers will be combined with lidar data to retrieve vertical profiles of individual aerosol species. Cloud layers will be detected and classified based mainly on the synergy of the lidar and the sky cameras and a realistic 3D representation of cloud conditions around the measurement site will be produced. Lidar profiles will be implemented as a priori information for radiative transfer purposes, that are necessary in order to obtain high quality trace gases profiles from the DOAS/MAX-DOAS spectrophotometer. Fast synergistic data processing will ensure that the algorithm can be applied for near real time public data dissemination in the future.

1. INTRODUCTION

Air quality forms an indisputable major global concern as in year 2014 only 12% of the Earth's urbanized population lived in cities whose air quality levels comply with the World Health Organization, WHO, guideline levels. Recent studies show that more people die because of air pollution than malaria and HIV combined, including 1.4 million people in China, 650000 in India and 180000 in Europe, per annum. Accurate information on air quality levels traditionally depends on local monitoring stations. Their task is to monitor multiple atmospheric components. These components can be summarized in four main categories: aerosol species, clouds, trace gases, and greenhouse gases. Since they are of quite different nature, variable vertical distribution, and also considering inherent difficulties in performing in situ measurements in high altitudes, it is rather challenging for a single station to provide a complete, spatially resolved dataset. While the use of remote sensing techniques overcomes difficulties such as targeting the elevated atmospheric components, the retrievals from single instruments typically leads to limited datasets. If a holistic approach is desired, synergies between remote sensing instrumental techniques are certainly required. The need for such datasets is also evident in radiative transfer modeling, considering that all atmospheric components interact with radiation.

This study proposes new potential synergies using a network of co-located remote sensing instruments installed in the facilities of the Laboratory of Atmospheric Physics, in Thessaloniki, Greece (40.63, 22.95, 60m): a depolarization Raman lidar, a visible and a thermal all-sky cameras, a Brewer spectrophotometer, and up to three mini DOAS/MAX-DOAS systems. The main goal is to provide the algorithmic means to reveal new information on atmospheric components. A prototype chain of processing algorithms that will operate with standard streams and perform automated inversions, will allow fast monitoring of the atmospheric components and their effect on radiation that is really hard to achieve otherwise.

2. INSTRUMENTATION

Multiple instruments that monitor the atmosphere are currently installed in the facilities of the Laboratory of Atmospheric Physics. They are constantly maintained and calibrated, and most of them have participated in international field campaigns. The current equipment includes:

• A double-monochromator Brewer spectrophotometer that performs parallel direct and global spectral measurements of the solar irradiance in the 290 - 363 nm wavelength range since 1997. Information on the calibration and QA/QC of the measured spectra can be found in [1] and [2]. It will provide aerosol properties such as the spectral UV Aerosol Optical Depth (AOD) and Single Scattering Albedo (SSA).

• A multiwavelength depolarization Raman lidar that is installed in LAP-AUTH since 2000. The current setup of the system includes three elastic channels at 355, 532 and 1064 nm, two nighttime only Raman channels at 387 and 607 nm and two polarization channels (parallel and cross) at 532 nm. A detailed description of the system can be found in [3]. It will provide aerosol properties such as the backscatter, depolarization, and extinction profiles, and also cloud properties such as the cloud boundaries, optical depth, and the ice/ water phase at zenith.

• Three mini DOAS/MAX-DOAS spectrophotometers that perform automated daytime measurements in the wavelength range 297 - 450 nm of direct solar irradiance and diffuse sky radiance since 2013. They consist of a cooled, miniature CCD spectrograph and a 2-axes tracker. They are described in [4]. They will provide trace gases' concentrations (NO_2, HCHO, SO_2, O_4, and O_3) and also aerosol properties such as the daytime spectral UV AOD, UV SSA and the Asymmetry Factor as well as the nighttime spectral UV AOD.

• A commercial digital all-sky camera with a fish-eye lens and a hemispheric dome that is installed since 2012. The addition of a thermal all-sky camera in the longwave infrared region to the current equipment would allow nighttime cloud detection (e.g. [5]), extending synergies with the lidar after sunset. They will provide cloud properties such as the cloud fraction, mask, classification, brightness temperature, and the cloud optical depth (COD).

3. METHODOLOGY

A synopsis of the implementation of the research idea is presented here. Initially, measurements from all relevant instruments will be gathered in a pilot dataset. They will be analyzed towards their potential to reveal information on the vertical and spatial distribution of aerosol species, trace gases, and clouds. The necessary algorithms that link the raw data with the final retrievals will be implemented first. This includes relevant routine inversion algorithms already adopted by the station, state of the art algorithms that are going to be adjusted to the current equipment and novel synergistic techniques that will be developed in the frame of this project. All the individual modules will be integrated into the umbrella algorithm: Synergistic Profiling Algorithm for the Retrieval of Trace Gases, Aerosol Species, and ClOUdS (SPARTACUS), which will automatically process the raw data from the instruments and apply the necessary inversions, proceed to the synergistic algorithms, and finally combine the retrievals to a common output. It will follow a standard input/output protocol in order to facilitate implementation in other stations in the future. Finally, an Operational Acceptance Testing will ensure that all processes are robust enough for near real time application. The final products of SPARTACUS algorithm will include:

• Aerosol Species Concentration Profiles: A new algorithm will be designed for the retrieval of aerosol species concentration profiles based on the synergy of lidar profiles and spectrophotometer columnar aerosol properties. Aerosols will be modeled as mixtures of up to 9 aerosol species with distinct size distributions and dry refractive indices, and variable total concentration similar to the OPAC database [6]. The additional parametrizations of [7] will also be considered. The hygroscopic growth will be calculated using humidity profiles from meteorological models and/or radiosondes and the absorbed water concentration per species will also be retrieved.

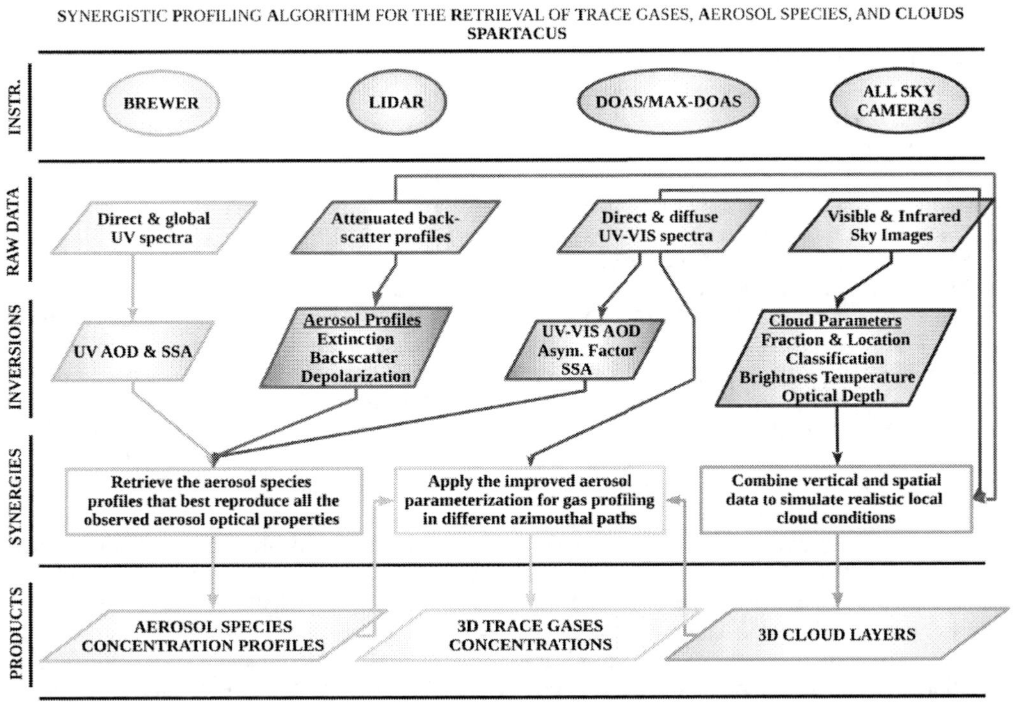

Figure 1: Flowchart of the umbrella algorithm SPARTACUS. Instruments are marked with round borders, algorithms with square boxes and products with parallelograms.

The species will include fine dust, coarse dust, fine sea salt, coarse sea salt, coarse volcanic ash, fine black carbon, fine carbonaceous, fine water soluble and potentially coarse pollen. During daytime, spectrophotometer columnar aerosol products will supplement lidar backscatter and depolarization profiles. During nighttime, the addition of the extinction profiles could suffice for a decent separation of some aerosol species (e.g. [7]). The potential to separate more species, though, by including lunar DOAS/MAX-DOAS spectral AOD will also be examined. Concerning the inversion method, look up table and optimal estimation approaches will be investigated. The first inversion category will focus on building a library of simulated aerosol mixtures and then apply machine learning techniques in order to identify the mixture that best describes the measurements. The latter will focus on building a regression algorithm (e.g. non linear least squares, total least squares) that will solve towards the concentration profiles combination that best reproduces the observations.

• 3D Trace Gases Concentration: Many techniques currently provide trace gas profiles from DOAS/MAX-DOAS systems (e.g. [8]) by applying an aerosol extinction profile, usually either from a climatology or from DOAS/MAX-DOAS O_4 measurements. In both cases, such a profile cannot compete with the high resolution and accuracy a dedicated lidar can offer. The application of the novel aerosol species concentration profiles will further improve the radiative transfer calculations that are required for the gas profiling due to the advanced aerosol parameterization which goes far beyond a simple extinction profile application. Both optimal estimation and lookup table approaches will be considered for the inversions. By measuring in different azimuthal paths, the possibility to derive the spatial distribution of traces gases around the instrument will also be examined. Finally, the novel cloud products can be applied in order to investigate the effect of partial cloud cover to the retrievals.

- 3D Cloud Layers: A new algorithm designed for realistic simulation of cloud conditions will be developed based on lidar, all sky cameras, and DOAS/MAX-DOAS measurements. While a lidar can accurately and diurnally monitor cloud properties at zenith, it lacks information about their spatial distribution that can be diurnally provided by sky cameras. The placement of the two cameras could allow the application of stereophotometric techniques and the inference of cloud base height information (e.g [9]). Cloud base trigonometry with the DOAS/MAX-DOAS systems can also be investigated during daytime as there are currently three of them available. As cloud optical depth, base brightness temperature, and classified type can also be indicative of the cloud base height and thickness, they can be applied in order to improve the stereophotometric retrievals, also in combination with the accurate lidar zenith cloud products. Finally, a pattern recognition algorithm will identify continuous cloud layers from sky images and cluster them in categories based on similarity. This will determine whether it is safe to expand the detailed lidar information in zenith to other parts of the sky with similar cloud conditions.

4. THE IMPACT

The ability to obtain the vertical distribution of individual aerosol species and trace gases is still quite fresh in the scientific community. The atmospheric modeling community for sure would benefit the most as such information can be invaluable for the evaluation of forecasting models. The same products could be utilized in the evaluation of satellite observations as well. As cloud formation depends largely on the existing aerosol load and some aerosol species originate from trace gases, applications in the fields of aerosol and cloud interactions and aerosol formation in elevated layers are also possible. Since both aerosols and trace gases are factors that negatively affect public health, the knowledge of their concentration levels inside the boundary layer is important in preventing detrimental exposure by timely warning the public. Likewise, a near real time detection of extreme aerosol outbreaks, such as dust and volcanic ash plumes, can be critical in avoiding aviation hazards. Applications could even focus towards agriculture as there is evidence that aerosol species and trace gases could damage crops directly by being absorbed by the plants.

ACKNOWLEDGMENTS

We acknowledge support of this work by the project "PANhellenic infrastructure for Atmospheric Composition and climatE change" (MIS 5021516) which is implemented under the Action "Reinforcement of the Research and Innovation Infrastructure", funded by the Operational Programme "Competitiveness, Entrepreneurship and Innovation" (NSRF 2014-2020). This research is co-financed through the Operational Program «Human Resources Development, Education and Lifelong Learning 2014-2020» in the context of the project "Optimization and application of methods for ground-based remote sensing of aerosols and ozone in the lower troposphere for investigating their variability" (MIS 5004486). Kalliopi A. Voudouri acknowledges the support of the General Secretariat for Research and Technology and the Hellenic Foundation for Research and Innovation. Dimitrios Karagkiozidis and Fani Gkertsi acknowledge the support of the Operational Programme «Human Resources Development, Education and Lifelong Learning» in the project "Strengthening Human Resources Research Potential via Doctorate Research" (MIS-5000432), implemented by the State Scholarships Foundation (IKY).

REFERENCES

[1] Kazadzis, et al. Applied Optics, 44(9), 1681-1690 (2005)
[2] Garane, et al. Ann. Geophys., 24(12), 3215-3228 (2006)
[3] Siomos, et al. Atmos. Chem. Phys., 18, 11885-11903 (2018)
[4] Gkertsi, et al. Atmospheric Environment, 180, 51-58 (2018)
[5] Aebi, et al. Atmos. Meas. Tech., 11, 5549-5563, (2018)
[6] Hess, et al. Bulletin of the American Meteorological Society, 79, 831-844 (1998)
[7] Hara, et al. Remote Sensing, 10 (2018)
[8] Frieß, et al. Atmos. Meas. Tech. Discuss., in review (2018)
[9] Crispel, P. and Roberts, G. Atmos. Meas. Tech., 11, 593-609 (2018)

EPJ Web Conferences **237**, 08024 (2020) https://doi.org/10.1051/epjconf/202023708024
ILRC 29

TWO PARAMETER-RETRIEVAL ALGORITHMS OF AIRCRAFT WAKE VORTEX WITH DOPPLER LIDAR IN CLEAR AIR

Chun Shen, Jianbing Li*, Hang Gao

State Key Laboratory of Complex Electromagnetic Effects on Electronics and Information System, National University of Defense Technology, Changsha 410073, China
**Email: jianbingli@nudt.edu.cn*

ABSTRACT

Aircraft wake is a pair of strong counter-rotating vortices generated behind an aircraft, which might be very hazardous to a flowing aircraft and the detection of which has attracted much attention in aviation safety field. This conference paper introduces two parameter-retrieval algorithms, i.e., Optimization method and Max-min method. They have been integrated into a toolbox and can retrieve the parameters of wake vortex efficiently and robustly.

1. INTRODUCTION

Aircraft wake vortex is a strong counter-rotating turbulence generated by a flying aircraft. It might be very dangerous for a following aircraft because it may cause the following aircraft to roll out of control. To avoid this hazard, a special flight separation rule was proposed by ICAO, but it is conservative and has limited the flight capacity of airports. Therefore, it is very important to detect the wake vortex in real time to make a compromise between the flight safety and capacity. The Single European Sky Air Traffic Management Research(SESAR)[1] plan in EU and the Next Generation Air Transportation System(NGATS)[2,3] plan in US have also delivered a lot of projects on this issue.

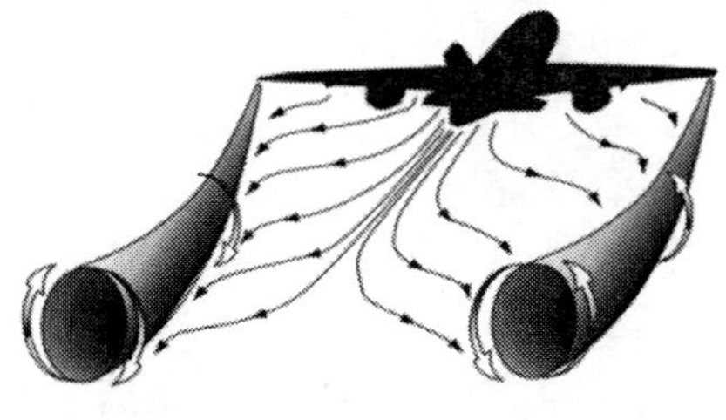

Fig. 1 Formation of Wake Vortex

Basically, there are two types of sensors for wake vortex detection: Radar and Lidar[4]. For wake vortex generated in clean air, the proposed sensor is Lidar, and the representative parameter-retrieval algorithms include the TV method[5] and the radial velocity method [6-8]. We have been studying on the radar characteristics and detection of wake vortex for many years and we recently proposed two methods to retrieve the parameters of aircraft wake vortex, say the Optimization method and the Max-Min method. In this conference paper, we briefly introduce these two methods, and whose performances are verified with simulated and field detection data.

2. TWO NEW ALGORITHMS

2.1 Optimization Method

For Lidar detection of aircraft wake vortex in clear air, the particles to be sensed are the aerosols, which are generally with weak inertia. In this sense, the obtained Doppler velocity is assumed to be approximately the same to that of the ambient velocity. According to this principle, a movement equation[9] is established to connect the Doppler velocity and the parameters of wake vortex, and an optimization method has been proposed to solve the equation well.

The radial velocity of the ith detection cell (equal to and opposite to the doppler velocity, $-V_{\mathrm{D}}^{(i)}$) represents the projection of the background velocity ($V^{\text{flow},(i)}$)on the lidar line of sight:

$$-\mathbf{V}_{\mathrm{D}}^{(i)} = \mathbf{V}^{\text{flow},(i)}\Big|_{\mathrm{r}} = \mathbf{V}^{\text{wind},(i)}\Big|_{\mathrm{r}} + \mathbf{V}^{\text{wake},(i)}\Big|_{\mathrm{r}} \quad (1)$$

where $\mathbf{V}^{\text{wind},(i)}$ and $\mathbf{V}^{\text{wake},(i)}$ are the background wind velocity and wake vortex velocity at the detection cell, $\mathbf{X}\big|_{\mathrm{r}} = \mathbf{k}\cdot\mathbf{X}$ are the velocity vector $\mathbf{X}$ project on the lidar beam, and $\mathbf{k}$ is the direction vector of the beam with an elevation Angle of α: $\mathbf{k} = [\cos\alpha \quad \sin\alpha]$

According to equation (1), the equation of wake vortex characteristic parameter-retrieval is composed of background wind velocity, wake velocity and doppler velocity:

$$\mathcal{F}^{(i)}(\Gamma_L,\Gamma_R,O_L,O_R;V_D^{(i)},x_i,y_i)=V^{\text{wind},(i)}\Big|_r+ V^{\text{wake},(i)}(\Gamma_L,\Gamma_R,O_L,O_R)\Big|_r+V_D^{(i)}=0 \quad (2)$$

The equation contains six unknowns (O_L, O_R respectively represent left vortex-core and right vortex-core position), so when we use M detection cells $\left[x_i, y_i, V_D^{(i)}\right]$ (M≥6), the equation can be solved:

$$[\hat{\Gamma}_L,\hat{\Gamma}_R,\hat{O}_L,\hat{O}_R]=\underset{\Gamma_L,\Gamma_R,O_L,O_R,r_c}{\arg\min}\sum_{i=1}^{M}\left|\mathcal{F}^{(i)}(\Gamma_L,\Gamma_R,O_L,O_R;V_D^{(i)},x_i,y_i)\right|^2 \quad (3)$$

The strengths of this method include:

1. The ground effect of wake vortex has been considered by introducing two imaging vortices in the velocity model.
2. The wake vortex moves while the Lidar beam scans up and down alternately, which leads to distortion of Doppler velocity on the RHI plane. This distortion has been fixed by taking into account the acceleration of wake vortex.
3. The fitting algorithm has been introduced to predict the vortex core positions, which could be very helpful to get robust and accurate retrieval results.

2.2 Max-min Method

The difference between the maximum and minimum doppler velocities at different elevation and the same radial distance R_k is called "doppler velocity range" [10]

$$\Delta V(R_k)=\max_{\varphi\in[\varphi_{\min},\varphi_{\max}]}\{V_D(R_k,\varphi)\}-\min_{\varphi\in[\varphi_{\min},\varphi_{\max}]}\{V_D(R_k,\varphi)\} \quad (4)$$

where $V_D(R_k,\varphi)$ represents the doppler velocity.

The "velocity range" at the center of the vortex-core should be significantly larger than that at other radial distances. Therefore, the "velocity range" at different radial distances is fitted as a double Gaussian curve, and two peak points corresponding to the left and right vortex cores can be obtained. The radial distance from the peak point to the lidar is the estimated value of left or right vortex radial distance(R_{c1} or R_{c2}).

The Doppler velocity along the elevation direction is used to determine the left and right vortices' elevation angles(φ_{c1} and φ_{c2}). If the maximum and minimum Doppler velocities are found at $(R_{ci},\varphi_{ci,\max})$ and $(R_{ci},\varphi_{ci,\min})$ ($i=1,2$), the elevation angle of a vortex-core can be approximated as $\varphi_{ci}=\frac{1}{2}\left(\varphi_{ci,\max}+\varphi_{ci,\min}\right)$ $(i=1,2)$.

From the Burnham-Hallock velocity model, the tangential velocity of a single vortex whose core located at (R_i,φ_{ci}) is:

$$V^{\alpha}=\frac{\Gamma_i}{2\pi R}\cdot\frac{R^2}{R^2+r_c^2} \quad (5)$$

where R is the distance from the detection cell to the vortex-core, Γ_i is the circulation, r_c is the vortex-core radius.

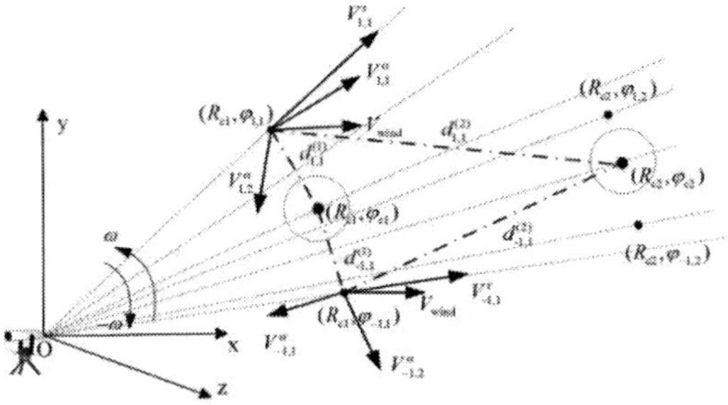

Fig. 2 Velocity Component of The Detection Cell

Fig 2 shows velocity component of the detection cell on the radial distance of the left vortex, V_1^r and V_{-1}^r respectively represent velocity above and below the vortex core:

$$V_1^r=\frac{\Gamma_1}{2\pi}\cdot\frac{R_{c1}}{d_{11}^2+r_c^2}\cdot\sin(\varphi_1-\varphi_{c1}) -\frac{\Gamma_2}{2\pi}\cdot\frac{R_{c2}}{d_{12}^2+r_c^2}\cdot\sin(\varphi_1-\varphi_{c2})+V_{\text{wind}}\cdot\cos\varphi_1 \quad (6)$$

$$V_{-1}^r=\frac{\Gamma_1}{2\pi}\cdot\frac{R_{c1}}{d_{-11}^2+r_c^2}\cdot\sin(\varphi_{-1}-\varphi_{c1}) -\frac{\Gamma_2}{2\pi}\cdot\frac{R_{c2}}{d_{-12}^2+r_c^2}\cdot\sin(\varphi_{-1}-\varphi_{c2})+V_{\text{wind}}\cdot\cos\varphi_{-1} \quad (7)$$

where d_{1i}, d_{-1i} $(i=1,2)$ are the distances from the detection cell (above and below the vortex core) to the ith vortex core. φ_1 (φ_{-1}) is the elevation

 https://doi.org/10.1051/epjconf/202023708024

angle of the extreme Doppler velocities above (below) the vortex core. Normally the Lidar set at a long distance to the wake vortex, we can assume $V_{\text{wind}} \cdot \cos\varphi_1 \approx V_{\text{wind}} \cdot \cos\varphi_{-1}$, thus the effect of background wind is mitigated by the subtraction of the two velocities:

$$\Delta V_1 = V_1^r - V_{-1}^r \approx \begin{bmatrix} \Psi_1 & \Psi_2 \end{bmatrix} \begin{bmatrix} \Gamma_1 \\ \Gamma_2 \end{bmatrix} \tag{8}$$

where

$$\Psi_1 = \frac{R_{c1}}{2\pi} \cdot \left[\frac{\sin(\varphi_1 - \varphi_{c1})}{d_{11}^{\ 2} + r_c^2} + \frac{\sin(\varphi_{c1} - \varphi_{-1})}{d_{-11}^{\ 2} + r_c^2} \right]$$

$$\Psi_2 = -\frac{R_{c2}}{2\pi} \cdot \left[\frac{\sin(\varphi_1 - \varphi_{c2})}{d_{12}^{\ 2} + r_c^2} + \frac{\sin(\varphi_{c2} - \varphi_{-1})}{d_{-12}^{\ 2} + r_c^2} \right]$$

with r_c=0.052b_0 being the vortex core size, and b_0 being the separation between the two vortex cores. $[\Gamma_1, \Gamma_2]$ can be easily solved by equation (8).

3. RESULTS

3.1 Wake-Vortex Parameter-Retrieval System

We have developed a comprehensive wake vortex simulation and parameter-retrieval system as shown in Fig. 3.

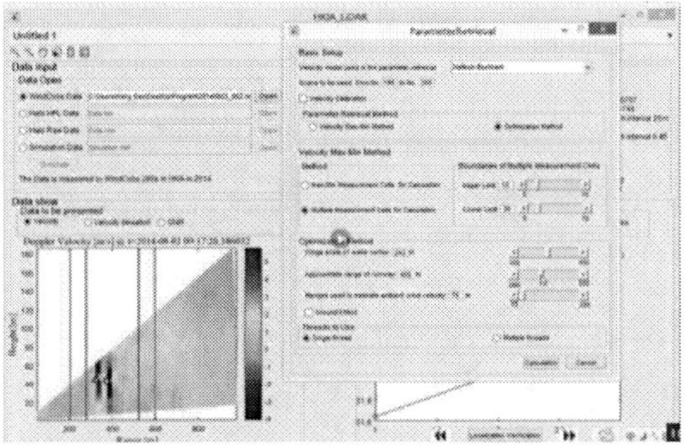

Fig. 3 Wake Vortex Parameter-Retrieval System

The system can well simulate the dynamics, scattering, echo signal, Doppler spectra, and etc. It can handle the Lidar detection data from a number of different Lidars, for example the Halo Streamline, Wind Cube 400S, WindTracer, etc. The two algorithms have been integrated into the processing system to retrieve the wake vortex parameters from simulated data or field detection data, and good performances have been well verified.

3.2 Retrieval Result of Circulation

While calculating on the same computer, the time costs of the RV method, Optimization method, Max-min method and TV method are several minutes,5.6s, 3.5s and 1.9s. For the purpose of real time, this conference paper does not show the result of RV method.

Fig 4 shows the circulation result of the three methods. It can be found that Max-min method and Optimization method can get more robust result than TV method. The TV method could overestimate the circulation, which is generally caused by locating the tangential velocity improperly. The Max-min method and Optimization methods have a slight underestimation, which is caused by spatial averaging error of the simulated data.

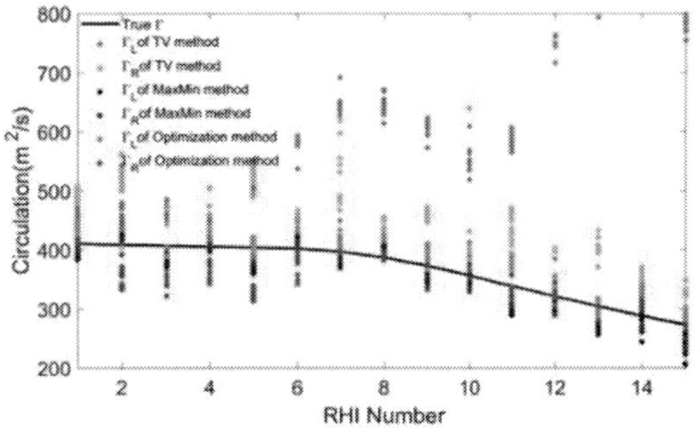

Fig. 4 Circulation Result on 15 Simulated RHIs

4. CONCLUSION

This paper introduces two parameter-retrieval method we recently developed, i.e., the optimization method and the Max-Min velocity method. They have been integrated into a software toolbox and the good performances have been well verified. The two methods can be very helpful for real time detection and prediction of wake vortex in air traffic control.

ACKNOWLEDGEMENTS

This work was supported by the National Natural Science Foundation of China (No. 61771479) and distinguished young scholar fund of Hunan Province (No.2018JJ1030).

REFERENCES

[1] Sesar Consortium. European Air Traffic Management Master Plan [R]. SESAR Consortium, 2009.

[2] J. Burden, P. A. Curry, D. Roby, F. Love. Introduction to the Next Generation Automatic Test System (NGATS) [R]. IEEE Autotestcon, 2005.

[3] P. A. Curry, J. Burden, G. A. Lundy. Next Generation Automatic Test System (NGATS) Update [R]. IEEE Autotestcon, 2006.

[4] S. Hannon, J. Alexthomson. Aircraft Wake Vortex Detection and Measurement with Pulsed Solid-State Coherent Laser Radar [J]. Optica Acta International Journal of Optics, 1994, 41(11): 2175-2196.

[5] F. Holzäpfel, T. Gerz, F. Köpp. Strategies for Circulation Evaluation of Aircraft Wake Vortices Measured by Lidar [J]. Journal of Atmospheric and Oceanic Technology, 2003, 20: 1183-1195.

[6] I. N. Smalikho, V. A. Banakh, F. Holzäpfel, S. Rahm. Method of Radial Velocities for the Estimation of Aircraft Wake Vortex Parameters from Data Measured by Coherent Doppler Lidar [J]. Optics Express, 2015, 23(19): A1194-A1207.

[7] I. N. Smalikho, V. A. Banakh. Estimation of Aircraft Wake Vortex Parameters from Data Measured by a Stream Line Lidar [C]// 21st International Symposium on Atmospheric and Ocean Optics: Atmospheric Physics, 2015.

[8] I. N. Smalikho, V. A. Banakh. Estimation of Aircraft Wake Vortex Parameters from Data Measured with a 1.5-Um Coherent Doppler Lidar [J]. Optics Letters, 2015, 40(14): 3408-3411.

[9] H. Gao, J. B. Li, P. W. Chan, K. K. Hon, X. S. Wang. Parameter-Retrieval of Dry-Air Wake Vortices with a Scanning Doppler Lidar [J]. Optics Express, 2018, 26(13):16377.

[10] H. Gao, J. B. Li, P. W. Chan, K. K. Hon. Parameter-Retrieval of Aircraft Wake Vortex Based on Its Max-Min Distribution of Doppler Velocities Measured by a Lidar [C]. IET International Radar Conference, Oct. 2018, Nanjing, China.

USING THE PHYSICAL OPTICS APPROXIMATION FOR ESTIMATING THE LIGHT SCATTERING PROPERTIES OF LARGE DUST PARTICLES FOR LIDAR APPLICATIONS

Alexander Konoshonkin*[,1], Natalia Kustova[1], Anatoli Borovoi[1], Alexandra Tsekeri[2], Josef Gasteiger[3]

[1]*V.E. Zuev Institute of Atmospheric Optics SB RAS, Tomsk 634055, Russia*
[2]*Institute for Astronomy Astrophysics Space Applications and Remote Sensing, Athens 15236, Greece*
[3]*Faculty of Physics, University of Vienna, Vienna 1010, Austria*
**Email: sasha_tvo@iao.ru*

ABSTRACT

The physical optics approximation was used to solve the light scattering problem of large dust particles with irregular shape. The comparison between the physical optics approximation and the DDA method was made for particles with volume-equivalent size parameters of 44.21. A full light scattering matrix (Mueller matrix) was calculated for particles with volume-equivalent size parameters from 44 up to 220 for two wavelengths: 0.532 and 1.064 µm. The dependences of the lidar and depolarization ratio, and extinction coefficient on the number of spatial orientations of the particle were obtained. It was shown that the elements of the light backscattering matrix obey power laws.

1. INTRODUCTION

Since the interpretation of the lidar signal is the inverse light scattering problem, it requires the calculation of the light scattering properties of the particles. In the case of large dust particles with realistic irregular shapes, the calculation of the scattering of light becomes a non-trivial task. This is due to the fact that the requirement for computational resources of numerical exact methods (FDTD, DDA) increases significantly with increasing the particle size. In this report, dust scattering properties are obtained within the framework of the physical optics approximation (PO), which has worked well for the solution of the light scattering of large atmospheric ice particles. The computational complexity of the PO does not increase with increasing the particle size, on the contrary, the larger the particle, the more accurate the approximation.

2. NUMERICAL RESULTS

Since the physical optics approximation provides satisfactory results for particles with size much larger than the wavelength, it suffices to show its applicability for particles with smaller sizes and thus define the lower limit of its applicability. Further, for any larger size accuracy will not be worse. In this report, the particle with the volume-equivalent size parameters x = 44.21 was chosen as the minimum size. There

$$x = 2\pi r_v / \lambda,\ r_v = \sqrt[3]{3V / 4\pi}\ ,$$

where r_v is the volume equivalent radius of the particle and V is volume of the particle. The PO calculations are performed for two lidar wavelengths of 0.532 and 1.064, and the refractive index was chosen as m = 1.48+i0.0 and m = 1.6+i0.0, respectively. The particle shape was taken from [1], shape *F*.

Comparison of the solutions using the ADDA [2] and the PO [3] is presented in Figs. 1 and 2. It can be seen that the solutions agree well with each other. However, in the exact backward scattering direction, the solution essentially depends on the number of orientations.

Due to the requirements for computing resources of the ADDA code, it was possible to obtain a solution only for 288 orientations. The physical optics approximation is much faster, so we analyzed the dependence of lidar characteristics on the number of particle orientations using the PO, see Fig. 3. Figure 3 shows that the lidar and depolarization ratios essentially depend on the number of orientations, and in this case there is no convergence even at 4608 orientations, despite this fact, 72 orientations are sufficient to correctly calculate the extinction coefficient. Nevertheless, based on the estimation of [4], we believe that 4608 orientations are quite enough. Here we mean special averaging over only two Euler angles β and γ, averaging over the third Euler angle α is performed analytically.

It should be noted that the PO is an iterative method. At the first iteration, the solution is found only for external specular reflection, then it is refined at each iteration, see [5] for details. The convergence of the method on the number of iterations is presented in Fig. 4.

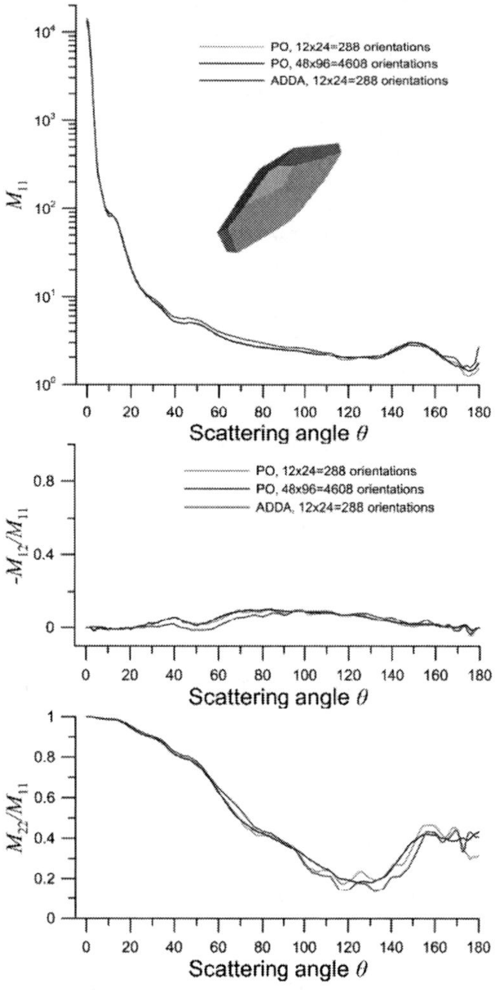

Figure 1. Elements of the Mueller matrix for a randomly oriented dust particle calculated within the physical-optics approximation and discrete dipole approximation for 0.532 μm.

For the calculations, we chose 7 iterations of the PO and 4608 special orientations. Figure 5 shows the dependence of the M_{11} element of the light scattering matrix in the vicinity of the backward scattering direction on the particle size. We see that the interference oscillations are not significant and can be neglected (see Fig. 6), which allow us to increase the step of the particle size incrementation. Solution for particles with sizes from 40 up to 220 is presented in Fig. 7.

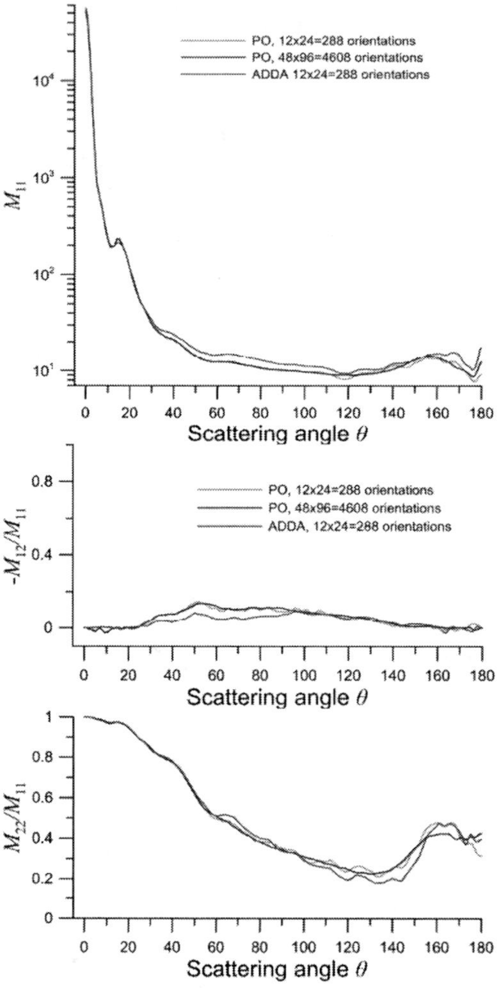

Figure 2. The same as Fig. 1, but for 1.064 μm.

Generally speaking, with a doubling in size it is necessary to increase the number of orientations four times. Therefore, the calculation for size $x = 200$ requires 115200 orientations.

However, Fig. 8 clearly shows that the solution in the exact backscattering direction, that is important for the lidar application, can be well

 https://doi.org/10.1051/epjconf/202023708025

approximated by power functions similar to ice particles, see [6] for details. Note that this is shown only if we neglect the absorption effect. When the absorption is taking into account the approximation can change.

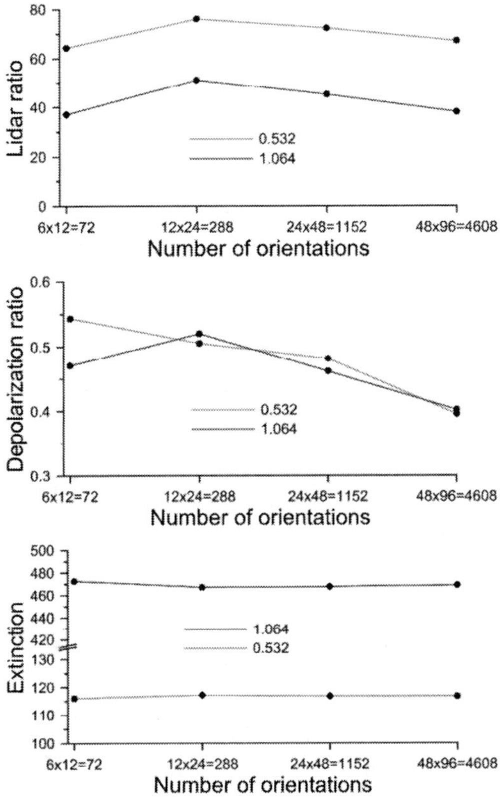

Figure 3. The lidar and depolarization ratio, and extinction coefficient vs. number orientations.

3. CONCLUSION

Thus, it is shown that the physical optics approximation can be efficiently used to solve the problem of light scattering of large dust particles with arbitrary shape. The only condition is that the particle has to be faceted. This solution does not take the absorption effect into account for simplicity. The physical optics approximation significantly exceeds the exact numerical methods in speed. The results presented in the report are obtained on a desktop computer. It is also shown that the interference oscillations are not significant in case of dust aerosol particles; and the solution in the exact backscattering direction obeys power laws, as was shown for atmospheric ice crystals, which significantly simplifies the solution of the problem.

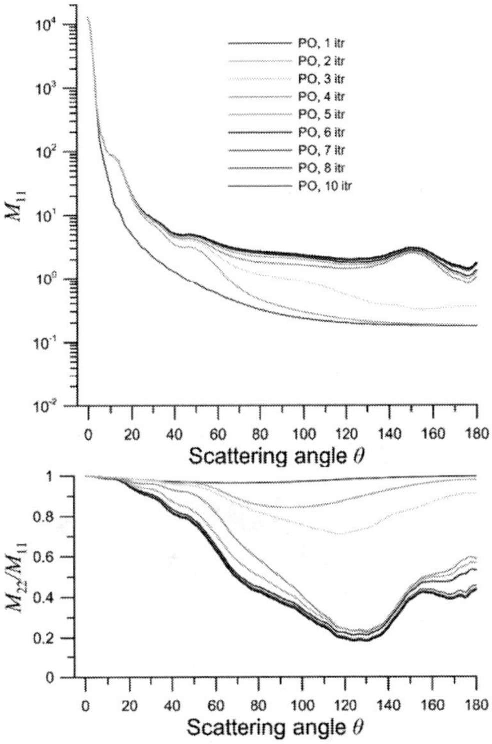

Figure 4. Element of the Mueller matrix depending on the number of iteration.

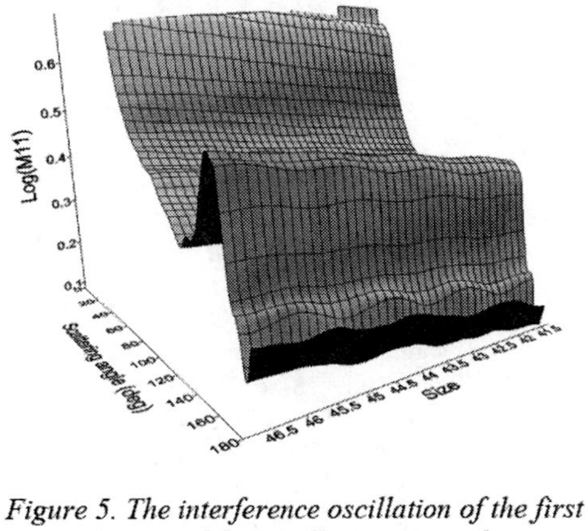

Figure 5. The interference oscillation of the first element of the Mueller matrix in the backscattering direction.

https://doi.org/10.1051/epjconf/202023708025

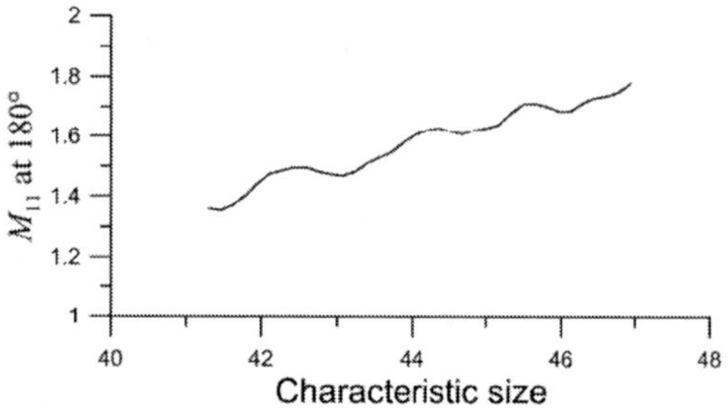

Figure 6. The interference oscillation

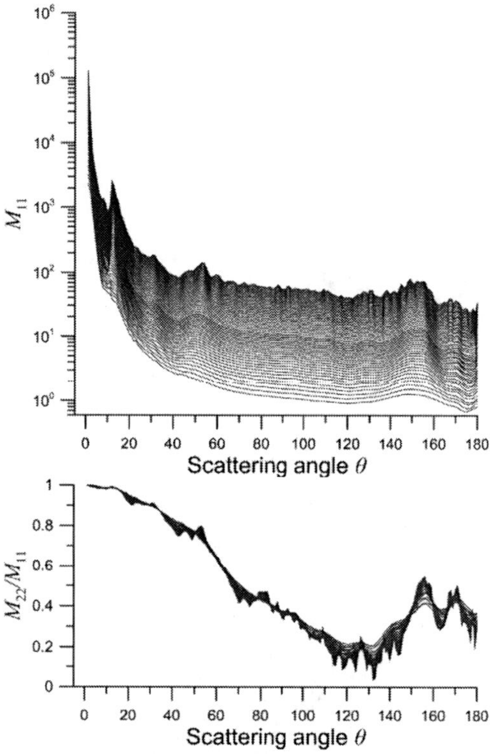

Figure 7. The elements of the Mueller matrix for a dust particle with size from 40 to 220.

ACKNOWLEDGEMENTS

This work is supported by the RFBR under Grants no. 18-05-00568, no. 18-55-53046, and budget project AAAA-A17-117021310145-6. Authors acknowledge PRACE for awarding access to MareNostrum at Barcelona Supercomputing Center (BSC), Spain, and GRNET for awarding computational time in the National HPC facility - ARIS - under project ID pa170906-ADDAPAS and pr005038-REMOD. Calculation of Mueller matrixes within the PO is supported only by Russian Science Foundation (18-77-10035).

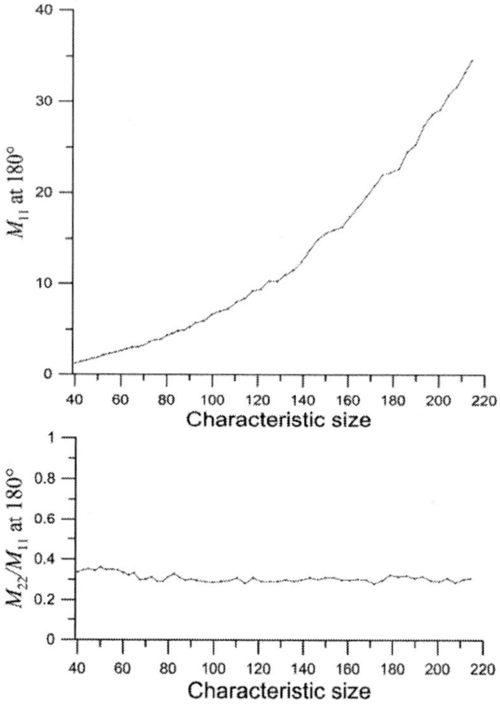

Figure 8. The elements of the Mueller matrix at exact backscattering direction.

REFERENCES

[1]J. Gasteiger, M. Wiegner, S. Groß, V. Freudenthaler, C. Toledano, M. Tesche, K. Kandler, Tellus B: Chemical and Physical Meteorology 63(4), 725-741 (2011)

[2] M. A. Yurkin, V. P. Maltsev, A. G. Hoekstra, J. Quant. Spectrosc. Radiat. Transfer 106, 546-557 (2007)

[3]A. Borovoi, A. Konoshonkin, N. Kustova, J. Quant. Spectrosc. Radiat. Transfer 146, 181-189 (2014)

[4]A. V. Konoshonkin, A. G. Borovoi, N. V. Kustova, H. Okamoto, H. Ishimoto, Y. Grynko, J. Förstner, J. Quant. Spectrosc. Radiat. Transfer 195, 132-140 (2017)

[5]A. V. Konoshonkin, N. V. Kustova, A. G. Borovoi, J. Quant. Spectrosc. Radiat. Transfer 164, 175-183 (2015)

[6] A. V. Konoshonkin, A. G. Borovoi, N. V. Kustova, J. Reichardt, Optics Express 25, 22341-22346 (2017)

Validation of the GOME-2 Absorbing Aerosol Height product using elevated layer top height obtained from Thessaloniki EARLINET station

Konstantinos Michailidis [1], Nikolaos Siomos[1], Dimitris S. Balis [1], Maria-Elissavet Koukouli[1] Kalliopi-Artemis Voudouri[1], Tuinder Olaf [2], Gijsbert Tilstra[2], Wang Ping[2]

[1]*Laboratory of Atmospheric Physics, Physics Department, Aristotle University of Thessaloniki, Greece*
[2]*Royal Netherlands Meteorological Institute (KNMI), De Bilt, the Netherlands*
komichai@physics.auth.gr

ABSTRACT

The purpose of this research is to investigate the ability of GOME-2 instrument on board on the MetOpA and MetOpB platforms, to deliver accurate geometrical features of lofted aerosol layers over the area of Thessaloniki. For this purpose, we use ground-based lidar data from the Thessaloniki lidar station that belongs to EARLINET network. Results of the validation between absorbing aerosol height (AAH) from GOME-2 sensor and THELISYS lidar dataset products will be presented.

1. INTRODUCTION

Aerosols affect the Earth's radiation budget directly by interacting with solar and terrestrial radiation and indirectly by affecting cloud processes, having a large impact on air quality. Such effects depend on the optical properties of the aerosols and their vertical distribution in the atmosphere. Knowledge of the aerosol layer height, ALH, is essential for understanding the impact of aerosols on the climate system. Ground-based observations are important to verify the accuracy and validity of satellite observations and model-based results. The purpose of validation is to detect biases in the comparison of space-borne absorbing aerosol height, AAH, products with independent ground based observations and establish the accuracy levels of such. Lidar instruments can provide aerosol profile information such as the backscatter and extinction coefficients at different wavelengths which are representative of the aerosol load, with a vertical resolution of a few meters. In this study we use data from satellite GOME-2 instrument and ground based lidar system located in Thessaloniki, Greece.

2. DATA AND METHODOLOGY

2.1 THELISYS LIDAR

Thessaloniki lidar system (THELISYS) is located at the Laboratory of Atmospheric Physics (LAP) (40.5°N, 22.9°E, 50 m above sea level) in Aristotle University of Thessaloniki (AUTH). THELISYS has been in operation since 2000 and is part of the European Aerosol Lidar Network (EARLINET). The European Aerosol Research Lidar Network has coordinated ground-based lidar activities on the European continent, and it holds a comprehensive database of European lidar data sets giving information on the horizontal, vertical and temporal distribution of aerosols on a continental scale [1]. The THELISYS lidar dataset used in our study includes vertical profiles of the aerosol extinction coefficient at 355nm and 532nm, and the aerosol backscatter coefficient at 355nm, 532nm, and 1064nm. Measurements are typically performed every Monday and Thursday according to the EARLINET schedule of data acquisition. All lidar profiles are publicly available on the EARLINET database [2] (www.earlinet.org). The time period between 2007 and 2018 is selected, so as to perform this validation study.

2.2 GOME-2 /METOP

In this study we use data from GOME-2 on board the MetOp-A and MetOp-B satellite platforms [3]. The GOME-2 instrument is a nadir looking and scanning UV-VIS spectrometer on board the METOP-A and METOP-B satellite platforms since 2006 and 2012 respectively that measures backscattered solar light. GOME-2 is comprised of four main optical channels of 1024 pixels giving in total 4096 spectral channels for a range of 240–790 nm and a spectral resolution of 0.26–0.51 nm, with a ground pixel size 80x40km, a total of 24 pixels in the forward scan, on a 1920km of swath width. MetOp satellites are flying in sunsynchronous orbits with an approximate equator crossing time, around 09:30LT (descending node) and repeat cycle of 29 days. Data from GOME-2 (MetOpA and MetOpB) have been processed by the Royal Netherlands Meteorological Institute (KNMI) as part of the EUMETSAT Satellite Application Facility on Atmopsheric Composition Monitoring (https://acsaf.org/)

The absorbing aerosol index (AAI) derived from UV reflectances derived from the 340/380nm wavelength pair, is an operational product of the GOME-2 instrument [6]. The AAI product indicates the presence of elevated absorbing aerosols in the Earth's atmosphere. Aerosol types that are mostly encountered in this region are desert dust and aerosols from biomass burning and volcanic eruption events. A positive value for the AAI can only be explained by the presence of absorbing aerosols. Whenever the AAI is greater or equal than 2, the amount of observing aerosol is considered sufficient to result in a reliable retrieval of the AAH from satellite data [7]. Furthermore, AAI is sensitive to clouds therefore it is derived for both cloud free and cloudy scenes [8]. Factors such as sunglint, solar eclipses, surface height, and scattering geometry have to be taken into account as they can lead to abnormal AAI and AAH values [5].

2.3 METHODOLOGY

The time period from January 2007 to August 2018 is selected for the comparison between satellite (GOME-2) and ground-based Absorbing Aerosol Height (AAH). Aerosol geometrical properties carry information about the structure of lidar profiles such as PBL height and lofted aerosol layers, including vertical profiles of the aerosol extinction coefficient at 355nm, and the aerosol backscatter coefficient at 355nm, 532nm and 1064nm. The THELISYS data analysis, for lidar dataset, is based on the methodology of [4] that applies the wavelet covariance transform (WCT) to the raw lidar data in order to extract geometrical characteristics like PBL and cloud boundaries. We used this method in the case of lofted layers and we have chosen to identify the three major lofted layers with the largest contribution to the vertical profile. The GOME2 AAH product is provided directly by KNMI. The collocation of AAH satellite product have been restricted to an area of 300km radius around Thessaloniki EARLINET station. The closest measurement in time has been selected for each overpass observation, within the limits set by the spatial collocation criteria. Since the satellite can detect only the top of the upper layer, the maximum value of the lidar features is selected per measurement for the comparison. In Figure 1, we present the spatial coverage for AAH collocated satellite product as derived from the comparison with ground based lidar measurements, at 6h temporal interval. The concentric red circles denote regions of 100, 200, 300km from the location of THELISYS lidar system.

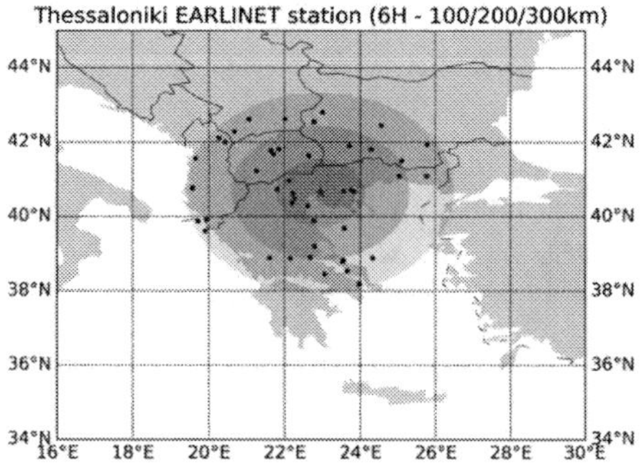

Figure 1: The concentric red circles denote regions of 100, 200, 300km from the location of THELISYS lidar system (red asterisk).

3. RESULTS

In Figure 2, we show the AAH timeseries from the GOME-2 satellite measurements in blue and THELISYS observations in orange. In our data analysis we set as spatial criterion for coincidence over Thessaloniki, the value of 300km. We notice that there are more satellite measurements than the ground lidar. This difference is due to the fact that satellite sampling is daily, while lidar measurements usually occur two days a week for a short time period. Gaps occurring in the time series in terms of lidar measurements are due to cases where lidar was unable to measure because of presence of clouds. Especially, during the winter months it seems there is a limitation of lidar measurements. Apart from cloudy conditions, due to hardware limitations it is not possible for the lidar to operate during days with strong winds. There are quite large values for the height products. For satellite observations the maximum values reach 15 km. This product is being retrieved can never be larger than 15 km. A value of 15km indicates that the real aerosol layer is at 15km or higher.

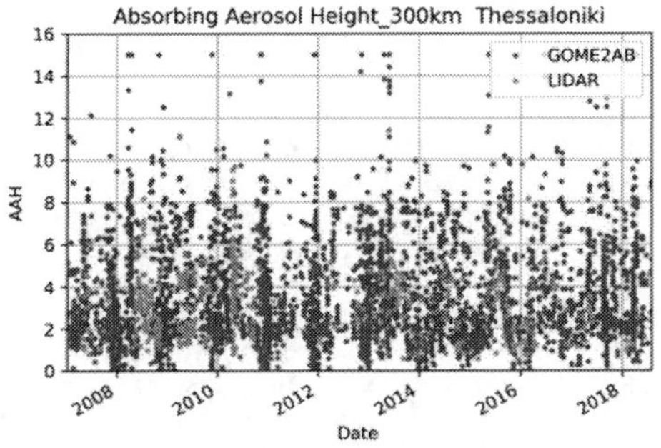

Figure 2: Timeseries of the GOME2 (blue) on MetOpA and MetOpB and the lidar (orange) dataset.

In Figure 3, a histogram depicting the absolute bias between of two AAH products is presented. 67 common cases for the period 2007 to 2018 were identified, with a mean bias value 0.28km, standard deviation of 2.8km and a rather symmetric distribution function.

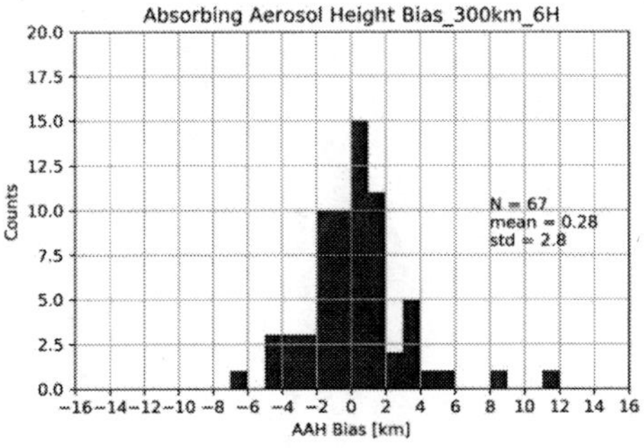

Figure 3: Histogram of the absolute bias between the satellite AAH and lidar upper layer top for the co-located cases .

Furthermore, a scatterplot of the collocated common layers (300km & 6H) is presented in Figure 4. The correlation coefficient for the collocated values is 0.197. We notice that for values up to 3km there is good agreement between the two datasets, while for measurements larger than 3km the dispersion increases. Cases that exhibit large absolute and relative biases will be studied separately in order to identify the source of the discrepancies.

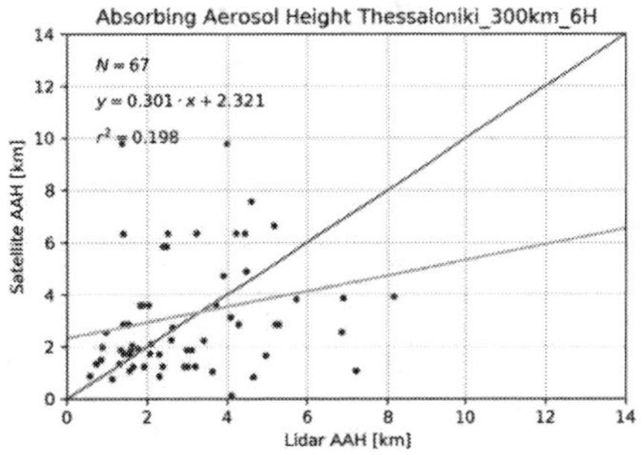

Figure 4: Scatterplot of the Lidar (THELISYS) AAH against the satellite AAH.

4. CONCLUSIONS

The time period from January 2007 to August 2018 is selected for the comparison between GOME-2 MetOpA and MetOpB and ground-based Lidar Absorbing Aerosol Height (AAH). Overall, the agreement between the lidar and satellite AAH product is quite promising, with a mean bias of 0.28±2.8km. In the future, the sensitivity of the bias to the spatial search radius, product quality flag are going to be examined. Also, studies that focus on the climatological circulation patterns will reveal information on the seasonal variations of the geometrical and optical properties that are observed. Study of cloud effects on the comparison of the two products will give answers about uncertainties that result in the calculation of the products. Furthermore, other lidar stations that are part of the EARLINET network will be included in the comparison for further validation.

ACKNOWLEDGEMENTS

We acknowledge support of this work by the project "PANhellenic infrastructure for Atmospheric Composition and climatE change"

https://doi.org/10.1051/epjconf/202023708026

(MIS 5021516) which is implemented under the Action "Reinforcement of the Research and Innovation Infrastructure", funded by the Operational Programme "Competitiveness, Entrepreneurship and Innovation" (NSRF 2014-2020) and co-financed by Greece and the European Union (European Regional Development Fund). The data of the GOME-2 Absorbing Aerosol Index are provided by KNMI in the framework of the EUMETSAT Satellite Application Facility on Atmospheric Composition (AC SAF).

REFERENCES

[1] Balis, D., Koukouli, M.-E., Siomos, N., Dimopoulos, S., Mona, L., Pappalardo, G., Marenco, F., Clarisse, L., Ventress, L. J., Carboni, E., Grainger, R. G., Wang, P., Tilstra, G., van der A, R., Theys, N., and Zehner, C.: Validation of ash optical depth and layer height retrieved from passive satellite sensors using EARLINET and airborne lidar data: the case of the Eyjafjallajökull eruption, Atmos. Chem. Phys., 16, 5705-5720, https://doi.org/10.5194/acp-16-5705- 2016, 2016

[2] Pappalardo, G., Amodeo, A., Apituley, A., Comeron, A., Freudenthaler, V., Linné, H., Ansmann, A., Bösenberg, J., D'Amico, G., Mattis, I., Mona, L., Wandinger, U., Amiridis, V., Alados-Arboledas, L., Nicolae, D., and Wiegner, M.: EARLINET: towards an advanced sustainable European aerosol lidar network, Atmos. Meas. Tech., 7, 2389-2409, https://doi.org/10.5194/amt-7-2389-2014, 2014.

[3] Hassinen, S., Balis, D., Bauer, H., Begoin, M., Delcloo, A., Eleftheratos, K., Gimeno Garcia, S., Granville, J., Grossi, M., Hao, N., Hedelt, P., Hendrick, F., Hess, M., Heue, K.-P., Hovila, J., Jønch-Sørensen, H., Kalakoski, N., Kauppi, A., Kiemle, S., Kins, L., Koukouli, M. E., Kujanpää, J., Lambert, J.-C., Lang, R., Lerot, C., Loyola, D., Pedergnana, M., Pinardi, G., Romahn, F., van Roozendael, M., Lutz, R., De Smedt, I., Stammes, P., Steinbrecht, W., Tamminen, J., Theys, N., Tilstra, L. G., Tuinder, O. N. E., Valks, P., Zerefos, C., Zimmer, W., and Zyrichidou, I.: Overview of the O3M SAF GOME-2 operational atmospheric composition and UV radiation data products and data availability, Atmos. Meas. Tech., 9, 383-407, https://doi.org/10.5194/amt-9-383-2016, 2016.

[4] Siomos, N., Balis, D. S., Voudouri, K. A., Giannakaki, E., Filioglou, M., Amiridis, V., Papayannis, A., and Fragkos, K.: Are EARLINET and AERONET climatologies consistent? The case of Thessaloniki, Greece, Atmos. Chem. Phys., 18, 11885-11903, https://doi.org/10.5194/acp-18-11885-2018, 2018.

[5] M. de Graaf, P. Stammes, O. Torres, and R.B.A. Koelemeijer, *Absorbing Aerosol Index: Sensitivity analysis, application to GOME and comparison with TOMS,* J. Geophys. Res. 110, D010201, doi:10.1029/2004JD005178, 2005.

[6] Tilstra, L. G., O. N. E. Tuinder, and P. Stammes (2010), GOME-2 Absorbing Aerosol Index: statistical analysis, comparison to GOME-1 and impact of instrument degradation, in Proceedings of the 2010 EUMETSAT Meteorological Satellite Conference, EUMETSAT P.57, ISBN 978-92-9110-089-7, Cordoba, Spain.

[7] Tuinder, O. N. E. and L. G. Tilstra (2016), Near Real-Time, Offline and Data Set Aerosol Products - Product User Manual, Doc. No. O3MSAF/KNMI/PUM/002, Issue 1.61, 11 January 2016, KNMI, De Bilt, The Netherlands.

[8] Wang, P., O. N. E. Tuinder, L. G. Tilstra, M. de Graaf, and P. Stammes (2012), Interpretation of FRESCO cloud retrievals in case of absorbing aerosol events, Atmos. Chem. Phys., 12(19), 9057–9077, doi:10.5194/acp-12-9057-2012.

Variation of ozone and PBL from the lidar observations and WRF-Chem model in NYC area during the 2018 summer LISTOS campaign

Kaihui Zhao [1*], Yonghua Wu[2,3], Jianping Huang [4], Rongsheng Jiang [5], Guillaume Gronoff [6], Timothy A. Berkoff [6], Fred Moshary [2,3]

[1] Yale-NUIST Center on Atmospheric Environment, Nanjing Univ. of Information Science & Technology, China. *Email: 645746215@qq.com

[2] the City College of New York (CCNY), New York, NY 10031, USA.

[3] NOAA–Cooperative Science Center for Earth System Sciences and Remote Sensing Technologies

[4] NOAA-NCEP Environmental Modeling Center and IM System Group Inc., College Park, MD 20740, USA

[5] School of Atmospheric Sciences, Nanjing University of Information Science and Technology, China

[6] NASA Langley Research Center, Hampton, VA 23681, USA

ABSTRACT

High ozone (O_3) episodes frequently occur in New York metropolitan and the downwind coastal area in summer. In this study, lidar/ceilometer are combined with WRF/Chem model to investigate an O_3 event on Aug. 27~30 2018. We examine the spatial-temporal variabilities of O_3 and planetary-boundary-layer height (PBLH) and assess the model performance on simulating surface O_3 during this episode. By comparing with the lidar observations, the WRF/Chem is able to capture high O_3 distribution in the PBL at noon and indicates consistent diurnal evolution for the ground O_3. Nevertheless, in the early morning and night, the model overestimates the ground O_3 and underestimates the PBLH.

1. INTRODUCTION

Surface ozone is a secondary pollutant formed by a series of photochemical reactions involving oxides of nitrogen (NO_x = NO + NO_2) and volatile organic compounds (VOCs) in the presence of solar ultraviolet radiation. Lidar and numerical models are powerful tools for understanding the mechanisms for the regional transport and local chemical reactions compared with the routine surface measurements [1]. The coastal areas in the northeast US and mid-Atlantic ocean are still suffering from high O_3 in summer through various processes, such as downwind transport from urban area [2], local photochemical production of O_3 and recirculation brought about by sea breeze [3]. However, most of the modeling studies of ozone exceedance at the coastal areas were limited to the regional chemistry models with relatively coarse resolution of land-use data, which may pose a large uncertainty to the O_3 forecast.

New York City is situated in the northeastern US with an estimated population more than 8 million distributed over a land area of about 784 km^2. Summer O_3 exceedances of US-EPA National Ambient Air Quality Standard (NAAQS) are often observed in the urban and coastal areas. Thus, it is important to investigate the formation mechanisms of high O_3 events and evaluate the model forecast skills. In this study, an ozone lidar and a ceilometer are used to characterize diurnal vertical distribution of ozone and PBLH on 27–30 August 2018 in

NYC area. The ranged-resolved O_3 and PBLHs are focused to investigate the formation mechanisms of O_3 during a heat-wave event.

2. METHODOLOGY

2.1 Model and configurations

The version 3.7.1 of the WRF/Chem model was used in this study [4]. The grid point dimensions from the outermost to the innermost domains are 85 × 75, 76 × 70, and 70 × 61 with horizontal resolutions of 12 km, 4 km and 1.3 km, respectively. Outputs from the whole atmosphere community climate model (WACCM) provide initial and lateral conditions of chemical species to drive the WRF/Chem simulations.

The details for the model application to NYC area can be seen in [5].

2.2 Observational data

Two ceilometers (Vaisala CL-51 & CL-31) make the PBLH measurements at CCNY campus (40.821° N ,74.0°W) and Eatons Neck (coastal site, 40.93 °N, 73.40 ° W), respectively [6]. The NASA-Langley mobile ozone lidar is operated for range-resolved ozone profiles at Westport site, CT (41.07 ° N, 73.2 ° W). The details of the lidar specifications and measurement analysis can be seen in [7-8]. In addition, the measurements of ground-level O_3 and other species are operated by the NYSDEC (New York State Department of Environment Conservation). The hourly mean raw data were used to validate the model predictions. Further, both surface meteorological and sounding data were used to evaluate the model performance on simulating meteorological variables. Vertical profiles of temperature, relative humidity and wind speed measured at Upton site (OKX, 40.87°N, -72.86°E, 20 m above sea level) and are utilized to assess the simulated vertical profiles.

3. RESULTS and DISCUSSION

3.1 O_3 spatial-temporal variations

Figure 1 shows the spatial distribution of WRF-Chem-simulated surface O_3 at 14:00 EST

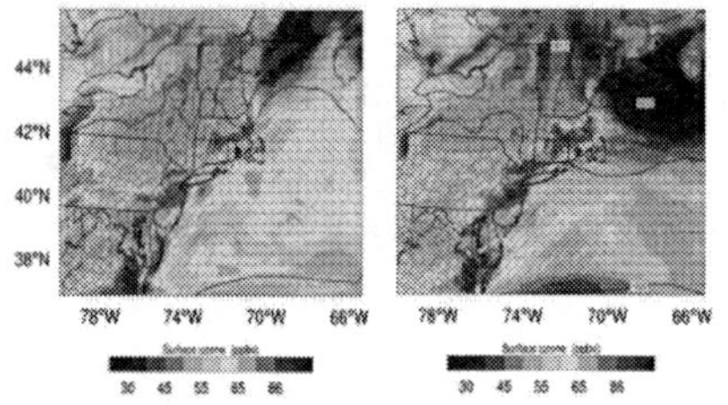

Fig.1 Surface O_3 by WRF/Chem model (shaded colors) and wind (black barb) at 14:00 EST on (left) Aug. 28, (right) Aug. 29, 2018

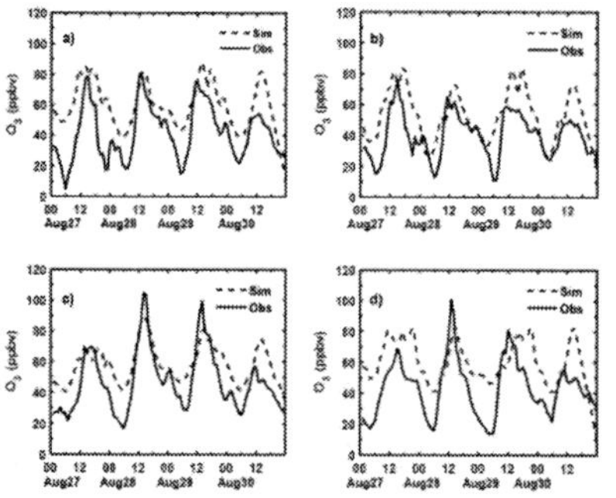

Fig.2 Simulated (dashed:) and observed (solid) surface O_3 at a) Queens College, b) CCNY (urban), c) Eatons Neck (coastal), and d) Babylon.

on Aug. 28 and 29, 2018, respectively. High O_3 can be seen in the NYC and the coastal area, along with the prevailing southwest winds. Figure 2 illustrates the time series of surface O_3 during 27-30 August 2018 at NYC urban and Long Island stations. The model captured the diurnal variations, the daily O_3 peak values, and an increasing trend at different sites quite well except Aug. 30, but surface O_3 concentrations are over-predicted in the morning and night time. With favorable meteorological conditions for O_3 formation, the hourly maximum ozone higher than 70-100 ppb was observed in the NYC and

https://doi.org/10.1051/epjconf/202023708027

surrounding regions, leading to violation of the (NAAQS) for surface O_3 (70-ppb for daily maximum of 8hr-ave). On August 30, the O_3 concentration decreased a little due to the effect of cloud. In addition, the nighttime over-prediction of O_3 is probably attributed to the underpredicted PBLH and underestimated NO concentration.

Figure 3 (a) and (c) illustrate the spatial and temporal distribution of O_3 mixing ratio measured with the NASA Langley O_3 lidar at Westport on 28-29 August 2018. The model results are shown in Fig.3 (b) and (d) for comparison. The WRF/Chem is able to reproduce the ozone diurnal variations and the O_3 structure in the PBL. On 28 August, O_3 concentration higher than 100-ppbv was observed at 20:00 UTC and extended to 1 km height in the PBL. One day later, the hourly-average maximum O_3 was reduced to about 80.0 ppbv. Meanwhile, a layer with O_3 concentration lower than 50.0 ppbv maintains at the height of 1-km above ground level during these two days. The vertical distribution of ozone indicates that the local emissions and chemical production play a critical role in the exceedance of NAAQS for O_3.

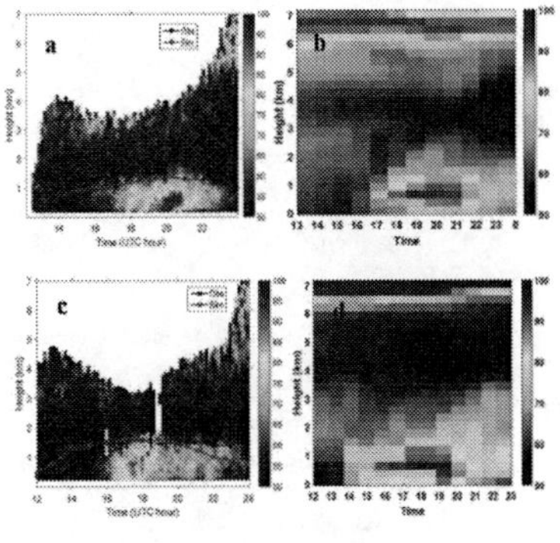

Fig. 3 Spatial and temporal distribution of O_3 mixing ratio (ppbv) at Westport (red line: simulated PBL height; black line: observed PBL height at CCNY on 28, 29 August 2018 (a and c: ozone lidar; b and d: WRF/Chem model) (UTC-4=local time)

3.2 PBL and meteorological profiles

Figure 4 shows the spatial and temporal variations of simulated (boundary layer height) PBLH and the ceilometer observation at Eatons Neck, a coastal site in Long Island Sound. The model shows higher PBLH in the land but much lower in the Long Island sound and marine area at 19:00 UTC (local noon time) Aug.29, 2018. Such large differences in the land and costal/marine area need to be validated from the observations. Near the coastal region, the simulated maximum PBLHs shows better agreement with the ceilometer observation, but the PBL development time was about 2-3 hours earlier than the observation and the PBLH is over-estimated during 8:00-12:00 UTC (UTC-4=local time, 4:00-8:00 early morning). The over-predictions of the PBL height could be due to the uncertainties of the PBL schemes and surfaces.

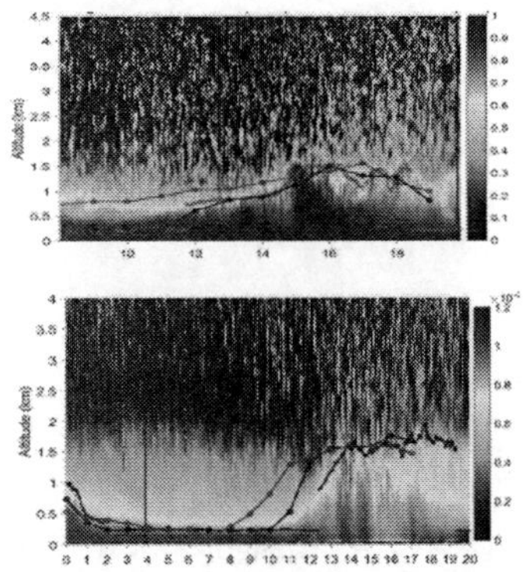

Fig 4. PBLH spatial distribution at 19:00 UTC on Aug. 29 by WRF-Chem model; ceilometer observation at Eatons Neck (coastal site) on Aug. 28 (top) and 29 (bottom). (red line: model; black: obs.; pink: 2-min ave for the obs. PBLH)

Figure 5 shows simulation-observation comparisons of vertical profiles of potential temperature, relative humidity, and wind speed in the morning (08:00 EDT) during 27-30 August 2018. Overall, the simulated vertical profiles agree well with the observations. The potential temperature significantly increased and the slope under 1 km became more stable during this event (figure5). The relative humidity obviously increased above 1-km which is consistent with the low ozone concentration layer. In contrast, the simulated wind profiles did not agree well with observations due to the complex topography. The wind speeds were persistently weak during this event. The lower wind speed is favorable for ozone photochemical accumulation of ozone and its precursors.

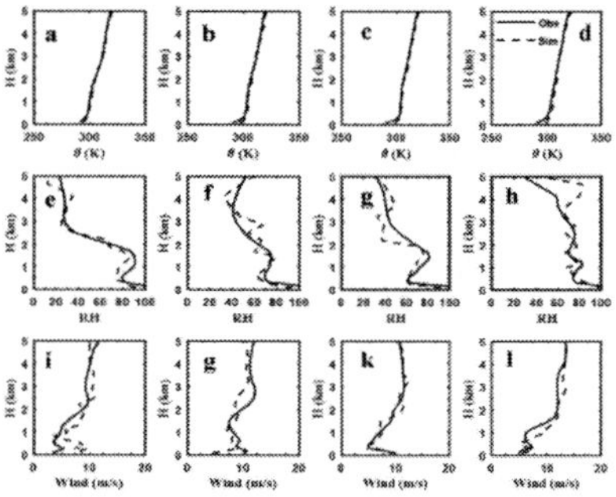

Fig.5 Comparison of WRF/Chem-simulated vertical profiles with radiosonde observed potential temperature, relative humidity (RH%), wind speed at 08:00 EDT, OKX Upton during the event (panels a), e) and i) for Aug. 27; b), f), and j) for Aug. 28; and c), g) and k) for Aug.29; d), h), l) for Aug. 30.

4. CONCLUSION

This extended abstract presents a case study of ozone exceedance event in NYC urban and its coastal area on August 27-30, 2018 with WRF/Chem simulations and observational data. The maximum hourly-averaged ozone concentrations higher than 70-100 ppb were observed at both the urban and rural or coastal areas. The diurnal variation of surface ozone, PBLH and ozone vertical distribution were well captured by the WRF/Chem model. The vertical distribution of ozone indicates that the local emissions and chemical production play a critical role in the exceedance event of NAAQS of O_3. The model was successful in reproducing the maximum PBLHs but the PBL development time with 2-3 hours earlier than the observations near the coastal areas.

ACKNOWLEDGEMENTS.

The New York State Energy Research and Development Authority (grant # 100415 and 137482), NESCAUM (grant # 2411) and the NOAA-Cooperative Science Center for Earth System Sciences and Remote Sensing Technologies under the Cooperative Agreement Grant #NA16SEC4810008, the National Science Foundation of China (Grant 41575009), the Xianyang major science and technology projects (2017YFC0210102) support this study.

REFERENCES

[1] Chaxel E, et al., 2009: Ozone production from Grenoble city during the August 2003 heat wave. Atmos. Environ., 43(31):4784-4792.

[2] Angevine, W.M., et al., 2004: Coastal Boundary Layer Influence on Pollutant Transport in New England. *J. Appl. Meteor.,* 43, 1425–1437.

[3] Banta, R M, et al., 2005: A Bad Air Day in Houston. Bull. Amer. Meteor. Soc., 86(5):657-669.

[4] Grell, G. A., et al., 2005: Fully coupled "online" chemistry within the WRF model, Atmos. Environ., 39, 6957–6975.

[5] Zhao, K., et al., 2019: A high-resolution modeling study of a heat wave-driven ozone exceedance event in New York City and surrounding regions, Atmos. Environ., 199, 368–379.

[6] Gan, C. M., et al., 2010: Statistical comparison between hysplit sounding and lidar observation of planetary boundary layer characteristics over New York City, SPIE Proc., vol. 7684-57

[7] Young, R. D. et al. 2017: Langley mobile ozone lidar: ozone and aerosol atmospheric profiling for air quality research, Appl. Opt. **56**, 721-730.

[8] Farris, B. Met al., 2019: Demonstration of an off-axis parabolic receiver for near-range retrieval of lidar ozone profiles, Atmos. Meas. Tech., 12, 363-370.

AUTHOR INDEX

AUTHOR INDEX

AUTHOR INDEX

AUTHOR INDEX

AUTHOR INDEX

AUTHOR INDEX

AUTHOR INDEX

AUTHOR INDEX